Lecture Notes in Civil Engineering

Volume 98

Lecture Notes in Civil Engineering (LNCE) publishes the latest developments in Civil Engineering - quickly, informally and in top quality. Though original research reported in proceedings and post-proceedings represents the core of LNCE, edited volumes of exceptionally high quality and interest may also be considered for publication. Volumes published in LNCE embrace all aspects and subfields of, as well as new challenges in, Civil Engineering. Topics in the series include:

- Construction and Structural Mechanics
- Building Materials
- Concrete, Steel and Timber Structures
- Geotechnical Engineering
- Earthquake Engineering
- Coastal Engineering
- Ocean and Offshore Engineering; Ships and Floating Structures
- Hydraulics, Hydrology and Water Resources Engineering
- Environmental Engineering and Sustainability
- Structural Health and Monitoring
- Surveying and Geographical Information Systems
- Indoor Environments
- Transportation and Traffic
- Risk Analysis
- Safety and Security

To submit a proposal or request further information, please contact the appropriate Springer Editor:

- Mr. Pierpaolo Riva at pierpaolo.riva@springer.com (Europe and Americas);
- Ms. Swati Meherishi at swati.meherishi@springer.com (Asia - except China, and Australia, New Zealand);
- Dr. Mengchu Huang at mengchu.huang@springer.com (China).

All books in the series now indexed by Scopus and EI Compendex database!

More information about this series at http://www.springer.com/series/15087

Eduardo Toledo Santos · Sergio Scheer
Editors

Proceedings of the 18th International Conference on Computing in Civil and Building Engineering

ICCCBE 2020

Springer

Editors
Eduardo Toledo Santos
Escola Politécnica
University of São Paulo
São Paulo, São Paulo, Brazil

Sergio Scheer
Federal University of Paraná
Curitiba, Paraná, Brazil

ISSN 2366-2557 ISSN 2366-2565 (electronic)
Lecture Notes in Civil Engineering
ISBN 978-3-030-51294-1 ISBN 978-3-030-51295-8 (eBook)
https://doi.org/10.1007/978-3-030-51295-8

This Springer imprint is published by the registered company Springer Nature Switzerland AG
The registered company address is: Gewerbestrasse 11, 6330 Cham, Switzerland

Eduardo Toledo Santos · Sergio Scheer
Editors

Proceedings of the 18th International Conference on Computing in Civil and Building Engineering

ICCCBE 2020

Springer

Editors
Eduardo Toledo Santos
Escola Politécnica
University of São Paulo
São Paulo, São Paulo, Brazil

Sergio Scheer
Federal University of Paraná
Curitiba, Paraná, Brazil

ISSN 2366-2557 ISSN 2366-2565 (electronic)
Lecture Notes in Civil Engineering
ISBN 978-3-030-51294-1 ISBN 978-3-030-51295-8 (eBook)
https://doi.org/10.1007/978-3-030-51295-8

This Springer imprint is published by the registered company Springer Nature Switzerland AG
The registered company address is: Gewerbestrasse 11, 6330 Cham, Switzerland

Editor's Preface to the Proceedings of the 18th International Conference on Computing in Civil and Building Engineering

Civil and Building Engineering have long been application areas for computing and other information and communication technologies (ICT). These proceedings are from the 18th International Conference on Computing in Civil and Building Engineering, a series started almost 40 years ago by the ISCCBE—International Society for Computing in Civil and Building Engineering. Along with the CIB W78—IT in Construction Working Group, these are two of the most important international institutions dedicated to promote computing in the construction sector and, in 2020, came together to promote the 2020 ICCCBE + CIBW78 Virtual Joint Conference.

Despite all research efforts, the construction industry until recently was reluctant to more fully adopt ICT on its processes. That status started swiftly to change with the widespread Building Information Modelling (BIM) adoption which spearheaded a cultural change on the sector, making some other related technologies worth attention in the eyes of this industry stakeholders.

So now we are at a very special moment when research in this important area can make a real difference, impacting productivity and quality of thousands of construction projects around the world, both in building and infrastructure projects, reducing or eliminating so common time and cost overruns.

The papers contained in this volume are representative of the ultimate research trends on this theme and are authored by some of the leading researchers in the world. Together, they provide an updated review of current scientific developments as well as practical applications of Computing in Civil and Building Engineering.

Some outstanding examples are the sections on machine learning, Internet of things, blockchain, drones and vision-based applications, as well as on the recent trends of BIM uses in infrastructure projects and during the construction phase.

We hope this book will be useful and informative both to the academic community and to practitioners in the construction industry looking for high quality and updated research on Computing for Civil and Building Engineering.

Eduardo Toledo Santos
Sergio Scheer

Contents

Rule-Based Systems Applications

Blockchain Applications for AEC

BIM and Internet of Things (IoT) Frameworks and Applications

Artificial Intelligence Applied to the Built Environment

Artificial Intelligence Techniques for Smart City Applications

Daniel Luckey, Henrieke Fritz, Dmitrii Legatiuk, Kosmas Dragos, and Kay Smarsly(✉)

Chair of Computing in Civil Engineering, Bauhaus University Weimar, Weimar, Germany
{daniel.luckey,kay.smarsly}@uni-weimar.de

Abstract. Recent developments in artificial intelligence (AI), in particular machine learning (ML), have been significantly advancing smart city applications. Smart infrastructure, which is an essential component of smart cities, is equipped with wireless sensor networks that autonomously collect, analyze, and communicate structural data, referred to as "smart monitoring". AI algorithms provide abilities to process large amounts of data and to detect patterns and features that would remain undetected using traditional approaches. Despite these capabilities, the application of AI algorithms to smart monitoring is still limited due to mistrust expressed by engineers towards the generally opaque AI inner processes. To enhance confidence in AI, the "black-box" nature of AI algorithms for smart monitoring needs to be explained to the engineers, resulting in so-called "explainable artificial intelligence" (XAI). However, when aiming at improving the explainability of AI algorithms through XAI for smart monitoring, the variety of AI algorithms requires proper categorization. Therefore, this review paper first identifies objectives of smart monitoring, serving as a basis to categorize AI algorithms or, more precisely, ML algorithms for smart monitoring. ML algorithms for smart monitoring are then reviewed and categorized. As a result, an overview of ML algorithms used for smart monitoring is presented, providing an overview of categories of ML algorithms for smart monitoring that may be modified to achieve explainable artificial intelligence in civil engineering.

Keywords: Artificial intelligence (AI) · Machine learning (ML) · Smart cities · Smart infrastructure · Smart monitoring · Explainable artificial intelligence (XAI)

1 Introduction

In the last decade, developments within the ongoing socioeconomic digitalization have created the vision of smart cities, which aspires to connect all aspects of urban life. The basis for connecting aspects of urban life in smart cities is being built around contemporary and emerging technologies, such as cloud computing, the Internet of Things and cyber-physical systems, representing the latest chain in industrial revolution, referred to as Industry 4.0 (Acatech 2015). A key aspect of the aforementioned technologies is adopting and advancing artificial intelligence (AI) techniques, which have

E. Toledo Santos and S. Scheer (Eds.): ICCCBE 2020, LNCE 98, pp. 3–15, 2021.
https://doi.org/10.1007/978-3-030-51295-8_1

proven their ability to process large amounts of data towards developing learning rules, e.g. via machine learning (ML), making complex associations, and predicting outcomes of complex physical processes. Although applications related to smart cities are expected to become a trillion-dollar market in the next five years (PWC 2019), the term "smart city" has not been officially defined (OECD 2019; Johnson et al. 2019). However, several key components of smart cities have already been well-established, such as smart living, smart governance, smart citizen (people), smart mobility, smart economy, and smart infrastructure (Mohanty et al. 2016).

Smart infrastructure is of particular importance for civil engineering and provides the foundation for key components of smart cities (UN Economic and Social Council 2016). Therefore, smart infrastructure is considered the backbone of smart cities. Smart infrastructure is realized via smart (wireless) structural health monitoring (SHM) systems, referred to as "smart monitoring", which enables timely detection of structural degradation, thus resulting in low maintenance, repair, and disruption costs (Ogie et al. 2017). Because of aging infrastructure, smart monitoring has been gaining increasing popularity for leveraging the aforementioned benefits of smart infrastructure.

Smart monitoring fosters automation in SHM; therefore, aspects of SHM are essential for defining objectives in smart monitoring. SHM is typically associated with structural condition assessment using structural response data and encompasses data acquisition, data communication, data analysis, data storage, and data retrieval. Specifically, data analysis leads to conclusions drawn from structural response data, with respect to damage detection, damage classification, damage localization, condition assessment, and life-time prediction (Kabalci and Kabalci 2019). Data analysis is usually performed using data-driven models that extract information from structural response data. While several data-driven models draw from statistical processing and experimental mechanics, the increasing amounts of data in long-term monitoring systems have fueled research in adopting AI algorithms for data analysis and processing. The intelligence inherent to AI algorithms is compatible with the automation necessary for smart monitoring, as part of smart infrastructure. Moreover, several AI algorithms used in smart monitoring are commonly referred to as "big data" algorithms and therefore serve a twofold purpose, (i) to detect patterns representing complex physical processes that otherwise would remain undetected, and (ii) to exploit, to the best possible extent, large amounts of data available in long-term SHM systems that are otherwise only partially utilized.

Smart monitoring, thus smart infrastructure, has taken advantage of distributed artificial intelligence, a subfield of artificial intelligence. In particular, multi-agent technology, representing a major branch of distributed artificial intelligence, has been deployed to advance different fields of smart monitoring, such as dam monitoring (Mittrup et al. 2003), wind turbine monitoring (Hartmann et al. 2011), and bridge monitoring (Smarsly et al. 2007). Multi-agent systems have also been reported as an enabling technology of self-managing smart monitoring systems (Smarsly et al. 2012) and process scheduling in smart infrastructure applications (Bilek et al. 2003). Facilitating wireless smart infrastructure, multi-agent technology has been extended towards mobile multi-agent systems, as reported in (Smarsly and Law 2013), proposed to enable agent-based software modules to autonomously migrate from one wireless sensor node to another in an attempt to analyze smart infrastructure on demand. As could be demonstrated in a study presented in (Smarsly et al. 2011), the mobile multi-agent approach leads to significantly reduced resource consumption in wireless smart monitoring systems, as compared to traditional approaches. Further artificial intelligence techniques, such as neural networks (Dragos and Smarsly 2016), support vector regression (Steiner et al. 2019) and evolutionary algorithms (Nguyen et al. 2007), have been implemented into smart monitoring systems in a decentralized manner.

Most recent approaches have as common ground machine learning algorithms, a subcategory of AI, that have been adopted for smart monitoring purposes (Smarsly et al. 2016). Generally, ML algorithms in civil engineering may be distinguished by their application into (i) ML algorithms used for so-called surrogate modeling, where ML algorithms substitute conventional algorithms to achieve higher computational efficiency and (ii) ML algorithms used to solve abstract problems pertaining to data analysis, such as pattern recognition or classification problems, in which ML algorithms are deployed to analyze large amounts of data to classify given signals (or pictures) with respect to predefined classes.

In general, artificial intelligence algorithms, and, by extension, machine learning algorithms, may be categorized into *symbolic AI*, which includes inference and search algorithms using explicit symbolic programming, and into *subsymbolic AI*, which is generally considered "black-box" in terms of internal mechanisms. Subsymbolic AI, such as deep learning neural networks, shows good performance in analyzing complex engineering problems that involve large data sets and is therefore widely used in smart monitoring. However, the widespread adoption of subsymbolic AI/ML algorithms in smart monitoring is still limited, due to mistrust expressed by engineers towards the opaque inner mechanisms of subsymbolic AI/ML algorithms, and, by extension, to the reasoning and reproducibility of the outputs. While an explanation of the algorithms is inherent in symbolic AI, there is a strong need to explain subsymbolic AI/ML algorithms. The need for explaining the reasoning behind decisions made by subsymbolic black-box AI/ML algorithms has led to the development of "explainable artificial intelligence (XAI)" (Gunning and Aha 2019; Barredo Arrieta et al. 2019). XAI is a technical discipline aiming to comprehensibly present AI systems and to clarify why and how AI systems generate certain outputs (Adadi and Berrada 2018).

Addressing the explainability of AI/ML algorithms for smart monitoring requires a concise overview of existing approaches using AI/ML algorithms in smart monitoring. From the broader perspective of smart cities, AI/ML algorithms for smart city applications have been reported in reviews and summary papers, for example by Guo et al. (2019) and Mohapatra (2019). Soomro et al. (2019) have reviewed big data analytics for smart cities and Martins (2018) has discussed the impact of ML algorithms on innovations in smart cities. Furthermore, Nosratabadi et al. (2019) have surveyed deep learning and ML models for smart cities. Regarding smart monitoring, Bao et al. (2019) have presented a review on data science approaches in SHM, and Joshuva et al. (2019) have reviewed machine learning algorithms for monitoring wind turbines. However, to the knowledge of the authors, no review has focused on categorizing AI/ML algorithms for highlighting the need for XAI in smart monitoring.

This paper essentially constitutes a preliminary step towards adapting XAI approaches for smart monitoring. By reviewing and categorizing AI/ML algorithms for smart monitoring and discussing general XAI concepts, an overview of which AI/ML algorithms used in SHM may be modified towards adopting XAI in smart monitoring is shown. Subsequently, the review presented herein is summarized, and a concise outlook on potential future work is provided.

2 Machine Learning Algorithms in Civil Engineering

Because of the ability to recognize and to classify patterns in large data sets, ML algorithms are of increasing interest in civil engineering. In the following subsections, a categorization of ML algorithms is provided, and ML algorithms of particular relevance to smart monitoring applications are reviewed and categorized.

2.1 Categorization of Machine Learning Algorithms

The term "intelligence" in "artificial intelligence" denotes the ability of an entity to capture, to process, and to respond to input of different kind (Legg and Hutter 2007). Extending the definition of intelligence, the term "artificial intelligence" describes the ability of an artificial entity (e.g., a software or computer system) to achieve specific goals under a variety of environmental conditions. However, to qualify as "intelligent", a system needs to possess the ability to respond to previously unknown (environmental) conditions through learning and adaption (Hutter 2005). In a broader sense, AI is the ability of a computer system to approximate the intellectuality of human beings. To mimic human behavior, Russel and Norvig (2014) have defined six categories of AI: machine learning, robotics, computer vision, natural language processing, knowledge representation, and automated reasoning.

In intelligent systems, ML helps adapt a system to new circumstances through processing and analyzing data, extrapolating patterns, and making predictions. By combining concepts of computer science with optimization and statistical concepts (Mohri et al. 2012), ML essentially represents the learning processes of AI, often described as converting experience into expertise or knowledge (Shalev-Shwartz and Ben-David 2014). In summary, ML algorithms show two distinct advantages, as compared to traditional algorithms (Russel and Norvig 2014; Shalev-Shwartz and Ben-David 2014):

1. ML algorithms operate with previously unknown (i.e., newly derived) data on which the system has not been trained, and
2. ML algorithms are adaptable to changes in the data.

However, ML algorithms need to learn from experience or knowledge of domain experts (Shalev-Shwartz and Ben-David 2014). Depending on the type of learning, ML algorithms may be categorized into

i. supervised learning,
ii. unsupervised learning, and
iii. reinforcement learning,

as shown in Fig. 1. In *supervised learning*, ML algorithms use labeled input-output pairs as training data, and the system learns based on given examples. Typical learning problems in supervised learning are *classification* and *regression*. In classification, data is sorted into predefined categories, while in regression, the outputs to corresponding input data are calculated. In contrast to supervised learning, the training data in *unsupervised learning* is not labeled. A typical problem of unsupervised learning is *clustering*, where data is grouped according to commonalities. In *reinforcement learning*, no training data is provided. Instead, the system develops a strategy to maximize a predefined cumulative reward. Figure 1 shows the categorization of AI and ML algorithms as well as the subcategories mentioned above. In addition, examples of ML algorithms, corresponding to the subcategories, are illustratively provided in Fig. 1.

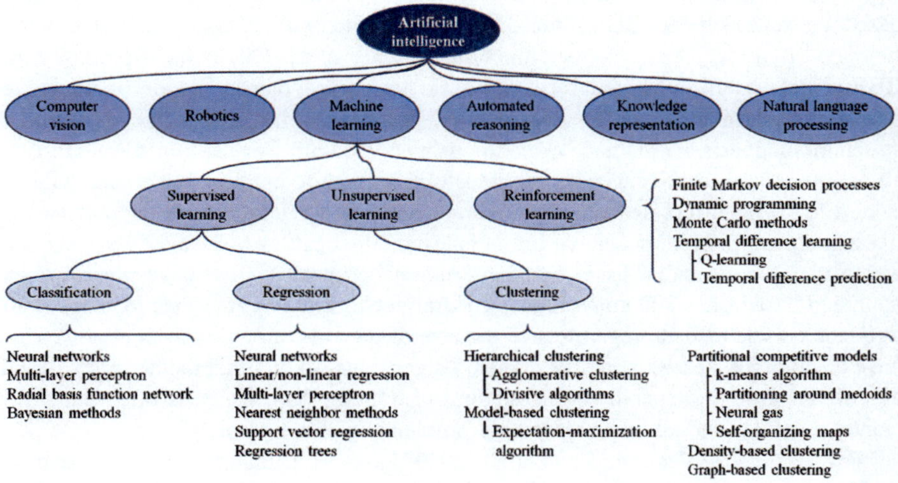

Fig. 1. Categorization of machine learning algorithms.

Depending on the category of ML algorithms, mixed forms of (i), (ii), and (iii) are likely to be used (Burkov 2019; Salehi and Burgueno 2018). For example, artificial neural networks are trained with different specifications and, depending on the purpose and structure of the artificial neural network (ANN), may fit into any of the three categories. Therefore, the categorization presented in Fig. 1 is regarded as a starting point to approach the basic concepts of machine learning but cannot be considered generally valid for any ML specification.

2.2 Machine Learning Algorithms for Smart Monitoring

The application areas of machine learning in smart monitoring are manifold. This paper focuses on ML algorithms applied to data analysis, which, according to Bisby and Briglio (2005), may pursue the following goals,

i. damage detection,
ii. damage classification,

iii. damage localization,
iv. condition assessment, and
v. life-time prediction,

to be achieved primarily by supervised and unsupervised ML algorithms as well as algorithms that combine both categories. In the remainder of this subsection, ML algorithms addressing the above goals are reviewed, distinguishing between supervised and unsupervised/hybrid ML algorithms.

Supervised Machine Learning Algorithms for Smart Monitoring
For *damage detection and damage classification*, support vector machines (SVM) are common. For example, Li et al. (2019) have identified damage based on SVMs and Lamb waves in smart monitoring. Gui et al. (2017) have compared different SVM-based optimization techniques for damage detection with a Gaussian radial basis function (RBF) chosen as kernel function. Gardner et al. (2016) have proposed an RBF-kernel based SVM, fed by a finite element-based damage model to generate output data, while Pan et al. (2018) have proposed a framework for data-driven structural diagnosis and damage detection using SVM with wavelet transform, Hilbert-Huang transform, and Teager-Huang transform as feature extraction methods. Ghiasi et al. (2016) have reported on a new kernel function for least square support vector machines using multidimensional orthogonal-modified Littlewood-Paley wavelets and a thin plate spline radial basis function. Abdeljaber et al. (2018) have presented an approach based on a 1-D convolutional neural network (CNN) to detect damage with two labeled sets of data, regardless of the size of the structure. Gunawan et al. (2018) have examined k-nearest neighbors (k-NN) algorithms, stating that the accuracy of the algorithms strongly depends on the amount of training data, which is often not sufficiently available for solving engineering problems in smart monitoring.

For *damage localization*, Zhao et al. (2019) have presented an algorithm based on ANN regression using acoustic emission sensors for carbon fiber reinforced polymer composite materials. The training data required for the artificial neural network has been obtained from a finite element model.

To advance *condition assessment* of smart structures, Nazarian et al. (2018) have combined SVMs, ANNs, and Gaussian naïve Bayes techniques to assess the condition of a masonry building with timber frames. The ML model has been trained by finite element model simulation data to relate the change of stiffness of different building components to intensity and location of the damage sources.

Aiming at *life-time prediction*, Sysyn et al. (2019) have addressed a railway crossing based on features extracted by principal component analysis and partial least square regression. Hoang et al. (2018) have predicted the scour depth at bridges by using support vector regression, for which several feature selection algorithms have been combined, with the variable neighborhood search feature selection method providing the best outcome.

A number of studies have been reported that aim at *combinations* of the data analysis goals within smart monitoring, for example pursuing damage detection and damage classification together. Vitola et al. (2016) have presented a combination of principal component analysis (PCA) with k-NN and PCA with bagged trees. Vitola et al. (2017a) have compared different k-NN algorithms to detect and to classify damage based on identical data sets, linked with the research conducted by Tibaduiza et al. (2018) and Vitola et al. (2017b), who combine PCA and k-NN components to detect und to classify damage of sandwich structures and composite plates. Joshuva and Sugumaran (2018) have compared classification and regression algorithms with respect to damage detection and damage classification, including a sequential minimal optimization classifier, a simple logistic algorithm classifier, a multilayer perceptron in terms of a feedforward artificial neural network, logistic regression, and an RBF network. The authors have been able to define five different damage classes. Vashisht et al. (2018) have compared Bayesian ANNs, CNNs, and long short-term memory ANNs to identify and to localize damage in a cantilever beam with training data for the ANNs provided by finite element simulations.

Unsupervised and Hybrid Machine Learning Algorithms for Smart Monitoring Studies applying unsupervised/hybrid ML algorithms to achieve the goals of data analysis in smart monitoring are less common than supervised learning approaches, because labeled training data is usually available in smart monitoring. For example, Sierra-Perez et al. (2017) have presented a multi-layer ANN-based damage detection methodology for strain field pattern recognition, using a hierarchical non-linear PCA dimensionality reduction technique. Santos et al. (2016) have improved Gaussian mixture models to detect and to classify damage of bridges. Senniappan et al. (2016) have applied fuzzy cognitive maps to categorize cracks in reinforced concrete columns.

Furthermore, Diez et al. (2016) have used a k-NN outlier detector for performing k-means clustering on data in an attempt to isolate and to localize damaged joints of a bridge. Das et al. (2019) have used Gaussian mixture models for clustering unlabeled data and for feature separation by an SVM-calculated hyperplane for crack mode classification.

3 Results and Discussion: Towards Explainable Artificial Intelligence

The result of the review presented in the previous section is shown in Fig. 2 in terms of an overview of ML algorithms for smart monitoring. As can be seen from Fig. 2, the ML algorithms are assigned to the goals of data analysis in smart monitoring, with the thickness of the lines connecting an ML algorithm and a data analysis goal denoting the quantity of papers found in literature. Regardless of the ML algorithm and the data analysis goal, it has been concluded that intransparency and mistrust in ML algorithms that are black-box in nature are hindering the widespread adoption of the algorithms in civil engineering practice. Particularly following the enforcement of the European data protection regulation, which requires comprehensible decision making in AI, the incomprehensibility of ML-based decision making further limits the distribution and implementation of ML algorithms.

XAI has the potential to overcome implementation obstacles and provide explanations as well as additional information regarding decision-making processes, hence offering more comprehensible ML algorithms. However, when designing XAI-based

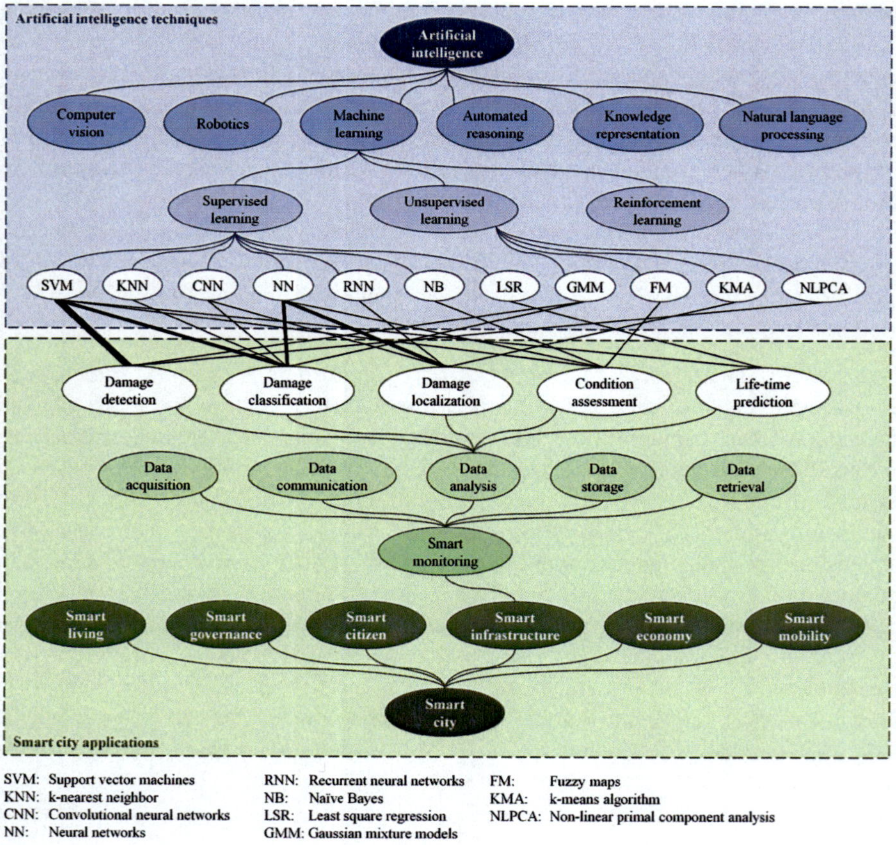

Fig. 2. Review of machine learning algorithms for smart monitoring.

ML algorithms, different levels of explanations must be considered, ranging from "comprehensive explanation" in case of complex subsymbolic ML algorithms to "no explanation" in case of symbolic ML algorithms, as implemented in expert systems that inherently explain themselves. Further distinctions must be made with respect to the experience and expertise of human individuals that are addressed by the explanations, such as technicians using ML algorithms in engineering practice or computer scientists implementing data analysis into smart monitoring systems.

In general, an explanation is considered a collection of human-interpretable features, relevant to decisions provided by ML algorithms. The explainability of ML algorithms is often referred to as interpretability, with "interpretation" denoting a mapping of abstract concepts that are comprehensible for human individuals (Montavon et al. 2018). Different efforts towards implementing XAI approaches have been reported. For example, LIME, a local interpretable explanation, presents a model-independent approach towards approximating black-box models around any classifier of interest and explaining the predictions of the classifier in an interpretable manner

(Ribeiro et al. 2016). The layer-wise relevance propagation (LRP) algorithm for image classification serves as another XAI implementation example. LRP decomposes the classifier and iterates the relevance of each layer of a network backwards, starting with the output prediction (Bach et al. 2015). Aiming to explain autonomous decisions made by smart monitoring systems with respect to sensor fault diagnosis, Fritz (2019) has implemented an XAI approach that extends deep learning NNs coupled with blockchain technology. In summary, representing an open research problem in smart city applications, it can be concluded that explanations must be adapted to the goal of data analysis, to the level of explainability, and to the target audience.

4 Summary and Conclusions

Smart infrastructure is a key component of smart cities and requires smart monitoring to achieve more reliable, durable, and cost-efficient infrastructure as compared to the past. Smart monitoring is a combination of SHM and AI algorithms. ML algorithms, a subcategory of AI algorithms, are used to automatically analyze sensor data. However, the black-box nature of ML algorithms typically used in smart monitoring, although efficient in analyzing sensor data, causes intransparency and mistrust expressed by engineers, thus hindering the exploitation of the ML full potential in engineering practice.

XAI is supposed to enhance the transparency, thus the confidence, in ML algorithms. Drawing from trends in current ML applications for smart monitoring, this paper has presented a preliminary step towards adapting XAI approaches in smart monitoring. ML algorithms commonly deployed to smart monitoring have been reviewed and XAI approaches have been presented, proposed to overcome the obstacles of incomprehensibility of ML algorithms. For smart monitoring, ML algorithms may require different levels of explanations based on their purpose and the human individuals addressed. In conclusion, the overview of ML algorithms in smart monitoring provided in this paper has demonstrated that an in-depth analysis of explainability and levels of explanation for ML algorithms is required to advance smart monitoring and smart city developments.

Acknowledgments. The authors gratefully acknowledge the support offered by the German Research Foundation (DFG) under grants SM 281/9-1, SM 281/14-1, and SM 281/15-1. This research is also partially supported by the German Federal Ministry of Transport and Digital Infrastructure (BMVI) under grant VB18F1022A. Any opinions, findings, conclusions or recommendations expressed in this paper are those of the authors and do not necessarily reflect the views of DFG or BMVI.

References

Acatech – National Academy of Science and Engineering: Industry 4.0, Urban development and German international development cooperation (Acatech position paper), Herbert Utz Verlag, Munich, Germany (2015)

Adadi, A., Berrada, M.: Peeking inside the black-box: a survey on explainable artificial intelligence (XAI). IEEE Access **6**(2018), 52138–52160 (2018)

Abdeljaber, O., Avci, O., Kiranyaz, S., Boashash, B., Sodano, H., Inman, D.: 1-D CNNs for structural damage detection: verification on a structural health monitoring benchmark data. Neurocomputing **275**(2018), 1308–1317 (2018)

Bach, S., Binder, A., Montavon, G., Klauschen, F., Müller, K.R., Samek, W.: On pixel-wise explanations for non-linear classifier decisions by layer-wise relevance propagation. PLoS ONE **10**(7), e0130140 (2015)

Bao, Y., Chen, Z., Wei, S., Xu, Y., Tang, Z., Li, H.: The state of the art of data science and engineering in structural health monitoring. Engineering **5**(2), 234–242 (2019)

Barredo Arrieta, A., Diaz Rodriguez, N., Del Ser, J., Bennetot, A., Tabik, S., et al.: Explainable artificial intelligence (XAI): concepts, taxonomies, opportunities and challenges toward responsible AI. Inf. Fusion **58**(2020), 82–115 (2019)

Bilek, J., Mittrup, I., Smarsly, K., Hartmann, D.: Agent-based concepts for the holistic modeling of concurrent processes in structural engineering. In: Proceedings of the 10th ISPE International Conference on Concurrent Engineering: Research and Applications, Madeira, Portugal, 26 July 2003

Bisby, L.A., Briglio, M.B.: ISIS educational module 5: an introduction to structural health monitoring. SAMCO Final Report. Winnipeg, Manitoba, Canada: ISIS Canada (2006)

Burkov, A.: The Hundred-Page Machine Learning Book (2019). ISBN-13: 978-1999579500

Cha, Y.-J., Choi, W., Büyüköztürk, O.: Deep learning-based crack damage detection using convolutional neural networks. Comput. Aided Civil Infrastruct. Eng. **32**(5), 361–378 (2017)

Das, A., Suthar, D., Leung, C.: Machine learning based crack mode classification from unlabeled acoustic emission waveform features. Cem. Concr. Res. **121**(2019), 42–57 (2019)

Diez, A., Khoa, N.L.D., Makki Alamdari, M., Wang, Y., Chen, F., Runcie, P.: A clustering approach for structural health monitoring on bridges. J. Civil Struct. Health Monit. **6**(2016), 1–17 (2016)

Dragos, K., Smarsly, K.: Distributed adaptive diagnosis of sensor faults using structural response data. Smart Mater. Struct. **25**(10), 105019 (2016)

Fritz, H.: An explainable artificial intelligence model coupling deep learning and blockchain technology. Bachelor thesis. Chair of Computing in Civil Engineering, Bauhaus University Weimar, Germany (2019)

Gardner, P., Barthorpe, R.J., Lord, C.: The development of a damage model for the use in machine learning driven SHM and comparison with conventional SHM methods. In: Proceedings of the International Conference on Noise and Vibration Engineering 2016 (ISMA 2016) and International Conference on Uncertainty in Structural Dynamics (USD 2016), Leuven, Belgium, 13 September 2016 (2016)

Ghiasi, R., Torkzadeh, P., Noori, M.: A machine-learning approach for structural damage detection using least square support vector machine based on a new combinational kernel function. Struct. Health Monit. **15**(3), 302–316 (2016)

Gunawan, F., Soewito, B., Surantha, N., Tuga, M.: One more reason to reject manuscript about machine learning for structural health monitoring. In: Proceedings of the 2018 Indonesian Association for Pattern Recognition (INAPR) International Conference, Jakarta, Indonesia, 7 September 2018 (2018)

Gui, G., Pan, H., Lin, Z., Li, Y., Yuan, Z.: Data-driven support vector machine with optimization techniques for structural health monitoring and damage detection. KSCE J. Civil Eng. **21**(2), 523–534 (2017)

Gunning, D., Aha, D.W.: DARPA's explainable artificial intelligence program. AI Mag. **40**(2), 44–58 (2019)

Guo, X., Shen, Z., Zhang, Y., Wu, T.: Review on the application of artificial intelligence in smart homes. Smart Cities **2**(3), 402–420 (2019)

Hartmann, D., Smarsly, K., Law, K.H.: Coupling sensor-based structural health monitoring with finite element model updating for probabilistic lifetime estimation of wind energy converter structures. In: Proceedings of the 8th International Workshop on Structural Health Monitoring, Stanford, CA, USA, 13 September 2011 (2011)

Haugeland, J.: Artificial Intelligence. The Very Idea. MIT Press, Cambridge (1987)

Hoang, N.-D., Liao, K.-W., Tran, X.-L.: Estimation of scour depth at bridges with complex pier foundations using support vector regression integrated with feature selection. J. Civil Struct. Health Monit. **8**(3), 431–442 (2018)

Hutter, M.: Universal Artificial Intelligence – Sequential Decisions Based on Algorithmic Probability. Springer-Verlag GmbH, Heidelberg (2005)

Johnson, P., Robinson, P., Philpot, S.: Type, tweet, tap, and pass: how smart city technology is creating a transactional citizen. Gov. Inf. Q. **37**(1), 101414 (2019)

Joshuva, A., Sugumaran, V.: A study of various blade fault conditions on a wind turbine using vibration signals through histogram features. J. Eng. Sci. Technol. **13**(1), 102–121 (2018)

Joshuva, A., Aslesh, A., Sugumaran, V.: State of the art of structural health monitoring of wind turbines. Int. J. Mech. Sci. **9**(5), 95–112 (2019)

Kabalci, E., Kabalci, Y.: From Smart Grid to Internet of Energy, 1st edn. Academic Press, London (2019)

Kelley, T.: Symbolic and sub-symbolic representations in computational models of human cognition: what can be learned from biology? Theor. Psychol. **13**(6), 847–860 (2003)

Langley, P.: The changing science of machine learning. Mach. Learn. **82**(3), 275–279 (2011)

Legg, S., Hutter, M.: Universal intelligence: a definition of machine intelligence. Mind. Mach. **17**(4), 391–444 (2007)

Li, R., Gu, H., Hu, B., She, Z.: Multi-feature fusion and damage identification of large generator stator insulation based on Lamb wave detection and SVM method. Sensors **19**(7), 3733 (2019)

Martins, J.: Towards smart city innovation under the perspective of software-defined networking, artificial intelligence and big data. RTIC – Revista de tecnologia da informação e comunicação **8**(2), 1–7 (2018)

Mittrup, I., Smarsly, K., Hartmann, D., Bettzieche, V.: An agent-based approach to dam monitoring. In: Proceedings of the 20th CIB W78 Conference on Information Technology in Construction, Auckland, New Zealand, 23 April 2003 (2003)

Mohanty, S.: Everything you wanted to know about smart cities. IEEE Consum. Electron. Mag. **5**(3), 60–70 (2016)

Mohapatra, B.: Machine learning applications to smart city. ACCENTS Trans. Image Proces. Comput. Vis. **5**(14), 1–6 (2019)

Mohri, M., Rostamizadeh, A., Talwalkar, A.: Foundations of Machine Learning. MIT Press, Cambridge (2012)

Montavon, G., Samek, W., Müller, K.-R.: Methods for interpreting and understanding deep neural networks. Digit. Signal Proc. **73**(2018), 1–15 (2018)

Nazarian, E., Taylor, T., Weifeng, T., Ansari, F.: Machine-learning-based approach for post event assessment of damage in a turn-of-the-century building structure. J. Civil Struct. Health Monit. **8**(2), 237–251 (2018)

Nguyen, V.V., Smarsly, K., Hartmann, D.: A computational steering approach towards sensor placement optimization for structural health monitoring using multi-agent technology and evolutionary algorithms. In: Proceedings of the 6th International Workshop on Structural Health Monitoring, Stanford, CA, USA, 11 September 2007 (2007)

Nomura, Y., Shigemura, K.: Development of real-time screening system for structural surface damage using object detection and generative model based on deep learning. J. Soc. Mater. Sci. **68**(3), 250–257 (2019)

Nosratabadi, S., Mosavi, A., Keivani, R., Ardabili, S., Aram, F.: State of the art survey of deep learning and machine learning models for smart cities and urban sustainability. In: Proceedings of the 18th International Conference on Global Research and Education Inter-Academia, Budapest, Hungary, 4 September 2019 (2019)

Ogie, R.I., Perez, P., Dignum, V.: Smart infrastructure: an emerging frontier for multidisciplinary research. ICE Smart Infrastruct. Constr. **170**(1), 8–16 (2017)

Organisation for Economic Co-operation and Development (OECD): Enhancing the contribution of digitalisation to the smart cities of the future (2019). https://one.oecd.org/document/CFE/RDPC/URB(2019)1/REV1/en/pdf. Accessed 20 Jan 2020

Pan, H., Azimi, M., Lin, Z., Yan, F.: Time-frequency based data-driven structural diagnosis and damage detection for cable-stayed bridges. Journal of Bridge Engineering **23**(6), 04018033 (2018)

PricewaterhouseCoopers: Creating the smart cities of the future (2019). https://www.pwc.com/gx/en/sustainability/assets/creating-the-smart-cities-of-the-future.pdf. Accessed 21 Jan 2020

Ribeiro, M.T., Singh, S., Guestrin, C.: Why should I trust you?: Explaining the predictions of any classifier. In: Proceedings of the 22nd ACM International Conference on Knowledge Discovery and Data Mining, San Francisco, CA, USA, 13 Aug 2016 (2016)

Russel, S.J., Norvig, P.: Artificial Intelligence: A Modern Approach, 3rd edn. Pearson Education Ltd., Harlow (2014)

Salehi, H., Burgueno, R.: Emerging artificial intelligence methods in structural engineering. Eng. Struct. **171**(2018), 170–189 (2018)

Santos, A., Figueiredo, E., Silva, M., Santos, R., Sales, C., Costa, J.: Genetic-based EM algorithm to improve the robustness of Gaussian mixture models for damage detection in bridges. Struct. Control Health Monit. **24**(3), e1886 (2016)

Senniappan, V., Subramanian, J., Papageorgiou, E., Mohan, S.: Application of fuzzy cognitive maps for crack categorization in columns of reinforced concrete structures. Neural Comput. Appl. **28**(1), 107–117 (2016)

Shalev-Shwartz, S., Ben-David, S.: Understanding Machine Learning. From Theory to Algorithms. Cambridge University Press, New York (2014)

Sierra-Perez, J., Torres Arredondo, M.A., Alvarez-Montoya, J.: Damage detection methodology under variable load conditions based on strain field pattern recognition using FBGs, nonlinear principal component analysis, and clustering techniques. Smart Mater. Struct. **27**(1), 015002 (2017)

Smarsly, K., Lehner, K., Hartmann, D.: Structural health monitoring based on artificial intelligence techniques. In: Proceedings of the International Workshop on Computing in Civil Engineering, Pittsburgh, PA, USA, 24 July 2007 (2007)

Smarsly, K., Law, K.H., König, M.: Resource-efficient wireless monitoring based on mobile agent migration. In: Proceedings of the SPIE (Vol. 7984): Health Monitoring of Structural and Biological Systems 2011, San Diego, CA, USA, 6 March 2011 (2011)

Smarsly, K., Law, K.H., Hartmann, D.: A multiagent-based collaborative framework for a self-managing structural health monitoring system. ASCE J. Comput. Civil Eng. **26**(1), 76–89 (2012)

Smarsly, K., Law, K.H.: A migration-based approach towards resource-efficient wireless structural health monitoring. Adv. Eng. Inform. **27**(4), 625–635 (2013)

Smarsly, K., Dragos, K., Wiggenbrock, J.: Machine learning techniques for structural health monitoring. In: Proceedings of the 8th European Workshop on Structural Health Monitoring (EWSHM), Bilbao, Spain, 5 July 2016 (2016)

Soomro, K., Bhutta, M., Khan, Z., Tahir, M.: Smart city big data analytics: an advanced review. Wiley Interdisc. Rev. Data Min. Knowl. Discovery **9**(5), e1319 (2019)

Steiner, M., Legatiuk, D., Smarsly, K.: A support vector regression-based approach towards decentralized fault diagnosis in wireless structural health monitoring systems. In: Proceedings of the 12th International Workshop on Structural Health Monitoring. Stanford, CA, USA, 10 September 2019 (2019)

Suleiman, A.R., Nehdi, M.L.: Modeling self-healing of concrete using hybrid genetic algorithm-artificial neural network. Materials **10**(2), 135 (2017)

Sysyn, M., Gerber, U., Nabochenko, O., Li, Y., Kovalchuk, V.: Indicators for common crossing structural health monitoring with track side inertial measurements. Acta Polytechnica **59**(2), 170–181 (2019)

Tibaduiza, D., Torres Arredondo, M.A., Oyaga, J., Anaya, M., Pozo, F.: A damage classification approach for structural health monitoring using machine learning. Complexity **2018**, 5081283 (2018)

United Nations Economic and Social Council: Smart cities and infrastructure (2016). https://unctad.org/meetings/en/SessionalDocuments/ecn162016d2_en.pdf. Accessed 25 Jan 2020

Vashisht, R., Viji, H., Sundararajan, T., Mohankumar, D., Sarada, S.: Structural health monitoring of cantilever beam, a case study – using Bayesian neural network and deep learning. In: Proceedings of the 13th International Conference on Systems, Athens, Greece, 22 April 2018 (2018)

Vitola, J., Tibaduiza, D., Anaya, M., Pozo, F.: Structural damage detection and classification based on machine learning algorithms. In: Proceedings of the 8th European Workshop On Structural Health Monitoring (EWSHM), Bilbao, Spain, 5 July 2016 (2016)

Vitola, J., Pozo, F., Tibaduiza, D., Anaya, M.: A sensor data fusion system based on k-nearest neighbor pattern classification for structural health monitoring applications. Sensors **17**(2), 417 (2017a)

Vitola, J., Pozo, F., Tibaduiza, D., Anaya, M.: Distributed piezoelectric sensor system for damage identification in structures subjected to temperature changes. Sensors **17**(6), 1252 (2017b)

Zhao, Z., Yua, M., Dong, S.: Damage location detection of the CFRP composite plate based on neural network regression. In: Proceedings of the 7th Asia-Pacific Workshop on Structural Health Monitoring, Hong Kong, China, 12 November 2018 (2019)

Check for updates

Use of Artificial Intelligence in a Regulated Design Environment – A Beam Design Example

Ebrahim Karan(✉), Mahdi Safa, and Min Jae Suh

Department of Engineering Technology, Sam Houston State University, Huntsville, TX 77340, USA
ebrahim.karan@shsu.edu

Abstract. The development of computer design tools in relation to construction management is perceived as a mainstream to automate several construction processes. Automation is not necessarily based on Artificial Intelligence (AI). But if we power it up with data, we can make a machine to automatically mimic and eventually supersede human intelligence. However, until now, not much evident research has been conducted to pursue successful applications of AI in the design. In using AI for design decision making, the intelligent agent should first gather information from the environment and then transform them into internal context. Last, the agent acts on its perceived environment in a way that maximizes its chances of success. To better assess the applicability of AI to the construction management, we classify the design environment into two categories; in a regulated design environment, a collection of written regulations governs the design process. In contrast, the needs and expectations of people are the basis for the input data in a human-controlled design environment. Text analysis is probably the most common way of interpreting design and construction related data. Choosing an appropriate AI technique depends largely on how much we know about the environment and the problem at hand. Search methods and optimization theories seem more suited to solve regulated design problems compared to learning methods. An example of the selection of correct beam size is used to describe the methods for choosing an alternative from a set of acceptable choices (those satisfy the regulations) with the highest possible reward.

Keywords: Artificial intelligence · Design · Construction · Search and optimization methods

1 Introduction

Lately, with the advancement in computing technology and a significant increase in the data volume, there is an added emphasis on the automation and formal reasoning. Automating manual tasks largely began with programming a software program and then creating connections between various software solutions. Artificial intelligence (AI) is taking the automation one step further, through performing high-volume, frequent, computerized tasks with very little human intervention. AI and automation have

E. Toledo Santos and S. Scheer (Eds.): ICCCBE 2020, LNCE 98, pp. 16–25, 2021.
https://doi.org/10.1007/978-3-030-51295-8_2

been used interchangeably in the construction management literature, yet there are different. Automation focuses on performing repetitive, monotonous tasks. Intelligent machines (compared to automated machines) are not driven by the manual configuration. Instead, they mimic human intelligence actions [1].

An intelligent agent has been generally defined as a machine capable of understanding its environment and mimics cognitive functions (e.g. learning, reasoning) of human mind to achieve its goals. Understanding the environment is most of the time done through sensory data, such as image and video (vision sensory modality input), audio (auditory system), and in near future smell (olfactory sensory modality input) and taste. However, conventional ways of gathering information such as questionnaire, surveys, user's assessment and prioritization are still considered of value. After this information is gained, the intelligent agent should transform them into internal context. This step consists of converting data into machine-readable format. Despite the difference among various data processing techniques (e.g. neural network, linear filtering, or feature matching for processing images), they ultimately transform the data in quantitative form (e.g. assigning greyscale values to an image) which allows quantitative analysis in a formal and rigorous fashion. Last, the agent acts on its perceived environment in a way that maximizes its chances of success. AI techniques and algorithms play a significant role in taking actions to achieve a predefined goal [2]. The authors have briefly explained these techniques in [3] and readers are referred to this study for details regarding AI techniques to support design and construction. The purpose of this introduction is to give an overview of AI. Some examples of AI in design and construction are discussed in the next section to elaborate AI definition.

2 Examples of AI Applications in Construction Management

AI applications in construction management are virtually unlimited, but only three examples for design, planning, construction are presented here.

In architectural design, the action of an intelligent agent is to provide the client with various design solutions and the goal is to modify or alter their design solutions incorporating the feedback until the client is satisfied. In this application, the design environment is the interaction between the client and the design. This interaction can be recorded conventionally, for instance by asking the client to evaluate each design using a Likert-style rating scale, or by using advanced sensors such as eye tracking, facial expressions, or electrocardiogram. Next, a decision-making technique shall be used for modeling the consequences of actions and solving architectural design problems. In a recent study, a window design scenario was formulated as a Markov decision process where participants evaluated the quality of each design solution based on their preference by giving it a score [4]. The goal of the intelligent agent in this scenario was to determine which decision to make so that the score differences between two consecutive designs is maximized.

Scheduling in construction management is to determine when activities will take place depending upon their durations and resource requirements. An intelligent agent can consider millions of alternatives for project delivery and continuously update durations, resource availability, and time constraints [5]. In this application, the planning environment contains dependencies between tasks, availability and productivity of the appropriate resources, updates on the project's development (e.g. daily reports from the field), and even weather conditions. This raw data should be turned into information a machine algorithm can use. The goal of the intelligent agent would be to reduce construction project delays or complete the project at minimum cost without violating any project rules or contract specification. Most likely a learning algorithm will be used to achieve the goal, by trying different alternative and learn what improved results (reduce time or cost), and what worsened results, in an effort to reduce error (differential outflow).

3D printing offers a quick tool for architects and construction companies. The addition of AI to 3D printing optimizes the printing process, corrects errors in real-time comprehensively, and predicts the behavior of materials and offers more accurate final products. For example, 3D concrete printing supported by AI can optimize and change the concrete mix in real time to meet the project conditions (e.g. temperature, concrete thickness) [6]. In this application, as-designed drawings and jobsite conditions (weather, surface condition, humidity) are considered the construction environment. The goal of the intelligent agent would be to reduce material waste and meet the design specifications (e.g. compressive strength of concrete). Search methods and optimization theories would be appropriate for such applications when the problem is solved by the selection of the best action (with regard to some constraints and criteria) from a set of alternatives. Of course AI applications in design and construction are not limited to these examples and many other applications (e.g. AI controls sensors in smart homes) can also be used. Salehi and Burgueno [7] provided a comprehensive review of emerging AI methods in structural engineering for purposes such as structural health monitoring and damage detection.

3 Research Objective and Scope

The objective of this research is to demonstrate the potential for the use of AI in design through the development of an intelligent agent capable of making design decisions on its own. The intelligent agent and its difference from automated design systems, as well as the work scope should be clearly defined. In order to be an intelligent, the proposed agent should be able to perceive its environment and act upon that environment through some mechanisms [8]. The two keywords here are "environment" and "mechanisms". Figure 1 describes a conceptual view of environments during a project lifecycle. The type of data perceived from an environment varies among different phases of the project (planning, designing, constructing and operating), but we can generally categorize an environment into design (when only the planning and design of the project is carried out) and construction (when physical construction begins and then continues to the operation and maintenance of the built facility). This study focuses only on the regulated design environments, and does not address other environments. In a regulated design environment, the actions (or decisions in this case) are governed by a collection of regulations, standards and guidelines (e.g., building codes, safety requirements, and loads). Therefore, the client or architect's attitude and desire toward functional clues and aesthetic forms are not considered at all unless they are specified in the regulations.

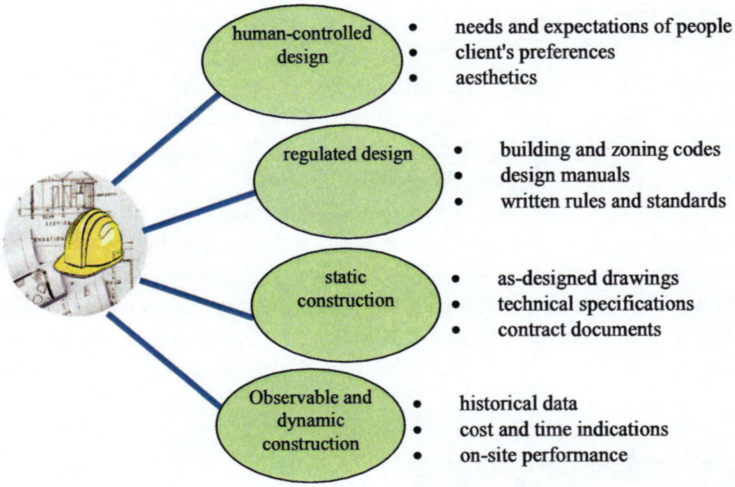

Fig. 1. Interrelated AI environments to support design and construction

It is essential to convert the data gained from the environment into a machine-readable data set. Generally speaking, data in a regulated environment may take the form of text, tables, or figures. A review of some of the most common technical standards such as the International Building Code (IBC) and those published by the American Association of State Highway and Transportation Officials (AASHTO) indicates that textual data is the predominant form of data presented in these standards and regulations. The raw to machine-readable data conversion reported herein is only limited to text analysis. An example of the selection of correct beam size is used to describe the methods for choosing an alternative from a set of acceptable choices (those satisfy the regulations). Using AI, computers can be trained to accomplish specific design tasks by processing large amounts of textual data without fatigue.

An intelligent agent is not necessarily an autonomous agent operating on a designer's behalf without any interference. Instead, an intelligent agent employs some knowledge or learns from experience to make some design decisions with some degree of independence or autonomy. The agent uses guidance mechanisms to change part of the environment in order to achieve its programmed goals. These mechanisms are AI techniques and algorithms when decisions about the designs are made and are actuators converting decisions into actions. Arms, lights, valves, or speakers are all actuators of a machine or a robot. This study focus on a design environment, thus the focus is on AI techniques and algorithms. To choose the right algorithm (or technique), we should first categorize the input and then understand the nature of the decision problem at hand. There are numerous mechanisms available for solving regulated design problems (supervised learning, reinforcement learning, etc.), a beam design example is used in this paper to show how to choose the right algorithm and categorize the problem.

4 Beam Design Scenario

The project consists of building an addition to an existing structure. We are focusing on the design of the beam for the new roof. The new gable roof will be perpendicular to the existing gable roof (see Fig. 2). Because the framing for the addition will overlay the existing roof, a structural engineer should design a beam (rafter or truss) to support the weight of the new roof. The top priority is not adding more weight to the exiting roof. The new beam or rafter will be set on the addition walls or columns (posts) up to the point of the eave of the existing roof. This structural element (shown in Fig. 2) is generally treated as a cantilever, which is supported on only one end and the other end is exposed beyond the support. The proposed intelligent agent comes into play when the architectural design is complete and the form and shape of the new roof is determined.

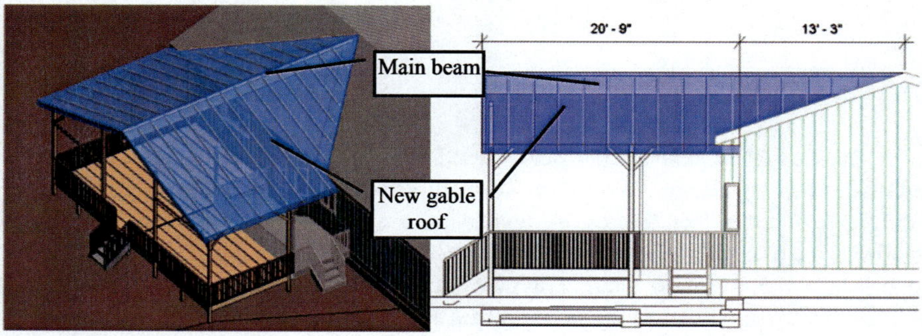

Fig. 2. 3D (left) and elevation (right) views of the beam design scenario

Design problems can be solved by an engineer with proper training. Similarly, AI can be used to design a structure based on known properties learned from the training data. Thus we shall train an intelligent agent at various stages to acquire data, discover its decision boundary, and make it clear about the goal. In the present scenario, the related 2018 International Residential Code (IRC) and the 2018 IBC requirements for residential roof and structural design are used. Because of the data type, a text analysis architecture is needed to learn the feature representation of the input data. In the next section, the use of text analysis is outlined. Since there are many ways to achieve the final design, it is important to train the agent such that it can clearly evaluate each design solution and determine whether the solution is improved or not (e.g. is the new solution lighter than the previous one?). With all these combinations of beam rafters and roof trusses, the model may be able to consider literally thousands of different design solutions that meet the code requirements. It is essential for the agent to learn which design solution is preferred and what actions to take to improve the design solution. The role of states (condition of the design at every stage) and possible actions that the agent can execute (i.e. list of decisions) are discussed in Sect. 6.

5 Text Analysis for the Proposed Scenario

Text analysis is the automated process of analyzing and interpreting text data such as those presented in building codes [9]. In this study we train the intelligent agent to understand how to analyze, understand, and derive meaning from building codes and standards by tagging examples of text. With enough samples of tagged text, the agent will be able to differentiate between topics and codes. Text extraction is first used to extract important data such as keywords. In the present scenario, we train the agent to index the keyword "roof" and generate a tag cloud. Only in IBC alone, more than 180 "roof" keywords were extracted. Next, a concordance tool is used to identify the preceding and following contexts in the concordance of the keyword "roof". The concordance tool is used along with a sentiment analysis to clean regular expressions and filter out most of the unwanted texts. Table 1 shows the concordance of the keyword "roof" in IBC 2018. Note that only frequent contexts are listed in Table 1.

Table 1. Result of the text analysis (concordance and sentiment analysis) for the keyword "roof" (an example).

Target	Preceding context	Count	Following context	Count
Roof	Flat	18	Live load	35
Roof	Live load	12	Snow load	14
Roof	Support/Members	8	Members	9
Roof	Pitched	7	Load	7
Roof	Area	6	Area	6

Once we have identified, extracted, and filtered out the contexts relevant to our project (e.g. roof), the next step is to have an understanding of that context. In the beam design scenario, the roof load seems to be the essential context as it appeared frequently with the keyword "roof" in the document. Because the IBC document is written in English (a natural language), we need to rely on text mining and natural language processing (NLP) techniques to find the relevant section and other clues for the design process. At this stage, we provide a general understanding of the phrases and words, such as roof type, ordinary roof, green roof, etc. from individual phrases or sentences in the relevant code section. Once the agent is able to get insights from the text, it can move on to the next design step, load analysis. Here, we should feed the agent with building's information such as roof surface, slope, location of the project to find the base snow and wind loads (or directly provide them). These steps are summarized in the flowchart shown in Fig. 3. The building's information is usually available in the CAD drawings, however, many architectural design tools also deliver textual data. For example, building information modeling (BIM) tools use an interoperable format for the uniform representation and exchange of project information in a plain text form called Industry Foundation Class (IFC). Text analysis can be used to capture the contextual information of each IFC element, as well as attributes and properties [10]. Now it is time to choose an appropriate AI technique to select the beam type.

6 Choosing the AI Technique for the Proposed Scenario

Choosing an appropriate AI technique depends largely on how much we know about the environment and the problem at hand. We categorized the problem in Sect. 4 and then showed how to categorize the input in Sect. 5. The input is a labeled data thus we are dealing with a supervised learning problem. After categorizing the problem and understanding the design environment, the next step is to choose the AI applicable and practical to implement in a reasonable time. Accuracy of the data, time needed to build, train, and test the agent, and problem complexity are some of the key elements affecting the choice of a technique.

Search methods and optimization theories seem more suited to solve regulated design problems compared to learning methods. Search methods can be generally divided into blind (exhaustive) methods and informed (heuristics) methods. Blind search is used when no information regarding the problem is available. In contrast, when we know how to direct search to seek a goal and domain-specific knowledge is available (like the beam design scenario), an informed search is used. Some of the most common informed search methods include bandwidth search, best first search, branch and bound algorithm, greedy search, and hill climbing. Regardless of the informed search methods, the following components should be determined or defined:

- States: description of the objective situation (e.g. design a beam in compliance with codes)
- Search Space: The set of possible situations, a beam that meets the size requirement and carry the design loads.
- Search Path: Sequence of states the intelligent agent actually visits.
- Solution: Design solutions that meet the requirements.
- Strategy: How to choose the next design (e.g. minimum cost, reduce weight, continue with a design solution or consider another one).
- Path cost: Assigning a numeric value to each path.

The beam design scenario can be modelled as a hierarchy of linked design solutions in a tree structure. Figure 3 shows the root node (type of the horizontal structural member) representing a particular state corresponding to the initial state. A greedy algorithm is chosen for the beam design scenario because it takes all of the data gained from the previous steps, and then sets a rule for which changes to make to the design solution at each step of the search. Basically it takes the closest node to the goal state and continues its searching from there [11]. For example, the 34 ft long span makes the *truss* members the closest node to the goal state. Thus, the search continues from there and two basic types of truss forms (i.e. pitched truss characterized by its triangular shape, and parallel chord truss, or flat truss) are evaluated.

Each design solution is evaluated based on the required material presented in board feet. Although this cost factor does not guarantee to find the solution with the minimum cost (e.g. the cost of flat truss can be higher because due to the use of steel bracing for joints), but the use of greedy search guarantees to find the best solution when several solutions exists. The advantage of a greedy algorithm over other informed search methods is its efficiency and simplicity. By dividing the design problem recursively based on the roof condition, there is no need to combine all design solutions. For instance, taking into account all possible values for overall truss depth, with, and chord (top or bottom) sizes only for flat trusses will result in around 15 × 6 × 16 = 1440 design solutions, not to mention the type of lumber used for the member (e.g. Spruce-Pine-Fir, Douglas Fir lumber) and a different size for the web. Instead of evaluating thousands of design solutions, a greedy algorithm can find design solutions with less than a thousand iterations. In the proposed intelligent agent, we determine the following search path, each criterion is placed on one node:

Pitch >> overall truss depth (four levels: 12–16 in, …, 24–30 in) >> width (two levels: 6 in or less,…) >> chord for top and bottom (two levels: similar or different size) >> web side >> material.

All these nodes are shown as one single block in Fig. 3. The proposed beam design for this project is found to be a 34 ft Douglas Fir flat truss, with overall truss depth of 18 in, and the width equals to 4 in, 2 × 6 lumbers for both top and bottom chords, and 2 × 4 lumbers for the web. Eighteen other design solutions were found after 576 iterations.

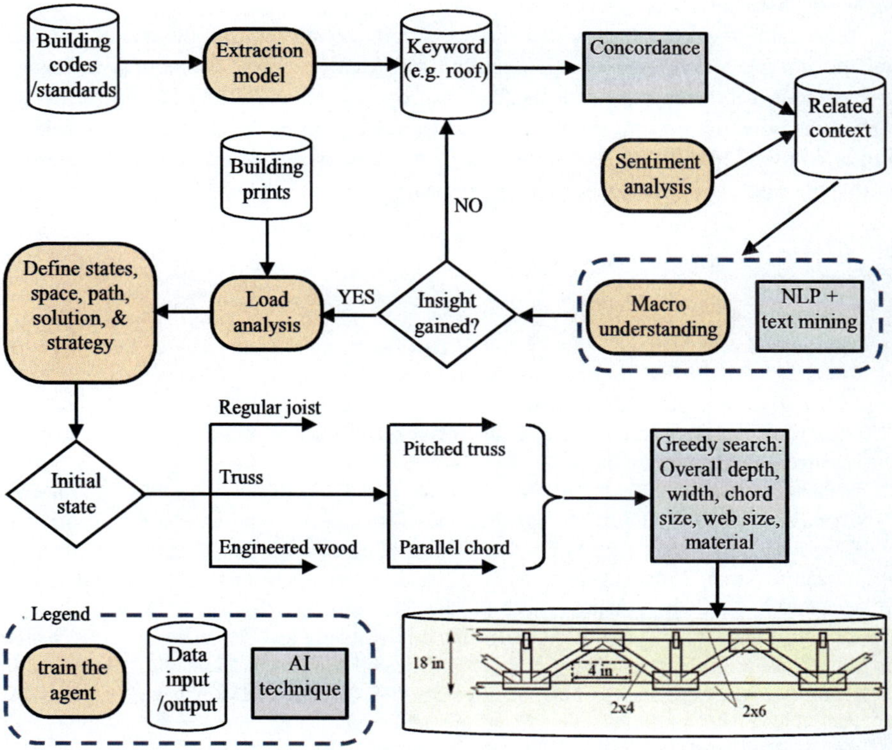

Fig. 3. Framework for the proposed intelligent agent (beam design scenario)

7 Conclusions and Future of AI in Construction Management

The analysis of a large amount of data and complexity of design decisions indicate the need for more intelligent tools. In this paper, an intelligent agent is explained through its interaction with a design and the way it changes the environment by doing something. We showed this interaction based on a perception-decision-action design scenario. Text analysis was used in the study for gathering information from the environment and transforming them into internal context. The actions (or design decisions) were formulated as a sequence of independent decisions using a greedy search method. Design solutions were observed (or evaluated) by the agent at consecutive intervals called states. Linking design and construction data together can build a more efficient approach, allowing broader context to be used within scenario decision support.

Despite the fear that AI systems are predicted to take over massive jobs, in the author's view, intelligent agents are unlikely to replace the human workforce. Instead, they will be able to reduce human errors and make design and construction operations more efficient. A secondary objective of this study is to educate managers at design and construction companies about the potential investment areas where AI has the greatest impact on their business's needs.

In addition to AI techniques, we should take into consideration the use of robotics and the internet of things to further reduce construction costs. In contrast to design environment, robots can use sensors (e.g. microphone, camera) to gather information from their construction environment as the project progresses. These robots can be supported by AI to track the real-time interactions of workers, equipment, and materials on the site and alert construction errors and safety uses.

References

1. Rouse, W.B., Spohrer, J.C.: Automating versus augmenting intelligence. J. Enterp. Transform., 1–21 (2018). https://www.tandfonline.com/doi/citedby/10.1080/19488289.2018.1424059?scroll=top&needAccess=true
2. Oprach, S., et al.: Building the future of the construction industry through artificial intelligence and platform thinking. Digitale Welt **3**(4), 40–44 (2019)
3. Mohammadpour, A., Karan, E., Asadi, S.: Artificial intelligence techniques to support design and construction. In: Proceedings of the International Symposium on Automation and Robotics in Construction (ISARC), Banff, Alberta, Canada. IAARC Publications (2019)
4. Karan, E., Asadi, S.: Intelligent designer: a computational approach to automating design of windows in buildings. Autom. Constr. **2019**(102), 160–169 (2019)
5. Liu, N., Kang, B.G, Zheng, Y.: Current trend in planning and scheduling of construction project using artificial intelligence. In: IET Doctoral Forum on Biomedical Engineering, Healthcare, Robotics and Artificial Intelligence 2018 (BRAIN 2018), Ningbo, China (2018)
6. Paul, S.C., et al.: An artificial intelligence model for computing optimum fly ash content for structural-grade concrete. Adv. Civil Eng. Mater. **8**(1), 56–70 (2019)
7. Salehi, H., Burgueno, R.: Emerging artificial intelligence methods in structural engineering. Eng. Struct. **2018**(171), 170–189 (2018)

8. Russell, S.J., Norvig, P.: Artificial Intelligence: A Modern Approach, 3rd edn. Pearson (2019)
9. Zhang, R., El-Gohary, N.: A machine learning approach for compliance checking-specific semantic role labeling of building code sentences. In: Proceedings of the 35th CIB W78 2018 Conference: IT in Design, Construction, and Management, Chicago, IL, USA. Springer (2018)
10. Karan, E., Irizarry, J., Haymaker, J.: Generating IFC models from heterogeneous data using semantic web. Constr. Innovation **15**(2), 219–235 (2015)
11. Chandel, A., Sood, M.: Searching and optimization techniques in artificial intelligence: a comparative study & complexity analysis. Int. J. Adv. Res. Comput. Eng. Technol. (IJARCET) **3**(3), 866–871 (2014)

An Interview-Based Method for Extracting Knowledge of Skilled Workers at Construction Sites Using Photographs and Deep Learning

Yuichi Yashiro[1(✉)], Rikio Ueda[1], Fumio Hatori[1], and Nobuyoshi Yabuki[2]

[1] Hitachi Plant Construction, Ltd., Toshima-Ku, Tokyo 170-8630, Japan
yuichi.yashiro.kp@hitachi.com
[2] Osaka University, Suita, Osaka 565-0871, Japan
yabuki@see.eng.osaka-u.ac.jp

Abstract. In the Japanese construction industry, the number of skilled workers has been decreasing year by year, and a large number of skilled workers will retire in the near future. Furthermore, at construction sites, education is carried out on-the-job-training (OJT) basis in the local environment of the site. Therefore, it is necessary to immediately establish a mechanism to effectively transfer knowledge before skilled workers retire. The first problem is the large amount of data that skilled workers have personally over many years, and it is difficult for them to organize manually. In this research, a system was developed for automatically classifying and extracting a large number of photographs. In this system, object detection with transfer learning is used. As a result of applying it to the special equipment of a nuclear power plant, the F-measure achieved 89%, and the time required for searching photographs was significantly reduced. The second issue is tacit knowledge in the brain of the expert. In general, it is possible to extract knowledge by interviewing experts. In this research, we developed a support system to extract tacit knowledge efficiently and adopted a method for conducting interviews effectively, which called the functional approach (FA) and the semi-structured interview (SSI). By applying FA, we can replace work-related things with "functions" and make many hypotheses for conducting SSI. As a result, the new method improved the time efficiency of interviews by 77.1% and increased the exhaustiveness (the number of knowledge/work step) by 2.9 times.

Keywords: Expert knowledge · Deep learning · Object detection · Transfer learning · Functional approach · Semi-structured interview

1 Introduction

The situation in the Japanese construction industry is severe with the number of skilled workers having decreased by 25% in the past 20 years, the number of elderly people having increased (age 55 or older account for 30%), and the number of young people having decreased (age 29 or younger account for 10%). Furthermore, in the future, the number of skilled workers is expected to decrease by 37% in 2025 (compared to 2014),

E. Toledo Santos and S. Scheer (Eds.): ICCCBE 2020, LNCE 98, pp. 26–40, 2021.
https://doi.org/10.1007/978-3-030-51295-8_3

and there is a shortage in the supply of human resources compared with other industries such as the manufacturing industry. In addition, following the 2007 problem with the retirement of the baby boomers, the mass recruitment generation (joining the company from around 1987 to 1991) is expected to retire in large numbers from 2026 to 2029.

Meanwhile, looking at the situation of the Japanese nuclear power plants covered in this research, due to the accident at the Fukushima Daiichi Nuclear Power Plant caused by the Great East Japan Earthquake in 2011, new construction projects in Japan have been suspended, existing plants are being checked for decommissioning, and compliance with new regulatory standards and periodic inspections have not been conducted for nine years. Therefore, the opportunity to transfer the know-how of experts involved in the construction and periodic inspections of nuclear power plants for a long period has been lost. Considering this market environment, while a large number of experts, who shall be forced to retire within a few years, work for companies, a framework for transferring the know-how cultivated so far to young people is urgently required. However, the effective transfer of know-how gives rise to various problems.

The first problem is that there is a lot of knowledge that exists personally in the brain of the expert, so-called "tacit knowledge," and it is difficult to extract such knowledge effectively. Education at the sites is mainly carried out on an on-the-job-training (OJT) basis, and the know-how is transferred from experts to young people by sharing knowledge through on-site work. Therefore, tacit knowledge is only shared within one site. A series of methods has not been established to externalize the tacit knowledge from the expert's brain, to organize it systematically as "explicit knowledge," to share the know-how among multiple organizations and to finally ensure that many on-site non-skilled workers master the know-how. An enormous amount of tacit knowledge exists in the brains of supervisors who were involved in the rush to build nuclear power plants between 1970 and 1980, and in the subsequent periodic inspections. It is urgently necessary to turn such tacit knowledge into explicit knowledge and accumulate it as digital data before the experts retire.

The second problem is the vast amounts of past data (photograph data, documents, drawings, etc.) possessed by experts. These data are full of know-how, but most of them are personally managed in a closed site environment and will be discarded when the experts retire. The vast amounts of data accumulated for many years are often left buried because it is a very time-consuming task to organize them, even for experts. Therefore, a mechanism that can efficiently extract the buried data is required.

In this research, we propose a method for efficiently and exhaustively extracting the tacit knowledge existing in the brains of experts using a semi-structured interview method and a functional approach, and a method for automatically and highly accurately extracting the vast amounts of past photograph data possessed by experts using object detection technology through deep learning and transfer learning.

2 Literature Review

The International Chamber of Commerce (ICC), in its "Know-how Protection Standard Clause" created in 1960, stated that "Know-how, alone or in combination, refers to the secret technical knowledge, experience, and their accumulation that are needed to complete and actually apply certain technologies useful for industrial purposes" [1]. In addition, Nonaka et al., focusing attention on the productivity of organizational knowledge in Japanese companies, stated that "the skills and techniques of organizational knowledge creation are the biggest success factors for Japanese companies," and summarized the advantages of Japanese companies in an easy-to-understand concept [2]. Figure 1 shows the "SECI" model, which is the basic theory of knowledge management proposed by Nonaka et al. SECI stands for "Socialization," "Externalization," "Combination," and "Internalization," which explains how knowledge-creating companies that have achieved excellent results by sharing and utilizing knowledge are generating organizational knowledge.

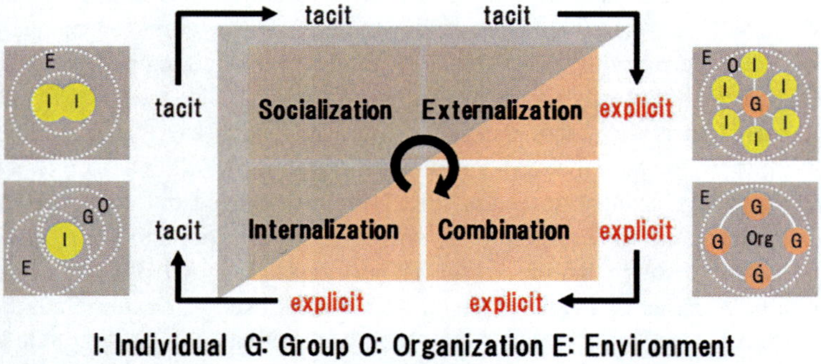

- Socialization: Acquires common tacit knowledge by sharing tacit knowledge within each individual.
- Externalization: Creates the concept and iconography of knowledge in an organization and turns them into explicit knowledge in a form that can be shared by groups.
- Combination: Combines explicit knowledge and systemizes it as knowledge that can be used between organizations.
- Internalization: Each individual acquires systemized knowledge through actions.

Fig. 1. SECI model

Generally, knowledge can be broadly divided into tacit and explicit knowledge. Polanyi defines tacit knowledge as "knowledge that cannot be explained in words" and explicit knowledge as "knowledge that can be explained in words" [3]. Tacit knowledge is an empirical rule derived from the knowledge and practice obtained from personal past experience and is accumulated personally in the brain. On the other hand, explicit knowledge is the documentation of the information and knowledge known only to individuals, and is standardized and expressed on paper or electronic media. As mentioned in the introduction, tacit knowledge often exists personally and subjectively in the brains of experts at construction sites, and thus the tacit knowledge must be turned into explicit knowledge to document the knowledge in a form that can be understood in a standard and objective manner.

There are two methods for extracting tacit knowledge in the brains of experts. One is a registration method by individuals. There is a mechanism (system) for registering tacit knowledge, and experts actively use the mechanism to register their own tacit knowledge. This is an effective method when there are many experts because tacit knowledge naturally accumulates by defining the system and operation method. Several studies have been reported on the methods of registering know-how by Web-based systems in ongoing construction projects [4–7]. Also, studies have been reported on efficient extraction by setting incentives and penalties for registrants [8]. However, when there is no enforcement, the registration is not actively performed, so that the method by individuals is disadvantageous in that the know-how cannot be exhaustively extracted.

The other is an interview method. A third person asks questions to the expert, and the expert answers the questions, thereby extracting the tacit knowledge existing in the brain of the expert. It is a method that is widely and generally used, and it is possible to extract tacit knowledge exhaustively, including deep knowledge outside the consciousness of the expert by interviewing dialogically over time. However, the disadvantage is that the interview needs to be conducted by a small number of people, and it takes a lot of time to extract tacit knowledge from many experts.

The interview method is broadly divided into a structured interview, unstructured interview, and semi-structured interview. The structured interview is a method of obtaining answers using a questionnaire sheet in which predetermined questions are structured. The unstructured interview is a method of asking questions freely and getting answers through dialogue. The semi-structured interview is an intermediate method, in which rough question items are determined beforehand, and further details are deepened according to the answers of the respondent. The method to be applied varies depending on the application, but there have been some reports of the use of semi-structured interviews to exhaustively extract the know-how from experts [9–11].

As a method of extracting data possessed by experts, there is a method of automatically extracting the data using machine learning. In particular, for photographs, a method of using Deep Learning (DL) is widely used. Until now, for photograph classification, a method has been generally used in which the feature amount detectors for the objects to be classified are designed and matched by humans, but by learning a huge number of annotation photographs using DL, it becomes possible to automatically create the feature amount detector, and recognition accuracy is significantly improved. This sparked the third AI boom, and at the Large Scale Visual Recognition Challenge (ILSVRC) in 2015, AI proved to be recognizable with accuracy beyond the human eye (AI misrecognition rate: 3.6%, Human misrecognition rate: 5.1%).

In photograph recognition by DL, a model called the Convolutional Neural Network (CNN) has become the mainstream at present. The convolution layer outputs a new feature map by performing convolution with filters on an arbitrary range of the input data, and the pooling layer compresses the data by performing sub-sampling to leave only statistical aggregate values. Finally, the output label is predicted by linear weighting using the fully connected layer. In the learning phase, a large number of prelabeled photographs are prepared and input to the CNN and the back-propagation process is performed to minimize the error between the output feature amount and the correct label. By repeating this, the weight of each neuron in the model is repeatedly updated. Finally, when an unknown photograph is input using the weight map obtained at the stage of learning until the loss function representing the error is minimized, the correct label can be predicted together with the correct probability based on the output feature amount.

In photographs taken at the actual site, various objects are reflected, and thus a method called Object Detection is widely used. This method makes it possible to identify where in the photograph and what is reflected in it, not in the entire photograph. In recent years, this technology has attracted attention, and many cases of its application to construction sites have been reported [12–14].

3 Proposed Method

3.1 Overview of the Proposed System Architecture

Figure 2 shows the overall architecture of the proposed system.

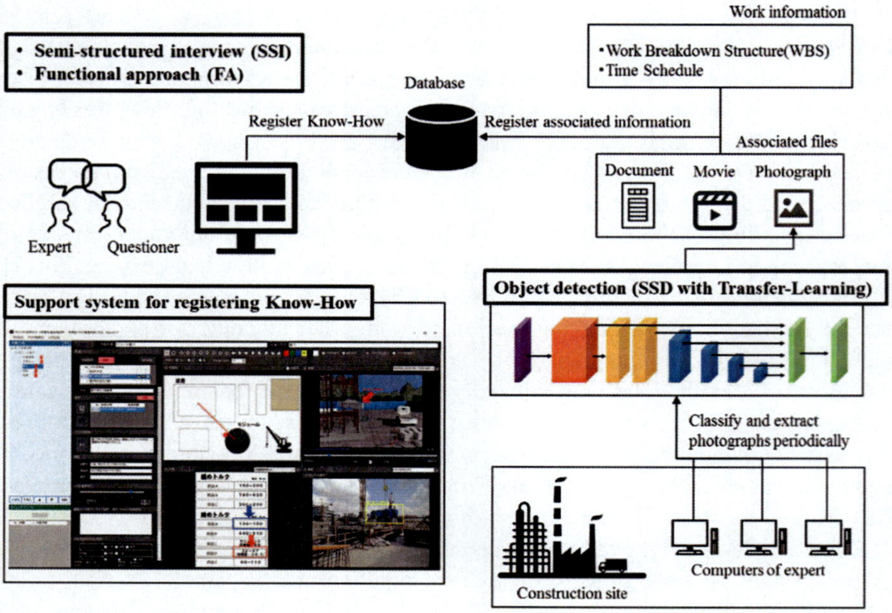

Fig. 2. System architecture

Centering on the database, the past associated files (documents, photographs, movies) and work information are linked to each other and registered before interviewing. In particular, for photographs, a method called SSD (Single Shot Multi-box Detector), which is one of the object detection methods, is adopted. In this method, a huge number of photographs stored in a computer of the expert at the construction site are automatically associated with each work step, classified, extracted, and registered. In SSD, recognition accuracy is being improved by using Transfer Learning. This method makes it possible to automate the work in which a large number of past photographs were classified by the human eye, thereby greatly reducing the work hours.

Next, during the interview, the questioner and the respondent browse the registered information associated with each work step on the support system for registering know-how and extract the know-how for each work step using the Functional Approach (FA) and Semi-Structured Interview (SSI). The extracted know-how is registered in the database from the system. By using the support system for registering know-how and methods such as the FA and SSI, the questioner can efficiently and exhaustively extract the know-how from the expert (respondent) as compared with the conventional interview.

3.2 Functional Approach (FA) and Semi-structured Interview (SSI)

The Functional Approach (FA) was developed in 1947 by Lawrence D. Miles of GE, and has been proposed as Value Engineering (VE) [15]. This is an approach of thinking by replacing objects and things with the concept of functions without being bound by preconceptions or stereotypes in solving problems. In FA, a technique called the Function Analysis System Technique (FAST) can decompose a form into functions that are represented by a functional system diagram systematizing the relationship between the purpose and means. Changing the factor of the problem to "for what?" instead of "why?" makes it possible to create a future-oriented means to achieve its true purpose. Figure 3 shows an example developed for a functional system diagram and SSI.

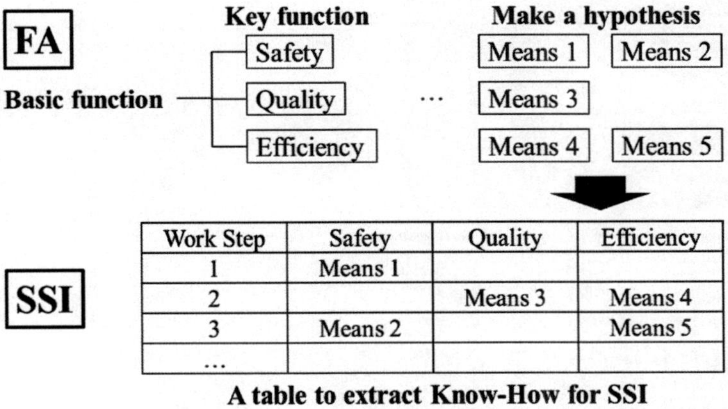

Fig. 3. Example of using FA and SSI

First, a basic function for accomplishing a certain work is defined based on the concept of FA, and then decomposed into key functions such as safety, quality, and efficiency. Furthermore, by digging into each key function, a hypothesis is made considering what are the means (assuming the know-how possessed by the expert) to satisfy each function. Then, each hypothesis is assigned to the corresponding work step of the target work, and a know-how extraction table for SSI is created. By performing

SSI in such a structured state, the know-how can be extracted efficiently and exhaustively without omission. In SSI, the answer to the hypothesis can be derived in a simple manner, but the result of the answer can be used to dig deeper and extract deeper tacit knowledge.

3.3 Support System for Registering Know-How

We have developed a support system for registering know-how for SSI (see Fig. 4). When the target project is opened, the work steps are displayed on the WBS (Work Breakdown Structure) panel, and when an arbitrary work step is selected, associated contents (maps, movies, photographs, documents) registered in advance are displayed. The questioner and expert perform SSI while browsing these contents. When the know-how has been extracted, it is registered on the know-how list panel. Necessary items such as the subject, details, importance, and experience are registered on the know-how list. Further, an icon associated with the know-how is registered on each content. For example, a place where know-how exists on a map, and a range or point where know-how exists in a photograph or a movie are registered with icons of various shapes. In this way, by linking the know-how with past data and recording it in a database as digital data, the know-how can be visualized in a form that anyone can easily understand and reuse.

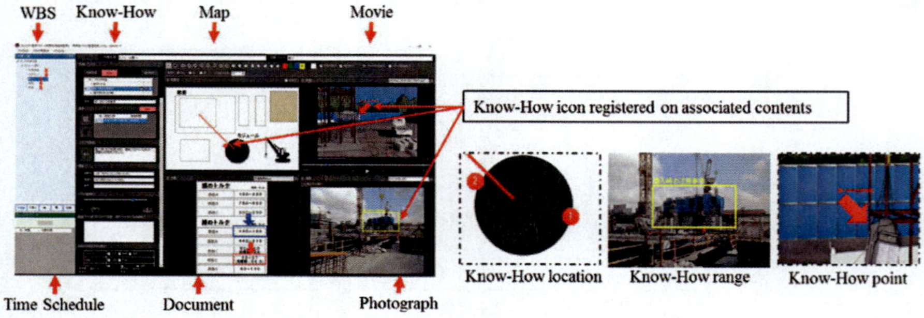

Fig. 4. Support system for registering know-how

3.4 Object Detection (SSD with Transfer Learning)

Figure 5 shows the outline of the SSD (Single Shot Multi-Box Detector) and transfer learning adopted this time. The CNN model of the base network uses VGG-16. The Extra Feature Layer after the base network can extract the feature amount by dividing the photograph step by step with boxes of various scales and aspect ratios, and the performance can be obtained even with a relatively low-resolution photograph. Also, since learning and object detection can be processed from input to output by the end-to-end deep learning, high-speed processing is possible.

First, based on many known photographs, many annotation data defining where (at which coordinates on photographs) and what (which labels) are present are created. All the annotation data are input by dividing the training data and the evaluation data at a

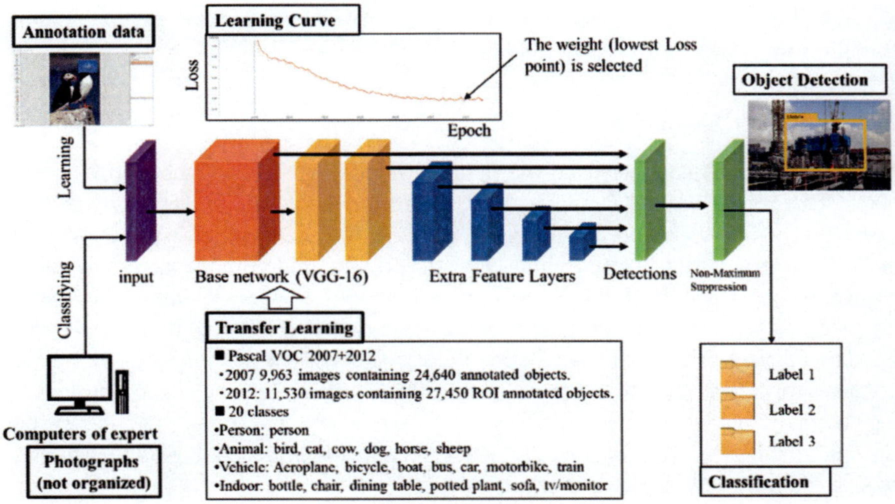

Fig. 5. SSD with transfer learning

certain ratio, and learning is repeatedly performed. At this time, in the base VGG-16, transfer learning is performed using weight data that have been learned in advance in 20 classes of Pascal VOC as initial values. As the learning proceeds normally, the loss value of the learning curve gradually decreases, and when the loss is minimized, weight data at that time are selected as the optimal learning state. By using a weight file that has already been learned by transfer learning, re-learning can be performed with various feature amounts already learned, so that learning progresses faster than learning without anything, and further, the loss can be reduced.

Once the optimal weight data have been obtained, the process is performed on a group of unorganized photographs on the computer of the expert that contains a huge number of unknown photographs (unorganized). This makes it possible to automatically classify the photographs containing the contents defined by each label by object detection. The classified photographs are extracted from each folder and registered in the database on the support system for registering know-how. Furthermore, the classification/extraction processing program is connected to each site in the network and periodically executed in the background, so that the latest classified photographs can be automatically obtained from all computers in the organization.

4 Results

Two types of experiments were performed. First, by using the support system for registering know-how, FA, and SSI, we evaluated how efficiently and how exhaustively the know-how could be extracted as compared with the conventional know-how extraction method by the interview. Second, by using the SSD with transfer learning, we evaluated the recognition rate at which the know-how could be classified, and how efficiently the classification was performed as compared with the conventional manual classification work.

4.1 Interview Experiment

The interview was conducted on the periodic inspection work of the nuclear power plant. Three special works (Work A, B, C) included in the inspection work were selected. The interview was conducted for the expert in each special work. In Work A, the interview was conducted without any preparation, and without using any of the system, FA, and SSI. In Work B, the system was used. In Work C, all of the system, FA, and SSI were used. Figure 6 shows the situation when SSI was performed (Work C). One monitor displays the simulation questions that were hypothesized by FA and prepared for SSI at each work step. The other monitor displays the support system for registering know-how. The interviewer asks the expert questions, and the system operator displays the past data and registers the know-how depending on the situation so that the interview can proceed smoothly.

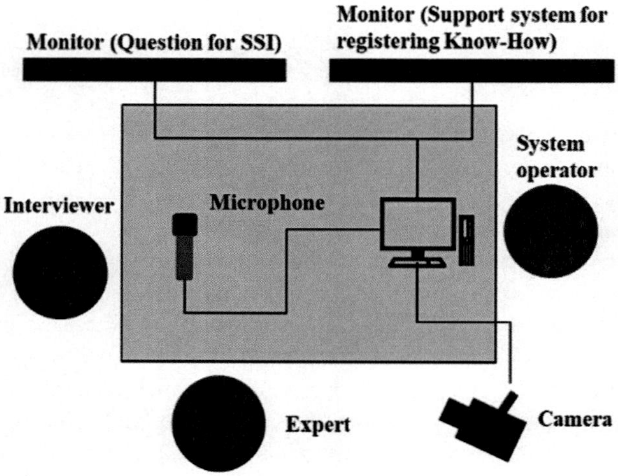

Fig. 6. The situation when SSI was performed

Two evaluation indexes were defined for the experiment. The first index was defined as "extraction time per know-how," which represents "time efficiency". The second index was defined as "the number of extraction know-hows per work step," which represents the "exhaustiveness".

Figure 7 shows the experimental result. The total interview time and the number of extractable know-hows vary depending on the work steps that differ for each Work, on whether the system was used, and on whether FA and SSI were applied. At this time, the interview time in Work C includes the preparation time for utilizing FA and SSI. Compared with the Work A using neither the system nor the interview method, Work B using the system reduced the know-how extraction time from 14.8 min (322 know-hows in 79.5 h) to 6.3 min (237 know-hows in 25 h). As a result, an efficiency improvement of 57.4% was obtained. Furthermore, compared with Work B, Work C using FA and SSI improved the number of extracted know-hows for one work step from 3.08 (237 know-hows for 77 work steps) to 4.31 (474 know-hows for 110 work steps). This index indicating the exhaustive extraction was improved 1.4 times. Ultimately, compared to Work A, Work C improved the time efficiency by 77.1% and increased the exhaustiveness by 2.9 times.

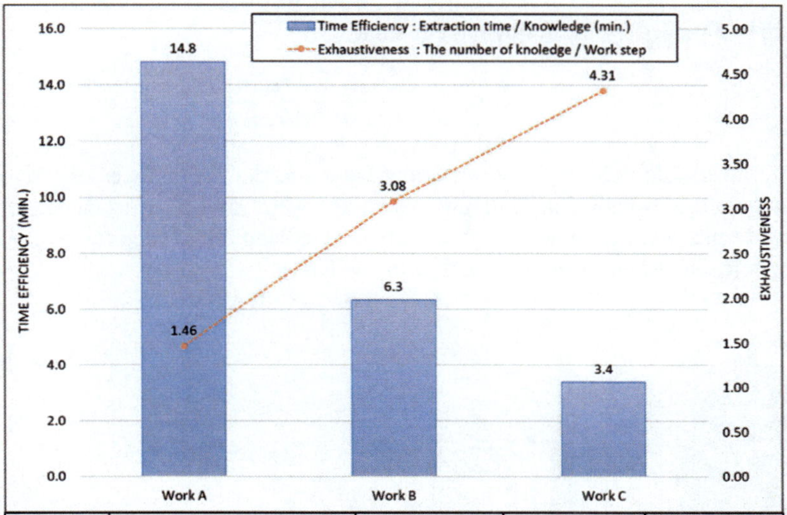

		Work A	Work B	Work C
Experimental Condition	The number of work step [A]	220	77	110
	Support System	False	True	True
	FA & SSI	False	False	True
Result	Interview time (min.) [B]	4,770	1,500	1,610
	The number of knowledge [C]	322	237	474
Evaluation	Time efficiency (min.) [B]/[C]	14.8	6.3	3.4
	Exhaustiveness [C]/[A]	1.46	3.08	4.31

Fig. 7. Effects of applying the support system, FA, and SSI in an interview

4.2 Photograph Extraction Experiment

In the automatic classification, seven special devices on the operating floor of the nuclear power plant were selected as classification labels (e.g., reactor pressure vessel, special hanging equipment, transfer equipment, special bolts, equipment in the reactor, etc.). From the past known photographs, the photographs with these labels were selected, and annotation data were created to randomly allocate them for learning and evaluation. The number of targeted photographs was 882 in total (707 for learning, and 175 for evaluation). In general, it is considered appropriate to learn about 80% of the total number of photographs for learning and the remaining 20% for evaluation. For the hyper-parameter at the time of learning, the number of epochs (the number of times of repeated learning) was set to 300, and the batch size (the number of data to be divided into some subsets) was set to 16. When the loss was minimized, the process was terminated early and the weight data at that time were obtained. In addition, the number of photographs for learning was small, and thus an image augmentation layer was inserted immediately after the input layer to randomly change the original photographs.

Recognition accuracy was evaluated by the F-measure based on the harmonic mean of the precision and recall that were obtained [16–18]. The precision is the ratio of data that were actually positive in the data predicted to be positive, and the recall is the ratio of data that were predicted to be positive in the data that were actually positive. The F-measure is obtained by the following equation.

$$\mathrm{F-measure} = \frac{2Recall \times Precision}{Recall + Precsion} \tag{1}$$

Also, in transfer learning, the relearning layer and the frozen layer (the layer that does not update weight data without relearning) were arbitrarily changed, and the learning results were compared. Figure 8 shows the conditions when the frozen layer and the relearning layer were changed in the VGG-16.

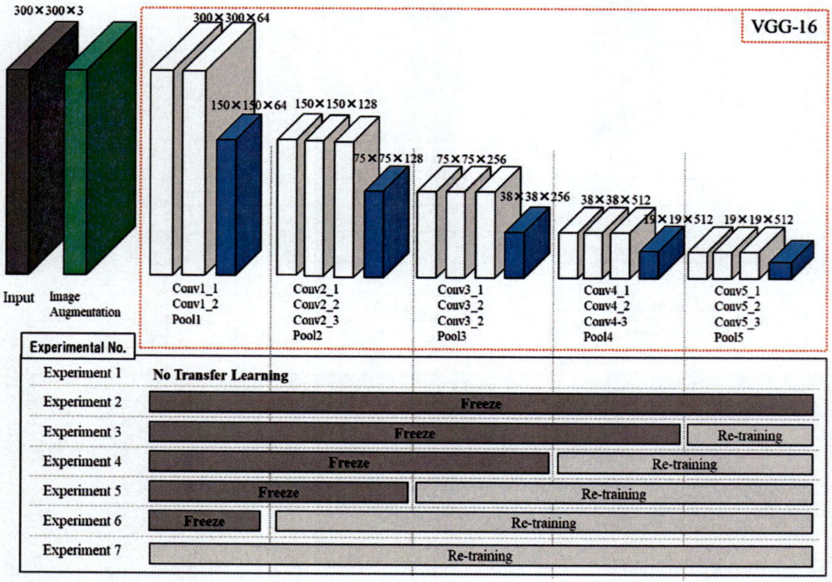

Fig. 8. The conditions when the frozen layer and the relearning layer

Figure 9 shows a comparison of F-measures according to the difference in the freezing conditions. When the transfer learning was not used (experiment No. 1), the F-measure was 66.9, whereas when the transfer learning was used (experiment No. 2–7), the F-measure was improved overall. When the number of frozen layers was relatively small, the F-measure tended to be high, and when the number of frozen layers was 3, a maximum F-measure of 90 was obtained.

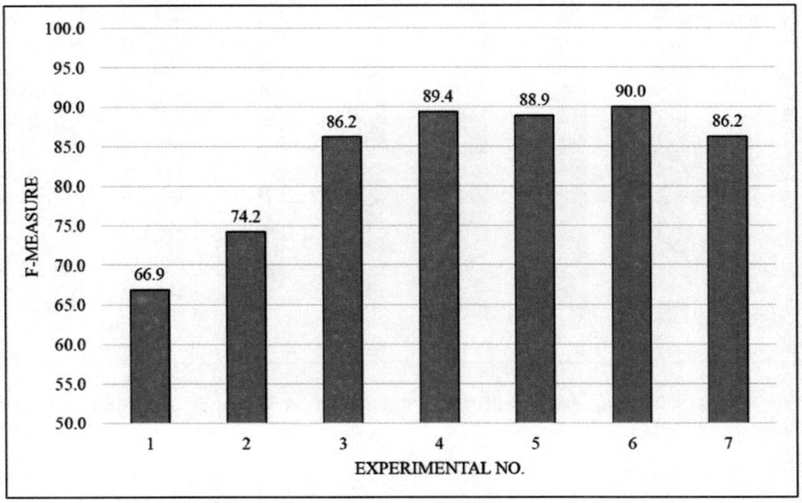

Fig. 9. F-measures when changing the number of frozen layers

Based on the above results, the dataset was divided into 5 groups (groups in which the learning photographs (80%) and the evaluation photographs (20%) were changed, respectively) using the transfer learning conditions when the number of frozen layers was 3 to perform Cross Validation and compare the F-measures with and without transfer learning. Figure 10 shows a comparison for each label, and Fig. 11 shows a comparison for each dataset group.

Although the F-measure differs for each label, the F-measures of all labels were improved by transfer learning (up to 29.2% for label 3). In addition, although the F-measure also differs for each data set, the F-measures of all data sets were improved by transfer learning (up to 23.1% for data set 1). On average for all dataset groups, the F-measure was 71.9% without transfer learning, and was 89.0% with transfer learning, resulting in an improvement of 17.1%.

Next, we verified the working hours when using the conventional method of manually searching and classifying a large number of past photographs, and the working hours when using the method of automatic classification by SSD (assuming F-measure: 89%). This time, we conducted an experiment on the 17,604 photographs provided by the on-site supervisors of the actual nuclear power plant. As the conventional manual method requires an enormous amount of time, it was determined by trial calculation. In the case of SSD, the actual result was measured. In the case of the manual method, although it was a trial calculation result, it took a total of 518 h for search/classification/extraction work, which required a very long time. On the other hand, in the case of SSD, Although it takes some time to create annotation data, the learning, classification, and extraction can be performed fully automatically (including

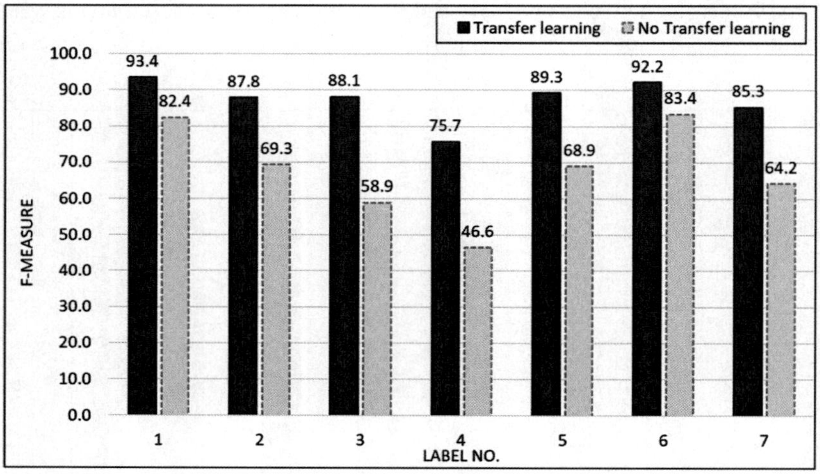

Fig. 10. Cross validation results with and without transfer learning (compared with labels)

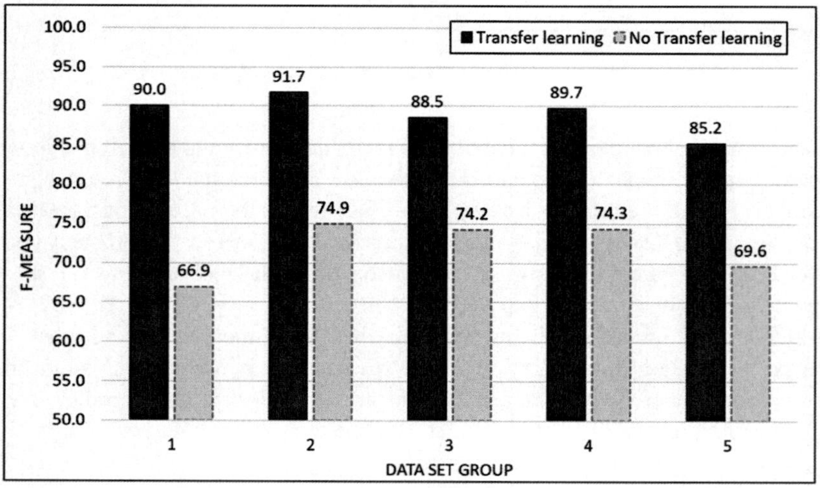

Fig. 11. Cross validation results with and without transfer learning (compared with data set)

manual reclassification in the case of misrecognition), so the process was able to be completed in a total of 38.3 h, and a large improvement in efficiency was obtained.

5 Discussion and Conclusion

We utilized the support system for registering know-how, the Functional Approach (FA), and the Semi-Structured Interview (SSI) to extract know-how from experts based on interviews. As a result, it was confirmed that the extraction time efficiency per know-how can be improved by 77.1% and the exhaustiveness of the extraction know-how for the work steps can be improved 2.9 times. This is considered to be because the system and the interview method enabled the interviews to proceed in a state where all information was organized systematically.

In the conventional method, it took a lot of time to search past data, to switch files, to switch screens, to think about questions, and to delve into the answers from experts, so the process could not proceed efficiently. As a result, each interview took a long time, and as a whole, the interviews took many days. However, Work B was able to smoothly extract tacit knowledge because experts could read the information organized on the system immediately and respond without stress. In addition, Work C could accurately obtain the desired results for the simulation questions (hypotheses), because the thinking on the part of the questioner was organized by making hypotheses with FA and preparing for SSI in advance.

Furthermore, for the huge number of past photographs stored in the system, the use of SSD and transfer learning made it possible to automatically classify and extract the huge number of past photographs that were possessed personally by experts. In addition, incorporating transfer learning (even though the number of learning photographs is as small as 882) made it possible to improve the F-measure by 17.1% compared with the case without transfer learning, and to classify the photographs with high accuracy of 89%. In general, there have been many reports that high accuracy can be obtained when most layers of VGG-16 are frozen and only the last layer is re-learned. However, in this research, more accurate results were obtained by reducing the number of frozen layers (up to three layers from the initial layer). This may be due to the fact that the special equipment (objects with special feature amount) of the nuclear power plant and the known learned weight files (objects with the general feature amount such as persons, animals, vehicles, etc.) are significantly different in the feature amount.

Also, it takes a lot of time to classify a huge number of past photographs. Conventionally, it was necessary for the expert to visually organize photographs, but it was difficult and impractical to classify these photographs, which had been accumulated for decades, in between the expert's daily work. In this research, the automation by SSD enabled the working hours to be greatly reduced. By executing these processes periodically on all the construction-site computers in the network, the latest classified photographs can always be centrally managed in an in-house central database, so interviews can be conducted effectively.

At present, this method is being rolled out to the special construction work of nuclear power plants, and the know-how of experts is gradually being accumulated. In the future, we will incorporate the accumulated know-hows into the contents of educational movies and Virtual Reality (VR) as transfer contents and will use them as one of the more effective methods of know-how transfer data.

References

1. International chamber of commerce. https://iccwbo.org/. Accessed 9 Dec 2019

2. Nonaka, I., Takeuchi, H.: The knowledge-Creating Company: How Japanese Companies Create the Dynamics of Innovation. Oxford University Press, Oxford (1995)
3. Polanyi, M.: The Tacit Dimension. Doubleday, Garden City (1966)
4. Chika, E.: Udeaja: a web-based prototype for live capture and reuse of construction project knowledge. Autom. Constr. **17**, 839–851 (2008)
5. Lin, Y.-C.: Enhancing knowledge exchange through web map-based knowledge management system in construction: lessons learned in Taiwan. Autom. Constr. **15**, 693–705 (2006)
6. Lin, Y.C.: Developing construction assistant experience management system using people-based maps. Autom. Constr. **17**, 975–982 (2008)
7. Lin, Y.C.: Developing project communities of practice-based knowledge management system in construction. Autom. Constr. **22**, 422–432 (2012)
8. Tserng, H.P.: Developing an activity-based knowledge management system for contractors. Autom. Constr. **13**, 781–802 (2004)
9. Babar, A.: BIM-based claims management system: a centralized information repository for extension of time claims. Autom. Constr. **110**, 102937 (2020)
10. Adams, W.C.: Conducting semi-structured interviews. In: Handbook of Practical Program Evaluation, pp. 492–505
11. Low, B.K.L.: The risk-taking propensity of construction workers—an application of quasi-expert interview. Int. J. Environ. Res. Public Health **15**, 2250 (2018)
12. Gil, D.: Classification of images from construction sites using a deep-learning algorithm. In: Proceeding of 35th ISARC (2018)
13. Kim, H.: Detecting construction equipment using a region-based fully convolutional network and transfer learning. J. Comput. Civ. Eng. **32**(2), 04017082 (2018)
14. Yabuki, N., Nishimura, N., Fukuda, T.: Automatic object detection from digital images by deep learning with transfer learning. In: Proceedings of 25th European Group for Intelligent Computing in Engineering (EG-ICE) International Workshop 2018, Lausanne, Switzerland, June 2018, pp. 3–15 (2018)
15. Miles, L.D.: Techniques of Value Analysis and Engineering (2015)
16. Koirala, A., Walsh, K.B., Wang, Z., McCarthy, C.: Deep learning – method overview and review of use for fruit detection and yield estimation. Comput. Electron. Agric. **162**, 219–234 (2019)
17. Zhong, L., Lina, H., Zhou, H.: Deep learning based multi-temporal crop classification. Rem. Sens. Environ. **221**, 430–443 (2019)
18. Heikal, M., Torki, M., El-Makky, N.: Sentiment analysis of arabic tweets using deep learning. Procedia Comput. Sci. **142**, 114–122 (2018)

Enriched and Discriminative Human Features for Person Re-Identification Based on Explainable Behaviors of Convolutional Neural Networks

Peter Kok-Yiu Wong[1], Han Luo[1], Mingzhu Wang[2], and Jack C. P. Cheng[1(✉)]

[1] Hong Kong University of Science and Technology, Hong Kong SAR, China
{kywongaz, han.luo}@connect.ust.hk, cejcheng@ust.hk
[2] Imperial College London, South Kensington, London, UK
m.wang@imperial.ac.uk

Abstract. Understanding pedestrian behaviors such as their movement patterns in urban areas could contribute to the design of pedestrian-friendly facilities. With the commonly deployed surveillance cameras, pedestrian movement in a wide region could be identified by the person re-identification (ReID) technique across multiple cameras. Convolutional neural networks (CNNs) have been widely studied to automate the ReID task. CNN models equipped with deep learning techniques could extract discriminative human features from images and show promising ReID performance. However, some common challenges such as occlusion and appearance variation are still unsolved. Specifically, our study infers that over-relying on discriminative features only may compromise ReID performance. Therefore, this paper proposes a new model that extracts enriched features, which is more reliable against those ReID challenges. By adding a feature dropping strategy during model training, our model learns to focus on rich human features from different body parts. Moreover, this paper presents an explainable approach of model design, by visualizing which human parts a deep learning model focuses on. Based on an intuitive interpretation of model behaviors that lead to inaccurate results, specific improvement of model architecture is inspired. Our improved results suggest that making existing models explainable could effectively shed light on designing more robust models.

Keywords: Computer vision · Deep learning · Enriched human features · Explainable convolutional neural networks · Person Re-identification

1 Introduction

Smart city planning has been strongly promoted, in response to the need of new area development and urban renewal. As citizens are the core entities in urban areas, understanding pedestrian behaviors is a key to revealing the infrastructural demand and encouraging human-centric design in an urban area [1]. For example, by identifying

E. Toledo Santos and S. Scheer (Eds.): ICCCBE 2020, LNCE 98, pp. 41–53, 2021.
https://doi.org/10.1007/978-3-030-51295-8_5

pedestrian movement patterns in a neighborhood, urban planners could propose more walking-friendly facilities to enhance pedestrian walkability [2, 3]. Traditionally, pedestrian behavioral analysis heavily relied on manual survey on street, such as conducting questionnaire or manual observation [2]. These approaches are labor intensive, requiring tedious data collection and processing.

More recently, such data acquisition could be enhanced by video analytics, since surveillance cameras are commonly deployed in different places, such as public estates and mass transit stations. Pedestrian movement could be studied by recording their traveling trajectories [3]. Yet, there could be many cameras scattering among a large district, which makes manually extracting pedestrian data very time-consuming, and possibly error-prone due to human fatigue. It is also tedious to manually identify long-term movement of each single person in a wide region, which requires precise identity mapping across multiple cameras.

Identification of pedestrian movement could be formulated into a problem of person re-identification (ReID). By definition, given a set of images capturing individual person, images of the same person across different cameras should be matched to establish the same identity [4]. To automate this process, computer vision techniques have been widely adopted. Particularly, recent studies are mainly based on convolutional neural network (CNN) equipped with deep learning, which shows promising performance in extracting features for generic object recognition [5], as well as human ReID [4]. Some typical categories of ReID models include part-by-part [6, 7] or attention-driven [8, 9] feature extraction, and multi-scale feature aggregation [10, 11]. A common objective among many feature extraction models is to obtain discriminative human features for retrieving the same person across cameras, and simultaneously rejecting any false target.

However, even the state-of-the-art feature extraction models may suffer from challenges such as partial occlusion and significant appearance variation across cameras [4]. This paper infers that, while newer ReID models are continuously published, there still lacks an explicit methodology to explain the behavior of existing model and the specific reason to their failure results. An explainable model may intuitively shed light on what aspect could be improved. There have been some techniques for interpreting behavior of generic CNNs. For example, the convolutional feature map could be extracted to compute a feature activation map (FAM) [12], which locates the attention of a CNN on an image, i.e. which part is more focused on. If put in ReID context, FAM could potentially reveal what features of a person are extracted by a model. Some studies attempted to visualize FAMs of their models [8–11]. Nevertheless, a detailed interpretation of the behaviors is still lacking, which may be necessary for developing more robust models.

Therefore, this paper proposes a new ReID model based on an explainable interpretation of how existing models behave. Particularly, the FAM visualization technique is utilized to explain the feature extraction behaviors of some baseline models chosen. Based on a detailed interpretation, a new model is designed to specifically address aforementioned challenges, including occlusion and appearance variation. Different from previous studies, this paper attempts to understand model behavior more thoroughly, in order to more intuitively explain the design rationale of new models. Such an approach of model design could be generalized to future studies. The improved performance in our study shows that an explainable design could effectively develop more robust ReID models.

2 Baseline Models

Several existing ReID models are studied, where ResNet50 [5] and OSNet [10] are particularly chosen as our baseline models. It is because, as illustrated in Sect. 3.1, they exhibit two opposite types of behaviors upon human feature extraction, which could intuitively justify the design rationale of our proposed model. Note that this paper aims at a more generic comparison among types of behavior.

ResNet50 is a commonly adopted feature extractor for various tasks, such as object classification and detection, as well as ReID. As shown in the model architecture in Fig. 1, the by-pass connections across layers in different depths create an effect of residual learning, which enables ResNet to learn very deep features effectively without suffering from problems such as vanishing gradient [5]. This ability of deep feature learning also contributes to the ReID task, where human identities could be more uniquely distinguished with deep features extracted. On the other hand, OSNet is a ReID-specific model published more recently, with competitive ReID performance achieved. The diagram in Fig. 2 shows the basic module of OSNet. In addition to using residual connections like ResNet, the output features are an aggregation of multiple sets of features obtained by convolutions with different receptive fields. Such a multi-scale feature aggregation makes OSNet highly flexible for extracting both coarse and fine features of human [10].

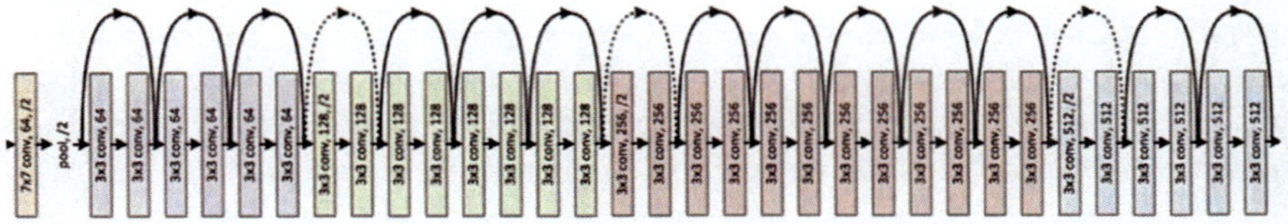

Fig. 1. Typical model architecture of ResNet, where the by-pass connections effectively enhance deep feature learning

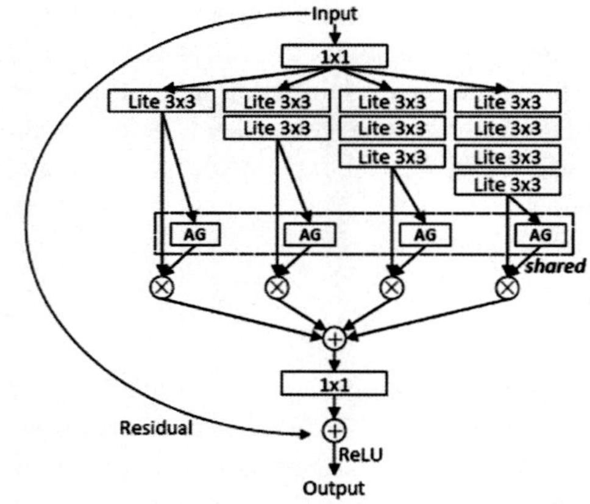

Fig. 2. Basic module of OSNet, which adaptively aggregates convolution feature maps from different receptive fields, along with a residual connection like ResNet. Lite: lightweight convolution layer. AG: aggregation gate

3 Proposed Methodology

This paper proposes a new CNN architecture for enhanced ReID robustness, based on an intuitively explainable strategy. Particularly, the technique of visualizing feature activation maps (FAMs) [12] is utilized to interpret which regions on an image that a CNN model focuses on to extract features for ReID. Intuitively speaking, such regions indicate which parts of a person that possibly make him/her discriminative from others. When a model processes an image, our study extracts its last convolution feature maps just before a global pooling operation. An activation map is then computed as the sum of absolute-valued feature maps along the channel dimension, followed by a spatial L2 normalization [12]. The computed FAM is then rescaled and overlaid onto the raw image, since these two are dimensionally aligned. The normalized value at each spatial position reflects how much the model focuses on that region during its feature extraction process. By analyzing the output FAMs, the behavior of a model is interpreted, e.g. whether the model wrongly focuses on any background clutter rather than a person. This would shed light on how to design an improved model.

3.1 Behaviors of Baseline Models

The behaviors of the two baseline models are analyzed, in terms of their FAMs produced from some images of pedestrians. The findings would then form the basis that justifies the design of our proposed model in Sect. 3.2. The DukeMTMC-reID dataset [13] obtained on a college campus is mainly used in our study. This dataset provides research values since the images include several challenging conditions such as occlusion and appearance variation across multiple cameras. All subsequent results throughout this paper are based on this dataset, unless otherwise specified. Codes and pre-trained models published under Torchreid [14] are utilized.

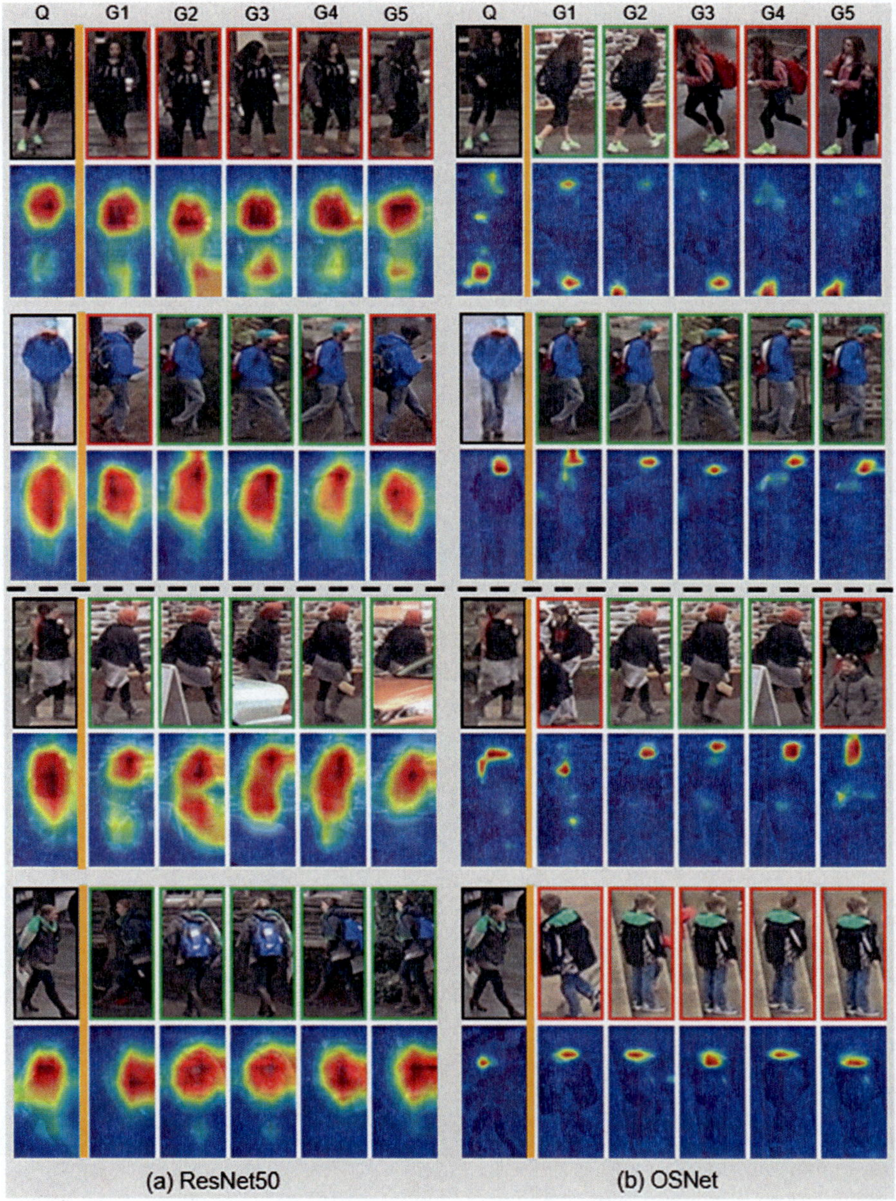

Fig. 3. Resulted FAMs from (a) ResNet50 and (b) OSNet. Q: query image. G1-G5: gallery images ranked by Euclidean distances among feature descriptors, with G1 as the closest match. Green bounding boxes denote correctly matched persons. Best viewed in color. Among upper two rows, OSNet performs better by locating very fine and discriminative features. Among lower two rows, ResNet50 performs better without being over-sensitive to eye-catching features.

The FAMs in Fig. 3 illustrate the behaviors of ResNet50 and OSNet respectively. Regions with darker red denote where a model focuses more on, while blue regions denote the opposite. It is observed that these models exhibit two extremes of ReID behaviors. As for ResNet50, the highly focused regions cover nearly the whole upper body of a person, while the legs in lower body are relatively less focused. The small parts such as the head and shoes are mostly ignored. As for OSNet, the focused regions squeeze into very fine parts that look eye-catching, such as the shiny hats, hair, T-shirt logos and shoes. The remaining portion of a human body are barely focused on. Based on the upper two rows of FAMs in Fig. 3, locating very discriminative features enhances ReID performance than having a too coarse focused region. Therefore, being more flexible and able to attend to some fine regions of a person is one factor for accurate ReID results.

Nevertheless, focusing only on some fine parts may make a ReID model over-sensitive to those eye-catching features, which may lead to inaccurate results. From the lower two rows of FAMs in Fig. 3, OSNet may have been misled by two different persons with only some small parts looking similar. For example, the red scarf of the woman may be wrongly matched with the red shirt logo of the man. Indeed, the two wrongly matched persons have quite different appearance in other body parts, on which OSNet focused less. It is inferred that relying only on fine regions for ReID may suffer from various problems, including occlusion and appearance variation. For instance, the focused parts may be occluded by surrounding objects, or the person him/herself when having the body turned around. Besides occlusion, the attended features may be lost in another camera due to appearance variation, such as having the scarf taken off. These may compromise the fine features extracted. Therefore, extracting rich features is possibly another factor for obtaining reliable ReID results.

3.2 Proposed Model

Regarding the shortcoming of being over-focused on small parts of a person, it is inferred that forcing the model to spread its attention over different regions on an image could possibly enrich the features extracted. Recently, [15] suggested an interesting technique named Batch DropBlock (BDB). The rationale behind BDB is that, by randomly hiding some features of a person during training, the model is forced to rely on the remaining parts. This trick prevents a model from always focusing on particular regions throughout its training process. Consequently, the model learns to explore richer features over an image. This aligns with our discussion in Sect. 3.1 about how to improve the baseline models. Therefore, this paper proposes to incorporate the BDB technique into OSNet, resulting in a new model for extracting both discriminative and enriched features. The model architecture is shown in Fig. 4.

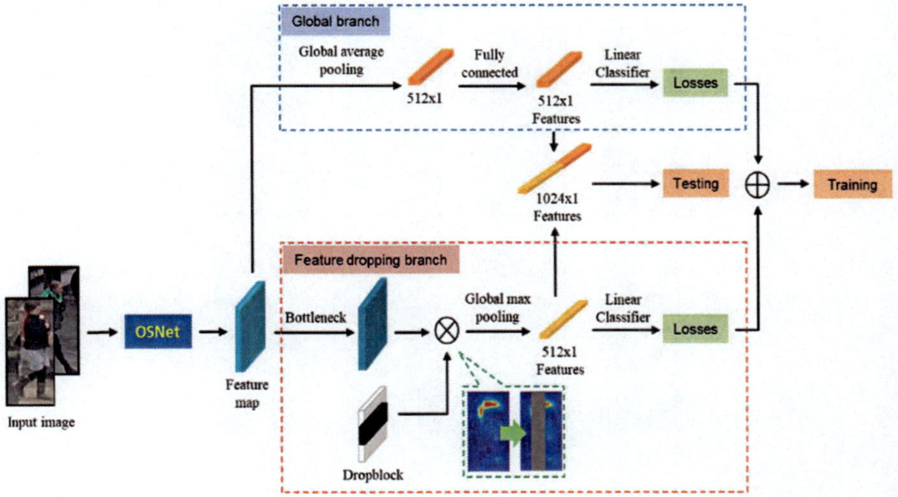

Fig. 4. Architecture of our proposed model OSNet + BDB, revised from [15]. The global branch alone forms the baseline OSNet. The feature dropping branch randomly places a rectangular dropblock to zero-out part of the bottlenecked feature map. The losses among two branches are summed for training. As for testing, the globally pooled features from both branches are concatenated as the feature descriptor of an image.

Our proposed model integrates OSNet [10] and BDB [15]. Readers are referred to their original publication for their detailed model design. There are two branches after obtaining the convolution feature map from OSNet. The global branch alone forms the baseline OSNet, with a global average pooling layer. Simultaneously, the feature dropping branch introduces a rectangular dropblock to zero-out part of the bottlenecked features, followed by a global max pooling layer. During testing, the globally pooled features from each branch are concatenated into a single feature descriptor of an image. As for training, each of them proceeds to the loss calculation and summation. It is inferred that both branches could supervise each other during training, which facilitates the model to extract both discriminative and enriched features.

Specifically, there are three highlights that differentiate our proposed model from original OSNet and BDB. Firstly, original BDB uses ResNet as the backbone feature extractor, while our model replaces it with OSNet to extract more discriminative features. Secondly, OSNet only has a single branch with global average pooling for discriminative features, while this paper adds a feature dropping branch for enriched features as well. Thirdly, original BDB slides a dropblock vertically over an image in training, while our study proposes a horizontally sliding dropblock. The difference among vertical and horizontal sliding is illustrated in Fig. 5. For comparison, our proposed model is divided into two variants, namely 'OSNet + BDB-vertical' and 'OSNet + BDB-horizontal'.

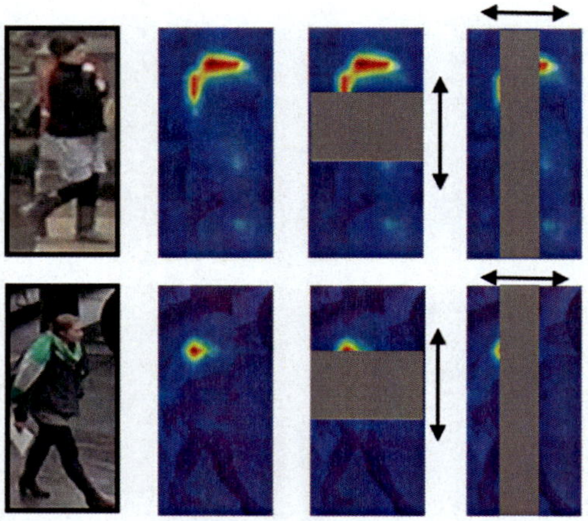

Fig. 5. Illustration of BDB process where a block of features is dropped during training, with the dropblock sliding vertically or horizontally to a random position. (a) raw image, (b) output FAM, (c) example of original BDB-vertical, (d) example of our proposed BDB-horizontal.

3.3 Experimental Design

Table 1. Configuration of training and testing our proposed OSNet + BDB

Item	Setting
Dropblock size for BDB-vertical	1/3 height of last convolutional map
Dropblock size for BDB-horizontal	1/3 width of last convolutional map
Distance metric for person matching	Cosine distance
Training dataset	DukeMTMC-reID [13]

Our proposed OSNet + BDB was trained with the configuration similar as that of the baseline OSNet [10]. Table 1 mainly summarizes the parameters that are specific to our proposed model. As for our added BDB technique, the dropblock size is chosen as one-third of the respective axis. Such a dropblock size was reported to be optimal in the original BDB [15]. Furthermore, a hybrid loss function combining cross-entropy and batch hard triplet losses is used for model training, whose positive effect was illustrated [16]. As for testing with the extracted feature descriptors, our study adopts the Cosine distance which yielded better performance than the typical Euclidean distance [17]. Both our training and testing utilized the DukeMTMC-reID dataset [13], which includes challenging conditions such as occlusion and appearance variation. For performance evaluation, two commonly used metrics in ReID are adopted, namely Rank-1 accuracy and mean average precision (mAP) [4]. These two metrics reflect how well a model could extract discriminative and reliable features respectively for ReID.

4 Results and Discussion

4.1 Quantitative Results

The quantitative results are summarized in Table 2, comparing among baseline models and variants of our proposed model. Based on the numerical results, it is inferred that adding BDB could enhance the performance of feature extraction. In terms of mAP, both variants outperform the baseline model, which means that our proposed model is more reliable in feature extraction for ReID. Our 'OSNet + BDB-horizontal' also outperforms the baseline model in rank-1 accuracy, suggesting its ability of extracting discriminative features as well.

Among the variants of our proposed model, 'OSNet + BDB-horizontal' achieves the highest rank-1 accuracy of 89.0% and mAP of 77.7%, with more than 1% advantage over 'OSNet + BDB-vertical'. Therefore, the BDB-horizontal strategy seems to bring more positive effect to the feature extraction process than BDB-vertical does. The underlying rationale would be explained in the subsequent qualitative analysis.

Table 2. Testing results among baseline and proposed models in DukeMTMC-ReID dataset

Model	Rank-1 accuracy (%)	mAP (%)
OSNet (baseline)	88.6	73.5
OSNet + BDB-vertical	87.6	76.5
OSNet + BDB-horizontal	**89.0**	**77.7**

4.2 Qualitative Results

The FAMs in Fig. 6 depict the behaviors of the baseline OSNet and two variants of our proposed OSNet + BDB. As discussed, focusing only on some eye-catching regions of a person may be easily misled under occlusion or appearance variation across cameras. Based on the FAMs, our proposed model addresses these problems by widely attending to different regions for extracting richer features of a person.

Proposed Model vs Baseline Models. As shown in Fig. 6, for the man in the 1st row query, OSNet relied on the very fine white logo on his red shirt, which was misled by another woman in red shirt and carrying a white item. Instead, our OSNet + BDB widely focused on the head, shoulder and shoes of the man, which successfully retrieved the same person even with the body turned around. It is inferred that the fine region (the logo on front body) was self-occluded when the man turned around. Intuitively, when the features a model relies entirely on are occluded, the matching robustness may be compromised. Regarding this problem, our model solicited features from different parts of the man, on top of just a fine region. Even when some features were occluded under pose variation or by other objects, the enriched features possibly help the model retrieve the same person.

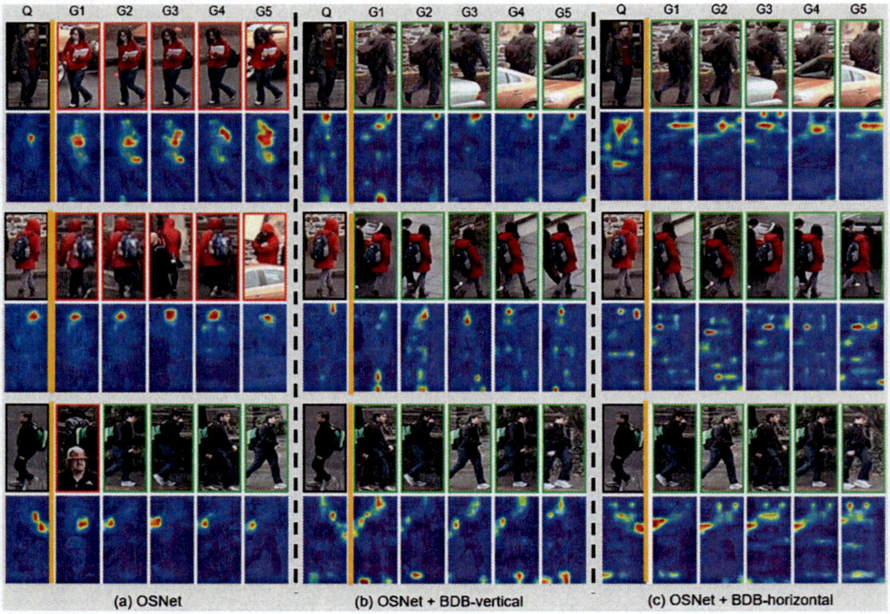

Fig. 6. Resulted FAMs from (a) OSNet, (b) OSNet + BDB-vertical and (c) OSNet + BDB-horizontal. Q: query image. G1-G5: gallery images ranked by cosine distances among feature descriptors, with G1 as the closest match. Green bounding boxes denote correctly matched persons. Best viewed in color. Our proposed model attends to different regions for extracting richer features of a person, for more robust matching.

As another example, for the woman in the 2nd row query, OSNet relied mainly on the red hood around her head, which was misled by another person with very similar clothing among upper body. There was a clear difference among features of their lower body, which was however not focused on. Interestingly, from the successful matching by our proposed model, the woman took off the red hood in another camera. Intuitively, relying entirely on that fine feature tends to be easily misled upon such appearance variation. Instead, our proposed model extracted various features from her head (including the red hood), backpack and shoes. Even having the hood took off, there were enriched features for matching. Overall, our proposed model outperforms the baseline one possibly because of the enhanced ability to extract richer features, as explained by the visualized FAMs.

BDB-Horizontal vs BDB-Vertical. Among the two variants of our proposed OSNet + BDB, the one with BDB-horizontal seems to outperform BDB-vertical. The FAMs in Fig. 6 show a clear difference among their behaviors of feature extraction. Specifically, BDB-vertical tends to separate the focused regions towards the top and bottom edges. In most of its output FAMs, the highly focused regions lied on either the head or shoes of a person. A possible explanation is that the main human body in the center covers most area of an image. When sliding the dropblock vertically, the main body has a high chance to be dropped, which forces the model to rely more on head and shoes. However, this apparently yields sub-optimal performance since some useful features on the main body are not fully utilized.

On the other hand, BDB-horizontal is more flexible in spreading its attention over arbitrary parts of a person. Its focused regions are not spatially limited to any particular position of an image, unlike BDB-vertical which mainly explores the top and bottom edges. Moreover, the attention of BDB-horizontal is not limited semantically, since it is able to capture the most eye-catching part as well as other useful features of a person. Such behavior may explain why BDB-horizontal outperforms BDB-vertical.

4.3 Potential Improvement

Our proposed OSNet + BDB could extract discriminative and enriched features of a person, leading to enhanced ReID performance. Nevertheless, it sometimes undesirably included features of another person on the same image. As shown in Fig. 7, the query target is the woman with yellow backpack, while our proposed model also attended to the woman with grey umbrella. Interestingly, both persons appeared together in most of their gallery images. Our model apparently mixed up their identities.

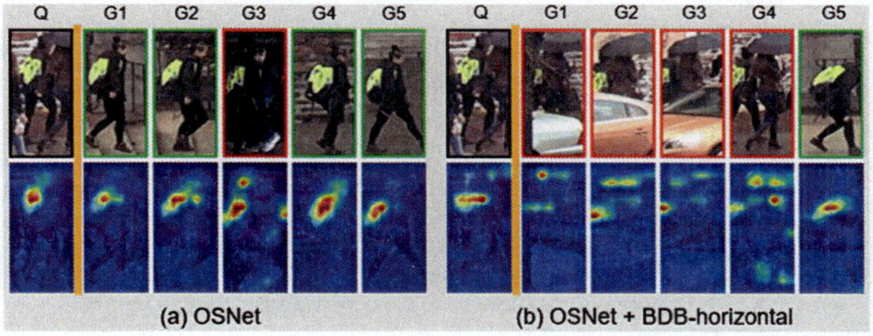

Fig. 7. Resulted FAMs from (a) OSNet and (b) OSNet + BDB-horizontal. Q: query image. G1-G5: gallery images ranked by cosine distances among feature descriptors, with G1 as the closest match. Red bounding boxes denote wrongly matched persons. Best viewed in color. Our proposed model may have included features of multiple persons on the same image.

Based on the discussion above, a potential improvement is to obtain images with accurately bounded targets before the subsequent feature extraction. Possible ways include image pre-processing for further alignment or segmenting out the exact human body boundary. As for the model architecture, the BDB technique may be further improved by making the dropblock shape deformable and adaptive to a human body, possibly with the help of human segmentation. This may further facilitate the model training for exploring more human-specific features. Overall, above are some directions for potential improvement of our proposed model.

5 Conclusion

This paper presents a new CNN architecture with enhanced ReID performance. Importantly, our design rationale is explainable by visualizing what features are focused on by a model. This strategy reveals that only extracting discriminative features from fine human parts may suffer from challenges such as occlusion and appearance variation. Our proposed model addresses these by integrating the baseline model with a Batch DropBlock technique, which enriches the human features extracted from different human parts for ReID. Such an explainable approach of model design is highly encouraged in future studies. For practical video analytics, a robust ReID model could effectively identify pedestrian movement in a neighborhood. This would greatly facilitate pedestrian behavioral analysis, which enables a human-centric urban design.

References

1. Ewing, R.: Eight Qualities of Pedestrian-and Transit-Oriented Design, Urban Land: The magazine of the Urban Land Institute (2013)
2. Sharifi, M.S., Christensen, K., Chen, A., Stuart, D., Kim, Y.S., Chen, Y.: A large-scale controlled experiment on pedestrian walking behavior involving individuals with disabilities. Travel Behav. Soc. **8**, 14–25 (2017)
3. Fu, L., Cao, S., Shi, Y., Chen, S., Yang, P., Fang, J.: Walking behavior of pedestrian social groups on stairs: a field study. Saf. Sci. **117**, 447–457 (2019)
4. Wu, D., Zheng, S., Zhang, X., Yuan, C., Cheng, F., Zhao, Y., Lin, Y., Zhao, Z., Jiang, Y., Huang, D.: Deep learning-based methods for person re-identification: a comprehensive review. Neurocomputing **337**, 354–371 (2019)
5. He, K., Zhang, X., Ren, S., Sun, J.: Deep residual learning for image recognition. In: Proceedings of the IEEE Conference on Computer Vision and Pattern Recognition, pp. 770–778 (2016)
6. Sun, Y., Zheng, L., Yang, Y., Tian, Q., Wang, S.: Beyond part models: person retrieval with refined part pooling (and a strong convolutional baseline). In: Proceedings of the European Conference on Computer Vision (ECCV), pp. 480–496 (2018)
7. Wong, P.K., Cheng, J.C.P.: Monitoring pedestrian flow on campus with multiple cameras using computer vision and deep learning techniques. In: CIGOS 2019, Innovation for Sustainable Infrastructure, pp. 1149–1154 (2020)
8. Li, W., Zhu, X., Gong, S.: Harmonious attention network for person re-identification. In: Proceedings of the IEEE Conference on Computer Vision and Pattern Recognition, pp. 2285–2294 (2018)
9. Quispe, R., Pedrini, H.: Improved person re-identification based on saliency and semantic parsing with deep neural network models. Image Vis. Comput. **92**, 103809 (2019)
10. Zhou, K., Yang, Y., Cavallaro, A., Xiang, T.: Omni-scale feature learning for person re-identification. In: Proceedings of the IEEE International Conference on Computer Vision, pp. 3702–3712 (2019)
11. Wang, G., Yuan, Y., Chen, X., Li, J., Zhou, X.: Learning discriminative features with multiple granularities for person re-identification. In: Proceedings of the 26th ACM International Conference on Multimedia, pp. 274–282 (2018)

12. Zagoruyko, S., Komodakis, N.: Paying more attention to attention: improving the performance of convolutional neural networks via attention transfer (2016). arXiv preprint arXiv:1612.03928
13. Ristani, E., Solera, F., Zou, R., Cucchiara, R., Tomasi, C.: Performance measures and a data set for multi-target, multi-camera tracking. In: European Conference on Computer Vision, pp. 17–35 (2016)
14. Zhou, K., Xiang, T.: Torchreid: A Library for Deep Learning Person Re-Identification in Pytorch (2019). arXiv preprint arXiv:1910.10093
15. Dai, Z., Chen, M., Gu, X., Zhu, S., Tan, P.: Batch DropBlock network for person re-identification and beyond. In: Proceedings of the IEEE International Conference on Computer Vision, pp. 3691–3701 (2019)
16. Almazan, J., Gajic, B., Murray, N., Larlus, D.: Re-id done right: towards good practices for person re-identification (2018). arXiv preprint arXiv:1801.05339
17. Luo, H., Gu, Y., Liao, X., Lai, S., Jiang, W.: Bag of tricks and a strong baseline for deep person re-identification. In: Proceedings of the IEEE Conference on Computer Vision and Pattern Recognition Workshops (2019)

Automating the Generation of 3D Multiple Pipe Layout Design Using BIM and Heuristic Search Methods

Jyoti Singh and Jack C. P. Cheng(✉)

Department of Civil and Environmental Engineering, The Hong Kong University of Science and Technology, Kowloon, Hong Kong
cejcheng@ust.hk

Abstract. The layout design and spatial coordination of multiple pipe systems is major challenge in the construction industry. The purpose of multiple pipe layout design is to find out an optimal layout for numerous individual pipes to route from a different start locations to different end locations in a 3D environment with no clashes under various kinds of constraints. Currently, pipe layout design is conducted manually by consultants, which is tedious, labor intensive, error-prone, and time-consuming. This paper proposes a BIM-based approach for layout design of multiple pipes using heuristic search methods. Algorithms are developed based on a directed weighted graph according to the physical, design, economical and installation requirements of pipe layout design. Clashes between pipes and with building components are considered and subsequently avoided in the layout optimization. Based on the developed algorithms, simulated annealing (SA) algorithm is used to approximate global optimization in a large search space for multiple pipe layout optimization. As for layout design, Dijkstra algorithm and two heuristic algorithms namely 3D A* and fruit fly optimization algorithm (FOA) are implemented and compared to obtain the multiple pipe system layout design. An example of a typical plant room with nine pipe routes is used to illustrate the developed approach on multiple pipe layout design. The result shows that the developed approach can generate optimal and clash-free multiple pipe system layout. Compared with the conventional method, the developed approach significantly reduces the time and cost for designing multiple pipe layout.

Keywords: Multiple pipe layout · Automatic design optimization · Building information modeling · Heuristic method

1 Introduction

Multiple pipe system layout design and spatial coordination plays an important and challenging role in construction industry. Efficient layout design of the pipe systems becomes more and more complicated due to numerous pipe systems involved in the layout space in compliance with various constraints such as physical, design, operational, and economical constraints [1]. The purpose of multiple pipe layout design is to find the most economical spatial arrangement of equipments and vessels and the route for interlinked pipes and accessories, as shown in Fig. 1, satisfying many constraints

E. Toledo Santos and S. Scheer (Eds.): ICCCBE 2020, LNCE 98, pp. 54–72, 2021.
https://doi.org/10.1007/978-3-030-51295-8_6

such as physical, construction, operation, maintenance, safety and other various code requirements [2]. The design of pipe system layout needs to consider factors such as building space geometry, system requirements, design code specifications, and the locations and configurations of relevant equipments [3].

Pipe systems aims to transfer fluids from a starting point to a destination point with a certain time interval through an appropriate route. Pipe layout is very popular and play an essential part in our daily lives, for example, in electric power plants, chemical plants, factories, buildings, sewage, automobiles, etc. [4], and its contribution to the overall capital cost is often nontrivial. Pipe layout design is one the most time consuming, expensive and complicated steps in any pipe system design since it can take over 50% of the total design man-hours and all other detail design depends on it [5]. One of the most exhausting aspect of pipe layout design is to conduct pipe route planning to connect start and end points. Especially, when large pipes network layout is involved in the 3D space. Pipe routing is usually determined after the positions of the major components are determined which is done by as per some industry guidelines, building requirements, and designer and consultant experience [1]. As detail functional design is characteristically less creative and more routine than the earlier route planning and designing steps [6], a complete automation and design optimization of multiple pipe route layout might offer an attractive way to cut down on tedious and irksome work, leading to saving in time and money. Effective layout design of the pipe system is constrained by various consideration, such as [4]:

- Physical constraints: Avoidance of clash with structural elements, architectural elements, and other obstacle around; clearance for entry and exit requirements.
- Design constraints: clashes among pipes; branching of pipes, minimum equipment clearance requirements; buffer for danger zones; accessibility requirement for valves and accessories.

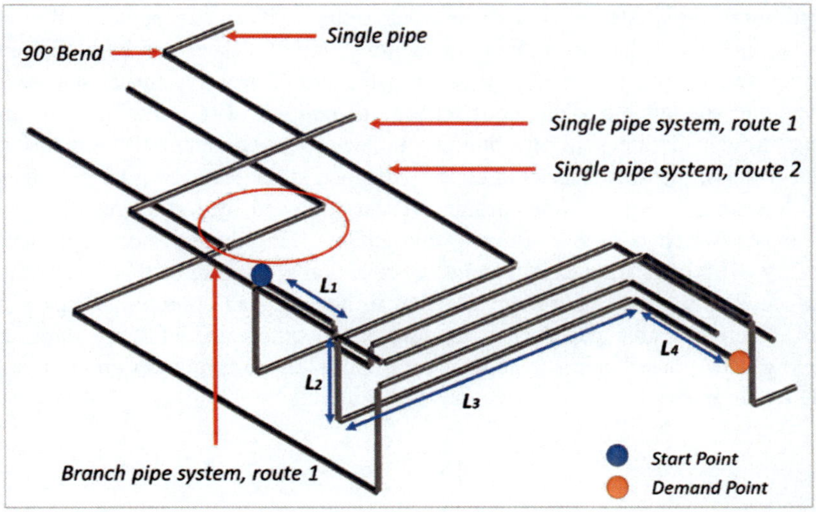

Fig. 1. Multiple pipe system layout

- Economical constraints: minimum pipe length and bends; installation cost (near to walls and obstacles for support); support sharing among pipelines.

Conventionally, engineers design the pipe route manually [7] or use physical model in which route planners interact with physical wires to find optimal route [8]. This conventional method requires huge time and cost with low accuracy. The use of computer-aided design (CAD) software has partially automated the manual process. Yet it requires human effort and decision making to plan, design and resolve the various design constraints. Also, different design aspects typically require different software's and the process is usually not well-integrated and hence making the design process expensive, tedious, labor intensive, error-prone, time-consuming and unadaptable to changes [9]. Moreover, the routing of pipes neither consider various pipe layout constraints nor considers all the possible solutions for the optimal layout that satisfy design requirements, thereby leading to uneconomical design. Physical pipe layout design is a complex spatial geometry problem. Since several solutions are possible for pipe routing design and checking each solution will be NP-Hard. Hence, a heuristic optimal layout technique for pipe system layout design is desired. Furthermore, the layout design gets more complicated and time consuming when the 3D space consists of multiple pipe systems. Therefore, a complete automation with heuristic design optimization approach is required to quickly and optimally plan multiple pipe layout in 3D environment.

With the advent of BIM technologies in the AEC industry, the conventional layout design process has been changed and improved with various potential design constraints addressed. BIM is used in this study because BIM can allow smooth integration of piping systems to architectural and structural envelope [10]. BIM employs digital representation of 3D parametric modeling of building to support the intelligent exchange of building information [11]. Furthermore, 3D BIM enables the designer to visualize and extract necessary geometry and semantic information of the space [12], location of equipments, start and end point of pipe system, and location of building (architectural and structural) components. Additionally, it can conceptualize routing of different pipe routes thus enabling quick identification of clashes and interference thus avoiding actual reworks on site. Besides, BIM provides a systematic approach to manage and support the intelligent exchange of building data information in digital format through the lifecycle of a building project [10]. Although BIM supports pipe design process, but still it lacks in clash avoidance at the stage of pipe routing, thus clash detection is required after design process followed by further modification, if clash exists, which is time consuming and tedious. Furthermore, pipe route as suggested by BIM application cannot guarantee selection of optimal path and still requires human intervention and consultant decision in choosing an appropriate pipe system layout. Therefore, this paper aims to develop an automated building information modeling (BIM) based heuristic approach for multiple pipe layout design optimization in 3D environment.

2 Literature Review

2.1 Multiple Pipe System Layout

Various sorts of automated strategies are developed and are widely used in present industrial practice, relatively less has been evolved on the layout routing problem in the literature. Several algorithms particle swarm optimization (PSO) [13], ant colony optimization (ACO) for ship building [14] and aircraft engines [3], genetic algorithm in plant layout (GA) [4], octree modeling and modified max-min ant system optimization algorithm in aero-engines [15] have been used for pipe routes planning. Yet, limited development has been made to pipe routing in built environment, as the constraints and requirements are distinct such as, objectives are of substantially different in nature, the constraints are often complex, and the routing is 3D. The researches mainly focus on single pipelines with two terminals with various imperceptive assumptions while developing algorithms. In addition, the optimization strategy only focused on clash-free pipe routes neglecting other factors such as economical, construction, design, operation, safety, etc. Siu and Niu [16] presented branch-pipe-routing approach for ships using improved genetic algorithm. However, the solutions obtained were only based on limited strategies thus could not be guaranteed as the optimal solution in terms industry practical constraints.

Although significant research efforts have been carried out to enhance the efficiency of pipe route planning in single or multiplanar, with fewer constraints, limited research have been focused on multiple pipe layout design in 3D. Also, the current researches do not focus on developing an automatic comprehensive strategy of extracting the details and information required to generate pipe layout planning, thus the process is still disintegrated and manual. Thereby leading to the process error prone due to lack of information flow. Therefore, this paper presents an integrated automated approach to generate optimal multiple pipe system layout in 3D environment.

2.2 Optimization Algorithm for Route Planning

Several algorithms have been developed to tackle NP-Hard optimization problems particularly when the environment is more complex and have shown better results in overcoming the limitation of classic methods [17]. Heuristic algorithms are most often employed when approximate solutions are enough and exact solutions are necessarily computationally expensive [18]. Global optimization algorithms have a goal of finding global optimum solutions. This generally involves more rigorous search than local optimization. However, the quality of the solutions obtained from the heuristic algorithms can be improved by spending additional time in the optimization process.

Some commonly used algorithms are Dijkstra's algorithm [19–21], Ant Colony Optimization has been used in path planning problems [22, 23], layout arrangement in marine engine room [24], pipe route design in ship design [14], aircraft engines [3], robot path planning [17]; Particle Swarm Optimization, it is a stochastic population-based optimization method, applied to solve many layout and routing problems such as cable routing optimization for offshore wind power plants [25], energy-competent routing protocol for MANETs [26], layout optimization for aero-engine [27], evacuation routing optimization [28], pipe routing [13, 29], layout arrangement in marine engine room [24]; Fruit Fly Algorithm is one of the recently proposed swarm intelligence optimization algorithm for solving shortest path problems for Mobile Ad-Hoc Network [30], travelling salesman problem [31, 32]; Genetic Algorithm has been used in various fields for shortest path and planning problems such as path planning for mobile robot planning [33], pipe routing [3], route planning [34–36], ship design [16], mobile ad-hoc networks [37]; A* Algorithm [38, 39], and Simulated Annealing (SA) for path optimization [40, 41]. SA is better than other techniques in finding global optima as it often accepts and check worst solution, thereby allowing more extensive search for the global optimal solutions [42].

Many of these heuristic methods are being widely used for route and path planning with various limitation such as requires more computational time, resulting in unavailable path, local optimized route, less accurate results, demands additional modification, requires lot of memory space. However, the combination of these heuristic algorithms can be effectively used to overcome their inherent limitation and can be efficiently used to plan out global optimized routes in a complex environment. Therefore, this paper presents an approach for multiple pipe layout design using combination of algorithms. Simulated annealing (SA) algorithm is used to approximate global optimization in a large search space for multiple pipe layout optimization. As for layout design, Dijkstra algorithm and two heuristic algorithms namely 3D A* and fruit fly optimization algorithm (FOA) are implemented and compared to obtain the multiple pipe system layout design. Dijkstra is used for comparison in this paper because of its ability to find the optimum route. Though, the time required to find the optimum path by Dijkstra becomes remarkably long when the search scope is broad, still it can provide better foundation for comparison with heuristic algorithms.

3 Proposed Approach

The objective of multiple pipe layout design approach is to find out an optimal clash-free layout of different pipe routes (paths) from their respective start location to an end location in a 3D environment with interference under various kinds of constraints. The approach includes large number (numerous) of pipe systems in a 3D layout space having single pipelines with two terminals and branch pipelines with multiple terminal. The proposed approach consists of four modules (1) BIM modeling, (2) Space and constraints modeling, (3) Fitness function, and (4) Optimization Algorithms.

3.1 BIM Modeling

To design a pipe system layout, geometric and semantic information about building layout space is needed and BIM technology allows smooth flow of these information such as, (1) information related to layout space are required to understand the confined working space for pipe system layout, (2) building components or objects information are needed to avoid interference of pipe routes with various obstacles and thereby, generating clash-free pipe layouts, (3) location of entrances, exits and gateways in order to provide accessible passage for user, (4) position and dimension of equipments to avoid interference with them and to provide necessary buffer for laying pipe routes around various equipments, (5) location coordinates of all the supply (start) and design (end) points as per the designer experience, (6) Selection of branch points as per the designer requirements. Branch points refers to the points adopting subdivision of pipe routes connecting one supply to many demands' points and many supplies to one demand point.

BIM is used to automatically extract and provide geometric and semantic information of 3D space for clash detection, spatial coordination, accessible routes and clearance, and multiple pipe system layout. In order to eliminate tedious human effort and errors, as well as to improve the efficiency of information flow, plug-ins for the BIM-platforms can be developed to automate the geometric and semantic information extraction. The information from BIM-model can be easily and automatically extracted and passed to an external file (e.g. spreadsheet, text document) for further design layout process.

3.2 Space and Constraint Modeling

Space modeling is a conceptual expression of space environment information. It is the key to obtain geometric information about the space environment quickly and accurately, which is important to generate a valid navigation graph. The graph, G = (N, E) is represented as combination of nodes (N) and edges (E) where, each node represents possible junction or joints 3D points and edge represents the segment running between two nodes. This method reduces our search space to a simple 3D array. Each node and edge in the graph are composed of weights which corresponds to a cost value reflecting any limitation as per the constraint modeling on total piping layout. In general, the cost function of total pipe layout is represented by cost of edge and nodes inhabited by the individual pipe routes. Thus, selecting the edges and nodes with lowest cost from the start node to end node will lead to generation of pipe layout with minimum cost.

Constraint modeling refers to the consideration and elimination of various constraints such as physical, design, operational and economical constraints as mentioned in Sect. 1. According to economical constraints, the pipe system layout should have minimum length, less bends, routed around the building elements for necessary supports and close enough to share common pipe supports, racks and hangers during installation. The economical constraints can be tackled by the formation of appropriate optimization function as mentioned in next section.

To eliminate physical constraints i.e. clashes with the building (architectural and structural) elements and other equipments and objects within the defined building space are examined. The nodes which intersects with various physical components are restricted for the pipe routes and are regarded as unwalkable for pipe layout. For example, in Fig. 2 the $node_{i+1}$ is restricted, as it clashes with the structural element and other pipe. In order to provide mandatory clearance for operational constraints, a buffer value based on piping code and requirements is selected and any nodes within that space is impeded and is restricted for routing of pipe layout. The clashes among various multiple pipe system, as shown in Fig. 2, are also tackled by a rule-based strategy and are discarded for pipe layout generation. Any clash of nodes and edges with already routed pipe systems during the layout generation is restricted as per the order of pipe system preference. For the branch pipe systems having common supply or end points can have a common route up to an appropriate location (node) depending upon the other constraints requirements.

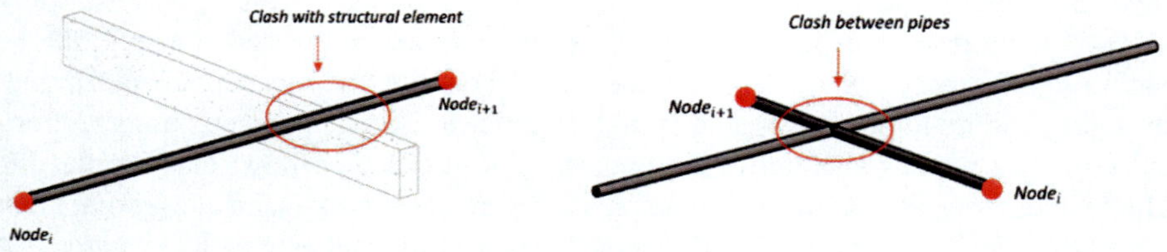

Fig. 2. Routing restriction due to clashes

3.3 Fitness Function

To evaluate the multiple pipe system layout for the optimal result, the total cost of the layout system is calculated in order to reduce its pressure loss by minimizing its length and bends, and installation cost, which is one of the major cost ignored while designing thereby leading to addition cost during installation or rework at the latter stage. The formulation of cost function for multiple pipe system layout is based on minimizing variables affecting pressure loss and installation cost.

Cost Due to Pressure Loss: The pressure loss suffered in the pipe system is caused due to the friction along the pipe length (straight path) and momentum exchanges results from the change in direction of the flow (bend) [44]. The friction along the pipe length is calculated as per the total distance travelled by the edges from the supply point to the demand point, following the pipe route. The distance travelled between the two nodes is calculated using Manhattan method [45], as shown in Eq. (1). The momentum exchanges are represented by the cost of nodes on which a bend is introduced. The cost of bend is calculated by the equivalent length method depending on the bend type. Equivalent method calculates the pressure loss through a bend as a length of straight pipe under the corresponding flow rate [44].

$$\mathrm{Li} = abs(X_i - X_{i+1}) + \mathrm{abs}(Y_i - Y_{i+1}) + \mathrm{abs}(Z_i - Z_{i+1}) \tag{1}$$

The cost function (Cp) based on pressure loss is calculated by the cost for movement for the start to end node (cost of edges and nodes), following the route, and is given as in Eq. (2):

$$C_P = \mathrm{C_m}\left[\sum\nolimits_{i=1}^{n} \mathrm{L_i} + \sum\nolimits_{j=1}^{x} \mathrm{NB_j}\right] \tag{2}$$

where, C_P = Cost for movement from the start to end node (US$); C_m = material cost of pipe per unit length (US$/m); L_i = total distance travelled among the nodes (m); NB_j = equivalent length for a bend (j) (m); n = total number of nodes; x = total number of bends. In this study, 90° curved elbow is considered and thus, NB_j = 30.

Installation Cost: The installation cost refers the cost incurred in fitting pipe system on site and depend on the location of pipe system routes. Nearer (with minimum clearance) the pipe system with a supporting component (architectural and structural elements) to hold the pipe system, lesser is the installation cost. Hence, the total installation cost of a pipe system is evaluated as per the perpendicular distance of pipe route nodes from the nearest supporting elements such as wall, column, beam etc., as shown in Fig. 3 and is given by Eq. (3):

$$C_I = I_C \sum\nolimits_{i=1}^{n} d_i \tag{3}$$

where, C_I = Installation cost of a pipe system; I_c = Installation Cost of pipe/unit length (US$/m); d_i = perpendicular distance of node from the nearest supporting elements (i) from supporting elements, (m); n = total number of nodes.

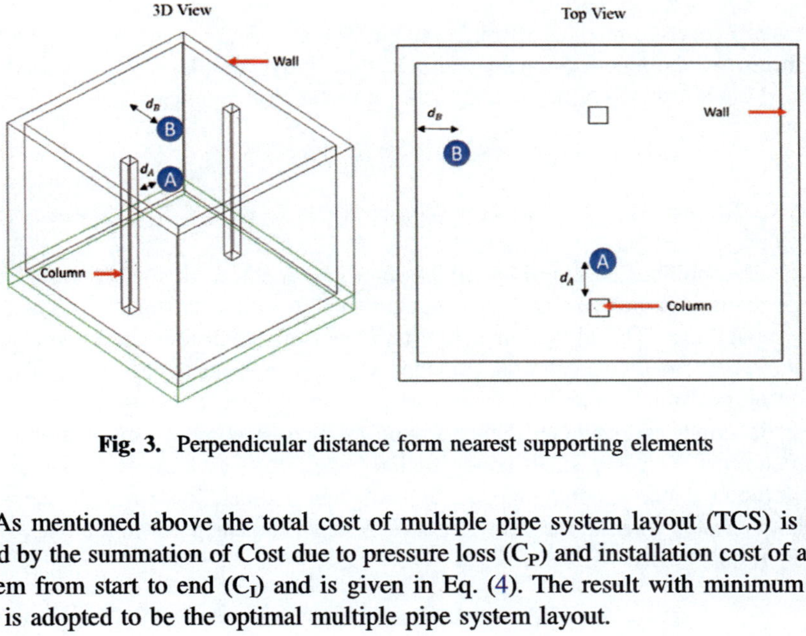

Fig. 3. Perpendicular distance form nearest supporting elements

As mentioned above the total cost of multiple pipe system layout (TCS) is evaluated by the summation of Cost due to pressure loss (C_P) and installation cost of a pipe system from start to end (C_I) and is given in Eq. (4). The result with minimum total cost is adopted to be the optimal multiple pipe system layout.

$$Min\,TCM = C_P + C_I = C_m\left[\sum\nolimits_{i=1}^{n} L_i + \sum\nolimits_{i=1}^{n} NB_j\right] + I_C \sum\nolimits_{i=1}^{n} d_i \tag{4}$$

3.4 Optimization Algorithms

This paper presents two sets of optimization approaches one for individual pipe route optimization and other for finding global optimum layout among large search spaces.

Pipe Route Optimization aims to find optimal individual pipe route connecting a single supply and demand point. The approach considerers two heuristic algorithm, namely A* and Fruit fly algorithm, and Dijkstra algorithm for pipe route optimization.

A* algorithm is a graph traversal greedy best-first-search algorithm and finds a least-cost path among possible solutions from start to end node using heuristic function [46]. As this paper demonstrate pipe system route in a 3D environment, the A* algorithm is modified to 3D A*. For the successful application of 3D A* algorithm, the 3D space is converted to weighted graph (nodes and edges). The nodes which are available for path is marked as walkable and the nodes which are not available for the path is marked as unwalkable.

As this study considers pipe route in the straight and orthogonal (90°) direction along the coordinated axis. During each search step in the 3D environment, there are six possible movement direction 2 in X-axis, 2 in Y-axis, and 2 in Z-axis. Path finding begin with the start node and moves to the other walkable neighbor nodes next to it until the end node is reached. The cost function of path as per A* algorithm fitness function as given in Eq. (4) and heuristic cost function. In this study, the Manhattan method [30] is used to determine the heuristic cost function shown in Eq. (5), as it is the standard heuristic method of square graph. The algorithm so far gives us shortest route based on the length (straight path). In this study, the 3D A* is modified and improved based on length, bends and the installation requirements.

$$\mathrm{h(n)} = abs(X_c - X_t) + \mathrm{abs}(Y_c - Y_t) + \mathrm{abs}(Z_c - Z_t) \tag{5}$$

where, X_c, Y_c, Z_c = X, Y, Z values of the current node; X_t, Y_t, Z_t = X, Y, Z values of the target node

Fruit fly optimization algorithm (FOA) is a new method for finding global optimization based on the food finding behaviour of the fruit fly using their smell and vision capabilities. The algorithm starts with position the fruit flies randomly and finding the best smell concentration (Smellc), which is given as per distance (Distc) of demand point from the current location as given in Eq. (6.1) and (6.2), among the different locations and using the same location for next iteration and calculation of the movement cost. The best smell concentration represents least cost of pipe route. In general, food location is not known, but in this layout optimization, we know the target location, and thus when the target location is reached the smell concentration will be highest and the search will be stopped.

$$\mathrm{Dist_c} = abs(X_c - X_t) + \mathrm{abs}(Y_c - Y_t) + \mathrm{abs}(Z_c - Z_t) \tag{6.1}$$

$$\mathrm{Smell_c} = \mathit{Fitness\,Function}\left(\frac{1}{Dist_c}\right) \tag{6.2}$$

Dijkstra algorithm is an exhaustive search algorithm to find optimum route. The implementation is same as A* algorithm only the heuristic function, considering the current node to target node, is eliminated. Path finding begin with the start node and moves to the other walkable neighbour nodes (six possible orthogonal movements) next to it until the end node is reached.

Finding an optimal individual pipe route based on pipe route optimization algorithm for a single system does not guarantee optimal multiple pipe system layout, when the 3D space consists of numerous pipe systems. Thus, a metaheuristic algorithm namely simulated annealing is combined with route optimization algorithm to obtain optimal clash-free layout of multiple pipe system in terms of minimum cost.

Global Layout Optimization is done using simulated annealing (SA). SA is a probabilistic technique for approximating the global optimum of a given objective function. It is basically composed of two stochastic processes: firstly, the generation of solutions and secondly the acceptance solutions. SA basically performs by generating a random solution and calculating the cost using the defined objective functions. Followed by generating a new neighboring solution and calculating new cost of the solution using the same objective function for all the multiple pipe system in the layout space as given by Eq. (4). The two results are then compared for the better result, if the better solution is found then it is accepted and marked as current solution otherwise the acceptance of the solution is done by probability function determined by Boltzmann distribution of the cost difference as calculated by Eq. (7). The acceptance criteria of the worse solution depend on the difference between the two solutions and the temperature parameter, as the worse solution are more likely to get accepted at high temperature. The process continues until the termination criteria is met i.e. final temperature is reached.

$$P = e^{\frac{(S_{old} - S_{new})}{K_b T}} \tag{7}$$

where, P = probability function of the acceptance criteria, S_{old} = initial (old) cost solution, S_{new} = new neighboring cost solution, K_b = Boltzmann constant, T = current temperature.

4 Illustrative Example

To illustrate the proposed approach, a typical BIM model of water plant room (see Fig. 4) is studied. Assuming that there is finite number of fixed obstacles and interferences such as architectural and structural elements, equipments, etc. in a 3D layout space. The purpose of multiple pipe layout planning is to obtain clash-free and optimized route for pipe system in the confined space. Dynamo was used to capture the 3D geometric, such as information related to plant room layout space, building components location and dimension as shown in Table 1, and semantic information, like location of supply and demand points, operational requirements and constraints information (branch pipe systems, buffer requirements, accessible clearance, etc.) from BIM model. Dynamo is a visual programming tool, which works as plug-in in Revit [48]. A total of nine routes were used for multiple pipe layout illustration. Supply and demand points of respective routes are shown in Table 2. All the information extracted from Dynamo is stored in excel file, which is further used for pipe layout optimization and determination. Optimization of pipe route depends on a minimum of Li, NBj, di variables see Eq. (4), resulting in more than a million possible solutions. Checking each solution will be NP-Hard and unserviceable, as it will increase time and space complexity. Hence, a two heuristic technique namely, 3D A* and FOA, and Dijkstra algorithm that is efficient and reliable to find optimal pipe route is combined with simulated annealing, due to its flexibility and its ability to approach global optimal solution of multiple pipe system layout within a reasonable amount of time. In this example, the developed 3D A*, FOA, Dijkstra - SA approach-based algorithm was implemented in Visual Basic Application in Excel.

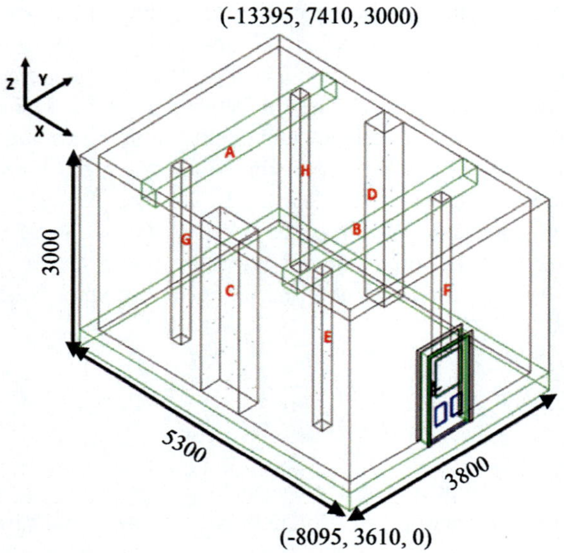

Fig. 4. Typical example of water plant room

Table 1. 3D geometric information of the layout space

S. No.	Object	(X, Y, Z) minimum coordinate value (mm)	(X, Y, Z) maximum coordinate value (mm)
1)	Plant room	(−13395, 3610, 0)	(−8095, 7410, 3000)
2)	Beam A	(−12497, 3610, 2700)	(−12197, 7410, 3000)
3)	Beam B	(−9520, 3610, 2700)	(−9220, 7410, 3000)
4)	Column C	(−11175, 7010, 0)	(−10775, 7410, 3000)
5)	Column D	(−11235, 3610, 0)	(−10435, 4010, 3000)
6)	Column E	(−9474, 6710, 0)	(−9274, 6910, 3000)
7)	Column F	(−9474, 4210, 0)	(−9274, 4410, 3000)
8)	Column G	(−12454, 6710, 0)	(−12254, 6910, 3000)
9)	Column H	(−12454, 4210, 0)	(−12254, 4410, 3000)

Table 2. Supply and demand points for pipe systems

Route. No.	Supply point (start)	Demand point (end)
R1.	(−9000, 3800, 2000)	(−13000, 3800, 2000)
R2.	(−9000, 3800, 2000)	(−13000, 3800, 1000)
R3.	(−13300, 4800, 2000)	(−8800, 3800, 2500)
R4.	(−8500, 7200, 1600)	(−9000, 3700, 2600)
R5.	(−9000, 3700, 2800)	(−10000, 7200, 800)
R6.	(−12000, 7200, 500)	(−11500, 6700, 2500)
R7.	(−10300, 7200, 2500)	(−13300, 5200, 500)
R8.	(−9800, 7200, 1500)	(−13300, 6700, 500)
R9.	(−9500, 7000, 500)	(−9100, 6600, 2500)

A flow chart depicting the developed approach for the multiple routing layout is given in Fig. 5. To determine the pipe route, the room layout space was converted to graph of 100 mm × 100 mm × 100 mm grid size. A random order of all the pipe routes was selected and used for multiple pipe layout generation. The pipe routing of each system starts in the order of the selected order based on the adopted pipe route optimization algorithm. The path finding starts from the start node, to find the next clash-free optimal node for the path, total six orthogonal neighbor nodes were checked, and their respective total movement cost as mentioned in Eq. (4). was calculated. The neighbor node with the lowest movement cost was chosen to be next potential search node. The search for the next potential node continues until the target node was reached. The nodes that obstruct the building objects or already exists in some other pipe routes are discarded and marked as unwalkable. A new strategy is proposed to avoid the clashes among different pipe routes. The nodes that clash with already existing pipe routes are restricted and discarded for the new pipe routes. Any node that clashes with the start and end points (nodes) of the other pipe routes are marked as unwalkable except if the multiple pipe routes have same start points or end points.

After finding all the pipe routes as per respective selected order. The total movement cost for the entire multiple pipe system layout is calculated and is marked as the current cost. Following which new random order of pipe route is generated and is selected to determine new pipe routes and the new multiple pipe system layout cost. Subsequently, metropolis cycle and the temperature cycle are also updated. The process continues until the stopping criteria (final temperature) is reached and resulting in the multiple pipe layout having minimum layout cost. The simulated annealing parameters used, and their respective values are presented in Table 3.

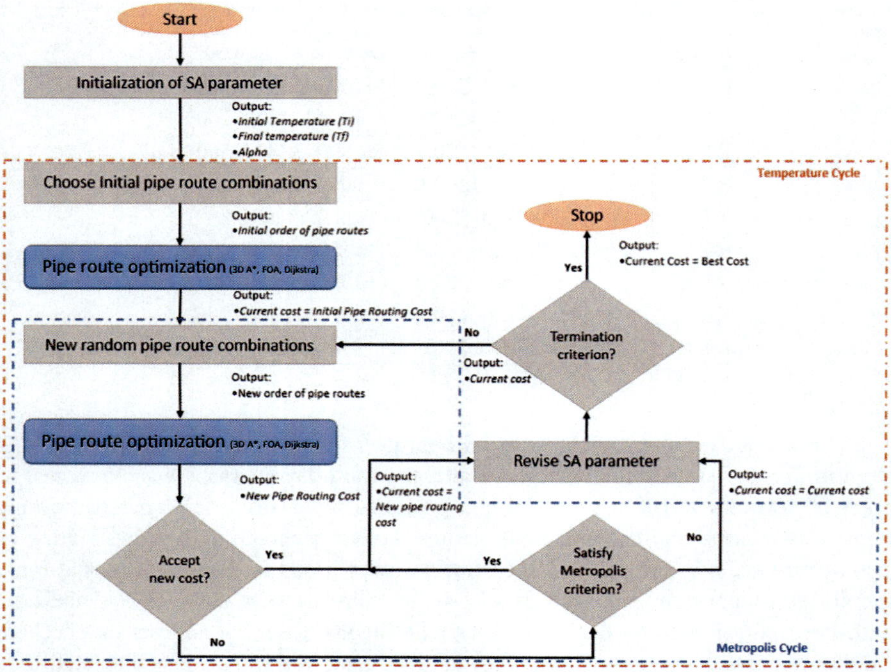

Fig. 5. The developed approach for the multiple routing layout optimization

Table 3. Simulated annealing parameter

S. No.	Parameter	Value
1.	Initial temperature (T_i)	0.1
2.	Final temperature (T_f)	10
3.	Alpha (α)	0.9
4.	Cooling schedule equation	T_{i+1} (current temperature) = $\alpha * T_i$

The output of the developed approach using all the above-mentioned path route optimization algorithms (3D A*, FOA, Dijkstra) - SA approach for multiple pipe system layout are compared (Table 4) among each other based on cost of pipe layout and time taken by the algorithm. The variation of cost curve and convergence curve of all the algorithms with each iteration is shown in Fig. 6 and Fig. 7. While implementation of 3D A* algorithm for pipe route optimization, four possibilities were considered to know the how cost is comprised with time, as represented in Table 4. For this purpose, the heuristic cost (h_C) function used in 3D A* algorithm was multiplied by factor of 10000, 1000, 100, 1 (original 3D A*). Higher the factor value, lesser is the time to run the algorithm. The 3D A* results are also compared with FOA and Dijkstra results as shown in Table 4. After comparing the respective algorithms results it was found that 3D A with $1 * h_C$ shows best result with minimum total layout cost which is same as Dijkstra algorithm i.e. $644450 and takes approximately 50% lesser time than the Dijkstra algorithm. The output of the multiple pipe system layout is shown in Fig. 8.

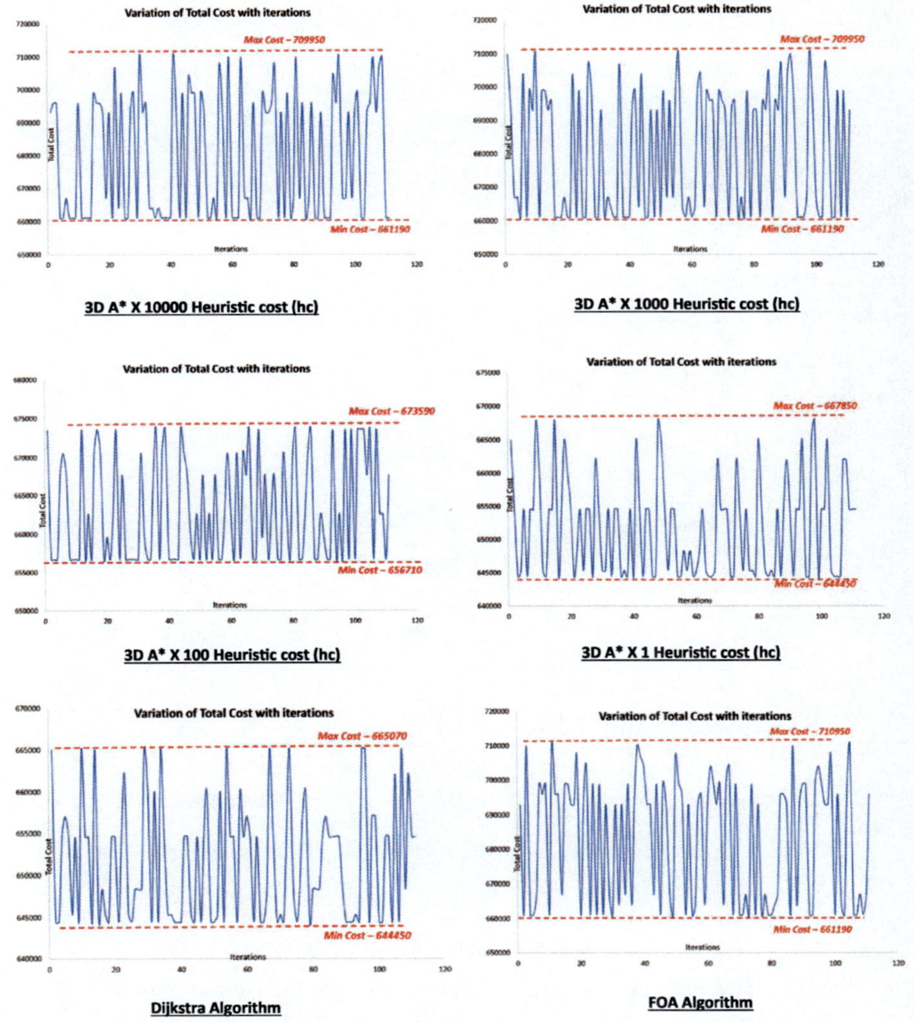

Fig. 6. Comparison of variation cost of multiple pipe system layout

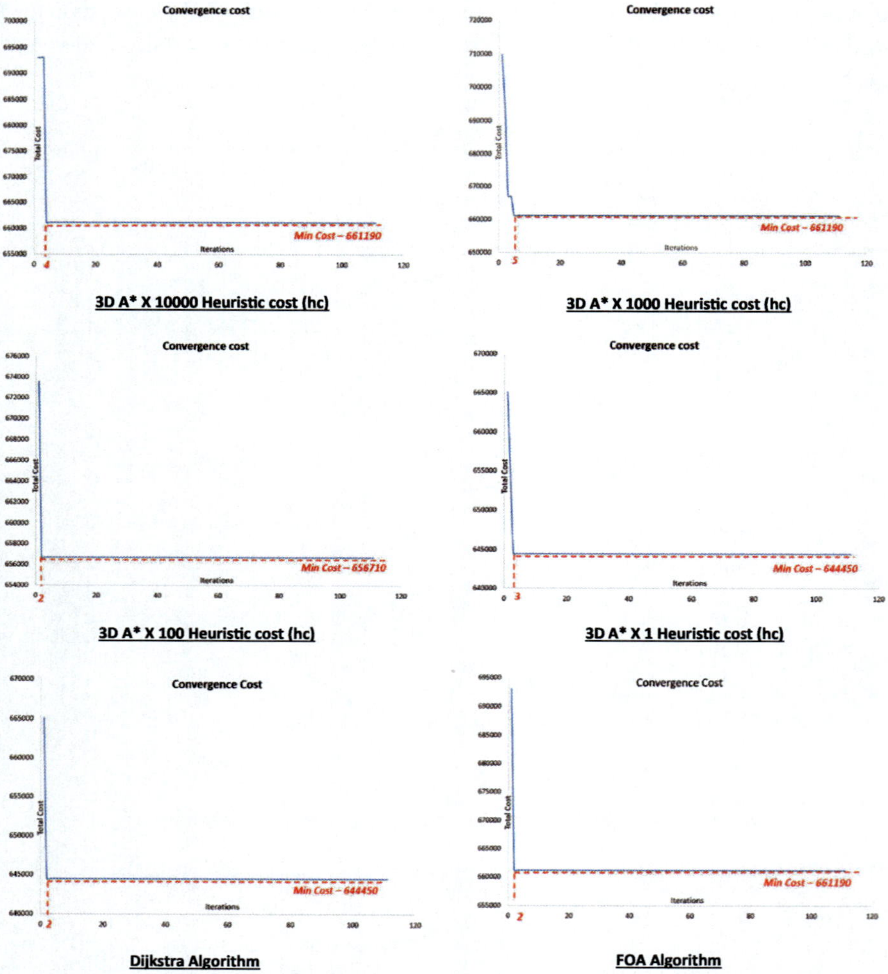

Fig. 7. Comparison of convergence cost of multiple pipe system layout

Table 4. Comparison of output of multiple pipe system layout

S. No.	Pipe route optimization algorithm - SA		Cost ($US)	Time (hh:mm:ss)
1.	3D A* - SA	10000 × heuristic cost	661190	00:37:18
2.		1000 × heuristic cost	661190	00:41:49
3.		100 × heuristic cost	656710	06:07:17
4.		1 × heuristic cost	**644450**	**13:29:31**
5.	Dijkstra		**644450**	28:07:27
6.	FOA		661190	00:33:20

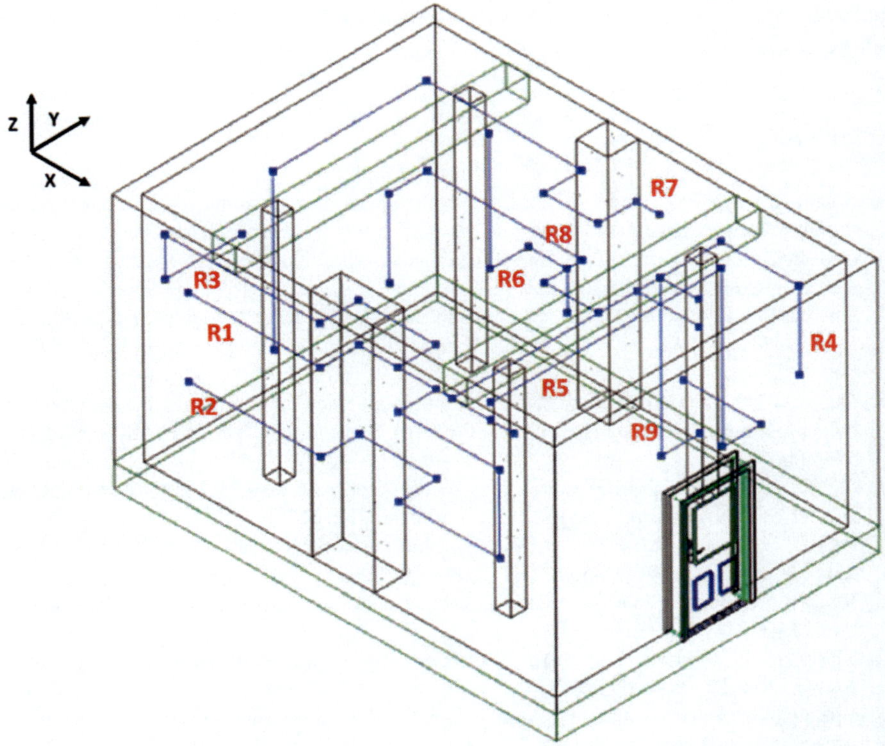

Fig. 8. Output of multiple pipe system layout

5 Conclusion and Future Work

Multiple pipe system layout design and spatial coordination plays an important and challenging role in construction industry. In this paper, a BIM-based approach is proposed to automate multiple pipe system layout optimization in 3D environment using combination of two algorithms. The approach compares pipe route optimization algorithms, Dijkstra algorithm and two heuristic methods (A* and FOA algorithm). The algorithms are modified and used in a directed weighted graph to obtain the multiple pipe system layout. Clashes among pipes and with building components are considered and avoided in the layout optimization. The approach also considers branch pipe routing connecting one supply to many demands points and many supplies to one demand point. The framework is also tested by an illustrative example to further explain the functionalities of the different modules in the framework. After comparing the respective algorithms results it was found that 3D A* shows best result with minimum total layout cost which is same as Dijkstra algorithm and takes approximately 50% lesser time than the Dijkstra algorithm. With the adoption of this framework, time and human effort for the optimized pipe systems design are significantly reduced. The developed framework can also be used in designing of other trades such as mechanical etc. with further modifications. Therefore, extending the framework for other trades will be considered in the future work.

References

1. Singh, J., Deng, M., Cheng, J.C.P.: Implementation of mass customization for MEP layout design to reduce manufacturing cost in one-off projects. In: Proceeding of the 26th Annual Conference of the International Group for Lean Construction, Chennai, India, pp. 625–635 (2018). http://iglc.net/Papers/Details/1587
2. Guirardello, R., Swaney, R.E.: Optimization of process plant layout with pipe routing. Comput. Chem. Eng. **30**, 99–114 (2005). https://doi.org/10.1016/j.compchemeng.2005.08.009
3. Qu, Y., Jiang, D., Gao, G., Huo, Y.: Pipe routing approach for aircraft engines based on ant colony optimization, J. Aerosp. Eng. **29**(3) (2016). https://doi.org/10.1061/(ASCE)AS.1943-5525.0000543
4. Ito, T.: A genetic algorithm approach to piping route path planning. J. Intell. Manuf. **10**, 103–114 (1999). https://doi.org/10.1023/A:1008924832167
5. Park, J.-H., Storch, R.L.: Pipe-routing algorithm development: case study of a ship engine room design. Expert Syst. Appl. **23**(3), 299–309 (2002)
6. Kang, S., Myung, S., Han, S.: A design expert system for auto-routing of ship pipes. J. Ship Prod. **15**(1), 1–9 (1999)
7. Blaschke, J.C., Jatzek Jr., H.A.: Electronic engine 'MockUp' shortens design time. Aerosp. Am. **23**, 98–100 (1985)
8. Bile, W., Ruxin, N., Jianhua, L.: Architecture of cable harness and tube assembly planning system in virtual environment. Comput. Integr. Manuf. Syst. **13**, 1579–1585 (2007)
9. Jebamalai, J.M., Marlein, K., Laverge, J., Vandevelde, L., van den Broek, M.: An automated GIS-based planning and design tool for district heating: scenarios for a Dutch city. Energy **183**, 487–496 (2019)
10. Qiqi, L., Wong, Y.-H.: A BIM-based approach to automate the design and coordination process of mechanical, electrical, and plumbing systems. HKIE Trans. **25**(4), 273–280 (2018). https://doi.org/10.1080/1023697x.2018.1537813
11. Eleftheriadis, S., Duffour, P., Stephenson, B., Mumovic, D.: Automated specification of steel reinforcement to support the optimization of RC floors. Autom. Constr. **96**, 366–377 (2018). https://doi.org/10.1016/j.autcon.2018.10.005
12. Wong, J.K.W., Zhou, J.: Enhancing environmental sustainability over building life cycles through green BIM: a review. Autom. Constr. **57**, 156–165 (2015). https://doi.org/10.1016/j.autcon.2015.06.003
13. Wang, C., Sun, X., Sun, L., Yuan, T.: A method based on PSO for pipe routing design. In: Proceeding of the 6th Annual IEEE International Conference on Cyber Technology in Automation, Control and Intelligent Systems, Chengdu, China (2016). https://doi.org/10.1109/CYBER.2016.7574862
14. Jiang, W.-Y., Chen, Y.L.M., Yu, Y.-Y.: A co-evolutionary improved multi-ant colony optimization for ship multiple and branch pipe route design. Ocean Eng. **102**, 63–70 (2015). https://doi.org/10.1016/j.oceaneng.2015.04.028
15. Qu, Y.-F., Jiang, D., Zhang, X.-L.: A new pipe routing approach for aero-engines by octree modeling and modified max-min ant system optimization algorithm. J. Mech. **34**(1), 11–19 (2018). https://doi.org/10.1017/jmech.2016.86

16. Sui, H., Niu, W.: Branch-pipe-routing approach for ships using improved genetic algorithm. Front. Mech. Eng. **11**(3), 316–323 (2016). https://doi.org/10.1007/s11465-016-0384-z
17. Xu, S., Ho, E.S.L., Shum, H.P.H.: A hybrid metaheuristic navigation algorithm for robot path rolling planning in an unknown environment. Mechatron. Syst. Control **47**(4), 216–224 (2019). https://doi.org/10.2316/J.2019.201-3000
18. Cook, S.A.: An overview of computational complexity. Commun. ACM **26**, 401–408 (1983)
19. Dijkstra, E.W.: A note on two problems in connexion with graphs. Numer. Math. **1**, 269–271 (1959)
20. Padhy, C.P., Sen, D., Bhaskaran, P.K.: Application of wave model for weather routing of ships in the North Indian Ocean. Nat. Hazards **44**(3), 373–385 (2008)
21. Wang, H., Mao, W., Eriksson, L.: A three-dimensional Dijkstra's algorithm for multi-objective ship voyage optimization. Ocean Eng. **186**, 106131 (2019)
22. Dorigo, M., Maniezzo, V., Colorni, A.: Ant system: optimization by a colony of cooperating agent. IEEE Trans. Syst. Man Cybern.-Part B Cybern. **26**(1), 29–41 (1996)
23. He, Y., Zeng, Q., Liu, J., Xu, G.: Path planning for indoor UAV based on ant colony optimization. In: Proceedings of the 25th Chinese Control and Decision Conference (CCDC), Guiyang, China, pp. 2919–2923 (2013)
24. Jiang, W.-Y., Lin, Y., Chen, M., Yu, Y.-Y.: An optimization approach based on particle swarm optimization and ant colony optimization for arrangement of marine engine room. J. Shanghai Jiaotong Univ. **48**(4), 502–507 (2014)
25. Jin, R., Hou, P., Yang, G., Qi, Y., Chen, C., Chen, Z.: Cable routing optimization for offshore wind power plants via wind scenarios considering power loss cost model. Appl. Energy **254** (2019). Article no. 113719. https://doi.org/10.1016/j.apenergy.2019.113719
26. Kumar, S.S., Manimegalai, P., Karthik, S.: An energy - competent routing protocol for MANETs: a particle swarm optimization approach. In: Proceedings of IEEE International Conference on Soft-Computing and Network Security (2018). Article no. 8573677. https://doi.org/10.1109/icsns.2018.8573677
27. Liu, Q., Mao, L.: Multi-objective layout optimization for branch pipe of aero-engine based on MOPSO. J. Mech. Eng. **54**(19), 197–203 (2018). https://doi.org/10.3901/JME.2018.19.197
28. Zhu, J.-L., Li, W., Li, H., Wu, Q., Zhang, L.: A novel swarm intelligence algorithm for the evacuation routing optimization problem. Int. Arab J. Inf. Technol. **14**(6), 880–889 (2017)
29. Liu, Q., Wang, C.: A discrete particle swarm optimization algorithm for rectilinear branch pipe routing. Assembly Autom. **31**(4), 363–368 (2011). https://doi.org/10.1108/01445151111172952
30. Darwish, S.M., Elmasry, A., Ibrahim, S.H.: Optimal shortest path in mobile ad-hoc network based on fruit fly optimization algorithm. In: The International Conference on Advanced Machine Learning Technologies and Application, pp. 91–101 (2019)
31. Jiang, Z.-B., Yang, Q.: A discrete fruit fly optimization algorithm for the traveling salesman problem. PLoS ONE **11**(11), e0165804 (2016). https://doi.org/10.1371/journal.pone.0165804
32. Iscan, H., Gunduz, M.: An application of fruit fly optimization algorithm for traveling salesman problem. Procedia Comput. Sci. **111**, 58–63 (2017). https://doi.org/10.1016/j.procs.2017.06.010
33. Shitagh, N.A., Jalal, L.D.: Path planning of intelligent mobile robot using modified genetic algorithm. Int. J. Soft Comput. Eng. (IJSCE) **3**(2), 31–36 (2013)
34. Mathias, H.D., Foley, S.S.: Improving a genetic algorithm for route planning using parallelism with speculative execution. In: ACM International Conference Proceeding Series (2019). Article no. 3333251. https://doi.org/10.1145/3332186.3333251

35. Xin, J., Zhong, J., Yang, F., Cui, Y., Sheng, J.: An improved genetic algorithm for path-planning of unmanned surface vehicle. Sensors (Switzerland) **19**(11) (2019). Article no. 2640. https://doi.org/10.3390/s19112640
36. Alves, R.M.F., Lopes, C.R.: Using genetic algorithms to minimize the distance and balance the routes for the multiple traveling salesman problem. In: IEEE Congress on Evolutionary Computation, CEC 2015 - Proceedings, pp. 3171–3178 (2015). Article no. 7257285. https://doi.org/10.1109/cec.2015.7257285
37. Hasdemir, S., Yilmaz, S., Sen, S.: A novel multi-featured metric for adaptive routing in mobile ad hoc networks. Appl. Intell. **49**(8), 2823–2841 (2019). https://doi.org/10.1007/s10489-018-01401-4
38. Kumar, S.S., Cheng, J.C.P.: A BIM-based automated site layout planning framework for congested construction sites. Autom. Constr. **59**, 24–37 (2015). https://doi.org/10.1016/j.autcon.2015.07.008
39. Tan, Y., Song, Y., Liu, Y., Wang, X., Cheng, J.C.P.: A BIM-based framework for lift planning in topsides disassembly of offshore oil and gas platforms. Autom. Constr. **79**, 19–30 (2017). https://doi.org/10.1016/j.autcon.2017.02.008
40. Hajad, M., Tangwarodomnukun, V., Jaturanonda, C., Dumkum, C.: Laser cutting path optimization using simulated annealing with an adaptive large neighborhood search. Int. J. Adv. Manuf. Technol. **103**(1–4), 781–792 (2019). https://doi.org/10.1007/s00170-019-03569-6
41. Daryanavard, H., Harifi, A.: UAV path planning for data gathering of IoT nodes: ant colony or simulated annealing optimization. In: Proceedings of 3rd International Conference on Internet of Things and Applications, IoT 2019 (2019). Article no. 8808834. https://doi.org/10.1109/iicita.2019.8808834
42. Kirkpatrick, S., Gelatt, C.D., Vecchi, M.P.: Optimization by simulated annealing. Science **220**(4598), 671–680 (1983). https://doi.org/10.1126/science.220.4598.671
43. Nayyar, M.L.: Piping Handbook. Mcgraw-Hill, New York (2002). ISBN 9780070471061
44. Black, P.E., Vreda, P.: Manhattan distance. In: Dictionary of Algorithms and Data Structures, pp. 16–54 (2006)
45. Hart, P.E.G., Nilsson, N.J., Raphael, B.: A formal basis for the heuristic determination of minimum cost paths. IEEE Trans. Syst. Sci. Cybern. **4**, 100–107 (1968)
46. Autodesk Inc., Dynamo. http://dynamobim.org/

Guidance System for Directional Control in Shield Tunneling Using Machine Learning Techniques

Kensuke Wada(✉), Hirokazu Sugiyama, Kojiro Nozawa, Makoto Honda, and Shinya Yamamoto

Shimizu Corporation, 3-4-17 Etchujima, Koto-Ku, Tokyo, Japan
wada@shimz.co.jp

Abstract. In shield tunneling, it is essential to control the shield jacks appropriately so that the shield machine follow a planned path, which depends on the position of the shield machine and the geological conditions. However, the quality of the directional control of the shield machine depends on the skill of its operator. Herein, a guidance system that provides a method for controlling shield jacks that is equivalent to the techniques utilized by a skillful operator is described. The developed guidance system uses machine learning models trained by gradient tree boosting with the operational data related to the actions of skilled operators. The models predict the optimal point at which the resultant force of shield jacks should be acted upon to control the propulsive direction of a shield machine. To validate the performance of the guidance system, a shield machine was driven according to the system at a site under construction. Deviation from the planned path and the attitude of the shield machine were within set tolerances. The results show that our guidance system has applicability in real environments, indicating the future possibility of self-driving shield machines.

Keywords: Shield tunneling machine · Guidance system · Machine learning

1 Introduction

In Japan, the number of construction workers continues to decrease because of a declining birth rate and an aging population. Employing skilled workers is becoming increasingly difficult. Therefore, cutting down on labor and assisting inexperienced workers are required.

The quality of shield tunneling depends on the skill of the operator who controls the shield machine because the operation method of the shield machine is largely left to the operator. Thus, the shortage of skilled operators leads to the deterioration of the shield tunnel's quality. Self-driving shield machines are effective in ensuring the quality of shield tunnels, but there have been several challenges in their development thus far. For example, the operations of a screw conveyor for discharging excavated soil and shield jacks for advancing a shield machine were automated using fuzzy models [1]. There has also been a study in which the shield jacks to be activated were selected automatically by using monitoring data to estimate the physical properties of the ground

E. Toledo Santos and S. Scheer (Eds.): ICCCBE 2020, LNCE 98, pp. 73–88, 2021.
https://doi.org/10.1007/978-3-030-51295-8_7

and calculating the moment required for excavation [2]. However, these methods did not lead to full automation of shield machines: differences in ground and construction conditions in various locations meant that it was difficult to predict the proper operations of shield machines.

Recently, machine learning has attracted attention. It has become possible to construct high-performance models easily. The realization of self-driving shield machines is feasible if machine learning techniques are applied to the monitoring data of shield tunneling. It has already been shown that models capable of making judgments equivalent to that of an operator can be constructed using machine learning [3–6]. These models predict how the rotational speed of a screw conveyor should be set and the optimal point at which the resultant force of shield jacks should be acted upon. As the first step toward self-driving shield machines, a guidance system for directional control in shield tunneling using machine learning techniques was proposed [7].

This study presents the results of investigating the applicability of the proposed guidance system. To investigate the system's feasibility, the machine learning models used by the guidance system were trained with the data obtained at an existing shield tunnel, and the models' accuracy was evaluated. After confirming that the outputs of the models were comparable with operation by experts, a shield machine was driven according to the guidance system at another site under construction. To validate the performance of the guidance system, deviation from the planned path and the attitude of the shield machine were assessed.

2 Shield Tunneling Method

2.1 Overview

Figure 1 shows a schematic diagram of the shield tunneling method. In this method, a tunnel is constructed while a tunneling shield supports the surrounding ground. It stabilizes the face using earth or slurry in a chamber to counter earth and hydraulic pressure. The shield tunneling method is often used for construction in soft ground because it makes possible the suppression of the deformation of the ground. The construction process of the shield tunneling method is as follows: excavating the ground in contact with the face using a cutter, discharging the excavated soil in a chamber using a screw conveyor, advancing a shield machine using shield jacks that are mounted along the circumference of a shield, and assembling the lining segments.

2.2 Operations of a Shield Machine

Figure 2 shows an example of the operational interface of a shield machine. Three types of action exist in the operator's manipulations: controlling the propulsive direction of a shield machine, making the pressure in the chamber equal to that of the ground, and injecting an additive agent to the chamber and a bulking agent to the tail voids. An operator must manipulate the shield machine while comparing the planned values with the values shown on the shield machine's sensors. The planned values are the shield machine's azimuth direction, its pitch, the extension lengths of the shield jacks' strokes, and so on. It is not easy to manipulate a shield machine to match all the planned values, so operators need to be skilled at their work.

Controlling the propulsive direction of a shield machine is the most important task in ensuring the accurate construction of a shield tunnel. When an operator activates only some shield jacks, the force acting on the shield machine is decentered, and a shield machine can veer to follow the planned path.

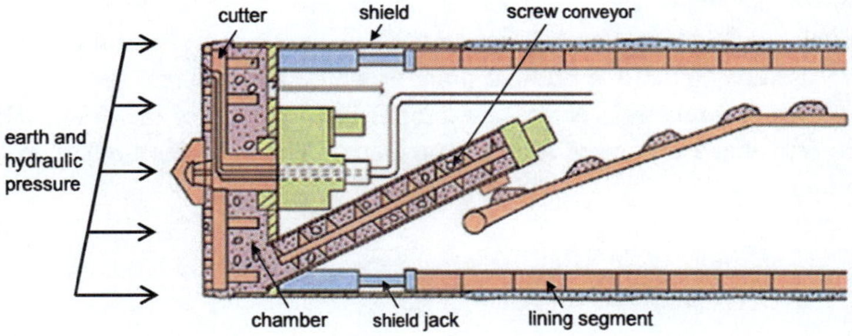

Fig. 1. Schematic of the shield tunneling method (the figure on the website of Shield Tunneling Association of Japan [8] was revised).

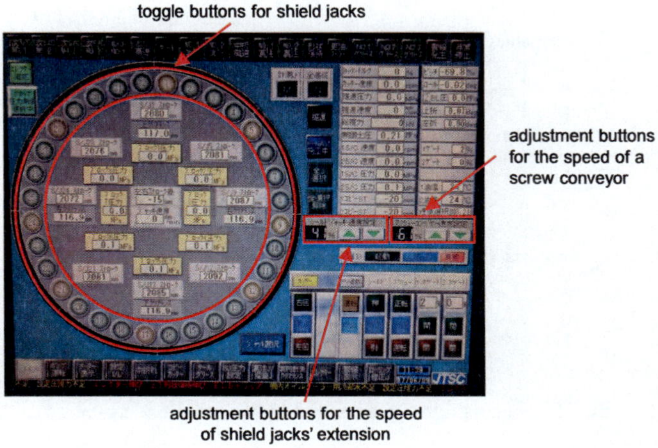

Fig. 2. Example of the operational interface of a shield machine.

3 Guidance System for Directional Control in Shield Tunneling

3.1 Outline of the Guidance System

The developed guidance system proposes the optimal way to control the shield jacks and to make a shield machine follow a planned path. Figure 3 shows the processing flow executed by the guidance system. In the first step, a supervisor inputs the planned values required for the day's excavation. When the shield machine starts to excavate, the guidance system obtains data from the shield machine and calculates the feature variables. The necessity for shield jacks' operations is predicted from the feature variables using a machine learning model. When necessary, other machine learning models are used to predict the optimal point at which the resultant force of the shield jacks should be acted upon to control the propulsive direction of the shield. The shield jacks to be activated are selected from the predicted point and presented to an operator, as shown in Fig. 4.

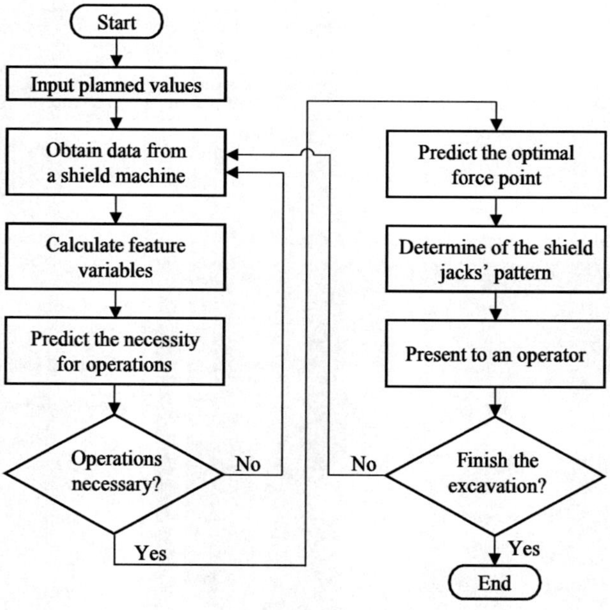

Fig. 3. Processing flow executed by the guidance system.

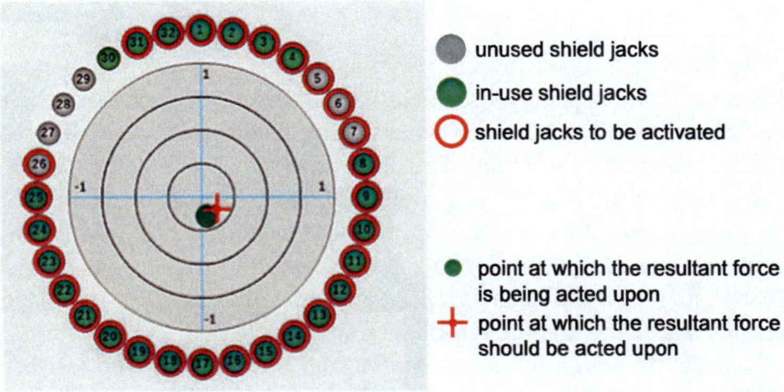

Fig. 4. Example of the guidance presented to an operator.

3.2 Gradient Boosting Decision Tree

A gradient boosting method [9] is one of the ensemble learning methods. A strong learner is built by combining multiple weak learners. When decision trees are used as weak learners, this method is called a "gradient boosting decision tree" (GBDT). The GBDT builds a strong learner through the following process. An initial strong learner $F_0(\boldsymbol{x})$ is given as

$$F_0(\boldsymbol{x}) = f_0(\boldsymbol{x}) \tag{1}$$

where $\boldsymbol{x}$ is a vector of feature variables and $f_0(\boldsymbol{x})$ is a weak learner of a decision tree. The root-mean-square error E_0 of $F_0(\boldsymbol{x})$ is calculated as

$$E_0 = \sqrt{(\boldsymbol{y} - F_o(\boldsymbol{x}))^2} \tag{2}$$

where $\boldsymbol{y}$ is a vector of the target values. A weak learner is $f_1(\boldsymbol{x})$ built with E_0 as the target value and connected to the strong learner:

$$F_1(\boldsymbol{x}) = f_0(\boldsymbol{x}) + f_1(\boldsymbol{x}) \tag{3}$$

When this process is repeated M times, the following strong learner is obtained:

$$F_M(\boldsymbol{x}) = f_0(\boldsymbol{x}) + f_1(\boldsymbol{x}) + \cdots + f_M(\boldsymbol{x}) \tag{4}$$

The GBDT can build a high-performance learner by connecting weak learners to reduce the error. In this study, XGBoost [10], an open-source GBDT library, was used to build the machine learning models.

3.3 Data for Machine Learning Models

Table 1 shows the data for three machine learning models used by the guidance system. One model predicts the necessity for shield jacks' operations, using Fc as the target variable (Fc-Model). Fc is given as 1 when in-use shield jacks were changed by an operator and as 0 when they were not. The other two models predict the horizontal and vertical coordinates of the point at which the resultant force of shield jacks should be acted upon, using Fx and Fy as the target variables (Fx-Model and Fy-Model). Fx and Fy are calculated from each thrust of the shield jacks. The variables other than Fc, Fx, and Fy are the monitoring data that an operator refers to during the manipulations of a shield machine. These include the information of the shield machine's states and the deviation from the planned values. In addition, the moving averages between 1 and 10 s, 11 and 20 s, 21 and 30 s, and 31 and 60 s are added to feature variables because an operator would consider temporal variations.

3.4 Predicting the Necessity for Shield Jacks' Operations

The Fc-Model predicts the necessity for shield jacks' operations to avoid excessive operations. These operations are deemed necessary when a few minutes have passed since the previous instruction of the guidance system and when the output of the Fc-Model is more than the threshold. Here, the waiting time is set at 2 min, which is enough time for the effect to be reflected after the operations of shield jacks.

Most of the Fc are 0 because an operator controls shield jacks once in several tens of seconds to several hundreds of seconds. If the Fc-Model is built using raw Fc as the target variable, almost all values outputted by the Fc-Model are 0. For this reason, the Fc-Model is trained with weighted data.

The threshold must be set to predict whether the operations of shield jacks are necessary. Figure 5 shows the concept of the probability distribution of the predicted values outputted by the Fc-Model. The numbers of true negatives (TN), false negatives (FN), true positives (TP), and false positives (FP) can be calculated if the threshold is fixed. Here, the threshold is set to minimize the balanced error rate (BER).

$$BER = 0.5 \times \left(\frac{FP}{TN + FP} + \frac{FN}{FN + TP} \right) \tag{5}$$

When the output value of the Fc-Model is larger than the threshold, the operations of shield jacks are necessary.

3.5 Prediction of the Optimal Force Point

The optimal point at which the resultant force of shield jacks should be acted upon is predicted by the Fx-Model and Fy-Model. When the operations were predicted to be necessary, these models outputted the horizontal and vertical coordinates of the point. The Fx-Model and Fy-Model are trained with the data classified as TP and FN in Fig. 5. It is considered that more accurate models can be built because only characteristic data that were predicted to require the operations are used.

Table 1. Data for machine learning models.

Variable	Unit	Target	Machine's state	Difference form plan
Binaray data of the operator's manipulations (*Fc*)	–	✓		
Horizontal coordinate of the resultant force point (*Fx*)	–	✓		
Vertical coordinate of the resultant force point (*Fy*)	–	✓		
Torque (*To*)	%		✓	
Velocity (*Ve*)	mm/min		✓	
Thrust (*Tr*)	kN		✓	
Face pressure (*Fp*)	MPa		✓	
Pitch (*Pt*)	deg		✓	
Roll (*Ro*)	deg		✓	
Deviation from the planned left-right difference of the extensions of shield jacks's strokes (*Dj*)	mm			✓
Deviation from the planned azimuth direction (*Dd*)	deg			✓
Deviation from the planned pitch (*Dp*)	deg			✓
Horizontal deviation from the planned path (*Dh*)	mm			✓
Vertical deviation from the planned path (*Dv*)	mm			✓

3.6 Determination of Shield Jacks' Pattern

The guidance system must present how shield jacks avoid applying excessive pressure to lining segments. Hence, the system chooses the pattern of shield jacks presented to an operator from those selected by an expert in the past. The system compares the resultant force point predicted by the machine learning models with that selected by an expert. The system presents the shield jacks' patterns with the shortest distance between the models' and expert's patterns.

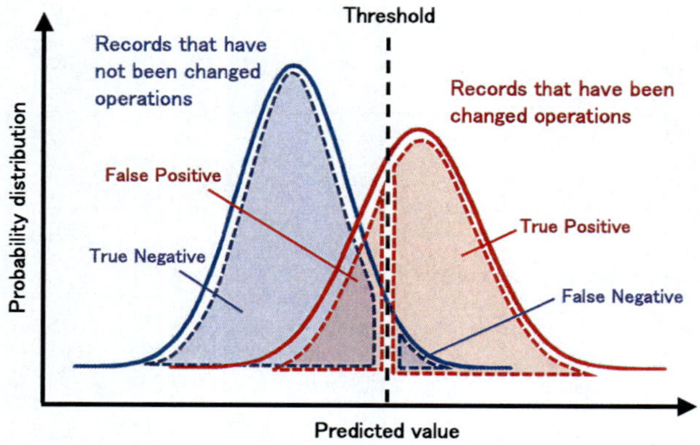

Fig. 5. Concept of the probability distribution of the predicted values outputted by the *Fc*-Model.

4 Results

4.1 Feasibility of the Machine Learning Models

Target Site. To confirm the feasibility of the machine learning models, their performance was evaluated using the data obtained in an existing shield tunnel. Figure 6 shows the target site, which is the ramp for a highway with a total length of 448 m, including downslopes between 5.9% to 7.6% and a steep curve with a radius of 50 m. It was constructed by the shield machine with a diameter of 10.8 m.

The machine learning models were trained and evaluated using the data of 10-s intervals obtained at the section displayed by the red line in Fig. 6. The total quantity of data was approximately 140,000. Approximately 70% of these data were used to build the models. The remaining data were used to evaluate the performance of the models. These were randomly divided for each segmental ring of the lining.

Performance Analysis of the Models. Figure 7 shows the comparative results between the values predicted by the models (shown on the vertical axis) and the operator's manipulations (shown on the horizontal axis). Only data that were predicted by the *Fc*-Model to require the operations of shield jacks are plotted in Fig. 7. The correlation coefficients are 0.749 for (a) and 0.730 for (b), which show positive correlations. However, the accuracy of the *Fx*-Model is lower in the area where the values of *Fx* are far from the origin.

The lift charts, which examine the extent to which the accuracy of the models deteriorates, are shown in Fig. 8. After the predicted values were arranged in descending order and the data block was divided into 60, the averages of each unit of the divided block are plotted in Fig. 8. The average errors of *Fx* within the range −0.2 to 0.2 is 0.0199, whereas the average errors out of the range is 0.0889. The performance of the *Fx*-Model drops when it deviates from the range of −0.2 to 0.2. The performance of the *Fy*-Model is good everywhere.

The trajectory of the point at which the resultant force of shield jacks is acted upon was plotted to confirm that the models can predict well regardless of the positions of *Fx* and *Fy*. Figure 9 shows this trajectory: a segmental ring with the point in each quadrant is shown from (a) to (d). Although the predicted points move more frequently than operation records, the difference between them is not so significant.

Figure 10 shows the influence rate that each feature variable has on the *Fx*-Model and *Fy*-Model. The rate of a feature variable's influence is calculated from the severity of the deterioration of the models' accuracy when the models predict the processed data. The processed data are obtained by randomly shuffling the values of one feature variable of raw data. The greater the feature variables' influence rate, the worse the models' accuracy becomes. The influence rate is given as 1 when the accuracy of the models deteriorates most. *Dh* and *Dd* have a large influence on the *Fx*-Model. *Dp* and *Dv* greatly impact the *Fy*-Model. The results imply that the models can make reasonable predictions because these feature variables correspond to the items that an operator regards as important.

These results show that the models can predict the point of the shield jacks' resultant force when *Fx* is within the range of −0.2 to 0.2. The accuracy deterioration of the *Fx*-Model outside of this range may be caused by the lack of training data and could be solved by collecting more data. In addition, operators rarely set *Fx* to anything beyond the range of −0.2 to 0.2. Thus, models that are applicable to actual environments can be built by using machine learning techniques.

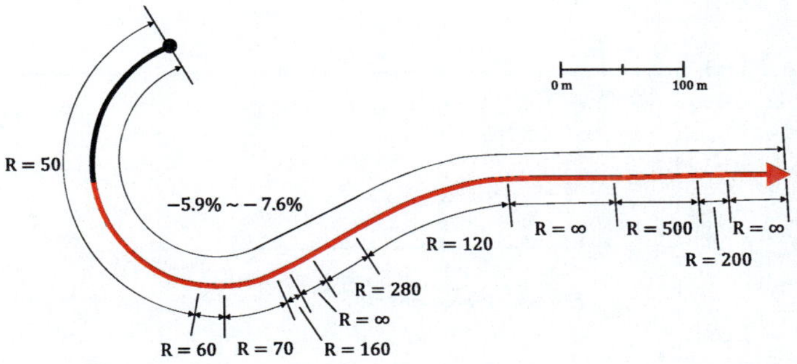

Fig. 6. Target shield tunnel for investigating the feasibility of the machine learning models. The red line indicates the section where the data were acquired for training and evaluating the models. R is the radius of the curve in meters.

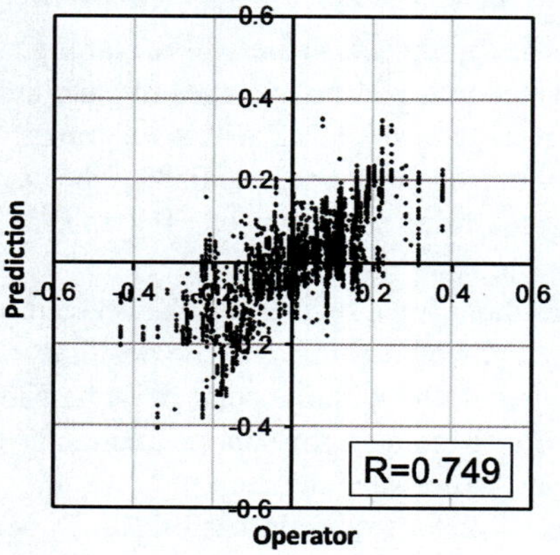

(a) *Fx*

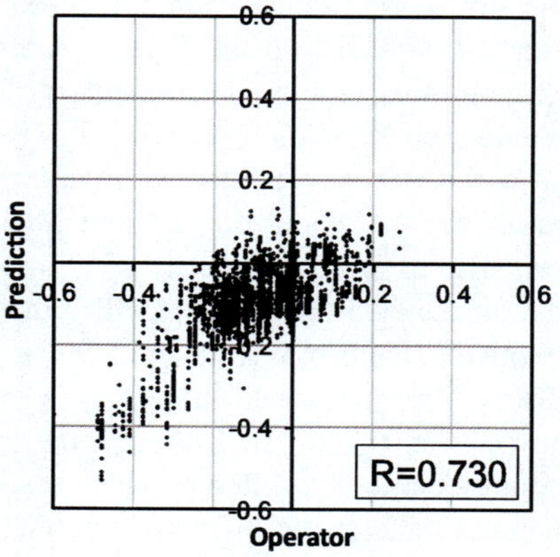

(b) *Fy*

Fig. 7. Comparative results between the values predicted by the models and the operator's manipulations for (a) *Fx* and (b) *Fy*. The horizontal axis indicates the operator's manipulations, and the vertical axis indicates the predicted values.

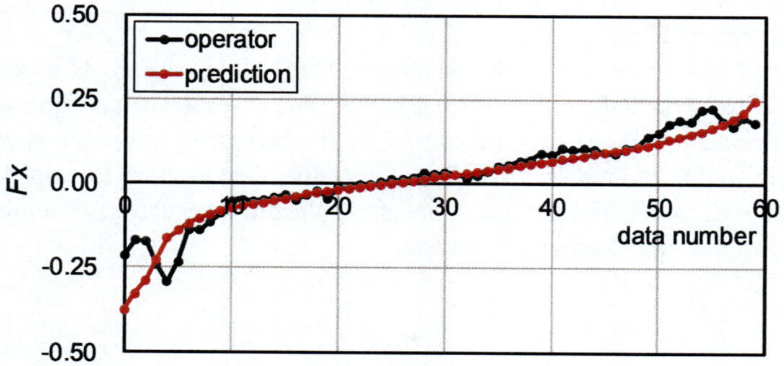

(a) *Fx*

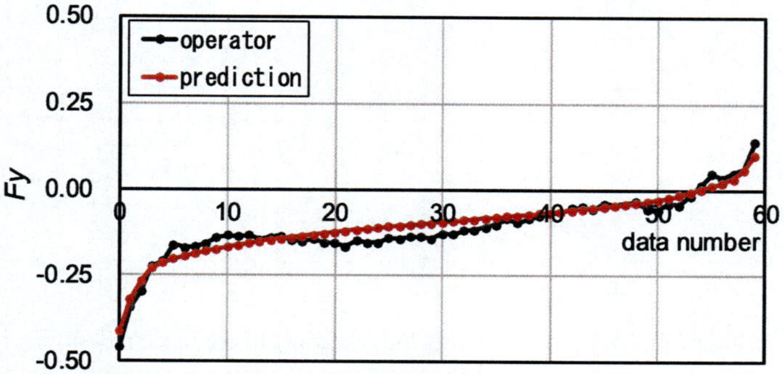

(b) *Fy*

Fig. 8. Lift charts of the values predicted by the models and the operator's manipulations for (a) *Fx* and (b) *Fy*. Black lines indicate the operator's manipulations, and red lines indicate the predicted values.

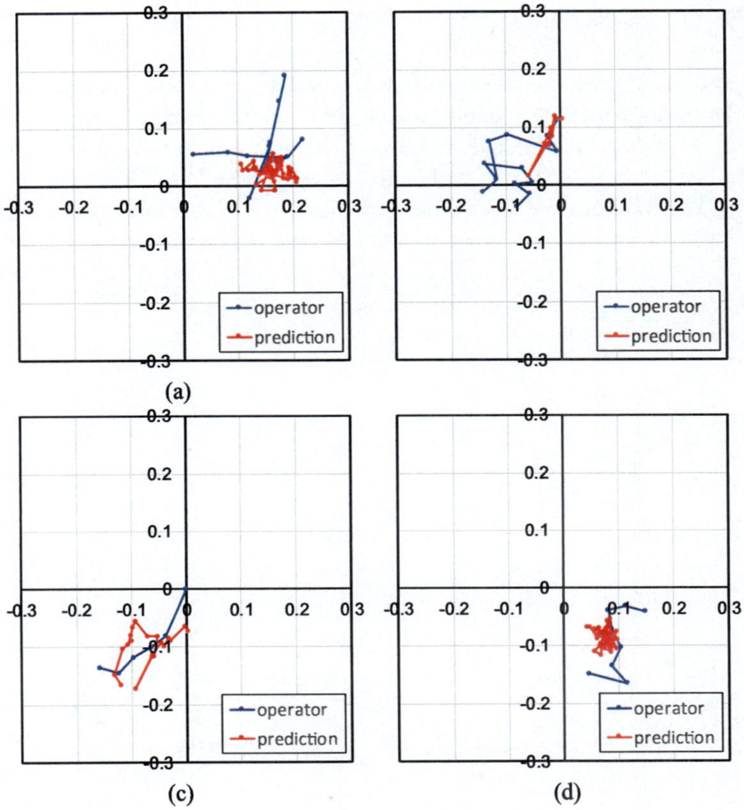

Fig. 9. Trajectory of the point at which the resultant force of the shield jacks is acted upon. A segmental ring with points in each quadrant is shown from (a) to (d). Blue lines indicate the operator's manipulations, and red lines indicate the predicted values.

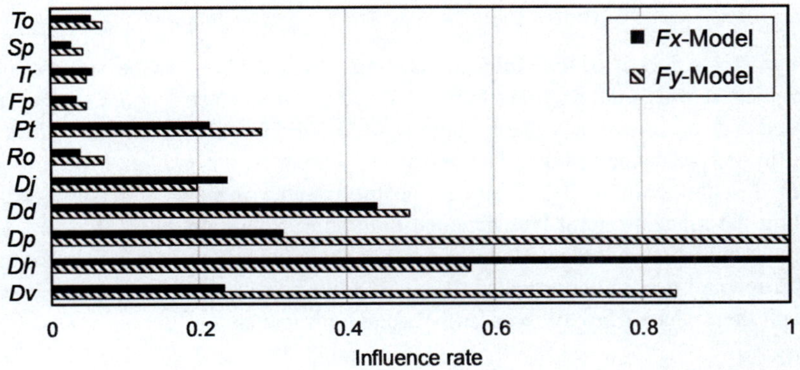

Fig. 10. Influence rate that each feature variable has on the *Fx*-Model and *Fy*-Model.

4.2 Applicability of the Guidance System

Target Site. The shield machine was driven according to the guidance system to validate its performance in a shield tunnel under construction. Figure 11 shows the target site, which has a total length of 1,845 m, including slopes of −5.0% to 5.0% and a curve with a radius of 220 m. It was constructed by a shield machine with a diameter of 11.9 m. The models were trained using the data of 1-s intervals obtained at the red line displayed in Fig. 11. The total quantity of data is approximately 1.2 million. In addition, the performance of the guidance system was evaluated at the location marked in Fig. 11.

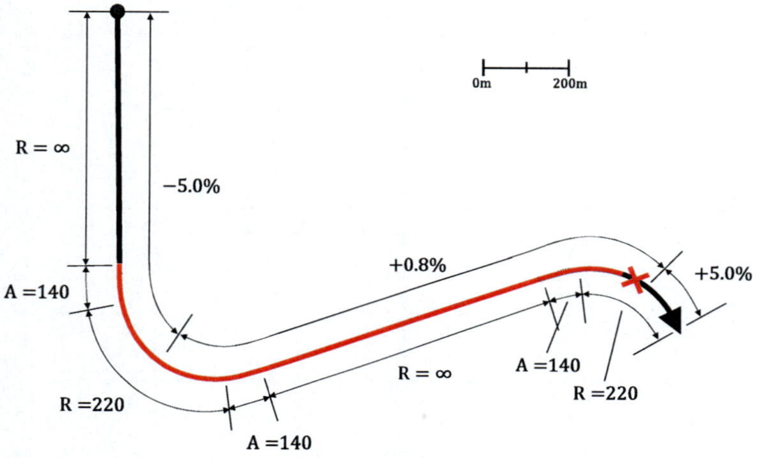

Fig. 11. Target shield tunnel for estimating the performance of the guidance system. The red line indicates the section at which the data were acquired for training the models, and the mark indicates the location where the shield machine was driven according to the guidance system. R and A are the radius of the curve and the parameter of the clothoid in meters, respectively.

Performance Analysis of the Guidance System. Figure 12 shows the time transitions of the point at which the resultant force of shield jacks was acted upon. Black dashed lines indicate the operator's manipulations. Red solid lines indicate the values presented by the guidance system. The reason the output of the guidance system is not presented at the start of excavation is that the moving average of feature variables inputting the models cannot be calculated until some data have been obtained. Comparing the operation records with the outputs of the guidance system, it was confirmed that the operator generally controlled the shield jacks according to the guidance system, although there were a few differences between both values.

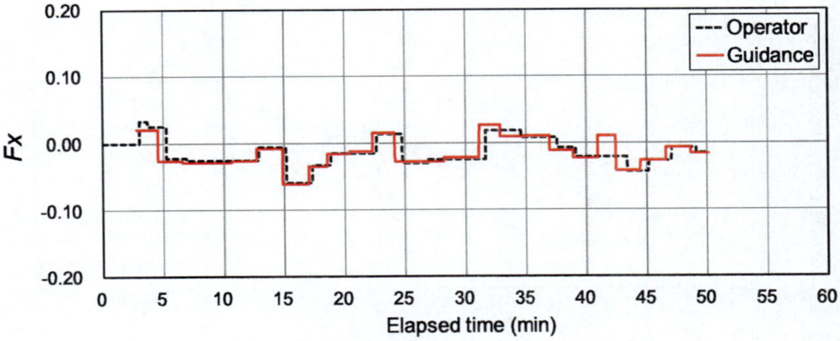

(a) *Fx*

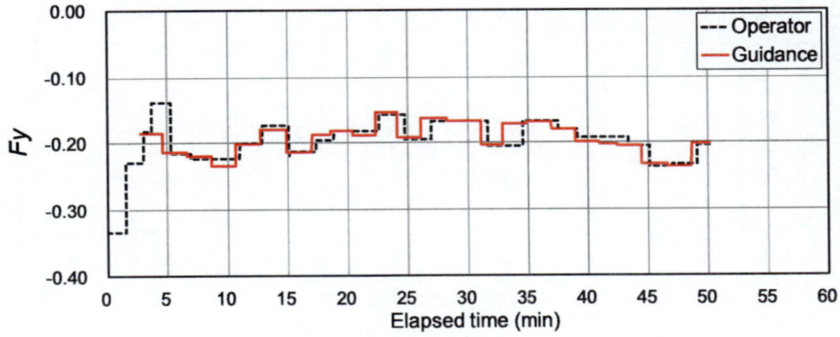

(b) *Fy*

Fig. 12. Time transitions of the point at which the resultant force of shield jacks was acted upon, which are (a) *Fx* and (b) *Fy*. Black dashed lines indicate the operator's manipulations, and red solid lines indicate the values presented by the guidance system.

Figure 13 shows the time transitions of the observed values of control items at the site. The control items are the left–right difference of the extension length of the shield jacks' strokes, the shield machine's azimuth angle, and its pitch. As for the left–right difference of the extension length of the shield jacks' strokes and the shield machine's azimuth, the observed values reached the target values at the end of the segmental ring. The observed value did not reach the target value for its pitch, although it became closer to the target value than it did at the start of excavation.

It was possible to excavate the length of one segmental ring within the tolerances according to the guidance system. In particular, the accuracy of the shield machine's control in the horizontal direction was high. However, in the vertical direction, the accuracy was insufficient. The location where the shield machine was driven by the guidance system is the section that switches from 0.8% to 5.0% upward slope. It is considered that the *Fy*-Model failed to learn the changes of the slopes' ratio. In addition, the output values of the guidance system were frequently changed. This result implies that the guidance included unnecessary operations. It is likely that there is room for improvement in setting the threshold for determining the necessity for the shield jacks' operations.

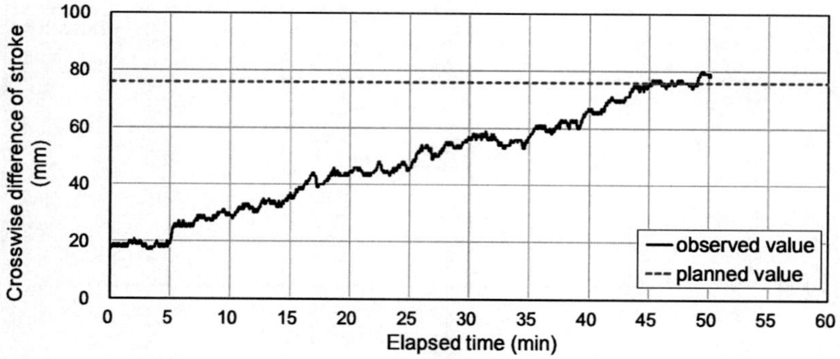

(a) Left–right difference of the extension length of shield jacks' strokes

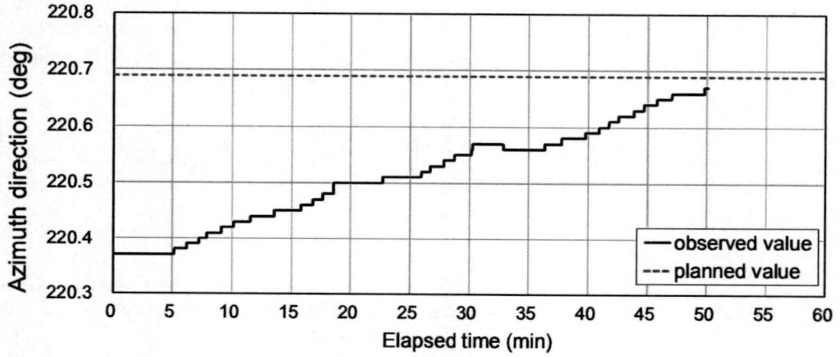

(b) Shield machine's azimuth angle

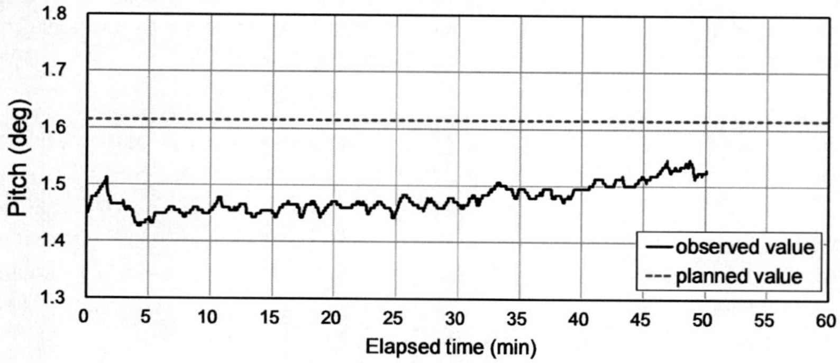

(c) Shield machine's pitch

Fig. 13. Time transition of the observed values of (a) the left–right difference of the extension length of shield jacks' strokes, (b) the shield machine's azimuth angle, and (c) its pitch when the shield machine was driven by the guidance system. Black solid lines indicate the observed values, and red dashed lines indicate the planned values.

5 Discussion

In this study, it was suggested that machine learning models can replace operators in the manipulation of shield machines. However, there are two challenges in the practical realization of the guidance system.

One is the deterioration of the accuracy of the models due to the lack of feature variables. An operator manipulates a shield machine while checking the tail clearance (the distance between the lining segments and the shield). When the tail clearance is small, operations inconsistent with the plan are often performed to avoid breaking the lining segments. Thus, it is necessary to add the information regarding tail clearance to the feature variables of machine learning models.

The other challenge is that the machine learning models must be built for each site. The operation methods of shield machines depend on the specifications of shield machines and the conditions of construction. In other words, the relationship between the target variables and the feature variables at one site will not be the same at another site. Generalization of the data used in the models is needed to make the system usable everywhere.

For self-driving shield machines, machine learning can be a useful tool. Enhancement of the acquisition of data explaining various situations at shield tunnels will allow improving the performance of the machine learning models. In future works, models predicting the operations of shield machines (excluding directional control) will be also built using machine learning techniques, and these will be added to the guidance system.

6 Conclusion

This study developed a guidance system for directional control in shield tunneling using machine learning techniques as the first step toward the realization of self-driving shield machines. By presenting how shield jacks operate, the guidance system makes it possible to ensure the quality of shield tunnels in a way that does not depend on the skill of an operator. The following conclusions were derived by evaluating the performance of the machine learning models and the guidance system.

The results reveal correlation between the values predicted by the machine learning models and the operator's manipulations. In addition, the difference between them was small. The models equivalent to a skilled operator's techniques were able to be built with the data obtained from the shield machine's sensors.

The control items were within the set tolerance levels when the shield machine was driven by the guidance system. No instruction that adversely affected the quality of the tunnel was outputted by the system. Although the system's outputs may have included some redundant instructions, this challenge can be solved by changing the threshold for determining the necessity for the shield jacks' operation. In addition, the performance of the guidance system deteriorated because the rate of the tunnel's slopes changed. Therefore, the machine learning models must be built with more data. Although there are some improvements to be made, the guidance system was able to be applied to the actual construction site.

References

1. Kuwahara, H., Harada, M., Seno, Y., Takeuchi, M.: Application of fuzzy reasoning to the control of shield tunneling. J. JSCE **391**, 169–178 (1988). (in Japanese)
2. Tateyama, K., Nishitake, S., Kazama, K.: Automatic advancing control of shield machine (in Japanese). J. Rob. Soc. Jpn. **12**, 928–932 (1994)
3. Sugiyama, H., Wada, K., Nakaya, T., Ogi, T.: Preliminary study on shield machine's operations by using artificial intelligence. In: 2017 JSCE Annual Meeting, vol. 6, pp. 675–676. Japan Society of Civil Engineers, Japan (2017). (in Japanese)
4. Wada, K., Sugiyama, H., Nozawa, K., Honda, M., Nakaya, T., Ogi, T.: Autonomous direction control of shield machine with AI. In: 2018 JSCE Annual Meeting, vol. 6, pp. 285–286. Japan Society of Civil Engineers, Japan (2018). (in Japanese)
5. Wada, K., Sugiyama, H., Nozawa, K., Honda, M.: Machine learning model considering the operating characteristics of a shield machine. In: 2019 JSCE Annual Meeting, vol. 6, no. 814. Japan Society of Civil Engineers, Japan (2019). (in Japanese)
6. Sugiyama, H., Wada, K., Nozawa, K., Honda, M.: Evaluation of the performance of the machine learning model for directional control of a shield machine. In: 2019 JSCE Annual Meeting, vol. 6, no. 813. Japan Society of Civil Engineers, Japan (2019). (in Japanese)
7. Sugiyama, H., Nozawa, K., Wada, K., Hokari, T.: Development of the support system for the shield machine's operations. In: 2018 JSCE Annual Meeting, vol. 6, pp. 287–288. Japan Society of Civil Engineers, Japan (2018). (in Japanese)
8. Shield Tunneling Association of Japan. http://shield-method.gr.jp. Accessed 10 Jan 2020
9. Jerome, F.: Greedy function approximation: a gradient boosting machine. Ann. Stat. **29**(5), 1189–1232 (2001)
10. Chen, T., Guestrin, C.: XGBoost: a scalable tree boosting system. arXiv, 1603.02754 (2016)

Classification of the Requirement Sentences of the US DOT Standard Specification Using Deep Learning Algorithms

Kahyun Jeon[1], Ghang Lee[1(✉)], and H. David Jeong[2]

[1] Yonsei University, Yonsei-Ro 50, Seodaemun-gu, Seoul 03722, South Korea
glee@yonsei.ac.kr
[2] Texas A&M, 334 Francis Hall, 3137 TAMU, College Station, TX 77843-3137, USA

Abstract. This aim of this study is to classify requirement sentences from the specifications of US DOT using natural language processing (NLP) and a deep neural network. At the contract phase of the project, the requirements analysis of contract documents is a significant task to prevent claims or disputes caused by ambiguous or missing clauses, but it is highly human-intensive and difficult to identify requirements within a given short period. In this article, the requirement sentences identification model was proposed based on deep-learning algorithms. First, the critical terms that define what the requirement sentence is were identified, and then all sentences were labeled using the pre-defined critical terms. Second, three vectorizing methods were used, including two pre-trained methods—GloVe and Word2Vec—and a self-trained method to produce word embedding. Third, the automated classification of requirements sentences was experimented using three deep-learning models: the convolutional neural network (CNN), the long-short-term memory (LSTM), and the combination of CNN+LSTM. In the evaluation of nine total experiments, the results showed that the F1 scores of the CNN model were the highest at 92.9% and 92.4% for both the Word2Vec model and the Glove model. This study provided a way to achieve a high level of classification accuracy with simple deep-learning models and pre-trained embedding models.

Keywords: Construction requirement · Natural language processing · Deep learning · Word embedding

1 Introduction

In the bid phase of a construction project, clients request a large number of requirements through a set of contract documents, which they expect to be delivered during a project. The contractors must precisely analyze all requirements and submit the project execution plan with a reasonable cost and schedule for a successful project. In general, the client's requirements are often described not in structured and standardized texts but in an unstructured natural language [1]. Because of this unstructured and unstandardized nature of a natural language, the same meaning of a requirement can be described in different ways. Thus, a heuristic or knowledge of the contractor is the only way to

E. Toledo Santos and S. Scheer (Eds.): ICCCBE 2020, LNCE 98, pp. 89–97, 2021.
https://doi.org/10.1007/978-3-030-51295-8_8

identify requirements in the bid documents, so this is very laborious, and error-prone. Most contractors have difficulty in finding the specific requirements in a short period of bidding time. If a contractor missed a requirements sentence, that might be led to claims, disputes or even serious lawsuits from the client [2]. Therefore, systemically identifying the requirements sentences is critical for managing potential risks.

This study aims to learn which word-embedding and deep-learning algorithms work best for identification of the requirements sentences in construction specification documents. In recent years, machine learning and deep learning have shown promising results in the field of natural language processing (NLP). Neural networks (NNs), such as convolutional neural networks (CNN) [3] and recurrent neural networks (RNNs), have been used for information retrieval, text classification, and document classification [4]. To input the text data into these kinds of deep network models, a word embedding —a method to group similar words through vectorization of words or phrases—is an essential step. The most popular word-embedding algorithms are Word2Vec [5] and Global Vectors for Word Representation (GloVe) [6]. These two pre-trained word-embedding model have shown highly accurate results in the deep learning tasks.

Using these two pre-trained word-embedding methods and a self-trained word-embedding method and the three popular deep learning methods—convolutional neural network (CNN), long-short-term memory (LSTM), and a combination of CNN+LSTM, this study analyzes and finds the requirements sentences from the standard specification of the US Department of Transport (DOT). The standard specification of the US DOT is one of the contract documents used in the bid phase of a US DOT project. This paper is structured as follows. First, related works using NLP and deep models in the construction field are briefly reviewed. Second, each step of research framework is described in detail. Third, the experiments are described to discuss the datasets, hyperparameters, and evaluation scores of each model's performance. The paper concludes with a discussion of the lessons learned for future work.

2 Related Work

Identifying the client's requirements and how to effectively satisfy all the requirements are the key to a project's success [7, 8]. Nevertheless, it is challenging to capture the client's requirements from hundreds-page-thick contract documents. To relieve this problem, several studies deployed NLP and the artificial intelligence (AI) algorithms for different phases of a project.

For the bid phase, Lee et al. in 2019 tried to find the malicious clauses, which could result in claims and disputes to detect potential bidding risks in contractual documents [9]. This study proposed a bidding risk prediction model using NLP. Previously in 2017, Lee et al. developed a bidding risk prediction algorithm using four machine learning algorithms: artificial neural networks (ANN), support vector machine (SVM), k-nearest neighbors (KNN), and Naïve Bayes classification models [1]. Son et al. in 2019 proposed a schedule delay risk index (SDRI) developed based on the frequency analysis of the pre-defined critical terms in order to estimating a schedule delay in a bid phase [10].

For the design and construction phase, Uhm et al. analyzed over 9,800 sentences compiled from 27 requests for proposal (RFPs) to define objects and methods required for automated design compliance checking using a grammatical analysis approach [12]. Zhu et al. used text analysis and statistical analysis to understand the patterns of the occurrences and cross-references between for building components, contract documents, and project management concepts in unstructured documents such as requests for information (RFIs) [13].

Despite these studies, the use of NLP and AI in construction requirements engineering is relatively new. This study deploys a total of nine cases—a combination of three word-embedding methods and three deep-learning models, in the requirement sentences classification of construction contract documents.

3 Research Methods

This study consists of three parts; 1) data preparation and preprocessing; 2) training and cross validation of deep models; 3) evaluation. The entire research framework is described in Fig. 1 and each process will be explained as follows.

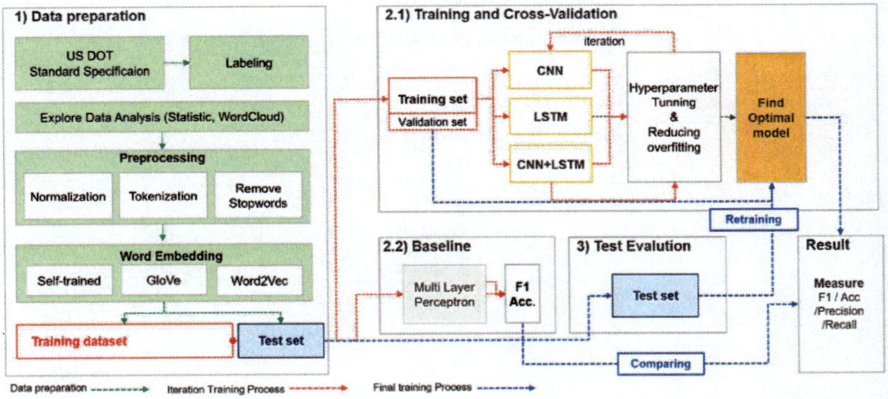

Fig. 1. Research framework

3.1 Data Preparation and Preprocessing

Data Collection. Since the US DOT specification was in a PDF format, it was necessary to convert the PDF file into a plain text file such as '*.txt' or '*.csv', then into individual sentences. After converting a specification file into individual sentences, all the tables and figures, which required additional studies to make them useful for text mining, and other unnecessary objects such as annotations and footnotes were removed. In case of sentences consisting of several sub-sentences, they were divided into unit sentences, which could be mapped to individual requirements.

Labeling Rule. In advance to labeling requirements in contracts, it is necessary to define what requirements are first. A couple of previous studies have identified that some specific words appeared more frequently than the other words in requirements sentences [11, 12]. Based on the studies, a category of critical terms was grouped into four: modality verbs, action verbs, prohibit words, and constraint words (Table 1). Based on the categories, the requirement types of each sentence were manually determined and labeled: 1) The modality verbs were "must", "shall", and "may". The modality verbs signified the level of obligation. For example, if "must" was used in a sentence, it meant a strong obligation. And if "shall" or "may" were used, it meant a recommended regulation. 2) The action verbs signified a specific task that the contractor should do. 3) The prohibit words such as "no" and "not" indicated a negative action. 4) The constraint words used to emphasize specific limitations, regulations, or instructions that the contractor should follow. If a sentence contained one of the above critical terms, it was considered a requirement sentence. For example, the sentence "Do not perform work related to submitted documents or drawings before approval of the client" included four critical terms: not, submit, before, and approval. Thus, this sentence was labeled '1' to indicate a requirement sentence. If not, the sentence was labeled '0' to indicate a non-requirement sentence.

Table 1. Categories of critical terms.

Prohibit	Modality verbs	Action verbs	Constraint verbs
No	Must	Approved	At least
Not	Shall	Meet	After
	May	Permit	Before
		Prevent	Exceed
		Notify	Less than
		Submit	More than
		Required	Prior to
			Until
			Within

Preprocessing. The typical preprocessing of NLP starts from the 'exploratory data analysis (EDA)', normalization, tokenization, and stopword removal. These steps were conducted in order (Fig. 2). The mean number of words that comprised a sentence was recorded around 18.44 words and the mean length of each sentence is 142.8 characters. After the initial round of preprocessing, the sentences composed of less than 50 characters were removed because the analysis revealed that they were titles or non-sentential phrases. The distribution of the final labeled dataset is shown in Table 2. The percentage of the requirement sentences (50.1%) was almost the same as that of the non-requirement sentences (49.9%).

Word Embedding. If a training dataset is relatively small like the one in this study, it is generally not recommended used the dataset-specific self-trained word embedding– a method to create embedding vectors solely relying on the training dataset. In this case, use of pre-trained word embedding vectors such a GloVe and Word2Vec is known to be more effective than the self-trained word embedding. This study used both self-

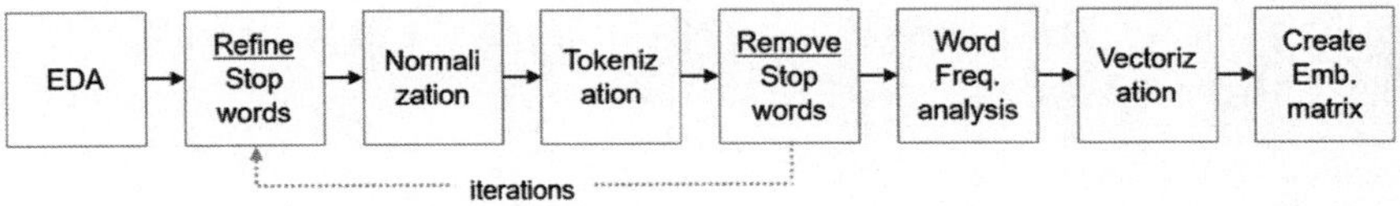

Fig. 2. A sequence of preprocessing

Table 2. Distribution of the refined datasets

	Total number of sentences	The number of requirements (percentage)	
Original datasets	11,733	Requirements	6,402 (55%)
		Non-requirements	5,227 (45%)
Refined datasets	11,060	Requirements	5,535 (50.1%)
		Non-requirements	5,525 (49.9%)

trained and pre-trained vectorization approaches. In this study, a total of three types of embedding methods were applied: 1) self-trained dataset embedding, 2) pre-trained GloVe (100 dimension), and 3) pre-trained Word2Vec (300 dimension). The final corpus was composed of 8,404 unique tokens (words), but only the top 1,000 frequently appeared words were used as features to focus on meaningful words following the general NLP practice.

3.2 Design the Deep Neural Networks

Baseline. Abaseline model provides a minimum performance threshold to compare experimental models. The Multi-Layer Perceptron (MLP), a fundamental architecture of neural model, was used as a baseline. An MLP has at least one hidden layer between the input and the output layers, and the other layers can be stacked continuously until reaching the output layer.

Convolutional Neural Network (CNN). Recently, the Conv1D network has shown promising results in the NLP field compared to the RNN model, and it is even much faster for simple tasks. The computational cost is less than that of RNN. The Conv1D layer is useful for extracting local features regardless of where the critical terms appear. It can extract the features considering the context.

Long-Short Term Memory (LSTM). A recurrent neural network (RNN) model can remembers the context and reflects the information in a time series; however, if some information is separated for a long distance, the network forgets the previous information. LSTM overcome this weakness using memory cells [13]. Selecting the number of hidden layers and the number of memory cells in LSTM depends on the application domain and the context of the dataset. The optimal number of hidden units can easily be smaller than the number of inputs, and sometimes just two hidden units work well with a small dataset. Thus, the more the training dataset is fed, the more multiple layers should be used (Fig. 3).

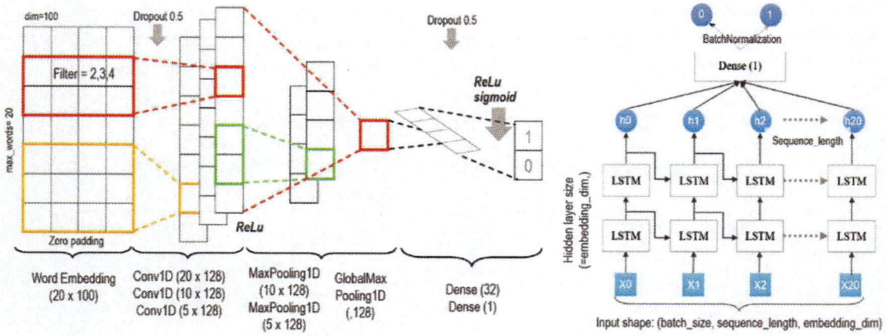

Fig. 3. CNN and LSTM architectures

CNN-LSTM Combination Model. While CNN can capture specific word regardless of where it appears in the context, LSTM can remember past words even if they are located far away. Thus, a combination model of these two models can be used to capture contextual information with the recurrent structure and to construct the representation of text in a convolutional neural network [14]. The detailed model architecture is described in Fig. 4.

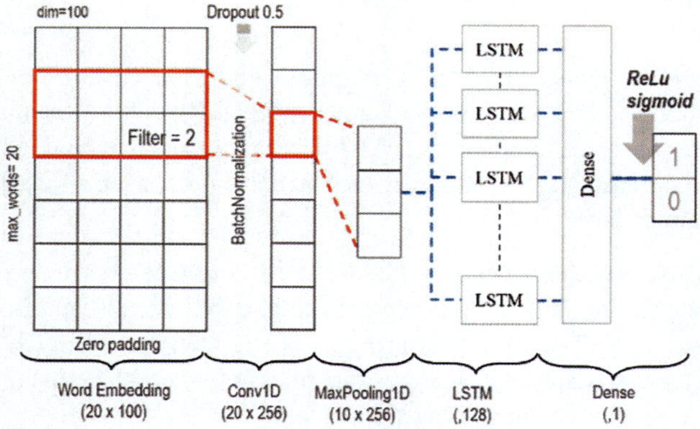

Fig. 4. CNN-LSTM combination model architecture

3.3 Training and K-Fold Cross Validation of Deep Models

K-Fold Cross Validation. A hold-out validation, which splits the dataset into "train" and "test", is only useful when the training and validation datasets are enough. If a dataset is small, the K-fold cross-validation is much reliable [15]. The variable K refers to the number of groups in which a given dataset is to be split. A given dataset is randomly shuffled and divided into K groups before every iteration of validation. Then, in each of the experiments of K, the one-fold used for validation and the other folds are concatenated together for training. Thus, every sample can be used in a training set 'K-1' times and in the test set once, so the average score of the validation accuracy can be guaranteed. The goals of a cross-validation are to predict the performance of unseen data and to find the optimal model chosen by hyperparameter tuning [16].

Evaluation. There is no perfect answer when searching for the optimal model because hyper-parameters that determine the model's performance are changeable depending on various factors, such as the size of the dataset, the network model architecture, and even the hardware environments where the experiments are conducted. In general, finding an optimal model begins with the overfitting model by changing the hyper-parameters one by one, which mitigates overfitting. Once the model's performance has improved sufficiently, this optimal model can be trained once again with the full training datasets, and the test dataset can be evaluated.

4 Results

The experiment was conducted for a total of nine experiments, applying three embedding methods to three neural network models and the results of each F1 score, precision, recall, and accuracy were described in Table 3. As compare to the baseline, the F1-scores of all three models were higher than the baseline, and even the lowest score recorded by LSTM with word2vec was higher than the highest score of the baseline (Fig. 5). CNN had the highest F1 score, and LSTM had the lowest. Although the learning curve of the baseline seemed to be well-trained, all scores of the baseline were under 90%.

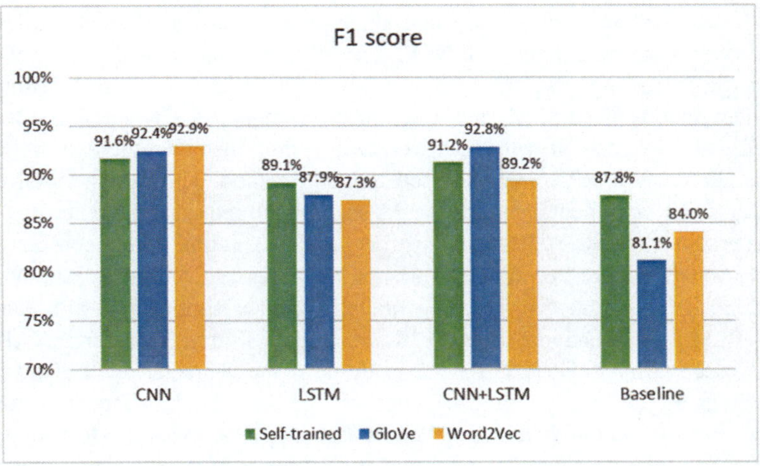

Fig. 5. Comparing F1 scores of test evaluation

Comparison Between Word Embedding Methods. Both pre-trained word embedding vectors were effective for CNN, and CNN showed the best performance recorded at 91.6% for self-trained, 92.4% for GloVe, and 92.9% for word2vec models. The self-trained dataset embedding seemed to be effective for LSTM and the baseline; however, the F1 scores of LSTM and the baseline with a self-trained model were 89.1% and 87.8%, respectively, which were lower than the that of CNN at 91.6%. Thus, CNN was more efficient with this dataset.

Impact of the Combination Model of CNN+LSTM. Contrary to expectations, the combination model did not significantly outperform the single-type network. For the GloVe embedding model, the CNN-LSTM model achieved 92.8%, which was slightly higher but almost the same as 92.4% for the CNN model; however, for word2vec embedding, 89.2% for the CNN-LSTM model was much lower than the best score at 92.9% for CNN.

Table 3. Test evaluation scores

(%)		CNN			LSTM			CNN+LSTM		
		Self-trained	Glv.	W2V	Self-trained	Glv.	W2V	Self-trained	Glv.	W2V
Val	F1	91.4	88.2	91.3	87.8	81.6	88.0	89.7	86.6	90.9
Test	Recall	88.5	91.5	93.3	85.2	84.6	84.1	93.1	92.2	93.9
	Precision	95.5	93.9	93.0	94.2	92.3	91.6	89.8	93.8	85.5
	F1	91.6	92.4	92.9	89.1	87.9	87.3	91.2	92.8	89.2

5 Conclusion

This study aimed to automatically identify construction requirements from the US DOT standard specification documents using three word embedding models and three popular deep learning models for NLP. The research started with data collection, labeling, preprocessing, and then the baseline model was trained first. Three deep neural networks, CNN, LSTM, and a combination model of CNN and LSTM, were designed and implemented with three word embedding methods: the two pre-training methods GloVe, Word2Vec, and the self-training method. Through the total of nine experiments, all deep neural models could classify the requirement sentences with a high accuracy (over 90%). This is a particularly meaningful result considering the relatively small dataset. In future works, it is necessary to apply state-of-the-art embedding methods and deep-learning models, such as Embedding with Language Model (ELMo), bidirectional Encoder Representations from Transformers (BERT), and very deep CNN (VDCNN) for the analysis of the character level of texts. This study is meaningful as a preliminary study on the text classification of requirement sentences in construction documents. The datasets could be expanded to various types of other documents that can be generated in the entire lifecycle of the construction project. The robust datasets can also be preprocessed based on knowledge of the construction domain in future works.

Acknowledgements. This work was supported by an Institute for Information & Communications Technology Promotion (IITP) Grant funded by the Korea Government (MSIT) (No. 2019-0-01559-001, Digitalizing Construction Project Requirements Using Artificial Intelligence and Natural Language Processing).

References

1. Lee, J., Yi, J.-S.: Predicting project's uncertainty risk in the bidding process by integrating unstructured text data and structured numerical data using text mining. Appl. Sci. **7**(11), 1141 (2017)
2. Walsh, K.P.: Identifying and mitigating the risks created by problematic clauses in construction contracts. J. Legal Aff. Dispute Resolut. Eng. Constr. **9**(3), 03717001 (2017)
3. Kim, Y.: Convolutional neural networks for sentence classification. arXiv preprint arXiv:1408.5882 (2014)
4. Le, H.T., Cerisara, C., Denis, A.: Do convolutional networks need to be deep for text classification? In: Workshops at the Thirty-Second AAAI Conference on Artificial Intelligence (2018)
5. Markovits, H., Brisson, J., de Chantal, P.-L., Thompson, V.A.: Interactions between inferential strategies and belief bias. Mem. Cogn. **45**(7), 1182–1192 (2017)
6. Pennington, J., Socher, R., Manning, C.: Glove: global vectors for word representation. In: Proceedings of the 2014 Conference on Empirical Methods in Natural Language Processing (EMNLP) (2014)
7. Sommerville, I., Sawyer, P.: Requirements Engineering: A Good Practice Guide. Wiley (1997)
8. Wilson, W.M., Rosenberg, L.H, Hyatt, L.E.: Automated analysis of requirement specifications. In: ICSE. Citeseer (1997)
9. Lee, J., Yi, J.-S., Son, J.: Development of automatic-extraction model of poisonous clauses in international construction contracts using rule-based NLP. J. Comput. Civil Eng. **33**(3), 04019003 (2019)
10. Son, B.-Y., Lee, E.-B.: Using text mining to estimate schedule delay risk of 13 offshore oil and gas EPC case studies during the bidding process. Energies **12**(10), 1956 (2019)
11. Sharma, A., Kushwaha, D.S.: Natural language based component extraction from requirement engineering document and its complexity analysis. ACM SIGSOFT Softw. Eng. Notes **36**(1), 1–14 (2011)
12. Vlas, R., Robinson, W.N.: A rule-based natural language technique for requirements discovery and classification in open-source software development projects. In: 2011 44th Hawaii International Conference on System Sciences (2011)
13. Hochreiter, S., Schmidhuber, J.: Long short-term memory. Neural Comput. **9**(8), 1735–1780 (1997)
14. Shuang, K., et al.: Combining word order and CNN-LSTM for sentence sentiment classification. In: Proceedings of the 2017 International Conference on Software and e-Business (2017)
15. Wong, T.-T.: Performance evaluation of classification algorithms by k-fold and leave-one-out cross validation. Pattern Recogn. **48**(9), 2839–2846 (2015)
16. Rodriguez, J.D., Perez, A., Lozano, J.A.: Sensitivity analysis of k-fold cross validation in prediction error estimation. IEEE Trans. Pattern Anal. Mach. Intell. **32**(3), 569–575 (2009)

Assessment of Effect of Strain Amplitude and Strain Ratio on Energy Dissipation Using Machine Learning

Jamal A. Abdalla(✉) and Rami A. Hawileh

American University of Sharjah, Sharjah, United Arab Emirates
{jabdalla, rhaweeleh}@aus.edu

Abstract. In this study Artificial Neural Networks (ANN) as a machine learning technique is used to predict and assess the effect of strain amplitude and strain ratio on energy dissipated in steel reinforcing bars in reinforced concrete members. The steel reinforcement bars were experimentally tested and were subjected to variable strain amplitudes beyond yield. The developed machine learning model, which is based on Back-Propagation ANN, accurately predicted the experimentally measured dissipated energy. The developed model is then used to deeply assess the effect of a range of strain amplitudes and strain ratios in the amount of energy dissipated at the first cycle, in an average of selected number of cycles and in all cycles, all at different levels of low-cycle fatigue loading of the reinforcement bars. It is concluded that the developed machine learning model can accurately predict the hysteresis energy dissipated in steel bars subjected to low-cycle fatigue load and more importantly it is a viable machine learning tool for deep assessment of the tested specimens with several parameter values that were not covered by the experimental program, but within the domain bounded by the maximum and minimum values of the training data. Based on the prediction and the deep assessment results, several conclusions were drawn.

Keywords: Machine learning · Artificial Neural Network · Dissipated hysteresis energy · Low-cycle · Fatigue life

1 Introduction

The main objective of reinforcement bars in concrete members is to carry the tensile force. However, for reinforced concrete buildings in seismic regions, reinforcement bars play another important role which is the absorption and dissipation of energy. Therefore the energy dissipated by reinforced concrete buildings subjected to seismic load depends, to a great extent, on the energy absorbed and dissipated by the steel reinforcing bars [1–4]. The amount of energy dissipated by reinforced concrete structures in the event of earthquake could be related to the type and amount of seismic damage. Therefore, the energy dissipated by the steel reinforcing bars is an important parameter in identifying and quantifying seismic damage in reinforced concrete buildings in the event of severe seismic loading [1, 2].

E. Toledo Santos and S. Scheer (Eds.): ICCCBE 2020, LNCE 98, pp. 98–108, 2021.
https://doi.org/10.1007/978-3-030-51295-8_9

In the event of severe seismic loading, plastic energy dominates. The plastic energy dissipated by steel reinforcing bars depends on many factors including strain amplitude (ε_a), the maximum strain ($\varepsilon_{s,max}$) and the minimum strain ($\varepsilon_{s,min}$), strain ratio (R), frequency of the applied dynamic load (low, intermediate, high), load type (periodic, random, etc.) and load duration. In this study the strain amplitude (ε_a) and the strain ratio (R) are used as input parameters to the ANN to predict the amount of energy dissipated by steel reinforcing bars at different stages of dynamic loading – first cycle, second through fifth cycles and total cycles to failure. These energy quantities can be related to fatigue life of the steel reinforcing bars and also to the level of damage in reinforced concrete buildings in the event of earthquake [1, 2].

ANN as a machine learning tool has been used extensively for prediction of energy demands and energy consumptions in several fields [5–7]. It has also been used for predicting energy in buildings. Kalogirou et al. [5] used ANN to predict the energy consumption of a passive solar building. The ANN thermal behavior of the building was proved to be very accurate and faster compared to evaluation of the dynamic thermal building model that was constructed on the basis of finite volumes and time marching. Kalogirou [8] also provided a survey of applications of artificial neural networks for energy systems.

With the advancement in soft computing, ANN as a machine learning technique, have been utilized effectively to solve many civil and building engineering problems related to behavior of building materials, structural members and building structures as a whole [9]. It used to predict shear resistance of reinforced concrete beams [10], optimum seismic design of un-bonded post-tensioned precast concrete walls [11], prediction of bond strength [12] and in assessing slope stability and predicting the factor of safety [13–15].

Recently, with the vast increase in the amount data, independent and dependent variables and the complex relationships among them, deep machine learning became necessary [16–18]. Oudeh et al. [16–18] investigated the viability of implementing ANN with Neural Implementation Diagram (NID) and Sequential Feature Selection (SFS) as deep learning mechanism, respectively, to accurately predict the fiber reinforced polymer shear resistance in externally strengthened RC [16, 17] and to predict the compressive strength of ultra-high performance concrete [18].

Although ANN has been used to predict the fatigue life of steel reinforcing bars and other metals [19–24], however, very few studies were carried out to predict the hysteresis energy dissipated on steel reinforcing bars subjected to low-cycle fatigue using ANN [25, 26]. Therefore, in this study ANN is used to predict the hysteresis energy dissipated in steel reinforcing bars subject to low-cycle fatigue that is based on the strain amplitude and the strain ratio which are used as input parameters for the ANN and also for assessing their effect in dissipated energy.

2 Hysteresis Energy in Low Cycle Fatigue

Over the years several researcher used energy as a parameter for predicting the fatigue life of metals [27–29], among others. For constant amplitude loading the total strain energy per cycle is the summation of elastic and plastic energy and is given by:

$$\Delta W = \Delta W_e + \Delta W_P \tag{1}$$

$$\Delta W = \oint_{cycle} \sigma d\varepsilon_e + \oint_{cycle} \sigma d\varepsilon_p \tag{2}$$

Where ΔW_e is the elastic strain energy and ΔW_p is the plastic strain energy; ε_e is the elastic strain and ε_p is the plastic strain and σ is the stress. As previously indicated, low-cycle fatigue life of steel reinforcing bars is a function of the amount of energy dissipated during the cyclic loading. Experiments were conducted and several samples of steel reinforcing bars have been subjected to low-cycle fatigue loading with variable maximum strain amplitude ranging from 3% to 10% and different strain ratios ranging from 0 to −1. The strain ratio (R) and the constant strain-amplitude (ε_a) are given by:

$$R = \varepsilon_{s,min} / \varepsilon_{s,max} \tag{3}$$

$$\varepsilon_a = \frac{\Delta\varepsilon}{2} = \frac{\varepsilon_{s,max} - \varepsilon_{s,min}}{2} \tag{4}$$

Where $\varepsilon_{s,max}$ is the maximum applied total (elastic + plastic) strain and $\varepsilon_{s,min}$ is the minimum applied total (elastic + plastic) strain. In low-cycle fatigue the elastic energy is usually small compared to plastic energy and therefore can be neglected [30, 31]. Figure 1 shows the elements of strain energy in cyclic loading for BS 460B subjected to a complete reversal maximum strain of 0.08 [30]. For each specimen, the hysteresis plastic energy is then calculated using numerical integration to compute the area enclosed within hysteresis loop for first cycle, average cycles and total cycles. Figure 1 also shows the hysteresis loop or the plastic energy dissipated for one specimen.

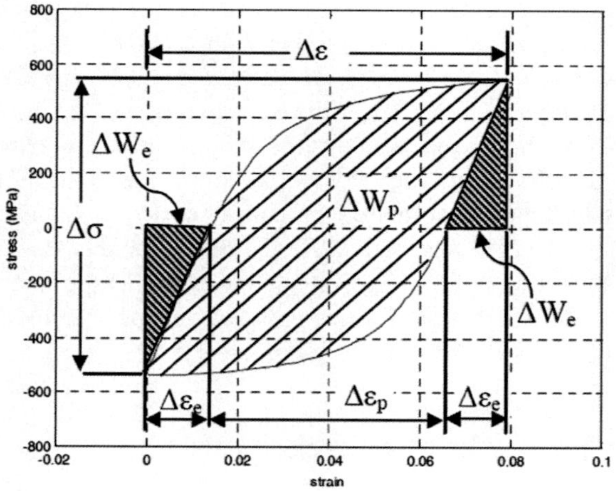

Fig. 1. One cycle hysteresis loop for 460B with $\varepsilon_{s,max}$ = 0.08, R = −1 [26].

3 Predictions of Dissipated Energy Using ANN

Employing a Universal Testing Machine (UTM) several low-cycle fatigue tests were conducted on BS 460B and BS B500B. Displacement-controlled, low-cycle fatigue tests involved applying sinusoidal axial strains ranging from 3% to 10% with different strain ratios ranging from zero to −1. The frequency of the sine function was 0.05 Hz (period of 20 s). The detailed description of the experimental program and all test results were given in a previous study [1].

Artificial Neural Networks are computational and information processing paradigms that consist of interconnected neurons which are used to model and relate complex, often nonlinear, relationships among parameters. They have wide range of applications in different fields including classification, prediction, clustering and function approximation, among others [26]. There are several ANN that are used for prediction, however, the most widely used ANN for prediction is Back-propagation feed-forward (BPFE) multi-layer perceptron (MLP) which has been used in this investigation [26, 32].

Figure 2 shows the neural network architecture used in this study. It has one input layer, one hidden layer and one output layer. The input layer consists of two nodes, mainly, the tensile strain amplitude (ε_a) and it is in the range of 0.02–0.08 and the strain ratio (R) is in the range of −1 and 0. The output layer of the ANN consists of three nodes which are the amount of energy dissipated in the first cycle (W_{P1}), average energy of cycles (W_{PA}) and total energy dissipated (W_{PT}) in all cycles as shown in Fig. 2 below. A supervised learning (associative learning) where the ANN is trained with a set of input parameters and target or desired output parameters are used. Initial random values were generated for the weight and the experimental training data set was used to train the ANN and the testing data set was used to test the ANN [26].

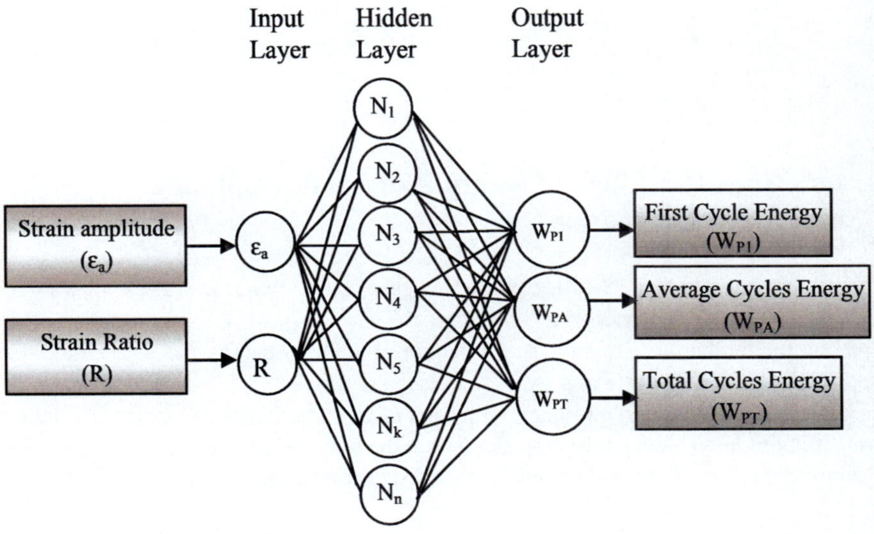

Fig. 2. Neural network architecture

4 Results and Discussions

Back-Propagation Feed-Forward (BPFF), Multi Layer Perceptron (MLP) ANN was subjected to a supervised training using some of the experimental data. The trained ANN is then used with the test data to predict energy values and the accuracy of the prediction of the trained ANN of the first cycle, average of cycles and total cycles are then compared with the actual measured or desired values.

Table 1 summarizes the measured and predicted energy parameters and the corresponding ratios. It indicated in Table 1 that the ANN gave better predictions of first cycle energy where most test samples are within ±10% of the measured first cycle energy values (W_{P1}), followed by the predictions of average cycles energy where most test samples are within ±15% of the measured average cycles energy values (W_{PA}). The predictions of the total cycles energy are the least accurate with most test samples are within ±30% of the total cycles energy values (W_{PT}).

Table 1. Summary of measured and predicted energy for test specimens

ε_a	R	First cycle (W_{ANN} /Wp_1)	Average cycle (W_{ANN} /Wp_A)	Total energy (W_{ANN} /W_{PT})
0.03	−1	1.093	1.056	1.015
0.05	−1	0.837	0.860	0.974
0.03	−0.5	1.100	1.071	1.261
0.06	−0.5	1.000	0.885	1.002
0.03	0	1.023	1.044	1.030
0.04	0	0.900	0.802	1.297
0.04	−1	1.057	1.126	1.078
0.06	−1	0.923	0.893	1.255
0.08	−1	0.954	0.919	1.162
0.035	0	1.082	1.583	0.792
Average		0.997	1.024	1.087

5 Assessment of Effect of Strain Ratio and Amplitude on Energy Dissipation

The performance of the ANN for predicting first cycle (W_{P1}), average cycles (W_{PA}) and total cycles (W_{PT}) energy is satisfactory. Table 2 shows the performance of the ANN on predicting these quantities. Prediction of the total energy is less accurate than prediction of the first cycle and average cycles energy.

The trained ANN were provided with a set of input variables for different values of strain amplitude (ε_a) and strain ratio (R) within the range of that of the training samples to study the effect of these parameters in energy prediction.

Table 2. Performance of the ANN on the test data

Performance Criterion	W_{P1}	W_{PA}	W_{PT}
Mean Square Error (MSE)	32.2960	56.0648	60703.2
Normalized Mean Square Error (NMSE)	0.03168	0.05677	0.23890
Mean Absolute Error (MAE)	4.63652	6.51717	176.463
Minimum Absolute Error (MinAE)	0.05160	1.04389	2.36970
Maximum Absolute Error (MaxAE)	11.7344	11.2677	490.368
Correlation Coefficient (r)	0.98920	0.98583	0.88445

Figure 3(a), (b) and (c) show the variation of the first cycle energy, average cycles energy and total cycles energy with strain amplitude for different strain ratios. It is observed from Fig. 3(a) that as the strain amplitude increases the energy dissipated in the first cycle (W_{P1}) increases. For small strain amplitudes (0.03–0.06) the energy dissipated in the first cycle is smaller for complete and semi-complete strain reversals as compared to energy dissipated in the tension only case (R = 0). At stain amplitude $\boldsymbol{\varepsilon_a} = 0.07$ all strain ratios yield the same first cycle energy. At large strain ratios (0.08–0.10) the energy

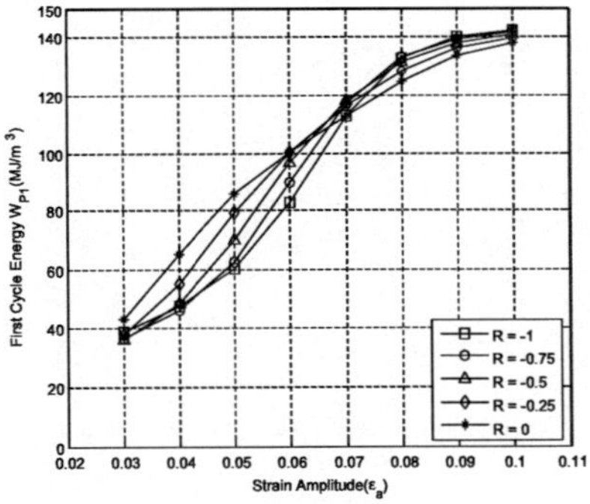

(a) First cycle energy (W_{P1})

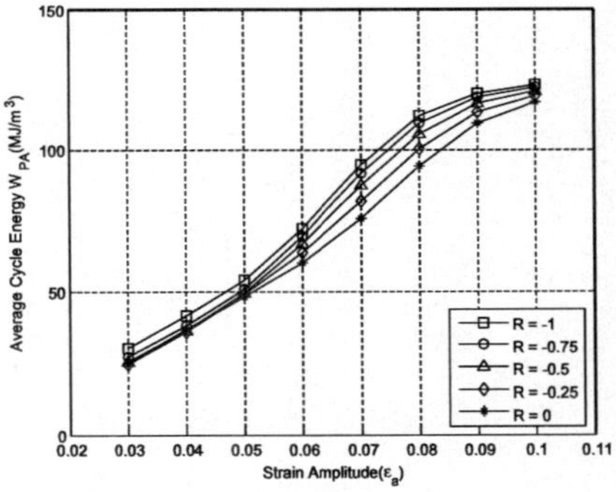

(b) Average cycles energy (W_{Pa})

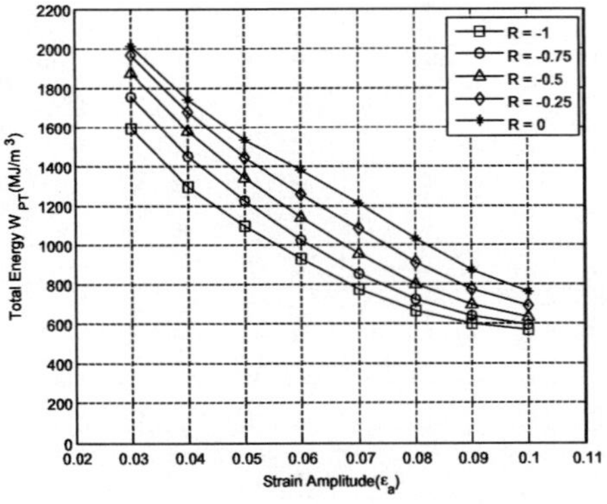

(c) Total cycles energy (W_{PT})

Fig. 3. Variations of energy (W_{P1}, W_{Pa}, W_{PT}) with strain amplitude dissipated in the first cycle (W_{P1}) is larger for complete and semi-complete strain reversals as compared to energy dissipated in the tension only case (R = 0). It is observed from Fig. 3(b) that as the strain amplitude increases the energy dissipated in the average cycles (W_{PA}) increases with R = −1 yielding the largest values, irrespective of the strain amplitude. It is observed from Fig. 3(c) that as the strain amplitude increases the energy dissipated in the total cycles (W_{PT}) decreases with R = −1 yielding the smallest values and R = 0 yielding the largest values for all strain amplitudes.

Figure 4(a), (b) and (c) show the variation of the first cycle energy, average cycles energy and total cycles energy with strain ratio for different strain amplitudes. It is observed from Fig. 4(a) that at low and intermediate strain amplitudes ($\varepsilon_a = 0.03-0.06$) the amount of energy dissipated in the first cycle (W_{P1}) is affect more by the strain ratio. At high strain amplitudes the effect of strain ratio is insignificant on the amount of energy dissipated in the first cycle. For energy dissipated in average cycles (W_{PA}), the effect of strain ratio is more apparent at intermediate and high strain amplitudes (ε_a = 0.05−0.10) as shown in Fig. 4(b). For energy dissipated on total cycles (W_{PT}), the effect of strain ratio is very much the same for all strain amplitudes as shown in Fig. 4(c).

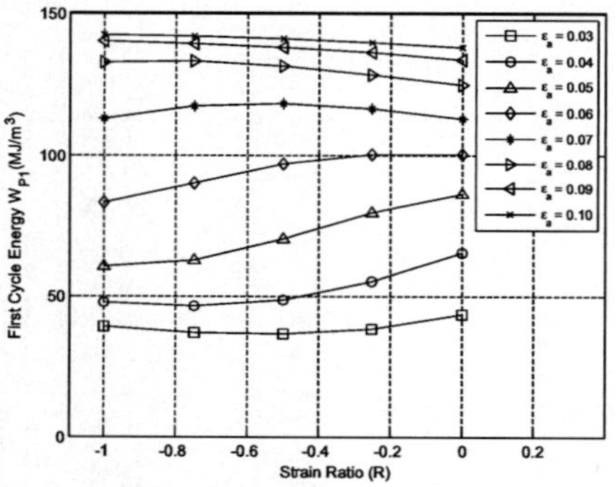

(a) First cycle energy (W_{P1})

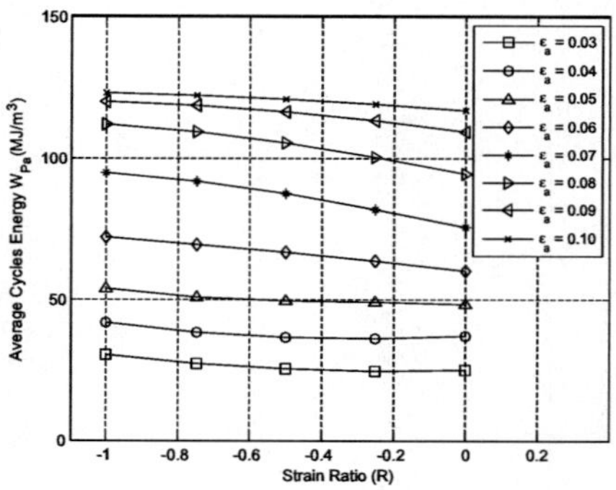

(b) Average cycles energy (W_{Pa})

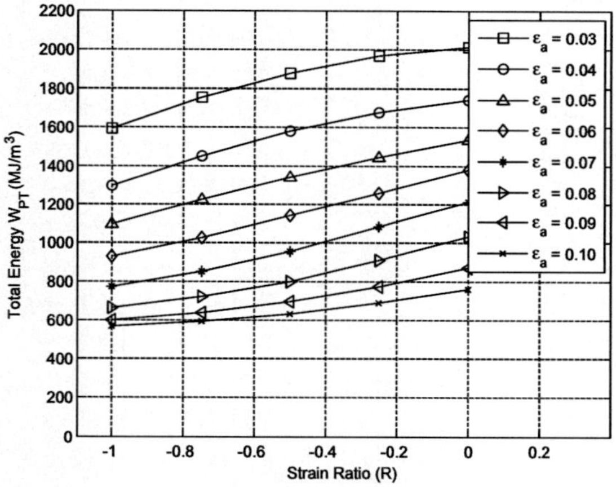

(c) Total cycles energy (W_{PT})

Fig. 4. Variations of energy (W_{P1}, W_{Pa}, W_{PT}) with strain ratio

Figure 5(a), (b) and (c) show the dissipated energy (W_{P1}, W_{Pa}, W_{PT}) and its variation with the two parameters – strain amplitude and strain ratio. It is observed from Fig. 5(a) that, the largest energy dissipated in the first cycle is at the maximum strain amplitude ($\varepsilon_a = 0.1$) and at complete load reversal, i.e., at strain ratio (R = −1) while the small energy dissipated in the first cycle is at the minimum strain amplitude ($\varepsilon_a = 0.03$) and at strain ratios approaching tension only (R = 0). It is observed from Fig. 5(b) that, the largest energy dissipated in the average cycles is at the maximum strain amplitude ($\varepsilon_a = 0.1$) and at complete load reversal, i.e., at strain ratio (R = –1) while the small energy dissipated in the average cycles is at the minimum strain amplitude ($\varepsilon_a = 0.03$) and at strain ratios approaching tension only (R = 0). The behavior of first cycle energy and average cycles energy are similar.

It is observed from Fig. 5(c) that, the largest energy dissipated in the total cycles is at the minimum strain amplitude ($\varepsilon_a = 0.03$) and at strain ratios approaching tension

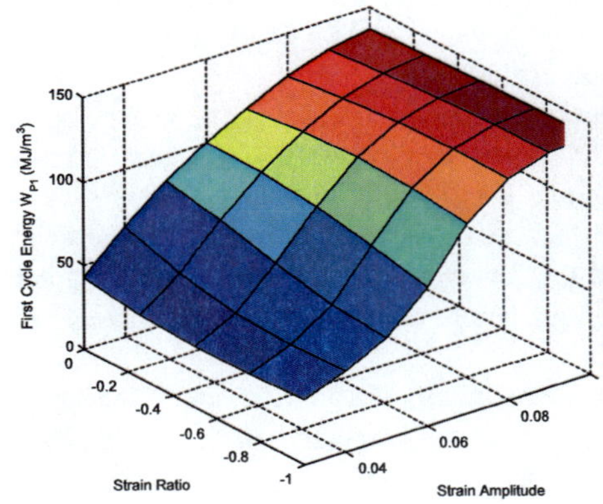

(a) First cycle energy (W_{P1})

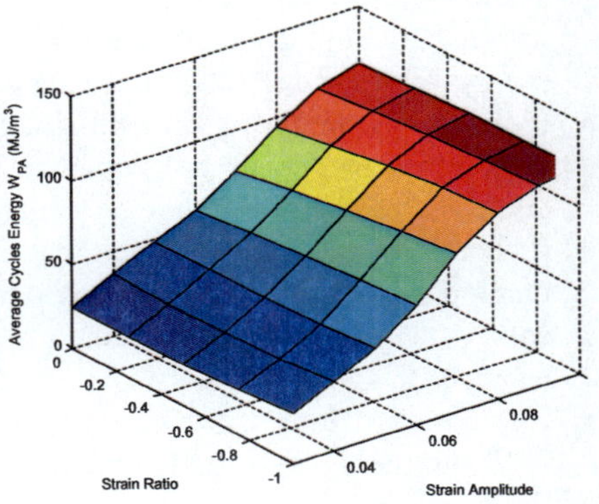

(b) Average cycles energy (W_{Pa})

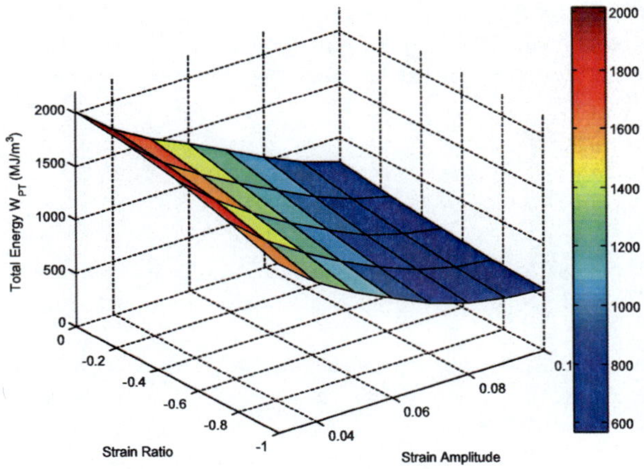

(c) Total cycles energy (W_{PT})

Fig. 5. Energy (W_{P1}, W_{Pa}, W_{PT}) surface variation with strain amplitude and strain ratio only (R = 0) while the smallest energy dissipated in the total cycles is at the maximum strain amplitude ($\varepsilon_a = 0.10$) and at complete load reversal, i.e., at strain ratio (R = −1). The behavior of total cycles energy is different from that of the first cycle energy and the average cycles energy.

6 Summary and Conclusions

This paper presented the results of prediction of dissipated energy of low-cycle fatigue test of steel reinforcing bars using Artificial Neural Network. Based on the trained ANN, an assessment was carried out to investigate the effect of a range of strain amplitudes and strain ratios on the amount of energy dissipated at different levels of low-cycle fatigue loading. It can be concluded from this study that:

- ANN accurately predicted the desired hysteresis energy dissipated in the first cycles, average cycles and total cycles. Predictions of energy for most test samples are within ±10% of the measured first cycle energy values, ±15% of the average cycles energy values and ±30% within the total cycles energy values.
- It observed from the assessment that the first and average cycles energies peak at large strain amplitude and complete load reversal (R = −1) while total energy dissipated peaks at low strain amplitude and tension only load (R = 0).
- It observed from the assessment that the first and average cycles energies are their minimum at small strain amplitude and tension only load (R = 0) while the total energy dissipated at its minimum at high strain amplitude and complete load reversal (R = –1).
- It is observed from the assessment that at an intermediate strain amplitude (ε_a = 0.07) the energy dissipated in the first cycle is same irrespective of the value of the strain ratio R.

Acknowledgement. The support for the experimental part of the research presented in this paper had been provided by the American University of Sharjah, Faculty Research Grant number FRG08-15. The support is gratefully acknowledged. The views and conclusions, expressed or implied, in this document are those of the authors and should not be interpreted as those of the sponsor.

References

1. Abdalla, J.A., Hawileh, R.A., Oudah, F., Abdelrahman, K.: Energy-based prediction of low-cycle fatigue life of BS 460B and BS B500B steel bars. Mater. Des. **30**(10), 4405–4413 (2009)
2. Hawileh, R.A., Abdalla, J.A., Oudah, F., Abdelrahman, K.: Low-cycle fatigue life behaviour of BS 460B and BS B500B steel reinforcing bars. Fatigue Fract. Eng. Mater. Struct. **33**(7), 397–407 (2010)
3. Chen, F., Yuan, X., Yi, W.: Towards a unified low-cycle fatigue model of steel rebars: a meta-analysis. Constr. Build. Mater. **216**, 564–575 (2019)

4. Hawileh, R.A., Abdalla, J.A., Al-Tamimi, A., Abdelrahman, K., Oudah, F.: Behavior of corroded steel reinforcing bars under monotonic and cyclic loadings. Mech. Adv. Mater. Struct. **18**(3), 218–224 (2011)
5. Kalogirou, S., Bojic, M.: Artificial neural networks for the prediction of the energy consumption of a passive solar building. Energy **25**, 479–491 (2000)
6. Yokoyama, R., Wakuia, T., Satake, R.: Prediction of energy demands using neural network with model identification by global optimization. Energy Convers. Manag. **50**(2), 319–327 (2009)
7. Ekici, B., Aksoy, U.: Prediction of building energy consumption by using artificial neural networks. Adv. Eng. Softw. **40**, 356–362 (2009)
8. Kalogirou, S.: Applications of artificial neural-networks for energy systems. Appl. Energy **67**, 17–35 (2000)
9. Adeli, H.: Neural networks in civil engineering: 1989–2000. Comput. Aided Civil Infrastruct. Eng. **16**(2), 126–142 (2001)
10. Abdalla, J.A., Elsanosi, A., Abdelwahab, A.: Modeling and simulation of shear resistance of R/C beams using artificial neural network. J. Franklin Inst. **344**(5), 741–756 (2007)
11. Abdalla, J.A., El Saqan, E.I., Hawileh, R.A.: Optimum seismic design of unbonded post-tensioned precast concrete walls using ANN. Comput. Concr. **13**(4), 547–567 (2014)
12. Abdalla, J.A., Hawileh, R.A., Al-Tamimi, A.: Prediction of FRP-concrete ultimate bond strength using Artificial Neural Network. In: Fourth International Conference on Modeling, Simulation and Applied (2011)
13. Abdalla, J.A., Attom, M.F., Hawileh, R.A.: Prediction of minimum factor of safety against slope failure in clayey soils using artificial neural network. Environ. Earth Sci. **73**(9), 5463–5477 (2015)
14. Kaveh, K., Hamze-Ziabari, S.M.: Soft computing-based slope stability assessment: a comparative study. Geomech. Eng. **14**(3), 257–269 (2018)
15. Abdalla, J.A., Attom, M.F. and Hawileh, R.A.: Artificial neural network prediction of factor of safety of slope stability of soils. In: Proceedings of the 14th International Conference on Computing in Civil and Building Engineering (2012)
16. Abuodeh, O.R., Abdalla, J.A., Hawileh, R.A.: Prediction of shear strength and behavior of RC beams strengthened with externally bonded FRP sheets using machine learning techniques. Compos. Struct. **234**, 111698 (2020)
17. Abuodeh, O.R., Abdalla, J.A., Hawileh, R.A.: Predicting the shear capacity of FRP in shear strengthened RC beams using ANN and NID. In: 8th International Conference on Modeling Simulation and Applied Optimization (ICMSAO2019) (2019)
18. Abuodeh, O.R., Abdalla, J.A., Hawileh, R.A.:, Prediction of compressive strength of ultra-high performance concrete using SFS and ANN. In: International Conference on Modeling Simulation and Applied Optimization (ICMSAO 2019) (2019)
19. Abdalla, J.A., Hawileh, R.A.: Predictions of low-cycle fatigue life of steel reinforcing bars using artificial neural network. In: Proceedings of the 3rd International Conference on Modeling Simulation and Applied Optimization (ICMSAO 2009), Sharjah, UAE (2009)
20. Pleune, T., Chopra, O.: Using artificial neural networks to predict the fatigue life of carbon and low-alloy steels. Nucl. Eng. Des. **197**, 1–12 (2000)
21. Genel, K.: Application of artificial neural network for predicting strain-life fatigue properties of steels on the basis of tensile tests. Int. J. Fatigue **26**, 1027–1035 (2004)
22. Abdalla, J.A., Hawileh, R.A.: Modeling and simulation of low-cycle fatigue life of steel reinforcing bars using artificial neural network. J. Franklin Inst. **348**(7), 1393–1403 (2011)
23. Abdalla, J.A., Hawileh, R.A.: Energy-based predictions of number of reversals to fatigue failure of steel bars using artificial neural network. In: The 13th International Conference on Computing in Civil and Building Engineering (2010)

24. Durodol, J.F., Ramachandra, S., Gerguri, S., Fellows, N.A.: Artificial neural network for random fatigue loading analysis including the effect of mean stress. Int. J. Fatigue **111**, 321–332 (2018)
25. Abdalla, J.A., Hawileh, R.A.: Artificial neural network predictions of fatigue life of steel bars based on hysteretic energy. J. Comput. Civil Eng. **27**(5), 489–496 (2013)
26. Abdalla, J.A., Hawileh, R.A.: Predictions of Hysteresis Energy Dissipation in Steel Reinforcing Bars using Artificial Neural Networks. In: Topping, B.H.V., Tsompanakis, Y. (eds). Proceedings of the First International Conference on Soft Computing Technology in Civil, Structural and Environmental Engineering. Civil-Comp Press, Stirlingshire, UK, Paper 32 (2009). https://doi.org/10.4203/ccp.92.32
27. Halford, G.: The energy required for fatigue. J. Mater. **1**(1), 3–18 (1966)
28. Ellyin, F., Kujawski, D.: Plastic strain energy in fatigue failure. ASME J. Press. Vessel Technol. **106**(4), 342–347 (1984)
29. Tchankov, D., Vesselinov, K.: Fatigue life prediction under random loading using total hysteresis energy. Int. J. Press. Vessels and Pip. **75**, 955–960 (1998)
30. Park, J., Nelson, D.: Evaluation of an energy-based approach and a critical plane approach for predicting constant amplitude multiaxial fatigue life. Int. J. Fatigue **22**, 23–39 (2000)
31. Lagoda, T.: Energy models for fatigue life estimation under uniaxial random loading Part I: the model elaboration. Int. J. Fatigue **23**(467–480), 31 (2001)
32. NeuroSolutions software version 5.0. Source. www.nd.com. Accessed May (2009)

Machine Learning for Whole-Building Life Cycle Assessment: A Systematic Literature Review

Natalia Nakamura Barros(✉) and Regina Coeli Ruschel

University of Campinas, Campinas, Brazil
natalianakamura.arq@gmail.com, ruschel@unicamp.br

Abstract. Life Cycle Assessment (LCA) is a methodology to systematically investigating impacts from interactions between environment and human activities. However, the number of parameters and uncertainty factors that characterize built impacts over their full-lifecycle, preclude a broader LCA adoption. This enable faster progress towards reducing building impacts by combining established environmental impact assessment methods with artificial intelligence approaches, such as machine learning (ML) and neural networks. This article will present previous research on ML for LCA of buildings. To achieve this goal, we perform a Systematic Literature Review (SLR). SLR was governed by the question "What are scientific research developed for Architecture, Engineering and Construction (AEC) industry in LCA and ML context?". This SLR was performed in three databases: Scopus, Engineering Village and Web of Science, using keywords: Life Cycle Assessment, Machine Learning, Learning, Building and Neural Network. From SLR, we identified best practices, acquired and developed by other studies, clarifying how to interpret large data sets monitored through advanced analysis to improve LCA. The results showed: (i) number of articles increase in recent years; (ii) the most searched environmental indicators are energy consumption and Global Warming Potential (GWP); (iii) machine learning is mainly used for prediction impacts and; (iv) the most used ML method is Artificial Neural Networks. Advances in LCA and ML field can contribute to calculation and analysis of buildings environmental indicators, as well as can develop and improve LCA methods. The combination of reliable data and ML will produce an unprecedented change in speed and accuracy of LCA.

Keywords: Life cycle assessment · Machine learning · Neural networks

1 Introduction

Buildings are the highest contributors to energy demand, greenhouse gas (GHG) emissions, resource consumption and waste generation. There is an opportunity to confront climate change, global warming and scarcity of resources, rethinking building design. Environmental impact assessment methods, such as life cycle assessment (LCA), are increasingly used. However, they could not obtaining expected benefits due to high number of parameters and uncertainty factors that characterize whole-building

E. Toledo Santos and S. Scheer (Eds.): ICCCBE 2020, LNCE 98, pp. 109–122, 2021.
https://doi.org/10.1007/978-3-030-51295-8_10

environmental impacts. In addition, effort and cost required for a reliable assessment preclude a broader LCA adoption. Therefore, it seems possible to make faster progress towards reducing building impacts, combining established environmental impact assessment methods with artificial intelligence (AI) approaches, such as machine learning (ML) and neural networks (NN) [1].

Energy and environment assessment of building's performance is necessary to propose measures to save energy and reduce building environmental impacts, such as greenhouse gases emission. Currently, solving this complex problem usually requires an interdisciplinary team, knowledge on specific software or algorithm, specialist user, large amount of data collection and long computational time. Common language lack often complicates the interpretation between these two areas significantly different, but highly connected [2].

A major challenge in obtaining a reliable LCA comes from an inadequate understanding of underlying activities related to each product's life cycle stages, based on expert's knowledge. Data-driven modeling, on the other hand, is an emerging approach that uses ML's benefits to build models that complement or replace knowledge-based models. Incorporating suitable data analysis models to use data product and process real-time can significantly improve LCA techniques [3].

This article will present previous research on machine learning for life cycle assessment of buildings. To achieve this goal, we perform a Systematic Literature Review (SLR).

2 Method

The systematic literature review (SLR), following Kitchenham and Charters [4] protocol, governed by question "What are scientific research developed for Architecture, Engineering and Construction (AEC) industry in LCA and ML context?". This SLR was performed in three relevant databases: Scopus, Engineering Village and Web of Science, using keywords: Life Cycle Assessment, Machine Learning, Learning, Building and Neural Network. Table 1 shows SLR protocol, Table 2 shows number of articles per database and Table 3 shows assessment criteria of the dimensions of studies quality.

In Step 1, we excluded duplicate articles, with 203 articles, or 46%, resulting in 236 articles. Step 2 resulted in 110 excluded articles. The main exclusion criterion was article deals with another study area. The recurrent area of study was education. This was due to choice keyword: 'Learning'. Other areas corresponded: Chemical, Biological, locally competitive algorithm, TI, Forestry. Many articles did not deal with Machine Learning and therefore were exclude. Finally, in step 3, we applied the assessment criteria of the dimensions of studies quality (Table 3), whose focus was to verify: (i) the study addresses precisely the subject of systematic review and; if (ii) the study was carried out in an identical context to that defined for this review. In this stage, 111 articles were exclude.

Table 1. Systematic literature review protocol.

Goal		Verify scientific research developed for Architecture, Engineering and Construction (AEC) industry in LCA and ML context
Intervention (Context/focus)		Articles related to LCA and ML in Architecture, Engineering and Construction (AEC)
Population		Studies published in English
Results		Benefits, results, context and application related to study
Research questions		How many artifacts were produced according to focus/context: LCA and ML? What is the goal study? What are Machine Learning techniques used? What are impact indicators analyzed? What is the study object? What are the benefits of each artifact in terms of optimization, prediction, decision-making support and new systems? What are research centers involved in research development?
Who might be interested in this research?		Researchers, specialists, architects and engineers interested in building life cycle assessment and machine learning
Data bases		Scopus, Engineering Village and Web of Science
Studies types		Scientific articles indexed in journals and peer-reviewed
Language		English
Search terms		Life Cycle Assessment AND Machine Learning/Life Cycle Assessment AND Learning AND Building/LCA AND Learning AND Building/Neural Network AND Life Cycle Assessment
Inclusion criteria	IC1	Articles must have been published in peer-reviewed journals or conference proceedings
	IC2	Studies that contain search terms in: Title, abstract or keywords
	IC3	Studies written in English
Exclusion criteria	EC1	Duplicated articles
	EC2	Full text not available
	EC3	Study not written in English
	EC4	Study rated LOW on the Quality Assessment criteria
	EC5	Study rated MEDIUM on the Quality Assessment criteria
Data Extraction		Description of study context; goal, method, application, research center, relevance and benefits
SLR steps	Step1	Exclusion of duplicated articles - EC1
	Step2	Exclusion of articles that do not meet inclusion criteria - IC1, IC2 and IC3 - and that meet exclusion criteria - EC2 and EC3
	Step3	Exclusion of articles that are classified as low quality - EC4 and medium quality - EC5

Source: adapted from [4].

Table 2. Number of articles per database.

	Data base	Search criteria	Search terms	Number of articles
Research date: 02/09/2019	Scopus	Article Title, Abstracts, Keywords	"Life Cycle Assessment" AND "Machine Learning"	14
			"Life Cycle Assessment" AND learning AND building	36
			LCA AND learning AND building	39
			"Neural Network" AND "Life Cycle Assessment"	82
	Engineering Village	Subject/Title/Abstract	"Life Cycle Assessment" AND "Machine Learning"	9
			"Life Cycle Assessment" AND learning AND building	32
			LCA AND learning AND building	35
			"Neural Network" AND "Life Cycle Assessment"	45
	Web of Science	Topic (title, abstract, author's keywords and Keywords Plus)	"Life Cycle Assessment" AND "Machine Learning"	14
			"Life Cycle Assessment" AND learning AND building	41
			LCA AND learning AND building	33
			"Neural Network" AND "Life Cycle Assessment"	59
TOTAL articles: 439				

Source: authors.

3 Results

The SLR resulted in 15 articles presented in Table 4. The number of published articles related to LCA and ML increased mainly in 2019 (Fig. 1). Most articles focused on new artifacts development; there was two-literature review article [1, 6]. The main researches benefits were optimizing building performance, predicting impacts, decision-making support and developing new systems, as shown in Fig. 2. Such classification took into account the main benefit listed by article's authors. The description of these articles in this topic will follow the benefits mentioned above.

Table 5 shows environmental impacts, objects of study and ML techniques used in selected articles by SLR.

3.1 Articles Classified as Literature Review

D'Amico et al. [1] briefly presented previous attempts to employ ML and LCA techniques in civil and structural engineering. They briefly reviewed the status of ML and NN applications in structural and civil engineering, and highlighted its potential use in research related to sustainable structural materials and structures. Finally, this brief communication aimed to launch a broad and open data call to support research [1].

Manfren, Caputo and Costa [6] presented a review of modeling tools developed to identify optimal solutions for district-wide energy systems. For example, Geographic Information Systems (GIS) are essential for storing, organizing and viewing spatial data and performing spatial calculations (preliminary phase). After that, accounting, simulation and optimization techniques can be employed to correctly design distributed generation (DG) systems. Finally, the environmental and economic impact can be addressed by calculating local dispersion of pollutants, external costs and carrying out LCA.

Table 3. Assessment criteria of the dimensions of quality of primary studies.

Evaluation	Quality of study performance	Relevance to the review issue	Relevance to the review focus
High	The proposed method meets the standards required for subject under study; the study strictly followed proposed method; and the results are supported by facts and data	The study addresses precisely the subject of systematic review	The study was conducted in a similar context to one defined for the review
Medium	The proposed method has gaps in relation to standards required for subject under study; or the study does not show that it followed the proposed method in its entirety; or the results are not entirely based on facts and data	The study partially addresses the subject of systematic review	The study was conducted in a context similar to that defined for the review
Low	The proposed method does not meet standards required for subject under study; or the study does not show that it followed proposed method; or the results are not based on facts and data	The study only addresses the subject of systematic review only superficially	The study was conducted in a different context from defined for the review

Source: adapted from (HARDEN e GOUGH, 2012 *apud* [5], p. 146).

Table 4. Year, classification, journal and author's affiliation of articles selected by SLR.

Article	Year	Classification	Journal	Author's affiliation
[1]	2019	Literature review	Structures	REBEL (Resource Efficient Built Environment Lab); Edinburgh Napier University; University of Edinburgh; Expedition Engineering, Useful Projects Ltd., London, UK; Cambridge Architectural Research (CAR), Cambridge, UK; World Green Building Council, London, UK
[6]	2011	Literature review	Applied Energy	Politecnico de Milano, Milano, Italy
[7]	2019	Optimize performance	Journal of Building Engineering	Concordia University, Montreal, Quebec, Canada
[8]	2016	Optimize performance	Energy and Buildings	University of Texas at San Antonio, United States; Islamic Azad University, Tehran, Iran; National Petrochemical Company, Tehran, Iran
[2]	2019	Impact prediction	Journal of Cleaner Production	University of Palermo, Italy; Aachen University, Germany
[3]	2018	Impact prediction	Proceedings - IEEE International Conference on Big Data, 2017	Syracuse University, New York, USA; National Institute of Standards and Technology, Gaithersburg, Maryland, USA
[9]	2019	Impact prediction	Proceedings of the 6th International Symposium on Life-Cycle Civil Engineering, IALCCE 2018	University of Nottingham, Nottingham, United Kingdom
[10]	2015	Impact prediction	Transportation research part d-transport and environment	Korea National Defense University, Seoul, Korea; University of Illinois at Urbana-Champaign, Urbana, IL 61801, USA
[11]	2019	Impact prediction	Sustainable Cities and Society	COMBO Solutions, Lyon, France Ecole Polytechnique Fédérale de Lausanne (EPFL), Fribourg, Switzerland
[12]	2013	Impact prediction	Journal of Civil Engineering and Management	University of Nebraska-Lincoln, NE 68588-0500, USA
[13]	2019	Impact prediction	International Journal of Life Cycle Assessment	University of Illinois at Urbana-Champaign, Urbana, IL 61801, USA
[14]	2009	Decision making support	International Conference on Artificial Intelligence and Computational Intelligence	Tongji University, Shangai 200092, China
[15]	2016	Decision making support	Journal of Environmental Accounting and Management	Luxembourg Institute of Science and Technology (LIST), Luxembourg
[16]	2016	Decision making support	Energy	University College London, London, UK
[17]	2013	New system	Applied Mechanics and Materials	Tianjin University of Technology, China

Source: Authors

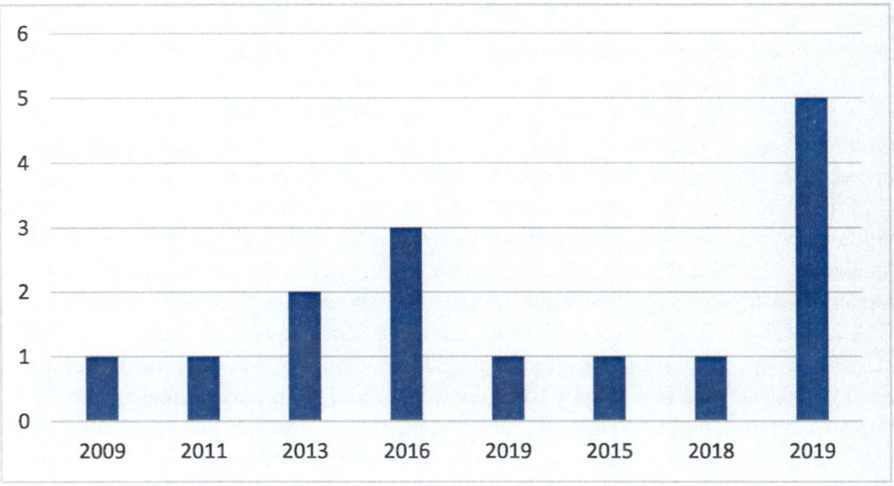

Fig. 1. Number of articles by year of publication. Source: authors.

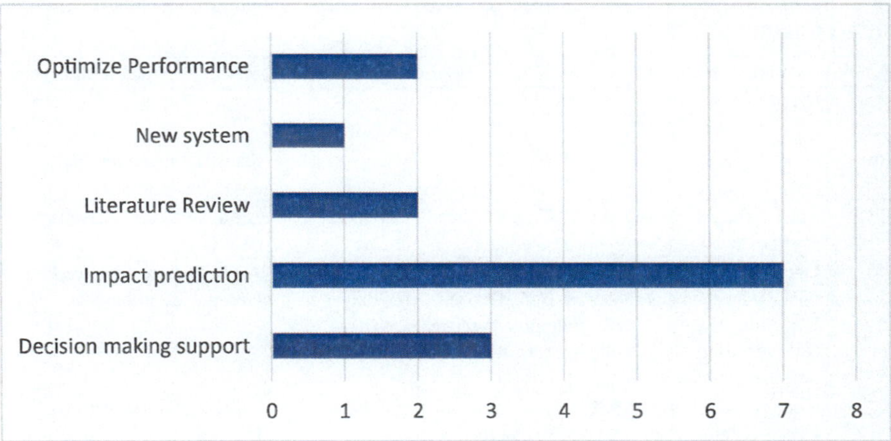

Fig. 2. Number of articles according to main benefit presented. Source: authors.

3.2 Articles Classified as Optimize Performance

Sharif and Hammad [7] developed a study focused on artificial neural network (ANN) to obtain renovation scenarios to minimize Total Energy Consumption (TEC), Life Cycle Cost (LCC) and environmental impacts. A set of data representative of renewal scenarios was developed from results obtained by Simulation-Based Multi-Objective Optimization (SBMO). The simulated data refer to existing buildings, related to several factors, including TEC, external temperature, building envelope components, Heating, Ventilation and Air-Conditioning (HVAC) and lighting systems. The simulation took about 180 s using SBMO model, while proposed ANNs can provide results in less than 1 s. ANNs were developed as surrogate models to emulate simulation models of real computationally expensive buildings. The computational time savings associated with proposed substitute models was significant [7].

Azari et al. [8] used a multi-objective optimization algorithm to explore ideal building envelope design in relation to energy use and life cycle environmental impacts of office building in Seattle, Washington. Design inputs include insulation material, window type, window frame material, wall thermal resistance and south and north window-to-wall ratios (WWR). Using these variables, they tried to find design combination that produces the smaller operational energy and the smaller environmental impact. The eQuest 3.65 simulation tool is used to assess operational energy, while LCA and Athena IE are used to estimate environmental impacts. In addition, an ANN and genetic algorithm (GA) approach is used to develop future populations/generations of combinations and find ideal design combination. The environmental impact categories of interest in LCA include global warming, acidification, eutrophication, formation of air pollution and ozone depletion. The results of this research have potential to meet architecture and construction research community to design guidelines that simultaneously reduce operational energy consumption and environmental impacts.

Table 5. Environmental impacts, object of study and machine learning technique used in articles selected by SLR.

Article	Environmental impacts	Object of study	Machine learning technique
[1]	None	Structure (building)	None
[6]		Energy systems	Statistical regression, NN, SVM, etc.
[7]	Total Energy Consumption (TEC), o LCC and environmental impacts	Buildings	Artificial Neural Network (ANN)
[8]	Operational energy, Global Warming Potential (GWP), Acidification Potential (AP), Ozone Depletion Potential (ODP), Eutrophic Potential (EP) and atmospheric pollution formation potential	Commercial building	Artificial Neural Network; Genetic Algorithm
[2]	Energy; GWP, ODP, AP, EP, Photochemical Ozone Creation Potential (POCP) and the Abiotic Depletion Potential (ADPFossil)	Buildings	Artificial Neural Network (ANN)
[3]	Energy	Injection molding	Bayesian Network (BA)
[9]	Fuel consumption	Road pavement	ANN; BA
[10]	Energy; fuel consumption	LCA method	Predictive usage mining for life cycle assessment (PUMLCA) algorithm
[11]	GWP	Residential and commercial building	ANN, Multiple Linear Regression and Support Vector Regression
[12]	Energy	Residential building	Stochastic Markov model and Neural Network

(*continued*)

Table 5. (*continued*)

Article	Environmental impacts	Object of study	Machine learning technique
[13]	Energy; GWP	LCA method	Artificial Neural Network (ANN); Monte Carlo, Bayesian network
[14]		Ashlar brick	Back propagation neural network (BPNN) and Genetic Algorithm (GA)
[15]		Construction materials	Agglomerative, clustering technique and self-organizing map
[16]	LCCF (life cycle carbon footprint) and LCC (life cycle cost)	Residential building	Multi objective genetic algorithms (MOGA)
[17]			Back propagation neural network (BPNN)

Source: Authors

3.3 Articles Classified as Impact Prediction

D'Amico et al. [2] developed a method to decision-making support during planning and design of buildings with high-energy performance, in order to provide an estimate of life cycle environmental and energy building performance. Different values of transmittance, construction materials and energy networks were chosen and simulated to represent Italian buildings for different climatic zones, climatic conditions and shape factors. The development of this approach allowed implementation of energy and environmental database. The database consists of 36 columns and 780 rows for a total matrix composed of 28,080 cells. The columns consist of thermo-physical and environmental inputs and outputs of representative environmental impacts, such as Global Warming Potential (GWP), Ozone Depletion Potential (ODP), Acidification Potential (AP), Eutrophic Potential (EP), photochemical ozone creation potential (POCP) and abiotic depletion potential fossil (ADPFossil). In the lines, there are 13 construction models located in three cities for five climatic zones, which were simulated for four scenarios of energy carriers (780 cases were analyzed).

After pre-processing phase, D'Amico et al. [2] explored various ANN topologies, changing the number of neurons, hidden layers, activation functions and all learning parameters to minimize Mean Squared Error (MSE) and identify ideal configuration. The results emphasized potential of ANN to predict energy demand and building environmental impacts. Thus, the authors demonstrated that ANN is a reliable alternative for simultaneously determining energy demand and environmental impacts.

Li et al. [3] used Bayesian Network (BN) to estimate energy consumption of injection molding process. BN learned from data presented problems such as: (i) arcs tend to indirectly connect related nodes; (ii) there are arcs being directed incorrectly. These problems are not errors that need to be corrected; differences are considered between structure learned and structure created by specialists. In this way, the learned model guided by specialists (mixture of specialized knowledge and knowledge of data mining) improves forecast accuracy.

Perrotta et al. [9] developed an approach based on application of Boruta Algorithm (BA) and NN, for evaluation and calculation of fleet of trucks fuel consumption, which can be used to estimate emissions and pavement roads use. The data includes geographical positional system (GPS) of vehicles, speed, torque and engine revolution, vehicle's weight and fuel consumption measured with nearest 0.001 L. BA is applied first to select most significant variables in order to avoid over adjustment. Then, an NN that includes all variables selected by BA was developed. The study showed that NN are suitable for analyzing large amounts of data, coming from fleet and road asset management databases, and the developed NN model is able to estimate impact of parameters related to rolling resistance (pavement roughness and macrotexture) in truck's fuel consumption.

Ma and Kim [10] proposed a time series modeling technique use, predictive usage mining for life cycle assessment (PUMLCA), as an alternative to conventional constant rate method. There are five stages of PUMLCA: data preprocessing for handling missing and abnormal values, seasonal period analysis, segmentation analysis, time series analysis and predictive LCA. The proposed algorithm captured patterns of sensor data use on large scale with automatic segmentation algorithm and time series analysis, and was able to assess environmental impact of complex system over a real time horizon. The results showed that predictive model of PUMLCA method can provide a similar level of forecast accuracy to constant rate method when data is constant, and a greater forecast accuracy when data has complex patterns. The automatic segmentation algorithm expanded important patterns and helped to predict future values more accurately.

Duprez et al. [11] developed a method to predict GWP of new design alternatives in a short period of time and with a high coefficient of determination, using ML. Three metamodels, Multiple Linear Regression (MLR), Support Vector Regression (SVR) and ANN, were chosen in order to find out if they are able to predict GWP accurately, in addition to being less computational than original model. As a result, ANN offered better results than MLR and SVR. When trained on approximately 25,000 samples, it is able to predict new GWP values with a determination coefficient greater than 0.9 and a low RMSE. MLR and SVR fail to accurately predict new values because they are not suitable for complex – MLR - or very slow - SVR models. In this way, it was possible to reduce computational time with high precision of results.

Wang and Shen [12] developed a stochastic Markov model to improve accuracy of life cycle energy consumption forecast, considering longitudinal uncertainties in building conditions, degree-days and useful life. The Markov building deterioration model was developed based on historical records of similar conditions, and is able to predict the useful life and expected condition of building at a specific time. The annual variation in energy consumption was simulated as joint process of deterioration of building and temperature change. To calculate annual energy consumption, probabilistic distribution of energy consumption was estimated by NN with available data set. The proposed stochastic approach can produce a much more restricted distribution and seems closer to the measured data, which indicates that the longitudinal uncertainty both in the thermal condition of the building and in the temperature, and can explain a lot of uncertainty in the variation of the residential energy performance [12].

Ziyadi and Al-Qadi [13] adopted quantitative uncertainty analysis methods to characterize, propagate and quantify uncertainties in Life Cycle Inventory Analysis (LCIA) model. Several data analysis methods were used to conduct a complete analysis of model's uncertainty. To propagate input variability through system using interval analysis, a substitute ANN model was trained and tested to replace simulation. Then, Monte Carlo (MC) sampling method was used as a direct way to propagate input uncertainty and was compared to indirect nonlinear optimization method that tries to maximize output range, given input domain. Assuming that model's parameters are random variables, probability distributions can be used to represent these parameters. So, a Bayesian inference, as a commonly used approach, can be adopted to quantify the uncertainty of parameters. Finally, the model correction method can be adopted as a method to characterize uncertainty in model form.

3.4 Articles Classified as Decision-Making Support

Shi and Xu [14] presented a systematic method based on LCA to analyze environmental performance of construction materials. For an additional step, back propagation neural network (BPNN) and the hybrid algorithm GA-BP are introduced to evaluate environmentally building materials, respectively. Compared with BPNN, the authors identified that hybrid GA-BP algorithm is much more suitable for selection of environmentally construction materials and has greater precision. This research is useful for construction professionals in selection of sustainable building materials.

Marvuglia et al. [15] tested two different grouping techniques to distinguish classes of materials based on their environmental performance. The first is an agglomerative clustering technique and the second is self-organizing map (SOM). A vector of six elements represented each material: five values indicating the midpoint of potential environmental impact of material and, in addition, its accumulated primary energy (non-renewable), all normalized in cubic meters and depending on material conductivity. The two grouping techniques produced coherent results and visual observation of obtained clusters allowed identification possible explanatory variables that could be used to determine limit values to distinguish materials classes based on their environmental performance.

Schwartz, Raslan and Mumovic [16] used multi objective genetic algorithms (MOGA) to find ideal designs for residential complex renovation, to minimize life cycle carbon footprint (LCCF) and LCC over an estimated useful life 60 years. The results showed MOGA use has potential to reduce LCCF and LCC. Finally, when comparing LCA with the most used performance-based decision-making design procedures, the study highlights that these different methods use can lead to different design solutions.

3.5 Articles Classified as New System

Xia and Liu [17] established a green building assessment index system, based on life cycle theory and BPNN use, through a comparative Chinese and international building classification system green analysis. In this article, BPNN and assessment tools, is a satisfactory solution for assessment of green buildings. With this rigor, scientific assessment will be basis for choosing the best plan for green building systems, which can also be used to assess green building that has been completed.

4 Final Considerations

SLR showed an increase in number of articles dealing with LCA and ML in recent years. Machine learning is mainly used to predict impacts, the most researched environmental indicators are energy consumption and Global Warming Potential (GWP), and the most used ML method is Artificial Neural Networks.

ANN was used mainly to optimize building performance and impact prediction [2, 7–9, 11]. Given a large set of variables, the articles demonstrated that ANN is a reliable alternative to achieve the goal of study. ANNs are advantageous due to their ability to overcome several limitations typical of traditional software, such as collecting environmental and energy data, knowledge of physical problem and software language, long computational time and need to calibrate model [2]. All of this, ML, and NN models in particular, offer a better and more reliable decision support tool for engineers and architects, reducing uncertainties in LCA field [9].

The ANN implementation in software can allow appropriate decision support tool development. The possibility of using an extremely consistent instrument to predict building performance enables decision makers to make more sustainable decisions when analyzing reliable energy and environmental assessments [2, 11].

Another advantage is the possibility of having several outputs from the same ANN, so that different problems in same case study can be solved, which would normally require several software tools and experienced users [2]. In addition, there is a significant reduction in simulation time with ANN use [7]. However, the validity of neural network solution is strongly linked to database reliability, which is generally difficult to implement [2].

Bayesian Network can also be used to estimate energy consumption. The indicator calculated using BN is comparable to real Eco-Indicator. The advantage is the calculated indicator can actually reflect actual energy consumption of process, considering almost all possible values for missing information based on the knowledge learned from data. However, BN structure learned from data presented wrong or missing connections, which needed to be corrected with specialist's knowledge [3].

Several advantages have been identified: (i) BN is suitable for small data sets; (ii) BN allows efficient different sources of knowledge use: knowledge provided by domain experts and the knowledge learned from data; (iii) BN can answer queries based on incomplete information, that is, it can provide an estimate for a query considering almost all possible values for this missing information based on knowledge learned from data [3].

Authors also used other algorithms. Boruta Algorithm was used to quickly and effectively identify the most significant variables to fuel consumption model of truck's fleet [9]. The BPNN and GA-BP were introduced to assist in choice of environmentally building materials [14]. The agglomerative clustering technique and self-organizing map (SOM) allowed the identification of possible explanatory variables that could be used to distinguish classes of materials based on their environmental performance [15]. MOGA can be used to assist designer's decision-making, in terms of LCCF and LCC [16].

Finally, Ma and Kim [10] developed an algorithm, called PUMLCA, which presented a higher forecasting accuracy when data has complex patterns. By modeling usage patterns such as trend, seasonality and level from a time series of usage information, predictive LCA can be performed over a real-time horizon, which can provide a more accurate estimate of environmental impact.

Advances in LCA and ML area can contribute to calculation and analysis building environmental indicators, as well as in development and improvement LCA methods. In short, Machine Learning algorithms and techniques have the potential to increase accuracy in LCA. The reliable data and AI combination will produce an unprecedented change in speed and accuracy of LCA.

References

1. D'Amico, B., Myers, R.J., Sykes, J., Voss, E., Cousins-Jenvey, B., Fawcett, W., Richardson, S., Kermani, A., Pomponi, F.: Machine learning for sustainable structures: a call for data. Structures **19**, 1–4 (2019). https://doi.org/10.1016/j.istruc.2018.11.013
2. D'Amico, A., Ciulla, G., Traverso, M., Lo Brano, V., Palumbo, E.: Artificial Neural Networks to assess energy and environmental performance of buildings: an Italian case study. J. Clean. Prod. **239**, 117993 (2019). https://doi.org/10.1016/J.JCLEPRO.2019.117993
3. Li, Y., Zhang, H., Roy, U., Lee, Y.T.: A data-driven approach for improving sustainability assessment in advanced manufacturing. In: Nie, J.Y., Obradovic, Z., Suzumura, T., Ghosh, R., Nambiar, R., Wang, C., Zang, H., BaezaYates, R., Hu, X., Kepner, J., Cuzzocrea, A., Tang, J., Toyoda, M. (eds.) IEEE International Conference on Big Data (Big Data). IEEE, Boston (2017)
4. Kitchenham, B., Charters, S.: Guidelines for performing Systematic Literature Reviews in Software Engineering, UK (2007)
5. Dresch, A., Lacerda, D.P., Antunes Junior, J.A.V.: Design Science Research: A Method for Science and Technology Advancement. Springer, Cham (2015)
6. Manfren, M., Caputo, P., Costa, G.: Paradigm shift in urban energy systems through distributed generation: methods and models. Appl. Energy **88**, 1032–1048 (2011). https://doi.org/10.1016/j.apenergy.2010.10.018
7. Sharif, S.A., Hammad, A.: Developing surrogate ANN for selecting near-optimal building energy renovation methods considering energy consumption, LCC and LCA. J. Build. Eng. **25**, 100790 (2019). https://doi.org/10.1016/J.JOBE.2019.100790
8. Azari, R., Garshasbi, S., Amini, P., Rashed-Ali, H., Mohammadi, Y.: Multi-objective optimization of building envelope design for life cycle environmental performance. Energy Build. **126**, 524–534 (2016). https://doi.org/10.1016/j.enbuild.2016.05.054
9. Perrotta, F., Parry, T., Neves, L.C., Mesgarpour, M.: A machine learning approach for the estimation of fuel consumption related to road pavement rolling resistance for large fleets of trucks. In: The Sixth International Symposium on Life-Cycle Civil Engineering (IALCCE 2018), Belgium (2018)
10. Ma, J., Kim, H.M.: Predictive usage mining for life cycle assessment. Transp. Res. Part D Transp. Environ. **38**, 125–143 (2015)

11. Duprez, S., Fouquet, M., Herreros, Q., Jusselme, T.: Improving life cycle-based exploration methods by coupling sensitivity analysis and metamodels. Sustain. Cities Soc. **44**, 70–84 (2019). https://doi.org/10.1016/j.scs.2018.09.032
12. Wang, E., Shen, Z.: Lifecycle energy consumption prediction of residential buildings by incorporating longitudinal uncertainties. J. Civ. Eng. Manage. **19**, S161–S171 (2013). https://doi.org/10.3846/13923730.2013.802744
13. Ziyadi, M., Al-Qadi, I.L.: Model uncertainty analysis using data analytics for life-cycle assessment (LCA) applications. Int. J. Life Cycle Assess. **24**, 945–959 (2018)
14. Shi, Q., Xu, Y.: The selection of green building materials using GA-BP hybrid algorithm. In: 2009 International Conference on Artificial Intelligence and Computational Intelligence, pp. 40–45 (2009)
15. Marvuglia, A., Kanevski, M., Benetto, E.: Machine learning for toxicity characterization of organic chemical emissions using USEtox database: learning the structure of the input space. Environ. Int. **83**, 72–85 (2015). https://doi.org/10.1016/j.envint.2015.05.011
16. Schwartz, Y., Raslan, R., Mumovic, D.: Implementing multi objective genetic algorithm for life cycle carbon footprint and life cycle cost minimisation: a building refurbishment case study. Energy **97**, 58–68 (2016). https://doi.org/10.1016/j.energy.2015.11.056
17. Xia, L., Liu, J.: Research on green building assessment system based on bp neural network and life cycle assessment (LCA). Appl. Mech. Mater. **357–360**, 508–514 (2013). https://doi.org/10.4028/www.scientific.net/AMM.357-360.508

Advanced BIM Platform Based on the Spoken Dialogue for End-User

Sangyun Shin[1], Chankyu Lee[2], and Raja R. A. Issa[2(✉)]

[1] M.E. Rinker, Sr. School of Construction Management, University of Florida, Box 115703, Gainesville, FL 32611, USA
sang87yun@ufl.edu

[2] M.E. Rinker, Sr. School of Construction Management, University of Florida, Box 115703, Gainesville, FL 32603, USA
raymond-issa@ufl.edu

Abstract. Since the advent of automatic speech recognition-based virtual assistants (e.g., Google Assistant and iOS Siri), it has become more common to use voice-based information retrieval systems in search engines. In addition, recently, such search engines have served not only to accept keyword-based commands from the human voice but also have evolved to recognize and understand more natural human language. As a result, various industries including the Architecture, Engineering, Construction and Operations (AECO) have been exploring the application of Automatic Speech Recognition (ASR) systems to their processes in order to improve work efficiency/ productivity. While other areas have somewhat succeeded in interacting with machines (computers) using natural language, the AECO field is still utilizing keyword-based speech recognition. and lags behind the research of other industries trying to apply natural language-based ASR systems. Thus, this paper aims to apply natural language-based voice recognition system to building information modeling (BIM) and be used to manipulate the building model. In order for the BIM software to understand natural language queries and show the appropriate results in response to it, connecting natural language with BIM data is the key effort. This is because, in general, the words and sentences people use on a daily basis (i.e., natural language) are not highly correlated with the BIM data. With natural language processing (NLP) and structured querying language (SQL), therefore, this paper tries to propose not only a way to connect natural language and BIM data but also a way to manipulate BIM itself.

Keywords: BIM · Natural Language and Structured Query Language

1 Introduction

Building Information Modeling (BIM) is not a new technology or concept in the Architecture, Engineering, Construction, and Operation (AECO) industry. According to the UK National BIM report (2018), the rate of BIM use has been growing compared to a decade ago (see Fig. 1). This has resulted in people trying to look for the methods to make better use of BIM by integrating other technologies into it. Also, in the last decade, computational power has improved exponentially which ended up putting

E. Toledo Santos and S. Scheer (Eds.): ICCCBE 2020, LNCE 98, pp. 123–132, 2021.
https://doi.org/10.1007/978-3-030-51295-8_11

Artificial Intelligence in the hands of the masses. Such progresses in the availability of Artificial Intelligence (AI) have also had a significant impact on the AECO industry. The deployment of BIM and the integration of computational power and technology throughout the life cycle of a buildings has made AI and its related technologies (e.g., speech recognition) more feasible to utilize within the AECO industry.

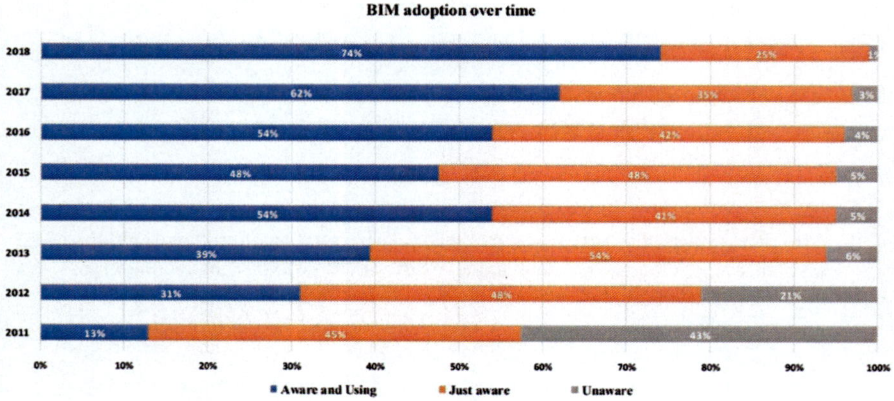

Fig. 1. Yearly use of BIM from 2011 to 2018 (National BIM Report 2018).

In recent years, new types of output devices such as VR and AR have begun to appear, and many studies have been attempted to apply them to BIM instead of using the existing output device (e.g., Monitors). According to Sun et al. (2018), VR and AR are usually used not in static spaces such as laboratories but in dynamic spaces such as construction sites. These environments make it difficult to use existing input devices, keyboards, and mouse. This study was started to find ways to better utilize BIM in these circumstances.

The aim of this study is to determine whether it is possible for users to manipulate building information model using their natural language without using existing input devices (e.g., keyboards and mouse). This research question is derived from the changing environments such as VR, AR, and holograms and advanced artificial intelligence technologies. According to Oh et al. (2017), the existing input devices such as keyboard and mouse are not suitable for using new output devices (e.g., VR, AR, and Holograms). In addition, with the rapid development of speech recognition systems as well as their accuracy rates, these systems have begun to require new types of input devices that can be controlled in the changed output environments.

2 Literature Reviews

Many fields have focused on applying speech recognition systems with a variety types (e.g., chatbot, and virtual assistants) to improve work efficiency (e.g., healthcare, information retrieval, language translate, and even manufacture industry). However, in the AEC field, there was not much effort to apply speech recognition. Some research has tried to apply it into the BIM environment as shown in Table 1.

Table 1. Previous studies applying speech recognition systems.

Parameter	Voice 360 (2015)	Motawa et al. (2017)	Kim et al. (2018)
ASR H/W interface	Computer microphones	Computer microphones	Virtual assistants (Cortana)
Input type	Keyword-based commands	Natural language queries	Keyword-based commands
Purpose of use	Information retrieval from BIM in Revit	IFC-based information retrieval	Information retrieval and modification for BIM data in Revit
Data approach methods	Keyword matching-based approach	Keyword matching-based approach	Keyword matching-based approach
Existence of a knowledge base	Does not apply	Does not apply	Does not apply
Ease of use	Requires mastery of keywords and less intuitive	Requires another system to link with IFC	Requires of mastery of keywords

Table 1 shows the limitations of previous studies in using speech recognition systems since these were not based on natural language queries but keyword-oriented commands. In addition, the keyword-based commands meant that there was no need to consider a knowledge base with which to interpret commands. However, the current study is focused on using natural language as a manipulation method in the BIM environment, the knowledge base will play a central role in interpreting natural language. In order to understand a query's exact meaning and its context, a knowledge base is a crucial part of the speech recognition system. According to Mahesh et al. (1996), to extract and manipulate text meanings, a natural language processing (NLP) system must have a significant amount of knowledge about the world and the domain of discourse available to it. The knowledge-based approach to NLP concerns itself with methods for acquiring and representing such knowledge and for applying the knowledge to solve well-known problems in NLP, such as ambiguity resolution (Chaudhary et al. 2014). Therefore, a knowledge base should be built around the BIM environment in order to better utilize NLQ-based voice recognition. As this was not done in previous studies, the current study seeks to establish and apply this knowledge base in the BIM environment.

The proposed framework in this study will be developed in AutodeskTM Dynamo in order to connect the main modules. Dynamo is an open-source, visual programming platform developed by Autodesk, which can work as a visual scripting interface with the Autodesk^{TM} Revit API to extend its parametric capabilities for various life cycle information management purposes. Dynamo enables users to create, customize, retrieve, and document data from a Revit file by scripting in a visual workspace (Autodesk 2017). Applications of such functions include, but are not limited to scripting behavior, generating both geometric and non-geometric information (e.g., enrichment of the semantic information levels for model elements in BIM), and information retrieval and validating existing models (Preidel et al. 2017). The visual programming interface is a significant advantage of Dynamo, as it allows users with little computer programming knowledge to carry out various project life-cycle analysis on models (Danhaive and Mueller 2015; Zibion 2018). According to Shin et al. (2017), visual programming tools were increasingly being utilized when implementing simulations in a BIM environment. Within Dynamo's workspace, data flow through "wires" to support input and output ports to "nodes," establishing a logical flow of data in the visual program. Nodes process and execute the input data by operating to create the output and represent the sequence of executed actions. Various functions are provided in the Dynamo library, grouped in different categories, available for use in composing an intended information process. Users also can extend Dynamo for their specific needs by creating custom nodes using existing core nodes and publish those for future applications (Autodesk 2017; Preidel et al. 2017).

3 Proposed Framework for Semantic-Based Building Information Retrieval Feasible

The goal of this research is to develop an algorithm for a question and answer system in order to interact between end-users and BIM utilizing their voice based on natural language queries. This study is based on the research related to the automatic speech recognition systems that has been conducted over the past decades for accepting natural language that consists of several sentences as well as for understanding the exact intention. Because of that, accepting human voice, even if it consists of long sentences, is not challenging anymore. However, trying to understand the intention of the voice that has a specific meaning or purpose is totally different from simply accepting it. In order to do that, there have been numerous attempts to better understand the human voice having a specific intention in the field of search engines such as in Google. And by being able to apply syntax and semantic analysis simultaneously as well as enabling linking everyday words and jargon for non-professionals to better understand their queries, it is possible for search engines to understand human questions more accurately. In the AECO field, however, since studies related to the ASR systems are at their initial phase, research such as on the connection between terms has not yet made a lot of progress.

So, for the computer to understand human spoken language in terms of building construction, it needs to understand BIM structure related to making 3D models and how building information is processed and stored in BIM. Compared to previous research, the proposed algorithm is designed to be more responsive to user-specific needs. This is accomplished by working efficiently when BIM is utilized in VR, AR, or even Holograhic environments, and even when end-users wish to look for the building data in the process of and after construction. The main difference between this study and previous research in the AECO field is that a Natural Language Query (NLQ) is changed automatically into a structured query language, which is a database language, for the purpose of manipulating BIM software (e.g., Revit).

To demonstrate the possibility of communication between end-users and BIM using natural human language, this research will process a set of hypotheticals (or scenario-based) test cases to validate its results. Figure 2 shows the overall workflow of this research. It is divided into total three main parts: scenario-set part, methodology part, and validation part. In the scenario-set part, a scenario is established for the implementation of the proposed framework. The reason for setting up the scenario is to reduce errors that can be caused by processing large data. The proposed scenario is

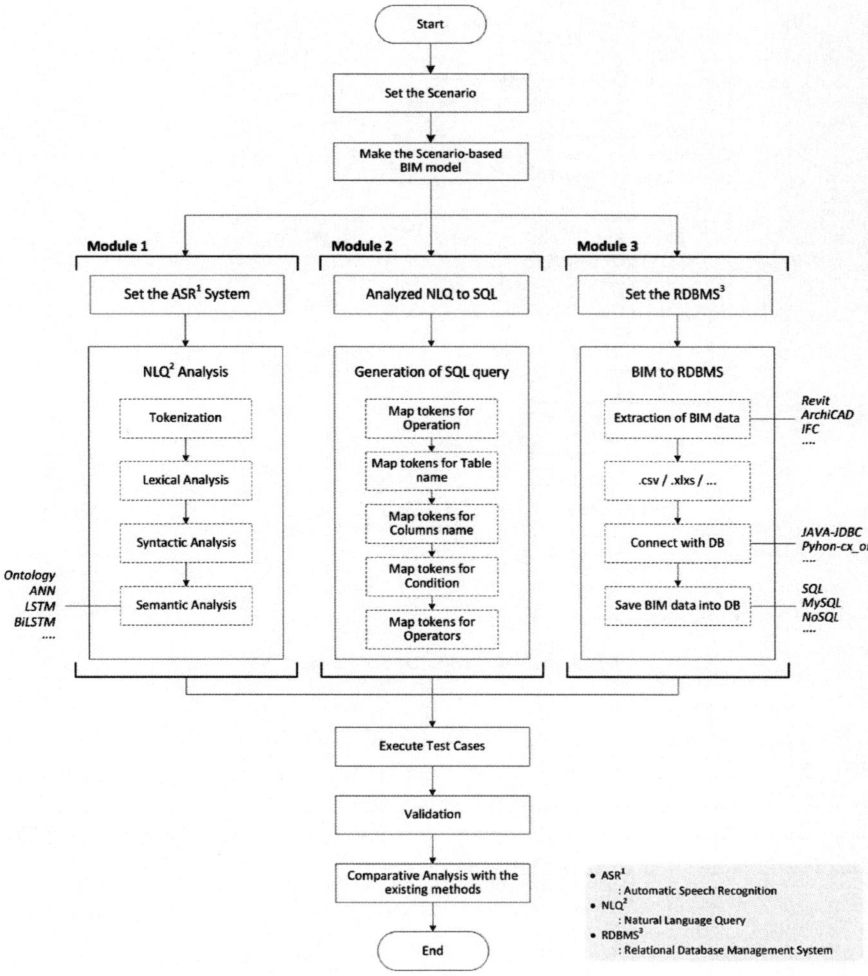

Fig. 2. Overall research process.

shown in Figs. 3 and 4. Figure 3 shows what process the scenario will take, and Fig. 4 shows the use case diagram for the proposed scenario. With using ASR devices like Alexa, end-users will ask several questions for the purpose of manipulating BIM software. Those questions are based on natural language.

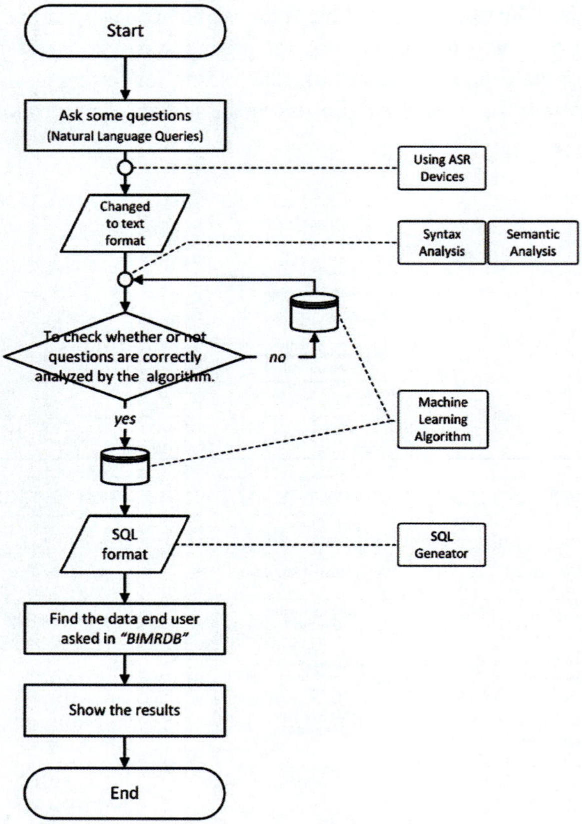

Fig. 3. The proposed scenario.

The proposed scenario is based on the Level of Development (LOD) by BIM Forum (2018). For the BIM model to be used in the field, it is usually created with LOD 400. However, the purpose of this study is to demonstrate the possibility of manipulating BIM with human voice, so there is no need to fully implement the BIM model. The implemented BIM model is shown in Fig. 5.

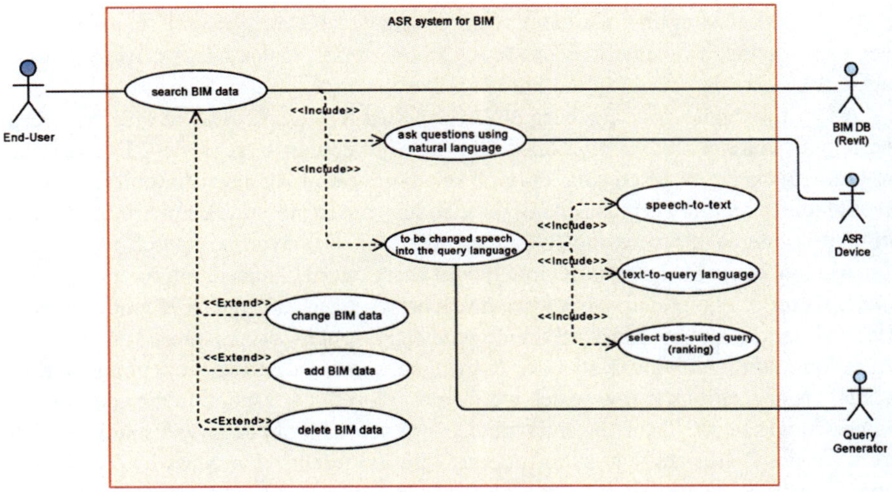

Fig. 4. The proposed scenario based on use case diagram.

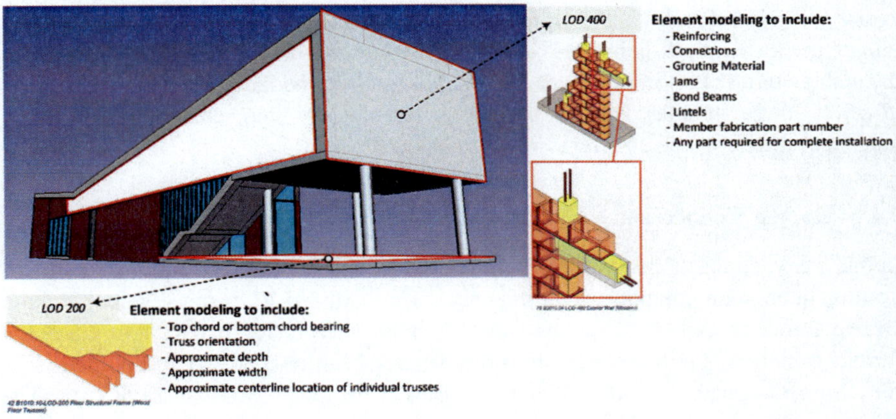

Fig. 5. The Level of Development (LOD) of the scenario-based BIM model (Adapted from BIM Forum 2018).

The most important part of the proposed methodology flow is how BIM accepts, understand and operates a natural human language when the end-user tries to manipulate BIM. Currently, BIM by itself does not understand natural human language such as the following sentences: "I want to change wall type that is located on the second floor into the different type, but I don't know exactly what kind of types there are, so could you please show me the list for that?". In order for BIM to understand the long sentences the user has said and to perform appropriate actions in response to it, not only it is important to properly accept the long sentence, but it is also important that it should be able to extract only the meaningful parts from the long sentences. Module 1 in Fig. 2

is dedicated to accepting natural human language and analyzing it to extract the meaningful parts. The smart devices (e.g., iPhone series, Galaxy series, Alexa, and so on) will be used in accepting the human voice.

A lot of technologies have been already applied to these devices in order to accept the human voice more exactly. According to Gonfalonieri's article in 2018, Agashe, program manager at Microsoft, current smart devices that have Automatic Speech Recognition function such as Siri, Bixby, Google Assistant, and Alexa are built based on natural language processing (NLP), a procedure of converting speech into words, sounds, and ideas. Such smart devices record users' words. Indeed, interpreting sounds take up a lot of computational power, the recording of users' speech is sent to a smart devices' servers to be analyzed more efficiently. Such servers break down users' "questions" into individual sounds. It then consults a database containing various words' pronunciations to find which words most closely correspond to the combination of individual sounds. Then, the smart device's servers send the analyzed sounds back to users' devices. This study aims to proceed with the module 1 process by using smart devices such as Alexa.

Module 2 serves to change the natural language queries in already transformed-text format via Module 1 to structured query language. Module 3 serves to transform the BIM data to CSV format data in order to be stored in the Relational Database Management System (RDBMS). This process will be conducted via using Dynamo which is one of the applications being provided by Autodesk for developing Revit plugins. In the architecture shown in Fig. 2, each module appears to be independent of each other, however, these modules interact with each other every step, changing the natural language queries to the structured query language.

3.1 Module Connection

In this study, connections between each module are the keys to verification. In order to do this, three main computer environments will be utilized: Dynamo, Python, and the Oracle Database (see Fig. 6). Dynamo not only extracts Revit data into CSV file format, which is able to be input into the database, but also takes the modified CSV files from the database and applies or updates them to the original Revit data. The Python environment plays an important role in connecting between Revit and the Oracle Database. Python enables natural language queries to be accepted into the computer using smart devices (e.g., iPhone's Siri or Amazon's Alexa) and analyzed using natural language processing. Then, analyzed queries can be converted to SQL (Ghosal et al. 2016), the database language, in Python. These processes will enable BIM to interact with the end-user. Lastly, Python has a role to enable the input of CSV files extracted by Dynamo into the Oracle database as well as to update the original CSV data in Revit. The Oracle Database works to save, modify, and delete data while communicating with natural language queries.

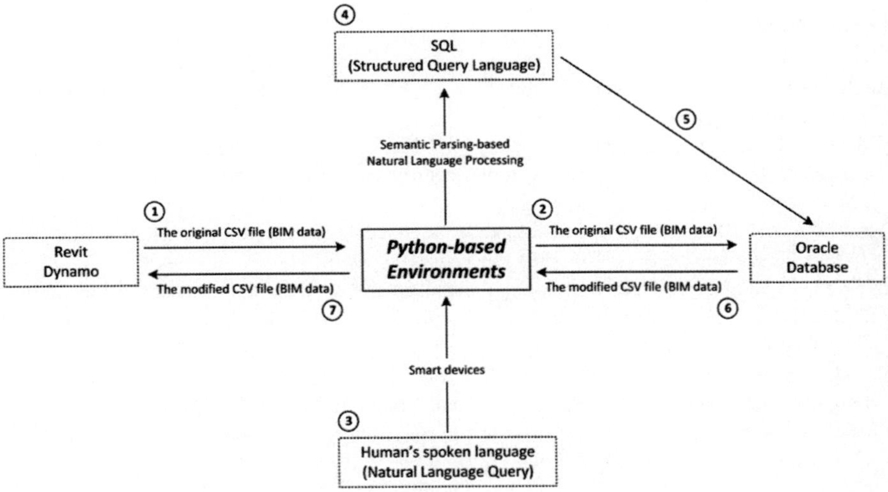

Fig. 6. Module connection.

4 Conclusion

The purpose of this study is to create an environment in which BIM is utilized not only through existing input devices but also the human voice through the use of a specific framework consisting of three main modules. A human's spoken word that is called a conversation can be used as the most convenient way to interact, even if the target is something like a computer. Issuing voice commands to BIM yields two main benefits. First, there would be no problem in using BIM even if the end-user is not a BIM expert because the proposed framework is based on accepting natural human language. Second, in case of having a device that can connect between user and BIM, users can modify and supplement the BIM model in real-time even if they are out in the field.

This study is mainly focused on how to use automatic speech recognition in BIM. The proposed framework consists of three modules. The first is a Voice-to-Text module that changes natural human language-based queries into text form-based queries. The second is a Query Analysis module that converts the modified text queries into SQL queries corresponding with it. The last is a BIM-to-RDBMS module that changes a BIM model (e.g., rvt format) into a Relational Database linking several tables. Its role is for the SQL query to find the corresponding data in the database.

This paper is meaningful in creating a framework that corresponds to the overall framework of research. It is also meaningful that we could confirm the feasibility of each module. However, it has a limitation that the three modules could not be implemented in one environment. Future research should look at implementing three modules in one environment in the next phase.

References

Autodesk Dynamo Studio: Dynamo primer for V1.3 (2017). https://primer.dynamobim.org/01_Introduction/1-2_what_is_dynamo.html

BIM Forum: LOD Specification 2018 Part I (2018). chrome-extension://oemmndcbldboiebfnladdacbdfmadadm/https://bimforum.org/wp-content/uploads/2018/09/BIMForum-LOD-2018_Spec-Part-1_and_Guide_2018-09.pdf

Chaudhary, A., Battan, A.: Natural language interface to databases-an implementation. Int. J. Adv. Res. Comput. Sci. **5**(6), 192–195 (2014)

Danhaive, R.A., Mueller, C.T.: Combining parametric modeling and interactive optimization for high-performance and creative structural design. In: International Association for Shell and Spatial Structures (IASS), no. 20, pp. 1–11 (2015)

Ghosal, D., Waghmare, T., Satam, V., Hajirnis, C.: SQL query formation using natural languageprocessing. Int. J. Adv. Res. Comput. Commun. Eng. **5**(3) (2016)

Ghosh, P.K., Dey, S., Sengupta, S.: Automatic SQL query formation from natural language query. Int. J. Comput. Appl. **975**, 8887 (2014)

Gonfalonieri, A.: How Amazon Alexa works? Your guide to Natural Language Processing (AI). Medium, 21 November 2018. https://towardsdatascience.com/how-amazon-alexa-works-your-guide-to-natural-language-processing-ai-7506004709d3

Kim, H., Jun, H.: Basic research on BIM application of artificial intelligence-based virtual assistant: based on algorithm based BIM platform utilizing speech recognition and artificial intelligence. Architectural Inst. Korea **38**(1), 96–99 (2018)

Motawa, I.: Spoken dialogue BIM systems-an application of big data in construction. Facilities (2017)

National NBS: National BIM report 2018. Royal Institute of British Architects (2018). chrome-extension://oemmndcbldboiebfnladdacbdfmadadm/https://www.thenbs.com/-/media/uk/files/pdf/nbs-national-bim-report-2018.pdf?la=en

Oh, J.Y., Lee, J., Lee, J.H., Park, J.H.: AnywhereTouch: finger tracking method on arbitrary surface using nailed-mounted IMU for mobile HMD. In International Conference on Human-Computer Interaction, pp. 185-191. Springer, Cham (2017)

Preidel, C., Daum, S., Borrmann, A.: Data retrieval from building information models based on visual programming. Visual. Eng. **5**(1), 18 (2017)

Shin, S., Jeong, S., Lee, J., Hong, S.W., Jung, S.: Pre-occupancy evaluation based on user behavior prediction in 3D virtual simulation. Autom. Constr. **74**, 55–65 (2017)

Sun, Q., Patney, A., Wei, L.Y., Shapira, O., Lu, J., Asente, P., Kaufman, A.: Towards virtual reality infinite walking: dynamic saccadic redirection. ACM Trans. Graph. (TOG) **37**(4), 1–13 (2018)

Zibion, D., Singh, D.P.V., Braun, M.S.A., Yalcinkaya, D.S.M.: Development of a BIM-enabled software tool for facility management using interactive floor plans, graph-based data management and granular information retrieval (2018)

Surface Scratch Detection of Monolithic Glass Panel Using Deep Learning Techniques

Zhufeng Pan[1,2], Jian Yang[1,2,3(✉)], Xing-er Wang[1,2], Junjin Liu[4], and Jianhui Li[4]

[1] State Key Laboratory of Ocean Engineering, Shanghai Jiao Tong University, Shanghai 200240, People's Republic of China
j.yang.1@sjtu.edu.cn

[2] Shanghai Key Laboratory for Digital Maintenance of Buildings and Infrastructure, School of Naval Architecture, Ocean and Civil Engineering, Shanghai Jiao Tong University, Shanghai 200240, People's Republic of China

[3] School of Civil Engineering, University of Birmingham, Birmingham B15 2TT, UK

[4] CABR Technology Co., Ltd., Beijing 100013, People's Republic of China

Abstract. Glass has been widely used in the construction sector with various kinds of applications in recent decades. However, the surface scratches generated from manufacturing process and service stage such as windborne debris impacts may lead to a strength degradation of glass material. The microscopic cracks propagation from such scratches may hence trigger glass facture unexpectedly and yield serious safety problems. In order to detect the glass damage due to such scratches, traditional manual inspection techniques have many limitations. The latest development of deep learning technology has rendered the possibility to automate such damage detection process. However, most detection methods use bounding box to roughly locate the damage in grid-cell level. To precisely describe the location of scratches, a pixel-level instance segmentation Mask R-CNN model is proposed. A total number of 1032 images with scratches are collected by a microscopic camera system to build the training and validation dataset, in which the scratches are annotated manually in pixel level. Data augmentation is adopted to improve the diversity of the dataset. During the training process, transfer learning strategy is applied to obtain the feature parameters for reducing the computation cost. Test is then performed in new architectural glass panels to evaluate the performance of the model. Test results demonstrate that the proposed trained network is satisfactory, achieving a mean average precision of 96.5% and the detection missing rate of 1.9%.

Keywords: Glass panel · Surface scratch detection · Deep learning · Mask R-CNN · Instance segmentation

1 Introduction

Glass has become a popular building material in the construction sector with a wide range of applications, such as glass façade and load bearing structural glass. According to the statistical data from the China Architectural and Industrial Glass Association, the

E. Toledo Santos and S. Scheer (Eds.): ICCCBE 2020, LNCE 98, pp. 133–143, 2021.
https://doi.org/10.1007/978-3-030-51295-8_12

annual architectural glass production of China has accounted for more than 50% that of the world. Glass has superior mechanical strength that can be ten times higher than that of concrete. However, the brittle nature of glass material is always a problematic issue, which can lead to the sudden fracture or failure. The surface scratches generated from polishing or grinding in glass production process, abrasion wear [1], hit by debris such as sand particles in the storm during service life, is one critical factor influencing the mechanical strength of the glass material. They could be triggering factor for a sudden glass fracture. Therefore, early stage detection of glass surface scratches with high precision becomes important for maintaining the safety and the serviceability of the glass members [2]. Previous inspection of the glass surface scratches was usually undertaken manually. Evaluation results were often determined by the experienced inspectors, which has problems of time consuming, labor intensive and error-prone.

To overcome these drawbacks, several non-contact vision-based inspection methods in combination with image processing techniques (IPTs) have drawn attention to many researchers. The advantage of IPTs comes from the identifiability for the majority of defects in appropriate environment. Abdel et al. [3] compared four types of edge detection methods. Results show that fast Haar transform can achieve the highest accuracy in monitoring multiple defects like concrete cracks. A study from Yeum and Dyke [4] revealed the feasibility of detecting steel cracks via integrating IPTs with sliding window techniques. However, IPTs may not be robust when testing in new datasets and it is quite sensitive to shadows or illumination. To address this, different denoising algorithms [5] have been proposed to optimize the performance. Nevertheless, the variety of image edges still limits its effectiveness in the real-world situations.

Machine learning-based technique has developed rapidly and shows the strength of multiple feature recognition attributed to its capability of resisting external interference. Within the category of this technique, the convolutional neural network (CNN) proposed by LeCun [6] shows its superiority after winning the ImageNet Large Scale Visual Recognition Challenge (ILSVRC) [7]. It therefore has regain attentions in dealing with computer vision-based problems using deep learning techniques [8]. CNN can handle with object classification and detection with remarkable adaptability. Many researchers proposed different efficient object detectors, for example, two stage detectors like region-based CNN (R-CNN) proposed by Girshick generate independent region proposals by selective search to localize and classify objects. Faster R-CNN [9] further improves the accuracy and step up the training process by applying Region Proposal Network (RPN). In addition, several single stage detectors such as You Only Look Once (YOLO) [10] and single shot multibox detector (SSD) [11] with much less computational cost also have excellent performances in recognizing objects. Accordingly, it has found various applications in civil engineering especially in structural health monitoring, such as capturing concrete cracks [12], spalling [13] and steel delamination [14]. However, these methods based on bounding box can only present grid-cell level detection and cannot describe the objects precisely and directly, so it cannot assess the damage properly.

Semantic segmentation is the countermeasure through making a pixel-level prediction with the class of candidate objects or a region. Fully convolutional neural network (FCN) [15], a pixel to pixel and end to end encoder-decoder network architecture has been proposed by replacing fully connected layers with deconvolutional layers. As a result, it provides the possibility to output the same image size as the input by adding a mask simultaneously and can also reflect the quantification information of specific object. Yang et al. [16] successfully exploited the FCN architecture to identify and quantitatively determine concrete cracks with ideal accuracy. Several other elegant structures like U-net [17] which needs much less numbers of training crack dataset is also found to be a promising approach.

Mask R-CNN [18] is deemed to be the combination and the extension of FCN and Faster R-CNN network architecture for instance segmentation. This framework can simultaneously achieve the goal of object detection, localization as well as segmentation precise to unique instance with high accuracy and fast test speed. Mask R-CNN has been used for the flood region analysis assisted with the drones, scene text detection, human pose estimation etc. A report from Wei's efforts [19] using Mask R-CNN for concrete bughole detection has confirmed its possibility of application in defect detection of the structures. However, no use has been found to detect surface scratches in monolithic glass panels. In this study, a novel surface scratch detection approach is proposed by applying deep learning-based Mask R-CNN model.

2 Network Architecture

2.1 Framework

Mask R-CNN is a simple and efficient pipeline for implementation. The overall process consists of two stages: 1) The first stage is a convolutional backbone for feature extraction followed by a RPN to generate candidate regions for further processing; 2) The second stage includes a quantization-free layer called RoIAlign to align the extracted features exactly with the original input image. Finally, the network head is added with three branches to properly achieve different tasks. The structural pipeline for Mask R-CNN is shown in Fig. 1.

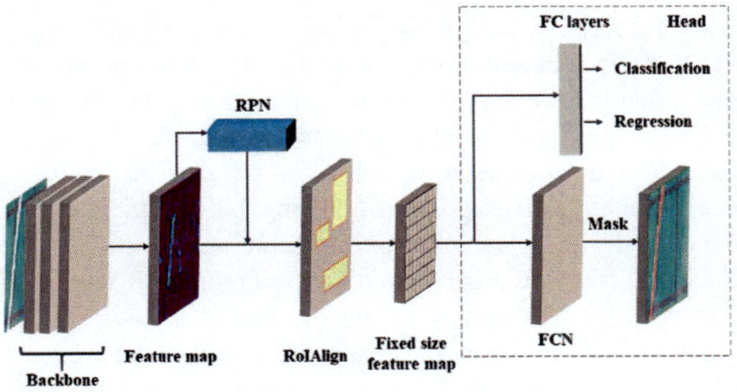

Fig. 1. Pipeline of mask R-CNN

In the first stage, the images with arbitrary sizes are served as the input into the network. Deep convolutional neural network acts as an encoder to extract different dimensions of the feature by convolution, pooling, activation and batch normalization operations. In this study, Resnet101-FPN is selected to generate the feature map for input images. Resnet [20] is a bottom-up deep learning network proposed in 2015 with up to 152-layer depth. It became well-known after winning the ILSVRC competition in 2015. With the introduction of residual blocks and shortcut connection concept, this deep network is able to avoid the degradation and gradient vanishing problem, which regularly happens in deep CNNs. Therefore, it outperforms the counterparts even in tasks with background interference. Feature pyramid Network (FPN) [21] is a top-down pathway with lateral connections hence it can detect objects across multiple scales, in particular, for small details and precisely locate the features, as depicted in Fig. 2. Compared to other methods like feature map hierarchy or featurized image pyramid, the output can have significant improvement for ensuring both high-level semantic and resolution level. This is also a generic framework so that it can be associated with different convolutional backbones.

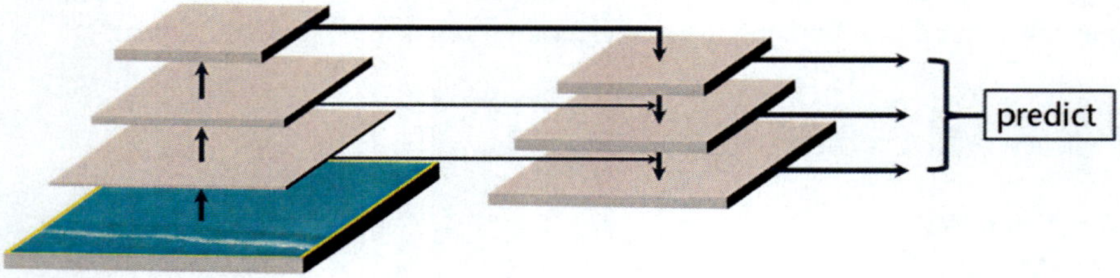

Fig. 2. Framework of feature pyramid network

RPN then takes the feature maps extracted by the shared convolutional backbone and produces a collection of rectangular region proposal candidates. RPN is a fully convolutional sub-network sliding through the feature map output by the last convolutional layer. In each sliding position, several types of rectangular boxes with pre-defined shapes and aspect ratios (anchors) are introduced to recognize and cover the objects with different sizes. Finally, these anchors will pass through two sibling layers. One will select positive anchors roughly based on the Intersection-over-Union (IoU). Those candidate anchors have the IoU overlap higher than 0.7 with a ground-truth box are considered to be positive [9], and negative for those IoU ratio lower than 0.3 otherwise. The additional one is responsible for refining the bounding box offset regression to get more accurate proposals. Following this operation, the RPN can detect the object bounds and present an "object-ness" score at each position, separating the desirable surface scratches foreground from the background. It will also calculate the position and the shape information of an anchor when it is thought to be foreground.

In the second stage, RoIAlign layer is exploited to collect the proposals selected by RPN and provides feature maps with fixed size to be projected to the features in the RoI, as depicted in Fig. 3. In faster R-CNN, as the spatial location of the proposals output by RPN is always expressed in real floating numbers, it does not align with the feature map generated by the convolutional backbone in integers. The previous RoI-Pool rounds it twice and results in position deviation between the post-processing proposal and the initial one. RoIAlign modifies it by utilizing bilinear interpolation, the value of the points at sample locations are computed accurately by nearby points in each RoI bins. It has been proved that RoIAlign can be an efficient way to solve the problem of pixel to pixel misalignment between the feature map and RoI in the original image. It is noted to be a crucial factor for improving the quality of pixel-level segmentation mask and output the spatial location of the objects more precisely.

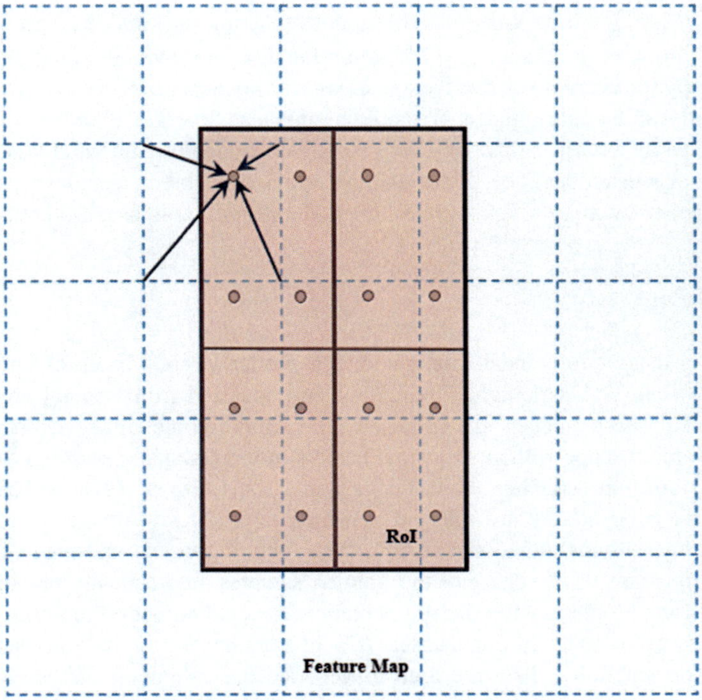

Fig. 3. Basic operation of RoIAlign

The output of RoIAlign is sent to the network head embracing three branches in the final step. The classification and the regression branch both encompass a couple of fully connected layers to determine the class and the location of bounding boxes. The innovation in proposed Mask R-CNN is the mask branch predicting accurate pixel-level mask in parallel with those tasks in Faster R-CNN. The mask branch is a fully convolutional neural network with both convolutional and deconvolutional layers. In this study, the number of classes in the last output layer is slightly modified from 80 in the original design to just 2 in our experiment, corresponding to the label of scratch or intact surface. Mask branch will generate a mask for the two classes individually in each sampled RoI, and only the mask associated with the predicted ground-truth class will be output.

2.2 Loss Function

During the training process, the main loss function of the Mask R-CNN is defined as the combination of the loss value in three independent tasks $\mathrm{L} = L_{cls} + L_{box} + L_{mask}$ on the sampled RoI. L_{cls}, L_{box} and L_{mask} are denoted as the losses in classification, regression and mask branch, respectively. L_{cls} is derived from $-\log P_u$ named as the log loss for the predicted true class u. L_{box} is calculated by the formula of $L_{box} = \mathrm{R}\left(p_i - p_i^*\right)$, where R denotes the smooth function illustrated in Fast R-CNN in detail [22]. In order to obtain L_{mask}, a sigmoid function is applied to obtain the average cross-entropy for every pixel, the average cross-entropy loss can therefore be calculated consequently for an entire image. By applying this loss function, it allows the network head to generate a mask output for every class independently without any interference from other classes unlike FCN, which couples both segmentation and classification task results together by a per-pixel softmax function and multinomial cross-entropy loss.

3 Training

To create a dataset for model training, images under various types of environment including strong and dark light illumination are acquired from several architectural glass panels. These images are collected by a microscopic image camera system including a microscope with an industrial camera to incrementally scan t glass panel to clearly capture thin scratches. Each image has a same size of 1920 × 1080 pixels. Images with good quality are selected and there are 720 raw images in total. Data augmentation methods including rotation, shear, flip, zoom and shift cnge are implemented to increase the diversity of the training samples automatically tavoid the possible overfitting problem. After the augmentation, the total number of training images is accumulated up to 1032. In this dataset, 70% of the images are randomly selected for training. The remaining 30% are used to generate the validation dataset. The VGG annotator app is utilized to annotate the ground truth data in pixel level for each image. The characteristics of surface scratches are labeled by polygon shape as the class label are added for each RoI. The annotation example is shown in Fig. 4. In order to reduce computation time and promote the accuracy of the result, transfer learning is employed. The Mask R-CNN framework is pre-trained on COCO, a well-known large-scale dataset developed by Microsoft with over 330k images to obtain the initial weights and biases. The experiment is conducted on a workstation embodied a Nvidia RTX 2080Ti graphic processing unit (GPU) with 11 GB graphics memory and is based on an open-source project implemented with Keras and Tensorflow [23] framework. Training hyper-parameters such as learning rate, weight decay and the momentum are set as 0.001, 0.0001 and 0.9, respectively.

Training process has altogether 200 joint epochs for both network head, the 4th and upper stage (C4+) of the backbone. It is finally determined by a trial-and-error method to fine tune the network and right after the validation step. Each mini batch has 2 images per GPU. The loss curve during the training and validation of the Mask R-CNN model is shown in Fig. 5. The loss decrease gradually and converge to around 0.3 in training and 0.6 in validation, respectively.

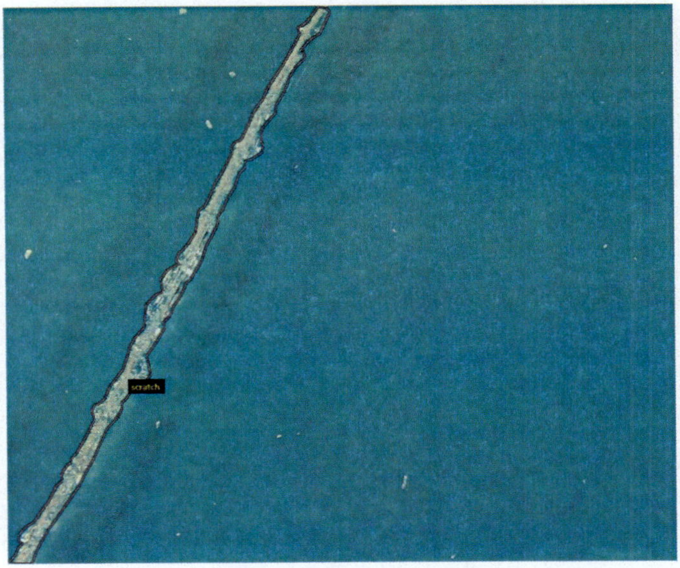

Fig. 4. Example of annotation in VGG annotator

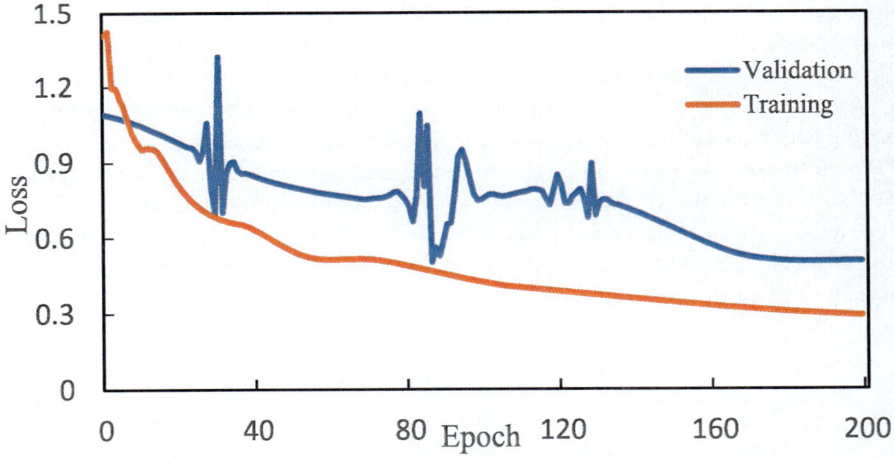

Fig. 5. Training and validation loss curve of Mask R-CNN

4 Experimental Test

4.1 Test Setup

To evaluate the performance of the proposed model in detecting scratches in new glass panel samples, several new architectural glass panels with mesoscale surface scratches are selected as test samples. Test images are also collected from the camera system mentioned in the Sect. 3. There are totally 72 test images obtained. It is noted that few grid lines from the background may act as crack-like feature interferences to influence the detection accuracy. Mean average precision (mAP) based on precision and recall value, as well as missing rate and false rate are used as the indicators to evaluate the performance of the proposed model. Precision is defined as the ratio of the true positive (TP) detection results in all the positive results indicating the prediction accuracy. Recall refers to the percentage of TP result among the sum of TP and the miss detected false negative (FN) results, as shown in Eq. (1) and (2). Whether the results are positive or not, are determined by the IoU threshold and the corresponding value is set as 0.5 and 0.75 hereon. mAP denotes the contribution of precision and recall in balance to best describe the evaluation performance of the proposed model.

$$\text{Precision} = \frac{TP}{TP + FP} \tag{1}$$

$$\text{Recall} = \frac{TP}{TP + FN} \tag{2}$$

4.2 Test Results

Test results for both glass panel with single surface scratch and multiple intersected scratches are shown in Fig. 6 and Fig. 7, respectively. The test result includes a bounding box, a mask overlay and a predicted score for a corresponding instance. The objects with higher score are likely to be surface scratches. It is clear that both types of scratches are well detected and extracted from the glass panel even under the complex texture background. In Table 1, the results of false rate and missing rate are 4.8% and 1.9%, respectively. It demonstrates that the surface scratches are almost completely detected, as the noise is also effectively avoided. Although few grid lines are falsely detected as scratches, they are tended to be neglected with much lower score. The test results reveal the robustness of the network architecture for detecting different characteristics of surface scratches.

Table 1. Test statistics of mask R-CNN

Class	Exact number	Correct detection	False detection	Missed detection	Missing rate	False rate
Scratch	105	98	5	2	1.9%	4.8%

Table 2. Results for various strategies in training process

Model	Training strategy			mAP (%)	
	Training epochs	Dataset size	Trainable layers	IoU50	IoU75
1	**200**	**1032**	**Head&C4+**	**96.5**	**65.3**
2	200	1032	Head	94.3	50.7
3	200	720	Head&C4+	86.7	54.4
4	300	1032	Head&C4+	97.0	65.8

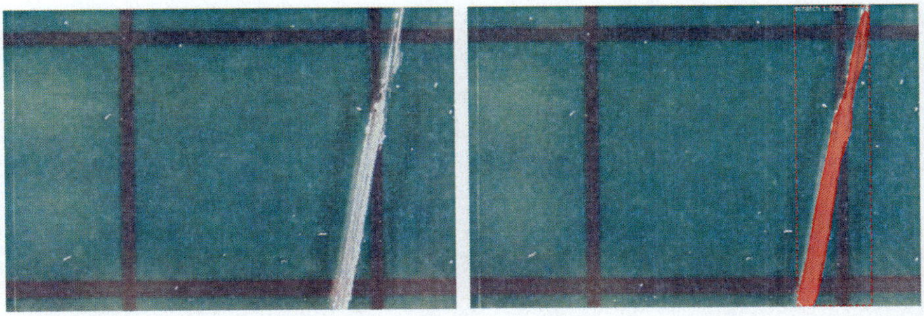

(a) Original image (b) Output with mask overlay

Fig. 6. Test result of single surface scratch

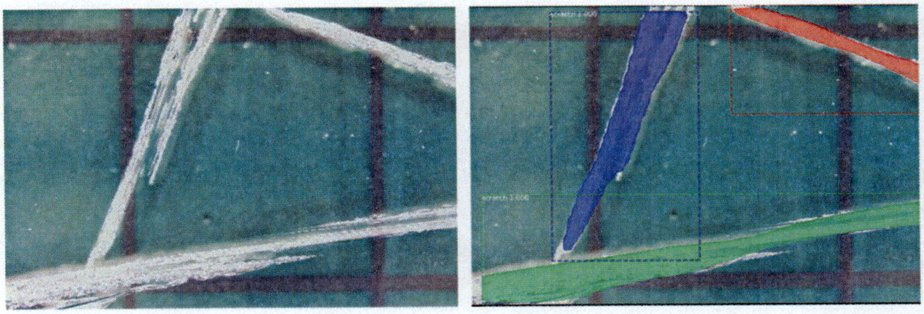

(a) Original image (b) Output with mask overlay

Fig. 7. Test result of multiple intersected surface scratches

An extensive study is also conducted to investigate the effect of various training strategies such as training epoch, data augmentation and trainable layers on the performance of the trained models. The results are given in Table 2. Take the results of IoU50 as an example, it manifests that 200 epochs are deemed to be the optimal epoch applied in model 1 when comparing with 300-epoch trained model 4. This is because the mAP only improves slightly from 96.5% to 97% whereas the computation time has an increase of 1.5 times more. It can also be concluded from the convergence of the loss curve during training. In addition, with the help of data augmentation method to increase the dataset size between model 1 and 3, the mAP gets promoted by 9.8%, indicating that the growth of dataset size has significant positive effect. Therefore, more images under complex scenes will be included in future works. In the proposed Mask R-CNN model, multi-level trainable layers are adopted and the according increase in mAP is 2.2% than model 2, which is expected to be cost-effective in application.

5 Conclusion

Surface scratches yield great potential threat to architectural glass as it will lead to significant strength reduction. However, traditional manual inspection techniques cannot meet the demand. In this study, a novel deep learning-based method is presented using Mask R-CNN to detect scratches of the monolithic glass panel automatically. The test result proves that such network architecture has good potential in capturing the scratch features without any pre-processing methods. It can achieve convincing performance, reaching a mAP value of 96.5%. The interference background of grid lines can influence the detection result, whereas the overall performance is robust with low false rate and missing rate. With the support of data augmentation and through multi-level training strategy, the mAP value can be enhanced by 9.8% and 2.2%, respectively.

Acknowledgements. This work was supported by the National Key Research and Development Program of China [Grant No. 2017YFC0806100], the National Natural Science Foundation of China [Grant No. 51908352] and the Science Research Plan of Shanghai Municipal Science and Technology Committee [Grant No. 18DZ1205603].

References

1. Petit, F., Ott, C., Cambier, F.: Multiple scratch tests and surface-related fatigue properties of monolithic ceramics and soda lime glass. J. Eur. Ceram. Soc. **29**(8), 1299–1307 (2009)
2. Schneider, J., Schula, S., Weinhold, W.: Characterisation of the scratch resistance of annealed and tempered architectural glass. Thin Solid Films **520**(12), 4190–4198 (2012)
3. Abdel-Qader, I., Abudayyeh, O., Kelly, M.E.: Analysis of edge-detection techniques for crack identification in bridges. J. Comput. Civil Eng. **17**(4), 255–263 (2003)
4. Yeum, C.M., Dyke, S.J.: Vision-based automated crack detection for bridge inspection. Comput. Aided Civil Infrastruct. Eng. **30**(10), 759–770 (2015)
5. Beck, A., Teboulle, M.: Fast gradient-based algorithms for constrained total variation image denoising and deblurring problems. IEEE Trans. Image Process. **18**(11), 2419–2434 (2009)
6. LeCun, Y., Bottou, L., Bengio, Y., Haffner, P.: Gradient-based learning applied to document recognition. In: Proceedings of the IEEE, vol. 86, pp. 2278–2324. IEEE (1998)
7. Krizhevsky, A., Sutskever, I., Hinton, G.E.: Imagenet classification with deep convolutional neural networks. In: Proceedings of the Neural Information Processing Systems Conference, Stateline, NV (2012)
8. LeCun, Y.A., Bengio, Y., Hinton, G.E.: Deep learning. Nature **521**, 436–444 (2015)
9. Ren, S., He, K., Girshick, R., Sun, J.: Faster R-CNN: towards real-time object detection with region proposal networks. IEEE Trans. Pattern Anal. Mach. Intell. **39**(6), 1137–1149 (2017)
10. Redmon, J., Farhadi, A.: Yolov3: an incremental improvement, arXiv:1804.02767 (2018)
11. Liu, W., Anguelov, D., Erhan, D., Szegedy, C., Reed, S., Fu, C.-Y., Berg, A.C.: SSD: single shot multibox detector. In: European Conference on Computer Vision, pp. 21–37. Springer (2016)
12. Cha, Y.J., Choi, W., Büyüköztürk, O.: Deep learning-based crack damage detection using convolutional neural networks. Comput.-Aided Civil Infrast. Eng. **32**(5), 361–378 (2017)
13. Beckman, G.H., Polyzois, D., Cha, Y.-J.: Deep learning-based automatic volumetric damage quantification using depth camera. Autom. Constr. **99**, 114–124 (2019)

14. Cha, Y.J., Choi, W., Suh, G., Mahmoudkhani, S., Büyüköztürk, O.: Autonomous structural visual inspection using region-based deep learning for detecting multiple damage types. Comput.-Aided Civil Infrastr. Eng. **33**(9), 731–747 (2018)
15. Long, J., Shelhamer, E., Darrell, T.: Fully convolutional networks for semantic segmentation. In: Proceedings of the IEEE Conference on Computer Vision and Pattern Recognition, pp. 3431–3440 (2015)
16. Yang, X., Li, H., Yu, Y., Luo, X., Huang, T., Yang, X.: Automatic pixel-level crack detection and measurement using fully convolutional network. Comput.-Aided Civil Infrastr. Eng. **33**(12), 1090–1109 (2018)
17. Liu, Z., Cao, Y., Wang, Y., Wang, W.: Computer vision-based concrete crack detection using U-net fully convolutional networks. Autom. Constr. **104**, 129–139 (2019)
18. He, K., Gkioxari, G., Dollar, P., Girshick. R.: Mask R-CNN. In: Proceedings of the IEEE International Conference on Computer Vision (ICCV), pp. 2980–2988 (2017)
19. Wei, F., Yao, G., Yang, Y., Sun, Y.: Instance-level recognition and quantification for concrete surface bughole based on deep learning. Autom. Constr. **107**, 102920 (2019)
20. He, K., Zhang, X., Ren, S., Sun, J.: Deep residual learning for image recognition. In: Proceedings of 2016 IEEE Conference on Computer Vision and Pattern Recognition (CVPR), Las Vegas, NV, 12 December 2016, pp. 770–778 (2016)
21. Lin, T.Y., Dollár, P., Girshick, R., He, K., Hariharan, B., Belongie, S.: Feature pyramid networks for object detection. In: Proceedings of 2017 IEEE Conference on Computer Vision and Pattern Recognition (CVPR), Honolulu, pp. 936–944 (2017)
22. Girshick, R.: Fast R-CNN. In: Proceedings of the IEEE International Conference on Computer Vision, Santiago, Chile, pp. 1440–48 (2015)
23. Abadi, M., Barham, P., Chen, J., Chen, Z., Davis, A., Dean, J., Devin, M., Ghemawat, S., Irving, G., Isard, M., et al.: Tensorflow: a system for largescale machine learning. OSDI **16**, 265–283 (2016)

BIM-enabled Design Tools, Information Management and Collaborative Environments

A Knowledge-Based Model for Constructability Assessment of Buildings Design Using BIM

Abdelaziz Fadoul(✉), Walid Tizani, and Carlos Arturo Osorio-Sandoval

Centre for Structural Engineering and Informatics, University of Nottingham, Nottingham, UK
abdelaziz.fadoul@nottingham.ac.uk

Abstract. Economic and time efficiency can be attained in the construction industry by applying the principles of constructability. Existing empirical studies demonstrate that incorporating these principles into initial stages of design maximise outcomes for all stakeholders including designers, contractors, and clients. Considering the complexity of current building design processes, there is a need to provide a decision support tool that can help designers in designing for constructability based on embedded information within the design model. Such a tool would be most beneficial at the conceptual design stage so that constructability is factored into the design solution starting from its inception. Therefore, this research investigates how contemporary process- and object-oriented models can be used to provide a mechanism that represents the subjectivity of design constructability to inform decision making. Consequently, it proposes a BIM-based model using embedded information within the design environment to conduct the assessment. The modelling framework is composed of three key parts: The *Constructability Model (CM)* which formulates user-based knowledge; the *BIM Design Model* which provides required data for the assessment; and the *Assessment Model (AM)* which reasons with the formulated knowledge and the BIM Design Model. The modelling framework is implemented in C#, using .NET Frameworks and Revit API. This paper demonstrates that using this framework, constructability related information can be captured and reasoned with to inform decisions at the early stages of the design process.

Keywords: Constructability assessment · Building information modelling · Building design

1 Introduction

The constructability concept aims to integrate engineering, construction and operation knowledge and experience to better achieve project objectives [1]. The term is defined by the Construction Industry Institute [2] as "the optimum use of construction knowledge and experience in planning, design, procurement, and field operations to achieve overall project objectives". Similarly, the Construction Industry Research and Information Association (CIRIA) defines the buildability as "the extent to which the

E. Toledo Santos and S. Scheer (Eds.): ICCCBE 2020, LNCE 98, pp. 147–159, 2021.
https://doi.org/10.1007/978-3-030-51295-8_13

design of a building facilitates ease of construction, subject to overall requirements for the completed building" [3].

The importance of deploying the constructability concept at the early design stage stems from the criticality of this phase in any architectural, engineering and construction (AEC) industry project. The building design is a fundamentally complex process, with the most influential design decisions being made in the early design stage [4]. This includes the consideration of design constructability, which is often ignored by designers and building clients until the commencement of the construction phase, when they are confronted by potential adverse and costly realities [5].

The significance of designing for constructability is globally recognised in the construction industry [6]. To date, several studies attempted to address the subject and accommodate its controversy aspects [7]. They adopted different approaches to benchmark design constructability and to enable the objective evaluation of abstract concepts. As a result, various techniques and methods have been developed to improve design constructability, including developing guidelines, checklists, expert systems, and empirical formulas [8–11]. However, barriers to implementing constructability are present in design practice. This is evidenced by the significant efforts, time, and human resources required to implement the concept within the design environment, which discourages many practitioners from considering constructability in their designs [12].

Researchers have investigated the employment of advanced ICT capabilities within the architecture, engineering, and construction (AEC) industry to address gaps. Building information modelling (BIM) is the most powerful technique available to effectively implement information modelling. BIM design tools offer great capabilities in managing vast amounts of information embedded in the building model, from initiation to demolition. This has enabled its adoption for assessing aspects such as cost, energy, functionality, aesthetics, and constructability [13]. In addition, the capability of implementing parametric design rules associated with building elements allows for dynamic changes during design development to explore various alternatives [14]. However, the exploitation of such technologies for implementing constructability has not been fully realised [15]. The study, therefore, investigates design-stage assessment of design constructability by examining contemporary process and object-oriented models.

This paper presents an original assessment framework to measure constructability of BIM-based design solutions. The paper discusses incentives to design for constructability, and the possible application of BIM technologies to facilitate such a process. The sections in this paper describe the proposed modelling framework and its components, and how they contribute to the overall assessment mechanism when modelling design constructability. An overview of the framework implementation in a prototype is also included in the form of a plug-in to BIM software Revit, to operate on its created design models.

2 Literature Review

2.1 Benefits of Designing for Constructability

The introduction of constructability principles in the construction industry has well-recognised benefits for owners, contractors and designers [2, 6]. While some of these benefits are tangible and manifest in the form of cost, time, quality and safety etc., others are more subjective in their nature and are observed in the sense of their physiological and psychological rewards for the project stakeholders, including client satisfaction [16].

With respect to the cost objective, identified benefits by previous studies included aspects such as savings in the total project cost [17–19]. This is estimated by 1–14% of capital cost [20]. Others expressed it as an increase in the cost-effectiveness [21], resulting in a lower cost of bidding [22], reduced site labour [23], and better usage of resources [24]. In terms of time, implementation of the concept improves the project productivity [21, 25] and reduces outage duration [24] for an early completion of the construction phase [16, 18, 19, 21, 24]. In respect of the project quality, application of constructability principles facilitates the delivery of higher quality of built products [9, 18, 21, 24]. With regards to safety, consideration of constructability results in a safer on-site environment [24, 26, 27].

However, those benefits may be extended further to cover the entire building process, including aspects such as: improving planning perception, materials acquisition, design solutions, construction approaches, site management, teamwork, job satisfaction, project performance, and stakeholder involvement and satisfaction [16].

2.2 Associated Challenges with Constructability Assessment

Constructability encounters many challenges in practical implementation [12]. One of the main obstacles is knowledge acquisition and representation to benchmark the extent of constructability principles application in design solutions [8]. This is mainly due to the nature of construction knowledge and its subjectivity from one constructor to another, given contrasting construction capabilities and needs. One way around this obstacle is to employ survey and interview approaches to extract human elements of construction knowledge and experience [7, 11, 15, 23, 28–34]. However, studies that adopt such approaches are only valid on the particular place of the study where the knowledge was originally collected [35]. Furthermore, the formulated knowledge is not captured from a user perspective and hence will not reflect their construction capabilities and preferences. Consequently, this would undermine the accuracy of obtained assessment results that are meant to advise on a constructor ability to build a design [36].

Another big challenge to design for constructability is the lack of a design tool to effectively apply the captured knowledge and map it onto design features. Current methods demand laborious efforts and resources to execute assessment calculations and interpret their outcomes. The dynamic design process and the need for ongoing modifications in designed products necessitate automation of the constructability assessment routine whenever design changes are introduced [37].

2.3 BIM for Constructability Assessment

BIM presents a substantial opportunity to improve design constructability using its rich information repository. It enhances the integration between the design and construction processes, enabling improved quality with savings in project cost and time [38]. Object-oriented models are capable of quantifying constructability, whereby designers can observe the effects of design decisions [15].

In addition, BIM-enabled tools can electronically model and manage the vast amount of a building's information throughout its entire lifecycle. Such information can be used to estimate, schedule, detail, automate fabrication drawing, and plan construction activities. Furthermore, incorporating the time dimension in the model allows different design and execution plans to be explored and tested for better design constructability [33].

Realising the benefits of BIM-enabled constructability assessment, many studies explored the subject for the full adoption in the construction industry [15, 33, 39–45]. However, the ASCE Constructability and Construction Research Council stated in one of its special journals that "*The potential of new technology-based tools such as 4D CAD or BIM have not been fully realized. This area could also include validation of new constructability software tools*" [46].

3 Current Limitations and Emerging Issues

Despite the recognition of constructability benefits and its potential to facilitate the construction process and meet set objectives, its implementation in a method or tool still stands as a challenge. In modern practice, evaluating design constructability paradigms is a complex process and demands more efforts, resources, and time than can usually be devoted to it in real construction projects. The design team has limited technical support to oversee and assess the possible consequences of decisions taken at various stages with respect to constructability during the design stage. Many constructability aspects are left out of consideration in favour of at a later stage, when it is too late to improve the design constructability performance. There is a much greater need to enhance and support the process using specialised tools during the conceptual design stage, where critical decisions are made, rather than during the later detailing stages where changes are more complex and costly [47, 48].

No current tools provide the necessary construction knowledge to inform design decisions based on constructability considerations. To date, research has tended to focus on formulating guidelines and measures for designers to follow rather than developing a mechanism or tool to support their application during the design process. The lack of a decision-support tool that quantifies design constructability is identified as the major cause of poor constructability performance in most projects [8].

BIM technologies have emerged as potential platforms for facilitating the design process of buildings. However, the potential use of their capabilities to design for constructability has not been fully realised [15]. Therefore, this research attempts to address the question of how to map and model design constructability with the employment of knowledge-based systems and data modelling techniques to inform design decisions.

4 A Proposed BIM-Based Constructability Assessment Framework

The proposed assessment mechanism separates between the formulation of construction knowledge and carrying out the assessment processes. The modelling framework is composed of three key parts: the *Constructability Model (CM)* which formulates user-based knowledge; the *BIM Design Model* which provides required data for the assessment; and the *Assessment Model (AM)* which reasons the formulated knowledge onto design features, as illustrated in Fig. 1.

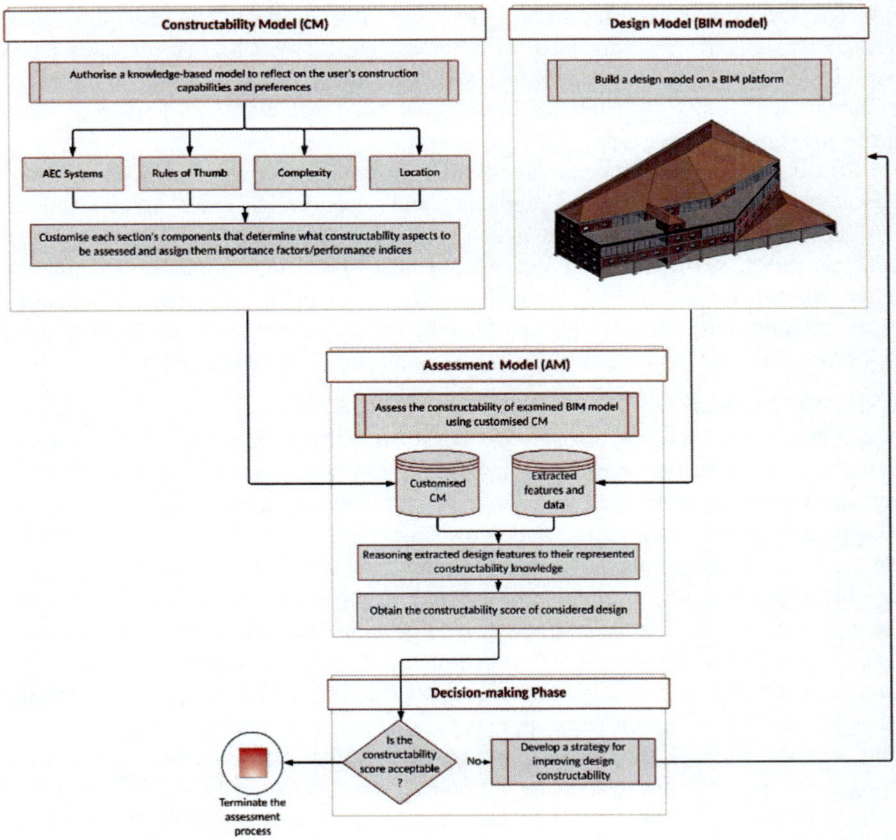

Fig. 1. Components of the proposed modelling framework

4.1 The BIM Design Model

The BIM design model refers to the digital building model that needs to be assessed for constructability. At this stage, designers build their design model using BIM authoring tool to a suitable level of details (LOD). Such information is an essential input for the assessment process, and consequently any delivered outcomes. Therefore, BIM-enabled tools should allow for the extraction of features extraction from their product models for the purpose of constructability analysis.

4.2 The Constructability Model (CM)

The CM is the knowledge-based model used for benchmarking the constructability of design solutions. Its information representations seek to model constructors' capabilities in terms of what they can build, and defining their preferences using various construction systems and methods. The represented knowledge is captured directly from users to impose their design objectives and meet particular project requirements. This is typically by enabling them to customise their own CMs. A specialised CM would typically be authored once for every type of project (e.g. multi-storey office buildings, multi-story car parks, and residential buildings, etc.), and is re-used many times for similar project types.

Figure 2 below illustrates the proposed CM. It consists of four main components: AEC Systems, Rules of Thumb, Complexity, and Location. The model components are designed to accommodate both quantitative and qualitative assessment of the design constructability, making ultimate use of embedded BIM data to model constructability in its multidimensional aspects. The contents of these components, extracted from users at the customisation stage, will guide the AM at the assessment stage in interpreting associated information with the BIM model and inform decision making.

CM: AEC Systems

The AEC systems module assesses the constructability of selected design elements based on available resources and imposed design constraints. With BIM-enabled, accurate estimation of design quantities, materials, and specifications, it is now possible to approximate demand for construction resources based on a scheduled program. This can be matched against accessible resources for design-builders to evaluate if they are capable to build it. Such evaluation of demand for resources with the available supply can be automated, having the first part accessible within the BIM model while building a resource database for the second part, which reflects users' construction resources. Consequently, the lack of a specific type of resources could be flagged at a specific stage of construction for the entire project period. Scores and indices may be assigned reflecting the intensity of observed shortage of resources. This is achieved by identifying the constructability factors and attributes to be considered during the assessment process based on the user's needs, and then calculating constructability indices of different design elements using analytical hierarchy process (AHP) method [49]. It ranks the elements based on their constructability from users' perspectives and hence enabling them to decide between alternative designs.

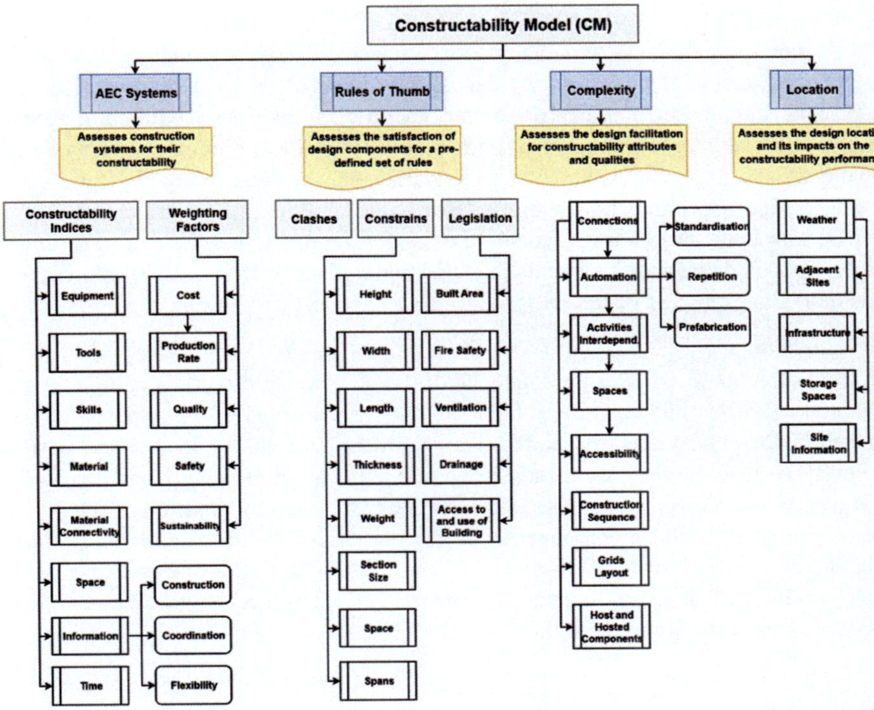

Fig. 2. Proposed CM hierarchy

CM: Rules of Thumb

This feature of the model allows users to assign a set of rules that need to be satisfied in their considered design. It takes advantage of a rule-based system to assess the design constructability based on available information in the design platform. These rules are applied to impose the design limitations and constraints in terms of spacing, layout or dimensions, which may later affect the construction process.

When customising this part of the assessment model, it is possible for users to enable rules that impose design constraints (restrictions of weight, height, length, width, etc.). Such restrictions could be applied based on the availability of elements, mode of transportation, site accessibility, available storage space, methods of constructions/installation and available working space. The rules to apply for a specific design can be activated when form the CM (not all rules need to be applied for all designs) to suit the given conditions. Although users could always opt to extend the package to include more rules, this might require some programming skills.

During the assessment process, the AM verifies the compliance of BIM model's components with enabled rules, assigns them weights (as specified by the user in the CM) and then determines a final score representing the constructability index based on these rules. By adding the design rules feature, the proposed framework is the first of its type to combine a numerical assessment system and a rule-based system, allowing for both quantitative and qualitative approaches when modelling design constructability.

CM: Complexity

The complexity module assesses the effects of selected design solutions in facilitating various constructability aspects during the construction process, such as the intricacy of the design, automation of the process and uncertainty associated with its different aspects. The CM introduces this feature to exploit BIM data in producing performance indicators that capture the existence of such qualities in a design solution. Features such as visualisation and interaction with the BIM model within a 3D environment facilitated by nowadays design tools, enable the users to better assess their designs interactively. Detailed construction simulations or simpler 4D animations also support in the decision-making process of the designer to improve constructability.

CM: Location

The location module assesses the influence of surrounding environment factors on the construction process and how that is factored in the considered design. Aspects such as weather in the region and site conditions should be catered for in selected design elements and the way they are installed. Additionally, site accessibility and proximity to delivery sources play a vital role in choosing construction methods (e.g., precast or *in situ* casting for concrete components). In the proposed CM, the assessment of these components is based on available information within the BIM model that can be employed for this part, with some user inputs. The high-level assessment of these aspects has been discussed in [37].

4.3 The Assessment Model (AM)

This process maps the customised CM on the actual BIM design model to benchmark its constructability. The design model is assessed based on selected AEC Systems and their suitability for its potential constructor's capabilities, satisfaction with enabled rules, complexity, and considerations of the project location. The AM extracts necessary information from the BIM model to process configured sections within the CM. The mechanism for calculating the Constructability score using the introduced model is as illustrated in Fig. 3.

5 Implementation of Constructability Assessment Prototype

The model was implemented in a prototype using object-oriented programming in a C# application. The prototype was developed using .NET Framework as a plug-in to BIM software Revit in order to operate on its created design models. The prototype was tested using typical design case studies, which has proved its usefulness in informing constructability decision-making. It also enabled the exploration and evaluation of what-if scenarios in design iterations, and construction methods.

Prototype development is based upon the elicited use-case to guide the programming direction, as demonstrated in Fig. 4. It shows the prototype functioning in four parts, namely: customising a new CM; modifying the customised CM for another use; interacting with the developed BIM model (initial analysis for its quantities), and assessing the design constructability (running the AM based on a selected CM).

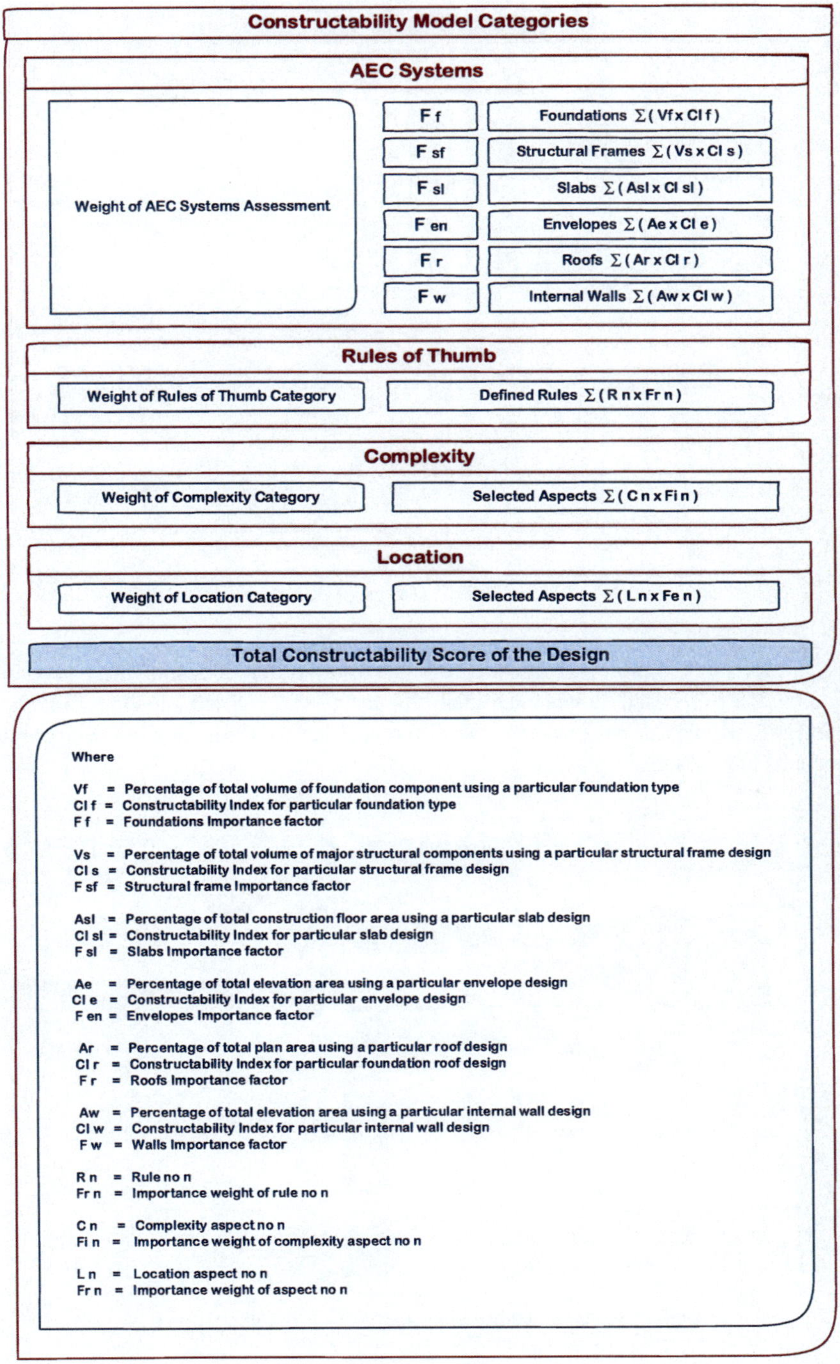

Fig. 3. Equation framework for calculating the Constructability score using a customised CM

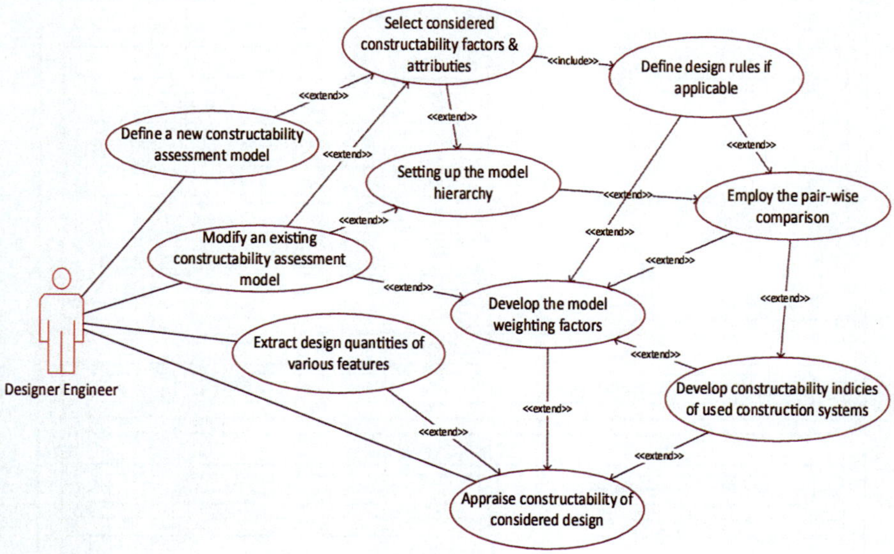

Fig. 4. Use case

The implemented prototype allows users to explore different design alternatives and decide on a design based on its constructability performance. Such a feature will enable the design optimisation and observe the difference due to using different construction systems as Fig. 5 shows.

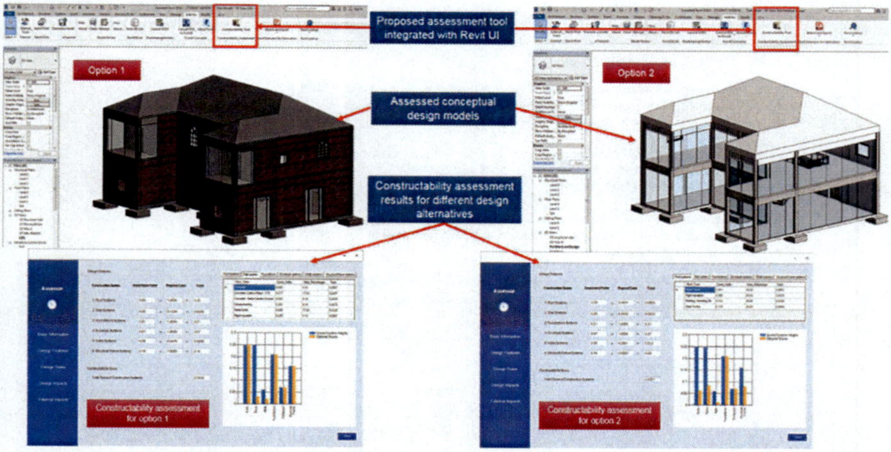

Fig. 5. Constructability assessment of design alternatives

6 Conclusion

Despite the recognised benefits of designing for constructability, it has been challenging to devise a tool that implements the concept. The paper identified the need for a decision-support tool that measures design constructability, as the major cause of poor constructability performance in most projects. Consequently, it investigated design-stage assessment of design constructability by examining contemporary process and object-oriented models.

The paper presented a framework to capture constructability related information while accommodating requirements of contrasting constructors' profiles. The framework is designed to exploit benefits of construction knowledge-based systems, object-based programming technology, and decision-making tools for modelling design constructability. It demonstrated the utilisation of BIM in constructability assessment of buildings design options through feature mapping and modelling technology.

This paper concludes that the devised modelling framework can be used to represent the subjectivity of constructability in its multidimensional aspects and inform decisions at the early stages. It enables the exploration of design alternatives for their constructability performances and hence an improvement to meet the design objectives.

References

1. Arditi, D., Elhassan, A., Toklu, Y.: Constructability analysis in the design firm. J. Constr. Eng. Manag. **128**(2) (2002)
2. CII, Constructability, a primer, publication 3-1, C.I.I. (CII), Editor. University of Texas at Austin, Austin, TX (1986)
3. Johnson, S.T., Jardine, F.M.: Construction Industry Research and Information Association (CIRIA) Programme on Remedial Treatment of Contaminated Land: Phase 3—Case Study Demonstrations. In: Contaminated Soil'95, pp. 1261–1262. Springer (1995)
4. Fadoul, A., Tizani, W.: Evaluation of current practice and associated challenges towards integrated design. Adv. Comput. Design **2**(2), 89–105 (2017)
5. CIRIA, Buildability: An Assessment, C.I.R.a.I. Association, Editor. London (1983)
6. McGeorge, D., Chen, S., Ostwald, M.: The development of a conceptual model of buildability which identifies user satisfaction as a major objective. In: CIB Conference, Rotterdam (1992)
7. Hei, W.W.: Developing and implementing an empirical system for scoring buildability of designs in the Hong Kong construction industry. In: Department of Building and Real Estate. The Hong Kong Polytechnic University (2007)
8. Fischer, M.A.: Using construction knowledge during preliminary design of reinforced concrete structures (1991)
9. Pheng Low, S.: Quantifying the relationships between buildability, structural quality and productivity in construction. Struct. Surv. **19**(2), 106–112 (2001)
10. Fox, S., Hietanen, J.: Interorganizational use of building information models: potential for automational, informational and transformational effects. Constr. Manag. Econ. **25**(3), 289–296 (2007)
11. Wong, F.W.H., et al.: A study of measures to improve constructability. Int. J. Q. Reliab. Manag. **24**(6), 586–601 (2007)

12. O'Connor, J.T., Miller, S.J.: Barriers to constructability implementation. J. Perform. Constr. Facil. **8**(2), 110–128 (1994)
13. Eastman, C., et al.: BIM handbook: A Guide to Building Information Modeling for Owners, Managers, Designers, Engineers and Contractors. Wiley, New York (2011)
14. Michael, T.: Current application of graphical programming in the design phase of a BIM project: development opportunities and future scenarios with 'Dynamo'. Northubria University (2016)
15. Hijazi, W., Alkass, S., Zayed, T.: Constructability Assessment Using BIM/4D CAD Simulation Model (2009)
16. Griffith, A., Sidwell, A.C.: Development of constructability concepts, principles and practices. Eng. Constr. Architectural Manag. **4**(4), 295–310 (1997)
17. Jergeas, G., Van der Put, J.: Benefits of constructability on construction projects. J. Constr. Eng. Manag. **127**(4), 281–290 (2001)
18. Elgohary, M., Fairclough, N., Ricciuti, R.: Constructability–from Qinshan to the ACR. Nuclear Plant J. **21**(5) (2003)
19. Trigunarsyah, B.: A review of current practice in constructability improvement: case studies on construction projects in Indonesia. Constr. Manag. Econ. **22**(6), 567–580 (2004)
20. Gray, C.: Buildability: The Construction Contribution. Chartered Institute of Building Ascot, UK (1983)
21. Pheng, L.S., Abeyegoonasekera, B.: Integrating buildability in ISO 9000 quality management systems: case study of a condominium project. Build. Environ. **36**(3), 299–312 (2001)
22. Gibson Jr., G., et al.: Constructability in public sector. J. Constr. Eng. Manag. **122**(3), 274–280 (1996)
23. Lam, P.: Buildability assessment: the Singapore approach. J. Build. Constr. Manag. **7**, 21–27 (2002)
24. Eldin, N.N.: Impact of employee, management, and process issues on constructability implementation. Constr. Manag. Econ. **17**(6), 711–720 (1999)
25. Poh, P.S., Chen, J.: The Singapore buildable design appraisal system: a preliminary review of the relationship between buildability, site productivity and cost. Constr. Manag. Econ. **16** (6), 681–692 (1998)
26. Francis, V., et al.: Constructability strategy for improved project performance. Architectural Sci. Rev. **42**(2), 133–138 (1999)
27. Trigunarsyah, B.: Project owners' role in improving constructability of construction projects: an example analysis for Indonesia. Constr. Manag. Econ. **22**(8), 861–876 (2004)
28. BCA. Annex A: Buildability and Constructability (2005) https://www.bca.gov.sg/newsroom/others/pr11032013_PA.pdf. 25 Jan 2017
29. Lam, P.T., Wong, F.W., Chan, A.P.: Contributions of designers to improving buildability and constructability. Des. Stud. **27**(4), 457–479 (2006)
30. Lam, P.T.I., Wong, F.W.H.: Implementing a Buildability Assessment Model for Buildability Improvement. Architectural Sci. Rev. **51**(2), 173–184 (2008)
31. Lam, P.T., Wong, F.W.: Improving building project performance: how buildability benchmarking can help. Constr. Manag. Econ. **27**(1), 41–52 (2009)
32. Lam, P.T., et al.: A scheme design buildability assessment model for building projects. Constr. Innov. **12**(2), 216–238 (2012)
33. Zhang, C., et al.: Quantitative assessment of building constructability using BIM and 4D simulation. Open J. Civil Eng. **06**(03), 21 (2016)
34. Zolfagharian, S., Irizarry, J.: Constructability assessment model for commercial building designs in the United States. J. Constr. Eng. Manag. **143**(8), 04017031 (2017)
35. Fadoul, A., Tizani, W., Koch, C.: Building information modelling for constructability assessment of buildings design. In: Conference Proceedings Civil Engineering 2018 (2018)

36. Fadoul, A., Tizani, W., Koch, C.: Constructability assessment model for buildings design. In: International Workshop on Intelligent Computing in Engineering (2017)
37. Fadoul, A., Tizani, W., Koch, C.: A BIM-based model for constructability assessment of conceptual design. Adv. Comput. Design **3**(4), 367–384 (2018)
38. Eastman, C., et al.: BIM Handbook: A Guide to Building Information Modeling for Owners, Managers, Designers, Engineers, and Contractors. Wiley, Hoboken (2008)
39. Autodesk White Paper, Improving Building Industry Results through Integrated Project Delivery and Building Information Modeling Report on integrated Practice (2008)
40. Eastman, C., et al.: Automatic rule-based checking of building designs. Autom. Constr. **18** (8), 1011–1033 (2009)
41. Yang, H.-H., et al.: Use of BIM for construtability analysis in construction. In: Proceedings of the Thirteenth East Asia-Pacific Conference on Structural Engineering and Construction (EASEC-13) (2013)
42. Tauriainen, M., et al.: The assessment of constructability: BIM cases. eWork and eBusiness in Architecture, Engineering and Construction, pp. 55–61 (2014)
43. Solihin, W., Eastman, C.: Classification of rules for automated BIM rule checking development. Autom. Constr. **53**, 69–82 (2015)
44. Bhatija, V.P., Thomas, N., Dawood, N.: A preliminary approach towards integrating knowledge management with building information modeling (K BIM) for the construction industry. Int. J. Innov. Manag. Technol. **8**(1), 64–70 (2017)
45. Fadoul, A., et al.: A constructability assessment framework for buildings design using Building Information Modelling (BIM). In: 17th International Conference on Computing in Civil and Building Engineering (ICCCBE). Tampere, Finland (2018)
46. Gambatese, J.A., Dunston, P.S., Pocock, J.B.: The way forward: recommendations for future constructability research and practice. In: Constructability Concepts and Practice, pp. 142–145. ASCE Publications (2007)
47. Stuart, K.: Modern Warfare 3 smashes records: $775 m in sales in five days (2011). http://www.theguardian.com/technology/2011/nov/18/modern-warfare-2-records-775m website. Accessed 2 December 2014
48. Aouad, G., Lee, A., Wu, S.: Constructing the Future: nD Modelling. Routledge, New York (2006)
49. Saaty, T.L.: Decision making with the analytic hierarchy process. Int. J. Serv. Sci. **1**(1), 83–98 (2008)

BIM to Develop Integrated, Incremental and Multiscale Methods to Assess Comfort and Quality of Public Spaces

Thibaut Delval[1], Brice Geffroy[2], Mehdi Rezoug[1(✉)], Alexandre Jolibois[1], Fabrice Oliveira[1], Samuel Carré[1], Mélanie Tual[1], and Julien Soula[1]

[1] Centre Scientifique et Technique du Bâtiment, 84 Avenue Jean Jaurès, 77420 Champs sur Marne, France
Mehdi.REZOUG@cstb.fr

[2] Paris La Défense, 110 Esplanade du Général-de-Gaulle, 92932 Paris La Défense, France

Abstract. Our approach applied BIM to the urban scale to feed multiscale simulations with input data and to collect results which can be used to create global indicators to evaluate current urban projects, in any field of interest. In our case the whole available data of the territory - including new and existing buildings and infrastructure - of the National Interest Operation of Paris La Défense (which is France's largest business district) has been collected and formatted into several databases. Our achieved interoperable BIM model has been adapted to perform complex multi-physical studies and simulations in several technical fields (including noise exposure, wind comfort, artificial and natural lighting, energy consumption, environmental impacts and global comfort) in order to understand and assess the quality of use of public spaces.

This paper shows a smart combination and integration of these databases and parameters which made possible to define new methodologies to evaluate urban comfort, as well as the quality of use of public spaces. It also presents an application of this research in the context of public commenting process, where results and methodologies were used to develop a tool to show end users the impact of land use decisions on the quality of use of public spaces.

Keywords: BIM · CIM · GIS · Knowledge management · Data science · Cooperative design · Smart cities · Simulation

1 Background: Setting up an Interoperable Simulation Ready Data Base

Paris La Défense, the first European business district, is the only vertical district in France with a high urban density. The territory hosts 70 high-rise buildings, 245.000 m^2 of shops and more than 3.6 million m^2 of offices. The urban planning of La Défense, based on a slab, has created a 31-ha pedestrian public space where 180,000 employees, 42,000 inhabitants and 45,000 students meet daily. The public establishment Paris La Défense is responsible for the development, management, animation and

E. Toledo Santos and S. Scheer (Eds.): ICCCBE 2020, LNCE 98, pp. 160–179, 2021.
https://doi.org/10.1007/978-3-030-51295-8_14

promotion of the business district. The establishment continues the urban transformation of the territory by promoting the diversity and by arranging the spaces in urban places. To meet its objectives and strengthen the attractiveness of the territory on a global scale, Paris La Défense also develops services and improves the customer experience by opening the territory to experimentation.

The use of an evaluation tool, developed by the CSTB, allows global and scientific vision of the urban comfort and its qualities of use. This tool meets the challenges of the planners by allowing the comparison between projects with objective criteria, in order to make informed decisions during studies. By appreciating the quality of the public spaces according to their uses over time, the establishment disposes of a solution to assess exploitation performances of the sites. Finally, this tool optimizes the layout and installations of ephemeral events to improve the level of satisfaction of users during animations.

1.1 Adaptation of the Model Geometry for Simulations

Multi-physical simulations models allow to study, understand and predict many complex applications such as urban comfort levels. Among the various categories of these models, some are based on 3D analysis to segment and identify the 3D environment elements (i.e. the building shapes and basic geometrical characteristics and dimensions), and their properties and governing physical laws [1].

Tremendous progress has been observed in computational sciences and engineering over the past decades, making it now possible to perform large scale computations based on geometric models spanning many spatial and temporal scales.

On the other hand, many important scientific and technological developments in the last few decades resulted in better Information Models, able to represent the 3D environment on different scales, and support semantic information attached to geometric models. For example, Geographic Information Systems (GIS) such as CityGML [2, 3] are a many-scale models with resolutions from continental down to building scale; and Building Information Models (BIM) [4] are starting to revolutionize the construction sector, both in buildings and infrastructure.

CityGML is the international standard of the Open Geospatial Consortium (OGC) for the representation and exchange of 3D city models. This format is not restricted to modelling buildings only, but also covers all relevant features one can find within urban areas. For each feature, a semantic definition, a set of attributes, relationships to other objects and a 3D spatial representation are provided.

In CityGML geometries, features are represented geometrically by the well-known Boundary Representation (B-rep): Volume features such as buildings are represented by solids, which are defined by their bounding surfaces. These surfaces must be mutually nonoverlapping, non-penetrating, and completely seal a volume without gaps. In general, the advantage of a solid representation of features is that the volume can be computed [5]. This is necessary, for example, in environmental applications for the computation of energy related performances.

When a solid representation for a building is not available, alternatively a representation by surfaces that do not completely seal the feature (multi surfaces) is possible.

Area or line features such as single walls, roads or energy system network are represented by surface or line geometries.

The CityGML features are organized into modules (building module, vegetation module, transportation module, etc.), which can be arbitrarily combined as needed for a specific simulation. To give just a few examples:

- The Building module aggregates several objects like RoofSurface, CeilingSurface, GroundSurface, FloorSurface, WallSurface. Each object can be enriched, for example with thermal characteristics of the surfaces (i.e. WallConductivity, WallThermalResistance, etc.) required for energy analysis.
- The Transportation module provides roads, railways and similar features. Physical and technical characteristics associated with these classes (ex: type of vehicles, traffic density, etc.) can be used to simulate road traffic noise in urban areas.
- Solitary vegetation objects like trees and plant covers are provided by the vegetation module. Physical properties of such objects (ex: foliar index, vegetation density, evapotranspiration, etc.) may be used for microclimate simulations.
- The City Furniture module is used to represent city and street furniture like streetlights. Such objects are relevant for street lighting simulation.

To prepare the necessary input data for each simulation model, we structured physical characteristics by parameter sets, in the appropriate modules, as a generic attribute. Figure 1 illustrates an example of how these attributes can be structured according to the type of application.

Thematic simulation	CityGML Thematic Module	CityGML Feature type	IFC Feature	Simulation attribut (Input/Output parameters)	Attribute type	Geometry Representaion	LOD
Urban Heat Island Simulation	Building	Building, BuildingPart	IfcBuilding	Building ID Usage Date of construction	integer string integer	MultiSurface, Solid	Lod2, Lod3, Lod4
		WallSurface	IfcWall, IfcCurtainWall	Wall ID Orientation of the wall Opening rate (in %)	integer Double integer	Surface	Lod2, Lod3, Lod4
		RoofSurface, OuterFloorSurface, OuterCeilingSurface	IfcRoof, IfcCovering,	Roof ID roof type Roof materials	integer string Liste of string	Surface	Lod1-4
	Relief	TINRelief, BreaklineRelief, MassPointRelief, RasterRelief	IfcTriangulatedFaceSet	Relief ID hydrous deficit (mm) Permeability index (%)	integer Double Double	TriangulateSurface, MultiPoint, MultiCurve, RectifiedGridCoverage	Lod0-4
	Transportation	Road, Track,TrafficArea,Square, AuxiliaryTrafficArea	IfcGeographicElement	Section ID Width of street road (space between buildings) (W in m) length of section (in m)	integer Double Double	MultiSurface	Lod0-4
	Vegetation	PlantCover, SolitaryVegetationObject	IfcGeographicElement	the plant or vegetated area ID Number of trees per built square meter (Tree/m2) Percentage of low vegetation in the vegetated area (%)	integer integer Double	MultiSurface	Lod2-4
	WaterBody	WaterSurface	IfcGeographicElement	WaterSurface ID Type of WaterSurface Surface (m2)	integer Liste of string Double	Surface	Lod2-4
	CityFurniture						
	LandUse	LandUse	IfcSpatialZone,IfcBuilding ElementProxy or IfcGeographicElement (No	LandUse ID FunctionType Surface (m2)	integer string Double	MultiSurface	Lod0 -4
	Generics (GenericCityObject)	(core:CityObject) GenericAttributeSet	IfcBuildingElementProxy (No matching semantic class is available)	Groupe ID Building Density (%) Road Coverage (%)	integer Double Double		Lod0 -4
	CityObjectGroup						
	Tunnel						
	Bridge						

Fig. 1. Extract of the used data dictionary

In this study, several simulations were performed using numerical tools that are directly interoperable with the CityGML format [6]. Some of the most important parameters required are directly derived from data such as surface geometry and orientation, material, etc. that exist in 3D CityGML models. In addition, global urban parameters like the roughness length in urban wind and airflow simulation, result "automatically" from 3D data about obstacle shape, dimensions and relative position, thus making the use of a CityGML extremely advantageous in comparison to previous works [6].

However, the geometric requirements needed for mesh-based 3D simulations using Computational Fluid Dynamics (CFD) models, such as simulation of wind load on buildings, pollution dispersion, flood propagation, etc., makes the use of the CityGML geometry much less efficient.

In order to carry out the CFD aeraulic simulations, aimed to quantify the pedestrian comfort regarding the wind, we developed and applied a geometric processing algorithm, using Rhino python API, to automatically simplify and adapt CityGML geometry.

Many projects such as CityDoctor[1] or CityGML Quality Interoperability Experiment[2] concentrate on creating automated procedures which validate and identify sources of errors in city models. However, mainly designed for engineering applications, commercial CFD tools require computer-aided design (CAD) data formats which are used in these disciplines. Hence, a conversion from the GIS data format (i.e. CityGML) to a CAD data format is necessary first [7].

For the application of the processing algorithm, the CityGML model LoD (Level of Detail) 1 and 2, are used as an input data. It is converted into a CAD data structure (STL Format). In this process, not only point objects are needed: additional geometrical and topological information, e.g. surface groups and solid representation, is necessary as well.

In most cases 3D city models still cannot be used in CFD simulations, even if they are transferred to a CAD format. The main reason for this is the presence of geometrical errors detected in the model and of many unnecessary geometric details e.g. short edges or small faces in a building's geometry, which may lead to an unsatisfactory CFD mesh, resulting in non-converging solutions or unacceptably high computing time.

In the first step, the algorithm iteratively treats building geometries (i.e. CAD models) one by one, by solving and eliminates the following errors and topological problems (some examples are shown in Fig. 2):

- self-intersections
- non-manifold geometries
- interior surfaces
- holes
- stray elements
- gaps between neighboring faces
- duplicate vertices and faces
- missing polygons
- multiply used edges
- wrong direction of face normal vectors

[1] https://www.citydoctor.eu/.

[2] https://www.coors-online.de/citygml-quality-interoperability-experiment-report-veroeffentlicht/.

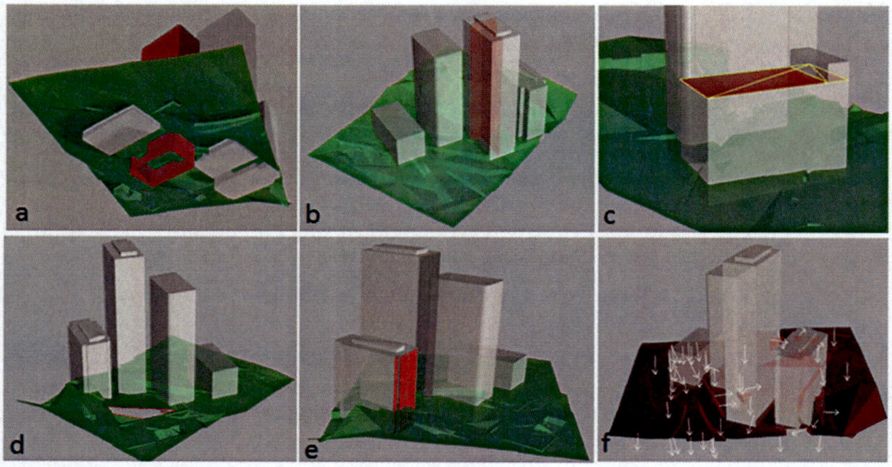

Fig. 2. Example of a building and terrain model with quality deficits a) missing ground surface, b) presence of interior surfaces, c) duplicate faces, d) missing polygons, e) gaps between neighboring faces, f) wrong direction of face normal.

Some errors, among the above list, were corrected automatically using our algorithm. For example, the elimination of separation surfaces inside buildings, missing polygons of the ground surfaces were automatically created. Other errors still need manual correction, e.g. repair holes and gaps in surfaces.

The second geometric specification that had to be addressed is the complexity of integrating buildings and terrain surface into the same model and identifying the valid intersection between them. This process makes additional requirements on the 3D city model, such as the fact that terrain and buildings must be watertight and without self-intersection.

Connected terrain and building may form a 2-manifold model when only a building exterior is present. This is a common representation of city models which focus on the external terrain and the shapes of buildings.

The basic operations of Boolean algebra are conjunction, disjunction and negation. For our purposes we only need conjunction. Rhino union-operator can be used to merge buildings and terrain that are connected (see Fig. 3). To do this, the building – terrain shell must be perfectly continuous (no holes) at the building/terrain junctions. There can be no creases, interruptions, folds or dangling edges, non-manifold conditions (redundant "T" surfaces) or faces in such junctures.

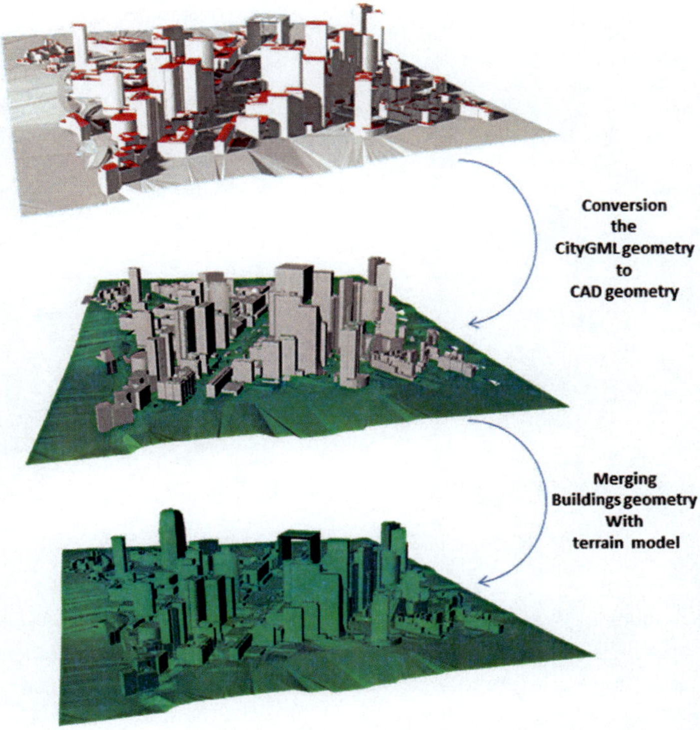

Fig. 3. Result of converting between data types and special union operation

1.2 Integration of New Projects in the Data Base, IFC to CityGML

CityGML had been chosen as the main format of the "La Defense" BIM database. But most of the building project (new construction or renovation) will be produced by CAD BIM Software (ArchiCAD, Revit, Allplan, …) and exported in IFC format. Most CAD tools cannot directly export to CityGML which is why a complete bi-directional conversion between IFC and CityGML was developped.

The GeoBIM project [8] shows how building envelopes in CityGML could be extracted form IFC Format. However, one of the conclusions of this work is that this conversion is very difficult, and that some specific geometric details cannot be handled automatically. For example, extracting glass surfaces was very difficult, whereas it is very important to have them in CityGML model, in order to address comfort simulation in public spaces.

The algorithm we developed is based on a quite simple mathematical assumption: if the IFC model is perfectly structured (with all building elements and spaces perfectly in contact), computing the union of all geometries should result with the exact external envelop of the building. This hypothesis has been checked on small synthetic models and turned out to work quite well (see Fig. 4).

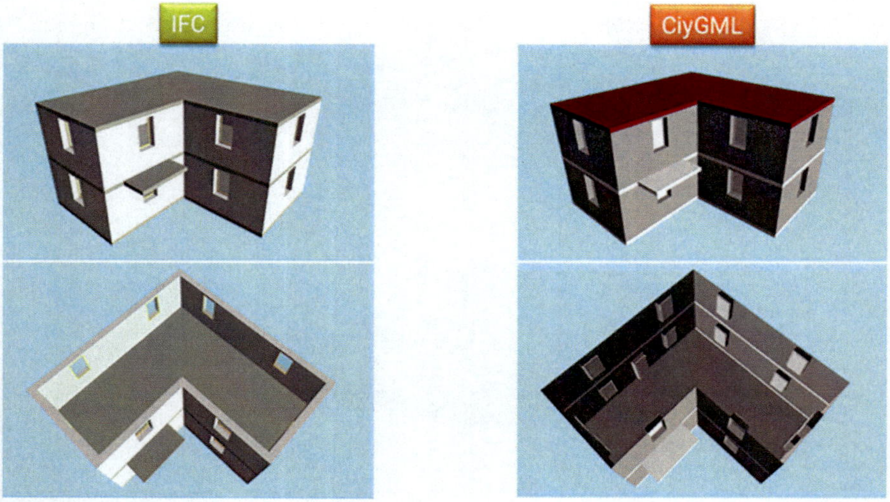

Fig. 4. Synthetic tests of IFC to CityGML conversion.

In order to handle larger and "real life" buildings, specific treatments were used to deal with openings (replacing window or door geometry by the geometry of the opening) and curtain walls (grouping all curtain wall part into a unique façade element). By tagging input geometrical elements with the corresponding IFC class, one can assign resulting geometric faces to corresponding CityGML elements (RoofSurface, WallSurface, Window). Finally, a LOD3 CityGML model containing external envelop geometry and all IFC information needed for exploitation or simulation is obtained (see Fig. 5).

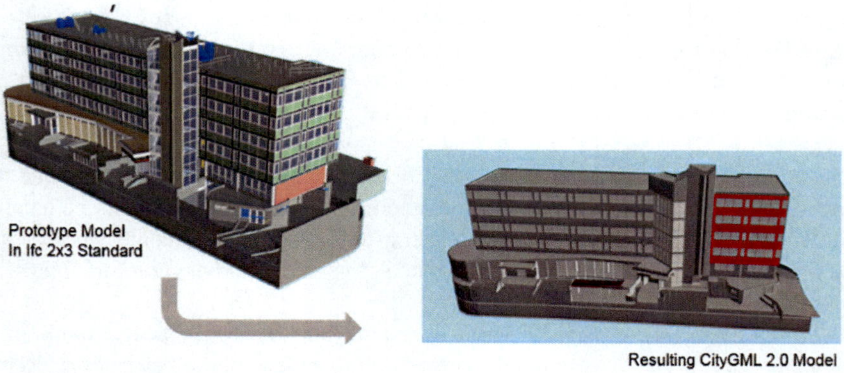

Fig. 5. Example of application of the conversion algorithm from IFC 2x3 to CityGML 2.0

1.3 Need for New Parameters: Time Dependence and Use

In order to properly address overall comfort in outdoor spaces, it is critical to consider not only the different physical parameters human senses are sensitive to (sound pressure, lighting level, outdoor wind speed, temperature and humidity…) but also their time variability. Indeed, these parameters first undergo strong variations from season to season, day to day, hour to hour and even to smaller time scales: for instance, the pass-by of a single car in a quiet environment will last maybe a few seconds but will dramatically change nearby sound levels and could possibly cause discomfort to someone sitting nearby.

Besides, the user of a public space experiences its surrounding environment at a specific time and day, and for a specific duration, from a few seconds to a few hours. Therefore, it is rather meaningless to assess the comfort one might experience based solely on yearly average values, which is nevertheless what is most commonly done in impact studies for instance.

Finally, the sensitivity one has to its environment – and therefore the sense of comfort – depends as well on the type of activity one is doing, and therefore on the use of the public space. As a result, threshold values corresponding to different comfort levels vary as well with the type of activity: reading a book, running, having lunch, etc.

Given those issues (the dependence with time of physical environment descriptors and comfort level, and the dependence with activity of comfort thresholds) and with the objective to more properly assess overall quality of urban spaces, it was found necessary to enrich the physical descriptors of the outdoor environment resulting from numerical simulation:

- include a sufficiently fine time-dependence to provide a meaningful information and at the same time limit data size,
- include a way to take into account the use of the public space.

This has been achieved by acquiring more accurate meteorological data (hourly time series over a year), re-run some simulations and apply specific post-treatment depending on the field.

1.4 Infrared and Lighting Exposure

Lighting and infrared radiative exposure severely affects visual and thermal comfort. To achieve accurate analysis of spatial and temporal variations of light exposure, meteorological data with a time step of less than 30 min has been used (see Fig. 6).

Depending on the nature of the available data, various models [9, 10] are used to reconstruct illumination values. Other models [11, 12] are used as well for spectral distributions (both in the visible and infrared domains) of direct and diffuse illumination as well as the type of sky (see Fig. 7).

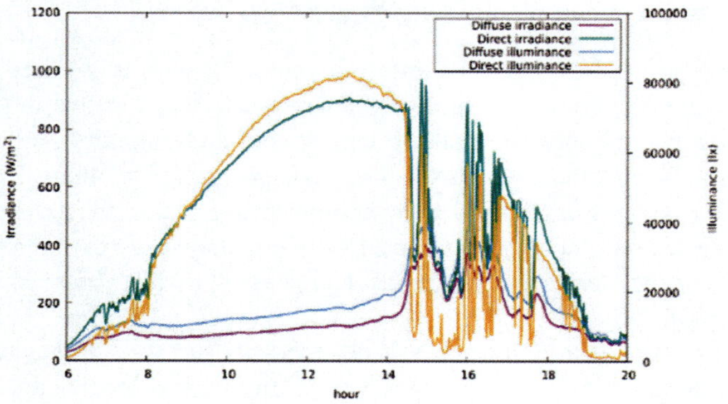

Fig. 6. Example of time history of lighting parameters over a day

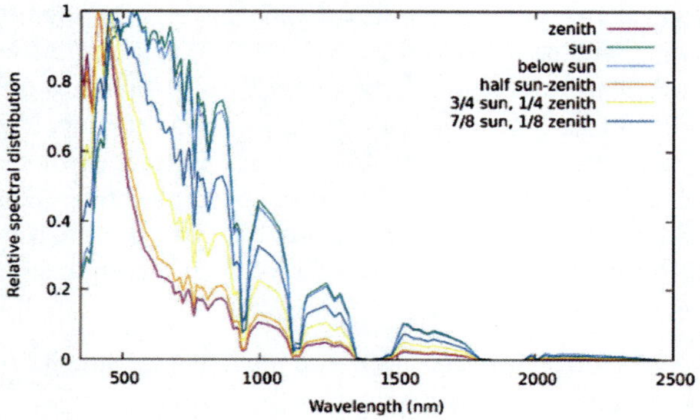

Fig. 7. Spectral decomposition of natural light components

Considering temporal variations during a representative year - with a sufficiently fine time step - corresponds to more than 8000 situations, which raises obvious problems of computation time. The use of a parametric sky model considerably reduces this complexity and allows for finer analyses. This classical approach in lighting simulation [13] breaks down the hemisphere into sectors and carries out the annual assessments in two steps: determination of extended sky factors, and then temporal analysis.

The determination of extended sky factors between the studied areas and the angular sectors of the sky is of a similar complexity to that of a lighting simulation. Each portion of the hemisphere can be assimilated to a radiative source. These factors consider direct and indirect contributions (via multiple inter-reflections) and are independent of climatic conditions. After this step, each virtual sensor associated with the areas of interest studied is therefore linked to the radiation conditions through its extended sky factor vector.

In this work, the Phanie[3] lighting simulation software uses a parametric sky model with irregular and adaptive decomposition depending on the solar gains of the site in question (see Fig. 8). This sky model also integrates a spectral dimension, for a more realistic consideration of colored inter-reflections.

Fig. 8. Example of an adaptive decomposition of the parametric sky model

This process [14] is very fast and allows for assessments to be carried out over much longer periods of time in order to obtain more realistic analyses. For each instant in a typical year, the radiative quantities associated with each virtual sensor are deduced from simple scalar products between the extended sky factors and the spectral irradiance associated with each sector. These spectral irradiances are deduced from the climatic conditions at each moment.

In this study, sky coefficients are used to reconstruct global and infrared radiative inputs.

1.5 Wind Exposure

In order to assess the influence of wind in thermal comfort, simulated acceleration factors referenced to the closest meteorological monitoring station have been used to calculate wind speed and direction at any location (see Fig. 9). Meteorological wind speed has also been tuned to be representative of the site roughness.

[3] http://www.cstb.fr/dae/fr/nos-produits-et-formations/outils-de-calcul/phanie.html.

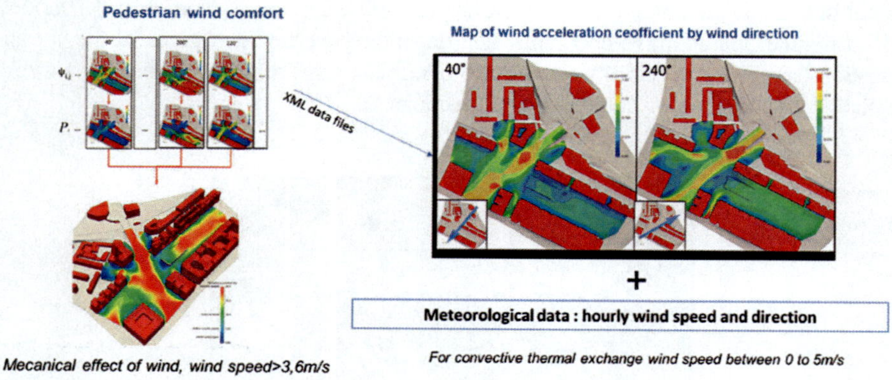

Fig. 9. Methodology to extrapolate wind speed any location and at each time step

1.6 Thermo-Aeraulic Comfort

The overall thermo-aeraulic comfort can be addressed through the Physical Equivalent Temperature (PET) [15] which essentially provides a way to do an energy balance between the human body and the outdoor environment, taking into account conduction (based on temperature difference), convection (based on relative wind speed), radiation (based on infrared exposure) and phase change latency (based on humidity). This calculation can be done at each time step based on meteorological data, simulated wind and infrared exposure results (see Fig. 10).

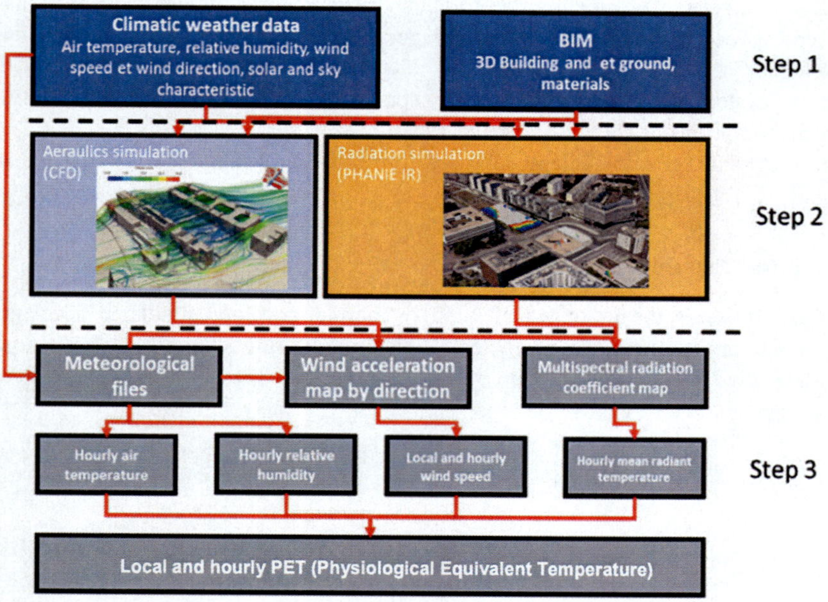

Fig. 10. Methodology to compute time varying PET based on simulation results

1.7 Noise Exposure

Time dependence of noise exposure essentially comes from time variability of urban noise sources which are here road and rail traffic, even though one could add that other unmodelled noise sources also induces time variability. On the other hand, usual noise exposure usually considers yearly average noise levels.

To enrich existing noise simulation data and add time variability, it has been decided to define a set of correction factors between yearly average levels and hourly levels, based on the evolution of the two types of traffic, and apply this set of factors to any calculation point in the area of interest. Existing studies on the evolution of road traffic in urban areas [16, 17] have shown that the largest variability happens between weekdays and weekends, and from hour to hour. This is also true regarding railway traffic, although the time evolution is specific to each line. Each value of traffic evolutions has then been converted into a correction term on noise level in dB (see Fig. 11).

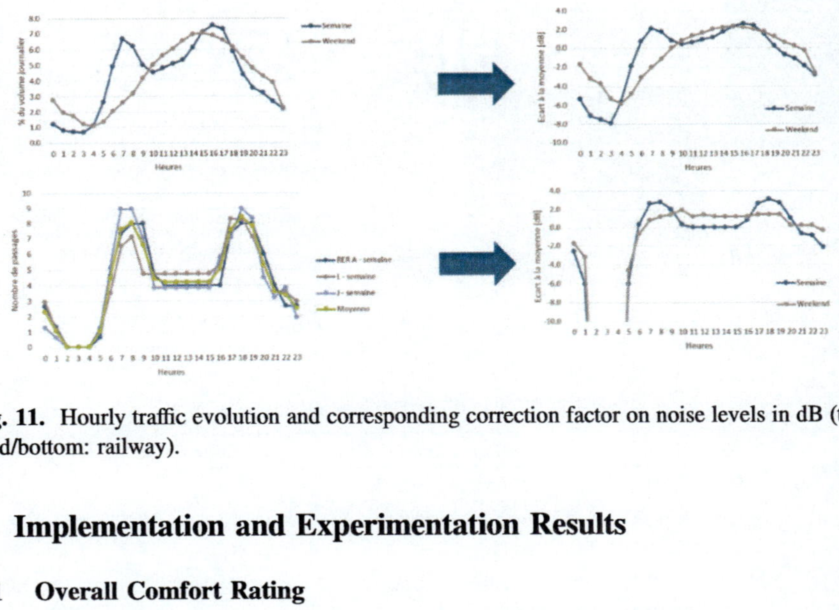

Fig. 11. Hourly traffic evolution and corresponding correction factor on noise levels in dB (top: road/bottom: railway).

2 Implementation and Experimentation Results

2.1 Overall Comfort Rating

The overall rating is carried out hourly based on physical thermo-aeraulic, radiative and acoustic inputs (see Fig. 12). With the climatic database, the time histories are extrapolated over a year and used for the assessment of overall outdoor comfort. The methodology of comfort rating is carried out to combine at each hour all physical input according an approach proposed by Cohen [18]. Each physical time history is coded in a scale 1 to 5 (see Table 1), from worst to best comfort level, according to pedestrian activities (such as standing, seating, walking, etc.). Each activity translates into a different metabolism value when calculating the PET.

After this step, the hourly summation function over all coded time history is carried out. Finally, an occurrence analysis according to two thresholds is carried out to classify environmental quality of each time step (see Table 2), which can be good, moderate or poor, corresponding respectively to a cumulative value over 10, between 10 and 5 and below 5. Finally, this occurrence analysis of environmental quality (comfort rating) time history provide urban planners important decisional information to manage urban space for example with a specific scenario focused on a user's activity, a time period, specific days of the week and for different seasons.

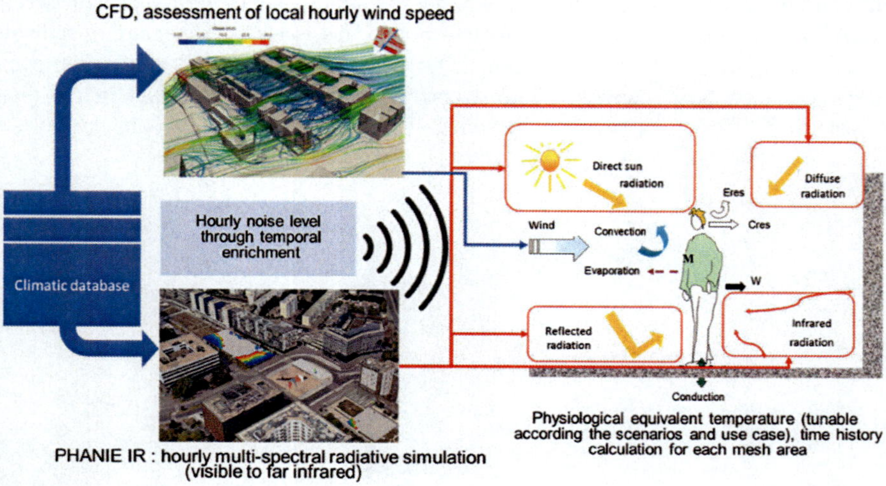

Fig. 12. Time history input for the overall comfort rating

Table 1. PET coding table for all activities

PET	Sensation	Code
<4 °C	Extremely cold	1
4 °C à 8 °C	Very cold	2
8 °C à 13 °C	Old	3
13 °C à 18 °C	Slightly froid	4
18 °C à 23 °C	Neutral	5
23 °C à 29 °C	Slightly warm	4
29 °C à 35 °C	Warm	3
35 °C à 41 °C	Very warm (health risk)	2
41 °C	Extremely warm (danger)	1

Table 2. Time history of multi-component comfort indexes, coding indexes, cumulative comfort and coding of environmental quality.

Year	Month	Day	Hour	Minute	Pet time history °C	Coding PET	Laeq time history dBA	Coding Laeq	Wind (m/s)	Coding wind comfort	Cummulative comfort	Environmental quality level
1996	6	8	14	22	29,5	2	67,6	1	4,2	0	3	1
1996	6	8	14	52	29,1	2	67,6	1	4,2	0	3	1
1996	6	8	15	22	28,9	4	67,7	1	4,2	0	5	1
1996	6	8	15	52	25,6	4	67,7	1	3,8	0	5	1
1996	6	8	16	22	24,7	4	67,6	1	3,8	0	5	1
1996	6	8	16	52	24,1	4	67,6	1	3,8	0	5	1
1996	6	8	17	22	23,0	5	67,4	1	3,8	0	6	2
1996	6	8	17	52	21,8	5	67,4	1	3,8	0	6	2
1996	6	8	18	22	21,0	5	67,1	1	3,8	0	6	2
1996	6	8	18	52	20,2	5	67,1	1	3,8	0	6	2
1996	6	8	19	22	19,2	5	66,6	1	3,8	0	6	2
1996	6	8	19	52	17,8	4	66,6	1	3,8	0	5	1
1996	6	8	20	22	17,1	4	66	1	3,8	0	5	1
1996	6	8	20	52	16,9	4	66	1	3,8	0	5	1
1996	6	9	5	52	12,9	2	60,6	2	0,6	5	9	2
1996	6	9	6	22	13,2	4	62,3	2	0,6	5	11	3
1996	6	9	6	52	13,1	4	62,3	2	0,6	5	11	3
1996	6	9	7	22	16,6	4	63,4	2	0,6	5	11	3
1996	6	9	7	52	18,5	5	63,4	2	0,6	5	12	3
1996	6	9	8	22	20,3	5	64,3	2	0,6	5	12	3
1996	6	9	8	52	18,7	5	64,3	2	0,6	5	12	3
1996	6	9	9	22	22,0	5	65,4	1	1,0	5	11	3
1996	6	9	9	52	25,4	4	65,4	1	1,0	5	10	2
1996	6	9	10	22	29,0	4	66,2	1	1,0	5	10	2
1996	6	9	10	52	30,5	2	66,2	1	1,0	5	8	2
1996	6	9	11	22	32,0	2	66,7	1	1,0	5	8	2
1996	6	9	11	52	32,5	2	66,7	1	1,0	5	8	2
1996	6	9	12	22	33,9	2	67	1	1,4	5	8	2
1996	6	9	12	52	34,0	2	67	1	1,4	5	8	2
1996	6	9	13	22	34,3	2	67,4	1	1,4	5	8	2
1996	6	9	13	52	34,2	2	67,4	1	1,4	5	8	2
1996	6	9	14	22	33,9	2	67,6	1	1,4	5	8	2

2.2 How to Build Indicators to Assess the Quality of Public Spaces from a User's Perspective?

Previous sections of this work focused on physical criteria in order to assess comfort in public spaces. However, when one is interested in urban quality, it is straightforward to realize that this is a notion that is simple to grasp but complex to analyze. If, with the blink of an eye, anyone can tell how good she/he feels in a public space, it is far more complicated to detail and objectivize what this quality is made of: parameters are numerous and there is a part of subjectivity. Thus, the first part of our approach has focused on the definition of indicators relevant to grasp the reality of urban quality in a broad sense. In order to do so, we built on existing methodologies on the assessment of public spaces and definition of urban quality. We then worked with Paris La Defense urban experts to adapt our methodology to the specificities of the territory: we asked them to confirm (or infirm) the relevance of indicators and define ponderation. We ended up with more than 200 indicators grouped in five categories of analysis as following:

- uses and activities,
- accessibility and integration in the urban environment,
- connection to the living,
- comfort and image,
- social interactions.

So that this methodology was user-centered, we elaborated a questionnaire submitted to users of the public space where we conducted the test. The results to this questionnaire on perceptions and representations of the public space were used to define how important the five categories would be. We also integrated perception as a key indicator (maximum ponderation) for each category of analysis.

2.3 Methodology to Assess Urban Quality from a User's Perspective

We wanted our methodology to assess the urban quality of public spaces to reflect a user-centered approach. Our idea was that the assessment of the urban quality of public spaces should take in account users' perception and representation at first. In order to do so, we built a methodology where the categories of analysis' importance would depend on the results of questionnaires conducted in the studied area on the perception and representation of the public space and where for each category of analysis, perception would be a key indicator.

In complement, we considered that indicators had to be weighed according to their relevance to the territory and asked professionals to do so.

So, our methodology is based on a broad definition of urban quality from a user's perspective. These two principles reflect in the large number of used indicators, the importance of users' perceptions as an indicator for every category of analyses and the ponderation's system based on the results of questionnaires. Our methodology also relies on professionals' expertise for the ponderation of indicators. Finally, the methodology allows to assess the consistency of analysis according to the availability of data.

The methodology has been applied to the studied area, Place Mandela in the area of interest of Paris La Defense. The results shown in Fig. 13 is a radar chart where 80 informed indicators are automatically evaluated according to the data collected in the BIM model. The user can select the categories of analysis as well as the parameters needed to evaluate the key indicator he wants to consider.

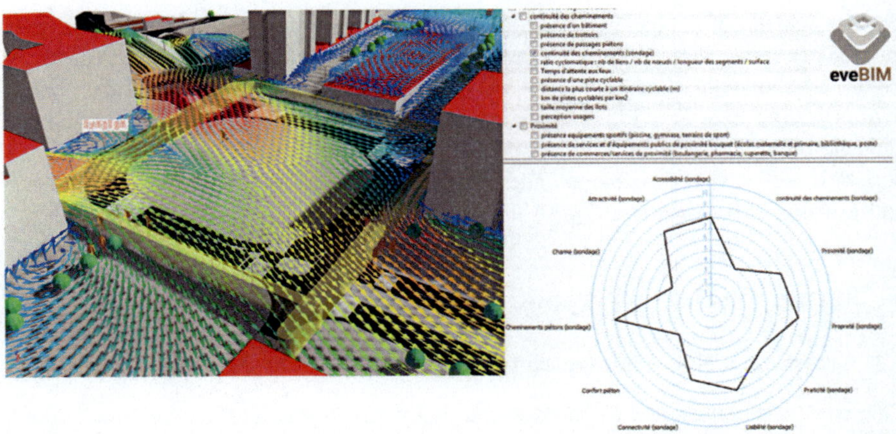

Fig. 13. Urban quality analysis on the area of interest (place Mandela)

2.4 Integrating Simulation Results into Urban Quality Analysis

The last step of the research was to build a common environment for both methodology urban comfort and quality of use. A specific implementation into the OpenBIM viewer eveBIM®[4] has been achieved in order to integrate the results of the different simulation needed. The program enables the user to draw specific zones on the simulation maps so as to define the most interesting area to study comfort results. The calculated comfort criteria are integrated into the corresponding use zones.

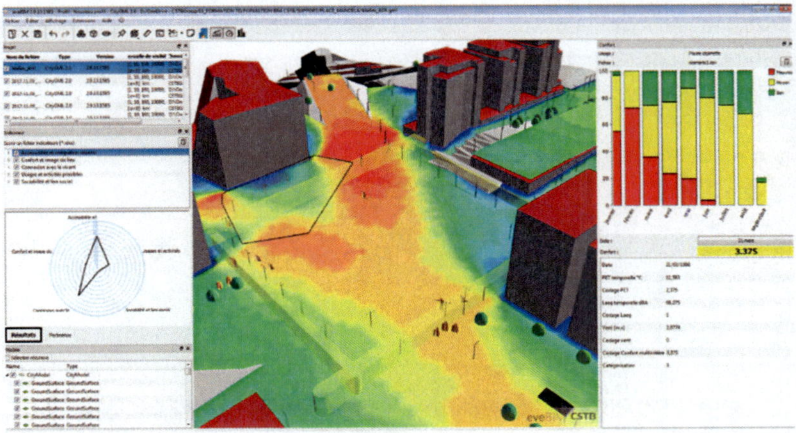

Fig. 14. Example of comfort rating analysis on the simulated area in the eveBIM® viewer

[4] https://logiciels.cstb.fr/bim-et-maquette-numerique/evebim-v2/.

The analysis of the quality of use then integrates a temporal dimension that the user must select to visualize the different results on the area and to grasp the overall comfort. Figure 14 shows the area of interest with comfort maps and the result of the overall comfort rating here at 3.375. Calculation details are shown as well, in particular the acoustic comfort and wind comfort corresponding to the given use and the selected period. It will be thus possible to understand why the comfort obtained a certain rating, and to reconsider urban strategy on the selected area.

3 Future Works and Applications

3.1 Towards a New Data Operating Approach

This work has contributed to generate multi-physics, spatial and temporal data which can be merged to data coming from other databases such as environmental sensors, open data (air quality, real time traffic information, quality dimension and availability of available services etc.), IOT, meteorological stations or field investigation (see Fig. 15).

This new large database can then be used to gain more accurate knowledge of urban spaces and therefore to manage urban spaces more efficiently. It would then be possible to use new disruptive tools such as machine learning in order to predict short-term environmental quality assessments (health and comfort) and identify new designs, services (for example pedestrian alert or recommendation to use urban space) or management strategies to adapt the urban space to daily life needs.

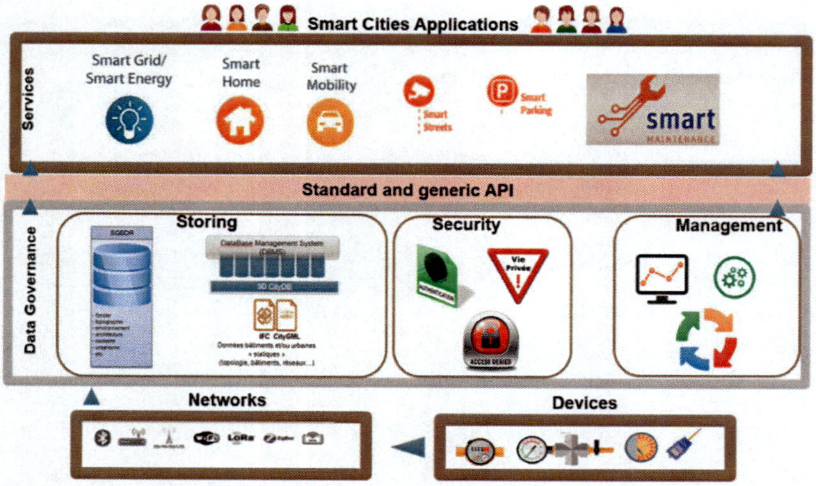

Fig. 15. IOT strategy for data management and storage

3.2 Experiment a Tool to Codesign Public Spaces with Users

Another ongoing work concerns the development of a co-design tool based on the research conducted to assess urban quality of public spaces (see Fig. 16). The requirements of this tool are based on the needs of Paris La Defense public consultation team. It aims at revealing to end users the complexity of making the right choices when designing public spaces with a high level of quality of use. The tool allows to define zones with different purposes (zones to relax, play, walk…), to implement urban furniture (benches, lights…) and vegetation. The next phase will be to test it with users to improve its ergonomics. In a first phase, this tool will be displayed during workshops with a CSTB professional using it according to participants' wishes. This first phase will be used to bring improvements to the overall functioning of our tool aiming at allowing autonomous use in a second phase.

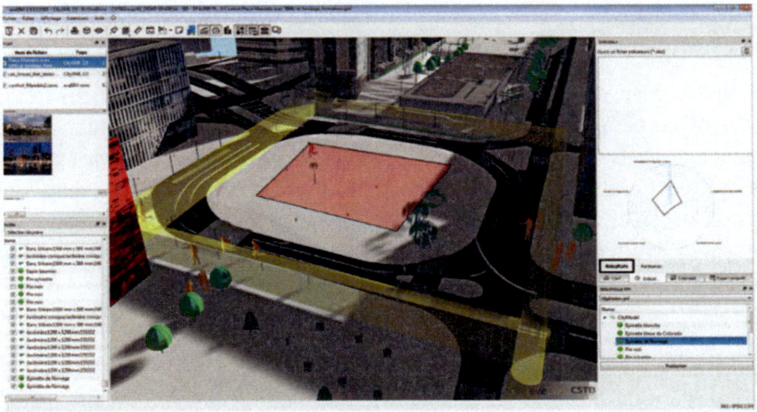

Fig. 16. View of the co-design tool under development (eveBIM viewer)

3.3 Integrate New Input Data from Climate Change

Future climate change can be predicted using the database provided by climate services such as the DRIAS database. Data are available for several time periods (e.g. between 1971 and 2100) and models (e.g. CNRM, IPSL, Eurocordex) at a fixed spatial (e.g. 1 km, 5 km or 8 km) and temporal (e.g. day, month) scale.

The anomaly map makes possible to visualize the different characteristics of climatic evolutions at the urban scale (i.e. frequency, duration, intensity) in average climate conditions but also extremes ones (i.e. heat waves, floods, storms), between the present and the future (Fig. 17).

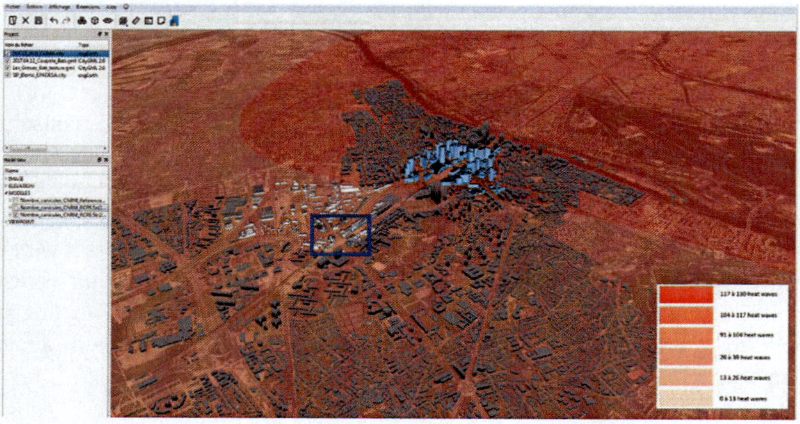

Fig. 17. Integration of climate change evolution in the database (eveBIM viewer)

This type of map was first integrated into our numerical model in order to visualize the impact of climate change between the current and the future situation of urban development projects. In a second time, it will be used as input data on our multi-exposure comfort simulations tool. A scale adaptation of the current map would be necessary in order to carry out these studies at the right scale. To this end, a down-scaling (spatial and temporal) would make it possible to refine and target, on areas at risk, impact studies.

4 Conclusion

The BIM applied to the urban scale has allowed implementation of territorial digital models in standard and open formats, including IFC and CityGML. New datasets are now available for multi-scale simulations whose results can lead to creating global indicators to evaluate the planned developments.

Comfort is a global concept on which several research studies have been already carried out at the building scale, but rarely at the large scale such as neighborhood and city levels. To be evaluated one of our main challenge was to work with a huge set of parameters that are often heterogeneous and multiscale (building, city and territory), multiformat and multi-actors. The interdependence of physical phenomena in the perception of comfort led us into new ways of defining use cases and time series for simulation.

A new interaction with the user, during public consultation, was tested based on our decision tool prototype to co-design public spaces.

The multi-scale BIM provides a first piece of answer in the centralization of exploitable data for the assessment of comfort and quality of public spaces, but still bring issues concerning data storing and management.

References

1. Biljecki, F., Stoter, J., Ledoux, H., Zlatanova, S., Çöltekin, A.: Applications of 3D city models: state of the art review. ISPRS Int. J. Geo-Inform. **4**, 2842–2889 (2015)
2. Kolbe, T.H., Gröger, G., Plümer, L.: CityGML – interoperable access to 3D city models. In: Proceedings of the First International Symposium on Geo-Information for Disaster Management (2005)
3. Kolbe, T.H.: CityGML–3D geospatial and semantic modelling of urban structures. In: Presentation on the GITA/OGC Emerging Technologies Summit in Washington (2007)
4. BuildingSMART. http://www.buildingsmart-tech.org/speci.cations/ifc-releases/ifc4-release
5. Cox, S., Lake, R., Portele, C., Whiteside, A.: Geography Markup Language GML 3.1. OGC Doc. No. 03–105r1 (2003)
6. Delval, T., Jolibois, A., Carré, S., Aguinaga, S., Mailhac, A., et al.: Building/City information model for simulation and data management. In: 12th European Conference on Product and Process Modelling (ECPPM 2018), September 2018, Denmark. https://doi.org/10.1201/9780429506215
7. Piepereit, R., Beuster, A., von der Gruen, M., Voß, U., Pries, M., Wagner, U.: Towards wind-simulation of virtual 3D city models. In: A Collaborative Vr Environment, Int. Arch. Photogramm. Remote Sens. Spatial Inf. Sci., XLII-4/W15, pp. 61–66 (2019). https://doi.org/10.5194/isprs-archives-xlii-4-w15-61-2019
8. Ohori, K.A., Krijnen, T., Diakité, A., Ledoux, H., Stoter, J.: Processing BIM and GIS models in practice: experiences and recommendations from a GeoBIM project in The Netherlands. ISPRS Int. J. Geo-Inform. **7**(8), 31 (2018)
9. Liu, B.Y., Jordan, R.C.: The long-term average performance of flat-plate solar-energy collectors: with design data for the U.S., its outlying possessions and Canada. Sol. Energy **7** (2), 53–74 (1963)
10. Brichambaut, C.P.D.: Rayonnement solaire et échanges radiatifs naturels. Gauthier-Villars (1963)
11. Chain, C.: Caractérisation spectrale et directionnelle de la lumière naturelle. Application à l'éclairage des bâtiments. Ph.D thesis, Institut National des Sciences Appliquées (INSA) de Lyon (2004)
12. Tourasse, G.: Mesure et Analyse Statistique Tout Temps du Spectre du Rayonnement Solaire. Ph.D thesis, Université de Lyon (2016)
13. Martinez, A., Perraudeau, M., Dumortier, D., Carré, S.: A new sky subdivision for climate-based daylight modelling. In: 27th Conference of the CIA in South Africa, vol. 1 (2011)
14. Perraudeau, M., Carré, S.: Analyse du bilan de l'environnement lumineux intérieur. In: International Building Performance Simulation Association (IBPSA) Conference, Lyon (2008)
15. Höppe, P.: Physiological equivalent temperature - a universal index for the biometeorological assessment of the thermal environment. Int. J. Biometeorol. **43**(2), 71–75 (1999)
16. Battermann, S., Cook, R., Justin, T.: Temporal variation of traffic on highways and the development of accurate temporal allocation factors for air pollution analyses. Atmos. Environ. **107**, 351–363 (2015)
17. Sampson, J.: Traffic flow variations in urban areas. In: Proceedings of the 36th Southern African Transport Conference (SATC 2017) (2017)
18. Cohen, P., Potchter, O., Schnell, I.: A methodological approach to the environmental quantitative assessment of urban parks. Appl. Geogr. **48**, 87–101 (2014)

Augmented BIM Workflow for Structural Design Through Data Visualization

Luiza C. Boechat(✉) and Fabiano Rogerio Corrêa

Universidade de São Paulo, São Paulo, SP 05508-220, Brazil
luizaboechat@usp.br

Abstract. Structural analysis of projects generates a large amount of data, especially when it involves complex or not modular geometries. By analyzing the generated data, it is possible to create optimized outputs for structures in an iterative process, until reaching a solution that satisfies the project constraints. Often, it is challenging to conciliate optimization with issuing of intermediate versions to be used for project coordination by other stakeholders. This is due to recurrent project changes, related to architectural or client requirements that arise during the design process. Moreover, structural design is generally carried out in a variety of software by different engineers within the design team. Therefore, it is rather challenging to organize, control and analyze all the information generated in structural design at each step, aiming to enhance both general team efficiency and structural optimization. The increasing popularity of BIM (Building Information Modeling) in this field could be a pivoting factor towards data-driven design and informed decision-making in all levels of a structural project team.

In this context, this paper presents a proposal and an application of a BIM data workflow to enhance the work of structural designers as well as the collaboration within the design team. The workflow is based on data extraction through visual programming, using Dynamo, and the creation of an automatically updatable dashboard within Power BI. Python is used to create custom-build operations that apply specifically to the structural assessment context. Discussion about the suitability of this process for other applications within decision-making activities in structural engineering is also given.

Keywords: BIM · Data visualization · Decision-making · Data-driven design · Structural engineering

1 Introduction

Structural engineering is a discipline that has been largely influenced by software. It has been many decades since the impact of Information Technology (IT) started to be discussed in this field. A survey from the 1990s in Great Britain pointed out that 68% of the respondent engineers already believed that IT was accelerating the traditional functions of engineering [1]. This happened thanks to the use of computer-aided drawings (CAD) and computer-aided engineering (CAE), especially finite element analysis (FEA).

E. Toledo Santos and S. Scheer (Eds.): ICCCBE 2020, LNCE 98, pp. 180–196, 2021.
https://doi.org/10.1007/978-3-030-51295-8_15

Since then, the use of computational tools in design has been intensified. Nowadays, most of the structural engineering offices use FEA and 2D or 3D CAD to develop projects. The growing access to computers along with the increasing complexity of projects has made it possible, and not rare, to analyze various load combinations for a given project. Additionally, parametric design has also substantially increased the number of geometric options to be investigated. All of this contributes to the generation of what was coined by Joyce et al. [2] as "big-design-data" - large amounts of data that have been generated during design and analysis of options, rather than passively collected. This data does not have the same collecting speed or size as general big data, but also presents the same issues of data volumes and complex relationships.

Aiming to enhance the work of structural designers, the following aspects about the structural engineering data should be considered:

- size and variability;
- usage for collaboration;
- usage for optimization;
- usage in the context of BIM (Building Information Modeling).

Given the aspects cited above, this article presents the visualization of structural data as a powerful tool to get meaningful insight and more efficient collaboration in structural engineering. In this article, BIM is considered as per the NBS (United Kingdom's National Building Specification) definition: "a process for creating and managing information on a construction project across the project lifecycle" [3]. One of its key outputs, the BIM model is investigated as the central platform to allow structural data integration.

More specifically, this research addresses the problem of complexity in the structural design process and proposes that moving towards a more data-driven workflow, with the support of BIM, can address this issue. To achieve this goal, firstly, a literature review was carried out to assess the latest advances in academic research in this area. Then, a constructive research method was taken, creating a construct (or artifact), which is presented as an Augmented BIM Workflow for structural engineering. The proposed workflow, in which designers' internal collaboration is supported by computational tools, allows for a new degree of automation and data-driven decision-making.

One specific implementation, applied to a cultural building project, is presented to illustrate the concepts developed. The output of structural data into a common BIM model along the development of the project is investigated as the solution for establishing a merged and organized structural database, aiming to support informed decision-making. The workflow is based on data extraction through programming, both traditional and visual, followed by the creation of an automatically updated dashboard.

2 Considerations on Structural Engineering Data

2.1 Size and Variability

"Big-design-data" is currently a reality in structural engineering offices, especially when dealing with complex projects. In the context of design, complexity is either related to randomness on the information description or to the size of the process [4]. Two complementary enhancements were the main enablers of engineering needing and being able to handle complexity: advances in the construction industry and advances in computing.

Regarding advances in the construction industry, Rahman [5] highlights that even though there are historical examples of complex buildings, the rate of change in many aspects of building design has increased complexity in this field in the last decades. Four sources of complexity are cited: sophisticated building components, sophisticated systems, multi-disciplinary integration, and novel forms. All of those have a significant impact on structural engineering.

Regarding computing, research progress in the finite element method and graphic computing coincided with the rapid growth of computer power [6]. From the time when the term "finite analysis" was coined and PRONTO (prototype of CAD software) was released, the cost of processing power has had a decrease of several orders of magnitude [6–9], as shown in the graph in Fig. 1. The consequent increase in the frequency of use of computation as support for engineering contributed to enable complex engineering at a much faster pace.

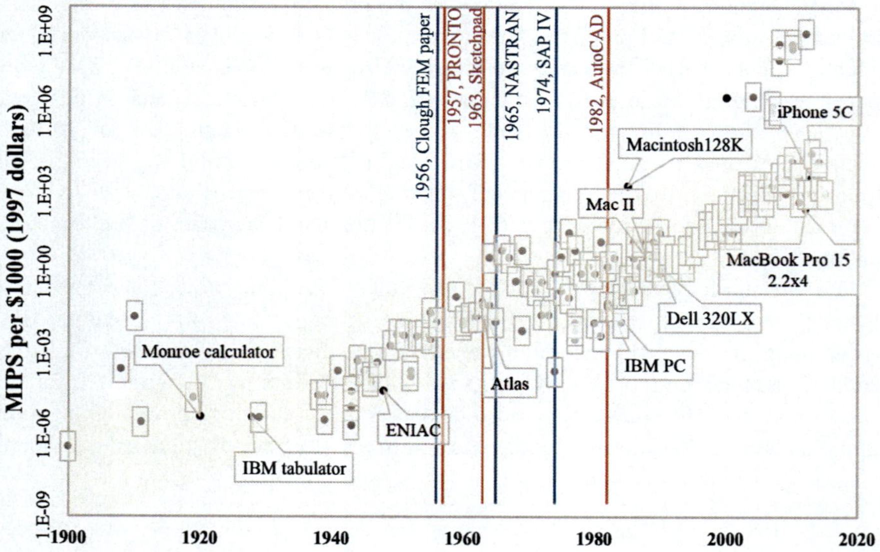

Fig. 1. Evolution of MIPS (million instructions per second) per cost of computers, and milestones in the initial development of CAD and FEA. Produced by the authors, data from [6–9].

These advances made it possible and not rare that structural engineers explore an amplified design space before reaching a final solution, mainly through parametric modeling combined with FEA. The novel forms previously cited as an example of construction advances, hinder the determination of which are the critical permutations of single load cases, from all possible ones. The result is that the designer must deal with hundreds of load combinations. Surely, nowadays computers can handle those analyses. But this rapidly evolves into the need of managing large datasets while exploring options. In terms of size, outputs from sophisticated FEA commonly reach some giga-bytes of size for one given geometry.

Another relevant aspect of structural data is variability: structural data often comes from different sources. Most analysis software is focused on the final part of the design process, for verification rather than for exploration [10]. Commonly, FEA software is complemented by custom-built spreadsheets, in which the designer performs the required code checks of structural elements. The advent of plug-ins for CAD that allow for structural analysis in a parametric workflow has enhanced the efficiency of analysis in the conceptual design phase. However, as these plug-ins are not as sophisticated as traditional FEA software, they are not often used as a solely sufficient tool, but rather complemented with other software.

2.2 Use for Collaboration

The construction industry is distinguished by its high fragmentation and complexity of its products and services. The fragmentation enforces the need for effective collaboration between different companies [11]. Complexity, mainly due to largeness (or as per previously defined, the size of the process), also increases the need for effective collaboration within the same company.

It is challenging to conciliate optimization with issuing of intermediate structure versions to be used for project coordination by other stakeholders. This is often due to many project changes, related to architectural or client requirements that arise during the design process. There is an industry perception that the ascending adoption of digital design tools and the competitive pressure have accelerated the pace of building design. The design team is now routinely challenged to propose original designs at an unparalleled velocity. Chock [12] elaborates that this compressed design environment will lead to better design only if there are more effective collaboration and a paradigm shift towards information-based design, with all stakeholders engaged.

In this context, structural data must be prepared for what Loos [10] denominates "integrated design". From the external collaboration point of view, data needs to provide rapid information to architecture at other disciplines, such as HVAC, plumbing and landscape architecture, as concurrent design gets more common in the construction industry. Automatic or semi-automatic data visualization tools become rather relevant in this context. Within the same specialty, complex projects and tight deadlines will often require different designers with various backgrounds and previous experiences to work in parallel towards a common final product. It is important that in this case, the same quality is ensured on their work, with key common targets agreed throughout the project. It is proposed that the use of structural data to establish key performance indicators (KPIs) in this case is a valued option for smooth and successful internal collaboration.

2.3 Use for Optimization

Through the analysis of the generated structural data, it is possible to create optimized outputs for structures and reiterate the process, until reaching a feasible solution that satisfies the given project constraints. Optimization is in itself a very relevant research field in structural engineering. In this context, optimization problems generally look for the structural system that most efficiently – or with the minimum cost – resists the imposed loads.

However, in most problems, there is no single optimum design. Moreover, not every design aspect that should be optimized is easily translated into numbers. Examples of that are aesthetics, degree of innovation in architecture history and impact on the use of the building.

As stated by Mueller [13] "Unlike the human design process, optimization on its own cannot handle unformulated objectives and constraints". Harding [14] also discuss the "illusion of optimal design", describing the situation in which automatic optimization acts over a very limited design space, resulting in a sub-optimal solution.

In this context, the investigation of structural data visualization as broader support for optimization seems promising, as it allows the designer to navigate through alternatives rather than compressing them in a single number. The informed decision-making process is then facilitated, since all structural and architectural information of different design proposals may be presented in a comprehensive and global overview [10].

2.4 Use in the Context of BIM

Building Information Modeling adoption in the construction industry is promising. Especially in the design phase, the impacts of BIM, such as a reduction in design and in drafting errors, along with increased responsiveness to clients, have been already investigated in the literature. In what concerns structural engineering, an increase in productivity of up to 41% was measured thanks to improvements in drawing production alone [15].

Besides the increase of productivity in drawings, which is inherent to BIM, it is also important to consider the opportunities that it offers for knowledge management in design and construction. In fact, among the researches and users in this field, it is a widely accepted idea that information is the most relevant part of BIM [16, 17].

BIM processes promote the integration of construction data in an information model that applies principles similar to object-oriented programming, which is very useful for data science [18]. In this context, BIM is suggested as an input source to a data warehouse which is the basis of applying data science to the construction industry – comprising data visualization, statistics, machine learning and data mining [19] (Fig. 2). Through an extract-transform-load (ETL) process, valuable BIM data could be transformed to be used as an input to a target data warehouse. Analyzing the stored data, knowledge and insights can be generated, which could further enhance productivity and collaboration in the design and engineering.

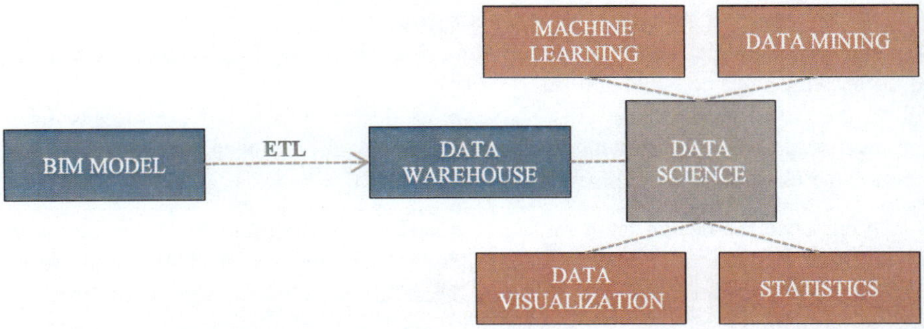

Fig. 2. BIM model and data science possible integration and applications. Adapted from [19].

3 Structural Project Workflow

3.1 Traditional Project Workflow

A sequence of different steps exists in the construction industry, from the beginning to the completion of a project. The design process is one of the first steps. It is in general divided into four successive phases: conceptual design (or predesign), schematic design, design development, and construction documents [20].

Traditionally, the structural engineering design workflow is based on a responsive approach [12]. The following items summarize its steps based on literature [10, 12, 15] and general industry experience.

1) Firstly, the base geometry and requirements are provided from architecture either through the desired grid, in case of frame structures, or a massing model, in case of more complex shapes;
2) In the conceptual design phase, a general check is carried out. For frame structures, this may be done with simple calculations, checking for instance if spans are reasonable for the structural height restriction and if suitable bracing systems are possible. More complex shapes could require a more sophisticated feasibility check. All those checks are generally made through limited examples and simple models. The conceptual design is achieved by an incremental learning process consisting of the generation and evaluation of a set of design alternatives. The nature of these processes is divergent and convergent respectively. In the end, it is expected that the designer can provide a final shape and topology, estimate the quantity of material that the project requires and its feasibility from the structural point of view;
3) When the project evolves to schematic design, the current architecture geometry is generally imported to a FEA software and:
   a. The compatibility between imported geometry and structural feasibility determined in conceptual phase is checked;
   b. An abstract structural model is created by the structural engineer, based on architecture and inputs such as section properties, boundary conditions and loads. This model is then analyzed for all required combinations.

c. The sizing of structural elements is performed.

4) In detailed design, the same steps of items 3.a, 3.b. and 3.c. are performed, but this time with more consolidated information;
5) In the construction documents phase, most design decisions have already been taken and compatibility with other disciplines has already been checked. The final documentation is then consolidated.

Often, once the design team finishes the analysis and communicate its outputs to the architect, there are several new changes, due to architecture and client requirements, that should be incorporated in the design. There are also changes resultant from coordination with projects of different disciplines, such as electrical, plumbing and HVAC engineering. If design modifications are made in the geometric modeling tools, those changes need to be remodeled in the engineering tools as well, an approach which is time-consuming and can lead to errors [21].

The inherent aspects of structural data (size and variability, use for optimization, internal and external collaboration), explored in Sect. 2, have variable impacts and potentials of exploration throughout the project phases and steps. In this context, impact is defined as a measure of the change a certain aspect may cause in the common structural engineering design workflow. Potential is defined as a measure of the possibility that the structural data generates relevant insight when used for a given aspect. The graph in Fig. 3 illustrates this variation.

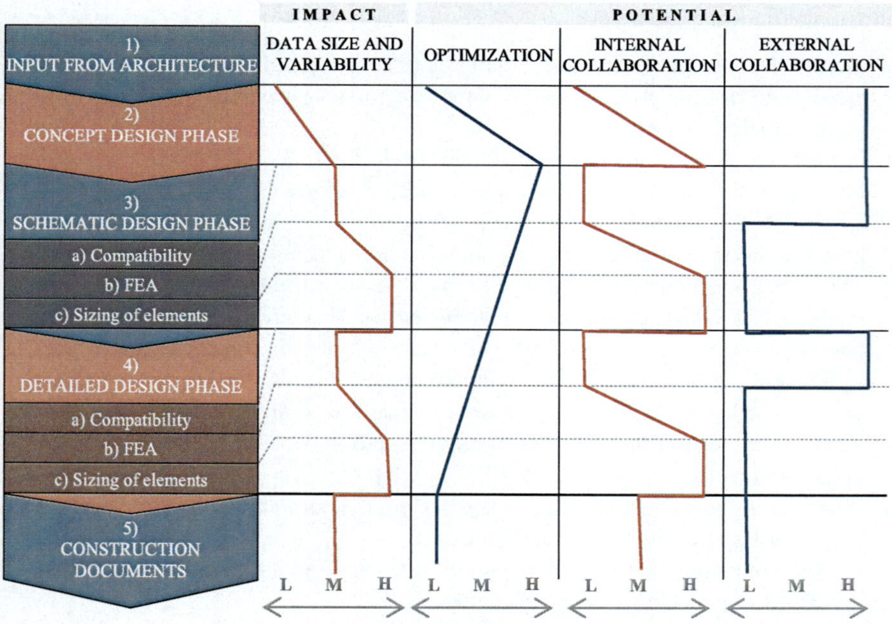

Fig. 3. Qualitative impact or potential of the use of structural data per aspect discussed during each project phase. Impact and potential may vary from project to project and niche of application. L: low; M: medium; H: high. Produced by the authors.

In the traditional project workflow, size, and variability of data impact increases during design space exploration, which marks the concept design phase, and in each iteration of FEA and sizing of elements. One relevant consequence of the high size of data and variability of sources is reduction in productivity when reacting to design change.

Although optimization may be performed up to the Construction Documents phase, it is much more likely to have an important impact on the project outcome in step 2, when there are a lot more topics to be defined. Therefore, the potential of using structural data for optimization reaches its higher point towards the end of this phase.

Internal collaboration is indispensable: if different engineers compose the design team, there is a need of aligning the project outputs to be provided to the client in a single approved output. This means agreeing on compatible design assumptions in terms of loads, boundary conditions and allowable final sections, keeping the same quality of project even if different engineers were responsible for it. Often, in this context, members of the design team need to produce reports for communicating with more senior members and with pairs, which is a time-consuming task. The potential of using structural data to support internal collaboration increases during the concept design phase and at each iteration of FEA and sizing of elements.

Moreover, collaboration with other companies is relevant to minimize the need for re-design and to increase overall design productivity. Structural data may be used for this purpose especially during the concept design phase, when there are a lot of open topics to be aligned with external parties and during each compatibility step.

The use of BIM software and methodology is an improvement on the whole process, although currently explored by structural engineers mostly for compatibility and generation of construction or end-of-phase submission documents. Consequently, there is an opportunity window in exploring the potential of the use of structural data through BIM models on the other steps. The establishment of a workflow that explores this potential is called in this paper an "Augmented BIM Workflow".

4 Implementation of an Augmented BIM Workflow

4.1 Proposed Workflow: Focus on Internal Collaboration

It is proposed that by enhancing the use of structural data in the traditional project workflow, a better outcome can be achieved, both in terms of quality of the final product and productivity of design. For that, BIM models can be used as a central platform, the main constituent of the proposed data warehouse.

This data warehouse should be explored based on agreed KPIs. As design constraints and objectives of data visualization may vary considerably from project to project, it is recommended that KPIs be defined within the design team during phases 1 to 3 of each project. Of course, once this approach is adopted in many projects, the experience of selecting KPIs may be adapted from one to the other, becoming rather a management standardized procedure of the company.

In this paper, the "Data and Control" step permeates phases 3 and 4. A similar procedure could also be explored in phases 1 and 2, as "Data and control for optimization", or in phase 5, named as "Data and control for documentation", but these are not explored in the present paper.

The objective of the present implementation is to allow that through the visualization of data in each step, the design team improves the outputs quality control, reaching a more efficient internal collaboration. Because of this focus, the steps of FEA and sizing of elements are the most relevant for data extraction, transformation, and load.

To include the Data and Control step, a joint base of structural and geometric information, extracted from all relevant sources used in the project, is created – the data warehouse. Structural data is explored through data visualization aiming to generate insight and support informed decision-making. The diagram in Fig. 4 outlines the present implementation process and its applicability to the relevant project phases.

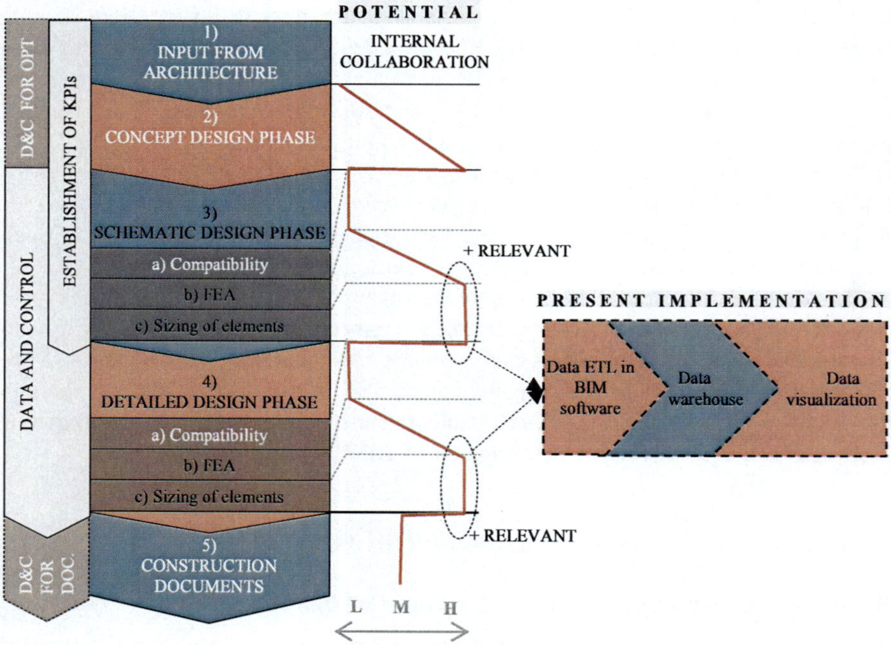

Fig. 4. Present implementation of the Data and Control step in the design process, focused on enhancing internal collaboration. L: low; M: medium; H: high. Produced by the authors.

In this article, the information sources are an Excel spreadsheet, with FEA resulting internal forces, and a Revit model, with geometric and coordination information. The data extraction is done through a script using Dynamo, a visual programming tool that runs within Revit. This tool is complemented by custom Python scripts that run in the same environment. The data is gathered in the BIM warehouse, and exported as a general table, in.xls format. The visualization of the information is done through Power BI, which allows for relatively flexible graphical outputs, enabled via compatibility with Python scripts. This implementation method is illustrated in Fig. 5.

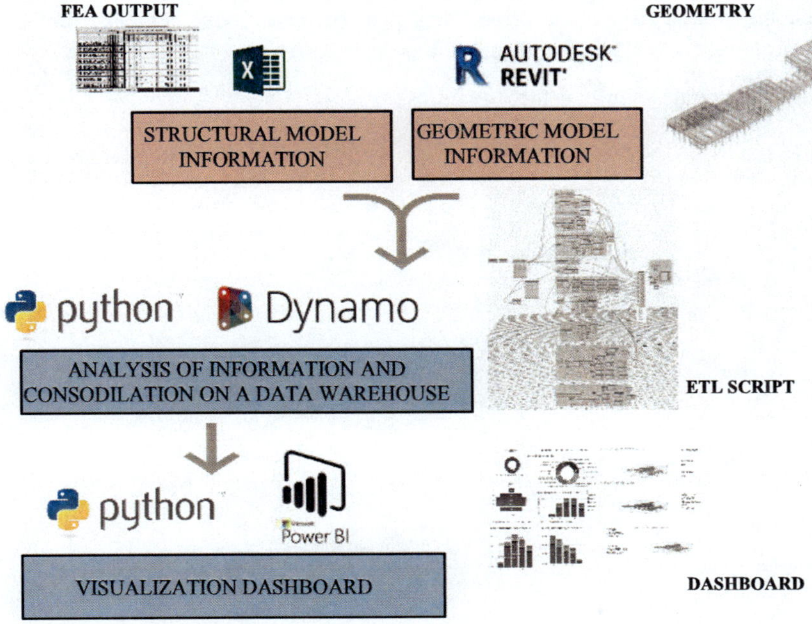

Fig. 5. Steps and software used in the present implementation of the Data and Control step. Produced by the authors.

4.2 Particularities of the Application Structural Project

The implementation of the proposed Augmented BIM Workflow is carried out for a frame structure of a cultural center building. The total built area is around 25000 m^2, divided into five different blocks. The building is formed by a steel frame structure with concrete slabs. Although this application was inspired by an existing project, a topologically similar model was generated for the illustrations in this article, forming a mischaracterized shape, not to compromise the confidentiality of the project.

Some features of this project that can make the inclusion of a Data and Control step especially interesting are:

- Important structural requirements due to long cantilevers (over 8 m), long spans (up to 20 m) and high live loads required by the client (up to 1000 kg/m^2);
- Low modularity, resulting in low element repetition and control and documentation difficulties – as per architectural concept, column alignments are inclined, and the building is "embedded" in the ground, resulting in many different levels despite low maximum height;
- Recurring design changes: architectural and complementary discipline projects were developed in parallel, and due to parallel coordination, architectural changes and client requests, several structural models were made on schematic and detailed design phases;

- Given the complexity of the project and possible separation in blocks, different structural engineers had to work on it in parallel.

Figure 6 illustrates some of these features.

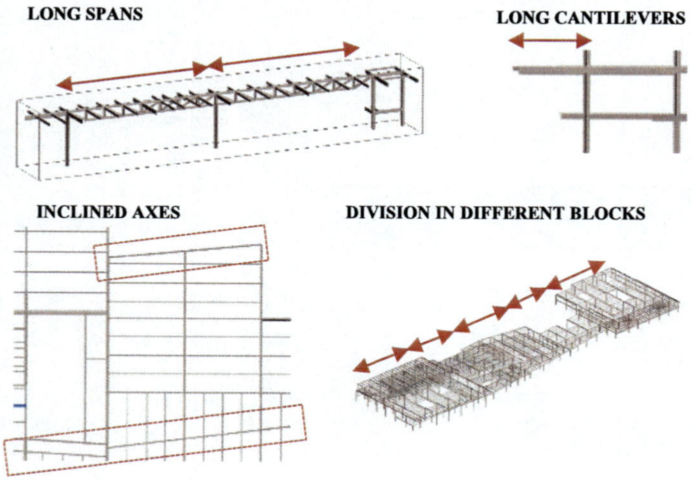

Fig. 6. Special features that make Data and Control step application more relevant: long spans, long cantilevers, inclined alignments, and separation in different blocks.

With those special features in mind, especially the requirement of concomitant design by different team members, it makes sense to analyze performance indicators that would examine the maintenance of project quality within the different blocks. In the example of this article, it is proposed to analyze:

- the occurrence of each type of element (beam, bracing, column, truss) per block;
- the predominance of each type of beam per block per % of total weight;
- the occurrence of each type of section (rolled W sections, welded rectangular hollow sections of given dimensions, etc.) per block;
- the predominance of each type of beam per block per % of total weight;
- the checking of alignment of columns;
- the maximum capacity ratio of each beam element as per NBR8800 composite beams checks;
- the number of connections at each end of a beam.

4.3 Dashboard Implementation

To implement the ETL process, a script was developed in the Dynamo plugin, using both visual and Python programming. Figure 7 shows a general scheme of this script.

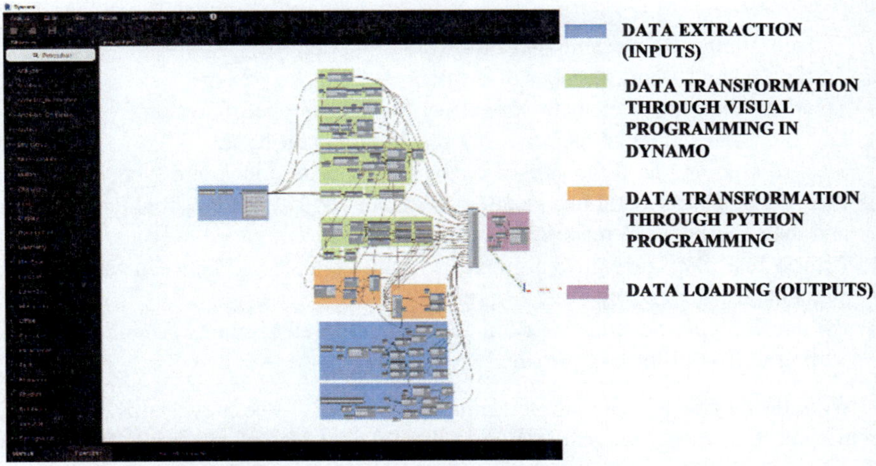

Fig. 7. General scheme of the script. Produced by the authors.

From the data in the geometric and structural bases, the following information was taken:

- Tag - unique tag per element from the structural model that was used to track information throughout the project in the Revit model;
- Block - block of the building to which the element belongs - A, B, C, D or E;
- Cross-section of element – the name of the type in the Revit model;
- Beginning and ending section - for variable sections;
- Type of element - classified into beams, columns, bracing or trusses;
- Element length;
- Element volume;
- Nodes - coordinates of the start, end, and center nodes of each element (to be able to reproduce geometry in the dashboard);
- Building floor to which the element belongs;
- Distributed area load in ULS (ultimate limit state) corresponding to the floor and block of each element;
- Vertical check function coded in Python to check if the columns are correctly vertical, given a certain tolerance;
- Connections check function coded in Python that tests the number of bars connected to each of the elements in its start and end node, given a certain tolerance;
- Utilization rate according to the structural sizing worksheet;
- Effective width according to the structural sizing worksheet;
- Linear weight according to a section table.

This base was imported into Power BI by creating a query. The choices of the graphs to be generated and the final layout of the dashboard were derived from an analysis of the practicality and ease of understanding of the data in the project. The graphics created were:

a) Donut chart: number of elements in the model;
b) Funnel: number of elements per block in the model;
c) Stacked column chart: number of elements per block and type;
d) Treemap: the sum of element weight per block and type of section;
e) Donut chart: number of elements per block, type, and floor;
f) Stacked column chart: the sum of element weight per block and type;
g) Python line plot: general line geometry with not vertical columns highlighted in red and their marks listed for easy identification;
h) Python line plot: general line geometry, with connections between more than 3 elements highlighted by a red marker;
i) Python line plot: general line geometry, with colored elements according to their scaling utilization rate as provided in Excel.

With those graphs, the design team could have a general visualization of the occurrence of element type and section in the different blocks, relating to the element utilization. In this way, different designers collaborating in the project could diminish the total amount of sections specified. If one type of section was used only in one block, it could be desired to change it to another one that is more common in the project, if the utilization allows so, reaching a more consistent final project. Moreover, this workflow gives the structural engineer flexibility to add any checks that can be coded as an automated check to the project, which is the case for example of graphs g) and h).

Graphs a, b, c, d, e, f can be generated directly in the data analysis software Power BI. On the other hand, graphs g, h, i require the python visual interface present in the software, as they aim to reconcile data and geometry, which is in principle out of the scope of standard Power BI options. Figure 8 and 9 shows the resultant dashboard for this implementation.

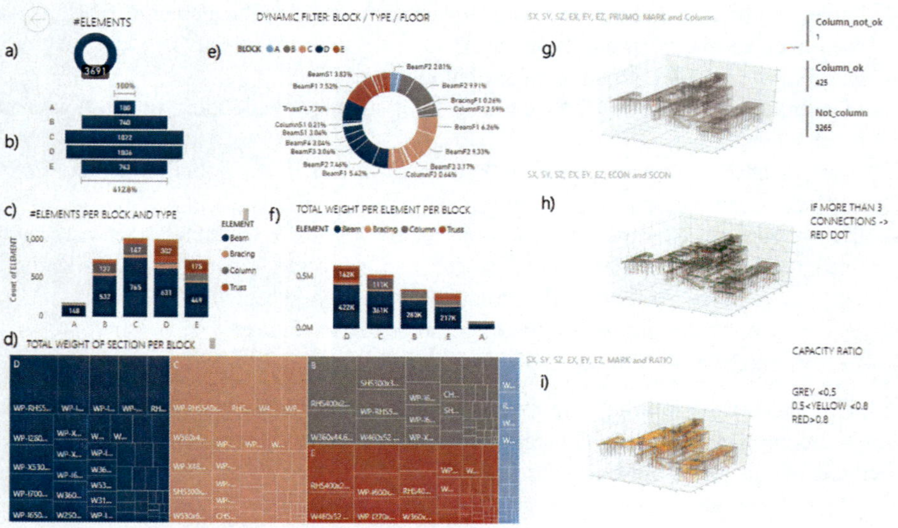

Fig. 8. General view of the resultant dashboard, with graphs a) to i). Produced by the authors.

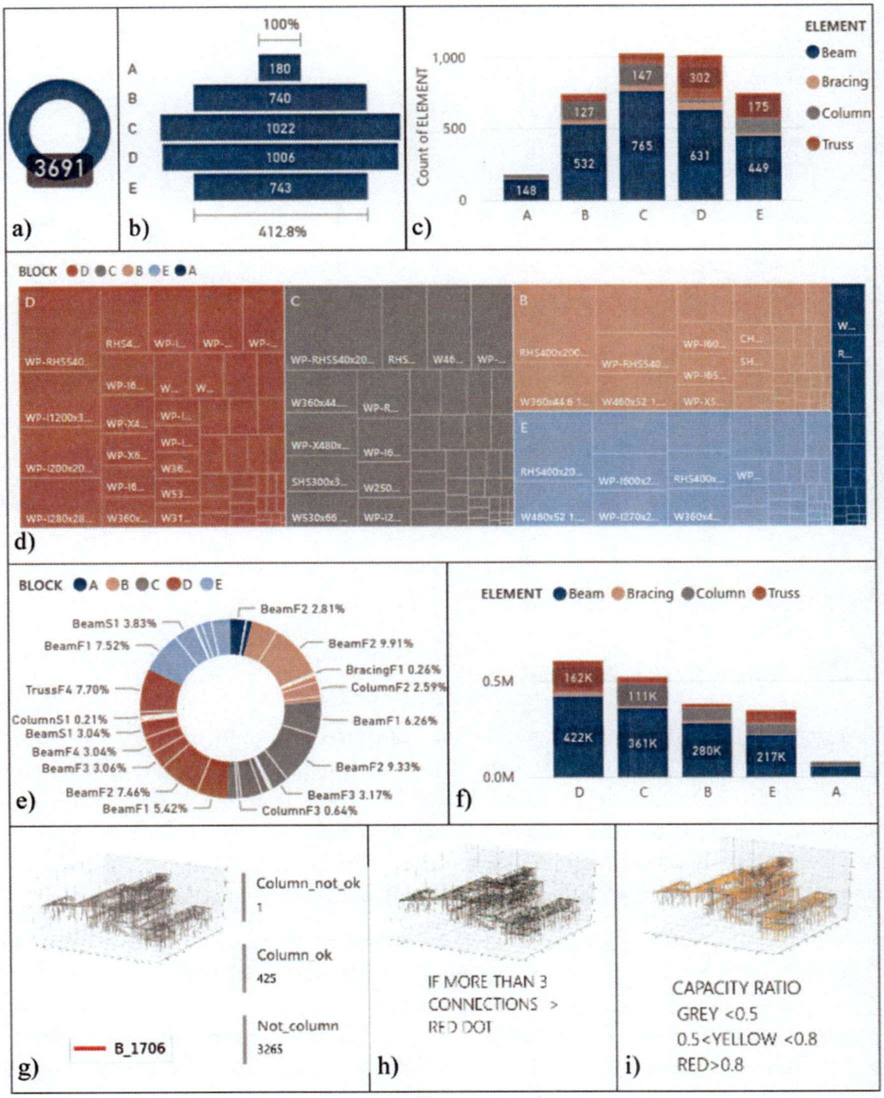

Fig. 9. Closer view of graphs a) to f). Produced by the authors.

In terms of ease of visualization, a very relevant feature that this workflow allows is the use of dynamic information and geometry filters. Since the base is structured, the data analysis software allows that, when selecting a particular group in one of these graphs, all other graphs highlight this same group, which facilitates relating the information on the different graphs. This is illustrated in Fig. 10.

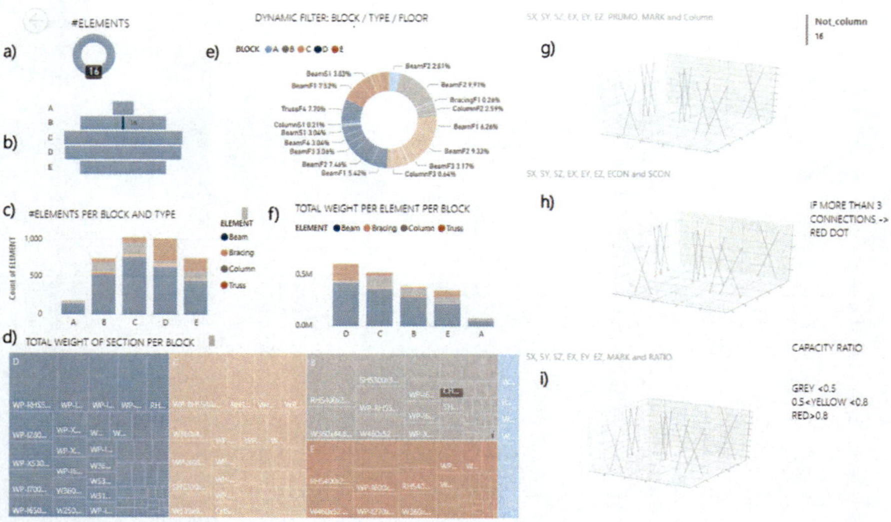

Fig. 10. Resultant dashboard with filtered information and geometry, showing bracings of a single block. Produced by the authors.

5 Discussion and Conclusion

The proposed uses of structural engineering data present a high potential for performance improvement and analysis optimization at structural engineering offices. The presented implementation of Augmented BIM Workflow is only a sample of how this idea could be applied to design and collaboration. Analyses to be performed and KPIs to be generated should be defined specifically by project. Possible applications and extensions of the present workflow, with a focus on internal collaboration, include:

a) Assisting the project engineer by identifying and analyzing discrepant areas in the project, such as regions with higher than average area steel rates;
b) Facilitating project performance monitoring by the senior engineer in charge of the project by defining KPIs for periodic verification online in real-time;
c) Possibility of adapting the project indicators for a macro analysis, creating a view for managing directors on various projects;
d) Possibility of adapting to vision with the main indicators for the client, who generally lay in engineering, allowing her/him to better understand the value of hiring a qualified structural engineering office;
e) Generation of an information base that can be used as input for the optimization of structures.

A wide range of applications of this methodology is possible. In this way, it should be adapted to each context of the project and the specific needs of the company, especially regarding applications a), b), c), and d). Analyzing application e) opens yet another possibility: could these structured data, generated from complex structural analyzes, be used to some extent as input to intermediate stages of the project?

Take the script application project as an example. Each of those building blocks had around 20 different versions of structural models due to geometric changes. In conversations with industry professionals, or taking the literature on structural project management, it is clear that this type of rework is relatively common in large building projects. This situation, combined with the potential described in e) above, raises the following question: What if the structural engineer could use the extracted and analyzed data to guide the next rounds of interaction with the architecture? This shall be the object of future research, as a continuation of the Augmented BIM Workflows.

References

1. Gardner, P.: Feature: Gardner Feature IT in structural engineering (results of the members' survey on IT and computers). Struct. Eng. **73**(21), 374–379 (1995)
2. Joyce, S., Guo, G., Dy, B., Ibrahim, N., Poorthuis, A.: Seeing numbers: considering the effect of presentation of engineering data in design. In: Proceedings of the IASS Annual Symposium 2019 – Structural Membranes 2019, pp. 1155–1162. IASS, Barcelona (2019)
3. NBS.: What is Building Information Modeling (BIM)? https://www.thenbs.com/knowledge/what-is-building-information-modelling-bim. Accessed 28 Jan 2020
4. De Biagi, V., Chiaia, B.: Complexity of structures: a possible measure and the role for robustness. In: Van Mier, J.G.M., Ruiz, G., Andrade, C., Yu, R.C., Zhang, X.X. (eds.) VIII International Conference on Fracture Mechanics of Concrete and Concrete Structures, pp. 1–9, Toledo (2013)
5. Rahman, M.: Complexity in building design. In: Proceedings of the 3rd International Holcim Forum for Sustainable Construction, Re-inventing Construction, pp. 14–17 (2010)
6. Fish, J., Belytschko, T.: A First Course in Finite Elements. 1st edn. Wiley, Location (2007). https://doi.org/10.1002/9780470510858
7. Field Robotics Center. Carnegie Mellon University. https://frc.ri.cmu.edu/~hpm/book97/ch3/processor.list.txt. Accessed 28 Jan 2020
8. Tornincasa, S., Di Monaco, F.: The future and the evolution of CAD. In: 14th International Research/Expert Conference, Trends in the Development of Machinery and Associated Technology, pp. II-1–II-18. Mediterranean cruise (2010)
9. Clough, R. W., Wilson, E.L.: Early finite element research at Berkeley. In: Fifth US National Conference on Computational Mechanics, pp. 1–35 (1999)
10. De Laet, L., Kawaguchi, K., Ohsaki, M., Takeuchi, T., Loos, L., Verbeeck, K.: Structurally informed decision-making by means of data visualizations during the conceptual design phase. In: Kawaguchi, K., Ohsaki, M., Takeuchi, T. (eds). Proceedings of the IASS Annual Symposium 2016 "Spatial Structures in the 21st Century", pp. 1–10. IASS, Tokyo (2016)
11. Miles, J.: IT-Supported Collaboration for Structural Engineering. Struct. Eng. Int. **15**(3), 139–144 (2005). https://doi.org/10.2749/101686605777963099
12. Chok, K., Donofrio, M.: Structure at the velocity of architecture. In: Proceedings of the 30th Annual Conference of the Association for Computer Aided Design in Architecture (ACADIA), pp. 218–226, New York (2010)
13. Mueller, C.T.: Computational exploration of the structural design space, Doctoral dissertation, Massachusetts Institute of Technology (2014)
14. Harding, J., Joyce, S., Shepherd P., Williams, C.: Thinking topologically at early stage parametric design. In: Hesselgren, L., Sharma, S., Wallner, J., Baldassini, N., Bompas, P., Raynaud, J. (eds) Advances in Architectural Geometry 2012. Springer, Vienna (2013). https://doi.org/10.1007/978-3-7091-1251-9_5

15. Sacks, R., Barak, R.: Impact of three-dimensional parametric modeling of buildings on productivity in structural engineering practice. Autom. Constr. **17**(4), 439–449 (2008). https://doi.org/10.1016/j.autcon.2007.08.003
16. Elinwa, U.K., Agboola, O.P.: Beyond BIM–a classroom approach to virtual design education. Procedia-Soc. Behav. Sci. **83**, 393–397 (2013). https://doi.org/10.1016/j.sbspro.2013.06.077
17. Lendvorským, T., Fredon, V.: BIM for infrastructure—manage your information. Autodesk University. https://www.autodesk.com/autodesk-university/class/BIM-Infrastructure-Manage-Your-information-2016#handout. Accessed 28 Jan 2020
18. Eastman, C.M.: Building Product Models: Computer Environments, Supporting Design and Construction. CRC Press, Boca Raton (2018). https://doi.org/10.1201/9781315138671
19. Siddiqui, R.: BIM and data science. http://www.aecbytes.com/feature/2018/BIM-DataScience.html. Accessed 28 Jan 2020
20. American Institute of Architects. Integrated Project Delivery: A Guide. http://info.aia.org/SiteObjects/files/IPD_Guide_2007.pdf. Accessed 28 Jan 2020
21. Verbeeck, K., Loos, L., De Laet, L.: How to re-open the black box in the structural design of complex geometries. In: Proceedings of the International Conference on Structures and Architecture, Portugal (2016)

Towards a BIM-Based Decision Support System for Integrating Whole Life Cost Estimation into Design Development

Mariangela Zanni[1,2](✉), Tim Sharpe[1], Philipp Lammers[2], Leo Arnold[2], and James Pickard[2]

[1] Department of Architecture, Strathclyde University, 16 Richmond Street, Glasgow G11XQ, UK
m.zanni@cartwrightpickard.com

[2] Cartwright Pickard, 1 Canal Side Studios, 8-14 St Pancras Way, London NW1 0QG, UK

Abstract. A common barrier to achieving design intent is the absence of information about operational performance during key design stages. This results in uninformed decision-making, which has negative impacts on actual building performance and Whole Life Cost (WLC). It is proposed that Building Information Modelling (BIM) has the potential to assist design decisions at initial stages, if the model can utilize reliable and robust cost and performance data from buildings in use. This paper describes the findings of a research project that has investigated the integration of WLC estimation into BIM processes. Firstly, the research attempts to gather tacit expert knowledge by reverse-engineering unstructured processes to clarify information requirements for WLC assessment. The Integrated DEFinition (IDEF) 3 structured diagramming modelling technique was utilised for process mapping of information flows. As a result, value adding information, and the chain of interdependencies, were determined. The model has informed the development of a decision-support tool that automates prescriptive tasks and streamlines information exchange procedures. The research findings support the idea that BIM can assist the accurate estimation of WLC of an asset, from the early design stages, through the implementation of a structured information exchange. However, source data is fragmented and in propriety formats and further work is needed to standardise and gather performance data and BIM model structures.

Keywords: Building Information Modelling (BIM) · Whole Life Cost (WLC) · Process modelling · Integrated DEFinition (IDEF) language · Decision support

1 Introduction

Successive UK construction industry reports have documented the fact that the sector has been suffering from low innovation and underperformance in terms of productivity and quality of the final product [1–3] and this has led to the UK Government's target of a 33% reduction in WLC [2, 4]. Research has shown that over a 30-year life span, the

E. Toledo Santos and S. Scheer (Eds.): ICCCBE 2020, LNCE 98, pp. 197–206, 2021.
https://doi.org/10.1007/978-3-030-51295-8_16

cost of operating the asset could be as much as four times the cost of designing and constructing the building, and that 80% of the operation, maintenance, and that replacement costs of a building can be influenced in the first 20% of the design process [5, 6]. However, buildings rarely perform as expected [7, 8]. This can impact on energy consumption, environmental quality and operational costs (OpEx) [5, 9]. It has been argued that the most common barrier to achieving design intent is the absence of reliable predictive information during design and construction stages, leading to poor decision-making, which impacts on performance and WLC [10]. Furthermore, researchers have identified the lack of standard methods and format for calculating WLC as the main barrier for its implementation [11].

As a consequence, the need for a more holistic approach to building design is gaining momentum. BIM has proven its potential to facilitate the building development process during planning, designing, construction, and operation [12]. Nevertheless, to date, BIM has focused on the production of construction information and although the use of BIM for calculating CapEx has become commonplace, its potential to be used for estimating WLC is not yet developed. This research implements a holistic approach for integrating WLC considerations throughout the design process. It considers trade-off relationships between economic cost and environmental performance issues. Several BIM software solutions have been developed with the aim of calculating WLC, but none of them offers a complete solution [13, 14].

This paper is based on a project that responds to the objectives of the UK Government's Transforming Infrastructure Performance (TIP) strategy [15] for adopting cost, schedule and performance benchmarks to support the selection, budgeting and design of projects through the development of disciplined definitions of required outcomes, ensuring they can be delivered in line with whole-life performance, cost and schedule targets. The project also addresses the objectives of the Construction Sector Deal (CSD) [4] by providing a common data framework and technology platform that support the design, delivery, and operation of higher performing buildings with lower carbon emissions and WLC. This can be achieved by harnessing the power of digital technology and data analytics [16]. The project focuses on the Build-to-Rent (BTR) housing sector [13, 17, 18] as an area in which WLC can impact on commercial viability. This is a critical issue for investors as they strive to maintain high performance standards for quality, which will lead to better return on investment and reduced risk.

2 Research Strategy and Methods

This research follows the pragmatic paradigm throughout the research design, data collection, and analysis. Mixed methods were utilized [19], and rationalized through abductive reasoning [20] for the development and validation of a structured standardised process for more accurate and timely WLC estimation through BIM. Interviews, workshops, and the IDEF3 structured diagramming technique were utilized to identify value-adding tasks and coordinate them into a holistic process. Triangulation of qualitative and quantitative methods served to increase the validity and reliability of the data collected. The research process followed the Technology Readiness Level

(TRL) framework [21]. Exploratory research served to satisfy the Discovery stage (Levels 1–2). Process mapping coordinated the components of the process into a structured framework and model (Levels 2–3). Computer programming and coding resulted in an automated, streamlined prototype tool that facilitated the standardised process (Levels 4–5). The next stages of development will be extrapolating lessons learnt into a large scale fully functioning software application that can be validated and marketed into a wide range of scenarios (Levels 6–9).

2.1 Conceptual Process Modelling

A conceptual model, in this study, is defined as a set of constructs, inferred by observable events, used to describe an event, object, or process. Thus, the identified propositions in a conceptual model are logical rather than epistemological. The chain of interdependencies was coordinated explicitly into a reverse-engineered systematic process that uses the IDEF3 structured diagramming modelling technique [22]. The IDEF3 method was selected due to its high descriptive power, which is considered appropriate for detailed processes that handle know-how knowledge. IDEF3 captures descriptions about sequences of activities, while also identifying critical decision points, or milestones, of the process from different perspectives [22, 23]. The main assumption of the systemic model is that it represents an open system, hence its functions (i.e. tasks) need to be descriptive rather that prescriptive (see Fig. 1). Nevertheless, information flows and ranges of values (i.e. rules) were prescribed.

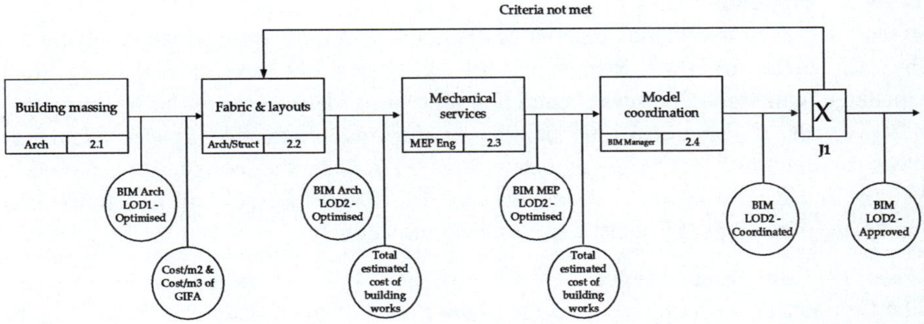

Fig. 1. IDEF3 decomposition of Stage 2 of design development

2.2 The Case Study – Proof of Concept

The research implemented a case study approach to validate the efficacy of the developed concept. Case study approach is recommended when an in depth exploration of a situation is required, with emphasis on understanding processes as they occur in their context [24]. The selected building was a BTR scheme situated in Northwest London. The design team followed the IDEF3 modelled process to create BIM models in Autodesk Revit. Three levels of design development were implemented (see Fig. 2).

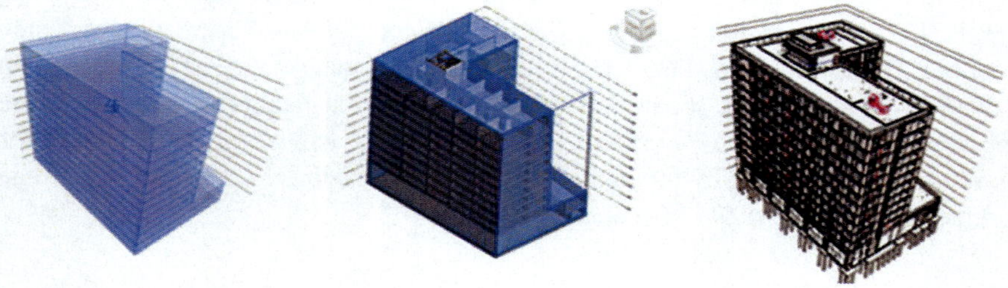

Fig. 2. Levels of model development (i.e. massing, functional units, and components)

Level 1 – Massing Model

The Proof of Concept model was developed in accordance with the IDEF3 definition using the Revit Conceptual Mass/Mass in Place tool. Mass Floors are activated automatically by defining Revit Levels in the BIM model. Roof level were excluded as this led to incorrect calculation of the Gross External Area (GEA). Furthermore, internal voids and double spaces were not considered at this stage. The number of apartment types and auxiliary spaces were calculated using a Unit Mix Tool. The Massing model contained a single Uniclass2015 code to determine the overall use of the building. In this case, Entity code Residential buildings (En_45_10) was selected. The Unit Mix Tool was aligned with more detailed SL (Spaces and Locations) codes for the different types of uses.

Level 2 - Functional Unit Model

In the Functional unit model, external walls, floors and roofs were added, applying the 'by face' option of Revit. Stair wells, lift and riser shafts were created using Shaft Openings with symbolic lines. Functional units were defined using the Rooms capability. Types of Functional unit include: Apartments, Circulation Spaces, Plant and other Auxiliary Spaces (Refuse and Bike Storage), Resident's Lounge, and Lobby. SL, EF, Ss, and Pr Uniclass2015 codes were used to specify Spaces and Locations, Elements and Functions, Systems, and Products respectively.

Level 3 – Component Model

The Component model represented the highest level of granularity and detail. It is the level where each product was specified as close to the as-built as possible. At pre-construction stage, the performance specifications were predicted. These values could be substituted with actual performance by performing post-occupancy evaluation studies and updating the digital twin accordingly. SL, EF, Ss, and Pr Uniclass2015 codes were used to specify Spaces and Locations, Elements and Functions, Systems, and Products respectively, as above. In addition to these, detailed specifications were selected for each product type using the naming of the National Building Specification (NBS) Product Data Templates (PDTs) [25].

3 Discussion of Research Findings

The research findings support the idea that BIM technologies can assist in the accurate estimation of WLC of an asset, from the early design stages, through the implementation of standardised information exchange procedures. Despite the potential, there remain many challenges to be faced before accomplishing fully automated, integrated, and interoperable data. This section discusses the challenges encountered for the standardisation and automation of the BIM-enabled WLC assessment process. It describes the principles that underlie BIM-enabled WLC estimation (Sect. 3.1) along with the development of a small-scale prototype application that validates the analytical predictions and tests the feasibility of the system in a relevant environment (Sect. 3.2).

3.1 Standardisation of the Design Process

The main challenge in the attempt to standardise the early stages of the design process is that the problem is not sufficiently defined yet [26, 27]. In addition to that, design problems are often idiosyncratic and difficult to understand [28]. Therefore, the process of design is highly iterative, and the experimentation and comparison of alternative solutions is necessary [29]. The IDEF3 model addresses this by incorporating design loops as part of the process, until certain conditions are satisfied. The tool has addressed this issue by reverse-engineering the process bottom-up by identifying key performance indicators in terms of functional performance. Standardisation of WLC information used the Uniclass2015 coding system [30] and data from several cost and carbon databases. For WLC assessment to become meaningful, material specifications, for example, k-values, and recycled content should become explicit for the calculations to be accurate. Otherwise, the range of possible costs is so broad that the committing to a decision cannot be made with confidence. For example, for aluminium double-glazed windows, the following parameters were considered: (i) thermal conductivity (U-values), (ii) sound transfer (dBs), (iii) fire rating, (iv) strength, (v) translucency, (vi) security, and (v) cleanability. Nevertheless, the level of information contained in BIM components is not comprehensive enough to enable the objects to be associated with WLC datasets. Key assumptions regarding several specifications (e.g. materials, properties) need to be made very early in the process.

3.2 Sources of Data - Data Availability

The accuracy of WLC substantially depends on the quality and availability of data from all phases of a building. It was found that historical databases contain several irregularities and ambiguities, and they do not offer neither the granularity nor the classification and clustering that is required to be used in an effective way. Datasets are not annotated consistently and the specifications and factors affecting WLC (such as intensity of use, environmental conditions) are not parametric. Moreover, annotations of datasets are not aligned with those in BIM entities. Therefore, direct alignment to model objects was not possible. Furthermore, accessibility to WLC information presented the biggest challenge, as it is either stored in static documents, or it is proprietary. Data from manufacturers, suppliers, and specialist subcontractors is also difficult to obtain due to concerns regarding liability and commercial competitiveness. Consequently, the level of detail of existing readily available datasets is not appropriate to be used in conjunction with BIM. It was inferred that collection of in-use data, which is fit-for-purpose, can be a viable solution however at present there is a paucity of longitudinal studies that collect data on the performance and reliability of systems. BIM-enabled WLC integration is also hindered by the fact that existing databases are static, are not open source, and contain insufficient and unproperly structured information. This is mainly due to failures of the industry to collect WLC information but also due to lack of understanding of what information is required. For a BIM-enabled WLC estimation process, data need to be structured and formatted in a way that communicates with existing systems and calculations tools to become useful.

3.3 System Architecture – Modules and Connections

The prototype software tool performs the following functions: (i) Extraction of area measurements, functional units, and component quantities out of BIM models of varying levels of design development (massing, concept, developed, and technical design); (ii) Uses WLC benchmarks from several cost and carbon databases to create data structures that combine New Rules of Measurement (NRM) codes with Uniclass2015 for varying levels of granularity; (iii) Performs WLC calculations using quantities extracted from the BIM model (through a Revit plugin combined with Excel spreadsheets); and (iv) Exports the result of calculations to Microsoft PowerBI for the creation of dashboards. It consists of the following modules, illustrated in Fig. 3.

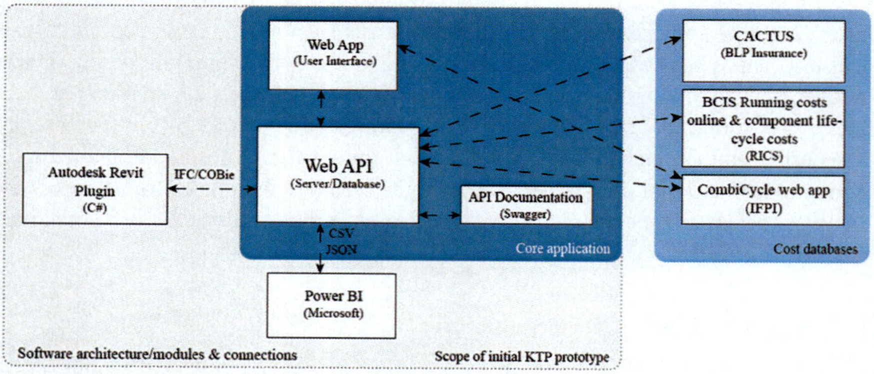

Fig. 3. Software architecture and connections

Autodesk Revit Plugin – Quantity Take-Off

As described in Sect. 3, quantity take-off for the Massing and Functional units' models (Levels 1 and 2) is the square meterage of useable areas. These quantities were then multiplied using benchmarks acquired by the Royal Institution of Chartered Surveyors (RICS) Running Cost Online subscription service. It needs to be stressed that, for the quantity extraction process to work, the BIM model needs to be built using a prescriptive method both for the geometric and non-geometric information. Apart from the example of the Roof and Rooms discussed above, the Uniclass2015 classifications of the components need to be specified in advanced as part of the Employer's Information Requirements (EIR). Several workarounds and bespoke solutions had to be developed for extracting the quantities of the architectural elements due to limitations of Revit. For example, Revit calculates the area of a window as a box, while the WLC calculations require the area of the opening. Inaccuracies in quantity extraction have also been noticed for the structural elements where there were intersections between columns and slabs. The solution that was implemented in this occasion was to use the "Join" tool so that the tool considers only one of the elements. This is a limitation of the process as well, since at early design stages it is not efficient to make these adjustments as they are time consuming. Another challenge was that the Mechanical, Electrical and Plumbing (MEP) BIM model was not comprehensive enough to be used as a system. For that reason, the proof of concept utilised parabolic curves for estimating cost, carbon, and energy consumption of extract ventilation systems for 50 years based on the ratio of wet areas to total area of the building, which differs between typologies. It is suggested that future work should incorporate the Industry Foundation Classes (IFC) data structure to overcome the inherent limitations of Revit, as it offers dedicated information containers for model components and information. Work is also needed to develop structured processes for model construction by designers.

Web Application – Server (Database and User Interface)
The web application contains the re-structured datasets of cost and carbon information. This data was pre-processed; adjusted for London and to represent totals for a 50-year calculation period. Some datasets were also annualised to be used for creating more detailed breakdown of cash flows. Discount factors were applied separately in PowerBI. All components were classified using Uniclass2015 codes and were linked to the BIM model through the identification of identical classifications. For the Proof of Concept, all component and database classifications were specified explicitly in advance. For future development, the software should be adjusted to find the closest match and present the alternative cost and carbon information available in the database. The database follows an object-oriented structure where components were defined by Pr (Product) codes and functional units (i.e. Revit Rooms) by SL (Space and Location) codes. Relationships between components and functional units are also built in. All entities have a list of properties attached to them and they are also defined by a Revision that is associated to stage of design development and a user of the system. The user interface of the web application is accessible via a browser. This solution was preferred to a desktop application as it offers better accessibility. Additionally, the web application is supplemented by Application Programming Interface (API) documentation developed using Swagger. This contains instructions on how to integrate the web application with external applications, offering the option to create connections to a multi-platform structure.

4 Conclusions and Future Work

To date, key decisions (relating to WLC) are often based on judgement rather than data, which leads to an increase in the level of risk and uncertainty. There is no comprehensive tool to assist designers commit to decisions in a timely manner and with confidence. The scope of the project was to integrate cost and carbon information related to maintenance and operation of a completed building throughout design delivery, from the early stages, as an integral part of the natural design development process. The Proof of Concept case study focused on the BTR housing sector, for whom accurate WLC can influence long term financial viability. Its main objectives were: (i) to investigate the emerging landscape of BTR housing, establish the need and opportunity for WLC assessment, and explore the capabilities of BIM to facilitate this purpose from the early design stages, and (ii) to develop and verify a structured standardised process for more accurate WLC estimation through BIM and assess the benefits of this approach. As a result, the project proved the potential of BIM to accelerate the process and reduce the tedious and repetitive tasks that are currently used to assess WLC. It demonstrated how BIM has the potential to facilitate more accurate prediction of WLC more efficiently in terms of time and effort involved so as to achieve quality assurance. As such, WLC can consider all significant and relevant initial and future costs and benefits of an asset, throughout its life cycle, while fulfilling environmental, and other, performance requirements. This holistic approach can result in high-performance buildings that are cost-effective, thus sustainable.

4.1 Challenges and Limitations

The project identified several issues that are barriers to BIM-enabled WLC estimation. These can be summarised as follows:

- Lack of a common standardised framework and method that provides appropriate definitions of information requirements and is aligned with BIM standards and protocols for delivering design information which is fit-for-purpose for WLC estimation.
- Alignment of modelled elements, annotations, and properties to information contained in cost databases. Models are constructed *ad hoc* at early design stages in order to produce quick 3D information and restructuring these later is time-consuming and generates inaccuracies.
- Availability, accuracy, reliability of cost and performance data as well as accessibility and structure of available data. WLC and performance data is not routinely collected from existing buildings. Consequently, data that exists is small-scale, fragmented, held in propriety formats and is not open source.
- Proprietary file formats of BIM authoring software are restricting the extraction of information from the models. Implementation of open standards has the potential to facilitate transparency and information exchange procedures.

4.2 Future Work

Based on the findings above, the next stage of the project will focus on:

- Facilitating standardised procedures for the collection, structuring, and analysis of cost and carbon data in order to comply with the granularity of the Uniclass2015 classifications. This approach will allow benchmarking at varying levels of detail for both areas and components, based on UK standards. Whilst government is mandating use of BIM, this is not extended to performance data, and this is vital to the accuracy and robustness of modelling tools.
- Development of structured and standardised approaches for model construction at design stages to enable increased granularity of data at later design stages. All levels of geometric development need to be populated with appropriate information that is useful for WLC assessment so as to enable association of objects to cost and carbon datasets.
- Developing a web-based Software-as-a-service (SaaS) application that links WLC calculations to BIM authoring tools for extracting quantities natively and performing calculations "on-the-fly" so as to streamline integrated cost and carbon assessment.
- Re-structuring datasets of existing historical databases that contain cost and carbon information by clustering and classification. These will be used to populate and maintain a dynamic WLC database and use it to identify patterns, dependent variables, and predict future performance.
- Testing the effectiveness of visualisation methods that assist the interpretation of WLC estimations and determining the ones that have the most significant effect on improving the final design product/solution.

References

1. Cabinet Office. Government Construction Strategy, May 2011. GOV.UK (2011)
2. HM Government: Construction 2025: strategy. https://www.gov.uk/government/publications/construction-2025-strategy
3. Farmer, M.: The Farmer Review of the UK Construction Labour Model (2016)
4. HM Government: Construction Sector Deal (2018)
5. BSI: ISO 15686-5:2017 - Buildings and Constructed Assets – Service Life Planning – Part 5: Life-Cycle Costing (2017)
6. White, G., Boyne, P.: Facilities management. In: Pittard, S., Sell, P. (eds.) BIM and Quantity Surveying. Routledge, London (2016)
7. BSRIA: BG 64/2016 - Building Performance Evaluation in Domestic Buildings – a BSRIA Guide (2016)
8. Sharpe, T., McGill, G., Gupta, R., Gregg, M., Mawditt, I.: Characteristics and Performance of MVHR Systems - A Meta Study of MVHR Systems Used in the Innovate UK Building Performance Evaluation Programme (2016)
9. RICS: RICS Professional Guidance: Life Cycle Costing, 1st edn. RICS (2016)

10. Zanni, M.A., Soetanto, R., Ruikar, K.: Towards a BIM-enabled sustainable building design process: roles, responsibilities, and requirements. Archit. Eng. Des. Manag. **13**(2), 101–129 (2017). https://doi.org/10.1080/17452007.2016.1213153
11. Opoku, A.: The application of whole life costing in the UK construction industry: benefits and barriers. Int. J. Archit. Eng. Constr. **2**(1), 35–42 (2013). https://doi.org/10.7492/IJAEC.2013.004
12. Eastman, C.C., Teicholz, P., Sacks, R., Liston, K.: BIM Handbook: A Guide to Building Information Modeling for Owners, Managers, Designers, Engineers and Contractors, 2nd edn. Wiley, Hoboken (2011)
13. Zanni, M., Sharpe, T., Lammers, P., Arnold, L., Pickard, J.: Developing a methodology for integration of whole life costs into BIM processes to assist design decision making. Buildings **9**(5), 114 (2019). https://doi.org/10.3390/buildings9050114
14. Kehily, D., Underwood, J.: Embedding life cycle costing in 5D BIM. J. Inf. Technol. Constr. **22**, 145–167 (2017)
15. HM Government: Transforming Infrastructure Performance (2017)
16. Centre for Digital Built Britain: Gemini Principles (2018)
17. Urban Land Institute (ULI): Build to Rent Best Practice Guide, 2nd edn. ULI UK Residential Council (2016)
18. Mayor of London: Homes for Londoners: Affordable Housing and Viability Supplementary Planning Guidance 2017 (2017)
19. Creswell, J.W.: Research Design: Qualitative, Quantitative, and Mixed Methods Approaches. Sage, Thousand Oaks (2009)
20. Dubois, A., Gadde, L.-E.: Systematic combining: an abductive approach to case research. J. Bus. Res. **55**(7), 553–560 (2002). https://doi.org/10.1016/S0148-2963(00)00195-8
21. Nuclear Decommissioning Authority: Guidance on Technology Readiness Levels (2014)
22. Mayer, R.J., Menzel, C.P., Painter, M.K., Dewitte, P.S., Blinn, T., Perakath, B.: Information integration for concurrent engineering (IICE) - IDEF3 process description capture method report, DTIC Document (1995)
23. Mayer, R.J.: IDEF family of methods for concurrent engineering and business re-engineering applications, College Station, Texas, USA (1992)
24. Yin, R.: Case Study Research: Design and Methods. SAGE Publications, Thousand Oaks (2013)
25. Jones, L.: Product Data Templates For Manufacturers. https://toolkit.thenbs.com/articles/pdts
26. Sebastian, R.: Managing Collaborative Design. Eburon Uitgeverij BV (2007)
27. Dorst, K., Cross, N.: Creativity in the design process: co-evolution of problem-solution. Des. Stud. **22**(5), 425–437 (2001)
28. Laseau, P.: Graphic Thinking for Architects & Designers. Wiley, Hoboken (2001)
29. Goldschmidt, G.: Linkography: Unfolding the Design Process. The MIT Press, Cambridge, London (2014). https://doi.org/10.1016/j.destud.2015.10.003
30. NBS: Uniclass 2015 - a universal classification system for the construction industry | NBS. https://www.thenbs.com/news/uniclass-2015-a-universal-classification-system-for-the-construction-industry. Accessed 05 Feb 2018

Value Diversity as a Driver for Renovation Design Support: A Clustering-Based Approach to Accelerate the Exploration of Design Space

Aliakbar Kamari(✉), Poul Henning Kirkegaard, and Carl Schultz

Department of Engineering, Aarhus University, 8000 Aarhus, Denmark
ak@eng.au.dk

Abstract. Sustainable building renovation design is a highly challenging and ever more pressing issue in the construction sector. Architects must search through a large number of potential designs while aiming to optimize a range of Key Performance Indicators (KPIs) that measure, for example, construction cost, energy efficiency, an occupant's sense of privacy, and so on.

Software tools for supporting this task often seek to automatically discover renovation solutions (or scenarios) through search procedures that optimize certain criteria, and deliver a small number of similar, optimal designs with no intermediate layers of abstraction, and no support in analyzing and exploring the diversity of the design space.

In contrast, we present a new approach that emphasizes the diversity of KPI outcome first, that combines both physics-based KPIs in addition to human-centered KPIs such as privacy, thermal comfort, and so on. The architect is presented with clusters of designs that have relatively similar KPI evaluations, and can then subsequently explore (i.e. fine-tune) their designs within each cluster via a semantic domain model of renovation, called NovaDM that we have previously developed. Thus, each cluster represents a qualitatively distinct combination of near-optimal KPI values, i.e. all lying close to the Pareto front, and exploring the design space within a cluster ensures that the design solution (or scenario) obtains the desired combination of values (e.g. medium construction cost, high privacy, etc.).

Keywords: Sustainable building renovation · Sampling · Clustering · Design space exploration

1 Introduction

Development of design scenarios for sustainable renovation of existing buildings is a profoundly challenging task. An important challenge for practitioners in this context is *complexity*, both in terms of climate change and the urgent need for a sustainable built environment, and in terms of its diversity of stakeholders [1]. According to the study in [2, 3], the construction community now has a better understanding and appreciation of the complexity of sustainability challenges, and has come to realize that **sustainable renovation** of existing buildings must include: a construction perspective, a social and economic perspective, as well as an eco-efficiency perspective. In this framework,

E. Toledo Santos and S. Scheer (Eds.): ICCCBE 2020, LNCE 98, pp. 207–227, 2021.
https://doi.org/10.1007/978-3-030-51295-8_17

architects (or the **design team**) must search through a large number of potential design scenarios while aiming to optimize a range of sustainability Key Performance Indicators - KPIs [4], that measure a range of different "soft" and "hard" KPIs [5], for example, investment cost, energy efficiency, an occupant's sense of privacy, and so on.

An effective way to cope with the explained complexity above is through the development and application of methodologies and methods that facilitate and support the decision-making process. Likewise, Jensen et al. [6] state that **in Denmark there is a lack of simple and holistic tools** to **support the involved decision-makers towards reaching informed decisions** (specifically) in the **early stages of building renovation** projects. Various Decision Support Systems (DSSs) and tools for building renovation already exist [7–9]. Forty-three decision support tools were considered in a literature review by the authors [10]. Some of the tools encompass the aspect of setting goals for sustainability, some focus explicitly on the discussion of objectives among the decision-makers, and the rest provide a fixed set of criteria but leaving the weighting of the criteria open to the decision-makers. However, the majority of the software tools in the design stage of renovation projects often seek to automatically discover renovation solutions (or scenarios) through search procedures that optimize certain criteria, and deliver a small number of similar, optimal designs **with no intermediate layers of abstraction**, and **no support in analyzing and exploring the diversity of the design space**.

Using "cluster analysis" we present a new approach that emphasizes the diversity of KPI outcome first, that combines both physics-based KPIs in addition to human-centered KPIs such as: degree of privacy, indoor thermal comfort, indoor air quality, and so on. The architect/designer is presented with clusters of designs that have relatively similar KPI evaluations, and can then subsequently explore (i.e. fine-tune) their designs within each cluster via a semantic domain model of renovation, called NovaDM [11] that we have previously developed. Thus, each cluster represents a qualitatively distinct combination of near-optimal KPI values, i.e. all lying close to the Pareto front, and exploring the design space within a cluster ensures that the design solution (or scenario) obtains the desired combination of values (i.e. medium construction cost, high privacy).

The research in this paper builds upon our previous work in developing a BIM-based DSS, named PARDIS, which generates optimal and sustainable renovation scenarios, and evaluates them for multiple KPIs tailored towards dwellings [12]. The key challenge is that the KPI evaluation of each design scenario requires the use of computationally expensive simulations, which severely limits the proportion of the design space that can be sampled beforehand to form clusters, particular in relation to the enormous size of the full design space. For example, the total number of renovation scenarios in the Danish residential dwelling case study that we use for empirical evaluation is in the order of 2.8×10^{26} scenarios. To this end, we investigate a range of sampling strategies via space mapping. Our key contributions are:

- We develop two complementary strategies for sampling the full renovation design space based on NovaDM: uniform random and *n*-ary aspect coverage
- We study and integrate two clustering techniques in a newer version of PARDIS, to clustering the KPI's simulation outcomes for given scenarios;

- We present a comprehensive set of results by evaluating our approach on four samples of renovation scenarios related to a single-family dwelling renovation project in Denmark.

2 PARDIS - Process integrAting Renovation DecIsion Support

PARDIS is a plug-in for Autodesk Revit that architects or engineers can use on an existing BIM-model to invent, animate, and assess the enormous number of potential renovation scenarios. The overall framework for the development of PARDIS including different phases and activates has been discussed in [12]. Figure 1 demonstrates the tool's user interface. PARDIS is being developed to support the renovation sector with the following features:

a) optimizing the selection process of renovation alternatives, and the generation of the renovation scenarios, through the use of specific renovation typologies and constraints,
b) evaluating standard sustainability Key Performance Indicators - KPIs (e.g. energy consumption, investment cost, indoor thermal comfort, air-quality etc.) over the generated renovation scenarios,
c) visualizing the generated renovation scenarios for ease of comprehension by the designers, and for enabling discussion between expert stakeholders (i.e. architects, engineers, contractors etc.) and non-expert stakeholders (i.e. building occupants, owners etc.) towards making an informed decision about the most appropriate and optimal renovation scenario in a short time frame.

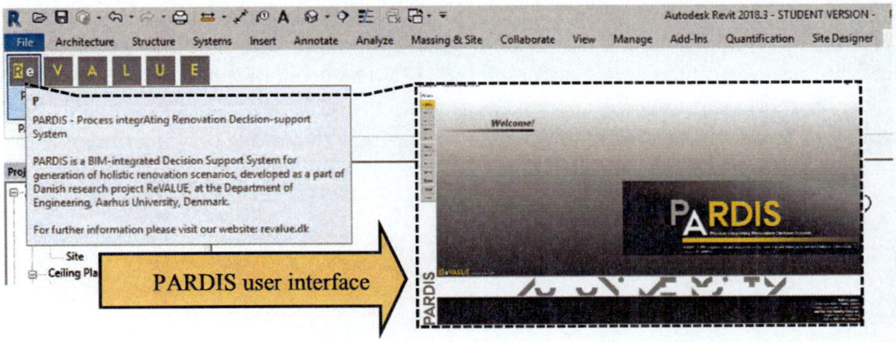

Fig. 1. The PARDIS front-end user interface (version 0.1.0) (A video of the PARDIS first prototype can be watched in the following link: https://www.youtube.com/watch?v=9uFbLQ9Z_H4)

Figure 2 illustrates the workflow of the scenario generation system that has been discussed and presented by the authors in [13, 14], on the development of a rapid constraint-based renovation design support framework through the application of a formal (logic-based) domain-specific renovation modeling language, named NovaDM [11]. First the design team specifies the (a) project-specific renovation alternatives in the form of a NovaDM action tree, (b) renovation scenario constraints, and (c) the BIM-model of the design to be renovated. The scenario generator parses the action tree into Answer Set Programming (ASP) facts and generates (optimal) scenarios that are consistent with the given constraints using the ASP reasoning engine Clingo [15].

The PARDIS system architecture enables different "KPI Simulator" components to be plugged in (Fig. 2), i.e. PARDIS itself does not restrict KPI evaluation to any particular KPI simulator, but allows different simulators to be employed. In this paper we are using the ICEbear KPI simulator [20].

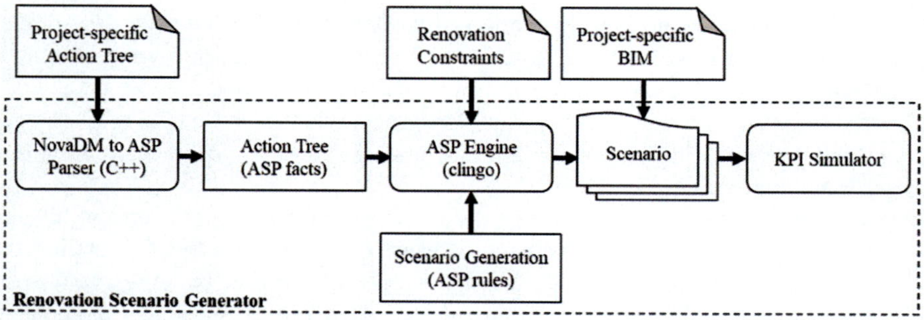

Fig. 2. Workflow of the renovation generation system

In addition, the scenarios are built on the basis of three different categories of **constraints**. The first category of constraints determines the renovation depths and level of interventions that illustrate the impact of different ambition levels concerning the European environment and economy [16, 17]. The depth of renovation refers to the extent and size of measures applied, as well as to the level of resulting energy and emissions reduction. It is used to generate scenarios with various depths of renovation, including Minor, Moderate, Deep, and Nearly Zero Energy Building – NZEB.

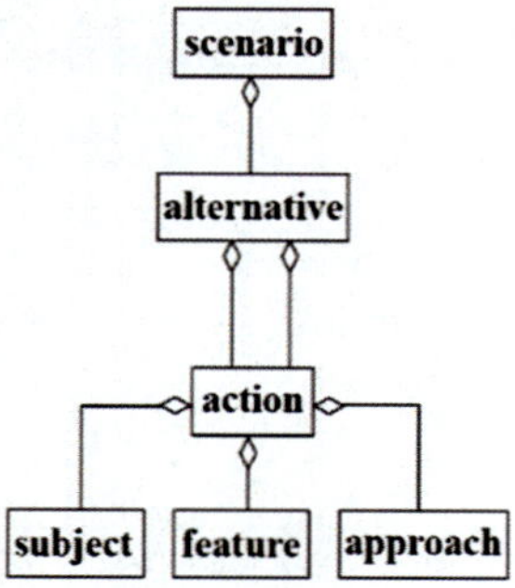

Fig. 3. The relationships between concepts in NovaDM

The second category of constraints determines the typologies of the existing building according to TABULA [18]. Likewise, the third category of constraints targets the architectural aspects and the appearance of the renovation scenario [19], especially related to the buildings' facades.

As illustrated in Fig. 3, the scenarios are structured through the use of various concepts including:

Renovation Subject: the object that is the target of renovation, e.g. windows, doors, etc.

Renovation Approach: the way that the subject will be renovated, e.g. repair, replace, remove, refurbish, or modify.

Renovation Feature: the aspect of the subject being renovated, e.g. window frame material. Features have types (e.g. material) and values (e.g. fiberglass, uPVC, etc.).

Renovation Aspect: subjects, approaches and features are collectively called **aspects** of renovation. This enables us to talk about subjects, approaches, and features together as a single (abstract) class of concepts, e.g. our *n*-ary aspect coverage sampling strategy refers to coverage of subjects, approaches and features.

Renovation Action: an instance of a subject, an approach, and a set of feature types and values, e.g. "window frames will be replaced with fiberglass material".

Renovation Alternative: the set of (mutually exclusive) actions for a given subject, e.g. all the ways that windows can be renovated according to the specified approaches and features.

Renovation Scenario: a set of actions such that each action belongs to a different alternative, e.g. one action for windows and one action for floors.

For each renovation project, the design team needs a way of specifying the particular subjects, features, and approaches that are available. For this, a formal **action tree language** for compactly specifying sets of actions is developed (adopted from [14]). Each action tree (see Fig. 4) corresponds to a renovation alternative. Traversing an action tree corresponds to generating actions in a scenario, starting from the root node. Each node assigns zero or more aspects (subject, feature or approach) to the action. Tree nodes can be either xor-nodes "–" (meaning that the current action must be built by traversing exactly one child), or and-nodes "+" (meaning that the current action must be built by traversing all children).

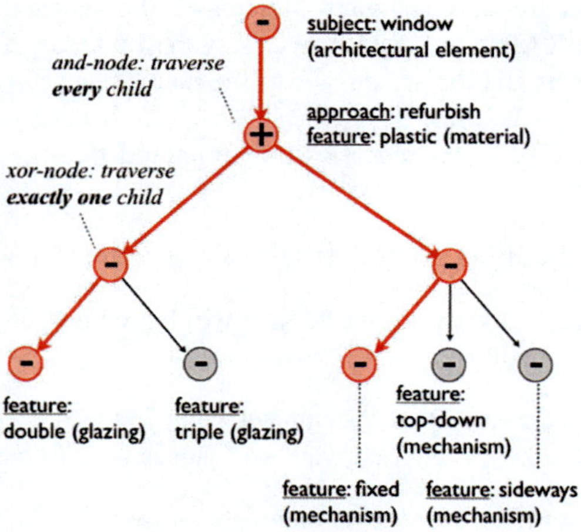

Fig. 4. A NovaDM action tree

An important shortcoming to the process demonstrated in Fig. 2 emerges when the user tries to generate a large number of scenarios (i.e. one million scenarios). What happens is that the scenarios are generated and simulated, then, the tool is forced to deliver and visualize a small number of (most likely) similar and optimal designs with no intermediate layers of abstraction, and no support in analyzing and exploring the diversity of the design space. More generally, it is significant challenge to give a designer a realistic, practical impression of the overall landscape of renovation design options given the vast size of the total number of renovation scenarios to choose from; as mentioned previously, the Danish residential dwelling project that we evaluate our approach on has 2.8×10^{26} possible renovation scenarios, and is quite typical of renovation projects.

Visualization dashboards have a limit to demonstrate a large number of simulation outcomes, and more importantly, it is not realistic for the user to be presented with a large number of generated scenarios and their simulation outcomes. This adds even more complexity to the decision-making process. The development of an appropriate solution to this issue is the main objective of the present paper.

3 Methodology

3.1 System Development Overview

Figure 5 illustrates the workflow for the integration of renovation scenario sampling and clustering. Rather than generating scenarios using the ASP engine as illustrated in Fig. 2, we instead generate scenarios using our two custom sampling strategies (described in detail in Sect. 3.2).

Each scenario is then evaluated according to user-specified KPIs using a simulator. For example, to calculate an estimate of “annual daylight exposure” (a useful KPI), a tool such as ICEbear runs a *365 day light simulation* based on the geometry of the building, its geographic location, properties and geometry of architectural components such as windows, etc. This is in contrast to solving a closed-form equation, which is typically much faster to compute than running a simulation; i.e. simulations typically are based on *iterative* calculations (e.g. nested *for* loops) rather than a single calculation. We refer to a simulation as being **high fidelity** if it seeks to make a relatively **precise estimate** of the KPI value, at the expensive of **increased computational runtime**.

In this paper we are particularly interested in working with such high fidelity simulations because they can generate very useful, precise information about a renovation option, but due to their relatively long runtimes they severely limit the number of scenarios that can be evaluated. This strongly motivates our need for *sampling* and *clustering* renovation scenarios. In this paper we use the ICEbear simulator [20] (like we have done so for PARADIS), but we are by no means restricted to ICEbear and, in principle, any high fidelity simulator can be "plugged in". We present ICEbear in detail in the following section.

The evaluated scenarios are then clustered as described in Sect. 3.3.

There are various benefits that can be achieved as the result of a holistic and sustainable renovation to higher energy performance standards. Many are tangible and possible to quantify, while others are less so and may be difficult to allocate a monetary value. These renovation objectives must be identified and targeted early in the design process while renovation scenarios are being developed. Regarding the full scope of this discussion, Kamari et al. [4] attempts to address this through a new holistic sustainability Value Map, which applies Checkland's Soft Systems Methodologies—SSM, together with Keeney's Value Focused Thinking—VFT. The Value Map consists of the three overall categories *Functionality*, *Accountability*, and *Feasibility*. Although the list can be extended in future, Table 1 summarizes the evaluating KPIs for generated renovation scenarios in this study according to the study in [4]. As mentioned previously, we use an extended version of the hourly dynamic simulation tool ICEbear [20] to simulate the generating renovation scenarios for the listed KPIs. ICEbear[1] is a tool that strives to facilitate design buildings for architects and engineers, evaluating the impact of the geometry on the indoor climate and energy demand at the same time. It is based upon algorithms for auto-generating hourly building performance data at a room level basis.

[1] Link to download: http://www.idbuild.dk/icebear.

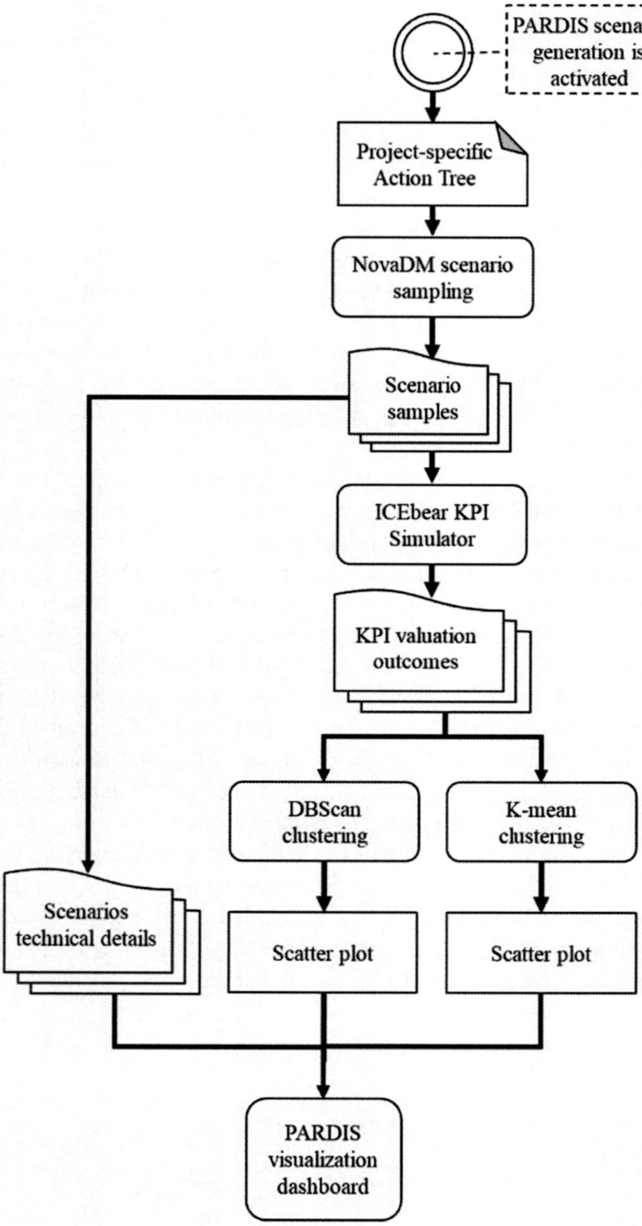

Fig. 5. Workflow for integrating clustering methods

Table 1. Outline for the selecting KPIs for evaluation of the generating renovation scenarios based on findings from [4]

KPIs	Evaluation [value measurement]
Energy consumption [21, 22]	Reduction of energy consumption for heating measured in kWh/m2/year [less better]
Discomfort hours above 27 and 28 (°C) [23, 24]	Number of hours [less better]
Indoor Air Quality – IAQ [22, 25]	% out of Class III according to EN 15251 [less better]
DF (daylight factor) [26]	0 < DF < 5 [bigger better]
Daylight requirements according to BR18 [22]	% > = 10 [bigger better]
View-out quality	% of openings area on façade regarding adjacent buildings [client dependent]
Degree of privacy	% of openings area on façade regarding adjacent buildings [client dependent]
Degree of Satisfaction [27]	% regarding indoor thermal comfort & air quality [bigger better]
Investment cost [28]	Price of the procurement in DKK (Danish Krone) [less better]

3.2 Sampling Strategies

Considering a complete renovation design space specified by a given action tree, our workflow samples a subset of this space used to approximate the full space. In this section we define two sampling strategies: **uniform random** sampling and ***n*-ary aspect coverage** sampling.

We have developed these particular two approaches with the intention of *complementarity*. With uniform random sampling we intend to generate a set of scenarios that spread out over the entire scenario space, irrespective of the structure of scenario space. By avoiding any "bias" in which scenarios are selected, we establish a tradeoff between (a) the *likelihood* of discovering high performing scenarios and (b) the time taken (i.e. the number of random scenarios generated): increasing computation time spent in generating scenarios increases the likelihood of discovering high performing scenarios.

With *n*-ary aspect coverage we intend to "map out" the structure of the scenario space in the sense that certain renovation options will outperform other renovation options with respect to certain KPI values, all other renovation aspects being left unchanged. We intend that *n*-ary aspect coverage will help designers identify promising *partial* renovation scenarios that make certain tradeoffs in the KPIs that they perform well in, and can then be further refined.

The empirical study presented in Sect. 4 sheds light on the efficacy of this combination of sampling strategies.

Uniform Random Sampling: Uniform random sampling has the property that any scenario defined by the set of action trees is equally likely to be chosen (with repeats).

Example: the action tree in Fig. 4 represents a total of six renovation scenarios. With uniform random sampling, each of these six scenarios must have an equal probability of (1/6) of being selected every time we "select" a new random scenario.

We have developed a recursive algorithm for uniform random sampling that incrementally constructs a scenario by stepping through each action tree:

- When an **And** node is visited, all children are visited.
- When an **Xor** node is visited, **exactly one** branch is chosen with probability in proportion to the sum of (partial) actions in each branch.

In order to calculate the probability of selecting a particular branch at an **Xor** node, we need to count the number of (partial) actions represented by each node. To illustrate this, Fig. 6 displays the example action tree from Fig. 4 now annotated with the number of scenarios represented by each subtree (bold numbers). The root node has 6 scenarios, as does its single child node. Next, the left branch has two options (either *double* or *triple* glazing) and the right branch has three options (either a *fixed*, *top-down* or *sideways* mechanism). Suppose the algorithm is currently visiting this left branch (Xor node with two glazing options in its children nodes); it needs to select exactly one branch to go down. The probability of picking either branch is (a) the number of partial actions of each child (b) divided by the sum of partial actions from all children. Each child has one action, and there are two children, so the probability of selecting either branch is ½ or 50%.

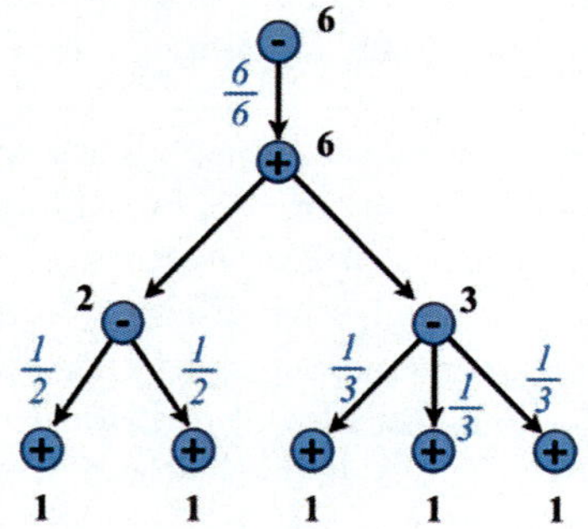

Fig. 6. Action tree nodes from the action tree in Fig. 4 annotated with partial actions count (bold numbers) and branch probabilities (blue fractions) in uniform random sampling.

Figure 7 illustrates the general case:

- Leaf nodes consist of one partial action
- **And** nodes consist of the *product* of children actions (because every partial action in one branch can be combined with every partial action in every other branch, i.e. via cross product), and when visited, every branch is selected with 100% probability when constructing a random scenario

– **Xor** nodes consist of the *sum* of children actions K (because every partial action in one branch is mutually exclusive with every other partial action in every other branch), and when visited, every branch is selected with probability in proportion to the number of partial actions of that branch divided by K

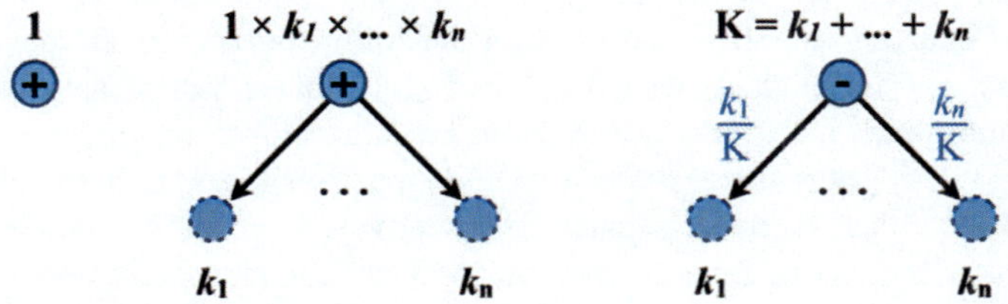

Fig. 7. Partial action counts and branch probabilities in for each type of node in an action tree.

***n*-ary Aspect Coverage Sampling:** Aspect coverage means that the set of sampled scenarios, together, ensures that certain combinations of aspects appear in at least one scenario. Unary aspect coverage (i.e. 1-ary) means that every aspect appears in at least one scenario. 2-ary coverage means that every pair of aspects (that are not mutually exclusive) appears in at least one scenario, and so on.

Example: The action tree in Fig. 4 consists of seven aspects (window subject, refurbish approach, two glazing features, and three mechanism features). 1-ary aspect coverage requires that every aspect appears in at least one scenario in the set of sampled scenarios. The following set satisfies this and thus would achieve 1-ary coverage:

Scenario 1: subject = window; approach = refurbish; glazing = double; mechanism = fixed
Scenario 2: subject = window; approach = refurbish; glazing = double; mechanism = top-down
Scenario 3: subject = window; approach = refurbish; glazing = double; mechanism = side-ways
Scenario 4: subject = window; approach = refurbish; glazing = triple; mechanism = fixed

2-ary coverage would require that every pair of aspects (that are not mutually exclusive) appear together in at least one scenario. The following set achieves this:

Scenario 1: subject = window; approach = refurbish; glazing = double; mechanism = fixed
Scenario 2: subject = window; approach = refurbish; glazing = double; mechanism = top-down
Scenario 3: subject = window; approach = refurbish; glazing = double; mechanism = side-ways
Scenario 4: subject = window; approach = refurbish; glazing = triple; mechanism = fixed

Scenario 5: subject = window; approach = refurbish; glazing = triple; mechanism = top-down
Scenario 6: subject = window; approach = refurbish; glazing = triple; mechanism = side-ways

Our algorithm creates an *n*-ary coverage sample set (for a given *n*) as follows. Scenarios are constructed by traversing the action tree in the usual way starting from the root and enumerating all scenarios, with the following additional constraint. The first time that an **Xor** node is visited, all its *t* children are visited as usual creating *t* partial actions, each belonging to a different scenario. However, the second time the **Xor** node is visited, only the first *n* of its children are visited. E.g. in the above 1-ary example, observe that the three mechanism aspect values are fully explored once (i.e. $t = 3$), and then the second time the mechanism **Xor** node is visited (i.e. when glazing value is triple) only the first mechanism value node is visited (i.e. $n = 1$).

We note that this algorithm does not compute a *minimal* set. That is, in the first example the minimal set is as follows:

Scenario 1: subject = window; approach = refurbish; glazing = double; mechanism = fixed
Scenario 2: subject = window; approach = refurbish; glazing = triple; mechanism = top-down
Scenario 3: subject = window; approach = refurbish; glazing = triple; mechanism = side-ways

We leave this topic for future research.

3.3 Cluster Analysis (Clustering)

Each generated scenario is evaluated according to *n* KPI numerical values. We can thus think of each scenario as an *n*-dimensional point in an *n*-dimensional space, i.e. a coordinate system with *n* orthogonal axes, where each axis represents a KPI value. Points that are "close together" in this "KPI space" have similar outcomes, even though the renovation actions that make up the scenarios themselves might be quite different.

We identify "clusters" of points (scenarios) in this KPI space to identify groups of scenarios that are similar in outcome. In general, clustering [29] is used to categorize a set of data in such a way that data in the same group (called a cluster) are more similar (in a well-defined sense, in our case based on KPI values) to each other than to those in other groups (clusters) [30]. Clustering can be achieved by various *clustering algorithms* that differ significantly in their interpretation of what constitutes a cluster and how to efficiently find them. Clustering algorithms can be categorized based on their cluster model [31]. In the present paper we apply two different clustering techniques due to making extra experiments as well as to compare the final outcomes. The clustering analysis presented are:

a) **Centroid-based clustering (or *k*-means clustering):**
One of the most common techniques for clustering numeric data is called the k-means algorithm [32]. *K*-means-type algorithms require the number of clusters – *k* – to be specified in advance. Clustering is then treated as an optimization

problem: find the k cluster centers and assign the objects to the nearest cluster center, such that the squared distances from the cluster are minimized [33].
In this work, we initialize the clusters using instances chosen at random from the data set (this is a common approach when using k-means). The data sets we used are composed solely of numeric features, and "distance" between points within the KPI value space is defined as the Euclidean distance. The final issue is how to choose k which is one possible drawback of the k-means method, i.e. that some "k" must be chosen *a priori*. Although a wrapper search can be used to locate the best value of k [34], we assign $k = 5$ in this paper to get an initial insight into how k-means performs with a relatively low number of clusters. We have adapted sample code in C# language is used from an online open-source[2].

b) **Density-based clustering:**
In density-based clustering [35], clusters are configured as areas of higher density than the remainder of the data set. Objects in these sparse areas - that are required to separate clusters - are usually considered to be noise and border points. The most popular density-based clustering method is DBSCAN [36]. It is one of the most common clustering algorithms as well as being the most cited in scientific literature. DBSCAN is a density-based clustering non-parametric algorithm: given a set of points in some space, it groups together points that are closely packed together (points with many nearby neighbors), marking as outliers points that lie alone in low-density regions (whose nearest neighbors are too far away). The key differences between DBSCAN and k-means are that: (1) DBSCAN clusters can be arbitrary, polygonal shapes whereas k-means clusters tend to be circular (owing to k-means cluster membership being the minimum straight-line distance to a cluster center); (2) DBSCAN does not define a fixed number of clusters k beforehand.

In our study we compared both clustering methods to determine which approach is more suitable in a renovation context.

4 Case Study

This section presents a case study where the system is empirically evaluated by applying it to real renovation tasks for a single-family dwelling located in Hjerting near the city of Esbjerg, Denmark. Figure 8 illustrates the Revit model of the dwelling that is developed based upon the obtained drawing material and pictures. The dwelling is typical in terms of its building typology, yellow brick finishes on the exterior walls, colorful building elements and dark color selections on the interior finishes. It has a heating floor area of approximately 92 m^2, basement excluded and uses district heating as its heat supply. Furthermore, it is seen that despite the regular and periodical appearance the building has strong architectural features. The family residents have decided to renovate the building, and their budget is about 200,000.00 DKK.

[2] Visual Studio Magazine: https://visualstudiomagazine.com/articles/2013/12/01/k-means-data-clustering-using-c.aspx.

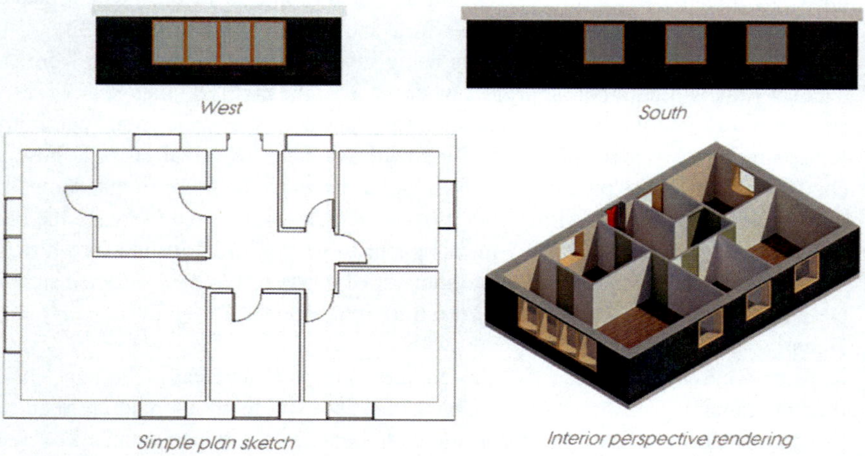

Fig. 8. The apartment's floor plan, west and south elevations in Revit

We generated 1-ary, 2-ary and 3-ary aspect coverage sample sets, and five uniform random samples (with 1000 scenarios per sample set). Figure 9 demonstrates the examples of data generated from sampling. Every scenario in a sample set was then evaluated using ICEbear and the values were normalized to a relative percentage of the range of values in the set (e.g. if the minimum and maximum scenario cost in a sample set is 100 k DKK and 200 k DKK, then a scenario with 150 k DKK cost has a relative value of 50%). While all eight KPIs are relevant in evaluating the impact of renovation, only two KPIs are actually modified by differing renovation options in this case study, i.e. construction cost and energy consumption.[3] Each sample set was then clustered using K-means (with k = 5) and DBSCAN.

Table 2 presents the statistics for the four sampling strategies[4]. Firstly, we observe that approximately between 2 to 3 scenarios can be evaluated per second using ICE-bear, which gives us a benchmark for the size of sample sets that are practical to evaluate:

- hundreds of scenarios takes between 1–5 min;
- a few thousand of scenarios takes between 5–10 min;
- from five to seven thousand scenarios takes between 10–30 min.

[3] This is because the renovation options do not alter the physical structure of the building. As part of future research we are currently developing an efficient approach for also incorporating spatial changes into renovation design.

[4] Experiments were run on an Intel Xeon(R) CPU E5-2620 v4 desktop computer running Windows 10, with 64 GB RAM and 2.10 GHz 2.10 GHz (2) processors.

Fig. 9. The examples of data generated from sampling

It is thus feasible to evaluate sample sizes up to around 100 k by leaving the system to run over night, and sample sizes of one million by leaving the system to run for a week.

Secondly, we observe that *n*-ary coverage sample sizes grow quickly as *n* increases, with 3-ary being the largest practical set that can be evaluated in a reasonable amount of time without leaving the system to run overnight.

Thirdly, scenario sample generation and clustering time for both k-means and DBSCAN (not shown) was relatively fast compared to evaluation time, taking less than a few seconds in all cases, and is thus not the bottleneck in the workflow of scenario sampling and evaluation.

Table 2. Number of scenarios generated, and total and average time to evaluate the KPI values of scenarios with different sampling strategies. Uniform random statistics are an average (mean) of five uniform random sets.

Sample strategy	Number of scenarios generated	Total time to evaluate KPIs of all scenarios using ICEbear (seconds, 1dp)	Mean time to evaluate KPIs per scenario using ICEbear (seconds, 2dp)
1-ary coverage	166	70.0	0.42
2-ary coverage	1657	500.0	0.30
3-ary coverage	7090	2011.0	0.28
Uniform random	1000	282.2	0.28

Figure 10 illustrates 8 scatterplots (only showing one of the five uniform random samples). The left column plots are the k-means clusters, the right column plots are the DBSCAN clusters. The plots in the first three rows are 1-, 2-, and 3-ary aspect coverage sample sets (respectively) and the last row is one of the uniform random sample sets. Different symbols and colors are used to represent different clusters. The smaller the KPIs values are (cost and energy consumption), the higher performing the scenario is, i.e. in the context of optimization, this could be framed as a *minimization* task. Thus, the Pareto front pushes towards the lower left corner of the scatterplots.

We note that the *uniform distribution* in our random sampling method refers to the *scenario* space and **not** the *KPI* space, i.e. every scenario that can be generated is equally likely to be selected. This method is **not** expected to generate scenarios that cover the KPI space uniformly; we instead expect that the density of scenarios in the very high (and very low) performing regions of the KPI space, near the pareto front, is much lower than in the central region. We can observe this pattern in the bottom row of Fig. 10, where the central area appears to be more densely populated than the edges of the circular region.

Comparing Coverage and Uniform Random Sampling: The results show that both sampling strategies are highly complementary. Aspect coverage tends to produce relatively high performing scenarios that approach the Pareto front, with a relatively small number of scenarios sampled. Uniform random scenarios instead cover the whole KPI space with a higher density in the central region and decreasing density away from the center.

Uniform random sampling produced the overall highest performing scenarios (i.e. with both minimum values in both cost and energy consumption), but as a percentage of the sampled set (i.e. 1000 scenarios) this was a relatively very small proportion, i.e. the average scenario performance was significantly lower than in 1-ary and 2-ary coverage.

The three coverage strategies show a trend in that increasing coverage class (i.e. increasing the n in n-ary) produces a sample set that approaches the Pareto front and “fills out” the corresponding clusters. The trend also suggests diminishing returns with increasing coverage class: 1-ary coverage already identifies some key clusters, 2-ary and 3-ary coverage expand these boundaries further towards the Pareto front and discovers some intermediate clusters. This suggests that 4-ary coverage and higher is not expected to reveal any new and drastically unique clusters, but would take a significantly longer amount of time to complete scenario KPI evaluation.

Thus our results suggest the following overall sampling strategy: coverage aspects up to 3-ary coverage are useful to identify a set of high performing clusters. This also provides a basis for approximating the Pareto front. Uniform random sampling may then be used to generate between 1000 and 7000 scenarios (depending on the computation time that the user is prepared to spend) with the aim of discovering very high performing scenarios that dominate scenarios in the coverage sets.

Comparing K-means and DBSCAN: DBSCAN produces significantly more intelligible clusters in the three coverage sample sets compared to *k*-means, where many clusters significantly. In the uniform random case, k-means separates the scenarios into equivalently sized vertical bands, which, while somewhat arbitrary, is more useful than DBSCAN placing all scenarios into the same cluster.

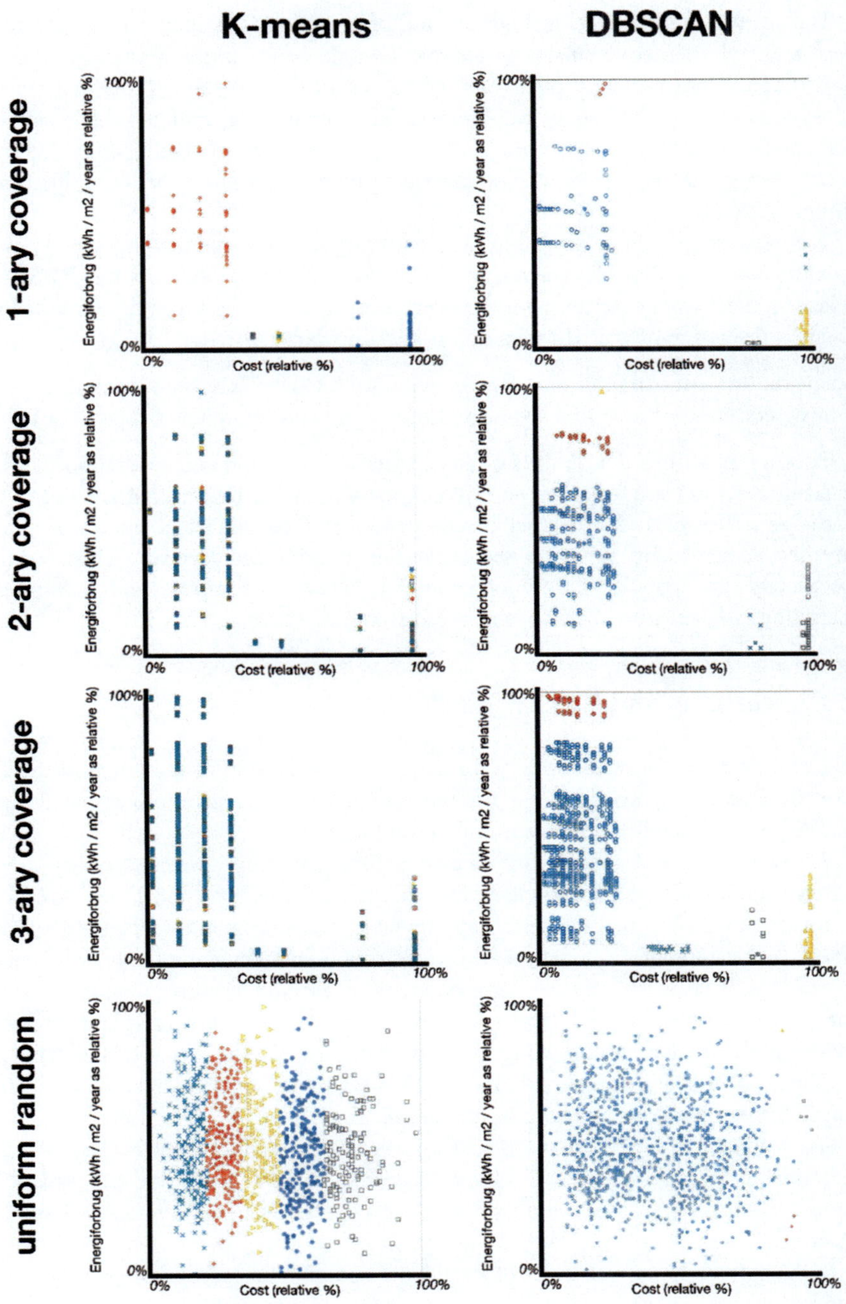

Fig. 10. Eight scatterplots presenting KPI evaluation of scenarios and generated clusters, across four sampling scenarios (1-, 2-, 3-ary aspect coverage and uniform random) and two clustering algorithms (K-means and DBSCAN).

The results suggest that, in both the uniform random case and coverage sample cases, a useful alternative clustering approach would be to simply uniformly partition the KPI space into n-dimensional cells, where each cell represents a different cluster. A partition of the full KPI space would be useful given that the random samples show a wide spread across the range of the KPI space, and a series of smaller cells could be used closer to the Pareto front to distinguish subtle differences between high performing scenarios.

The next stage in our workflow is identifying and communicating the types of renovation actions that characterize each discovered cluster. E.g. in the DBSCAN clusters in the 1-ary coverage case, the user needs to know the types of actions in the scenarios in the red cluster (for example). In particular:

- actions that are common to scenarios within the cluster (cluster cohesion)
- those common actions that are also rare in other clusters (cluster separation)

This set of actions that defines a cluster (via the cohesion and separation metrics) are used to further refine and explore scenarios within the cluster that are not in the sample set. We leave the detailed investigation of this next stage as future work; currently we are using cohesion and separation metrics and Decision Tree learning (specifically the popular C4.5 algorithm) to discover compact and informative descriptions of scenarios that characterize discovered clusters.

5 Conclusions and Future Work

We have developed a methodology for sampling and clustering scenarios in the full renovation design space, in order for renovation designers to explore real designs alternatives in a value-drive, outcome-focused manner.

We developed and empirically evaluated four sampling strategies (*n*-ary aspect coverage) and investigated two clustering approaches (K-means and DBSCAN) on a real renovation project of a residential apartment block in Denmark. Our results suggest that the four sampling strategies should be used in a complementary way, and can be gracefully scaled up to generate larger sample sizes with a correspondingly steady increase (with diminishing returns) in design space "decision-making information" revealed in the clusters, according to the amount of time that the user is prepared to take on scenario KPI evaluation. Approximately 2–3 scenarios can be evaluated per second using ICEBear, giving a benchmark for sample sizes.

Our results also suggest that DBSCAN is significantly more effective in producing intelligible clusters compared to k-means. Moreover, a simple uniform cellular partition of the KPI space may produce a useful set of clusters, and this approach should be explored.

In the current study, we have not included renovation decisions that alter the actual physical structure of the building, which is the focus of our current research efforts. The primary challenge is efficiently pruning combinations of physical alternations that are geometrically or spatially invalid [37–41], e.g. the dimension of two windows growing so far that they overlap. Moreover, the values of six of our primary KPIs only change when the physical structure of the building changes. Thus, in the current study we only evaluated two KPIs, which can be easily illustrated on a 2D scatterplot. Moving up to eight KPIs raises the issue of how to visualize the clusters that have been discovered (i.e. in an 8-dimensional KPI space) – for this we are currently using Principle Component Analysis (PCA) [42] as a means to incrementally reduce the dimensions from 8D to 2D/3D for visualization. A detailed account of the efficacy of PCA in this way is thus future research.

Acknowledgements. The authors of the paper would like to show their gratitude to the Danish Innovation Foundation for financial support through the ReVALUE[5] research project.

References

1. Kamari, A., Jensen, S.R., Corrao, R., Kirkegaard, P.H.: A holistic multi-methodology for sustainable renovation. Int. J. Strateg. Property Manag. **23**(1), 50–64 (2019)
2. Kamari, A., Corrao, R., Petersen, S., Kirkegaard, P.H.: Sustainable renovation framework: introducing three levels of integrated design process implementation and evaluation. In: PLEA 2017 Conference, Edinburgh, UK, pp. 748–755 (2017)
3. Kamari, A., Jensen, S.R., Corrao, R., Kirkegaard, P.H.: Towards a holistic methodology in sustainable retrofitting: theory, implementation and application. In: WSBE 2017 (World Sustainable Built Environment) Conference, Hong Kong, China, pp. 702–708 (2017)
4. Kamari, A., Corrao, R., Kirkegaard, P.H.: Sustainability focused decision-making in building renovation. Int. J. Sustain. Built Environ. **6**(2), 330–350 (2017)
5. Kamari, A., Corrao, R., Petersen, S., Kirkegaard, P.H.: Tectonic Sustainable Building Design for the development of renovation scenarios – Analysis of ten European renovation research projects. In: SER4SE 2018 (seismic and Energy Renovation for Sustainable Cities) Conference, Catania, Italy, pp. 645–656 (2018)
6. Jensen, P.A., Maslesa, E.: Value based building renovation – a tool for decision making and evaluation. Build. Environ. **92**, 1–9 (2015)
7. Ferreira, J., Pinheiro, M.D., Brito, J.D.: Refurbishment decision support tools: a review from a Portuguese user's perspective. Constr. Build. Mater. **49**, 425–447 (2013)
8. Kamari, A., Jensen, S., Christensen, M.L., Petersen, S., Kirkegaard, P.H.: A hybrid Decision Support System (DSS) for generation of holistic renovation scenarios—case of energy consumption, investment cost, and thermal indoor comfort. Sustainability **10**(4), 1255 (2018)
9. Kamari, A., Laustsen, C., Petersen, S., Kirkegaard, P.H.: A BIM-based decision support system for the evaluation of holistic renovation scenarios. J. Inf. Technol. Constr. **23**(1), 354–380 (2018)
10. Nielsen, A.N., Jensen, R.L., Larsen, T.S., Nissen, S.B.: Early stage decision support for sustainable building renovation: a review. Build. Environ. **103**, 165–181 (2016)
11. Kamari, A., Schultz, C., Kirkegaard, P.H.: NovaDM: towards a formal, unified renovation domain model for the generation of holistic renovation scenarios. In: ECPPM 2018 (12th European Conference on Product & Process Modelling) Conference, Copenhagen, Denmark, pp. 197–205 (2018)
12. Kamari, A., Schultz, C., Kirkegaard, P.H.: Towards a BIM-based Decision Support System for rapid generation and evaluation of holistic renovation scenarios. In: CIBW78 2019, pp. 244–254. Northumbria University, Newcastle (2019)

[5] Website: www.revalue.dk.

13. Kamari, A., Schultz, C., Kirkegaard, P.H.: Unleashing the diversity of conceptual building renovation design: integrating high-fidelity simulation with rapid constraint-based scenario generation. In: SimAUD 2019 (10th Annual Symposium on Simulation for Architecture and Urban Design) Conference, Atlanta, USA, pp. 29–36 (2019)
14. Kamari, A., Schultz, C., Kirkegaard, P.H.: Constraint-based renovation design support through the renovation domain model. Autom. Constr. **104**, 265–280 (2019)
15. Gebser, M., Kaminski, R., Kaufmann, B., Ostrowski, M., Schaub, T., Wanko, P.: Theory solving made easy with clingo 5. In: OASIcs-OpenAccess Series, Informatics 52, Schloss Dagstuhl-Leibniz-Zentrum fuer Informatik (2016)
16. TECNALIA: Intervention criteria & packaged solutions for buildings renovation towards a NZEBR (2015). http://www.nezer-project.eu/download/18.343dc99d14e8bb0f58bb3b/1440579936965/NeZeR_D2_3_NZEBR%20criteria.pdf
17. EU [European Union]: Boosting Building Renovation: What potential and value for Europe? (2016). www.europarl.europa.eu/RegData/etudes/STUD/…/IPOL_STU(2016)587326_EN.pdf
18. TABULA: Typology Approach for Building Stock Energy Assessment (2015). http://episcope.eu/building-typology
19. Boeri, A., Antonin, E., Gaspari, J., Longo, D.: Energy Design Strategies for Retrofitting: Methodology, Technologies, Renovation Options and Applications. WIT Press, Southampton (2014)
20. Purup, P.B., Petersen, S.: Rapid simulation of various types of HVAC systems in the early design stage. Energy Procedia **122**, 469–474 (2017)
21. Danish Energy Agency: 'Energimærkning af huse', Energimærkning boliger (2017). https://sparenergi.dk/forbruger/boligen/energimaerkning-boliger/huse
22. Danish Building Research Institute (2017), 'Be18'. https://be18.sbi.dk/be/
23. Dansk Standard: DS/EN 15251 Input Parameters for Design and Assessment of Energy Performance of Buildings – Addressing Indoor air Quality, Thermal Environment, Lighting and Acoustics (2007). https://webshop.ds.dk/en-gb/standard/ds-en-152512007
24. Dansk Standard: DS/EN ISO 7730 Ergonomi inden for termisk miljø – Analytisk bestemmelse og fortolkning af termisk komfort ved beregning af PMV- og PPD-indekser og lokale termisk komfortkriterier (2006)
25. Dansk Standard: DS 447 Ventilation i bygninger - Makaniske, naturlige og hybride ventilationssystemerDeursch, R. (2011). BIM and Integrated Design. Hoboken: Wiley & Sons (2013)
26. VELUX: Architecture for wellbeing and health|the daylight site, pp. 1–19 (2016). http://thedaylightsite.com/architecture-for-well-being-andhealth/. Accessed February 2018
27. Norback, D., Lampa, E., Engvall, K.: Asthma, allergy and eczema among adults in multifamily houses in Stockholm (3-HEStudy) - associations with building characteristics, home environment and energy use for heating, PLoS One **9**(12) (2014)
28. Molio: Molio Price data (2016). https://molio.dk/molio-prisdata/prisdata-footer/brug-m
29. Jain, A.K., Dubes, R.C.: Algorithms for Clustering Data. Prentice Hall, New Jersey (1988)
30. Everitt, B.: Cluster Analysis. Wiley, Chichester (2011)
31. Murty, M.N., Jain, A.K., Flynn, P.J.: Data clustering: a review. ACM Comput. Surv. **31**(3), 264–323 (1999)
32. MacQueen, J.B.: Some methods for classification and analysis of multivariate observations. In: Fifth Symposium on Math, Statistics, and Probability, pp. 281–297. University of California Press, Berkeley (1967)
33. Lozano, J.A., Pena, J.M., Larranaga, P.: An empirical comparison of four initialization methods for the k-means algorithm. Pattern Recogn. Lett. **20**, 1027–1040 (1999)

34. Milligan, G.W., Cooper, M.C.: An examination of procedures for determining the number of clusters in a data set. Psychometrika **50**, 159–179 (1985)
35. Kriegel, H.P., Kröger, P., Sander, J., Zimek, A.: Density-based clustering. WIREs Data Mining Knowl. Discovery **1**(3), 231–240 (2011)
36. Ester, M., Kriegel, H.P., Sander, J., Xu, X.: A density-based algorithm for discovering clusters in large spatial databases with noise. In: Second International Conference on Knowledge Discovery and Data Mining (KDD 1996), pp. 226–231. AAAI Press (1996)
37. Wałęga, P.A., Schultz, C., Bhatt, M.: Non-monotonic spatial reasoning with answer set programming modulo theories. Theory Pract. Logic Program. **17**(2), 205–225 (2017)
38. Wałęga, P.A., Bhatt, M., Schultz, C.: ASPMT (QS): non-monotonic spatial reasoning with answer set programming modulo theories. In: International Conference on Logic Programming and Nonmonotonic Reasoning, pp. 488–501. Springer, Cham (2015)
39. Bhatt, M., Lee, J.H., Schultz, C.: CLP (QS): a declarative spatial reasoning framework. In: International Conference on Spatial Information Theory, pp. 210–230. Springer, Heidelberg (2011)
40. Schultz, C., Bhatt, M.: Spatial symmetry driven pruning strategies for efficient declarative spatial reasoning. In: International Conference on Spatial Information Theory, pp. 331–353. Springer, Cham (2015)
41. Schultz, C., Bhatt, M., Borrmann, A.: Bridging qualitative spatial constraints and feature-based parametric modelling: expressing visibility and movement constraints. Adv. Eng. Inform. **31**, 2–17 (2017)
42. Pearson, K.: On lines and planes of closest fit to systems of points in space. Phil. Mag. **2**(11), 559–572 (1901)

Collaborative Workflows and Version Control Through Open-Source and Distributed Common Data Environment

Paul Poinet(✉), Dimitrie Stefanescu, and Eleni Papadonikolaki

Bartlett School of Construction and Project Management, University College London, London WC1E 7HB, UK
p.poinet@ucl.ac.uk

Abstract. Collaboration in architectural design becomes an increasingly complex task involving various actors working distributed in different locations. This complexity is even more hindered by the fact that the various actors involved in a project operate on different software environments and need to access accurate and up to date data at any time. Consequently, managing and keeping track of design changes throughout the workflow still remains a challenge for all actors involved in the design. This is a review paper that presents the state of the art in advanced collaborative design workflows, both in academia and industry, and introduces Speckle, a distributed Common Data Environment (CDE) and open-source data platform for Architecture, Engineering and Construction (AEC), as well as its version control capabilities.

Keywords: Collaborative workflows · Distributed common data environment · Version control

1 Introduction

Current modelling practices in the Architecture, Engineering and Construction (AEC) sector involve the sharing of large building models, through entire Revit or IFC (Industry Foundation Classes) files. Such practices and ways of sharing information are challenging for coordination and usually create unnecessarily large data payloads: *"Because the IFC payload normally includes a high ratio of such definitions, there is the need for specific investigation."* [1]. Therefore, tracking changes throughout the design process becomes very problematic, creating inefficiencies for design communication, as each party needs to save and send back the complete modified model if any change needs to be communicated. These existing frictions in the design process have been criticized and foreseen by Peters [2], who argues for the need to shift from centralized, file-based processes to a more decentralized design process within which many different software environments need to be connected: *"Traditionally, CAD and BIM systems have been monolithic applications, but increasingly there is pressure to connect. The promise of a single piece of software to carry out all design tasks and be used from conception through to operation is still not realised, and may not be the best*

E. Toledo Santos and S. Scheer (Eds.): ICCCBE 2020, LNCE 98, pp. 228–247, 2021.
https://doi.org/10.1007/978-3-030-51295-8_18

strategy anyway. [...] now many pieces of software with many different commands and sub-programs are used, and these form complex design processes.". Facing such challenges, existing efforts conducted in academia through the Linked Building Data (LBD) community group have been focusing on the use of the Semantic Web standards, as an open, decentralized alternative to the existing centralized design processes of storing and sharing data. Parallel efforts conducted in industry have been focusing on the development of version controlled technological platforms (such as 3D Repo [3] and Konstru [4]) aiming at collecting data from different existing Computer-Aided Design (CAD) and Building Information Modelling (BIM) software environments, allowing users to manage change and access the complete history of their data models uploaded online.

Considering the above, this paper reviews the state of the art in emerging alternative modelling practices both in academia and in industry, and introduces Speckle, an open-source Common Data Environment (CDE) for AEC, aiming at facilitating collaborative workflows. These new ways of working aim at enabling clearer coordination, minimizing data payloads by transferring only the design changes, and tracking clear authorship and history throughout the design process.

The paper is divided into seven sections. After introducing the current challenges and posing the problem statement by raising the lack of collaborative frameworks in the context of computational modelling in AEC, in this first section, the second describes the state of the art in existing innovative modelling practices (developed both in academia and in industry), acting as alternatives to the more traditional, centralized and file-based BIM approaches. The section also highlights the converging directions taken in both parties. Subsequently, the third section introduces Speckle and discusses both the technological similarities and differences between this platform and the use of the more traditional Semantic Web standards. The fourth section focuses on version control systems and change management strategies for the AEC sector. As those are fairly new concepts in this domain, the section firstly looks at the existing version control systems used in software engineering and introduces existing version controlled technological platforms for AEC practices. In the same section, the existing version control capabilities available in Speckle are described. The fifth section describes a design to fabrication workshop conducted at the Centro de Estudios Superiores de Diseño de Monterrey (CEDIM), during which the Speckle platform has been deployed throughout the design process. Then, the sixth section discusses the current efforts in the development of open-standards in both academia and industry, and discusses the limitations of Speckle as well as future work that still needs to be undertaken. Finally, the seventh section concludes the present research paper.

2 Existing Alternatives to the Centralized BIM Approach

2.1 Existing Research in Academia

Scholars actively propose standards and develop solutions to move from the traditional, centralized file-based approach to a more decentralized, web-based paradigm. The Linked Building Data (LBD) Community Group, part of the World Wide Web

Consortium (W3C), proposes alignment with the existing technological standards from the Semantic Web. Such research efforts led to tangible implementations and web interfaces enabling users to collaborate through Common Data Environments (CDEs) [5] or explore building data through RDF (Resource Description Framework) queries [6]. Other research efforts led to the development of *Drumbeat*, a Representational State Transfer Application Programming Interface (REST API): *Drumbeat* is *"a Web of Building Data concept and software for publishing and linking BIM information on the Web in construction projects and facility management."* [7]. Ultimately, the purpose of such REST API is to make use of the existing Semantic Web standards in order to connect to multiple CAD and BIM software platforms [8].

Apart from following the Semantic Web technology stack set by the World Wide Web Consortium, the AEC has independently utilized its Research & Development (R&D) efforts to create specific custom in-house interfaces which solve in the short-term local workflow frictions, interoperability issues and other technological problems. The next section describes some of these efforts.

2.2 Custom Modelling Practices and Collaboration Workflows in Industry

Various AEC firms, ranging from architecture and engineering practices to consultancies and construction firms, have been independently developing their own in-house custom processes to leverage and deliver fabrication data to external trades [9]. These self-driven ways of improving workflows can take different shapes, depending on each company's specific activities. The current section describes three initiatives (1) within the engineering practice of BuroHappold Engineering through the open-source Building Habitat object Model (BHoM) platform [10], (2) within the consultancy practice Proving Ground through the development of custom interoperability workflows for architectural design practices such as Conveyor [11], and (3) within the software company Robert McNeel & Associates through the Rhino.Inside technologies [12].

Engineering Practice: BuroHappold Engineering. As a very large engineering company occupying 23 offices around the world and bringing together more than 1800 employees, BuroHappold Engineering faces both internal and external interoperability challenges. Because the company operates within different sectors of the AEC, BuroHappold employs engineers specialized in multiple fields; structural engineers, MEP (Mechanical, Electrical, and Plumbing) engineers, façade consultants and architects are very often working altogether on the same architectural projects. Since each specialization uses specific software for unique purposes, extra work-load is generated to communicate and translate data to other specialists who operate within another set of software platforms. To tackle such issues, the BuroHappold Engineering's Computational Collective has undertaken for the past few years an R&D agenda which led to the co-creation of the open-source Buildings and Habitats object Model (BHoM) platform [10], enabling seamless data exchange across the different used software packages. BHoM has been designed as a hybrid model for code architecture, integrating a number of concepts from across existing languages/platforms. Specifically, BHoM has a data structure and data manipulation strategy that is directly compatible with both visual flow-based programming and text-based imperative code. Practically, BHoM offers a neutral schema to define design objects that can be converted to and from various software platforms.

Consulting for Architectural Design Practices: Proving Ground. Proving Ground is a small-sized and leading digital design consultancy practice that delivers custom interoperability workflows and tools for diverse AEC clients, enabling *"digital transformation with creative data-driven solutions to the building industry"* [13]. Nathan Miller, Chief Executive Officer of Proving Ground, states that *"[...] inventing customized workflows are a necessary part of the design process and have a profound impact on the delivery of the architecture."* [14]. Proving Ground always tries to abstract and generalize these specific workflows, from which are further developed open-source or commercial plug-ins that could serve more generic purposes and be reused for future collaborations.

Latest R&D by Proving Ground focused on mesh-based workflows for handling complex geometry in Building Information Modelling (BIM), aiming at reconciling parametric modelling processes with BIM-related software. Starting from the initial observation that friction still exists between the two – *"[...] BIM's inadequacy in supporting these data structures produces both friction and waste during production, including painful remodeling processes and incomplete or low-fidelity documentation for complex geometries. [...] a time-consuming process that directly contradicts BIM's fundamental utility for project delivery: to simultaneously manage design geometry across multiple, interdependent, multi-media and on-demand representations."* [15] – Proving Ground developed a custom workflow based on a communication pipeline between the Application Programming Interfaces (APIs) of Rhino3D and Revit, to process complex mesh-based geometries and translate them into clean IFC objects through Revit families, enabling the end-user to *"automate mesh importing by directly accessing geometry contained within a Rhino file."* This feature allows users to *"[...] both control mesh edge visibility and then parameterize them as Revit families."* [15]. Such custom workflow led to the development of Conveyor, enabling end-users to seamlessly share complex geometries between Rhino3D and Revit.

Research and Development in the Software Industry. Realizing the needs manifested by both engineers (e.g. BuroHappold Engineering) and architects (e.g. Proving Ground), Robert McNeel & Associates has developed Rhino.Inside® [12], an open source Rhino Work in Progress (WIP) project which allows Rhino3D [16] and Grasshopper3D [17] to run inside other 64-bit Windows applications such as Revit [18], AutoCAD [19], and other CAD Software application which offer a .NET [45] Software Development Kit (SDK).

2.3 Comparison of Alternatives to the Centralized BIM Approach

Within all the existing platforms described above, firms resort to developing their own custom tools and workflows, as the currently available software platforms in AEC do not meet their specific needs. Generally, the different proposed solutions aim at minimizing the technological framework helping in transporting information, and share the following goals and features:

- Improving interoperability through custom development enabling the transfer of data-rich geometrical objects.
- Keeping geometrical objects as lightweight as possible for serialization and further reconstruction.
- Using extensible neutral formats to facilitate both object customization (scaling and nesting of data-rich information) and interoperability.

Even though the custom solutions and tailored software development described in the previous sections are very valuable in solving immediate and critical issues related to the conception and construction phases of different building projects, they are being developed in isolation. These approaches are naturally narrow in scope and do not meet wider related challenges of data interchange, collaboration and inefficiency across the supply chain. Therefore, there is no common agreement for setting up an underlying data-exchange infrastructure that would enable better linkage among different practices, which constantly need to build custom applications and workflows for unrevealed design problems. Ramsgaard Thomsen [20] foresaw this particular need and consequent modelling paradigm shift: *"Rather than building common standards and libraries for known practices, we need to develop the fundamental infrastructures for yet unknown practices. This position fundamentally challenges some of the cornerstones of present modelling paradigms."* Therefore, there is a need to move away from efforts in design and code experimentation to solve local problems towards the development of solutions for increased transparency and participation at every stage.

AEC needs open-source platforms for combining all these efforts of software developers through to computational designers, engineers, architects in an accessible computational design and coding ecosystem. Similarly, the Linked Building Data (LBD) Community Group has directed towards considering open data models, as it is recognized that the AEC still uses a lot of different models, software, and ontologies. Standardization has still a long way and it is noticed that large software vendors are not keen to support open data models but rather develop and promote closed platform solutions. A common work around is that these parties use own models by creating new data models with custom attributes. Therefore, the LBD Community Group promotes the use of open data models, aiming at developing *"a concept and workflow of a web application for an open, consensual development of a common data model"* [21].

3 Speckle: An Open-Source Common Data Environment (CDE) for AEC

3.1 What Is Speckle?

Speckle [22] differentiates itself from other commercial web-based interoperability platforms by proposing a complete open-source data framework for AEC [23]. Speckle was originally developed at UCL in 2016 by Dimitrie Stefanescu as part of the InnoChain project, a H2020 Marie Curie European Training Network. Speckle does not enforce a predefined topology of communication patterns, but rather allows for the emergence (and analysis) of meaningful data-driven dialogue amongst the stakeholders involved in the design process. Regarding schemas, Speckle, contrary to the existing industry standard IFC, promotes composability over completeness and provides a programmatic infrastructure for end-users to define their own, domain-, company-, or even project-specific, object models. Furthermore, Speckle can support pre-existing object models (such as BHoM, and even subsets of IFC) "out of the box", provided that they exhibit certain technical characteristics [24].

Data transfer to and from each end-user is orchestrated via "Senders" and "Receivers" and by a given Speckle Server, which ensures its availability in the case the original source is offline. Furthermore, the server allows for efficient updates by leveraging several mechanisms, such as caching, object immutability and partial, differential updates. Via its Admin web app, Speckle implements a Discretionary Access Control (DAC) model, which gives full control to data authors on how accessible their information is, and with whom. The Speckle Admin web app is similar to other existing CDE in the sense that it presents a *"single source of information used to collect, manage and disseminate documentation, the graphical model and non-graphical data for the whole project team (i.e. all project information whether created in a BIM environment or in a conventional data format)."* [25]. However, Speckle's CDE differs from other existing CDEs at it is not file-based and does not offer any Information Container Data Drop (ICDD) functionality [26]. Instead, resources in Speckle are dynamically sent and received from the different CAD software integrations (or clients), and are represented as JSON objects which can be accessed through Speckle's CDE directly via their respective URLs/URIs. In [27], the authors describe a *"distributed common data environment"* as an evolved version of a traditional CDE, which *"links information artefacts without the need to convert from one format to the other"*. Following such definition, Speckle can be defined as a distributed CDE.

Speckle's distributed CDE allows for either fully public or private resources (Streams, objects, projects), but as well for granular privacy and security settings customised to the roles and needs of each stakeholder a particular resource is shared with. Furthermore, Speckle allows for resources to be enriched with extra metadata such as description, tags, comments, so as to be able to respond to the project's needs and allow for diagonal queries. Finally, Speckle also offers a version control interface allowing the end user to trace back the history of the created *Streams* (or collections of objects). This particular aspect is discussed in more details in Sect. 5.

Since its inception in 2016, it has been adopted by a large number of progressive AEC companies, e.g. Arup, Royal HaskoningDHV, HOK, Bjarke Ingels Group, BVN, HENN, Woods Bagot, Aurecon, Grimshaw and KPF, to name a few, as a key piece in their digital transformation efforts. Some of the Speckle Project Contributors have been involved in the development of specific web applications which answer their employer's needs. For example, Arup Carbon – an embedded carbon estimation tool – has been using Speckle, which calculates the embodied carbon of a model sent from Revit [18], based on the objects' attached material. Woods Bagot has also contributed to the Speckle platform by developing more specific features to its Admin interface web app [28].

As the first section of the present paper introduced numerous efforts from the World Wide Web Consortium (W3C) [29] and the Linked Building Data (LBD) community group [30] in promoting the use of the Semantic Web technology stack for the future development of CDEs in AEC, the next sub-section compares Speckle and Semantic Web technology stacks by highlighting respectively their differences and similarities.

3.2 Speckle/Semantic Web Ontologies Comparison

Whereas Speckle does not apply the same technological standards set by the World Wide Web Consortium (W3C) and the Linked Building Data community group, and developed through the dedicated Linked Data for Architecture and Construction (LDAC) workshops, its technology stack is constituted of the same "components" that compose the Semantic Web Stack. The main difference lies in the technical implementation of these components, listed in Table 1 below, comparing their technical implementations within both Speckle and the Semantic Web framework.

Table 1. Speckle Tech Stack/Semantic Web Tech Stack.

Components	Identifiers	Syntax	Data interchange	Taxonomies (schemas)	Querying	Trust/auth
Semantic web	URI/URL	XML (RDF/XML) JSON-LD	RDF	RDF-S	SPARQL	–
Speckle	URI/URL (hashes)	JSON	Speckle object	Speckle schemas	MongoDB queries	Passport.js

Identifiers. Both frameworks make use of Uniform Resource Identifiers (URIs) and Uniform Resource Locators (URLs) to identify and access unique resources. Although the main syntax used within the Semantic Web paradigm is the Extensible Markup Language (XML), Speckle relies on the JavaScript Object Notation (JSON) to define and query objects.

In regards with object identity, Speckle relies on an immutable object identifier (hash) as well as a client provided id. The hash represents a unique string that is specific to a given object's sate. For example, a point object defined by "[0, 0, 0]" (hash = 13e81f6567b656a19c629377c7f5a698) will have a different hash if its coordinates are changed to, for example, "[1, 0, 1]" (hash = 53ed8181875a36311d34ba1b5b46ff29). The hash is generated by passing the object's byte array footprint from the computer's memory to a standard hashing function, namely MD5. The resulting byte array is then converted to a hexadecimal string. This property is used as the unique key of the object's URI.

Data Interchange and Taxonomies. Another noticeable difference between the two technology stacks (see Fig. 1) lies in the chosen data interchange format and its related schema. While the Semantic Web and Linked Building Data community strongly encourage the use of the Resource Description Framework (RDF) [31] and its related

Schema (RDF-S), Speckle uses its own data interchange model and related taxonomy: the SpeckleObject, a simple base class that adds a set of simple properties related to authorization and commenting to all applicable resources.

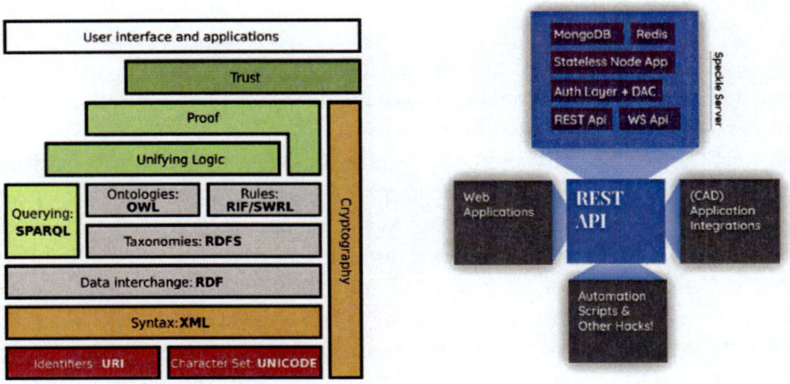

Fig. 1. The Semantic Web Stack (left) and the Speckle Technology Stack (right)

The main reason why Speckle privileges a JSON-based serialisation syntax over an XML-based one is purely out of convenience and cross-platform development ease. The Speckle framework exposes a Representational State Transfer Application Programming Interface (REST API) enabled by a Stateless Node Server App which is consumed by multiple software platforms and clients which benefit from the wealth of JSON serialising libraries.

A JSON-based syntax does not necessarily conflict with the Semantic Web paradigm, as the W3C community has introduced JSON-LD (a syntax designed to easily integrate into deployed systems that already use JSON, providing a smooth upgrade path from JSON to JSON-LD) and created a specific working group, the JSON-LD Working Group, dedicated in developing and maintaining the JSON-LD syntax. Such effort is *"intended to be a way to use Linked Data in Web-based programming environments, to build interoperable Web services, and to store Linked Data in JSON-based storage engines."* [32].

In fact, JSON-LD has been created so that the conversion from a JSON-LD document to RDF triples can be operated seamlessly: *"Conversion of a JSON-LD document, especially one in the expanded form, to RDF triples is straightforward. A subject, which could also be used as an object in another triple, is defined by @id. All other JSON-LD properties are converted to predicates."* The similarity between the Terse RDF Triple Language (Turtle) and JSON has been even more emphasized by the authors: *"JSON-LD looked like a more or less direct translation of Turtle to JSON, the syntax was changed dramatically in the latest versions and allows now data to be serialized in a way that is often indistinguishable from traditional JSON"* [33]. Therefore, it can be argued that the JSON syntax chosen for the Speckle REST API does not conflict with the standards promoted by WC3.

The main Speckle clients are developed and maintained for both Rhinoceros [16] and Autodesk Revit [18], respectively a Computer-Aided Design (CAD) and a Building Information Modelling (BIM) software widely used in Architecture, Engineering and Construction (AEC). Because Autodesk Revit and Rhino3D provide rich . NET [45] APIs, it was necessary to enable expert users to bootstrap their own object models built upon those APIs. Therefore, Speckle provides a "SpeckleKit" solution based on its SpeckleObject data interchange format. At its essence, a SpeckleKit is a taxonomy (a set of object definitions) grouped together with their translation routines to and from existing CAD software [24].

One might argue that privileging the use of a custom object model (such as the SpeckleObject class) instead of the recommended RDF data model might miss the concept of triples (the subject-predicate-object linkage) enabling relationship between multiple resources. Nonetheless, Speckle provides a similar quality which can be deployed within more classical conceptual modeling approaches - such as entity–relationship or class diagrams. At the level of SpeckleObjects, users can add semantic triples [31] to the basic set of predefined objects defined in a SpeckleKit. A semantic triple codifies the relationship between two entities in the form of "subject – predicate – object" expressions, thus expressing informational relationships in a machine and human readable way. Using dot notation, one can define such a triple as: myBeam. startPoint = myStartPoint. In this example, myBeam is the subject, startPoint is the predicate and myStartPoint is the object. Dynamically typed languages, such as JavaScript or Python, allow for the dynamic creation of such triples on any object without them being defined on its base class. Nevertheless, statically typed languages, such as C#, do not easily allow for this behavior. Consequently, in the specification and subsequent implementation of the base geometry and primitive classes, the technical implementation opted for a compromise whereby custom properties can be defined inside a designated field named properties. This specific field can then be implemented as a specific dynamic structure native to coding framework itself (e.g., in JavaScript, it will be simply another default Object, as all objects are key-value pairs, whereas in C#, it will be implemented as a Dictionary <string, object>, which is one of the key-value pair primitives offered by the language. At the level of a SpeckleStream class, "children" and "ancestors" fields enable expert users to specify hierarchical relationships between objects. Moreover, the clients of each created SpeckleStream can be accessed, enabling the mapping of multiple senders and receivers. Very similarly to RDF, the linking structure provided by Speckle forms a directed, labeled graph, where the edges represent a directional/hierarchical link between two resources, represented by graph nodes.

Querying. To query resources, W3C recommends to use SPARQL, which is a specific query language for RDF. Because Speckle does not use RDF data models and employs MongoDB to store data, it uses the query language developed specifically for MongoDB, as well as query-to-mongo, a Node.js package to convert query parameters into a mongo query criteria and options. A query tester has been implemented on the Speckle website to demonstrate the latter (see Fig. 2) [34].

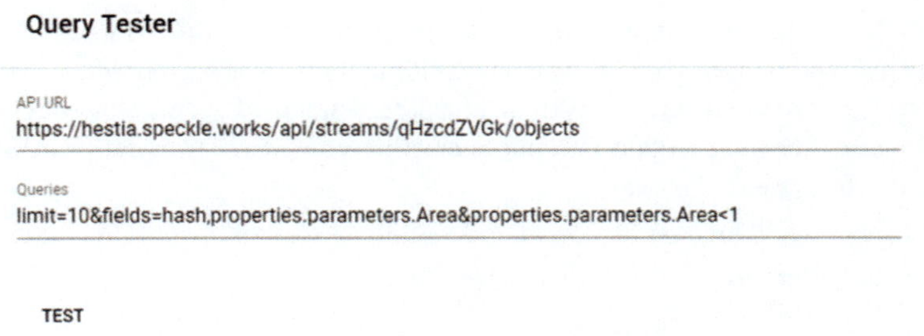

Fig. 2. Speckle's query tester interface [34]

4 Towards a Version Control Platform for AEC

4.1 Version Control Systems in Software Engineering

In software engineering, the terminologies "revision control" or "version control" encompasses any kind of practice that keeps track and provides control over changes to source code. Today, Git [35] and Github [36] are the most popular version control platforms. These platforms enable programmers to manage their code in a more fluent way, and separate their concerns through the creation of multiple parallel branches (see Fig. 3): *"a branch is the fundamental means of launching a separate line of development within a software project. A branch is a split from a kind of unified, primal state, allowing development to continue in multiple directions simultaneously and, potentially, to produce different versions of the project. Often, a branch is reconciled and merged with other branches to reunite disparate efforts."* [36] To understand similarities and differences between two different branches, version control systems make use of the "diff" utility, a data comparison tool that calculates and displays the differences between two resources. *Diffing* is a function that takes two input data sets (generally text) and outputs the changes between them. "Git diff" [35] is a multi-use Git command that when executed runs a diff function on Git data sources. These data sources can be commits, branches, files and more.

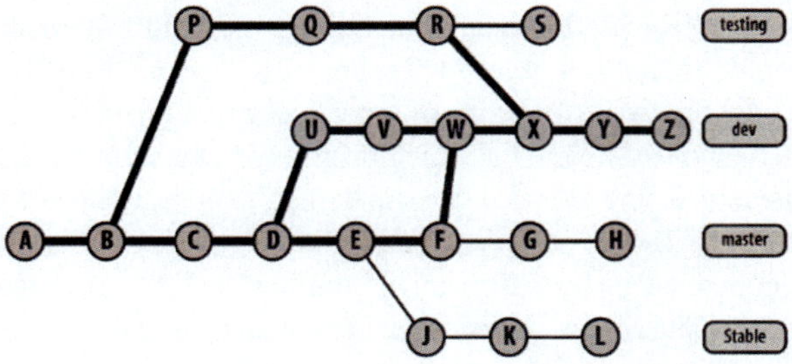

Fig. 3. Commits and branches reachable from the dev branch [36]

Similarly, to software development, design processes in the AEC are complex, decentralized, and happen parallelly across multiple trades and stakeholders. Therefore, it becomes necessary to have a system that keeps track of changes throughout the design process. The next section introduces existing version control platforms acting as version control systems for AEC.

4.2 Existing Version Control Platforms for AEC

3D Repo. 3D Repo [37] is an *"open-source version control framework that enables coordinated management of large-scale engineering 3D data over the Internet"* [3], within which users are able to gather and centralize various BIM models coming from multiple software platforms. This open-source platform offers today multiple features such as "3D diff", issue tracker, change and clash detection, enhancing collaboration among different stakeholders involved in a project. The *3D diff* technology enables the comparison between two sets of objects (or files) as opposed to the classic diffing technology introduced above and used across existing version control systems and is a patented technology allowing the detection of similarities and differences between two 3D models [38]. 3D Repo also provides a file-format-independent version-controlled repository for all common 3D assets such as meshes, materials, textures, animations, engineering assemblies, etc.

Konstru. Similarly to 3D Repo, Konstru [4] enables interoperability among multiple software platforms and offers version control capabilities such as *Merge & Compare* which highlights the differences between multiple models, and *Change Management*, which tracks the model history and creates reports on updates, changes and user activity.

4.3 Version Control in Speckle

Stream Revision History. More than a simple REST API, Speckle offers a web-based management interface within which the end-user can keep track of project data history. Through StreamDiff [39], a *diffing* mechanism comparable to the *Git diff* command in Git and GitHub, it is possible to observe the similarities and differences between two "Streams" (or two collections of objects). Technically, StreamDiff compares two JSON strings that represent the two collections of objects in a text format. The StreamDiff end point produces a ResponseStreamDiff which informs on: the number of SpeckleObjects contained exclusively in the original Stream ("inA"), the number of SpeckleObjects contained exclusively in the newly created Stream ("inB"), and the number of SpeckleObjects contained in both Streams ("common"). This information enables the description of a complete Stream Version History which can be accessed and explored by the end user.

From existing clients and CAD Software integrations supported by Speckle, users send Streams to a Speckle Server. If a Stream is modified, it can be registered as a "child" of the newly created Stream by the user. By registering these "child" versions of the working Stream, the end user can access the version history of a specific Stream via the Speckle Admin web app. Each version displays how many objects have been added ("Added"), removed ("Removed") and conserved ("Common"), compared to the previously created child Stream. This way, Speckle keeps track of the overall version history per Stream (see Fig. 4).

Fig. 4. The stream revision history available via the Speckle admin interface.

Project Timeline History. The Stream version history described above informs users on how a particular Stream has evolved over time. However, the design process of a construction project involves the creation, management and curation of multiple Streams. Therefore, one also needs to keep track of design transaction and interaction between multiple stakeholders who share multiple Streams. This has been addressed with *SpeckleViz*, an interactive User Interface developed within the Speckle admin web app, helping Speckle users to get a better understanding of the data flow of a specific Speckle Project across users, Streams and documents. Figure 5 illustrates the main User Interface, within which circle nodes represent Speckle Senders (S) and Speckle Receivers (R), square nodes represent Streams. Arrows (or graph edges) represent either data that has been shared to a Stream by the user (Receiver to Stream) or data that has been retrieved by a user from a Stream (Stream to Sender). Generally, both nodes and edges are coloured according to their respective timestamp: dark blue for the newest created, and light grey for the oldest (see Fig. 5).

The presence of both a Stream revision history (a version control system per Stream) and a Project timeline history (that keeps track of the dataflow across multiple Streams) within the same Project, or CDE, has proven to be a very powerful feature that helps the end users to curate data coming from different sources, and thus collaboratively throughout an entire design process. The next section illustrates such feature by describing a design to fabrication workshop within which the Speckle's CDE has been used and deployed.

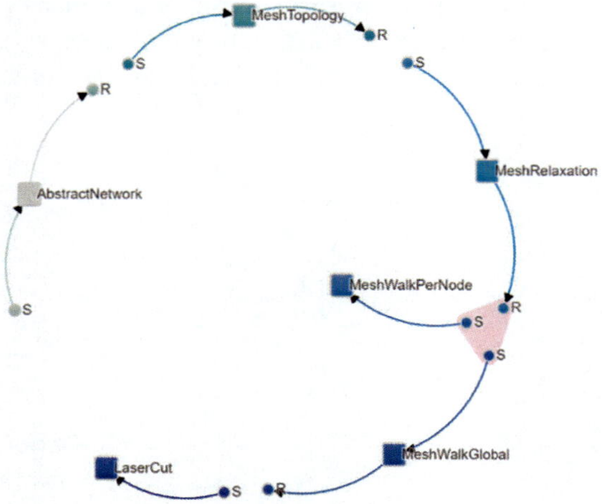

Fig. 5. The SpeckleViz interface available via the Speckle Admin web app. Here, circle nodes represent Speckle Senders (S) and Speckle Receivers (R), square nodes represent Streams. Arrows (or graph edges) represent either data that has been shared to a Stream by a user (Receiver to Stream) or data that has been retrieved by a user from a Stream (Stream to Sender). The content of each Stream is displayed in Fig. 6.

5 Case Study at the Centro de Estudios Superiores de Diseño de Monterrey (CEDIM)

5.1 Case Study Brief and Set-Up

SpeckleViz has been tested against data generated from design to fabrication during the "Piped Assemblies" Workshop conducted at the Centro de Estudios Superiores de Diseño de Monterrey (CEDIM) from the 26th of November to the 7th of December 2018. As part of the Architectural Design module, this workshop addressed the challenge of introducing state of the art web-based collaborative computational design workflows to 11 undergraduate students in architecture, through the design and digital fabrication of a free-form networked structure made of laser-cut polypropylene plastic strips. The main brief of this workshop was to introduce architecture students to understanding the production process of their design, based on Computer Numerical Control (CNC) technologies. The teaching objectives also consisted of raising awareness of digital workflows and interoperability concerns, which are an integral part of today's architectural practice. Speckle has been chosen as main data exchange platform throughout the design process of the project, as its learning curve is very low once basic knowledge of Rhino3D and its Visual Programming interface – Grasshopper3D – is acquired by the students. The main learning outcome for the students consisted in gaining a holistic understanding of a complete design to fabrication process, from the earliest design phases to the latest production stages, and through the use of a data exchange platform – Speckle.

During the workshop, the first author acted as a facilitator through the deployment of a Speckle server and the setup of a Speckle project within the main Admin interface, used for gathering geometrical data shared by the students. This allowed more flexibility in the design and fabrication process as the students were not required to work in the same place. At the beginning of the workshop, two main groups were formed. While the first one mainly focused on the overall design of the structure, the second concentrated on the fabrication of first prototypes in order to test the material, connection details and design to fabrication workflow. Once that the overall design was fixed, the students were divided again into four different groups that were spread across three different locations in which different parts of the production process took place: (1) the laser cutter at STM Robotics for fabrication, (2) university's workshop for assembly and (3) CEDIM's exhibition space for installation. In (1) and (2), the students were subdivided again into two different groups, so that they could focus parallelly on different parts of the structure during production. Overall, the students were able to collaborate and work remotely from these three different locations by sharing their data through Speckle.

5.2 Case Study Data Collection and Analysis

Through the Grasshopper Speckle client, both Speckle Senders and Receivers were used to seamlessly share design data across all the phases of the design process. The data collected originated directly from the different Rhino-Grasshopper sessions manipulated by the students, and could serve three different purposes:

- **Exchanging design ideas:** data could be shared in the sole purpose of exchanging design ideas. This way, students could always log in to the admin interface and explore and be inspired by the different models shared by their classmates.
- **Keeping track of a Stream's design history:** data could be collected in order to keep track of the design history of a particular Stream, through the Stream Revision History interface described in Sect. 4.3.
- **Keeping track of the project's timeline history:** Finally, data could also be gathered in order to keep track of the chronological evolution of the design process, from the lowest level of detailing to the highest, as described in Sect. 4.3.

For the sake of conciseness, the next section focuses primarily on the data collected in regards to the project's timeline history.

5.3 Case Study Outcomes

At each design transaction, the users were able to enrich geometric information with the necessary metadata that was needed for their specific subtasks. For example, each generated strip was given an approximated length, area and label, which were crucial information for both the unrolling, nesting and assembly processes. The enrichment of the shared geometrical objects with their related metadata allowed to maintain a high degree of information efficiency and little overhead in term of data payload, as the minimum required information could be seamlessly shared via Speckle Streams, without the need of converting and exporting entire files. Multiple informational

Streams were subsequently created, aggregated and queried at various steps in the design process (see Fig. 6). Throughout the design process of this specific case study, each Stream geometrically depends on each other, and the overall workflows goes from the lowest level of detailing to the highest:

- 1. Definition of the overall abstract network (generated by the design group).
- 2. Base topological mesh design generation (generated by the design group).
- 3. Mesh relaxation (generated by the design group).
- 4. Strip segments generation (inspected by the fabrication group).
- 5. Unrolling and nesting (generated and inspected by the different production groups responsible respectively for one particular node of the structure).

Here, the design Streams in Speckle maintained their relevance and usefulness both in the later fabrication stages, during laser cutting and labelling processes, as well as during assembly, by being able to visually relate parts to the whole, interactively, and on-site.

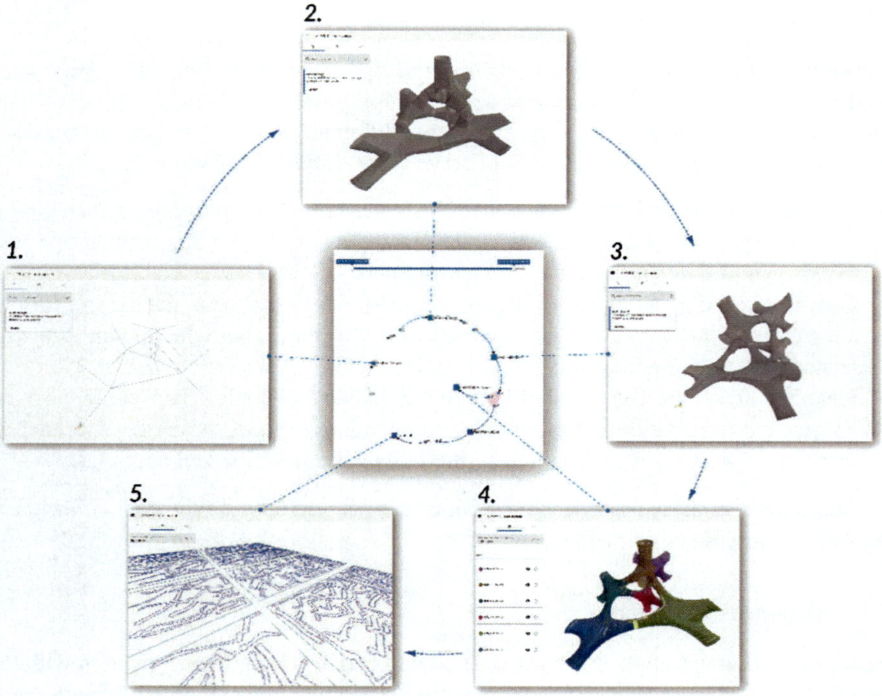

Fig. 6. The overall workflow of the Piped Assembly Workshop has been deployed through the SpeckleViz interface. The user can visualize the data exchanges and access each Stream within the Speckle viewer environment, from the design to the fabrication stages. Here, the SpeckleViz interface refers to Fig. 5.

Whereas the present case study happened in the context of an academic workshop for undergraduate students in architecture, the authors believe that these particular design methodologies and new ways of sharing data through the Speckle platform could also be deployed in practice, as they present great potential for seamless collaboration strategies. For example, the *Building Information Generation* [40] workflow, a modelling strategy deployed by Front Inc., that consists of a strategic alternation between the generation of objects in Grasshopper and their subsequent storage and classification within staged Rhino3D models, could largely benefit of the distributed Speckle's CDE which would enable a live connection between the different deployed Rhino3D and Grasshopper clients.

6 Discussion

Whereas the idea of introducing open standards – to combat the existing monopoly of software vendors' monolithic applications and related proprietary file formats – is commonly shared by both academia and AEC practitioners, there exists a divergence of views regarding what this open standard should be, and how it should be implemented. While academia is promoting the use of Semantic Web standards in the long term through the research efforts [5, 6] undertaken by the Linked Building Data community group [30], the industry is developing its own data transfer and version control solutions (e.g. Konstru [4], 3D Repo [3], Speckle [22]) in order to solve crucial issues (e.g. the existing interoperability gaps between software platforms) in the short term [9], through Research & Development efforts undertaken by BuroHappold Engineering through BHoM [10] and by Proving Ground through Conveyor [11], for example. In other words, while academia is reflecting on what should be the best ontology to be adopted by the AEC industry, the latter is developing its own solutions, using data formats that are the most suitable for their needs and the easiest to implement and deploy – as seen in Sect. 3.2.

This existing discrepancy between the slow release cycle of open standards suggested by academia or consortiums, and the parallel fast pace software development existing in industry, has been already pointed out by Fabian Scheurer in a similar context, criticizing the slow release cycle for the IFC standard compared to the fast pace development of CAD software: *"Although there has been an ongoing attempt since 1995 with the "Industry Foundation Classes" (IFC), an open standard for so called "Building Information Modeling" (BIM), there have been long delays between releases of new versions. This timeframe between releases poses an interesting dilemma: while new versions of software packages with new functionality appear on the market every year, a four year innovation cycle for the underlying data format seems rather unhurried. Can the IFC standard keep up with the development?"* [41]. Therefore, it seems reasonable to take into consideration the latest technological developments from the software industry when defining an open standard for decentralized data exchange in AEC.

The academic case study described in Sect. 5 demonstrated the potential of the distributed CDE provided by Speckle, which enabled the data collection from different Rhino3D and Grasshopper sessions at any time during the design process. More than providing a neutral data format, which – as discussed in Sect. 2.3 – is the current converging trend in industry, Speckle starts acting as an online data management platform for AEC with strong versioning capabilities. From such platform, the authors believe that the data generated throughout the design to fabrication stages of a large-scale architectural project could be seamlessly stored, inspected and retrieved – through the Stream revision history described in Sect. 4.3 – and thus at any time during the process – through the project timeline history described in Sect. 4.3.

6.1 Limitations

Although Speckle is useful for interoperability, data management and collaborative workflows across platforms, it presents some limitations, both in terms of integration capabilities and maintenance.

- **Limited integration capabilities:** Currently, Speckle provides application integrations for the following software clients: Rhinoceros [16]. Grasshopper3D [17], Autodesk Revit [18], Dynamo [42], Blender [43] and GSA [44]. Although Speckle is able to support the development of application integrations for software packages which have an Application Programming Interface (API) or Software Development Kit (SDK) built within the .NET Framework [45], the open-source platform is less capable of integrating within software packages which do not offer .NET APIs or SDKs.
- **Maintenance:** As mentioned above, Speckle provides five different application integrations, and it is expected that this number increases over time as the Speckle community continuously grows and therefore might gain interest in integrating the open-source platform within other specific software packages. Speckle already requires high maintenance as each software package that integrates a Speckle client could modify its .NET API or SDK unexpectedly. Consequently, the contributors to the Speckle platform would need to revisit the affected Github [36] repositories and rewrite specific object model conversions.

6.2 Future Research

In this paper, the main concepts and principles of the Speckle platform have been discussed, such as its version control capabilities and its ability to facilitate collaborative workflows. Although the latter has been demonstrated through a case study in Sect. 5, it consisted of a relatively small-scale artefact which enabled the facilitator to contain and highly curate the overall workflow via the Speckle platform throughout the whole design to fabrication process. Future work will test the Speckle platform against larger and more complex construction projects in order to assess its robustness against bigger and more intricate datasets. The authors are currently collaborating with Grimshaw Architects who provided relevant building data from a large-scale and complex infrastructure project that is being undertaken in United Kingdom. Speckle, and more particularly its project timeline history feature *SpeckleViz* (described in Sect. 4.3), are currently being tested against Grimshaw's datasets. This will enable the authors to better understand both the strengths and the weaknesses of the open-source platform.

Parallel work is also investigating the application of the Speckle platform within other domains than the building industry, such as the naval architecture industry.

7 Conclusion

This paper introduced the current challenges in collaborative design workflows within the AEC sector and highlighted a general trend from both academia and the industry to converge towards the use of neutral object models and open-source platforms. Whereas academia looks specifically at Web Standards in order to implement software agnostic Common Data Environments (CDEs), the AEC industry is developing specific versioning control systems to give more control over the produced data history to the end user. Speckle aims at combining the best of both worlds: a neutral object model which abstracts data originated from the various connected software clients, and a version control system allowing the user to explore the data history of each created data Stream. Consequently, Speckle facilitates information exchange and collaboration across software packages in a modular, rather than monolithic, way; it also describes a bottom-up approach to the problem of interoperability. In this way, Speckle offers an alternative to the traditional centralized design processes generating large data payloads [1], by proposing an open-source and distributed CDE within which data can be seamlessly exchanged across different users and software platforms.

Acknowledgements. The present research has received funding from InnovateUK under the competition “Increase Productivity, Performance and Quality in UK Construction” (proj. no. 104799). Speckle was originally developed at The Bartlett School of Architecture as part of the InnoChain project, which received funding from the European Union’s Horizon 2020 research and innovation programme under the Marie-Sklodowska-Curie grant agreement No 642877.

References

1. Vonthron, A., Koch, C., König, M.: Removing duplicated geometries in IFC models using rigid body transformation estimation and flyweight design pattern. Visual. Eng. **6**(1), 1–15 (2018)
2. Peters, B., Peters, T.: Computing the Environment: Digital Design Tools for Simulation and Visualisation of Sustainable Architecture, 1st edn. Wiley, Chichester (2018)
3. Doboš, J.: 3D Repo: version controlled 3D repository. Poster presented at the Advances in Architectural Geometry (AAG) Conference, London, United Kingdom (2014)
4. Konstru. https://konstru.com/. Accessed 14 Jan 2019
5. Werbrouck, J., Pauwels, P., Beetz, J., Van Berlo, L.: Towards a decentralised common data environment using linked building data and the solid ecosystem. In: 36th CIB W78 2019 Conference Proceedings, pp. 113–123. Northumbria University, Newcastle (2019)
6. Rasmussen, M.H., Hviid, C.A., Karlshøj, J.: Web-based topology queries on a BIM model. Paper presented at LDAC2017 – 5th Linked Data in Architecture and Construction Workshop, Dijon, France (2017)
7. Drumbeat. http://www.drumbeat.fi/concept.htm. Accessed 14 Jan 2019
8. VisuaLynk. https://visualynk.com/. Accessed 14 Jan 2019
9. Deutsch, R.: Convergence: The Redesign of Design, 1st edn. Wiley, Chichester (2017)

10. The Buildings and Habitats object Model (BHoM). https://bhom.xyz/. Accessed 14 Jan 2019
11. The Proving Ground's Conveyor Plug-in. https://provingground.io/tools/conveyor/. Accessed 14 Jan 2019
12. Rhino.Inside. https://www.rhino3d.com/inside. Accessed 14 Jan 2019
13. Proving Ground. https://provingground.io/about/. Accessed 14 Jan 2019
14. Miller, N.: [make]SHIFT: information exchange and collaborative design workflows. In: ACADIA 2010: LIFE in:formation on Responsive Information and Variations in Architecture, Proceedings of the 30th Annual Conference of the Association for Computer Aided Design in Architecture (ACADIA), New York, United States, pp. 139–144 (2010)
15. Miller, N., Stasiuk, D.: A novel mesh-based workflow for complex geometry in BIM. In: ACADIA 2017: DISCIPLINES & DISRUPTION, Proceedings of the 37th Annual Conference of the Association for Computer Aided Design in Architecture (ACADIA), Cambridge, MA, United States, pp. 404–413 (2017)
16. Rhinoceros (typically abbreviated Rhino, or Rhino3D) is a commercial 3D computer graphics and computer-aided design (CAD) application software developed by Robert McNeel & Associates. https://www.rhino3d.com/. Accessed 14 Jan 2019
17. Grasshopper3D (typically abbreviated Grasshopper) is a visual programming language and environment that runs within Rhino3D
18. Autodesk Revit is a Building Information Modelling (BIM) software for architects, landscape architects, structural engineers, mechanical, electrical, and plumbing (MEP) engineers, designers and contractors
19. AutoCAD is a commercial computer-aided design and drafting software application developed by Autodesk
20. Thomsen, M.R.: Complex modelling: questioning the infrastructures of information modelling. In: Herneoja, A., Österlund, T., Markkanen, P. (eds.) Proceedings of the 34th International Conference on Education and Research in Computer Aided Architectural Design in Europe (eCAADe): Complexity & Simplicity, pp. 33–42. eCAADe (Education and Research in Computer Aided Architectural Design in Europe) and ITU/YTU (2016)
21. Open Data Model for AEC Industry's LDAC Workshop GitHub repository. https://github.com/linkedbuildingdata/SummerSchoolOfLDAC/blob/master/Notebooks/02-05-Open-Datamodel-Coding-Challenge.ipynb. Accessed 14 Jan 2019
22. Speckle. https://speckle.systems/. Accessed 14 Jan 2019
23. Stefanescu, D.: Alternate means of digital design communication. Ph.D. thesis, UCL, London (2020)
24. Speckle's Schemas and Object Models (.NET). https://speckle.systems/docs/developers/object-models. Accessed 14 Jan 2019
25. Common Data Environment (CDE). https://www.designingbuildings.co.uk/wiki/Common_data_environment_CDE. Accessed 14 Jan 2019
26. Formerly known as Information Container for Data Drop (ICDD), the standard has been recently renamed as Information Container for Linked Document Delivery. https://www.iso.org/standard/74389.html. Accessed 14 Jan 2019
27. Bradley, A., Li, H., Lark, R., Dunn, S.: BIM for infrastructure: an overall review and constructor perspective. Autom. Constr. **71**, 139–152 (2016)
28. Speckle Community Meetup. https://www.youtube.com/watch?v=J8XCVL1ihp4. Accessed 14 Jan 2019
29. World Wide Web Consortium (W3C). https://www.w3.org/. Accessed 14 Jan 2019
30. Linked Building Data (LBD) Community Group. https://www.w3.org/community/lbd/. Accessed 14 Jan 2019
31. Cyganiak, R., Wood, D., Lanthaler, M., Klyne, G., Caroll, J.J., McBride, B.: RDF 1.1 concepts and abstract syntax. W3C Recommendation **25**(02), 1–22 (2014)

32. JSON-LD Primer. https://www.w3.org/TR/json-ld11/. Accessed 14 Jan 2019
33. Lanthaler, M., Gütl, C.: Complex modelling: on using JSON-LD to create evolvable RESTful services. In: Proceedings of the Third International Workshop on RESTful Design. ACM International Conference Proceeding Series, pp. 25–32. Association of Computing Machinery
34. Speckle's Query Tester interface. https://speckle.systems/docs/developers/api-specs. Accessed 14 Jan 2019
35. Chacon, S., Straub, B.: Pro Git. Apress, New York (2014)
36. Loeliger, J., McCullough, M.: Version Control with Git, 2nd edn. O'Reilly Media, Sebastopol (2012)
37. 3D Repo. https://3drepo.com/. Accessed 14 Jan 2019
38. 3D Repo's 3D Diff Technology. https://3drepo.com/3d-diff-model-change-detection-made-simple/. Accessed 14 Jan 2019
39. Speckle's StreamDiff endpoint. https://speckleworks.github.io/SpeckleSpecs/#streamdiff. Accessed 14 Jan 2019
40. Van Der Heijden, R., Levelle, E., Reise, M.: Parametric building information generation for design and construction. In: Proceedings of the 35th Annual Conference of the Association for Computer Aided Design in Architecture – Computational Ecologies, Design in the Anthropocene (ACADIA 2015), Cincinnati, Ohio, United States, pp. 417–430 (2015)
41. Scheurer, F.: Digital craftsmanship: from thinking to modeling to building. In: Digital Workflows in Architecture, pp. 110–129. Birkhäuser, Basel (2012)
42. Dynamo is a graphical programming interface within Revit. Dynamo is similar to and takes inspiration from the Grasshopper visual programming language and environment for Rhino3D
43. Blender is a free and open-source 3D computer graphics software toolset used for creating animated films, visual effects, art, 3D printed models, motion graphics, interactive 3D applications, and computer games
44. GSA is a structural analysis package software developed by Oasys for advanced analysis and design of buildings, bridges and tensile structures
45. .NET Framework (pronounced as "dot net") is a software framework developed by Microsoft that runs primarily on Microsoft Windows

Check for updates

Using BIM and GIS Interoperability to Create CIM Model for USW Collection Analysis

Carolina Midori Oquendo Yosino(✉) and Sergio Leal Ferreira

University of São Paulo, São Paulo, SP, Brazil
{carolinayosino, sergio.leal}@usp.br

Abstract. Computational tools based on the CIM (City Information Modeling) concept have been developed to support the analysis and project of the urban environment. These tools demonstrate how information modeling technology contributes to the visualization and handling of data in urban management context. A devised solution to get a CIM model is through the interoperability between BIM (Building Information Modeling) and GIS (Geographic Information System) applications. This research proposes this integration to be applied and generate a CIM modeling. The aim is to present an efficient way to take advantage of the BIM modeling data, which have a real probability to be sent to city hall in future as part of the project approving process. Hence, this turns possible the calculation of USW (Urban Solid Waste) production and the planning of their collection with an immediate visual feedback. In this research, BIM and GIS data were integrated through interoperability between .RVT and .GdB files. All data were used together to get a result analysis of the USW production and respective collection in a specific urban region. Also, it was possible to analyze data accuracy per USW collection points from the provided data for each BIM modeling. The simulated CIM modeling in this work allowed the creation of a three-dimensional environment for the visualization of the USW production. Also, it gave support to the analysis of waste collection routes in order to maximize the use of collectors and allowed a preliminary verification of how a building affects the waste production and collection routes in a city.

Keywords: CIM · BIM · GIS · USW

1 Introduction

BIM concept (Building Information Modeling) has influenced the change of project management in vertical constructions. The changes are present in how the models are projected and parametrized and in the building construction control. Likewise, the CIM concept (City Information Modeling) comes with an analogous objective to BIM, however applied to the urban environment. The CIM proposal is to join data from urban subsystems to obtain a global view of the functioning of the urban environment, thus improving the management and the city planning.

Therefore, technological tools based on the CIM concept are necessary to support an integrated and complete urban management. Lima [1] highlights that before technological evolution, the data to develop urban planning were obtained through

E. Toledo Santos and S. Scheer (Eds.): ICCCBE 2020, LNCE 98, pp. 248–271, 2021.
https://doi.org/10.1007/978-3-030-51295-8_19

socioeconomic analysis, without any type of data processing. Today, on the other hand, technologies applied to information modeling allows the use of open digital data in an overlap of several types of urban parameters, which lead to better data visualization and decision making related to urban planning.

CIM takes the benefit of the immense amount of data generated daily on the urban environment and presents itself as a plausible way to support the management and the use of these data for urban environment planning. Authors such as Beirão [2] and Xu et al. [3] have developed computational tools based on the CIM concept to support the analysis of the urban environment and its systems. These researches have demonstrated how information technology modeling can contribute to the visualization and handling of urban data. Later, this will enable the creation of an integrated and informative urban management available for all users in the environment.

However, there are still some technical barriers that blocks the design of platforms based on the CIM concept such as the interoperability between systems that supports the development of technology tools, and the lack of open data to support the parameter models. In this way, the present research has as the main focus to study the interoperability between BIM and GIS models to conceive a CIM environment with the capacity to manage data related to the Urban Solid Waste (USW) collection, through an analysis made from the individual inputs of the quantity of waste generated by each building.

Through the development of a CIM simulation, this work seeks to deepen the understanding on how the CIM concept, applying BIM and GIS interoperability, can contribute to the management of urban assets in a planned and integrated way.

2 Urban Solid Waste Management and Logistics

The management of USW in Brazil is owned by the public city administration. If there is negligence in the service of this system, the accumulation of waste generated and its incorrect packaging can cause several direct damages to the population served, such as the silting up of rivers, clogging of manholes, destruction of green areas, in addition to the proliferation of various diseases [4]. On the other hand, if USW management is well executed through a comprehensive and organized collection and with the correct disposal of waste, the served population has, beyond the multiple direct benefits, a sense of efficiency in public management [5].

However, USW management is shown to be a highly expensive urban subsystem in Brazil. In small municipalities the USW management can represent 7 to 15% of the city hall budget and they still do not serve the population adequately [6].

In the report released by IPEA (Institute for Applied Economic Research) [7] it's presented that only with an USW management improvement, it would be possible to minimize expenses in this area without impact the services negatively, endorsing the fact that optimizing the management processes helps to save significant budgets for local governments.

Within the scope of urban cleaning, worldwide data on USW generation by macro-regions are used to infer the volume of collection in each region to create collection routes. However, the individual data for each building isn't used to achieve a better precision of the contribution of each constructive model to the USW logistical collection system. The lack of precision in the USW production data just generates both overestimated and underestimated USW collection networks [5].

In order to further analyze the effectiveness of the USW management system in Brazil, it's interesting to investigate the related data to the scope of the USW collection service. According to the Abrelpe [8] report, the coverage rate for USW collection services in Brazil was 91.24%. The Southeast region achieved the highest collection rate in the country and the Northeast region was lowest one (see Fig. 1).

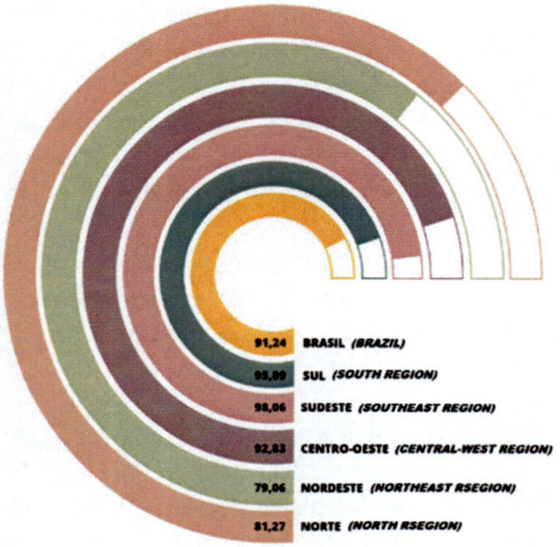

Fig. 1. Index of USW collection coverage in Brazil (%) (Abrelpe [8]).

It's noted with such data that the waste collection service in the country is close to cover the entire population, demonstrating the effectiveness in universal service. Even so, in order to develop USW management in Brazil in a financially effective basis, it's necessary to increase the efficiency of the system through investment in new management strategies. González and Leal [9] report that only with the implementation of new strategies - in the short, medium and long term - it will be possible to control and manage the USW collection system properly.

CIM concept is an interesting approach to seek efficiency in the management of USW. According to Almeida and Andrade [10] CIM can assist as a proper technology of urban infrastructure through the analysis of the collection of solid urban waste, having the ability to use data from buildings to measure their individual contributions in the waste generation.

3 CIM Model from BIM and GIS Interoperability

As it's a relatively new topic, in the last few years several authors still discuss about the CIM concept and where it might be applied. Both nationally and internationally, the CIM has been identified as "one of the ways for a city to achieve the status of Smart City" [11].

In this context, Thompson et al. [12] portrays the CIM as a "holistic and interdisciplinary approach to the generation of spatial data models through the integration, application and visualization of city data". Dantas, Sousa and Melo [13] complement that the CIM concept is composed not only by information about urban infrastructure, but it also creates connections with relevant information to administration and human activity. CIM has the power to facilitate visualization, analysis, monitoring of the urban environment and it also provides a global view of a region.

Cavalcanti and Souza [14] reported that the current technology used for urban management does not process various engineering subsystems concomitantly. That is why it isn't possible to manage the cities considering all its aspects. For this reason, academic researchers have been moving towards the study of the CIM concept as a technological tool to support city management.

According to Bertei et al. [15], to enable an infrastructure system in the city scope, it's necessary to include all possible public infrastructure works in this scope. From road, waste collection, water and sewage supply, electricity and gas infrastructure, to health care, security and recreation buildings.

However, for CIM-based tools to become a reality, it's necessary not only access to urban data sets, but also the investment of time and efforts to create ways to store, decode and manipulate this data, so that buildings have information crossed with the elements of its surroundings to make possible the reproduction of the natural movement of the city. From the well elaborated computerization of city planning, it will be possible to create urban models closer to reality, thus helping in the reliable management of urban environments.

Certainly, the development of CIM technological tools will rise the urban planning to a new level of detailing capability, so that becomes possible to cross several sets of urban data. Those tools will enable a deeper understanding of cities functioning and how to adjust urban environment without impact negatively its daily dynamics and functionality.

For that, it's necessary to cross data from the urban environment and constructive data, and a proposal to merge these two types of information is the interoperability between BIM and GIS systems.

According to Kymmell [16], interoperability is the ability of different file formats to be integrated to each other and to transfer relevant information as well, without data loss. In order to create a CIM environment, interoperability is essential in the process of joining data from the urban environment with building environments.

The best way to determine interoperability is to try to link several files that contain examples from different areas of the project and place them in a single test file, in order to observe all the expected processes [16].

To find a software that works in perfect interoperability is important for the success of simulation based on the CIM concept. On the other hand, when the interoperability between systems doesn't operate in order to achieve its optimal functioning, there's a loss of important information, preventing any substantial use of the resulting model.

Therefore, it's extremely important to find BIM and GIS systems that have a quite good interoperability, so that it's possible to create a strong bridge between these two systems, which allows the information crossing without loss or data change.

4 Methodology

This work was based on the Design Science Research (DSR) methodology. It's a methodological process of elaborating artifacts to achieve better results, that is, producing innovative systems that modify existing situations [17]. The developed artifact in this study was a simulated urban model that follows the CIM concept premises, through interoperability between BIM and GIS systems (See Fig. 2).

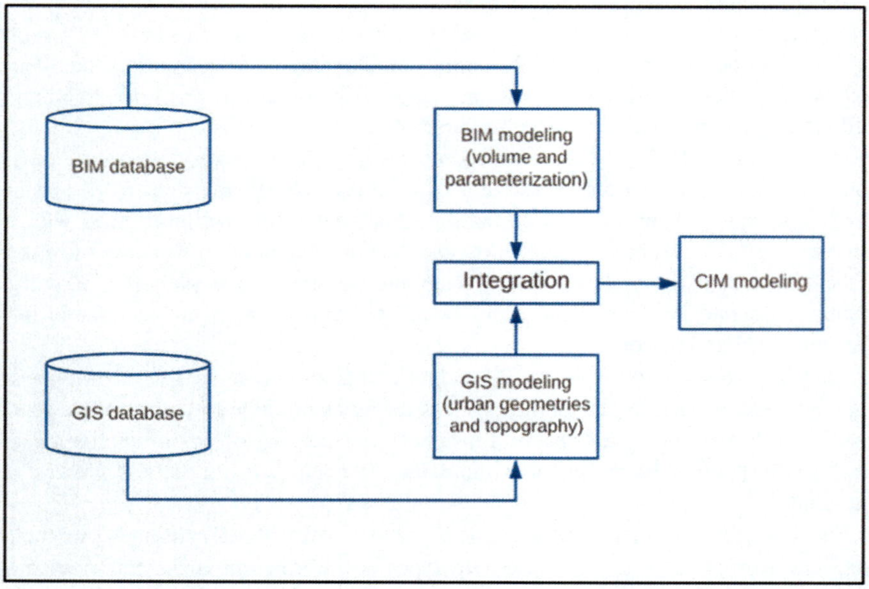

Fig. 2. Artifact design workflow, representing the interoperability between the BIM and GIS systems for the development of the CIM model.

The evaluation process is experimental. According to Lacerda [17], the experimental proposed analysis of the artifact is "to study the artifact in a controlled environment to verify its qualities". Within this classification, the simulation model presented in this research will follow four main stages of project development:

1. BIM modeling;
2. GIS modeling;
3. BIM and GIS data integration;
4. Data analysis in a CIM environment.

Finally, with results of the design of this artifact it's expected to highlight the importance of CIM modeling for the visualization, analysis and data management of urban subsystems. The development of each stage of the project is presented in the next sub-chapters.

4.1 BIM Modeling

This step was mainly focused on the modeling of the built volume and its proper software-based BIM parameterization. Thus, in this first stage, the BIM modeling was carried out using four steps: 1. Definition of the basic prerequisites; 2. Software selection; 3. Building modeling; and 4. GIS platform file importing. The Fig. 3 presents the workflow of the BIM modeling.

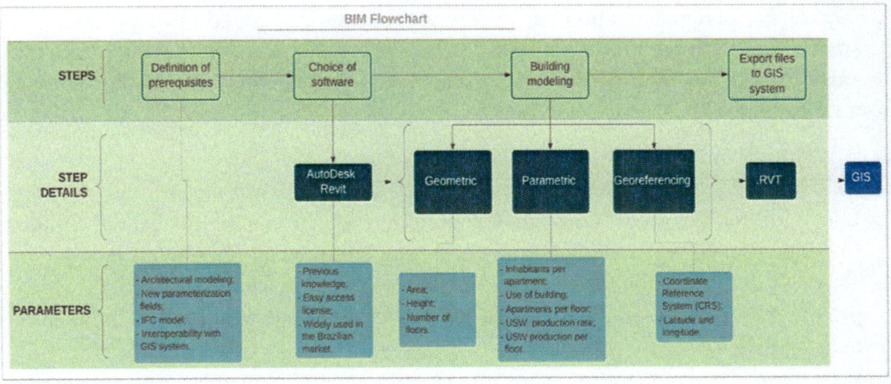

Fig. 3. BIM modeling stage workflow.

First, it was identified the basic prerequisites that a BIM platform must have to ensure the development of BIM objects with the required characteristics to integrate with the GIS environment and to make a CIM environment. Thus, to obtain the expected result, the BIM platform must meet the following requirements:

a. Three-dimensional architectural BIM modeling software. As constructive data wasn't used for the development of the artifact and the BIM models was designed at the 100 LoD (Level of Development) [18] level, the BIM platform that meets the three-dimensional architectural modeling was considered enough for this work.
b. Software must allow the insertion of new parametric fields. USW data was added to the BIM modeling through object parameterization by fields that weren't part of the Default BIM environment fields.

c. GIS interoperability. The chosen BIM platform must guarantee interoperability with the GIS system with no loss of critical information when transferring data.

From the data obtained in the previous step, a survey of platforms based on the BIM concept available on the market that meets these prerequisites was carried out. After this analysis, it was decided to use the BIM Revit 2018 software, from Autodesk [19].

Revit was selected because it's a three-dimensional architectural modeling focused platform, it presents specific tools to create new parameters and has interoperability with the most used GIS platforms in the AEC (Architecture, Engineering and Construction) market.

The building modeling step was divided into: i. Geometric; ii. Parametric; and iii. Georeferencing.

The Geometry (i) was developed through a modeling of low-level geometric information. So that, the generated files would not be too heavy, which would make it difficult to insert the models in the GIS system. Thus, it was decided to create volumes that presented only the required basic structures to create the building, composed only by the Floors and Walls layers. Through the geometric modeling, it was obtained the basic volume data of BIM object, such as area and number of floors, the necessary information to parameterize the model.

The object parameterization (ii) in Revit 2018 was developed through the Project Parameter tool. The parameterization rules for the BIM models were made to identify the quantity of USW that each building produces per day. Thus, five parameters were inserted in the models:

a. Apt/Floor: Number of dwelling units per floor;
b. Hab/Apt: Defines the average number of residents per dwelling;
c. Uso_edif: Indicates the usage of the building, which can be residential building (medium and high standard, and houses), and commercial (small and large);
d. Tx_usw: Indicates the production of solid waste per capita, according to data provided by Abrelpe [8]. The rate varies according to the use of the building and is measured in kg/inhabitants.day;
e. Prod_usw: Total production of solid urban waste per floor per day, measured in Kg/day and calculated by:

$$Prod_{USW} = Tx_{USW} \times Numer\,of\,floors \times \frac{Apt}{Floor} \times \frac{Hab}{Apt} \tag{1}$$

Finally, the GIS interoperability (iii) has a relevant role within BIM modeling, since it's through the interoperability that will be possible to insertion of BIM models into the GIS environment. For that, a key topic to perform the integration with the GIS system is to specify the Coordinate Reference System (CRS) of the BIM modeling. It was defined that for this project the CRS followed was EPSG: 3857 - WGS84 Web Mercator (Auxiliary Sphere). This data was inserted in the BIM modeling, together with parameterized latitude and longitude coordinates of each building, so that, each model is inserted in the correct position and scale (Fig. 4).

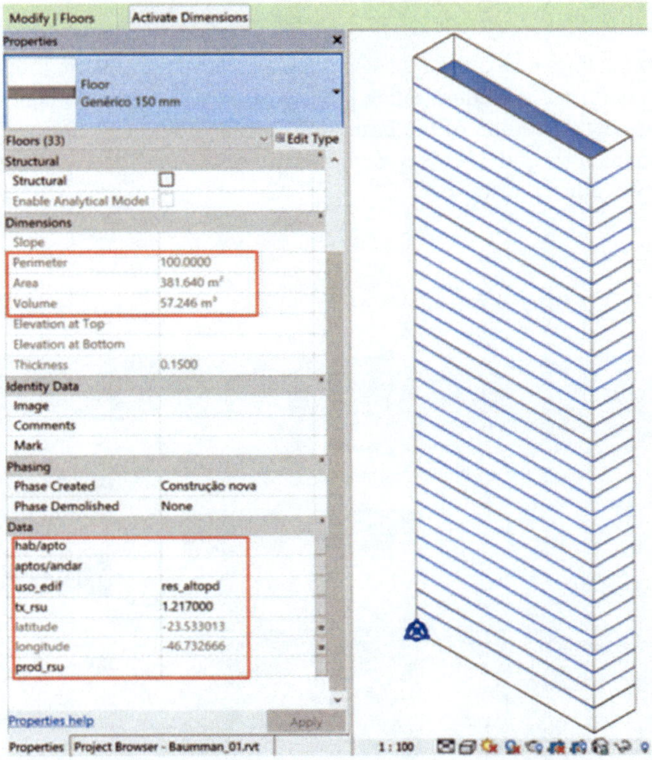

Fig. 4. Volumetric, parametric and georeferenced data in BIM software, Revit.

Altogether, six buildings were modeled in BIM, all located in the city of São Paulo, Brazil. The summary of BIM modeling information is in Table 1.

Table 1. Summary of BIM modeling data.

Summary – BIM modeling									
Modeling ID	Quantity units	Housing units		Area (m^2)		Inhabitants /unit (average)	Use of building	USW production (Kg/day)	
		/Floor	Total	/Floor	Total			Per capita	Total
Froben 01	16	4	64	488.4	7814.4	3.6	Standard	1.217	280.4
Froben 02	16	4	64	488.4	7814.4	3.8	Standard	1.217	295.97
Froben 03	16	4	64	488.4	7814.4	3.1	Standard	1.217	241.45
Baumman 01	32	2	64	381.64	12212.48	3.1	High	1.217	241.45
Baumman 02	32	2	64	381.64	12212.48	2.5	High	1.217	194.72
Baumman 03	32	2	64	381.64	12212.48	2.9	High	1.217	225.88

4.2 GIS Modeling

In the GIS modeling stage, the focus was on the development of urban modeling through GIS platforms. The construction process of the urban environment was divided into three steps: 1. Definition of the basic prerequisites for the platform; 2. Choice of software for modeling; and 3. Urban environment modeling. The Fig. 5 represents the GIS modeling summary workflow.

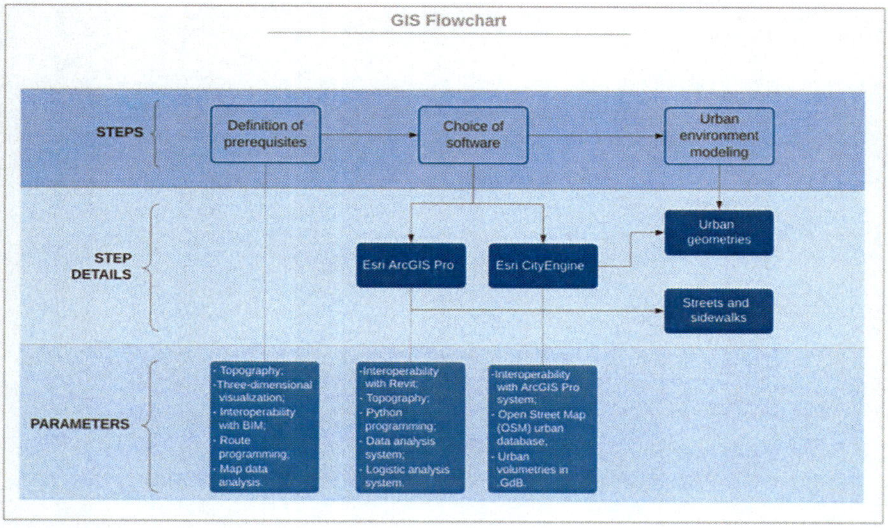

Fig. 5. GIS modeling stage workflow.

As it occurred in the BIM modeling stage, in order to select the GIS platform that was used in this research, prerequisites were established to guarantee the modeling of the parameterized urban environment and the BIM data integration:

a. GIS platform should allow topographic visualization of the region and three-dimensional objects;
b. Interoperability with BIM platform;
c. Modeling of volumetric representative of urban environment (buildings);
d. Georeferenced data analysis;
e. Logistic analysis.

It is worth mentioning that building volumes are created in GIS to compose the urban environment where BIM models are inserted. In a simulation that has all buildings in a BIM model, it isn't necessary building models in GIS.

To compose the GIS modeling, two platforms were selected, ArcGIS Pro and CityEngine, both from Esri. The ArcGIS Pro platform is a tool based on the GIS concept, whose main aspects are 2D and 3D mapping, image processing, data analysis, data management and integration through a layered system [20]. Those tools also ensure interoperability with Autodesk platforms.

However, ArcGIS Pro does not have the ability to model urban geometries, which culminated in the need to add a tool capable of filling this gap. Therefore, the tool CityEngine [21] was inserted into the project scope in order to develop the modeling volume of the urban environment (see Fig. 6).

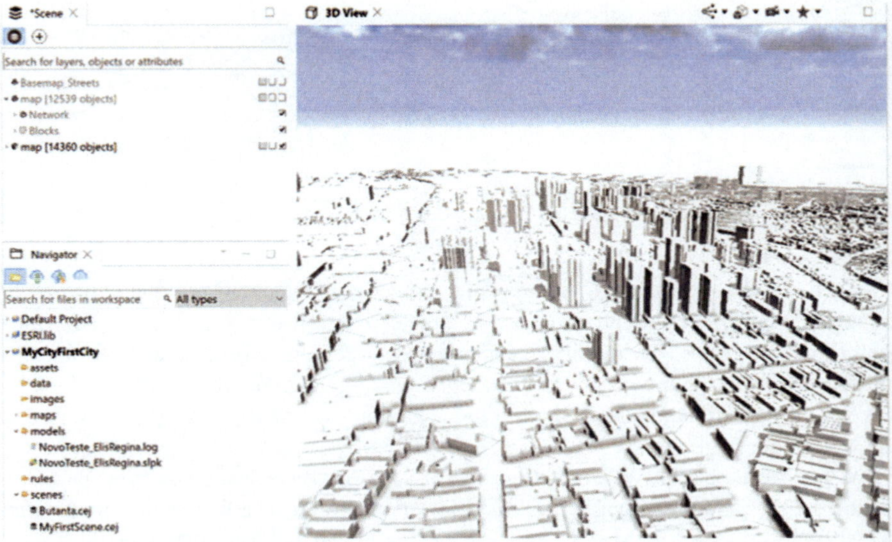

Fig. 6. CityEngine platform interface, from Esri Company.

In the modeling of the urban environment, the streets and sidewalks of the city are relevant to the success of simulation, as it makes up the environment in which the BIM models were inserted. It provides information about the urban environment through the parameterization performed on the GIS platform.

In this sense, the CityEngine software was used for the volumetric modeling of the urban environment through the open data access of the Open Street Maps (OSM) platform. The developed model in CityEngine was imported into ArcGIS Pro software in the Geodatabase format (.GDB) and inserted in the GIS model as a new layer of three-dimensional information.

In the ArcGIS Pro platform, the three-dimensional data obtained from the CityEngine are superimposed to the topography layer. This creates the scene of the urban environment (see Fig. 7) and ends the urban areas volumetric modeling phase.

Fig. 7. Topographic and volumetric data combination in ArcGIS Pro, from Esri Company.

Then, it was necessary to parameterize objects. Unlike BIM, urban geometries created in the GIS environment were parameterized following a single rule, which established an USW production average in the region. This parameterization is important to insert in the urban environment the average USW production provided by Abrelpe [8], and to compare them with the actual production data that were inserted in the BIM modeling.

While Abrelpe's USW production data [8] are estimated averages from data collection on the amount of waste deposited in landfills, that is, a calculation made after the USW production, BIM data is based on the actual number of habitants of each building, which can be updated at any time and are calculated before the production of waste.

As the set of volumes created in the GIS system changes based on different parameters, it was necessary to establish rules that defines useful building areas, building usage type and the number of users of each building. ArcGIS Pro does not have tools that can create these rules automatically. To solve this problem, Python language algorithms were coded so that, when running them in the GIS system, it was possible to categorize the volumetries according to their shapes and, finally, assign the production of USW for each modeled building.

The parameters created in Python language were number of floors, inhabitants, building usage type and rate of USW production. With that, it was possible to establish the USW production per represented volume. Next, it will be explained the logic to code each algorithm to calculate the listed parameters.

Building Usage Type. It's the parameter created from the data of the projected area and height and aims to establish rules that define the use of the building according to the volume size. The parameter is divided into three main categories: shed (for areas smaller than 20 m^2), residential (areas between 20 and 1500 m^2) and commercial (areas above 1500 m^2). The residential category will be further subdivided into three others, through height analysis: house (up to 9 m height), medium standard building (between 9 and 45 m height) and high standard building (45 m height). The shed category will not be used in any subsequent calculation, as they are buildings with minimal dimensions, and it isn't feasible to use them as dwelling, but as a deposit. The algorithm below was developed to calculate this parameter.

```
uso_edif=uso(!area_proj!,!height!)
def uso(area,altura):
  if (area<40):
    return 'comercial_peq'
  elif (area>40 and area<1500):
    if(altura<9):
      return 'res_casa'
    elif(altura>9 and altura<45):
      return 'res_mediopd'
    elif(altura>45):
      return 'res_altopd'
  elif (area>1500):
    return 'comercial'
```

[Algorithm for calculating building usage type. Carolina Yosino, São Paulo (2020)]

Floors. This parameter identifies the number of floors in each building using the building's height and building usage type. For this study, it was considered that medium-standard residential buildings have a ceiling height of 2.8 m and high standard buildings have 3.2 m. To not consider the hall areas in the buildings, the parameters does not consider 4 m height in the calculation. For commercial buildings it was considered 4 m ceiling height. The algorithm below represents the floors quantity parameter.

```
pav_edif=pavimentos(!height!,!uso_edif!)
def pavimentos(altura,uso):
  if (uso=='comercial_peq'):
    return 1
  elif (uso=='res_casa'):
    return 1
  elif (uso=='res_mediopd'):
    return ((altura-4)/2.8)
  elif (uso=='res_altopd'):
    return ((altura-4)/3.2)
  elif (uso=='comercial'):
    if (altura<=4):
      return 1
    elif (altura>4):
      return (altura/4)
```

[Algorithm for calculating the number of floors. Carolina Yosino, São Paulo (2020)]

Inhabitants. This parameter determines the number of inhabitants per volume. The used inputs to determine this parameter are area, floors and building usage type. Based on a study made by Ferreira and Castiñeiras [22], an average of 3.2 inhabitants per dwelling unit was established. In medium standard buildings, it was considered 4 dwelling units per floor, and 2 units per floor in for high standard buildings. In the case of commercial units, there is no statistic that determines how many employees each company has, since this data will depend on the commercial segment. Thus, it was assumed that all businesses in the region have 1 employee for each 80 m^2 of commercial area. The algorithm to calculate the number of inhabitants per volume is shown below.

```
hab_edif=hab(!area_proj!,!pav_edif!,!uso_edif!)
def hab(area,pavimentos,uso):
  if (uso==‘res_casa’):
    return 3.2
  elif (uso==‘res_mediopd’):
    return (pavimentos*4*3.2)
  elif (uso==‘res_altopd’):
    return (pavimentos*2*3.2)
  elif (uso==‘comercial_peq’):
    return 1
  elif (uso==‘comercial’):
    return ((area*pavimentos)/80)
```

[Algorithm for calculating the number of inhabitants. Carolina Yosino, São Paulo (2020)]

USW Production Rate. Parameter that defines the production of urban solid waste per capita (kg/inhab.day). The rate of USW production varies according to the building usage type. For residential type buildings, in the Southeast region in Brazil in 2017, it was considered the average production per capita of 1.217 kg/inhab.day [8]. For commercial buildings, it isn't possible to accurately measure the production of USW, since this value depends on the type of business. Therefore, for commercial buildings, an average above the residential per capita was stipulated, only as a way of differentiating contribution according to the use of the building. Then, the average used for this type of volume was 1.362 kg/inhab.day, presented by Abrelpe's annual report [8]. The production of USW for large commercial buildings varies too much and it also has specific collection process. So, the large commercial buildings were excluded from the study. The algorithm below shows the calculation of the USW production rate.

```
TX_RSU=taxa(!uso_edif!)
def taxa(uso):
  if (uso==‘res_casa’):
    return 1.217
  elif (uso==‘res_mediopd’):
    return 1.217
  elif (uso==‘res_altopd’):
```

```
    return 1.217
  elif (uso=='comercial_peq'):
    return 1.362
  elif (uso=='comercial'):
    return 0
```

[Algorithm for calculating the USW production rate. Carolina Yosino, São Paulo (2020)]

USW Production. The last parameter is related to the calculation of the USW production generated by each building per day. This parameter was obtained by multiplying the number of inhabitants with the USW production rate, and it was represented in kg/day. This parameter was used to compose the USW production data within BIM data to evaluate the USW collection in the region. The production of USW is calculated by the following algorithm.

```
prod_rsu = (!hab_edif!*!tx_rsu!)
```

[Algorithm for calculating USW production. Carolina Yosino, São Paulo (2020)]

Summary of GIS Data. Through the parameterization rules coded on Python, all the set of GIS volumes are parameterized. Altogether, 1,766 volumes were modeled in the GIS environment, which represents residential and commercial buildings. The Table 2 displays the summary of the GIS data.

Table 2. Summary of GIS modeling data.

Summary – GIS modeling								
Use of building	Modeled units	Área (m^2)		Height (m)		Inhabitants /unit (average)	USW production (Kg/day)	
		Minimum	Maximum	Minimum	Maximum		Per capita	Total
comercial_peq	117	0	40	–	–	1.0/unid	1362	159.35
commercial	165	1500	–	–	–	0	0	0
res_casa	1355	40	1500	0	9	3.2	1217	5276.91
res_mediopd	87	40	1500	9	45	3.2	1217	9701.36
res_altopd	42	40	1500	45	–	3.2	1217	5659.97

4.3 BIM and GIS Data Integration

This stage was focused on the interoperability of BIM and GIS models to enable the modeled to be integrated and both analyzed together.

To promote data interoperability, three ways of integrating BIM volumes with GIS data was studied:

a. IFC and CityGML: The first proposal was to export BIM data to IFC format, convert the IFC format to CityGML and then, insert the file in the GIS environment.

The study of this type of data transformation was made by using the FME Workbench tool, which is a visual workflow editor for data conversion. Through FME Workbench, it was noted that lots of information about BIM modeling had lost when converting .IFC data to .GML. This occurs because several parameters contained in the IFC such as data related to installations, floors and cladding standard, do not correspond to the CityGML standard. Since for the development of this artifact, the definition of pavements with their parameters is relevant and the conversion of IFC data to CityGML is complex, it was decided not to follow this path for BIM and GIS interoperability.

b. Export Revit in .RVT + database formats: This form of data integration seeks to export geometric data in .RVT format to be inserted in the GIS environment as a three-dimensional data layer, and the parametric data was exported to a database system, in this case, the Microsoft Access format (.ACCDB). The dissociation of geometric data from parametric data brings as benefits (i) the power to select which parameters will be inserted in the CIM model and (ii) a cleaner file to be imported into the GIS system. However, when BIM data from Revit was exported in a database format, there was no georeferencing associated with this data, so, it was necessary to assign location coordinates for each object properly. Also, it was needed to convert the data from the .ACCDB format to the Shapefile .SHP standard, again through the FME Workbench tool.
c. Export BIM to .RVT format only: The last analyzed way to insert BIM data in GIS environment was through a single file of the type .RVT. ArcGIS Pro can understand the information contained in the .RVT file when it's inserted in Revit's Layer Floors. Therefore, it was necessary to institute all BIM parameterization in the Floors layer.

Finally, it was concluded that export information through database system with volumetric data through .RVT file (b), and export BIM to .RVT format only (c) are both viable solutions in the simulation. However, for the development of the artifact, preference was given to data integration through the single import of data in the .RVT format, since the ArcGIS Pro platform has good interoperability with Revit.

However, it's important to mention that the integration of GIS data from a database (b) is also interesting for the formation of a CIM model, since it allows greater flexibility in the choice of data to be exported to GIS and enables to make a faster and simpler analysis into a large volume of data in the database.

4.4 Data Analysis in CIM Environment

The last stage of the project was intended to analyze the integrated data from GIS and BIM to create an CIM environment where urban and building data are presented and used together. Data related to the production of USW based on BIM and GIS modeling were analyzed together to obtain route of USW collections that meet the demand of the pre-defined area. The workflow in Fig. 8 presents the steps to get the data analysis in a USW route collection.

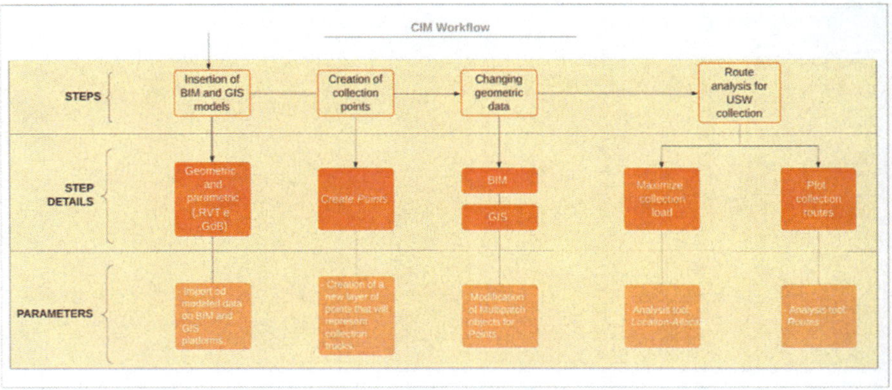

Fig. 8. CIM modeling stage workflow.

First, BIM and GIS data were integrated by importing the BIM file in .RVT format into the GIS environment. The .RVT file is inserted in the GIS platform as a new layer of three-dimensional information, and together with urban data available in GIS environment. Then, a simulated urban environment is obtained, characterized by topographic, geographic data, construction volumes with their parameters (see Fig. 9).

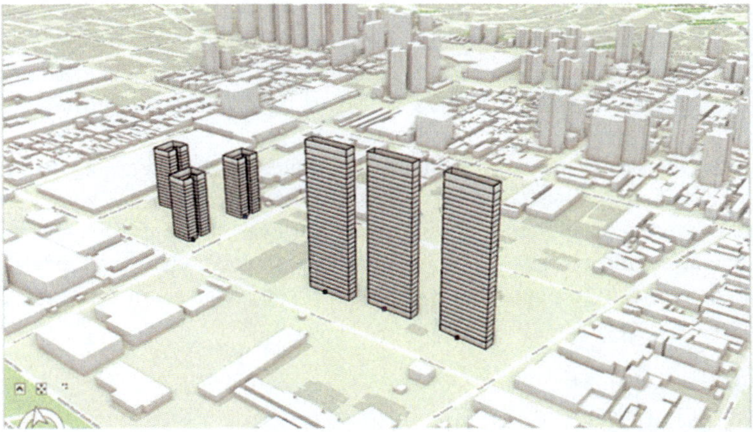

Fig. 9. Volumetric BIM and GIS data integrated into ArcGIS Pro platform.

With the BIM and GIS data, it was found that the studied area generated 22,278.32 kg of USW per day. To identify the collector points in CIM modeling, georeferenced points were created using the Create Points tool from ArcGIS Pro. Each created point was assigned to a collector which name and maximum capacity parameters set for each one, as shown in Fig. 10. In this study, 22 collection points were created with a capacity of 1.0 ton each.

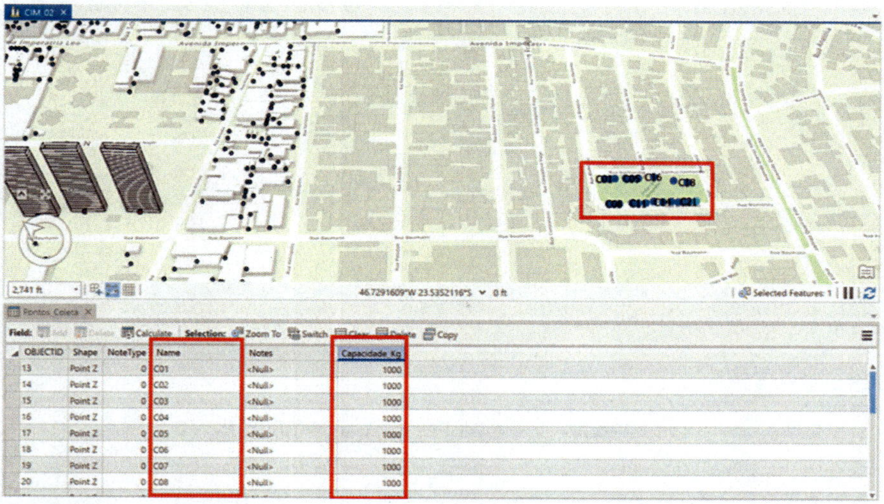

Fig. 10. Creation of collectors in CIM modeling.

The third identified step in the CIM workflow was the transformation of BIM and GIS Multipatch objects type to Points objects type. Esri [23] explains that the Multipatch is applied to GIS objects that store a collection of snippets that together represent a three-dimensional object. According this study, it could be attested that to develop a data analysis where each building will represent a parameter in a georeferenced space, it's recommended that each volumetry is represented by a single geolocated point, and not a set of snippets. Therefore, it was necessary to transform Multipatch objects to Points.

The ArcGIS Pro platform provides a tool called Feature Vertices to Points to perform this transformation. Through this tool, each volumetry generate a number of points that represent each vertex of the object, however, the tools also generate, as a result, several duplicated data. To eliminate those duplications, it was developed an algorithm in Python to recognize the duplicated data and retains only one representative point of each volumetry.

```
import arcpy
  fc = r' <building points file directory>
  selected_old = []
  osm_id_aux = `` ''
  fc = arcpy.MakeFeatureLayer_management(fc,<building demand point>)
  with arcpy.da.SearchCursor(fc, ['OBJECTID_1', 'ObjectId']) as cur:
    for i, row in enumerate(cur):
      if not (str(row[1])) == osm_id_aux:
        osm_id_aux = str(row[1])
        selected_old.append(str(row[0]))
  where = 'OBJECTID_1 in ({})'.format(','.join(selected_old))
```

```
arcpy.SelectLayerByAttribute_management(fc, "NEW_SELECTION", where,
'INVERT')
arcpy.DeleteRows_management(<building demand points>)
```

[Algorithm to recognize the duplicated data. Carolina Yosino, São Paulo (2020)]

The Fig. 11 presents the result of the duplication algorithm. On this figure, the modeled volumetries have only one georeferenced point to represent themselves.

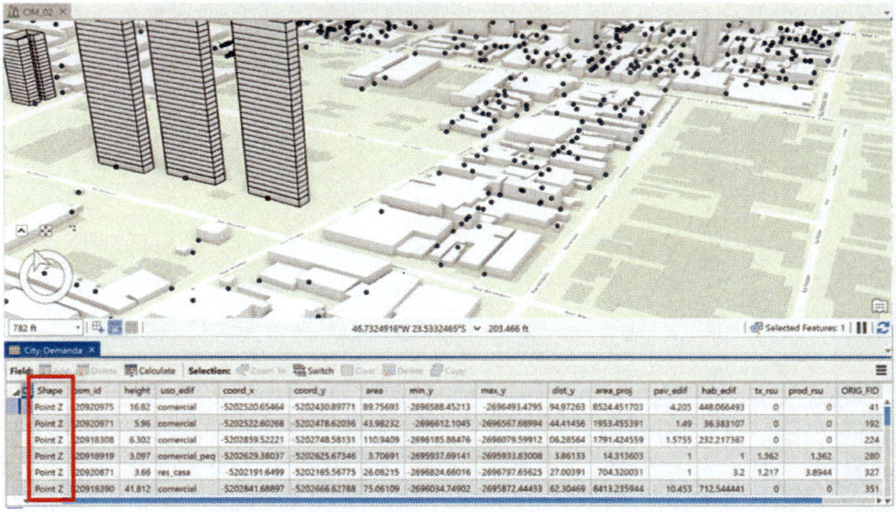

Fig. 11. Volumetries represented by a single georeferenced point.

With this transformation of data format, the artifact displays BIM and GIS volumetric in the proper format to develop the analysis of the collection of USW through the Network Analysis tool. In this study, two types of analysis were performed: Maximize collection load and Define Routes.

The Maximize Capacitated Coverage tool has the main function of allocate demand for each defined facility to maximize the full use of the permissible load. For the development of this analysis, the demand will be represented by the production of USW for each building, while the facilities will be represented by the collectors.

To define the data collectors and facilities in the Maximize Capacitated Coverage tool, it was necessary to link the relevant data collectors with the default facilities model. Thus, maximum truck collection quantity will be linked with the data of the Capacity tool. It was also determined that the maximum range for each facility was 10 km.

After running the analysis in the Maximize Capacitated Coverage tool, it returned buildings which collector will serve each building in order to maximize the collect load usage. In the Fig. 12, each blue line represents a demand met by the collector. The column *DemandWeight* set the quantity (in kilograms) for each collector load, while column *DemandCount* input how many buildings are being served by each collector.

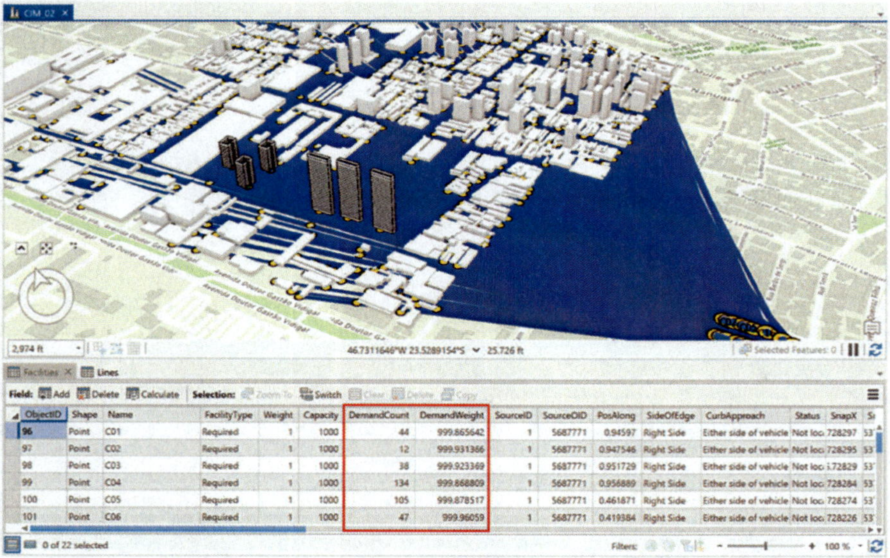

Fig. 12. Analysis maximize capacitated coverage in ArcGIS Pro.

Through the analysis of the generated data, it's noted that all collectors presented the load usage above 97%. That met the initial study premise of create routes that seeks to use the maximum capacity of each collector. However, the tool does not search for demand points close to each other to create more efficient routes.

After generating from maximize collectors load usage process, it was necessary to trace the routes for each collector in order to meet all the served demand points. For this, the data obtained through the Maximize Capacitated Coverage analysis was used in the Create Routes tool.

The Create Routes tool from ArcGIS Pro is used to trace the shortest path between predetermined points, called Stops. The obtained data in the previous analysis to relate the served points with their collectors will be used Stops in the Create Routes tools. Thus, routes were drawn for each collector, identifying the necessary routes for the total use of their loads.

When combining the data from the Maximize Capacitated Coverage analysis with the Create Routes tool, a limitation was noted in the data analysis of the ArcGIS Pro system, which does not allow more than 150 points to be connected on the same route. It was noted that, due to the limitation of the system itself, for the general service of the region it would be necessary to use an additional collector. The 22 previously sized collectors meet the demand presented, however, for this purpose, some routes would pass through more than 150 demand points, causing the tool to malfunction. Therefore, for the correct operation of the route analysis, it was decided to add one new collector in the region. Thus, it was necessary to create an algorithm in order to identify routes with more than 150 points and transfer additional demand points to the path of a collector with available load capacity. The algorithm below presents this function.

```
#Select and delete all stops with RouteName equal to Null
arcpy.SelectLayerByAttribute_management('CIM Network Analysis\Route
\Stops', 'NEW_SELECTION', 'RouteName is Null', 'NON_INVERT')
arcpy.DeleteRows_management('CIM Network Analysis\Route\Stops')
#Select all stops with a sequence starting at 150
arcpy.SelectLayerByAttribute_management('CIM Network Analysis\Route
\Stops', 'NEW_SELECTION', 'Sequence >= 150', 'NON_INVERT')
#Select all facilities with DemandCount less than 149
arcpy.SelectLayerByAttribute_management('CIM Network Analysis\Location
-Allocation\Facilities', 'NEW_SELECTION', 'DemandCount < 149 and Facil-
ityType <> 0', 'NON_INVERT')
#Based on the selected stops and facilities, the algorithm makes the
adjustment so that all sequences reach up to a maximum of 149
fieldsStops = ['ObjectID','RouteName','Sequence','prod_RSU']
fcStops = 'CIM Network Analysis\Route\Stops'
fcFacilites = 'CIM Network Analysis\Location-Allocation\Facilities'
cursorStops = arcpy.da.UpdateCursor(fcStops, fieldsStops)
for rowStops in cursorStops :
  allocated = False
  cursorFacilities = arcpy.UpdateCursor(fcFacilites)
  for rowFacilities in cursorFacilities :
  if ((rowStops[3] <= (1000 - rowFacilities.getValue('DemandWeight'))) and
(not allocated) and (int(rowFacilities.getValue('DemandCount')) < 149)) :
 rowFacilities.setValue('DemandCount', (rowFacilities.getValue('Demand-
Count')+1))
    rowStops[2] = rowFacilities.getValue('DemandCount')+1
    rowStops[1] = rowFacilities.getValue('ObjectID')
 rowFacilities.setValue('DemandWeight',rowFacilities.getValue('Demand-
Weight')+rowStops[3])
    cursorStops.updateRow(rowStops)
    cursorFacilities.updateRow(rowFacilities)
    allocated = True
#Recalculate data from facilities that had more than 149 DemandCount
fcStops = 'CIM Network Analysis\Route\Stops'
fcFacilites = 'CIM Network Analysis\Location-Allocation\Facilities'
arcpy.SelectLayerByAttribute_management(fcFacilites, 'NEW_SELECTION',
'DemandCount <= 149 and FacilityType <> 0', 'INVERT')
cursorFacilities = arcpy.UpdateCursor(fcFacilites)
arcpy.SelectLayerByAttribute_management(fcStops, 'NEW_SELECTION')
for rowFacilities in cursorFacilities :
   weightSum = 0.0
   arcpy.SelectLayerByAttribute_management(fcStops, 'NEW_SELECTION',
'RouteName = \" + str(rowFacilities.getValue('ObjectID')) + '\"')
   demandCountFc = int(arcpy.GetCount_management(fcStops)[0])-1
   rowFacilities.setValue('DemandCount', demandCountFc)
```

```
cursorStops = arcpy.UpdateCursor(fcStops)
for rowStops in cursorStops :
 if not (rowStops.isNull('prod_RSU')) :
  weightSum = float(rowStops.getValue('prod_RSU')) + weightSum
rowFacilities.setValue('DemandWeight', weightSum)
cursorFacilities.updateRow(rowFacilities)
```

[Algorithm for the analysis of Maximize Capacitated Coverage. Carolina Yosino, São Paulo (2020)]

The developed algorithm allowed that each route represented by a collector had no more than 150 collection points. Thus, from the use of the Routes Create tool, it was set the smaller distances routes for each collector to meet all served points (see Fig. 13).

Fig. 13. Route analysis in ArcGIS Pro.

As can be seen in Fig. 13, the route of each collector is represented by a different color in ArcGIS Pro for the purpose of visual identification of the traced route by each collector.

5 Results and Discussions

Through this research it was possible to develop a computational modeling based on the concept of City Information Modeling to analyze the logistics of the collection of USW in a pre-defined urban region. To achieve that purpose, it was used data from parameterized buildings in a BIM environment.

For that, the Design Science Research methodology was applied in this research with the purpose of developing the simulated CIM environment as an artifact. The main objective with the development of the artifact was to make possible the union of information from two areas of knowledge: Urban Engineering and Building Construction Engineering, in order to present a new way of visualizing the production of USW in a given urban area, through the analytical study of a simulated environment.

With the artifact in functional state, it's possible to analyze USW collection routes in a given region from data provided by parameterized construction elements in a BIM environment, thus obtaining a simulation with more accurate data on the USW production of each residence. It's also feasible to modify parameters such as housing density of each building, USW production rate and load volume of each collection truck according to the reality and changing of the region, making the planning phase more accurate and closer to the reality.

For the development of the artifact, it was necessary to join parameterized data in BIM and GIS modeling. To this end, it used the technologies such as Revit, from the Autodesk developer company, City Engine and ArcGIS Pro, from Esri developer company. The integration of data from these modeling tools allowed the integration of urban and constructive data, which culminated in the creation of a simulated CIM environment.

From the simulation in its functional state it was possible to verify the positive points of the artifact:

a. Construction data and urban data were integrated into the GIS platform and the data were read together;
b. The BIM parameters inserted in the GIS environment were recognized and the ArcGIS Pro platform was able used them for the define the USW production in the region;
c. The artifact was able to generate collection routes from the integration of BIM and GIS data;
d. The generated routes had as principle the optimization of the load of the collectors;
e. In the development of this artifact, it was possible to use complete BIM models (volume and parameters), since the interoperability between Revit and ArcGIS Pro systems proved to serve the purpose in this scenario. However, it's important to note that in cases where it isn't possible to send three-dimensional data in a GIS environment, the integration of BIM parameters through a database is also valid for the global analysis of the urban environment.

On the other hand, some negative points of the proposed artifact were also noted:

a. Since the ArcGIS Pro platform wasn't able to generate three-dimensional objects from the urban environment, it was necessary to add one more technological tool for the development of the artifact such as the City Engine. It was used only to model three-dimensional elements. This fact implies in a greater difficulty to create simulated urban models, since the use of many technological tools makes the process of data integration more difficult;

b. It's interesting to import BIM models in GIS with as little information as possible to make the final simulation lighter in the sense of data quantity, which facilitates data processing on the GIS platform, but limits the use of BIM models in LoD1;
c. The artifact wasn't able to generate routes that optimized both the load of the collectors and the shortest routes. For that, it was necessary to create an algorithm that covers these two variables in the logistical analysis.

Finally, through the application of BIM and GIS interoperability, CIM modeling presents itself as a viable solution for planning and controlling the urban environment. In the same way, this work used the solution to study the collection of USW, CIM modeling can also be applied to other urban subsystems, such as energy, sewage and drainage control. With the application of current technologies, it's possible to work with large databases that allow the intersection of information on different areas of Urban Engineering. This will generate a complete CIM model with the ability to have a general perception about the city functioning.

References

1. Lima, M.Q.: Limites e Possibilidades do City Information Modeling (CIM) em Planejamento Urbano. In: IV Encontro Da Associação Nacional De Pesquisa E Pós-Graduação Em Arquitetura E Urbanismo, pp. 1–16. Enanparq, Porto Alegre (2016)
2. Beirão, J.N.: CItyMaker: Designing Grammars for Urban Design. Architecture and the Built Environment, 1st edn, Lisbon, Portugal (2012)
3. Xu, X., et al.: From building information modeling to city information modeling. J. Inf. Technol. Constr. **19**, 292–307 (2014)
4. Jacobi, P.R., Besen, G.R.: Gestão de resíduos sólidos em São Paulo: Desafios da Sustentabilidade. Estudos Avançados **25**(71), 135–158 (2011)
5. Lima, R.S., Lima, J.P., Silva, T.V.V.: Roteirização em arcos com um sistema de informações geográficas para transportes: aplicação em coleta de resíduos sólidos urbanos. J. Transp. Lit. **6**(2), 180–196 (2012)
6. Cunha, V.: Gerenciamento da Coleta de Resíduos Sólidos Urbanos: Estruturação e Aplicação de Modelo Não-Linear de Programação por Metas, 1st edn. University of São Paulo, São Paulo (2001)
7. IPEA – Instituto de Pesquisa Econômica Aplicada: Diagnóstico dos Resíduos Sólidos Urbanos: Relatório de Pesquisa. 1st edn. IPEA, Brasília - Brazil (2012)
8. Abrelpe, C.: Panorama dos Resíduos Sólidos no Brasil 2016, 14th edn. Abrelpe, Brasil (2017)
9. González, G.A., Leal, J.E.: Planejamento das Rotas de Coleta de Resíduos Sólidos Domiciliares Especiais por Meio de Um Sistema de Informações Geográficas para Transporte: Aplicação do Algoritmo do Problema de Roteirização dos Veículos com Janela de Tempo. In: XVIII Simpósio de Pesquisa Operacional & Logística da Marinha, pp. 1–11. Edgard Blücher, Brazil (2016)

10. Almeida, F., Andrade, M.: A integração entre BIM e GIS como ferramenta de gestão urbana. In: 7h Encontro de Tecnologia de Informação e Comunicação na Construção - Edificações, Infra-estrutura e Cidade: Do BIM ao CIM, pp. 371–383. Antac, Porto Alegre (2015)
11. Amorim, A.L.: Cidades inteligentes e city information modeling smart cities and city information modeling. In: XX Congress of the Iberoamerican Society of Digital Graphics, pp. 481–488. SIGraDi, Argentina (2016)
12. Thompson, E.M., et al.: Planners in the future city: using city information modelling to support planners as market actors. Urban Plan. **1**(1), 79–94 (2016)
13. Dantas, H.S., Sousa, J.M., Melo, H.C.: The importance of city information modeling (CIM) for cities' sustainability. In: IOP Publishing, IOP Conference Series: Earth and Environmental Science, vol. 225, pp. 1–8 (2019)
14. Cavalcanti, A., De Souza, F.: O Uso do CIM e a Difusão das Ideias no Campo das Políticas Públicas no Setor do Gerenciamento das Cidades. In: 7h Encontro de Tecnologia de Informação e Comunicação na Construção - Edificações, Infra-estrutura e Cidade: Do BIM ao CIM, pp. 400–410. Antac, Porto Alegre (2015)
15. Bertei, R.M., et al.: Desenvolvimento de um Sistema de Informação para o Gerenciamento de Redes de Infraestrutura Urbana. RAIMED - Revista de Administração IMED **4**(3), 300–313 (2014)
16. Kymmell, W.: Building Information Modeling: Planning and Managing Construction Projects With 4D CAD and Simulations, 1st edn. The Mcgraw-hill, New York (2008)
17. Lacerda, D.P., et al.: Design science research: método de pesquisa para a engenharia de produção. Gestão Produção **20**(4), 741–761 (2013)
18. BIM Forum: LOD Spec 2019: For Building Information Models. 1st edn. BIM Forum, EUA (2019)
19. Revit Overview Homepage. https://www.autodesk.com/products/revit/overview. Accessed 18 Jan 2020
20. ArcGIS Pro Overview Homepage. https://pro.arcgis.com/en/pro-app/get-started/get-started.htm. Accessed 10 Jan 2020
21. CityEngine Overview Homepage. https://www.esri.com/en-us/arcgis/products/esri-cityengine/overview. Accessed 12 Jan 2020
22. Ferreira, C.E., Castiñeiras, L.L.: Transição Demográfica e Demanda Por Moradias: projeção de domicílios do Estado de São Paulo até 2050. In: SP Demográfico: Resenha de Estatísticas Vitais do Estado de São Paulo, vol. 1, no, 18, pp. 1–14. São Paulo (2018)
23. Esri's 3D analysis webpage. https://desktop.arcgis.com/en/arcmap/latest/extensions/3d-analyst/multipatches.htm. Accessed 20 Jan 2020

Information Management in AEC Projects: A Study of Applied Research Approaches

David Fürstenberg[1,2(✉)]

[1] Norwegian University of Science and Technology (NTNU), 7491 Trondheim, Norway
david.furstenberg@ntnu.no
[2] COWI Norway, 5058 Bergen, Norway

Abstract. Using the potential that lies within digital transformation seems to be a promising response to reverse the traditional low productivity within the AEC (Architecture, Engineering and Construction) industry. One possible approach to the digital transformation of this industry has been identified as increased and effective use of information in projects. Since only structured information can create value, information management is necessary. This paper provides a systematic overview of the available research on information management in AEC projects with the focus on research approaches. 27 papers were thoroughly analyzed and showed that researchers often give little attention to a detailed description of their research approach. An interpretative/constructivist and pragmatic worldview seems to dominate. The observed research purpose was mainly explorative-descriptive which suggests that information management in AEC projects is a rather young field of research. The most common observed research methods are qualitative and design. Data is mainly collected by case studies and software prototypes are presented as artefacts. Most analyzed papers were conceptual and therefore no data analysis was needed. Whereas the most used method of data analysis for empirical papers was statistical. More empirical evidence through case studies of information management and especially the testing of proposed frameworks and prototypes in real-world projects seems urgent.

Keywords: Information management · AEC projects · Systematic literature review

1 Introduction

Over the last decades, productivity in the AEC (Architecture, Engineering and Construction) industry has not increased as much as in other industries [1, 2]. Using the potential that lies within digital transformation to accomplish faster planned, designed and built projects seems a promising response to reverse this trend. While the AEC industry traditionally is not an early adopter of new trends [3, 4], new demands from public clients are pushing the industry towards digital transformation.

One possible approach to the digital transformation of the AEC industry has been identified as increased and effective use of information in projects [2, 5]. Software development has made it possible to produce information-rich Building Information

E. Toledo Santos and S. Scheer (Eds.): ICCCBE 2020, LNCE 98, pp. 272–284, 2021.
https://doi.org/10.1007/978-3-030-51295-8_20

Models (BIM) for both buildings and infrastructure projects [6, 7]. Geospatial data contain information about the environment and existing assets. Using geospatial data and Geographic Information Systems (GIS) seems therefore like a promising solution to enhance BIMs even more. Beside graphical information provided in BIM or GIS there is also non-graphical information that is important in AEC projects like e.g. quality assurance documentation. Designers, however, rarely use the available information efficiently since it is unstructured and provided in different silos. Information management is therefore necessary to create value from the available unstructured information.

Since all research should be well grounded in theory and should provide a sound research design [8], this paper aims to provide a systematic overview of the available research on information management in AEC projects. Therefore, the focus of this paper is on the research methodology by addressing the following research questions:

RQ1: What is the most common research approach in information management in AEC projects?

RQ1.1: Which philosophical worldviews is the research on information management in AEC projects mainly based on?

RQ1.2: What is the most common research purpose in information management in AEC projects?

RQ2: What are the most common research methods in information management in AEC projects?

RQ3: What is the most common form of data analysis in information management in AEC projects?

AEC projects can be classified into different phases. In this paper the Norwegian system "Next step" as outlined by Knotten et al. [9] was used. Only papers describing the design and construction phases were included. The author limited the investigation to these phases to focus on the collaboration between design consultants and contractors. According to Fürstenberg and Lædre [10] information management in these phases needs more investigation. Thus, all papers only describing information management in the asset management phase were excluded. Another limitation is on the acronym "AEC" as a search term. While the acronym includes "construction", using it as a search term excludes papers that only use "construction" and not the acronym "AEC" to describe this industry. Including both "AEC" and "construction" would return too many hits and make a qualitative assessment impossible.

2 Methodology

A literature review was carried out according to the steps described by Blumberg et al. [11]: 1) build information pool, 2) apply filter to reduce pool size, 3) rough assessment of sources to further reduce pool size, 4) analyze literature in pool and 5) refine filters or stop search.

Firstly, the information pool was built by searching in the following databases: Scopus and Oria – a Norwegian academic library database. Due to limited functionalities, Oria was only used for double checking the number of hits returned from

Scopus. According to Blumberg et al. [11], words that appear in the working title or theoretical concepts presented in the paper are important keywords. Therefore, the databases were searched for both "information management" and "AEC" in all fields using the following expression: ALL ("information management" AND aec). This returned around 1,000 hits.

Secondly, filters were applied to only include peer-reviewed journal articles. This was done by ticking off for document type "article". The reviewing process of journal articles secures a high standard of quality at the expense of actuality. Conference papers have either a shorter or non-existing review process. That makes them more actual, but an individual quality assessment of the conferences is necessary. Due to limited resources, such a quality assessment was not possible, and quality was prioritized before actuality. Furthermore, books were excluded because they can consist of assembled publications based on several research approaches and philosophies. This returned 590 articles on Scopus and 526 articles on Oria. Another filter was applied to only include papers that have "information management" in the abstract. This returned 73 articles from Scopus – in Oria this filter was not available.

Thirdly, the titles and abstracts were roughly assessed to check for relevance for this study. Only articles in English describing the design and construction phases were considered. In total 29 articles were found to be relevant for information management in AEC projects.

The steps 4 and 5 involve a full-text assessment of each hits and to either stop the search or refine the filters if the pool is still too large. Since the number of the hits was rather small, no further filters were applied, and the search was stopped. Two papers were regarded irrelevant and excluded after the full-text assessment. Thus, the final sample consisted of 27 unique peer-reviewed journal articles (Fig. 1).

Lastly, all the papers of the final sample were analyzed regarding the underlying research approach, research methods and analyses of the collected data. For this purpose, an Excel spreadsheet containing their attributes was created. Title, author name(s) and publication year were exported from a reference handling tool. The following attributes were extracted from the papers: research approach, philosophical worldview, research purpose, research method(s), instrument(s) of data collection and type of data analysis.

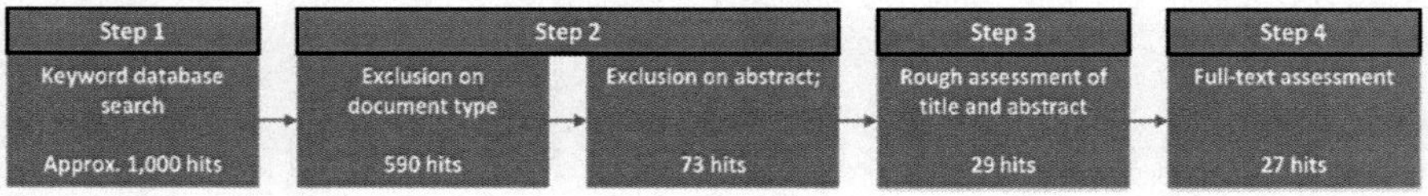

Fig. 1. Filtering process to obtain the final sample

3 Theoretical Background

3.1 Information Management

Information management is a term used in different fields of research (e.g. business and management, information and communication technology (ICT), or librarianship [12]). Detlor [13] describes information management as "the management of the processes and systems that create, acquire, organize, store, distribute, and use the information". He outlines three major perspectives: organizational, library and personal. From the library perspective, information management concerns the collection of information and providing it to users. From the organizational perspective, information management serves the goal to archive business objectives and strategies. From the personal perspective it serves the goal to archive objectives of the individual. Information management of AEC projects has the organizational perspective.

With the ever-increasing number of data files created in our society, the (unstructured) information increases as well. AEC projects are no exception and produce different forms of information, both graphical (e.g. 3D models, drawings, visualizations) and non-graphical (e.g. technical guidelines, technical reports or QA documentation). While the importance of information management of AEC projects was reported on earlier [14, 15] it was the digital transformation that raised its importance even more. For example, it is possible for an experienced user to find relevant information on a drawing even if it is unstructured. Whereas it is difficult to almost impossible to find relevant information in digitally transformed projects where information is only stored in digital models. A systematic management of information is therefore necessary. "Better Information Management" as an alternative description of the acronym BIM or the designated role of an "information manager" in the UK are other indicators for the increased importance of information management in AEC projects. Furthermore, Fischer et al. [16] report that especially new delivery forms within AEC projects, namely Integrated Project Delivery (IPD) could benefit from "integrated information" by information management.

For the author, information management is a managerial task supported by IT systems that aims to improve the workflow within AEC projects. Therefore, the overall goal of information management is to provide the right information at the right time for the right users.

3.2 Digital Transformation

Practitioners and researchers use different definitions of the digital transformation process in AEC projects, some of the definitions are frequently used interchangeably. I-scoop [17] distinguishes "digitization", "digitalization" and "digital transformation".

- **digitization:** "transformation from analog to digital (…) with the goal to digitize and automate processes or workflows".
- **digitalization:** "use of digital technologies and of data (…) in order to create revenue, improve business, replace/transform business processes and create an

environment for digital business, whereby digital information is at the core." In other words, using the digitized data to create an improved product.

- **digital transformation:** builds upon digitization and digitalization and "encompasses all aspects of business, regardless of whether it concerns a digital business or not, … ultimately leading to a new economy."

3.3 Classification of Research Approaches

To set research in a broader context, especially for research within social sciences, it is important to know its theoretical perspective or "philosophical worldview" [8]. This worldview influences the researcher and should therefore always be addressed in advance. Several authors use different classifications. While Neuman [18] describes five types of social sciences – *positivistic, interpretative, critical, feminist, and post-modern*, Creswell and Creswell [8] describe four worldviews – *post-positivism, constructivism, transformative and pragmatism*, Saunders et al. [19] describe four slightly different categories – *positivism, interpretivism, realism, and pragmatism* and Blumberg et al. [11] describe three categories – *positivism, interpretivism and realism*. Based on these different classifications, the author regarded the following three categories most appropriate for the purpose of this paper.

(Post-)positivism: Positivists think that all knowledge already exists, and theory needs to be deducted from it. Post-positivists however agree that knowledge is not the absolute truth for studying human behavior [8]. (Post)positivist researchers are objective, believe only in "real facts" [20] and have the explanation of causality as their overall goal. This worldview is rooted in natural sciences and was introduced in the 19th century. Quantitative methods are used for data collection.

Interpretivism/Constructivism: Constructionists think that knowledge depends on the social context and needs to be created inductively by interpreting observations [21]. Alvesson and Sköldberg [20] describe an "aha experience" as central in many publications. Interpretivist/constructivist researchers are not objective and often they "want to criticize, change, or destroy some X that they dislike in the established order of things" [22]. Creswell and Creswell [8] state that constructivist researchers rely "as much as possible on the participants' view of the situation studied" and that they recognize that their own background influences the interpretation of the observed phenomena. This worldview was introduced in the 1960ies. Qualitative methods are used for data collection.

Pragmatism: Pragmatists center their research around the research question [19]. It is also important to build a rationale [8]. Pragmatist researchers can be objective or subjective, but they acknowledge that research occurs in social contexts. This worldview was introduced in 1980ies. Pragmatists use mixed methods for data collection. The use of methods is not strict, they shall serve to understand the problem [23]. Knowledge is created both deductively, inductively and abductively.

The above described worldviews imply different forms of knowledge creation; *deductive, inductive and abductive*. When deducing knowledge, researchers formulate hypotheses based on existing theory. The hypotheses need to be verified or falsified by testing. When inducing knowledge, researchers turn this process upside down. Theory does not exist yet; it needs to be created inductively by observing real life phenomena. Research questions are used instead of hypotheses. Abduction can be seen as a combination of these two forms. Theory does not exist and needs to be created by observation. Since generalizing from cases may lead to wrong conclusions, the theory needs to be verified deductively by testing at a later stage.

Research can also be approached by the existence of data collection. On the one hand, researchers do collect data and rely on testing before considering ideas as knowledge [24]. In this case they conduct *empirical research*. Normally, empirical studies create knowledge by deduction. However, this depends on the number of the investigated sample. With small samples the knowledge is rather created by induction. On the other hand, researchers do not collect data (or only to a limited extend) and rely on their experience for creating theoretical frameworks. In this case they conduct *conceptual research*. Therefore, conceptual studies create knowledge by induction. However, conceptual research often develops models that are tested empirically at a later stage. In this case, the knowledge is created by abduction.

Yet another research approach is by its purpose, like described by Yin [25]. He uses the following three categories:

- **descriptive:** describes "a phenomenon in its real-world context"
- **exploratory:** identifies research questions or procedures for subsequent studies
- **explanatory:** explains *"how or why some conditions came to be"*

3.4 Classification of Research Methods

Creswell and Creswell report on [8] three research methods for empirical studies within social sciences: quantitative, qualitative and mixed methods.

Quantitative Methods: based on quantitative information which can be collected by surveys or experiments.

Qualitative Methods: based on qualitative information (text, audio, video) which can be collected through case studies, document analyses, observations or interviews.

Mixed Methods: a combination of quantitative and qualitative methods.

Vaishnavi and Kuechler [26] report on **design as a research method**, mainly applied for conceptual studies within Information Sciences. Software prototypes or "artefacts" [26] are the result of such studies.

3.5 Classification of Data Analysis

Data gathered by quantitative methods consist of numbers and can therefore be analyzed by statistical means. Specialized data programs like SPSS are most commonly used. According to Creswell and Creswell [8] it is important to inform the reader about the types of statistical analyses used.

Blumberg et al. [11] describe content analysis as the most common form of qualitative data analysis. In short, this means coding transcripts, documents, audio and video material. Creswell and Creswell [8] describe this process as making sense out of the collected data by taking them apart and putting them back together.

4 Findings

This paper investigated the most commonly applied research approaches, research methods and data analyses in the field of information management in AEC projects.

4.1 Observed Research Approaches

15 of 27 papers of the final sample do not have a method chapter. Information about the research approach and methods is sometimes only briefly mentioned in the abstract, introduction and/or conclusion chapter. Knowledge is equally created by abduction (11 of 27) and induction (13 of 27), while only three of 27 papers deduce knowledge. Most of the papers investigated are conceptual (19 of 27). 13 of these conceptual papers do not have a method chapter (Table 1).

Table 1. Papers of the final sample.

	Has method chapter	No method chapter	Share
Conceptual	[27–32]	[33–45]	70%
Empirical	[46–51]	[52, 53]	30%
Share	56%	44%	

Only two of the 15 papers that do have a method chapter mention their worldview. One states a constructivist and one both a positivist and an interpretivist worldview which Creswell and Creswell [8] would define as pragmatism. Based on the definitions of the worldviews presented earlier and the research methods used, the remaining papers were classified by the author. Only three papers seem to have a positivist approach (using quantitative methods). Both interpretivist/constructivist and pragmatist worldviews seem to be equally represented in the final sample (Fig. 2).

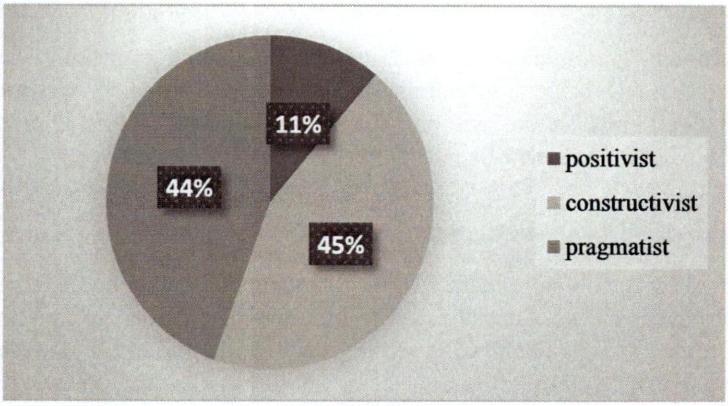

Fig. 2. Philosophical worldview

The research purpose like outlined in [25] was mainly exploratory and to a smaller extend descriptive and explanatory (see Table 2).

Table 2. Research purpose

Purpose	Quantity
Descriptive	7
Exploratory	19
Explanatory	1
Total	**27**

4.2 Observed Research Methods

Table 3 shows that the majority of the investigated papers used qualitative methods (13 of 27), followed by design as a research method, like outlined in [26]. Quantitative and mixed methods are rarely used but equally shared in the final sample.

Table 3. Research methods

Method	Quantity
Quantitative	3
Qualitative	13
Mixed methods	3
Design	8
Total	**27**

In Table 4 the instruments of data collection used in the papers are shown. Although all papers have a literature review, only 12 of 27 papers state that they did such a review to set the research in the context of the existing knowledge. Case studies were mainly used to collect qualitative data. Within the case studies there was an equal share between interviews, observations, document analyses and questionnaires. One paper analyzed typical construction documents like bill of quantities, logfiles or change orders. One paper interviewed engineers from an engineering company about their workflow for data integration. Surveys, logfiles and experiments were equally used to collect quantitative data. In 14 of 19 conceptual papers a software prototype was produced. Four of these prototypes were based on empirical data. 9 of the 14 prototypes were tested under simulated surroundings, none of the prototypes were tested in real projects.

Table 4. Instruments of data collection

Method of data collection	Main method	Second method	Third method	Fourth method
Case study (interviews)	1		1	
Case study (observations)	2			
Case study (questionnaire)	1			
Case study (document analysis)	2			
Literature review	15	12		
Document analysis	1			
Interview	1			
Survey	2			
Logfiles	1			
Experiments	1		1	
Prototype		10	4	
Prototype test (lab)			6	3
Total	**27**	**22**	**12**	**3**

4.3 Observed Data Analysis

Most of the papers investigated (19 of 27) did not explain how they analyzed the data. However, it must be mentioned that 18 of these papers were conceptual. This means only one empirical paper did not describe how the data was analyzed. The other empirical papers analyzed the data either by statistical means or by content analysis like described in [11].

5 Discussion

The findings showed that about half of the investigated papers did not have a chapter explaining the research approach, methods or data analyses used. On the one hand, it must be mentioned that most of these papers are conceptual. Conceptual papers usually describe frameworks or develop new methods or prototypes. Strict use of methods is therefore not as important as in empirical papers. On the other hand, like mentioned by Creswell and Creswell [8] all researchers should have a clear standing towards their philosophical worldview before the research is executed as it will influence them. Since only two author groups stated their theoretical perspective only subjective interpretations of the dominating philosophical worldview could be made. It seems like interpretivist/constructivist and pragmatist approaches dominate and these worldviews represent equal shares of the papers of the final sample.

The high number of conceptual papers presenting software prototypes suggest that the main interest of research within information management in AEC projects is on technical aspects rather than processual. Moreover, this high share of conceptual papers indicates that information management in AEC projects lacks a proper theoretical model.

The research purpose of most of the assessed papers is exploratory or descriptive. This suggests that the field of information management in AEC projects is a rather young field of research. Blumberg et al. [11] explain that exploratory studies work well within fields that are new or little researched.

Qualitative methods are the most common research methods. When design is used as a research method, usually software prototypes are produced and tested under theoretical conditions. The most common instruments of data collection are case studies.

Like mentioned earlier, most of the assessed papers are conceptual ones presenting frameworks, methods or software prototypes. In conceptual studies, data is normally only collected to a limited extend. Therefore, there is limited need for data analyses. Apart from one, all the empirical papers stated how they analyzed their data. Depending on the form of data collection, this was either a statistical analysis or a content analysis.

6 Conclusion

This paper set out to find the most common 1) research approach, 2) research methods and 3) form of data analysis in information management in AEC projects. Interestingly, it appears that researchers concentrate on technical aspects rather than processual aspects, despite the importance of the latter. An overweight of conceptual papers indicates that information management in AEC projects represents a novel field of research without empirically based theory. To put it short, a knowledge gap was identified.

Although most authors describe their applied methods to some extent, they are often not clearly marked and thus making it harder for the reader to validate them. Researchers need to pay more attention to properly displaying their methodology. Moreover, this paper revealed that researchers often do not mention their philosophical worldview explicitly. Researchers need to be more aware of their theoretical foundation or more likely, pay more attention to conveying it in the paper.

While different researchers have pointed out the importance of information management in AEC projects, there seems to be little empirical research on how information management is applied in real-world projects. The mainly explorative-descriptive research purpose of the investigated papers suggests that information management in AEC projects is a rather young field of research. Different theoretical frameworks and software prototypes are proposed, but they are rarely tested in real life projects. More empirical evidence through case studies of information management and especially the testing of proposed frameworks and prototypes in real-world projects seems therefore urgent.

Since most assessed papers lacked a thorough description of the research approach the author had to interpret the papers. Therefore, it might be that some of the subjective interpretations are not aligned with the original intentions of the authors of the papers.

References

1. Hampson, K.D., Kraatz, J.A., Sanchez, A.X.: R&D Investment and Impact in the Global Construction Industry. Routledge, Oxfordshire (2014)
2. Rodrigues De Almeida, P., Bühler, M., Gerbert, P., Castagnino, S., Rothballer, C.: Shaping the future of construction – a breakthrough in mindset and technology. World Economic Forum, Cologny/Geneva (2016)
3. Koskela, L., Vrijhoef, R.: Is the current theory of construction a hindrance to innovation? Build. Res. Inf. **29**(3), 197–207 (2001)
4. Reichstein, T., Salter, A.J., Gann, D.M.: Last among equals: a comparison of innovation in construction, services and manufacturing in the UK. Constr. Manag. Econ. **23**(6), 631–644 (2005)
5. Schober, K.-S., Hoff, P.: Digitization of the construction industry - building Europe's road to "Construction 4.0". Roland Berger GMBH, Civil Economics, Energy & Infrastructure Competence Center, Munich (2016)
6. Eastman, C.M., Teicholz, P., Sacks, R.: BIM Handbook: A Guide to Building Information Modeling for Owners, Managers, Designers, Engineers, and Contractors, 2nd edn. Wiley, Hoboken (2011)
7. Borrmann, A., König, M., Koch, C.: Building Information Modeling. Springer, Heidelberg (2018)
8. Creswell, J.W., Creswell, J.D.: Research Design: Qualitative, Quantitative & Mixed Methods Approaches, 5th edn. Sage, Los Angeles (2018)
9. Knotten, V., Hosseini, A., Klakegg, O.J.: "Next step": a new systematic approach to plan and execute AEC projects. In: Saari, A., Huovinen, P. (eds.) CIB World Building Congress 2016, vol. 3, pp. 484–495 (2016)
10. Fürstenberg, D., Lædre, O.: Application of BIM design manuals: a case study. In: Proceedings of the 27th Annual Conference of the International Group for Lean Construction (IGLC), Dublin, Ireland, pp. 145–156 (2019)
11. Blumberg, B., Cooper, D.R., Schindler, P.S.: Business Research Methods, 4th edn. McGraw-Hill Higher Education, London (2014)
12. Macevičiūtė, E., Wilson, T.D.: The development of the information management research area. Inf. Res. **7**(3), 7-3 (2002)
13. Detlor, B.: Information management. Int. J. Inf. Manage. **30**(2), 103–108 (2010)
14. Howell, G.A.: What is lean construction - 1999. In: 7th Annual Conference of the International Group for Lean Construction, Berkeley, USA, pp. 1–10 (1999)
15. Hicks, B.J., Culley, S.J., Mcmahon, C.A.: A study of issues relating to information management across engineering SMEs. Int. J. Inf. Manage. **26**(4), 267–289 (2006)
16. Fischer, M., Ashcraft, H., Reed, D., Khanzode, A.: Integrating Project Delivery. Wiley, Hoboken (2017)
17. I-Scoop. https://www.i-scoop.eu/digitization-digitalization-digital-transformation-disruption/ . Accessed 10 Jan 2019
18. Neuman, W.L.: Social Research Methods: Qualitative and Quantitative Approaches, 7th edn. Pearson, Harlow (2014)
19. Saunders, M.N.K., Lewis, P., Thornhill, A.: Research Methods for Business Students, 7th edn. Pearson, Harlow (2016)
20. Alvesson, M., Sköldberg, K.: Reflexive Methodology: New Vistas for Qualitative Research, 2nd edn. Sage, London (2009)
21. Crotty, M.: The Foundations of Social Research: Meaning and Perspective in the Research Process. Sage, London (1998)

22. Hacking, I.: The Social Construction of What? Harvard University Press, Cambridge (1999)
23. Rossman, G.B., Wilson, B.L.: Numbers and words: combining quantitative and qualitative methods in a single large-scale evaluation study. Eval. Rev. **9**(5), 627–643 (1985)
24. Bryman, A.: Social Research Methods, 5th edn. Oxford University Press, Oxford (2016)
25. Yin, R.K.: Case Study Research: Design and Methods, 6th edn. Sage, Thousand Oaks (2018)
26. Vaishnavi, V.K., Kuechler, W.: Design Science Research Methods and Patterns: Innovating Information and Communication Technology. CRC Press, Boca Raton (2015)
27. Li, D., Lu, M.: Integrating geometric models, site images and GIS based on Google Earth and Keyhole Markup Language. Autom. Constr. **89**, 317–331 (2018)
28. Yalcinkaya, M., Singh, V.: Exploring the use of Gestalt's principles in improving the visualization, user experience and comprehension of COBie data extension. Eng. Constr. Archit. Manag. **26**(6), 1024–1046 (2019)
29. Olawumi, T.O., Chan, D.W.M.: Building information modelling and project information management framework for construction projects. J. Civ. Eng. Manag. **25**(1), 53–75 (2018)
30. Senescu, R.R., Aranda-Mena, G., Haymaker, J.R.: Relationships between project complexity and communication. J. Manag. Eng. **29**(2), 183–197 (2013)
31. Zou, Y., Kiviniemi, A., Jones, S.W., Walsh, J.: Risk information management for bridges by integrating risk breakdown structure into 3D/4D BIM. KSCE J. Civ. Eng. **23**(2), 467–480 (2019)
32. Akinyemi, A., Sun, M., Gray, A.J.G.: An ontology-based data integration framework for construction information management. Proce. Inst. Civ. Eng. Manag. Procure. Law **171**(3), 111–125 (2018)
33. Lee, D.G., Park, J.Y., Song, S.H.: BIM-based construction information management framework for site information management. Adv. Civ. Eng. **2018**, 1–15 (2018)
34. Jeong, Y.: A study on the BIM evaluation, analytics, and prediction (EAP) framework and platform in linked building ontologies and reasoners with clouds. Adv. Civ. Eng. **2018**, 1–15 (2018)
35. Caldas, C.H., Soibelman, L.: Automating hierarchical document classification for construction management information systems. Autom. Constr. **12**(4), 395–406 (2003)
36. Shahid, S., Froese, T.: Project management information control systems. Can. J. Civ. Eng. **25**(4), 735–754 (1998)
37. Lee, K., Chin, S., Kim, J.: A core system for design information management using industry foundation classes. Comput. Aided Civ. Infrastruct. Eng. **18**(4), 286–298 (2003)
38. Brown, A., Rezgui, Y., Cooper, G., Yip, J., Brandon, P.: Promoting computer integrated construction through the use of distribution technology. Electron. J. Inf. Technol. Constr. **1**, 51–67 (1996)
39. Xu, X., Ma, L., Ding, L.: A framework for BIM-enabled life-cycle information management of construction project. Int. J. Adv. Rob. Syst. **11**(1), 126 (2014)
40. Ding, L., Xu, X.: Application of cloud storage on BIM life-cycle management. Int. J. Adv. Rob. Syst. **11**(1), 129 (2014)
41. Cheng, E.W.L., Li, H., Love, P.E.D., Irani, Z.: A generic conceptual framework for a Total Information Networking System in construction. Int. J. Serv. Technol. Manage. **3**(1), 125–138 (2002)
42. Caldas, C.H., Soibelman, L., Han, J.: Automated classification of construction project documents. J. Comput. Civ. Eng. **16**(4), 234–243 (2002)
43. Ramaji, I.J., Memari, A.M.: Product architecture model for multistory modular buildings. J. Constr. Eng. Manag. **142**(10), 04016047 (2016)
44. Jiang, S., Zhang, H., Dalian, J.Z.: Research on BIM-based construction domain text information management. J. Netw. **8**(6), 1455–1464 (2013)

45. Zheng, R., Jiang, J., Hao, X., Ren, W., Xiong, F., Ren, Y.: BcBIM: a blockchain-based big data model for BIM modification audit and provenance in mobile cloud. Math. Probl. Eng. **2019**, 1–14 (2019)
46. Noordin, M.F., Burhanuddin, L.A., Kanaa, A.: The current state of information management and knowledge management in the Malaysian construction industry. Aust. J. Basic Appl. Sci. **6**(6), 138–145 (2012)
47. Mohamed, S., Stewart, R.A.: An empirical investigation of users' perceptions of web-based communication on a construction project. Autom. Constr. **12**(1), 43–53 (2003)
48. Demian, P., Walters, D.: The advantages of information management through building information modelling. Constr. Manag. Econ. **32**(12), 1153–1165 (2014)
49. Xu, S., Luo, H.: The information-related time loss on construction sites: a case study on two sites. Int. J. Adv. Rob. Syst. **11**(1), 128 (2014)
50. Wu, Y., Wang, Y.: A study of smart construction and information management models of AEC projects in China. Int. J. Simul. Syst. Sci. Technol. **17**(21), 2.1–2.8 (2016)
51. Kang, J.H., Anderson, S.D., Clayton, M.J.: Empirical study on the merit of web-based 4D visualization in collaborative construction planning and scheduling. J. Constr. Eng. Manag. **133**(6), 447–461 (2007)
52. Tsai, M.K.: Improving communication barriers for on-site information flow: an exploratory study. Adv. Eng. Inform. **23**(3), 323–331 (2009)
53. Anumba, C.J., Issa, R.R.A., Pan, J., Mutis, I.: Ontology-based information and knowledge management in construction. Constr. Innov. **8**(3), 218–239 (2008)

Discrete-Event Simulation and Building Information Modelling Based Animation of Construction Activities

Carlos Arturo Osorio-Sandoval[1(✉)], Walid Tizani[1], Estacio Pereira[2], Christian Koch[3], and Abdelaziz Fadoul[1]

[1] Centre for Structural Engineering and Informatics, University of Nottingham, Nottingham, UK
carlos.osoriosandoval@nottingham.ac.uk
[2] Department of Civil Engineering, University of Nottingham, Nottingham, UK
[3] Department of Intelligent Technical Design, Bauhaus-Universität Weimar, Weimar, Germany

Abstract. Discrete-event simulation (DES) technology has been widely used in the construction context, particularly in the academic community for research purposes. However, it has been neglected by industry for various reasons, including the usual way in which its results are presented, which is abstract and confusing. In this regard, animation is viewed as an appropriate tool to reduce the complexity in which simulation outputs are presented to stakeholders. In this paper, we present a brief overview of a framework to integrate DES and building information modelling (BIM) within a game engine. This framework consists of four components, the environment, the user input, the simulation module and the animation module. The simulation module allows users to import their own BIM models; input project parameters related to resource constraints; and simulate the construction of the BIM model with the selected parameters by linking generic DES models of construction activities to the corresponding BIM objects. The animation module allows the user to visualise an animation of the simulation results in a 3D environment or using mixed reality technologies. The resulting animation is useful to verify and validate the simulation model, to analyse the construction activities and as a supporting tool to plan construction activities under resource constrains. The animation can also be used to present simulation results to domain experts unfamiliar with simulation techniques.

Keywords: Building information modelling · Discrete-event simulation · Game engine · Animation

1 Introduction

In the construction context, simulation is the science of modelling construction systems to understand their underlying behaviour and experimenting with the resulting model on a computer [1]. It has become the method of choice for comparative research of construction systems [2], as it provides appropriate means to test what-if scenarios in a timely manner [3]. The prevalent approach for simulating construction systems has

E. Toledo Santos and S. Scheer (Eds.): ICCCBE 2020, LNCE 98, pp. 285–294, 2021.
https://doi.org/10.1007/978-3-030-51295-8_21

traditionally been discrete-event simulation (DES) [4]. DES concerns the modelling of a system as it evolves over time by a representation in which the state variables change instantaneously at separate points in time. These points are the ones at which an event occurs, where an event is defined as an instantaneous occurrence that may change the state of the system [5].

Despite its potential, the use of simulation in construction has been mostly restricted to the academic community for research purposes [1] with limited use in planning and designing construction operations [6, 7]. Researchers in the area have attributed the lack of implementation of simulation in industry to several reasons. Firstly, building and maintaining simulation models requires significant time and effort [8, 9], as well as large amount of data [10]. Secondly, specifying and maintaining the interdependencies between activities and resources needs a significant amount of manual work [10]. Thirdly, the creation of a simulation model is mostly project-specific and reusing models for different projects is generally not possible [1, 11]. Moreover, researchers have identified that the results from simulation models are usually abstract and confusing [12] or shown in statistical and text formats [13], which leads to a lack of trust by industry practitioners toward their effectiveness [7].

In this regard, the field of simulation in construction is moving toward its integration with other tools [4] with a special interest of researchers on enhancing visualisation [6, 13–18]. The gaming industry is one of the leaders in utilizing and advancing visualization technologies [13]. Gaming technologies, like game engines, have drawn the attention of researchers in the field, as they provide ways to achieve immersion in visualisation due to their compatibility with mixed reality technologies, ubiquity as they offer multiplatform deployment, high-quality graphics and network capabilities that allow for multiple simultaneous users, among other characteristics [19].

On the other hand, building information modelling (BIM) is one of the most promising recent developments in the architecture, engineering, and construction (AEC) industry. With BIM technology, an accurate virtual model of a building is digitally constructed [20]. This model contains precise geometry and relevant data needed to support the design, procurement, fabrication, and construction activities required to realize the building [21].

In contrast with simulation in construction, industry practitioners have quickly adopted BIM technologies, as a wide variety of stakeholders are demanding BIM and changing contract terms to enable it [20]. In 2018, 74% of the AEC companies in the UK were aware and using BIM and it is projected that this number will increase to 90% within three to five years [22].

In recent years, researchers have attempted to integrate BIM and construction simulation techniques in an effort to overcome the latter shortcomings [8–12, 23, 24]. However, most attempts have relied on using BIM generated material take-offs as input for the simulation models, which on its own is not sufficient for detailed construction simulations. Moreover, few of these attempts have focused on the visualisation of simulation results.

Game engines have also been integrated with BIM technologies with many proposed applications, from mixed reality architecture visualisation to utilisation of BIM models as game content, which brings the possibility of using real world-based game scenarios able to take advantage of all the information contained in them [19].

In this paper, we present a brief overview of a framework to integrate DES and BIM within a game engine. This framework consists of four components, the environment, the user input, the simulation module and the animation module. The simulation module allows users to import their own BIM models; input project parameters related to resource constraints (e.g. number of workers and their skills, location of raw materials and available tools and equipment); and simulate the construction of the BIM model with the selected parameters by linking generic DES models of construction activities to the corresponding BIM objects. The animation module, which is the focus of this paper, allows the user to visualise an animation of the simulation results in a 3D environment to verify and validate the simulation model and analyse the construction activities. The resulting animation can also be visualised using mixed-reality technology as the game engine in which the framework was developed supports this feature

2 Related Work

2.1 Integration of BIM and DES

Researchers have attempted to integrate BIM with DES in an effort to overcome the shortcomings of the former in terms of the time and effort required to build and maintain them, the manual work required to specify interdependencies between tasks, and the fact that reusing them is generally not possible. Table 1 compares some of the most relevant studies in the area with our proposed framework. The characteristics shown in Table 1 are (1) uses BIM data as input for the simulation model, (2) reduces manual work required to specify and maintain interdependencies between activities, (3) allows reusing of simulation models for different projects, (4) presents a 3D animation of simulation results, (5) presents an animation that includes resources utilisation, and (6) supports mixed reality visualisation.

Table 1. Comparison of research integrating BIM and DES

Reference	Purpose	Characteristics					
		(1)	(2)	(3)	(4)	(5)	(6)
König et al. [11]	Scheduling	x	x	x	x		
Wang et al. [8]	Improve 4D applications	x			x		
Lu and Olofsson [10]	Evaluate construction performance in terms of time	x	x	x	x	x	
Liu et al. [23, 25]	Scheduling	x	x	x			
Chang et al. [9]	Planning	x	x				
Wu et al. [12]	Integration of BIM, DES and ABS	x			x		
Barkokebas et al. [24]	Modular construction manufacturing performance improvement	x	x				
This study	Integrate DES & BIM in a game engine	x	x	x	x	x	x

As can be seen in Table 1, researchers have integrated BIM with DES mainly to take advantage of the data contained in BIM models, which can be used as input for the simulation models in different ways. In [8, 10, 11, 24] the quantity take-offs generated in BIM were used as input for the simulation models. In [23, 25] the entities that move through the simulation model are guided by enriched information extracted from BIM and a database with WBS information. In [9] BIM data, such as materials, geometry and size were translated into resource data for the simulation. In [12] a BIM database was utilised as input for the simulation model. In the proposed framework, the input for the simulation model is made through enriched entities, similar to the ones presented in [23], consisting of C# scripts attached to the BIM objects within the environment of the framework.

2.2 Enhancement of Simulation Visualisation Through Animation

In an animation, a simulation model's operational behaviour is displayed graphically as the model moves through time. Researchers have used this technique to verify and validate simulation models [26]. Additionally, integration of animation techniques and DES have been made in an effort to reduce the complexity in which simulation results are presented to stakeholders unfamiliar with simulation techniques. Table 2 compares some of the most relevant studies in the area with our proposed framework. The characteristics shown in Table 1 are (1) uses BIM data, (2) the system can be used in different types of projects, (3) supports mixed reality visualisation, (4) allows interaction with the animation, (5) developed in a game engine.

Table 2. Comparison of research enhancing simulation visualisation

Reference	Purpose	(1)	(2)	(3)	(4)	(5)
Kamat and Martinez [6, 14]	Visualisation of DES modelled construction operation					
Al-Hussein et al. [15]	Visualisation of DES modelled tower crane operations					
Rekapalli and Martinez [16]	Running animations and simulations concurrently				x	
ElNimr and Mohamed [13, 27]	Integration of 3D visualisation and DES for modelling crane movement				x	x
Chen and Huang [17]	Visualising transport operations in construction			x		
This study	Integrate DES & BIM in a game engine	x	x	x		x

Most systems utilise post-processing visualisation, which does not allow the users to interact with the simulation during the visualisation stage. This type of visualisation relies on trace files generated during the simulation as it runs. The animation uses the trace files to replicate what happened during the simulation after it finishes. In our framework, the animation module utilises this concept to animate the simulation results with the initial input parameters from the user. However, users can simulate a short period of time (for example, a day or a week) and visualise the progress of the project in that period. After visualising the animation, users can change the parameters and continue to simulate the project with the new parameters.

3 Overview of the Framework

The proposed framework to integrate BIM and DES consists of four components, as depicted in Fig. 1, the environment, the user input, the simulation module and the animation module.

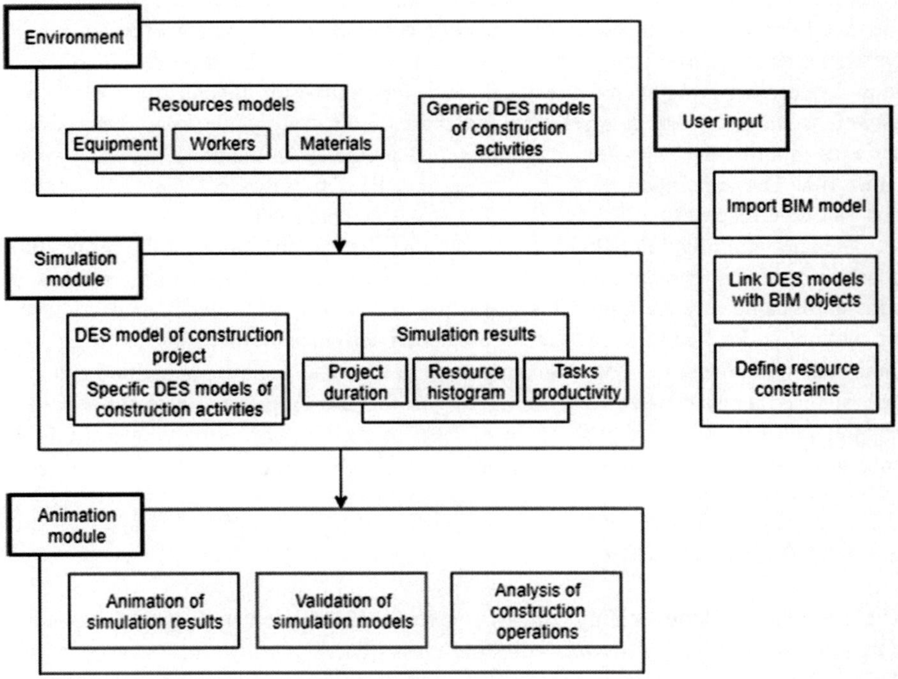

Fig. 1. Framework overview

The environment, developed in the Unity game engine, is preloaded with models that play different roles in the construction project that the user wishes to simulate. On the one hand, there are models that represent different types of resources, such as materials or workers. These models consist of the combination of two components, namely, a physical model of the resource and a script containing its behaviour and properties. On the other hand, there are generic DES models that represent different construction activities. These models are contained in Unity scripts developed in the C# programming language using the SharpSim library [28].

Users import a BIM model into the environment. Within the environment, users link the generic DES models to each BIM object that represents the product of the corresponding construction activity. For example, an instance of the DES model that represents the construction activity "masonry walls with concrete blocks" is linked to each BIM object that represents a wall built with concrete blocks. Once a generic DES model is linked with a BIM object, it becomes a specific DES model for that particular element. Finally, users define project parameters related to resource constraints (e.g. number of workers and their skills, location of raw materials, and available tools and equipment) in the user interface.

The simulation module creates a DES model of a construction project based on the building information model imported by the user. This DES model consists of the sum of the specific DES models of the construction activities attached to each BIM object in the environment. The simulation considers resource constraints according to the parameters that the user input. The user can simulate the developed model for as many runs as necessary and obtain the resulting completion times of the construction project, as well as statistics regarding resource utilisation and tasks productivity. These results serve as input to the animation module, which produces an animation of each simulation run. The simulation module suggests that BIM provides sufficient information that can be utilised to build DES models accurately and quickly.

The animation module allows the user to visualise an animation of the results of a chosen run of the simulation in a 3D environment that domain experts unfamiliar with simulation can easily interpret. This animation can be utilised to verify or validate the simulation model, and to analyse the construction activities involved in the project and their interactions with each other to support planning. The resulting animation can also be visualised using mixed-reality technology as the chosen game engine supports this feature. Research shows that the field of DES in construction is moving towards an enhancement in visualisation. This module was designed to be in line with this trend.

4 The Animation Module

The animation module mainly utilises two classes, depicted in Fig. 2, the *AnimationState* class and the *SaveableStateNow* class. There is only one instance of the *AnimationState* class in the environment provide by the game engine. This object contains a list of multiple instances of the *SaveableStateNow* class, which are generated during the simulation.

A new instance of the *SaveableStateNow* class is added to the list in the *AnimationState* object whenever an event that changes the state of a game object occurs during the simulation. The game object can represent a human or equipment resource or a material, and its state can change in four different ways, namely "created", "destroyed", "resized" or "moved".

The *SaveableStateNow* instance stores information regarding the game object that changed, the type of change that took place and the time in which it occurred. This information includes the parameters that changed if the object was "resized" or "moved".

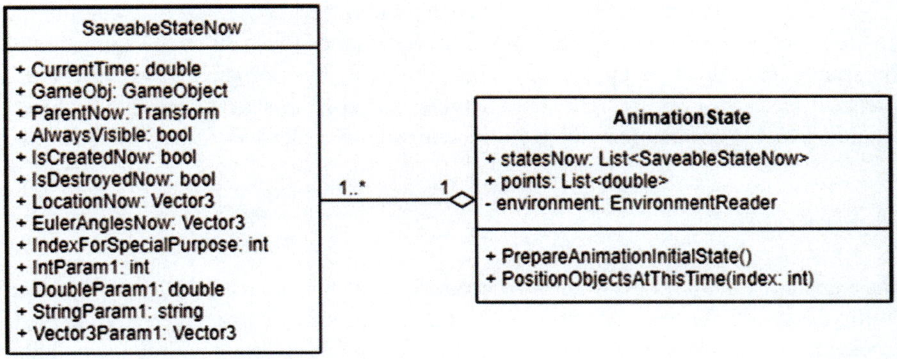

Fig. 2. Animation module class diagram

For example, during the simulation of the construction of a masonry wall with concrete blocks, several game objects representing a concrete block each will be "created" when the materials are delivered to the construction site. Each concrete block will be "moved" from the storage area to the site in which it is needed during the construction process. A block may need to be "resized" if its position within its corresponding row in the wall is equal to the first or last column of blocks as illustrated by the highlighted block in Fig. 3. The "destroyed" status is reserved to temporary game objects that are not part of the finished product of the simulation, such as formwork molds, concrete batches in liquid state, etc.

Fig. 3. A concrete block that needs to be "resized"

At the end of the simulation, all the instances of the *SaveableStateNow* class in the list are sorted by time for each simulation run. The user can visualise a 3D animation of the simulation results of any chosen simulation run by reproducing its *SaveableStateNow* instances in order. The game engine changes the state of the game objects according to the information stored in the corresponding *SaveableStateNow* instance.

5 Limitations

The animation module of the framework produces and replays a record of what happened during the simulation exclusively for visualisation purposes. This is referred to as post-processing visualisation and it does not allow the users to interact with the simulation during the visualisation stage. However, the framework allows users to simulate a period of time (for example, a day or a week) and visualise the animation of the project progress in that period. After visualising the animation, the user can change the parameters of the simulation and continue to simulate the project with the new parameters.

At the moment, the resulting animations only show the discrete points in time in which events happened, but they do not show the continuous motion of the game objects.

6 Conclusions and Future Research

This paper described the animation module of a framework to integrate BIM and DES within a game engine. The animation module allows users to visualise a 3D animation of the results of construction activities simulation. The resulting animation can be used to verify or validate simulation models and as a supporting tool to plan construction activities under resource constrains. The animations can also be used to present simulation results to domain experts unfamiliar with simulation techniques.

The next step to enhance the animation module of the proposed framework is to implement the animation of the continuous motion of the game objects during the simulation.

Acknowledgments. The author gratefully acknowledges the funding from the National Council of Science and Technology (Conacyt, Mexico), granted to carry out this research work.

References

1. AbouRizk, S., Halpin, D., Mohamed, Y., Hermann, U.: Research in modeling and simulation for improving construction engineering operations. J. Constr. Eng. Manag. **137**(10), 843–852 (2011)
2. Ben-Alon, L., Sacks, R.: Simulating the behavior of trade crews in construction using agents and building information modeling. Autom. Constr. **74**, 12–27 (2017)
3. Hamdan, S.B., Barkokebas, B., Manrique, J.D., Al-Hussein, M.: A BIM-based simulation model for inventory management in panelized construction. In: ISARC. Proceedings of the International Symposium on Automation and Robotics in Construction (2015)

4. AbouRizk, S.: Role of simulation in construction engineering and management. J. Constr. Eng. Manag. **136**(10), 1140–1153 (2010)
5. Law, A.M., Kelton, W.D.: Simulation Modeling and Analysis, vol. 3. McGraw-Hill, New York (2000)
6. Kamat, V.R., Martinez, J.C.: Validating complex construction simulation models using 3D visualization. Syst. Anal. Modell. Simul. **43**(4), 455–467 (2003)
7. Leite, F., Cho, Y., Behzadan, A.H., Lee, S., Choe, S., Fang, Y., Akhavian, R., Hwang, S.: Visualization, information modeling, and simulation: Grand challenges in the construction industry. J. Comput. Civil Eng. **30**(6), 04016035 (2016)
8. Wang, W.-C., Weng, S.-W., Wang, S.-H., Chen, C.-Y.: Integrating building information models with construction process simulations for project scheduling support. Autom. Constr. **37**, 68–80 (2014)
9. Chang, S., Son, J., Jeong, W., Yi, J.-S.: BIM-integrated simulation of construction operation simulation for lean production management. In: Modular and Offsite Construction [MOC] Summit & 1st International Conference on the Industrialization of Construction [ICIC] (2015)
10. Lu, W., Olofsson, T.: Building information modeling and discrete event simulation: towards an integrated framework. Autom. Constr. **44**, 73–83 (2014)
11. König, M., Koch, C., Habenicht, I., Spieckermann, S.: Intelligent BIM-based construction scheduling using discrete event simulation. In: Proceedings of the Winter Simulation Conference (2012)
12. Wu, C., Jiang, R., Li, X.: Integration of BIM and computer simulations in modular construction, A case study. Modular and Offsite Construction (MOC) Summit Proceedings, vol. 1, no. 1 (2016)
13. ElNimr, A.A., Mohamed, Y.: Application of gaming engines in simulation driven visualization of construction operations. J. Inf. Technol. Constr. (ITcon) **16**, 23–38 (2011)
14. Kamat, V.R., Martinez, J.C.: Visualizing simulated construction operations in 3D. J. Comput. Civil Eng. **15**(4), 329–337 (2001)
15. Al-Hussein, M., Niaz, M.A., Yu, H., Kim, H.: Integrating 3D visualization and simulation for tower crane operations on construction sites. Autom. Constr. **15**(5), 554–562 (2006)
16. Rekapalli, P.V., Martinez, J.C.: Discrete-event simulation-based virtual reality environments for construction operations: technology introduction. J. Constr. Eng. Manag. **137**(3), 214–224 (2010)
17. Chen, H.-M., Huang, P.-H.: 3D AR-based modeling for discrete-event simulation of transport operations in construction. Autom. Constr. **33**, 123–136 (2013)
18. Sacks, R., Gurevich, U., Belaciano, B.: Hybrid discrete event simulation and virtual reality experimental setup for construction management research. J. Comput. Civil Eng. **29**(1), 04014029 (2013)
19. Osorio-Sandoval, C.A., Tizani, W., Koch, C.: A method for discrete event simulation and building information modelling integration using a game engine. Ad. Comput. Design **3**(4), 405–418 (2018)
20. Eastman, C., Teicholz, P., Sacks, R., Liston, K.: BIM handbook: a guide to building information modeling for owners, managers, designers, engineers and contractors, Wiley (2011)
21. Azhar, S.: Building information modeling (BIM): Trends, benefits, risks, and challenges for the AEC industry. Leadersh. Manag. Eng.g **11**(3), 241–252 (2011)
22. "The National BIM Report 2018," NBS, 2018. https://www.thenbs.com/knowledge/the-national-bim-report-2018. Accessed 17 May 2019
23. Liu, H., Al-Hussein, M., Lu, M.: BIM-based integrated approach for detailed construction scheduling under resource constraints. Automation in Construction **53**, 29–43 (2015)

24. Barkokebas, B., Zhang, Y., Ritter, C., Al-Hussein, M.: Building information modelling and simulation integration for modular construction manufacturing performance improvement. In: Proceedings of the European Modeling and Simulation Symposium, Barcelona (2017)
25. Liu, H., Altaf, M., Lei, Z., Lu, M., Al-Hussein, M.: Automated production planning in panelized construction enabled by integrating discrete-event simulation and BIM. In: 5th International/11th Construction Specialty Conference (2015)
26. Sargent, R.G.: Verification and validation of simulation models. In: Proceedings of the 2005 Winter Simulation Conference (2005)
27. ElNimr, A., Fagiar, M., Mohamed, Y.: Two-way integration of 3D visualization and discrete event simulation for modeling mobile crane movement under dynamically changing site layout. Autom. Constr. **68**, 235–248 (2016)
28. Ceylan, A., Gunal, M.M.: A methodology for developing DES models: event graphs and SharpSim. In: Proceedings of the 2011 European Modelling and Simulation Symposium, Rome (2011)

BIM for Infrastructure Projects

Implementation, Performance and Waste Management Analysis of Decentralized Wastewater Treatment Systems Using BIM Technology

Matheus Alves Dariva(✉) and André Araujo

The Federal University of Uberlândia, Uberlândia, MG, Brazil
matheusalvesdariva@gmail.com, andre.araujo@ufu.br

Abstract. Brazil has more than 23 million inhabitants in rural areas without adequate sewage collection or treatment, which corresponds to 75% of the rural population. The absence of adequate sanitary conditions can cause disease and harm the environment. Decentralized wastewater treatment systems are alternatives to ensure the health of the population and at the same time maintain the environmental integrity of these locations. However, several projects in this area do not include important data for the construction, good performance, and operation of these systems. This partially nullifies the benefits of collecting and treating effluents. Currently, the Building Information Modeling (BIM) methodology fits into this scenario as a possibility for technological innovation and progress in productivity and project presentation. However, there are two barriers to the implementation of this technology in the production of hydraulic and sanitary projects: the scarcity of studies related to the management of the design process of these systems and the lack of BIM components that represent decentralized wastewater treatment systems. This work presents the development of a technological artifact with the potential to fill these gaps: a template created in Autodesk Revit software with several components of decentralized wastewater treatment systems available for use in projects. It is not simply a catalog in which the designer can choose the equipment to be used, but components that provide productivity in the project design, learning and flexibility in the construction of the models, fitting to the market standards.

Keywords: Decentralized wastewater treatment system · BIM · Sustainability

1 Introduction

Basic sanitation is defined by law No. 11,445 of January 5, 2007, as the set of services, infrastructures and operational facilities in four areas: drinking water supply, sanitary sewage, solid waste management and rainwater management [1]. In addition to providing important definitions, the law also establishes national guidelines and fundamental principles for universal access to sanitation, a proposal that necessarily involves resolving the situation of isolated communities. On this document, two of the objectives of the Federal Basic Sanitation Policy are: 1) prioritize plans, programs, and

E. Toledo Santos and S. Scheer (Eds.): ICCCBE 2020, LNCE 98, pp. 297–319, 2021.
https://doi.org/10.1007/978-3-030-51295-8_22

projects aimed at implementing and expanding basic sanitation services and actions in areas occupied by low-income populations; 2) provide adequate conditions of environmental health to rural populations and isolated communities.

The term "isolated communities" was disseminated by the Brazilian Association of Sanitary and Environmental Engineering (ABES-SP) to designate housing units that are not connected to public basic sanitation services. This isolation can occur due to several factors: technical, economic or political infeasibility, a long distance from the municipality's headquarters, difficult access, low population density, large dispersion among households or irregular land tenure. In these locations, the water distribution and sewage collection networks do not exist or are insufficient, leading to the adoption of local, single-family or semi-collective solutions [2]. In this regard, it is essential that different strategies have adopted that respect the cultural and environmental diversity of the communities.

1.1 The Universalization of Basic Sanitation, a Distant Goal

Good sanitation practices are essential not only to prevent diseases, especially those of water transmission, but also to promote health, protect the environment and increase the population's quality of life [3]. The sewer sanitary system can be defined as the set of activities, infrastructures and operational facilities for the collection, transportation, treatment and final disposal of wastewater, from building connections to its final release into the environment [1]. Despite being an essential basic sanitation service for the promotion of the population's health and for environmental protection, it is estimated that about 2.4 billion people worldwide still live without access to adequate sanitation practices, which represent about 32% of the global population [4].

This is also the reality in Brazil, since, among the national basic sanitation services, the sewer sanitary system is the one with the lowest municipal coverage [5], just over half of Brazilian municipalities have this service (55.2%) [6]. In 2013, a study was carried out by the IBGE, which included 65130 households, in which it indicated that only 63.5% of these are served by wastewater collection systems, with huge disparities between the Southeast Region (88.3%) and North (19, 3%), for example [7].

In Brazilian municipalities, the most common destination for domestic sewage is still rudimentary cesspools (53.17%), with septic tanks (8.03%) and open ditches, direct disposal in water bodies and other forms of incorrect disposal (3.35%) [5]. This situation is more common in small municipalities and districts, with less than 50,000 inhabitants, predominantly rural locations with a more dispersed population [6]. Therefore, if the universalization of sewage services in Brazilian municipalities is still a distant goal, the situation in isolated communities is even more serious.

Currently, about 30 million people live in the Brazilian countryside, which is equivalent to 16% of the country's population [6]. The IBGE data indicate that only 8% of the rural households surveyed are served by a wastewater collection system and that 13% do not have any type of wastewater treatment system [8]. Among households that have some form of treatment, 58% adopt solutions considered unsuitable for sanitary sewage, such as the use of improperly constructed absorbent pits and other forms of treatment and final disposals such as the discharge of sewage into ditches, water bodies and other locations.

1.2 Decentralized Wastewater Treatment Systems

Despite the low coverage rate of rural areas in Brazil by sewage collection networks, this in itself is not an aggravating factor for sanitary conditions [3, 9]. Decentralized wastewater treatment systems, if well designed, built and operated, are good alternatives to ensure the health of the population and at the same time maintain the environmental integrity of these locations [10]. The decentralization strategy is increasingly complementary and not opposed to the centralization of wastewater treatment in the quest for the universalization of basic sanitation services [11].

In general, decentralized systems are those that collect, treat and make the final disposal or reuse of wastewater in a place close to its generation, unlike what occurs in traditional centralized systems, where wastewater in cities is collected and treated in large sewage treatment plants. Two terms are used to classify decentralized systems: single-family and semi-collective. Single-family systems refer to those that serve a family that lives in one or two very close households. Semi-collective systems are those that treat the wastewater of a small group of houses or spaces destined to commerce or provision of services such as villages, churches, small schools, grocery stores, shops, and rural warehouses [2].

Decentralized systems are gaining more and more attention as they present several benefits widely discussed in the literature, such as the demand for less financial resources in implementation, the contribution to local sustainability [12] and the opportunity to reuse treated sewage in agriculture [13]. Although decentralized systems have some advantages over centralized systems, there are limitations and many challenges that need to be overcome to make their implementation efficient, viable and safe for isolated communities. These points are summarized in a table adapted from Bueno [14] and Tonetti et al. [2], which divides them into four areas: social, economic, environmental and operational (Table 1).

The first challenge in implementing this type of system is choosing the most appropriate sewage treatment technologies for each situation. This is a complex task, which involves evaluating many variables simultaneously. There is a wide variety of technologies available for rural and isolated areas, but there is no consensus on which would be the most technically appropriate. The decision must, therefore, consider local specificities, given that there are significant differences between Brazilian regions regarding their environmental, socioeconomic and cultural characteristics [2].

Therefore, there must be a greater concern in studying and working efficiently with the information related to this type of system in the design stage. The design in civil construction is one of the main stages of work and has a fundamental role in the quality and production of buildings. It is in this phase that the concepts of space organization are defined, as well as the technology to be used, costs, benefits, etc. The project's solutions directly reflect on the entire construction process and on the quality of the product to be delivered to the customer [15, 16].

It is known that sewage treatment is essential for reducing pollution of water resources and improving the health of the population. However, several projects do not include important data for the construction, good performance, and operation of the wastewater treatment system. Fundamental information, such as materials used for construction, destination of the sludge produced, type of treated sewage, type of system, rate of removal of organic matter, the number of people served and the frequency

Table 1. Possibilities, limitations and challenges of using decentralized wastewater treatment systems. Source: Adapted from Bueno [14] and Tonetti et al. [2].

Area	Possibilities	Limitations	Challenges
Social	Contribute to improving the health of the local population	Requires the interest of residents, family and neighbors to understand the functionality of the system	To guarantee the acceptance of the population and their involvement in the technology choice process
	They can generate work and income		
	Can help produce food, contributing to food security	There is not always acceptance by the population in the implementation of the system	
	Adaptable to customs and culture		To guarantee the help of technicians in clarifying doubts and presenting viable options
	They are usually well accepted by the population and inspection bodies	Needs the effective participation of the population served	
	Can help compose the local landscaping		
Economic	The simplest systems have low installation costs	The most appropriate type of system does not always correspond to the economic reality of the population served	To reduce costs by using alternative materials for the construction of the systems (construction rubble, bamboo, green coconut, etc.) or materials that were already available on site and that can be reused (pipes, connections, water tanks, concrete rings, bricks, tires, etc.)
	Consume little energy and external inputs		
	Some by-products of the system have commercial value and can generate income (food, biogas, ornamental plants, etc.)		
	There is savings in fertilizers when using treated wastewater in agriculture		To reduce costs and generate jobs with the use of local labor
Environmental	Single-family systems can be compact	Most systems need a large area for their implementation	To avoid overload and contamination of soils with low infiltration rate or soils with good drainage in very dense places
	Use little inputs and energy in construction and operation	Factors such as climate and soil type can generate large variations in the final quality of treated sewage in some systems	To avoid contamination of water by nitrates, phosphates and pathogens from systems located very close to surface or underground water reserves
	Reduce pollution of soil and local water bodies		
	They can improve local ecological conditions	Very impermeable soils, with groundwater close to the surface, or areas that are frequently flooded do not allow the infiltration of treated sewage into the soil	To avoid eutrophication of surface water bodies
	Promote the reuse of water and nutrients locally		To ensure correct disposal of sludge produced in the system
Operational	No need to build a local wastewater collection system and pumping stations	Some systems have a high frequency of operation and maintenance, which often must be performed by residents	To maintain regularity in the operation and maintenance of the systems (filter material exchange, constant plants pruning, sludge final disposal, etc.)
	Have good operational flexibility		
	Can be extended over time		
	Have low consumption of materials and energy		
	In most cases, treatment is free		To monitor treated wastewater and receiving water bodies and to create an information system, when possible
	Do not require specialized labor		
	Can treat gray and black waters separately		
	Are little influenced by natural disasters		

of maintenance [13], are neglected and replaced by standard components [15, 16], which is often not the best solution for the building or community in question, when it comes to a decentralized process. As a result, the benefits of collecting and treating effluents are partially canceled. The community needs to take this problem very seriously, and, with the help of scientific and technological research, develop safe and feasible alternatives so that sewage does not become a new environmental problem, taking instead the environmental advantages of its disposal [13].

1.3 About BIM Technology

Currently, study methods and standards are being applied to improve civil construction projects production. Strategies of constructive conceptions are researched that present approximation with the reality of the construction sites, or even control over the entire planning, favoring mainly the construction progress [17]. The BIM methodology fits into this scenario as a possibility for technological innovation and progress in the project's presentations productivity. One of the pioneers of the concept was Chuck Eastman [18], a professor at the Georgia Institute of Technology in the United States. According to him, BIM is a work philosophy that integrates architects, engineers, and builders (AEC) in the elaboration of a precise virtual model, which generates a database with topological information and necessary subsidies for budget, energy calculation, and forecast of inputs and actions at all stages of construction.

Developed by Autodesk, Inc., Revit is one of the software with BIM technology available on the market and includes resources for architectural design, mechanical, electrical and hydraulic systems engineering, structural engineering and construction. It also supports a collaborative and multidisciplinary design process, allowing everyone involved to keep up to date on the stages of work [19]. In this software, every virtual model of construction is designed with the modeling and use of families. According to Autodesk [20], a family is a group of elements with a set of parameters (semantic and/or geometric information) and a related graphic representation, they constitute all components of a project.

Elements modeled on BIM platforms are based on representations of the physical and functional characteristics of a component or set of components to be used in the construction of the building. One aspect of BIM is the availability to contain building information, these data are defined in relation to the geometric and non-geometric attributes of the elements [21] (Table 2).

Table 2. Examples of geometric and non-geometric attributes of elements in BIM. Source: Adapted from the Government of SC - Notebook for presentation of projects in BIM [21].

Geometric attributes	Non-geometric attributes
Size	Product performance data
Volume	Technical specifications and standards compliance
Form	Cost
Density	Color, brand, technical details
Area	Classification system (DEINFRA, Uniformat, SINAPI (…))

The Level of Development (LOD), developed by BIMForum, is a specification to clearly articulate the content and reliability of the BIM model in different phases of a project. It is related to the amount of information and details that are present in a component of a virtual design [22]. According to BIMForum [23], a BIM model element is associated with a specific LOD, in which it evolves from LOD 100 (preliminary basic model) to LOD 400, which corresponds to the completed model with all necessary information for the construction. Finally, there is the LOD 500, which characterizes what has been accomplished. In this evolution, each component (family) is increased with geometric and non-geometric parameters and are categorized as follows:

- LOD 100: The model element is represented graphically with generic symbols;
- LOD 200: The model element is converted into an object or set of generic objects with the definition of its basic dimensions;
- LOD 300: The model element is graphically represented as a specific system, object or set in terms of quantity, size, shape, location and orientation;
- LOD 350: The model element presents the vision of the construction and the identification of the interfaces between the specialties;
- LOD 400: It includes the final details of the model element, in order to generate enough information for the perfect characterization of the services to be performed;
- LOD 500: The model element is a field-verified representation in terms of size, shape, location, quantity and orientation (Fig. 1).

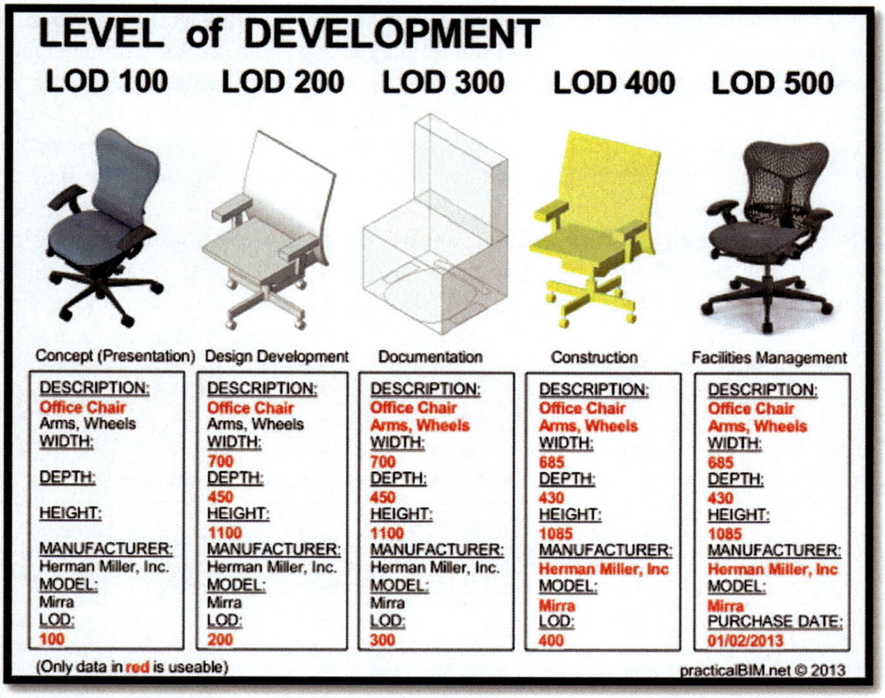

Fig. 1. Level of Development applied to an office chair. Source: PracticalBIM [24].

LOD is sometimes interpreted as a Level of Detail, rather than Level of Development, but there are important differences. The Level of Detail is essentially the amount of detail included in the model element (visual design). The Level of Development is the degree to which the element's geometry and the information attached to it were thought, that is, it is the Level of Detail of the component associated with its Level of Information. That is, the Level of Detail can be considered as an input for the element, while the Level of Development is an output with reliable information attached [23] (Fig. 2).

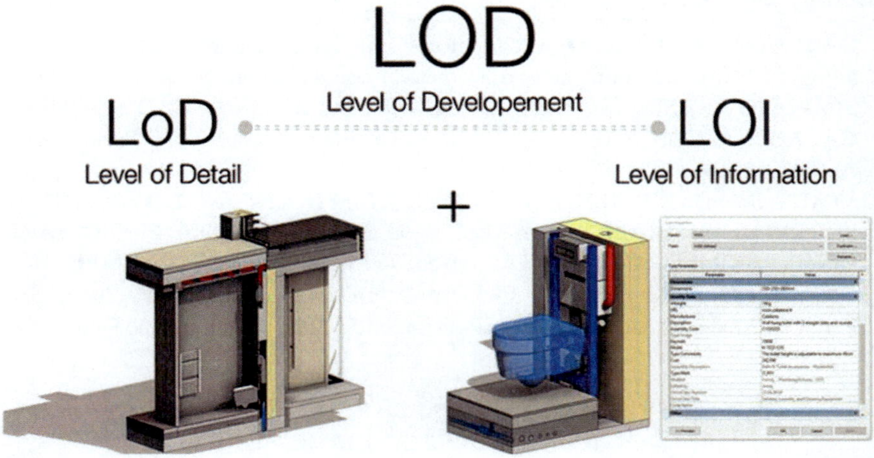

Fig. 2. Level of Development in relation to the Level of Detail and Level of Information. Source: Authors (2020).

Despite recent efforts by software vendors, civil construction organizations, professionals and the government to promote the BIM as the way to define buildings, most projects are still carried out in a traditional way, based on several two-dimensional drawings of each disciplines. Today, although technology exists that can make professional work in most disciplines in the AEC sector more efficient, most professionals are resisting efforts to move towards BIM [15–25].

Bazjanac [25] highlights that after trying briefly to model information and under pressure to meet tight deadlines and without significant additional resources to carry out their work on a given project using the BIM methodology, engineers, and architects come to the conclusion that it is not worth the effort and end up opting for the traditional way of projecting. In this way, professionals already have a well-defined workflow in their offices with ready components to be used in their projects, saving time and money. The lack of BIM components in the market and the difficulty in creating these families, also indicated by Memon et al. [26] and Santos [27], is one of the major barriers to implementing this concept in the AEC sector.

Another issue to be highlighted is that although there are several kinds of research regarding BIM technology, there are still few works that focus their attention on the process of hydraulic and sanitary projects from the management point of view, that is, the relationship of this system with the other project disciplines, the moment this project is elaborated, the pertinent standards, how it will be executed on-site and how the operation and maintenance process works [28]. Ywashima and Ilha [29] also report that this context is reflected in the market, where most offices still work in a segmented manner and the solutions adopted do not always meet the needs of users and the maintenance of the building.

A systematic survey of the articles published in Brazilian scientific congresses focused on the built environment (Fig. 3) revealed the scarcity in research that relates project-based technologies to hydraulic and sanitary systems, where in only two events an article with this approach was published:

1. DARIVA, M. A.; ARAUJO, A. L. Concepção de projetos hidrossanitários com tecnologia BIM: estudo comparativo ao método conduzido em plataforma CAD. In: SIMPÓSIO BRASILEIRO DE TECNOLOGIA DA INFORMAÇÃO E COMUNICAÇÃO NA CONSTRUÇÃO, 2., 2019, Campinas. Anais [...]. Porto Alegre: ANTAC, 2019.
2. STAUT, Solange L. S.; ILHA, Marina. S. de O. BIM NO ENSINO DO PROJETO DE SISTEMAS PREDIAIS HIDRÁULICOS E SANITÁRIOS NA ENGENHARIA CIVIL. In: SIMPÓSIO NACIONAL DE SISTEMAS PREDIAIS DESEMPENHO E INOVAÇÃO DE SISTEMAS PREDIAIS HIDRÁULICOS, XIII., 2019, SÃO PAULO. Anais... São Paulo: Escola Politécnica da Universidade de São Paulo, 2019.

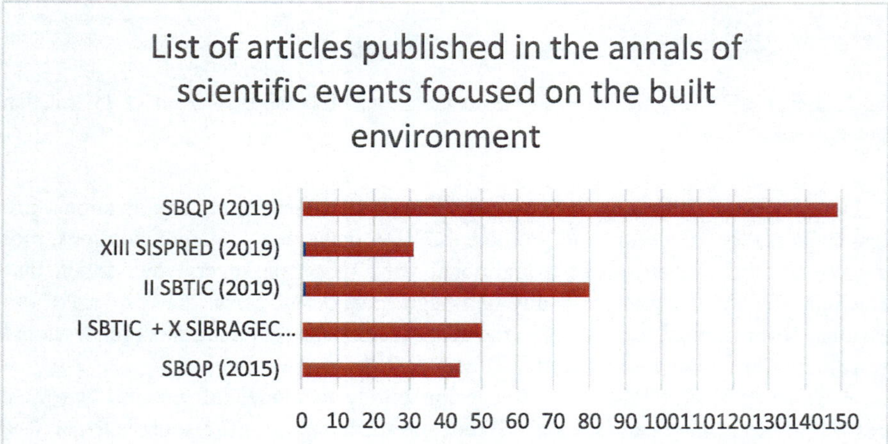

	SBQP (2015)	I SBTIC + X SIBRAGEC (2017)	II SBTIC (2019)	XIII SISPRED (2019)	SBQP (2019)
Articles related to the process of hydraulic and sanitary projects	0	0	1	1	0
Other articles	44	50	79	30	149

Fig. 3. List of articles published in the annals of scientific events focused on the built environment (comparison between the number of articles related to the design process of Building Hydrosanitary Systems and others from other disciplines). Source: Authors (2020).

In this current academic, market and environmental context, briefly outlined above, some issues challenge the development of decentralized wastewater treatment systems projects by professionals in the AEC area as an alternative to the universalization of basic sanitation:

1. The adequacy to sustainability parameters;
2. The incorporation of the ICT (Information and Communication Technology) concepts;
3. The treatment of adequate and enough information for the correct implementation of the systems;
4. Ensuring learning about the many types of decentralized sewage treatment, its challenges, and its limitations;
5. The easy access to the different components and the possibility to study and choose the best system in the elaboration and presentation of projects.

2 Objective

This work aims to confirm these issues highlighted above by creating families of decentralized sewage treatment systems at a High Level of Development (LOD 400) that enable the study, manipulation, learning and the best choice of the numerous options of this equipment.

3 Methodology

For the development of this work, a strategy of Design Science Research was adopted, whose main characteristic is the production of innovative artifacts aimed at solving problems found in the real world, contributing to the development of the theory of the disciplines in which it is applied, these artifacts can be models, methods, concepts or implementations [30]. Table 3 summarizes the general instructions for the conduct and evaluation of this type of methodology.

Table 3. General instructions for the Design Science Research Conduction/Evaluation. Source: Adapted from Hevner, March and Park [30, p. 83].

Instruction	Description
The relevance of the problem	The objective of Design Science based research is to develop technology-based solutions to important and relevant management problems
Design as artefact	Research based on Design Science should produce a viable artifact in the form of a construct, model, method and/or an instantiation
Design evaluation	The usefulness, quality and efficacy of the artifact must be rigorously demonstrated by means of well executed evaluation methods

(*continued*)

Table 3. (*continued*)

Instruction	Description
Design contributions	Research based on Design Science should provide clear and verifiable contributions in the specific areas of the artifacts developed and present a clear grounding in design fundamentals and/or design methodologies
Design as a research process	The search for an effective and efficient artifact requires the use of means that are available, to achieve the desired ends, while satisfying the laws governing the environment in which the problem is being studied
Research communication	Research in Design Science should be presented to both the more technology-oriented and the more management-oriented audience

3.1 The Relevance of the Problem and Design as an Artifact

First, through the bibliographic survey, the relevance of problems associated with decentralized wastewater treatment systems were verified. It was found that:

- Only 63.5% of households in Brazil are served by a sewage collection system [7];
- The most common destination for domestic sewage in Brazilian municipalities are still considered inadequate solutions, such as rudimentary cesspools (53.17%), septic tanks (8.03%), open ditches and direct disposal in water bodies (3.35%) [5];
- Currently, about 30 million people live in rural Brazil (16% of the country's population) [6].
- Only 8% of rural households are connected to the sewage collection system and 13% do not have any kind of wastewater treatment system [8];
- Among the rural households that have some form of treatment, 58% adopt solutions considered inadequate [8];

From so many alarming data it was conjectured that this problem could also be associated with the issue related to the production of projects, where, through a new bibliographic the survey, it was verified that:

- The use of AutoCAD standard blocks in different projects is a common practice in the market [15, 16]. This factor can represent productivity gains in the project issue [16], however, when it comes to the design of a decentralized wastewater treatment the system, a "fad" in using the same block in different cases can be created, which often does not represent the best solution for the building or the community in question.
- Several projects do not include important data for the construction, good performance and operation of the wastewater treatment system [13].

- The BIM methodology fits into the current scenario as a possibility of technological innovation and progress in productivity and project presentation [18]. However, the lack of BIM components in the market and the difficulty in creating these families are one of the great barriers to implement this concept in the AEC sector [25–27].
- Technological solutions in the area of hydraulics and sanitation have not been consistently studied from a management point of view, that is, related to the project process [28].
- Although there is a wide range of systems available, many professionals in the AEC field are not aware of these possibilities nor of the functionality of some equipment [2].

In view of this, thought was given to the development of an artifact that could resolve these issues. The idea was to create a template in Revit that would include families of decentralized wastewater treatment systems with LOD 400. From these families, it would be possible to extract sufficient information for the characterization and construction at the time of the work, and not only that, but it would also be possible to verify maintenance and operation data of the equipment, final destination of the sludge produced, type of sewage treated, type of system, rate of organic matter removal of and amount of people served. In addition to the non-geometric information, the proposal also aimed to treat geometric information automatically and with a certain degree of flexibility that would allow the professional to choose the best system for the building, but would not allow the use of diameters/sizes outside the market standards. That is, the sizing of the equipment would have to be done automatically from the entry of some information (number of people served, the volume of sewage produced, room temperature, etc.), ensuring productivity in the preparation of the project. Finally, we would have components of decentralized wastewater treatment systems that are self-dimensioned according to commercial diameters/sizes, with sufficient information for construction, operation and maintenance, and that enable not only the best choice but also the learning of the designers.

The first step was to choose the systems, from the literature, that would be modeled. It was based on the systems proposed in the book "Tratamento de esgotos domésticos em comunidades isoladas: Referencial para a escolha de soluções" by Tonetti et al. (2018). Among the several options highlighted in the book, the 7 considered most efficient in the organic matter removal and one for the final disposal of the treated sewage was chosen. These are them:

- Biodigester Septic Tank
- Evapotranspiration Tank
- Wetlands/Garden Filter
- Banana Circle
- Anaerobic Filter
- Sand Filter
- Vermifilter
- Soak Pit

The chosen systems were modeled in the Revit Family Editor, where all parameters were configured very rigorously to guarantee all expected contributions. Each family was configured in order to guarantee the self-dimensioning, flexibility, and analysis of all the information mentioned above. For this, several algorithms were created to be applied to several parameters of the families, as shown, for example, in Fig. 4.

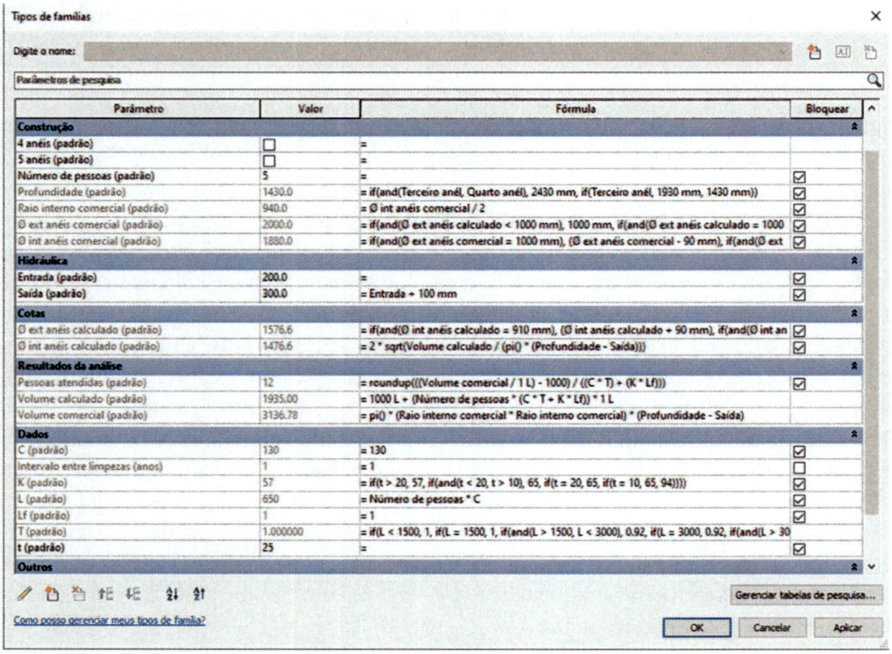

Fig. 4. Table of parameters for the "Biodigester Septic Tank" component in the Revit Family Editor. Source: Authors (2020).

All families were made to this standard of rigor. It was also concerned not to overload the archives with families of the vegetation type, making them heavy. For this purpose, light textures and RPC (Rich Photorealistic Content) type vegetations were added for the Evapotranspiration Tank and Banana Circle systems (Fig. 5). Only in the Filter Garden component was modeled vegetation used in the Revit family editor (Fig. 6). As smaller vegetations are used in this type of system, it would not be interesting to place numerous RPC vegetation families, but only one family representing the type of plant used for sewage treatment.

Fig. 5. Views of the Banana Circle and the Evapotranspiration Tank. Source: Authors (2020).

Fig. 6. Perspective and frontal section of the Filter Garden. Source: Authors (2020).

Another concern was the choice of materials, it was chosen to use concrete rings in circular systems because of it is a cheap material and easily found in building material stores. This was the case of the filters (Anaerobic Filter, Sand Filter, and Vermifilter), the Septic Tank and the Soak Pit (Fig. 7).

Fig. 7. Views of Vermifilter, Sand Filter, Biodigester Septic Tank, Anaerobic Filter and Soak Pit. Source: Authors (2020).

Finally, all the created families were grouped in a Revit template that makes it possible to evaluate and study each one of them, to self-size the systems from the input of some data (Fig. 8 and Fig. 9), automatically create tables of quantities (Fig. 10), to analyze information by positioning the mouse cursor on some parameter (Fig. 11) and to identify each of the materials that compose each component (Fig. 12).

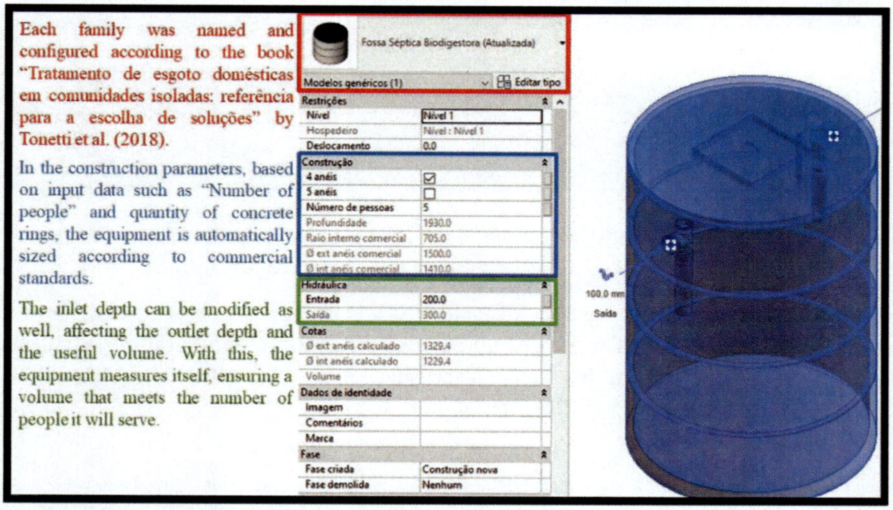

Fig. 8. Procedure for self-sizing of the Biodigester Septic Tank. Source: Authors (2020).

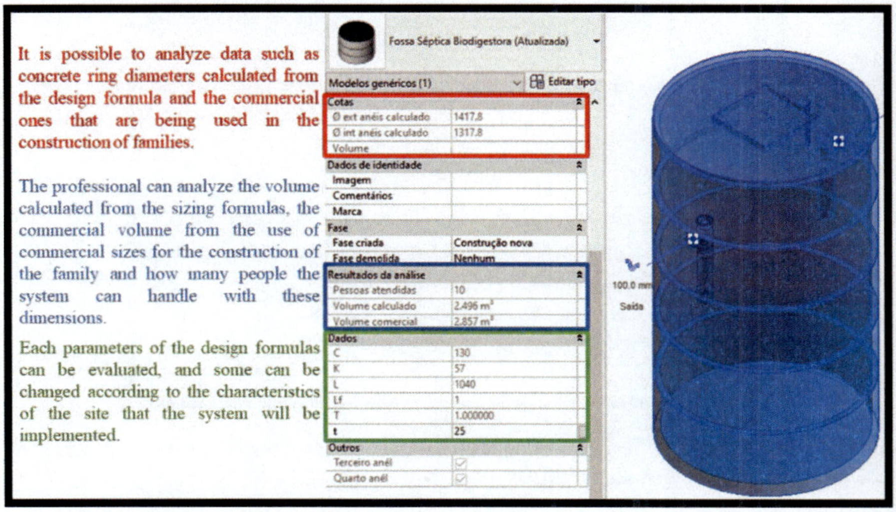

Fig. 9. Parameters assigned for the self-sizing of the Biodigester Septic Tank. Source: Authors (2020).

Kit do Sistema de Tratamento de Esgoto Descentralizado	
Quantidade	**Descrição**
3	Anel de concreto LISO 1.20 (Ø 1200 mm)
2	Anel de concreto LISO 1.20 (Ø 1200 mm) com FUNDO (70 mm)
1	Anel de concreto LISO 1.20 (Ø 1200 mm) com FUNDO PERFURADO (70 mm)
2	Anel de concreto LISO PERFURADO 1.20 (Ø 1200 mm)
1	Anel de concreto LISO PERFURADO 1.20 (Ø 1200 mm) com FUNDO (70 mm)
2	Camada de Brita N° 4
1	Cap 100mm, Esgoto Série Normal - TIGRE
2	Cap Soldável PERFURADO 20mm, PVC Marrom, Água Fria - TIGRE
4	Luva Simples 100mm, Esgoto Série Normal - TIGRE
1	Tampa circular (Ø 1200 mm) c/ inspeção quadrada (500x500mm)
2	Tampa circular (Ø 1200 mm) c/ inspeção quadrada (500x500mm) - Superfície pintada com tinta preta
4	Tubo de PVC - Série Normal - DN 100
1	Tubo de PVC - Série Normal - DN 100 PERFURADO
2	Tubo de PVC Marrom - Soldável - DN 25 com CAP perfurado de PVC soldável na ponta (Respiro)
4	Tê 100 x 100mm, Esgoto Série Normal - TIGRE

Dimensões				
Quantidade	**Descrição**	**Ø ext anéis (m)**	**Ø int anéis (m)**	**Volume Útil**
1	Filtro Anaeróbio	1.20	1.11	895.11 L
1	Fossa Séptica Biodigestora	1.20	1.11	1112.84 L
1	Sumidouro	1.20	1.11	870.92 L

Fig. 10. Quantitative tables of materials extracted from the Biodigester Septic Tank. Source: Authors (2020).

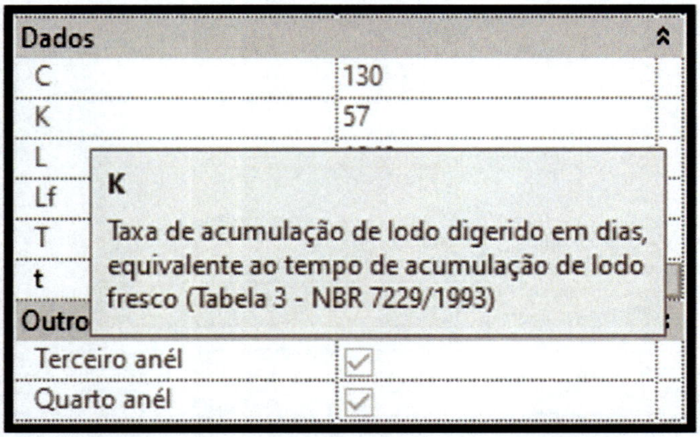

Fig. 11. Demonstration of the information analysis from the positioning of the mouse cursor on some parameters. Source: Authors (2020).

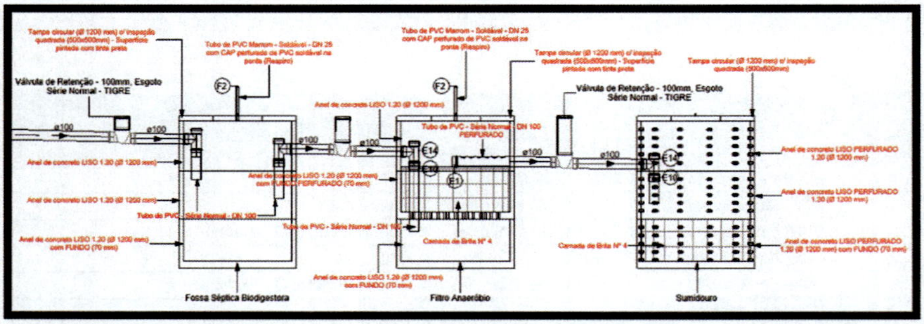

Fig. 12. Example of identification of the materials that compose the families of Biodigester Septic Tank, Anaerobic Filter and Soak Pit. Source: Authors (2020).

3.2 Design Evaluation and Contributions

After the preparation of the families and the template in Revit, it was made available to some professionals in the area of AEC from various locations in Brazil with the families of Biodigester Septic Tank, Anaerobic Filter and Soak Pit available for use in projects. With these families, it was already possible to create an integrated system (Fig. 13) to treat the black water coming from the toilet and to guarantee adequate final disposal for the grey water coming from sinks, washbasins, showers, etc. Some months after that the template was shared, a questionnaire was elaborated in Google Forms with questions that clarify the points highlighted in this work, and it was sent to all professionals who acquired the material (25 in total). This way, it was possible to evaluate the possible contributions that the artifact in question could provide.

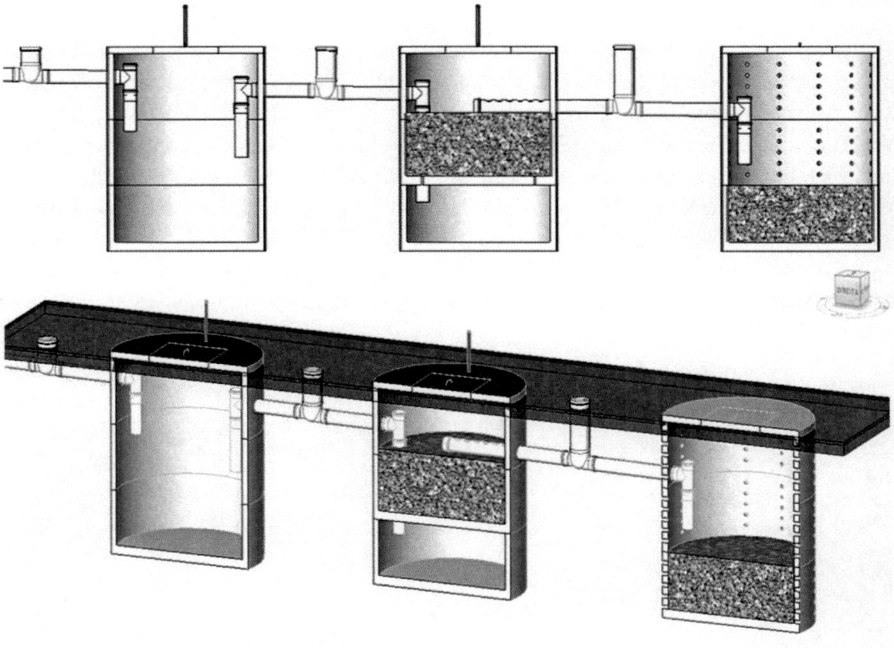

Fig. 13. Frontal cut and perspective of the decentralized sewage treatment system integrated with the families of Biodigester Septic Tank, Anaerobic Filter and Soak Pit. Source: Authors (2020).

Five questions were developed to compose the questionnaire, which had the following configuration:

1. Knowing that in the project stage the Development Level (LOD) evolves from LOD 100 (preliminary basic model) to LOD 400, which corresponds to the model completed with enough information for the construction, how do you classify the Development Level of the families?
2. How do you rate the control of the information available in families?
3. Have the acquired families contributed in any way to learning about decentralized wastewater treatment systems?
4. Before using these families, what other methods were used to represent a decentralized wastewater treatment system?
5. In your opinion, does the lack of components (families) harm, or create resistance, to the implementation of BIM in the AEC (Architecture, Engineering, and Construction) sector?

The results of this research will be presented and discussed in the next topic (Results and Discussions).

3.3 Research and Communication Process

The template created with the families were made available at (www.darivabim.com) where any professional in the AEC area can download the material. Information videos were recorded about each of the components available in the template and posted on YouTube in the channel "DarivaBIM" (Fig. 14), which can be accessed for free through the link: (https://www.youtube.com/channel/UCUyry49-3fD9RGUAJ_xDdcA).

The videos were divided into episodes and present guidelines on how families work, their possibilities and limitations. Thus, the contribution to the participation of professionals in the elaboration of hydraulic and sanitary projects can be more effective when related to situations that require the implementation of wastewater treatment systems.

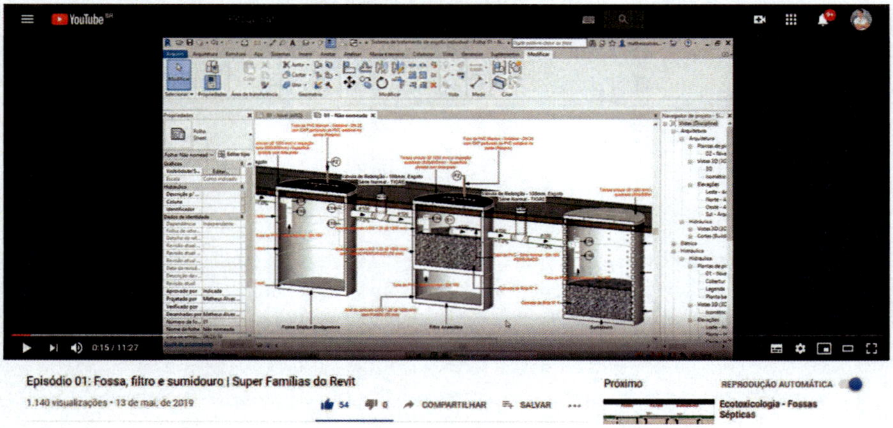

Fig. 14. Episode 01, available on YouTube, of the video series dealing with the components of a decentralized wastewater treatment system. Source: Authors (2020).

4 Results and Discussions

With the evaluation made from the Google Forms questionnaire, it was possible to get an idea of the current context in the market regarding wastewater treatment system projects. It was also possible to obtain qualitative results regarding the level of development of families, i.e. opinions of professionals regarding the details and information contained in the components and whether these data were enough for the construction, good performance and operation of the equipment. Of the 25 professionals requested, 19 answered the questionnaire. The results can be observed below.

In relation to the first question, all professionals replied that the families acquired are designed at a high level of development (LOD 350/LOD 400). For most of them, the information contained in the families is enough for a perfect characterization of the services to be performed (LOD 400), as shown in Fig. 15.

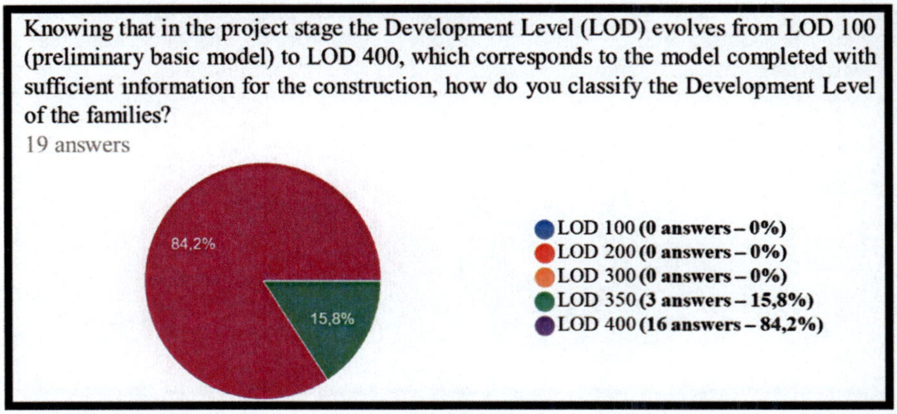

Fig. 15. Result of the first question in the questionnaire. Source: Authors (2020).

The result of the second question (Fig. 16) shows that, for most professionals, the control of the information available in families is great. This means that these professionals have adapted with excellence to the workflow established in the families. As previously mentioned, the flexibility of the components guarantees the automatic sizing of the equipment according to commercial standards, not allowing the use of sizes/diameters outside the market context. In the opinion of the interviewees, this is a factor that guarantees productivity in project development.

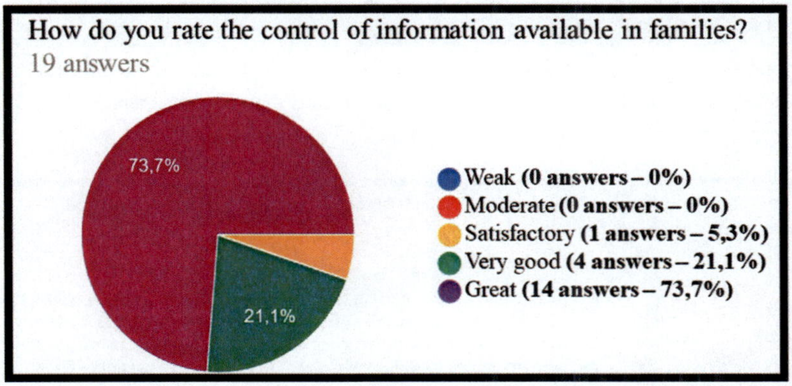

Fig. 16. Result of the second question in the questionnaire. Source: Authors (2020).

For all interviewees, the families contributed in some way to learn about wastewater treatment systems, as shown in Fig. 17. This may represent a paradigm shift, where the project process is also a learning opportunity for the professional, and not only a sophisticated tool for the selection of the best solution for those professionals who hold the heuristic knowledge.

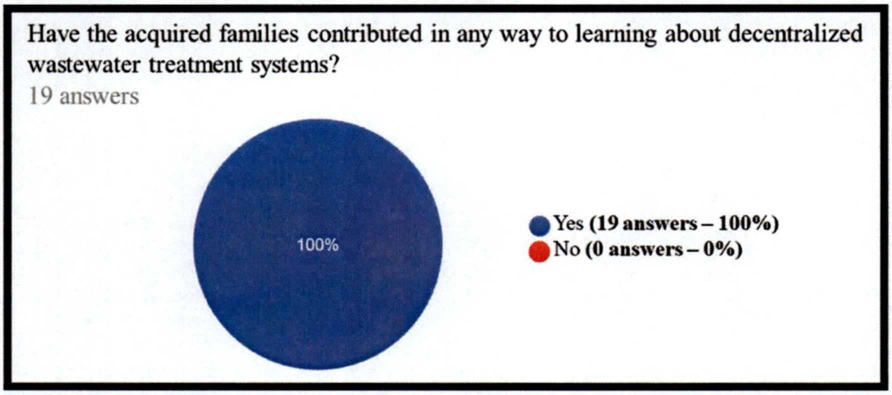

Fig. 17. Result of the third question in the questionnaire. Source: Authors (2020).

As some authors have found [15, 16], the use of AutoCAD standard blocks in different projects are common practice in the market. Observing the result of the fourth question (Fig. 18), this method, besides being common in the market, also extends the use of low LOD families when working with BIM technology. This is a factor that hinders the representation of projects since not enough information is guaranteed for the construction of the systems in question. It is also concluded that this factor can be linked to the difficulty of professionals in creating BIM components or acquiring them in high LOD since there were no interviewees who selected these options. It is interesting to point out that 10.5% of them answered that they did not use this type of system in projects. Some reasons can be deduced from this: the difficulty in finding high LOD components; the lack of knowledge about the possibilities of wastewater treatment systems and the impacts caused by their incorrect disposal; the lack of interest in using this type of system in hydraulic and sanitary projects.

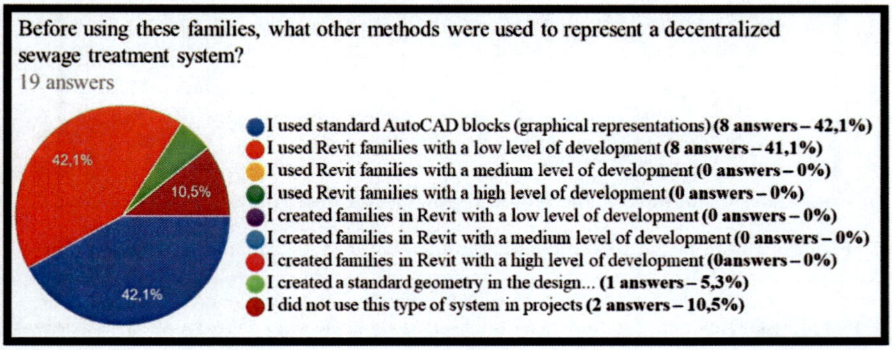

Fig. 18. Result of the fourth question in the questionnaire. Source: Authors (2020).

Finally, as some authors have already observed [15–26], the lack of BIM components available in the market creates a resistance to the implementation of the concept in the AEC sector. This is shown by the results of the fifth question (Fig. 19), where all professionals agree or fully agree with this statement.

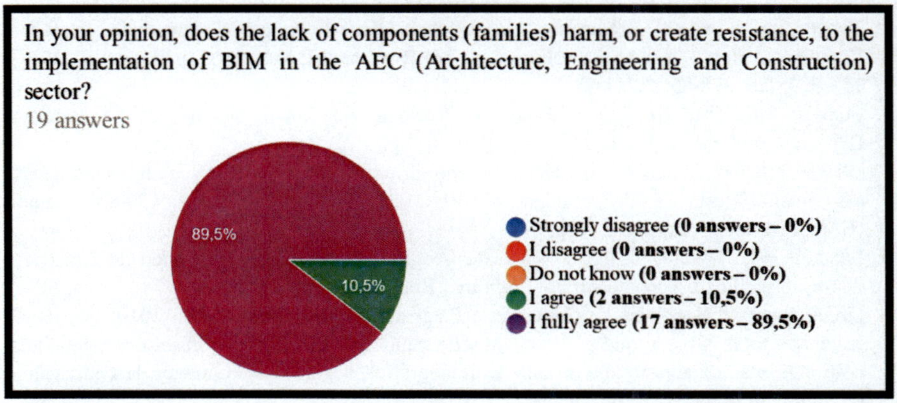

Fig. 19. Result of the fifth question in the questionnaire. Source: Authors (2020).

All questionnaire results are available on the link: (https://docs.google.com/forms/d/e/1FAIpQLScYKvbFiTWAshAk_-s5CWPVH3kIt_35MaznrOVgdL3RUpWVVA/viewanalytics).

5 Conclusion

From the current environmental and market context related to the final disposal of sewage, it was found that the lack of BIM components that represent decentralized systems of wastewater treatment hinders the presentation of these projects. This is a problem that causes damage to the proper functioning of these types of equipment and, consequently, to the environment and the health of the population. The preliminary results indicate that the artifact produced in this work has the potential to fill this gap, providing possibilities of choice by the designers, productivity in the design, learning and a certain flexibility in the construction of the models, which fits the market standards. This may be, in another perspective, the path to a new concept of sustainability in the AEC sector: the ability to accommodate all design disciplines harmonically to produce efficient construction systems from their parameters.

References

1. Brasil: Presidência da República Federativa do Brasil. Casa Civil. Decreto nº 7.217, de 21 de junho de 2010. Regulamenta a Lei nº 11.445, de 5 de janeiro de 2007, que estabelece diretrizes nacionais para o saneamento básico, e dá outras providências (2010). http://www.planalto.gov.br/ccivil_03/_ato2007-2010/2010/Decreto/D7217.htm. Accessed 05 Jan 2020
2. Tonetti, A.L., et al.: Tratamento de Esgotos Domésticos em Comunidades Isoladas: Referencial para a escolha de soluções. 1. ed. Biblioteca Unicamp, Campinas, SP (2018). 153 p. ISBN 978-85-85783-94-5
3. Funasa: Ministério da Saúde. Fundação Nacional de Saúde. Manual de Saneamento: Orientações técnicas. 3. ed. Brasília (2007). 408 p.
4. WHO/UNICEF: Progress on sanitation and drinking water – 2015 update and MDG assessment. World Health Organization (WHO) and The United Nations Children's Fund (UNICEF), Geneva (2015). 90 p.
5. Landau, E.C., Moura, L. (eds.): Variação geográfica do saneamento básico no Brasil em 2010: domicílios urbanos e rurais. Embrapa, Brasília (2016). 975 p.
6. IBGE: Instituto Brasileiro de Geografia e Estatística. Censo demográfico 2010: População residente, total, urbana total e urbana na sede municipal, em números absolutos e relativos, com indicação da área total e densidade demográfica, segundo as Unidades da Federação e os municípios. Ibge, Rio de Janeiro (2010)
7. PNAD: Pesquisa nacional por amostra de domicílios: síntese de indicadores, 2. ed. Ibge, Rio de Janeiro (2015). 296 p.
8. PNAD: Pesquisa nacional por amostra de domicílios, vol. 33. Ibge, Rio de Janeiro (2013). 133 p.
9. de Sousa, J.T., et al.: Utilização de wetland construído no pós-tratamento de esgotos domésticos pré-tratados em reator UASB. Engenharia Sanitaria e Ambiental [s.l.] **9**(4), 285–290 (2004). FapUNIFESP (SciELO). http://dx.doi.org/10.1590/s1413-41522004000400004
10. Massoud, M.A., Tarhini, A., Nasr, J.A.: Decentralized approaches to wastewater treatment and management: applicability in developing countries. J. Environ. Manage. [s.l.] **90**(1) 652–659 (2009). http://dx.doi.org/10.1016/j.jenvman.2008.07.001. Accessed 05 Nov 2019
11. Libralato, G., Ghirardini, A.V., Avezzù, F.: To centralise or to decentralise: an overview of the most recent trends in wastewater treatment management. J. Environ. Manage. [s.l.] **94**(1), 61–68 (2012). http://dx.doi.org/10.1016/j.jenvman.2011.07.010. Accessed 05 Nov 2019
12. Metcalf & Eddy, Inc., Tchobanoglous, G., Burton, F.L., Stensel, D.H.: Wastewater Engineering: Treatment, Disposal, and Reuse, 4th edn. Mcgraw-Hill, Boston (2003). 1819 p.
13. Bettiol, W., et al.: Lodo de esgoto: Impactos ambientais na Agricultura. 1. ed. Embrapa Meio Ambiente, Jaguariúna, SP (2006). 349 p. ISBN 85-85771-37-2
14. Bueno, D.A.C.: Filtros anaeróbios com pós-tratamento em filtros de areia intermitentes: Desempenho em operação crítica. 466 f. Dissertation (Master's Degree) - Civil Engineering Course, Sanitation and Environment Area, State University of Campinas, Campinas (2017)
15. Crotty, R.: The Impact of Building Information Modelling: Transforming Construction. SPON/Routledge, London (2012). 72 p. ISBN 9781136860560
16. de Castro Panizza, A.: Colaboração CAD no projeto de Arquitetura, Engenharia e Construção: Estudo de caso. Dissertation (Master's in Civil Engineering) - Faculty of Civil Engineering, State University of Campinas, Campinas, SP (2004)
17. Magalhães, R.M., Mello, L.C.B.B., Bandeira, R.A.M.: Planejamento e controle de obras civis: estudo de caso múltiplo em construtoras no Rio de Janeiro. Revista Gestão e Produção, Rio de Janeiro, RJ, Brasil, ano 2018, vol. 25, no. 1, pp. 44–55 (2018). http://www.scielo.br/pdf/gp/v25n1/0104-530X-gp-0104-530X2079-15.pdf. Accessed 15 Nov 2019

18. Eastman, C., et al.: Manual de Bim: Um guia de modelagem da informação da construção para arquitetos, engenheiros, gerentes, construtores e incorporadores. Bookman, Porto Alegre (2014). 483 p.
19. Autodesk: Revit: Developed for Building Information Modeling, Brazil (2019). https://www.autodesk.com.br/products/revit/overview. Accessed 12 Jan 2020
20. Autodesk: About the families, Brazil (2019). 13. Accessed 12 Jan 2020
21. Santa Catarina Government: Caderno de apresentação de projetos de Santa Catarina. State Planning Secretariat, Florianópolis, SC, Brazil (2018). http://www.spg.sc.gov.br/visualizar-biblioteca/acoes/comite-de-obras-publicas/427-caderno-de-projetos-bim/file. Accessed 02 Nov 2019
22. Eastman, C., Solihin, W.: Classification of rules for automated BIM rule checking development. Georgia Institute of Technology, North Ave NW, Atlanta, EUA (2015)
23. Bimforum: Level of Development (LOD) Specification Part I & Commentary: For Building Information Models and Data. US: BimForum (2019). 256 p. https://bimforum.org/lod/. Accessed 15 Jan 2020
24. Practicalbim: What is this thing called LOD [S. l.], 1 March 2013. http://practicalbim.blogspot.com/2013/03/what-is-this-thing-called-lod.html. Accessed 02 Nov 2019
25. Bazjanac, V.: Virtual building environments (VBE) - applying information modeling to buildings. United States: N. p. (2004). Web
26. Memon, A., Ismail, A.R., Irfana, M., Azman, N.: BIM in Malaysian construction industry: status, advantages, barriers and strategies to enhance the implementation level. Res. J. Appl. Sci. Eng. Technol. **8**, 606–614 (2014). https://doi.org/10.19026/rjaset.8.1012
27. dos Santos, W.R.: Estudo de caso de Implementação de Modelagem de Informação da Construção em Microescritórios de Arquitetura. Dissertation (Master in Housing: Planning and Technology) - Institute of Technological Research of the State of São Paulo, São Paulo (2016)
28. Emmitt, S.: Gestão do processo de projeto em arquitetura, engenharia e construção: origem e tendências. Gestão & Tecnologia de Projetos, Brasil, vol. 5, no. 3, pp. 27–38, December 2010. ISSN 1981-1543. https://doi.org/10.4237/gtp.v5i3.173
29. Ywashima, L.A., de O. Ilha, M.S.: Concepção de projeto dos sistemas hidráulicos sanitários prediais: mudanças no processo de projeto com a utilização de Building Information Modeling (BIM). In: Encontro Nacional de Tecnologia no Ambiente Construído, 13 (2010). Canela. Anais … Porto Alegre: ANTAC (2010)
30. Dresch, A., Lacerda, D.P., Júnior, J.A.V.A.: Design Science Research: Método de Pesquisa para Avanço da Ciência e Tecnologia. Bookman Editora, Porto Alegre, RS (2015). 204 p. ISBN 8582602995, 9788582602997

Integrated Platform for Interactive and Collaborative Exploration of Tunnel Alignments

Marcel Stepien(✉), Andre Vonthron, and Markus König

Ruhr-Universität Bochum, Chair of Computing in Engineering,
Universitätsstraße 150, 44801 Bochum, Germany
marcel.stepien@ruhr-uni-bochum.de

Abstract. The planning of alignments in tunneling projects is a very complex and crucial process. It involves the consideration of numerous conditions and studies of different variants. Essential information depends upon the existing environment, i.e. on existing building structures and ground conditions. The collected information and results quantify metrics and parameters, which are used to define necessary constraints for alignment definition.

Because of the large and varied scope of these constraints, the application of optimization methods is unpractical. Usually, multiple alignment variants are created and compared manually by using specific and standalone software solutions for modeling alignments and involving experts for each criterion to find a near optimal solution. On the one hand, the modeling of typical alignment curves requires the consideration of many discontinuities along the alignment, which limits local modification. On the other hand, it is not possible so far to modify and evaluate variants directly, for example in team discussions. Such evaluations are often associated with a significant involvement of additional resources.

An approach is presented to create an integrated platform to support the planning process. This includes the strategic use of techniques and systems that allow an interactive exploration of tunnel alignments. Geometric modeling methods are utilized to provide local modeling and successive creation and manipulation of alignments without being limited to transitions of conventional alignment modeling methods. Technical equipment, such as touch tables, are part of the concept to add a collaborative value that allows simultaneous actions of multiple persons, especially when discussing and comparing variants in project teams.

Keywords: Building information modeling · Tunnel information modeling · Interactive exploration · Alignment planning

1 Introduction

The planning of tunneling projects requires enormous effort. The most important task is to find a valid alignment. For this purpose, a significant amount of data needs to be gathered and evaluated. That includes the exploration of ground and groundwater

E. Toledo Santos and S. Scheer (Eds.): ICCCBE 2020, LNCE 98, pp. 320–334, 2021.
https://doi.org/10.1007/978-3-030-51295-8_23

conditions, field studies, operational and technical requirements, cadastral maps, surveys or the status of building environments at the construction site. These are extensively collected by drilling boreholes, building shafts, mining and surveying [1, pp. 20–39]. By combining and comparing the gathered data, the requirements for creating a tunnel alignment are derived. Such an alignment is forming a strategy for a cost-efficient route. Usually multiple strategies will be explored at once, since the consideration of necessary constraints and conditions leads to alternative options [2]. In current approaches, multiple variants of alignments are created and made available for selection. Accordingly, the evaluation of a projects constraints and conditions results in the proposal and selection of an alignment. Evaluating a near optimal solution involves a multitude of empirical, analytical and numerical methods [1, pp. 98–114]. To be able to process the data entirely and draw consistent conclusions, building information modeling is applied. The method usually requires the usage of specific tools to complete particular challenges, such as modeling curves, planning schedules, analyzing constraints, performing simulation procedures and estimating costs. Additionally, there are applications that are considered multidisciplinary and used to distribute and manage whole projects. Such applications are also used to evaluate created alignments.

As stated by Rudolf et al. [2], one major driver in the decision process of selecting an alignment is project costs. Ideally, an alignment should describe a cost-efficient strategy, but not necessarily the cheapest solution. Many more aspects have to be taken into consideration, such as safety, funding, risks and the environmental impact. A universal approach for evaluating possible alignments currently does not exist, resulting in assessments that are difficult to investigate. Usually, the applicable alignment is selected by exclusion, using classifications or grades and comparing the variants based on a ranking resulting from a weighted-criteria analysis [3]. However, because the scope of the collected data is large and varies drastically it is difficult to express and compare tunnel alignments with simple metric values.

Furthermore, the method of creating multiple static variants of alignments does not take into account the potential solutions in-between the preselection. An approach for an ongoing and interactive modeling, planning and evaluation of alignments could provide a wider range of possible solutions. The CAD applications developed to create complex alignments are currently used by experts or trained individuals only. A collaborative approach between all decision-makers is currently not part of the design concept. However, it could provide a useful method for an ongoing creation and evaluation of alignments. In an article by Kang et al. [3] the applicability of optimization methods for alignments in highway construction were tested and applied. While the optimization can successfully be performed, the computation time can exceed multiple hours. For tunnel alignments many more parameters have to be considered, which only increase the already time-consuming optimization. Therefore, the use of optimization methods is considered unpractical for estimating an optimal tunnel alignment, especially in a collaborative scenario where decisions are discussed and applied immediately.

The following Sect. 2 summarizes the most relevant research, such as the method of tunnel information modeling, alignment modeling and known techniques for touch interaction on touch devices. In Sect. 3 a concept for an integrated platform for interactive collaboration and exploration of tunnel alignments is introduced, improving the planning process by enabling interactive and local modeling of alignments on collaborative touch devices. The presented approach concentrates primarily on horizontal alignments, since vertical alignments depend heavily on a detailed exploration of soil and groundwater conditions, which exceeds the scope of this paper. In Sect. 4 the introduced concept of an integrated platform is tested by implementing a prototype.

2 Related Work

This paper builds upon the method of tunnel information modeling, the conventional approach for alignment modeling and required touch interaction. The integrated platform expands on the presented methods and applies them to a collaborative context, aiming to improve the modeling of alignments and decision making.

2.1 Tunnel Information Modeling

In the research by Koch et al. [4] a tunnel information model (TIM) has been introduced. The main feature of the TIM is to gather and link different models to analyze and visualize project contents. Collected models are joined into a multi-model container, representing a BIM model with enriched information. The intended use of the TIM concept is primarily data management. As illustrated in Fig. 1, Koch et al. embedded the TIM into a framework that enabled a visualization and time based simulation. Analytical features provided further depth in exploring the data.

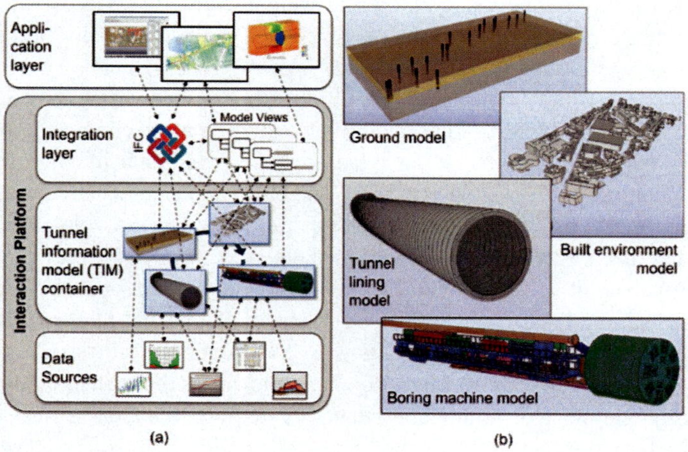

Fig. 1. Framework with application layers (a) and the four sub-domain models containing in TIM (b) [4]

The interaction platform of TIM is composed of several layers of components, which are the Integration-layer, TIM-container and Data-sources. The Data-sources component contains data that is linked on-demand, such as a series of time-dependent measurements or simulation results. The TIM-container component upholds all data needed for planning, construction and maintenance of tunnels. They are linked into a holistic and object-oriented multi-model container. In this case, the models are distinguished between four sub-domains. Those are the ground model, the built environment model, the tunnel lining model and the boring machine model. Each domain is referred to as significant data for the tunneling design and construction process. The Integration-layer describes the used model data schema, which in this case is primarily the Industry Foundation Classes (IFC). The IFC schema is developed and documented by buildingSMART, distributing IFC 4.1 since 2018 and standardized in DIN EN ISO 16739 [5]. The geometric representation can occur at different levels of detail and is augmented by semantic and topological information. Also, model views are specified to create subsets of information required for analytic tasks. Additionally, an Application-layer is conceived to perform basic user interaction. The Application-layer is serving as an entry point to the application. Results are visualized in a 3D viewer.

2.2 Procedure for Modeling Tunnel Alignments

In general, 3D alignments are segmented and split into horizontal and vertical parts. Using a combination of circular arcs, straight lines and clothoid arcs for creating the horizontal alignment part is a well know and often used approach [6–8]. The vertical part is usually a combination of straight lines and parabolic curves. Both horizontal and vertical alignment segments are defined as a consecutive sequence, in order to calculate paths that are without discontinuity and free of kinks. Accordingly, an alignment must comply particularly with geometric constraints and conditions. In addition, a project environment using BIM based models specifies visual boundaries for alignment creation, such as roads, building structures and topological information.

As described in Sect. 2.1, the TIM uses the IFC to store and provide BIM compliant model data. One of the additions in recent years is the IFC Alignment extension, which has been initiated by Amann et al. in 2013 [8]. The extension provides entities to store alignments in a segmented form. By using the extension, an alignment is created by defining an IfcAlignment entity. The axis of an IfcAlignment requires the definition of an IfcAlignmentCurve, which again requires defining horizontal and vertical alignment segments, recreating approaches commonly used to design alignments. The extension enables the introduction of alignments into the concept of TIM.

The conventional method is applied in a number of state-of-the-art applications designed to create alignments. The creation process is usually assisted by software solutions and methods that make it easier to handle alignments, resulting in vastly different approaches for the planning process. One the one hand, creating alignments in AutoCAD Civil3D [9] uses straight lines for initial input, which are converted into an alignment by automatically inserting circular arcs and manually modeling the transitions as spiral curves. Parameters of specific curve types, such as the radius of circular arcs segments, can either be adjusted manually or by interacting with control points, as shown in Fig. 2. To ensure continuity, exclusively global modification is performed. There are five different types of control points, each performing a different modification requiring high precision for interaction. On the other hand, Card/1 [10] uses a more traditional approach, where alignment segments are modelled separately and are ultimately merged by connecting start- and endpoints of each section. The application requires modeling alignments sequentially, meaning one segment at the time. Subsequently, the continuity and length of segments are constantly evaluated.

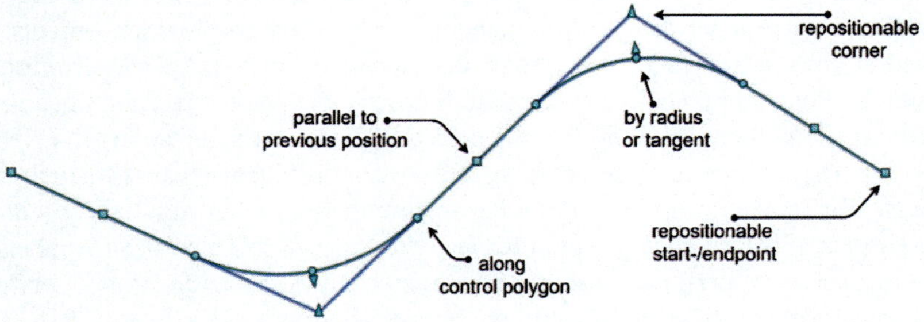

Fig. 2. Alignment exploration in Civil3D

2.3 Touch Interaction

Currently, a collaborative approach between decision-makers is not part of the design concept of commonly used CAD applications. Embedding the method of creating alignments to touch tables establishes a collaborative environment. By including all decision-makers into the planning and modeling process an ongoing and interactive exploration of alignments is performed.

The exploration of alignments on touch tables requires the definition of strategies for precise touch interaction. On a computational device the precision of interactions are examined by comparing touch input to mouse input, using Fitts' Law for measuring accuracy [11, 12]. The approaches usually focus on the pixel precision provided by default on the device hardware. Forlines et al. [11] reached the conclusion that when working with single interactions, a mouse is more suitable due to accuracy, while when working with multiple interactions at once, or on a large display, touch interactions turned out to be more suited.

In an early article by Potter et al. [13], that got further expanded on by Albinsson and Zhai [14], advanced techniques for high precision touch interactions are exemplified, such as Take-Off, Land-On, Cross-Key and Cross-Lever. By performing gestures and using software assisting tools provided by the application, such as virtual levers, visual feedback or buttons, the accuracy of interactions increased. These techniques provide templates to specify needed strategies for touch interactions for an integrated platform.

3 Concept

A general concept based on the known TIM is explained, extending the design for creating tunnel alignments. The proposed method of alignment planning and modeling will be described in detail and the collaborative value will be described. The methods will be applied to the integrated platform and general procedures for user interactions will be provided. The creation of horizontal alignment elevates the need for a general method of creating alignments that operates with simplified touch interactions. Accordingly, a method for stepwise creation of alignments and a set of interaction strategies for touch input have to be examined.

3.1 Integrated Platform

To improve the current practice to plan and evaluate alignment variants in tunneling projects, an integrated platform is proposed. The concept of an integrated platform extends the approach of the previously described TIM [4] by introducing planned alignment models. Our approach concentrates primarily on planning and modeling horizontal alignments. Vertical alignments require the consideration of an in-depth exploration of soil and ground-water conditions, which exceeds the scope of this paper. The platform is conceived to be a web-service with the TIM assigned to be located on the server and handling a repository of model data, as shown in Fig. 3. The client is conceived to handle user interactions for modeling and planning. By outsourcing analytical and evaluating mechanisms, we aim to reduce the computational complexity of modeling and planning processes.

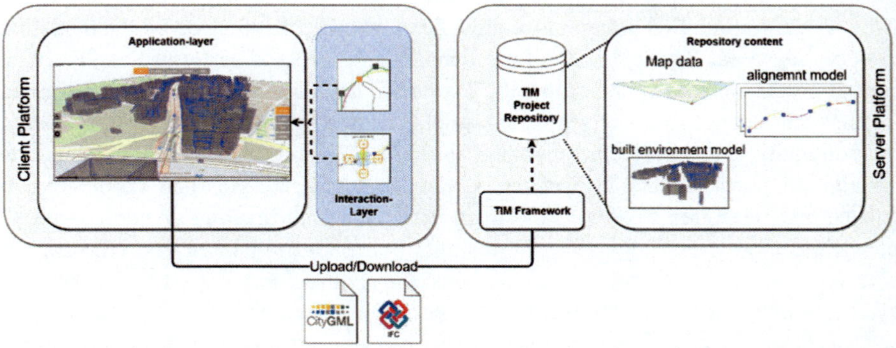

Fig. 3. Integrated platform embedded in TIM framework

In addition to the layers known by TIM, we introduce the Interaction-layer into the concept of the client. The Interaction-layer is usually part of the Application-Layer. In the context of the integrated platform, the interaction components are externalized to be able to specify essential interactions and touch techniques. Thus, techniques for creating and modeling alignments are provided more clearly. Associated interaction strategies and techniques should be documented in detail, since dependencies between interactions require them to be transparent and unambiguous at all times.

To deal with the large scope of significant planning data, the server is designed to handle different types of data and to link them into a holistic project. Data and files, such as created alignments and existing building structures, are stored in a model repository managed by the TIM. Our approach primarily uses XML-based data as input to create a project environment. In particular, the City Geography Markup Language (CityGML) is used, since GML can be easily retrieved and translated in a web-environment and model data occurs in different level of detail. Overlaying CityGML and 2D maps using geo-localization, enables the creation of project environments with sufficient information for modeling horizontal alignments. The project environment serves as visual boundary. Details about buildings can be made accessible for the planning process.

3.2 Method for Creating Alignments

Handling different curve types of alignment segments challenges local modification. As discovered in Sect. 2.2, existing modeling approaches often imply global modification effects to maintain continuities after editing. In most cases the global modifications affect segments along the alignment and perform a modification on a larger scale than desired. A method should be applied that only performs local modification and harmonizes the different curve types used to create segmented alignments.

For that matter, the advantages of spline curves are utilized, which are curves that are comprehensively adjustable by manipulating parameters and enable local modification. Provided by the definitions of Piegl and Tiller [15, pp. 117–139], this paper specifically uses Non-Uniform Rational B-Splines (NURBS). By keeping modifications within a controlled polygon, we are able to restrict the influence on the modeling of curve segments. The interaction for the modeling process is performed explicitly on control points connecting the segments. Therefore, the modification is only performed on the start- and endpoints of segments, simplifying the creation process by calculating the continuity and curves corresponding to the modification automatically. Thus, the remaining control points and parameters, such as degree, weights and knots, are calculated and set corresponding to the placement, type and direction of bordering segments and their own restrictions. By restricting and setting parameters, NURBS are able to represent straight-lines, circular-arcs and clothoid-arcs [16, 17]. Henceforth, such a NURBS-segment is referred to as specialized, since the resulting curve represents a specific type. If dependencies between segments are considered, alignments can be extended by inserting and removing specialized NURBS-segments at start- and endpoints. A local modification can be achieved by transforming both segments associated with the currently edited control point. An approach for sequential creation of horizontal alignments by inserting and modifying specialized NURBS-segments is introduced and illustrated in Fig. 4. In the following subsections the stepwise procedure is explained in more detail.

Step 1: Creating a Control Polygon. Starting with a collection of known conditions and constraints derived from planning data, the first step is to create a basic path of

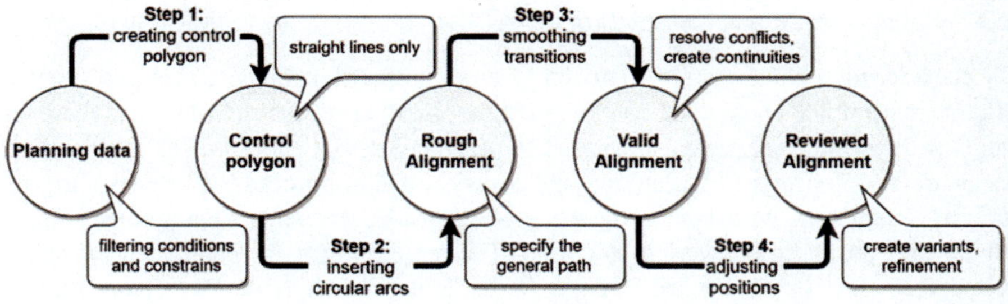

Fig. 4. General method for creating a horizontal alignment

straight-line segments, matching the control polygon for the alignment. The conditions and constraints can be, for example, the position of roads, buildings and detected soil like sand or clay that needs to be handled strategically. Tunneling alignments are usually planned along streets, minimizing the occurrence of terrestrial settlements. Using NURBS, a straight-line is created by either setting the weight of control points between start- and endpoint to zero or by setting the degree to one.

Step 2: Inserting Circular-Arcs. By inserting circular-arc segments, a rough alignment is created. The alignment currently resembles a Dubins Path [18], because the sudden change in the continuity between straight-line and circular-arc segments needs to be resolved. Provided by the definition from Piegl and Tiller [15, pp. 298–310], circular-arc segments are created by calculating the position and weight automatically, forming an isosceles triangle for the control polygon, using initially the start- and endpoints, as well as the direction vector generated by the previous segment.

Step 3: Smoothing Transitions. By adding a transition curve between straight-lines and circular-arc segments, the curvature will be smoothed, resolving the sudden change in continuity between straight-lines and circular-arcs. Modeling tunneling alignments usually use clothoid curves for transition between segments, since its curvature increases linearly with its curve length [7, 19]. The resulting alignment can be called valid, because it includes all required geometric properties, such as a valid composition of segment types and the smooth continuity. By applying the method of controlled clothoid spline curves provided by Walton and Meek [17], the NURBS segment represents a pair of clothoid curves. Since the method is applicable to B-Splines, only the position of control points needs to be calculated to compile with the theorem. The applied weight is therefore holistically and equally distributed.

Step 4: Adjusting Positions. Finally, the positions and curves of alignments need to be adjusted and transformed, since step 1 to 3 use an algorithmic approach to generate a valid alignment that usually diverges from the intended path. The adjustments are performed by interacting and transforming control points. This is basically a refinement process for the alignment, therefore calling them a reviewed alignment. By creating duplicates of the reviewed alignment, variants can be derived. The variants can differ, not only in the positioning of control points, but also in the composition of used segments, enabling a visual comparison.

3.3 Collaboration and Interaction

By considering a collaborative approach where multiple stakeholders plan and discuss alignments simultaneously and continuously we aim to create an environment where an ongoing evaluation of variants is possible. A collaborative environment enables live feedback and improves the planning and decision making process. The possibility of directly interacting with control points enables the adjustments of alignments intuitively and creates more varied solutions. This environment is established by using room-ware devices to our advantage, in particular touch tables with multi-touch capabilities. A touch table is a stationary device that can be easily surrounded and viewed by a group of people.

The multi-touch capabilities enable to perform simultaneous interactions. Contradictions during an interaction, such as editing and navigating at the same time, causes unexpected influence on interactions and at worst resulting in undesired changes in the alignment model. To prevent such events our approach uses a task oriented method that requires the selection of the current task at hand. This is further expended on in Sect. 3.4. In general, multiple significant but different tasks should not be performed concurrently, but rather be processed sequentially.

Furthermore, performing an interaction requires the consideration of certain interaction requirements. Accordingly, for defining an interaction on touch devices four leading questions are considered. Those questions are structured as objective, strategy, flaws and accuracy:

- What is the intended result produced by the interaction? (objective)
- How should the interaction strategy behave? (strategy)
- Which problems could occur and how can they be solved? (flaws)
- How intensive or accurate should the response be? (accuracy)

A single touch interaction is partitioned into the initial touch event, a performed gesture and the release touch event, while using multi-touch enables to express more complex interactions by grouping multiple interactions into a single response. Based on known approaches, described in Sect. 2.3, a set of basic touch strategies will be derived, which are able to perform interactions for planning and modeling tasks. The strategies used in this paper are called Lock-On and Area-Group.

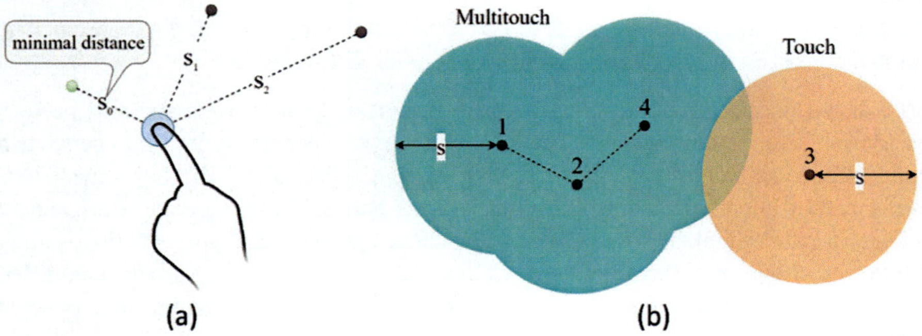

Fig. 5. The Lock-On strategy used for single interaction (a) and the Area-Group strategy used for grouping multi-touch interactions (b)

Lock-On Strategy. The task of selecting and editing needs to consider even the smallest objects. Depending on the perspective, distance and position of objects, interactions can be difficult. For example, an alignment can be hidden behind buildings or ground models. An interaction needs to be performed even if the conditions are disadvantageous.

A strategy called Lock-On is developed, based on the Take-Off strategy [14], which aims to improve the precision by performing distance checks and mapping the interaction to the closest object, as shown in Fig. 5 (a). In a 3D environment, the distance check is performed between the center points of objects and the interaction ray. The interaction ray is the vector between absolute camera position and interaction detected on the screen. The strategy can be adjusted by applying a threshold criterion for the distance. Setting the threshold to value S, so that $S \geq S_i$ with $S \in \mathbb{R} > 0$, results in fixed areas to detect interaction for each object. Mapping gestures to a selected object by Lock-On enables tasks to perform with an offset, which is a very useful property, due to the physical size and distances of a touch table.

Using this method ensures a more robust interaction. However, it is required to noticeably visualize interactable objects. Especially for complex geometry, finding the center point is usually a time-consuming estimation. Depending on the number and density of objects present, this method can get inefficient and inaccurate, since the number of checks increases simultaneously and a high density can lead to unintended interactions.

Area-Group Strategy. Considering the collaborative use, the implementation of separated multi-touch interactions needs to be considered, because multiple users will perform interactions simultaneously. To differentiate these interactions the Area-Group strategy is introduced, as shown in Fig. 5 (b). The Area-Group strategy is designed to group interactions based on a threshold distance value S, similar to the Lock-On strategy. A new interaction will be inserted into a group if it is within the distance of touch points performing an interaction already. However, applying this strategy requires multi-touch interactions to devolve. Depending on the number of touch points, different strategies will be executed, initially starting with one touch point and increasing the number by grouping points of interactions.

3.4 Modeling Process

The modeling process is designed to be performed task oriented. By switching between different modes of specialized interactions, a focused approach for the current task is presented. The modes are Navigate, Create, Edit and Extend, with Navigate as the default state. Since the navigation mode is not explicitly part of the creation process, the priority is on the latter modes. A general overview is provided in the activity diagram shown in Fig. 6.

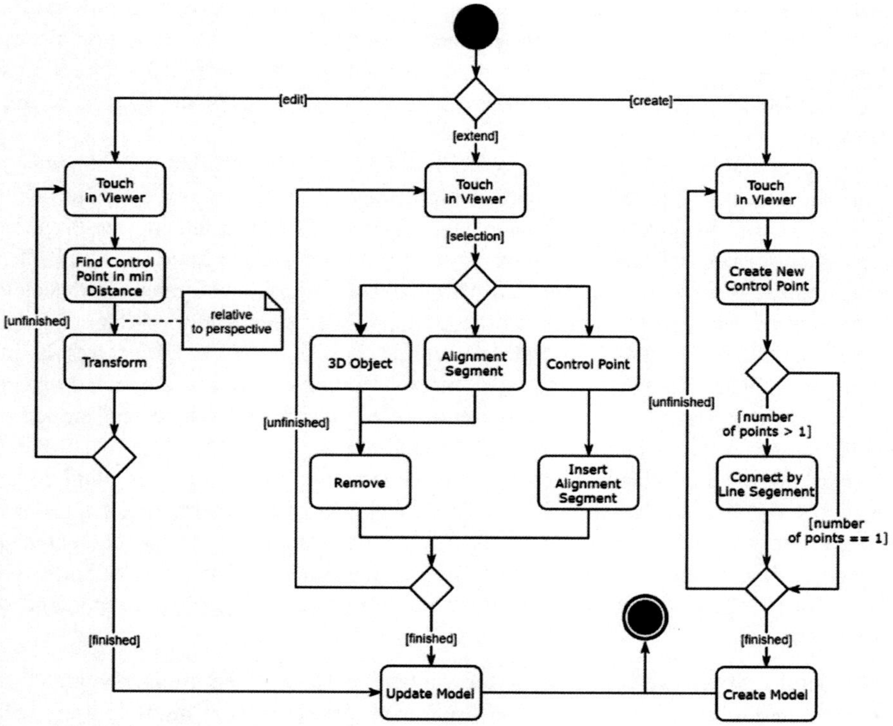

Fig. 6. Activity diagram of the task oriented modeling processes

The Create mode performs the creation of a control-polygon to initially work with. Each touch interaction generates a new control-point that will be connected with the previous control-point by a straight-line segment. Continuing the touch interaction generates more points and segments. Once sufficient segments are created, the segments are merged into a new alignment model containing the control-polygon.

Segments are inserted or removed by performing interactions in Extend mode. Depending on the configured sub-mode, a selection of control-points, alignment segments or environmental 3D objects is performed. New segments are inserted only by selecting control-points and replacing them with a new segment. Disused segments can be removed by selecting a segment instead of control-points and performing a remove command. Finally, related alignment models are updated.

By using the Edit mode, the alignments can be adjusted or transformed. This mode operates only on control-points. For each touch interaction the Lock-On strategy is performed, as described in Sect. 3.3. A touch gesture performs the transformation relative to the perspective of the 3D viewer. Once the transformation is performed, the alignment model is updated. Containing a reviewed alignment at best.

3.5 Comparing Methods

By using NURBS-segments we aim to reduce the complexity of using multiple curve types for creating alignments. The conventional method for creating alignments, which is still present and used in state-of-the-art applications, requires knowledge about explicit input of parameters and the resulting effect on the alignment. To be more precise, the challenge of using the conventional method is situated in handling multiple different parameters that are also different per curve type. This results in a combination of curve segments that are difficult to synchronize and to update if changes occur along the alignment. NURBS-segments, instead, use fundamentally the same function and require the same parameters as input for every curve type. Furthermore, by automatically calculating and setting most of the parameters needed to define NURBS, such as curve degree, knots and weights, we are reducing the parameter input only to control points and a type hint. By utilizing NURBS, the curve segments gain the ability of local modification for the alignment creation, through locally aligning the control polygon corresponding to the modification.

4 Implementation

A case study is conducted by implementing a prototype that will be embedded in a collaborative environment. The presented method for creating alignments will be evaluated by implementing tools to create segmented alignments using NURBS. The stepwise creation of alignments is exemplified by performing the creation in exemplary project environment.

4.1 Prototype

The client-platform can be accessed through the web-browser. By creating a project and loading planning data, the user can establish an environment, as shown in Fig. 7. We used open accessable city models and map sources to generate visual boundaries like roads, forests and special buildings for the planning of alignments. The plain visualization and menu interactions corresponds to the Application-Layer, as contextualized in Sect. 3.1. The user interface is subdivided into a navigation for different perspectives at the top, a menu with settings at the left, a menu for switching tasks at the right, a navigation intensity slider at the bottom and a 3D view for modeling at the center.

4.2 Creating Alignments

The creation of an alignment is exemplified in Fig. 8, applying the method described in Sect. 3.2 and 3.4. Using map data and building environments enables the creation of alignments within defined boundaries. The initial control polygon (Step 1) is created by successively picking the general path in the 3D view. By inserting new segments (Step 2 and 3), the alignment is created along the defined general path, replacing old control points in the process. Adding a function for duplicating existing alignments enables a quick and efficient creation of variants. By utilizing the touch-strategy Lock-On, as

Fig. 7. The client implementation of the prototype

described in Sect. 3.3, adjustments are performed (Step 4). These adjustments are performed by interacting with the control points and transforming the segments in the process. The control points are illustrated as blue rectangles in figure Fig. 8.

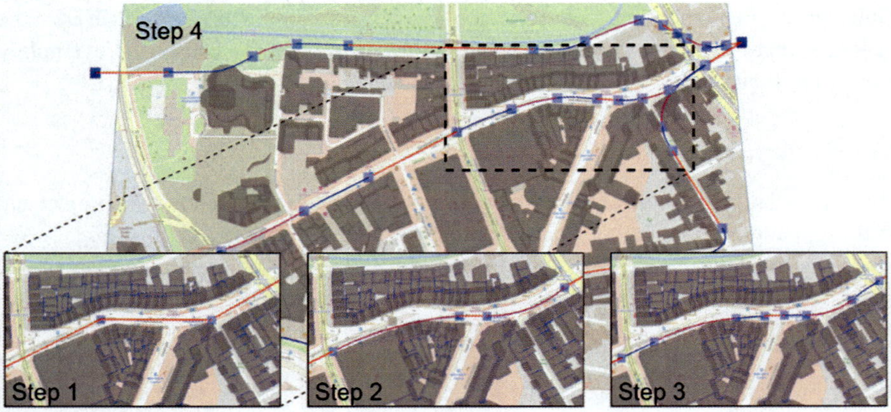

Fig. 8. Stepwise creation of alignments performed in prototype

4.3 Evaluation

To concentrate on the issued challenges, the prototype will be evaluated based on usability and practical proximity.

Usability. The prototype cannot be called intuitively applicable. We assume that the experience varies and leads frequently to undesired changes, since it depends on the user's knowledge and expectations. Those difficulties are eliminated by properly introducing the implemented interaction strategies and techniques to the user. The reduced and simplified user interface provides a focused project environment for collaborative planning and modeling, as well as encouraging team discussions. Switching between navigation and editing limits the capabilities for user interactions to the task at hand, as well as preventing unexpected changes. The prototype presents a highly efficient tool to create alignments by utilizing previously defined interaction strategies and techniques. It is recommended to appoint a moderator or coordinator among users, who is able to manage the workflow, switch tasks smoothly and streamline the systematic creation of alignments.

Practical Proximity. Using the concept of an integrated platform to create a prototype for planning and modeling alignments provides a beneficial use case of collaborative web-applications in tunnel construction. Improving the creation of alignments and variants by enabling an ongoing, collaborative and interactive planning and modeling are seen to be favorable. Using specialized NURBS-segments exemplifies an alternative to creating alignments, which in turn enables local modification. Applying multi-touch strategies and techniques provides a useful and interactive concept for planning in team discussions.

5 Conclusion

Building upon known tools like TIM, the integrated platform conceptualizes an application for interactive exploration of tunnel alignments. We presented a concept for enabling local modification and reducing the complexity of using multiple curve types by creating horizontal alignments systematically using NURBS-segments. Additionally, by comparing the method for creating alignments in state-of-the-art applications a method for stepwise alignment creation has been derived. By utilizing the potentials provided by touch tables, a collaborative approach between users has been examined and the required touch-interaction strategies has been defined. Adding the Interaction-Layer into the concept of the integrated platform provides an interface for implementing the required touch interactions. A prototype implementation shows that the integrated platform is a usable approach for planning and modeling alignments. A procedure for stepwise creation of horizontal alignments is also provided and implemented in the prototype implementation.

We expect to advance this approach by introducing further constraints, by considering more planning data and by generating more detailed models. Processing more data, such as ground and ground-water conditions, enables the consideration and integration of vertical alignments into the concept of the integrated platform. The introduction of rule-based checking into the integrated platform would generate a responsive environment with live feedback. By linking the planning and modeling process with simulation methods, we hope to increase the precision even further and improve the decision-making. Building upon our approach of using web-development technologies for implementation, a collaborative approach between applications is promising. The introduction of the integrated platform provides an additional example for digital construction and collaborative modeling in BIM based tunneling.

References

1. British Tunnelling Society and others: Tunnel lining design guide, Thomas Telford (2004)
2. Rudolf, J., Gall, V., Nitschke, A.: Selection of alignment and tunneling methods in urban settings. In: Proceedings of the World Tunnel Congress 2008—Underground Facilities for Better Environment and Safety, pp. 22–24 (2008)
3. Kang, M.-W., Jha, M.K., Schonfeld, P.: Applicability of highway alignment optimization models. Transp. Res. Part C Emerg. Technol. **21**(1), 257–286 (2012)
4. Koch, C., Vonthron, A., König, M.: A tunnel information modelling framework to support management, simulations and visualisations in mechanised tunnelling projects. Autom. Constr. **83**, 78–90 (2017)
5. Industry Foundation Classes (IFC), DIN EN ISO 16739
6. Shanmugavel, M., Tsourdos, A., White, B., Żbikowski, R.: Co-operative path planning of multiple UAVs using Dubins paths with clothoid arcs. Control Eng. Pract. **18**(9), 1084–1092 (2010)
7. Baass, K.G.: Use of clothoid templates in highway design (1982)
8. Amann, J., et al.: A refined product model for shield tunnels based on a generalized approach for alignment representation. In: Proceedings of the ICCBEI (2013)
9. Civil3D: Civil infrastructure design and documentation software: Autodesk (2019)
10. card_1: Application for Civil infrastructure design: IB&T GmbH (2019)
11. Forlines, C., Wigdor, D., Shen, C., Balakrishnan, R.: Direct-touch vs. mouse input for tabletop displays. In: Proceedings of the SIGCHI Conference on Human Factors in Computing Systems, pp. 647–656 (2007)
12. Sears, A., Shneiderman, B.: High precision touchscreens: design strategies and comparisons with a mouse. Int. J. Man-Mach. Stud. **34**(4), 593–613 (1991)
13. Potter, R.L., Weldon, L.J., Shneiderman, B.: Improving the accuracy of touch screens: an experimental evaluation of three strategies. In: Proceedings of the SIGCHI Conference on Human Factors in Computing Systems, pp. 27–32 (1988)
14. Albinsson, P.-A., Zhai, S.: High precision touch screen interaction. In: Proceedings of the SIGCHI Conference on Human Factors in Computing Systems, pp. 105–112 (2003)
15. Piegl, L., Tiller, W.: The NURBS Book. Springer Science & Business Media (2012)
16. Wang, L.Z., Miura, K.T., Nakamae, E., Yamamoto, T., Wang, T.J.: An approximation approach of the clothoid curve defined in the interval [0, pi/2] and its offset by free-form curves. Comput. Aided Des. **33**(14), 1049–1058 (2001)
17. Walton, D.J., Meek, D.S.: A controlled clothoid spline. Comput. Graph. **29**(3), 353–363 (2005)
18. Dubins, L.E.: On curves of minimal length with a constraint on average curvature, and with prescribed initial and terminal positions and tangents. Am. J. Math. **79**(3), 497–516 (1957)
19. Amann, J., Singer, D., Borrmann, A.: Extension of the upcoming IFC alignment standard with cross sections for road design. In: Proceedings of the ICCBEI, pp. 22–24 (2015)

BIM Component Library for Subway Public Works

Sarah Cardoso Nunes(✉), Sergio Leal Ferreira, and Jéssica Tamires Silva Brito

University of São Paulo, São Paulo, Brazil
{sarahnunes, sergio.leal, jessicatami}@usp.br

Abstract. The aim of this paper is to discuss and propose solutions for the introduction of BIM component library in the Brazilian context of contracting public works. Currently, public companies contract engineering and construction services through a bidding process, which seeks to ensure the most advantageous proposal for the government and to avoid overpricing in operations. However, transparency and active public accountability are still poorly used, creating misinformation and opportunities to misuse public resources. In light of this context, some public initiatives of BIM (Building Information Modeling) adoption for public construction aim to increase the quality of projects and budgets, resulting in greater transparency and information flow control. The present work focuses on the contracting process of "Companhia do Metropolitano de São Paulo - CMSP", the São Paulo state subway company, which are usually made under a semi-integrated regime. It is necessary that the CMSP develop or contract a called basic project. From this project a reference budget is calculated and used as a basis for the request for proposal value. In order to increase the speed of the contracting process and provide a better project development, it is proposed a BIM component library, which includes complete modules of subway station. Subway station designs have a high number of standardized solutions, which is closer to manufacturing processes. The research seeks to combine the BIM methodology with the construction standardization process, providing elements to assist the development and control of the subway works.

Keywords: BIM · Subway · Bidding process · Public works

1 Introduction

In the last few decades, the construction industry has failed to keep up with technological development and productivity gains generated by other industrial segments, such as the automobile industry and agribusiness. Fact that impacts the growth of national economies, the quality of life of the population and environmental issues.

According to The Economist magazine [1], civil construction has the lowest productivity rates compared to other industries. In the last 20 years, Germany and Japan, models in industrial efficiency, have shown virtually no growth in construction

E. Toledo Santos and S. Scheer (Eds.): ICCCBE 2020, LNCE 98, pp. 335–361, 2021.
https://doi.org/10.1007/978-3-030-51295-8_24

productivity, and in the United States the number has halved since the late 1960s, as shown in Fig. 1.

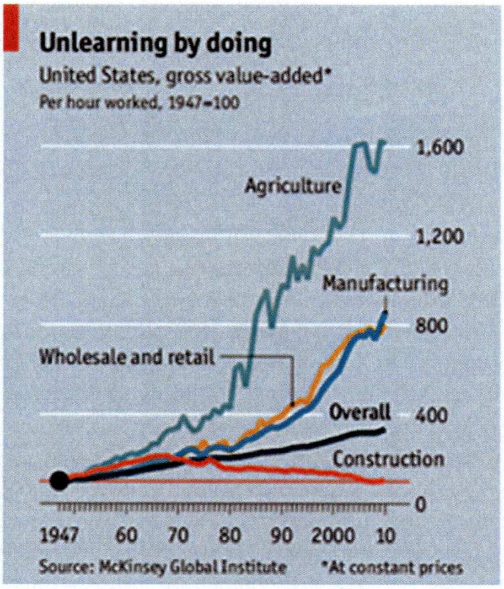

Fig. 1. Industry productivity index (Added value per hour worked) [1]

In Brazil, following the world trend, a study by CBIC - Câmara Brasileira da Indústria da Construção in partnership with FGV - Fundação Getúlio Vargas between the period 2007 to 2012, reveals that the Total Factor Productivity (TFP) of construction companies civilian sector presented an average decline of 0.4% per year [2]. The data are shown in Table 1.

Table 1. Average productivity growth rate (% per year) [3]

Period	Labor productivity	Capital productivity	Total factor Productivity (TFP)
2007/2012	−0,2%	−0,5%	−0,4%
2007/2010	−1,2%	1,4%	0,0%
2010/2012	1,3%	−3,3%	−1,0%

In addition, the difficulty that the Brazilian Public works administration has to carry out the works within the desired standards of quality, cost and terms is notorious [3]. Problems in resource management are generally associated with corruption, which results in losses of approximately R $ 200 billion per year, about 3% of national GDP [4].

The low performance of the civil construction sector, added to the lack of technological innovations and of transparency in the administration of public projects point to the need for research, modernization and restructuring of the infrastructure and urban development industry.

Digital technologies, such as 3D printing, prefabrication, intelligent construction equipment, mobile devices can help to avoid waste, increase agility and quality of projects and works. BIM (Building Information Modeling) platform has also the potential to provide and manipulate the precise and consistent data needed for such technological innovations [5].

According to the EU BIM task group [6], a strategic factor for enabling technology is its adoption by the public sector. In Brazil, on May 17, 2018, BIM's National Dissemination Strategy was instituted through Decree No. 9.377, with the purpose of developing measures to regulate, train and disseminate technology nationally. For the public sector, the implementation of BIM represents the control and management of a large volume of data, better understanding of projects, transparency in budgets, with less risk of burdensome works, resulting in more construction of urban equipment, with the same or less investment.

The challenges for the public sector are innumerable, however, the research will be directed to the infrastructure industry, especially in the area of urban mobility, with the objective of offering an improvement to the process of contracting and development of subway building projects, under the perspective of using BIM technology.

Metro works, due to their nature, are extremely complex, and involve exceptional efforts and costs. According to Mainardi and Saccomano [8], the BIM methodology presents itself as an efficient management tool for the São Paulo Metro company, as it breaks with old paradigms, and offers several integration and development tools that are in line with the most modern technologies.

The subway station projects have a high number of standardized solutions, elements and materials, a fact that awoke to the investigation of the construction industry's approach to the manufacturing processes. According to Johnston [9], the way out of the construction chain is in changing processes, in which construction becomes more like manufacturing. From integrated solutions and standardized components, a new way of working will be produced, reducing the fragmentation of the system and bringing together the various agents of construction.

The standardization of construction makes it possible to increase the quality and reduce the cost of buildings. According to Gabarra [10], infrastructure projects, due to their very nature of effort and investment by the community, require the industrialization of their construction in order to optimize the large volume of invested resources.

The research seeks to combine the BIM technology with the standardization process of components of the construction industry, proposing the development of a collection of BIM components, with the purpose of increasing the quality of the projects and reducing the contracting and development deadlines of the subway station projects.

2 Objectives

The objectives of the study can be divided into general and specific terms, as follows:

2.1 General Objectives

- To study and characterize the current functions of BIM object libraries, including their different types and stages of development, in order to understand their impacts on the digital construction industry.
- Complement the knowledge about the Building Information Modeling paradigm, through the study of new techniques and processes that can contribute to the efficiency, quality and productivity of the infrastructure and urban development industry.

2.2 Specific Objectives

- Understand the process of contracting and developing projects for the construction of metro stations.
- Map and survey data about BIM component libraries for generic and public projects.
- Study the standardization process of components of the construction industry, seeking to apply to subway works.

3 Methodology

In order to establish the framework of known empirical solutions, a literature review was prepared, covering the following themes: process of contracting public works, project process on the São Paulo Subway, industrialization and standardization in stations subway, and finally generic and public BIM libraries. From this study it was possible to confirm the proposed artifact: library of BIM components to support the process of developing bidding projects. The scheme developed for the execution of Design Science Research is illustrated in Fig. 2.

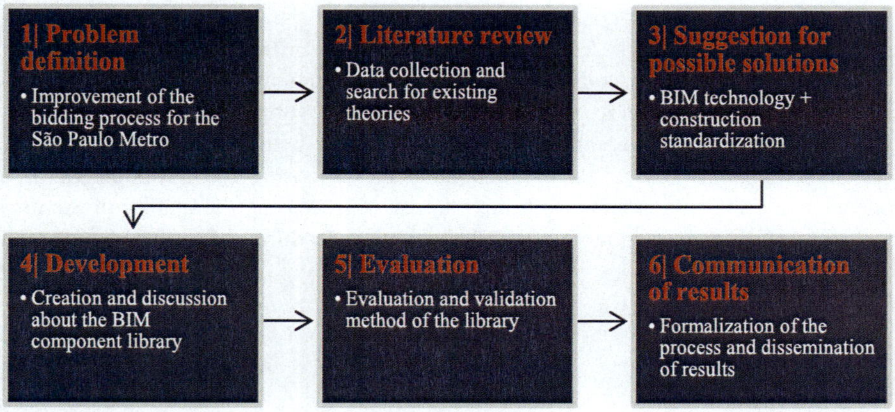

Fig. 2. DSR execution scheme [11]

The case study was used to understand the initiatives practiced, in order to expand the knowledge that will allow the design and construction of an artifact that covers quality and applicability.

The research is in the development stage of the BIM component library for the subway stations. After completing this step, a method for evaluating and validating the artifact will be developed, followed by the formalization and dissemination of the research.

4 Literature Review

4.1 Public Works Contracting Process: Bids and Contracts

Given that the objective of the research is to study and develop a library of BIM components for the subway works, it is necessary to understand the context in which type of work is inserted. According to Altounian [12], public works have prominence in the investment group made by public sector, both for its materiality and for its social importance. However, it is not uncommon to find serious irregularities in the management of resources channeled to this area, such as overpricing, targeting, delay in completion or even the abandonment of projects.

Given this scenario, improving the management of projects and public works is fundamental for the country's development, and for that it is necessary to identify and solve the problems of the current system [13]. This research item focuses on understanding the process contracting public works, with the aim of examining which points can be improved with the use of the BIM methodology associated with the standardization process of construction elements.

In Brazil, the contracting of engineering works and services is carried out through bids, which aims to guarantee a more advantageous proposal for the government (avoiding overpricing in operations), guaranteeing equal conditions to all who want to contract with the Government and promote sustainable national development [14].

There are several legal structures that govern public contracts in Brazil, the main one being Law 8,666/1993, which imposes the "rules for public tenders and contracts", the main one for contracting civil construction [15]. The document that regulates public and mixed companies of the Union, the States, the Federal District and the Municipalities is Law No. 13,303, of June 30, 2016. From this Law, all public companies can follow their own bidding procedures, provided the precepts of Law No. 13,303/2016 are guaranteed [16].

The Metropolitan Company of São Paulo - METRÔ, has its own regulation, which establishes the company's bidding and contracting procedure, in addition to its relations with suppliers, contractors, partners and donors [17]. The document presents several contracting regimes: contract for unit price, contract for global price, task, full contract, semi-integrated and integrated contracting. The regimes are chosen according to the scale of the work or the engineering service to be provided, the precision of the quantity and budget inherent in the basic project, the time available for operation of the project and the intellectual and technological nature of the bid object.

Art. 42 of Law No. 13,303/2016, establishes that public or mixed companies must use semi-integrated contracting, in the case of engineering works and services. In this contract regime, the contracted companies are responsible for the development of the executive project, execution of the work and for all the operations necessary for the final delivery of the object. Therefore, in order to call for tenders in this type of contract and in others it is necessary to prepare the "basic project", which is an intermediate phase between what is defined by the American Institute of Architects (AIA) as the "design development" phase and the "construction document" phase.

Law 8.666/1993 defines the "basic project" as a "set of necessary and sufficient elements, with an adequate level of precision, to characterize the work or service, or complex of works or services object of the bidding process, elaborated based on the indications of the preliminary technical studies, which ensure the technical feasibility and the appropriate treatment of the environmental impact of the enterprise, and which makes it possible to assess the cost of the work and define the methods and the execution time" [15]. However, the law does not establish the level of accuracy of the basic project, and as a result of this degree of uncertainty, it allows an increase or decrease of 25% in cost, in cases of new buildings, and 50% in renovations.

The flaws in the elaboration of the projects are highly related to the main attributes that cause the contractual amendments of term and value in the works of public entities [3]. In the case of infrastructure works, object of analysis of this research, the use of the BIM methodology is pointed out as one of the solutions to the problem. According to the Paraná Department of Infrastructure and Logistics (SEIL) [13] - BIM provides significant improvements in the technical quality of projects, in sustainability and in the control and transparency of public works.

Considering the importance of initial studies and projects, the research in question studies ways to improve the development process of basic projects, giving them greater quality and information control through the BIM methodology and public project standardization processes, with a focus in the subway stations. In the following item, the process of designing the stations and the bidding project for the São Paulo Subway will be described.

4.2 São Paulo Subway

According to the São Paulo Metro Expansion Management methodology [18], the feasibility of an enterprise in the company has 5 phases, as shown in Fig. 3.

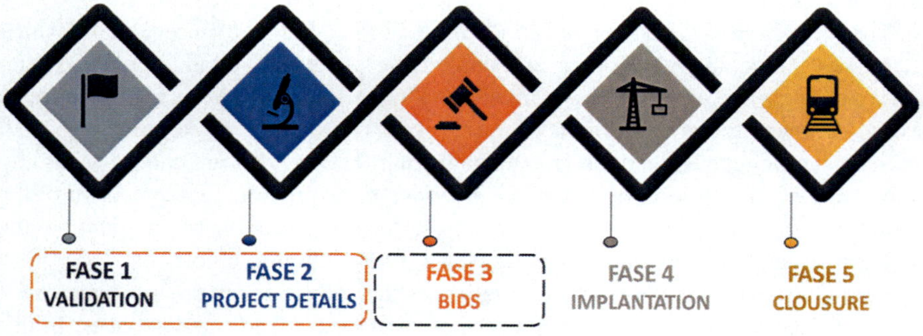

Fig. 3. Phases of the enterprise in the CMSP [18]

The process of designing the lines and developing the bidding projects is extensive. To elucidate the issue of time in the METRÔ works contracting process, Fig. 4 presents a timeline from the conception of the extension of Line 2-Green to the completion of the basic project for one of the stations on the line.

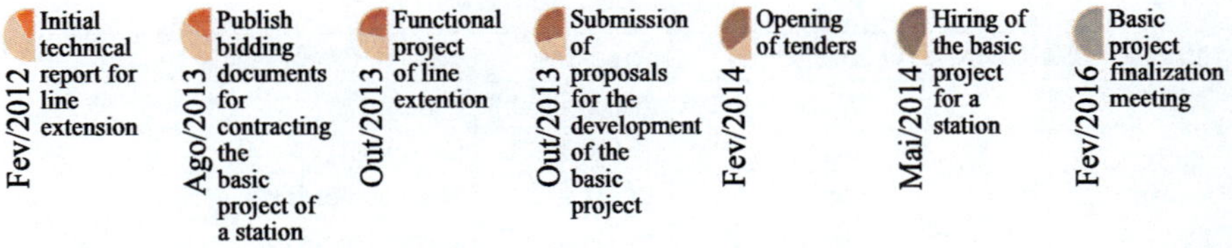

Fig. 4. Design and contracting process for Ponte Grande station - Line 2-Green [11]

As presented in item 4.1, the initial design phase, comprising the preliminary studies and the basic project, is essential for the bidding process, as well as for the smooth progress and execution of the work. The following item will explore the design process of subway stations, in order to analyze how the library of BIM components, object of the study, can optimize the design process of subway stations.

4.3 Project Process on the São Paulo Subway

The design guidelines for the São Paulo Metro stations were established by the architect Marcello Fragelli, in the 1960s, for the construction of the first metro line in Brazil, the North South Line – currently Line 1-Blue. Based in these guidelines, the process was shaped, which, even today, guides the design exercise of subway installations [19].

Functional Project. The functional project is the first phase of the project for the implementation of a new metro line. In this stage, the main issues are defined, such as: the line layout, the location of its stations, the integration terminals, the maintenance yards and the train parking. Complex planning studies analyze the consequences of implementing a line, through its impacts on public transport, land use and occupation, the environment and urban infrastructure.

The elaboration of the Functional Project presupposes the carrying out of studies of [20]:

- regional characterization;
- reorganization of public transportation;
- demand estimate;
- alternatives and route definition;
- details of the recommended layout alternative;
- preliminary studies that define the necessary and sufficient elements for the development of the basic project;

The estimated demand of each station will guide the dimensioning of the project, in order to avoid reaching high rates in relation to the comfort level, above 5 passengers/m^2. According to Santos [26], the technical standards of comfort level used by the PBMC are based on the work of Jonh J. Fruin. This study establishes six comfort levels, depending on the area available per individual, shown in Table 2.

Table 2. Comfort level (m^2/person) [21]

NÍVEL DE CONFORTO	m^2/pessoa	pessoas/m^2	Fatores Psicológicos
A	mais de 1,2	menos de 0,8	nenhum
B	de 1,2 a 0,9	de 0,8 a 1,1	nenhum
C	de 0,9 a 0,7	de 1,1 a 1,4	zona de conforto
D	de 0,7 a 0,3	de 1,4 a 3,3	contato evitável
E	de 0,3 a 0,2	de 3,3 a 5,0	contato inevitável
F	menos de 0,2	mais de 5,0	desconforto fisico e psicológico

Based on the technological definition of the line - which consists of the type of train, type of track, power supply, electronic signaling, communication, etc. - the geometric parameters of the permanent track are established and, adding the conditions determined in the regional characterization, it is then it is possible to draw the alternatives of layout in plan and profile. The maximum slope that the composition can overcome is the factor that determines the profile of the line [22].

After studies carried out on a macro scale, detailed in-depth investigations are initiated to support the basic design process [19]. From the definitions of the functional design, the basic architectural and civil engineering projects of the stations of the proposed line are elaborated. Generally, the basic design of a subway line is bid for stretches, in which different designers develop the projects of the different subway buildings that make up the line.

Basic Project. The basic architectural design defines the architectural part of the ventilation shafts, of the stations and their accesses, as well as of the construction elements, considering the complementary projects (structure, ventilation, electrical, electronics, hydraulics and roads), as well as the facilities and equipment necessary for the operation, comfort and safety of users. The project must contemplate the legislation related to accessibility, firefighting and escape route, land use and occupation (setbacks, occupancy rate and permeability rate), in addition to complying with municipal and state legislation regarding the environment.

The critical design analysis, performed by the CMSP technicians, assesses the capacity of a project to meet the proposed requirements, seeking to identify problems and propose the development of solutions [23]. Each discipline, or project phase, is analyzed by specific teams or professionals, and usually the same project is analyzed by different technicians and departments.

The São Paulo Subway has several guidelines and instructions for preparing the basic project. The specification of the materials and equipment for the stations is carried out based on the technical documents that describe and list the codes, units and geometric data for each product. The lists also indicate reference drawings, called "standard design" or "projeto padrão" - PP in Portuguese. From the standard design, each designer must represent geometrically and indicate through codes the elements used in the design. PP's will be studied in item 4.4 - Industrialization and standardization in subway stations.

4.4 Industrialization and Standardization in Subway Stations

According to Gabarra [10], the adoption of standardization and industrialization of construction allows a lower proportional cost, as it dilutes equipment costs, generates cheaper elements, beyond the increase of reliability and control of deadlines. The public department that manages the railways in England - Network Rail, encourages a standardized approach to create and maintain its buildings, pointing out some benefits of this approach [24]:

- Project effort and cost reduction;
- Consistency of appearance (brand);
- Maintenance facilitated by the use of known components;
- Cost savings through the management of bulk purchases;
- Potential for feedback and product improvement;
- Greater quality control and reliability;
- Reduction of human dangers and errors through familiarity.

The CMSP also has standardized elements that are specified and detailed in the standard projects (PP's). The PP's for architectural and furniture elements is an important tool for the development and analysis of basic projects. These drawings set design and installation guidelines, and are used in the design of stations, ventilation shafts and maneuvering yards, among other buildings in the subway network [25]. The development and use of these standard projects demonstrate the concern of the CMSP in adopting standardized solutions with proven quality.

The proposal of the BIM component library is to group both the standardized elements already used by the company, as well as offering complete building modules approved by the CMSP, considering different construction methods, so that the designers develop only the particular parts of each station, such as, for example, access and surroundings.

This research item is intended to understand the composition of a station, from the construction methods available to the parts and elements that can be standardized.

Typology of Subway Stations. In a survey of the existing subway stations of the Metropolitan Region of São Paulo - RMSP, the underground stations appear in greater number. As the conditions necessary for the construction of elevated and surface roads are increasingly restricted in the city, this item of research will study underground subway stations, in order to present some of the elements and construction methods that can compose the library of BIM components for the CMSP. Figure 5 shows the types of existing station in the RMSP and the respective opening years.

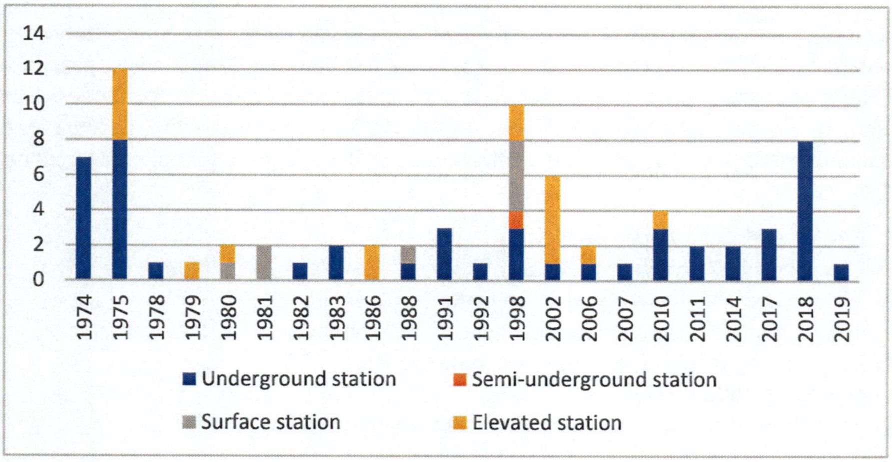

Fig. 5. Subway stations: type and year of inauguration [11]

According to the studies made by Carvalho [7] the organization of underground subway stations can be described from five main areas:

- External accesses: located on expropriated lots or sidewalks.
- Passenger lounge - usually in the form of a mezzanine located on the first buried level. In this space there is communication between the accesses and the paid areas, therefore it is the place where the ticket offices and the blockades are installed.
- Areas of loading and unloading platforms.
- Operating areas - with offices and support spaces for employees.
- Technical rooms - equipment rooms, maintenance areas, reservoirs.

The flowchart, with the layout of the areas mentioned above, is shown in Fig. 6.

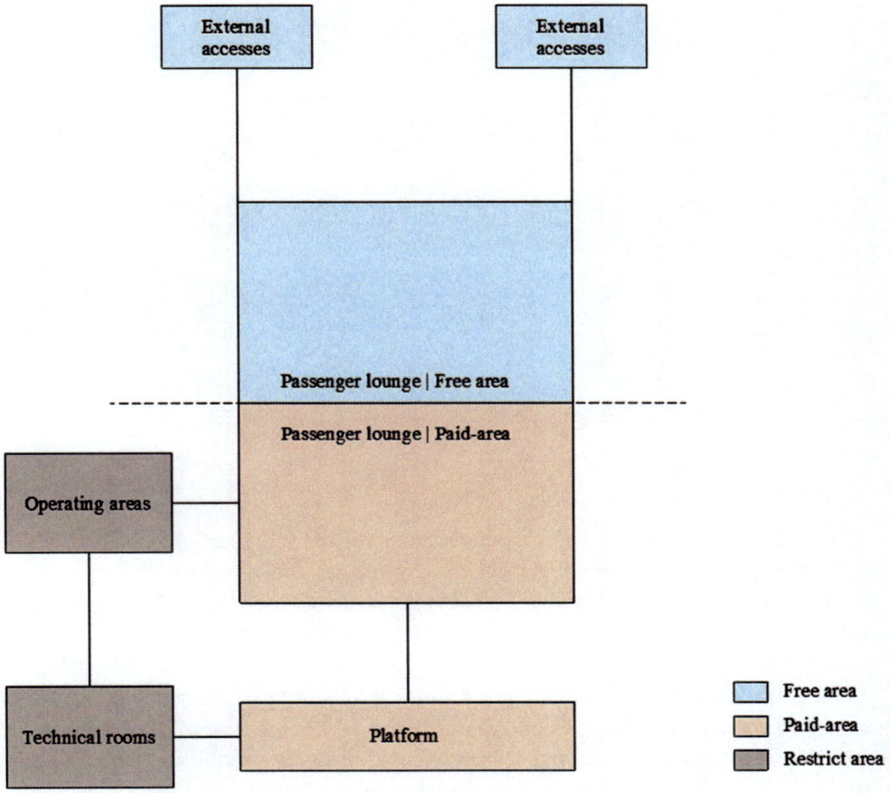

Fig. 6. Plan of stations in plan [7]

Gabarra [10] studied the standardization and industrialization in the construction of the São Paulo subway, presenting an analysis of the expansion of Line 5-Lilac. According to the author, an entire stretch of line may present the same season pattern, according to its characteristics of constructive density, urban pattern and relief of the region where it is located. After defining the premises of the modal and consolidating the network planning, Gabarra [10] proposes that the standard treatment of the stations be considered according to the type of section that the station will be implemented, dividing them into three main groups, according to Fig. 7.

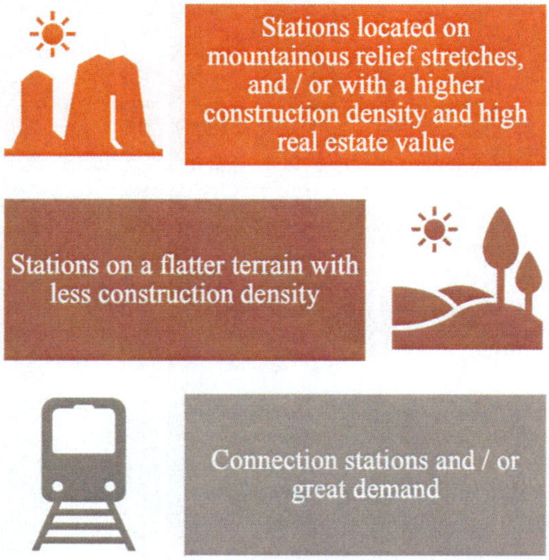

Fig. 7. Station groups [11]

The characteristics of stations located on mountainous relief stretches, and/or with a higher construction density and high real estate value are: **NATM tunnel**[1] for platforms and distribution mezzanine, together with **drying wells**[2] and connection tunnel for passenger access and necessary features [10]. This standard scheme has some variations depending on the position and size of the drying wells, according to the scheme in Fig. 8.

When the center of the well is aligned with the axis of the track, it is not necessary to execute a connecting tunnel, a condition that also favors the flow of passengers at the station. This type of station has a very lean circulation system, requiring that part of the needs program be kept in external constructions or in shallow widening of the ditch.

The NATM tunnel has a lateral platform and a mezzanine over the rail, where passengers are distributed. According to Gabarra [10] this tunnel can also be executed using the **Tunnel Boring Machines – TBM**[3] methodology, considering the typology of the station with displaced well, and making some adjustments, such as the overlapping of the platforms. Considering this configuration, stations made with a tunnel platform can be standardized, adding reliability to projects, costs and deadlines.

[1] New Austrian Tunneling Method - NATM consists of a sequential excavation of the massif using shotcrete as a support, associated with other elements such as an-chors and fibers in the concrete, depending on the massif's self-supporting capacity (CMSP, 2019) [19].

[2] Drying wells are excavations, built by the vertical NATM method, consisting of circular wells, which can have varying numbers, housing the entire program of a station or only part of it [19].

[3] The mechanized method (TBM - TUNNEL BORING MACHINES) uses machines that, under the protection of a metallic armor, perform the excavation and assemble the precast concrete (or metallic) segmented coating. The advance of the machine is due to the reaction of jacks against the mounting rings already assembled. In the case of TBM for rock, without armor, the advance occurs through shoes anchored on the side walls of the tunnel [19].

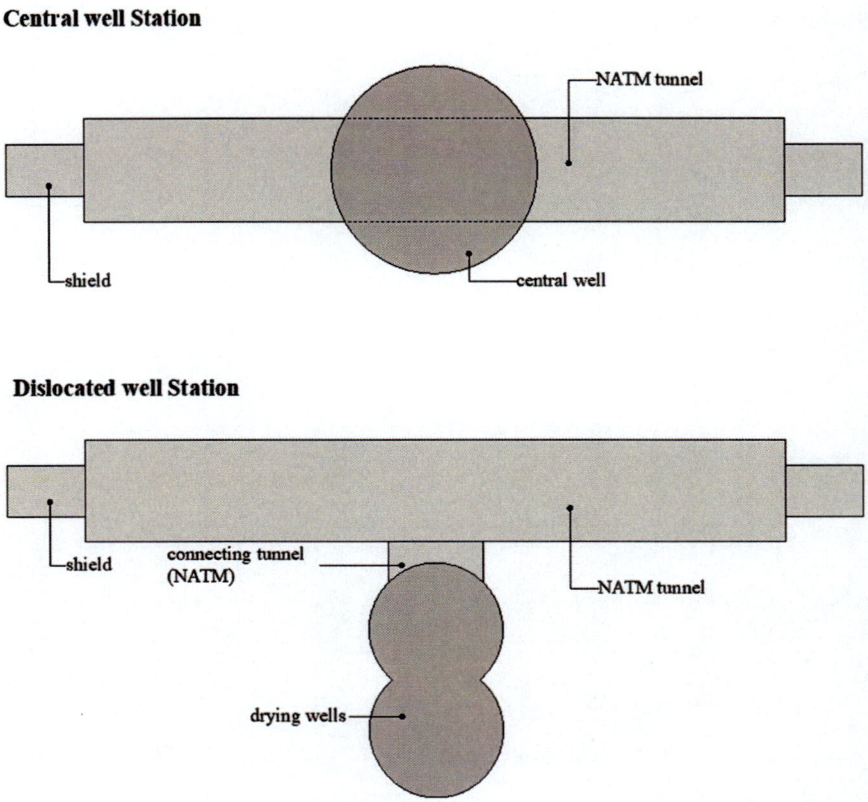

Fig. 8. Scheme of stations with platform in NATM tunnel [11]

In the other group of stations on a flatter terrain with less construction density, stations with a **trench or VCA**[4] platform are viable, being able to use the rectangular containment system or five drying wells, depending on the conditions of the site, as shown in Fig. 9.

The drying well system, due to the geometry, allows for greater wall extensions without shoring. As it is a determining element for the layout and for the escalators, the modulation of the shoring by struts should be standardized in the seasons. In this type of station, Gabarra [10] suggests the location of the technical rooms either inside the ditch, since the excavated space is used and the cables are distributed more directly.

In the case of connection stations and/or great demand, standardization is compromised due to the specific legal requirements that they must meet and also so that they are provided with comfort for their users. In these cases, Gabarra [10] suggests the

[4] Trench or VCA (Open Pit Ditch) is a construction system that consists of opening a ditch along the entire length of the tunnels and stations and, after its containment and waterproofing, the roof slab is made, on which is reconstituted the urban fabric of streets, squares, etc. [10].

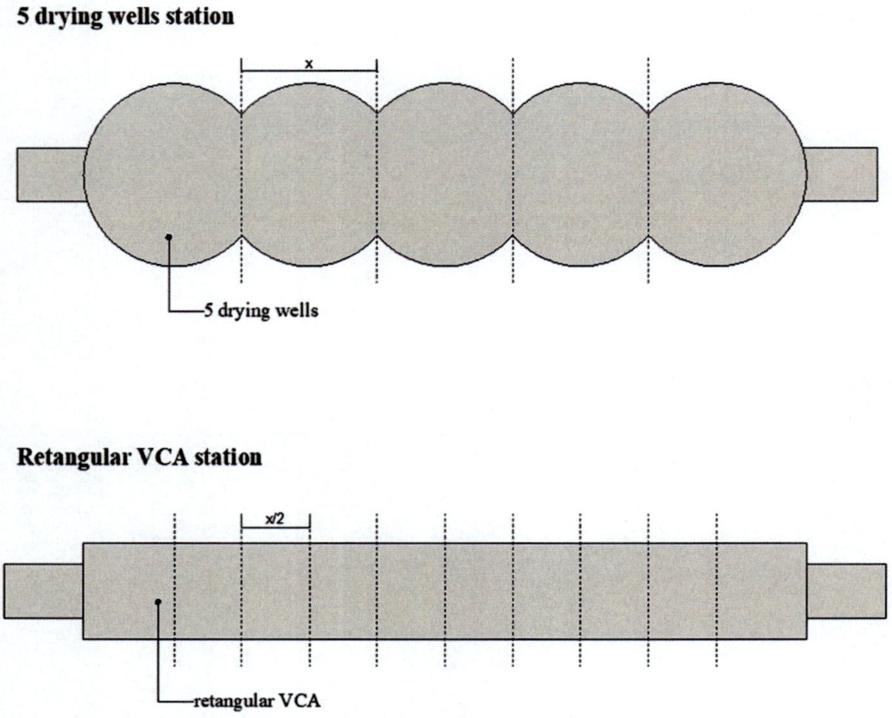

Fig. 9. Scheme of stations with VCA platform [11]

elaboration of specific projects that consider the characteristics of the lines, connections and the surroundings of these stations.

The production of the macro modules of the BIM component library for CMSP will be based on the studies by Gabarra [10] on the standardization of São Paulo subway stations, as presented in this item. For the development of the library, surveys of generic and public BIM libraries were carried out, with the purpose of observing their characteristics, typologies and stages of development.

4.5 BIM Libraries

The organization and classification of project information is essential for the work, highlighting the importance of well-structured libraries that lead to savings, through the reuse of information and reduction of rework.

BIM libraries store basic blocks from which BIM models are built. These blocks represent all types of objects that make up a building: beams, columns, slabs, floors, walls and ceilings, doors and windows, pipes, connections and ducts, furniture and equipment, etc. [26]. In this sense, the library of BIM components for the CMSP, seeks through the compilation of elements and construction systems to offer macro modules of the stations, with several components, in order to facilitate the design process of the stations.

The study of BIM libraries is essential for the structuring of a component library for the subway works that presents quality and functionality. The next item of this paper will present an analysis of two generic and one public BIM library, as listed in Fig. 10.

Fig. 10. Component libraries analyzed

BIM Object. BIM Object is a library started with the Autodesk Seek, created by the Autodesk with the purpose of promoting objects for its BIM software (Revit). However, due to the need for object compatibility for other BIM software, the platform was sold to BIM Object, a Swedish company that disseminated the components and increased the collection for various compatibilities. The library is built in a shared way, which means that any user can create an object and publish it on the page.

BIM Object is the most widely used library available on the internet. According to data from the platform itself, there are more than 1.4 million registered architects, engineers and designers, allowing them to produce, share and use components of the platform.

In addition, BIMObject [27] presents more than 400 thousand parametric BIM objects available in more than 1400 different brands for download - although not all of them are parametric. There are companies that manufacture building components that offer digital versions of their products, like Saint-Globain, Schneider, Roca, Kingspan and others. More than 43 million object downloads have been made and more than 360 million views have already taken place.

Among the objects available, there are: construction, electronics, engineering and infrastructure objects, electrical installations, sanitary installations, windows, building materials, furniture and others. The Fig. 11 is a screenshot of the main screen of the BIM Object library.

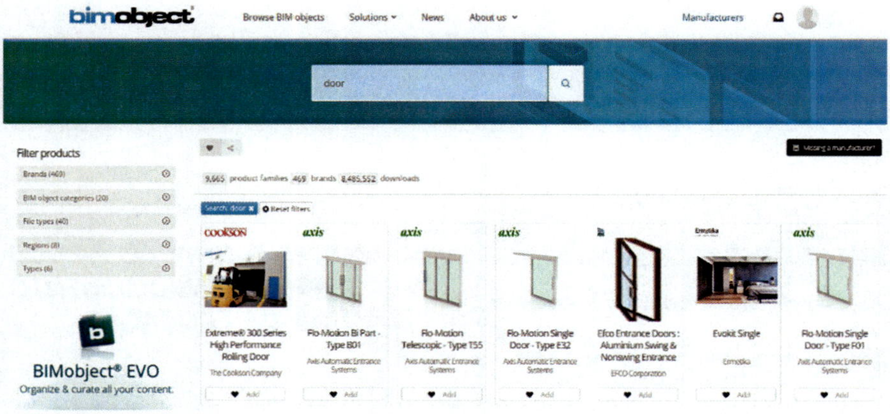

Fig. 11. Product search page, with details for the available filters [28]

Basically, like most free BIM library platforms, its operation is based on the sharing of objects between users. To be able to download the files from the platform the user must register. Although it is widely used, BIM Object does not have its own certification. That is, there are no pre-established criteria to which users' productions have to comply. The lack of certification must be considered as a deficiency, since without control the shared BIM objects may be of poor quality, poorly sized, and may not respect the specific technical standards of each country. On the other hand, as any user can share his production, BIM's knowledge becomes more widespread, democratized and accessible.

NBS National Library. The original UK library belonged to the Royal Institute of British Architects (RIBA) and features an extensive collection of generic and manufacturer BIM objects, ranging from fabric construction systems to mechanical and electrical objects - despite a smaller collection than the BIM Object library. According to the website NBS [29], all objects meet the requirements of the internationally recognized NBS BIM Object Standard.

The certification of the quality is the main advantage associated with BIM objects on this platform, which guarantees safety and workability for users. The NBS standard requires all objects to take a consistent approach to classification, apply a standard naming convention to facilitate use, and standardize approaches to the level of information and presentation of objects. In the meantime, the NBS BIM Object Standard outlines the requirements that productions must meet in order to be certified and operate in the Common Data Environment (CDE). There are general, information, geometry, functional and metadata requirements. According to the website, NBS [29] values a consistent approach to classification that follows the new international standard ISO 19650, complemented by ISO 12006-2.

Among these certified objects, there are doors, paving systems, electronic accessories, insulation, coverings, roofs, panels, sanitary accessories, walls, stairs, windows, furniture, installations and others, as shown in Fig. 12:

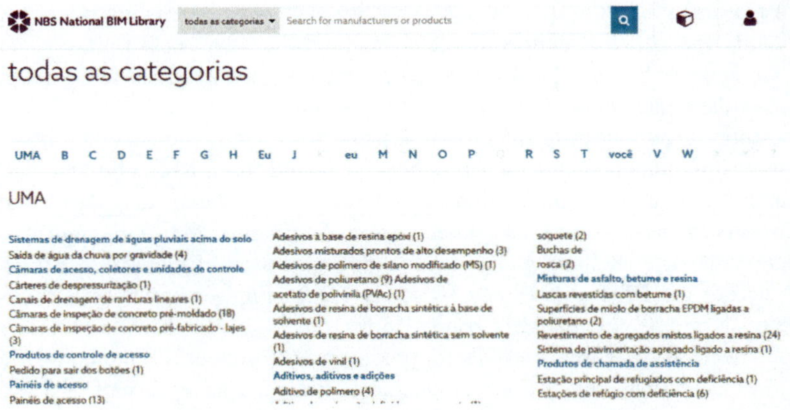

Fig. 12. Product search page, with details of the categories [30]

The platform also features some objects produced by companies. Among the main producers of BIM objects that are registered on the platform, we can mention: Mitsubishi Electric UK, Trovex Innovations, Movins Designs and Premier Loft Ladders. Most of them are UK manufacturers. NBS also seeks to enable the dissemination of BIM and, therefore, the objects are made available in formats compatible with the following software: REVIT, ArchiCAD, Vectorworks, AECOsim and IFC.

FDE Library. The Foundation for the Development of Education (FDE) created in 1987 is an institution responsible for implementing the policies defined by the Secretary of Education of the State of São Paulo. The production model of school buildings managed by FDE follows a systematization characterized by the existence of standardized modules, components and services. The standardization of the constructions of this institution makes the initiative conducive to the application of Building Information Modeling - just as it happens with subway stations. In addition, by serving the State Secretariat, FDE's hiring must also be through bids, highlighting one more common point between the buildings of the FDE and those of the São Paulo subway.

The similarities between the process of contracting and standardizing the construction of the subway and the FDE justify the study of its BIM components library.

In 2013, FDE started a BIM implementation, with the objective of making the processes of design, construction, renovation, restoration and expansion of schools more efficient, productive and transparent, meeting a very important public demand - in addition to allowing construction managers to have greater control and accuracy of project phases. The implementation took place with the modeling of two schools that served as pilots.

An important fact that differentiates the FDE library from the others studied so far, is that the institution has implemented BIM so that all its components follow the premises of "adherence to standardization and methods already established in the FDE, as well as compatibility in quantitative extraction with the price listing rules" [32]. In other words, it is a well-structured implementation project that seeks to take advantage of all the benefits brought by the BIM paradigm - a good example to be taken into account in the construction of the library for the subway.

The foundation's website offers families of components, furniture and own catalogs, as well as templates for the disciplines of architecture, hydraulic and electrical installations - including configurations, standardization of sheets, annotative families, tags, models of views, tables and other resources that meet the FDE standard - and technical manuals containing guidelines for the preparation of projects in BIM [32].

To access the BIM objects developed by FDE, simply register as a student or professional on the foundation's page. Unlike other generic libraries that contain various objects made by several different producers and without a common goal for all components, the FDE library features categories designed for each part of a school's construction, renovation or expansion project.

Figure 13 presents a piece of furniture modeled by FDE, in 3D visualization - something that did not exist before with so many details and features in the traditional dynamics of design.

Fig. 13. 3D visualization of some living room furniture modeled by FDE [33]

The FDE library, unlike the other libraries studied so far, is not a shared platform or is sold as a service. The functioning of the library is promoted by FDE, responsible - together with other actors - for the development, updating and availability of the archives to the public. It is, therefore, a compilation that has a specific objective - to provide contributions to students and professionals for school projects. Figure 14 shows the most recent version of the technical catalogs, manuals and families made available by the FDE in July 2019.

Fig. 14. Latest FDE technical catalogs updated [33]

The BIM methodology also allows the study of project costs in a fast, safe and connected way, increasing the potential use of the FDE's quantitative and budgetary research area, which deals with many projects at the same time (Table 3).

Table 3. Quantitative table for some blocks of the FDE family [33]

A	B	C	D	E
Nota-chave	Descrição	Código de listagem	Área	Count
QE-02	Poste para rede de voleibol - fundação direta (descoberta)	16.04.001		3
QE-03	Trave de futebol de salão - fundação direta (descoberta)	16.04.002		6
QE-12	Quadra de esportes / piso de concreto armado / fundação direta 600m² (descoberta)	16.04.007	600.00 m²	2
QE-23	Espaço multiesportivo / piso de concreto armado / fundação direta 160m² (descoberto)	16.04.043	160.00 m²	1
QE-26	Quadra de esportes / piso de concreto / laje alveolar (com pintura)	13.02.010	600.00 m²	1
QE-27	Quadra de esportes / piso de concreto / pré-laje treliçada (com pintura)	13.02.011	600.00 m²	1
QE-28	Quadra de esportes / piso com proteção acústica sobre laje (com pintura)	13.02.012	600.00 m²	4

An issue raised by Esteves [34] is the interoperability between software. This is a huge challenge, but one that must be worked on so that the components are accessible to a larger number of people and, more than that, so that the components are used in each specific program, according to the needs of each one. each process. Until the last update in July 2019, the FDE component library featured BIM objects compatible only with the REVIT parametric software, a fact that was discussed by FDE. In September

2016, at a meeting of the BIM working group, problems of interoperability between software were addressed, particularly related to the ARCHICAD software. At this meeting, the general director of GRAPHISOFT stated that the company proposes to transcribe the existing content of the FDE to the ARCHICAD format – extension .pln -, but the amounts of information to be transposed in this first experience were not defined at the meeting.

This effort is beneficial, if carried out with caution and quality - since for use in an additional format to be efficient, it is necessary that the components are correctly transcribed, and that the data be like those existing in the components that already exist. The benefits of this quality translation are the increase in the range of objects - as other designers can start using the files, which are now compatible with other software - and the guarantee that the FDE is not practicing market reserve. As of January 2020, the new transcribed files were not yet available for users to use.

According to Esteves [34], the FDE aims to model the 5300 schools in its custody using the BIM paradigm. To do this, it is thinking to work with drones and scanners, which allow you to view buildings quickly and easily. In the InovaDay lecture, the FDE representative also points out that the implementation of this proposal also depends on the future of public policies in the State of São Paulo.

5 BIM Library of Components for CMSP

The library of BIM components proposed for the São Paulo Subway is a collection of macro modules of subway buildings. In addition to providing guidance on the issues of standardization and modulation of constructions, the development of the library will discuss organizational issues, such as parameterization of modules, finishing materials, graphic representation, and also criteria for standardizing the model, such as formats and classifications to be used.

In order to streamline the production of basic CMSP projects and ensure accurate budgets, the basis for the company's bids, the library of BIM components for subway stations will present different construction methods, to suit different types of terrain and depth of stations. However, they should reserve greater autonomy for the development of the part on the ground of the buildings, since each location has particularities that must be studied to meet the public interest and the dynamics of the city.

Each macro module in the library will consist of standard objects approved by the CMSP and connected with reference tables for budget. In this way, the study will provide elements that pave the way for making the construction industry more apt to use modern and precise tools to assist the development and control of subway works.

BIM has been a reality at the CMSP since 2011 with the creation of the BIM working group - GT BIM. Besides bidding the basic design of the Ponte Grande station (2014) and the basic and executive project of Line 15-Silver [36], the CMSP has developed a series of routines and standards, which contribute to the organization and workflow in BIM.

According to the competition notice for Line 15-Silver [35], the objective of BIM for the São Paulo Subway is to improve collaboration in the development of projects, enable greater clarity in resolutions, improve the extraction and traceability of information during the development of the project, work and post-work.

5.1 Library Structure

After the surveys and studies carried out, the division of the building's macro modules is the first question to be structured. These modules can be divided according to the type of section that the station will be implemented, as pointed out in the studies by Gabarra [10] and described in the literature review of this paper, or by sectors of the building that have greater individuality.

Considering the deep underground parts of the stations, since the areas located in the shallow contention require specific adjustments according to the local urban parcel and the special conditions of each station, the library would be divided into four main groups, described in the following items.

Group A - Station with Tunneled Platforms. The implantation of these stations must be in mountainous relief stretches, and/or with greater construction density and high real estate value. The macro module of this type of station is composed of the underground part of the station, excluding the passenger lounge, which is usually located on the first buried level, that is, it is part of the shallow containment structure. Therefore, the macro module would consist of the drying well (s), in which the intermediate mezzanines are located, and the NATM/TBM tunnel, in which the distribution mezzanine and platforms are positioned. The lower levels of the station, where the cable hold is located and some technical parts are also part of the macro module. This group of stations has some variations of macro modules, which will be presented through diagrams in Fig. 15. It is important to consider these variations, as the option of each depends on the location of the land in relation to the rail, and also on the geotechnical parameters of the area. There may also be different tunnel options for the platforms, considering the option of tunnels executed in the TBM methodology.

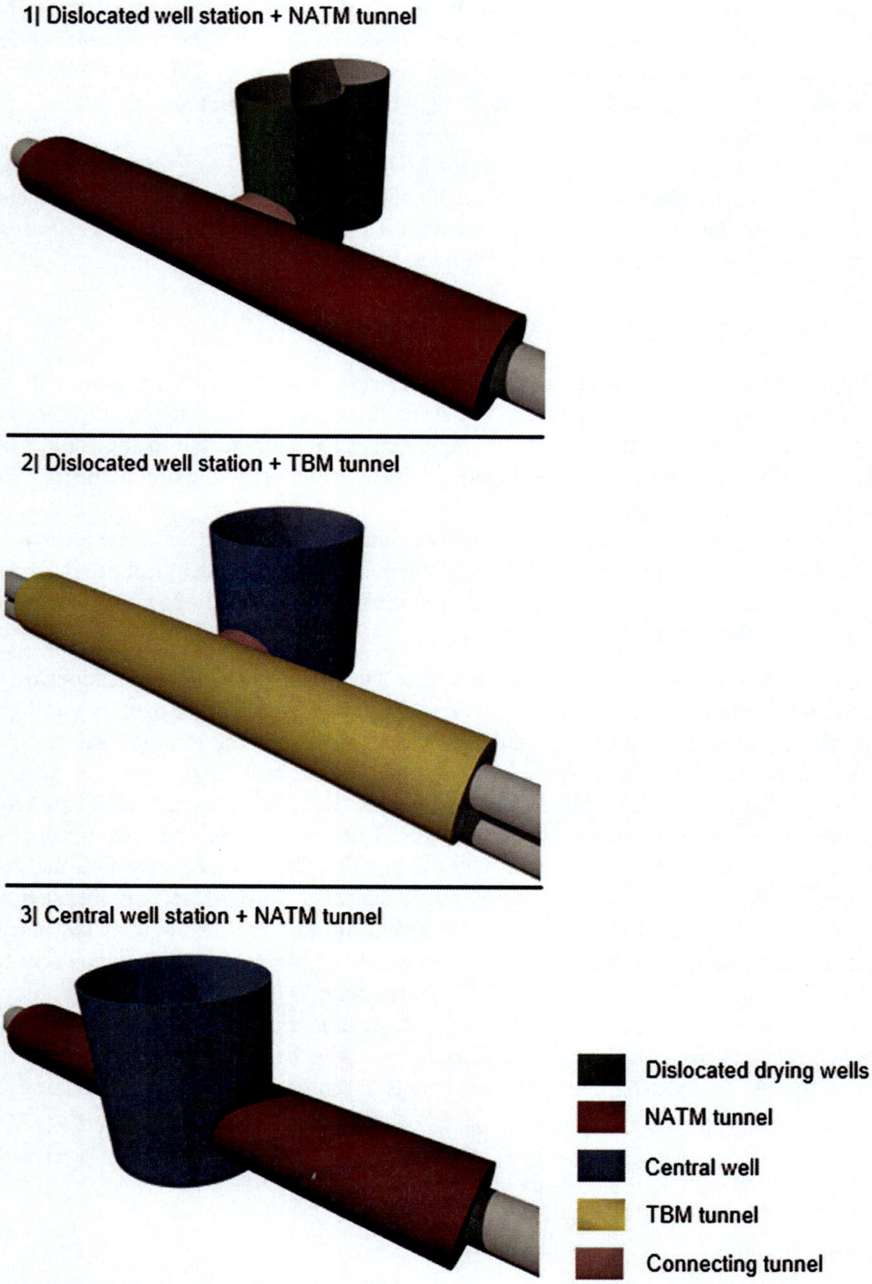

Fig. 15. Group A macro modules [11]

Group B - Station with VCA Platforms. This type of station is ideal for stretches of flatter relief and areas with less construction density. This group of stations contains macro modules with a rectangular containment system and five drying wells, which can be used depending on the condition of the soil at the site to be implanted. Due to the greater volume of excavation, these modules include practically all parts of the station, such as platforms, the distribution mezzanine, the passenger lounge, as well as the technical and operational rooms, with only access and urbanization of the part over the ditch to be projected later. The dimensioning of the ditch is relevant in this type of station, and depends on the passenger estimate of each station, therefore the parameterization of the module must be related to the estimate mentioned. The types of modules in this station group are illustrated in Fig. 16.

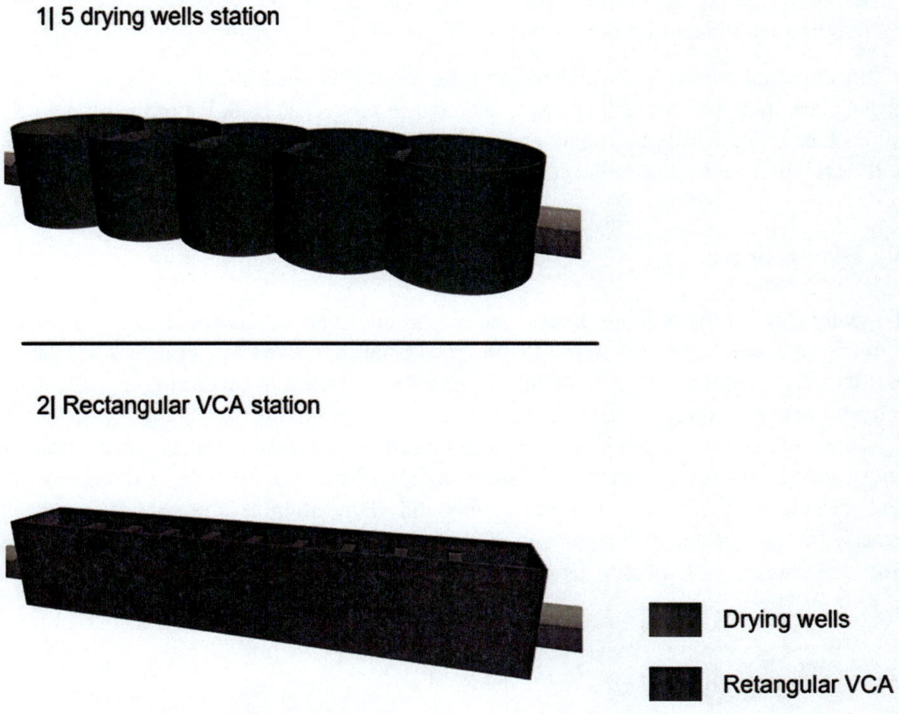

Fig. 16. Group B macro module types [11]

Group C - Buildings on the Ground. Another macro module of the proposed library is the building of technical rooms, which is usually located above ground in the station's grounds.

It is worth mentioning that this type of division of the macro modules – groups A, B and C – considers more complete solutions for subway buildings, incorporating, for example, the ventilation system. As these buildings have a very complex and large ventilation system, standardizing this solution and including it in the macro module would reduce the time for discussions and formalize the proposal for greater quality and efficiency.

Group D - Individualized Sectors. Another way to organize the library is to separate the buildings by sectors, being able to fit the different modules according to the needs of each project. Some examples of possible station sectors that can configure a macro module if this group are listed below:

- Operating room block;
- Sanitary block;
- Accesses. There are some types of accesses that were used in several stations of a line, as is the case of the domes of Line 5-Lílas. This type of access may become an item in the proposed component library.
- Blocks of technical rooms, separated by theme if it is not possible to deploy the building as a whole on the ground.

In this configuration, the macro modules would be used at the beginning of the project, advancing the modeling stage, and also serving as a check list of the guidelines for each area of the station. The parameterization of the macro modules must be studied and developed according to the instructions and technical guidelines of the CMSP.

6 Discussion

To really support the bidding process, an important aspect of the component library is to relate each element and material to reference tables approved by public entities, such as SINAPI. Questions about updating the library are raised at this point, as entry and service costs are adjusted periodically.

Modeling influences labor productivity, and allows innovations in the production of construction elements, so that their manufacture becomes more mechanized, controlled and scalable. Currently, through prototyping and 3D printing, it is possible to produce exactly the component parameterized in the software. This already occurs with items from the studied FDE library, as shown in Fig. 17.

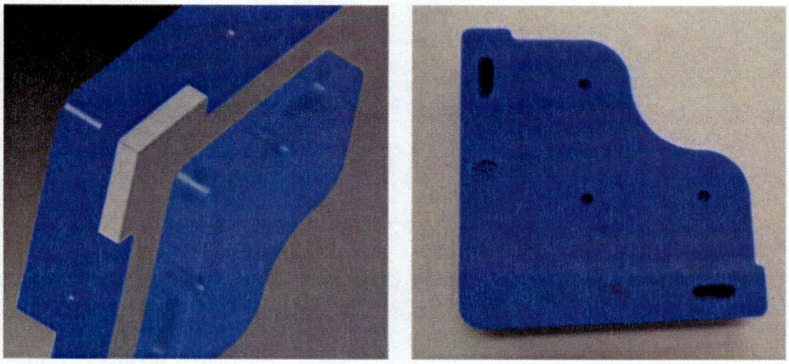

Fig. 17. 3D printing of a part of an FDE object, from its prototype [31]

This fact raises some questions about the component library to be developed, such as what the level of detail of the objects in the library should be, in the sense that different details are necessary for each purpose. In the case of prototyping and 3D printing, as shown above, the objects require information for manufacture, translating into ultra-detailed models, being able to load the models excessively.

Another discussion raised is about the participation of objects developed by manufacturers in the CMSP library. In the case of specific elements, produced especially for subway works, if the manufacturer models the object within the standardization adopted by the library and the CMSP approves the item, can it be absorbed by the library? In this case, the proposed library would house elements developed by manufacturers, as long as these objects do not have the name of the producer, so that they do not offer legal complications in relation to tenders.

Regarding the classification of the model, a document that must be analyzed is the "IFC Infra Overall Architecture Project", developed by Building Smart, which deals with a common basis for data on infrastructure projects, in addition to the guidelines developed by the CMSP for its library of objects.

7 Results

The main result of this research is the library of BIM components for subway station bidding projects, with due consideration learned in the development process and the evaluation of the artifact.

It is also expected to identify solutions and technological innovations, related to the BIM platform and the standardization of components, which contribute to the improvement and reduction of contracting and development deadlines for subway station projects.

Currently the research is focused on the development of a first macro module of group A - station with central well + tunnel in NATM.

References

1. Groves, D.: Efficiency eludes the construction industry - Least-improved. The Economist 1 (1), 1–2, 17 (2017). https://www.economist.com/business/2017/08/17/efficiency-eludes-the-construction-industry. Accessed 22 Aug 2018
2. CBIC/FGV: A Produtividade da Construção Civil Brasileira, 40 p. (2014). (CE No2418/14 - Projeto 054/14). http://www.cbicdados.com.br/media/anexos/070.pdf. Accessed 19 July 2018
3. Santos, H.P.: Diagnóstico e análise das Causas de Aditivos Contratuais de Prazo e Valor em Obras de Edificações em Uma Instituição Pública. Master's thesis, Escola de Engenharia, Universidade Federal de Minas Gerais (2015)
4. Portugal, M.A.: Corrupção em obras públicas: Uma análise dos procedimentos licitatórios no papel de combate à corrupção. 2017. 236 f. Dissertação (Mestrado) - Curso de Administração, Centro Universitário Fei, São Paulo (2017)

5. World Economic Forum: London: An Action Plan to Accelerate BIM Adoption. Geneva: World Economic Forum, 20 p. (2018). http://www3.weforum.org/docs/WEF_Accelerating_BIM_Adoption_Action_Plan.pdf. Accessed 18 Aug 2018
6. EUBIM Task Group: Manual relativo à aplicação da Modelação da Informação da Construção (BIM) no Setor Público Europeu: Ação estratégica para o desempenho do setor da construção. União Européia, 84 p. (2017). http://www.eubim.eu/handboo. Accessed 10 July 2018
7. Oakley, T.C.: Arquitetura das estações de conexão na rede de metrô de São Paulo. 2017, 209 p. Dissertação (Mestrado) – Faculdade de Arquitetura e Urbanismo, Universidade de São Paulo, São Paulo (2014)
8. Mainardi, I.S., Mozarte, D.: BIM aplicado ao Metrô: porque o metrô investe neste novo processo. In: SEMANA DE TECNOLOGIA METROFERROVIÁRIA, vol. 19, São Paulo, Anais, pp. 1–9. Aeamesp, São Paulo (2013)
9. Johnston, J.: Platforms: bridging the gap between construction and manufacturing. In: NBS (Comp.). National BIM Report 2018, pp. 4–9. Riba Enterprises (2018) http://www.bimacademy.global/wp-content/uploads/2018/05/NBS0850-BIM-Report-2018-LR-.pdf. Accessed 3 Sept 2018
10. Gabarra, M.M.: Industrialização e padronização para expansão da rede de metrô de São Paulo, 138 f. Dissertação (Mestrado) - Curso de Arquitetura, FAUUSP, São Paulo (2016)
11. Self elaboration (2019)
12. Altounian, C.S.: Obras públicas: licitação, contratação, fiscalização e utilização. Belo Horizonte: Fórum (2016)
13. Paraná: Secretaria de Estado de Infraestrutura e Logística (SEIL). http://www.infraestrutura.pr.gov.br/modules/conteudo/conteudo.php?conteudo=123. Accessed 10 Oct 2018
14. Brasil: Controladoria-Geral da União. Licitações e contratações. http://www.portaltransparencia.gov.br/entenda-a-gestao-publica/licitacoes-e-contratacoes. Accessed 25 Sept 2019
15. Brasil: Presidência da República. Casa Civil. Lei nº 8.666, de 21 de junho de 1993. Regulamenta o art. 37, inciso XXI, da Constituição Federal, institui normas para licitações e contratos da Administração Pública e dá outras providências. Diário Oficial [da] República Federativa do Brasil, Poder Legislativo, Brasília, DF, Seção 1, p. 8269, 22 June 1993
16. Brasil: Presidência da República. Casa Civil. Lei nº 13.303, de 30 de junho de 2016. Dispõe sobre o estatuto jurídico da empresa pública, da sociedade de economia mista e de suas subsidiárias, no âmbito da União, dos Estados, do Distrito Federal e dos Municípios. Diário Oficial [da] República Federativa do Brasil, Poder Legislativo, Brasília, DF, Seção 1, p. 1, 1 July 2016
17. COMPANHIA DO METROPOLITANO DE SÃO PAULO (CMSP). Regulamento de Licitações, Contratos e demais ajustes. São Paulo: 2019. 88 p.
18. Companhia do Metropolitano de São Paulo: Relatório de empreendimentos. São Paulo, out/2019, 19 p. (2019)
19. Companhia do Metropolitano de São Paulo (CMSP). Subterrâneo – Tipos de construção. http://www.metro.sp.gov.br/tecnologia/construcao/subterraneo.aspx. Accessed 19 Dec 2019
20. Companhia do Metropolitano de São Paulo (CMSP): Projeto Funcional da Conexão da Linha 2-Verde com a Linha 3-Vermelha – Trecho Vila Prudente-Dutra. São Paulo, 150 p. (2011)
21. dos Santos, MA.M.: Companhia do Metropolitano de São Paulo - METRÔ/ S. P. A DISTRIBUIÇÃO DA DEMANDA DE VIAGENS NO METRÔ E SUA APLICAÇÃO NO DIMENSIONAMENTO DA OFERTA DE TRENS. 11º Congresso de transporte e trânsito. Belo Horizonte, 11 p. (1997)
22. Tupinambá, N.S.: Diretrizes técnicas para o projeto de uma linha de metrô, no Metrô-SP - Análise de um caso real 2007. 177p. Dissertação (Mestrado) - Faculdade de Arquitetura e Urbanismo, Universidade de São Paulo, São Paulo (2007)

23. Companhia do Metropolitano de São Paulo (CMSP): MAN-10-201 – Instrumento normativo: Elaboração e fornecimento da documentação técnica de engenharia civil, arquitetura, geotecnia, topografia, desapropriações e via permanente. São Paulo (2011)
24. Network Rail. Station Design Strategy for Network Rail, London, 43 p. (2015)
25. Mainardi, I., et al.: Processo de desenvolvimento de objetos BIM de elementos padronizados de arquitetura do Metrô-SP. In: Semana de Tecnologia Metroferroviária, vol. 24, São Paulo. Anais. AEAMESP, São Paulo, pp. 1–32 (2018)
26. AGÊNCIA BRASILEIRA DE DESENVOLVIMENTO INDUSTRIAL – MINISTÉRIO DA INDÍSTRIA, COMÉRCIO EXTERIOR E SERVIÇOS (ABDI-MDIC). BIM na Quantificação, orçamentação, planejamento e gestão de serviços da construção. Brasília-DF: ABDI-MDIC, 67 p. 3 v. (Coletânea Guias BIM) (2017)
27. BIM Object. https://www.bimobject.com/pt-br. Accessed 12 Aug 2019
28. BIM Object. https://www.bimobject.com/pt-br/product. Accessed 14 Aug 2019
29. NBS National BIM Library. https://www.nationalbimlibrary.com/en/nbs-bim-object-standard/. Accessed 15 Aug 2019
30. NBS National BIM library. https://www.nationalbimlibrary.com/en/find-bim-objects/. Accessed 15 Aug 2019
31. FUNDAÇÃO PARA O DESENVOLVIMENTO DA EDUCAÇÃO (FDE). http://arquivo.fde.sp.gov.br/fde.portal/PermanentFile/File/Audi%C3%AAncia_P%C3%BAblica_BIM_FDE_14_06_2016.pdf. Accessed 5 Jan 2020
32. FUNDAÇÃO PARA O DESENVOLVIMENTO DA EDUCAÇÃO (FDE) (2016). http://arquivo.fde.sp.gov.br/fde.portal/PermanentFile/File/Projeto_Basico_BIM_FDE%2025-05-16.pdf. Accessed 5 Jan 2020
33. FUNDAÇÃO PARA O DESENVOLVIMENTO DA EDUCAÇÃO (FDE). https://produtostecnicos.fde.sp.gov.br/Login.aspx
34. Esteves, R.G.: Implantação BIM na FDE. In: INOVADAY 2016, São Paulo (2016). http://inovaday.com.br/br/bim-building-information-modeling/. Accessed 5 Jan 2020
35. Secretaria dos Transportes Metropolitanos, CONCORRÊNCIA INTERNACIONAL Nº 01/2017, PROCESSO STM Nº 816/2017, Concessão Linha 15 – Prata (2017)
36. Mainardi, I.: O BIM chegou à primeira estação: Ponte Grande. In: Semana de Tecnologia Metroferroviária, vol. 21, pp. 1–22 Anais, São Paulo. Aeamesp, São Paulo (2015)

Strategy for Defining an Interoperability Layer for Linear Infrastructure

Robin Drogemuller[1(✉)], Sara Omrani[1], Fereshteh Banakar[1], and Russell Kenley[2]

[1] Queensland University of Technology, Brisbane, QLD 4000, Australia
robin.drogemuller@qut.edu.au
[2] Swinburne University, Melbourne, VIC 3122, Australia

Abstract. The structure of the IFC model originally focused on buildings – vertical construction. The containment hierarchy of project, site(s) – building(s) – building storeys, reflects this. There are a number of activities within buildingSMART working on infrastructure extensions – horizontal construction - to the IFC model, including roads, bridges, tunnels, ports and waterways. Each of these activities is defining the necessary extensions to the IFC model to support their particular domain of application.

The IFC model has a four layer architecture, with the resource layer defining base concepts that are necessary for the higher levels, the core layer defines central concepts around product, process and control, the interoperability layer contains concepts that shared across a number of domains and is the main mechanism to support interoperability and finally the domain layer containing the concepts that are needed in a particular domain of application. Each of the existing buildingSMART activities in the infrastructure area are defining their own domain models. However, there is a need to define concepts in the interoperability layer to support information exchange between the various domains.

This paper describes the initial considerations around the development of this interoperability layer for infrastructure. The paper is informed by the current work within buildingSMART in defining the various domain models, as well as current project work by the authors on modelling roads and stormwater/flooding.

Keywords: Infrastructure · Interoperability · IFC road · IFC rail · IFC bridge

1 Introduction

The first version of the IFC model was released in 1996 (buildingSMART 2019), concentrating on the representation of buildings. The IFC model defined a containment hierarchy that supported the concept of storeys and a strict cartesian, right angled coordinate system. Once the IFC project became more widely known, engineers working on infrastructure projects became interested in the IFC developments. One of the earliest significant infrastructure projects, starting in 2002, was a French initiative, IFC-Bridge (Lebegue 2005). There was also collaboration with Japanese researchers (Yabuki and Li 2006). One example of an IFC roads project was undertaken in Korea (Lee and Kim 2011). The long history of infrastructure projects using the IFCs as a

E. Toledo Santos and S. Scheer (Eds.): ICCCBE 2020, LNCE 98, pp. 362–371, 2021.
https://doi.org/10.1007/978-3-030-51295-8_25

basis for their data modelling and the proposals for extensions to the IFC model to cover various types of infrastructure leads one to ask why it has taken so long for these to be explicitly added to the official IFC model. There are a number of technical reasons for this delay – the lack of a containment structure in infrastructure projects, issues with location referencing and the need for geospatial coordinates.

The buildingSMART developments that support infrastructure need to be consistent:

- with each other;
- with the entire IFC model;
- future extensions to other types of infrastructure; and
- the processes that rely on this information throughout the lifecycle of these assets.

The final point provides the major motivation for this paper, since the authors are currently working on a project to extend the COBie concept to road infrastructure projects. This requires the identification of objects and attributes that are needed by asset managers, together with who will provide this information and at what stage of the design/construction process this information should be entered.

2 Background

The IFC 1.0 model to support interoperability between building design software was first released in 1996 (buildingSMART 2019). Its development grew out of work on the STEP (ISO10303) standard and the Building Construction Core Model (ISO10303:106) (Froese 1996). While the IFC model used the EXPRESS data definition language and the ISO10303:21 file exchange format, there was a major difference in that the IFC approach did not support multiple inheritance within the object model as was supported in the ISO10303 parts. This was mainly due to issues that the software implementors had with implementation against the STEP models. There was parallel development of a STEP Application Protocol (AP225: Building elements using explicit shape representation) (Haas 1997). However, the wider support of the IFC initiative meant that there was no further significant buildings related work within the STEP developments.

While the various IFC infrastructure projects are undertaken separately, the buildingSMART development processes provide an environment that supports coordination between the groups (Plume and Erismann 2019). This is essential because of the high degree of interrelatedness when infrastructure is built (Fig. 1).

Fig. 1. Proposed road-rail grade separation at Toorak, Melbourne, Australia. https://www.abc.net.au/news/2019-02-13/government-announces-sky-rail-for-toorak-road-level-crossing/10806336

Coordination becomes even more important when road and rail meet at level crossings (Fig. 2), although this interface is currently out of scope for IFC Road (Plume and Erismann 2019).

Fig. 2. Level crossing, Werribee, Australia. https://your.levelcrossings.vic.gov.au/werribee?tool=qanda

Other directly relevant buildingSMART initiatives are the IFC Infra Overall Architecture project (André Borrmann et al. 2017) and the IFC Alignment project (Fig. 3). IFC Alignment is an important precursor to the other IFC Infra projects as it provides common locational and relationship constructs that are needed by the other projects. Another buildingSMART report addresses asset management issues (Jackson 2018).

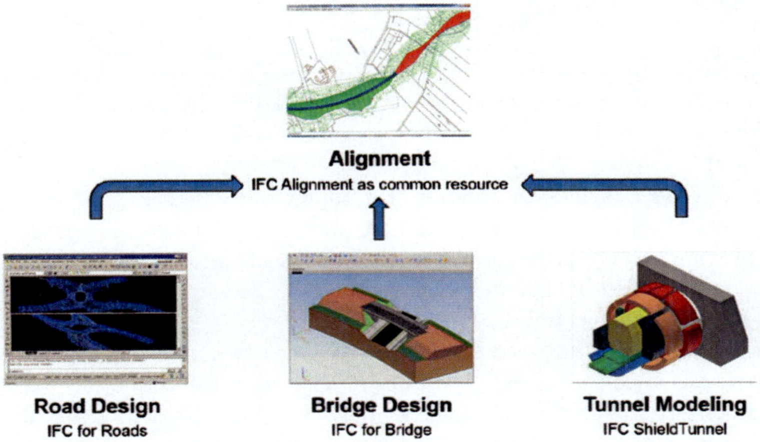

Fig. 3. IFC Infra projects and IFC Alignment (Amann and Borrmann 2015)

Lifecycle issues have been addressed with projects that have examined how non-BIM technologies, such as image recognition, can assist in asset management (Koch et al. 2012). IFC-Bridge also considered these through a reference process map (Borrmann 2019)

From the above review it can be seen that there has been significant work on infrastructure within the IFC community. What are the contributions that are expected from the work proposed in this paper? Firstly, the authors have been and are involved in a number of infrastructure projects over time including assessment of the possibility of supporting the IFC model in string-based infrastructure software, BIM-based asset management (Hampson et al. 2018a; Hampson et al. 2018b; Perumpilly et al. 2019), support for road and rail projects including supporting ancillary services such as power, hydraulics and communications () and the use of asset management covering road, rail and hydraulic assets to support flood modelling. This will lead to:

1. Review of commonalities between the extensions from both a software implementation perspective and user perspectives, i.e. what information needs to be communicated between the users of the extensions;
2. Checking how objects are added through the project/asset lifecycle;
3. Whether there are issues with the proposed IFC extensions when dealing with information that is currently out of scope;

4. Consideration of issues around different representations required to present information to a range of stakeholders;
5. External review of the proposed extensions.

The overall goal is to ensure that a wide range of considerations are checked to provide feedback to the various IFC extension project teams.

3 Review of Commonalities

The IFC model has a layered architecture (Fig. 4). The resource layer contains the definitions for many general concepts that are needed in any project, such as dates, measurements, geometry, etc. Most of these were adapted from the STEP standards (Pratt 2001).

The core layer defines the objects that support the modelling of the physical components of a structure, the processes needed to manage construction processes and the controls that constrain the products and processes. The object definitions at this level are not detailed enough for general use. The detailed, usable definitions are defined higher in the hierarchy.

The interoperability layer is where the objects become usable and is where most of the activity proposed in this paper will occur. This layer defines objects that will be shared amongst the various stakeholders. They may be applicable across the entire project, such as start dates and end dates of particular activities, or they may only be shared between two different pieces of software, such as a structural analysis of an entire structure and detailed analyses of the connections within that structure. This second case is significant as there may be a need to iterate back and forth between the two types of analyses. This is also the error where serious errors can be made. The raft of issues around the Grenfell Tower disaster are examples of failure in decision making between disciplines.

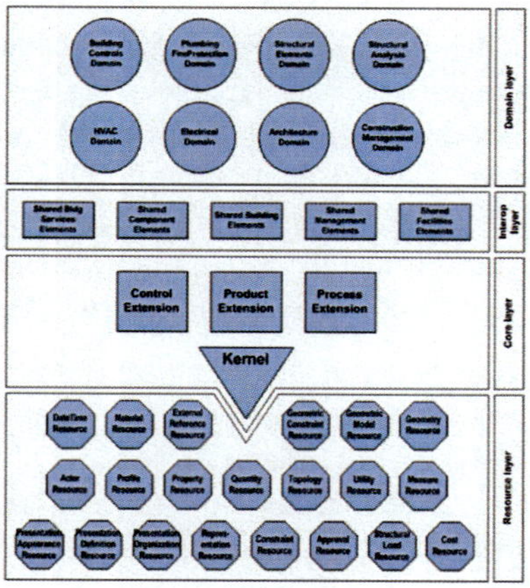

Fig. 4. IFC 4.1 model architecture (buildingSMART 2019)

The domain layer contains the objects that are specific to a particular discipline and contain all of the information required by that discipline at particular stages of a project.

Model View Definitions (MVDs) (https://technical.buildingsmart.org/standards/mvd/) define which objects and attributes will be exchanged between two parties at a defined stage in the project lifecycle.

The IFC 4.0 containment hierarchy consists of project → site(s) → building (s) → building storey(s) → space(s). The concept IfcSpatialzone was added in IFC 4.1 to support infrastructure projects for which IfcBuilding is not suitable. An interesting issue for exploration is the situation where a building, such as a railway station, the railway line and the roads servicing (Fig. 5) it need to be represented. How will the containment hierarchies interact?

Fig. 5. Cross River Rail project - proposed Fairfield station upgrade - https://www.couriermail.com.au/news/queensland/brisbane-southside-train-stations-in-60m-upgrade-for-cross-river-rail/news-story/10adffda2f36d23c3e6fc997330e74e5

4 Project/Asset Lifecycle

It is necessary to understand the flow of information throughout the life cycle of infrastructure to assess whether a representation is sufficient or not. Defining a process map is a common activity within IFC projects (Fig. 6).

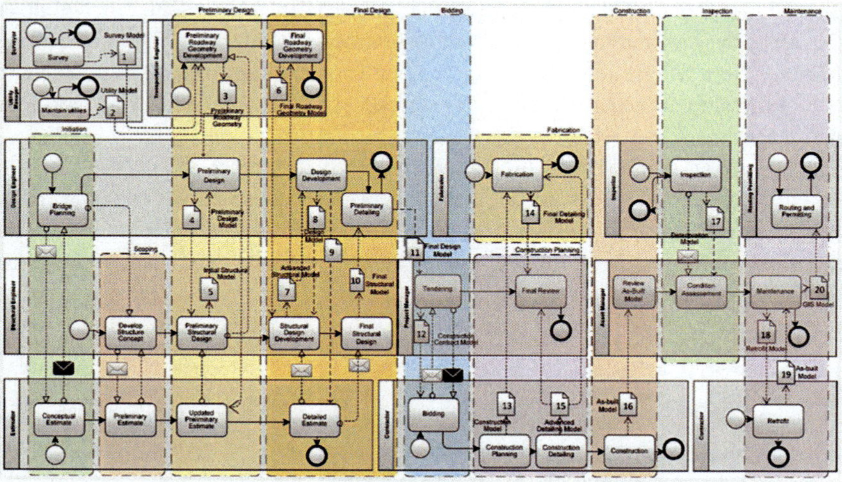

Fig. 6. Process map from IFC-Bridge (Andre Borrmann 2019)

When examining interoperability between IFC Road and IFC Rail, for railway level crossings for example (Fig. 2), the process maps for each discipline need to be checked to ensure that there is appropriate synchronization between the two disciplines.

5 Scope of Infrastructure Extensions

The process for defining extensions to the IFC model is rigorous and requires extensive validation through building examples of the types of structure that are within scope. Figure 7 is an example of just one of the test models developed for the IFC Bridge extension. No one could complain that it was not comprehensive. However, there can be issues where a road or rail line abuts a bridge. In the Logan Enhancement Project (https://loganenhancementproject.com.au) ArcGIS was used to model the entire project, with Revit models of the bridges, overpasses, retaining walls, etc. embedded in ArcGIS. Care was needed in embedding the Revit models in ArcGIS to ensure that the strings representing the roads and road structures aligned appropriately with the matching features on the Revit models. Ideally, this process should be automatic. This would require another user case around the interface between the differently sourced models.

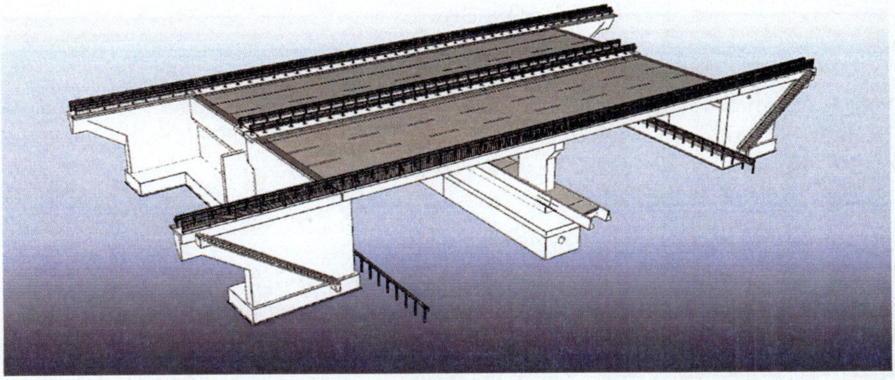

Fig. 7. IFC Bridge example (WTM Engineers GmbH, Hamburg; http://ifcinfra.de/ifc-bridge/bridge-abschluss/)

6 Multiple Representations

Different stakeholders in projects often require explicit representations of the project to suit specific needs (Fig. 8).

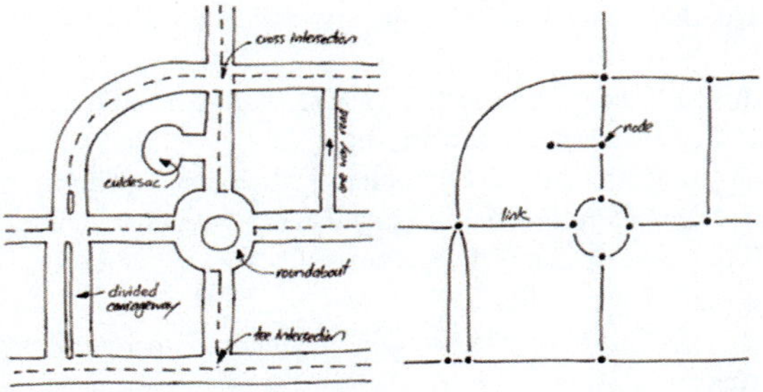

Fig. 8. Sketched local road configuration (L) and network representation (R) (André Borrmann et al. 2017)

Where possible, these multiple representations should be derivable from the same underlying model. This is especially important if the two representations support user interaction and editing of the model.

7 Conclusions

The release of the IFC Infra extensions to the IFC model is having significant repercussions on the scope and structure of the IFC model itself and potentially on the projects for which the IFC model is used. Consideration of interoperability between the IFC Infra extensions and with the appropriate existing system models will be essential for the successful uptake of the next version of the IFC model.

The following issues have been identified:

1. Identifying commonalities;
2. Modelling interaction between disciplines over the asset lifecycle;
3. Handling current and future scope of infrastructure extensions;
4. Supporting multiple representations; and
5. Review of the extensions by third parties.

The authors are engaged in several live projects which will provide an opportunity to consider the above issues and to provide feedback to the Infra extension project teams within the timeframe defined in the programs of work defined for each extension project.

References

Amann, J., Borrmann, A.: Creating a 3D-BIM-compliant road design based on IFC alignment originating from an OKSTRA-accordant 2D road design using the TUM Open Infra Platform and the OKSTRA class library (2015). https://publications.cms.bgu.tum.de/reports/2015_Amann_TR1.pdf

Borrmann, A.: IFC Bridge Project Status Overview (2019). http://ifcinfra.de/wp-content/uploads/2019/04/2019-03-26_IFC-Bridge_Overview_.pdf

Borrmann, A., Amann, J., Chipman, T., Hyvärinen, J., Liebich, T., Muhič, S., Scarponcini, P.: IFC Infra Overall Architecture Project - Documentation and Guidelines (2017)

buildingSMART: IFC Specifications Database (2019). https://technical.buildingsmart.org/standards/ifc/ifc-schema-specifications/

Froese, T.: STEP and the building construction core model. In: Paper presented at the Computing in Civil Engineering, Anaheim, California, United States (1996)

Haas, W.R.: ISO/STEP AP225 for the Exchange of 3D CAD-data and building models. Int. J. Constr. IT **5**(2), 113–122 (1997)

Hampson, K., Wang, J., Drogemuller, R., Wu, P., Omrani, S., Shemery, A.: Digital asset information management (DAIM): case studies (2018a). https://sbenrc.com.au/app/uploads/2019/05/SBE0281.pdf

Hampson, K., Wang, J., Wu, P., Drogemuller, R., Omrani, S., Shemery, A.: digital asset information management (DAIM): a guide and manual (2018b)

Jackson, P.: Infrastructure Asset Managers BIM Requirements (TR 1010) (2018). https://www.buildingsmart.org/wp-content/uploads/2018/01/18-01-09-AM-TR1010.pdf

Koch, C., Paal, S.G., Rashidi, A., Zhu, Z., König, M., Brilakis, I.: Achievements and challenges in vision-based inspection of large concrete structures. Paper presented at the 1st International Conference on Performance-Based and Life-Cycle Structural Engineering (2012)

Lebegue, E.: IFC-BRIDGE V2 data model (2005). http://www.mediaconstruct.fr/sites/all/Documents/Bridge/ifc-bridge-v2-data-model.pdf

Lee, S.-H., Kim, B.-G.: IFC extension for road structures and digital modeling. Procedia Eng. **14** (2011)

Perumpilly, R., Kenley, R., Harfield, T.: CONie is not COBie: information exchange differences between network and building handover to asset management. Paper presented at the IOP Conf. Series: Materials Science and Engineering, Malaysia (2019). https://iopscience.iop.org/article/10.1088/1757-899X/512/1/012041/meta

Plume, J., Erismann, C.: Delivery Plan for Infra Extensions to IFC - draft 3 (2019). https://forums.buildingsmart.org/

https://discourse-cloud-file-uploads.s3.dualstack.us-west-2.amazonaws.com/business6/uploads/buildingsmart1/original/1X/6606662ad78ac5405621d85a4014603ac608bf38.pdf

Pratt, M.J.: Introduction to ISO 10303 - the STEP standard for product data exchange (2001). https://tsapps.nist.gov/publication/get_pdf.cfm?pub_id=821600

Yabuki, N., Li, Z.: Development of new IFC-BRIDGE data model and a concrete bridge design system using multi-agents. In: Corchado, E., Yin, H., Botti, V., Fyfe, C. (eds.) Intelligent Data Engineering and Automated Learning – IDEAL 2006. Springer, Heidelberg (2006)

Study of Building Information Modelling Implementation on Railway Infrastructure

Ali Aryo Bawono[1,2](✉), Christian Maximilian von Schumann[2], and Bernhard Lechner[2]

[1] Rapid Road Transport Department, TUMCREATE, Singapore 138602, Singapore
ali.aryo@tum-create.edu.sg

[2] Chair and Institute of Road, Railway and Airfield Construction, Technical University of Munich, 81245 Munich, Germany

Abstract. In the very recent, Building Information Modelling (BIM) technology is implemented in many aspects in the construction industry such as buildings, bridges, dams, tunnels, as well as in railways. McGraw Hill reported that 12% of the contractors in France, Germany and the UK have been using BIM for at least six years or more. Similarly, in the US and Canada, where more than a third of the BIM-using contractors have experience for six or more years. South Korean contractors show a 65% BIM adoption rate [1]. In Singapore, the Building and Construction Authority has implemented the BIM roadmap since 2010 with the aim that 80% of the industry will implement BIM by 2015 [11]. By now, Singapore is one of the leading authorities with its guidelines and implementation [12, 13].

Despite a notable implementation of BIM in those industries, many companies are facing challenges when it comes to BIM implementation in railway infrastructure. The phenomenon is however, not studied extensively. Most of the existing studies were investigating issues in the context of bridges and tunnelling projects, while studies on BIM implementation in railway projects were found less [2]. In infrastructure and especially in railway projects the implementation is lacking in comparison to the broad implementation of BIM in building projects [1]. Not only are the implementation rates of BIM in railway infrastructure far lower compared to those in buildings, but also little specific scientific publications are available on the implementation of BIM in railway infrastructure [3].

This paper presents the study of BIM implementation in railway infrastructure based on a survey, with a semi-structured interview method, done on BIM-using railway contractors/users. The questionnaire was carefully designed to understand better the perspective of contractors/users in using BIM. Key challenges, including technical, personal, and process aspects are addressed. Point of view of managers and engineers who are responsible for BIM are discussed. Lastly, expectations on improvements for the features and solutions from the users were revealed so BIM can be used in a better way.

Keywords: Building Information Modelling (BIM) · Railway · Industry 4.0

E. Toledo Santos and S. Scheer (Eds.): ICCCBE 2020, LNCE 98, pp. 372–382, 2021.
https://doi.org/10.1007/978-3-030-51295-8_26

1 Introduction

In the very recent, Building Information Modelling (BIM) technology is implemented in many aspects in the construction industry such as buildings, bridges, dams, tunnels, as well as in railways [2, 4, 5]. McGraw Hill reported that 12% of the contractors in France, Germany and the UK have been using BIM for at least six years or more. Similarly, in the US and Canada, where more than a third of the BIM-using contractors have experience for six or more years. South Korean contractors show a 65% BIM adoption rate [1]. In Singapore, the Building and Construction Authority has implemented the BIM roadmap since 2010 with the aim that 80% of the industry will implement BIM by 2015 [11]. By now, Singapore is one of the leading authorities with its guidelines and implementation [12, 13].

Despite a notable implementation of BIM in those industries, many companies are facing challenges when it comes to BIM implementation in railway infrastructure [6, 7]. The implementation of BIM in railway projects is lacking in comparison to the broad implementation of BIM in building projects [1]. Not only are the implementation rates of BIM in railway infrastructure far lower compared to those in buildings, but also little specific scientific publications are available on the implementation of BIM in railway infrastructure [3]. Most of the existing studies were investigating issues in the context of buildings, bridges and tunnelling projects, while studies on BIM implementation in railway projects were found less [7].

Challenges within BIM implementation in railway infrastructure consists of but not limited to: high environmental requirements, complex network topology, very high requirements on the track geometry and different track gauge, a big number of different supply divisions, requirement on the interaction of the network and the rolling stock, and compliance to international as well as national standards [8]. Furthermore, railway infrastructure has very high needs in terms of the complexity of certain alignment elements as well as the requirement to allow deviations from the design. In contrast to road alignment, there are several different transition curves combined with a cant that works slightly different compared to the superelevation of roads. The requirements to the track geometry are high with only little allowed deviations. As an example, the gauge design only allowed to have deviations less than 3.0 mm [9]. Not to mention, the high requirement on the environmental aspect in combination with necessity in large of land use often causes a lot of interfaces with diverse third parties. For this reason, effective communication of the project to the public is essential for railway projects. Yet, a precise schedule for railway track projects is highly important, as delays cause more expenses to the projects [15].

This paper presents the study of BIM implementation in railway infrastructure based on a survey, with a semi-structured interview method, done on BIM-using railway contractors/users. The questionnaire was carefully designed to understand better the perspective of contractors/users in using BIM. Key challenges, including technical, personal, and process aspects, are addressed, which are mainly coming from the point of view of managers and engineers who are responsible for BIM. Lastly, expectations on improvements for the features and solutions from the users were revealed so BIM can be used in a better way.

2 Methodology

2.1 Survey

Surveys were conducted to get reliable information and data. The survey was designed according to the methods and practices developed by Statistics Canada [10]. The method was chosen due to its well-fitting content and its clear structure. The development of the survey consists of four steps: defining the objectives. Selecting a survey frame, determining the sample design, designing the questionnaire, collecting and processing the data, analyzing and disseminating the data, and lastly documenting the survey. The simplified methodology for the survey can be seen in Fig. 1.

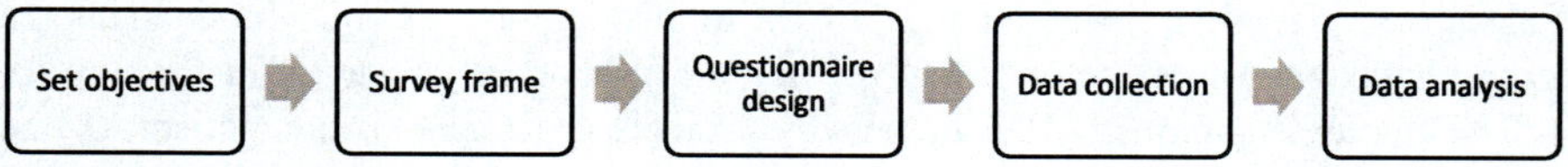

Fig. 1. Methodology of survey used in this study.

The questions posed conform to the survey's statement of objectives. It was designed well, so they are easily understood and can be accurately answered by respondents. The questionnaire was tested and then revised accordingly before the final implementation. The survey frame was set based on users experience by using BIM in railways project related.

Two types of questions were used: closed and open. The closed questions are those where answers are listed. They were used when there is a need to restrict the respondent to a predetermined set of responses. The open questions were also used as this is becoming more important when examining an issue that is not well understood [10].

The questionnaire was structured into six sections: (1) general, (2) BIM usage, (3) BIM standardization, (4) Software, (5) Benefits/Challenges, and (6) opinion. Firstly, the "General" part consists of general questions related, such as company size, railway department size, and experience of using BIM. The four mid-sections are gathering concrete answers for the diverse topics connected to the BIM implementation within railway engineering. Lastly, "Opinion" section is elaborated to gather understanding from respondents on their overview of the application of BIM in railway infrastructure.

The data was collected with an interviewer assisted questionnaire or rather a semi-structured interview. A semi-structured interview method, which based on prepared questions and topics, was chosen. The advantage of this mode of data collection is that the questions can be prepared ahead, and it provides reliable and qualitative data for comparison without limitation for alternative discussions if they evolve [14]. Furthermore, it is flexible with the option to skip questions if necessary. Prior to the conduction of the interviews, the questionnaire was sent to the participants to give them an idea about the scope of the discussion. Lastly, the data collected were analysed.

2.2 Respondents

Case studies on BIM implementation in railway projects found to be rare. Yet, five different companies have participated in the study. All considered as big companies as they have more than 250 employees within the company. However, their railway department was running by less number, varies from less than five to more than 25 engineers. One company does not have a specific railway department. Four out five companies have at least four years' experience of using BIM for railway projects, while only one company has only one-year experience. All stated their clients mandate to use BIM with Employer's Information Requirement (EIR). The characteristics of the companies (survey participants) are summarized in Table 1.

Table 1. Characteristics of the survey participants.

Company	Number of employees in		Years of BIM expertise	Client mandates BIM use with EIR (Yes/No)
	Company	Railway department		
A	>250	>25	1	Y
B	>250	5–25	4	Y
C	>250	N.A.	4	Y
D	>250	<5	4	Y
E	>250	5–25	5	Y

3 Results and Discussions

3.1 Survey Results

The survey results are summarized in Table 2 and Table 3, which give detailed information from five different companies (company A, B, C, D, and E). Table 1 gives the insight of issues when the users use BIM for their railway projects, what are their satisfaction level, what are their challenges, and related usage information.

Table 2 elaborates the opinion of each user from five different companies, on what benefits they have received for the implementation of BIM on their railway projects, how BIM can be improved for future implementation.

Company A explains there are at least two benefits by using BIM in their railway projects: (1) the visualization effect which allows more accessible communication with the public, (2) the better process of continuous design that can produce the better accessible data, lesser loss data, and better-preserved data for the next projects. Company A mentioned that the biggest challenges were the need for motivating and accompanying the employees on the BIM technology implementation itself. It was emphasized that it was not technical issues the limiting factor.

Company B, which has experiences in handling station and open track projects with the BIM method, stated the importance of the willingness from their clients to implement BIM for their projects. Only when the client of a project is completely convinced by BIM and wants to implement it, there will be a success in terms of BIM.

Company C mentioned the benefits of BIM implementation for their railway projects: the better product quality due to the reduction of design errors, and the better visualization of the project development. Similar to Company A, this company has also mentioned the importance of people's mindset and motivation for implementing BIM. They also emphasized the importance of open file formats that can be used as proper design tools.

Company D mentioned that one of the most significant benefits of implementing BIM in their projects is better visualization. Furthermore, better product quality in comparison to a traditional 2D design is resulted due to easier and better design process of the whole project. Besides the technical challenge of rapidly growing file sizes and required hardware, Company D also stated their opinion that another challenge is to change the mindset of the people to use BIM.

Company E stated one significant benefit by implementing BIM in their projects is the capability of BIM to handle the whole project as a 3D model, including all the interfaces. This becomes important as they often need to see if all interfaces are working properly prior to the construction process. Further, BIM provides the calculation of volumes, which are characteristic for railway projects, by much faster, more accurate and easier in comparison to the conventional method. They have the opinion that the bigger and more complex a project is the bigger benefits will be achieved through the usage of BIM. However, they believe that the whole process needs to be standardized. Yet, to have the same understanding of BIM implementation within the whole project team is rather important.

Satisfaction on the time consumption in developing BIM Model is arguable. Three out of five companies state that it requires a lot of time to develop the 3D BIM model. Only one company (Company A) states differently. Two of the reasons for dissatisfaction are firstly they need to make a big effort for combining model into different software. Secondly, as there is no clear of Employer's Information Requirement (EIR) applied to the project causing a lot of re-work.

Table 2. Issues and satisfaction on BIM implementation for railway projects.

Issue	Company A	Company B	Company C	Company D	Company E
Percentage of model type	80% 3D models 20% 2D	50% 3D models 50% 2D	100% 3D models	90% 3D models 10% 2D	95% 3D models 5% 2D
Future plan for 3D model usage	3D information models with extensive embedded data	N.A.	N.A.	N.A.	3D information models with extensive embedded data
Satisfaction on software (++) Very Satisfied; (+) Satisfied; (−) Unsatisfied; (− −) Very Unsatisfied	Buildings: Revit (−) Alignment: ProVi (+) not really working in a sense of BIM Review: Navisworks (++)	Buildings: Revit (−) Alignment: ProVi (− − reg. BIM) Card/1 (− − reg. BIM) Review: Autodesk Navisworks (++)	Modeling: Autodesk Infrastructure Design Suite (++) Alignment: Bentley InRoad (++) For Review: Bentley Navigator (++) Autodesk Navisworks (++)	Autodesk Revit (+) For review: Autodesk Navisworks (N.A.)	Bentley AecoSim (++) Bentley InRail (++) Bentley InRoad (++) For review: Bentley Navigator (++)
Satisfaction of time consumption for modeling/drawing	The time consumption of 3D BIM was very unsatisfying	N.A.	The time consumption for 3D BIM was not satisfying due to high efforts put into combining models of Bentley and Autodesk software	The time consumption for 3D BIM was very satisfying	The time consumption for 3D BIM would have been very satisfying if the EIR of the client would have been clearer. A lot of rework emerged because of that

(continued)

Table 2. (*continued*)

Issue	Company A	Company B	Company C	Company D	Company E
Variation of modeling by disciplines	BIM is only used within the discipline responsible for stations and track works, other disciplines are not following yet	Due to missing software BIM is only used for architecture related structures like stations Only very limited modeling of railway tracks with integrated railway specific discipline design possible	The BIM usage is not varying between the disciplines	No drainage, no track works, no railway specific disciplines and no alignment is modeled in a sense of BIM	All railway specific disciplines but signaling and electrification are modeled (signaling and electrification not even CAD is applied)
Challenges of different software publishers	No mature interface solution is existing	• Huge interface loss • No possibility to integrate software solutions • Many stand-alone solutions	Existing interfaces were not well developed	N.A.	N.A.
Data exchange	PDF and DWG as standard exchange format within the design process	PDF as standard exchange format within the design process	Information is exchanged with discipline models, 3D PDFs, Navisworks, DWFX or DWG files depending on the purpose	Data exchange via . PDF file format	PDF and 3D PDF for exchange between design and project management
Future plan for data exchange	Exchange via disciplinary models	Exchange information via discipline models if it would be technically feasible	N.A.	Exchange via disciplinary models but to many lacking behind disciplines	Exchange via disciplinary models at least for reviewing with all participants
Open file formats	IFC (only for delivery, not for internal processes) Opinion: open file formats are very important for the implementation of BIM in railway infrastructure	CPIXML (− −)	IFC, DWFX, iModel (−) Unsatisfied due to huge data loss Opinion: open file formats are extremely important, to be able to use the proper tool for the different tasks	iModel (N.A.) Open file formats are important for the implementation of BIM in railway infrastructure	XML (+) Opinion: open file formats are important for the implementation of BIM in railway infrastructure
Data exchange platform	Internal servers	Central servers	Internal servers	Internal servers	Internal servers
Future plan for data exchange platform	Common Data Environment accessible to all project stakeholders	Development to a CDE can be observed	Common Data Environment accessible to all project stakeholders	Common Data Environment accessible to all project stakeholders	Common Data Environment accessible by all stakeholders
Assistance for the BIM implementation	• BIM guideline provided by the client • A clear guideline for BIM in railway engineering would be desired • Definitions for level of detail for railway specific objects would be desired	• BIM guideline provided by the client • A clear guideline for BIM in railway engineering would be desired • Definitions for level of detail for railway specific objects would be desired	• Usage of the guideline proposed in the EIR; if nothing is specified the Singapore BIM Guide Version 2 is used	• Clients standards created with the help of the Singapore BIM Guide Version 2 • A clear Guideline for BIM in railway engineering would be desired • Definitions for level of detail for railway specific objects are provided by NBS	• Local's guideline for buildings is existing approach • A clear guideline for BIM in railway engineering would be desired • Definitions for level of detail for railway specific objects are provided by the client

Table 3. Opinion from BIM-railway projects users.

Issue	Company A	Company B	Company C	Company D	Company E
Benefits especially for railway infrastructure	• Visualization for public communication purposes • Continuous design process • Data accessible for everyone, less data loss, better data conservation for the future	• Clash detection of the different disciplines • Integration of the discipline design	• Better visualization of the project development • Reducing design errors • Better product quality	• Better product compared to a 2D design as it is easier to understand the project as a whole • Visualization • Clearance from the track can be derived more easily	• Capability to see the whole project as a 3D model including the interfaces between disciplines • Possibility to see if interfaces are working properly • Faster and easier volume calculation
Most beneficial lifecycle phase for BIM in railway infrastructure	N.A.	Design Operation/Maintenance	N.A.	All lifecycle phases are equal beneficial	N.A.
Characteristics that influence the benefits delivered through BIM	The more objects a project contains, which might be standardized the bigger are the benefits delivered through BIM	No difference of the benefits concerning project size and complexity	No certain characteristics that would influence the benefits of BIM were mentioned, as long as the right tools are used for the right things	The larger and more complex the project is the bigger are the benefits	• The larger and more complex the project is the bigger are the benefits • With established workflows also smaller projects might be more beneficial • The more objects a project contains, which might be standardized the bigger are the benefits delivered through BIM
Toughest challenge when introducing BIM in railway infrastructure	• To motivate employees to work with BIM • Technical solutions are not the limiting factor	• BIM is only working if the client is standing completely supporting • Currently the absence of standardized data exchange is a huge barrier	• The transition from the 2D to the 3D design characteristics • The change of the people's mindset and their motivation	• Changing the attitude and the mindset of the people • Handling the rapidly growing file size within BIM models	To have the same understanding of BIM within all project participants
BIM should be used for railway projects	Agreed	Strongly agreed	Strongly agreed	Strongly agreed	Agreed - depending on the project
BIM is already good to solve typical railway issues	Disagreed	Strongly disagreed	Strongly agreed	N.A.	Agreed

(*continued*)

Table 3. (*continued*)

Issue	Company A	Company B	Company C	Company D	Company E
BIM is expected to be great to solve typical railway issues	N.A.	Disagreed	N.A.	Strongly agreed	N.A.
BIM in the design phase is already delivering benefits	N.A.	Strongly agreed	Strongly agreed	Strongly agreed	Strongly agreed
BIM is only for large companies	Strongly Disagreed	Strongly disagreed	Strongly disagreed	Strongly disagreed	Strongly disagreed
BIM is not sufficiently standardized for railway projects	N.A.	Strongly agreed	Agreed	N.A.	Strongly agreed

3.2 Key Challenges for BIM Implementation on Railway Projects

There are few challenges within BIM implementation in railway infrastructure. They may include: high environmental requirements (noise protection, vibrations and land consumption), complex network topology, very high requirements on the track geometry and different track gauge, a big number of different supply divisions is necessary within the railway system (signaling, power supply for high and low voltage, telecommunications), interaction of the network and the rolling stock and therefore different requirements for different rolling stock, and compliance to international as well as national standards [8].

According to the survey, there are three typical issues, namely: technical, personal, and process (summarized in Table 4). The technical challenges consist of handling of big files, lack of standardized data exchange, and lack of proper design software. The last two challenges mentioned created a big hassle. As many times, the engineers have to convert several files before they can access it. Not to mention that there were data/information losses when the file needs to be converted/exchanged.

In terms of personal issues, the main challenge is to adjust the mindset of any stakeholders to the new technologies offered by BIM. Internally, companies also have to make an effort to motivate their engineers to take part and drive the implementation of BIM for their projects. Similarly, another study reported that cultural issues, which include many aspects such as ones are resistance to change, and the lack of cooperation between stakeholders, are playing big parts of issues when it comes to BIM implementation [16].

It is understood that many companies used to do 2D design for their projects. It requires significant effort for the company as well as their employees to improve the design process into the 3D design model. The transition from the 2D to 3D design needs to be managed well. Currently, the absence of standardized data exchange is a huge barrier for the transformation of the implementation of BIM in railway infrastructure. Finally, the whole process with BIM has to be established. The process will not well work when one stakeholder has a different view on how the project should be designed and processed.

Table 4. Challenges according to the survey participants.

Issues	Challenges
Technical	Handling of growing file sizes
	Lack of standardized data exchange
	Lack of proper design software
Personal	Change attitude and mindset of people
	Motivate people
	To have the same understanding of BIM within the whole project team
Process	Transition from 2D to 3D design characteristics
	Absence of standardized data exchange
	BIM is only working if the client is completely convinced to use BIM

3.3 Opinion and Expectation from Respondents on BIM Implementation in Railway Projects

It can be seen from Fig. 2, that all respondents agree that BIM offers many benefits, particularly in the design phase. Therefore, all respondents agree that BIM should be used for railway projects. However, it depends on the type, scale, and phase of the projects. Company A and B doubt that BIM can be implemented in all typical railway projects. Yet, most of the companies expect that BIM eventually can be a great software solution for railway projects.

It is noticeable that most companies are complaining that currently, there is no sufficient standard applied for railway projects. It was also noted from the interview results, that in the future, companies are expecting for more usage of 3D BIM Models with extensive embedded data. Further, companies using BIM for railway projects are also expecting a better and more accessible Common Data Environment (CDE) features to all project stakeholders. When these issues can be well managed, many companies considered that BIM could be a great tool to be used for railway projects.

According to those company's experience, BIM can be used by companies regardless of their company size. This brings valuable insight that BIM can offer benefits for even a small company.

The responses were recorded in five different levels: strongly agree (++), agree (+), neutral (0), disagree (−), and strongly disagree (− −). As can be seen from Fig. 2, when respondents have a strong response, regardless of positive or negative, it was symbolized with a thicker bar (double size).

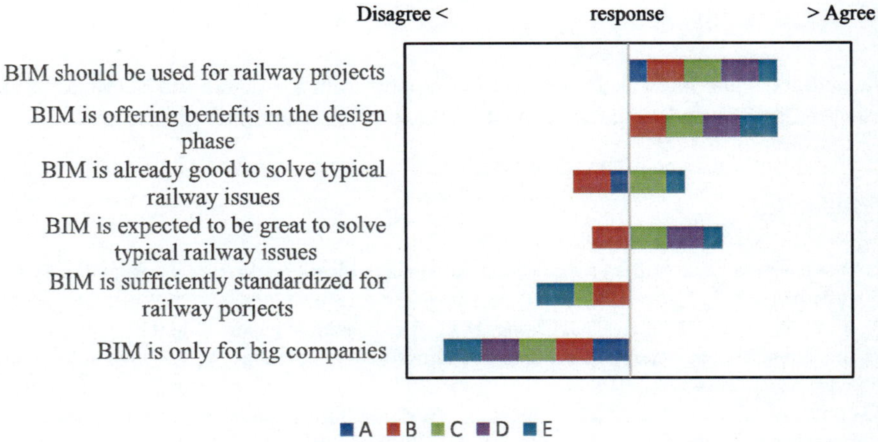

Fig. 2. Opinion from different companies towards BIM implementation in railway projects.

4 Conclusions

This paper presents the study of BIM implementation in railway infrastructure based on a survey with a semi-structured interview method. The surveys were conducted to BIM-using railway contractors/users. The survey was completed with a questionnaire which was carefully designed to understand better the perspective of contractors/users in using BIM. Concluding remarks were drawn, which include key issues and challenges, satisfaction, and opinion on BIM implementation for railway projects.

There are three types of challenges in BIM implementation for railway projects, including technical, personal, and process aspects. The technical challenges consist of handling of big files, lack of standardized data exchange, and lack of proper design software. In a personal level, motivation on engineers for using BIM needs to be well addressed within the company. Meanwhile, the whole (technical and managerial) transition process from the 2D to 3D design cannot be ignored, and they need to be managed well.

Satisfaction on the time consumption in developing BIM Model is arguable. Three out of five companies state that it requires a lot of time to develop the 3D BIM model. Only one company (Company A) states differently. Two of the reasons for dissatisfaction are firstly they need to make a big effort for combining model into different software. Secondly, it is because there is no clear of Employer's Information Requirement (EIR) applied to the project causing a lot of re-work.

All respondents agree that BIM offers benefits and it should be used for railway projects. However, it might depend on the type, scale, and phase of the projects. It was noted from the interview results, that in the future, companies are expecting for more usage of 3D BIM Models with extensive embedded data. Further, companies using BIM for railway projects are also expecting a better and more accessible Common Data Environment (CDE) features to all project stakeholders.

5 Acknowledgement

The authors would like to acknowledge the support from TUMAsia and TUMCREATE (Campus for Research Excellence And Technological Enterprise) programme.

References

1. The business value of BIM for construction on major global markets. McGraw Hill, Bedford (2014)
2. Shou, W., Wang, J., Wang, X., Chong, H.Y.: A comparative review of building information modelling implementation in building and infrastructure industries. Arch. Comput. Methods Eng. **22**(2), 291–308 (2014)
3. Bradley, A., Li, H., Lark, R., Dunn, S.: BIM for infrastructure: an overall review and constructor perspective. Autom. Constr. **71**, 139–152 (2016)
4. RIBA Entreprises: National BIM report 2016. RIBA Entreprises, Newcastle (2016)
5. Craven, P.: BIM Survey and Specification and Reference Guide. Plowman Craven Limited, Harpenden (2017)
6. Neves, J., Sampaio, Z., Vilela, M.: A case study of BIM implementation in rail track rehabilitation. Infrastructures **4**(1), 8 (2019)
7. Kapogiannis, G., Mlilo, A. (eds.): Digital Construction Strategies Digital Construction Strategies and BIM in Railway Tunnelling Engineering. IntechOpen (2019)
8. Ehrbar, H.: Building information modelling - a new tool for the successful implementation of major projects of German railways/Building information modelling - Ein neues Werkzeug zur erfolgreichen Realisierung von Großprojekten der Deutschen Bahn. Geomechanik Tunnelbau **9**(6), 659–673 (2016)
9. Lichtberger, B.: Track Compendium: Formation, Permanent Way, Maintenance, Economics. Eurailpress, Hamburg (2005)
10. Statistics Canada: Survey Methods and Practices. Statistics Canada, Ottawa 12-587-X (2003)
11. Kaneta, T., Furusaka, S., Tamura, A., Deng, N.: Overview of BIM implementation in Singapore and Japan. JCEA **10**(12), 1305–1312 (2016)
12. Building and Construction Authority: Singapore BIM Guide: Version 2. Building and Construction Authority, Singapore (2013)
13. Building and Construction Authority: BIM Essential Guide for BIM Adoption in an Organization. Building and Construction Authority, Singapore (2012)
14. Cohen, D., Crabtree, B.: Qualitative Research Guidelines Project. RWJF (2006)
15. Love, P.E., Zhou, J., Edwards, D.J., Irani, Z., Sing, C.-P.: Off the rails: the cost performance of infrastructure rail projects. Transp. Res. Part A Policy Pract. **99**, 14–29 (2017)
16. Sardroud, J.M., Mehdizadehtavasani, M., Khorramabadi, A., Ranjbardar, A.: Barriers analysis to effective implementation of BIM in the construction industry (2018)

BIM Support in the Tendering Phase of Infrastructure Projects

Stefania Limp Muniz Correa(✉) and Eduardo Toledo Santos

University of São Paulo, São Paulo, SP, Brazil
{stefania.correa, etoledo}@usp.br

Abstract. Although the construction industry has been adopting Building Information Modelling (BIM) for at least two decades, many of its concepts, practices and standards were created for and are now consolidated on building projects. Therefore, in order to use BIM methodology in infrastructure projects, it is frequently necessary to resort to some adaptation. Recently, there has been an increasing interest on the application of BIM to infrastructure, which has its own complexities and particularities. Infrastructure may imply very diverse kinds of structures like tunnels, dams, airports, highways, among others, each potentially requiring a different BIM-based approach, as the challenges arising on each one are not the same. On the other hand, even for building projects, studies related to the application of BIM in the bidding stage are still few. The aim of this article is to clarify, through a literature review and analysis, how BIM can be used to support the tendering process of infrastructure projects. The findings show that BIM can be used in the bidding stage of infrastructure projects to improve communication between stakeholders, reduce (or even eliminate) error-prone tasks and rework, generate accurate quantity take-offs and automate time-consuming activities during this usually time-pressed phase. BIM implementation can also be of great value as it can support risk identification and reduction by minimizing misinterpretation of schedules and design, as well as contributing to better constructability analysis, sub-contractor coordination, facilitating communication and supporting the risk analysis, leading to more reliable and accurate offers to the owner.

Keywords: BIM · Infrastructure · Tendering · Bidding · Risk management

1 Introduction

The Construction industry is responsible for an annual revenue of 10 trillion USD, representing 6% of the global GDP [1], having a direct relation with people day-to-day living and with the global economy, mainly in urban areas. Urbanization induces the growth in infrastructure investments to provide mobility, communication, fueling, etc. A 2017 McKinsey and Company survey [2] states that, by 2025, construction expenditures is projected to a total of 14 trillion USD, highlighting the need of improving construction productivity to a better outcome. A different survey [3] suggests that an increase of 1% of GDP in infrastructure investment would represent an additional 1.3 million direct and indirect jobs in Brazil, as an example of the impact globalization and a greater revenue could represent.

E. Toledo Santos and S. Scheer (Eds.): ICCCBE 2020, LNCE 98, pp. 383–392, 2021.
https://doi.org/10.1007/978-3-030-51295-8_27

In a scenario where it is common sense that productivity needs to be boosted, new technologies and methodologies are gaining relevance. Building Information Modelling (BIM) is regarded as an opportunity of improving not only the productivity of construction itself, but the whole supply chain methods, from the conception to the operation of the facility.

Sacks et al. [4] define BIM as a "modeling technology and associated set of processes to produce, communicate, and analyze building models."

A study by Kassem and Succar [5] on the macro adoption of the methodology shows that countries are increasingly embracing BIM concepts, being the Netherlands, China, United States and United Kingdom recognized as being at the most advanced stage among nations. In the literature, numerous are the studies describing BIM adoption during the design, construction or operation phases, but there is a lack of studies focusing on the tendering phase of project's lifecycle.

Cost estimation, a central task on the tendering process, can be divided in three major processes: classifying construction products into items, computing their quantities, and pricing them [6]. The bidding stage is characterized by a high level of uncertainty, which can be interpreted not only as a hazard but as an opportunity, considering the risks involved [7].

This article will present a literature review on the adoption of BIM in the tendering phase, aiming to clarify the potential benefits of the use of the methodology in this stage of infrastructure projects.

2 Method

A literature review was used to achieve the objectives of this research. Aspects of BIM adoption on the tendering phase of the projects were studied, focusing on planning and risk management, as those are critical to increase the tender competitiveness.

A set of three query strings was prepared and processed on three different publication databases (Scopus, Web of Science and ScienceDirect):

- (bidding **OR** tendering) **AND** BIM;
- (4D planning) **AND** BIM;
- "Risk management" **AND** BIM.

The results were then analyzed by query and grouped by relevant categories, where important topics were highlighted and discussed. Finally, the applicability and coverage of the items are presented, generalizing the potential benefits of BIM adoption in the tendering phases.

3 Results and Analysis

3.1 BIM and Tendering

The tendering stage of a project can be a challenging phase in the project lifecycle. Estimators frequently have resort to the interpretation of 2D representations of the design, manually treating the information needed [8]. These experts often rely on their judgement and experience for this task, which can result in a manual, time consuming and error-prone process [9]. Aslesen et al. [7] state that rather than searching for the optimal price, it would be appropriate to find ways to reduce the uncertainty in the bids offered. The authors also affirm that a BIM model can be used for cost estimation and quantity take-offs (an opinion shared by Aram, Eastman and Sacks [9]), as well as to detect potential conflicts, interferences and collisions. Moreover, BIM has the potential for structuring the delivery and avoiding information loss during the project development, ensuring the coordinated delivery of the clients' requirements, and anticipating the impact of client's decisions on the operational performance of buildings at early project stages [10]. Hence, BIM could help collaboration and communication in the bidding process, which could reduce the uncertainty of the bid, therefore, increasing the competitiveness of the bidders.

Quantity Take-Off and Cost Estimating. When bidding, estimators must complete the following sequence of activities: divide the project into cost-items, calculate the quantity of each of them, calculate resource consumption and estimate the cost of each item, summarizing them to determine the total cost of the project [11]. As quantity take-off is a pre-requisite to a sequence of activities in the tendering processes (planning and budgeting are examples), reducing the subjectivity, uncertainty and time to deliver the results is highly recommended [6, 9, 12, 13]. However, to identify all the relevant factors, including them in the BIM model, to define their relationships with cost and to build a flexible and automated cost model is a hard and complex task [12]. In order to achieve good results, quantity take-off software solutions assume the model was developed satisfying a specific set of conditions [9, 13]. Even if the models comply with the required conditions, quantity take-off is still a challenge due to the following factors: design and construction models differ from one another as the items are modelled for different purposes, softwares are prepared to extract a rigid set of quantities (volume, surface area, etc.), not all of the properties are quantitative (some may have qualitative aspects), and the detailed design (which can provide accurate and precise information) is developed late in the process [9].

Taboada and Garrido-Lecca [14] state that the creation of quantity take-off standards promotes prompt, flexible and reliable quantity estimates. The standards comprised the use of templates, naming conventions, parameters definition, modelling criteria, model checkings and table of quantities' structures.

Using the IFC (Industry Foundation Classes) standard to do cost estimation was the topic of the study of Zhiliang et al. [11]. By using the IFC schema, the authors could input information to the model regarding the building products, the division items, the cost items and schedule, resources and price information. However, they highlight that IFC has limitations related to construction temporary elements (as scaffolding, for example).

Using a standard does not imply non-flexible results. Al-Mashta and Alkass [15] used the connection with external relational databases in order to show the results in different Work Breakdown Structures (WBS), enabling not only a streamlined process for generating detailed cost estimates, but also testing various scenarios to meet the client budget during the tendering phase.

Improving Communication. A sensitive point in tendering phases resides on the communication between the different stakeholders. During the lifecycle of a facility, design and construction teams exchange project information with different trades. The translation of this information from one trade to another is often manual, tedious and error-prone [16].

Data transferring is the core of interoperability, a concept that can be defined as "the ability to exchange data between applications, which smooths workflows and sometimes facilitates their automation" [4]. The IFC schema was developed to define an extensible set of consistent data representations of building information for exchange between AEC software applications [4].

IFC use turns out to be advantageous when it comes to the intention of employing computers to perform cost estimation, since interpreting the specifications written in human language to computer-processable formats is crucial [17].

Information Anticipation. Besides defining standards and classification systems, when automating cost estimation, it is important the design to be accurate, as it will be interpreted by computers, not by humans. Collins and Genthry [16] state that designers are frequently unaware of fabrication and assembling details of building components. Therefore they proposed the use of a Knowledge Base for Architectural Detailing (KBAD) since the early stages of the design conception as fundamental to provide designers with the right information from subcontractors.

Another form of integrating component information between stakeholders is suggested by Berard and Karlshoej [18], who studied the use of Information Delivery Manuals (IDM) as a way to develop information systems that support digital collaboration. It was noticed that IDM process requires a considerable effort, being a time-consuming process. However, they recommend its application in the development of BIM components, as a way of assuring the right information will be considered.

Model Checking. In order to check if the information needed was considered during model development, one of the important BIM uses is the automated rule checking to assure the design is compliant with pre-requisites. Despite the fact that BIM adoption in the tendering stages is still incipient, model checking tools can help estimators, bidders and the jury to verify the compliance with requirements and guidelines [19].

3.2 BIM and Risk Management

In the early stages of project development, there is a high level of uncertainty in terms of project cost. This uncertainty tends to decrease as time passes by and the project is detailed, determining its solutions. Risk analysis support the tracking these uncertainties and provides a range of possible final costs [20].

Zou, Kiviniemi and Jones [21], in their study on BIM relationship with risk management, state that not identifying properly the risks in the beginning of a project can lead to higher risks in the forthcoming phases. A strategy that general contractors commonly adopt is to do contingency planning. A 2017 Dodge Data & Analytics report [22] shows that 77% of general contractors use contingency planning when bidding. It is mentioned that the main reason for that is that contractors usually begin to consider risk mitigation after construction has started, not in the early stages of a project. This could be a consequence of the fact that general contractors are usually involved too late in the process, so more collaborative delivery methods would promote an earlier risk mitigation strategy. The report also states that 57% of general contractors agree that BIM could reduce the risk on their projects, and highlights that this number rises to 68% when only large contractors (those with revenues of $500 million or more) are considered. According to the report, 57% of the respondents say that the most effective strategy for risk mitigation is the occurrence of regular meetings with full project teams focusing on risks, highlighting the importance of more collaborative approaches. 91% of the general contractors seem to agree with the relevance of collaboration, being the three most important benefits of it: to increase the reliability in overall project performance (47%), to improve project safety (38%) and to improve project schedule (35%). When only large general contractors are considered, these three benefits are: increased reliability of overall project performance (53%), improved project safety (27%) and reduced rework (12%).

With the same approach, Chang [23] states that the focus of risk management seems to be more on the contingency planning to unexpected costs, instead of an active management of risks. According to a 2013 Infrastructure Risk Group report [20], in general terms, the optimistic bias seems to be a current practice in cost estimation of projects, leading to contingency plans that are spent even when they are not needed.

BIM Influence on Risk Management. There are several studies that mention BIM relationship with risk management, not only on the risk reduction of projects, but also on the creation on new ones once the methodology is adopted [24–26].

The most relevant risk examples that can be mitigated by the use of BIM refer mainly to a better project understanding and coordination, eliminating object clashes and providing a collaborative work environment [26]. Zou, Kiviniemi and Jones [27] still point out that, on the early stages of a project, the main risks involve project feasibility, securing budget and establishing a governance regime. For all these risks, BIM presents mitigation potential, be it by visualization of 3D or 4D models or walkthroughs, or by improving interoperability between applications (when IFC is used), reducing the possibility of errors due to manual processing of data.

Barriers to BIM Implementation on Risk Management. IRG report [20] points out that the main obstacles mentioned by general contractors for an increased BIM implementation for risk management resides in the fact that the construction supply chain is very fragmented, making it difficult to have a collaborative approach for information and cooperation flow. Competition among bidders is also mentioned as an important barrier to a fully integrated adoption.

Tomek and Matejka [24] also mention that the absence of client demand for using BIM can make companies struggle to find arguments for BIM implementation. However, Sacks et al. [4] mention that the BIM model can be used as a risk-mitigation tool by general contractors once it reduces the uncertainty associated with material quantities.

BIM adoption can also be responsible for the initiation of new risks. These risks regard the different software and standards, the necessary changes in data security, and legal liability aspects [25]. The performance of the new methodology can also be impaired when the risks are not identified properly. Chien et al. [28] state that having inadequate project experience, lack of software interoperability, difficulties in model management, and inefficient data interoperability can pose significant challenges to BIM implementation.

3.3 BIM and Planning

When bidding, engineers commonly have to present a preliminary version of a master construction schedule. Construction planning and the resulting schedules are often based in the experience of engineers. In addition, revising a schedule is expensive and time-consuming, and integration of schedules with 4D planning applications is still limited [29]. As a result, the final schedule, by relying in static information, tacit knowledge, norm, policies and 2D drawings, lacks specific, spatial and dynamic information (for example, dangerous zones, who and when would be exposed to hazards, etc.) [30]. This subjectivity in planning construction activities is also mentioned by Bansal [31], who states that the interpretation of schedules and its link to the BIM model can vary according to the project team member experience. The author also mentions that understanding, communicating and discussing a construction schedule can be difficult when the schedule does not include spatial aspects. Choi et al. [32] agree and also extend the risks of a inadequate workspace planning, mentioning probable productivity losses, altogether with safety hazards and issues of poor quality. Shang and Shen [33] point the increasing of struck hazards generated by the traditional planning process, which, by being based on 2D drawings, lacks in addressing the dynamic nature of a construction jobsite.

Moon et al. [34] state that when "schedule-workspace interference" ('when workspaces that share parallel schedules and are physically adjacent to one another exist simultaneously') occurs, assuring the performance of the schedules is difficult and it could represent a threat to constructability due to the increased collision risk between resources.

4D Planning Benefits. Candelario-Garrido et al. [35] state that integrating BIM to the project execution plan can ensure a better control and prevention of time and cost inefficiencies during the project. Although the traditional planning method is 20% faster than the 4D planning method, the authors state that 4D planning can be up to 40% more efficient in terms of visual progress monitoring and another 40% in terms of updating the data process. According to Swallow and Zulu [36], the most perceived benefits from the use of 4D planning are the value addition through visualization and better communication of project outputs, which would contribute to a better health and safety management, considering site planning and logistics.

Rolfsen and Merschbrock [37] advocate that 4D BIM provides clarity when conveying a masterplan, being perceived as most useful in early project stages.

Barriers to 4D Implementation. According to a recent Torres-Calderon et al. [38] study, only 35–40% of US top companies embraced 4D planning, and only in a small fraction of their projects. The interviews made by the authors have shown that there is the perception that linking the 3D elements to the schedule outweighs the benefits of 4D planning use. Moreover, to maintain the simulation up-to-date is still a challenge, which is given to a third party. Yun et al. [39] add that the use of the methodology brings obstacles regarding the efficiency of the data produced, once it can only be used on animations and visual presentations. Besides that, although a wide variety of simulation applications are available today, they present difficulties in terms of performance.

Haiati et al. [40] also state that the data-entry process is still time-consuming, making it cost-prohibitive. They also mention that 4D planning needs to be simplified so it can be wider adopted by project teams.

The lack of standards for the representation of building elements and tasks can also be an impeditive to the adoption of 4D BIM. As each application has its own standards for viewing and navigating the model and the schedule, collaboration and iteration can be difficult due to the increasing complexity of construction projects [41].

Crowther and Ajayi [42] point out difficulties regarding cultural aspects of the project can represent an obstacle to 4D implementation: lack of shared responsibility outside of the planner and BIM coordinator, severe lack of understanding and training regarding 4D BIM and complexity of carrying out the process effectively.

4 Conclusion

The research showed that knowing how to deal with uncertainty is crucial when bidding to a project. The traditional process, based in 2D drawings, require from estimators previous experience to interpret and to formulate hypothesis on how to build the facility, resulting in a very subjective process. The manual aspects of the tendering process are of great importance, as they can cause errors and culminate with a very labor-intensive process. Finding ways to reduce the risks of misinterpretation in the bidding stage can be of great value to the future phases of a project lifecycle, once it can ensure project feasibility and overall performance.

The literature review showed that BIM can be used in the early stages of a project, providing more accurate quantity take-offs and cost estimations, as well as enabling model checking. BIM can also provide a more feasible and safer schedule, as it enables the consideration of spatial aspects, as well as the detection of dynamic conflicts, favoring workspace planning. The adoption of the methodology can be useful to risk management during the bidding phase, reducing the risks of the project, as it can also, by its nature, improve the communication and collaboration between the teams involved in the tendering process (not only between general contractors and designers, but also between them and subcontractors).

Being the potential benefits of BIM adoption stated, there are obstacles that need to be correctly addressed. Interoperability is still a challenge when it comes to BIM adoption. The lack of standards can pose difficulties not only to quantity take-offs and cost estimates, but also to developing 4D animations, as linking the schedule with the 3D objects (and updating it) can still be a time-consuming and error-prone task. Culture aspects must also be considered during BIM adoption, once the sharing of responsibilities and training can greatly affect the implementation of the new methods.

The literature review showed very little research done specifically to address the bidding phase and even less if only infrastructure projects are considered. Therefore, especially considering the demand for and the impact of infrastructure projects, it is necessary to target this theme in future research initiatives, as BIM adoption on infrastructure projects is increasing in recent years.

References

1. The Boston Consulting Group: Digital in Engineering and Construction - The Transformative Power of Building Information Modeling (2016)
2. McKinsey & Company: Reinventing Construction: A Route to Higher Productivity (2017)
3. McKinsey & Company: Bridging global infrastructure gaps. McKinsey Glob. Inst., June 2016
4. Sacks, R., et al.: BIM Handbook: A Guide to Building Information Modeling for Owners, Designers, Engineers, Contractors, and Facility Managers, 3rd edn. Wiley (2018)
5. Kassem, M., Succar, B.: Macro BIM adoption: comparative market analysis. Autom. Constr. **81**, 286–299 (2017)
6. Ma, Z., Liu, Z.: BIM-based intelligent acquisition of construction information for cost estimation of building projects. Procedia Eng. **85**, 358–367 (2014)
7. Aslesen, S., et al.: Winning the bid – a step-wise approach using BIM to reduce uncertainty in construction bidding. In: Proceedings of the 26th Annual Conference International Group for Lean Construction, Chennai, India, pp. 68–78 (2018)
8. Ma, Z., Wei, Z., Zhang, X.: Semi-automatic and specification-compliant cost estimation for tendering of building projects based on IFC data of design model. Autom. Constr. **30**, 126–135 (2013)
9. Aram, S., Eastman, C., Sacks, R.: A knowledge-based framework for quantity takeoff and cost estimation in the AEC industry using BIM. In: 31st International Symposium on Automation and Robotics in Construction and Mining, ISARC 2014 - Proceedings, pp. 434–442 (2014)
10. Rodriguez-Trejo, S., et al.: Hierarchy based information requirements for sustainable operations of buildings in Qatar. Sustain. CITIES Soc. **32**, 435–448 (2017)
11. Zhiliang, M., Zhenhua, W., Wu, S., Zhe, L.: Application and extension of the IFC standard in construction cost estimating for tendering in China. Autom. Constr. **20**(2), 196–204 (2011)
12. Aram, S., Eastman, C., Beetz, J.: Qualitative and quantitative cost estimation: a methodology analysis, in computing in civil and building engineering. In: Proceedings of the 2014 International Conference on Computing in Civil and Building Engineering, pp. 381–389 (2014)
13. Olatunji, O.A., Sher, W.: Perspectives on modelling BIM-enabled estimating practices. Australas. J. Constr. Econ. Build. **14**(4), 32–53 (2014)
14. Taboada, J.A., Garrido-Lecca, A.: Case study on the use of BIM at the bidding stage of a building project. In: 22nd Annual Conference of the International Group for Lean Construction Understanding and Improving Project Based Production, IGLC 2014, pp. 1473–1482 (2014)
15. Al-Mashta, S., Alkass, S.: Integrated cost budgeting and estimating model for building projects. AACE Int. Trans. **1**, 22–36 (2010)
16. Collins, J., Gentry, R.: KBAD knowledge base for architectural detailing. In: Proceedings of the 35th eCAADe Conference, vol. 2, pp. 657–666 (2017)
17. Eastman, C., Min Lee, J., Suk Jeong, Y., Kook Lee, J.: Automatic rule-based checking of building designs. Autom. Constr. **18**(8), 1011–1033 (2009)

18. Berard, O., Karlshoej, J.: Information delivery manuals to integrate building product information into design. Electron. J. Inf. Technol. Constr. **17**, 64–74 (2012)
19. Ciribini, A.L.C., Bolpagni, M., Oliveri, E.: An innovative approach to e-public tendering based on model checking. Procedia Econ. Financ. **21**, 32–39 (2015)
20. IRG: Managing Cost Risk and Uncertainty in Infrastructure Projects (2013)
21. Zou, Y., Kiviniemi, A., Jones, S.W.: BIM-based risk management: challenges and opportunities. In: Proceedings of the 32nd CIB W78 Conference, pp. 847–855 (2015)
22. Dodge Data & Analytics: Managing Risk in the Construction Industry (2017)
23. Chang, C.Y.: Harnessing BIM data in the management of project risks: the Bayesian risk-bearing capacity approach. In: Transforming the Future of Infrastructure through Smarter Information - Proceedings of the International Conference on Smart Infrastructure and Construction, ICSIC 2016, pp. 537–542 (2016)
24. Tomek, P., Matejka, A.: The impact of BIM on risk management as an argument for its implementation in a construction company. Procedia Eng. **85**, 501–509 (2014)
25. Ganbat, T., Chong, H., Liao, P., Wu, Y.: A Bibliometric review on risk management and building information modeling for international construction. Adv. Civ. Eng. **2018**, 1–13 (2018)
26. Ahmad, Z., Thaheem, M.J., Maqsoom, A.: Building information modeling as a risk transformer: an evolutionary insight into the project uncertainty. Autom. Constr. **92**(March), 103–119 (2018)
27. Zou, Y., Kiviniemi, A., Jones, S.W.: A review of risk management through BIM and BIM-related technologies. Saf. Sci. **97**, 88–98 (2017)
28. Chien, K.F., Wu, Z.H., Huang, S.C.: Identifying and assessing critical risk factors for BIM projects: empirical study. Autom. Constr. **45**, 1–15 (2014)
29. Tauscher, E., Smarsly, K., König, M., Beucke, K.: Automated generation of construction sequences using building information models, in computing in civil and building engineering. In: Proceedings of the 2014 International Conference on Computing in Civil and Building Engineering, pp. 745–752 (2014)
30. Choe, S., Leite, F.: Construction safety planning: site-specific temporal and spatial information integration. Autom. Constr. **84**, 335–344 (2017)
31. Bansal, V.K., Pal, M.: Construction projects scheduling using GIS tools. Int. J. Constr. Manag. **11**(1), 1–18 (2011)
32. Choi, B., Lee, H.S., Park, M., Cho, Y.K., Kim, H.: Framework for work-space planning using four-dimensional BIM in construction projects. J. Constr. Eng. Manag. **140**(9) (2014)
33. Shang, Z., Shen, Z.: A framework for a site safety assessment model using statistical 4D BIM-based spatial-temporal collision detection. In: Construction Research Congress 2016: Old and New Construction Technologies Converge In Historic San Juan, pp. 2187–2196 (2016)
34. Moon, H., Kim, H., Kim, C., Kang, L.: Development of a schedule-workspace interference management system simultaneously considering the overlap level of parallel schedules and workspaces. Autom. Constr. **39**, 93–105 (2014)
35. Candelario-Garrido, A., García-Sanz-Calcedo, J., Rodríguez, A.M.R.: A quantitative analysis on the feasibility of 4D planning graphic systems versus conventional systems in building projects. Sustain. Cities Soc. **35**, 378–384 (2017)
36. Swallow, M., Zulu, S.: Benefits and barriers to the adoption of 4D modeling for site health and safety management. Front. Built Environ. **4** (2019)
37. Rolfsen, C.N., Merschbrock, C.: Acceptance of construction scheduling visualizations: bar-charts, flowline-charts, or perhaps BIM? Procedia Eng. **164**, 558–566 (2016)

38. Torres-Calderon, W., Chi, Y., Amer, F., Golparvar-Fard, M.: Automated mining of construction schedules for easy and quick assembly of 4D BIM simulations. In: Computing in Civil Engineering 2019: Visualization, Information Modeling, and Simulation - Selected Papers from the ASCE International Conference on Computing in Civil Engineering 2019, pp. 432–438 (2019)
39. Yun, S., Jun, K., Son, C., Kim, S.: Preliminary study for performance analysis of BIM-based building construction simulation system. KSCE J. Civ. Eng. **18**(2), 531–540 (2014)
40. Haiati, O., Von Heyl, J., Schmalz, S.: BIM and sequence simulation in structural work - development of a procedure for automation. In: IGLC 2016 - 24th Annual Conference of the International Group for Lean Construction, pp. 73–82 (2016)
41. Castronovo, F., Lee, S., Nikolic, D., Messner, J.I.: Visualization in 4D construction management software: a review of standards and guidelines. In: Computing in Civil and Building Engineering - Proceedings of the 2014 International Conference on Computing in Civil and Building Engineering, pp. 315–322 (2014)
42. Crowther, J., Ajayi, S.O.: Impacts of 4D BIM on construction project performance. Int. J. Constr. Manag. (2019)

BIM Implementation, Current Status and Practice

An Investigation of Contractual Requirements for BIM Adoption in the Brazilian Public Sector

Douglas Malheiro Brito(✉), Emerson de Andrade Marques Ferreira, and Dayana Bastos Costa

Federal University of Bahia, Salvador, Brazil
douglas_ssa@hotmail.com, ferreira.eam@gmail.com, dayanabcosta@ufba.br

Abstract. Previous research has shown that Building Information Modeling (BIM) can reduce errors, rework, time and cost in project design and construction. Besides, it provides greater value delivered at the maintenance and operation stage, reducing owner-operating expenses. To maximize BIM use benefits, owners, such as the public sector, can directly affect the enterprise viability by selecting the project delivery system. That has led to the search for alternatives contractual requirements to traditional models. The main objective of this study is to investigate contractual requirements that influence BIM adoption process in public organizations. The research strategy adopted was a survey. Sixty-nine professionals from Brazilian Architecture, Engineering, Construction and Operation (AECO) sector answered a digital questionnaire. The target public was BIM adoption experts, academic professionals and public managers from fourteen public organizations that have already begun BIM implementation. The investigation revealed that project delivery systems considered with the greatest potential to maximize the benefits of BIM use were those less adopted and disseminated in the Brazilian public sector. The results also show that contractual requirements for BIM adoption might consider aspects, such as, technical qualification in public procurement, methods and tools for BIM deliverables quality checking, items that must compose a BIM mandate, and favorable cost-benefit factors for the development of BIM component virtual libraries by public organizations. The main contribution of this research is to provide a better understanding of contractual requirements that could be incorporated into project delivery systems to maximize BIM benefits in the public sector.

Keywords: BIM adoption · Project delivery system · Survey · Public organization · Contractual requirements

1 Introduction

BIM is a modeling technology associated with a set of processes to produce, communicate and analyze building models (Eastman et al. 2011). For Hardin and Mccool (2015), BIM is also an initial catalyst for a period of innovation and technology revival,

E. Toledo Santos and S. Scheer (Eds.): ICCCBE 2020, LNCE 98, pp. 395–408, 2021.
https://doi.org/10.1007/978-3-030-51295-8_28

generating innovations in many areas and changes in support processes, since new technologies do not fit in the previous processes.

For Sacks and Gurevich (2016), BIM adoption is a complex process that requires the development of a strategy that understands organization maturity, industry capabilities, national policies and regulations, education, hardware and software acquisition, changes in the project delivery systems, among other critical factors.

According to Bernstein et al. (2014), reduction of errors, rework, costs and time are some of the BIM use benefits for owners. Among the biggest beneficiaries are customers, maintenance and operation managers, as well as owners, according to Eadie et al. (2013).

Porwal and Hewage (2013) proposed a structure for public construction project delivery based on collaborative BIM processes and recommended studies in areas related to national guidelines, legal structure and specific contractual issues for implementation.

For Eastman et al. (2011), governments are usually owners and it also has control of the service provider's selection, contract type and delivery processes as well as general specifications and requirements. Even so, some public sector organizations do not realize their potential to change or control how a project is conceived, being unaware of the contractual requirements that are influenced by BIM adoption.

The main objective of this study is to investigate contractual requirements that influence BIM adoption process in public organizations, contributing to a greater understanding of this process and successful BIM implementations in the public sector.

2 Contractual Aspects in BIM Adoption

Brazil has unbalanced BIM diffusion areas and low levels of modeling, collaboration and integration capacity. Concerning maturity, the greatest performance was in the technological infrastructure component, while in measures and benchmarking; education; objectives and milestones; regulatory structure; and notable publications, had the lowest levels. Brazilian political actions were characterized as to raise awareness, encourage and observe, in a completely passive pattern (Kassem and Succar 2017).

Formal adoption in the Brazilian governmental sphere is still recent, having started from some states and federal agencies. However, in 2018, the Brazilian federal government launched BIM National Dissemination Strategy and an implementation committee responsible for various government actions, such as the development of company accreditation programs; BIM platform with national object library; stimulating demand in hiring; and vocational education.

For Eastman et al. (2011), some of the actions that owners can take as a way of encouraging BIM are: include specific pre-qualification criteria in the selection of contractors, such as knowledge and experience in BIM; build and educate a qualified network of BIM service providers; change requirements for deliveries, incorporating scope, detail and organization of model information and its uses; and propose performance-based contracts and shared incentive plans.

In this way, the choice of the project delivery system is one of the fundamental ways in which owners can affect in efficiency, productivity and profitability. Concerns about the sector's inefficiency in recent decades have led to the search for alternatives to the traditional Design-Bid-Build (DBB) model, in which the design, contracting and construction stages are segregated and generally performed by different agents. Another system that was established was the Design-Build (DB), in which the contractor is responsible for project design and construction (BSI 2013).

In recent years, another delivery systems have emerged and brought about more radical changes, such as the Design-Build-Operate/Maintain (DBO/M) that expands the role of the project contractor up to operation and maintenance phase. Integrated Project Delivery (IPD) is another emerging delivery system that proposes risks and responsibilities sharing between parties since the project beginning, integrating people, systems, business structures and practices in a process that leverages everyone's talents to reduce waste and optimize efficiency (BSI 2013).

Bernstein and Laquidara-Carr (2014) investigated AECO sector perception regarding the selection of these delivery systems, verifying that there is no absolute consensus on the benefits and obstacles of the most used systems such as DBB and DB. As an exception, there is a greater perception of a positive impact on time, risks and responsibilities sharing for DB use, as well as search for cost reduction with a greater tendency in the DBB selection.

Emerging systems such as DBO/M and IPD tend to have an increasing use since the latter was considered the best delivery system to achieve improved communication between those involved, productivity and increased efficiency. This makes most owners who seek better control of the schedule, costs and increased quality, analyze its use (Bernstein and Laquidara-Carr 2014).

The main public project delivery systems in Brazil are Law no. 8.666/2013; Public-Private Partnerships (PPPs) law; and the law of the Differential Contracting Regime (DCR). They are similar to the following delivery systems: Law no. 8.666/1993 with the DBB system; DCR with the DB system in the integrated contracting regime for design, construction and pre-operation stages; PPP law with the DBO/M system, encompassing stages of design, construction, operation and transfer to the public sector.

The project delivery system selection by the owner usually correlates his level of control with the risks assumed. The DBB method is at the intermediate level of risk, offering a high degree of control over the project, reliable information on the cost before construction and clearly defined parts roles, although it may be a choice that does not favor for schedule and creates a non-collaborative culture between the parties. In the low level of risk are the IPD and DB systems that significantly reduce the owner risks, however, if their ability to generate value for the investment or the building performance is reduced, it can result in increased operating costs (Bernstein and Laquidara-Carr 2014).

Porwal and Hewage (2013) proposed a structure for contracting public construction projects based on collaborative BIM processes, aiming to maximize the value delivered by the Canadian construction industry. A case study was conducted in a public project contracted at the lowest price, with the DBB system, to prove the viability of the proposed method.

The results indicated that guidance is needed from where to start, what tools are available, how to face legal, contractual and cultural challenges, among others.

Compared to the traditional DBB contracting method, the proposed BIM collaborative partnership was considered suitable for the public sector because the selection process was open, objective, fair and free from political influence, with significant improvements in terms of cost, value delivered and performance in carbon emissions Porwal and Hewage (2013).

According to Eastman et al. (2011), another central issue for the public sector involves the current processes based on the lowest price, which usually diminish the importance of criteria such as time and quality, without guaranteeing the return of maximum value. With BIM adoption, the sooner the model can be developed and shared, the greater its usefulness. For this reason, DB contracting models provide a great opportunity to explore the benefits, as the contractor is responsible for the design and construction, and both can participate in the design phase. Other systems can also make progress with BIM, however, achieving partial benefits if there is no collaboration during the design (Eastman et al. 2011).

Becerik-Gerber and Rice (2010) investigated the BIM use in the United States with a sample of 424 respondents and identified that although the traditional DBB system was still the most used, it was already possible to notice that many of the projects adopted delivery methods latest and most collaborative, including DB, IPD, performance-based contracts and alliance. IPD was cited as the most effective in facilitating the BIM use, as it seeks to create a more comprehensive collaborative atmosphere, aligned with the objectives of all members, encouraging them to work together throughout all phases.

In addition to selecting the project delivery system, public organizations should consider the influence of the BIM adoption process on other contractual aspects. Kassem et al. (2015) analyzed guides, protocols and mandates, documents that are published for a wide audience and that aim to promote understanding, regulate the implementation or demand requirements for BIM use.

BIM guides are characterized as descriptive and optional documents, clarifying goals, reporting research and simplifying complex topics. Protocols are also optional, however, they are prescriptive, providing, in detail, steps or conditions to achieve a measurable goal or result. In contrast, mandates are prescriptive and dictated by an authority, identifying what should be delivered and, sometimes, how, when and by whom (Kassem et al. 2015).

Sacks and Gurevich (2016) consider that the preparation of adequate guidelines on how to use BIM, to offer value, to be demanded in contracts is one of the main steps that can be exercised by large clients in the construction sector (companies, authorities and public departments) for a guided and systemic implementation. Based on the analysis of fifteen mandates, a set of requirements was recommended for organizations to develop, evaluate or update their documents. Of the ten main topics, interoperability, BIM manager role, collaboration modes, requirements for operation and maintenance, BIM Execution Plan (BEP) and simulations showed similarities in the documents, although content, frequency, specificity and level resolution have varied.

Faria et al. (2016) also analyzed mandates of pioneering countries in adopting, as the United States, United Kingdom, Finland, checking that have: BEP; definition of file formats for exchanging information; expected uses; roles and responsibilities; intellectual property of deliverables; Level of Development (LOD) of the components; and quality control of models.

Regarding delivery systems, although the DBB system is still the most traditional, some documents address the DB and the IPD trend towards increased collaboration and integration, even though there are many practical, legal and logistical obstacles, especially in organizations that have their procurement processes and legislation (Sacks and Gurevich 2016).

Planning for BIM adoption involves BEP preparation, which is prepared by the client to explain how the information modeling aspects of a project will be carried out (BSI 2013). According to Sacks and Gurevich (2016), BEP is an essential adoption component, as it contains the intended modes of collaboration and information sharing, roles and responsibilities of those involved, software to be used, scope and LOD to be achieved, quality control procedures, object composition and naming conventions.

The procedures for quality control of BIM models must be performed before integration with the federated model and made available as a reference for other designers. This verification involves aspects of both the model itself and its components and aims to ensure information consistency and conformity of the components and their use.

The following aspects must be verified: objects with correct typologies defined; duplicate, very small or overlapping objects; LOD established for a given type of component in the corresponding milestone or phase; and objects following intended uses (Eastman et al. 2011).

Gurevich et al. (2017) conducted action research and case studies in five public agencies in the United Kingdom that adopt BIM to propose an Adoption Impact Map, which lists possible actions taken by the agencies, intermediate results and possible social impacts in the value delivered to the occupants and in its functionality.

The results showed that the artifact of BIM objects library would have an impact on standardization implementation by the organization, generating positive effects on time, cost and maintenance. The five agencies studied emphasized the importance of the asset information value obtained at the end of a BIM design and construction process, bringing benefits to operation and maintenance activities. (Gurevich et al. 2017).

3 Method

The research strategy adopted in this research involved the following steps:

- Literature review on contractual aspects that are influenced by BIM adoption according to the literature;
- Development of a digital questionnaire on Google Docs on BIM adoption by public organizations, having, as part of it, five questions on the evaluation of contractual requirements that are influenced by BIM use in organizations;
- Surveying by sending questionnaires to a sample of Brazilian professionals from the AECO sector, including public managers, academics and specialists in BIM adoption;
- Analysis of the responses to the questionnaire by calculating agreement percentages and degree of importance average in questions with Likert scale.

The questionnaire comprised a section to identify the respondent's profile and another with evaluation containing the five questions on contractual aspects. In two questions, participants could select items considered to be priorities to compose a mandate and choose one or more existing Brazilian public procurement systems with the potential to maximize BIM benefits. The other questions evaluated by a Likert scale covered the degree of importance of requirements for technical qualification in hiring, methods and tools to verify BIM models quality and factors that generate a favorable cost for the development of a BIM component library. The five-level Likert Scale ranged from 1 - Very Low; 2 - Low; 3 - Medium; 4 - High; up to 5 - Very High.

A pilot questionnaire previously assessed the survey questions through two BIM implementation specialists. Subsequently, the questionnaire was sent by email to approximately 300 Brazilian professionals, obtaining a sample of 69 respondents, between May and July 2018. Those who did not respond to the survey in the first three weeks were reminded of the invitation via a new email.

4 Contractual Requirements for BIM Adoption in the Brazilian Public Sector

The initial part of the questionnaire consisted of identifying the survey participants, followed by the evaluation questions.

The respondents' profile involved aspects such as age, gender, education, areas of expertise in the AECO industry and the time of personal experience (studies and applications) with BIM, as shown in Figs. 1 and 2.

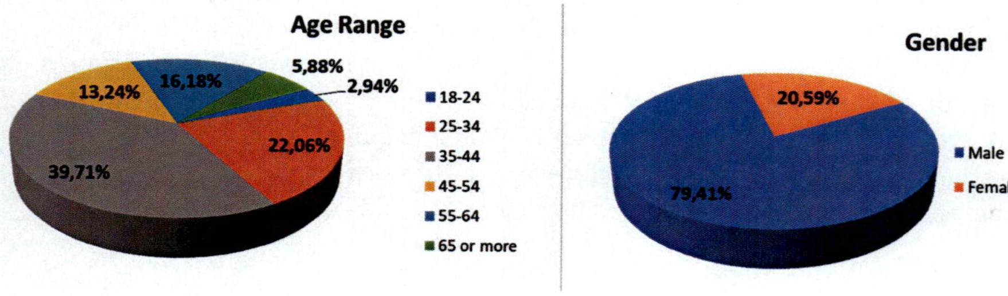

Fig. 1. Age range and gender of participants

About half of the professionals are architects, followed by civil engineers. Regarding the education level, 42.65% have a master's degree, followed by 25% who have a specialization/MBA. The sample covers ten states in four Brazilian regions. Also, 64.71% of respondents have at least 5 years of personal experience with BIM, with 48.53% reporting 8 years or more of experience.

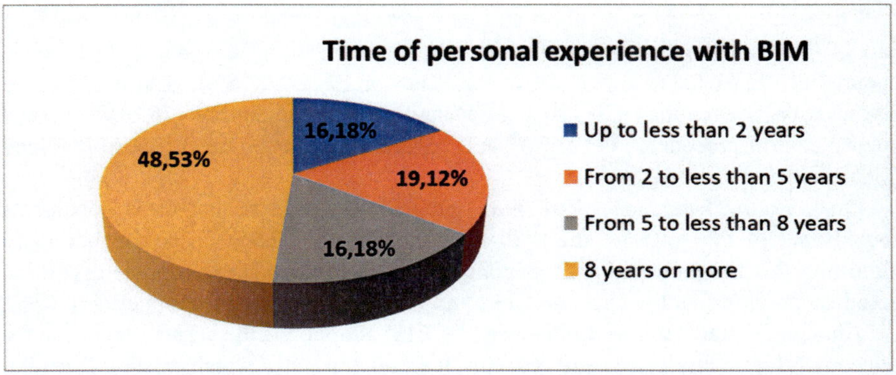

Fig. 2. Time of personal experience with BIM of participants.

The first contractual aspect evaluated was the organization's BIM adoption policy through the mandate publication, a prescriptive and mandatory document that guides implementation planning and execution. A set of fourteen items present in these documents were identified, according to the literature review, so that those considered priority by the respondents were chosen, as shown in Table 1.

Table 1. Evaluation of priority items in a mandate

Ranking	Priority items in a mandate	% Agreement
1	Interoperability for information exchange	82,61%
2	Requirements or template for BIM execution plan	81,16%
3	Data naming and organization conventions	79,71%
4	Levels of development (LOD) and models structuring	76,81%
5	Model quality control procedures	72,46%
6	Roles and responsibilities of those involved	69,57%
7	Project delivery systems and qualification	63,77%
7	Patterns of graphic representation and components	63,77%
9	Desired collaboration modes	62,32%
10	Uses, functions and analysis/simulations expected by life cycle phases	60,87%
11	Payment schedule (especially with definition of the design phase remuneration)	53,62%
11	Operation and maintenance requirements (COBie)	53,62%
11	Intellectual property rights of models and their components	53,62%
14	Software specification	30,43%
15	Outro	2,90%

The results revealed that almost all items were considered necessary by more than half of the evaluators, with five of them being chosen by at least about 70%: interoperability (82.61%); requirements or template for BEP (81.16%); data naming and organization conventions (79.71%); LOD and models structuring (76.81%); model quality control procedures (72.46%); and roles and responsibilities of those involved (69.57%).

On the other hand, only 30.43% of the sample considers that these documents should specify the software that will be used and only 2.9% chose another item, indicating that the proposed set was comprehensive for mandates scope composition, based on previous studies such as Sacks and Gurevich (2016) and Faria et al. (2016).

Another contractual aspect influenced by BIM adoption is the criteria definition for selecting project delivery systems. For this, it was investigated which existing Brazilian public delivery systems for construction projects would have the greatest potential to maximize BIM benefits, among them, Law no. 8,666/1993, DCR and PPP regimes, or otherwise not yet regulated.

Although this question had a higher percentage of respondents (11.59%) who indicated that they did not know how to answer, the systems considered most appropriate were those that are still less used and disseminated in Brazil. DCR with variable remuneration for performance, which has variations that can be associated with the DB or DBB system (49.28%) and PPP encompassing design, construction and operation with variable remuneration, according to performance, equivalent to the DBO/M system (42.03%), as shown in Fig. 3.

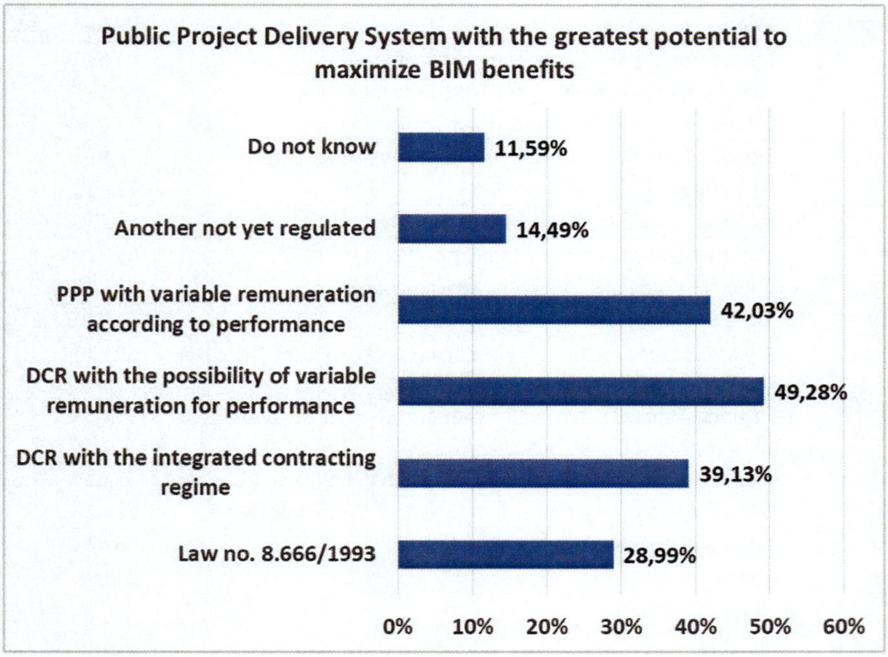

Fig. 3. Public project delivery system to maximize BIM benefits.

The percentage of participants who selected another form not yet regulated (14.49%) and who cited the IPD corroborates the growing trend in the valuation of new delivery systems that use alternative mechanisms, as previously identified in other markets as in the survey by Becerik-Gerber and Rice (2010) in the US. Similarly, one of the comments suggests that the Brazilian public sector should study the modalities known in the US as IPD and DB, reducing the current segregation and disintegration between project stages and results.

It is noteworthy that the DCR regime with the integrated contracting of the design, construction and pre-operation stages, the closest to the American DB system, obtained 39.13% of the answers. On the other hand, Law 8.666/1993, with separation of the stages of project design, contracting, construction and operation, equivalent to DBB system, reached the lowest percentages with 28.99%, reinforcing the existing criticisms about their deficiencies regarding integration, cost and time control and low generation of value delivered in the operation and maintenance stages.

In addition to the appropriate delivery system selection in public projects, another important action to be taken is the criteria definition for qualifying in BIM of interested bidders, maintaining equality and ensuring that the contractor provides the expected performance.

Among the four requirements assessed through the degree of importance in a Likert scale, the highest averages were obtained by team qualification and professional experience in BIM use (4.20); and assessment of the quality of the deliverables in previous projects (4,19), both considered highly or very highly important for more than 80% of the professionals, as shown in Table 2. These results highlight the need for public organizations to develop objective criteria for analyzing the qualification of professionals and the quality of BIM models.

Table 2. Evaluation of requirements for technical qualification

Requirements for technical qualification in contracting for BIM use	% (4) or (5)	Average	Standard deviation	Ranking
Team qualification and professional experience in BIM use	87,0%	**4,20**	0,70	**1**
Assessment of the quality of the deliverables in previous projects	82,6%	**4,19**	0,83	**2**
Availability of company adequate technological infrastructure	71,0%	**3,87**	0,84	**3**
Company experience in similar projects	62,3%	**3,78**	0,78	**4**

Among the comments of the participants, there is the importance of companies having managers or coordinators of BIM processes, or even a group that plays this role with proven experience, not only based on documentation. Another way of guaranteeing professional qualification would be to create certification processes for professionals or companies in skills necessary for the use of BIM.

The other requirements were also considered highly important, according to the following sample averages: availability of company adequate technological infrastructure (3.87); and company experience in similar projects (3.78), which indicates that they should not be discarded by public managers and can be evaluated as complementary criteria in the qualification process.

The creation of objective criteria for assessing BIM models quality is not only strategic for the initial contracting phase, but it is also during the design phases until the project's operations.

To verify BIM models quality, the evaluation of the methods and tools that obtained the highest averages of importance have some degree of automation and validation software support and specific electronic spreadsheets, such as automatic verification of duplicate, overlapping or inconsistent objects (4.44); and automatic verification of specific programmable rules (4.44). Then, specific checklists to control the quality of data and projects were evaluated with a sample mean of 4.06, as can be seen in Table 3. As proposed by Sacks and Gurevich (2016), these criteria may be included in the project's BEP.

The manual visual verification of specific views and manual verification with filters use were evaluated with average degrees of importance of 3.31 and 3.52, respectively, which reinforces the tendency that they are methods with less productivity, being useful in certain situations that require qualitative analysis. One of the respondent's comments emphasizes the importance of these qualitative analyses to meet functionality, flows, aesthetics, budget, among others, which would depend on a holistic interpretation of the project coordinator for decision making.

Table 3. Evaluation of methods and tools for checking BIM models quality

Methods and tools for checking BIM models quality	% (4) or (5)	Average	Standard deviation	Ranking
Automatic verification of duplicate, overlapping or inconsistent objects	93,9%	**4,44**	0,61	**1**
Automatic verification of specific programmable rules	89,4%	**4,44**	0,68	**2**
Specific checklists to control the quality of data and projects	81,8%	**4,06**	0,74	**3**
Manual visual verification of specific views	54,7%	**3,52**	0,94	**4**
Manual verification with filters	43,1%	**3,31**	0,98	**5**

Other requirements to be verified, according to some participants, including checking of the modeled objects, LOD compatibility among disciplines and identification of possible information and elements that still need to be modeled. Furthermore, it was mentioned the need to review some technical standards that have subjectivities in the interpretation of their requirements so that they can program objective rules for measuring compliance.

Still, regarding the requirements and guidelines for receiving deliverables, another potential action investigated was the cost-benefit analysis in the development of BIM components libraries. Bearing in mind that the elaboration, maintenance and updating of a specific BIM components library by a public institution is not a decision widely accepted by the Brazilian sector, professionals were questioned about the agreement with this issue, as shown in Fig. 4.

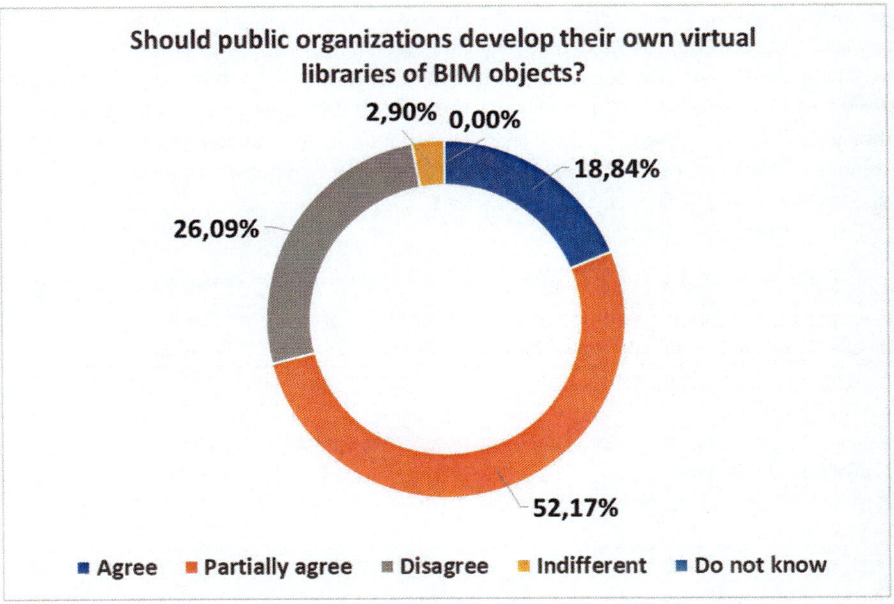

Fig. 4. Development of BIM objects libraries by public organizations.

The results showed that just over half of the professionals partially agree, while 26.09% disagree and 18.84% agree, confirming the existence of different perceptions on the topic. Among the participant's comments, some consider that public agencies should develop, as well as integrate, libraries developed by the industry, or that they should develop for their items with greater specificity, while for components in common use a more comprehensive library should be created from a federal agency or association.

Other observations of respondents mention that libraries should be restricted to specific situations, such as when the institution has projects with standardized components. Among those who disagree, it is argued that this is an attribution of the market, materials manufacturers with software companies, and that library management needs qualified professionals, elaboration covering the diversity of existing competing software and updating to be continued.

Adopting the definition of the 71.01% who agree, even if only partially, with the development of virtual libraries by public organizations, seven organizational factors were assessed as to the degree of importance in generating a favorable cost-benefit.

The factors considered with the highest degrees of importance were: existence of requirements for projects modulation and/or standardization (3.96); personnel and technological infrastructure availability for development (3.94); and repeatability in projects typology and specifications (3.91), with emphasis on the latter, which obtained 76.6% of responses with a high or very high degree of importance, as shown in Table 4. It is also observed that these three factors bring together aspects related to the processes, people, technologies and characteristics of organizations, considered important for cost-benefit analysis.

Then, the factors considered the fourth and fifth most important were: high complexity of the projects (3.74); and existence of a BIM objects national repository or library (3,67), comprising a factor that corresponds to a characteristic of the organization and another that involves external environment, which reveals a diversity of aspects that would influence the cost-benefit.

Table 4. Evaluation of generating cost-benefit factors to BIM component libraries

Generating cost-benefit factors favorable to the development of BIM component libraries	% (4) or (5)	Average	Standard deviation	Ranking
Existence of requirements for projects modulation and/or	70,2%	**3,96**	0,81	**1**
Personnel and technological infrastructure availability for development	66,7%	**3,94**	0,95	**2**
Repeatability in projects typology and specifications	76,6%	**3,91**	0,75	**3**
High complexity of the projects	59,6%	**3,74**	1,05	**4**
Existence of a BIM objects national repository or library	56,3%	**3,67**	0,95	**5**
Variability in the projects typology and specifications	44,7%	**3,32**	0,93	**6**
Low complexity of the projects	29,8%	**3,00**	1,10	**7**

On the other hand, the factors that obtained an average degree of importance in generating favorable cost-benefit were: variability in the projects typology and specifications (3, 32); and low complexity of the projects (3.00). These characteristics are opposed to factors that obtained high importance related to projects repeatability and high complexity, possibly indicating greater relevance of repeatability and potential for standardization in the use of own libraries. These findings reinforce the results obtained by Gurevich et al. (2017) that associated the creation of a BIM objects library with standardization implementation by an organization, generating positive effects on time, cost and maintenance.

Among the comments made by the participants, the gain obtained by standardization was reinforced, which would be reduced when there is no significant volume of repeat projects. In these cases and, when the organization does not design, it was mentioned that the public organization would simply be concerned with defining what information is necessary for contract fulfillment.

5 Conclusions

The study main objective was to investigate contractual requirements that influence the BIM adoption process in public organizations. The contribution of this research is to provide a greater understanding of the contractual aspects of this process and to promote implementations that achieve the objectives established by the organizations.

A survey using digital questionnaires was used to evaluate these contractual aspects, based on the responses of a sample of 69 Brazilian professionals involved with BIM use in public organizations, among public managers, academics and experts.

The results showed a set of items considered a priority in the mandate composition and indication of the Brazilian public project delivery systems considered with the greatest potential for maximizing BIM benefits. Emerging delivery systems that are still less used and disseminated in Brazil were considered the most appropriate.

Besides, the participants highlighted as technical qualification requirements the need for public organizations to develop objective criteria for analyzing qualification of professionals and BIM models quality. The quality verification methods and tools considered to be of the greatest importance were those that have some degree of automation, support of validation software and specific spreadsheets.

The research also evaluated the factors that generate cost-benefit favorable to the development of a BIM components library, especially those related to the project repeatability and high complexity. The findings may indicate the greater relevance of repeatability and the potential for standardization in the use of own libraries.

As for recommendations for future research, it is expected that these contractual requirements assessed according to the professional's perception involved in the BIM adoption can be investigated as to their influence on the implementation process by public organizations.

References

Becerik-Gerber, B., Rice, S.: The perceived value of building information modeling in the U.S. building industry. J. Inf. Technol. Constr. (ITcon) **15**(15), 185–201 (2010)

Bernstein, H.M., et al.: The Business Value of BIM for Construction in Major Global Markets. McGraw Hill Construction, Bedford (2014)

Bernstein, H.M., Laquidara-Carr, D.: Project Delivery Systems: How they Impact Efficiency and Profitability in the Buildings Sector. McGraw Hill Construction, Bedford (2014)

BSI. PAS 1192-2:2013: Specification for information management for the capital and delivery phase of construction projects using BIM, London W4 4AL, British Standards Institution (2013)

Eadie, R., Browne, M., Odeyinka, H., Mckeown, C., Mcniff, S.: BIM implementation throughout the UK construction project lifecycle: an analysis. Autom. Constr. **36**, 145–151 (2013)

Eastman, C., Teicholz, P., Sacks, R., Liston, K.: BIM Handbook: A Guide to Building Information Modeling for Owners, Managers, Designers, Engineers, and Contractors, 2nd edn. Wiley, Hoboken (2011)

Faria, D.R.G., Barros, M.M.S.B., Santos, E.T.: Proposição de um protocolo para contratação de projetos em BIM para o mercado da construção civil nacional. In: Encontro Nacional de Tecnologia do Ambiente Construído, 16, São Paulo (2016)

Gurevich, U., Sacks, R., Shrestha, P.: BIM adoption by public facility agencies: impacts on occupant value. Build. Res. Inf. **45**(6), 610–630 (2017)

Hardin, B., Mccool, D.: BIM and Construction Management: Proven Tools, Methods, and Workflows, 2nd edn. Wiley, Indianopolis (2015)

Kassem, M., Succar, B., Dawood, N.: Building information modeling: analyzing noteworthy publications of eight countries using a knowledge content taxonomy. In: Issa, R.R.A., Olbina, S. (ed.) Building Information Modeling: Applications and Practices, pp. 329–371. American Society of Civil Engineers, Reston (2015)

Kassem, M., Succar, B.: Macro BIM adoption: comparative market analysis. Autom. Constr. **81**, 286–299 (2017)

Porwal, A., Hewage, K.N.: Building information modeling (BIM) partnering framework for public construction projects. Autom. Constr. **31**, 204–214 (2013)

Sacks, R., Gurevich, U., Shrestha, P.: A review of building information modeling protocols, guides and standards for large construction clients. J. Inf. Technol. Constr. (ITcon) **21**, 479–503 (2016)

A Practice-Based Conceptual Model on Building Information Modelling (BIM) Benefits Realisation

Thayla Zomer[1,2](✉), Andy Neely[1,2], Rafael Sacks[1,3], and Ajith Parlikad[1,2]

[1] Centre for Digital Built Britain, University of Cambridge, Cambridge, UK
ttdsz2@cam.ac.uk
[2] Institute for Manufacturing, University of Cambridge, Cambridge, UK
[3] Technion - Israel Institute of Technology, Haifa, Israel

Abstract. Most of the existing research on BIM implementation and benefits realisation has departed from a technocentric perspective and instead promoted the belief that the benefits of using BIM technologies are an inevitable outcome of adoption in and of itself. Recently, scholars have acknowledged that important links between implementation practices and realisation of benefits have received little attention in the literature. There have been calls for more research taking a more detailed account of how outcomes can be achieved. Thus, by adopting a practice-based perspective as a theoretical lens, in this paper we investigate the 'what', 'who' and 'how' of successful BIM processes implementation. We propose a conceptual framework on the underlying conditions of successful implementation and assert that the achievement of benefits is dependent on the interaction of those three aspects. Building on qualitative data from nine construction projects from three client organisations in the UK, we show that the relationship between BIM implementation as a set of technologies and processes and performance cannot be understood without taking into account not only 'what' new processes exist in a BIM project and its interdependencies but also 'who' implements and engages with them and 'how' existing structures are reconfigured when enacting those practices. In alignment with recent research challenging the perceptions of BIM enactment as a linear process, our findings provide new insights into why the proclaimed benefits of BIM have not always been realised as an outcome of a 'symbolic' implementation of information management processes and lack of reconfiguration of existing institutions.

Keywords: Building Information Modelling · Benefits · Project performance

1 Introduction

Building Information Modelling (BIM) has become a subject of intense study in construction management research, as its use in construction projects has become increasingly popular (Papadonikolaki et al. 2019). In the last decade, there has been growing attention from both academia and industry towards the potential benefits of

E. Toledo Santos and S. Scheer (Eds.): ICCCBE 2020, LNCE 98, pp. 409–424, 2021.
https://doi.org/10.1007/978-3-030-51295-8_29

BIM (Ahankoob et al. 2019). Extant research has investigated the perception of industry practitioners regarding the impact of BIM on project performance (Zuppa et al. 2009; Smits et al. 2017), analysed the benefits of BIM in specific projects and/or compared BIM and non-BIM projects (Barlish and Sullivan 2012; Bryde et al. 2013; Poirier et al. 2015), or quantitatively analysed the impact of BIM uses (technologies) on project performance (Franz and Messner 2019). Additionally, another line of research has focused on proposing approaches for measuring those benefits (e.g., Yang and Chou 2019). Indeed, the realisation of BIM benefits has become an extensively researched area (Dowsett and Harty 2019).

Existing impact literature, however, while beneficial in promoting technological change, also promotes the belief that the benefits of using BIM technologies are an inevitable outcome of adoption in and of itself (Dowsett and Harty 2019). Moreover, most of the existing research on benefits realisation has been designed to test macro-level relationships and omits many of the details about the specific processes or implementation aspects Franz and Messner (2019). Important links between implementation practices and realisation of benefits have received little focus in the existing literature (Dowsett and Harty 2019). Indeed, studies on benefits realisation tend to adopt a normative perspective, and recent studies have recognised that those normative frameworks need to be complemented with more practically oriented approaches, which would allow learning across organisational and project boundaries and provide a more detailed account of how outcomes are achieved (Miettinen and Paavola 2014).

Therefore, by drawing on practice-based theory (Jarzabkowski et al. 2016), we argue that partial models that focus only on isolated, technocentric and macro-level aspects of BIM implementation, such as BIM uses (technology uses), are liable to misattribute performance effects. As proposed by practice theory, we suggest that understanding performance effects of BIM requires the adoption of a micro-level perspective to explore 'how' BIM processes are enacted in practice and in context (Jarzabkowski et al. 2016)—that is, how the new workflows associated with information management are enacted by project practitioners (who engage in those practices). Thus, in this paper, we investigate the 'what', 'who' and 'how' of BIM implementation and propose a conceptual model of the underlying conditions of successful BIM implementation. We assert that the achievement of benefits is dependent on the interaction of those three aspects.

By adopting a practice lens to investigate BIM benefits realisation, this paper contributes to both project management and BIM literature. First, although there have been some studies that adopt a practice-based lens and then transcend the dominant normative perspective (Kokkonen and Alin 2016; Nørkjaer et al. 2019), contributing to the calls for more research adopting a more realistic view of BIM implementation (Miettinen and Paavola 2014), there is scarce practice-based research so far on performance. The adoption of a practice-based perspective represents an important theme for rethinking project management research, and there have been calls for more studies investigating the link between practice and project performance (Svejvig and Andersen 2015). Second, while there has been considerable focus on 'why' construction firms need to adopt digital technologies and thus deliver assets more effectively and efficiently, there has been limited attention paid to 'how' they can realise their expected benefits and generate value (Love and Matthews 2019). In this paper, we build an integrative model in which the complex links between new processes associated with BIM implementation—the ways in which they are engaged (how), who engages with them and their potential outcomes—can be fully recognised. By exploring those links, we offer new insights into the successful implementation of BIM.

The paper is structured as follows. First, an overview of the benefits assessment literature is presented, summarising the main research approaches and existing gaps. The theoretical approach is then described, introducing Jarzabkowski et al.'s (2016) model from management research as the analytical lens used as the basis to develop the conceptual model proposed in this paper. Afterwards, the research method is presented, followed by an analysis of the entangled elements proposed in the conceptual model. The paper ends with a discussion on its contributions and limitations as well as suggestions for further research.

2 Background

BIM can be defined as a set of interacting policies, processes and technologies to manage project data in digital format throughout a project's life cycle (Succar 2009). BIM is often perceived as a solution for many inefficiencies in the construction industry (Zuppa et al. 2009). Accurately measuring the influence of BIM implementation on project performance, however, is a difficult task (Franz and Messner 2019). Most of the existing studies on BIM implementation and performance have adopted either surveys with project practitioners and their perceptions on the benefits and/or conducted case studies on benefits for individual projects. Some examples of existing empirical research include Smits et al. (2017), who conducted a survey of Dutch organisations and analysed the impact of BIM maturity (strategy, BIM uses, process, information, infrastructure and personnel) on key project performance indicators (time, cost and quality performance). Their findings demonstrated that only the maturity of the BIM strategy (i.e., mission, vision, goals and objectives, along with management support, BIM champions and BIM planning committee) is a predictor of time, cost and quality performance.

Examples of case study research include Barlish and Sullivan (2012), for example, who developed a framework calculation model to determine the value of BIM and applied it in case studies within a large industrial setting where similar projects could be evaluated (some projects implementing BIM and some using traditional approaches). Another example includes Bryde et al. (2013), who explored the extent to which the use of BIM has resulted in reported benefits on a cross section of 35 construction projects. There have also been case studies on non-achievement of expected benefits (Love et al. 2017).

Despite the contributions of these studies in advancing the understanding of the potential of BIM to improve industry productivity, overall, they tend to adopt technocentric conceptions of BIM (Dowsett and Harty 2019). In general, studies espousing the gains that can be acquired from digital technologies tend to overplay the expected benefits and downplay the change that is needed for them to be fully realised (Love and Matthews 2019). Successful BIM implementation and the realisation of its full potential require changes in work practices (Kokkonen and Alin 2016; Sacks et al.

2016). Past research has posited that intra- and inter-organisational changes (e.g., new workflows and new roles, and changes to procurement, contracts and team organisation) are required to successfully implement BIM. However, critics argue that such claims lack a grounding in theory, pointing out that existing studies only vaguely explore the real nature of the change required for BIM implementation (Çidik et al. 2017; Akintola et al. 2019). Changing working practices in construction projects is challenging (Kokkonen and Alin 2016), so it is therefore important to understand the practical aspects related to BIM and its successful implementation (Nørkjaer Gade et al. 2019).

As previously described, in this research we adopt a practice-based view of BIM process implementation, a more granular approach recently addressed in management research to explain performance heterogeneity among firms (Jarzabkowski et al. 2016). From a practice-based perspective, one could argue that, currently, attention in the BIM literature has been mostly placed on the 'what' of technologies and information management processes implementation. Indeed, the 'how' of implementation has been overlooked, and the micro-level changes that are often required have largely been ignored (Love and Matthews 2019). By adopting a practice-based framework, we posit in this study that closer attention should also be paid to the micro-level enactment of processes and practices in BIM projects, in order to understand the conditions under which the full potential of BIM can be realised.

2.1 A Practice Perspective

BIM is a holistic concept and comprises not only technologies but also processes to manage project data in digital format throughout a project's life cycle (Succar 2009). The implementation of new BIM processes represents a change at the inter-organisational level, impacting inter-firm processes and requiring those processes to realign (Alin et al. 2013).

From a practice lens, performance differentials related to the adoption of new processes or activities, such as the information management process of BIM, can only be understood if three related elements are considered: the 'what', 'how' and 'who' of practices (Jarzabkowski et al. 2016). The 'what' of BIM represents the technologies and information management processes embedded in projects and established, for instance, by policy mandates and frameworks such as the BIM Level 2 suite of standards as established in the UK (BSI 2019). The implementation of new information management processes and practices, however, does not occur automatically and unproblematically. Rather, they are enacted in context and require the consideration of 'how' they are enacted (Jarzabkowski et al. 2016). Information management processes and practices are also developed, transferred and enacted by practitioners, thus requiring the consideration of who is engaged with the practices. Such a practice perspective also emphasises that outcomes depend on the interaction of those three elements (Jarzabkowski et al. 2016). This leads us to propose an integrated model that can explain performance heterogeneity of BIM projects, built on real BIM project data, as described in more detail in the next section.

3 Methods

Given the purpose of analysing the links between implementation of BIM as a set of new processes and technologies and outcomes, we adopted a qualitative method, which is also strongly associated with the practice-turn lens (van der Hoorn and Whitty 2017).

Rich and detailed data is usually required when adopting a practice-based perspective. We conducted nine case studies on BIM implementation in projects of three different client organisations—herein called organisation A (six projects), organisation B (one project) and organisation C (two projects). The context of analysis is the UK, as the UK has been recognised as one of the leading countries on BIM adoption (Georgiadou 2019) and provides a representative context to explore the implementation of new processes in projects and the wider industry transformation.

The 2011 UK government construction strategy set a target to introduce a progressive programme of mandated use of fully collaborative BIM (BIM Level 2) for government projects by 2016, a policy shift that other countries have also begun to emulate (Dainty et al. 2017). As part of the policy shift, a framework (PAS 1192 documents) was developed, setting out the requirements for achieving BIM Level 2 by establishing a framework for collaborative work and information requirements. The PAS 1192 establishes the requirements for the level of model detail (the graphical content), model information (non-graphical content, such as specification data), model definition (its meaning) and model information exchanges (NBS 2019). The three analysed client organisations have implemented BIM Level 2 and the related PAS 1192 standards and were identified as representative projects to investigate the implementation of new BIM processes.

We selected institutional construction projects (schools and university buildings), as BIM implementation is mature and disseminated among institutional projects, given clients' requirements (government mandate). We also selected similar projects in terms of complexity, aiming to compare the implementation approach. The selected projects were at different stages of their life cycles at the time of data collection. Exploratory interviews (open-ended questions), attendance and observations of BIM meetings and workshops, document analysis (project documentation) and longitudinal investigation of the projects were the data collection methods applied to learn from BIM implementation within the projects. All data collected in the form of interviews and notes taken while observing meetings and workshops were transcribed. An overview of the interviews conducted with project stakeholders is presented in Table 1.

The collected data was triangulated, and a content analysis was carried out to understand the enactment of the new practices related to the BIM Level 2 along with the use of new technologies. Following the practice-based framework proposed by Jarzabkowski et al. (2016), we analysed the 'what', 'how' and 'who' of BIM processes enactment as recommended by the PAS 1192 suite of standards. The content analysis consisted of three main stages (Gioia et al. 2012): i) open coding of the transcribed data, ii) creation of second-order themes, and iii) development of aggregated conceptual themes.

In the data analysis, we coded elements of processes' enactment that were related to the achievement of improved performance, linking the 'what', 'how' and 'who' of enactment. In the first-order analysis, we tried to adhere to informant and data terms, allowing key themes to emerge from the data. We then looked at similarities and differences among the themes and looked for similarities and differences across the cases. We clustered the categories that were similar and for which there were cross-case patterns into second-order themes. Then, in the second-order analysis, we moved to the theoretical realm and looked at distilling the emergent second-order themes into aggregate dimensions suggesting concepts that help explain the phenomena, specially the 'how' of new information management processes and how they should be enacted in order to achieve the envisaged outcomes. When exploring the theoretical realm, we looked at the management literature and adopted concepts from institutional theory (Scott 2014) to link the change process in projects following BIM implementation to practitioners' enactment. To illustrate how conclusions have been reached when analysing the data, categorising the data and developing the conceptual model, we used vignettes from the data, similar to previous research (Jacobsson et al. 2017).

Table 1. Interviews conducted with project stakeholders.

Organisation	Project	Interviews	Life cycle stage (at the moment of data collection)	Interview duration (total)
Organisation A	Project 1	Project manager (client), project manager (contractor), facilities manager, information manager (client)	Technical design	6 h 40 m
	Project 2	Project manager (client), project manager (contractor), information manager (contractor)	Handover	2 h 25 m
	Project 3	Project manager (client), project manager (contractor), information manager (contractor), architects and workshop with all stakeholders	Construction	6 h 40 m
	Project 4	Project manager (client), BIM coordinator (contractor)	Operation (project completed)	2 h 20 m
	Project 5	Digital engineer (contractor)	Handover	45 m
	Project 6	Project managers (client), information manager (client), facilities managers	Concept design	3 h 55 m
Organisation B	Project 1	Client's technical advisor (TA), project manager (client), project manager and information manager (contractor), facilities operator, CAFM provider and individual responsible for the digital transformation strategy (client)	Construction	11 h

(*continued*)

Table 1. (*continued*)

Organisation	Project	Interviews	Life cycle stage (at the moment of data collection)	Interview duration (total)
Organisation C	Project 1	Project manager (client), information manager (client), architect, facilities manager (client), BIM coordinator (contractor)	Operation (project completed)	2 h 49 m
	Project 2	Project manager (client), information manager (client), facilities manager (client), BIM coordinator (contractor)	Construction	3 h

4 An Integrated Practice Perspective on Successful BIM Implementation

An integrated practice-based approach to BIM implementation is presented in Fig. 1. As previously mentioned, BIM can be seen as a combination of processes and technologies, and the implementation of new information management processes and practices is considered in this study. Our case study data revealed that information management processes/practices, the practitioners involved in the implementation of those processes/practices and how the implementation takes place (the enactment of new processes and reconfiguration of existing institutional aspects) are entangled and related to the achievement of the envisaged goals.

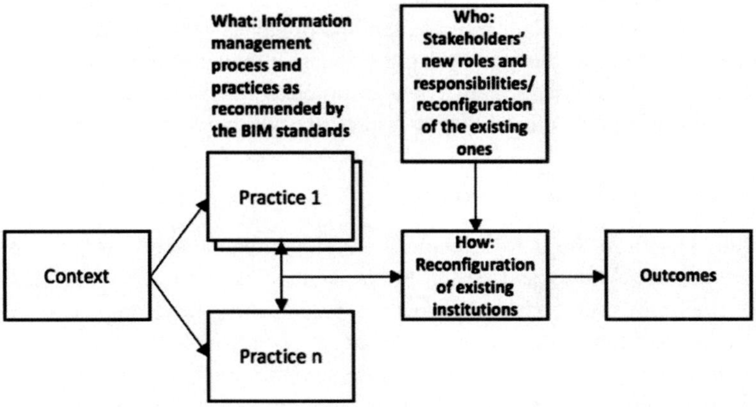

Fig. 1. A practice-based schematic model of successful BIM implementation.

4.1 What BIM Practices?

As mentioned earlier, in the context under analysis (UK), the 'what' aspect refers to the processes/practices that are taken up within BIM implementation as defined by the PAS 1192 suite of standards. Those practices include, for example, the definition of the information exchange and collaborative working requirements at early stages of the project that should be undertaken in parallel with other procurement and project definition activities. The context of the project/delivery approach provides the basis for which the information management processes and practices are enacted.

The first aspect observed regarding the adoption of information management processes and practices is that there are interdependencies between practices, as shown in Fig. 1. Examples of such interdependency and combined effect on performance could be observed, for instance, regarding the definition of the employer information requirements (EIR) by the client in the early stages of the project and the handover of information at the end of the project. Project stakeholders highlighted, for example, that a poor definition of a handover process in the early stages (as part of the definition of information requirements) led to issues later on when the information models had to be handed over:

'We weren't given a handover strategy in the EIR, and we still haven't been given one. So it's a bit of a... I wouldn't like to say a free-for-all, but it's very different, in places, using characters and stuff, or spaces or underscores. It's still being used very haphazardly, so it can be a bit random why some things are named differently than others. The designers might have tagged things in the first instance, and that's just been carried all the way through. It ended up that there was a lot of tagging to be done right at the end, when, really, it should've been done further back, and then you can just carry it through'.

By considering that there is interdependency between processes and practices, misattribution of the performance effects of any single practice, as it has been considered in the existing literature regarding the use of isolated technologies, should be considered. Moreover, the data revealed that the practices are also strongly shaped by the practitioners who perform and advocate them. Inferences about the relationships between practices and performance are incomplete, therefore, if the practitioners involved and the varying effects they can have are not taken into account.

4.2 Who Engages in BIM Practices?

It is acknowledged in the BIM literature that BIM implementation requires reconfiguration of power relations and new roles (Aksenova et al. 2019). The practice perspective emphasises that characteristics of practitioners, such as their roles and organisational positions, have strong implications for practices use and implementation (Jarzabkowski et al. 2016). The case study data confirmed that not only do new roles need to be established in a BIM project for practitioners to engage in practices, such as the appointment of a party to undertake the information management function, but project stakeholders' existing roles and responsibilities also need to be modified and updated. The data has shown that the engagement of the right stakeholders in each activity is essential to achieve the envisaged outcomes.

From an organisational perspective, the client's authority system and internal division of labour in organisation A, for instance, led to the involvement of only some information users in the definition of information requirements, as noted by a digital engineer for one of the projects when explaining what went wrong:

'You need to get the whole team agreeing on things. When I say the team, I mean the whole team, not just, maybe, the people on the project team. I think you have to think about the building as a whole life cycle, and someone at the project level might have a completely different opinion on what they want someone on the facilities side, and maybe even someone in the technical aspects'.

Despite having an information manager in place and having defined the information requirements (enacted the activity as recommended by the standards), the non-involvement of all information users influenced the outcomes at the end of the project, as highlighted by a project manager for another organisation's A project:

'Towards the end, the facilities team got more involved, and their requirements were different. What they've asked for has changed, because more people have become involved. If we had, maybe, been stricter with asking those questions at the beginning, we maybe wouldn't have had a lot of change towards the end'.

Traditionally, the facility manager's involvement in the asset data-management process commences at the handover phase (Alnaggar and Pitt 2019), and the existing roles have not been modified in the analysed projects, impacting the enactment of the practices and the achievement of the envisaged outcomes.

The client's information manager for organisation C's projects also highlighted that stakeholders' existing roles and responsibilities need to change—for instance, in the case of facilities managers who need to get involved in setting up the correct data format structure to consequently avoid producing information in the wrong format. This aligns with previous research that has identified that BIM implementation involves the creation of new work tasks (and consequently the establishment of new roles) and changing specialists' roles to become, for instance, responsible for a work task that is new to that specialist (Alin et al. 2013).

4.3 How Should Information Management Processes Be Practiced?

As Fig. 1 shows, the data revealed that what project stakeholders actually do in practice (i.e., how they perform activities) mediates the achievement of the envisaged outcomes following the implementation of information management processes and practices according to the PAS 1192 standards.

The findings revealed that when implementing BIM, project members used their existing knowledge to make sense of the policy message and the new processes that needed to be implemented, which led them to assimilate new knowledge of the standards and processes into existing ways of doing things and reproduction of habitual dispositions, routines, procedures, power systems, etc. The findings revealed that a reconfiguration of existing institutions (i.e., regulative, cognitive and normative) (Scott 2014) is necessary when enacting BIM processes and that the reproduction of existing institutions has impacted the achievement of envisaged outcomes, as explained next.

Reconfiguration of Cultural-Cognitive Rules

The analysis of how project stakeholders enacted the new process as recommended by the PAS 1192 suite of standards revealed that project team members have repeated habits and procedures based on existing knowledge and beliefs. Previous research on innovation in construction projects has already identified that the socio-cognitive environment or the broader belief system within the architecture, engineering and construction (AEC) industry shapes the beliefs and ideas of individual actors and influences their actions and the adoption and use of new technologies (Jacobsson et al. 2017). Indeed, as pointed out by Scott (2014), the internal interpretive processes of project stakeholders are shaped by external cultural frameworks, and a lack of reconfiguration of organisational and industry frameworks was identified as a cause of decoupling between the enactment of new practices and achievement of the envisaged outcomes.

The data revealed that existing knowledge, for instance, was replicated. In an interview with the lead contractor for one of organisation C's projects, for example, the contractor noted that the architect relied heavily on traditional ways of working and knowledge, despite having the skills to work with new BIM tools, which created issues with consistency during the project and led to rework:

'We need drawings for our sign-off, for example, and they are usually extracted from the BIM model. Sometimes, after it's been extracted from the model, the architect does some changes in the previously used software as well, but they are not reflected on the model. This is a huge issue for us because we need everything to be consistent. We couldn't really rely on the model but then we were still having to double check the drawings to make sure that everything was working okay. It did slow the process down a hell of a lot. Rather than just producing the drawings from the new software and things being accurate, they were still exporting those views from the new software, going into the old software'.

Indeed, as pointed out in the institutional theory literature, it is often difficult for established firms and organisations to implement new knowledge, and learning often builds upon existing knowledge (Geels 2004).

Moreover, the cultural-cognitive conception of institutions also stresses the role played by the socially mediated construction of a common framework of meanings and emphasises the role of shared understandings and professional ideologies in conditioning how organisational actors interpret and take actions (Scott 2014). For organisation A's projects, for example, contractors noted that they had to amend the models to be able to use the models. The lead contractors in the interviews highlighted that designers have their design 'ethos' or their dominant assumptions and practices, and that they rely on their general knowledge and replicating those practices in a BIM project, without considering further project stages and the wider implications to the project.

In summary, shared belief systems, expectations about the physical asset and deeply engrained habits and procedures based on existing knowledge were found to orient perceptions and steer actions of project stakeholders. Deeply engrained habits were also followed when exchanging information. According to the main contractor for one of organisation A's projects, although subcontractors were using a common data environment for information exchange in the analysed project and in projects in general, they relied on their general habits and exchanging information in both ways:

'I'm finding this across the industry as a whole. It's difficult to get subconsultants out of the habit of just sharing drawings before uploading onto a system'.

If the new information management processes and practices are not enacted in the way they are supposed to be enacted and by the person who should be engaged in its enactment, the expected outcomes cannot be achieved. The previously discussed examples of reproduction of taken-for-granted habits as part of the stable social system when enacting BIM have also been reinforced because they are normatively endorsed and backed by existing normative and regulative rules, which also needs to be reconfigured, as described next.

Reconfiguration of Normative and Regulative Rules

Existing organisational and industry's normative elements, such as authority systems and codes of conduct, were also being reproduced and influencing the achievement of the expected outcomes. In organisation B, for instance, given the established codes of conduct when selecting contractors, BIM capabilities were not taken as criteria for selection of the supply chain, as noted by the person responsible for the implementation of the BIM strategy within the client organisation:

'We get criticised if we don't allow smaller contractors into our frameworks. We didn't want BIM to be a pass/fail kind of thing'.

The non-consideration of BIM capabilities when appointing the lead contractors increased the risk of not having the capabilities necessary to develop information models with appropriate quality and suitable to support decision-making across the whole life cycle of the asset.

Moreover, the findings have also shown that existing regulative elements—reward and cost structures, governance systems, power systems, procedures, etc.—have been reproduced when enacting new information management processes. In terms of rewards and cost structures, for example, contractors noted that because of the dominant cost and reward structure in the AEC industry and the way that subcontractors get paid on projects, they tend to perform work as quickly as possible, and consequently have not changed their work practices to use the information models to support decision-making on site:

'A lot of the actual decisions as to exactly where, in a particular space, something is installed is down to the tradesmen on the day. And they really want to do it as quick and simple as possible. Because that is how they get paid'.

Because of the traditional cost and reward structures, lead appointed parties also tend to be involved only in the specific life cycle stage for which they provide services. Keeping the current reward and cost structure of contracts in a BIM project limits the adoption of an integrated and holistic approach, which may result in rework, as pointed out by a contractor:

'We need to have the designers more involved, but I'm not sure how you would achieve that, because they tend to only get money up to a certain point. That's just the way the contract works. So naturally, they're less inclined to continue with the same involvement that they have previously done'.

The current workflows and formal rules at an organisational level have also not been changed for organisation A's projects. The workflow for change management, for instance, was followed, and multiple changes had to be made in the 3D models as a consequence, representing rework to the main contractor:

'We found the user change on this job, the design cost was four or five times what it was on this project. Some of that was probably because there was a lot more change, it was a lot more detailed, we went through the loop several times. It might just have been to do with the process'.

Another example of organisational workflows and procedures that have not been reconfigured and thus resulted in the enactment of processes that did not necessarily lead to the envisaged outcomes includes organisation B's requirements specification procedure, which has been followed in the definition of information requirements as well, as pointed out by one of the building's final user:

'I suspect that one of the lessons learned from all of this would be how the client and the end user in this type of construction project can work together earlier to look at what potential future asset management requirements the final user might need, and how that could feed into the client's decision on the specification of the building, because a lot of the time, just speaking from the experience I'm going through at the moment, it's very much organisation B is the client, we are the end user and the stakeholder; organisation B makes the decisions on the specifications, not us'.

Moreover, the governance system employed by organisation B when selecting contractors was not updated to consider the appointment of parties taking into account their BIM capabilities and assessment of those capabilities, as noted by the person responsible for the implementation of the digital transformation strategy in organisation B:

'When we put a BIM question into our evaluation matrix, are we actually asking people who don't have the skill sets to actually evaluate the answers to that question? As mentioned earlier, there are an awful lot of people who are evaluating those questions. You have to expect a level of competence for somebody who is scoring that, and I don't think that's necessarily there'.

In summary, project stakeholders and the way they enact the information management process and the necessary reconfiguration of existing structure mediates the implementation of BIM processes and the realisation of the envisaged outcomes. The practice perspective, therefore, sheds light on a more realistic view on the conditions of BIM implementation, building on extant literature, as discussed next.

5 Discussion

One of the key findings of this study is that BIM (i.e., the respective information management processes and technologies) are rarely blueprints that will be implemented in a project context unproblematically, leading to improved project performance. Project stakeholders mediate the effect of the implementation of new information management processes and technologies by constructing and reconstructing the practical implementation through the lens of their pre-existing practices and knowledge.

Previous research examining the impact of embedded contexts on innovation and BIM implementation has already identified that the institutional context (defined as the practices, policies and procedures implemented by the various stakeholders in the AEC supply chain) hinder the BIM adoption and implementation process (Poirier et al. 2015). Poirier et al. (2015) also posit that education and training, as well as the establishment of codes of practice by professional associations, are the primary means through which the influence of the institutional context can be mitigated. Our findings actually show that even when codes of practice are established, such as in the case under analysis and the PAS 1192 standards, project stakeholders might still draw upon existing normative, regulative and cultural-cognitive institutions and reproduce those institutions, influencing the achievement of outcomes envisaged by the implementation of the standards. Other recent studies have also shown that the implementation of technologies in construction firms is facilitated by a mutually constitutive relationship between user, firm and institution (Morgan 2019), but have not revealed the mechanisms through which the existing structure might influence the implementation and respective outcomes.

Moreover, while there have been studies taking into account specific aspects and dimensions of those institutions, such as the socio-cognitive environment (Jacobsson et al. 2017), those have focused, for instance, on the influence of the socio-cognitive environment in the adoption and acceptance of technologies (Jacobsson et al. 2017). Jacobsson et al.'s (2017) findings show that when ICT applications are interpreted in the socio-cognitive environment of the construction industry, some of their features align better than others in the way that the industry current functions. Our findings extend the existing research in two ways. First, they show that existing institutions not only influence the acceptance of new technologies and practices but also how the implementation takes place. Second, our findings show that project stakeholders mediate the implementation of practices and realisation of outcomes in the sense that they can enact the practices and reconfigure existing institutions—such as the socio-cognitive environment—or not, impacting the achievement of the envisaged outcomes. We also take a more holistic view of institutions than previous studies by considering normative, regulative and cultural-cognitive institutions.

Furthermore, the extant literature on the changes driven by BIM has mostly adopted a deterministic approach (Dowsett and Harty 2019). The deterministic change agendas that have surrounded the industry often fail to account for the structural challenges related to practical implementation. There are still only limited studies examining the 'real world' implementation of BIM in organisations and projects (Dainty et al. 2017). In alignment with recent research challenging the perceptions of BIM enactment as a linear process (Dainty et al. 2017), our findings from a practice-based and institutional theories perspective connect the macro- (policy and mandate on BIM standards implementation) and micro-levels (implementation in projects) and provide new insights on the complex conception of a mandate impact and wider transformation within the industry. Our findings also offer new insights into why the proclaimed benefits of BIM have not always been realised, as an outcome of a 'symbolic' implementation of information management processes and lack of reconfiguration of existing institutions, including existing norms, codes of conduct and procedures.

6 Conclusion

Since scholars have acknowledged the need for more research with more realistic views on the conditions for BIM implementation, studies applying social theories have started to emerge in the literature. Less focus, however, has been placed on understanding successful implementation of innovations in construction projects through the lens of social theories. The practice-based approach adopted in this study revealed that achievement of the expected outcomes envisaged by the implementation of BIM as an innovation depends on the stakeholders' engagement with the new implemented processes and uses of technologies and how those enact the transformation of existing cultural-cognitive, regulative and normative rules. The practice-based lens adopted in this research, therefore, sheds light on other aspects of successful implementation not yet discussed in the literature, or found as not relating to performance. Previous research, for instance, has not found a relationship between having processes in place and achievement of outcomes, but has posited that process maturity is important and deserves more attention. Our findings reveal practical aspects of process implementation that mediate the implementation of those processes and realisation of benefits.

This study, however, is limited to the investigation of a small sample of projects. The causal relationship between enactment of BIM processes and respective reconfiguration or non-reconfiguration of existing institutions and improved performance would need to be quantitatively tested in a larger sample of BIM projects, which is the next step of this research.

References

Ahankoob, A., Manley, K., Abbasnejad, B.: The role of contractors' building information modelling (BIM) experience in realising the potential values of BIM. Int. J. Constr. Manag. (2019, in press). https://doi.org/10.1080/15623599.2019.1639126

Akintola, A., Venkatachalam, S., Root, D.: Understanding BIM's impact on professional work practices using activity theory. Constr. Manag. Econ. (2019, in press)

Aksenova, G., Kiviniemi, A., Kocaturk, T., Lejeune, A.: From finnish AEC knowledge ecosystem to business ecosystem lessons learned from the national deployment of BIM. Constr. Manag. Econ. **37**(6), 317–335 (2019)

Alnaggar, A., Pitt, M.: Towards a conceptual framework to manage BIM/COBie asset data using a standard project management methodology. J. Facil. Manag. **17**(2), 175–187 (2019)

Alin, P., Maunula, A.O., Taylor, J.E., Smeds, R.: Aligning misaligned systemic innovations probing inter-firm effects development in project networks. Proj. Manag. J. **44**(1), 77–93 (2013)

Barlish, K., Sullivan, K.: How to measure the benefits of BIM - a case study approach. Autom. Constr. **24**, 149–159 (2012)

Bryde, D., Broquetas, M., Volm, J.M.: The project benefits of building information modelling (BIM). Int. J. Proj. Manag. **31**(7), 971–980 (2013)

British Standards Institutions. BIM Level 2 (2019). https://bim-level2.org/en/. Accessed 20 Feb 2020

Çıdık, M.S., Boyd, D., Thurairajah, N.: Ordering in disguise: digital integration in built-environment practices. Build. Res. Inf. **45**(6), 665–680 (2017)

Dainty, A., Leiringer, R., Fernie, S., Harty, C.: BIM and the small construction firm a critical perspective. Build. Res. Inf. **45**(6), 696–709 (2017)
Dowsett, R.M., Harty, C.F.: Assessing the implementation of BIM – an information systems approach. Constr. Manag. Econ. **37**(10), 551–566 (2019)
Franz, B., Messner, J.: Evaluating the impact of building information modelling on project performance. J. Comput. Civil Eng. **33**(3), 04019015 (2019)
Geels, F.W.: From sectoral systems of innovation to socio-technical systems insights about dynamics and change from sociology and institutional theory. Res. Policy **33**(6), 897–920 (2004)
Georgiadou, M.C.: An overview of benefits and challenges of building information modelling (BIM) adoption in UK residential projects. Constr. Innov. **19**(3), 298–320 (2019)
Gioia, D.A., Corley, K.G., Hamilton, A.L.: Seeking qualitative rigor in inductive research: notes on the Gioia methodology. Organ. Res. Methods **16**(1), 15–31 (2012)
Jarzabkowski, P., Kaplan, S., Seidl, D., Whittington, R.: On the risk of studying practices in isolation linking what, who, and how in strategy research. Strateg. Organ. **14**(3), 248–259 (2016)
Jacobsson, M., Linderoth, H.C.J., Rowlinson, S.: The role of industry: an analytical framework to understand ICT transformation within the AEC industry. Constr. Manag. Econ. **35**(10), 611–626 (2017)
Kokkonen, A., Alin, P.: Practitioners deconstructing and reconstructing practices when responding to the implementation of BIM. Constr. Manag. Econ. **34**(7–8), 578–591 (2016)
Love, P.E.D., Zhou, J., Matthews, J., Sing, M.C.P., Edwards, D.J.: System information modelling in practice: analysis of tender documentation quality in a mining mega-project. Autom. Constr. **84**, 176–183 (2017)
Love, P.E.D., Matthews, J.: The 'how' of benefits management for digital technology from engineering to asset management. Autom. Constr. **107**, 1–15 (2019)
Miettinen, R., Paavola, S.: Beyond the BIM utopia approaches to the development and implementation of building information modelling. Autom. Constr. **43**, 84–91 (2014)
Morgan, B.: Organizing for digitalization through mutual constitution: the case of a design firm. Constr. Manag. Econ. **37**(7), 400–417 (2019)
National Building Specification (2019). https://www.thenbs.com/knowledge/what-is-the-pas-1192-framework. Accessed 22 Jan 2020
Nørkjaer Gade, P., Nørkjaer Gade, A., Otrel-Cass, K., Svidt, K.: A holistic analysis of a BIM-mediated building design process using activity theory. Constr. Manag. Econ. **37**(6), 336–350 (2019)
Papadonikolaki, E., van Oel, C., Kagioglou, M.: Organising and managing boundaries a structurational view of collaboration with building information modelling (BIM). Int. J. Project Manag. **37**(3), 378–394 (2019)
Poirier, E., Staub-French, S., Forgues, D.: Embedded contexts of innovation: BIM adoption and implementation for a specialty contracting SME. Constr. Innov. **15**(1), 42–65 (2015)
Sacks, R., Gurevich, U., Shrestha, P.: Review of national standards and organizational guides for BIM adoption. J. Inf. Technol. Constr. (ITCon) **21**, 479–503 (2016)
Scott, W.R.: Institutions and Organisations, 4th edn. Sage, California (2014)
Smits, W., van Buiten, M., Hartmann, T.: Yield-to-BIM: impacts of BIM maturity on project performance. Build. Res. Inf. **45**(3), 336–346 (2017)
Succar, B.: Building information modelling framework a research and delivery foundation for industry stakeholders. Autom. Constr. **18**(3), 357–375 (2009)
Svejvig, P., Andersen, P.: Rethinking project management a structured literature review with a critical look at the brave new world. Int. J. Proj. Manag. **33**(2), 278–290 (2015)

van der Hoorn, B., Whitty, S.J.: The praxis of 'alignment seeking' in project work. Int. J. Proj. Manag. **35**(6), 978–993 (2017)

Yang, J.B., Chou, H.Y.: Subjective benefit evaluation model for immature BIM-enabled stakeholders. Autom. Constr. **106**, 1–12 (2019)

Zuppa, D., Issa, R.R.A., Suermann, P.C.: BIM's impact on the success measures of construction projects. In: Computing in Civil Engineering, pp. 503–512 (2009)

BIM in Latin American Countries: An Analysis of Regulation Evolution

Fernanda Almeida Machado[1(✉)], Joyce Paula Martín Delatorre[1], and Regina Coeli Ruschel[2]

[1] Rua James Joule, 65, Cidade Monções, São Paulo, SP, Brazil
[2] University of Campinas, Avenida Albert Einstein, 951, Campinas, SP, Brazil
ruschel@unicamp.br

Abstract. Digital transformation in the construction industry has been encouraged by initiatives for macro adoption of information and communication technologies promoted by governments and non-profit institutions in several countries. One of the trends targeted by these initiatives involves the regulation of Building Information Modeling, which already crosses cultural and organizational barriers. Previous macro adoption studies have mapped BIM forecasts and implementations globally, comprising BIM-led digital transformation in more than twenty countries. Although some of these initiatives have already been evaluated, these studies covered countries mainly in North America, Europe, Asia and Australasia. Still, there is a gap in approaches with a focus on the recent BIM macro adoption movement in Latin American countries. This study aims to map and analyze BIM macro adoption initiatives in the region of Latin America, covering the leading public sector roles of each country which BIM is already part of a national strategy. It is an exploratory study based on the survey research method for characterizing BIM macro adoption initiatives in Latin America for the last fifteen years. The authors' research strategy considered the accessible publications available in government and specific institutions official websites in the Portuguese, Spanish, and English languages. These publications were grouped by country and organized in a timeline for highlighting temporal evolution. For content analysis, they were qualified as guides, protocols and mandates. A content matrix was structured. The statuses and main roles of the public sector for each country are presented as well as the characterization of Latin America a bloc.

Keywords: Digital transformation · BIM · Macro adoption · Latin America

1 Introduction

1.1 BIM Macro Adoption

Digital transformation in the construction industry has been encouraged by initiatives for macro adoption of Information and Communication Technologies (ICTs) promoted by governments and nonprofit institutions in several countries [1]. One of the trends targeted by these initiatives involves the regulation of Building Information Modeling (BIM), which already crosses cultural and organizational barriers [2].

E. Toledo Santos and S. Scheer (Eds.): ICCCBE 2020, LNCE 98, pp. 425–451, 2021.
https://doi.org/10.1007/978-3-030-51295-8_30

Beyond the set of tools, processes and policies [3] subsidized by a backbone of information models [4], BIM is characterized as an organizational innovation [5]. This is due to its potential to promote new ideas, processes, products or services [6] within the construction industry. Previous studies recognize that government support and leadership is a critical driver to elevate BIM adoption in n-scales [1, 7–9]. According to [8], BIM usage is accelerating powerfully and globally, driven by government and major private owners. However, such initiatives that ultimately influence BIM adoption at country-level or market-wide scale vary significantly from territory to territory [10].

[11] defines macro adoption as the implementation and diffusion of BIM within a country or across countries denoting a large collection of organizational adopters operating within a defined national border. A set of efforts including academic and industry surveys has been tracking forecasts and macro adoption comprising BIM-led digital transformation in more than twenty countries [12]. Although some of these initiatives have already been evaluated, these studies covered countries mainly in Europe, North America, Asia, and Australasia. Most of the analyses involved macro maturity components [3] or similar such as: objectives, stages and milestones; champions and drivers; regulatory framework and noteworthy publications; learning and education; measurements and benchmark; standardized parts and deliverables; and finally, technology infrastructure.

Still, there is a gap in approaches focused on the recent BIM macro adoption movement in Latin American (Latam) countries, even though we can find overview and/or individual analyses about, e.g., Brazil, Chile, Mexico, and Peru initiatives [13–15]. Due to this context, it is relevant to investigate what is the current state of BIM macro adoption in Latin America and its evolution in the last fifteen years. In this sense, this research aims to map and analyze BIM macro adoption initiatives in the region of Latin America, by covering the leading public sector roles of each country where BIM is already part of a national strategy.

1.2 Similar Studies

There are both previous academic and industry studies that mention individual efforts from Latin American countries in the lens of macro adoption.

BICP Global Studies [13] presented a comprehensive literature review of significant global BIM adoption initiatives in more than twenty countries, with a focus on successful enablers in international regions. Through a well-known world map indicating the status of openBIM standards, BIM mandates and programmes, the study highlighted that at that time Mexico, Peru, and Chile had fixed future mandates: (i) Mexico defined standards for BIM projects starting in 2017; (ii) Chile focused on BIM obligation for government projects starting in 2020; (iii) as well as Peru starting in 2022. In 2017, Brazil had no BIM federal mandate planned. An overview of Brazil and Chile comprising macro adoption components such as regulation, key champions, and noteworthy publications demonstrates that deeper research is necessary in terms of aligning BIM initiatives with the Latin American context.

2 Method

2.1 Research Strategy

This is an exploratory study based on survey research method [16], for characterizing BIM macro adoption initiatives in Latin America in the last fifteen years (2005–2019).

For the sampling of the survey eighteen Latin American countries that integrate the Inter-American Construction Industry Federation (FIIC) were analyzed. These countries are Argentina, Bolivia, Brazil, Chile, Colombia, Costa Rica, Ecuador, Guatemala, Honduras, Mexico, Nicaragua, Panama, Paraguay, Peru, Dominican Republic, El Salvador, Uruguay, and Venezuela (Fig. 1). FIIC is a nonprofit integrating institution with the aim to promote the growth and development of the construction industry in Latin America, as well as of the Chambers and Federations that compose it and therefore of the companies affiliated to them [17].

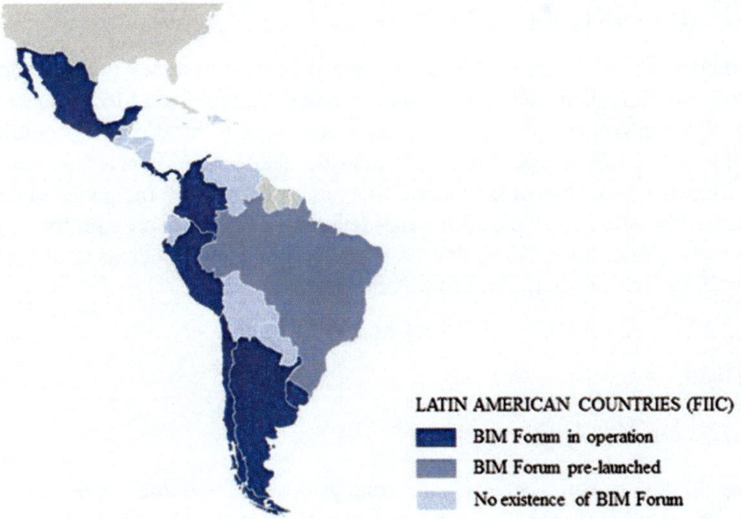

Fig. 1. Latin American bloc for BIM macro adoption analysis.

Therefore, the basis for analysis was defined as the collection of accessible publications available in official websites of (i) governments; (ii) national standards institutions; (iii) chambers for the construction industry; (iv) and BIM Forum or similar entities. Portuguese, Spanish, and English languages were considered. The keywords for searching the documents were: BIM, MIC, Building Information Modeling, Modelagem da Informação da Construção, Modelado de la Información para la Construcción, Modelado de la Información para la Edificación, Modelamiento de la

Información de Edifícios. The searches were carried out between from November/2019 to January/2020. Documents related to stakeholders' presentations only were excluded.

The publications were grouped by country in a content matrix and their data organized in graphics for highlighting key information, e.g., temporal evolution. For the content analysis, they were coded according to:

a. identity data: code; year; title; and language of the published document;
b. primary source: Government Body (G); National Standard Institution (NS); Chamber for Construction Industry (C); and BIM Forum or similar (BF);
c. category for content qualification according to BIM Knowledge Content (BKC) clusters taxonomy [18]: Guides (G), Protocols (P), Mandates (M), and Others (O);
d. category for content penetration: BIM-Specific Document (B); and Public Document with BIM mentioned (C);
e. content items inclusion: Modeling Standards (MS); Component Guidelines (CG); LoX Specification (LOX); Communication and Collaboration Protocols (CCP); Target Model Uses (MU); BIM-based Deliverables (BD); Information Exchange Formats (IXF); and Classification Systems (CSys).

The matrixes with the main information are presented in annex tables (criteria a to d), and those with content items in charts (criteria e). It is relevant to mention that the selection of these content items for checking is related to how many documents can drive industry to BIM-based project standards. The authors excluded the content analysis from national standards because they are not available for public access. The outcomes of these matrixes presented the detailed status by each country. Then, the graphics presented in Sect. 3.3 enabled a visual comparison in a cross-country level to synthesize BIM initiatives in the Latin America bloc covered.

3 Results

3.1 Latam Overview

According to the research strategy, the analysis comprised the following eighteen countries presented in Table 1. Among these, only eight countries are active members of the BIM Network of Latin American Governments. This network, originally known as *Red BIM de Gobiernos Latinoamericanos*, is an initiative created in 2018 for collaboration and joint learning to accelerate individual efforts. The network is financially supported by the Inter-American Development Bank (*Banco Interamericano di Desarollo* – BID) which is interested in promoting BIM macro adoption in the region. Since 2018, the network promoted four official meetings with representatives from Argentina, Brazil, Chile, Colombia, Costa Rica, Mexico, Peru, and Uruguay [19–22].

Table 1. Countries of Latam with macro BIM adoption efforts (2005–2019)

Country	BIM network	Chamber for the construction industry	Chamber acronym	BIM forum status
Argentina	Current member	*Cámara Argentina de la Construcción*	CAMARCO	In operation
Bolivia	–	*Cámara Boliviana de la Construcción*	CABOCO	–
Brazil	Current member	*Câmara Brasileira da Industria da Construção*	CBIC	Pre-Launched
Chile	Current member	*Cámara Chilena de la Construcción*	CChC	In operation
Colombia	Current member	*Cámara Colombiana de la Construcción* *Cámara Colombiana de la Infraestrutuctura*	CAMACOL CCI	In operation
Costa Rica	Current member	*Cámara Costarricense de la Construcción*	CCC	In operation
Ecuador	–	*Cámara de la Construcción de Guayaquil*	CCONSTRUCCIÓN	–
Guatemala	–	*Cámara Guatemalteca de la Construcción*	CONSTRUGUATE	–
Honduras	–	*Cámara Hondureña de la Industria de la Construcción*	CHICO	–
Mexico	Current member	*Cámara Mexicana de la Industria de la Construcción*	CMIC	In operation
Nicaragua	–	*Cámara Nicaraguense de la Construcción*	CNC	–
Panama	–	*Cámara Panameña de la Construcción*	CAPAC	In operation
Paraguay	–	*Cámara Paraguaya de la Industria de la Construcción*	CAPACO	–
Peru	Current member	*Cámara Peruana de la Construcción*	CAPECO	In operation*
Dominican Republic	–	*Cámara Dominicana de la Construcción*	CADOCON	–
El Salvador	–	*Cámara Salvadoreña de la Industria de la Construcción*	CASALCO	–
Uruguay	Current member	*Cámara de la Construcción del Uruguay*	CCU	In operation
Venezuela	–	*Cámara Venezolana de la Construcción*	CVC	–

In the lenses of the Chambers for the Construction Industry, which represent each country in the FIIC, it is relevant to observe that Colombia is the only one that has a chamber focused on infrastructure. Also, we have noticed that in those countries where an entity organization (e.g., BIM Forum, or similar) is responsible for coordinating all macro adoption activities, this entity is directly co-founded by the indicated chambers. Among the eight current member countries of BIM Network of Latam Governments, all of them already have an entity such as BIM Forum in operation, with exception of Brazil where BIM Forum Brazil was pre-launched in 2019. In Peru, this entity is named *Comite BIM Peru*. This correlation is supported by FIIC, which established the Latin American Network of Technological Innovation Centers (INCONET) in 2010. INCONET is responsible for Research, Development and Innovation in the construction sector; and from its ICT strategic area emerged BIM Forum Latam in 2015 [23]. The BIM Forum Latam was created to promote collaboration and associative work between BIM entities in different countries in order to facilitate internationalization to different markets and generate instances of coordination and collaboration of multinational projects with the use of BIM [24].

There are also external influences on BIM macro adoption in Latin America countries observed through the common presence of Central for Digital Built Britain (CDBB) from UK [25]. The international programme of CDBB supported the methodological transfer to Latin American countries of the implementation of UK BIM in public projects. CDBB formalized such initiative through a Memorandum of Understanding (MoU) signed by UK and Chile (2016), Brazil (2016), and Mexico (2019) committing to the exchange of information on national strategies for the consistent diffusion of BIM. Also, CDBB initiated collaboration approaches (e.g., workshops) with Colombia, Uruguay, and Peru. CDBB is a partnership between the Department of Business, Energy and Industrial Strategy from UK and the University of Cambridge. Therefore, the Latam bloc is heavily influenced by the UK in the development of BIM strategies.

Considering this overview, efforts from governments (i.e., through BIM Network of Latam Governments, international programme of CDBB) and market (i.e., represented in a high-level by Chambers, Federations, BIM Forums or similar) are aligned aiming at accelerating BIM macro adoption mainly in Argentina, Brazil, Chile, Colombia, Costa Rica, Mexico, Peru, and Uruguay. Beyond, we can highlight Panama, which already has a BIM Forum Panama with the aim at supporting government initiatives in terms of standardization as well as BIM diffusion.

3.2 Integrated Analysis of BIM Macro-adoption in Latam

This section presents an integrated analysis of macro adoption by each country, highlighting those who are leading the bloc and the recent initiatives. The analysis is based on the list of publications found per country (tables in the Annex) and characterization of content of publications.

Argentina (Annex 1). Argentina is a current member country of the BIM Network aforementioned. It presents 7 BIM publications (Fig. 2 left), all in Spanish, from 2017 to 2019, mainly developed by its chamber for construction industry CAMARCO, with small participation of government and BIM Forum Argentina. The official documents from Argentina aiming at BIM diffusion and acceleration were mainly produced as guides and are BIM specific documents (Fig. 3). In 2019, *the Ministerio del Interior,*

Obras Publicas y Viviendas launched the first government body' set of documents, which is qualified as protocol. It is relevant to mention that in 2019, the Argentinian public sector presented the BIM Implementation Plan through the SIBIM programme, in which such set of documents was generated [26].

Concerning the content items, the *Estándares* from BIM Forum Argentina and the SIBIM set of documents can support Argentinian industry in terms of BIM-based project standards, because they cover most of the content with the potential to drive BIM adoption (Fig. 2 right). The other published documents are related to reports from industry studies or conceptual approaches in a high-level degree, e.g., information management and disruptive innovation.

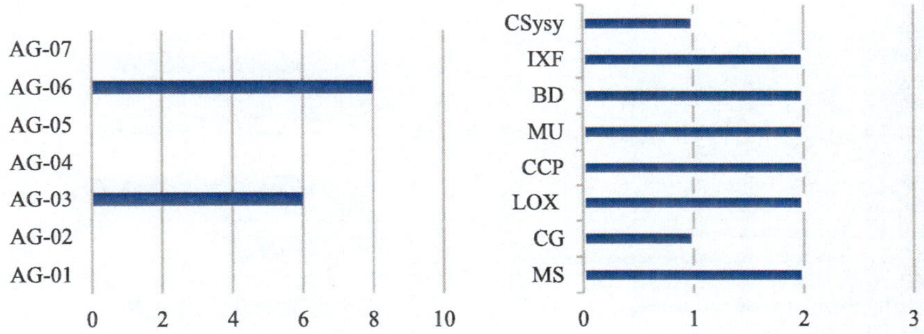

Fig. 2. Argentinian BIM document characterization: (left) documents with project-based content and (right) project-based content observed

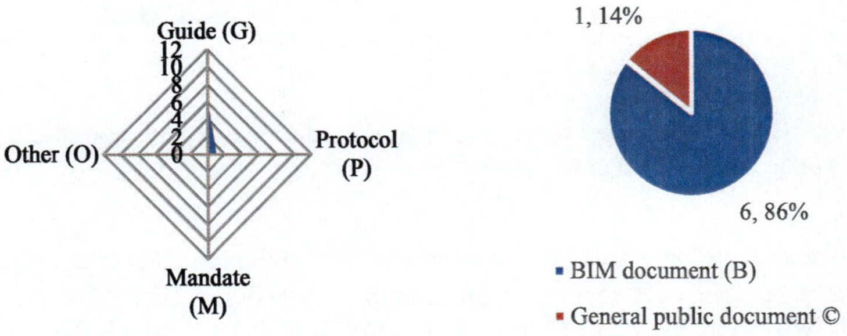

Fig. 3. Argentinian category of BIM documents

Brazil (Annex 2). Brazil is a current member country of the BIM Network aforementioned. It presents thirty-six BIM publications (Fig. 4 left), all in Portuguese, from 2010 to 2019, mainly developed by the government (federal or state) and its national standard institution, with small participation of non-governmental entities. Despite the amount of documentation, only a small portion supports project activities, covering most of the content with the potential to drive BIM adoption (Fig. 4 right). In addition, Brazilian BIM documentation is diverse, however BIM specific (Fig. 5).

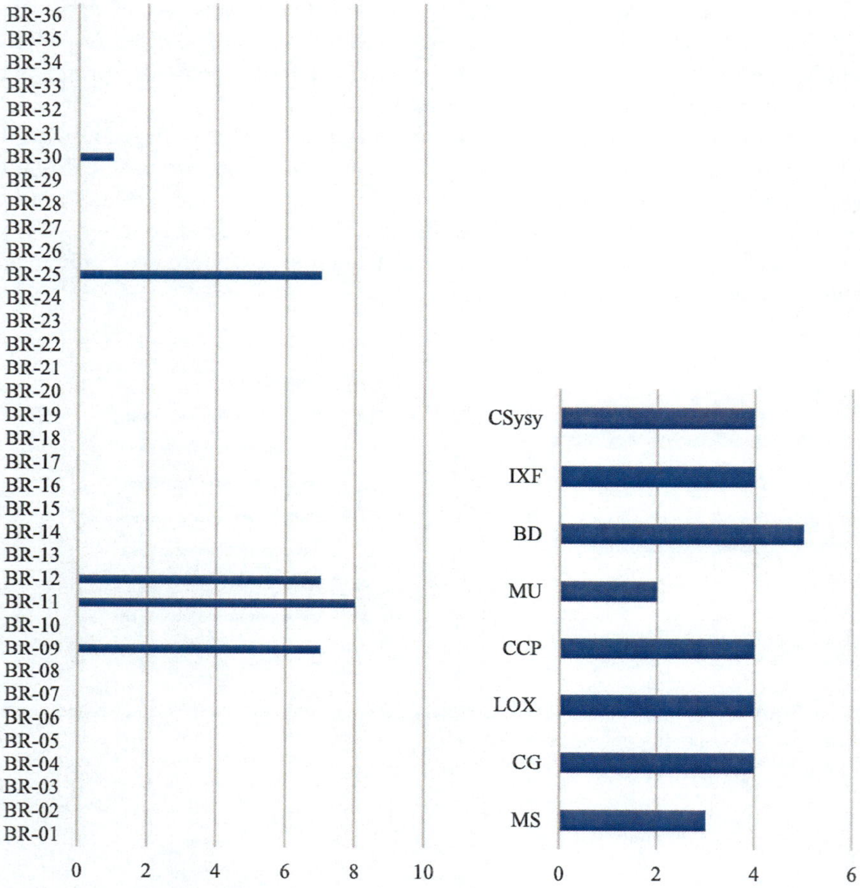

Fig. 4. Brazilian BIM document characterization: (left) documents with project-based content and (right) project-based content observed

The official documents from Brazil aiming at BIM diffusion and acceleration started with four protocols published as standards by *Associação Brasileira de Normas Técnicas* (ABNT) in 2010, 2011, 2012, 2014, 2015 and 2018. In 2015, the government supported the project Sector Dialogues for BIM, which presented a survey summarizing the current framework for the application and dissemination of BIM in Brazil and in the European Union. Such publication provided recommendations to government entities and other institutions to strengthen and speed up BIM macro adoption in Brazil. In the same year, the state government of Santa Catarina was the first to include BIM as a requirement in the public projects and launched three more guides in 2018. In the sequence, the chamber for construction industry CBIC, in 2016, and the ABDI-MDIC, in 2017, published guides with good practices in BIM for the public or private sectors.

A milestone in Brazil occurred in 2018 with the publication of a federal decree. The decree has established the Strategic Committee for the Implementation of Building Information Modeling. Due to this initiative, in 2018, MDIC published the National Dissemination Strategy for BIM in Brazil. The BIMBR Strategy comprises actions of induction to structure BIM macro adoption through its application in pilot programmes; i.e., the development of design projects from 2021 on, design and execution projects from 2024 on, and including post-execution projects from 2028 on. Other actions include, for instance, the promotion of a National BIM Library, already launched in 2018. Finally, in 2019, a technical cooperation agreement between the government of seven Brazilian states was signed. In addition, the BIM Forum Brazil was pre-launched.

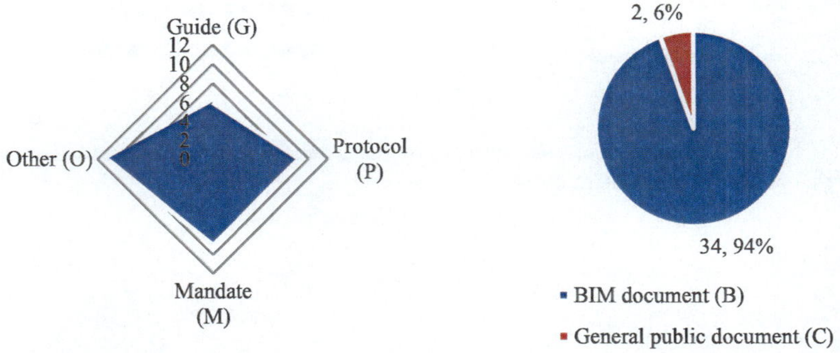

Fig. 5. Brazilian category of BIM documents.

Chile (Annex 3). Chile is a current member country of the BIM Network aforementioned. It presents thirteen BIM publications (Fig. 6 left), mostly in Spanish and specifically by PlanBIM also in English, from 2013 to 2019. Documents were outstandingly developed by the government, with small participation of national standard institution and chamber for construction industry. There is a greater number of documents that can drive industry to BIM-based project standards in comparison with Brazil (Fig. 6 right). In this context, the publication that stands out is BIM Standard for Public Projects: Information Exchange between Appointing and Appointed Parties. The Chilean BIM documentation is mostly concentrated in mandates, followed by protocols and guides (Fig. 7).

MOP published the first Chilean official document in 2013 to define concepts, requirements, methods, and examples about the use of BIM technologies in new public building projects and terms of development in the country. This document aimed at the creation of a Chilean BIM Standard, which directly benefits the MOP. In 2017, the *Cámara Chilena de la Construction* (CChC) published guidelines and best practices related to the use of BIM in projects and guidance to the implementation of BIM in organizations. In 2018, MOP published four terms of reference describing the deliverables linked to BIM required in public bidding. The *Ministerio de la Salud* quoted the use of BIM in 2018 for greater control over project design in the institutional and management model for the development of the country's health infrastructure. In 2019, *Ministerio de la Salud* also published a guide for the design of hospital establishments.

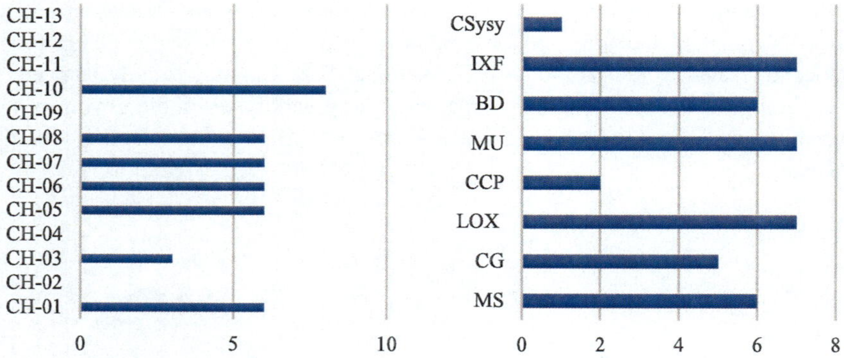

Fig. 6. Chilean BIM document characterization: (left) documents with project-based content and (right) project-based content observed.

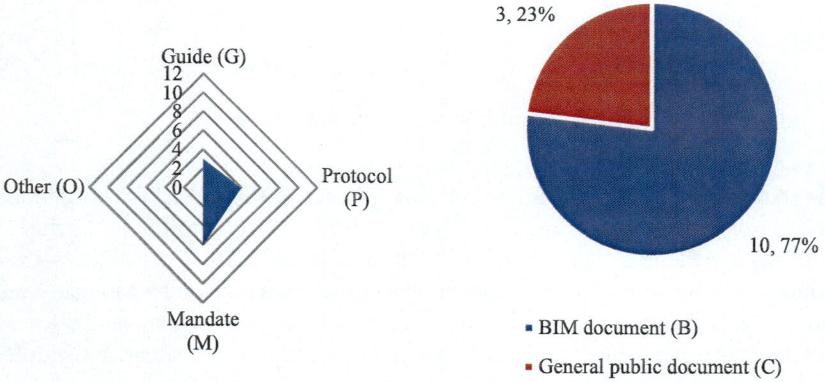

Fig. 7. Chilean category of BIM documents.

The year of 2019 was also a milestone for Chile with the launch of BIM standard for public projects by Planbim, a 10-year programme promoted by *Corporación de Fomento a la Producción* (CORFO), which emerged within the framework of the Strategic Program for Productivity and Sustainable Construction, Build 2025. Since 2017, it is housed in the CORFO Digital Transformation Committee. Chile also published two national standards in 2019: an adaptation of the International Standard (ISO) 19650 series [27]. Planbim and BIM Forum Chile are introducing high-level conceptual approaches and initial guidance.

Colombia (Annex 4). Colombia is a current member country of the BIM Network aforementioned. It presents three BIM publications punctually covering all content with the potential to drive BIM adoption (Fig. 8). These documents are all in Spanish and were developed from 2018 to 2019, two by the government, and one by the chamber for construction industry CAMACOL. Colombian BIM documentation are concentrated in guides and mandates (Fig. 9).

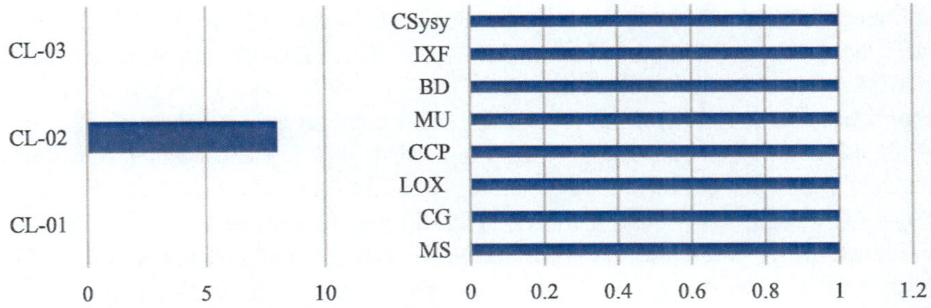

Fig. 8. Colombian BIM document characterization: (left) documents with project-based content and (right) project-based content observed.

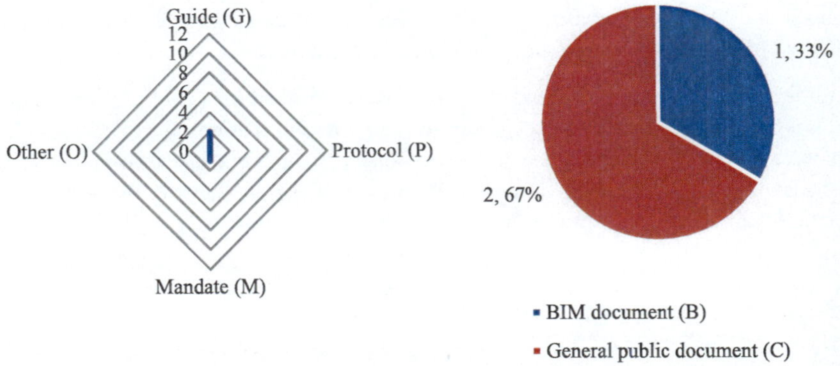

Fig. 9. Colombian category of BIM documents.

The first official BIM guide was published by CAMACOL in 2019 and focuses on guidelines for the adoption of BIM in the companies. This document gives detailed guidance and best practices for the use of BIM in projects, referencing the ISO 19650 series [27]. In 2019, the Colombian Government also included the promotion of BIM by the public sector in its sector strategy for the generation of new sources of growth. The presence of a BIM Forum is a sign that this construction industry is organizing a structured initiative in the next years.

Costa Rica (Annex 5). Costa Rica is a current member country of the BIM Network aforementioned. It presents one BIM publication, in Spanish, developed by both the chamber for construction industry CCC and BIM Forum Costa Rica. This only official document is a conceptual guide for BIM implementation in organizations. Therefore, the documental collection and analysis presented indicate that the country is starting its path to macro adoption. It is relevant to mention that this document does not match with the criteria e for structured content analysis.

Mexico (Annex 6). Mexico is a current member country of the BIM Network aforementioned. It presents three BIM publications, all in Spanish, from 2015 to 2019, developed by the national standard institution as well as by the government. Mexican BIM documentation is concentrated in two protocols and one mandate. None of these documents match with the criteria for our structured content analysis.

Official documents from Mexico aiming at BIM diffusion and acceleration started with two protocols published as standards by the *Sistema Integral de Normas y Evaluación de la Conformidad* (SINEC) in 2015 (already archived) and 2017. In 2019, the *Secretaria de Hacienda y Crédito Público* (SHCP) launched the Mexican BIM Mandate with objectives and strategies to improve processes of development of public infrastructure as well as the main stakeholders involved. The plan comprises efforts concerning of preparation (e.g., planning, pilot programmes, BIM diffusion) since 2019; adaptation (e.g., guides, processes and technologies, training) from 2020 to 2021; and implementation (e.g., application in public projects) from 2022 to 2026. It is relevant to mention that in 2015, an unofficial document co-authored by Arcadis and financed by the UK Embassy presented recommendations to the development of strategies for BIM macro adoption initiatives in Mexico [28].

Panama (Annex 7). Panama has only one BIM publication (Table 2), in Spanish, which is an official guide by the chamber for construction industry CAPAC about BIM principles focused on introducing initial approaches to the construction industry. The presence of a BIM Forum is a sign that this construction industry is organizing a structured initiative in the next years.

Table 2. Panamá content

Code	Category	MS	CG	LOX	CCP	MU	BD	IXF	CSys
PN-01	G/B	–	–	Y	–	Y	–	–	–

Peru (Annex 8). Peru is a current member country of the BIM Network aforementioned. It presents seven BIM publications, all in Spanish, from 2018 to 2019, mainly developed by the government, and its national standard institution. Only one of the documents, which is a mandate, covers 62.5% of the content associated with project-based BIM content (Fig. 10). Peruvian BIM documentation are scares and concentrated in protocols and a mandate (Fig. 11).

Official documents from Peru aiming at BIM diffusion and acceleration started with two protocols published as standards in 2018. These two standards are the adoption or adaptation of ISO 19650 series [27]. In 2019, six Mandates related to BIM were published by government entities. VIVIENDA published three resolutions and reports associated with the definition of a technical team and the definition of general guidelines for the use of BIM in construction projects. The National Competitiveness and Formalization Council included the progressive adoption of BIM throughout the public sector for investment in infrastructure, in any of its modalities of execution. It was an initiative to bring significant improvements in the productivity of infrastructure investment, as well as in its social profitability. In 2019, the Decree Nº 289-2019-EF was published in the *Diario Official*, in which provisions are approved for the progressive incorporation of BIM into public investment. The presence of an entity named *Comite BIM Peru*, with the same characteristics of a BIM Forum, is a sign that this construction industry is organizing a structured initiative in the next years.

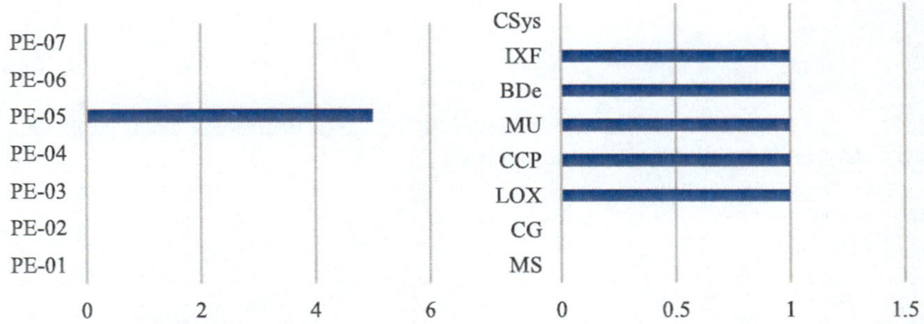

Fig. 10. Peruvian BIM document characterization: (left) documents with project-based content and (right) project-based content observed.

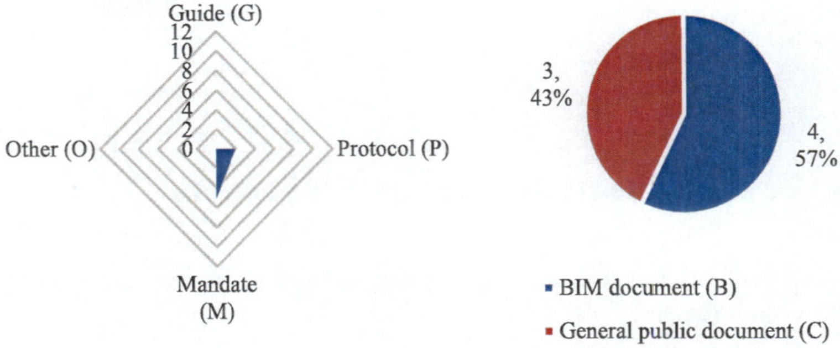

Fig. 11. Peruvian category of BIM documents.

Uruguay (Annex 9). Uruguay is a current member country of the BIM Network aforementioned. It presents five BIM publications (Fig. 12 left), all in Spanish, from 2019, developed by the government, the chamber for construction industry CCU and the BIM Forum Uruguay. BIM-based project content is partially covered, approaching only communication and collaboration protocols and target model uses (Fig. 12 right).

Official documents aiming at BIM diffusion and acceleration have been published since 2019, indicating initial efforts. Both national plans published by *Transforma Uruguay* concerning productive transformation mention BIM. CCU and BIM Forum Uruguay have introduced high-level conceptual approaches and initial guidance (Fig. 13).

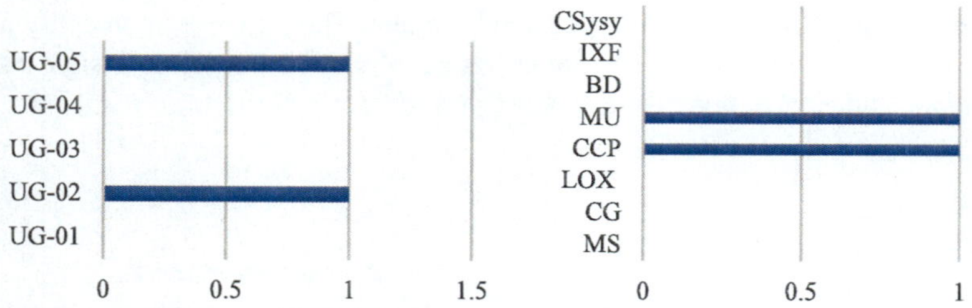

Fig. 12. Uruguayan BIM document characterization: (left) documents with project-based content and (right) project-based content observed.

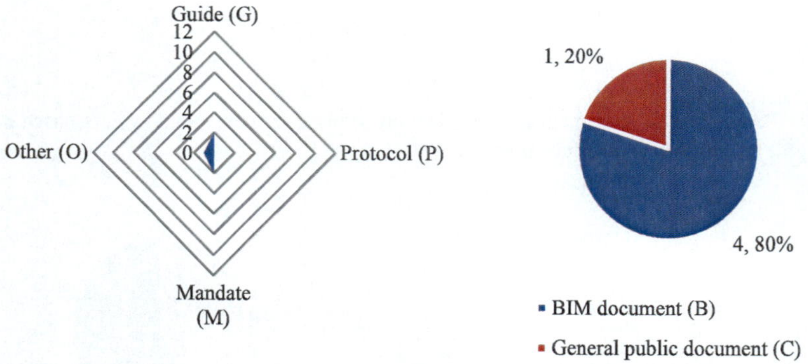

Fig. 13. Uruguayan category of BIM documents.

Others. The authors did not find any relevant official document considering established research criteria.

3.3 Synthesis

Considering the survey, we can highlight initial production of BIM publications in terms of macro adoption since 2010. From 2016 on, there has been a growth movement towards BIM macro adoption in Latin America. The period of 2015–2016 is aligned to the first MuOs between the UK and Latam countries aiming at knowledge transfer from the UK BIM experience. It is noticed that every country that collaborated with the UK has published documents to support macro adoption and is part of BIM Network of Latin American Governments. Also, in 2015 FIIC established BIM Latam Forum with the intention to strength the efforts in the bloc through the chambers for construction industry and prospect financial supports from banks such BID. A peak in 2018 is noticed, much of it due to Brazil's influence. Brazil has published the BIMBR Strategy with a series of protocols to drive both the public sector and the construction industry (Fig. 14).

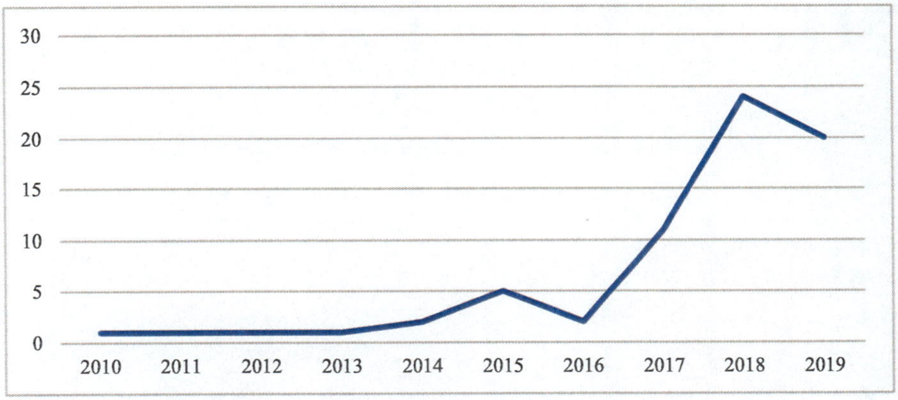

Fig. 14. Latam bloc for BIM macro adoption analysis: number of Latam publications per year.

The greater volume of publications from Latin America belongs to Brazil. Most of it is classified as mandates. Chile takes the second place in this context. In addition to them, Colombia, Mexico, Peru, and Uruguay also own mandates. In addition to the quantitative analysis, we can analyze the documents by category (Fig. 15).

Brazil was pioneer in the adoption of BIM in Latin America with the effort to create protocols, i.e., standards. The same path of initially creating protocols is identified in Mexico and Peru. Chile, in sequence to Brazil, contributes by publishing mandates that require BIM usage in public projects. When we temporarily record BIM documentation in Latin America, we noticed that protocols emerged first, then mandates and more recently guides (Fig. 16). In all graphics presented, the authors did not consider documents without date of publication.

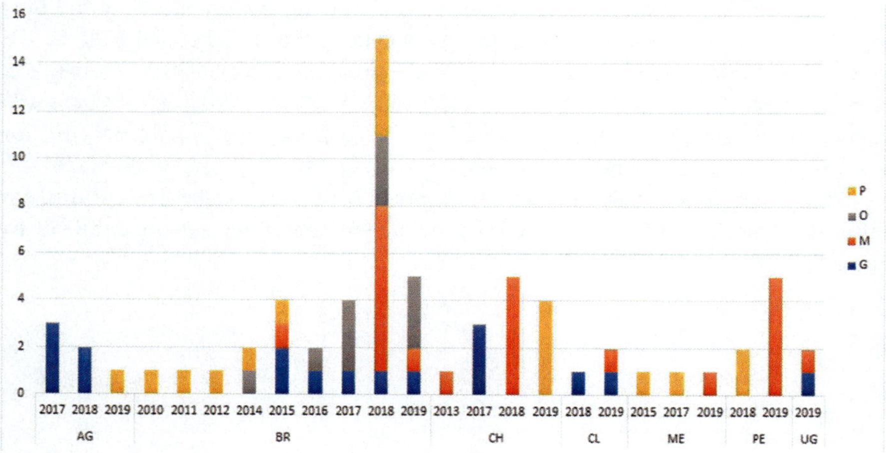

Fig. 15. Latam bloc category of documents by country and year.

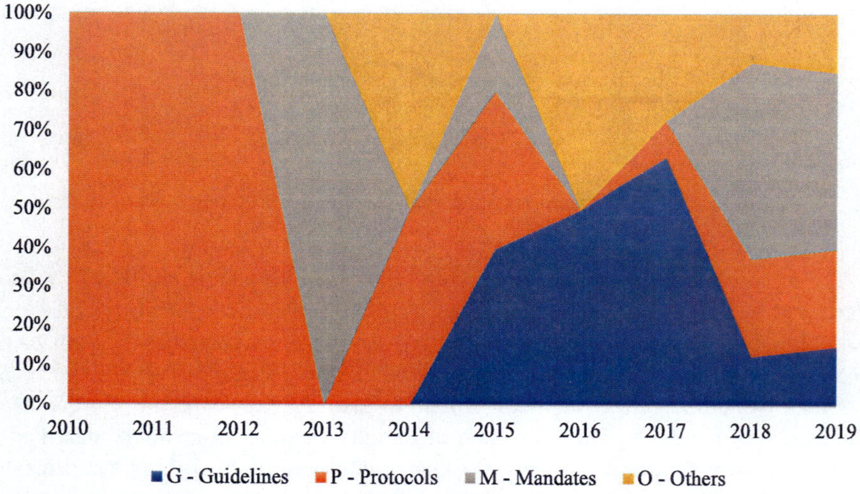

Fig. 16. Latam bloc category of documents per year.

4 Considerations

The UK has great influence on the macro adoption of BIM in Latin America, through the CDBB international programme and MoUs. Another strength in the bloc, which is also connected to CDBB is the FIIC and its chambers for construction industry. In a collaborative effort, Chile brought the issue of creating a BIM Network of Latam Governments together with FIIC, encouraged by CDBB/UK, and supported by BID.

BIM evolution in each country, even though the bloc has common efforts, had different dynamics. Considering the initiatives from the beginning, Brazil was a pioneer in producing documents for macro adoption, with an initial effort in developing protocols, e.g., standards. In this sense, it is noticed that Brazil, and Chile, have the greatest expression in terms of BIM documentation and project-based content. In addition, in terms of document diversity, Brazil is the leader, covering multiple instances that directly generate documents. In Chile, there is a centralized effort of this production in Planbim. Brazil and Chile are followed by Mexico, which has initiated its efforts since 2015. Therefore, Brazil and Chile are the leaders of BIM macro adoption in Latam bloc. Countries such as Mexico, Argentina, Peru, Uruguay and Colombia had later expertise development in this sense and were involved by the BIM Network of Latin American Governments. It is relevant to mention that Colombia has few official publications, however, with higher maturity content to drive BIM-based project standards. Therefore, quantity is not an indicator for quality considering the content analysis. A great number of documents analyzed focused on conceptual approaches and initial steps of a BIM implementation. Countries with mandates, and more complete protocols or guides in terms of project-based BIM content have greater potential for success in adopting BIM. This is the case of Brazil, Chile, and Colombia.

This survey is limited to the research criteria and does not include academic publications or documents from nonprofit entities which are not officially related with the government. The next steps are positioning Latam bloc in comparison to global BIM strategies.

Annexes

Annexes 1. Argentina BIM Documents

Code	Year	Category	Document title	Language	Source type	Source name
AR-01	2017	G/B	*Encuesta Nacional BIM 2016: Adopción de Building Information Modeling en Argentina*	Spanish	C	CAMARCO
AR-02	2017	G/B	*Adopción de BIM en Argentina Resumen Ejecutivo*	Spanish	C	CAMARCO
AR-03	2017	G/B	*Estándares BIM Forum Argentina Versión 1.0*	Spanish	BF	BIM FORUM ARGENTINA
AR-04	2018	G/B		Spanish	C	CAMARCO

(*continued*)

(*continued*)

Code	Year	Category	Document title	Language	Source type	Source name
			BIM: de la competitividad a la innovación radical: la gestion de la información en las organizaciones			
AR-05	2018	G/C	*Gestíon Integral de Proyectos. Procesos colaborativos en la industria de la construcción*	Spanish	C	CAMARCO
AR-06	2019	P/B	*SIBIM - Sistema de Implementación BIM*	Spanish	G	Ministerio del Interior, Obras Públicas y Viviendas
AR-07	N/M	G/B	*BIM. Alternativa o tendencia?*	Spanish	C	CAMARCO

Annexes 2. Brazil BIM Documents

Code	Year	Category	Document title	Language	Source type	Source name
BR-01	2010	P/B	*(Archived) - ABNT NBR ISO 12006-2:2010 Construção de edificação — Organização de informação da construção*	Portuguese	NS	ABNT
BR-02	2011	P/B	*ABNT NBR 15965-1:2011 Sistema de classificação da informação da construção Parte 1: Terminologia e estrutura*	Portuguese	NS	ABNT
BR-03	2012	P/B	*ABNT NBR 15965-2:2012 Sistema de classificação da informação da construção Parte 2: Características dos objetos da construção*	Portuguese	NS	ABNT
BR-04	2014	P/B	*ABNT NBR 15965-3:2014 Sistema de classificação da informação da construção Parte 3: Processos da construção*	Portuguese	NS	ABNT
BR-05	2014	O/C	*Plano de Metas para a gestão 2015–2018*	Portuguese	G	Beto Richa Governador*
BR-06	2015	P/B	*ABNT NBR 15965-7:2015 Sistema de classificação da informação da construção Parte 7: Informação da*	Portuguese	NS	ABNT

(*continued*)

(*continued*)

Code	Year	Category	Document title	Language	Source type	Source name
BR-07	2015	G/C	*MANUAL DA CONSTRUÇÃO INDUSTRIALIZADA Conceitos e Etapas Volume 1: Estrutura e Vedação. ABDI. Agência Brasileira de Desenvolvimento Industrial*	Portuguese	G	ABDI
BR-08	2015	R/B	*DIÁLOGOS SETORIAIS PARA BIM - BUILDING INFORMATION MODELING NO BRASIL E NA UNIãO EUROPEIA*	Portuguese	G	Governo
BR-09	2015	M/B	*Caderno BIM: Apresentação de Projetos de Edificações em BIM*	Portuguese	G	Governo de Santa Catarina
BR-10	2016	O/B	*Anúncio público sobre a utilização da tecnologia BIM (Building Information Modeling) pela Fundação para o Desenvolvimento da Educação – FDE. Apresentação dos objetivos a serem perseguidos, e da metologia a ser utilizada na implantação de novos processos*	Portuguese	G	FDE
BR-11	2016	G/B	*Coletânea Implementação do BIM: Building Information Modeling - CBIC*	Portuguese	C	CBIC
BR-12	2017	G/B	*Coletânea Guias BIM ABDI-MDIC*	Portuguese	G	ABDI-MDIC
BR-13	2017	O/B	*DECRETO DE 05 DE JULHO DE 2017: Institui o Comitê Estratégico de Implementação do Building Information Modeling*	Portuguese	G	Governo
BR-14	2017	O/B	*Portaria nº 1734-SEI, de 8 de setembro de 2017*	Portuguese	G	Governo
BR-15	(2017a 2018)	O/B	*Conjunto de Resoluções CE-BIM (2017–2018)*	Portuguese	G	Governo
BR-16	2018	M/B	*Livreto Estratégia: BIM BR Construção Inteligente*	Portuguese	G	MDIC
BR-17	2018	M/B	*ESTRATÉGIA BIM BR: Estratégia Nacional de Disseminação do Building Information Modelling – BIM*	Portuguese	G	MDIC
BR-18	2018	G/B	*Mapeamento Internacional de Bibliotecas BIM*	Portuguese	G	ABDI
BR-19	2018	P/B	*ABNT NBR ISO 16354:2018 Diretrizes para as bibliotecas de conhecimento e bibliotecas de objetos*	Portuguese	NS	ABNT

(*continued*)

(*continued*)

Code	Year	Category	Document title	Language	Source type	Source name
BR-20	2018	P/B	*ABNT NBR ISO 16757-1:2018 Estruturas de dados para catálogos eletrônicos de produtos para sistemas prediais Parte 1: Conceitos, arquitetura e modelo*	Portuguese	NS	ABNT
BR-21	2018	P/B	*ABNT NBR ISO 16757-2:2018 Estruturas de dados para catálogos eletrônicos de produtos para sistemas prediais Parte 2: Geometria*	Portuguese	NS	ABNT
BR-22	2018	M/B	*Caderno BIM: Coletânea Cadernos Orientadores. Caderno de especificações Técnicas para Contratação de Projetos em BIM - Edificações. SECRETARIA DE INFRAESTRUTURA E LOGÍSTICA DEPARTAMENTO DE GESTÃO DE PROJETOS E OBRAS. Governo do Estado Paraná*	Portuguese	G	Secretaria de Estado de Infraestrutura e Logística
BR-23	2018	M/B	*Decreto nº 9.377, de 17 de maio de 2018*	Portuguese	G	Governo
BR-24	2018	P/B	*GUIA BÁSICO TEKLA BIM SIGHT. Governo de Santa Catarina.*	Portuguese	G	Governo de Santa Catarina
BR-25	2018	M/B	*Caderno de Apresentação de Projetos em BIM de Santa Catarina. TERMO DE REFERÊNCIA DE APRESENTAÇÃO DAS DIRETRIZES PROJETUAIS. VERSÃO 2.0*	Portuguese	G	Governo de Santa Catarina
BR-26	2018	M/B	*Estratégia de Disseminação do Building Information Modelling no Rio de Janeiro - Estratégia BIM-RJ*	Portuguese	G	Governo do Rio Janeiro
BR-27	2018	O/B	*Resolução CG-BIM 1/2018*	Portuguese	G	Governo
BR-28	2018	O/B	*Resolução CG-BIM 2/2018*	Portuguese	G	Governo
BR-29	2018	O/B	*Resolução CG-BIM 3/2018*	Portuguese	G	Governo
BR-30	2018	M/B	*Caderno de Encargo de Projetos em BIM e CAD: TERMO DE REFERÊNCIA DE APRESENTAÇÃO DAS DIRETRIZES PROJETUAIS. V2.0*	Portuguese	G	Governo de Santa Catarina

(*continued*)

(*continued*)

Code	Year	Category	Document title	Language	Source type	Source name
BR-31	2019	O/B	*Edital n 3/2019: Termo de colaboração para execução de ações para promover ganho de produtividade e competitividade do setor de construção civil. Ministério da Economia.*	Portuguese	G	SDIC
BR-32	2019	O/B	*REQ 2.742/2019 - Requer o registro da Frente Parlamentar em Defesa do Sistema de Modelagem da Informação de Construção - BIM perante a Mesa de Câmara dos Deputados*	Portuguese	G	Governo
BR-33	2019	G/B	*Manual do Objeto BIM: Autodesk Revit - Plataforma BIM BR*	Portuguese	G	ABDI
BR-34	2019	M/B	*DECRETO Nº 9.983, DE 22 DE AGOSTO DE 2019*	Portuguese	G	Governo
BR-35	2019	O/B	*Acordo de cooperação técnica que entre si celebram o Governo do Estado de Santa Catarina, o Governo do Estado do Espírito Santo, o Governo do Estado de Minas Gerais, o Governo do Estado do Paraná, o Governo do Estado do Rio de Janeiro, o Governo do Estado do Rio Grande do Sul e o Governo do Estado de São Paulo*	Portuguese	G	COSUD
BR-36	N/M	G/B	*Road Show BIM resultados da pesquisa e desdobramentos*	Portuguese	C	CBIC

Annexes 3. Chile BIM Documents

Code	Year	Category	*Document name*	Language	Source type	Source name
CL-01	2013	M/B	*(Archived) Términos de Referencia Uso de modelos BIM. Dirección de Arquitectura MOP (v1.15_oct2013)*	Spanish	G	MOP
CL-02	2017	G/C	*Orientaciones Técnicas Para Diseño de Anteproyectos de Hospitales Complejos*	Spanish	G	Ministerio de la Salud

(*continued*)

(*continued*)

Code	Year	Category	*Document name*	Language	Source type	Source name
CL-03	2017	G/B	*Guía inicial para implementar bim en las organizaciones*	Spanish	C	CChC
CL-04	2017	G/B	*Recomendaciones prácticas para el uso de herramientas bim en la coordinación de proyectos desarrollados en forma tradicional. Bim forum chile corporación de desarrollo tecnológico - cámara chilena de la construcción*	Spanish	C	CChC
CL-05	2018	M/B	*Términos de referencia BIM. Factibilidad. Diseño Anteproyecto. División de edificación pública – dirección de arquitectura mop. Versión 1.18*	Spanish	G	MOP
CL-06	2018	M/B	*Términos de referencia BIM etapa: factibilidad contrato: consultoría diseño. División de edificación pública – dirección de arquitectura mop. Versión 1.18*	Spanish	G	MOP
CL-07	2018	M/B	*Términos de referencia BIM etapa: ejecución contrato: obra según bases: DS. MOP N° 75/2009. Versión 1.16*	Spanish	G	MOP
CL-08	2018	M/B	*Términos de referencia BIM etapa: ejecución contrato: obra - pago contra recepción según bases: DS. MOP N° 108/2009. Versión 1.19*	Spanish	G	MOP
CL-09	2018	G/C	*Nuevo modelo institucional y de gestión para el desarrollo de la infraestructura sanitaria del país*	Spanish	G	Ministerio de la Salud
CL-10	2019	G/B	*BIM standard for public projects: Information Exchange between Appointing and Appointed Parties*	Spanish English	G	PLANBIM
CL-11	2019	G/B	*Guía de diseño para establecimientos hospitalarios de mediana complejidad 2019*	Spanish	G	Ministerio de la Salud

(*continued*)

(*continued*)

Code	Year	Category	*Document name*	Language	Source type	Source name
CL-12	2019	M/B	*NCh-ISO 19650/1- Organización y digitalización de la información en obras de edificación e ingeniería civil, incluyendo modelamiento de la información de edificios (BIM) - Gestión de la información utilizando modelamiento de la información de edificios - Parte 1: Conceptos y principios*	Spanish	NS	INN
CL-13	2019	M/B	*NCh-ISO 19650/2 - Organización y digitalización de la información en obras de edificación e ingeniería civil, incluyendo modelamiento de la información de edificios (BIM) - Gestión de la información utilizando modelamiento de la información de edificios - Parte 2: Fase de desarrollo de los activos*	Spanish	NS	INN

Annexes 4. Colombia BIM Documents

Code	Year	Category	Document name	Language	Source type	Source name
CO-01	2018	G/C	Ministerio de Vivienda, Ciudad y Territorio Informe de Rendicíon de Cuentas. Vigencia 2018	Spanish	G	MINVIVIENDA
CO-02	2019	G/B	BIM KIT. Guías para la adopcíon BIM em las organizaciones	Spanish	C	CAMACOL
CO-03	2019	M/C	Pacto por el crecimiento y para la generación de empleo del sector de industrias para la construcción. Estrategia sectorial para la generación de nuevas fuentes de crecimiento	Spanish	G	Gobierno de Colombia

Annexes 5. Costa Rica BIM Documents

Code	Year	Category	Document title	Language	Source type	Source name
CR-01	N/M	G/B	*Guía de implementación BIM para las empresas. BIM Forum Costa Rica. Cámara Costarricense de la Construcción*	Spanish	C	CCC

Annexes 6. Mexico BIM Documents

Code	Year	Category	Document title	Language	Source type	Source name
MX-01	2015	P/B	*NMX-000-11-2015: Modelado de Información para la Construcción*	Spanish	NS	SINEC
MX-02	2017	P/B	*NMX-C-527-1-ONNCCE-2017: Industria de la Construcción-Modelado de Información de la Construcción-Especificaciones-Parte 1: Plan de Ejecución para Proyectos*	Spanish	NS	SINEC
MX-03	2019	M/B	*Estrategia para la implementación del Modelado de Información de la Construcción (MIC) en México*	Spanish	G	Secretaria de Hacienda y Crédito Público

Annexes 7. Panama BIM Documents

Code	Year	Category	Document name	Language	Source type	Source name
PN-01	N/M	G/B	*Los Principales Términos B.I.M. en la industria de la construcción*	Spanish	C	CAPAC

Annexes 8. Peru BIM Documents

Code	Year	Category	Document name	Language	Source type	Source name
PE-01	2018	P/B	*NTP-ISO 29481-2:2018. Modelado de la información de los edificios. Manual de entrega de la información. Parte 2: Marco de trabajo para la interacción. 1a Edición*	Spanish	NS	ISO
PE-02	2018	P/B	*ETP-ISO/TS 12911:2018. Guía marco para el modelado de información de la edificación (BIM). 1a. Edición*	Spanish	NS	ISO
PE-03	2019	M/C	*Informe N' 849-2019-VIVIENDA/IVCS-DGPRCS-DC*	Spanish	G	VIVIENDA
PE-04	2019	M/C	*Resolución Directoral N° 053-2019-VIVIENDA/VMCS-DGPRCS*	Spanish	G	VIVIENDA
PE-05	2019	M/B	*Resolución Ministerial Nº 242-2019-VIVIENDA*	Spanish	G	VIVIENDA
PE-06	2019	M/C	*Plan Nacional de COMPETITIVIDAD Y PRODUCTIVIDAD 2019-2030. DOCUMENTO RESUMEN. Govierno del Perú*	Spanish	G	Consejo Nacional de Competitividad y Formalización
PE-07	2019	M/B	*DECRETO SUPREMO Nº 289-2019-EF*	Spanish	G	Diario Oficial

Annexes 9. Uruguay BIM Documents

Code	Year	Category	Document title	Language	Source type	Source name
UY-01	2019	M/C	*Plan Nacional de Transformación Productiva y Competitividad. Rendición de cuentas y actualización 2019*	Spanish	G	Transforma Uruguay

(*continued*)

(*continued*)

Code	Year	Category	Document title	Language	Source type	Source name
UY-02	2019	G/B	*Introducción al BIM Versión 1.0 | setembre 2019*	Spanish	BF	BIM Forum Uruguay
UY-03	N/M	M/B	*Estrategia de implementación de BIM en Uruguay. Sistema Nacional de Transformación Productiva y Competitividade - Transforma Uruguay*	Spanish	G	Transforma Uruguay
UY-04	N/M	O/B	*Proyecto: Realización de diagnóstico de la situación de Uso del BIM en Uruguay y elaboración de requerimientos de los TDR's – BIM*	Spanish	C	CCU
UY-05	N/M	G/B	*Guia de Primeros Pasos BIM: BIM Forum Uruguay*	Spanish	BF	BIM Forum Uruguay

References

1. Cheng, J., Lu, O.: A review of the efforts and roles of the public sector for BIM adoption worldwide. J. Inf. Technol. Constr. (ITcon) **20**(27), 442–478 (2015)
2. Ahmed, A.L., Kassem, M.: A unified BIM adoption taxonomy: conceptual development, empirical validation and application. Autom. Constr. **96**, 103–127 (2018). https://doi.org/10.1016/j.autcon.2018.08.017
3. Succar, B., Kassem, M.: Macro-BIM adoption: conceptual structures. Autom. Constr. **57**, 64–79 (2015). https://doi.org/10.1016/j.autcon.2015.04.018
4. Isikdag, U.: Enhanced Building Information Models. SpringerBriefs in Computer Science (2015). https://doi.org/10.1007/978-3-319-21825-0
5. Succar, B., Kassem, M.: Building information modelling: point of adoption. In: CIB World Conference Proceedings, vol. 1 (2016)
6. OECD: Oslo manual: guidelines for collecting and interpreting innovation data. Organisation for Economic Co-operation Development: OECD Publishing (2005)
7. Wong, A., Wong, F., Nadeem, A.: Comparative roles of major stakeholders for the implementation of BIM in various countries. In: Proceedings of the International Conference on Changing Roles: New Roles, New Challenges, Noordwijk Aan Zee, The Netherlands, vol. 5 (2009)
8. Smith, P.: BIM implementation–global strategies. Procedia Eng. **85**, 482–492 (2014)
9. Yang, J., Chou, H.: Mixed approach to government BIM implementation policy: an empirical study of Taiwan. J. Build. Eng. **20**, 337–343 (2018)
10. Edirisinghe, R., London, K.: Comparative analysis of international and national level BIM standardization efforts and BIM adoption. In: 32nd CIB W78 Conference Proceedings (2015)
11. Kassem, M., Succar, B.: Macro BIM adoption: comparative market analysis. Autom. Constr. **81**, 286–299 (2017)

12. Charef, R., Emmitt, S., Alaka, H., Fouchal, F.: Building information modelling adoption in the European union: an overview. J. Build. Eng. **25**, 100777 (2019)
13. McAuley, B., Hore, A., West, R.: BICP Global BIM Study - Lessons for Ireland's BIM Programme Published by Construction IT Alliance (CitA) Limited (2017). https://doi.org/10.21427/d7m049
14. BIMe Initiative Homepage, Macro BIM Adoption in Peru. https://bimexcellence.org/projects/macro-adoption/completed-adoption-studies/macro-bim-adoption-in-peru/. Accessed 13 Mar 2020
15. Loyola, M., López, F.: An evaluation of the macro-scale adoption of Building Information Modeling in Chile: 2013–2016. Rev. de la Constr. **17**, 158–171 (2018)
16. Floyd Jr., F.J.: Survey Research Methods. Sage Publications, Thousand Oaks (2013)
17. FIIC. https://www.fiic.la/index.php/nosotros. Accessed 02 Jan 2020
18. Kassem, M., Succar, B., Dawood, N.: A proposed approach to comparing the BIM maturity of countries. In: 30th International Conference on Applications of IT in the AEC Industry Proceedings (2013)
19. PlanBIM. https://planbim.cl/planbim-impulsa-el-1er-encuentro-bim-de-gobiernos-latinoamericanos-bim-en-la-construccion-publica/. Accessed 02 Jan 2020
20. DNIT. http://www.dnit.gov.br/noticias/brasil-e-sede-do-2o-encontro-bim-de-governos-latino-americanos. Accessed 02 Jan 2020
21. CND. https://www.cnd.org.uy/es/tercer-encuentro-bim-de-gobiernos-latinoamericanos. Accessed 02 Jan 2020
22. Argentina. https://www.argentina.gob.ar/noticias/cuarto-encuentro-bim-de-gobiernos-latinoamericanos-en-buenos-aires. Accessed 02 Jan 2020
23. FIIC-INCONET. https://fiic-inconet.org/#!/-inicio/. Accessed 02 Jan 2020
24. BIM Forum Latam. http://www.bimforum.la/. Accessed 02 Jan 2020
25. CDBB. https://www.cdbb.cam.ac.uk/AboutDBB/International. Accessed 13 Mar 2020
26. Argentina. https://www.argentina.gob.ar/noticias/plan-de-implementacion-de-metodologia-bim-para-obra-publica. Accessed 02 Jan 2020
27. ISO, ISO 19650-1:2018. https://www.iso.org/standard/68078.html. Accessed 13 Mar 2020
28. BIM Forum Mexico. http://bimforummexico.mx/wp-content/uploads/2016/09/EbimMEXICO.pdf. Accessed 02 Jan 2020

Germany's Governmental BIM Initiative – The BIM4INFRA2020 Project Implementing the BIM Roadmap

André Borrmann[1](✉), Christian Forster[2], Thomas Liebich[3], Markus König[4], and Jan Tulke[5]

[1] Chair of Computational Modeling and Simulation, Technical University of Munich, Munich, Germany
andre.borrmann@tum.de
[2] HOCHTIEF ViCon GmbH, Essen, Germany
[3] AEC3 Deutschland GmbH, Munich, Germany
[4] Chair of Computing in Engineering, Ruhr University Bochum, Bochum, Germany
[5] Planen Bauen 4.0 GmbH, Berlin, Germany

Abstract. The technology of Building Information Modeling (BIM) promises a significant increase in productivity in the design, construction and operation of buildings and infrastructure facilities. In the entire world, the AEC industry is starting to transform itself by moving from 2D drawings to digital building models that do not only represent 3D geometry of the building components, but also all the non-geometric data required throughout the building's lifecycle. As this technological change has an impact on all stakeholders of the AEC in a fundamental way, a carefully planned transition is necessary to avoid economical damage. In many countries around the world, government-driven initiatives are underway to initiate the necessary changes. Also Germany is preparing the transition to BIM-based workflows in its AEC industry. In 2015, the German Ministry of Transport and Digital Infrastructure defined a comprehensive roadmap for the stepwise introduction of BIM methods with the goal to make BIM mandatory for federal infrastructure projects in 2020. The authors present the outcomes of the large-scale project BIM4INFRA2020 initiated by the government that helped to pave the way towards the German BIM mandate through a number of measures, including the precise definition of the BIM implementation on the basis of well-defined use cases, the guidance and supervision of a number of carefully selected pilot projects, the creation of extensive guidelines and handouts, and the formulation of measures to be taken by industry and government to overcome yet existing hurdles.

Keywords: BIM roadmap · BIM use cases · BIM pilot projects · BIM mandate

E. Toledo Santos and S. Scheer (Eds.): ICCCBE 2020, LNCE 98, pp. 452–465, 2021.
https://doi.org/10.1007/978-3-030-51295-8_31

1 Introduction

The technology of Building information modeling (BIM) promises a significant increase in productivity in the design, construction and operation of buildings and infrastructure facilities (Borrmann et al. 2018). In the entire world, the AEC industry is under transformation - overcoming the conventional inefficient and error-prone practices based on 2D drawings by moving to the work with digital building models. These digital models that do not only represent 3D geometry of the building components, but also all the non-geometric data required throughout the building's lifecycle. As this technological shift has a significant impact on all stakeholders of the AEC industry in a fundamental way, a carefully planned transition is necessary to avoid economical damage.

In many countries around the world, government-driven initiatives are underway to initiate the necessary standards, guidelines and contract templates. The most prominent example is the United Kingdom where in 2010 the Cabinet office has initiated the BIM initiative with April 2016 demarking the official starting point for mandatory use of BIM in all public projects (Cabinet Office 2011). During the preparation process, a large number of carefully elaborated guidelines and standards defining BIM processes, contents and handover have been developed. Other examples of government-driven national BIM initiatives include Finland (Senate Properties 2007), Sweden (Statsbygg 2013), Denmark (BIPS 2013), the Netherlands (Rgd 2013) and Singapore (BCA 2013).

2 German Roadmap for Establishing BIM

Like many other industrialized countries, also the German government is pushing forward the transition to BIM-based workflows in its AEC industry. In 2014, after analyzing the reasons of failure in a number of large-scale construction projects, a national commission recommended as one measure for reducing projects risks to make extensive use of digital methods such as BIM in order to make the complexity of large projects manageable (BMVI 2013).

In consequence, the German Ministry of Transport and Digital Infrastructure has developed a strategic plan for the stepwise introduction of BIM methods in public construction projects. The resulting BIM roadmap ("BIM-Stufenplan") has been published in December 2015 (BMVI 2015).

It defines the following general goals:

- Increase in planning accuracy and reduction in cost overruns
  - visualization of planning alternatives
  - less design errors by collision detection and enforced collaboration between stakeholders
  - precise assessments of cost increases caused by owners' change requests
  - improved reliability of construction processes by simulation of construction sequences

- Improved communication with the public
  - comprehensible visualization of construction projects for maximum transparency
- Optimization of life-cycle costs
  - simulation of life-cycle costs (including operation and maintenance costs)
  - provision of the digital model to the owner as a basis for facility management

The BIM roadmap provides a rough definition of "Niveau 1" for BIM-based project delivery encompassing the following key features:

- project procurement and delivery based on ISO 19650
- defining employer's information requirements (EIR) in each project providing the specification of use cases, model extent, as well as object and attribute catalogues
- creation of a BIM Execution Plan (BEP) by the supplier detailing the way the EIR are proposed to be satisfied
- implementation of the principle of federated BIM model management according to ISO 19650
- usage of a common data environment according to ISO 19650
- use of open data exchange standards (mainly IFC, but also OKSTRA and GAEB)
- formal checking of BIM models for fulfilment of the EIRs

Apart from these characteristics, the general execution of construction projects is kept as-is to the largest extent possible. This includes the legal frameworks as well as the remuneration regulations (Honorarordnung für Architekten und Ingenieure – HOAI). These decisions were taken to allow an easy and hurdle-free transition to BIM-based project execution. For a future Niveau 2, changes in laws and remuneration regulations might be required.

The defined Niveau 1 is expected to be implemented in federal infrastructure projects starting in December 2020. To reach this goal, three phases have been defined:

- Phase 1: 2015 – 2017: Preparation Phase
- Phase 2: 2017 – 2020: Extended Pilot Phase
- Phase 3: from 2020: BIM Niveau 1 for new projects

During the preparation phase (Phase 1), pioneer pilot projects were conducted (Borrmann et al. 2016) BMVI (2014), standardization activities were started and programs for professional training were initiated. This paper reports on the outcomes of Phase 2, which was mainly driven by the large-scale project BIM4INFRA2020.

3 The BIM4INFRA2020 Project

To implement phase 2 of the BIM roadmap, the German Ministry of Transport tendered a large-scale project, which was won by the BIM4INFRA2020 consortium. The consortium consisted of Planen Bauen 4.0, Technical University of Munich, Ruhr-University Bochum, HOCHTIEF ViCon, AEC Deutschland, Obermeyer Planen + Beraten, WTM Engineers, Max Bögl, and Kapellmann & Partner, thus bringing together a broad range of backgrounds, views and expertise from academia, BIM consulting, design offices, construction companies and lawyers. The consortium's work started in January 2017 and ended in August 2019.

The workplan of BIM4INFRA2020 consisted of the following main items:

- Definition of the BIM 2020 scenario
- Legal compliance on BIM project execution
- Guidance and supervision of BIM pilot projects
- Publication of guidelines for BIM implementation
- Development of a database strategy

The following sections report on the main outcomes of the individual work packages. The legal opinions are excluded as they are of limited value for the international audience. The interested reader is referred to (BIM4INFRA 2018a).

4 Definition of the BIM 2020 Scenario

To allow the market to prepare for the upcoming BIM mandate, a clear specification of the expected BIM execution of construction projects was required. The German Ministry of Transport asked BIM4INFRA2020 for the specification of three different scenarios named "Entry", "Raise", and "High performance", with both, increasing implementation effort, but also increasing benefits (Fig. 1).

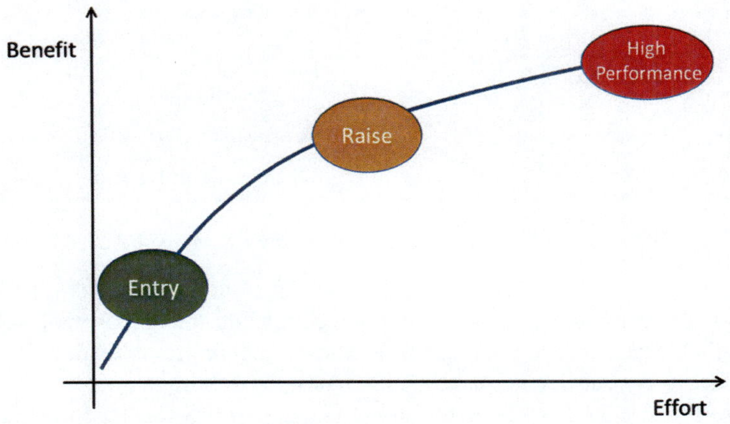

Fig. 1. The three scenarios "Entry", "Raise" and "High performance" with increasing effort, but also increasing benefit.

BIM4INFRA2020 decided to design the scenarios by defining on the one hand a framework of fixed boundary conditions that hold for each of the three scenarios, and on the other hand a flexible part based on a selection of BIM use cases that is specific for each scenario. The proposed rigid framework consists of the following boundary conditions that follow the aforementioned definition of BIM Niveau 1:

- Use of Common Data Environment
- Provision of Employer's Information Requirements
- Provision of BIM Execution Plan
- Consistent use of Classification Systems
- Procurement and project execution according to ISO 19650

These boundary conditions provide the basis for any sound BIM implementation and are thus integral part of all three scenarios.

The three scenarios are distinguished by the selection of mandatory BIM use cases. The concept of BIM use cases as the primary way of describing the purpose of apply the methodology has been developed at PennState (Kreider and Meisner 2013) and is adopted widely in the entire world. BIM4INFRA2020 defined 20 standard use cases and provided a detailed description for each of them. The complete list of use cases is depicted in Fig. 2.

1. Initial State Modeling	11. Bill of quantities / Tendering
2. Design option analysis	12. 4D Execution Planning
3. Visualization	13. Logistics planning
4. Simulation and dimensioning	14. Drawing Generation: Detailed Design
5. Coordination	15. Construction Progress Monitoring
6. Design Progress Control	16. Change Management
7. Drawing Generation: Preliminary Design	17. Billing
8. Safety Design	18. Issue Management
9. Design Approval	19. As-built documentation
10. Cost estimation	20. Operation and Maintenance

Fig. 2. List of standard use cases defined by BIM4INFRA2020

The use cases are associated with dedicated design/construction/operation phases of a building's lifecycle, thus providing a link to the current project execution phases, also with respect to contractual and honorarium issues.

The selection of the use cases for an individual scenario was based on an intense effort-benefit analysis. A panel of 40 experts consisting of owners, operators, designers and contractors was involved in a survey to assess the benefit of each use case as well as the expected effort that is required, both in terms of investment (including organizational changes, training, investment in hard- and software etc.) as well as in terms of increased effort in project execution. The scoring scale ranged from 0 to 3. The aggregated results of the survey are depicted in Fig. 3.

As shown in Fig. 3, the ratio between benefit and effort was calculated for each use case. Subsequently, thresholds were defined for associating a use case with a certain scenario. Use cases with a benefit-effort ratio over 1.5 were assigned to the "Entry" scenario, uses cases with a benefit-effort ratio between 1.2 and 1.5 were assigned to the

No	Use case	Investment Effort	Project Effort	Weighted Effort	Benefit	Benefit-Effort Ratio
1	Initial state modeling	1,7	1,1	1,5	2,5	1,67
2	Design option analysis	1,7	0,9	1,5	2,3	1,53
3	Visualization	1,0	0,7	0,9	2,5	2,73
4	Structural analysis and dimensioning	2,5	1,4	2,2	1,9	0,86
5	Coordination	1,5	0,8	1,3	2,6	1,99
6	Design progress control	1,8	0,7	1,5	2,0	1,35
7	Drawing generation: Preliminary design	1,8	0,8	1,5	2,3	1,53
8	Safety design	1,8	1,3	1,7	1,7	1,01
9	Design approval	2,0	0,8	1,7	2,1	1,26
10	Cost estimation	1,9	0,6	1,5	2,5	1,71
11	Bill of quantities	2,1	0,9	1,8	2,6	1,47
12	4D Execution planning	1,8	0,9	1,5	2,3	1,56
13	Logistics planning	2,0	1,3	1,8	1,8	1,01
14	Drawing generation: Detailed design	1,8	1,0	1,6	2,2	1,39
15	Construction progress monitoring	1,9	0,8	1,6	2,1	1,34
16	Change management	1,8	0,9	1,5	2,3	1,46
17	Billing	2,2	1,0	1,8	2,2	1,23
18	Issue management	1,6	0,9	1,4	2,4	1,73
19	As-built documentation	2,3	1,5	2,0	2,6	1,31
20	Operation and maintenance	2,4	1,3	2,1	2,7	1,30

Fig. 3. Aggregated results of effort-benefit assessment of the individual use cases conduct by 40 German BIM experts

"Raise" scenario, and the use cases with benefit-effort ratio between 1.0 and 1.2 were assigned to the "High performance" scenario. It must be noted that the higher scenario includes all use cases of the lower one. The use case "Calculation and dimensioning" has not been included due to its very low effort-benefit ratio of 0.86.

Figure 4 depicts each of the 20 use case in an effort-benefit diagram. In addition, it maps colored areas representing the individual scenarios "Entry", "Raise" and "High performance". The boundaries of the areas are determined by the effort-benefit ratio thresholds as defined above.

Based on a careful tradeoff analysis and taking into account the synergies between the use cases, BIM4INFRA2020 recommended to make the "Raise" scenario mandatory for public construction projects. At time of writing, the government has not yet taken a decision in this regard, but is expected to do so very soon.

For more details, the reader is referred to the final report on the definition of the scenario that has been published on the BIM4INFRA2020 website (BIM4INFRA 2018b).

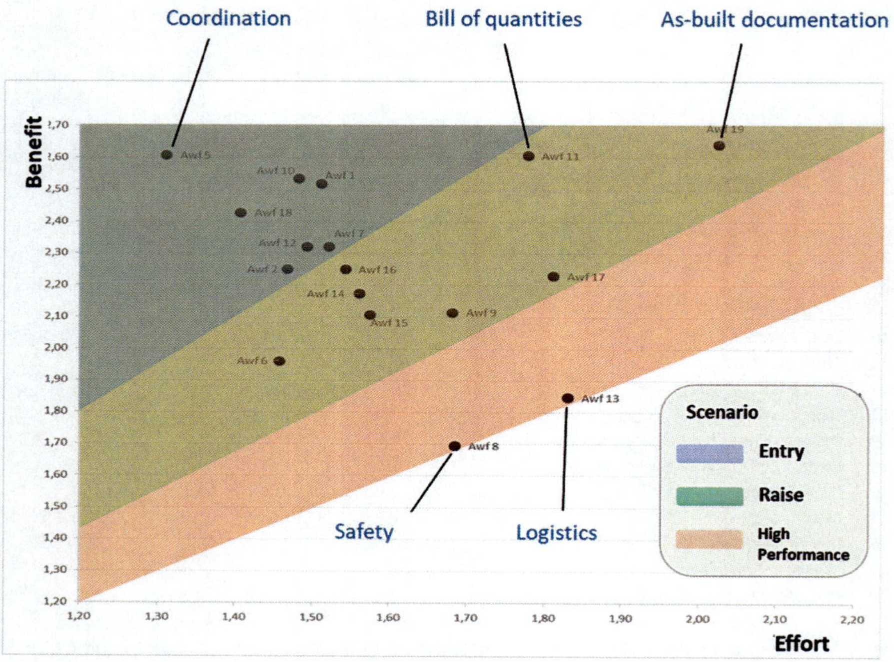

Fig. 4. Localization of the individual use cases in the effort-benefit diagram. The colored areas visualize the individual scenarios "Entry", "Raise" and "High performance".

5 Guidance and Supervision of BIM Pilot Projects

In the frame of the BIM4INFRA2020 project, seven BIM pilot projects have been conducted and intensively supervised in order to identify potential shortcomings and technological gaps that must be addressed before installing the mandatory use of BIM for all infrastructure projects. These six pilot projects were conducted:

- 3 bridges
  - Highway A99, Bridge 27/1
  - Highway A1, Bridge 533
  - Federal Road B31, Gauchachtalbrücke
- 1 tunnel
  - Highway A44: Eastern Tunnel Chain
- 1 resting place
  - Highway A7: resting place
- 1 watergate
  - Watergate Wedtlenstedt

Figure 5 depicts the BIM models of four of these six BIM pilot project.

In all projects, the focus was on the interaction between the public clients and the contracted partner. Specifically, BIM4INFRA2020 advised the projects in defining

Fig. 5. Visualization of the BIM models of four of the six pilot projects supervised and supported by BIM4INFRA2020

appropriate Employer's Information Requirements (EIR), helped in analyzing the BIM Execution Plans and provided support in checking the received models for completeness and consistency. In order to investigate different aspects and learn about potential pitfalls, the construction projects applied the BIM methodology in different setups and in different phases of the building lifecycle. The lessons learned were documented in detail and published in respective project reports.

The most comprehensive project in terms of cross-phase coverage was the A99 BW27/1 project where the early design, the detailed design and the construction phase were covered (Fig. 6, Fig. 7). In all phases, BIM models were created and handed over to the client. Particularly challenging was the fact that all three phases were conducted by different companies using different BIM authoring tools, which required the extensive use of the vendor-neutral format Industry Foundation Classes (IFC) to hand over the models between the different design phases. The respective lessons learned on the proper configuration of import and export modules were published in a comprehensive technical report (Trzeciak and Borrmann 2018).

While the pilot projects were conducted without major problems, the required effort was extraordinarily high due to insufficient knowledge and experience on both, the client and suppliers side. Accordingly, it became clear that precise guidelines were required to enable the market at large to conduct BIM projects on a broad basis with reasonable effort. At the same, the awareness was risen that the industry's shift to BIM-based working practices required a transition phase with necessary investments (organizational and financial) on all sides, before the payoff in terms of more efficient project execution can be delivered.

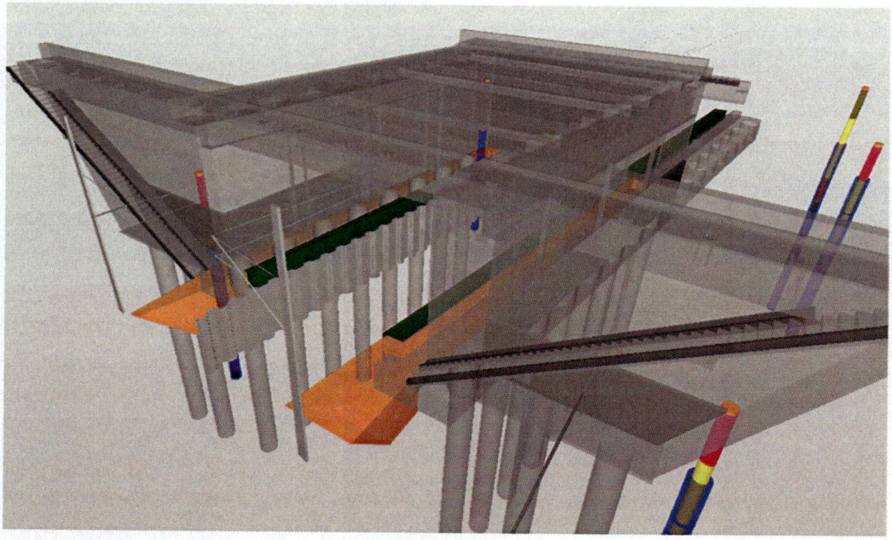

Fig. 6. Detailed model of the BIM pilot project Highway A99 Bridge 27-1

Fig. 7. Photogrammetrically created point cloud of the final bridge after construction

6 BIM Guidelines

In order to document the knowledge gathered in various pilot projects (including those mentioned in Sect. 5) and make it available to the public, BIM4INFRA2020 created 10 comprehensive guideline documents addressing the fundamental aspects of BIM project execution. In detail, the guideline documents address:

1. Basics and overall process
2. Employers Information Requirements (including templates and examples)
3. BIM Execution Plan (including templates and examples)
4. Specification of BIM services (related to contract and honorarium)

5. Contract templates
6. BIM use cases
7. BIM discipline models and Levels of Development
8. Vendor-neutral data exchange
9. Data exchange using IFC
10. BIM Technology

The guidelines add up to 250 pages in total and are available online free of charge (BIM4INFRA 2019). The target audience is AECO professionals on both client as well as supplier/contractor side.

An example from Guideline 6 is provided in Fig. 8.

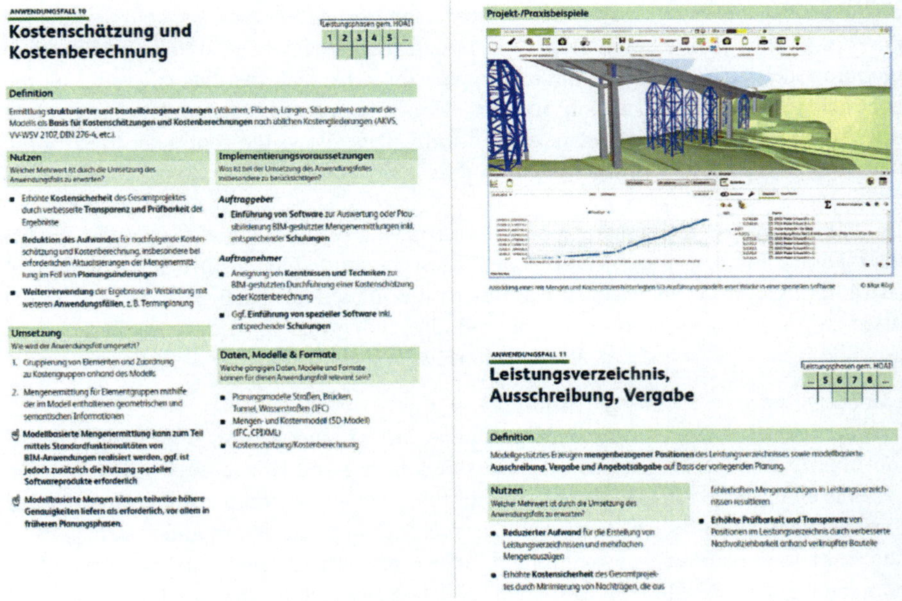

ANWENDUNGSFALL 10

Kostenschätzung und Kostenberechnung

Leistungsphasen gem. HOAI: 1 | 2 | 3 | 4 | 5 | ...

Definition

Ermittlung **strukturierter und bauteilbezogener Mengen** (Volumen, Flächen, Längen, Stückzahlen) anhand des Modells als **Basis für Kostenschätzungen und Kostenberechnungen** nach üblichen Kostengliederungen (AKVS, VV-WSV 2107, DIN 276-4, etc.).

Nutzen

Welcher Mehrwert ist durch die Umsetzung des Anwendungsfalls zu erwarten?

- Erhöhte **Kostensicherheit** des Gesamtprojektes durch verbesserte **Transparenz und Prüfbarkeit** der Ergebnisse
- **Reduktion des Aufwandes** für nachfolgende Kostenschätzung und Kostenberechnung, insbesondere bei erforderlichen Aktualisierungen der Mengenermittlung im Fall von **Planungsänderungen**
- **Weiterverwendung** der Ergebnisse in Verbindung mit weiteren **Anwendungsfällen**, z. B. Terminplanung

Umsetzung

Wie wird der Anwendungsfall umgesetzt?

1. Gruppierung von Elementen und Zuordnung zu Kostengruppen anhand des Modells
2. Mengenermittlung für Elementgruppen mithilfe der im Modell enthaltenen geometrischen und semantischen Informationen

Modellbasierte Mengenermittlung kann zum Teil mittels Standardfunktionalitäten von BIM-Anwendungen realisiert werden, ggf. ist jedoch zusätzlich die Nutzung spezieller Softwareprodukte erforderlich

Modellbasierte Mengen können teilweise höhere Genauigkeiten liefern als erforderlich, vor allem in früheren Planungsphasen.

Implementierungsvoraussetzungen

Was ist bei der Umsetzung des Anwendungsfalles insbesondere zu berücksichtigen?

Auftraggeber

- **Einführung von Software** zur Auswertung oder Plausibilisierung BIM-gestützter Mengenermittlungen inkl. entsprechender **Schulungen**

Auftragnehmer

- Aneignung von **Kenntnissen und Techniken** zur BIM-gestützten Durchführung einer Kostenschätzung oder Kostenberechnung
- Ggf. **Einführung von spezieller Software** inkl. entsprechender **Schulungen**

Daten, Modelle & Formate

Welche gängigen Daten, Modelle und Formate können für diesen Anwendungsfall relevant sein?

- Planungsmodelle Straßen, Brücken, Tunnel, Wasserstraßen (IFC)
- Mengen- und Kostenmodell (5D-Modell) (IFC, CPIXML)
- Kostenschätzung/Kostenberechnung

Projekt-/Praxisbeispiele

ANWENDUNGSFALL 11

Leistungsverzeichnis, Ausschreibung, Vergabe

Leistungsphasen gem. HOAI: ... | 5 | 6 | 7 | 8 | ...

Definition

Modellgestütztes Erzeugen **mengenbezogener Positionen** des Leistungsverzeichnisses sowie modellbasierte **Ausschreibung, Vergabe und Angebotsabgabe** auf Basis der vorliegenden Planung.

Nutzen

Welcher Mehrwert ist durch die Umsetzung des Anwendungsfalls zu erwarten?

- **Reduzierter Aufwand** für die Erstellung von Leistungsverzeichnissen und mehrfachen Mengenauszügen
- Erhöhte **Kostensicherheit** des Gesamtprojektes durch Minimierung von Nachträgen, die aus fehlerhaften Mengenauszügen in Leistungsverzeichnissen resultieren
- **Erhöhte Prüfbarkeit und Transparenz** von Positionen im Leistungsverzeichnis durch verbesserte Nachvollziehbarkeit anhand verknüpfter Bauteile

Fig. 8. Detailed description of the BIM use cases specified by BIM4INFRA2020 in Guideline 6

7 Database Concept

As a result of the development of BIM4INFRA2020, it was identified that, in addition to the preparation of guidelines, a technical support of the public clients and contractors for BIM-based project management is necessary.

For public clients, it is essential that the ordering, checking and acceptance of the relevant building information can be done in a standardized, uncomplicated and well-organized way. The public clients should be able to define and check the information requirements with the help of suitable tools. This provides the contractors with clear and consistent guidelines to enable them to implement the contracted information deliveries in a proper manner. This can improve the overall efficiency of the public sector and the productivity of the construction industry. Clients need digital support in the definition of project-specific information requirements. This also offers an advantage for the contractors, as the internal processes for creating bids and digital models can be easily adapted to the standardized specifications.

Contractors should be supported in the creation of building models based on the defined Employers Information Requirements (AIA) by appropriate software systems. Currently, the contractors must adapt the objects of the building models to the specified AIA in each case according to the specific project. If AIA-compliant templates were available for certain model objects, the model creation could be simplified and made more efficient. There are also advantages for the client, since the delivered models have a standardized structure and contain fewer errors.

The handover of results during and at the end of project phases is carried out in digital model-based form. For this purpose, appropriate systems for submission and checking by the public client must be provided. Public clients need tools for checking digital models in order to be able to compare the AIA with the data delivered by the contractors. Automated checking reduces errors and ensures high model quality. Test rules should be provided in a standardized form. This allows the contractor to carry out a preliminary check of his digital models to be delivered in order to avoid rework at an early stage.

The following recommendations for actions are based on the analysis of the international and national status quo. The corresponding report is available at (BIM4INFRA 2017). The functionalities of the individual databases or modules are listed in the following. Access to the data and individual functions should be made possible using web interfaces and web portals.

Classification and Property Database

In order to enable a seamless exchange and consistent reuse of digital building models, the objects must be described and classified in a standardized way. Classification systems and property databases are used for this purpose. Therefore, it is recommended to implement databases and corresponding user interfaces for the creation, management and use of classification systems and object properties. High quality, adapted to national regulations, and high availability must be ensured. In addition, an international coordination and linking should be promoted. The definition of the content must be carried out in close coordination with all stakeholders. Classifications and properties are maintained by individual expert groups. To ensure a consistent structure and to avoid duplication or ambiguity, overall coordination and quality management is necessary.

Employers Information Requirements Module

Classifications and properties form the basis for the definition of the AIA. Configurable templates should be made available to the public clients to create the precise information requirements for different phases and use cases. The AIA database maps the information requirements for the digital building models, i.e. information requirements can be defined for individual BIM use cases for the building models and objects to be included. Extensible templates for standard tendering procedures can be published in the AIA database, depending on the public client, the type of building and the size of the project. The database is also used to store project-specific AIA that can be re-used in similar projects. As a rule, project-specific AIA should be created based on AIA templates; however, they can also be customized or enhanced. The AIA are provided in machine-readable form. Thus, the requirements for the model objects can be transferred to existing tools. In addition, machine-readable checking rules are also generated, so that a quick check of the submitted models with regard to the AIA is possible.

Object Template Database
Configurable BIM object templates should be used for model creation. The object templates should satisfy the AIA so that subsequent modifications of a model object are not necessary. To enable contractors to ensure that they use suitable BIM object templates, a certification database with an associated quality assurance concept should be implemented. The configuration of BIM object templates using project-specific AIA should be as automatic as possible. The AIA for model objects should be provided in an evaluable and uniform format. With the help of software tools that can also be used by BIM object template providers and integrated into their own products, it should be possible to adopt, test and configure a BIM object template.

The interaction of the individual databases and modules is shown in Fig. 9. The described functionality is currently under development by the German Centre for the Digitisation of the Construction Industry (BIM Germany).

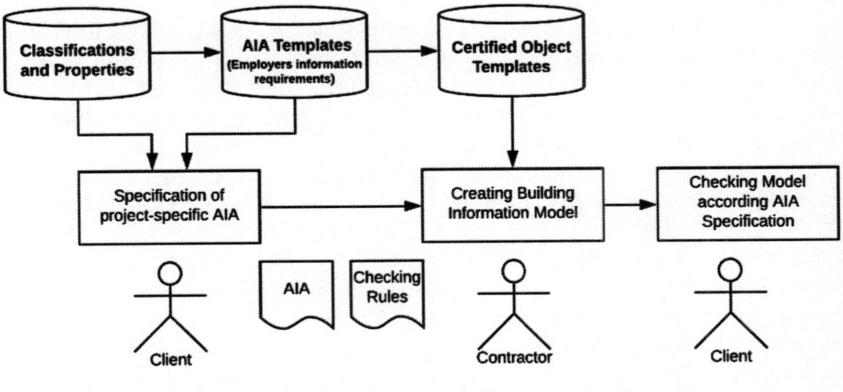

Fig. 9. Databases and modules of the future BIM portal for Germany

8 Conclusion

The BIM4INFRA2020 project provided an important corner stone for the broad implementation of BIM in Germany. By supervision of the BIM pilot projects and the intense consultation with various stakeholder groups, the most critical hurdles were identified. It became apparent that the BIM implementation process is at risk to be slowed down by the limited experience and capabilities of both, the public administration as well as the supplier side. The starting date for the mandatory use of BIM in federal infrastructure projects has been accordingly shifted to December 2020. The exact scope of the BIM mandate has not been specified but the groundwork conducted by BIM4INFRA2020 provides a solid basis for decision making.

The most important results of the BIM4INFRA2020 project were (1) the definition of the BIM 2020 scenarios by assessing the BIM use cases, (2) the publication of comprehensive BIM guidelines, and (3) the database concept. All these results provide crucial building blocks for implementing the German BIM roadmap.

Recently, the German Centre for the Digitisation of the Construction Industry (BIM Germany) was launched by the German government (BMVI 2020). It covers both the infrastructure as well as the housebuilding sector and will continue the work of BIM4INFRA2020 by pushing forward standardization and harmonization, developing a comprehensive training concept, providing consultancy and support for BIM projects and creating a vision for the future digitalization of the German construction sector.

References

BCA Singapore: Singapore BIM Guide V2, Building and Construction Authority, Singapore (2013). https://www.corenet.gov.sg/integrated_submission/bim/BIM_Guide.htm. Accessed 30 Jan 2020

BIM4INFRA: Umsetzung des Stufenplans "Digitales Planen und Bauen": Konzept für Datenbanken, Project report (2017). https://bim4infra.de/wp-content/uploads/2018/08/BIM4INFRA2020_AP5_Datenbankkonzept_FINAL.pdf

BIM4INFRA: Umsetzung des Stufenplans "Digitales Planen und Bauen": Klärung von Rechtsfragen, Project report (2018a). https://bim4infra.de/wp-content/uploads/2019/09/BIM4INFRA_AP3.1.pdf. Accessed 30 Jan 2020

BIM4INFRA: Umsetzung des Stufenplans "Digitales Planen und Bauen": Szenariendefinition und Empfehlung, Project report (2018b). https://bim4infra.de/wp-content/uploads/2018/09/AP1.2-AP1.3_BIM4INFRA_Bericht-Stufenplan.pdf

BIM4INFRA: Umsetzung des Stufenplans "Digitales Planen und Bauen": Handreichungen, Project report (2019). https://bim4infra.de/handreichungen/. Accessed 30 Jan 2020

BIPS: Danish BIM Guide (2013). http://bips.dk. Accessed 30 Jan 2020

BMVI: Reformkommission Bau von Großprojekten. Federal Ministry of Transport and Digital Infrastructure (2013). https://www.bmvi.de/SharedDocs/DE/Artikel/G/reformkommission-bau-von-grossprojekten.html. Accessed 30 Jan 2020

BMVI: BIM-Pilotprojekte des BMVI (2014). http://www.bmvi.de/SharedDocs/DE/Pressemitteilungen/2014/099-dobrindt-reformkommission-grossprojekte.html. Accessed 30 Jan 2020

BMVI: Digitales Planen und Bauen – Stufenplan zur Einführung von Building Information Modeling (BIM). Federal Ministry of Transport and Digital Infrastructure (2015). http://www.bmvi.de/SharedDocs/DE/Artikel/DG/digitales-bauen.html. Accessed 30 Jan 2020

BMVI: Nationales Zentrum für die Digitalisierung des Bauwesens "BIM Deutschland" nimmt Arbeit auf, Press Release, German Ministry of Transport (2020). https://www.bmvi.de/SharedDocs/DE/Pressemitteilungen/2020/002-zieschang-bim-kompetenzzentrum.html

Borrmann, A., König, M., Koch, C., Beetz, J.: Building information modeling: Why? What? How?, In: Borrmann, A., König, M., Koch, C., Beetz, J. (eds.) Building Information Modeling. Springer, Cham (2018)

Cabinet Office: Government Construction Strategy, Cabinet Office, London, UK (2011). https://www.gov.uk/government/publications/government-construction-strategy. Accessed 30 Jan 2020

Eastman, C., Teicholz, P., Sacks, R., Liston, K.: BIM Handbook: A Guide to Building Information Modeling for Owners, Managers, Designers, Engineers, and Contractors, 2nd edn. Wiley, Hoboken (2011)

ICE: BIM Maturity Measurement Tool, Institution of Civil Engineers, UK (2015). https://www.ice.org.uk/disciplines-and-resources/best-practice/bim-maturity-measurement-tool

Kam, C., Senaratna, D., Xiao Y., McKinney, B.: The VDC Scorecard: Evaluation of AEC Projects and Industry Trends, CIFE Working Paper #WP136, Standford University (2013)

Kam, C.: Managing BIM projects, organizations, and policies: turning aspirations into quantitative measures of success, In: Kensek, N. (ed.): Building Information Modeling: BIM in Current and Future Practice. Wiley, Hoboken (2015)

Kreider, R.G., Messner, J.I.: The uses of BIM. Classifying and Selecting BIM, Pennsylvania State University (9th version) (2013)

Rgd: Rgd BIM Standard, Rijksgebouwendients, Niederlande, Verfügbar unter (2013). http://www.rijksvastgoedbedrijf.nl/english/expertise-and-services/b/building-information-modelling-bim

Senate Properties: Senate Properties' BIM Requirements 2007 - Finland (2007)

Statsbygg: BIM Manual V 1.2.1, Statsbygg, Norway (2013). http://www.statsbygg.no/Files/publikasjoner/manualer/StatsbyggBIM-manual-ver1-2-1eng-2013-12-17.pdf. Accessed 30 Jan 2020

Trzeciak, M., Borrmann, A.: Design-to-design exchange of bridge models using IFC: a case study with Revit and Allplan, In: 12th European Conference on Product and Process Modelling, Copenhagen, Denmark (2018). https://publications.cms.bgu.tum.de/2018_TRZECIAK_ECPPM.pdf. Accessed 30 Jan 2020

Improving the Design Process Quality Using BIM: A Case Study

Luiz Fernando Domingues[1](✉) and Eduardo Toledo Santos[2]

[1] Modo Projetos Engenharia, São Paulo, Brazil
luiz@modoprojetos.com.br
[2] University of São Paulo, São Paulo, Brazil
etoledo@usp.br

Abstract. Given the constant search for increasing accuracy in construction projects as well as expanding efficiency on the implementation of real estate projects and meeting high quality requirements, design companies have been exploring new ways to enhance their processes, focusing on improving information flow, conceptualization and consistency in the development of their products. As a result, companies have sought BIM (Building Information Modeling) to improve their processes, but do not always decide to implement it, partly due to the limited availability of metrics that prove its efficiency. Aiming to improve this scenario, this paper presents metrics obtained before and after the implementation of BIM in a real case. Focusing on the quality improvement of design and coordination processes on projects using BIM methodology, this paper has, as object of study, a multidisciplinary design company that serves mainly the mixed use and medium-sized buildings construction market. The research method applied was the action-research, with a case study comparing the quality indicators of similar projects under 3 different scenarios: the first, elaborated using a 2D CAD tool; the second, using BIM platforms and; the third one, using BIM supplemented with the help of BIMserver, Insight and Dynamo applications. The data for analyses was gathered through the company's management database, focusing on the analysis of the resources spent on design development and on rework due to lack of quality. It was verified that the implementation of BIM, with the proposed tools, has reduced the rework due to design errors by 42.5%.

Keywords: Building Information Modeling · BIM tool · Quality · Project management · Construction · Metrics

1 Introduction

The performance of the civil construction sector, in terms of productivity, is low, having unsatisfactory design quality among its main causes. With the advancement of information technology, new tools are emerging to assist designers to obtain more consistent results that may increase efficiency and precision in the execution of building designs.

It has been known for a long time that design quality is a major factor in the performance and efficiency of construction projects [1], and that the use of Information

E. Toledo Santos and S. Scheer (Eds.): ICCCBE 2020, LNCE 98, pp. 466–482, 2021.
https://doi.org/10.1007/978-3-030-51295-8_32

Technology (IT) tools has helped to improve it [2]. More recently, Building Information Modeling tools and processes [3] enabled increased information sharing and collaborative decision capabilities, making it easier and more efficient reaching quality on building designs [4].

The quality of the design processes supported by BIM may be optimized by automating repetitive design activities, assisting multidisciplinary collaboration and supporting technical decision making of great impact for the project development. With BIM support, professionals can focus more on engineering processes and less on producing documentation.

Even with much evidence available on the positive return on BIM implementation, BIM adoption in many countries and markets are still in early stages. Being a very conservative sector by its own nature, construction stakeholders can benefit from successful case studies with hard evidence focusing on BIM implementation on different markets and facility types, that matches their own contexts.

This work aims to demonstrate, with the analysis of data collected during a transition process from CAD technology to BIM methodology, and later with the introduction of BIMserver, Insight and Dynamo software tools in the design process, the benefits achieved by changes in the design platform and processes of an engineering design company.

2 Method

The research method adopted was the action-research [5] which is somewhat similar to a case study, but one where the researcher is no longer just an observer, but performs an active role, interacting with the studied object. In this case, the first author, as a manager in the studied company, acted on the implementation of BIM methods on its design process and, later on, introduced additional BIM tools as a way to improve detected quality gaps.

The main objective of this research was to test the hypothesis that BIM adoption improves design quality, reducing rework on the design phase. Also, for this specific case, it was verified if additional implementation of specific BIM tools, targeted at identified design performance problems, were effective to mitigating them, and further improving quality and efficiency of the design process. These tools were:

- BIMserver [6] is an open platform, based on the IFC [7] standard, which enables storing and managing Building Information Model data. In this research, it was used for model management and for comparing model versions for identifying changes;
- Autodesk Insight [8] is a BIM enabled-tool for energy analysis, used for studying building energy performance through simulation. Integrated with Autodesk Revit, it allowed the company designers to better understand the effects of their design decisions on energy performance and, particularly, to assess architect's proposed solution before detailing energy-related systems.
- Dynamo [9] is an open source tool used for expanding Revit native functions, enabling the implementation of algorithms through a visual language. Its is manly used for automating repetitive tasks, interacting with BIM models on Revit or

model development based on complex parameters or external data. In this case study, it was used for checking and guarantying consistency of related parameters on different BIM models.

3 Case Study

In this section, the studied company and its market context are presented. Then, its design process is detailed, especially the BIM-related tasks.

It is important to present such details here so that the precise context in which the results were obtained are clearly stated. It is a common error by novice BIM adopters not to pay enough attention to BIM implementation planning, jeopardizing the intended BIM benefits the company was expecting to obtain.

3.1 Company Presentation

The company that served as the basis for the preparation of this study is an engineering design office located in São Paulo City, Brazil, that provides solutions for the disciplines of concrete and metal structures, HVAC, and electrical and plumbing building systems.

This engineering design office is specialized in the development of mixed-use projects, usually containing auditoriums, administrative rooms, classrooms and apartments (e.g., large church facilities), having developed designs for more than 10 projects of this kind in the last 2 years. It also has, in its portfolio, designs for commercial, residential and hotel developments, hospitals, conference centers and supermarkets. The services provided cover all design and preconstruction tasks, including survey of existing installations, preliminary design, schematic design and legal and construction design documents, as well as 4D planning, technical assistance to construction and consultancy. Currently, the organization uses Building Information Modeling (BIM) methodology on 100% of its contracts.

3.2 Design Process

At the beginning of every design contract, all participants of the new project are brought together to define various issues related to its development. To ensure that all parties have the same understanding of what will be produced, a first kick-off meeting is held to define objectives, resources, restrictions, deadlines and schedules for the project, in addition to presenting all guidelines and standards used by the company, and any other specific request from the client. This procedure is performed considering the importance of objective directions to the design participants who need to understand what should or should not be included in the BIM model, and what information can or cannot be used from the model at each project milestone. All consolidated items are recorded in the BIM Execution Plan (BEP), which establishes processes and needs according to the respective client, with the following main contents being covered:

Preliminary BIM definitions:

- Selection of BIM uses for the project;
- Definition of teams and their respective responsibilities;
- Delimitation of the intellectual property of models and their components;
- Definition of the Work Breakdown Structure (WBS) and the milestones for BIM uses;
- Modeling guidelines;
- Definition of the software, including versions, to be used.

Design and coordination guidelines:

- Definition of the design phases;
- Definition of the level of development of elements in the model;
- Definition of collaboration procedures;
- Definition of model control and quality procedures.

Project delivery guidelines:

- Definition of schedule and delivery dates;
- Definition of delivery file formats;
- Definition of deliverables.

This information is necessary for the development of the project and it is defined with the client during the kick-off meeting and included in the BEP. Based on the definition of the levels of detail for each stage of the design work, the schedules are analyzed and revised.

At the beginning of each design stage, three tasks are developed simultaneously for all disciplines, with the intention of making the design process more collaborative. At the end of each of these activities, which normally lasts one day, a meeting is held to present the solutions developed by each participating engineering group, together with a report by each discipline.

The first task is the organization of the spaces, where each engineering trade studies its needs for technical areas and spaces above the ceiling and for shafts throughout the building, according to the solutions that are being proposed, modeling projected infrastructure and equipment. The second task is for coordination among building systems: each discipline studies the technical needs of the other disciplines presented on the previous day, with the aid of clash reports and virtual tours, identifying conflict spots or which are inadequate in relation to the needs of the project. In the last task, coordination includes, besides building systems, also architecture and structures. The team valuates improvements and more efficient solutions for the design as a whole.

After the initial conceptualization tasks, the design stage is developed with all disciplines working in parallel, with demands being organized through meetings every morning between the coordinators and the architect responsible for planning. A general production schedule, subdivided by designer, is made available daily to everyone by e-mail (Fig. 1).

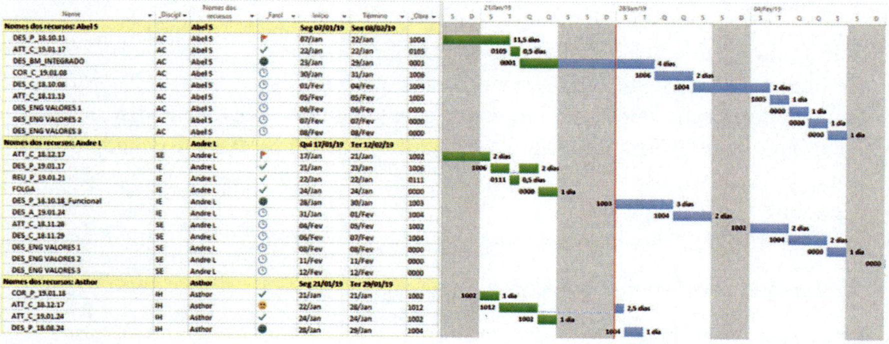

Fig. 1. Example of production schedule

As a way of guaranteeing the quality of all developed designs, the design coordinators carry out some checks for internal control of deliverables in order to ensure that everything is in accordance with the expected standards, ensuring that the requirements documented on the BEP are being met.

Weekly analyzes are carried out on the designs that had tasks executed in the current week, where the coordinators take a tour of the models, following all the installations routes and their interactions with the structure and architecture, making sure that the needs exposed by the client are being met, verifying whether duplicate or unintentionally inserted elements exist in the model and ensuring that the inserted elements correspond to the level of detail defined previously.

The communication of technical problems found in the internal analyses is done with DWF and BCF (BIM Collaboration Format) files: the first file type is used for storing comments on 2D documents to be sent to customers and the second one for documenting problems detected in 3D models.

At the end of each design phase, reports are generated with the Navisworks software to detect interferences between elements in the models. In these reports, soft clashes (elements not within a minimum required distance in relation to another element or system) and hard clashes (overlapping elements) are specified. These clash reports, along with the detected interferences resolved in the models, are delivered to the client for assuring him that the design is free from errors and there will be no coordination problems during the work on site.

After finishing construction documents phase, the company offers 4D Planning services [10] to the client, which consists in the development of a schedule of construction activities in the form of an animation for the purposes of validation and optimization of the construction sequence. This BIM use enables the visualization of which elements are on time, advanced or delayed, thus enabling the comparison of planned vs actual schedule, as long as the 4D model is updated during the execution phase.

The combination of the 3D model with the established schedule for construction favors the monitoring of deadlines and a better understanding of work progress, making it easier to identify unexpected delays or planning problems. This service, when contracted, provides a suggestion for work sequencing on a file open for editing, allowing the client to update it according to the progress of the work and, thus, guarantee control of the activities that are being or will be performed.

The sequencing provided is a suggestion based on studies carried out by the company and aimed at productivity and time and cost effectiveness, which can be modified according to the specific construction process of each client. The tool is developed using Microsoft's MS Project timeline software, in conjunction with Autodesk's Navisworks visualization platform.

All requests for adjustment (change order) or clarification (request for information) are documented in the company's organizational database aiming to provide accurate and proper service to each issue, in addition to feeding its database of lessons learned.

All project demands are classified with an indicator called "Demand Classification", which is a characterization of the demand source for all services performed in each project. It can be a planned work or it may be originated due to error (from the technical team, from partners, from management, from client decisions, etc.). This is done with the main objective of monitoring the performance and quality of developed projects.

4 Results and Discussion

Through the analysis of the company's design process during its operation, it was possible to observe points of failure, which needed treatment in order to improve the quality of the designs and achieve greater customer satisfaction. Among the greatest difficulties encountered, the complexity in the flow of information stood out, the multidisciplinary of designs and coordination problems and technical consistency, which were subsequently dealt with, respectively, with the implementation of the BIMserver, Insight and Dynamo software tools.

During the development of some designs using the company's process, information flow problems were found. Small adjustments in one discipline generated the need for adjustments in others, consequently creating a demand for storage, organization and distribution of information during internal production interactions. This need led to the search for tools that would allow data and communication management through a model sharing platform.

Additionally, postponing the multidisciplinary analysis of the designs proved to be a great generator of reworks, due to the low interaction between architecture and building systems in the design phase. It was found that the lack of deeper conceptual studies in early stages proved to be a major obstacle in the development on later stages since, in the traditional production process, the impact of poorly conceived architectural solutions with regard to energy efficiency are only perceived in later stages of building systems design, requiring rework from engineering teams.

Another problem encountered was the difficulty of ensuring technical compatibility between disciplines. Through the use of integration software, already incorporated in the company process, a gap was found in the process of trade coordination and in the consistency of building systems designs, since these tools analyze physical, temporal and dynamic interferences, but not technical compatibility between the disciplines. For example, is the electrical power supply for equipment shown on the air conditioning design the same as that shown in the electrical design? (it may have changed and not updated in the model of the other trade).

As presented in Sect. 2, the company has a production schedule focused on the organization of tasks, controlled by the coordination team and operated by the planning team. When allocating a task in the schedule, the coordinators indicate, using an index called "Demand Control", which was the cause that originated that task, allowing to gauge the number of days spent on the normal development of each discipline and on failures from different sources, which are classified as follows:

- CC - Failures caused by errors in Coordination and Compatibility
- DP - Design Development
- EC - Client Error
- EG - Management error
- EI - Internal Team Error
- EP - Partner Error (Architecture Office, for example)

Using fixed approximate percentage values, coordinators can allocate up to 3 types of errors per task, dividing them into percentages of 50%, 30% or 20% that add up, if they are from the same source, tracing the origin of the demands. In completing the example shown in Table 1, the task had a profile of 80% of time spent for its execution in design development (useful work) and 20% of its time spent for correcting internal errors (rework), caused by employee failures during design.

The use of indexes in fixed bands for filling the records was decided so that it became easier and faster to fill, since each coordinator could have a slightly different interpretation of the same demand if the value could vary freely, leading to greater inconsistencies in data collection. The information is used to aid in the identification of the demand sources, thus facilitating the classification of hours worked due to internal, coordination, partner, manager and customer errors, helping in the analysis of the company's performance and, in some cases, serving as a basis for requesting contractual amendments.

Table 1. Example of demand classification

CD 1–50%	CD 2–30%	CD 3–20%
DP	DP	EI

The company implemented this task control system prior to the implementation of BIM. This allowed the recording of project data that were prepared on a CAD platform, as well as another one using BIM and yet another one using BIM together with tools to aid the additional quality improvement proposed in this work. Thus, for the comparative analysis of the results obtained, both with the implementation of BIM, and with the implementation of BIM quality improvement tools, three mixed-use building projects designed for the same client were chosen, each developed with one of the cited platforms.

The first design, developed entirely on a CAD platform, shown in Fig. 2, is a 4,500 m^2 building project, located in the city of Praia Grande, south coast of the state of São Paulo. The building is divided into Block A, Central Block and Block B, with Blocks A and B consisting of three floors each, that house classrooms, administrative rooms and apartments, while the Central Block is ground floor and consists of only one auditorium, with capacity for 2,100 people.

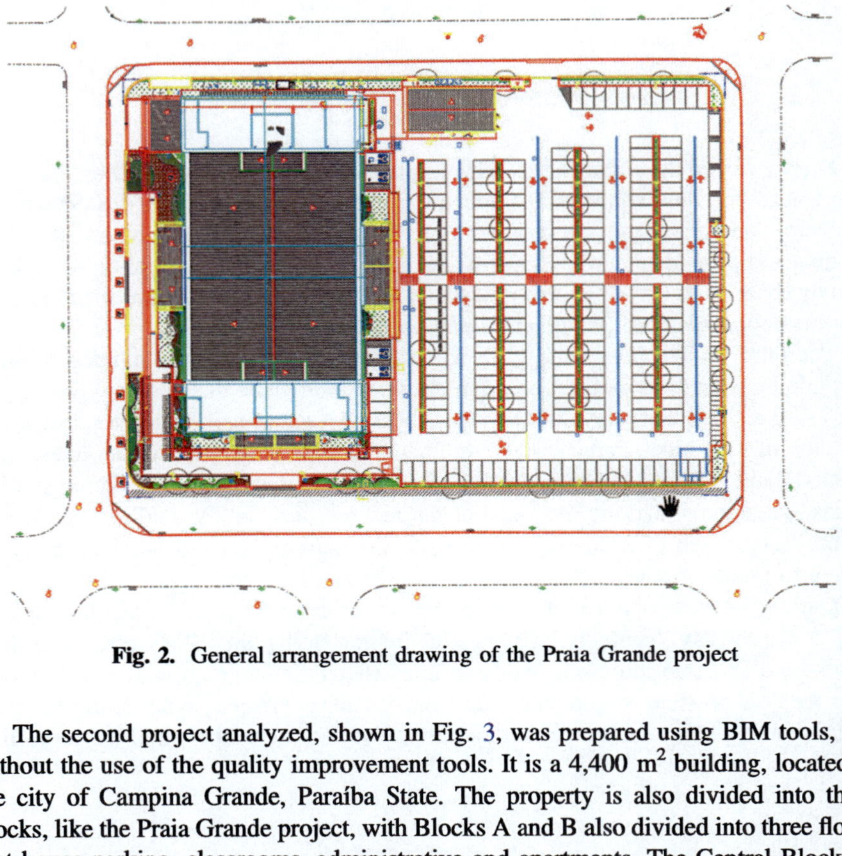

Fig. 2. General arrangement drawing of the Praia Grande project

The second project analyzed, shown in Fig. 3, was prepared using BIM tools, but without the use of the quality improvement tools. It is a 4,400 m^2 building, located in the city of Campina Grande, Paraíba State. The property is also divided into three blocks, like the Praia Grande project, with Blocks A and B also divided into three floors that house parking, classrooms, administrative and apartments. The Central Block, in this project, has an underground parking lot and a ground floor with an auditorium for 1,200 people.

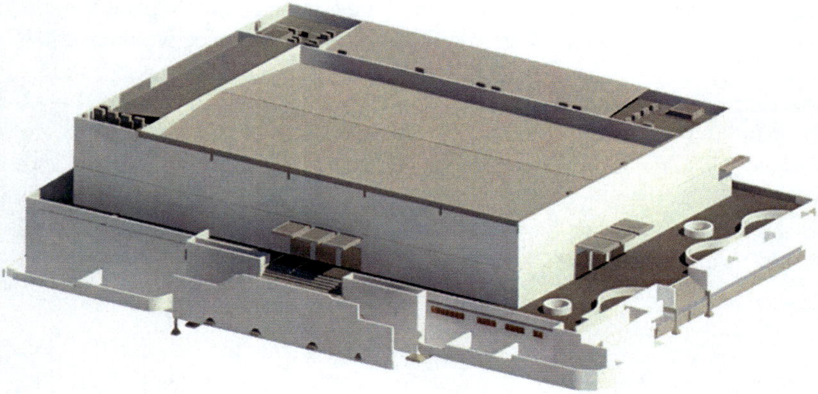

Fig. 3. Perspective of the Campina Grande project

During design development, all BIM implementation tasks developed by designers (creation of libraries, templates and learning of tools) were separated and allocated in a schedule as training and internal work. In this way, time consumption related to learning and development of design aid accessories was segregated from the work time directly spent on design. With this procedure, efforts made in the implementation of the new methodologies did not influence recorded times.

The third project (Fig. 4) to have its results examined used, in addition to BIM methodology, the three tools proposed in this work to improve the quality of projects. It is a 7,300 m^2 building, also divided into Blocks A, Central and B and it is located in the city of Londrina, Paraná State. In this building, unlike the others presented, Blocks A and B have 5 floors that include an auditorium for 240 people, administrative rooms, classrooms, parking lots, apartments and a radio and TV studio. The Central Block is composed of 3 floors, two of which are parking lots and one with an auditorium for 1,800 people.

During the development of this project, as in the case of the Campina Grande project, every day spent on learning and implementing the BIMserver, Autodesk Insight and Dynamo tools was allocated in a schedule as training and internal work. This made it possible to compare data from the three projects, considering the proportion of time dedicated in relation to the total days of development of the respective design.

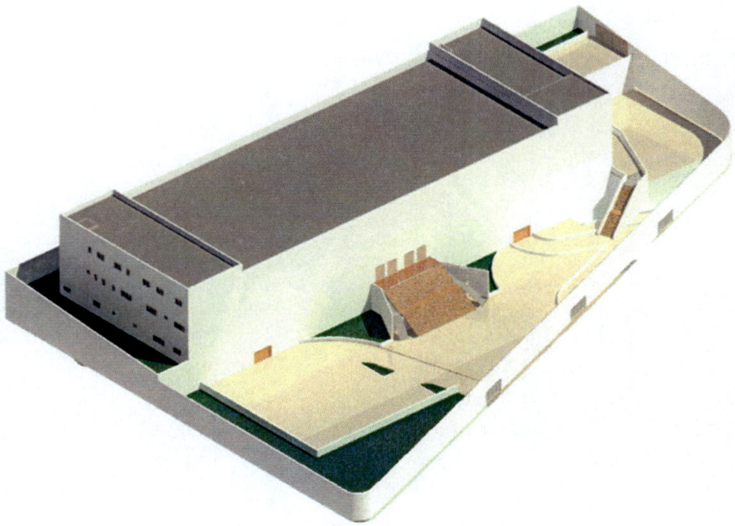

Fig. 4. Perspective of the Londrina project

Using buildings of the same typology and similar complexity added greater reliability to the results.

For the analysis of the project designs, reports were generated presenting the worked days in each project with the time spent in each discipline classified by designer. By separating the total time spent per employee in days worked due to each type of demand control, it was possible to characterize the results based on time spent on design development and time spent on correcting failures from different sources (rework).

In all three projects, the studied company provided designs to the (single) contractor of the same disciplines, allowing the comparison between results obtained from similar tasks. The designs prepared by the company for the three buildings were as follows:

- AC - Air Conditioning
- EC - Concrete Structure
- IE - Electrical Installations
- IG - Combustible Gas Installations
- IH - Hydraulic Installations
- II - Fire Protection Installations
- IS - Sanitary Installations
- SE - Electronic Systems
- ENE - Electrical Substation
- FU - Structural Foundations

The reports extracted from the task schedule for the three designs are presented in Tables 2, 3 and 4. It can be seen that the majority of time was spent on the design development and on rework caused by internal errors and in the correction of failures caused by coordination and compatibility errors. Therefore, it is understood that, to improve the efficiency of the design process, techniques and tools that provide increased consistency and coordination must be implemented.

Table 2. Results obtained from the Praia Grande project (CAD)

Project (Method)	Discipline	Designer	Worked Days	CC	DP	EC	EG	EI	EP
Praia Grande (CAD)	AC	Carlos Lima	32.75	–	26.25	2.50	–	4.00	–
	EC	Heloisa	55.25	2.50	48.25	–	2.50	2.00	–
	IE	Andre L	7.75	–	6.50	–	0.75	–	0.50
		Ednelson	20.25	1.75	10.25	3.75	–	4.50	–
		João G	16.00	–	8.50	3.50	–	4.00	–
		Karina	2.75	–	2.75	–	–	–	–
	IG	Alessandro	4.25	–	3.00	–	0.25	1.00	–
	IH	Alessandro	18.25	2.00	9.25	3.50	–	3.50	–
		Claudia	6.00	2.50	–	–	1.50	2.00	–
	II	Fernando	10.25	–	9.00	–	–	1.25	–
	IS	Alessandro	12.50	–	10.25	–	–	2.25	–
	SE	Andre L	21.00	2.00	16.00	–	–	3.00	–
		Ednelson	4.50	0.75	–	–	2.25	1.50	–
		João G	3.25	–	2.00	0.50	–	0.75	–
	ENE	Andre L	12.00	–	8.50	–	2.50	1.00	–
	FU	Heloisa	9.75	–	9.75	–	–	–	–

Table 3. Results obtained in the Campina Grande project (BIM)

Project (Method)	Discipline	Designer	Worked Days	CC	DP	EC	EG	EI	EP
Campina Grande (BIM)	AC	Abel S	38.00	–	32.50	–	3.50	2.00	–
	EC	Heloisa	25.25	2.50	19.25	–	–	2.25	1.25
		Leonardo C	33.25	–	24.50	3.50	–	5.25	–
	IE	Ednelson	42.25	–	31.00	3.75	–	7.50	–
		Carlos Crispin	8.25	1.25	6.75	–	–	0.25	–
	IG	Alessandro	3.50	–	3.50	–	–	–	–
	IH	Alessandro	12.50	–	12.50	–	–	–	–
		Matheus	9.50	3.25	3.75	–	–	2.50	–
	II	Fernando	5.00	–	5.00	–	–	–	–
		Matheus	3.00	–	1.80	–	–	1.20	–
	IS	Alessandro	7.50	–	7.50	–	–	–	–
		Gilvan	4.00	–	2.00	–	–	2.00	–
		Matheus	2.00	–	–	1.50	–	0.50	–
	SE	Andre L	15.00	1.50	10.00	–	3.50	–	–
		Ednelson	9.50	–	9.50	–	–	–	–
	ENE	Andre L	10.25	–	9.75	–	–	0.50	–
		Ednelson	0.50	–	–	–	0.50	–	–
	FU	Heloisa	8.00	–	7.75	–	–	–	0.25
	AQ	Luciana Vieira	13.00	–	13.00	–	–	–	–

Table 4. Results obtained in the Londrina project (BIM + Tools).

Project (Method)	Discipline	Designer	Worked Days	CC	DP	EC	EG	EI	EP
Londrina (BIM + Tools)	AC	Abel S	47.75	1.75	40.50	2.00	–	3.50	–
		Carlos Lima	1.00	–	1.00	–	–	–	–
	EC	Heloisa	27.50	–	24.75	–	2.75	–	–
		Leonardo C	43.00	2.00	36.50	1.00	–	3.50	–
	IE	Andre L	5.50	–	5.50	–	–	-	–
		Ednelson	30.75	0.25	26.25	2.00	–	2.25	–
		João G	4.00	–	4.00	–	–	–	–
		Carlos Crispin	12.00	–	8.25	1.25	–	2.50	–
	IG	Alessandro	3.25	–	2.50	–	–	0.75	–
		Matheus	2.00	–	2.00	–	–	–	–
	IH	Alessandro	8.25	–	7.75	–	–	0.50	–
		Asthor	7.50	–	7.50	–	–	–	–
		Andre M	1.00	–	–	–	–	1.00	–
		Matheus	11.75	–	7.75	2.00	–	2.00	–
	II	Fernando	7.75	–	5.00	–	–	2.75	–
		Gilvan	3.50	–	3.50	–	–	–	–
		Matheus	2.00	2.00	–	–	–	–	–
	IS	Claudia	8.50	–	6.50	–	–	2.00	–
		Gilvan	12.50	1.25	11.25	–	–	–	–
	SE	Andre L	23.25	–	20.50	–	–	2.75	–
		João G	9.50	–	6.25	–	–	2.50	0.75
	ENE	Ednelson	7.50	–	6.50	–	–	1.00	–
		Karina	2.00	–	2.00	–	–	–	–
	FU	Leonardo C	12.75	–	10.50	0.75	1.50	–	–
	AQ	Luciana Vieira	17.00	–	15.00	2.00	–	–	–

For a more effective comparative analysis of the results obtained, a graph was created, shown in Fig. 5, which compares the time spent correcting errors from each source to the total design development time in each of the three projects studied.

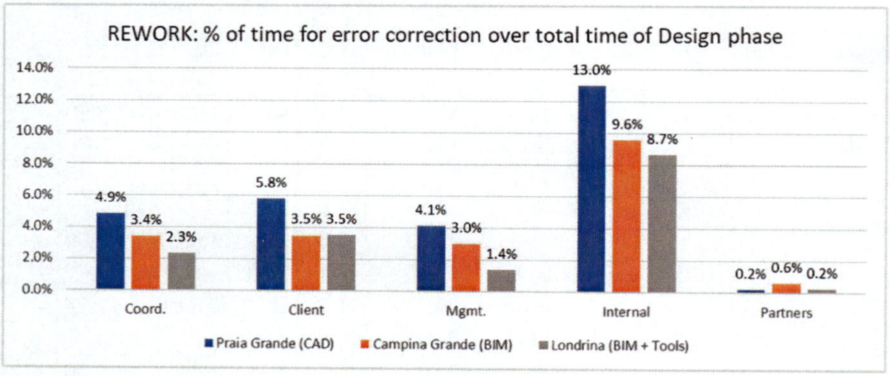

Fig. 5. Time spent correcting errors (rework) from each source, by project

It is noticed that there was a reduction of 30.6% in coordination errors with the implementation of BIM and later a further reduction of 32.3% with the implementation of the tools proposed in this work, totaling a reduction of 53% of the amount of time spent on rework due to coordination errors, even though the design executed with BIM + Tools had a significantly larger area than that developed in the CAD-based project.

Likewise, internal errors dropped by 26.1% with the implementation of BIM, and subsequently 9.4% with the implementation of the quality improvement tools. In total, in relation to the process using CAD, the reduction in rework time caused by internal errors was 33%. It is believed that the reductions in rework caused by process problems, such as coordination and internal errors, are an indicator of the greater accuracy that the implementation of the BIM platform and quality improvement tools brought.

On the other hand, it is understood that, despite the significant impact that the implementation of BIM and the proposed tools have caused on the number of errors from Client, Manager and Partners, these results were obtained indirectly, since these interested parties did not use BIM as a work platform. The only direct impact on their processes was the improved clarity that three-dimensional models bring to designs. Hence, data from these types of errors were more susceptible to particularities that occurred during the development of each project, unknown to us.

With the reduction of a large part of the errors that occur in the design process, Fig. 6 shows that with the implementation of BIM there was an increase of 11.1% in the portion of time spent on effective design development and another increase of 5.0% after the implementation of the proposed new tools.

In total, with the adoption of BIM and auxiliary tools, there was an increase of 16.7% in the proportion of design time over the total time spent by designers, which demonstrates an increase in the time available for activities that really add value to the design, as study of solutions, details, modeling, calculations and specifications.

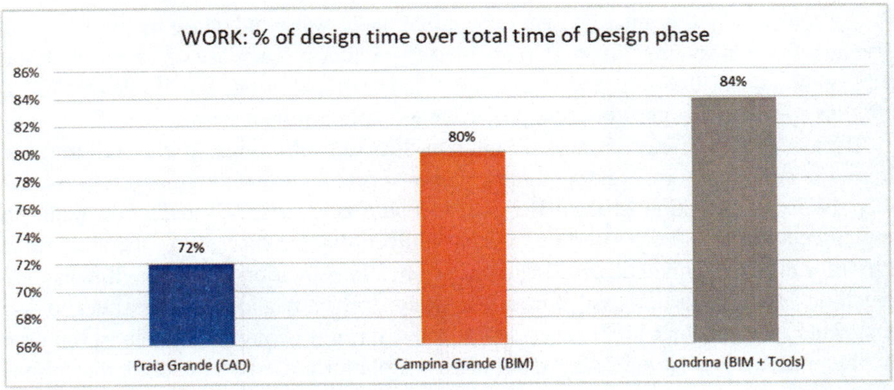

Fig. 6. Proportion of total time spent on effective design, by project

A detailed analysis of the impact of BIM adoption, as well of the implementation of the three additional tools, was done for each discipline and will be reported elsewhere. The main results of this analysis by discipline are shown on Fig. 7.

When evaluating the individual results from each of the disciplines, only the data related to internal errors were analyzed because these are the failures that are most likely to be impacted by the change of platform and improvement on tools. Also, it is understood that coordination errors are related to the design process as a whole and, therefore, should not be evaluated by discipline.

As can be seen on Fig. 7, the portion of time spent on rework due to internal errors (relative to total engineering design time) has decreased with BIM adoption for most of disciplines. The main exceptions are analyzed below.

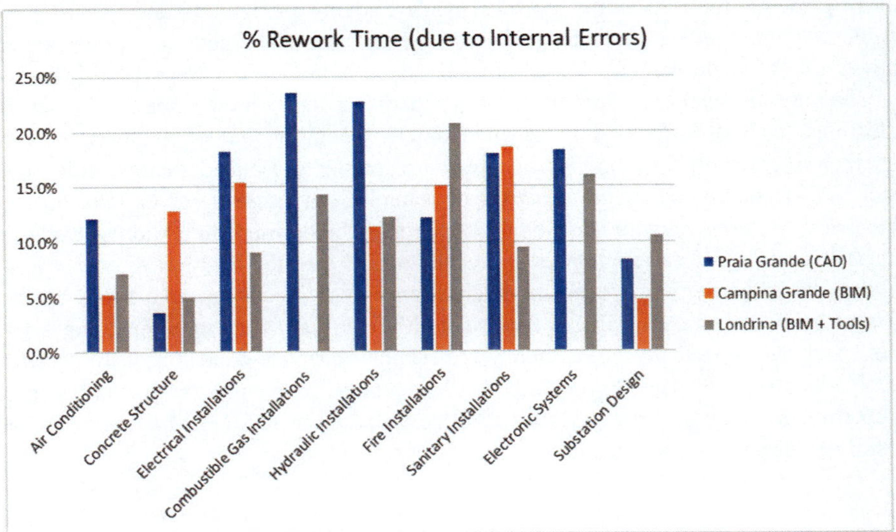

Fig. 7. Proportion of time spent on rework due to internal errors, by discipline and project

For the Air Conditioning (HVAC) discipline, there was a reduction by 56.6% in the time spent due to internal errors (EI) and, subsequently, an increase of 35.8% with the implementation of the proposed tools. The greater complexity of the HVAC technical solution specified for the Londrina project may explain this increase.

By analyzing the data from the Concrete Structure discipline, it can be seen that there was an increase of 255.5% in the amount of time linked to internal errors shortly after the implementation of the BIM platform and, later, a 43.7% reduction with the implementation of the proposed tools. It is believed that the increase in the amount of errors for this discipline occurred because, prior to implantation, the three-dimensional structural model was not used during the design process, allowing designers to use representative modeling approaches analytically, but not geometrically, as in the case of stairs that were modeled as flat slabs when the platform used by the other disciplines was CAD.

Through the study of the results obtained in the discipline of Fire Protection Installations it is noticed that, after the implementation of BIM, there was an increase of 22.9% in the amount of internal errors, and later, with the implementation of the tools proposals, there was another increase of 38.7%. It is believed that the increase in the number of errors possibly was caused by the incorporation of the approval of the legal project in the fire department, a scope that was not executed in the design elaborated in CAD platform (Praia Grande project). It is also understood that the increase in internal errors in the Londrina design in relation to the Campina Grande design, was caused by the greater complexity of the firefighting system design in the Radio and TV Studio area, where a more complex solution was required to prevent damage to equipment.

For the elaboration of the Substation Design, it can be seen that initially there was a reduction of 43.4% in the amount of internal errors shortly after the implantation of BIM and an increase of 123.4% after the implementation of the proposed tools. It is believed that the increase in internal errors in the Londrina design has occurred because of the greater complexity of the projected system, among the three designs studied, this is the only one that has generator and UPS systems, due to the need for uninterrupted power for the Radio and TV Studio area.

Through the analysis of performance comparisons between the same disciplines in the three studied designs, it is understood that the individual observations of the internal errors of the disciplines is more susceptible to isolated factors, since the increase in complexity of the design, for example, can cause great impacts on the number of failures, regardless of the tool or process being used. In addition, mistakes made in one discipline can impact the results obtained in the analysis of other disciplines, since the definition used for internal errors covers any mistake made by the design team during the design of the project. Finally, variations in engineering solutions adopted for each design, as well as the scope of hiring and the size of the system, certainly have a relevant importance in the demand for rework for internal errors, impairing the more accurate assessment of the isolated impact of BIM adoption in the building design in each discipline.

5 Conclusions

The implementation of BIM methodology and tools that helped to improve the quality and accuracy of designs, drew the attention of designers to the many possibilities that the use of BIM methodology allows. In general, the implementation and operation of the tools occurred with just a few obstacles, since the professionals had been working with BIM methodology for some time before starting to use the proposed software tools, which greatly facilitated the understanding and application of the platforms within the company's process.

All tools applied were aimed at improving the company's design process, since the authoring software already provides a great increase in the presentation of drawings in relation to products made on two-dimensional platforms. The cost-benefit of these deployments should also be stressed, since two of the tools used, Autodesk Insight and Dynamo, are already included in the Revit authoring software package and the third tool is a free platform, remaining only the cost of training.

In terms of process management, a substantial improvement can be seen, since the use of BIM brought up problems that existed in the process, making the strategy for solving them clearer. From the organization process, it was possible to extract data that allows assessing the performance of the company from the perspective of the quality and efficiency of its operations, both in terms of design, in relation to the flow of internal information and trade coordination.

By analyzing the results obtained with the use of additional BIM tools and comparing with a similar project that were elaborated using only BIM authoring tool and one using only 2D-CAD technology, one can perceive, through metrics, decrease in time for mitigating internal errors and coordination failures during design development, showing that the improvements made in the process through the implemented tools brought about a reduction in rework, meeting the specific objectives of the research.

It should be noted that the individual analysis of disciplines in relation to internal errors is very susceptible to variations in complexity and failures on other disciplines, as well as coordination errors can be understood as procedural failures, which are not well-structured or are not running properly.

Although the design exposed in the case study represents only a limited sample of the benefits brought by the deployment of the platforms, it is important as a reference for engineering organizations that aim to solve problems in their design process with the help of BIM tools, improving the quality of their projects. The description of the adopted strategies allowed the documentation of methods used in view of the obstacles encountered during its elaboration and production, and presented the ways the company found to extract benefits from the BIM methodology, congruent with its production process.

It is hoped that this presented case study can stimulate further research in the field covered in this contribution, stimulating increased BIM adoption by the more skeptic members of the construction sector.

References

1. Tilley, P.A., McFallan, S.L., Tucker, S.N.: Design and documentation quality and its impact on the construction process. In: Bowen, P., Hindle, R. (eds.) Proceedings of CIB W55 & W65 Joint Triennial Symposium (1999)
2. Kalay, Y.: The impact of information technology on design methods, products and practices. Des. Stud. **27**(3), 357–380 (2006)
3. Sacks, R., et al.: BIM Handbook: A Guide to Building Information Modeling for Owners, Designers, Engineers, Contractors, and Facility Managers, 3rd edn. Wiley, Hoboken (2018)
4. Sadek, K., El-Bastawissi, I., Raslan, R., Sayary, S.: Impact of BIM on building design quality. BAU J. Creative Sustain. Dev. **1**(1), 1–10 (2019)
5. Coughlan, P., Coughlan, D.: Action research for operations management. Int. J. Oper. Prod. Manag. **22**(2), 220–240 (2002)
6. BIMserver. https://github.com/opensourceBIM/BIMserver. Accessed 01 Feb 2020
7. BuildingSMART. https://www.buildingsmart.org/standards/bsi-standards/industry-foundation-classes/. Accessed 01 Feb 2020
8. Autodesk Insight. https://www.autodesk.com/products/insight/overview. Accessed 01 Feb 2020
9. Dynamo site. https://dynamobim.org/. Accessed 01 Feb 2020
10. BIM Dictionary. https://bimdictionary.com/en/4d/1. Accessed 31 Jan 2020

Challenges of District Information Modeling (DIM) Applied for Heritage Preservation

Eloisa Dezen-Kempter[1(✉)], Vitor E. Molina Jr[1], Leonardo H.G. Silva[1], Luiz P.D. Mendes[1], Maxwell F. Campos[1], Isabel A. Custodio[1], Lucas Alegretti[1], Vivian F.W. Rodrigues[1], Aleteia C.P.M. Pascual[2], Fernando B. Lima[2], Gisele Martins[2], Veruska B. Custodio[3], and Tatiane M.S. Alves[1]

[1] School of Technology, University of Campinas, Paschoal Marmo 1888, Limeira, Brazil
{elo,molina}@ft.unicamp.br, {l211885,l230315,m230316, i227118,l143541,v262987}@dac.unicamp.br, tatmeire@yahoo.com.br

[2] Institute of Architecture and Urbanism, University of São Paulo, São Carlos, Brazil
aleteiapedroso@sc.usp.br, birello@unemat.br, gisele.gmarquitetura@gmail.com

[3] University of the Triangulo Mineiro, Uberaba, Brazil
veruskabichuette@gmail.com

Abstract. The urbanization is a complex and continuous process requiring updated data over time. Information and Communication Technologies (ICT) are essential to assist in the planning process. However, distinct computational environments are necessary to use all available data. Faced with these challenges and aiming to generate technical inputs to foster a better understanding and management of urban spaces, this paper seeks to analyze the potentiality of the District Information Modeling (DIM) applied to a distinguished historical urban scenario. In this paper, we show that the Scan2-BIM2GIS process represents an essential technology for management at the city level. We reported the results of a study developed by postgraduate students in a course on Special Topics on Computer Graphics. The scope was to understand the steps for creating the DIM model of the Monte Alegre District, a former industrial area located in the city of Piracicaba, state of São Paulo. The starting point of the proposed approach was the District 3D geometric reconstruction obtained from an image-based surveying campaign with a Unmanned Aerial System (UAS), allowing to make a point cloud with excellent accuracy from the distinctive buildings, the roofs, the street system, and the open spaces. Building Information Modeling (BIM) tools permitted the creation of component-based structured 3D semantic models of the buildings with the desired Levels of Detail (LOD), and the Geographic Information System (GIS) environment enables indexing quantitative and qualitative attributes to these buildings. Furthermore, interoperability between software becomes a crucial factor, and improvement of data sharing will be relevant for future developments.

E. Toledo Santos and S. Scheer (Eds.): ICCCBE 2020, LNCE 98, pp. 483–495, 2021.
https://doi.org/10.1007/978-3-030-51295-8_34

Keywords: BIM · GIS · Point-to-BIM · Scanning · Heritage Management

1 Introduction

The non-existence and inadequacy of cartographic documentation tools at the scale of cities with a multidisciplinary approach have been impediments to the strengthening of the concept of urban heritage in the last century.

In Brazil, there is a patrimonial vision focused on the individual monument and not on the group of the historic built environment. The institutions of heritage protection lack instruments for a global understanding of the issues related to heritage in addressing historic cities and districts, such as the resulting cultural landscape, land uses, the scale of buildings, states of conservation, build changing of the style, and materiality, among others.

Regarding Urban Heritage, a wide-ranging updated and reliable geo-referenced multidisciplinary and multiscale data (e.g., from building to the entire district) is a critical pillar for its management improvement, including monitoring, conservation, and restoration process.

Scanning, BIM, GIS, and City Information Modeling (CIM) are among the technologies developed in the context of buildings, sites, and city management over the last decades. All of these ICTs linked with Information aim at structuring and organizing the data. What differentiates them, in a simplified way, are the scales of application and users' interests. [1] pointed out that in the CIM scenario, the ICTs application addresses the search for the most appropriate model to capture the objects' semantics and geometric representation relevant to the scope of a city. Semantic is a relevant part, as it allows the application of the model to different types of analysis and simulations.

However, issues related to integration and interoperability between these different ICTs still exist. This particular lead to combining content associated with different scales of representation, generalizing them and establishing processes, without any continuity solution, from an information system at the regional level (GIS) to the urban scale until reaching the building scale (BIM) [2]. Overcoming issues related to the integration of semantics and geometries without significant loss of content is the great challenge of researchers and software developers in recent years. The AEC industry has recognized the need to integrate BIM and GIS, and design software companies are looking forward to further collaboration to help simplify workflows and improve outcomes for both GIS and BIM users [3].

Our approach aims to investigate the integration of 3D data from geographical information about assets and districts, and information-rich building models. This integration can subsidize decision making based on knowledge and management of the cultural landscape, on the micro (building), mid (historical sets), and macro (district/city) scales. Thus, this work focuses on the potential application of BIM and GIS to the documentation and maintenance of historical sites. This article reports an academic experience for the integration of different computational tools for data processing and modeling, including digital scanning technologies, applied to the Monte Alegre District, located in Piracicaba, SP, Brazil. The contribution of this work refers to the integration of GIS and Historic BIM (HBIM) towards an innovative and more efficient way for the management of historical sites: the Historic District Information Model (HDIM). Certainly, the obtained results can contribute to leverage the process of documentation and management of large areas of historical and cultural interest by municipal bodies.

2 Background

According to [4], the integration of GIS and BIM deals with numerous challenges. The authors cite some efforts developed by [5], such as a CityGML extension, to get semantic IFC data into a GIS context. Other applications included conversion, translation, and extension of existing Standards that works as a bridge between BIM and GIS, such as InfraGML, IndoorGML, model-Unified Building Model-UBM, QUASY Project with QuVariants [6]. The authors [4] also stated that in previous works, [7] classified the integration methods into three different levels: **data**, **process**, and **application**. Based on these levels, [4] defined parameter criteria named EEEF (effectiveness, extensibility, effort, and flexibility), which list the pros and cons of each integration solution (see Table 1).

Table 1. Comparison of integration solution by EEEF criteria.

Level	Integration methods	Effectiveness	Extensibility	Effort	Flexibility
Data	New standards and models	Case-by-case	Case-by-case	Case-by-case	Case-by-case
	Conversion, translation and extension of existing standards (manual)	Medium	High	High	Medium
	Conversion, translation and extension of existing standards (semi-automatic)	Medium	Medium	Medium	Medium
Process	Semantic web technologies	High	High	High	Medium
	Services-based methods	High	Low	High	Low
Application	Application focused methods	Case-by-case	Low	Low	Low

The **effectiveness** means less information loss, while **extensibility** relates to the high degree of openness. The **effort** reflects the time, labor, and money cost involved in the method, and **flexibility** represents the possibility of one result be applied by other studies [4].

The application focused methods are implemented to work a specific use and can employ various tools for such integration of information between the two systems as a customized plug-in available for BIM software and a central database [8]. Another option used by [9] is a schema to work as an integration medium adopting an intermediate data format between GIS and BIM. In this case, the gbXML file was converted to KML and COLLADA. The advantage of this type of method is that it is not costly in terms of time, labor, and money.

Applications of GIS and BIM Integration in Heritage Management according to [6], included efficient information management in three dimensions: (a) segmentation; (b) structuring the hierarchical relationships; and (c) semantic enrichment (Fig. 1).

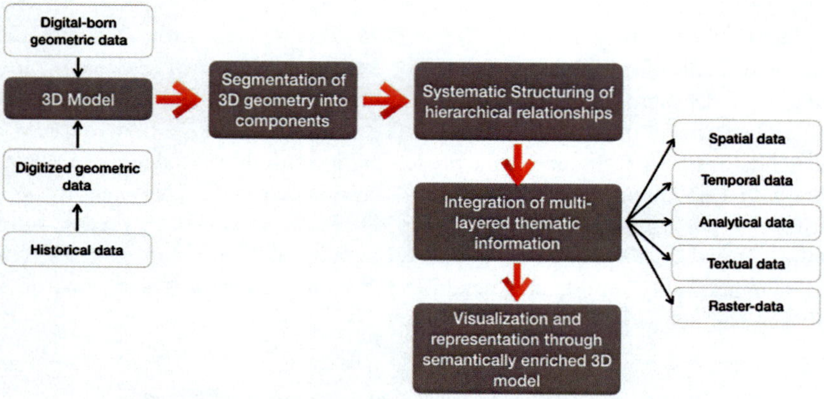

Fig. 1. The workflow diagram for semantically enriched 3D heritage models

GIS is the most common approach to managing multi-layered thematic information on a larger scale (city or district), including land use, construction typology, transportation, vegetation data. The main functions of GIS are to capture, store, manipulate, analyze, manage, and present all types of spatial or geographic data. This capability has facilitated its use in various applications, including heritage documentation.

Historic Building Information Modeling (HBIM) focuses on the integration of digitized geometric and historical data [10]. Accurate surveys data gathered by laser scanning or photogrammetry will result in segmented point clouds assisting in 3D modeling in BIM and also in the acquisition of orthophotos for data records in GIS.

Applied to urban heritage management, the GIS enables the storage, documentation and analysis of data assigned to preservation objects and surrounding spaces. It allows the construction of visualization products of spatial characteristics and topological relations - digital models and thematic cartographies - with varied levels of information and details linked to simple and complex geometries in 2D and/or 3D. It also makes possible the approximation between documental heritage GIS, other urban GIS and those with complex parametric urban models in BIM logic expanding the possibilities of documentation, modeling, understanding and simulation of the real world in digital environment [e.g. 11, 12]. For this effective and collaborative purpose, it is important that all information on an urban site is stored together, including data on territories and individual buildings on the site of interest.

Thus, 3D urban models approximate the real world, where features are modeled to a certain degree, and certain elements are simplified or omitted. The amount of detail in a 3D urban model, both in terms of geometry and attributes, constitutes its level of detail (LOD), indicating how thoroughly a spatial extension was modeled. For [13], the quantity and combination of contents (geometry and attributes) are driven by several factors, such as the intended use of the model, source of the primary data, acquisition techniques, spatial scale, financial, and human resources invested.

3 The Monte Alegre District

The Monte Alegre District (Fig. 2) was chosen as a test site for the proposed Scan2GIS2BIM-approach. The Municipal Preservation Body listed the District as a heritage site, and the protected-area perimeter covers 650 thousand square meters. Despite its historical and cultural importance for the state of São Paulo, the lack of adequate and comprehensive documentation that is required to carry out further preservation, conservation, and rebuilt actions, has led to a process of deconfiguration and even ruination of significant buildings in the complex.

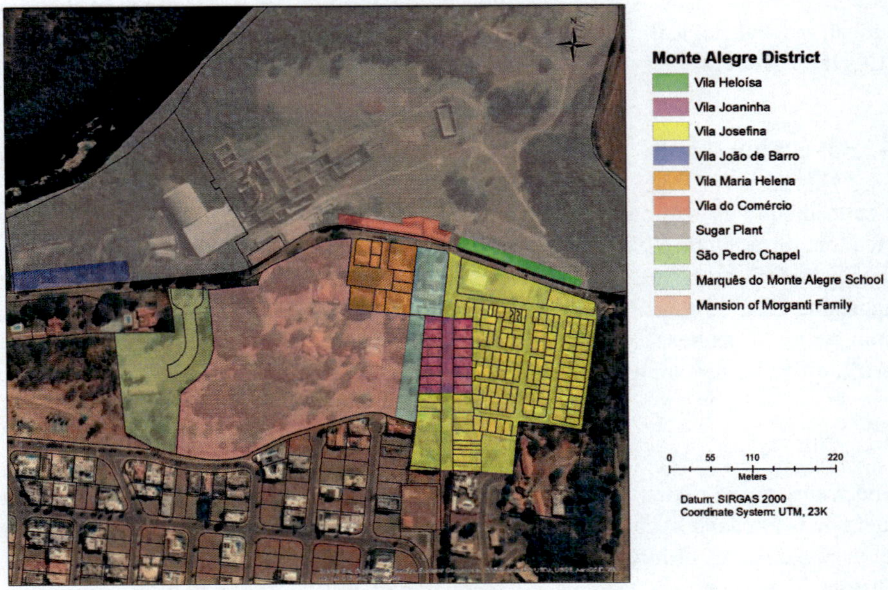

Fig. 2. The sectors of Monte Alegre District

The Case Study is a former industrial area located in the city of Piracicaba, state of São Paulo. The District has regional historical relevance, and represents one of the largest manufacturing centers in the State, formed from the culture of sugarcane and the sugar and alcohol production of the Monte Alegre Sugar Mill. Located on the banks of the Piracicaba River, the Monte Alegre District was built to meet the workers' demands of the Sugar Mill in the early twentieth century. The District development was becoming a relevant historical, architectural, and socio-cultural urban scenario for the city.

The Sugar Mill Assembly, which drove the emergence of the District, is the closest building to the Piracicaba River, and its construction started in the 19th century, together with the construction of the employer area where the Mansion of Morganti Family is located. The Sugar Mill operated until the 1980s, passing through some changes during its life cycle.

The 906 dwellings grouped in Vilas were organized according to various criteria, such as implantation dates, construction characteristics, morphology, scale, and the function of the worker at the plant. Vila Heloísa, the oldest and most modest, was implanted in 1920, followed by Vila Maria Helena and Vila Josefina between 1920 and 1930. Eight different houses arranged to form a square inside the court composed the Vila Maria Helena, intended for higher-ranking employees. Vila Marisa and Vila Joaninha were built between 1940 and 1950 and, finally, Vila Renata, established in 1950.

The Monte Alegre District contains reasonably well-preserved buildings types, like the old school inaugurated in 1927, and the Chapel of São Pedro built-in 1937. The Chapel stands out for its 600 m^2 of frescoes painted by Alfredo Volpi. All of these assets are under analysis for registration by the Council for the Defense of the Historical, Archaeological, Artistic, and Tourist Heritage of the State of São Paulo (CONDEPHAAT).

4 Overview of the Approach

The methodology was outlined in four steps: (i) aerial **survey** by UAS, aiming at the creation of a Digital Surface Model (DSM) and the orthophoto of the district; (ii) **historical and cadastral survey** in the archives of the city of Piracicaba to aggregate data as attributes in the **GIS model**; (iii) **BIM** modeling of significant community structures at different levels of detail **(LOD)**; (iv) **GIS and BIM integration** for the development of HDIM.

4.1 Surveying

The reality-based digital model, generated through UAS, had a double purpose. The first one is to enable accurate 3D documentation of the district as-is condition, framing an accurate source of information for interpreting the urban structure of Monte Alegre District. The second one is to allow the creation of other mapping products such as the site orthophoto, the start point from the digital interpretation in a GIS environment.

For the district DSM, 2294 images were acquired in two survey campaigns, completed on different days (Fig. 3). On the first day, the UAV survey was performed with an amateur UAS model DJI Mavic Pro with a 12.7 MP camera. Three flights were performed in a double grid at 70 m from the lowest point of the area, the first at nadir, and the second with the camera at 60° to capture the facades. These flights generated 645 images. On the second day, two flights in a double grid were performed with a professional UAS model DJI Inspire 2 with a 20 MP camera.

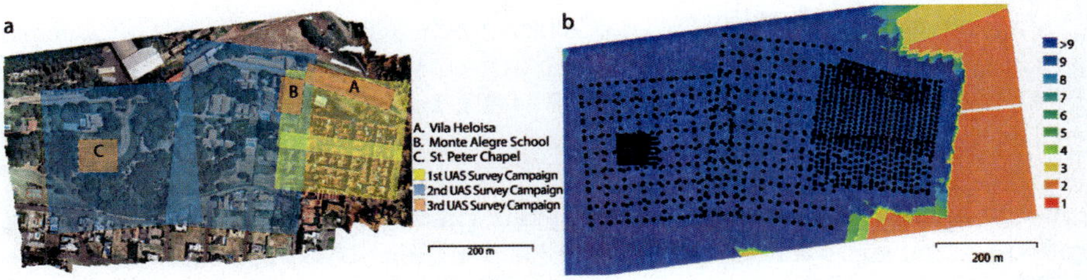

Fig. 3. (a) UAS Survey campaigns and (b) double grid of each flight in the area, the color schema represents pictures overlap

For this survey campaign, the UAS flight at 120 m from the site lowest point with the camera at nadir, generating 1649 images. On the third day specific flights were carried out for the buildings of interest (Chapel, School), in which, in addition to the double grid, a circular flight was made over the building with the camera at 45° (see Fig. 4c). For the Phototriangulation process, 8 control points were captured by Ashtec Geodetic GNSS receivers.

All 2294 photographs of the different surveys were processed in the same batch using Agisoft's Metashape software. Due to the high number of pictures, the process was set in low alignment parameters resulting in a dense point cloud with 23.5 million points and a point density of 31.9 points/m^2 (see Fig. 4a).

Fig. 4. (a) DEM of the total area, (b) DEM of the first UAS campaign, and (c) DEM of the School showing the grid of capture

Interior surveying in the São Pedro Chapel was driven using three types of cameras: a DSLR (Digital Single Lens Reflex) with a fixed super zoom lens (Nikon Coolpix P1000); a compact Mirrorless (Sony ILCE 6000); and cell phones (SM-A750G). The capture process with the DSLR equipment took place with the aid of a tripod and programmed shooting. With the other ones, the process was manual. All 498 photographs of these cameras were processed in the same batch using Agisoft's Metashape software, resulting in a dense point cloud (54.5 million points) and with high texture quality (see Fig. 5). The Baptistery section was captured with the Sony ILCE 6000 camera, and the 49 photos were processed separately, generating a TIN (Triangulated Irregular Network) mesh to be published on the Sketchfab platform[1].

Fig. 5. (a) Negative point cloud from interior of St. Peter Chapel, and (b) TIN mesh of the Baptistery

4.2 GIS

The data were incorporated into the software ArcMAP from the platform ArcGIS (ESRI) based on a SHP digital file of the neighborhood cartography obtained from municipal administration of Piracicaba-SP. The orthophoto was imported in a matrix format for the software environment, being kept in a data layer parallel to the vector concerning the geometry of the lots, blocks, and streets. From the layers overlapping, attributes such as area and perimeter were subsequently indexed, as well as other attributes defined as singular and measurable for the characterization of the study area such as construction year (Historical Evolution), buildings conservation status, characterization degree, and land use (see Fig. 6). Once the database constituted from in the lots related geometry, buildings projection, streets, blocks, and attributes, it was possible to obtain cartographic visualizations of the isolated and combined information, considered relevant for the understanding of this landscape and its management as a site of historical and cultural interest.

---

[1] Avaiable in https://sketchfab.com/3d-models/igreja-de-sao-pedro-f932c6c7d283435d8ac63bf447b2e765.

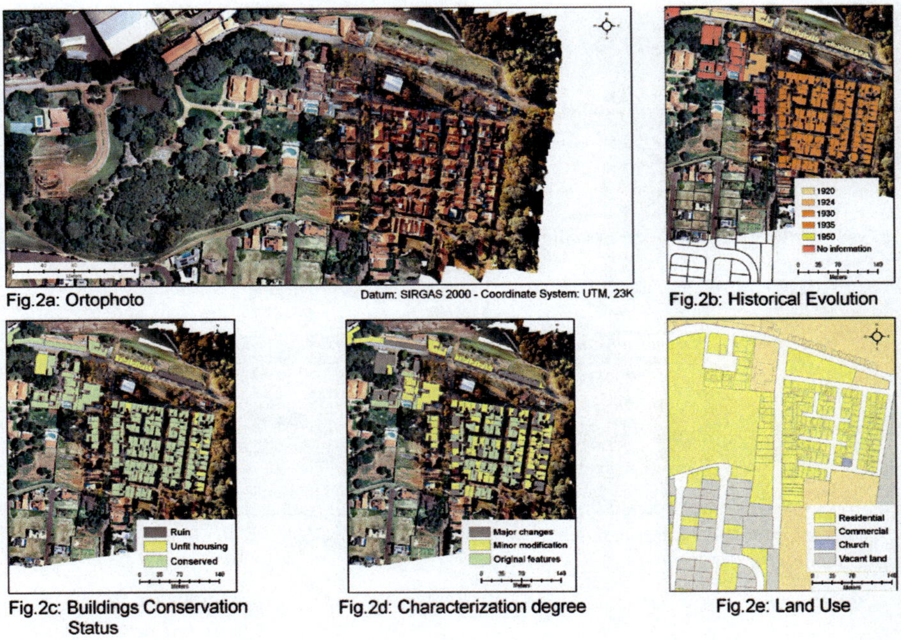

Fig. 6. Monte Alegre Cartography developed in ArcGIS Pro

4.3 HBIM

It is well-known the importance and revolution that BIM brought for the building's design and construction processes. In the last decade, efforts have been made to extend the BIM benefits (knowledge database), combined with scanning and computer vision, for the historic buildings. It is due, mainly, to the potential of this technology for the conservation of historic structures and sites (e.g., point-to-BIM, tangible and intangible information, building conditions assessment, ontology-based semantic information organization).

A different dimension of object characterization, regarding not only its geometric accuracy but also reliable information, defines the scope, purpose, and complexity of a 3D knowledge-based Model. The geometric and semantic specialization that this model can achieve requires the definition of the level of complexity to be reached. Around this general idea, different denominations have been created: Level of Detail (GIS environment) and Level of Development (BIM environment).

The geometric and semantic specialization, as well as the information and Knowledge enhancement that each model in different levels should be present, is depicted in Fig. 7.

Fig. 7. Correlation between different Levels of Geometric and semantic specialization, and information and knowledge enhancement in different technologies that use a 3D knowledge-based models as database.

Based on CityGML LoD 1, the geometric model of the District was developed from the vector maps and the point cloud. The Models of the most significant buildings, such as the Chapel, Vila Heloisa, Vila Josefina and the School was created in LoD 3 (detailed only the exterior) and LoD4 (detailed exterior and interior) (see Fig. 8). The software Revit (Autodesk) and Archicad (Graphisoft) were used for modeling, and the point cloud was the basis for creating the parametric components and details, also defining the height reference of the buildings.

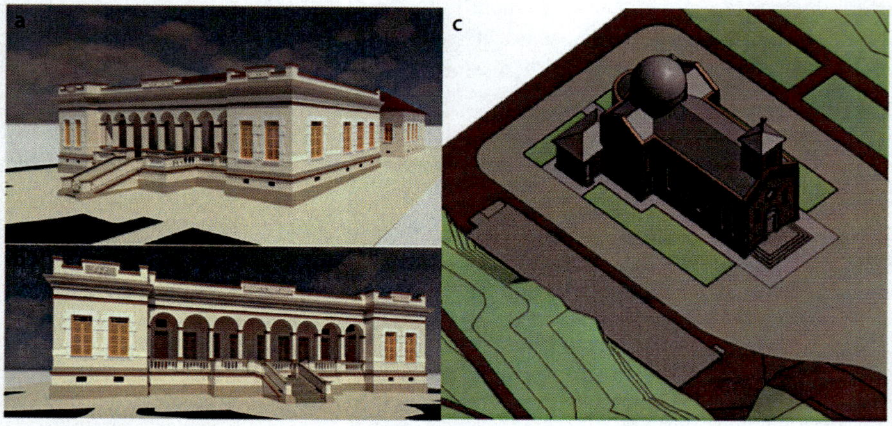

Fig. 8. (a) and (b) Render views from BIM model of Monte Alegre School in LoD 4, (c) BIM Model of St. Peter Chapel in LoD 3.

The other buildings were modeled in LoD 1 with a mass creation tool. Shared parameters tools in the Revit environment were created to define the building´s degree of characterization, state of conservation, and classes of use (see Fig. 9).

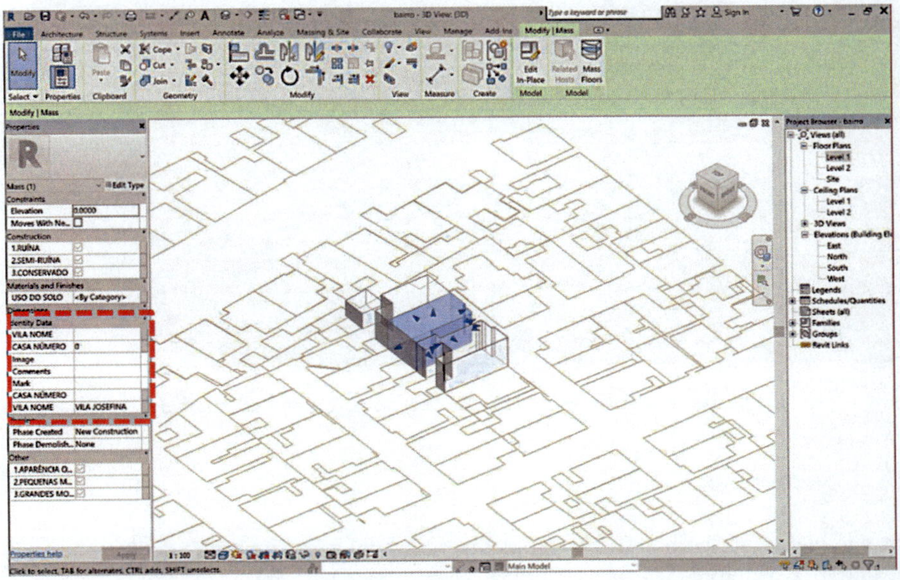

Fig. 9. Defined shared parameters in Revit environment for the mass representing the buildings in LoD1.

4.4 HDIM: GIS and BIM Integration

The integration method used in this study was the Application Focused Method defined by [4]. This method presents low extensibility; therefore, a small effort is required, employing the tools already known by the authors (e.g., Revit, Archicad, ArcMAP, ArcScene).

The point cloud, 2D technical drawings made in AutoCAD software, and the shapes produced in ArcGIS software were linked in the Revit software. In the Revit environment, shapes were converted into masses representing the heights of the buildings. Subsequently, materials created with the attribute's parameters (preservation degree and occupancy types) were applied to each building. An attempt to integrate the BIM models in the GIS does not show a satisfactory result due to the difficulty in integrating the IFC or RVT data into ArcScene. As this case study was developed in a time-limited discipline, it was not possible to explore the software to define a data integration workflow.

The Fig. 10 shows the LoD 1 model of the macroscale according to the classes of use show the microscale models (buildings) with LoD 3 and 4, Vila Heloisa, Vila Josefina, School respectively, inserted in the urban model.

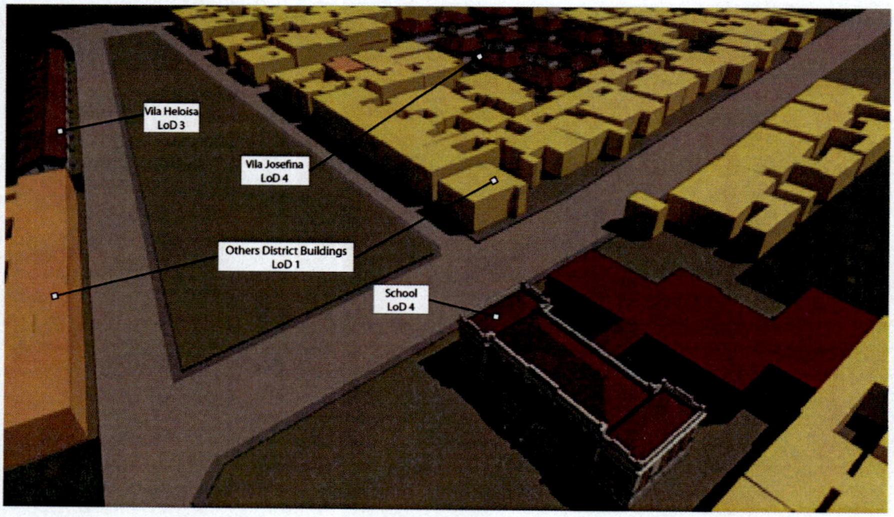

Fig. 10. Different levels of detail represented in the urban model

5 Conclusions

In this paper, we proposed and detailed a methodological workflow to implement a scan2GIS2BIM-approach to a Historical District Management.

The use of photogrammetry enabled the creation of an unstructured 3D geometric reconstruction of the district, also orthophotos, which were used in the GIS platform, and in the BIM environment for the data spatialization. BIM tools allowed the creation of component-based structured 3D semantic models of artifacts and buildings with the desired levels of detail (LOD).

The proposed workflow demonstrates that a 3D cadaster for Historical sites can benefit from the information that is maintained by the public administration and the Heritage Preservation Council and in other databases, creating an easily accessed and shared single information repository. In this case, the low cost of UAS-based context capture allows the continuous updating of the actual situation of the study area.

Despite the quality of the products generated, the proposed workflow encountered some difficulties, highlighting the lack of interoperability between software, which made it impossible to connect the BIM and GIS platforms more closely. The interoperability between the software, the definition of the attributes, and which LOD are the most suitable are some of the challenges found in this research.

The manipulation of acquired data is a method that requires higher human effort, especially when feeding the GIS system for the generation of Thematic Cartography and BIM modeling based on the point cloud. The use of GIS for applications in historical sites allows the characterization of the territory, generating information to understand how the district is inserted in its surroundings. Thus, interventions would not be limited only to the scale of the building but would extrapolate to the local scale.

Exporting the georeferenced mapping from the GIS platform to the DXF format to work it in Revit and Archicad environments (there is not a direct and simultaneous link between the software), only the georeferencing coordinates were maintained, losing all other attributes previously indexed in GIS software. Another challenge is to understand the modeling on each platform so that information is not lost in interoperability.

Thus, the integration of BIM and GIS semantics and instances in a single platform was complex and unsatisfactory, and it is part of future work to seek solutions to this problem.

References

1. Corrêa F.R., Santos E.T.: Na direção de uma modelagem da informação da cidade (CIM). In: VII Encontro de Tecnologia de Informação e Comunicação na Construção (TIC 2015), pp. 344–357. Blucher Engineering Proceedings, Recife (2015)
2. Ruffino, P.A., Bocconcino, M.M., Del Giudice, M., Osello, A.: From the elaboration process of point cloud to information systems both for planning and design management of cultural heritage. In: 2nd International Conference of Geomatics and Restoration. The International Archives of the Photogrammetry, Remote Sensing and Spatial Information Sciences, vol. XLII-2/W11, pp. 1031–1038 Italy (2019)
3. Andrews, C.: GIS and BIM Integration Leads to Smart Communities. ArcUser Newsroom (2018)
4. Liu, X., Wang, X., Wright, G., Cheng, J.C.P., Li, X., Liu, R.: A State-of-the-art review on the integration of building information modeling (BIM) and geographic information system (GIS). ISPRS Int. J. Geo-Inf. **6**(2), 1–21 (2017)
5. De Laat, R., van Berlo, L.: Integration of BIM and GIS: the development of the citygml GeoBIM extension. In: Advances in 3D Geo-Information Sciences. Springer, New York (2011)
6. Saygi, G., Remondino, F.: Management of architectural heritage information in BIM and GIS: State-of-the-art and future perspectives. Int. J. Herit. Digit. Era **2**, 695–714 (2013)
7. Amirebrahimi, S., Rajabifard, A., Mendis, P., Ngo, T.: A data model for integrating GIS and BIM for assessment and 3D visualisation of flood damage to building. Locate **15**, 10–12 (2015)
8. Irizarry, J., Karan, E.P., Jalaei, F.: Integrating BIM and GIS to improve the visual monitoring of construction supply chain management. Autom. Constr. **31**, 241–254 (2013)
9. Niu, S., Pan, W., Zhao, Y.: A BIM-GIS integrated web-based visualization system for low energy building design. Procedia Eng. **121**, 2184–2192 (2015)
10. Murphy, M., Mcgovern, E., Pavia, S.: Historic building information modelling – adding intelligence to laser and image-based surveys of European classical architecture. ISPRS J. Photogram. Remote Sens. **76**, 89–102 (2013)
11. Dore, C., Murphy, M.: Integration of HBIM and 3D GIS for digital heritage modelling. In: Digital Documentation, Edinburgh, Scotland, pp. 1–5 (2012)
12. Vacca, G., Quaquero, E., Pili, D., Brandolini, M.: GIS-HBIM integration for the management of historical buildings. Int. Arch. Photogram. Remote Sens. Spat. Inf. Sci. **42** (2), 1129–1135 (2018)
13. Biljecki, F.: Level of Detail in 3D City Models. TU Delft, Delft (2017)

Integrated Data Model and Mapping for Interoperable Information Exchange Between BIM and Energy Simulation Tools

Weiwei Chen(✉), Moumita Das, Vincent J. L. Gan, and Jack C. P. Cheng

The Department of Civil and Environmental Engineering, The Hong Kong University of Science and Technology, Hong Kong, China
wchenau@connect.ust.hk

Abstract. Energy efficiency is an important aspect in building design nowadays. During the design phase, energy simulation and analysis can enhance the overall building energy efficiency at a lower cost and less delay to the project schedule. Currently, energy analysis is performed digitally in energy simulation software applications which require users to draw a virtual model and define the parameters in the application platform before running the simulations. As building information modeling (BIM) technology becomes increasingly adopted in the architecture, engineering and construction/facility management (AEC/FM) industry, BIM models are often used in construction projects. However, these BIM models are not compatible with the existing energy simulation applications, and the required model conversion often results in information loss and distortion. One reason is that the current BIM data schemas, such as the Industry Foundation Classes (IFC), do not fully support data for energy simulation and analysis. This paper presents a data model to extend BIM schemas for supporting energy simulation and analysis. The data model was developed by studying the four most widely used energy simulation software, namely eQUEST and EnergyPlus. The proposed data model was then used to extend the latest IFC4 schema. Based on the results, a mapping between the extended IFC4 schema and gbXML was also developed to allow information exchange between BIM and energy simulation software. An example scenario is presented to illustrate the potential of the extended IFC4 schema for supporting energy analysis directly.

Keywords: Building information modeling · Energy simulation · Industry Foundation Classes · Interoperability · gbXML

1 Introduction

The energy consumption of buildings in most metropolitan cities around the world accounts for a large proportion of the overall energy consumption of the city. For instance, energy consumption of buildings accounted for 28% and 40% of the total energy use in United States and Hong Kong, for the years 2010 and 2011 respectively [1, 2]. However, if well devised energy conservation strategies and the best available

E. Toledo Santos and S. Scheer (Eds.): ICCCBE 2020, LNCE 98, pp. 496–506, 2021.
https://doi.org/10.1007/978-3-030-51295-8_35

technologies are implemented, the energy consumption of buildings may be significantly reduced by as much as 11.3%, by the year 2035 [1]. Energy performance analysis of buildings is one of the efficient approaches which provide decision support to building designers during the early design phases, and in turn can reduce the energy consumption during the construction and operation phase [3]. Due to this reason, energy simulation tools, such as IES VE, Ecotect, EnergyPlus and eQUEST, have gained significant popularity in the architecture, engineering, and construction/facility management (AEC/FM) industry. However, as complex simulation tools comprising improved user friendly interfaces and integrated multi-domain modeling concepts have increased, the need for precise design data from different domains has also arisen, leading to the problems of data interoperability [4]. Also, due to inaccurate input or insufficient information available to such simulation engines, the final building design tends to consume 20% more energy than its design intent [5]. Furthermore, in some of the energy simulation tools, the analysis models have to be drawn from scratch in proprietary format. Integration between BIM and the energy simulation tools through a neutral standard like IFC (Industry Foundation Classes) can address information requirement and interoperability issues. Building Information Modeling (BIM) is the process of generating, storing and managing information of a building throughout its life cycle [6]. BIM captures information related to a building which is necessary for energy related predictions. Such information includes building geometry like building orientation, building surfaces and openings related information, and building materials and equipment, HVAC, and lighting related information.

However, currently, the information exchange between BIM and energy simulation tools is realized through standardized data schemas like gbXML (Green Building Extensible Markup Language). However, the gbXML schema is designed only to cater to the energy simulation domain and is incapable of storing information from different BIM domains. Therefore, a universal BIM standard like IFC is more suitable for facilitating information exchange between energy simulation and other domains during the design phase. However, at present, the IFC standard does not fully cover the information requirement in the energy simulation domain [7]. Garcia and Zhu [8] studied an automated solution through the design of a novel Extensible Style Sheet Language Transformation (XSLT), and proposed series of instructions to facilitate the information exchange between building design and energy modeling fields. Therefore, in this paper, we present a comprehensive data model for facilitating the storage and exchange of energy simulation information. This data model has been designed by studying the information requirements of the four most popular energy simulation tools, each of which have different simulation engines. This data model is used to extend the latest version of IFC, which is IFC4. The result of the gap analysis between the proposed data model and IFC4 is also presented in order to demonstrate the importance of the proposed extension. Furthermore, considering the fact that the existing energy simulation tools are more compatible to gbXML as compared to IFC, a mapping between the gbXML schema and the extended IFC4 schema is presented.

2 Methodology for Selecting Entities for the Proposed Energy Simulation Data Model

(1) The first entity category is lighting, defining the attributes required for modeling the input parameters for lighting related energy simulation. For example, the electrical power consumed by a lighting device, type of lamps, distance of lighting device from the floor and the ceiling, and the total luminous flux per unit area incident to a surface. (2) The second entity category in the data model defines the interior and exterior surfaces of a building. These entities define attributes to accommodate information like the boundary condition of an internal floor, the construction layer of an external surface, the thermal conductivity and density of material used for each construction layer, and U-value of an internal surface (e.g., glass of a window). (3) The third entity category defines the ventilation related parameters such as the air flow rate of the outside air into the building, the types of ventilation, and the method of calculation of the air flow for energy simulation software. (4) The fourth entity category models both human and non-human occupancy related input parameters such as heat gain from the occupants and the electrical load of the equipment for the extended IFC schema. (5) The fifth entity category defines the information regarding the control and type of HVAC equipment, and the mechanism for supplying the facility with resources.

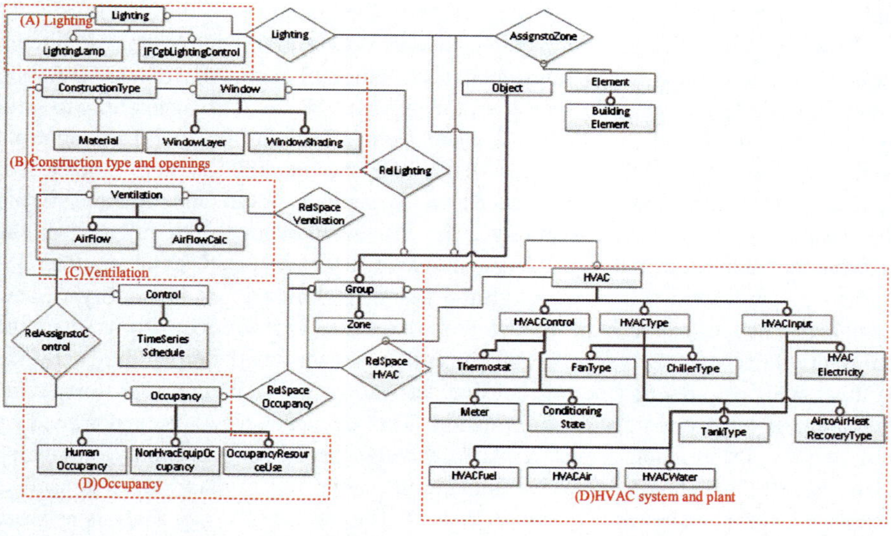

Fig. 1. The proposed data model for energy simulation

The actual information for energy simulation is stored in the attributes of the entities of the data model shown in Fig. 1. For example, Table 1 shows a detailed example of the parameters contained in the occupancy category. These parameters form the attributes of the entities under which they are grouped. For example, the bolded parameters in Table 1 describe information related to human occupants and therefore form the attributes of the entity, "Human Occupancy". The relationships between the entities are also defined in the proposed data model through a category of relationship entities. For example, the relationship entity "RelSpaceHVAC" defines the relationship between a thermal zone and the HVAC information associated with it. Defining relationship entities allows the reuse of entity definitions which would make the data model concise and easy to understand. For example, defining the relationship entity "RelAssignstoControl" links the time series schedule of the control devices with the occupancy, lighting ventilation and HVAC systems.

Table 1. Summary of requisite parameter of occupancy in energy analysis

	Parameters	eQUEST	IESVE	Ecotect	EnergyPlus
D1	People schedule	✓			✓
D2	Occupancy usage type	✓	✓		
D3	Type of use for given season	✓			
D4	Percentage of area for each type	✓			
D5	Tailored vent air flow for occupancy type	✓			
D6	Designed occupancy	✓			
D7	Heat gain from occupiers	✓	✓		
D8	Sensible heat gain from occupiers	✓	✓		
D9	Latent heat gain from occupiers	✓	✓		
D10	Equipment schedule	✓			
D11	Equipment input power density	✓			
D12	Equipment output power density	✓			
D13	Equipment sensible heat gain	✓	✓		
D14	Equipment latent heat gain	✓	✓		
D15	Equipment energy source e.g. gas	✓			
D16	Electrical load	✓		✓	
D17	Electricity sensible	✓		✓	
D18	Natural gas load	✓		✓	
D19	Natural gas sensible	✓		✓	

3 Developing the IFC4 Extension for Energy Simulation

The data model for energy simulation was developed from the energy simulation kernels with the aim of facilitating interoperability between BIM and energy simulation tools. Gudnason, G., et al. [9] summarized three approaches for implementation of domain data models: (1) to extend the BIM schema(s) with new concepts, attributes and relationships to accommodate the external information resources; (2) to extend existing BIM schema(s) with existing interfaces, for example, by using IfcRelationship subclasses and property sets; (3) to use a "link model" as a bridge between BIM and non-BIM data to provide generality, modularity and implementation scope. The first approach, which is to extend the existing BIM data schema, has been found to be the most efficient way to establish interoperability between BIM and other domain data models [9]. The IFC is identified as most comprehensive data model for building and architecture [6] and has a wider scope than other data standards, like gbXML [10]. Along with the energy simulation tools, IFC is compatible with the whole building lifecycle simulation and estimation tools. Therefore, we choose to extend the schema of IFC4 based on the data model for energy simulation. In the past, several efforts have been made for extending IFC in the areas of cost estimating and tendering [11] human comfort management [12], BIM-based facility maintenance management [12, 13] attempted to develop the IFC extension based on the data requirements in the Excel-based low energy design tool Passive House Planning Package (PHPP). However, since they only considered one data model contained in PHPP, the universality of the developed IFC extension is questionable.

The extension of IFC4 is developed for energy simulation, as shown in Fig. 2. The model is an extension to the domain layer and also includes some entities added to the interoperability layer. The extended schema can be classified into IFC object classes and IFC relation classes which define new entities (and attributes) for energy simulation and the relationship between two entities. The object classes can be divided into five main entities (as shown in Fig. 2) – (1) IFCgbLighting (three entities), (2) IFCgbConstructionType (five entities), (3) IFCgbVentillation (three entities), (4) IFCgbOccupancy (four entities), and (5) IFCgbHVAC (15 entities). The entities which are marked with an asterisk (Fig. 2) are existing entities in the IFC4 schema so that the extension could be linked to the existing schema. The geometry or space being modeled for energy analysis is associated with the entities of the extended IFC schema through relational entities like IFCRelLighting, IFCRelSpaceVentillation, IFCRelSpaceOccupancy, and IFCRelSpaceHVAC.

As discussed, Fig. 2 shows the entity-relationship diagram of the proposed IFC4 extension. The prototyped extension was developed in EXPRESS modelling language, which is the standard modelling language used by IFC schemas. An example of the EXPRESS model of the respective entities in the lighting category is shown in Fig. 2. The formalized schema of the IFC4 extension for energy simulation can be parsed by Java programs through JSDAI [14], which is a standardized API for parsing and manipulating EXPRESS schemas. The application of the IFC4 extension in facilitating interoperability between energy simulation engines and BIM will be discussed in detail in Sect. 4.

Due to the limited space, the IFCgbLight is selected as example to illustrate the methodology. The object classes and their attributes of IFCgbLight are discussed as follows. IFCgbLighting and its sub-entities, namely IFCgbLightingLamp and IFCgbLightingControl implement lighting related parameters such as luminaire type, method for lighting control (through occupancy, sensors, and time clocks), and lamp type respectively. The lighting extension forms part of the interoperability layer of the IFC schema as the attributes can be used across many of the domains. For instance, the luminaire type can be used in the facilities, electrical, and architectural domains.

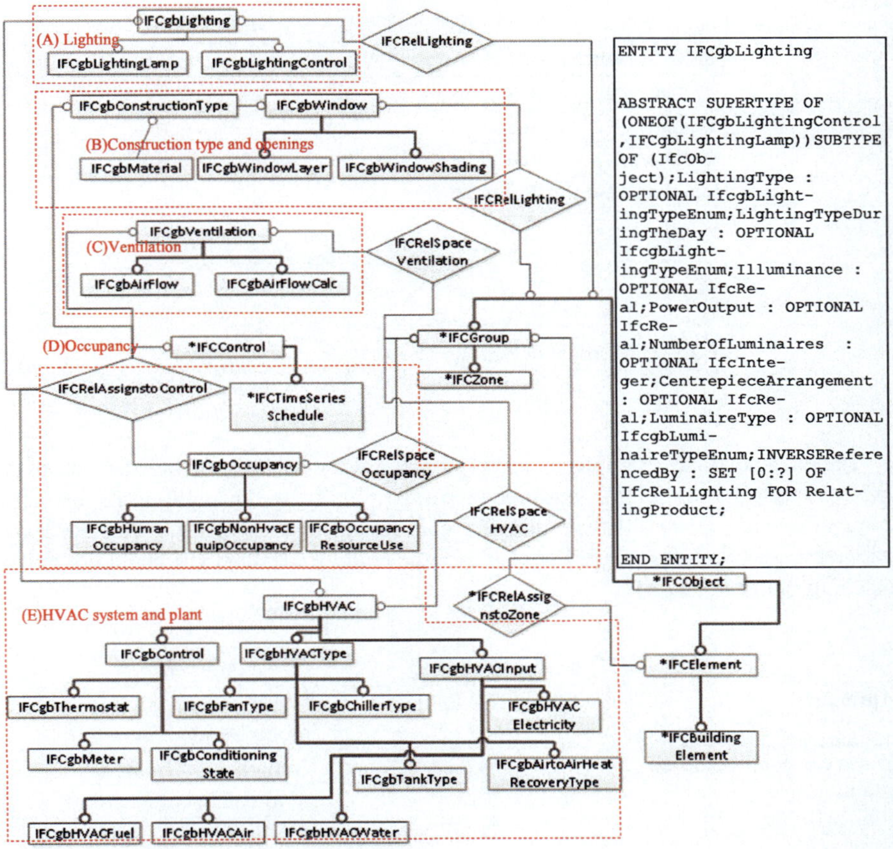

Fig. 2. The proposed extension of IFC4 for energy simulation

3.1 Methodology for Developing the Mapping Between IFC and gbXML

The methodology for deriving the mapping rules between IFC and gbXML entities is by comparing entity definitions and instances of similar elements in both formats. For example, analyzing the document segments that represent walls in IFC and gbXML gives insights about the geometry and semantic mapping of information inside these two schemas. The mapping rules can be divided into two main types: (1) mappings between elements and (2) mappings between attribute values. Table 2 shows a list of mapping rules that are most common for regular building models exported from Autodesk Revit or ArchiCAD. Most mapping between gbXML and IFC is 1-to-N mapping, where a gbXML element may find several matching elements in IFC. For example, all the IFC entities defining building surfaces, such as IfcWall, IfcSlab, and IfcFloor, could be the matching entity for Surface entity in gbXML, depending on the context of the mapping.

Table 2. Mapping gbXML and IFC elements

gbXML element	IFC element
gbXML	IfcProject
Campus	IfcSite
Building	IfcBuilding
BuildingStorey	IfcBuildingStorey
Space	IfcSpace
Surface	IfcWall, IfcSlab, IfcFloor, IfcCeiling
Opening	IfcWindow, IfcDoor
PolyLoop	IfcPolyLoop, IfcLoop, IfcFace
CartesianPoint	IfcCartesianPoint

Table 3 shows an example of mapping IFC attributes (some from the proposed IFC4 extension) with gbXML attribute or entities. Unlike the mapping in the element level, in which most gbXML entities have matching entities in IFC, attribute level mapping may not always exist. For example, the entity IfcChillerType could not find a matching entity in gbXML.

Table 3. Mapping IFC attibutes with gbXML entity/attribute

Attributes	Linked IFC Entity	Matching gbXML Entity/Attribute
Water use hourly schedule	*IFCTimeScheduleSeries	WaterUsePerCycle
Chiller type	*IFCChillerType	HydronicLoopEquipment
Efficiency specification for heater	*IFCgbHVACType	Efficiency
Heater fuel (hot water)	IFCgbHVACFuel	FuelLoad
HVAC schedule	*IFCTimeScheduleSeries	DaySchedule
Air supply temperature	IFCgbHVACAir	AirTemperature
Power and meter efficiency	IFCgbMeter	Efficiency
Cycle fans	*IFCFanType	AirLoopEquipment

Note: The entities which are marked with an asterisk are existing entities in the IFC4 schema

The mapping rules are supported by a customized JSDAI based parsers which aims to efficiently extract information from the IFC and gbXML files. The IFC parser considers the extended IFC4 schema. The gbXML parser is built based on JDOM for file input and output. Another issue in the mapping between IFC and gbXML is the geometry processing to find the building envelope. Unlike the detailed geometry in BIM, gbXML represents simplified geometry information used for energy simulation. Usually, the geometry in gbXML is comprised of several rooms (spaces) as thermal zones, and a bounding building envelope that defines the boundary of the building. The geometry in gbXML requires derivation of the exterior surfaces from the complex geometry in BIM. As pointed out in Bazjanac and Kiviniemi [15], BIM authoring software does not consider the distinction between exterior and interior surfaces and do not include such information in the output IFC file. Therefore, an exterior shell extraction algorithm developed by Cheng et al. [16] is used to derive the building envelope and transform IfcRoom to Space entity in gbXML. The translator developed for IFC and gbXML can deal with complex shapes and can derive building envelope accurately.

4 Implementation of the Interoperabiltiy Framework Between BIM and Energy Simulation Tools

In this section, we present a prototype framework that implements the proposed IFC extension for energy simulation (as discussed in Sect. 3). This prototype framework demonstrates the significance of the proposed IFC extension for integrating the BIM and energy simulation software. As shown in Fig. 3, the prototype framework consists of a Java based converter called the BIM-Energy simulation convertor, which translates the information stored in the extended IFC file to the input file of the respective energy simulation software. In this example, the input for the BIM-Energy simulation converter has been created by manually populating an IFC4 file with lighting simulation information according to the proposed IFC extension (as shown in Fig. 3).

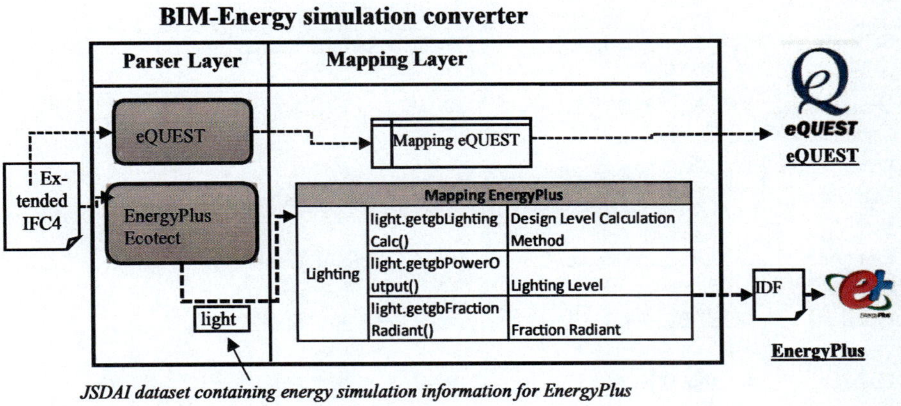

Fig. 3. The prototype of framework for integration of BIM with energy simulation software

In order to validate the proposed data mapping framework and the extended ifcgbXML, we performed the energy simulation using one lecture theater in HKUST campus. After building the BIM model using Autodesk Revit, we import the 3D model into the eQUEST and Energy Plus software and configure the simulation setting of energy consumption, as shown in Fig. 4 and Fig. 5.

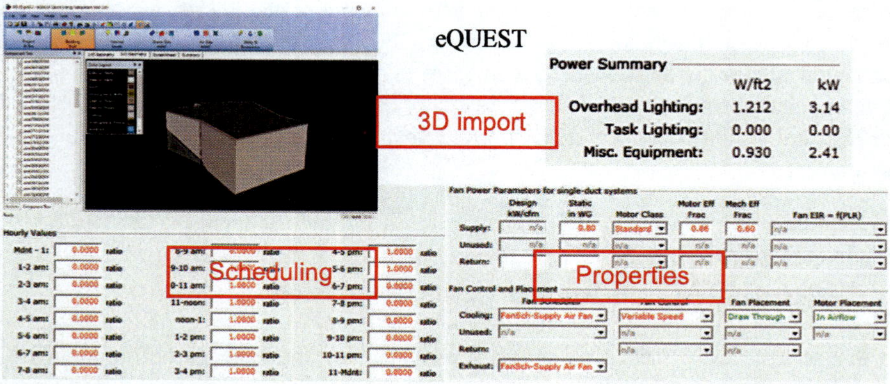

Fig. 4. The scheduling and properties in eQUEST

Table 4. The parameter names and their respective entity names in the extended IFC schema for day-lighting simulation in EnergyPlus and eQUEST

Parameter in name EnergyPlus	Data type and example	Entity name in the extended IFC schema	eQUEST	EnergyPlus
Zone	STRING; Example: "SPACE1-1"	IFCZone	Y	Y
Schedule name	STRING; Example: "LIGHTS-1"	IFCTimeSeriesSchedule	Y	Y
Design level calculation method	STRING ENUM; Example: LightingLevel, Watts/Area, Watts/Person	IFCgbLightingCalc	Y	Y
Lighting level	INTEGER; Example: 1584	IFCgbPowerOutput	Y	Y
Fraction radiant	DOUBLE; Example: 0.59	IFCgbFractionRadiant	Y	Y
Fraction visible	DOUBLE; Example: 0.2	IFCgbFractionVisible	Y	Y
Fraction replaceable	A value 0 or 1	IFCgbDaylightControl	Y	Y

The BIM-Energy simulation converter using IFC parser layer. This layer consists of parsers based on Java for extracting information related to energy simulation from the IFC file. For example, the EnergyPlus and eQUEST module (as shown in Fig. 3) would read all the entities shown in Table 4. As shown in Fig. 3, the EnergyPlus module extracts the information related to day-lighting simulation from the IFC file and converts it into a dataset, called "light" in this case, which can be parsed through pre-defined JSDAI functions.

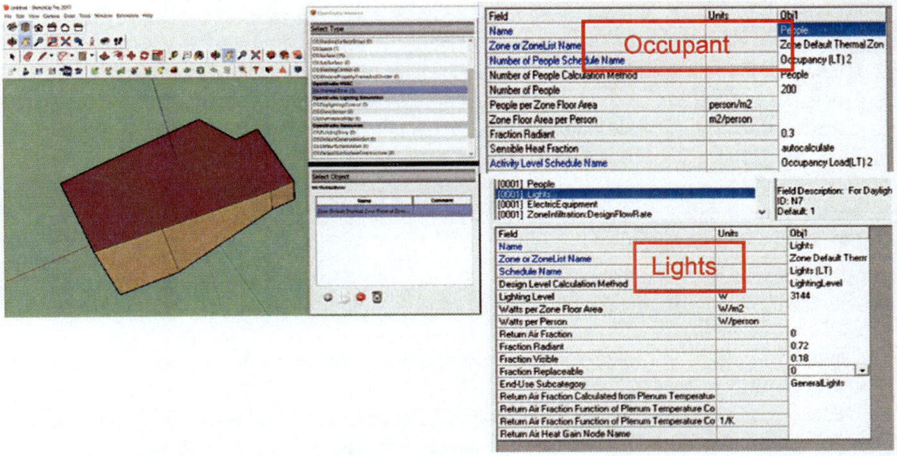

Fig. 5. The occupant, light, and equipment schedule in Energy Plus

5 Conclusions

In this paper, we present three components that aim to facilitate the interoperability between BIM and energy simulation software: (1) a comprehensive data model derived from four representative energy simulation tools, (2) an extension to the de-facto BIM data standard IFC for energy simulation, and (3) mapping between IFC and gbXML to facilitate data exchange between BIM software and energy simulation tools. The proposed data model was validated and is demonstrated through an example scenario which shows the improvement of seamless data exchange between BIM and energy simulation tools. The big challenge in this study is to identify the requirement information and extend the IFC to map data into gbXML data schema. The limitation of this study is that the IFC extension covers several entities, such as lighting, construction type, ventilation, occupancy, and HVAC, however, the rest of entities related to energy are needed to study. In future, extending the scope to cover more energy analysis applications and using ontology to extend IFC schema and gbXML schema including information requirements would be a positive development. Overall, the study has laid the foundation for energy analysis applications to the IFC model. It has also set out the future work necessary to bring the extension model from a concept to a working model, thereby furthering efforts to create a universal BIM standard.

References

1. Energy Information Administration (US) (ed.). Annual Energy Outlook 2012: With Projections to 2035. Government Printing Office (2012)
2. Electrical and Mechanical Services Department, Hong Kong Energy End-use Data 2013, Hong Kong (2013)

3. Al-Homoud, M.S.: Computer-aided building energy analysis techniques. Build. Environ. **36**(4), 421–433 (2001)
4. Clarke, J.: Energy Simulation in Building Design. Routledge, Abingdon (2007)
5. Dong, B., O'Neill, Z., Li, Z., Luo, D., Madhusudana, S., Ahuja, S., Bailey, T.: An integrated infrastructure for real-time building energy modeling and fault detection and diagnostics. Proc. SimBuild **5**(1), 448–455 (2012)
6. Eastman, C.M., Eastman, C., Teicholz, P., Sacks, R., Liston, K.: BIM Handbook: A Guide to Building Information Modeling for Owners, Managers, Designers, Engineers and Contractors. Wiley, Hoboken (2011)
7. Cheng, J.C., Deng, Y., Das, M., Anumba, C.: Evaluation of IFC4 for the GIS and Green Building Domains. In: Computing in Civil and Building Engineering (2014), pp. 2216–2223 (2014)
8. Garcia, E.G., Zhu, Z.: Interoperability from building design to building energy modeling. J. Build. Eng. **1**, 33–41 (2015)
9. Gudnason, G., Katranuschkov, P., Balaras, C., Scherer, R.J.: Framework for interoperability of information resources in the building energy simulation domain. In: Computing in Civil and Building Engineering, pp. 617–624 (2014)
10. Cemesova, A., Hopfe, C.J., Rezgui, Y.: An approach to facilitating data exchange between BIM environments and a low energy design tool. In: Proceedings of the BS2013, 13th Conference of International Building Performance Simulation Association, Chambery, France, pp. 25–30 (2013)
11. Ma, Z., Wei, Z., Wu, S., Zhe, L.: Application and extension of the IFC standard in construction cost estimating for tendering in China. Autom. Constr. **20**(2), 196–204 (2011)
12. Chen, W., Chen, K., Cheng, J.C.: Towards an ontology-based approach for information interoperability between BIM and facility management. In: Workshop of the European Group for Intelligent Computing in Engineering, pp. 447–469. Springer, Cham (2018)
13. Cemesova, A., Hopfe, C.J., Rezgui, Y.: PassivBIM–a new approach for low energy simulation using BIM. eWork and eBusiness in architecture, engineering and construction. In: ECPPM, Reykjavik, Iceland, p. 29 (2012)
14. LKSoftWare GmbH "JSDAI" (2015). http://www.jsdai.net/download
15. Bazjanac, V., Kiviniemi, A.: Reduction, simplification, translation and interpretation in the exchange of model data. In: CIBW, Maribor, Slovenia, vol. 78, pp. 163–168 (2007)
16. Cheng, J., Deng, Y., Du, Q.: Mapping between BIM models and 3D GIS city models of different levels of detail. In: 13th International Conference on Construction Applications of Virtual Reality, London, pp. 30–31 (2013)

Modeling Physical Damages Using the Industry Foundation Classes – A Software Evaluation

Mathias Artus(✉) and Christian Koch

Bauhaus-Universität Weimar, Marienstraße 13a,
99423 Weimar, Germany
mathias.artus@uni-weimar.de

Abstract. Bridges are designed to last for more than 50 years and consume up to 50% of their life cycle costs during their operation phase. Several inspections and maintenance actions take place within this period, during which bridge and damage information is collected, digitized and exchanged between different actors. Currently, the inspection and maintenance practice relies on paper based data collection and exchange, which is time consuming and error prone. A way of storing and exchanging damage data in a digital format would lower the costs for inspection and maintenance as well as support future needs, i.e. immediate simulations, automated maintenance planning and mixed reality supported inspections and maintenance. For this purpose, the paper presents and tests concepts for modeling damage information by the means of the Industry Foundation Classes (IFC) without extending or changing the existing IFC Schema. First, multiple use cases and related requirements for damage information are defined. This information is used to develop multiple concepts to model damage information by using existing IFC entities. The implementation and tests of the concepts will show that the IFC provide several meaningful ways to model damage information. However, existing IFC software does not fully support the concepts of IFC, particularly when visualizing specific entities.

Keywords: Building information modeling · Damages · Damage Information Modeling · Industry Foundation classes · Inspection and maintenance · Bridges

1 Introduction

1.1 Goal of a Damage Information Model

Bridges are designed to last for more than 50 years and consume up to 50% of their life cycle costs during its operation phase [1, 2]. Damage registration on site is done paper based. Whereby, processing the information, e.g. to decide about maintenance or to plan it, is done digitally. So, an engineer must digitize all information after an inspection respectively after retrieving the analog data [3]. This practice is error prone and leads to information loss [4]. A Damage Information Model (DIM), which supports a digital collection and exchange of damage data, could lower the costs for inspection and maintenance, reduce the loss of information on exchanging the data and reduce errors. Further, a higher automation of the inspection, rating and maintenance circle would be possible. Such an information model must cover current requirements, for

E. Toledo Santos and S. Scheer (Eds.): ICCCBE 2020, LNCE 98, pp. 507–518, 2021.
https://doi.org/10.1007/978-3-030-51295-8_36

example inspection planning, on-site inspections and bridge rating as well as future requirements, like instant load simulation, damage propagation or inspection in virtual reality.

1.2 Related Work

Because, the BIM concept is extended to new areas like infrastructure, Structural Health Monitoring (SHM) and simulation, new concepts of storing data are researched. In general three modeling concepts exist in the research of extending BIM. First, there are proprietary and new developed data formats to store additional data. Some examples for this are the usage of Excel spreadsheets [5] or Revit file extensions [6]. Completely new formats, either binary or text, are imaginable, too. A drawback of self-designed data formats is the lack of tools to work with these data. Many and well developed software tools for processing proprietary file formats exist, which allow to manipulate or visualize the data. However, proprietary data need to be transformed to use the data in third party applications. This practice is common in the Architecture, Engineering and Construction (AEC) sector. However, it is error prone and time-consuming.

IFC itself is suitable to model damages, too. Sacks et al. and Hamdan et al. have published an Information Delivery Manual (IDM) and a Model View Definition (MVD) for bridge inspection [7, 11]. These papers contain the processes, damage parameters, several damage types and an approach to model damages at the surface using the current IFC schema. Details will be discussed later in Sect. 5.2

Another approach is to store data using the linked data approach via ontologies. Data about the building is stored in the IFC format and additional data like damage data is linked and stored in another file or online with links to the building model [8]. This concept results in a multi-model approach.

Databases can build up upon the IFC schema or independent from it [9, 10]. This approach is as flexible as the ontology concept. However, it lacks a standardization. Even if the IFC schema is considered, specialized software is necessary to exchange the data.

Finally, the last approach is an IFC extension to support new features for data exchange [11]. Ma et al. [12] have extended the IFC model, in order to represent damages caused by earthquakes. IFC offers a standard for the data exchange. Therefore, every software vendor has to implement an IFC interface only. This paper focuses on the question in which way the existing IFC can be used to exchange damage data.

Beside the approaches above, there are many papers that deal with damage recognition. For example Torok et al. [13] have used a 3D point cloud for damage recognition. Their damage model consists of points, which are marked as crack or not. The points are not related to each other. Hence, a damage itself is only visible for an engineer and not for machines. A merging of all points to a single damage entity would help overcome this issue. Other work in the field of damage recognition has similar disadvantages and could be improved by an extensive DIM.

2 Methodology

Figure 1 shows the methodology, which was applied to define DIM's by the means of IFC and test them. Starting from practice, significant damages and damage parameters are identified. Upon this data, modeling concepts were defined. For later testing, software tests were defined and IFC files are generated. Finally, these IFC files have been imported into several BIM applications to check the quality of IFC implementation in this software.

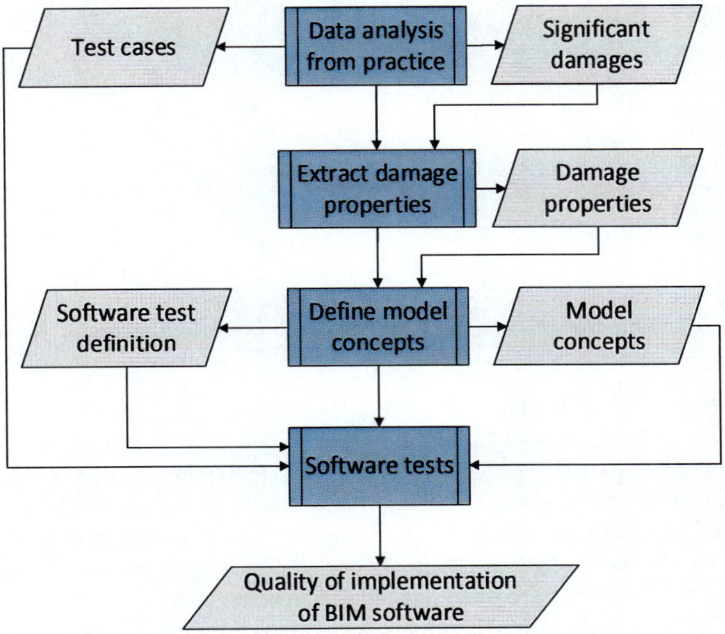

Fig. 1. Methodology for exploration of IFC abilities and IFC software

3 Damages in Practice

Artus and Koch [14] have pointed out 23 damage categories. Every damage gets a rating between 1 and 5 in Germany; 5 means the damage is crucial and 1 means it has no impact. It has been observed that cracks are the most frequent and the most significant damages. Figure 2 shows all damages with a rating of 2 or worse. Beside cracks, material changes have high ratings and hence, are significant for the final bridge rating.

In conclusion, a first attempt should be made to model cracks. Whereby, cracks can be understood as long, small and thin voids in the surface. Spallings are voids at the surface, too. Hence, it is possible to model both damage types, cracks and spallings, within the same model approach.

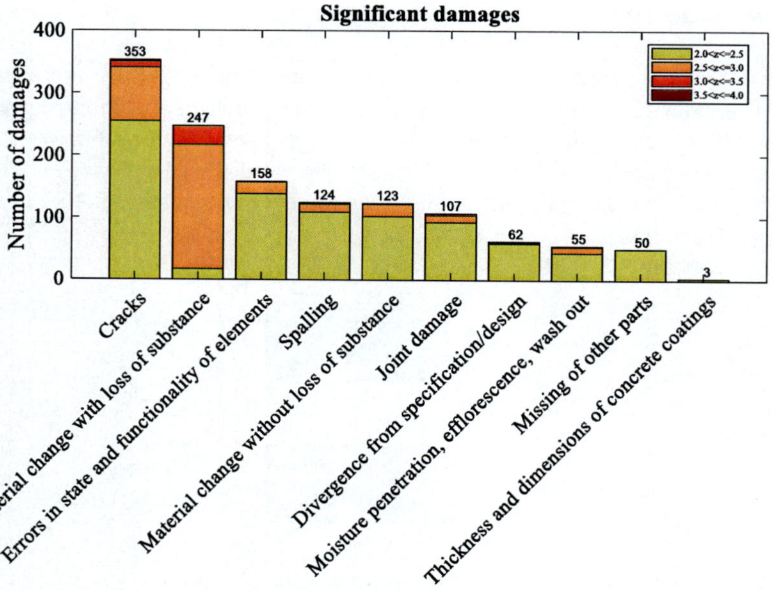

Fig. 2. Significant damages with condition ratings >2. Ordered by number of occurrence.

4 Damage Parameters

The authors have previously identified several parameters for damages [14]. Table 1 shows an overview of the parameters for a single damage. The parameters are sorted by 4 categories: visualization, relationships, inspection data and maintenance data. Visualization means data, which is necessary for visualization purposes. Relationships contain information about semantic relations between entities.

Relationships can lead to other effects, e.g. subtracting the geometry of a crack or spalling from the geometry of the building element. Inspection respectively maintenance data is the data from inspections respectively for maintenance. The geometries of damages are most important for the modeling. 2D geometry is mostly used in the paper-based workflow until now. For future needs, e.g. automatic maintenance planning, virtual reality inspection and automatic maintenance cost calculation, 3D geometries are necessary. Furthermore, images, text and audio recordings are necessary to review damages in the office later. Inspector positions and orientations can help to improve the inspection process later by revising this data. Additionally, data for the inspection would be measurements of chloride containment, x-ray or supersonic measurements. Data about material volume, prerequisites or actors are necessary for maintenance automation as well.

Table 1. Parameters for local damages

Visualization	Relationships	Inspection data	Maintenance data
Images	Related inspections	Audio recordings	Repair recommendation
2D geometry	Influences on material parameters	Text	Repair material
3D geometry	Influences on building element geometry	Inspector position	Material volume
Point Cloud	Related damages	Inspector orientation	Actors for repair
Mesh	Linked components, component groups and bridges	Measurements of physical parameters	Repair equipment
Texture	Other semantics	Damage type	Prerequisites for repair
		Rating	Repair process
			Repair approval criteria

5 Concepts for Geometrical Damage Models of Cracks and Spalling

All model approaches make usage of the final standard IFC 4 [15].

5.1 Modeling Damages as *IfcProxy* or *IfcBuildingElementProxy*

The most obvious approach to model a damage would be to use a generic element, like *IfcProxy* or *IfcBuildingElementProxy*. *IfcProxy* is deprecated since IFC version 4.0.2.1 [15]. Hence, *IfcBuildingElementProxy* is used for further illustration.

Figure 3 shows an approach to model a damage via an *IfcBuildingElementProxy*. All yellow boxes are entities, which represent objects. Black boxes represent relationships and the orange box is the damage, respectively the *IfcBuildingElementProxy*.

IfcBuildingElementProxy is a subclass of *IfcProduct* and hence, it has representations. These representations can be 2D or 3D. *IfcRelAgregates* represents the relation from the damage to the afflicted building element.

Measurements and ratings can be stored in an *IfcPropertySet*, which is linked to the damage. Hüthwohl et. al [16] have shown how the condition rating can be stored in the IFC. This approach uses the definition the Pset_Condition property set of the buildingSmart Data Dictionary (bSDD). Unfortunately, there are no further definitions for damage related data within the bSDD.

In addition, the *IfcElementAssembly* can be used to represent damages as assemblies of other damages, e.g. a spalling with corroded reinforcement. In the figure, the spalling and the corrosion would be an instance of *IfcBuildingElementProxy*. Hamdan et al. have shown an approach lto model damages assembled by individual damages

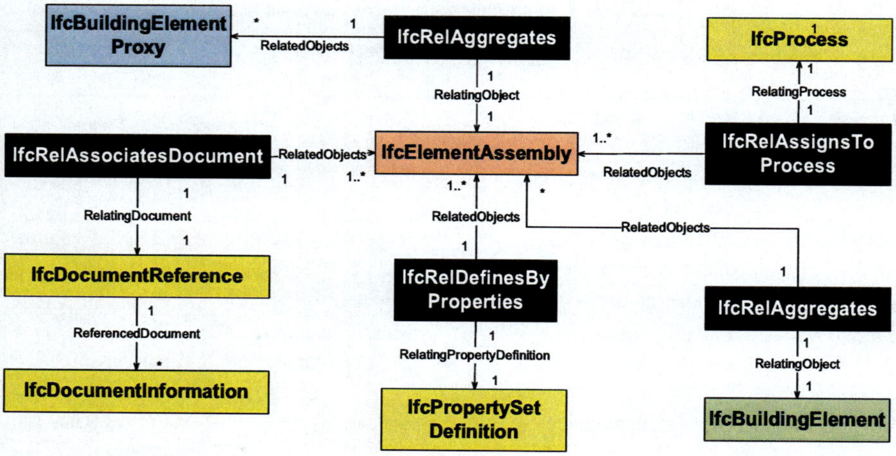

Fig. 3. Data model by the usage of *IfcBuildingElementProxy* as entity for the damage. The damage is orange, damage parts are blue, relationships are black, the damaged building element is green and yellow are other entities. Multiple superclasses, subclasses and selects are omitted.

[11]. However, their model has extensions of the IFC Schema, which is avoided by using the existing entity *IfcElementAssembly*.

Related documents, like pictures, sketches, audio recordings or other files, can be linked with the *IfcRelAssociatesDocument* relationship. Last but not least, processes, e.g. an inspection or maintenance action, can be connected by the *IfcRelAssignsToProcess* relationship or similar [16] and further, information about persons and time can be added then.

With this approach, many data from Table 1 can be stored. Disadvantageous is the lack of textures, the influences on material parameters and the influences on the building element geometry. Further, there is no possibility to store data for prerequisites or approval. Following concepts can make usage of all IFC relationships (black entities) from this concept. Hence, all advantages of this concept are valid for subsequent models.

5.2 Modeling Damages as IfcSurfaceFeature

Hüthwohl et al. [16] and Sacks et al. [16] have proposed an approach to model damages as assembly of multiple *IfcSurfaceFeatures*. An *IfcElementAssembly* represents a defect with a rating and consists of several element defects, which are modeled as *IfcSurfaceFeatures*. Figure 4 shows the usage of a surface feature, to model and visualize an element defect at the surface of a component. The result is visualizing a crack or corrosion via a texture on the 3D model. It is still disadvantageous that influences on the building element geometry cannot be modeled.

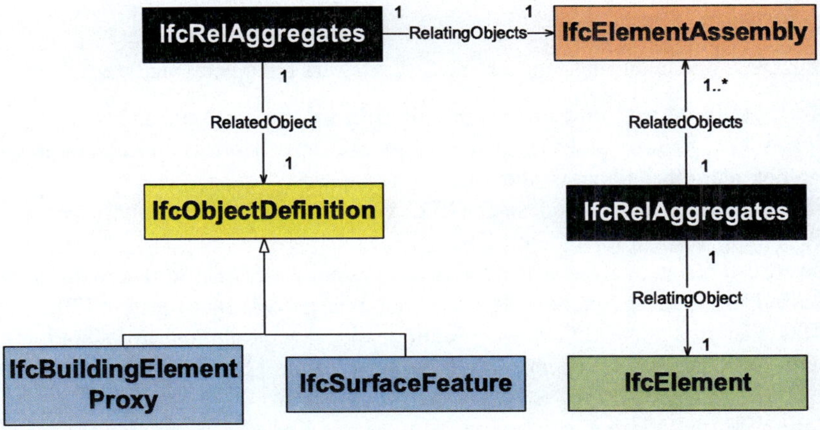

Fig. 4. Damage model with 2D texture using *IfcSurfaceFeature* and related elements. The damage is shown in orange, black are relationships, the damaged building element is green and yellow color shows other entities. Multiple superclasses, subclasses and selects are omitted for simplicity.

6 IFC Software Tests

6.1 Test Case Definition

The tests cover three different use cases.

A) **Usage of an *IfcBuildingElementProxy***: The test file contains an *IfcBuildingElementProxy* to model damages. It is a three dimensional damage, e.g. a spalling, at a beam. Additional data is linked to the IfcBuilingElementProxy, like the condition rating, assigned maintenance and documents. Further, the damage is modeled as an *IfcElementAssembly*, which offers the possibility to connect a spalling with a corrosion. This information should be visualized by the applications, too.

B) **Usage of an *IfcSurfaceFeature* to model a crack:** The test file contains an *IfcSurfaceFeature* to model damages. Hüthwohl et al [16] has developed this representation. All software should display a patch with a texture on the surface of a beam.

C) **Extending the *IfcSurfaceFeature* with additional data:** This is a combination of test case A and B. An *IfcSurfaceFeature* is used to model the spalling. Further, an *IfcBuildingElementProxy* is used to model the corrosion. All additional data is linked to the *IfcElementAssembly*. All additional data shall be visualized, like in test case A.

All test files are accessible at https://bitbucket.org/Noranius/dim-ifc-test-files-icccbe/src.

6.2 Software

For the tests, one authoring tool and six IFC viewers have been analyzed.

- **Revit** is a BIM authoring tool from Autodesk for the AEC sector [17].
- **apstex IFC Viewer** is a free Java IFC viewer. apstex offers an open source IFC toolbox and other libraries, too [18].
- **xBIM Xplorer** is an open source IFC viewer. It offers additional open source libraries to work with IFC, too [19].
- **Desite MD** is an analysis and information system for BIM. It offers multiple tools, i.e. model checking, quantity determination and process management [20].
- **BIM Vision** is a free IFC viewer. Additionally, several plugins are offered, i.e. for clash detection, model comparison or report creation [21].
- **usBIM** is a free IFC viewer [22].
- **Solibri Model Viewer** is a free model viewer and model check viewer [23].

6.3 Test Results

An overview about all test results can be found in Table 2019.

Tests with *IfcBuildingElementProxy* (A). Revit does not show the damage. Figure 5a shows this output. Revit should be capable of visualizing *IfcBuildingElementProxy*. This has been tested by exporting a custom family as building element proxy. Further, problems regarding the element assembly and the belonging to a spatial structure could be excluded, too. If there are any recommendations on the test files, feel free to contact.

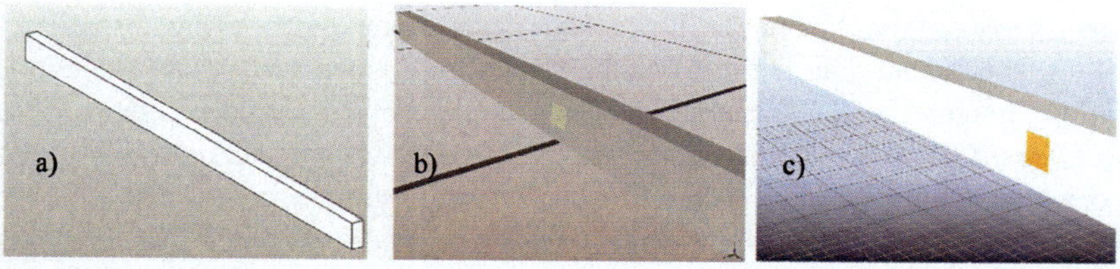

Fig. 5. a) Output of Autodesk Revit 2019 without any damage geometry. b) Output of Xbim with the damage geometry. c) Similar output of the apstex IFC Viewer.

DesitMD and the Solibri Model Viewer behave similarly. Finally, a relationship to different spatial elements has been done, i.e. site, story and building. In addition, this does not lead to a successful import. Because Revit is IFC certified, it should be possible to import proxy elements.

usBim is capable of the geometry, but does not show the related information like processes or documents. Bim Vision reveals better capabilities, because it visualizes the properties at least. Best results deliver apstex and Xbim. They show the geometry and related information. Figure 5b shows a proper output produced by Xbim.

Tests with *IfcSurfaceFeature* (B). Revit does not import the test file with the *IfcSurfaceFeature* properly. The same happened at Desite MD, which is shown by Fig. 6a. Other software, like apstex or xBIM, shows only an entry in the hierarchy of the project. Further investigation revealed that an *IfcRepresentationContext* must be set to display elements in xBIM. However, xBIM crashed after adjusting the test file in that way. BIM Vision, usBIM and Solibri have been able to show at least a colored patch on the surface without a texture. One of the best results by usBIM is shown in Fig. 6c.

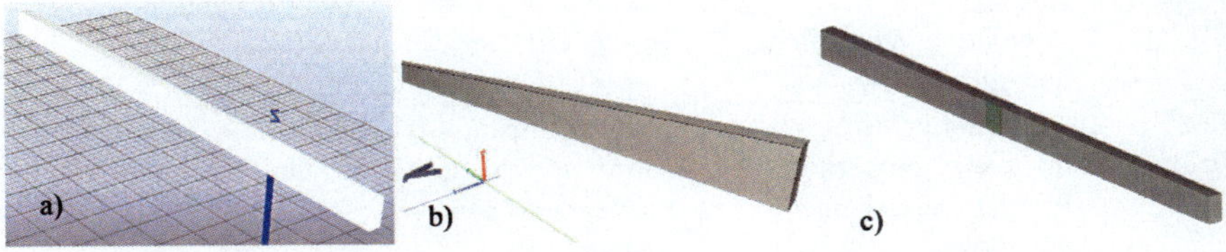

Fig. 6. a) Beam without a patch in apstex b) Output of Desite MD without any damage geometry, c) Output of usBIM with the patch on the beam

Tests with IfcSurfaceFeature and related information (C). The results of this test case can be understood as a combination of the results by case A and B. For example, Xbim and apstex are capable to visualize the relate damage data, but not the surface feature. It is similar with the other software. If in test case A the related data is displayed, it is displayed in case C. And if the surface feature is shown in case B, it is shown in case C, too.

Test Result Overview. In conclusion, the existing authoring tools and IFC viewers support the modeling concepts in different quality. Table 2 shows an overview of the software tests. xBIM and apstex have the most positive results. However, both lack visualizing textures. Problematic in all applications is the visualization of the *IfcSurfaceFeature*. None of the selected software was able to show the surface feature with texture.

Table 2. Overview over different software and the quality of handling the model concepts from section.

Software	Test A	Test B	Test C
Revit 2019	No damage no related data	No damage no related data	No damage no related data
apstex IFC viewer	Shows damage with geometry and related data	*IfcSurfaceFeature* in hierarchy only	IfcSurfaceFeature visible in hierarchy only. Related data visible
xBIM Xplorer	Shows damage with geometry and related data	*IfcSurfaceFeature* in hierarchy only	*IfcSurfaceFeature* without texture. Related data visible
Desite MD	Neither *IfcBuildingElementProxy* nor *IfcProxy*	Does not show the *IfcSurfaceFeature*	Neither damage geometry nor related data
BIM Vision	Geometry is visible and the related property set; Process and documents not shown	*IfcSurfaceFeature* is shown as patch on the beam without texture	*IfcSurfaceFeature* without texture; condition rating visible; No process or document
usBIM	Geometry shown; process and related documents not visible	*IfcSurfaceFeature* is shown as patch on the beam without texture	*IfcSurfaceFeature* without texture; No process or document
Solibri Model Viewer	Neither geometry nor process and related documents visible	*IfcSurfaceFeature* is shown as patch at the beam without texture	*IfcSurfaceFeature* without texture; No related data visible

7 Conclusion

During the life cycle of a bridge, data is generated during inspection and maintenance. The current practice, which relies on analog or abstract data, is error prone and time consuming. Getting to the next level of bridge data management, requires a data exchange format for damage information. Beside other methods, the IFC standard offers possibilities to exchange data. From an analysis of practically relevant data, it emerged that cracks and spalling occur most frequently. Hence, three approaches were presented to model cracks and spalling with existing IFC entities. Finally, the software tests showed that none of the tested programs supports all model concepts.

This work covers cracks and spalling only. As they are the most frequent an relevant damages. However, bridges are afflicted by other damages, too. Further work should cover other damage types. Furthermore, questions arise, whether the semantic significance of some entities for damage modeling is correct. Thus, an *IfcRelAggregates* represents a part whole relationship. This is incorrect for a damage. A new relationship would be more appropriate, such as *IfcRelDamaged*.

Finally, IFC is a powerful and flexible standard. The current standard offers many possibilities. Even much data, which are related to damages, can be stored within an

IFC file. However, this leads to a high complexity in implementation and hence different approaches and qualities in software. xBIM passes most of the test cases successfully. Regardless of this positive result, no software application was able to pass all three test cases. Future work about Damage Information Modeling will focus on prerequisites and approval criteria, because these are two parameters, which are not covered by the models yet. Moreover, modeling other damage types and distinguish different damage types in a single model is a goal for ongoing work.

Acknowledgments. We thank the "Thüringer Landesamt für Bau und Verkehr" for their support with knowledge from practice in bridge inspection and statistical data about bridges in federal state of Thuringia in Germany. Further, we thank Mr. Mohamed Said Helmy Alabassy for creating the IFC test files during his student project.

References

1. Schach, R., Otto, J., Häupel, H., Fritzsche, M.: Lebenszykluskosten von Brücken. Bauingenieur **81**, 7–14 (2006)
2. Wettlauf gegen den Verfall. http://www.welt.de/politik/interaktiv/bruecken/deutschlands-bruecken-wettlauf-gegen-den-verfall.html. Accessed 27 Sept 2018
3. Hammad, A., Zhang, C., Hu, Y., Mozaffari, E.: Mobile model-based bridge lifecycle management system. Comput.-Aided Civil Infrastruct. Eng. **7**, 530–547 (2006)
4. Borrmann, A., König, M., Koch, C., Beetz, J.: Building Information Modeling, 2nd edn. Springer, Cham (2018)
5. Marzouk, M.M., Hisham, M.: Bridge information modeling in sustainable bridge management. In: Chong, W.K.O., Hermreck, C. (eds.) ICSDC 2011, pp. 457–466. American Society of Civil Engineers (2012)
6. McGuire, B., Atadero, R., Clevenger, C., Ozbek, M.: Bridge information modeling for inspection and evaluation. J. Bridge Eng. **21**(4), 4015076 (2016)
7. Sacks, R., Kedar, A., Borrmann, A., Ma, L., Singer, D., Kattel, U.: See bridge information delivery manual (IDM) for next generation bridge inspection. In: University of Huddersfield (ed.) International Symposium on Automation and Robotics in Construction, vol. 33 (2016). http://eprints.hud.ac.uk/id/eprint/31380/1/2%20Sacks-2016-SeeBridge%20Information%20Delivery%20Manu.pdf
8. Lee, D.-Y., Chi, H.-L., Wang, J., Wang, X., Park, C.-S.: A linked data system framework for sharing construction defect information using ontologies and BIM environments. Autom. Constr. **68**, 102–113 (2016)
9. Jeong, S., Zhang, Y., O'Connor, S., Lynch, J.P., Sohn, H., Law, K.H.: A NoSQL data management infrastructure for bridge monitoring. Smart Struct. Syst. **4**, 669–690 (2016)
10. Sung, Y.-C., Chen, C.-C., Chang, K.-C.: Extended bridge inspection module of a life-cycle based bridge management system. In: ICCCBE2016 Organizing Committee (ed.) 16th International Conference on Computing in Civil and Building Engineering (2016). http://www.see.eng.osaka-u.ac.jp/seeit/icccbe2016/Proceedings/Full_Papers/215-092.pdf. Accessed 3 Sept 2018
11. Hamdan, A.-H., Scherer, J.R.: A generic model for the digitalization of structural damage. In: Caspeele, R., Taerwe, L., Frangopol, D.M. (eds.) Life Cycle Analysis and Assessment in Civil Engineering, vol. v.5. Chapman and Hall/CRC (2018)
12. Ma, L., Sacks, R., Zeibak-Shini, R.: Information modeling of earthquake-damaged reinforced concrete structures. Adv. Eng. Inform. **3**, 396–407 (2015)

13. Torok, M.M., Golparvar-Fard, M., Kochersberger, K.B.: Image-based automated 3D crack detection for post-disaster building assessment. J. Comput. Civil Eng. **28**(5), A4014004 (2014)
14. Artus, M., Koch, C.: State of the art in damage information modeling for bridges. In: Geyer, P., Allacker, K., Schevenels, M., de Troyer, F., Pauwels, P. (eds.) EG-ICE 2019, vol. 2394, CEUR Workshop Proceedings (2019). http://ceur-ws.org/Vol-2394/paper18.pdf. Accessed 17 Sept 2019
15. IFC4 Add2. https://standards.buildingsmart.org/IFC/RELEASE/IFC4/ADD2_TC1/HTML/. 1 June 2018
16. Hüthwohl, P., Brilakis, I., Borrmann, A., Sacks, R.: Integrating RC bridge defect information into bim models. J. Comput. Civil Eng. **32**, 1–14 (2018)
17. Revit. https://www.autodesk.de/products/revit/overview. Accessed 14 Oct 2019
18. Java IFC Viewer. http://www.apstex.com. Accessed 14 Oct 2019
19. xBIM XbimWindowsUI. https://github.com/xBimTeam/XbimWindowsUI. Accessed 15 Oct 2019
20. DESITE MD/ DESITE MD PRO. https://www.ceapoint.com/desite-md-md-pro/. Accessed 14 Oct 2019
21. BIM Vision. https://bimvision.eu/en/download/. Accessed 14 Oct 2019
22. usBIM.viewer+. https://www.accasoftware.com/en/downloadfreeware?prog=usbim_viewer. Accessed 14 Oct 2019
23. Solibri Model Viewer. https://www.idc.ch/solibri/produkte/solibri-model-viewer/. Accessed 14 Oct 2019

An IFC Representation for Process-Based Cost Modeling

Eduardo Luís Isatto(✉)

Universidade Federal do Rio Grande do Sul, Porto Alegre, RS 90035-190, Brazil
isatto@ufrgs.br

Abstract. One of the applications of BIM is for cost estimating in construction projects. However, the usual practices regarding the integration between cost and product models with BIM have been limited to the quantity takeoff of physical production measures from a digital model. The cost estimates of the building and its components usually happen outside the BIM model, through the same cost estimate method that has been traditionally used in the construction industry. Production costs are estimated based on previously known consumption rates of resources and theirs acquisition costs, which are commonly stored in external databases using "ad hoc" representation schemas, without any further information exchange with the BIM model. Nonetheless, the most recent version of the ISO 16739 standard (the Industry Foundation Classes, addendum 2, IFC4-add2) has included a number of new classes and properties aimed to extend BIM scope much beyond the product model. However, up to the moment, no specific schema using these new classes has been proposed in order to incorporate cost model information into the BIM model. This study was aimed to investigate the possibility of representing process-based cost information using IFC 4 objects. The research method approach used was case study. Some use cases of processes cost models were selected from real cost modeling cases, and generalize to design a process-based cost modeling schema in IFC. The resulting schema was then instantiated in a prototype, used to validate the logic and output of the representation against manual calculations. The representation of process-based cost modeling proved to be capable of successfully merging the cost modeling schema within a BIM model. It opens the possibility of close integration between product and cost models with BIM technology, as well as the storage and dissemination of cost models and their related information using open and interoperable representation.

Keywords: Building Information Modeling (BIM) · IFC · Cost modeling · Data representation · Cost estimating

1 Introduction

Effective cost management is a key factor for the success of a construction project [14]. It involves much more than just estimating the monetary effort, requiring an active posture of the managers in planning and influencing the behavior of project costs to achieve the desired results.

E. Toledo Santos and S. Scheer (Eds.): ICCCBE 2020, LNCE 98, pp. 519–528, 2021.
https://doi.org/10.1007/978-3-030-51295-8_37

Cost estimation of construction projects faces many difficulties particularly due to the unique nature of the construction product, among other factors. However, along the last decade the BIM technology has proven to be a valuable approach to perform quantity take-offs (QTO) and cost estimates [10].

Due to the possibility of automatically calculate and assess production quantities based on the geometric characteristics of building elements, BIM software is particularly capable of tracking the effects in costs due to changes in design, thus providing an extreme valuable support to target-costing strategies as those implied in Integrated Project Delivery (IPD) [1].

However, the use of BIM technology for cost estimation rarely goes beyond the QTO procedure. Cost estimates are also dependent on production costs per unit (unit costs), which are usually calculated based on previously known consumption rates of resources and their acquisition costs. These are commonly stored in external databases using "ad hoc" representation schema, without any further information exchange with the BIM model. As a consequence, the cost estimating process results highly dependent on the particular software being used, and the exchange of resulting data is made extremely limited.

Since the most recent version of the ISO 16739 standard, the Industry Foundation Classes schema (the IFC4) includes several new classes and properties aimed to extend BIM scope much beyond the product model, particularly in terms of resources and task types. Nonetheless, except for some references in the buildingSMART documentation of IFC4-add2 [2], publications reporting the use of these new classes for cost management are still rare, and approaches to incorporate cost model information based on resource usage into the BIM model are still to be proposed.

There are many advantages in representing cost information as a part of the BIM model. Firstly, it extends the scope of the BIM model from Product Data to Project Management Data, as suggested by Froese [7]. As a consequence, project data can be stored in a single place, or segmented in many places without losing its integrity. Secondly, it can help managers to deal with trade-offs related to the project "iron triangle"—cost, scope, and time—by providing a transparent and solid integration among cost objects, product elements, and project tasks. Thirdly, adopting ISO 16739 standard (the Industry Foundation Classes) can assure strong compatibility between different software and across building life-cycle stages. This facilitates, for example, the exchange of project information among distinct applications dedicated to specialized tasks, as project-scheduling and cost estimating, or even between these applications and ERP systems. Fourthly, an IFC-based cost model can enable cost information to be consolidated and exchanged with other project members through a federated model. Finally, process-based cost information can be stored as part of process libraries to be used in the context of a specific project, company, or even as public processes libraries (e.g., the Brazilian National System of Construction Costs and Indexes Survey—SINAPI) [3].

Considering the relevance of the research problem, this study aims to contribute to reducing the knowledge gap by proposing a process-based representation for cost information using IFC 4 objects. With such representation, it is expected to further incorporate resource-usage cost information directly into IFC projects files, or as member-of an IFC federated model.

2 Background

2.1 Process-Based Cost Modeling

Cost estimates methods vary depending on information availability and the intended purpose. They can range from approximate values early in the design to more precise values after the design is complete [6]. During the early design phase, *conceptual estimating methods* are used, as rough-order-of-magnitude (ROM), based on cost per unity or the production capacity, or even associated with the area or volume of the building (e.g., cost per square foot). As the design matures, *detailed estimating methods* can be used to develop cost estimates, based on cost per component (unit costs of assemblies) or per unit of work result (unit cost of processes). While conceptual estimating methods are based on historical cost patterns, detailed methods usually rely on the causes of the costs, the resource usage patterns.

All these methods are acceptable if the intended purpose is product pricing, but only the latter can provide adequate support for cost management during project execution. The reason is simple: product-based cost models are structured to represent building components or finished building and are thus concerned more on ends than on means [12]. Conversely, process-based models are naturally coupled with the project progress as they are structured to represent tasks as the cost objects.

As process-based costing method relies on process patterns, they are much more flexible to model production costs in contexts where product patterns do not exist or are difficult to identify, as in Engineering-To-Order (ETO) environments like construction. For example, all cut-and-bent reinforced bar configurations employ up to three processes: cut, bend and tie.

For the sake of clarity, in this article is assumed that tasks are the means to achieve a desired product as a result, and that these tasks demand the use of resources like material, labor, equipment, and other products (Fig. 1). Therefore, costs of the resources are allocated primarily to the processes, and then to tasks and products, in that order.

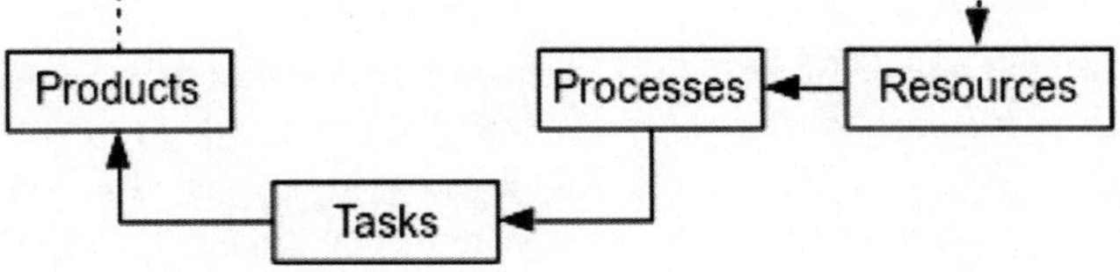

Fig. 1. Conceptual model for process-based costing

Process-based cost compositions are commonly used in construction industry. The Brazilian National System of Construction Costs and Indexes Survey (SINAPI) [3] uses these cost compositions to assess the unit price of all the public construction project, and its use is mandatory to all contractors. An example adapted from those cost compositions is presented in Fig. 2.

(96360) Internal drywall partition with metal studs and two-sided plasterboard lining, without voids (m²)					
Id	Resource description	Un	Qty	Unit cost	Item cost
37586	Steel rivet 27mm, with 23mm washer	pc	4.860	0.32	1.53
39413	Gypsum board 1200 x 2400 mm, e=12.5 mm,	m²	2.106	24.47	51.53
39419	Steel profile UW e=0.5mm, 70x300mm	m	1.521	5.37	8.16
39422	Steel profile CW e=0.5mm, 70x300mm	m	3.982	6.10	24.28
39431	Sealing tape 50x150mm	m	2.503	0.28	0.70
39432	Corner reinforcement tape	m	1.482	3.62	5.36
39434	Drywall compound	kg	1.033	4.87	5.02
39435	Steel metal screws 25mm	pc	20.008	0.05	1.00
39443	13mm screw for drywall	pc	0.808	0.13	0.10
88278	Drywall installer	h	0.690	19.26	13.29
88316	Assistant	h	0.173	16.56	2.85
Material:					97.68
Labor:					16.14
Total:					113.82

Fig. 2. Example of cost composition, adapted from Brazilian National System of Construction Costs and Indexes Survey (SINAPI)

The example in Fig. 2 illustrates the cost estimate per square meter for a process of wall partition installation that includes both the steel framing assembly and plasterboard installation in one single process. This cost composition may be part of the project cost database as it may be used to estimate the cost to any drywall of this type that is installed in a similar context to the one used to assess quantities and costs of the resources.

2.2 IFC Representation

Most of the research on BIM-based cost estimating approaches have been based on the product model, particularly the quantity take-off (QTO) [5, 10, 11, 15]. Some researches as Lee et al. [9] and Niknam and Karshenas [13] suggested an ontology-based approach for cost modeling that included the representation of product and process models. Monteiro and Martins [11] alert that the quality of BIM-based quantity take-offs are very dependent on the authoring, and requires rules to be followed in order to guarantee the consistency of the quantities obtained. Using IFC could avoid software dependency for QTO but there are very few examples of IFC-based cost estimating alternatives. One of the first studies on the topic was from Staub-French et al. [15], who proposed a cost estimating system that directly uses IFC files and external prices databases to accomplish the cost estimation. More recently, Lawrence et al. [8] proposed the use of flexible mappings between BIM models in IFC and cost information. Nonetheless, all of these studies were purely based on the product model. Among them, the only exception is Lee et al. [9] and Niknam and Karshenas [13], which partially included process information but still reside in a higher level of abstraction as a result of the ontology-based approach chosen for the representation.

The problem approached in this research is how to represent these resources costs and allocate them to the processes entities using the IFC4 schema. The IfcTask class is part of the IFC schema since version 1.0, with the main purpose of combining BIM model and project schedules (also know as BIM 4D modeling) [2, 6]. Nonetheless, examples of IfcTask usage for scheduling as the example E.13 provided by buildingSMART [2] are very few, and most of the existing BIM 4D modeling applications seem to not take advantage of this IFC class, favoring proprietary solutions to combine the BIM product model and project schedules to create 4D models.

More recently, project management capabilities of the IFC were greatly extended in the version 4, with several new classes in the process extension schema, as well as in the shared and specific construction management domains. It is noteworthy the class IfcTaskType, included as a way to generalize task definitions without the concern with task dependencies, duration or schedule. BuildingSMART [2] describes IfcTaskType as follows:

> *"A IfcTaskType is a reference definition for a unit of work that may be broken down into (a sequence of) subtasks. [...] Usage of IfcTaskType defines the parameters for one or more occurrences of IfcTask. [...] Fig. 145 shows the definition of a task type that is part of a task template library."*

The buildingSMART [2] also gives some indication on possible strategies to create a task template library and to combine task types with construction resources types. Those strategies, however, are all limited to very simple examples or hypothetical situations, and do not provide a comprehensive and complete approach on how to represent process-based cost compositions in the IFC file.

3 Research Approach

The first step taken was to design a preliminary approach to represent process-based cost compositions using the more recent IFC schema (IFC-add2). This was based on literature review and the strategies suggested by buildingSMART [2] for task template libraries and construction resource types entities.

To test the representation, the schema was instantiated in a prototype, used to extend the original IFC file in the EXPRESS notation with tasks, resources, and cost information. This file was then used to calculate the cost of the task and the associated wall, based on resources unit cost and usage, as well as the size of the wall.

The process-based cost composition data used for the test was the one shown in Fig. 2, associated with the installation of drywall internal partitions. This particular composition was chosen due the number and diversity of its resources, and also because it is a widely adopted and well-documented process in many countries.

That specific cost composition from the Brazilian National System of Construction Costs and Indexes Survey (SINAPI) is part of a public database and freely available without any cost.

The BIM model used for the test was chosen among well-document examples on the use of the software ifcopenshell and IFC4 schema. The chosen model is the resulting output of an example created by Kianwee Chen and Patrick Janssen [4] to demonstrate how to generate IFC files using ifcopenshell and python. The example is named "Creating a simple wall with property set and quantity information", and freely available at the IfcOpenShell Academy (http://academy.ifcopenshell.org/).

The test used ifcopenshell-python 0.60 and python 3.6.9 for creating the initial file, to include the necessary tasks, resources and cost information, and to perform calculation. The example was slightly modified for the sake of compatibility of the resulting file ('hello_wall.ifc') with the chosen software.

The test consisted in two steps. In the first, the original IFC file 'hello_wall.ifc' was read using a python script and ifcopenshell, and then the task, resources and cost information from the cost composition were inserted into the IFC model, which was them written as a new file 'wall_with_cost.ifc', in the EXPRESS notation.

Ifcopenshell was used in this step also to assure a minimum validity of the modified file against the IFC 4x2 schema. A python script was developed to read and modify the IFC EXPRESS file.

The second steps consisted in reading the file 'wall_with_cost.ifc' and performing the cost calculation of the unit cost of the process, and the cost of the task and the resulting product (the wall). In this test, only one task was associated with the wall.

4 Results

Figure 3 presents the strategy chosen in this research to represent process-based costs in the IFC schema. Basically, costs are allocated to tasks through the unit cost of the resources, which are represented as generic resources (resource types). These entities do not represent the resources themselves but their use for some purpose. Thus, both the unit costs and the base quantities can be represented as attributes of the resource entity. The unit costs can be associated with existing entities in the IFC project or library that play the roles of these resources, or may originate from external references as databases directly to the resources without the intervention of an existing entity. In this study, all the information about resources costs and quantities to be used in the chosen task were hard-coded in the python script, assuming that they do not result from any other entity of the model.

The processes were modeled as task types, as processes can be seen as generic tasks. The resource types and their attributes may be associated with the processes using the objectified relationship 'IfcRelAssignsToProcess'. With these entities, all the needed elements to calculate the cost of the composition are then defined.

It is noteworthy that processes (task types) may need to be associated with generic products (product types) when these products are needed as resources for other processes. For example, the process of preparing mortar results in a generic product (mortar), that may be further used as a resource for other processes (e.g. to execute brickwork).

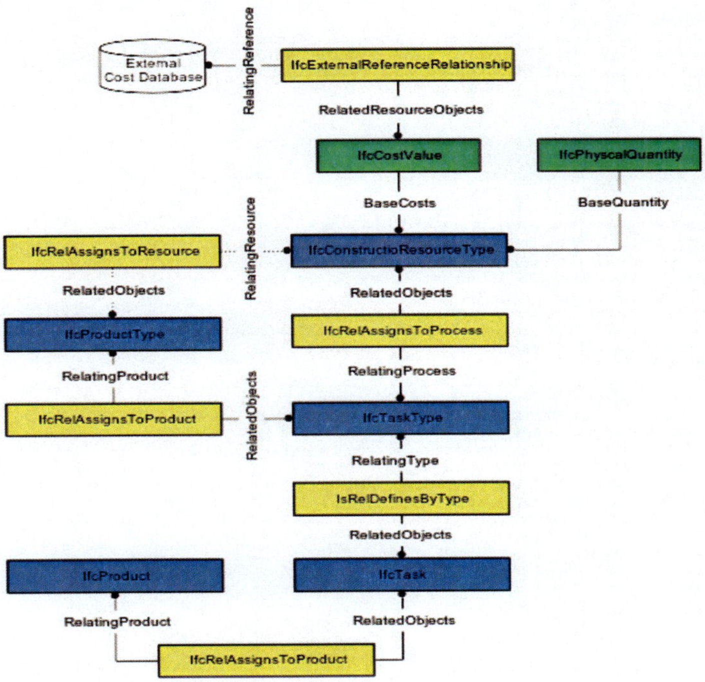

Fig. 3. Proposed IFC representation for process-based cost modeling

In a further step, the cost of the associated tasks and products can be calculated by multiplying the composition unit cost by the number of physical quantities in the associated product. In this study, the relevant attribute of the wall is its area, which is multiplied by the cost per square-meter of the composition, resulting cost of the task 'Installing the wall'. This task cost is then added to the cost of the wall. In this study only one task was applied to the product. However, it is very important to differentiate the cost of the task from the product cost as a single product can be associated with more than one task (e.g. to clean the surface, to sand, and to paint). This is particularly relevant when tasks do not create (or destroy) product entities.

The proposed schema does not include the instantiation of the IfcProductType in the IfcProduct. That instantiation is possible and may simplify significantly the effort in associating tasks and products instances, but would require that all products were represented by a resource type, which may not be feasible in many circumstances.

The first step was to include task, resource and cost data into the 'hello_wall.ifc' file. A python script was created to read the file, to create IFC entities representing tasks, resources, costs and the relationship entities shown in Fig. 3, using the data presented in Fig. 2. The file was then written to disk with a new name ('wall_with_-cost.ifc'). An excerpt of that file presenting the additional entities is shown in Fig. 4.

The second step consisted in reading the new IFC file, extracting task, resource and cost information, to calculate the unit cost of the process (task type). The cost of the task 'Install partition wall' was then derived from the unit cost of the process used

```
#128=IFCMONETARYUNIT('BRL');
#129=IFCTASKTYPE('0C5V8g_aD7hvY6ZAvRTda0',#5,'96360','Internal drywall partition with metal
     studs and two-sided plasterboard lining, without voids (m2)',$,$,$,$,$,.CONSTRUCTION.,$);
#130=IFCCOSTVALUE($,$,IFCMONETARYMEASURE(0.32),$,$,$,$,$,$,$);
#131=IFCQUANTITYCOUNT('*',$,$,4.86,$);
#132=IFCCONSTRUCTIONMATERIALRESOURCETYPE('0m9w5nnA961vcFDQmAMtQO',#5,'37586','Steel rivet 27mm,
     with 23mm washer (pc)',$,$,$,$,$,(#130),#131,.NOTDEFINED.);
#133=IFCCOSTVALUE($,$,IFCMONETARYMEASURE(24.47),$,$,$,$,$,$,$);
#134=IFCQUANTITYCOUNT('*',$,$,2.106,$);
#135=IFCCONSTRUCTIONMATERIALRESOURCETYPE('0wUJ9ZeAL6TBIFfmN0cStK',#5,'39413','Gypsum board 1200
     x 2400 mm, e=12.5 mm (m2)',$,$,$,$,$,(#133),#134,.NOTDEFINED.);
#136=IFCCOSTVALUE($,$,IFCMONETARYMEASURE(5.37),$,$,$,$,$,$,$);
#137=IFCQUANTITYCOUNT('*',$,$,1.521,$);
#138=IFCCONSTRUCTIONMATERIALRESOURCETYPE('3ZWv3peU93TR5f1iJrdRP2',#5,'39419','Steel profile UW
     e=0.5mm, 70x300mm (m)',$,$,$,$,$,(#136),#137,.NOTDEFINED.);
#139=IFCCOSTVALUE($,$,IFCMONETARYMEASURE(6.1),$,$,$,$,$,$,$);
#140=IFCQUANTITYCOUNT('*',$,$,3.982,$);
#141=IFCCONSTRUCTIONMATERIALRESOURCETYPE('0LhMs3lI18hB6zIYQFKNQ_',#5,'39422','Steel profile CW
     e=0.5mm, 70x300mm (m)',$,$,$,$,$,(#139),#140,.NOTDEFINED.);
#142=IFCCOSTVALUE($,$,IFCMONETARYMEASURE(0.28),$,$,$,$,$,$,$);
#143=IFCQUANTITYCOUNT('*',$,$,2.503,$);
#144=IFCCONSTRUCTIONMATERIALRESOURCETYPE('2A47nsYcj6GBNfFZ4nq3UJ',#5,'39431','Sealing tape
     50x150mm (m)',$,$,$,$,$,(#142),#143,.NOTDEFINED.);
#145=IFCCOSTVALUE($,$,IFCMONETARYMEASURE(3.62),$,$,$,$,$,$,$);
#146=IFCQUANTITYCOUNT('*',$,$,1.482,$);
#147=IFCCONSTRUCTIONMATERIALRESOURCETYPE('29yTWkxcn9aRIBIaDhIyAs',#5,'39432','Corner reinforce-
     ment tape (m)',$,$,$,$,$,(#145),#146,.NOTDEFINED.);
#148=IFCCOSTVALUE($,$,IFCMONETARYMEASURE(4.87),$,$,$,$,$,$,$);
#149=IFCQUANTITYCOUNT('*',$,$,1.033,$);
#150=IFCCONSTRUCTIONMATERIALRESOURCETYPE('08LRSU__bBhvciUfFGRhZQ',#5,'39434','Drywall compound
     (kg)',$,$,$,$,$,(#148),#149,.NOTDEFINED.);
#151=IFCCOSTVALUE($,$,IFCMONETARYMEASURE(0.05),$,$,$,$,$,$,$);
#152=IFCQUANTITYCOUNT('*',$,$,20.008,$);
#153=IFCCONSTRUCTIONMATERIALRESOURCETYPE('0wB$xLMcj2Fv7n9922LpHe',#5,'39435','Steel metal screws
     25mm (pc)',$,$,$,$,$,(#151),#152,.NOTDEFINED.);
#154=IFCCOSTVALUE($,$,IFCMONETARYMEASURE(0.13),$,$,$,$,$,$,$);
#155=IFCQUANTITYCOUNT('*',$,$,0.808,$);
#156=IFCCONSTRUCTIONMATERIALRESOURCETYPE('0V$h_yJuH45f_X8hGIk0ss',#5,'39443','13mm screw for
     drywall (pc)',$,$,$,$,$,(#154),#155,.NOTDEFINED.);
#157=IFCCOSTVALUE($,$,IFCMONETARYMEASURE(19.26),$,$,$,$,$,$,$);
#158=IFCQUANTITYCOUNT('*',$,$,0.69,$);
#159=IFCCONSTRUCTIONMATERIALRESOURCETYPE('1$rUFnYlbFFfxjoiQBtEzJ',#5,'88278','Drywall installer
     (h)',$,$,$,$,$,(#157),#158,.NOTDEFINED.);
#160=IFCCOSTVALUE($,$,IFCMONETARYMEASURE(16.56),$,$,$,$,$,$,$);
#161=IFCQUANTITYCOUNT('*',$,$,0.173,$);
#162=IFCCONSTRUCTIONMATERIALRESOURCETYPE('0xtuEhyo98QuVIxiXGy$8N',#5,'88316','Assistant
     (h)',$,$,$,$,$,(#160),#161,.NOTDEFINED.);
#163=IFCRELASSIGNSTOPROCESS('0eW3obryH8HRSJfkvMqB9y',#5,$,$,(#132,#135,#138,#141,#144,#147,#150,
     #153,#156,#159,#162),$,#129,$);
#164=IFCTASK('27l7nTkmrAd8WEPkypnkFP',#5,'0','Construction',$,$,$,$,$,$,$,$,$);
#165=IFCTASK('0iRNnNJyz448v9vZoLXTGg',#5,'1','Install partition wall',$,$,$,$,$,$,$,$,$);
#166=IFCRELDEFINESBYTYPE('1Kq088T6jAwOJB7JmASwZC',#5,$,$,(#165),#129);
#167=IFCRELNESTS('1BnZkzUEvDuOekA_S65OlB',#5,$,$,#164,(#165));
#168=IFCRELASSIGNSTOPRODUCT('0EJmNphm93qu5B5JzVYv1Z',#5,$,$,(#165),$,#66);
```

Fig. 4. An excerpt of the new IFC EXPRESS file with the new additional entities created to represent tasks, resources and cost information

(entity #129, 'Internal drywall partition with metal studs and two-sided plasterboard lining, without voids (m2)') and the area of the associated wall (entity #66, 'An awsome wall'). The resulting information is presented in Fig. 5.

A comparison between Figs. 2 and 5 shows that most of the values as identical, demonstrating that the cost composition information was effectively stored and retrieved using the IFC schema and EXPRESS notation.

Description		qty	cost	item cost
Steel rivet 27mm, with 23mm washer	pc	4.860	0.32	1.55
Gypsum board 1200 x 2400 mm, e=12.5 mm,	m²	2.106	24.47	51.53
Steel profile UW e=0.5mm, 70x300mm	m	1.521	5.37	8.16
Steel profile CW e=0.5mm, 70x300mm	m	3.982	6.10	24.29
Sealing tape 50x150mm	m	2.503	0.28	0.70
Corner reinforcement tape	m	1.482	3.62	5.36
Drywall compound	kg	1.033	4.87	5.03
Steel metal screws 25mm	pc	20.008	0.05	1.00
13mm screw for drywall	pc	0.808	0.13	0.10
Drywall installer	h	0.690	19.26	13.29
Assistant	h	0.173	16.56	2.86
Total cost:				113.90

Fig. 5. Calculation results for the cost of the task 'Install partition wall' based on information retrieved from the IFC file.

5 Conclusion

The article presents an initial study to represent process-based cost models using IFC4x2 schema and EXPRESS notation. Based on literature review and examples from buildingSMART, an IFC schema for representing process-based cost models was proposed. The proposed schema was instantiated in an IFC file based on a publicly available BIM model developed for demonstration purposes, which was populated with real cost data from the Brazilian National System of Construction Costs and Indexes Survey (SINAPI). The resulting IFC file was used to retrieve cost model information, and to calculate the cost of process, task and product instances. The results confirmed the effectiveness of the proposed representation to be capable of successfully integrating the proposed cost modeling schema within a BIM model. It opens the possibility of close integration between product and cost models with BIM technology, as well as the storage and dissemination of cost models and their related information using open and interoperable representation.

References

1. Ballard, G., Reiser, P.: The St. Olaf college fieldhouse project: a case study in designing to target cost. In: Bertelsen, S., Formoso, C. (eds.) 12th Annual Conference of the International Group for Lean Construction. Elsineor, Denmark (2004)
2. buildingSMART: Industry Foundation Classes -Version 4.2 bSI Draft Standard (2019). https://standards.buildingsmart.org/IFC/DEV/IFC4_2/FINAL/HTML. Accessed 06 Feb 2020
3. CAIXA (Caixa Econômica Federal): Sistema Nacional de Pesquisa de Custos e Índices da Construção (SINAPI), Relatório de Insumos e Composições – Dez/19 - Sem Desoneração

(2019). http://www.caixa.gov.br/site/paginas/downloads.aspx#categoria_556. Accessed 06 Feb 2019
4. Chen, K., Janssen, P.: Creating a simple wall with property set and quantity information (2015). http://academy.ifcopenshell.org/creating-a-simple-wall-with-property-set-and-quantity-information. Accessed 06 Feb 2020
5. Choi, J., Kim, H., Kim, I.: Open BIM-Based quantity take-off system for schematic estimation of building frame in early design stage. J. Comput. Des. Eng. **2**(1), 16–25 (2015)
6. Eastman, C., Teicholz, P., Sacks, R., Liston, K.: BIM Handbook: A Guide to Building Information Modeling for Owners, Managers, Designers, Engineers, and Contractors, 2nd edn. John Wiley, Hoboken (2011)
7. Froese, T.: Future directions for model-based interoperability. In: Construction Research Congress, p. 1–8. American Society of Civil Engineers, Reston (2003)
8. Lawrence, M., Pottinger, R., Staub-French, S., Nepal, M.: Creating flexible mappings between building information models and cost information. Autom. Constr. **45**, 107–118 (2014)
9. Lee, S., Kim, K., Yu, J.: BIM and ontology-based approach for building cost estimation. Autom. Constr. **41**, 96–105 (2014)
10. Mattern, H., Scheffer, M., König, M.: BIM-Based quantity take-off. In: Borrmann, A., König, M., Koch, C., Beetz, J. (eds.) Building Information Modeling. Springer International Publishing (2018)
11. Monteiro, A., Martins, J.: A survey on modeling guidelines for quantity takeoff-oriented BIM-Based design. Autom. Constr. **35**, 238–253 (2013)
12. Nguyen, H., Tommelein, I., Ballard, G.: Process-Based Cost Modeling to Support Lean Project Delivery. In: Tzortzopoulos, P., Kagioglou, M. (eds.): 16th Annual Conference of the International Group for Lean Construction, p. 577–588. Manchester, UK. (2008)
13. Niknam, M., Karshenas, S.: Integrating distributed sources of information for construction cost estimating using semantic web and semantic web service technologies. Autom. Constr. **57**, 222–238 (2015)
14. Pollack, J., Helm, J., Adler, D.: What is the iron triangle, and how has it changed? Int. J. Managing Projects Bus. **11**(2), 527–547 (2018)
15. Staub-French, S., Fischer, M., Kunz, J., Paulson, B.: A generic feature-driven activity-based cost estimation process. Adv. Eng. Inform. **17**(1), 23–39 (2003)

An Approach for Data Extraction, Validation and Correction Using Geometrical Algorithms and Model View Definitions on Building Models

Johan Luttun[1(✉)] and Thomas Krijnen[2]

[1] AECgeeks, 5616TW Eindhoven, The Netherlands
johan@aecgeeks.com
[2] Delft University of Technology, 2628BL Delft, The Netherlands

Abstract. The Industry Foundation Classes (IFC) cover a wide variety of subdomains in the construction industry. Model View Definitions (MVD) enable to specify a subset of the IFC schema to assess the content of a model for specific use cases and information exchanges.

However, IFC and MVD paradoxically complexify the workflow since it requires a deep understanding of the schema combined with construction knowledge to carry out simple use cases such as quantity checking or data export. This gap between domain specific queries and their expression in a computer-readable language weakens the opportunities provided to the building industry by Building Information Modeling.

Our research consists in the implementation of MVDs in a high-level programming language to extract data from building models, an assessment of the extraction results and geometrical processing algorithms to correct the explicit quantities and properties that are supplied as metadata alongside the elements in IFC building models.

Geometrical processing can be used to reduce and eventually correct errors on property values. We use a generic geometrical representation of IFC entity instances and apply geometrical transformations on those to obtain geometrical shapes. Boolean operations are used to identify relationships between elements. Eventually, incorrect data values are corrected directly in the IFC models accordingly to the IFC schema.

For instance, we authored an MVD to extract data pertaining to external walls from different IFC models and corrected the value of the IsExternal property of the models' IfcWall entities. This use case is of great importance for the cost estimation of a thermal renovation on a building as it gives a good estimate of the outer surface area of the building envelope.

Keywords: BIM · Checking · MVD · IFC · Geometry

E. Toledo Santos and S. Scheer (Eds.): ICCCBE 2020, LNCE 98, pp. 529–543, 2021.
https://doi.org/10.1007/978-3-030-51295-8_38

1 Introduction

1.1 Overview of the Difficulties Related to the IFC Open-Standard

The technical complexity of the IFC schema and the lack of reliable data in IFC files form a barrier towards a broader adoption of BIM-based collaborative methods that could be beneficial to the construction industry. First, there is a lack of open-access tools to efficiently extract data from an IFC file to a commonly used format, such as a spreadsheet. Furthermore, there are few formal ways used in practice to exchange definitions of data extraction or data validation processes, which compromises collaboration between actors and reusability of use cases throughout construction projects. Thus, domain knowledge experts cannot universally rely on the IFC models to understand the project, run analysis with accurate information and detect collisions, in order to save time and increase the overall quality of the built deliverables. Second, most of the information present in the IFC models is not reliable, and its use in calculations leads to uncertain and unverifiable results. Coherence between the value of the attributes in the IFC model and the physical descriptions and measures derived from the geometrical definition of the elements are rarely checked in practice, which weakens the opportunities offered by BIM, to understand the construction both visually and in terms of data.

1.2 Industry Foundation Classes

The Industry Foundation Class (IFC) standard, developed by buildingSMART, a worldwide community of actors from the construction industry, aims to provide interoperability between software applications while describing specific domain-related products. Indeed, the density of information conveyed between the project actors and used by different software applications requires a neutral format for import and export.

However, the IFC models used in the industry contain several errors [1] regarding to data accuracy, structure, or existence, which compromises the benefits brought by a BIM workflow: saving time and increasing project quality thanks to the storing of reliable data. The multiple translations between software applications, the heterogeneity of domain activities, and the diversity of use cases make digital exchanges error prone. In addition to data related errors, geometry errors are also responsible for unusable models and depreciating BIM added value.

Elements in a building element model have one or more geometrical representations, a placement that locates the element in 3D space and a set of associated semantic constructs such as material information and property sets, in which the latter allow for more user- or application defined key-value pairs. A subset of the property names in a model follows standardized patterns by buildingSMART, for example for IfcWall elements information, a field IsExternal is typically available to document whether the element is part of the external façade of the building. There are also numeric properties such as volume and surface area available in quantity sets. The geometry definitions in IFC constitute a large part of the IFC schema and a large part of the considerable implementation effort for applications to fully support IFC. These constructs, called representation items, can be simple explicit polyhedra or fairly complex sweeps and revolutions with boolean operations applied to them. The implicit geometry definitions and hierarchical placements make discovering simple geometric relationships between elements such as containment and adjacency, computationally expensive. Therefore, these relationships such as spatial containment in the project decomposition structure, connectivity information between walls and space boundaries are explicitly provided in the model as objectified relationships.

One of the most investigated topics in the construction industry is the reduction of the energy consumption of buildings and the limitation of their environmental hazards, particularly the carbon footprint, at each phase of their whole life cycle [2]. Thanks to IFC specifications, forecasting the CO2 released by a building throughout these phases, which requires an accurate modeling of quantities and materials used, is more affordable. The data stored in an IFC file offers a real opportunity to carry out such use cases, that depend on specific and accurate data. For example, the capacity to extract information about the external walls to calculate the façade area of a building is an important step towards the appropriation of BIM by end-users because it can be incorporated into decision-making processes, for instance at design or at earlier phases. The estimation of physical quantities allows for use cases that answer to other than purely technical rules, such as the financial ones. Thermal losses are determining variables for the calculation of energy consumption, which is a main part of a building's CO2 emissions [2]. Since conductive heat transfer takes place between heated and non-heated volumes through their common surface [3], insulating the building façade with less permissive materials reduces this transfer rate, and thus mitigates the amount of energy needed to maintain the thermal comfort temperature. In addition, façade geometry and materials intervene in the comfort, financial, and esthetic metrics of a building, and has been the object of optimization research [3, 4]. Therefore, using IFC to model external walls in terms of geometry, composition, quantities, and extract information about them brings several benefits to the construction industry.

1.3 Model View Definitions

The mvdXML format [5] is an open-standard developed by buildingSMART, whose function is to define a subset of an IFC model. This subset is called a Model View Definition (MVD) and represents a certain part of the IFC data available in the model. The objective of an MVD is to ensure the delivery of the right and valid IFC data in the exchanges of the project actors, throughout the digital chain.

In order to do so, the MVD standard uses Concepts objects, which are the implementation of Concept Templates. Concept Templates represent a general notion such as the relationship between an element and associated properties, that can be applied to one or many entities in the model. This notion can be represented as a tree structure. For example, the notion of externality for an IfcWall element can be modelled as shown in Fig. 1.

The nodes of the Concept Template tree are Rule or Constraint objects. The AttributeRule nodes impose the name of attributes while the EntityRule nodes define the type that should be used for a value. Moreover, the nodes can be constrained by adding Constraint nodes. Those rules can also be expressed as logical expression, as detailed in [6, 7].

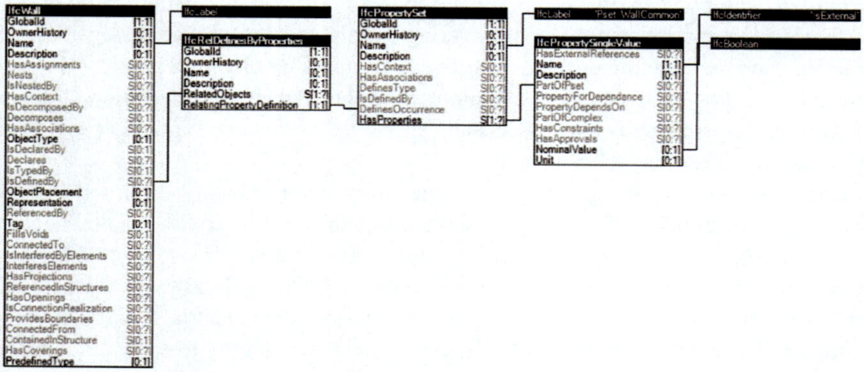

Fig. 1. The externality of IfcWall concept edited in the IfcDoc tool [8]

Overall an mvdXML file is a hierarchical structure forming a tree graph, with a root node mvdxml:root, and two main branches: the mvd:Templates, and the mvd:Views. In the templates branch, the Concept Templates are defined. In the Views branch, there can be one or several Concept Root nodes respectively representing an IFC entity towards which the concepts defined in the Concept Template nodes are applied. Moreover, Constraint nodes can be applied on the Rules nodes to impose a value of entity or attribute.

2 Literature Review

2.1 Checking and Validation

The diversity of disciplines and stakeholders involved in a construction project leads to an important number of rules to comply with [7, 9]. Indeed, checking whether a building project respects regulations require a set of numerous, precise, and reliable pieces of information about specific activities. BIM workflow allows to store, access, modify and share an unlimited amount of data about a construction project. The data enables to derive knowledge about the project and use it for rule checking, among other use cases [10]. However, the time saved will be only transferred from on operator to another if attributes values need to be analytically derived and then fulfilled to the model. The challenge is to write algorithms that can retrieve the implied knowledge of a building model by limiting the time-consuming tasks of data input and preprocessing [11]. To fulfill this objective, an assurance of the reliability of data is necessary. It depends on a strict formalization of the structure of the IFC information as well as compliance with minimal requirements on the model, such as minimal space requirements or absence of geometric collision [12].

The IFC model needs to respect storing requirements, that include the needs of application programs, in order to be useful for other applications than the one it has been originally edited on. In addition, rules are imposed on the data by regulatory conventions. Those rules have been acknowledged in previous research [6, 7] and concern the data existence and cardinality, the data content, the data uniqueness, and conditional dependency of data. The data existence imposes the presence of certain entities in the model, or as attributes of other entities. The data content requires a certain type or a certain value for data. The data uniqueness necessitates that for example only one GlobalId or one Name attributes should exist for any instance of an IfcRoot subclass. Finally, since there could be several ways to store data, the conditional dependency can be used in some cases to check the coherence of linked information.

2.2 State of the Art of Model View Definition Uses

As stated in [7], an MVD is an IFC subschema used for checking data compliance with diverse requirements. While is it mostly used to check the storing requirements, there are differences in the way MVD are embedded in a whole checking system. Moreover, the output of the validation process is often a viewer displaying with highlighting the geometric representation of non-compliant IFC instances.

In the state of the art section of [6], the different MVD methods are listed and categorized according to various parameters: IDM, Model subset extraction, Validation Rules, Implementation of the validation function, OpenStandard, Edit tool. Among the five methods listed, none of them implements a validation function. In this work, the authors check the data storing structure of the model as described earlier. They give detailed examples of MVDs structure as well as a logical definition of the rules. The elements checked are part of BIM regulations. The output consists in BCF files that report the non-compliant entities in a shareable way.

In [13], mvdXML files are used to check space requirements for the design of a hospital. This work illustrates how MVDs can be implemented at the project scope, for a construction project. For example, this study mentions how the MVDs are defined with the building experts, to ensure interoperability between software applications. In addition, an effort is made to involve the building end-users in the MVD definition process, by fulfilling their requirements through CSV spreadsheets. The validation output is rendered in a viewer.

In [7] the MVD presented is used to comply with the Precast Concrete Industry (PCI) requirements, which shows the reusability of the method for a specific types of building elements, as in [6]. This research also provides a logical notation of the MVDs. Indeed, the set of rules defined by the PCI are registered into what the authors called a PCI MVD. The MVD is edited into the IfcDoc tool and a building project is checked against it. The results of the validation are shown within the IfcDoc tool.

More recently, Eastman [10] used a BACnet MVD that offers promising perspectives for data communication between IFC files and Building Automation System (BAS). This work allows the integration of a BAS in the design process, until now excluded from BIM workflow, and enables its operation by creating new IFC modeling types. A prototype is tested, which shows how BAS object families can be modeled in the authoring tool Revit. The BACnet MVD is used to export these Revit objects content to populate IFC files and import the data in a web-based application, considered as more accessible than the IfcDoc tool. This study demonstrates how an MVD formalization can fix the interoperability problem on use cases that had not been tackled until now and thus expand BIM benefits.

2.3 Conclusion

Regarding checking processes, the distinction between data and geometric checking is well marked but there are few works which attempt to fix semantic-geometric coherence of models [14, 15]. The identification of this issue is of great importance to accurately deliver high quality models, and it is where BIM offers real opportunities, when it enables to detect implicit knowledge as a domain expert would [16].

In cited research [13], MVDs are used as validation tools for data structure compliance with Exchange Requirements. Generally, studies agree on the benefits of mvdXML standards, attributed to metrics such as simplicity of edition and reusability [6]. Current validation processes using MVDs consists in the edition of an mvdXML file, use a program to extract data from the IFC file, and eventually execute the mvdXML file on the IFC file. However, while there is a uniform way to define the requirements, the data extraction depends on a specific programming language, and set of tools. While it shows the strength of open standards, which is to read and edit data from any tools it also compromises reusability of such use cases.

In terms of the limitation of MVDs for data validation, [7] acknowledges that there is neither an easy nor convenient way for domain users to carry out the validation process by themselves. The IFC schema and mvdXML data model represent a barrier towards the adoption of this validation method. It is indeed necessary to provide user-friendly tools, as IfcDoc, to reduce computer science proficiency requirements from of the end-users.

Most of the information exchanges are hardly definable sequentially at the very beginning of a construction project, and it is therefore a complex task to impose a validation order. Efforts are concentrated on designing a system of exchange which can be hard to implement in practice. There are few mentions of an asynchronous approach instead, a tool or method that would allow any user in the chain to make the extraction and validation for its own use case. Generally, research assumes models have been validated throughout the phases of a project. However, these assumptions are valid when the collaboration is effective and BIM proficiency is advanced among the construction actors. In most situations, the model is wrong and there are few ways that give unfamiliar users to take advantage of the model it receives [7].

3 Contribution

3.1 Approach Overview

We propose a set of open-source tools that use MVDs to extract, filter, and eventually export data from an IFC model coupled with corrective geometrical algorithms, assuring coherence between geometric and semantic information in the model. Our approach enables a wide variety of use cases such as quality assurance and code checking thanks to powerful extractive features. It relies on a formal and shareable definition of IFC data extraction, through mvdXML Concept Templates, and offers the possibility to export the data in an easily readable format such as a spreadsheet.

The method is based on the definition of one or multiple queries on the IFC file through an easily generable but formal and shareable specification in the form of MVDs. The gap existing between the information needed and the data formulation relies on knowledge about buildings and IFC format. For example, if the total surface of the façade is to be known, building practitioners must extract the surface quantity of all the external walls, assuming there are no curtain walls, and sum them up. Once that formulation is stated and transformed in an MVD, data can be extracted from the IFC file for domain specific activities.

As a drawback of the freedom offered by a neutral format, the data is not stored uniformly in all IFC files [6] and makes it necessary to explore the model first in order to find how the hierarchical layout is articulated. For instance, for two different building projects, the information about the area of walls is stored in an IfcQuantitySet entity type whereas in the Duplex file it is stored in an IfcPropertySet entity type. In our approach, it obliges the user to build a new Concept Template per feature value. Therefore, we believe in a potential learning effect from this framework, enabled through the habit of editing trees of MVD to extract data.

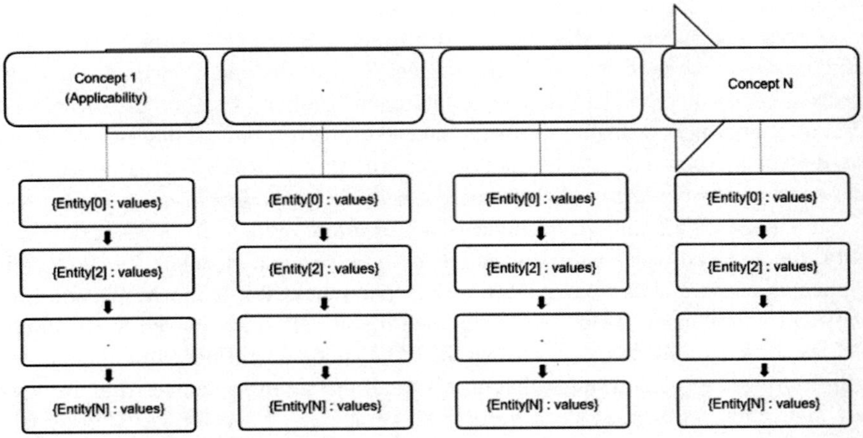

Fig. 2. The framework used for our approach. IFC values are collected throughout Concepts. The process shows the iteration over the Concepts, and the retaining of values mapped to entities. Those concepts can filter the entities as an mvdXML Applicability. For example, acts as a filter and only IfcWall instances pertaining to external walls will be extracted, also for each of the remaining concepts.

As shown in Fig. 2, the program developed through this research extracts data from Concept Templates units applied in series on IFC entities, whose type is determined through a Concept Root. Two functionalities exist in the program. First, as the entities pass the Concept Template units, the data required is collected. In addition, some Concept Templates can be made restrictive by constraining some of the values and thus excluding entities that do not comply with those constraints. Therefore, the data structure validation as in [6, 7] is also embedded in our extraction process, since data that do not comply will not be extracted.

3.2 Description of the Extraction and Validation Algorithm for Quantities of External Walls

The extractive functionality of our program relies on a recursive function that extracts attribute values from IFC entity instances while traversing the MVD Concept Templates. First, the program maps an mvdXML file to a high-level programming language to access its underlying structure. The mapping enables to store the Concept Template nodes applied on the ConceptRoot we want to extract data from in a variable to iterate over them. The program extracts the entities referenced by a ConceptRoot. Then, for each iteration over the Concept Template nodes, a traverse function is called, taking as input the entities and an mvdXML node, root of the considered MVD Concept Template tree, and will return a collection of EntityRule nodes mapping to literal value in IFC or literal values. Therefore, for each iteration, the results of the function correspond to one column of the spreadsheet export.

As presented in Fig. 3, the function recursively traverses the Concept tree while storing the data required by the AttributeRule, EntityRule, and Constraint mvdXML objects along the nodes. The behavior of the function differs based on the nature of its MVD node argument. Usually, for the initial call of the function all the IFC entities are passed with the root node of the Concept tree, an AttributeRule. The base case of the recursion occurs when the mvdXML node considered by the function is a leaf or when the child node of an EntityRule node is a Constraint. When the MVD node is an EntityRule the subset of currently supplied instances that matches the IFC entity supplied is taken and the recursive traversal continues. When the MVD node is a Constraint or leaf node the currently supplied literal values are passed to the higher-level invocation in the call stack. When the MVD node is an EntityRule there can be multiple AttributeRules meaning that at this point the results obtained from the lower level invocation on the call stack need to be combined. When the MVD node is an AttributeRule the different IFC instances can result in different attribute valuations so the size of the result set grows. For more details, we refer the reader to the open source repository containing the tools at https://github.com/opensourceBIM/python-mvdxml.

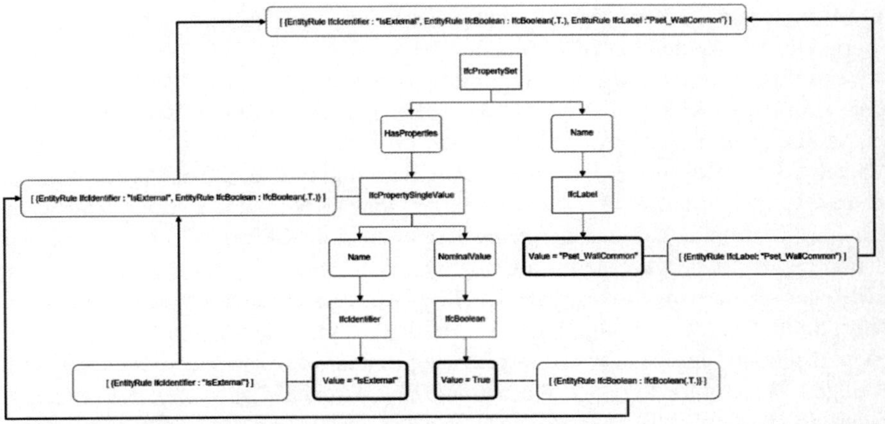

Fig. 3. Example of the extractive function applied on an IfcPropertySet EntityRule node. EntityRule, AttributeRule, Constraint are respectively represented by rectangular, slightly rounded edges, and bolded edges boxes. Output values are attached to the leaves by dotted lines and bolded arrows show the output process.

3.3 Limits of BIM and Proposition of a Corrective Approach

MVDs are efficient tools to extract data from specific entities but reliability of information is necessary to bring a real value to the construction industry. Indeed, as stated previously, many models delivered in the professional sector lack coherence and comport errors. For example, there can be errors on the quantities directly calculated in the attribute values of the entities. Moreover, non-trivial information must also be verified, as the externality of elements such as the walls. The main use example of our work aims to show that it is possible to use easily edited MVD to export data to a well-known format, to gather the necessary quantity information necessary to carry out the cost estimation of façade insulation or run a thermal analysis. Nevertheless, this simple use case in theory is unfeasible without the assurance that the IFC data exported is reliable.

The models shared in practice are not reliable when received by a project actor. The exchange that appears throughout the digital chain can be a cause because each export uses its own export module and the accumulated errors result in a flawed model. Moreover, there can be modeling errors, such as not defining the spaces correctly. Finally, one of the most important sources of errors is the absence of systematic checking on project. And this is more globally due to an absence of BIM execution plan at the project scope. The use cases not defined before the beginning of the design phase results in uncontrolled exchanges.

Hopefully, primary -or low-level errors- can be fixed thanks to geometrical algorithms. Open source tools such as IfcOpenShell [17] enable access to the geometry description of IFC entities. Moreover, it also provides functionalities to access the schema at runtime, to delete the wrong entities values and their relationships, to eventually construct new entity instances and write them onto a newly corrected file. Almost all the geometric and data related can be fixed using such methods, but the difficulty is that there is no finite set of errors to check. It is therefore hard to fully automate data fixing for every case, especially when all the error cases are unknown.

However, it can be done for frequently used data such as quantities or elements externality. The adoption of BIM by end-users relies on the confidence they have in the tool and their capacity to use it to fulfill their needs. It is preferable to be able to have basic information reliable than a lot of derived information based on false primary data.

As discussed in the Introduction section, the IFC schema offers a wide variety of geometry definitions that are potentially complicated to evaluate. Existing software can be used to transform the implicit engineering constructs in IFC into an explicit and universal description called a Boundary Representation (BRep). It is not an approximation (as would be the case for a triangle mesh for example) but has full geometric fidelity for arbitrarily curved surfaces. A BRep is a hierarchical structure of topological items: Solid, Shell, Face, Loop, Edge, Vertex. A Face for example contains one or more Loops that bound an area over a surface. A Loop contains a pairwise connected sequence of Edges. Three of the topological constructs have associated geometry. A Face has an underlying Surface, an Edge is associated to a Curve and Vertex is located at a Point. In this research IfcOpenShell is used to convert the IFC geometry into the BRep format of Open CASCADE [18], the latter implements a wide variety of modeling and analysis functionality, such as Boolean operations and the calculation of area and volume.

We consider an external wall as a wall that is not separating two internal spaces. For the simplicity of our example, we did not add additional constraints on the definition of externality for a wall. Checking the externality of a wall demonstrates the complexity of assigning a degree of trust to an IFC model and reinforces the need for the systematic verification of such cases. The IfcWall and IfcSpace representations enable to store the faces of those objects. Thus, we spot the internal walls as the one which have two of their biggest vertical faces intersecting shell representations of spaces. If a wall shares two common faces with a space, then it is not external (Fig. 4).

name	surface	external
Basic Wall:Party Wall - CMU Residential Unit Dimising Wall:139234	12	TRUE
Basic Wall:Party Wall - CMU Residential Unit Dimising Wall:139347	12	TRUE
Basic Wall:Party Wall - CMU Residential Unit Dimising Wall:139374	0	TRUE
Basic Wall:Foundation - Concrete (417mm):140479	11	TRUE
Basic Wall:Foundation - Concrete (417mm):140520	22	TRUE
Basic Wall:Foundation - Concrete (417mm):140554	10	TRUE
Basic Wall:Foundation - Concrete (417mm):140602	21	TRUE
Basic Wall:Foundation - Concrete (435mm):140913	5	TRUE
Basic Wall:Foundation - Concrete (435mm):140987	5	TRUE
Basic Wall:Foundation - Concrete (435mm):141018	2	TRUE
Basic Wall:Party Wall - CMU Residential Unit Dimising Wall:143239	38	TRUE
Basic Wall:Exterior - Brick on Block:138157	50	TRUE
Basic Wall:Exterior - Brick on Block:138310	48	TRUE
Basic Wall:Exterior - Brick on Block:143410	17	TRUE
Basic Wall:Exterior - Brick on Block:143534	17	TRUE
Basic Wall:Exterior - Brick on Block:138062	14	TRUE
Basic Wall:Exterior - Brick on Block:138237	12	TRUE
Basic Wall:Exterior - Brick on Block:143478	41	TRUE
Basic Wall:Exterior - Brick on Block:143590	41	TRUE
Basic Wall:Exterior - Brick on Block:184944	5	TRUE
Basic Wall:Exterior - Brick on Block:185014	11	TRUE
Basic Wall:Exterior - Brick on Block:185064	5	TRUE
Basic Wall:Exterior - Brick on Block:185101	10	TRUE
	503	

Fig. 4. Illustration of the spreadsheet output resulting from our program which has successfully exported the Name (name column) and the Area (surface column) values of IfcWall instances whose IsExternal attribute value is set to true. The summation of the area demonstrates the simplicity of the use case.

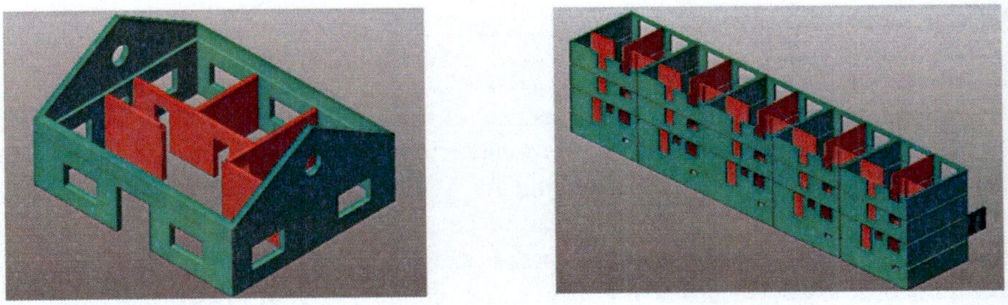

Fig. 5. Visualization of the filtered entities on Haus Elite and Smiley-West files. Source: [19]

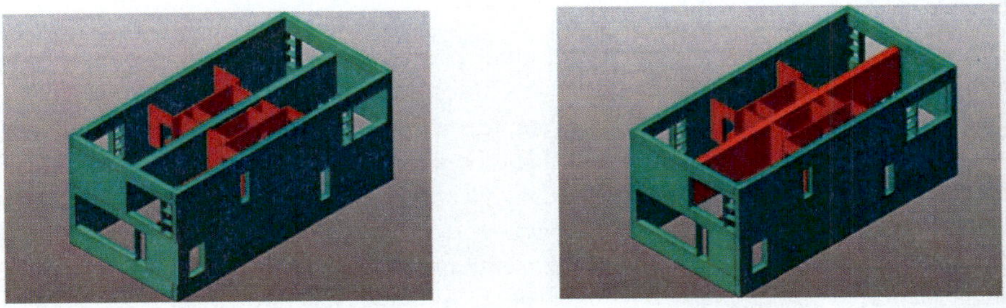

Fig. 6. Visualization of the filtered entities (the internal walls) for the Duplex file before and after applying the geometrical fixing. The wall on the first floor was incorrectly classified as external before the fixing.

4 Discussion

4.1 The Importance of Data Extraction

The ability to read IFC values is a fundamental part of checking a model or a project [9]. Indeed, the diversity of use cases and software applications has not delivered a unique way to validate data yet. We state that any validation operation requires a data extraction first. Before there could be any uniform way to validate data, extracting data could be formalized with the mvdXML format.

The use of a formal and shareable extraction through mvdXML files encourages BIM adoption as different methods for querying make it a complex task for end-users. In several works, a different approach has been used to query the information of IFC files in order to check the rule compliance, whereas extraction is at the core of BIM use cases. However, MVD was often used for software use cases, which was essentially checking the data structure, uniqueness, or existence. With our approach, mvdXML files can be used to directly extract information about the model for end-users in an easily accessible format. Therefore, we think of our approach as a method reducing the gap between end-users programming proficiency needs and the ability to exploit IFC data, while getting familiar with the mvdXML open-standard.

4.2 Results Discussion

Extracting the data and executing a validation regarding an MVD offers a real advantage for a whole range of users. Indeed, our approach can also be used by more experienced users who wish to reduce complexity of extraction and validation. We therefore hope to provide a robust support for MVD validation of IFC data.

The most common uses of MVDs can be extended to data extraction, for example in an easily readable way such as a spreadsheet for end-users. Also, MVDs can be constrained in a way such that IFC entities not complying with mvdXML Rules will not be extracted, thus marked as not conform. Eventually, filtered entities can be rendered in a viewer application.

Results are successfully extracted from the MVD and displayed in Fig. 5 and Fig. 6.

The results from Fig. 6 demonstrate that some external walls been have correctly fulfilled by the model's authors. However, there are remaining errors that cannot be totally spotted by our algorithm. For example, on Fig. 6, the IsExternal value of a wall was not correctly classified and could lead to estimation errors if repeated on bigger projects. The geometric algorithm used to identify external walls relies on a proper definition of entities and a correct modeling of spaces. Without these conditions, it misclassifies the walls regarding to their externality. For example, Fig. 7 shows the walls identified as external by the algorithm. Although the result may seem correct, there are still errors, especially small elements which are IfcWall instances, but which does not really represent a wall. Regarding this topic, Krijnen [16] implemented an unsupervised machine learning algorithm to carry out anomaly detection on the Duplex model, resulting in the discovery of misclassified elements.

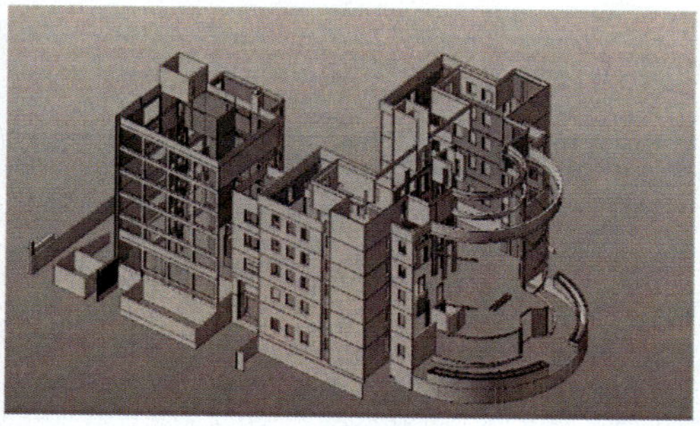

Fig. 7. Visualization of external walls on a model from the construction industry after applying the fixing algorithm. The model was delivered without any IsExternal classification, since all the walls were marked as external.

4.3 Validation Levels and Follow-up Works

IFC models must be validated, particularly regarding their semantic-geometric coherence, before using them to check project compliance with building codes. There is still a vast majority of models containing lacking, unverified, wrong, or misleading information. A model that contains errors, or whose data quality has not been checked appropriately, should not represent a trusty basis for real-life checking use cases such as structural analysis or cost estimation [14].

In the same way [9] proposes a hierarchy of rules, we believe in the necessity of the construction of a hierarchical level of trust for IFC models. This level of trust should start by checking that IFC instances representation correspond to their nomination, such as what has been initiated by Krijnen [16]. Then, the model should be observed and corrected. Our approach wishes to accomplish those use cases. Nevertheless, the identification of geometric-semantic potential issues should be made, as we did for instance for walls externality and elements quantities. Finally, the structuration of the data in the model, and the real-life checking will be properly be carried out.

5 Conclusion

This research seeks to highlight the opportunities offered by mvdXML files to extract data destinated to domain specific use cases from IFC files, as well as to show that IFC values can be corrected by processing the underlying description of their geometry. In the Introduction part, we demonstrated the role of IFC files and how we can use them to ease completion of use cases that would bring significant advances in the construction industry. The repository containing the set of tools used in our approach is available on https://github.com/opensourceBIM/python-mvdxml. We also showed MVD use to validate and filter IFC values and how it provides value to the digital transmission chain, improves models' quality, and offers promising perspectives through new use cases.

However, we noticed that while previous studies are encouraging, difficulties persist:

- In the industry, most of the models contain errors or are incoherent regarding essential information values such as the externality of elements or the physical quantity values, which does not allow for more complex checking based on derived values.
- Current approaches seek to forecast all the use cases rather than to give every project member the possibility to extract and check data independently. MVDs rely upon a project-based organization responsible for the BIM workflow processes, such as data exchanges. However, all the use cases cannot be anticipated.
- The end-users often cannot use the IFC data in an independent way. They need a viewer application or an authoring tool, which is an intermediary between themselves and the IFC data, that can hide the powerful information extraction and validation features offered by IFC and MVD standards.
- The data extraction and checking are usually made within software applications from a GUI, without the possibility to share and reuse rules from one application to another.

Assessing these difficulties and aware of their impacts, we developed an approach seeking to:

- Acknowledge the existence of incoherence between attribute values and geometrical description of elements through the example of the walls pertaining to the façade of a building and prove that it can be corrected with geometrical algorithms.
- Give the ability to end-users to extract data in a familiar format to gain independence and show how a query can simply be translated into an mvdXML Concept tree.
- Give the ability for experienced users to build upon our work to develop more complex parsing programs and explore elaborated uses of MVDs.
- Provide a flexible way to work with IFC data, that allows to process it digitally rather than depending on viewers or authoring tools, and share the extraction and validation formalization of this processed data.

Regarding the first point, we are open to collaborate on verifying the basic data of IFC files, which can be wrongly identified, such as the space definition or the IFC entity types, in order to ensure a high level of confidence for the most used IFC entity types.

References

1. Solihin, W., Eastman, C., Lee, Y.-C.: Toward robust and quantifiable automated IFC quality validation. Adv. Eng. Inform. **29**, 739–756 (2015). https://doi.org/10.1016/j.aei.2015.07.006
2. Wang, W., Zmeureanu, R., Rivard, H.: Applying multi-objective genetic algorithms in green building design optimization. Build. Environ. **40**, 1512–1525 (2005). https://doi.org/10.1016/j.buildenv.2004.11.017
3. Zemella, G., Faraguna, A.: Evolutionary Optimisation of Façade Design: A New Approach for the Design of Building Envelopes. Springer, London (2014)
4. Evins, R.: A review of computational optimisation methods applied to sustainable building design. Renew. Sustain. Energy Rev. **22**, 230–245 (2013). https://doi.org/10.1016/j.rser.2013.02.004
5. Chipman, T., Liebich, T., Weise, M.: Specification of a standardized format to define and exchange Model View Definitions with Exchange Requirements and Validation Rules
6. Zhang, C., Beetz, J., Weise, M.: Model view checking: automated validation for IFC building models. In: Mahdavi, A., Martens, B., Scherer, R. (eds.) eWork and eBusiness in Architecture, Engineering and Construction, pp. 123–128. CRC Press (2014)
7. Lee, Y.-C., Eastman, C.M., Solihin, W.: Logic for ensuring the data exchange integrity of building information models. Autom. Constr. **85**, 249–262 (2018). https://doi.org/10.1016/j.autcon.2017.08.010
8. IfcDoc. https://technical.buildingsmart.org/resources/ifcdoc/
9. Solihin, W., Eastman, C.: Classification of rules for automated BIM rule checking development. Autom. Constr. **53**, 69–82 (2015). https://doi.org/10.1016/j.autcon.2015.03.003
10. Tang, S., Shelden, D.R., Eastman, C.M., Pishdad-Bozorgi, P., Gao, X.: BIM assisted building automation system information exchange using BACnet and IFC. Autom. Constr. **110**, 103049 (2020). https://doi.org/10.1016/j.autcon.2019.103049
11. Gu, N., London, K.: Understanding and facilitating BIM adoption in the AEC industry. Autom. Constr. **19**, 988–999 (2010). https://doi.org/10.1016/j.autcon.2010.09.002
12. Azhar, S.: Building Information Modeling (BIM): trends, benefits, risks, and challenges for the AEC industry. Leadersh. Manage. Eng. **11**, 241–252 (2011). https://doi.org/10.1061/(ASCE)LM.1943-5630.0000127

13. Weise, M., Nisbet, N., Liebich, T., Benghi, C.: IFC model checking based on mvdXML, vol. 1, p. 9 (2016)
14. Häußler, M., Borrmann, A.: Model-based quality assurance in railway infrastructure planning. Autom. Constr. **109**, 102971 (2020). https://doi.org/10.1016/j.autcon.2019.102971
15. Daum, S., Borrmann, A.: Checking spatio-semantic consistency of building information models by means of a query language. Presented at the (2013)
16. Krijnen, T., Tamke, M.: Assessing implicit knowledge in BIM models with machine learning. In: Thomsen, M.R., Tamke, M., Gengnagel, C., Faircloth, B., Scheurer, F. (eds.) Modelling Behaviour, pp. 397–406. Springer International Publishing, Cham (2015)
17. IfcOpenShell. http://ifcopenshell.org/
18. OPEN CASCADE. https://www.opencascade.com/
19. Open IFC Model Repository. http://openifcmodel.cs.auckland.ac.nz/

BIM for Planning and Cost Estimating

Development of BIM-Based 4D Simulation System for Construction Schedule Planning

Fumio Hatori(✉), Kouji Satou, Joji Onodera, and Yuichi Yashiro

Hitachi Plant Construction, Ltd., Toshima-ku, Tokyo 170-8630, Japan
fumio.hatori.zq@hitachi.com

Abstract. Construction schedule planning is an important duty for carrying out efficiently construction projects. Recently, 4D simulation technology linked to the BIM model has made it possible to visually check the installation procedure of equipment and piping. However, the task of setting construction schedule information requires a lot of time, especially in large-scale plant construction BIM model. In addition, the actual installation work includes temporary scaffolding assembly and disassembly. Therefore, construction schedule planners should consider an appropriate schedule that takes into account these temporary scaffolding works. In order to conduct 4D simulation including scaffolding, it is necessary to create and add 3D scaffolding model that is not included in the original BIM model. It also required a lot of time. This paper proposes the method for automatically setting installation procedures based on rules using BIM information and an automatic creation system for scaffolding. The first is a function that automatically sets the installation order using the location information of the equipment and piping in the BIM model and the rules of the expert. The other is the interactive creation of 3D models of scaffolding using BIM spatial information. This system automatically creates 3D models of scaffolding by calculating the required number of standard scaffolding by instructing the spatial position in the BIM model. With these functions, computer system automatically sets up a installation schedule and significantly reduces time for 4D simulation setup.

Keywords: Construction schedule planning · BIM · 4D simulation · Scaffolding

1 Introduction

There are high expectations for the use of ICT from various viewpoints such as safety, quality improvement, and productivity improvement at construction sites. In particular, the use of building information modeling (BIM) is a major technological innovation, and is said to change the future of construction sites [1, 2]. However, in Japan, the fact is that construction planning and management are still conducted mainly with 2D drawings, depending on the experience of skilled workers. While skilled worker presence still remains within companies, we can respond in the traditional way. However, the retirement of skilled workers is approaching, and it is an urgent task for Japan to improve the efficiency of the planning expertise and construction management

E. Toledo Santos and S. Scheer (Eds.): ICCCBE 2020, LNCE 98, pp. 547–560, 2021.
https://doi.org/10.1007/978-3-030-51295-8_39

possessed by these experts. With the spread of BIM, the software functions have been enhanced. However, to make full use of them, it is necessary not only to become accustomed to the operation of software, but also to have the process planning knowledge and operational knowledge to manage the site [3].

The plant construction in which our company is involved is mainly equipment installation work, and the construction flow ranges from carrying in equipment and piping to installing and inspecting them. To proceed with construction efficiently, the process planning that determines the sequence in which equipment and piping are installed is important task that particularly requires the experience of site management. Building construction and equipment construction may be performed in parallel. In this case, the area where equipment can be installed is handed over from the building contractor before the work starts. The construction schedule will be decided in consideration of various matters such as the time to hand over the work area, the delivery time of equipment and piping, and the adjustment of workers and construction machinery such as cranes. The process planner creates an appropriate process by imaging the future site conditions in advance and considering various matters such as interference with other areas, arrangement of construction machinery, worker's flow line, etc. To support these, it is effective to simulate the construction visually.

As for the function to support visualization, it have been developed on BIM software, and a function (referred to as 4D simulation) that visualizes the on-site construction flow over time has been implemented [4]. However, to visualize the process planning by this function, it is necessary to set the start and end dates of work for a huge number of BIM objects, and it takes a great deal of manpower to perform this task manually. For this reason, the software has a visualization function, but its practical use is limited. Experts who have experienced on-site work can imagine the planned process and on-site conditions in their mind, but if they have little experience, this process is difficult for them. If BIM can visualize them efficiently, it will be very effective in managing construction work.

Therefore, in this study, we propose a method to automatically set the start and end dates of work on BIM objects, based on the spatial information obtained from BIM and rules which are established to formulate the knowledge from the process planning of experts. And we develop the tool that facilitates the identification of risks such as work interference by prior check of the construction site and supports the management work of unskilled workers by visually performing 4D simulation of the site situation that changes over time.

2 Literature Review

In some cases, visualization of process planning has been studied in conjunction with CAD tools even when the term BIM was not commonplace [5]. With the advent of BIM, various computer-aided scheduling tools using BIM have been developed [6]. In recent years, the function of visualizing the flow of building construction by 4D simulation has been a standard function for BIM software [7]. The 4D simulation function of the BIM software can display or hide objects in response to changes over time, and change the color settings of objects according to the actual status by setting the scheduled start and end dates of work and actual process information in BIM objects. However, in the process planning, it is important how to set the start and end dates of work, and there is no function for automatically setting them.

The methods of the process planning itself include the critical path method (CPM) and the program evaluation and review technique (PERT), and these are methods of manually setting work relationships. Methods to express the process specializing in construction were also studied [8, 9], and Wei-Chih, et al. [10] proposed a method of grasping the quantity of construction objects by BIM and integrating them with a process planning separately studied by the conventional process planning method. David, et al. [11] proposed a method to automate construction sequencing and scheduling based on functional requirements, and also studied the handling of temporary installation equipments such as scaffolding. Hyunjoo, et al. [12] proposed a method to generate a schedule using geometry data and material data described in the industry foundation class (IFC) used in BIM. Management information such as schedules may be stored in a database in another system, and methods to integrate these DBs with the BIM studied [13]. In addition, some studies were conducted on algorithms for optimizing process planning and project schedules. There have been reports of studies on optimal construction planning using the genetic algorithm for BIM models [14] and studies on improving planning using the tabu-search algorithm and 4D models [15]. Kateryna, et al. [16] studied a method to enable the management of more detailed tasks, including sub-schedules, using graph-based methods. In recent years, study has been conducted [17] to increase productivity by dividing projects into work zones and visualizing them from large processes to weekly processes, with the last planner system (LPS), which was developed by Glenn and Greg [18]. However, many of these studies have been conducted on general buildings for which the construction sequence was predetermined to some extent, and not on plants where piping are installed in large quantities during construction. In the case of plant construction, the specifications and equipment arrangement conditions in the building vary from project to project, and the site manager determines the procedure each time while considering the workability, etc. In addition, when installing piping, the work is often performed with scaffolding installed widely in the work space, and visualization including scaffolding is indispensable for studying workability; however, no studies from this perspective have yet been reported.

3 Proposed Method

3.1 Main Concept

In this study, we propose a method of automatically setting the order of work taking into account spatial constraints, especially for large amount of the product to be installed, such as piping work of plant construction including temporary scaffolds in the work space. The expert planned the process while imagining the spatial information of the construction area, including temporary information, in their mind. The expert understood that it was more efficient to carry the piping installed at a higher setting position farther from the entrance, considering the spatial constraints, and made a plan. Figure 1 shows the concept of a process planning system based on the proposed method. BIM makes it possible to acquire spatial three-dimensional information such as the installation location and installation height of equipment and piping. By including the construction information such as entrance information and the scaffold information in the BIM, the spatial information imagined by the expert can be obtained from the BIM. By using the knowledge of a skilled person who determines the order of work as a rule and acquiring necessary spatial information from BIM, the same construction plan as a skilled person can be automated.

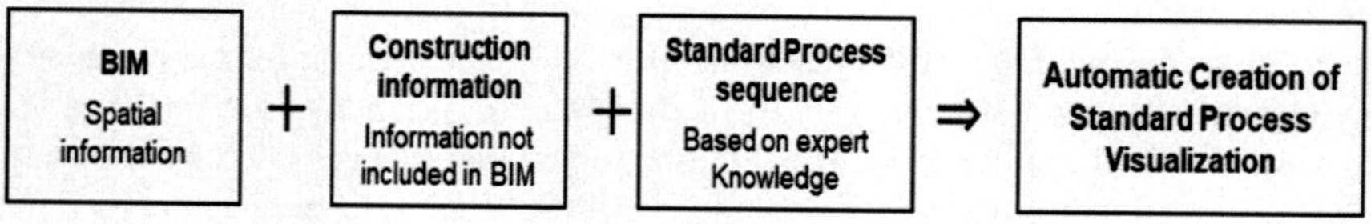

Fig. 1. Conceptual diagram of the proposed process planning system

3.2 Creating Standard Rules Based on Expert Knowledge

For establishing the rules to formulate the expert knowledge, as an example, we conducted a knowledge interview with planner for the construction inside the turbine building of a power plant. Figure 2 shows the standard sequence of work created by on-site experienced of expert as an example. Table 1 shows the examples of contents for each work sequence shown in Fig. 2 and information required for process planning.

The following is a characteristic rule for this process planning.

Generally, components to be installed at a higher position are first carried in and temporarily fixed. Next, common scaffolding for assembling the components at the higher position is set. After that, the equipment is carried in and installed. Next, piping to be installed at a position lower than the scaffolding are carried in, and finally piping are permanently installed.

In this example, the common scaffolding installation height is 2.5 m, and of the pipes to be installed at a position higher than 2.5 m, the piping to be installed farther from the carry-in entrance are carried in from the back to the entrance in order before being temporarily suspended. This has been established as a standard process planning rule based on expert knowledge. As in this example, the experienced expert consider the spatial information such as the installation height at the time of construction and the entrance location and route for carrying products, so we considered that if the spatial information can be acquired from BIM, the process planning can be automated.

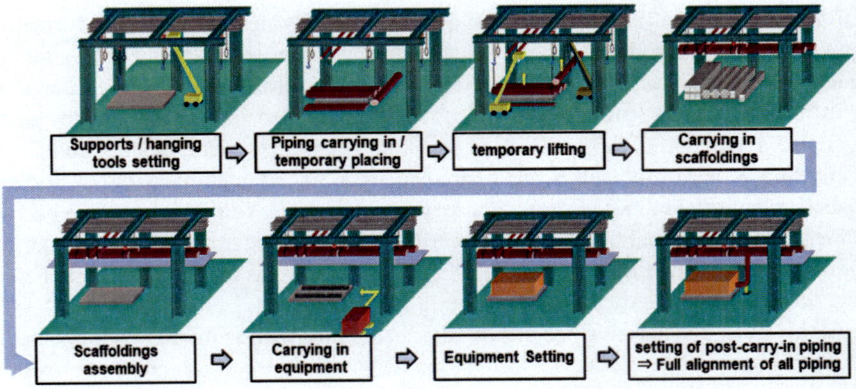

Fig. 2. Standard sequence for in-area work as an example in plant construction

Table 1. Contents for each work sequence and information required for planning

Work Sequence	Contents	Information for process planning
Support/hanging tool setting	The supports for piping and tools for temporarily hanging the piping are set	–
Piping earning in/temporary placing	The carry-in entrance is determined in construction area The piping to be installed father from entrance and upper position should be carried in first, and temporarily placed	Position of the carry-in entrance Height of installing point of the piping, (upper than 2.5 m)
Temporary lifting	The carried-in piping is lifted near the set position and temporarily fixed	–
Carrying in scaffoldings	Scaffoldings are carried in for setting the upper part of the piping	Installation height of scaffoldings (2.5 m)
Scaffoldings assembly	Scaffoldings for common use are set	Scaffoldings type
Carrying in equipment	Equipment is carried in And the equipment to be installed farther from the entrance should be carried in first	Position of the carry-in entrance
Equipment setting	Equipment is installed at the set position	–
Setting of post- carry-in piping => Full alignment of all piping	The pipes below the common scaffolding are carried in, and all piping installed at the set position	Height of installing point of the piping, (less than 2.5 m)

3.3 Acquisition of Spatial Information from BIM Data

From Table 1, we can see that the installation sequence of equipment and piping is determined by the installation height of piping in the area, the installation height of scaffolding, and the position information on the carry-in entrance as information required for planning. This is used as the standard process planning rule, and the sequence is schematically shown in Fig. 3. The common scaffolding is set at a height of 2.5 m from the floor, so that the working space is divided into an upper area and a lower area at a height of 2.5 m. First, of the products in the upper area, the products to be installed at higher positions are carried in from top to bottom in order, and the products to be installed farther from the carry-in entrance are carried in from the back to the entrance in order before being temporarily suspended or temporarily set. Next, the products in the lower area are carried in, in the same order as in the upper area, before being temporarily set. After that, they are fully set.

At this time, it is necessary to acquire the spatial information on the working area as the information on the scaffolding installation height and the carry-in entrance position, which is obtained from BIM. Equipment and piping are included in the BIM model, and set position information is acquired. Specifically, the center coordinate position of the object can be obtained from the 3D model of BIM. In addition, the scaffolding model and the carry-in entrance are determined temporarily at the time of construction and are not included in the original BIM, so these must be set.

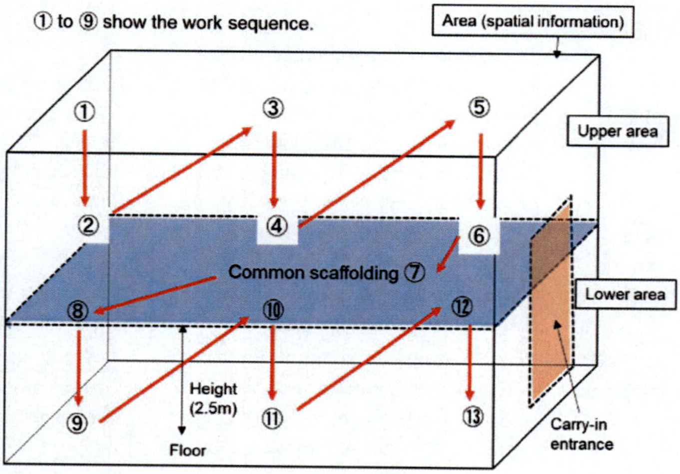

Fig. 3. Work sequence in the virtual area

3.4 Automatic Scaffolding Model Creation Function

Table 2 shows the types of scaffolding that are automatically drawn. Scaffolding specifications may vary from country to country, but in this case, they are roughly categorized into two types. One is the general built-up scaffolding that is set on a floor, using the specified scaffolding, and the framework consists of single tubes and clamps. The other is hanging scaffolding that is suspended from the ceiling, which has an advantage that have a space can work under the scaffolding floor.

Table 2. Types of scaffolding that are automatically drawn

Scaffolding	Built-up scaffolding	Hanging scaffolding
Image		
Characteristics	Assembled with specified scaffolding. Framework consists of single tubes and clamps.	Specification of hanging the scaffolding from the ceiling
Merit	· Easy assembly · Can be assembled on a stepped floor	· Can work under scaffolding floor

Figure 4 shows the setting conditions required to create each of the two types of scaffolding models. Based on the structure of each scaffolding, the input value of the height of the support point was set. For single tubes/wedges/built-up scaffolding, the height was set below the work floor, and for hanging scaffolding/staging, the height was set above the work floor. The maximum span was selected based on the laws and in-house assembly standards, and the adjustments of the span and the middle floor height were designed to limit the input values based on the structure of each scaffolding.

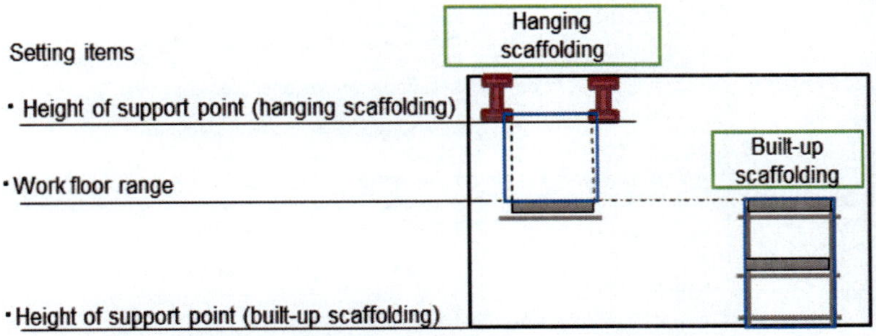

Fig. 4. Setting items required to create scaffolding models (elevation)

Table 3 shows the automatic drawing procedure for each scaffolding. First, the surrounding area spatial data is obtained to set the above setting items, and the position of the column or the hanging point is selected. The distance between columns and the distance between hanging points are set according to the scaffolding standards. After that, the positions of the beams and the work floor are selected according to the above positions, and the columns, beams, and the work floor are created using the standard scaffolding models as a framework model of scaffolding. Finally, base boards, handrails, suspension chains, etc. are created at predetermined positions.

Figure 5 shows an example of creating a scaffolding model using standard products. This is a relatively complicated column, and the procedure for creating the column is described below.

a. Extend the column height from work floor to handrail height.
b. Divide the column by standard single tubes (length: 1 to 4 m), and insert a joint material (length: 30 mm) between them.
c. Insert base material (15 mm height) between the column and floor.
d. For the case where the height of the divided column does not match the height of the handrail, that ties a 1.5 m long single tube with a free clamp.

These program has been developed using Autodesk's "AutoCAD". The scaffolding model can be created by reading the spatial information to set scaffolding from the BIM model and by interactively inputting conditions, such as selecting the range and type of scaffolding to be set.

Figure 6 shows a state in which a scaffolding model is added to the original BIM model obtained from a design without scaffolding. In addition, the scaffolding model automatically drawn by the system makes it possible to know the number of components required to create the scaffolding, so that the approximate amount of the scaffolding used in the work place can also be grasped (Fig. 7).

Table 3. Automatic drawing procedure of scaffolding

<table>
<tr><th>Types of scaffolding / Drawing procedure</th><th>Built-up scaffolding</th><th>Hanging scaffolding</th></tr>
<tr><td rowspan="3">Creating a skeleton</td><td colspan="2">Acquisition of surrounding area spatial data</td></tr>
<tr><td colspan="2">Position selection of columns or suspension points</td></tr>
<tr><td colspan="2">Selection of beams and work floor positions</td></tr>
<tr><td rowspan="3">Creating a skeleton with the standard scaffolding model</td><td>Creating columns with the standard model</td><td></td></tr>
<tr><td colspan="2">Creating beams with the standard model</td></tr>
<tr><td colspan="2">Creating work floors with the standard model</td></tr>
<tr><td rowspan="3">Other parts</td><td colspan="2">Drawing baseboards (created with the standard model)</td></tr>
<tr><td colspan="2">Drawing handrails (created with the standard model)</td></tr>
<tr><td></td><td>Creating the hanging chain</td></tr>
</table>

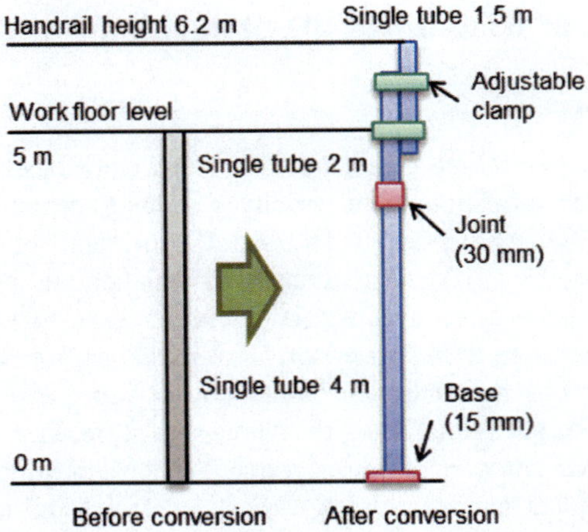

Fig. 5. Image of conversion of standard products to scaffolding model (column)

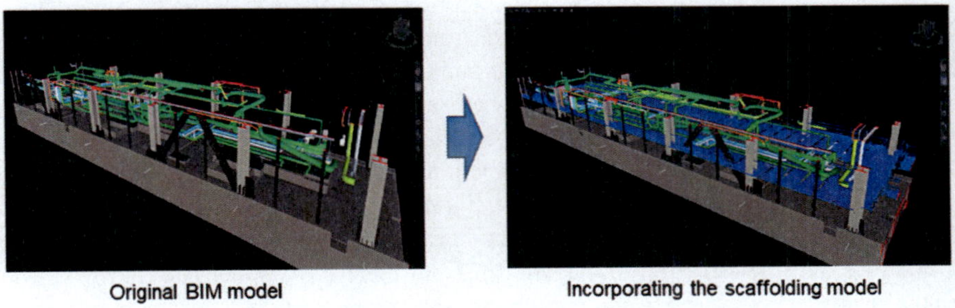

Fig. 6. Incorporation of scaffolding model into BIM model

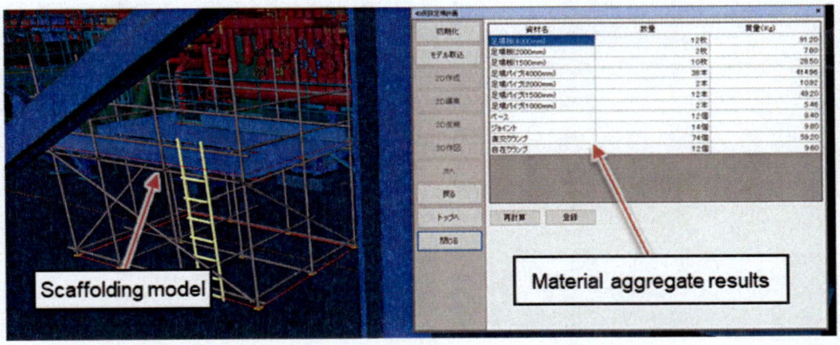

Fig. 7. Aggregation screen of scaffolding quantity

4 Development of BIM-Based 4D Simulation System

4.1 Platform System

Buildings, including plants, are constructed by separately designing the structure, equipment, piping, air conditioning, electricity and instrumentation, etc., and combining them. There are various design tools, and at present, the mainstream design software is standardized by BIM, so that shape information and attribute information can be integrated. At the time of construction, information that has been designed upstream is integrated using BIM viewer software, so that the interference between the structure and piping can be confirmed before construction, and the 4D simulation described above can be performed using this viewer software. This software supports a variety of design CAD software and uses Autodesk's "Navisworks", which has been used extensively in plant construction. We have developed and added a function to automatically set the process to this software as an add-on program.

4.2 BIM-Based 4D Simulation Functions

Using the developed program, the construction process (from carrying in equipment and piping to temporary setting) was automatically created for sample work when conducting the expert knowledge. The target equipment and piping are automatically selected as objects included in the preset construction area by selecting the construction area. Next, it is determined from which direction in the construction area the equipment and piping should be carried in, and the position of the carry-in entrance is designated. Figure 8 shows a setting screen of the carry-in entrance. Here, the position of the carry-in entrance and the direction of carrying in from there are displayed by arrows. This makes it possible to identify the equipment and piping installed farther from the set carry-in entrance. In this example, the carry-in entrance is set on the right side of the figure, and the product is carried in from here. Therefore, the piping and equipment to be installed on the left side (a place farther from the carry-in entrance) are first carried in. In addition, in this rule, the scaffolding is set at a height of 2.5 m from the floor. Therefore, the pipes to be installed at a position higher than the scaffolding are first carried in before being suspended temporarily on the ceiling, and then the scaffolding is installed. After that, the equipment and piping to be installed at a position lower than the scaffolding are carried in. The above sequence is applied as a standard rule. By setting the carry-in entrance and floor of the area as necessary information for construction, the information for determining the standard rule can be identified from the BIM spatial information. As for the standard rules, we plan to build and extend some rules in the future.

After the above setting, the start and end dates of work are automatically set for each object by equally dividing them so that all equipment and piping included in the designated area may be completed within the set process period. Actually, working hours differ for each object, but as of now, the schedule will be adjusted by the planner. Figure 9 shows an example in which the process set by the automatic creation is visualized by the 4D simulation using the BIM function.

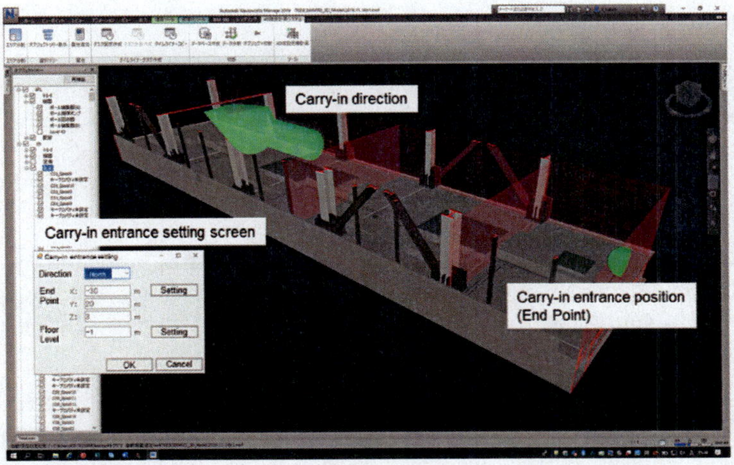

Fig. 8. Example of the carry-in entrance position setting screen

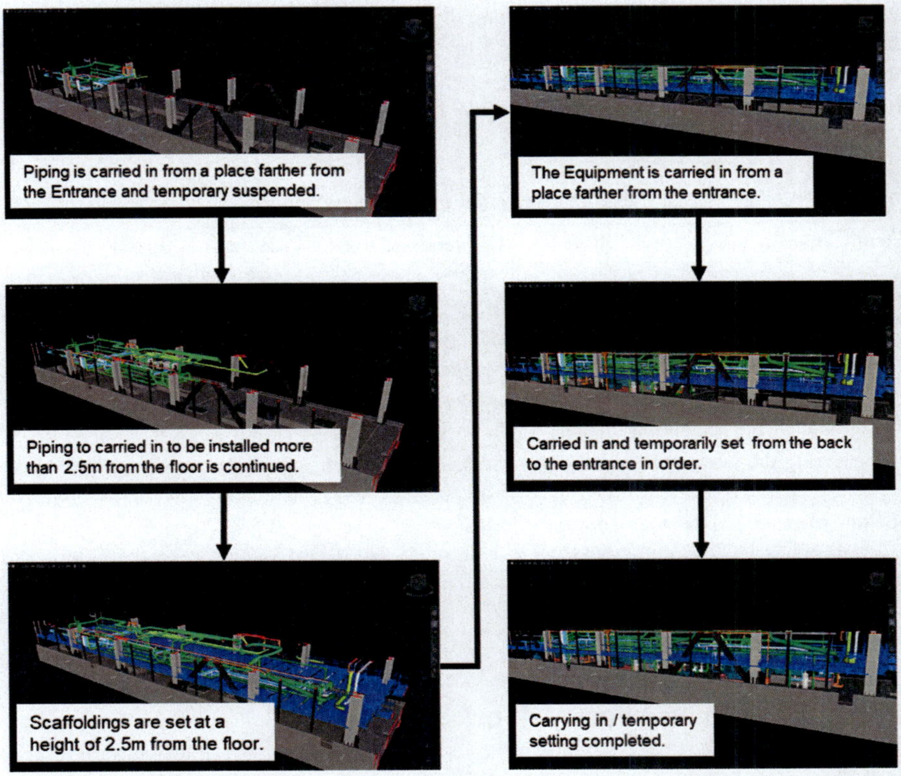

Fig. 9. Automatic creation of the area process and the visualization result screen

5 System Evaluation

Regarding the setting time of the 4D simulation using Navisworks, a comparison was made between the conventional method, which is completely set manually, and the method that is set using the system developed by proposed method, with the approximate values of the setting time. The volume conditions of the model used for evaluation are as follows.

a. Target plant: Equipment and piping in one construction area inside the power plant turbine building
b. Quantity in construction area: 19 equipment units, and around 2,000 number of piping objects

Table 4 shows the evaluation results of the system applied. In the past, it took 4 h to classify the BIM model according to the construction area and to select the target piping and equipment, etc., but the newly developed function enables setting in about 0.6 h. Scaffolding could not be considered in the past, but can be added with about 0.5 h of effort. The work sequence was assumed by experts, but can now be set automatically according to the standard rules. As a result, the start and end dates of work, which took about 4 h to set manually, can be set in about 0.5 h. In total, this makes it possible to reduce the setting time by 6.4 h, including the additional time of the scaffolding model, and the setting time with the 4D simulation function of Navisworks is expected to be reduced by 80% compared with the case of manual work.

Table 4. Evaluation of the system applied

Work Item	Conventional method		System Use	Effect
Setting of construction area	Manual setting	2 h	0.1 h	−1.9 h
Selection and classification of process planning objects	Manual setting	2 h	0.5 h	−1.5 h
Incorporating the scaffolding model	Impossible	–	0.5 h	+ 0.5 h
Considering the work sequence	Created by experts	4 h	Using rules 0.5 h	−3.5 h
Setting the start/end dates of work for each object	Manual setting			
Total		8 h	1.6 h	−6.4 h (80% reduction)

6 Conclusions and Future Work

In the process planning of large-scale plant construction, the number of products such as piping is large, and when performing the visualization by the 4D simulation of BIM software, many efforts are required for setting. In addition, some scaffolding installation work is required in construction work, so that the process planner needs to consider the work sequence including the scaffolding. However, to perform the 4D simulation including the scaffolding, it was necessary to create a new 3D model of the scaffolding that was not included in the original BIM model, which was also a time-consuming task. For this reason, in this study, we developed the systems for addressing these two problems, and confirmed the effects. The first is a function that uses the setting procedure based on the knowledge of expert as a standard rule, reads the installation position and arrangement information on piping and equipment included in the BIM model, and automatically sets the installation sequence. As a result, we have been able to reduce the setting time required to set the process information by automatically setting the installation sequence with the computer. The other is a function that interactively creates the necessary 3D model of scaffolding using the spatial information of BIM. We have made it possible to automatically calculate the required quantity of scaffolding and create the 3D model by specifying the spatial position to set scaffolding in the BIM model and by selecting the types of scaffolding. With these functions, the time required to create the scaffolding model has been reduced, and the quantity of scaffolding required for construction can be checked.

References

1. Ruben, S., Antonio, A.C., Jose, D.S., Lincy, P.: Infrometric analysis and review of literature on the role of BIM in sustainable construction. Autom. Constr. **103**, 221–234 (2019)
2. Hao, W., Xianhai, M.: Transformation from IT-based knowledge management into BIM-supported knowledge management: a literature review. Exp. Syst. Appl. **121**, 170–187 (2019)
3. Daniel, W.M.C., Timothy, O.O., Alfred, M.L.H.: Perceived benefits of and barriers to Building information modeling (BIM) implementation in construction: the case of Hong Kong. J. Build. Eng. **25**, 100764 (2019)
4. Xue, L., Jing, X., Qun, Z.: Research on construction schedule management based on BIM Technology. Proc. Eng. **174**, 657–667 (2017). 13th Global Congress on Manufacturing and Management, GCMM 2016
5. Kathleen, M., Martin, F.: Generating, evaluating and visualizing construction schedules with CAD tools. Autom. Constr. **7**, 433–447 (1998)
6. Alexandre, D.L., Naia, L., Robert, B., Guillaume, D.L.: Computer-assisted scheduling tools in the construction industry: a systematic literature review. IFAC PaperesOnLine **52–13**, 1843–1848 (2019)
7. Chuck, E., Paul, T., Rafael, S., Ghang, L.: BIM Handbook: A Guide to Building Information Modeling for Owners, Managers, Designers, Engineers, and Contractors, 2nd edn. John Wiley, New Jersey (2011)
8. Julio, C.M.: STROBOSCOPE State and Resource Based Simulation of Construction Processes, A dissertation submitted in partial fulfillment of the requirements for the degree of Doctor of Philosophy, University of Michigan (1996)
9. Adel, F.: Graphical modelling classification for construction project scheduling. Proc. Eng. **123**, 162–168 (2015). Creative Construction Conference 2015 (CCC2015)
10. Wang, W.C., Weng, S.W., Wang, S.H., Chen, C.Y.: Integrating building information models with construction process simulations for project scheduling support. Autom. Constr. **37**, 68–80 (2014)

11. Chua, D.K., Nguyen, T.Q., Yeoh, K.W.: Automated construction sequencing and scheduling from functional requirements. Autom. Constr. **35**, 79–88 (2013)
12. Hyunjoo, K., Kyle, A., SangHyun, L., John, H.: Generating construction schedules through automatic data extraction using open BIM (Building Information Modeling) technology. Autom. Constr. **35**, 285–295 (2013)
13. Shih-Ming, C., (Bud) Griffis, F.H., Po-Han, C., Luh-Maan, C.: A framework for an automated and integrated project scheduling and management system. Autom. Constr. **35**, 89–110 (2013)
14. Vahid, F., Kenneth, F.R., Julian, H.K.: Construction scheduling using genetic algorithm based on building information model. Exp. Syst. Appl. **41**, 7565–7578 (2014)
15. Borja, G.S., Anton, R., Jochen, R., Qian, C., Bryan, T.A.: Using a Tabu-search algorithm and 4D models to improve construction project schedules. Proc. Eng. **196**, 698–705 (2017). Creative Construction Conference 2017, CCC2017
16. Kateryna, S., Markus, K.: Recognition of process patterns for BIM-based construction schedules. Adv. Eng. Infrom. **33**, 456–472 (2017)
17. Daniel, H., Borja, G.S., Ernest, L.S.A.: BIM-based last planner system tool for improving construction project management. Autom. Constr. **104**, 246–254 (2019)
18. Farook, R.H., Glenn, B., Iris, D.T. :Improving construction work flow –the connective role of lookahead planning. In: Proceedings for the 16th Annual Conference of the International Group for Lean Construction, pp. 635–646 (2008)

A 4D BIM System Architecture for the Semantic Web

Calin Boje(✉), Sylvain Kubicki, and Annie Guerriero

Luxembourg Institute of Science and Technology,
5 Avenue des Hauts-Fourneaux, 4362 Esch-sur-Alzette, Luxembourg
calin.boje@list.lu

Abstract. The tedious process of linking product and process models, particularly at higher levels of detail makes the exploration of multiple construction management strategies prohibitive, especially when looking at the collaboration and social aspects of 4D review meetings (decision making features of support systems, taking into account the behaviour of professionals and the natural user interfaces that should support it). We consider the use of ontologies to be better suited for representing such complex systems, paving the way toward semantic digital twins of the construction site by including concepts beyond traditional 3D model data, such as actors, sensors, management workflows, information over the web, etc. Most importantly, access to web distributed resources can be used for intelligent processing and logic reasoning, bypassing the BIM legacy tools and formats which are becoming more and more limited in an increasingly interconnected world. To address these challenges, we propose a 4D BIM platform architecture that is able to encompass semantic web models representing the building and the construction scheduling information, which are further enriched by the semantics of collaboration planning meeting contexts. The system architecture is comprised of several layers, each of which are presented and discussed.

Keywords: BIM · 4D · System architecture · Semantics · Linked data

1 Introduction

The complexity of BIM data models at pre-construction and construction stages reaches its peak on one hand due to the involvement of several new actors (contractors, sub-contractors, suppliers, etc.) and on the other hand due to an increased application of BIM uses (scheduling, costing, site safety management, construction simulations, construction site logistics etc.). This presents challenges on two fronts: information cohesion and synchronicity. The first is caused by the necessity to ensure interoperability between modelling application domains, with the second to ensure models are up-to-date with actual site status and vice-versa. At the current stage, BIM relies on the Industry Foundation Classes (IFC) standard for exchanging information between different tools using the STEP format. However, due to multitude of application domains during construction stages, not all information can be federated within the same IFC model. Additionally, this static file-based information exchange is not compatible with

E. Toledo Santos and S. Scheer (Eds.): ICCCBE 2020, LNCE 98, pp. 561–573, 2021.
https://doi.org/10.1007/978-3-030-51295-8_40

the semantic web. There is a potential to be unlocked in BIM via a semantic web world view [1–3], where semantics play a key role in understanding, automating and keeping structured information valid and up-to-date. We believe this to be the logical transition towards a cyber-physical integration system and Digital Twins.

As the necessity for higher degree of integration with site sensing equipment becomes prevalent, linked data across the web begins to accumulate more and more value to construction practitioners. However, there is still a research gap concerning the usability of RDF/OWL models when in the context of capturing and representing human input when they interact with the digital building model data [4]. In addressing this gap, the system proposed within this article integrates heterogeneous data sources in RDF with the use of several OWL models. The system interface is capable of capturing human input on a collaborative device and keep track of external items (such as digital documents in various formats) for improving collaboration methods by using meta-data about these collaboration artifacts.

This research was developed under the 4DCollab project [5], which aims to deliver a novel, more holistic and collaborative 4D system, while considering a user-centric approach. The scope of this article is focused on the architecture of a prototype system under development. This is complemented by previous work from the authors on collaboration meetings using 4D BIMs [6, 7] and the development of the 4D Collab ontology [8] which acts as an underpinning semantic model within the developed system presented here. The objectives of this research are to outline a system architecture, describe the data workflows and highlight the key benefits of adopting a semantic web layer for integrating the different domains, thus enabling the fast development of a heterogeneous data environment. Additionally, the limitations of such an approach are highlighted from a technical perspective on data modelling.

2 Related Work

Reviewing the status of semantic web applications within the built environment, [9] and [4] note that ontology supported systems remain niche developments in research and especially industry, mainly due to the complexity of an additional semantic layer – which needs to be developed and maintained. [10] describe a framework for designing ontology supported systems by considering common open tools, libraries and formats – such as IfcOwl [11]. The use of IFC concepts in a semantic web environment bypasses legacy STEP file limitations, but it also means there is a need to convert IFC models to an RDF version [12]. The IFC-to-RDF converter is widely used to facilitate this step [13], allowing model information to be expressed in RDF and OWL under the IFC schema. The same IFC model data can therefore be used in a web-native language, linked with other resources and OWL models, whilst also allowing the data to be used for rules reasoning. The data can be accessed using the SPARQL query language or BimSPARQL [14] – a BIM domain adapted query language, which can be used to traverse RDF and OWL graphs. The BIM at a semantic web stage is not without limitations, being a subject of research which tries to deal with making ontology models more efficient and easier to use [4].

One of the more common uses for semantic web within the domain of design and construction is for rules checking, usually using domain specific ontology models with embedded rules [15–18]. Integration of heterogeneous web resources and sensor networks within one domain applied system is common all across the board from energy management [19–21], to construction cost estimation [22] or quantity take-off [23]. In all the cases mentioned above, the semantic layer (usually supported by an OWL ontology or RDF graphs) provides an integration canvas from digital models to virtual and physical sensors on site.

Since the advent of semantic web technologies within the architecture and construction fields [3], one of the first envisaged challenges was to tackle the multitude of native BIM data formats [2], as well as their mapping to IFCs [24]. A number of studies have described potential architecture of ontology-supported systems, such as [25] on mapping Revit models to the Building Topology Ontology (BOT) [26], thereby translating proprietary model data to a web native language. Similarly, [27] describe a system using a core of ontologies to support building evacuation design, which consider the BIM, as well as additional captured human input.

On the subject of 4D modelling - although 4D BIMs are commonly used during preconstruction and construction phases, the use of ontology models and even the IFC standard remain very limited. The linking of 2/3D objects to planning processes has been explored in research on automatically updating site scans to an IFC file [28], which would benefit more from a web interoperable environment. Several ontologies relevant to the 4D BIM domain are highlighted by [8], but most remain conceptual and not explored in multi-user contexts, such as synchronous collaborative sessions during project construction and management.

3 System Architecture

We propose a typical Client-Server architecture with additional layers to ensure integration with the semantic web-based languages, as well as BIM interoperability, as outlined in Fig. 1. The 4D Collab prototype system is able to store and link data in RDF (under RDFS and OWL schemas). As RDF was designed to act as common computer interpretable language (written in XML), and not as a programming language, the development of ontology-supported systems still has to use traditional programming environments to bridge the RDF data with dynamic in-memory data of the running applications. In this case, the central component in Fig. 1, the Server Application Layer (1) was programmed as a .NET WebAPI application. This can read and write directly in RDF using the dotNetRDF libraries which can also be used to interface with third party triples stores such as Jena Fuseki, Sesame, Stardog, Virtuoso, etc. The Client Application Layer (2) is envisaged to host different interfaces, according to user and development needs, while the layers at 3) and 4) in Fig. 1 act as means of storage and integration.

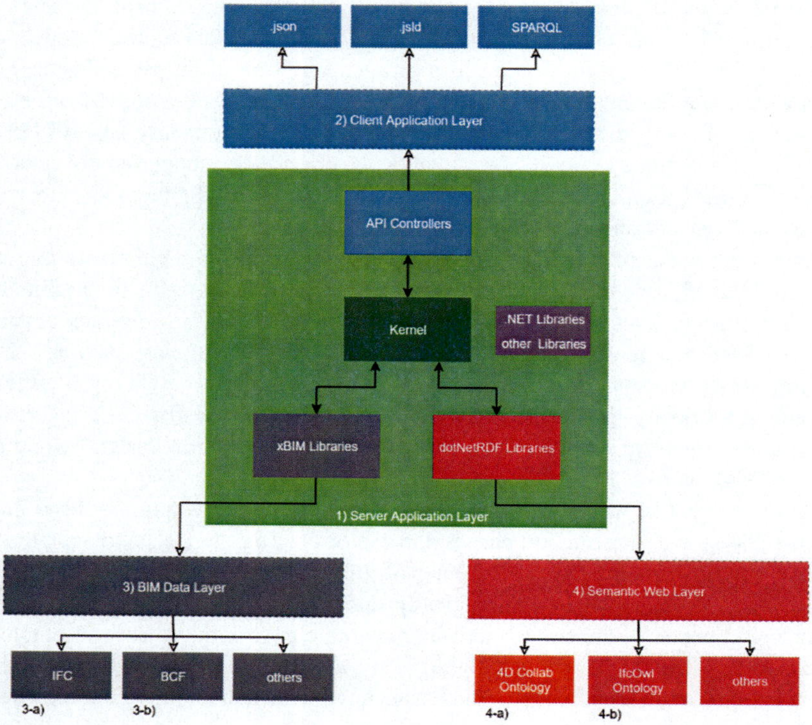

Fig. 1. Multi-layer architecture for the 4D Collab prototype system

3.1 Components by Layer

As outlined in Fig. 1 above, the four main layers and their components are as follows:

1) Server Application Layer – programmed on top of a .NET Framework, this embodies the main system application, that manages the logical workflow of the entire system, and deals with the overall information governance. The "Kernel" of the application consists of several developed classes and methods which facilitate the communication between the several layers. These classes use several imported libraries, such as the .NET Web API to build the skeleton framework for a Server-Client application, and other additional ones to integrate with 3) and 4) below.
2) Client Application Layer – this represents the client-side implementations which are able to consume the API from 1). As separate software implementations, these are not only independent from device or operating system, but also from demanding hardware specifications. Due to the integrated layers, the RDF and IFC stored information is parsed to several web compatible formats, such as ".json" (JavaScript Object Notation), ".jsld" (JSON for Linking Data), but also supports a SPARQL [29] endpoint for accessing the data directly in RDF triples.

3) BIM Data Layer – implements several features using the xBIM libraries in order to handle models and files directly in IFC (3-b), or BCF (BIM Collaboration Format) when necessary. This allows efficient model parsing (reading/writing), model querying, model element creation/modification - according to multiple schemas (IFC2x3, IFC4, BCF 1.0, BCF 2.0);
4) Semantic Web Layer – this consists of the semantic representation of the 4D Collab Model [8], which has been mapped to the IfcOwl ontology. An external IFC-to-RDF converter [13] is used to facilitate the conversion from initial IFC model files. Thus, the connected graphs allow the traversal of the information from multiple domains. Reasoning and rules injection is also supported, allowing a more comprehensive mapping between concepts and domains, wherever necessary.

3.2 4D Modelling Workflow in a Collaborative Setting

The initial prototype specifications were designed to support co-located collaboration between several end-users, as was previously investigated as part of the same research project [6–8]. However, a semantic layer was adopted to ensure extensibility and integration with the semantic web and Internet of Things (IoT) devices. This greatly expands the canvas of information which can be gathered from the internet, as well as construction site devices which track and keep the BIM up-to-date with the latest developments. This in turn would allow construction stakeholders to collaborate using an accurate 4D model, augmented by external resources and documents which are often required to support decision-making in such contexts [6].

In a typical collaborative setting, the proposed prototype system described in the previous section was considered for the workflow outlined in Fig. 2. Two main parallel threads (a and b) can be distinguished across a three stage workflow. The first thread (a) depicts the user interaction level, whilst the second (b) represents the system information flow, as follows:

1-a) The users start a collaborative session with regards to 4D modelling, by uploading the latest relevant documents which are used for specific use case decision-making. Due to various model views imposed by different user roles (architect, engineers, client, planner, etc.), the typical 4D model is often divided in a 2/3D model (in IFC) and a separate planning document (typically Gantt charts in various formats). The 3D models from different actors are usually merged beforehand if possible (when relying on IFC), but additional digital drawings and supporting documentation remains unlinked to specific BIM objects.

1-b) The imported documents are parsed by the system back-end, checked and integrated with the session workflow. The relevant information within imported documents is extracted and registered internally in multiple formats (in-memory objects, RDF graphs, native IFC files) to allow efficient manipulation and access to the data. It is at this stage that the 4D Collab prototype system registers and begins to keep track of meeting resources (digital documents) and its participants;

2-a) During the collaborative meeting, the users interact with the 4D model at object and document level. The interface needs to respect certain collaborative and ergonomic requirements to facilitate natural user interaction with the model [7]. The users typically annotate the model and raise issues for future correction by specialized technicians using typical BIM tools, hence the need to use common BIM formats for successive management stages;

2-b) The back-end server side of the system manages the ad-hoc API requests which are done by the client-end, ensuring correct communication of the data in various formats, as explained in the previous section. The system can handle IFC objects directly, but has to rely on the imbedded ontology models to use and store data about concepts which are out of scope for the IFC schema, such as 4D Collab ontology model for collaboration aspects and external documents. For example, external BIM documents need to be visualised from different cloud-based BIM environments (such as the Autodesk Forge Cloud, BIMServer, etc.)

3-a) Once the collaboration session is concluded, the end-users typically need to transfer BIM information from a federated model (4D) to other tools and formats so they can be corrected and modified using conventional BIM tools. The 4D Collab prototype system is able to generate complete 4D IFC models from the initial 3D, along with merged planning information according to specifications. An additional step is the exporting of annotations and issues as fully semantic models – hence the need to rely on the BCF format for complete communication of the change requests cross-platform;

3-b) As the parallel task of the previous, the back-end system is able to generate new files on demand based on existing documents (IFC, planning, screenshots, annotations, etc.) and based on the internal 4D Collab ontology model. This therefore completes the feedback and provides a method for end-users to export structured data from a complex multi-faceted collaborative environment. However, the full model view is stored in a triple store, which offers a holistic view of the collaboration process that can be used for knowledge processing.

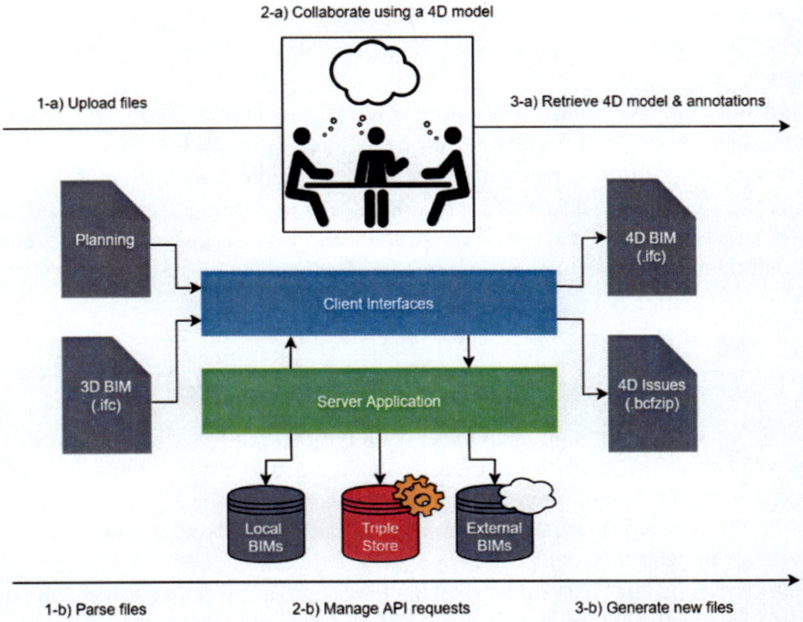

Fig. 2. Information workflow within the 4D Collab prototype system

4 Integrating Data Across Domains

One of the main challenges of the construction industry today is in integrating different models views, often from different application domains. The 4D modelling use alone is known to have several uses, such as scheduling, costing, construction simulation, safety management, etc. – all relying on spatio-temporal information with additional aspects about the site (resources, constraints, temporary works, weather, etc.). For the purposes of this research, we focus on collaboration aspects of the data models involved, namely the IFC, BCF and additional concepts of the 4D Collab ontology introduced in [8].

4.1 Domains

As the system was envisaged to support synchronous collaboration sessions between multiple users of BIM in a 4D context, the internal data is stored using the schema provided by the 4DCollab ontology model. This represents the domain of collaboration between actors, their digital assets (BIM models) and the events which take place during coordination meetings. The 4D model sits at the core of this ontology (mapped from the IFC schema), augmented by several other concepts which aim to formalize decision making aspects, external documentation, as well as the people which collaborate and the actions (events) they make on the multi-touch interface devices in use.

The BIM domain is broadly represented by the IfcOwl ontology (following the IFC 2x3 TC1, and the IFC4 ADD2 TC1 versions of the schemas), which is attached and mapped with the 4D Collab ontology. These store the actual IFC model data, which can be queried for more precise information. However, as its structure is quite heavy and recursive, its contents are not used to query complex geometric representations for example.

The formalization of the decision-making during collaboration can in part be accomplished by the BCF standard. However, there is no RDF/OWL format of the schema to be used directly. Thus, an extension to the initial 4D Collab model was used to capture BCF relevant data fields and relate them to existing concepts. This showcases the flexibility and extensibility of semantic web language models and how traversing domains can bypass file-based data transfers.

4.2 Data End Points

The main data types returned by the 4D Collab prototype system are:

1) JSON – the data used to communicate with typical web-based clients, usually on a web browser. The JSON format is very flexible and lightweight; however the data needs interpretation on the receiving end; Although the data transfer is very efficient, the data set is usually incomplete, as the data is broken down into several parts of the whole model graph;
2) RDF triples (in ".rdf" or ".ttl" formats) – the data transferred to a triple store, where complete graphs are uploaded/downloaded on demand. The full model semantics are presents, explicit or implicit, along with web accessible links to the domains and the resources that are used; there is no data expressivity loss during transfers, however information can also be broken down to logical units for more manageable chunks of information;
3) JSLD (JSON Linked Data) – is a hybrid approach which sends data in JSON format from 1), but with additional information on semantics by providing RDF like properties and web links to the concepts used (links to the ontologies and dictionaries which define each concept). This allows the client side applications to receive meta-data about the sent data properties;

The vast majority of web client libraries rely on light-weight languages for transferring information, usually XML or JSON, with reduced semantics. Although the data models in RDF/OWL can represent a high level of semantics, in order for them to be used by client-side applications, these need to be converted to different formats, thus implicitly reducing their scope and level of expressivity. This means that certain things are not present explicitly, and therefore parts of the information they carry is lost. Considering the various formats above, a semantic web system design needs to consider data transfer efficiency on the one hand, and semantics and extensibility on the other.

5 Use Case in a Collaborative Working Environment

A prototype web client was developed to act as a user interface which allows access to various data from the back-end layer. This makes the entire 4D Collab prototype system fully functional and accessible over the web on any device. To simulate a dynamic working environment with multiple users, the testing was carried out on a cutting edge collaborative device as shown in Fig. 3. To facilitate the capture of multiple touch events from several users, the Shariiing platform [30] and its imbedded web browser was used, with the testing interface shown in Fig. 4.

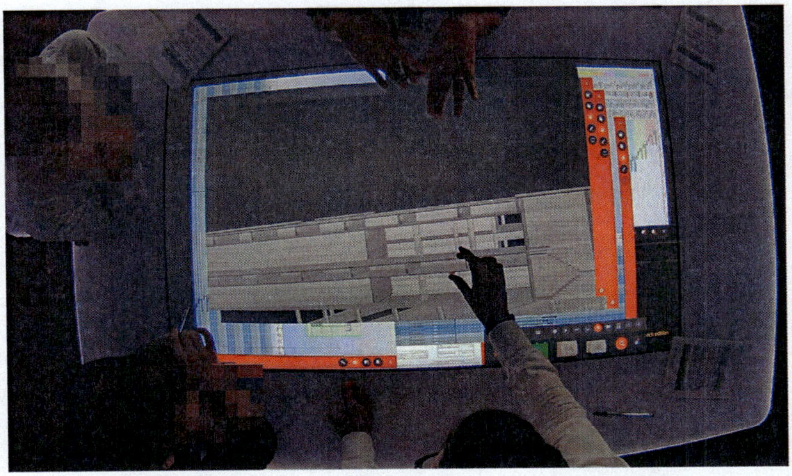

Fig. 3. Multi-user touch table used for collaborative sessions

The testing use case followed the described workflow in Sect. 3.2, using a sample IFC model. A sample gantt chart was imported to simulate a construction planning 4D use case. The IFC model and planning were not initially connected, being sourced from different modelling tools. As shown in Fig. 4, the main features of interest concerning a 4D modelling use case are the following:

1) Visualising the 3D model – this was achieved via an embedded ForgeViewer on the client interface, thus permitting the end-users to manipulate the model at BIM object level.
2) Creating/Modifying the 4D links – similar to the above, another window shows the planning in a Gantt chart format, allowing users to create/modify and link/unlink tasks with one or more 3D objects from 1).
3) Raising issues – problems which occur during collaborative meetings can be captured in a structured way by completing a form which is linked to the planning and the 3D view of the model, thus allowing the end-users to describe the problem which needs to be solved before the next coordination meeting;

Once the session was closed, the system was able to output several documents in IFC and BCF, which were imported and verified using eveBIM [31] – an academia sourced BIM platform which is both IFC (with planning view at task level) and BCF compliant, following strict schema restrictions.

From the initial system testing several limitations were observed, from both the technical and user interaction side:

1) The embedded 3D web viewer was not intended for a multi-touch device, thus the interaction remains limited and can often cause unexpected camera movements;
2) Visualising and manipulating the scheduling information can become difficult when the number of tasks increases. Additional automatic features on adjusting size and chart movement need to be explored;
3) Creating 4D related issues requires manual input within several fields according to BCF specifications. This is an error prone and time consuming procedure. More automatic ways of capturing model data and user input need to be explored for a more efficient process;

Fig. 4. 4D Collab interface within the Shariiing platform

4) Imported BCF annotations with references to a 4D model (referencing 3D objects and tasks at the same time) is not yet supported by the vast majority of BIM tools, hence a fragmentation of the model (both IFC and BCF) might be necessary at the end of a 4D BIM session;
5) Although the system supports multiple formats for communicating data over the web, most front-end tools rely on the ".json" format to visualize and exchange information. This makes it difficult to explore the semantic models in RDF/OWL in detail, and therefore difficult to take advantage of the enriched web semantics embedded within the model.

6 Conclusion and Future Work

A semantic web and linked data paradigm can empower the design and construction industries by providing enhanced interoperability and model extensibility with nearby domains and web devices. This article has introduced a novel architecture of a system prototype under development which focuses on linking data across multiple models using semantic web standards and tools. The Semantic Web Layer of the system introduced in Sect. 3 represents a means for integration, storage and reasoning capabilities. However, the lack of existing tools in industry which use such web standards makes the valorisation and exploration of semantic models prohibitive. Additionally, while linked data can empower tools and systems which post and use information on the web, this can be a limiting factor for the construction industry, due to data security and trust issues. Thus, the system was adapted to fit current tools used by industry practitioners both on and off the web, in order to provide a recursive loop for 4D modelling by relying on common formats (IFC, BCF). The use case in the previous section demonstrated some of the prototype features and highlighted several limitations concerning the user interaction, and a lack of automation means on the front-end. These limitations can be tackled by leveraging the imbedded semantics of the model.

Future work is expected to diverge into two main threads, with the first being on improving and validating the user interaction with the 4D model on multi-touch devices, whilst the second being focused on integration with site sensing capabilities to provide live data feeds to the system models and thus migrating towards a Digital Twin of the construction site.

Acknowledgements. The authors acknowledge the financial support from Fonds National de la Recherche (FNR) Luxembourg, and Agence Nationale de la Recherche (ANR) France to the 4DCollab grant reference: 11237662 (LU)/ ANR-16- CE10-0006-01(FR).

References

1. Howell, S., Rezgui, Y.: Beyond BIM: Knowledge Management for a Smarter Built Environment (2018)
2. Törmä, S.: Semantic linking of building information models. In: Proceedings - 2013 IEEE 7th International Conference on Semantic Computing, ICSC 2013 (2013)

3. Pauwels, P., Törmä, S., Beetz, J., Weise, M., Liebich, T.: Linked data in architecture and construction. Autom. Constr. **57**, 175–177 (2015)
4. Pauwels, P., Zhang, S., Lee, Y.-C.: Semantic web technologies in AEC industry: a literature overview. Autom. Constr. **73**, 145–165 (2017)
5. LIST: "4DCollab-project" (2019). http://www.4dcollab-project.eu/. Accessed 19 Mar 2019
6. Bolshakova, V., Guerriero, A., Halin, G.: Identification of relevant project documents to 4D BIM uses for a synchronous collaborative decision support. In: Creative Construction Conference 2018, CCC 2018, 30 June–3 July 2018, Ljubljana, Slovenia Identification (2018)
7. Kubicki, S., Guerriero, A., Schwartz, L., Daher, E., Idris, B.: Assessment of synchronous interactive devices for BIM project coordination: prospective ergonomics approach. Autom. Constr. **101**, 160–178 (2019)
8. Boje, C., Bolshakova, V., Guerriero, A., Kubickia, S., Halin, G.: Ontology assisted collaboration sessions on 4D BIM. In: Creative Construction Conference 2019, June–2 July 2019, pp. 591–596 (2019)
9. Abanda, F.H., Tah, J.H.M., Keivani, R.: Trends in built environment semantic Web applications: where are we today? Expert Syst. Appl. **40**(14), 5563–5577 (2013)
10. Ma, Z., Liu, Z.: Ontology- and freeware-based platform for rapid development of BIM applications with reasoning support. Autom. Constr. **90**(June), 1–8 (2018)
11. Beetz, J., van Leeuwen, J., de Vries, B.: IfcOWL: a case of transforming EXPRESS schemas into ontologies. Artif. Intell. Eng. Design Anal. Manuf. **23**(1), 89–101 (2009)
12. Pauwels, P., Terkaj, W.: EXPRESS to OWL for construction industry: towards a recommendable and usable ifcOWL ontology. Autom. Constr. **63**, 100–133 (2016)
13. Pauwels, P.: IFC repository (2020). https://github.com/pipauwel/IFCtoRDF. Accessed 22 Jan 2020
14. Zhang, C., Beetz, J., de Vries, B.: BimSPARQL: domain-specific functional SPARQL extensions for querying RDF building data. Semant. Web **9**(6), 829–855 (2018)
15. Pauwels, P., et al.: A semantic rule checking environment for building performance checking. Autom. Constr. **20**(5), 506–518 (2011)
16. Zhong, B.T., Ding, L.Y., Luo, H.B., Zhou, Y., Hu, Y.Z., Hu, H.M.: Ontology-based semantic modeling of regulation constraint for automated construction quality compliance checking. Autom. Constr. **28**, 58–70 (2012)
17. Lu, Y., Li, Q., Zhou, Z., Deng, Y.: Ontology-based knowledge modeling for automated construction safety checking. Saf. Sci. **79**, 11–18 (2015)
18. Zhang, S., Boukamp, F., Teizer, J.: Ontology-based semantic modeling of construction safety knowledge: towards automated safety planning for job hazard analysis (JHA). Autom. Constr. **52**, 29–41 (2015)
19. Dibley, M.J.: An intelligent system for facility management (2011)
20. Curry, E., O'Donnell, J., Corry, E., Hasan, S., Keane, M., O'Riain, S.: Linking building data in the cloud: integrating cross-domain building data using linked data. Adv. Eng. Inform. **27** (2), 206–219 (2013)
21. Hernández, J.L., et al.: An IFC-based interoperable framework for building linked-data. In: 6th Linked Data in Architecture and Construction Workshop (LDAC), CEUR Workshop Proceedings, pp. 44–51 (2018)
22. Abanda, F.H., Kamsu-Foguem, B., Tah, J.H.M.: Towards an intelligent ontology construction cost estimation system: using BIM and new rules of measurement techniques. Int. J. Comput. Electr. Autom. Control Inf. Eng. **9**(1), 294–299 (2015)
23. Liu, H., Lu, M., Al-Hussein, M.: Ontology-based semantic approach for construction-oriented quantity take-off from BIM models in the light-frame building industry. Adv. Eng. Inform. **30**(2), 190–207 (2016)

24. Törmä, S.: Web of building data - Integrating IFC with the web of data. In: eWork and eBusiness in Architecture, Engineering and Construction - Proceedings of the 10th European Conference on Product and Process Modelling, ECPPM 2014 (2015)
25. Rasmussen, M.H., Hviid, C.A., Karlshøj, J.: Web-based topology queries on a BIM model. In: 5th LDAC Workshop, November 2017
26. Schneider, G.F.: Towards aligning domain ontologies with the building topology ontology. In: 5th LDAC workshop, 13–15 November 2017, April 2017
27. Boje, C., Li, H.: Crowd simulation-based knowledge mining supporting building evacuation design. Adv. Eng. Inform. **37**(April), 103–118 (2018)
28. Hamledari, H., McCabe, B., Davari, S., Shahi, A.: Automated schedule and progress updating of IFC-based 4D BIMs. J. Comput. Civil Eng. **31**(4), 04017012 (2017)
29. Prud, E., Seaborne, A.: SPARQL query language for RDF (2006)
30. Immersion: Shariiing Platform (2020). https://www.shariiing.com/. Accessed 28 Jan 2020
31. CSTB: eveBIM (2020). https://www.evebim.fr/. Accessed 28 Jan 2020

A Metaheuristic Procedure Combined with 4D Simulation as an Alternative for the Scheduling Process of Housing Complexes

Pedro Bezerra and Sergio Scheer(✉)

Universidade Federal do Paraná, Curitiba, Brazil
bezerra.pedro92@gmail.com, scheer@ufpr.br

Abstract. The present work focuses on the application of a metaheuristic procedure, represented by a Genetic Algorithm, in the search of better alternatives for the organization of activities in the construction of housing complexes. Therefore, the algorithm is implemented in order to solve the Resource-Constrained Project Scheduling Problem – RCPSP. The problem modeling is adapted to the type of project in question, where, at first, the activity network of a single housing unit is defined through an activity-on-the-node graph and afterwards the Line of Balance method is used in order to determine the start and finish dates of all the other repetitive activities of the project. In order to test the efficiency of the algorithm, numerous experiments are performed with project instances from the PSPLIB, a specific benchmark database for the RCPSP, available online. Finally, in order to better visualize the schedules obtained in the optimization phase, a 4D simulation of a schedule for the execution of a housing complex, generated with the metaheuristic procedure, is performed with the software *Autodesk Navisworks*. The simulation phase allows the visualization of the projects constructability, makes it possible to confirm if the algorithm-generated schedule follows a logical sequence, as well as it supports the decision-making about the best execution alternatives.

Keywords: Operations research · Planning · Decision-making · Metaheuristics · 4D simulation

1 Introduction

The use of computational techniques to improve the planning process of construction projects has been addressed in research for several decades, being nowadays increasingly used in conjunction with project design and process simulation software, as a way to subsidize the decision-making both in pre-operational phases and during the construction execution. It's possible to observe a predominant use of meta-heuristic algorithms and discrete event simulation in conjunction with BIM models, what shows an increasing adoption of operational research concepts in the construction sector. These techniques are capable of addressing several specific objectives when planning activities, like the organization of activities, the management of resources and even the configuration of construction sites (Bezerra and Scheer 2018).

E. Toledo Santos and S. Scheer (Eds.): ICCCBE 2020, LNCE 98, pp. 574–593, 2021.
https://doi.org/10.1007/978-3-030-51295-8_41

In the context of public projects, especially in social housing, there is an even greater potential for the application of operational research, due to the repetitive nature of these projects, similar to an industrial production line. With the housing deficit currently seen in underdeveloped countries, alternatives that make the process of designing and building social housing faster are very necessary (Bezerra et al. 2018). In this sense, the increasing adoption of BIM software and the study of ways to integrate this technology with decision-making procedures from operational research represents a very current research agenda.

In the area of operational research, however, there are numerous techniques used to improve project execution strategies. Studies show greater relevance and knowledge about the application of genetic algorithms in the context of planning construction works. It is possible to observe the great potential of this technique in obtaining good results for activity scheduling and in the resource allocation process, being successfully applied in several studies on the search for solutions to optimization problems such as Resource Leveling Problem (RLP) and the Resource-Constrained Project Scheduling Problem (RCPSP) (Amiri et al. 2017; Bettemir and Sonmez 2015; Cheng et al. 2016; Faghihi et al. 2014).

Social housing projects, in general, are repetitive, that is, they are composed of a large number of activities that are carried out several times during execution. Therefore, traditional planning methods such as CPM (Critical Path Method) and Gantt charts are not effective when planning projects with such characteristics. Even after the advent of specific methods for this kind of project, such as Line of Balance, there is still a difficulty in considering uncertainties in the execution and the interaction between the resources involved. Thus, due to their repetitive nature, and also because their size is generally very large, social housing projects demand more complex management techniques (Marzouk et al. 2014).

This article aims to investigate new approaches to decision making in the construction sector that would assist in the planning of housing complexes of social interest. New ways of planning may be opportune for this type of project in order to allow faster and more rational construction, in addition to raising awareness among the agents involved about the importance of pre-operational phases on the project life cycle.

2 The Scheduling Process of Repetitive Projects

Traditionally, network techniques, such as CPM, have been used in the construction industry to schedule and control any type of project. These techniques, however, have great disadvantages when applied to repetitive projects.

The Line of Balance is the most used method in the construction sector for planning repetitive projects. Although it is already widespread, its application still requires a lot of work time. Thus, research on computational optimization methods becomes relevant, since there is a great demand for automation in the industry (Mathew et al. 2016).

Although studies have contributed to the evolution of scheduling techniques, there is not much research that focuses on applying them in the context of social housing. One of the biggest gaps is the need for an approach capable of evaluating multiple objectives, improving organization and costs, while guaranteeing the structural stability of the project (Faghihi et al. 2015; Marzouk et al. 2010).

In repetitive projects, also called linear projects, maintaining the continuity of the crews' work can generate improvements in objectives such as increasing productivity and reducing labor costs, even if this means an increase in the duration of the project. Traditional methods of planning repetitive projects are not very efficient when considering several objectives, since they require a lot of work and are capable of generating only one solution for the problem (Hyari and El-Rayes 2006). In this sense, operational research techniques, such as genetic algorithms, offer the opportunity to promote improvements in the automation of the decision-making process.

In the literature, most of the techniques identified are applied to the process of organizing or sequencing activities, using the Resource-Constrained Project Scheduling Problem (RCPSP) as a mathematical model for programming. Thus, it is necessary to conceptualize it in more detail.

3 The Resource-Constrained Project Scheduling Problem

The RCPSP can be formulated as follows (Kolisch and Hartmann 1999):

> "A single project consists of a set J = {0, 1,..., n, n + 1} of activities which have to be processed. Fictious activities 0 and n + 1 correspond to the 'project start' and to the 'project end', respectively. The activities are inter-related by two kinds of constraints. First, precedence constraints force activity j not to be started before all its immediate predecessor activities comprised in the set PJ have been finished. Second, performing the activities requires resources with limited capacities. We have k resource types, given by the set K = {1,..., k}. While being processed, activity j requires rj, k units of resource type k during every period of its non-preemptable duration pj. Resource type k has a limited capacity of Rk at any point in time. The parameters pj, rj, k and Rk are assumed to be deterministic; for the project start and end activities we have pj = 0 and rj, k = 0. The objective of the RCPSP is to find precedence and resource feasible completion times for all activities such that the makespan of the project is minimized."

Thus, if we call the completion time of activity *j* F_J, the project schedule can be represented by a vector of completion times (F_1, F_2,..., F_n). Let A (t) = $\{j \in J \mid F_J - p_J \leq t < F_J\}$ be the set of activities being processed at instant *t*. It is then possible to formulate the problem mathematically as follows (Kolisch and Hartmann 1999):

$$\text{Minimize}\, F_{n+1} \quad (1)$$

Subject to:

$$F_i \leq F_j - d_j \quad j = 0, 1, \ldots, n+1;\ i \in P_J \quad (2)$$

$$\sum\nolimits_{J \in A(t)} r_{j,k} \leq R_k \quad k \in K;\ t \geq 0 \quad (3)$$

$$F_j \geq 0 \quad j = 1, \ldots, n+1 \quad (4)$$

Equation 1 then represents the objective function of the problem, which consists of minimizing the time to complete the final project activity, thus reducing its duration. The constraints described in Eq. 2 enforce the precedence relationships between two activities. Equation 3, in turn, describes the limitation on the amount of resources used in each instant of time. Finally, Eq. 4 defines the decision variables (Kolisch and Hartmann 1999).

The RCPSP belongs to the class of combinatorial problems, which is characterized by the factorial growth of the computational time required to solve all possible solutions, as the problem size increases. To find solutions to these types of problems, researchers develop heuristic procedures based mainly on evolutionary algorithms (Ghoddousi et al. 2013; Bettemir and Sonmez 2015; Liu et al. 2015; Giran et al. 2017), integer linear programming (Menesi and Hegazy 2014) and network optimization (Potgieter and van Rooyen 2016).

Most of these heuristics use priority rules to schedule activities. They are able to solve resource conflicts by scheduling activities with higher priority, in the shortest time in which there are available resources. These techniques, in conjunction with evolutionary algorithms (Goldbarg et al. 2016), are combinatorial optimization tools widely used in the search for solutions for the RCPSP, what makes it important to conceptualize them more deeply.

4 Genetic Algorithms

Genetic Algorithms (GAs) refer to a class of adaptive search procedures, inspired by the genetic evolutionary principles of populations in nature. The algorithm mimics the course of natural evolution, where populations change over generations according to the principles of natural selection and the survival of the fittest. This means that the genes of highly capable individuals will influence an increasing number of individuals each generation, and their species evolves to become increasingly adapted to their environment (Ichihara 1999).

A GA simulates this process, taking an initial population of individuals and applying artificial genetic operators in each generation. Computationally, each individual is encoded in a string or chromosome, which represents a possible solution to a given problem. The individuals' adaptation is assessed according to a fitness function. Basically, highly fit individuals (better solutions) are given a better chance of reproducing, exchanging parts of genetic information, in a procedure called Crossover. There is also a Mutation operator, which is used to alter some genes on the chromosomes, giving diversity to the population. The offspring, in turn, can replace the entire current population or only individuals with lower fitness values. The cycle of evaluation, selection and reproduction is repeated until a satisfactory solution is found (Hillier and Ieberman 2010; Ichihara 1999). The process can computationally represented by the following pseudocode:

```
Genetic Algorithm
   (1)Create initial population
   (2)While termination condition is not satisfied:
         Calculate chromosome fitness
         Select individuals for crossover
         Apply crossover operator
         Apply mutation operator
         Update population
   (3)Return best solution
```

4.1 Crossover Operators

As highlighted earlier, the evolutionary process described by the genetic algorithm begins with the definition of an initial population. After its creation and the definition fitness function, it is possible to evaluate all individuals and apply a selection mechanism to choose a subset for reproduction. From this stage, genetic operators are implemented, starting with the Crossover operator. At this point, the genetic code of the chromosomes must be mixed so that new chromosomes are formed, called child chromosomes.

4.2 Mutation Operators

The mutation is a stochastic process that consists in altering the genetic material of the chromosomes, in a mechanism that produces new individuals in an "asexual" way. This type of operator can be applied to any chromosome in the population, parents or children, according to the parameters defined for the algorithm. The strategy aims to increase the diversification of the population.

4.3 Selection Operators

An artificial evolutionary process must define criteria that will control the population's reproduction. Initially, reproduction should be an opportunity offered with a greater chance to chromosomes with a higher fitness. On the other hand, if the population size is kept constant, it will be necessary to choose criteria for chromosome removal, which of the current individuals will survive and participate in the next generation. Here, two operators relevant to the present work are conceptualized:

- **Tournament:** the selection of chromosomes occurs through a kind of competition, where the winners will be selected to survive. In the tournament called binary, two chromosomes are chosen at random, advancing in the tournament the one with the higher fitness.
- **Elitism:** chromosomes are ordered and chosen according to their classification. For example, a number of k vacancies will be occupied by the k chromosomes with the higher fitness or through a percentage of them.

5 4D BIM Simulation

The planning process traditionally adopted by construction companies is widely criticized for its inability to take into account spatial and location aspects when activities are sequenced. In addition, production control during execution is often prioritized over pre-operational planning (Gledson and Greenwood 2016).

One of the obstacles faced in the construction industry is the difficulty in correctly visualizing the construction planning in space, generating schedules that are difficult to be interpreted. 4D models, in turn, intimately connect spatial and temporal aspects of the project, improving schedule reliability and minimizing communication problems through the generation of 4D animations, which consist of videos or virtual simulations of the schedule (Biotto et al. 2013; Brito and Ferreira 2015).

The main benefits of using 4D models for planning and control are: visualization of sequencing errors and of the space distribution conditions at the construction site, as well as support in the formulation of workflows and the logistical definition of equipment. The ability to communicate information about progress and discrepancies in relation to what was planned is strategic for project management, as it allows taking corrective actions in a timely manner (Brito and Ferreira 2015).

6 Method

In this sub-section, the proposed approach is presented in three steps: the problem modeling, the optimization module, where the genetic algorithm is implemented, and the simulation module, where the schedule generated by the algorithm is visualized through 4D BIM simulation.

6.1 Problem Modeling

The type of project considered consists of housing complexes composed of a large number of residential units with standardized design. The assumptions that describe the problem come from the Resource-Constrained Project Scheduling Problem (RCPSP) with some adaptations.

Assumptions. The following assumptions were considered:

- A project consists of several activities that can be represented in activity-on-the-node format, in other words, an acyclic graph where the nodes represent the activities and the arcs represent the precedence relations. To the set of activities, two more fictitious activities are added, one at the beginning of the project, being, therefore, direct or indirect predecessor of all the others, and a final activity, representing the completion of the construction.
- Activities are linked through start-finish precedence relationships. Therefore, no activity can be started until all of its predecessors have been completed.
- All known information about the project (precedence, duration, resource requirements) is deterministic (does not change over time).
- Only one mode of execution is considered for each activity.

- All activities are considered to be repetitive, performed obligatorily in all housing units of the project.
- The production rate of the activities is considered constant.
- Preemption, that is, the interruption of an activity after it has been started, is not permitted.
- There is a resource limitation to be considered throughout the project, which is also deterministic.

Project Caracteristics. According to the assumptions presented, the logical structuring of the type of project addressed in this work is obtained by means of an oriented graph, where the nodes correspond to the activities and the arcs represent the interrelationships between them. However, as highlighted in Sect. 2, this type of representation is not the most appropriate when it comes to repetitive projects.

Thus, through an adaptation of the method proposed by Ichihara (1999), the present study will seek to use the oriented graph technique to model the relationships between the activities of a single housing unit, along with the Line of Balance method, used specifically to schedule linear projects, in order to predict more efficiently the start-finish relationships between the housing units.

The Line of Balance method adopted here will take into account scheduling based on a maximum duration for the project (D_T). Thus, the execution time, which corresponds to the maximum time for the conclusion of all repetitive activities, can be calculated as follows (Mendes 1999):

$$T_R = D_T - (T_B + T_{MOB}) \quad (5)$$

Where:

T_R = Execution time.
D_T = Desired duration of the project.
T_{MOB} = Time to conclude the preliminary services in the site.

Then, the desired production rate can be calculated according to Eq. 6 (Mendes 1999):

$$R = \frac{T_R}{(N-1)} \quad (6)$$

Where:

R = Desired production rate.
N = Number of housing units.

With the calculated production rate, it is then possible to measure the of number of crews required for each activity (Eq. 7) (Mendes 1999). This calculation results, in general, in a decimal number, which must be rounded to the nearest whole number (Limmer 1997).

$$N_E = \frac{D_U}{R} \tag{7}$$

Where:

N_E = Number of crews.
D_U = Duration required for the activity completion in one housing unit.

With the approximation made for the crew size, it is then possible to calculate the specific production of each activity, by dividing the task duration by the corrected number of crews (Limmer 1997).

6.2 Optimization Module

The optimization module consists of the implementation of the genetic algorithm, composed by several steps, in the following order: 1. Generation of the initial population; 2. Assessment of the fitness of individuals; 3. Selection of individuals; 4. Application of the crossover operator; and 5. Application of the mutation operator (Fig. 1). The algorithms were implemented using the Python 2.7 programming language.

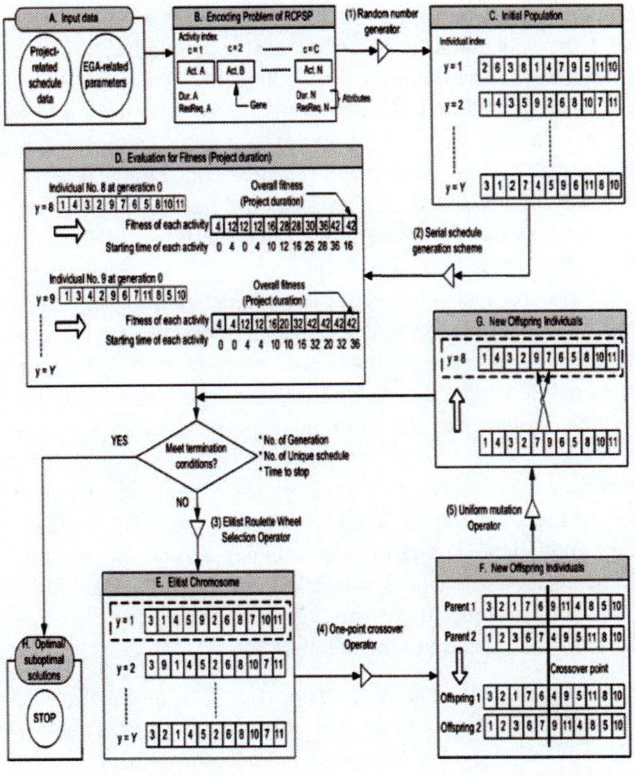

Fig. 1. Genetic algorithm operation (Kim and Ellis 2008)

In the next subsections, the implemented parameters are specified, as well as their operating mechanisms.

Solution Encoding and Initial Population. The algorithm proposed in the present work adopts a permutation-based encoding, characterized by an Activity List $\lambda = (j_1, j_2, j_3, \ldots, j_n)$, suggested in the literature as a relevant representation for the solutions in the RCPSP (Kolisch and Hartmann 1999; Kim and Ellis 2008).

Therefore, the solutions are represented here as a chromosome that corresponds to a sequence of activities. Each gene on the chromosome, in turn, is characterized by a number that represents an activity. A task has a lower priority than all of its predecessors and a higher priority than its successors. In this way, the precedence relationships are respected, as a given activity cannot appear on the list after the position where one of its successors is located (Kim and Ellis 2008).

Fitness Function. The objective function for the RCPSP, as evidenced by Eq. 1, is to minimize the duration of the project. Therefore, an individual's fitness is represented by the date of completion of the last project activity. Thus, starting from a list of activities of a project and knowing their durations and precedence relationships, it is possible to determine the end dates of each activity through a heuristic programming procedure based on priority rules. In the present approach, the Serial Method is adopted.

The serial method is done in stages. The number of stages is the same as the number of activities of the project. Associated with each stage, two sets are built: the decision set and the scheduled set. The already scheduled activities are part of the second set, being part of a partial project schedule. The decision set contains activities that have not yet been scheduled. An activity is selected according to the order of the activity list (representation of the individual) at each stage, and scheduled for the earliest date that meets the precedence and resource constraints. When all the activities of an individual are already scheduled, its general fitness is obtained by the maximum value between the end date of the last activity and the fitness value of the activity just before the last activity (Kim and Ellis 2008).

However, the above description applies to projects whose logical structure can be organized by PERT/CPM networks. As the present study deals with repetitive projects, it becomes necessary to take into account more appropriate scheduling methods. Thus, it is important to define new mathematical relationships to calculate the start dates of repetitive activities, which take into account their production rates, as evidenced in Subsect. 6.1.

For this, it is necessary to present the concept of critical unit (Ichihara 1999), which consists of the unit that needs to be verified to schedule activity j. When analyzing this unit, it becomes redundant to analyze the others. In other words, the start date of activity j is the first date that makes it possible to preserve the start-finish synchronism in the critical unit. Considering the construction of housing complexes, it is possible to define two cases for the calculation of the start date:

Case 1. The production rate of activity j is greater than or equal to that of activity i. The critical unit, in this case, is the last unit of activity i, as shown in Fig. 2. The start date of activity j can be calculated according to Eq. 8.

$$\text{Start}_j = \text{Finish}_i - \frac{(\text{UF}_j - \text{UI}_j)}{\text{R}_j} + 1 \quad (8)$$

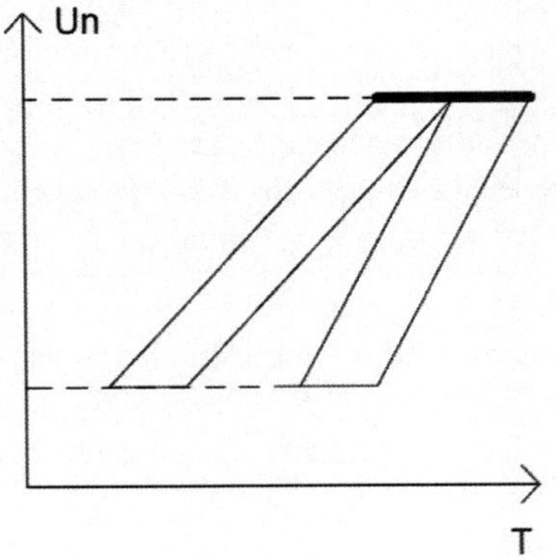

Fig. 2. Critical unit analysis for Case 1.

Where:

Start_j = Activity start date j.
Final_i = Date of completion of activity i.
UI_j = Housing unit where activity j is started.
UF_j = Housing unit where activity j is completed.
R_j = Activity *j* production rate.

Case 2. The production rate of activity *j* is less than that of activity *i*.
The critical unit, in this case, is the first unit of activity i, as seen in Fig. 3. The start date of activity j can then be calculated using Eq. 9.

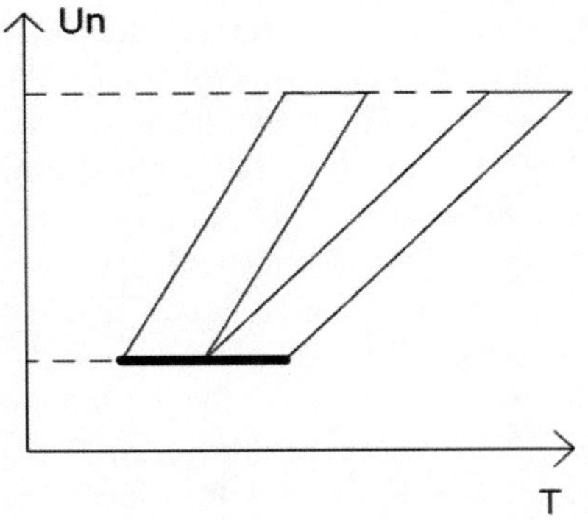

Fig. 3. Critical unit analysis for Case 2.

$$\text{In}\acute{i}textcio_j = \text{Final}_i - \frac{(\text{UF}_i - \text{UI}_i)}{\text{R}_i} - \frac{(\text{UI}_i - \text{UI}_j)}{\text{R}_j} + 1 \quad (9)$$

Selection Operators. In the implemented algorithm, first a selection by Elitism is used in the initial population, choosing a percentage of individuals with higher fitness values, which will necessarily pass to the next generation (Kim and Ellis 2008). Then, to select the individuals for crossover, the strategy of selection by tournament is adopted, where two pairs of individuals are formed, and each pair will then "compete" based on their fitness value, being chosen the best chromosomes from each pair. Then, these individuals will exchange genetic information through the crossover operator, generating new offspring.

Crossover Operator. The crossover operator implemented in the proposed algorithm is the One Point operator. The choice of this operator is due to the fact that it preserves parts of the "parent" chromosomes, maintaining the precedence relations (Hartmann 1997).

Mutation Operator. The Inversion mutation operator is implemented in the proposed algorithm. The operator swaps positions of adjacent elements. In order to do so, before performing the operation, the operator examines which elements in the list can be swapped with the ones right next to them without violating the precedence relationships. Then, one of these elements is chosen at random for the operation to be performed (Klimek 2010).

Termination Conditions. Three termination conditions are considered: maximum number of generations, processing time limit and number of unique individuals. It is possible that several individuals, by the end of a given iteration, have the same fitness value, even if the start dates of the activities are completely different. Thus, these unique individuals, in the search space, can be related to the same schedule, which represents the best solution obtained for the RCPSP (Kim and Ellis 2008).

6.3 Simulation Module

In the simulation module, the construction execution can be viewed virtually through the algorithm-generated schedules. This step is important because the solutions determined by the algorithm are abstract and do not take into account spatial or constructive conditions. Thus, with the 4D simulation, it is be possible to assess whether the schedules can be followed in the way they were created by the algorithm or if additional modifications are necessary.

The *Python 2.7* programming language offers the possibility to import data to .csv files. This information can then be directly imported into the 4D BIM simulation software Autodesk Navisworks, which already comes with the option to read and interpret data from files in .csv format, without the need to install plug-ins or save data extracted from the optimization module in some other format. The project, modeled with the BIM software Autodesk Revit, will then be connected to the schedule to perform the simulations.

7 Results

The results of the experiments carried out with the implemented algorithm and the application of the simulation module earlier described are presented below.

7.1 Computational Experiments

In this section, the results of the computational experiments with the implemented algorithm will be presented. The tests were done on project instances taken from PSPLIB (Kolisch and Sprecher 1996). 15 instances with 30 activities, 15 with 60 activities and 15 with 90 activities were randomly chosen for the tests. For instances of 30 activities, the optimal solutions are known and provided in PSPLIB. For instances with a greater number of activities, there are only lower bound values, which are the best values found for the instances so far. For each of the 45 instances, ten rounds were carried out, registering the best solution found and the computational time spent, totaling 450 experiments. The following configuration was used for the algorithm parameters:

- Population size: 50 individuals.
- Crossover rate: 0.5.
- Elitism size: 2 individuals.
- Maximum number of generations: 100.
- Number of unique individuals: 25.
- Termination time: 4200 s.

The mutation rate if calculated by the inverse of the individual's size, that is, if the individual is composed of 30 activities, the mutation rate will be 1/30 (0.033) (Bäck et al. 2018). The other parameters were defined according to Kim and Ellis (2008), due to the good results presented by the authors.

The algorithm and the genetic operators were programmed using the *Python 2.7* programming language and the experiments were performed on an Intel Core i5 computer, 1.7 GHz, with 8 GB of RAM (Fig. 4).

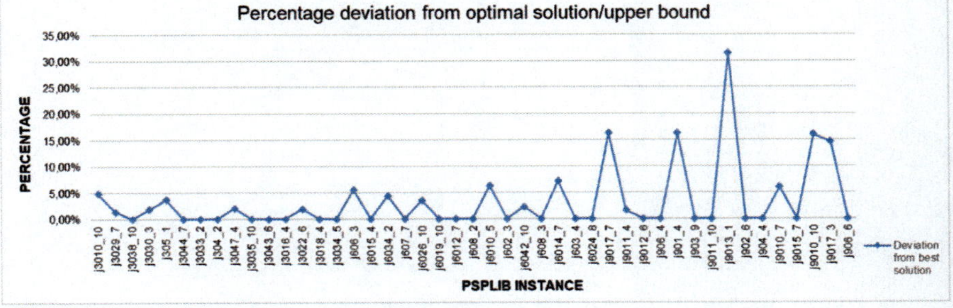

Fig. 4. Percentage deviation between the best solution found by GA and the optimal/upper bound solutions given in PSPLIB.

It was observed with the experiments that the algorithm is capable of generating better solutions over the generations. For instances of 30 activities, as expected, solutions close to optimal were found more easily, with a worse performance for instances of 90 activities, as expected. In Fig. 5, it is possible to observe that as the size of the project increases, so does the difficulty of finding good solutions and the necessary computational time.

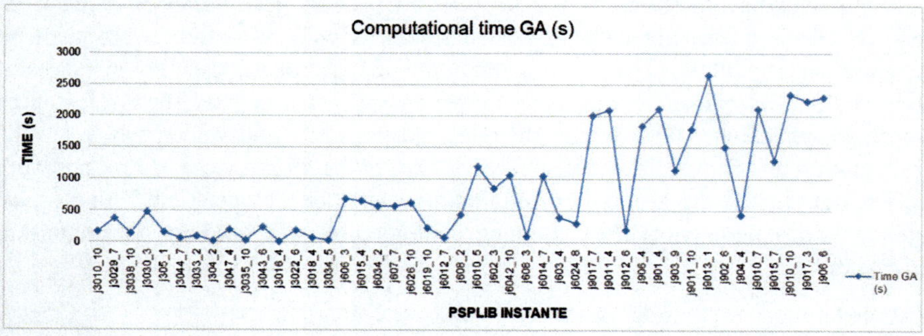

Fig. 5. Time required to find best solutions.

Three groups of three instances were also chosen, each with a number of activities, to verify the behavior of the algorithm until the last generation, in order to better evaluate both its efficiency and the ability to improve solutions over the generations. For this stage, instances were chosen based on their slower convergence behavior.

Figures 6, 7 and 8 show the behavior of instances j3029_7, j3030_3, j3047_4, j606_3, j6026_10, j6010_5, j9011_4, j906_4 and j9011_10. For all instances, it was possible to observe the ability of the algorithm to improve solutions over the generations, which is mainly due to the choice of selection by elitism, that prevents the best solutions found from being discarded.

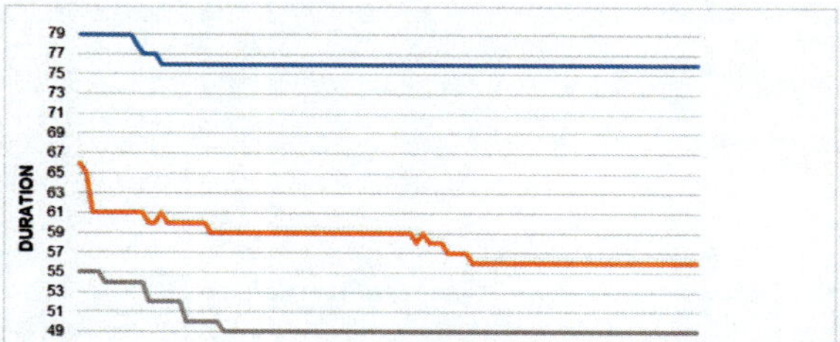

Fig. 6. Convergence behavior – 30 activities.

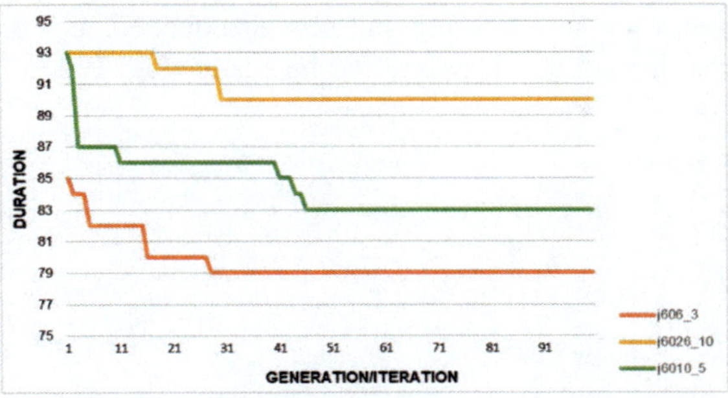

Fig. 7. Convergence behavior – 60 activities.

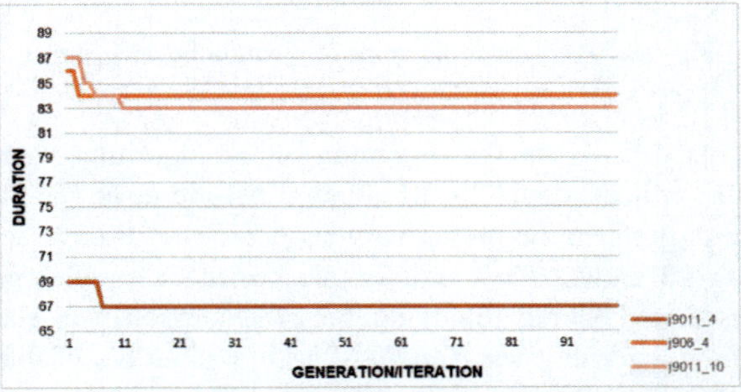

Fig. 8. Convergence behavior – 90 activities.

8 Simulation Module Application Example

As a way of exemplifying how the Line of Balance calculations presented previously can be applied in the optimization process, a simple standard project from the Brazilian government's housing programme "Minha Casa Minha Vida" was used. Altogether, 100 housing units were modelled for the simulation (Fig. 9).

Fig. 9. Housing complex project used for simulation.

A precedence network for a housing unit was also defined, as it would be done if the Line of Balance method was been applied manually (Fig. 10).

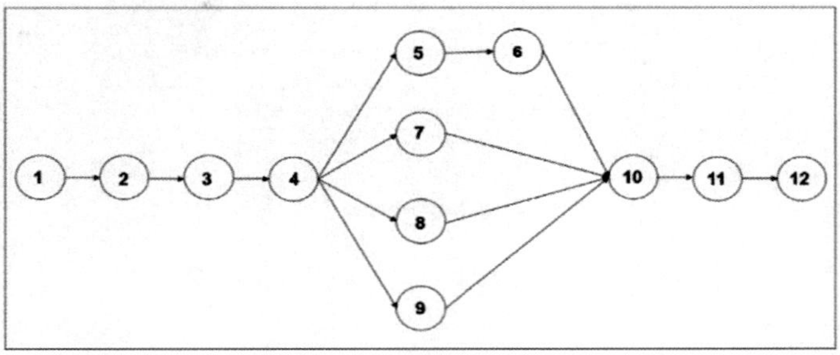

Fig. 10. Precedence network of a single housing unit.

Other input data defined for the execution of the algorithm were the maximum duration Dt of the project, equal to 30 months, or 660 days (22 days worked per month). The time to perform the preliminary services was 10 days and resource constraints was set to 20 crews. As it is a simple project, a smaller population of 10 individuals was used, and the maximum number of generations was also reduced to 20. The execution of the algorithm was very fast, lasting about 6 s (Table 1).

Table 1. Activity network for the second project.

Activity Number	Activity Name	Duration	Successors
1	Start	0	2
2	Excavation	2	3
3	Foundations	1	4
4	Masonry	4	5, 7, 8, 9
5	Flooring	1	6
6	Doors	1	10
7	Windows	1	10
8	Ceiling – Service area	1	10
9	Roof	2	10
10	Painting	2	11
11	Final services	1	12
12	Finish	0	

When scheduling the activities, the algorithm considers the duration to complete each activity in all housing units. Thus, if in one of the activities the resource limitation is exceeded, all similar ones will be reprogrammed, as well as their successors. The number of teams is also determined by the algorithm, according to the calculations presented previously.

The files in.csv format, generated by the algorithm, can be directly imported to *Autodesk Navisworks* (Fig. 12), generating a Gantt chart (Fig. 13). Thus, it is possible to affirm that the implemented process can significantly reduce the working time when scheduling a project with repetitive units, since it is able to determine start and end dates for all the activities of the project, following the Line of Balance logic, maintaining the continuity of the work and, at the same time, leveling the resources to be used during the project (Fig. 11).

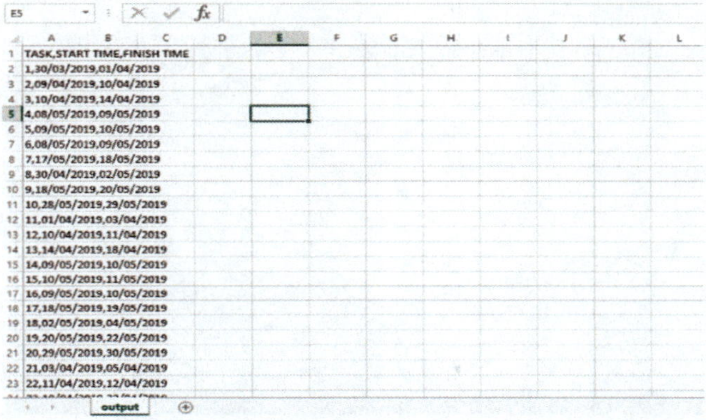

Fig. 11. .csv file.

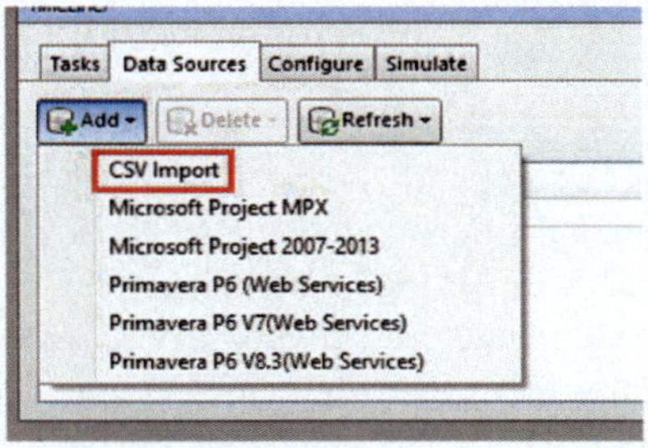

Fig. 12. Importing file to *Autodesk Navisworks.*

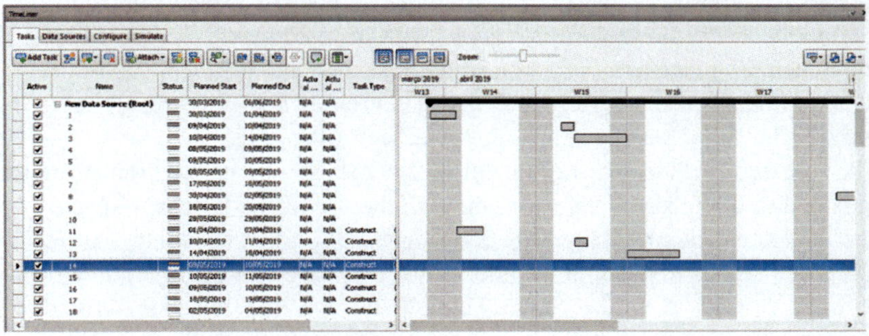

Fig. 13. Gantt chart in *Autodesk Navisworks.*

The simulation, as seen in Fig. 14, could show that the generated schedule respects the precedence relations between the activities and also demonstrates that the line of balance logic was well implemented in the schedule, since the execution of activities in a housing unit does not depend on the total completion of activities in the units that started to be built before.

Fig. 14. Simulation of the housing complex.

9 Conclusion

This article intended to address the functioning and relevance of applying a meta-heuristic procedure coupled with BIM 4D simulation when planning housing projects. The focus on housing developments was defined by the lack of studies, both national and international, that apply these techniques to horizontal projects, which in most cases are the solution found by governments in developing countries to promote housing policies.

Activity scheduling is the most discussed topic when it comes to the application of meta-heuristic procedures in civil construction. The mathematical modeling for this activity, in the vast majority of publications, was inspired by the Resource-Constrained Project Scheduling Problem (RCPSP), which suited well to the theme addressed in this work, since housing projects of social interest constantly deal with scarcity of resources and need to be delivered as soon as possible.

The research method was developed in three stages. First, the problem modeling was adapted to the focus of the work, through the use of the Line of Balance method. For the implementation of the genetic algorithm, the solution encoding was defined, as well as the fitness assessment procedure, which was based on a programming heuristic based on a priority rule, the Serial Method. Finally, the selection, crossover and mutation operators and the stopping criteria were also determined.

After the implementation of the algorithm, numerous experiments were carried out with project instances available in the database for RSPSP, the PSPLIB. Both the best and the worst solutions found were analyzed, as well as the time used by the algorithm to find them. Analyzing the behavior of the algorithms over the generations, it was concluded that the use of selection by Elitism is crucial so that good solutions are not lost during the search process.

After the experimental phase, the simulation module was performed. In this phase, the execution of a housing complex of 100 houses was simulated in the Autodesk Navisworks software, with a schedule generated by GA. First, the precedence relations for the execution of a housing unit were defined, the duration of the activities was determined and a resource constraint and a maximum duration for the project were set. The algorithm used this data to generate a schedule that included all 1000 activities necessary to complete the construction. The result of the optimization could then be read by Autodesk Navisworks directly, and the activities were connected to the previously designed BIM model.

The system created is relevant mainly as a way of reducing the human effort used when scheduling repetitive projects, since, for simple projects such as social housing units, in a few seconds it is possible to obtain a schedule that considers both work continuity and resource limitations. Simulation comes here as another tool to assist in decision making, since several possibilities generated by the algorithm can be tested and even modified through the thousands of functionalities in the BIM-based construction management software. The present paper also seeks to draw attention to the need for greater use this kind of software, since they are developed specifically for the construction market and allow a much wider visualization of the project execution, different from other commercial programs such as *Microsoft Project*, *Primareva P6* or even *Microsoft Excel.*

In this way, it is possible to consider the objectives of the work were achieved, since it was possible to implement the genetic algorithm, test its efficiency and also perform a 4D simulations of the optimized schedules without any compatibility issue. The system can, however, still be improved, since for projects with large numbers of activities the computational effort required proved to be very high. In addition, issues other than duration could still be included in the implementation, which could serve as a trigger for the development of future research related to simulation, optimization and Operational Research in the construction sector.

References

Amiri, R., Sardroud, J.M., de Soto, B.G.: BIM-based applications of metaheuristic algorithms to support the decision-making process: uses in the planning of construction site layout. Procedia Eng. **196**, 558–564 (2017)

Bäck, T., Fogel, D.B., Michalewicz, Z. (eds.): Evolutionary Computation 1: Basic Algorithms and Operators. CRC Press, Boca Raton (2018)

Bettemir, Ö.H., Sonmez, R.: Hybrid genetic algorithm with simulated annealing for resource-constrained project scheduling. J. Manag. Eng. **31**, 04014082 (2015)

Bezerra, P.H.P., Scheer, S.: Uso de *Building Information Modeling* para o aprimoramento e automatização do planejamento de obras: revisão sistemática da literatura. In: Encontro Nacional De Tecnologia Do Ambiente Construído. ANTAC, Porto Alegre (2018)
Bezerra, P.H.P., de Paula Lacerda Santos, A., Scheer, S.: BIM no planejamento de empreendimentos com sistemas de painéis pré-fabricados: uma alternativa para obras de habitação social. PARC Pesquisa em Arquitetura e Construção (2018)
Biotto, C.N., Formoso, C.T., Isatto, E.L.: Método para o uso da Modelagem BIM 4D na Gestão da Produção em Empreendimentos de Construção. In: Simpósio Brasileiro de Qualidade do Projeto No Ambiente Construído; Encontro de Tecnologia de Informação E Comunicação Na Construção, Campinas. ANTAC, Porto Alegre (2013)
de Brito, D.M., de Andrade Marques Ferreira, E.: Avaliação de estratégias para representação e análise do planejamento e controle de obras utilizando modelos BIM 4D. Ambiente Construído, Porto Alegre **15**, 203–223 (2015)
Cheng, M.-Y., Prayogo, D., Tran, D.-H.: Optimizing multiple-resources leveling in multiple projects using discrete symbiotic organisms search. J. Comput. Civil Eng. **30**, 04015036 (2016)
Faghihi, V., Reinschmidt, K.F., Kang, J.H.: Construction scheduling using genetic algorithm based on building information model. Expert Syst. Appl. **41**, 7565–7578 (2014)
Faghihi, V., et al.: Automation in construction scheduling: A review of the literature. Int. J. Adv. Manufact. Technol. **81**, 1845–1856 (2015)
Ghoddousi, P., et al.: Multi-mode resource-constrained discrete time–cost-resource optimization in project scheduling using non-dominated sorting genetic algorithm. Autom. Constr. **30**, 216–227 (2013)
Giran, O., Temur, R., Bekdaş, G.: Resource constrained project scheduling by harmony search algorithm. KSCE J. Civil Eng. **21**, 479–487 (2017)
Gledson, B.J., Greenwood, D.J.: Surveying the extent and use of 4D BIM in the UK. J. Inf. Technol. Constr. (ITcon) **21**, 57–71 (2016)
Goldbarg, M.C., Goldbarg, E.G., Luna, H.P.L.: Otimização combinatória e metaheurísticas, 1st edn. Elsevier, Rio de Janeiro (2016)
Hillier, F.S., Lieberman, G.J.: Introdução à pesquisa operacional. AMGH, Porto Alegre (2010)
Hyari, K., El-Rayes, K.: Optimal planning and scheduling for repetitive construction projects. J. Manage. Eng. **22**, 11–19 (2006)
Ichihara, J.A.: Um método de solução heurístico para a programação de edifícios dotados de múltiplos pavimentos-tipo. Universidade Federal de Santa Catarina, Florianópolis (1999)
Kim, J.-L., Ellis Jr., R.D.: Permutation-based elitist genetic algorithm for optimization of large-sized resource-constrained project scheduling. J. Constr. Eng. Manage. **134**, 904–913 (2008)
Klimek, M.: A genetic algorithm for the project scheduling with the resource constraints. Annales Universitatis Mariae Curie-Sklodowska, Sectio AI-Informatica **10**, 117–130 (2010)
Kolisch, R., Hartmann, S.: Heuristic algorithms for the resource-constrained project scheduling problem: Classification and computational analysis. In: Węglarz, J. (ed.) Project Scheduling. International Series in Operations Research & Management Science, vol. 14, pp. 147–178. Springer, Boston (1999)
Kolisch, R., Sprecher, A.: PSPLIB – a project scheduling problem library. Eur. J. Oper. Res. **96**, 205–216 (1996)
Limmer, C.V.: Planejamento, orçamentação e controle de projetos e obras. LTC, Rio de Janeiro (1997)
Liu, H., Al-Hussein, M., Lu, M.: BIM-based integrated approach for detailed construction scheduling under resource constraints. Autom. Constr. **53**, 29–43 (2015)
Marzouk, M., et al.: A simulation optimisation tool for planning of low-income housing projects. Civ. Eng. Environ. Syst. **31**, 51–63 (2014)

Marzouk, M., et al.: An optimization algorithm for simulation-based planning of low-income housing projects. J. Adv. Res. **1**, 291–300 (2010)
Mathew, J., et al.: Multi objective optimization for scheduling repetitive projects using GA. Procedia Technol. **25**, 1072–1079 (2016)
Mendes Jr., R.: Programação da produção na construção de edifícios de múltiplos pavimentos. Universidade Federal de Santa Catarina, Florianópolis (1999)
Potgieter, I.J., van Rooyen, G.C.: Maximizing slack in resource-constrained schedules: a heuristic approach. J. Comput. Civ. Eng. **30**, 04016018 (2016)

Strategic Planning of Work and the Use of 4D BIM for Multiple Floor Buildings

Luiz Reynaldo de Azevedo Cardoso[1(✉)], Thalyta de Miranda Lanna Rios[1], Tiely Zurlo Mognhol[2], and Alberto Vinicius Marostica[3]

[1] Department of Civil Construction Engineering, Polytechnic School, University of São Paulo – USP, São Paulo, SP, Brazil
{luizcardoso, thalyta.rios}@usp.br
[2] Master in Civil and Urban Construction Engineering, University of São Paulo – USP, São Paulo, Brazil
tielyzm@usp.br
[3] Mauá Institute of Technology, São Caetano do Sul, SP, Brazil
vinicius.alberto@maua.br

Abstract. The use of 4D BIM has been spreading, with significant scientific production in the area. The advantages of using it are notorious and well documented, highlighting the visualization of the work execution, increasing the comprehension and reliability of the planning. However, the need for 4D BIM support in the correct use of usual planning methods and techniques has been little emphasized and demonstrated in the literature. The study of 4D BIM applications in strategic construction planning has also been poorly documented, with simpler modelling and without the need for significant changes in the usual hiring processes for projects. This article aims to: demonstrate that the use of 4D BIM requires a consistent and reliable planning basis; to propose, based on the context of the Brazilian residential real estate market, a 4D BIM modelling method to support strategic construction planning decisions. In order to meet these objectives, a literature review was made, as well as basic concepts and the presentation of a case study. The case study is the planning of executing a residential complex with three towers of multiple floors, typical of real estate developments in the city of São Paulo. Two plans of attack regarding the execution of the work are simulated and visualized. The results are analysed and discussed based on the literature and conceptual considerations presented. The conclusion was by the feasibility and utility of applying the 4D BIM, considering the adopted assumptions, method and tools used. The main difficulty is the modelling time. This stems not only from the project as a whole, but also from the increase resulting from the modelling of complementary activities, such as wood forms, scaffolding, protection screens, and others. Limitations of the application used were identified. Even so, the expenditure of work is compatible with the final quality of the visualization and the benefits provided to planning and decision making. The method could be used without the need for significant changes in the usual project and planning contracting processes in the real estate market in Brazil. It was confirmed that 4D BIM requires the correct use of the usual planning techniques and tools However, it requires training the planner in 3D modelling and 4D BIM. Recommendations regarding the improvement of the efficiency and quality of modelling are made as well as regarding further studies in the area.

E. Toledo Santos and S. Scheer (Eds.): ICCCBE 2020, LNCE 98, pp. 594–612, 2021.
https://doi.org/10.1007/978-3-030-51295-8_42

1 Introduction

The use of BIM has been widespread, as well as scientific production in the area. Regarding 4D BIM, the literature in general emphasizes its advantages but not the fact that its use depends on knowledge and techniques and usual tools for planning works, and that most of the applications used in 4D BIM do not have the intelligence to plan.. This aspect is not omitted in literature, as it can be seen from works such as the ones of Biotto et al. [1], Vargas and Formoso [2], Brito and Ferreira [3], besides renowned authors such as Eastman et al. [4]. However, the little emphasis with which it is treated, generally, even with important exceptions, can contribute to induce inappropriate uses, on the assumption that 4D BIM solves problems that cannot solve.

Toledo [5] points out several causes of failures using BIM, among them are incorrect diagnosis of the needs of a company and, in consequence, of the processes and BIM tools to be used. He also highlights that there are several possible uses of BIM and that each company must try to identify which one is the most indicated to its needs, considering its processes, organizational profile, technologic culture, type of enterprise that it executes, partners and others. Among those is added the correct diagnosis of problems and planning deficiencies.

Researches about the causes of work delays [6] and time and cost deviations [7], show that the main causes of those problems are related to lacks of planning and managing (flawed and ineffective planning, programming failures and acquisition of materials, lack of communication, coordination and oversight of teams, quality problems and others). It is necessary to deepen those diagnoses because it is possible that those lacks are due to management and planning fails and not due to the usual methods and techniques, but to the bad use of them [8], which is not solved using the 4D BIM. On the contrary, 4D BIM demands consistent planning, made with the usual methods and techniques, as it will be demonstrated in this paper.

On the other hand, the possibility of a mistaken view of the use of 4D BIM can lead to the loss of the opportunity to use the real benefits that its use can provide to the planning and performance of works and companies. One of the main benefits, not the only one, is the visualization of the construction execution. Eastman et al. [4] points out the limitation of traditional methods in capturing the spatial and logistical impacts of activities and the difficulty that entrepreneurs have in understanding their programming, visualized in Gantt diagrams, which is the most common representation. Brazilian researchers already mentioned also point out the lack of visualization and abstraction of the representations of the schedules made with traditional techniques, such as precedence networks and balance lines. In this regard, the use of 4D BIM has had successful experiences, both internationally and nationally, as it has been reported, and will be demonstrated in this article.

The application of 4D BIM must also consider aspects of an organizational nature and the context in which the projects are carried out, particularly in the real estate market in Brazil.

In this market, projects are of long term, in high-risk and uncertain scenarios, sequenced implantation stages, with many and several agents acting in different moments of the process [9]. The predominant mode of contracting is Design-Bid-Built. This makes it difficult and may even prevent the integration of designers, planners, builders and suppliers in the construction of unique digital models of design, planning and construction, as recommended by the BIM methodology.

Most construction companies operating in this market are small and medium-sized, which occurs even in developed countries [10]. For them, the introduction of more radical changes in the processes of development, in the sense of significantly increasing the integration between the agents, as generally proposed by BIM approaches, tend to encounter resistance and implementation difficulties. Even in an international context, this aspect is also pointed out, as in Tsai et al. [11].

For these companies, which make up the majority of the market, the use of 4D BIM can be done without significant changes in their processes, especially in the initial stages of the project, in which strategic decisions related to the execution of the work are made, with greater responsibility. Strategic decision makers (developers and investors) are not usually familiar with concepts and techniques of usual schedule representations and may have difficulty understanding planning. The visualization of the construction and different construction execution strategies, provided by 4D BIM, becomes a powerful tool to increase understanding, reliability in planning and the quality of decisions to be made. For construction companies that compete in the construction market and participate in bids, the submission of proposals using this resource can become an important competitive factor [5]. In addition to being useful, the use of 4D BIM in this context is feasible, as the modelling in this stage can be simplified, since it basically requires the critical sequence of the work (relatively small number of activities, with simplified modelling), and not the complete and detailed one [4]. The planner themselves can do it, without the need to change the usual contracting and project preparation, planning and construction processes.

It is recognized that the traditional business process is very fragmented, little integrated and that it is necessary to overcome these deficiencies over time. The use of BIM can drive this integration, and the more integrated the process, the greater the benefits of its use. However, it is understood that the use of 4D BIM techniques and tools should not be disregarded in a more timely manner, which can provide a significant improvement in the usual planning processes.

This article aims to: demonstrate that the use of 4D BIM requires a consistent and reliable planning basis; to propose, based on the context of the Brazilian residential real estate market, a 4D BIM modelling method to support strategic construction planning decisions.

To meet these objectives, a literature review, a basic conceptualization and the presentation of a case study were carried out. The project object of the case study consists of a residential complex with 3 multi-storey towers. Two plans of attack are simulated in 4D BIM according to possible scenarios for the execution of the work.

The project and the scenarios were built based on real situations and are very representative of the Brazilian context and the real estate market in the city of São Paulo.

The intention is to contribute to the knowledge and discussion about the theme in the academic scope and also to subsidize professionals and companies regarding the use of the 4D BIM that are acting in this market.

The paper is structured in 5 topics: introduction, literature, concepts, presentation of case study and conclusions.

2 Literature

2.1 Sample

Searches were carried out in the main international databases Web of Science and Scopus, using the keywords "BIM", "building" and "plan". The search resulted in a total of 1,059 articles, 468 in the Web of Science database and 591 in the Scopus database. After repetition analysis, 301 articles were excluded from the sample and, after analysing the title and abstract, 677 articles were excluded, which had no direct interface with the theme of this work. The final sample, totalling 81 articles, specifically addresses the theme of 4D BIM applied to building works.

Based on this sample, we started to analyse the content of the most relevant articles, taking as parameter the Article Impact Index (AIF), which results from the product of the average annual citation and the journal impact index (JCRIF), according to Formula (1), below [12, 13]. There were 15 articles analysed, which correspond to the number shown in Table 1.

$$A\_IF = \text{Average annual citation} * (JCR\_IF + 1) \quad (1)$$

To these, 7 more articles were added for analysis. The number 16 is international and was chosen due to the proximity of the theme and approach to this article. The rest, totalling 6 (17 to 22 in Table 1), are by Brazilian authors, published in the *Ambiente Construído* Magazine, the most important Brazilian journal in the area.

These articles were selected because they were considered important for the purposes of this article to add to the analysis relevant works that show the vision of Brazilian authors and the reality of the country.

The production set is recent. Of the total of 22 analysed articles, 15 were published from 2015, and of these, 11 (half of the sample) between 2017 and 2020. The oldest are from 2014 (3 articles), 2013 (1 article) and 2010 (2 articles).

Among the countries, Brazil is the one with the largest number of articles in the sample (7 in total, 1 among the 15 most cited), followed by the USA (5 articles, all among the most cited). Then South Korea, Taiwan and Germany with 2 articles, finally China, United Kingdom, Slovenia and Canada with 1 article.

Table 1. Analysed papers. Source: authors

Number	Refrence	Authors	Title	AIF	Country
1	[14]	Pishdad-Bozorgi et al., 2018	Planning and developing facility management-enabled building information model (FM-enabled BIM)	106,26	EUA
2	[15]	Kropp et al., 2018	Interior construction state recognition with 4D BIM registered image sequences	79,70	China
3	[16]	Kim et al., 2013	Automated construction progress measurement using a 4D building information model and 3D data	77,92	Coréia do Sul
4	[17]	Sacks et al., 2010	Requirements tor building information modeling based lean production management systems for construction	59,62	Reino Unido
5	[18]	Golparvar-Fard et al., 2015	Automated Progress Monitoring Using Unordered Daily Construction Photographs and IFC-Based Building Information Models	56,86	EUA
6	[19]	Golparvar-Fard et al., 2011	Integrated sequential as-built and as-planned representation with D 4AR tools in support of decision-making tasks in the AEC/FM industry	42,65	EUA
7	[20]	Son et al., 2017	Automated Schedule Updates Using As-Buüt Data and a 4D Building Information Model	36,29	Coréia do Sul
8	[21]	Wang et al., 2014	Integrating building information models with construction process simulations for project scheduling support	36,13	Taiwan
9	[22]	Faghihi et al., 2014	Construction scheduling using Genetic Algorithm based on Building Information Model	34,93	EUA
10	[23]	Rebolj et al., 2017	Point cloud quality requirements for Scan-vs-BIM based automated construction progress monitoring	34,53	Eslovenia
11	[24]	Liu et al., 2018	BIM-based automated design and planning for boarding of light-frame residential buildings	31,88	Canadá
12	[25]	Sigalov; Konig, 2017	Recognition of process patterns for BIM-based construction schedules	21,47	Alemanha
13	[26]	Schwabe et al., 2019	Applying rule-based model-checking to construction site layout planning tasks	21,25	Alemanha
14	[27]	Bortolini et al., 2015	Site logistics planning and control for engineer-to- order prefabricated building systems using BIM 4D modeling	21,25	Brasil

(*continued*)

Table 1. *(continued)*

Number	Refrence	Authors	Title	AIF	Country
15	[28]	Choi et al., 2014	Framework for Work-Space Planning Using Four-Dimensional BIM in Construction Projects	20,61	EUA
16	[11]	Tsai et al., 2010	^eriences using building information modeling for a construction project		Taiwan
17	[1]	Biotto et al., 2015	The use of 4D modeling and Building Information Modeling for the management of production systems in construction projects		Brasil
18	[2]	Vargas; Formoso, 2020	Method for location-based production planning and control supported by BIM		Brasil
19	[29]	Manenti et al., 2020	BIM ^cution Plan: Proposal of guidelines for contractors and designers		Brasil
20	[30]	Rodrigues et al., 2018	A proposal to integrate the BIM model with the Last Planner system		Brasil
21	[31]	Bataglin et al., 2018	4D BIM applied to logistics management: implementation in the assembly of Engineer-to-order prefabricated concrete systems		Brasil
22	[3]	Brito; Ferreira, 2015	Evaluation of strategies for representation and analysis of construction planning and control using 4D BIM models		Brasil

2.2 Analysis

In view of the objectives and approach of this article, presented in the introduction, the content of the articles were analysed in terms of the following aspects: themes, context, diagnosis and objectives, types of works, methods and tools, results, difficulties and recommendations.

A summary of each of the aspects analysed will be presented.

2.2.1 Themes

All articles cover 4D BIM for building works, that is, the time area of project management. Thus, the themes were classified according to the processes: planning and control.

Initially, the high number of articles that focus on control draws attention, that is, the monitoring of the execution of the work and its comparison with the forecast. This topic is covered in 8 articles, or 36% of the total. On the other hand, Planning, which is carried out before the execution of the work - object of this article - is the focus of most articles (12 out of 22, or 55% of the total). Two articles have themes related to Planning, but with specific and distant focuses from the object of this article: FM (Facilities Management) with BIM (article 1) and contracts aimed at the use of BIM (article 19).

2.2.2 Context, Diagnosis and Objectives

The articles have varied contexts and objectives, sometimes more than one in the same article.

In general, little emphasis is given and information related to the more general context, that is, the organizational environment, the development and even the work itself. The motivation and objectives of the work are almost always focused on overcoming the technical barriers caused by the usual processes, which hinder the improvement of the planning and control performance. The exceptions are article 16, which proposes a method considering the context of project management (also approach to this article) and article 19, which focuses on the drafting of contracts, which is not related to the purpose of this article.

It is also noted that no article diagnoses failures in the use of traditional tools. The deficiencies stem from the methods and tools and not from their misuse.

Most articles (6 in total) are diagnosed and motivated by the need to automate, visualize and improve the precision and quality of works control, since the usual methods of control present one or more of the following problems: manual, slow, expensive, subjective and ineffective. There are two more articles related to this, which aim to increase the automation of Planning.

The second biggest problem diagnosed is the need to visualize the execution of the work in the Planning stage, considering that the usual tools do not enable it. Associated with this is the possibility of visually analysing different construction execution strategies. Planners and decision makers are cited as beneficiaries. Still related to this, the need to add physical space to planning is also mentioned.

Another aspect raised with relative importance is the need to improve the quality of information for planning. This is related to the deficiency of flows and the quality of information in the usual processes. This aspect is mentioned in the applications for industrialized works (flow of information from the project and work to the factory, aiming at production planning and sending components for the work). The integration of lean methods and tools with 4D BIM is also found as motivation (2 articles).

2.2.3 Types of Works

The articles focus, in general, on the objectives, methods and results. As already mentioned, little information is given about the projects and works chosen. There are 5 articles in which this information is not given, or the type of building chosen is not clear. In two of these, the focus is on site planning, regardless of the type of work. In 3 articles the buildings are of multiple floors without information on use. In the articles in which information is available (14), most works are residential (6 articles), 3 are university buildings, 2 are commercial buildings and 3 are industrial buildings. The construction types, usage and size of the works are varied and no relationship was identified between the type of work chosen and the objectives of the work. The exceptions are the 3 articles that aim to improve information for industrialized buildings, whose works are of this type. Even in these, the types, uses and size of buildings are varied. In this respect, this article differs, since the objectives and methods are applicable to a certain type of development and works (multi-storey residential buildings with conventional technologies) that needs to be well contextualized, as it will be done in the case study.

2.2.4 Methods and Tools

Not all articles explain the methods and tools used. Among those who present them, there are different methods and tools for similar objectives and there are uses of the same tools for different purposes and methods. Considering these aspects, we sought to identify patterns of methods and tools used. Articles 1 and 19 were excluded from the analysis for the reasons already mentioned.

Of the works related to Control (8 in total), 6 propose the use of methods and tools essentially aimed at automated image capture and transmission (high resolution cameras, laser scanners, point cloud models, MVS (Multi-View Stereo), RFI (radio frequency identification sensors) integrated with 3D and 4D BIM models. The focus of these works is the capture of the services performed and their comparison with the forecast. However, it is not reported or it is not clear how the replanning is done, due to deviations from what was planned. As already stated, the scope of these works is not the scope of the present article.

The other significant set, also in number 6, refers to Planning and proposes, essentially, the use of traditional planning tools and applications (MSProject, balance sheets, labour histograms, Excel), integrated with 3D and 4D BIM modelling, mostly with Revit and Navisworks. One of these (article 16) makes the 3D modelling of the building from 2D drawings, with Revit, from which the traditional planning tools (MSProject) and 4D BIM are integrated with Navisworks. This method is also the one used in this article. However, in article 16 [11] the building is completely modelled, with more than 4,000 elements, while in this article a simplified modelling was done, as it will be seen in the case study.

A third set, also of Planning (articles 8, 9 and 13) presents more complex solutions, with the creation of systems, models and algorithms, integrating traditional tools and 3D and 4DBIM modelling, aiming to automate the planning and integration of information. In one of these (article 13), for automation of jobsite planning, the Business Rule Management Systems (BRMS) and Drools tools for identifying rules were cited, combined with Industry Foundation Classes (IFC).

2.2.5 Results, Difficulties and Recommendations

In all articles, the proposed objectives were considered achieved. The good results are generally attributed to overcoming the limitations of the usual processes. In articles 17 and 18 the detection of planning problems from the visualization provided by 4D BIM is also reported as a good result. However, in the examples reported, it is noted that some problems could have been detected with the usual tools (line of balance in article 17) and perhaps even avoided with proper elaboration of the operational planning and construction project, in article 18, as the authors pointed out.

Even with the good results obtained, difficulties are reported and several recommendations are made. Those that appear most frequently and that are related to the scope of this article are:

- difficulties with the complexity of 3D modelling, which is cited in 6 articles. These difficulties stem from both the method used, the complexity of the building and the objectives proposed with the 4D BIM modelling;
- need for manual processes, cited in 6 articles;

- resistance and difficulties to change the usual processes, mentioned in three articles;
- need for professionals trained in BIM, design and planning (2 articles);
- difficulty in visualizing internal activities and activities with different durations (2 articles).

As a result of these difficulties, the following recommendations are made:

- a prior definition, as clear as possible, about what should be modelled, to avoid unnecessary work (article 14);
- prior definition of the information systems integrated between model and planning (codifications, descriptions and information added to the elements - articles 16 and 20);
- creation of libraries (article 16);
- use of 4D BIM at the beginning of the project, for general planning with little detailed models, when there is not much information. For operational planning, more detailed models are needed, focused on specific uses (article 14).

3 Concepts

Next is a quick conceptualization on the theme approached, restricted to the objectives of this paper.

3.1 Planning, Project and Project Management

Project planning will be defined based on classic authors on the subject, such as Ahuja et al. [32], Limmer [33], Kerzner [34], and Slack et al. [35], among others, as a decision-making process in that actions are chosen and organized, anticipating expected results, aiming to achieve, in the best possible way, pre-defined objectives.

Project is a temporary effort to achieve objectives and requires planning. The project is unique, made with limited resources and requires a plan (which we will call Planning) and a check on real progress, which we will call Control [35]. Planning is done before the beginning of the execution of the project and guides all its realization. Control, carried out during execution, detects deviations in relation to what was planned, generating the need for re-planning and re-checking, constituting a dynamic and continuous planning/ control cycle. The work to be planned and controlled is a project.

Planning and control, together with additional processes (coordination and others) are project management activities. The areas of knowledge of project management can be summarized, based on the same authors already mentioned in: scope, times, costs, quality, human resources, contracts, communication and risks. In the case of construction projects, we could also add technology, security and logistics as areas of knowledge, the latter involving the construction site.

The planning of the work results in the elaboration of products that can be generically called plans. The plans are expressed in specific documents, according to the area: scope (WBS-Work Breakdown Structure and EAP - Project Analytical Structure), time (attack plan, schedules and others), costs (budget and others), quality (work plan), security (security plan), construction site (construction site and logistics) and so on. The plans must also contain the planned control procedures in their respective area.

3.2 Hierarchical Planning Levels – Strategic Planning

Planning in general and timing in particular, which is what interests us in this article, has different characteristics and products, depending on the moment and objectives of the project.

Based on Halpin and Riggs [36] and also in contributions by Assumpção [37], time planning and its products can be synthesized hierarchically and defined as follows.

a) Strategic planning, which we will also call in this article as the Attack Plan: it defines a general plan for the execution of the work, containing the main stages of execution, the definition of the technologies of the main subsystems, definition of the sequences, trajectories and rhythms of production for the critical activities. It defines the execution deadlines; final and intermediary. This plan can be expressed in a basic schedule containing essentially the critical activities. More than one attack plan can be developed, with different strategies, for analysis and definition. It is the plan exemplified in the case study presented in this article.
b) Work schedule: it is based on the Attack Plan, containing all activities, with their predecessions and estimated durations and definition of the critical path, which is the same as the attack plan. The activities do not all need to have the same level of detail, as this can be done during the execution of the work.
c) Short term operating schedules, based on the previous ones, detailing the day-to-day activities of the work.

The plan of attack and the work schedule are done in general by planning professionals, which we will call Planners. For them to be effective, they must be done with the participation and validation of strategic decision makers (management of the construction company, developers, investors and others in the case of the Attack Plan) and the management of the work and those responsible for production in the case of work schedule (construction engineer, construction master, contractors and people in charge of production).

The operational schedules are usually drawn up by the construction management (construction engineer, masters, contractors and supervisors), with the participation of the Planner.

It is considered that the Planning, correctly executed, as well as the other management processes, is absolutely indispensable for the execution of a work. There are several techniques for this. The main and most used, widely known, are: Gantt Chart, precedence networks and balance lines, summarized below, based on the same authors and also in Mattos [38].

3.3 Planning Techniques

The main advantage of the Gantt Chart is the ease of communication. So it is the form of presentation usually used for network schedules. However, the serious limitation is the inability to express, algorithmically, the dependency relationships between activities. Its preparation is manual. This makes it impossible to identify the critical path, simulate different planning strategies and permanently re-plan according to the real progress of the work.

The technique that enables complete planning, identification of the critical path, clearance calculation and automatic re-planning is the precedence network, or PERT/CPM technique. The 4D BIM modelling requires the development of the network schedule, as it will be seen below.

The balance line has as main advantage the ease of elaboration and visualization of sequences, trajectories and rhythms of production, for many repetitive cycles. For this reason, they are suitable for works with many repetitive activities, such as buildings with many floors. However, its representation is graphical and not algorithmic, it is done manually, making it difficult to use and monitor for a large number of activities. They are useful for strategic planning for few activities. Thus, they can and should be used in addition to networks in buildings with multiple floors and repetitive works in general. In the case to be presented in this article, only the network technique will be used.

3.4 Planning Modelling with 4D BIM

The usual planning techniques mentioned, especially the precedence network, despite its limitations, have been the main planning tool used over the past decades [39]. However, one of its main limitations is the difficulty of interpreting the schedules, as they are abstract models and representations [3]. Even the representation in bars or in line of balance does not overcome this limitation.

4D BIM allows a three-dimensional visualization of the building construction. However, the most used applications do not have the intelligence to plan and need the information from the planning previously defined through the usual techniques. For this reason, 4D BIM does not replace the usual techniques. It needs to be connected to the usual applications, such as MSProject and others, to capture planning information and simulate the execution of the work.

It is also necessary, especially in multi-storey buildings, to synchronize the execution of the work with the installation and operation of vertical transport equipment (cranes and elevators), and also with safety equipment (trays, scaffolding and facade protection, among others). The usual planning and logistics techniques could also be used for this. However, they would increase the size and complexity of the schedule, requiring even more effort to understand it. The visualization of these activities provided by 4D BIM brings another quality and reliability to the planning. This aspect is particularly important for strategic planning, in which decision makers (developers, investors, directors of construction companies and even designers and suppliers), usually do not know planning concepts and are not familiar with the representations of the schedules. Strategic decisions are made at the beginning of the projects. You can, at this point, use simplified models. It is possible that there is not yet enough information to build the complete model. This is recommended by [4] - and confirmed in - the literature review presented above [27]. In addition, strategic planning, as already seen, consists of a relatively small number of activities, basically critical activities, further simplifying the modelling.

However, it is necessary that the 3D model, even simplified, contains the elements compatible with the construction processes and sequences to be simulated, which is defined by the Planner, not by the designers. It is also necessary that the elements of temporary buildings or transport and safety equipment be modelled, which is also not part of the designers' scope.

This problem can be worked around in two ways. One is the elaboration of these models by the designers with the support of the planner, which may require contractual arrangements different from the usual ones, increases in costs and time and difficulty of acceptance. The other alternative, which was considered in the case study to be presented, is for the planner to assume this task, inverting the relationship of the work to be done: the planner does the necessary additional modelling, with the support of the designers.

4 Case Study

4.1 General Presentation and Problem to Be Solved

The present case study is based on a real situation. It is a medium standard residential development, composed of 3 towers, each with 10, 15 and 20 SF (standard floors) respectively. The SFs are the same in all buildings, with 6 apartments per floor, with 2 or 3 bedrooms, totalling 270 apartments. The private areas of the apartments vary between 57 and 72 m2. Each tower has a ground floor with hall, party room, lounge and gym. In the common areas of the ground floor there is a swimming pool, playground, running track, barbecue place, tennis courts, football and gardens. There are two common basements with parking lots. The total built area is 38,596 m2.

The land has an area of 13,000 m2, and is crossed by a green area to be preserved, of 3,828 m2, dividing it into two parts. The smaller tower (T1), of 10 SF, is located in the smaller part of the land and the other larger towers (T2 - 15 PT and T3 - 20 PT) are in the larger part, as shown in the following implantation (Fig. 1). The T1 is fully sold.

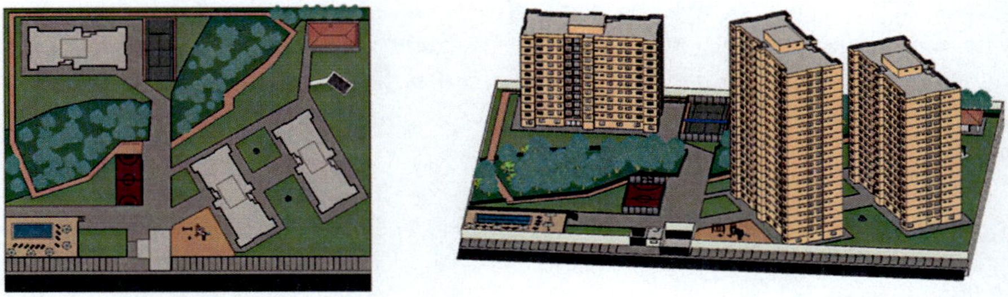

Fig. 1. Implementation of the complex, observing T1 (10 SF) located on the smallest part of the land and the others (T3-20SF and T2-15SF) on the largest part of the land. Source: Romanello et al. [40]), adapted by the authors.

The initial stage of the work will be carried out for the whole group and will consist of: implantation of the construction site, mechanized excavation, temporary containment and execution of the foundation piles.

After this common initial step, the execution of each tower begins, starting with the structure. The three towers will be built simultaneously. However, it is not possible to start them at the same time, due to the impossibility of simultaneously mobilizing teams, equipment and materials for the three towers. It was then adopted a gap of 30 days (one month) between the beginning of each tower, starting from the structure, being necessary to define a construction sequence.

The question then arises as to which tower to start and what construction sequence to adopt.

Considering that the biggest tower (T3-20 SF) is critical, the sequence that results in the shortest time is the one that starts with T3, then T2-20 SF and finally T1-10SF. This option is apparently the most advantageous, because in addition to resulting in the shortest term it also results in the lowest cost, since the indirect cost depends essentially on the time.

In the real case, the schedule was made with this sequence in the PERT-CPM network, on the MSProject and presented to the entrepreneurs (developers and investors) for decision. However, it was not evident to them that this was the best sequence, given their unfamiliarity with the concept of critical path and the lack of familiarity with usual representations of schedules. In addition, they understood that as the smallest tower (T1-10SF) was already sold, it was better to start with this one, of faster construction, providing quick visibility to the project, which would facilitate the sales of the other towers. However, this strategy, although better from the point of view of lower sales risk, was the worst from the point of view of time and cost, which caused an impasse. This impasse could be resolved by simulating the two alternatives for sequencing the towers in 4D BIM. Visualizing the construction of the work would provide entrepreneurs with greater confidence for decision making. These two simulations and visualizations were made for the case study and named respectively: Plan 1-shorter term and Plan 2 - less risk.

For these simulations, 3D modelling was necessary and the existing project was that of the basic architecture of a tower, in 2D. Then, for this case study, 3D modelling and simulation in 4D BIM were developed by undergraduate students of civil engineering during course completion work [40], guided by the first author of this article, who was the construction Planner. This case study reproduced a real and common situation in this type of development, which is that the Planner performs the 4D BIM simulation assuming responsibility for the 3D modelling work, as seen in the previous item.

4.2 Method and Tools

In order to make the 4D BIM modelling, considering the context of the case study work and the conceptualizations made in the previous item, the following method and tools were used.

4.2.1 Housing Development Model

The modelling of the towers was made from the 2D CAD files of a tower. The following subsystems were modelled: foundations in continuous helix piles and reinforced concrete blocks, provisional containments of the sides of the basements excavated in steel profiles with wooden planks, reinforced concrete reticulated structure executed on site, masonry fences, coated externally in roughcast, single mortar and paint, waterproofed slab cover, aluminium frames and glass. All structural elements were pre-designed and pre-dimensioned.

In external ground were modelled: swimming pool, playground, racing track, barbecue place, tennis and soccer courts; floors, gardens, parking for visitors, security cabin, walls and sidewalks based on parameters and in the usual specifications for this type of residential complex. Modelling was made in Revit and the result may be seen in Fig. 1.

4.2.2 Modelling of Complementary Activities for Simulation

The complementary activities, also modelled in Revit were: closing siding of the construction site, removal of ground and contentions with steel profiles and planks, wood forms and false work, protection trays (fixed and mobile), light suspended rockers and façade protection screen, racks elevators.

The forms, rockers, protection trays, protection screens and elevators racks were acquired from libraries and adjusted according to the conditions of the work.

4.3 Schedules of Attack Plans

The schedules of the attack plans Plan 1 - shorter term and Plan 2 - less risk were prepared in PERT/ CPM network, in MSProject and adopting the general sequence already mentioned in item 4.1. The activities are essentially critical: excavations and retainers, foundations, structure, masonry and facade of the towers, ground floor finishes and common basements. Predecessions, durations, production cycles were defined for repetitive activities and trajectories.

4.4 Schedules with Complementary Activities

These schedules also elaborated in MSProject, contain the activities of the attack plans plus the complementary ones for visualization, linked in the first ones. The activities mentioned in item 4.2.2 were inserted.

4.5 Simulation of Attack Plans

The simulation work with Navisworks consisted, briefly, in the import of the model files made in Revit, the creation of the sets - which are the elements of the model that will be associated with the planning - the import of the schedules made in MSProject for the Navisworks environment and linking the sets with the activities of the MSProject. Once this work is done, generating the simulations is simple. The result of the visualizations is seen in Fig. 2.

4.6 Results and Discussions

Attack plans 1 and 2 resulted in 22 and 24 months of work, respectively.

The modelling of the construction presented good general quality and a good degree of realism, although changes in zoom, speed and viewing angle are necessary to understand it in all details. The activities of shorter duration are little noticed with normal simulation speeds, as is the case with preparation and roughcasting for facades, as also reported in the literature review. The adoption of colours of elements close to the real ones, such as rockers, impairs the visualization as they are confused with the facade substrate. In any case, these deficiencies do not affect the overall quality and can be overcome with changes in colours, zooms, angles and simulation speed. The possibility of viewing the construction and, on the same screen, keeping track of the time is emphasized, which provides high reliability in planning and increased security in decisions.

The possibility of visualizing the construction and, on the same screen, of following the time counting, which propitiates high reliability in planning and increasing security in decisions should be noted.

The main difficulty is the modelling time. This stems not only from the project as a whole, but also from the increase resulting from the modelling of complementary activities. There is a need to model them on all floors - whenever you want to visualize trajectories - as is the case with all facade activities. The insertion of complementary activities also increased considerably the schedules, which went from approximately 150 to approximately 600 activities each plan. It was also necessary to model all the activities of the facades (roughcast, cast, paint and frames), and their visualization could be done by changing colours.

The application, however, does not offer this feature with proper viewing. The high modelling time is also reported in the literature review as one of the main difficulties encountered. This results in the need not only to simplify the models and uses of libraries – also recommended by the revised literature - but also the need to improve the resources of the applications used. It is also necessary to use coding and structure description of the modelled elements, integrated with the planning, as also recommended the literature review.

Even with the difficulties mentioned, the total expenditure of work can be considered compatible with the required speed, the final quality of the visualization and the benefits provided to planning and decision making. The method proved to be viable, in the sense that it could be used without the need for significant changes in the usual project and planning contracting processes. However, it requires training the planner in 3D modelling and 4D BIM, as also mentioned in the literature review.

It became clear that using 4D BIM brings advantages of visualization but demands knowledge and proper use of the usual methods and planning tools.

The method is applicable for construction works of buildings with high repeatability and verticality, as is the case with towers with multiple floors, object of this case study. It could also be used, in principle, for construction works of buildings with high repeatability and vertical and horizontal extension in general, as residential sets of houses in real estate developments and social housing estates. For that, there is a need for studies of application of the same approach for these types of works.

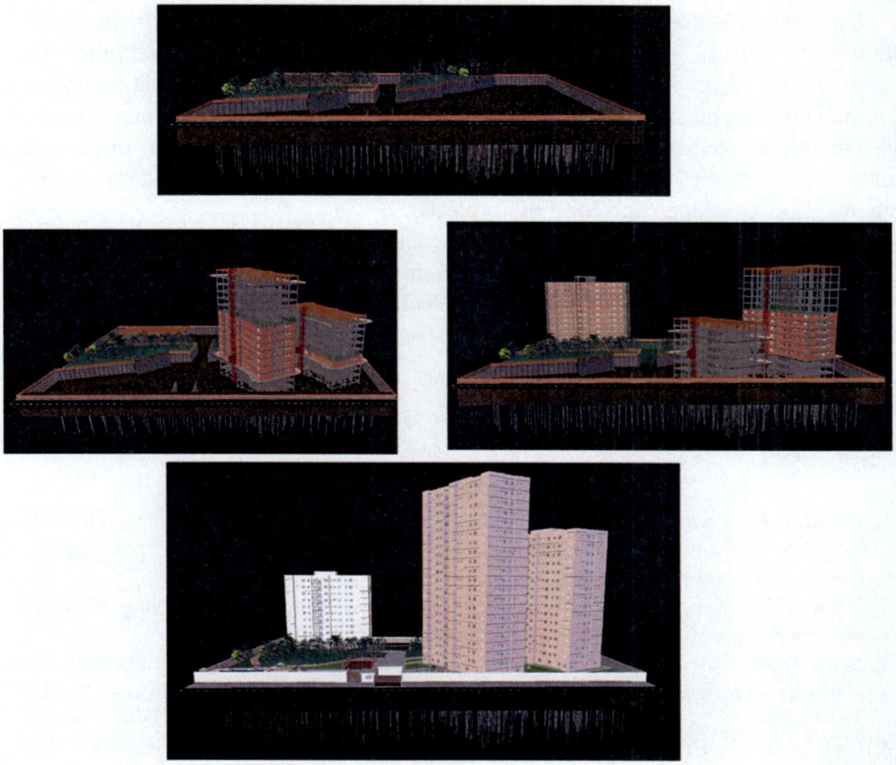

Fig. 2. Visualizations of the simulations of the two attack plans. Above, the initial stage of the work (construction site, retainers, excavations and foundations) common to both plans; In the middle: Attack plan-1, on the left, started by the largest tower (T3-20SF), with the shortest run and Plan and attack-2, on the right, started by the smallest tower (T1-10SF), with the longest term. Below the preview of the completed work. Source: Romanello et al. [40]), adapted by the authors.

5 Conclusions and Recommendations

The literature review and the case study aimed to demonstrate that 4D BIM requires the correct use of the usual planning techniques and tools. It also requires that the Planner to also be able to model 3D and use the 4D BIM tools. It was tried to demonstrate that it is possible and feasible to use 4D BIM without the need for major changes in the usual design and planning processes in residential real estate projects, such as are usually done in Brazil. In this context, and with the conceptual and practical premises presented, it is highly recommended to use 4D BIM for the strategic planning of this type of work. The method formulated for this purpose and used successfully in the case study presented is intended to be a contribution in this regard, and may have academic and technical-professional usefulness.

The method is applicable for construction works of buildings with high repeatability and verticality, as is the case with towers with multiple floors, object of this case study. It can be used, in principle, for construction works of buildings with high repeatability in general, such as residential housing and social housing units, including the use of more industrialized technologies, which is a trend observed in Brazilian constructions for these types of works. For that, there is a need for studies of application of the same approach for these projects.

In order to keep on works in the area, it is also recommended to continue the development of 4D BIM software resources to increase the simplification and efficiency of modelling. These resources and their needs have been described in the case study presented in this article. Further studies are recommended to additional diagnose the causes of planning deficiencies.

References

1. Biotto, C.N., Formoso, C.T., Isatto, E.L.: Use of 4D modeling and building information modeling in the management of production systems in construction projects. Ambiente Construído, Porto Alegre **15**(2), 79–96 (2015)
2. Vargas, F.B., Formoso, C.T.: Method for location-based production planning and control supported by BIM. Ambient. constr, Porto Alegre **20**(1) (2020)
3. Brito, D.M., Ferreira, E.A.M.: Evaluation of strategies for representation and analysis of construction planning and control using 4D BIM models. Procedia Econ. Finance **21**, 374–382 (2015)
4. Eastman, C., Teicholz, P., Sacks, R., Liston, K.: Manual de BIM: um guia de modelagem da informação da construção para arquitetos, engenheiros, gerentes, construtores e incorporadores. Porto Alegre, Bookman (2014)
5. Toledo, E.S.: BIM well done. Structure Magazine, April 2017
6. Filippi, G.A., Melhado, S.B.: A study on the causes of delays in the construction of real estate projects in the metropolitan region of São Paulo. Ambiente Construído, Porto Alegre **15**(3), 61–173 (2015)
7. Muianga, E.A.D., Granja, A.D., de Andrade Ruiz, J.: Deviations of costs and deadlines in construction projects: categorization and influencing factors. Ambiente Construído, Porto Alegre **15**(1), 79–97 (2015)
8. Mattos, A.D.: Why do schedules "stick"? Mundo Project Management (2007)
9. Rezende, M.A.P., Barros, M.M.S.B, Abiko, A.K.: Barriers and facilitators of technological innovation in the production of popular housing. In: ENTAC 2002. XIX Encontro Nacional de Tecnologia do Ambiente Construído, Foz do Iguaçu, Anais (2002)
10. de Brito Mello, L.C.B., de Amorim, S.R.L.: The civil construction building subsector in Brazil: a comparative analysis in relation to the European Union and the United States. In: Prod, São Paulo, vol. 19 no. 2 (2009)
11. Tsai, M.H., Wu, C.H., Matin, Md.A., Fan, S.L., Kang, S.C., Hsieh, S.H.: Experiences using building information modeling for a construction project. In: Conference: Proceedings of the International Conference on Computing in Civil and Building Engineering (ICCCBE), Nottingham, UK (2010)
12. Homrich, A.S., et al.: The circular economy umbrella: trends and gaps on integrating pathways. J. Cleaner Prod. **175**(20), 525–543 (2018)

13. Carvalho, M.M., Fleury, A., Lopes, A.P.: An overview of the literature on technology roadmapping (TRM): contributions and trends. Technol. Forecast. Soc. Change **80**, 1418–1437 (2013)
14. Pishdad-Bozorgi, P., Gao, X.H., Eastman, C., Self, A.P.: Planning and developing facility management-enabled building information model (FM-enabled BIM). Autom. Constr. **87**, 22–38 (2018)
15. Kropp, C., Koch, C., Konig, M.: Interior construction state recognition with 4D BIM registered image sequences. Autom. Constr. **86**, 11–32 (2018)
16. Kim, C., Son, H., Kim, C.: Automated construction progress measurement using a 4D building information model and 3D data. Autom. Constr. **31**, 75–82 (2013)
17. Sacks, R., Radosavljevic, M., Barak, R.: Requirements for building information modeling based lean production management systems for construction. Autom. Constr. **19**(5), 641–655 (2010)
18. Golparvar-Fard, M., Pena-Mora, F., Savarese, S.: Automated progress monitoring using unordered daily construction photographs and IFC-based building information models. J. Comput. Civil Eng. **29**(1), 04014025 (2015)
19. Golparvar-Fard, M., Peña-Mora, F., Savarese, S.: Integrated sequential as-built and as-planned representation with D 4AR tools in support of decision-making tasks in the AEC/FM industry. J. Constr. Eng. Manag. **137**(12), 1099–1116 (2011)
20. Son, H., Kim, C., Kwon Cho, Y.: Automated schedule updates using as-built data and a 4D building information model. J. Manag. Eng. **33**(4), 04017012 (2017)
21. Wang, W.C., Weng, S.W., Wang, S.H., Chen, C.Y.: Integrating building information models with construction process simulations for project scheduling support. Autom. Constr. **37**, 68–80 (2014)
22. Faghihi, V., Reinschmidt, K.F., Kang, J.H.: Construction scheduling using genetic algorithm based on building information model. Expert Syst. Appl. **41**(16), 7565–7578 (2015)
23. Rebolj, D., Pučko, Z., Babič, N.Č., Bizjak, M., Mongus, D.: Point cloud quality requirements for Scan-vs-BIM based automated construction progress monitoring. Automation in Construction **84**, 323–334 (2017)
24. Liu, H.X., Singh, G., Lu, M., Bouferguene, A., Al-Hussein, M.: BIM-based automated design and planning for boarding of light-frame residential buildings. Autom. Constr. **89**, 235–249 (2018)
25. Sigalov, K., Konig, M.: Recognition of process patterns for BIM-based construction schedules. Adv. Eng. Inform. **33**, 456–472 (2017)
26. Schwabe, K., Teizer, J., Konig, M.: Applying rule-based model-checking to construction site layout planning tasks. Autom. Constr. **97**, 205–219 (2019)
27. Bortolini, R., Formoso, C.T., Viana, D.D.: Site logistics planning and control for engineer-to-order prefabricated building systems using BIM 4D modeling. Autom. Constr. **98**(2019), 248–264 (2015/2019)
28. Choi, B., Lee, H.S., Park, M., Cho, Y.K., Kim, H.: Framework for work-space planning using four-dimensional BIM in construction projects. J. Constr. Eng. Manag. **140**(9), 04014041 (2014)
29. Manenti, E.M., Marchiori, F.F., Correa, L.A.: BIM execution plan: proposal of guidelines for contractors and designers Ambiente. constr. (conectados) **20**(1), 65–85 (2020)
30. Rodrigues, P.B.F., Machado, R.L., Mendes, J.R., Romagnoli, L.D.S.C.: A proposal to integrate the BIM model with the last planner system. Ambient. constr. Porto Alegre **18**(4) (2018)
31. Bataglin, F.S., Viana, D.D., Formoso, C.T., Bulhões, I.R.: 4D BIM applied to logistics management: implementation in the assembly of Engineer-to-order prefabricated concrete systems. Ambiente. constr. (conectados) **18**(1), 173–192 (2018)

32. Ahuja, H.A., Dozzi, S.P., Abourizk, S.M.: Project Management: Techniques in Planning and Controlling Construction Projects, 2nd edn. Wiley, Hoboken (1994)
33. Limmer, C.V.: Planejamento, orçamentação e controle de obras. Livros Técnicos e Científicos Editora, Rio de Janeiro (1996)
34. Kerzner, H.: A systems approach to planning, scheduling, and controlling, 10th edn. Wiley, Hoboken (2009)
35. Slack, N., Chambers, S., Harland, C., Harrison, A., Johnston, R.: Production Management. Atlas, São Paulo (1996)
36. Halpin, D.W., Riggs, L.S.: Planning and Analysis of Construction Operations. Willey, New York (1992)
37. Assumpção, J.F.P.: Project management in civil construction: model for strategic planning of building production. Doctorate thesis. Polytechnic School, University of São Paulo (1996)
38. Mattos, A.D.: Construction Planning and Control. PINI, São Paulo (2010)
39. Mattos, A.D.: Os 50 anos do PERT/CPM. Mundo Project Management, Ano 5, no. 25 (2009)
40. Romanello, A.P., Pelege, G.S., Silva, I.M.B., Ferreira, L.R.: BIM modeling in physical planning of works. Conclusion of Civil Engineering Course. Mauá Engineering School of the University Center of the Mauá de Tecnologia Institute. Sao Caetano do Sul (2019)

Conceptual Framework for Integrating Cost Estimating and Scheduling with BIM

Mírian Caroline Farias Santos[1(✉)], Dayana Bastos Costa[2], and Emerson de Andrade Marques Ferreira[2]

[1] Post-Graduation Program in Civil Engineering, School of Engineering, UFBA, Aristides Novis, 02 Federação, Salvador, Bahia, Brazil
miriancfarias@gmail.com

[2] School of Engineering, UFBA, Salvador, Brazil
dayanabcosta@ufba.br, ferreira.eam@gmail.com

Abstract. Many studies have emphasized the benefits of integration using different methodologies for reducing fragmentation. Modern simulation tools provide an excellent way to integrate information from a variety of work environments and disciplines. Building Information Modeling (BIM) can enable this communication more efficiently due to its rich capacity to store and manage project information. This work proposes a conceptual framework with assumptions for integrating the workflow into 3D BIM Modeling, Cost Estimating and Scheduling/Production Monitoring processes from project design to field production control. This study also provides a set of requirements to evaluate the integration between these processes. The research took place in two projects of a small-to-medium-sized enterprise and the research was developed using a case study approach, involving a literature review and a study of the workflow among the processes in two construction projects, Project 1and Project 2. This study provides a better understanding of the workflow for integrating 3D BIM Modeling, Cost Estimating, Scheduling, and Production Monitoring processes, identifying activities and information among them. The conceptual framework, with four assumptions of integration and a set of requirements, can help construction companies that wish to carry out an initial diagnosis of integration at the level of processes, people and technologies, as well as assessing whether there have been improvements at the end of an implemented integration.

Keywords: Integration · Cost Estimating · Scheduling · Workflow · BIM

1 Introduction

The construction industry is a complex and dynamic sector that involves many people in different stages of construction, covering functions such as planning, design, construction, and maintenance [1]. Due to its dynamic characteristics, many disciplines and project participants and different stages, the construction industry tends to suffer from poor communication among stakeholders and fragmentation represents a common problem [2, 3].

E. Toledo Santos and S. Scheer (Eds.): ICCCBE 2020, LNCE 98, pp. 613–625, 2021.
https://doi.org/10.1007/978-3-030-51295-8_43

Building Information Modeling (BIM) can store any information created for the execution of the project and, thus represent a comprehensive digital description of the building to be built [3]. For Jrade and Lessard [4], BIM can also improve technical work already in the design phase, creating 3D models that integrate all the resources of a building, linking schedule (BIM 4D) and costs data (BIM 5D).

Several studies have used BIM to facilitate the integration between planning and budgeting processes [4–9]. Some of these have obtained some successful results by solving integration issues and defining some requirements, such as the use of a standard work breakdown structure (WBS) for a work package-integrated model and the suggestion of a single platform for integration. However, some works fails due to not meeting the requirements of the main users, remained the linkages of cost and planning data to the BIM model laborious and manual, and did not consider field feedback data.

Moreover, due to the diversity of enterprises, companies and information technologies used, there is no consensus on the typical workflow between the different processes for an integrated control system using BIM. There are several methods and tools used by different authors and few works include the workflow among the people responsible for those processes, the activities and technologies involved. Therefore, this study aims to propose a conceptual framework to promoting integration between 3D BIM Modeling, Cost Estimating, Scheduling and Production Monitoring processes and identify requirements for the development of BIM models and BIM-related activities that result in the integration among those processes.

2 Integrated Scheduling and Cost Estimating Approach with BIM

2.1 An Integrated Workflow Among People, Processes and Technology

Workflows allow integrating not only the information of several business processes but also the information flow inherent in it [10]. Managing the flow of information within the organization makes knowledge management possible [11]. For Edwards [12], knowledge management in the organization is often described as the comprehension of three basic elements: people, processes, and technology.

People generate and apply knowledge, encourage sharing, and design and operate processes. Processes refer to the methods for obtaining, generating, organizing, sharing and transferring knowledge, defining the roles of each person in the organization. The technology involves mechanisms that store and provide access to databases and knowledge generated by people in different locations. [12, 13]. Thus, each person involved in the processes throughout the construction lifecycle needs specific information to develop their work. This information, besides being transferred and shared between different people, is also shared between these different processes. Such processes use different technologies that need to communicate in such a way that the information is not disaggregated or confusing but with enough quality to allow integration at all stages of construction.

Some advantages of adopting an integrated approach are access to the latest data and information, improving coordination, better decision-making and effective feedback [14]. Figure 1 summarizes the concepts involving an integrated workflow among people, processes and technology.

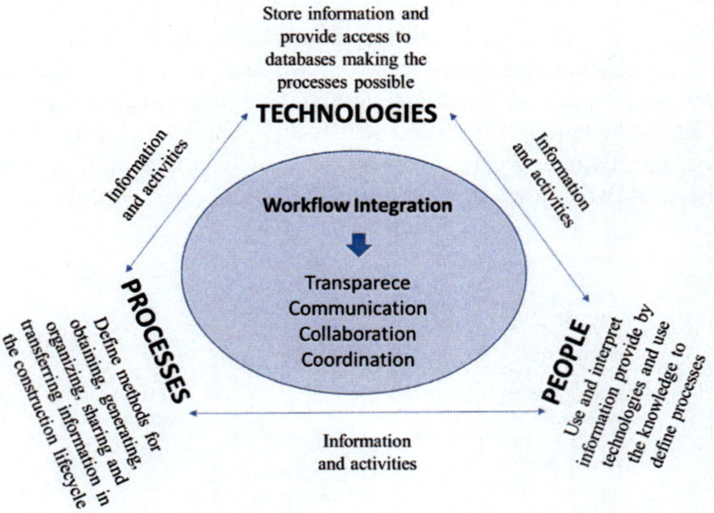

Fig. 1. Integrated workflow among people, processes and technology

2.2 Requirements for Integrated Construction Scheduling and Cost Estimating Systems Using BIM

The integration of cost and planning information can be understood as an effort to control cost and scheduling information simultaneously through a single controllable parameter [15]. Although these processes share common data, such as resources, quantitative and cost [8, 16], in practice, this information is tracked separately, mainly because this information is generated and maintained by different people. As a consequence, there are inconsistencies and inaccuracies in determining the cost and schedule status [17].

The Work-packaging Model (WPM) has been accepted as the broadest way to integrate cost and schedule data. This type of model eliminates the need for a cost breakdown structure (CBS) and adds cost data to the WBS [15]. The WBS is at its lowest level made up of individual work packages or task groups, which can be assigned schedules and cost estimates that can be monitored, controlled, tracked and measured separately [18, 19]. As long as work packages are prepared early in project planning, the information available is limited, but an analysis of BIM model data can be used to consider project-specific characteristics [20].

The BIM 4D model can facilitate the generation of information for the team, such as start and end dates of the elements and their criticality and visualization of the construction project over time [21]. Also, BIM 5D can be used to give designers more agile and accurate feedback on the cost of a project [22] by extracting component counts, space area and volume, material quantities directly from a model [23].

To guarantee integration and synchronization among computer management tools, especially in production planning and cost control system using BIM, Jrade and Lessard [4] also highlight the following specific requirements: (a) use of a single WBS standard, (b) construction of the model according to the WBS, (c) integration of different software, in which most software connections should be automated whenever possible, (d) association of the time and cost data to each of the components of the model, allowing for quick and easy modification through automation, (e) visualization as a communication tool among project stakeholders, and (f) budgeting and planning linked to a single platform. Figure 2 summarizes the connections and requirements for an integrated BIM workflow.

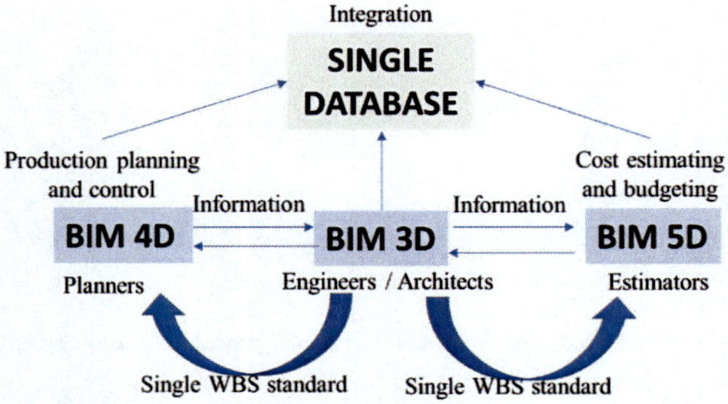

Fig. 2. Requirements for an integrated BIM workflow

3 Research Method

The case study research strategy adopted in this study aimed to investigate procedures and processes related to the understanding of the workflow between the 3D BIM Modelling, Cost Estimating, Scheduling, and Production monitoring. For this, the following questions were raised: How is the workflow between these processes? What problems commonly difficult integration? And how to improve and evaluate the effects of integration?

The study involved the literature review; the selection of the construction company and projects; the identification of the critical workflow concerning the four main processes: (1) BIM 3D Modeling, (2) Construction Scheduling, (3) Cost estimating, and (4) Production Monitoring in two construction project; and the development of a conceptual framework.

The literature review was carried out using the combination of keywords integration, BIM, BIM 4D/5D, cost/estimation, production planning and control on a Brazilian and an international database, CAPES and SCOPUS, respectively.

The criteria for selecting the construction company were the prior use of BIM with a minimum level of development of 300 (LOD 300) and the previously successful participation in research projects with the university involved in this research.

Company A used Revit for the BIM 3D modeling, Navisworks for 4D simulation, both from Autodesk, a commercial software program called TOTVS for cost estimating, Microsoft Project and Microsoft Excel were used for scheduling, and a customized software called CPObras was used for production task monitoring on-site. The study was carried out in two construction projects of Campany A.

Project 1 was a multi-residential housing complex with 500 apartment units distributed in 30 blocks, built using cast-in-place concrete walls. Project 2 was composed of 367 semidetached and isolated single-family houses also built using cast-in-place concrete walls. Both projects were located in Salvador, Brazil.

For the workflow study, the data collected with the involvement of the Manager, Intern (BIM Modeler), Planning Engineer and Field Engineer, were: staff meetings, project visits for direct observation and document analysis (2D CAD projects, reports from production monitoring software and BIM models), see Table 1.

Table 1. Description of data collection in Project 1 and Project 2

Project	Participant	Data collected	Environment	Duration of interview (h)
Project 1	Manager	Interview about routines and processes (Budgeting and Production Monitoring) developed by the company	Office	2
	Intern (modeler)	Interview about routines and processes (modeler) developed by the company		4
	Field engineer	Interview about routines and processes (Production Monitoring) developed by the Company/observation of field activities	Field	8
	Planning engineer	Interview about routines and processes (Production Planning) developed by the company		3
Project 2	Manager	Interview about routines and processes (Budgeting and Production monitoring) developed by the company	Office	3
	Intern (modeler)	Interview about routines and processes (modeling) developed by the company		6
	Field engineer	Interview about routines and processes (Production Monitoring) developed by the Company/observation of field activities	Field	20
	Planning engineer	Interview about routines and processes (Production Planning) developed by the company		3

A data flow diagram was used as the main tool to represent the information through the workflow. The study of the workflow aimed to identify the main parties responsible, software, type of information passed along the flow, and main obstacles to integrating the information among the main processes in Project 1 and Project 2.

From the data collected and literature review, the practical problem identified was the lack of process integration, poor communications and inefficient workflow among activities and information related to BIM modeling, cost estimating, scheduling and production monitoring processes.

Thus, based on the problems identified in the workflow study of Project 1 and Project 2, a conceptual framework for integrating the activities concerning the four main processes studied (3D BIM Modelling, Cost Estimating, Scheduling, and Production monitoring) was developed, as it will be presented in the next section.

4 Results

This section presents the workflow of Projects 1 and 2 and the conceptual framework developed.

4.1 Workflow Study of Project 1 and Project 2

The study of Project 1 and Project 2 workflow aimed to identify and define requirements for supporting integration between the studied processes from office to field.

The workflow study of Project 1 (Fig. 3) indicated that although the project had BIM 3D models and a standard WBS, the planning, estimating and modeling teamwork in an isolated manner because the WBS was only being used in the BIM model.

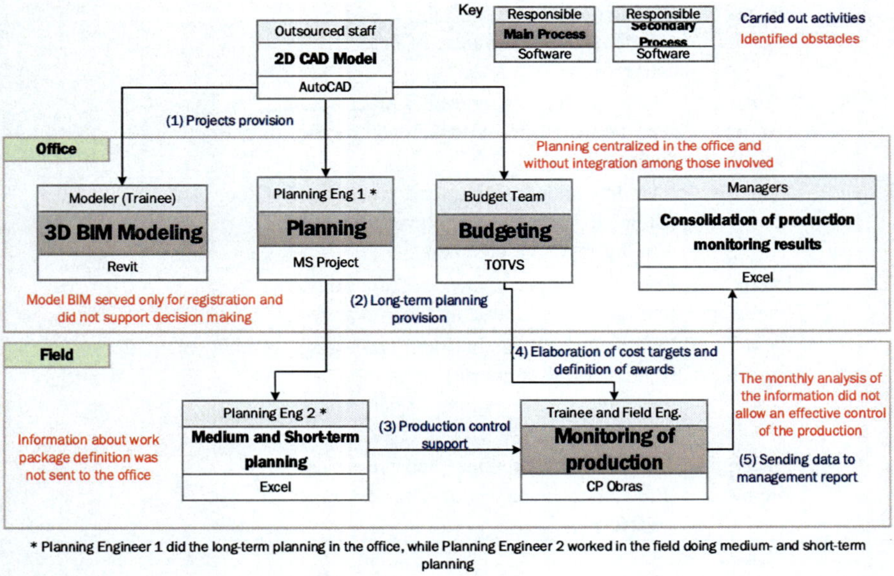

Fig. 3. The workflow of planning and cost processes – Project 1

Moreover, the BIM model was only useful to maintain information about the project and did not aid decision making. The medium-term and short-term planning was carried out in the field without taking into consideration the long-term planning and the cost information provided by the office teams. Besides, the production monitoring CPObras (customized company's software) did not promote effective feedback on the information because a different WBS was used. There were monthly reports of production. A brief description of the activities is presented in Table 2.

Table 2. Description of workflow through activities – Project 1

No	Activity	Description
1	Project provision	The 2D CAD projects were made available to the planning, budget and modeling teams
2	Long-term planning provision	Long-term planning has been made available to the field team for the preparation of medium and short-term planning
3	Production control Support	The medium- and short-term planning information served to support the production control defined in CPObras
4	Elaboration of target cost and definition of awards	The information about the labor costs of each service was included in the production monitoring software
5	Sending data to the management report	Production tracking data returned in the form of an Excel spreadsheet to the office

In Project 2 (Fig. 4), even with the support of the budget team, it was found that the use of the BIM 3D model was restricted to the office environment (quantity take-off), without being made available to the field team, and did not include the real cost and planning data.

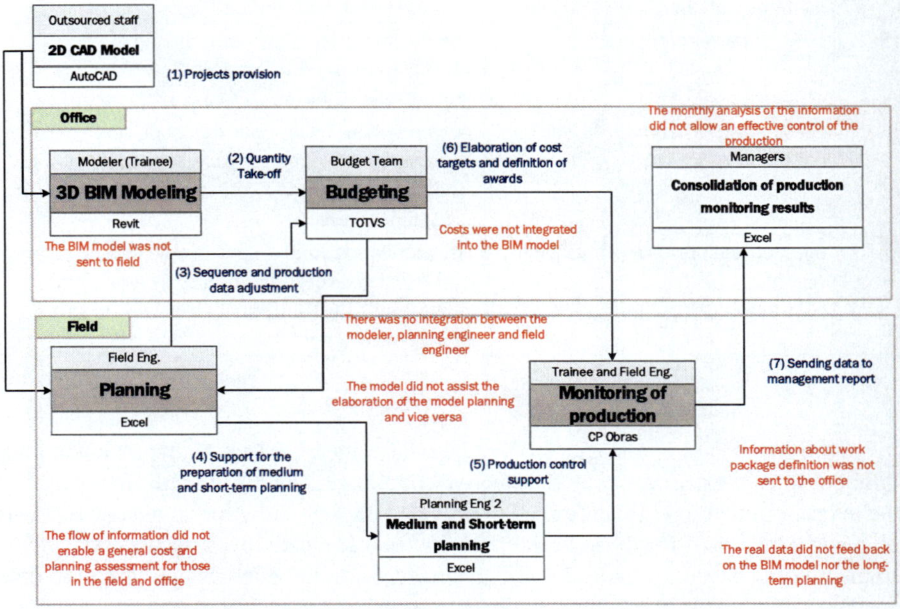

Fig. 4. The workflow of planning and cost processes – Project 2

There was no long-term planning to help define the execution goals, and the production planning was only based on a sequence of activities. The medium- and long-term planning were developed in the field based solely on the sequence of activities, independent of the cost estimating data, the elements modeled, and the activities described in the CPObras software. Although there were interactions between the estimating team and the field engineer to define the work sequence, when the work packages were defined in CPObras software, many activities were inserted and grouped, disrupting the integration cost estimating and production monitoring. A brief description of the activities is presented in Table 3.

Table 3. Description of workflow through activities – Project 2

No	Activities	Description
1	Project provision	The 2D CAD projects were made available to the field team to develop a sequence of activities and to the modeling team to develop the BIM model
2	Quantity take-off	The quantitative take-off was performed based on the BIM 3D model and the spreadsheet with the quantitative ones was sent to the budget team
3	Sequence and production data adjustment	The sequence of activities developed by the field engineer was sent to the budget team to be made compatible with the service data, so the activities specified in the field could match those provided by the budget team
4	Support for the preparation of medium and short-term planning	Based on the activity sequence, the field engineer has developed medium and short-term planning
5	Production control support	The medium- and short-term planning information served to support the production control defined in the CPObras software
6	Elaboration of target cost and definition of awards	The budgeting team input the labor cost information provided for each service in the production monitoring software to set target cost and labor awards
7	Sending data to the management report	The production follow-up data was returned in an Excel spreadsheet to the office, so monthly monitoring of this information was done in a management report

There was no use of a single WBS standard. Besides, the lack of integration among different software does not provide effective coordination and communication between the teams. Furthermore, the information provided did not allow for an overall cost and planning assessment for those in the field and office, mostly because there was not a single process or tool to integrate all the construction professionals and make the data available.

When the real project data returned to the office, the information was fragmented or grouped, so the team had to spend time compiling and understanding the information. This is mostly due to a lack of a defining work package based on field-planned activities that could be used among cost estimating, BIM modeling and production planning processes.

Due to the excessive manual activities required to make the production reports and due to their frequency of elaboration (monthly), the communication between the field and office teams was compromised and did not allow for rapid intervention if there were any deviations from the planned goals of the production and costs.

4.2 Conceptual Framework for Integrating 3D BIM Modelling, Cost Estimating, Scheduling and Production Monitoring

Based on the literature review and the workflow of Projects 1 and 2, a conceptual framework, including a set of assumptions was established (Fig. 5). The four assumptions established were:

a) 3D BIM modeling/Cost estimating Integration: the method should result in the integration of information among the activities, software, and teams in the Cost Estimating and 3D BIM modeling processes through the office and field environments.
b) Scheduling/Production Monitoring Integration: the method should provide the integration of information among the activities, software and teams in the Schedule and Production Monitoring processes through the office and field environments.
c) 3D BIM modeling/Scheduling/Production Monitoring Integration: the method should make possible the integration of information among the activities, software and teams in the BIM model and Schedule and Monitoring processes through the office and field environments.
d) Cost estimating/Scheduling/Production Monitoring Integration. the method should allow for the integration of information among the activities, software and teams in the Cost estimating and Production Planning and Monitoring processes through office and field environments.

For the elaboration of those assumptions, it was assumed the integration of processes one by one, so the integration could then be accomplished fully. For this, the pillars of knowledge management (people, technology and process) were used with a WBS as a common parameter of integration, as suggested by literature review.

Thus, for 3D BIM modeling/Cost estimating Integration, the 3D BIM model must be developed to provide value-added information to users of the cost estimation process. One must consider the numbers of modeled elements, units of measurement, as well as how the estimation program captures and uses the data from the 3D BIM model.

For Scheduling/Production Monitoring Integration, one must consider how the work packages will be structured and how the actual production data will be collected in the field to be used by the planning team. The focus is to make the feedback more agile and with enough quality to allow decision making in real-time.

In terms of 3D BIM modeling/Scheduling/Production Monitoring Integration, one should connect as hard as possible the modeled elements with the structure of the work packages made by the planning team. Thus, it will be possible to use the 4D simulation to view the planned activities, the activities performed, as well as to compare them.

Finally, for Cost estimating/Scheduling/Production Monitoring Integration, one must consider how cost data can impact the schedule and vice versa. The way work packages are organized influences financial disbursement and the availability of resources affects the planning of work packages. However, the key to this integration is to structure WBS thinking also of costs in terms of planned costs as well as real costs, to be possible to control them.

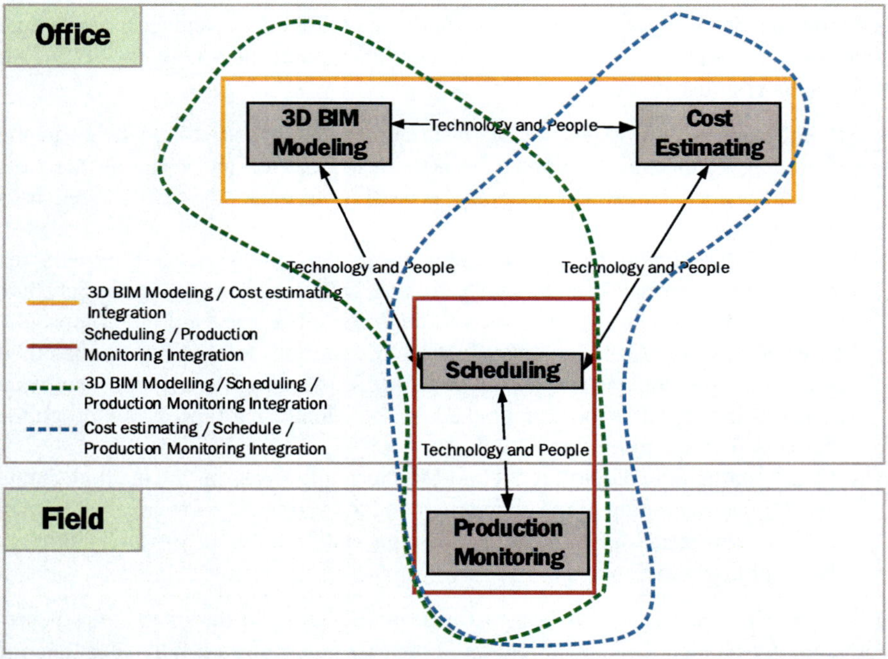

Fig. 5. Conceptual framework and assumptions

In addition to the assumptions, a set of requirements, based on Jrade and Lessard [4], was considered essential to integrate the workflow between Production, Budgeting, BIM Modeling, and Production Monitoring processes between field and office.

- Integration of different software and tools - generally, specific software is used for each process, so interoperability between them is necessary. This interoperability can be achieved using open file formats or using plugins that can transfer information between software. The important thing is that the information is not lost or changed.

- Use of a standard WBS among the processes - the use of a WBS common to all processes facilitates communication, both between software and between people. This is because different teams can exchange information more quickly and with greater precision, since they are speaking the same language.
- Communication between office and field – once the teams belong to different environments, distant or not, they are subject to poor communication. Integrating the field and the office allows decision making based on reality and considering the opinion of both teams.
- Review and update of information - for integration to occur, it is necessary to provide teams with updated information, preferably in real-time. Thus, costs, planning and the model can be revised and adjusted, when necessary.
- Automation and simplification of the processes - the simplicity of executing the processes must be considered when promoting integration. The reduction in the number of activities, people and technologies involved allows faster decision making. This simplification can sometimes be achieved through the automation of processes, that is, through the reduction of manual work, also increasing the precision of the results.
- People's responsibilities - when everyone involved in the processes knows what their duties are, it is easier to assign tasks and request feedback about the progress of the processes
- Single database and easy access to information - with a single database it possible to centralize information and it can be consulted more easily. But in addition to being centralized, this information must be made available to everyone involved in the integrated processes.

5 Conclusions

For the development of this research work, a study of the workflow among the processes of 3D BIM Modeling, Scheduling, Cost estimating, and Production Monitoring was carried out. The data flow diagrams obtained in the diagnostics allowed the visualization of obstacles and requirements for the integration between these processes from the people, environments and technologies' point of view.

Among the obstacles, the following stand out: (a) the centralization of planning in the office environment without considering information on the definition of work packages to be carried out in the field; (b) lack of periodicity in the evaluation the planning and cost information; (c) the BIM model serving only for registration and without practical applications in the company; and (d) lack of a common communication language, be it coding or nomenclature, between people, programs and processes.

To solve these problems, it was necessary to improve communication between field and office teams; define a standard communication language between these teams; involve the modeling, cost estimation and planning teams in the definition of the work packages, to represent more faithfully the way the services will be performed in the field; reduce the field data feed-back time; and expand the use of the BIM model.

The main contribution of this study is a conceptual framework for better understanding the types of activities is necessary to integrate the processes of 3D BIM Modeling, Scheduling, Cost estimating and Production Monitoring, for this four assumptions were established: (1) 3D BIM modeling/Cost estimating Integration; (2) Scheduling/Production Monitoring Integration; (3) D BIM modeling/Scheduling/ Production Monitoring Integration; and (4) Cost estimating/Scheduling/Production Monitoring Integration.

Besides, seven requirements to analyze the integration of the workflow between the studied processes were presented, which are: (a) Integration of different software and tools; (b) Use of a standard WBS among the processes; (c) Communication between office and field; (d) Review and update of information; (e) Automation and simplification of the processes; (f) People's responsibilities, and (g) Single database and easy access to information.

Based on this conceptual framework, the construction companies that wish to implement the integration of their processes can rely on the four assumptions and their requirements to carry out an initial diagnosis of integration at the level of processes, people and technologies, as well as assessing whether there have been improvements at the end of an implemented integration. This assessment can be carried out both for processes related to the company's internal environment and for those developed by companies outside the organization.

References

1. Nawi, M., Baluch, N., Bahauddin, A.: Impact of fragmentation issue in construction industry: an overview. In: MATEC Web of Conferences, vol. 9, pp. 1–8 (2014)
2. Alashwal, A., Fong, P.: Empirical study to determine fragmentation of construction projects. J. Constr. Eng. Manag. **141**(7), 04015016 (2015)
3. Preidel, C., et al.: Seamless integration of common data environment access into BIM authoring applications: the BIM integration framework. In: eWork and eBusiness in Architecture, Engineering and Construction: ECPPM 2016: Proceedings of the 11th European Conference on Product and Process Modelling (ECPPM 2016), Limassol, Cyprus, 7–9 September 2016, p. 119 (2017)
4. Jrade, A., Lessard, J.: An integrated BIM system to track the time and cost of construction projects: a case study. J. Constr. Eng. **2015**, 1–10 (2015)
5. Feng, C., Chen, Y., Huang, J.: Using the MD CAD model to develop the time–cost integrated schedule for construction projects. Autom. Constr. **19**(3), 347–356 (2010)
6. Chen, S., et al.: A framework for an automated and integrated project scheduling and management system. Autom. Constr. **35**, 89–110 (2013)
7. Fan, S., Wu, C., Hun, C.: Integration of cost and schedule using BIM. J. Appl. Sci. Eng. **13** (3), 223–232 (2015)
8. Fan, S., et al.: Cost-based scheduling method using object-oriented approach. Autom. Constr. **65**, 6577 (2016)
9. Wang, K., et al.: Applying building information modeling to integrate schedule and cost for establishing construction progress curves. Autom. Constr. **72**, 397–410 (2016)
10. Barros, A., Hofstede, A.: Towards the construction of workflow-suitable conceptual modelling techniques. Info. Syst. **8**, 313–337 (1998)

11. Choo, C.W.: The knowing organisation as learning organisation. Educ. Train. **43**(4/5), 197–205 (2001)
12. Edwards, J.: A process view of knowledge management: it ain't what you do, it's the way that you do it. Electr. J. Knowl. Manag. **9**(4), 297–306 (2011)
13. Hosseini, M.: The impact of people, process and technology on knowledge management. Eur. J. Bus. Manag. **6**(28), 230–241 (2014)
14. Khalfan, M., Raja, N.: Improving construction process through integration and concurrent engineering. Aust. J. Constr. Econ. Build. **5**(1), 58–66 (2012)
15. Cho, D., Russell, J., Choi, J.: Database framework for cost, schedule, and performance data integration. J. Comput. Civ. Eng. **10**(1061), 719–731 (2012)
16. Carr, R.I.: Cost, schedule, and time variances and integration. J. Constr. Eng. Manag. **119**(2), 245–265 (1993)
17. Staub, S., Fischer, M.: The practical need of integrating scope, cost, and time, durability of building materials and components. In: Lacasse, M.A., Vanier, D.J. (eds.) Institute for Research in Construction, pp. 2888–2898. Canada, Ottawa (1999)
18. Wayne, J.: Project Control: Integrating Cost and Schedule in Construction. 1st edn. Wiley, Canadá (2013)
19. Siami-Irdemoosa, E., Dindarloo, S.R., Sharifzadeh, M.: Work breakdown structure (WBS) development for underground construction. Autom. Constr. **58**, 85–94 (2015)
20. Isaac, S., Curreli, M., Stoliar, Y.: Work packaging with BIM. Autom. Constr. **83**, 121–133 (2017)
21. Umar, U., et al.: 4D BIM application in AEC industry: impact on integrated project delivery. Res. J. Appl. Sci. Eng. Technol. **10**, 547–552 (2015)
22. Sattineni, A., Macdonald, J.: 5D-BIM: a case study of an implementation strategy in the construction industry. In: 31ST International Symposium on Automation and Robotics in Construction and Mining, pp. 1–7 (2014)
23. Eastman, C., Teicholz, P., Sacks, R., Liston, K.: BIM Handbook: A Guide to Building Information Modeling For Owners, Managers, Designers, Engineers and Contractors. Wiley, New York (2008)

SINAPI and CPOS Review Proposal to Effective BIM Incorporation of These Measurement Criteria in Public Works

Rodolfo Pereira Silva(✉), Sérgio Leal Ferreira, and Luiz Reynaldo de Azevedo Cardoso

Universidade de São Paulo, São Paulo, Brazil
{rodolfosilva, sergio.leal, luizcardoso}@usp.br

Abstract. Building Information Modeling has been stimulated by the government in Brazil through decrees such as n° 9,983/2019, the creation of technical standards (ABNT NBR 15,965) and also concrete actions by the private sector. One of the characteristics of this technology that is most interesting at first is the possibility of performing a kind of virtual construction in line with a set of information (time, cost, specifications) for each modeled element. In this sense, one of the most common difficulties is the optimization regarding the automatic extraction of quantitative in order to realize a budget in harmony with the existing service standardization. This paper details the challenges inherent in modeling aiming the automatic quantities extraction from a public work in Brazil and demonstrates the difficulties in the relationship of the modeled objects and the SINAPI (Sistema Nacional de Pesquisa de Custos e Índices da Construção Civil) and of CPOS (Companhia Paulista de Obras e Serviços) measurement criteria. It was noted that the multiplicity of options for the same service, especially the presented by SINAPI measurement criteria, demands a huge modeling effort to assure a reliable extraction of quantitative data. It was also noted that the need to create a specific model for budgeting in the bidding phase, in order to make the necessary modifications to meet the measurement criteria. The study tested different ways of obtaining quantitative, comparing them in their advantages and disadvantages.

Keywords: BIM · Measurement criteria · Public works

1 Introduction

1.1 BIM and Public Works in Brazil

Public works are essential in order to supply to the population a wide range of services. Generally, their necessities comes from the demand of each locality or through a political issue. The weight of the public works in Brazilian finances can be pictured in Table 1.

Then, new technologies that helps to decrease execution time, to bring economy to the public vaults and to promote greater quality products will cause a huge positive

E. Toledo Santos and S. Scheer (Eds.): ICCCBE 2020, LNCE 98, pp. 626–642, 2021.
https://doi.org/10.1007/978-3-030-51295-8_44

Table 1. Public Works in Brazil (Adapted from [1])

Quantity of works	Origin	Last update	Amount	Total
93,001	Plataforma + Brasil	23/01/2020	80,189	R$ 1.27 trillion (1US$ = R$ 5,10)
	PAC	30/06/2019	12,812	

impact. Thereby, BIM (Building Information Modeling) approach, and its consequent technology, emerge as a promising alternative to the current stablished workflow.

MDIC [2] consider BIM as a series of procedures and technologies capable to produces digital models for construction, facilitating collaboration between diverse agents involved in all work phases, even the in posterior building operation.

To Jernigan [3], BIM is a system mainly based in construction information, capable to generate multiple types of innovation, resulting in benefits to projects and also to the environment, and to the people that will use the building space.

Eastman [4] states that a BIM model has the potential to generate multiple benefits, all over the building lifecycle, including initial studies of conception, of viability, including interference analysis, graphic representation and quantity generation, time and cost analysis, simulation to study sustainability (environment impact and energy consumption), building operation and even its demolition.

As the technology in BIM application and its use possibilities advances, its adoption becomes a wide reality, and currently is a fact in countries like United Kingdom, Netherlands, Denmark, France and Norway [5].

In Brazil, beyond private sector and some institutional initiatives, in federal and state ambit there are incentive politics toward the BIM technology adoption, and to the technical standardization production. Examples of this initiatives are the publication of the BIM Project Contract Workbook to public works of the Santa Catarina State in 2014, updated in 2018, and also the Paraná State BIM Project Contract Workbook published in 2018.

In the Federal level, the movement of BIM promotion started institutionally by the publication of the nr 9,377 decree (may 17, 2018) stablishing the BIM National Spreading Strategy and the Strategic Management Commission. In August 22, 2019 the nr 9,983 decree replaced the nr 9,377 decree adding some new directions to the strategy and to the Commission.

In private and institutional sector, the ABDI (Agência Brasileira de Desenvolvimento Industrial) publication of its five volumes of BIM Guides in 2017 can be cited as a milestone. AsBEA (Associação Brasileira do Escritórios de Arquitetura) also played an important role in BIM diffusion when in 2013 publish its chapters showing benefits of BIM adoption.

The Special Studies Commission nr 134 of the ABNT (Associação Brasileira de Normas Técnicas), the Brazilian technical standardization association, is responsible for the elaboration of BIM related normalization. The main standard is the ABNT ISO NBR 15,965, which is formed by 7 parts and intend to stablish a classification system of the building information. Only parts 1, 2 and 7 has been published, and the other ones are under elaboration [6].

1.2 Measurement Criteria

In Brazil, like some other countries, in a public work it is necessary to define which service prices will be adopted in the budget spreadsheet. To achieve the final price for the services, the composition of material, workforce and equipment, the measurement and remuneration criteria must be defined as much detailed as possible. All these information are compilated in bulletins that are periodically updated.

The CPOS (Companhia Paulista de Obras e Serviços) elaborated criteria of measurement and remuneration, and also the composition of services for the use in São Paulo State public works contracts. Since 1995 there is the periodical publication of its bulletins. Figure 1 shows the details of a roughcast service.

17.02.020 **Roughcast**				
1) It will be measured by the area covered with a hatter, not separating up to 2.00 m^2 and not considering chips. Spans above 2.00 m^2 can be deducted at the opening and the manufactured chips (m^2). 2) The item removes the supply of cement, sand and labor required for execution rough hat.				
17.02.020	**Roughcast**	m^2		
010139	Bricklayer	h		0.1000
010146	Servant	h		0.1125
020503	medium washed sand	m^3		0.0052
020508	CPII-E-32 Cement (50kg bags)	kg		2.2650
080125	concrete mixer capacity of 320 liters	h		0.0022

Fig. 1. Roughcast service from criteria CPOS.

An eventually wider used bulletin is published by SINAPI (Sistema Nacional de Pesquisa de Custos e Índices) which is, in its turn, a Caixa Econômica Federal (a Brazilian federal public bank) initiative that became to be mandatory in works that use federal resources or in contracts that involves public companies or mixed companies (with public capital participation) [7] e [8].

Since 2013, The SINAPI system is under a review process, and implanted a tree process system in order to identify the more types of services as possible, rising considerably the number of services in the compilation [9]. Below, there is a list of topics presented beyond the composition:

- Items and characteristics
- Equipment
- Quantification criteria
- Gauging criteria
- Execution
- Complementary information

- Pendencies

Figure 2 shows the process tree of the roughcast service. In this figure one can realize that there is thirty-two possible ways to do this service and, consequently, to describe it and associate to a specific cost.

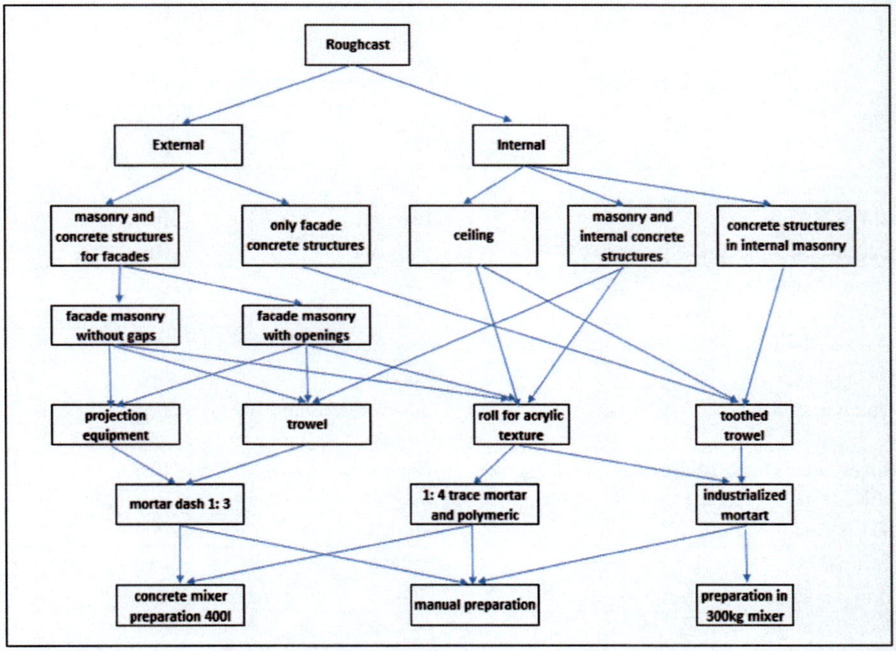

Fig. 2. Tree process systems of the roughcast from SINAPI.

Figure 3 illustrates one type of the SINAPI roughcast composition.

Code		composition description		unity
01.REVE.CHAP 012/02		ROUGHCAST APPLIED IN MASONRY (WITH THE PRESENCE OF VACES) AND FACADE CONCRETE STRUCTURES, WITH TROWEL MORTAR TRACE 1: 3 WITH		M²
Code SINAPI				
87905				
		COMPOSITION		
Item	Code	Description	Unity	Coefficient
C	88309	MASON WITH COMPLEMENTARY CHARGES	H	0.1830
C	88316	SERVANT WITH COMPLEMENTARY CHARGES	H	0.0910
C	87313	MORTAR TRACE 1: 3 (CEMENT AND THICK SAND)	M³	0.0042

Fig. 3. One roughcast composition from SINAPI.

1.3 BIM Model Element Classification

To elaborate the bidding spreadsheet it is necessary to classify elements according to some measurement or classification system. Generally, each country develops its own classification system, and it is used to standardize terminologies, to organize and import information, structure documentation and cost calculations [10].

Chart 1 summarizes some known classification systems.

Chart 1. Construction Information Classification Systems (CICS) [11].

CICS	Origin	Year of publication
Aus-Spec	Australia	–
BSAB 2.0	Sweden	2016
Cuneco Classification Systema (CCS)	Denmark	2016
Japan Construction Classification System (JCCS)	Japan	2006
Co-ordinated Building Information (CBI)	New Zealand	1997
NS 3451:2009	Norway	2009
OmniClass	United States and Canada	2006
Singapore Standard Industrial Classification SSIC	Singapore	2015
STLB-Bau	Germany	2004
Talo 2000	Finland	2010
Uniclass 2015	UK	2015

Some researchers already done application tentative of SINAPI system in the BIM context. Sakamori [12] showed a way to do the integration with Autodesk Revit when he assign the SINAPI code in the keynote field of each element together with an WBS code (Work Breakdown Structure) in order to organize the extracted spreadsheet that afterwards was used linked to SINAPI compositions. Felisberto [13] used Autodesk Revit and Autodesk Navisworks to do the classification with help of filtering to aid the user to choose the correct SINAPI code. Also made an WBS, but now in Navisworks and finalized the budget process via spreadsheet. He also found a difference in quantification results caused by distinct classification approaches (CEF and AD HOC) illustrated in the Table 2.

Table 2. Differences between results [13]

SINAPI	Description	Local	CEF method	AD HOC method
87266	Ceramic Half Height < 5 m^2	Internal	192.58	516.01
87267	Ceramic Half Height > 5 m^2	Internal	552.01	126.24
87527	Plaster for ceramics < 5 m^2	Internal	71.84	516.01
87529	Single grease for painting	Internal	974.50	436.09
87531	Plaster for ceramics 5 a 10 m^2	Internal	46.83	126.24
87536	Plaster for ceramics > 10 m^2	Internal	442.13	–
87878	Roughcast	Internal	1,110.70	1,078.34
87779	Plaster or single mass with gaps	External	244.83	422.49
87797	Plaster or single mass without gaps	External	684.92	302.32
87894	Roughcast without gaps	External	1,017.98	302.32
87905	Roughcast with gaps	External	336.35	422.49

Differences between results explicit that despite classification technology are functional, during the process of choose the right code there are moments of indecision, crossroads, that can be due to the modeling complexity, the high detail level of the measurement criteria and the also high quantity of options of services in the SINAPI system.

Another classification way found in the literature is through the Autodesk Revit tool called Classification Manager. It seems to be very practical but no real application was found or experienced yet by the authors.

1.4 Classification of Irregularities in Public Works

In Brazil, the Federal Constitution Court (TCU) publishes the Fiscobras report annually, which indicates the result of a series of inspections of public works according to some selection criteria. Thus, each inspected work is classified into three types of irregularities. Some of these irregularities come from deficiencies in the bidding spreadsheet, due to overestimated quantities, making possible undue payments (overbilling).

However, the TCU classification is based on the attitude that the public manager must take regarding the work. Thus, the first type is irregularity with recommendation of stoppage (called IGP), when the deviated value is considerable in relation to the contractual value, thus there is an immediate indication of stoppage of the work or cancellation of the bidding or contract. The second type is irregularities with partial retention of values (called IGR), which occurs when there is a financial loss for the public manager, however, it is possible to reverse this situation with reversal of payment or retention of values, without the work must be stopped. The third type is irregularities that do not affect the continuity of the work (called IGC), in which there are indications of some irregularities that must be explained to the manager, and fines may occur in some cases, but there is no need to stop the work [14].

In 2019, of the 77 works inspected by TCU, 59 presented serious irregularities (IGP or IGR or IGC). Figure 4 illustrates a work classified as IGP [14].

Work

Implementation of the new stretch of BR-040 RJ to climb the Serra de Petrópolis

Overall estimated cost of the work

R$ 870 milhões (US$ 170.58 millions)

Percentage executed

35.13%

Findings

overpricing in the construction budget

How much should be reimbursed to public coffers

R$ 870 milhões (US$ 54.11 millions)

Fig. 4. Adapted from Fiscobras 2019 reports [14].

Thus, there is no adoption of a system for classifying inconsistencies in the bidding spreadsheet, which could be used to review it prior to bidding, or as a way to point out corrections during the execution of the work, without necessarily having such inconsistent services already been measured, causing financial losses for the public manager. Such classification of inconsistencies system in the bidding spreadsheet could thus make it difficult to overbilling public works.

2 Justification

The present research is justified by the importance of studies that contribute to the development of improvements in the process of classifying the elements of a model in BIM, mainly for public works. Also, for setting out to create a tool for classifying inconsistencies in budgets, thus collaborating for inspection during the works or even for corrections in the bidding spreadsheet prior to the effective bidding.

3 Objectives

The main objective of the study is to demonstrate improvements that can be made in the SINAPI and CPOS measurement criteria in order to provide a suitable integration with the BIM model and a consequent greater flexibility, as well as facilitating execution by the construction company.

As a secondary objective is to develop a system to classify the inconsistencies of a budget, as an additional tool to help revision of the bidding spreadsheet or even a check tool to be used during work execution.

4 Methodology

4.1 Design Science Research (DSR)

According to Lacerda [15], the DSR proposes to build a prescriptive artifact, that is, something that have the capacity to solve a real problem, which causes a positive interference. There are four types of artifacts, summarized in Chart 2.

Chart 2. Types of artifacts in the DSR [15].

Description		
Types of artifacts	Constructs	Constructs or concepts form the vocabulary of a domain. They constitute a concept used to describe problems within the domain and to specify the respective solutions. Conceptualizations are extremely important in both natural and design sciences. They define the terms used to describe and think about the tasks. They can be extremely valuable to designers and researchers
	Models	A model is a set of propositions or statements that express the relationships between constructs. In design activities, models represent situations such as problem and solution. It can be seen as a description, that is, as a representation of how things are. Natural scientists often use the term 'Model' as a synonym for 'theory', or 'models' as theories that are still incipient. At Design Science, however, the concern is the usefulness of models, not the adherence of their representation to the Truth. Nevertheless, although it tends to be inaccurate about details, a model must always capture the structure of reality to be a useful representation
	Methods	A method is a set of steps (something-rhythm or orientation) used to perform a task. Methods are based on a set of underlying constructs (language) and a representation (model) in a solution space. The methods can be linked to the models, in which the steps of the method can use parts of the model as an input that composes it. In addition, methods are often used to translate a model or representation in a course for solving a problem. The methods are typical creations of research in Design Science
	Instances	An instantiation is the realization of an art-fact in its environment. Instances operationalize constructs, models and methods. However, an instantiation can, in practice, precede the complete articulation of its constructs, models and methods. Instances demonstrate the feasibility and effectiveness of the models and methods they contemplate

For the creation of an artifact, the research is developed from a BIM model of a penitentiary, of the Riversul-SP project, a work that is in the bidding phase. The model was designed from the 2D plans of the CAD and the analysis of the bidding spreadsheet was available in the bidding material. The bidding was based on the SINAPI (november 2018) and CPOS measurement criteria (november 2018).

5 Results and Discussion

5.1 Proposal for an Inconsistency Classification System for a Bidding Spreadsheet

In this research, according to inconsistencies in the classification that includes a bid planning action under study that was presented, an association of a type of inconsistency was made in a generic way, so that it can be used for other worksheets, or as a complementary way in classification of irregularities used by TCU.

Four main types of inconsistencies classification were elaborated, described below:

- type 1: such inconsistency occurs by mistakes in the classification of the elements, this directly impacts the extraction of quantitative, that is, it generates erroneous quantitative for the bidding spreadsheet;
- type 2: such inconsistency is observed due to the lack of interpretation of the service or a more accurate analysis of the composition of the remuneration for that service, generating the need to include additional services and thus causing double remuneration.
- type 3: this inconsistency occurs through the global analysis of the bidding, that is, individually the classification is correct, but according to its interaction with other classifications, it generates the need for a non-applicable execution process, and that, consequently, the spreadsheet will need to be corrected. It can be said that the type 3 inconsistency is more subtle, as it is necessary to know the compositions and their interrelations in the overall executive process of the work.
- type 4: this inconsistency is based on the lack of further information on the services measurement criteria, including their compositions. Therefore, when using more than one measurement criteria, the budgeter often choose the lowest cost service, but without knowing exactly the differences in remunerations for similar services and different criteria, due to the lack of information in some of their compositions.

5.2 Review of the SINAPI and CPOS Criteria for Some Modeled Items

Below, some modeled items are presented, such as understanding about the classification choice process, questioning this decision and, as it was possible, defining a relationship with the type of inconsistency present in the classification process, according to the types proposed.

Concrete Ballast for Foundation Blocks. SINAPI classification has three alternative, using the Ballast Notebook - Lot 01, with the following type tree as shown in Fig. 5.

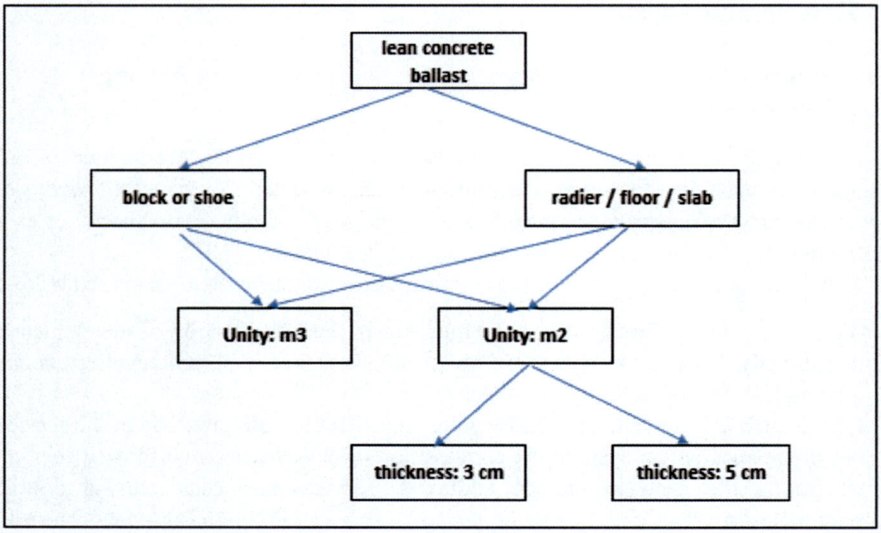

Fig. 5. Service tree for lean concrete ballast

What can be generally noted, and also by the compositions that are identical, is that the same service was split into three, simply to create services with a different measurement unit, from cubic meter (m^3) to square meter (m^2), this one with two alternatives of thickness already determined. As in the composition, in addition to the lean concrete to be used, there is also labor, it is understood that this remuneration is due to the launching service of the same.

Considering the CPOS, there are six other alternatives, three for non-structural machined concrete and three for non-structural concrete executed on site. It would still be necessary to add the launch service for the six cases of non-structural concrete.

The project only contains thin concrete ballast 5 cm thick. For the bidding spreadsheet, the SINAPI 96616 service - SLIM CONCRETE LASTRO, APPLIED TO CROWN BLOCKS OR SHOES was chosen. AF_08/2017. And a CPOS service is used to pay for the launch, being CPOS 11.16.020 - Launching, spreading and thickening of concrete or ballast mass and/or filling. Hence an error occurs, since the SINAPI 96616 service for lean concrete ballast differs from the SINAPI 94968 service for lean concrete precisely because it pays for its execution. Therefore, the remuneration apart from the launch service by the CPOS code 11.16.020 for the execution of the ballast is undue. Thus, there is a lack of analysis, not only of the services, but of their compositions, so that the bidding worksheet is prepared in the most correct way possible. In the event of payment for the launch of the ballast, it could be called invisible overpricing, since none of the parties (budgetist, fiscalization and construction company) became aware of this. In this way, it is also suggested to include in the title of the ballast service "including launch", thus facilitating the understanding of remuneration (Fig. 6).

Code	Criteria	Description	Unity	Quantity	Price	Total
96616	sinapi	lean concrete ballast applied to blocks	m3	5.33	424.56	2,262.90
94968	sinapi	lean concrete for ballast trace 1:4,5:4,5	m3	10.29	231.23	2,262.90
11.16.020	cpos	launching, spreading and thickening of concrete or ballast or filling mass	m3	15.62	54.78	855.66

Fig. 6. Part of the worksheet illustrating the inconsistency.

Thus, it can be classified as a type 2 inconsistency, as there was a lack of knowledge that the ballast service already pays for the launch service.

In Autodesk® Revit, it is not possible to include in addition to the classification, through the Classification Manager tool, also the composition of each service so that there is greater support in the decision of the budgetist.

Thus, it can be concluded that SINAPI services could be reduced to just one measurement criteria, preferably with a measurement unit in m^3, as it offers greater flexibility in the inspection of the work, as it pays for what is actually performed, up to the limit the thickness of the project, making it possible to punish the construction company, as long as the service does not want to be redone, for technical reasons. Unfortunately, ballast services are usually overlooked by construction companies, or performed in smaller thicknesses, making this service an opportunity for overpricing.

Regularization in Lean Concrete of Cyclopean Concrete. According to the project, this 3 cm layer separates the cyclopean concrete from the floor to be executed. In the author's practice, construction companies execute the structural concrete of the final layer directly on the cyclopean concrete, due to the practicality and non-uniformity of the stones, but usually measure the two services, regularization and structural concrete.

In this case, the budgetist also has the option of six types of non-structural concrete from CPOS and the addition of the launch of CPOS. By SINAPI, the lean concrete ballast floor service, code 99620, measured in m^3 or code 995240, measured in m^2 with a thickness of 3 cm, could be chosen. Another option would be to choose from SINAPI the lean concrete service for ballast, with three options that differ by the method of preparation (in 400 l, 600 l or manual mixer). The budgetist's choice was thin concrete in a 600 l concrete mixer, with the addition of CPOS's ballast launching service. The author found no specifics regarding the 600 l concrete mixer requirement, so it was probably an arbitrary choice by the budgetist.

As there is very little financial difference (approximately R$ 4.00 per m^3), between the unit values of the services in a concrete mixer, it could be reduced to just one type of preparation in a concrete mixer and another manually. Thus, it would bring flexibility regarding the use of the type of concrete mixer on site, considering that the same worksheet contains other types of concrete produced in a 400 L concrete mixer, which would be unfeasible for the performer to strictly follow each measurement criteria to have the correct remuneration.

In this way, there is the occurrence of type 3 inconsistency, as it is unlikely that the performer will strictly follow the manufacture of lean concrete in a 600 l mixer and in another nearby location, the execution of concrete will be in a 400 l mixer, just so that the remuneration is as faithful as possible. Therefore, it is clear that the best solution is to standardize the type of concrete mixer.

Structural Concrete - Floor, Blocks and Baldrame Beams. The structural concrete layer of fck = 40 Mpa, has a 20 cm thick design and was modeled as the first layer of the entire floor. As this service is currently performed by pumping, there is only one option at CPOS that was chosen in the bidding spreadsheet. At SINAPI there is no such concrete for pumping.

However, the launch has its separate remuneration. At CPOS, despite having a pumping launch service, the budgetist indicated the SINAPI pumping launch service, due to the considerable difference to less than the unit price (something around R $ 100.00 per m^3). When checking the composition of this service at CPOS, it is not in a detailed form, thus making an immediate comparative analysis of the measurement criteria impossible. The same concrete is used for the blocks, which were modeled with the shoe tool and baldrame beams, modeled as beams.

Masonry. Next to the cell sector, there is the multiple use sector, which includes bathrooms, divided into masonry, for the use of visitors. The author did not find any further specifications on this masonry. The budget spreadsheet contains the CPOS code 14.10.101 – 9 × 19 × 39 cm sealing concrete block masonry - class C. Due to the composition of this service, it is not possible to know which form of mortar preparation is being paid for, whether by hand or by a concrete mixer.

For SINAPI, there are several service variables with the use of this masonry, and it is necessary to choose, in addition to the type of masonry, which is the net area to be applied (greater or less than 6 m^2), whether it has openings or not, and finally, how to prepare the mortar for laying, whether it is by hand or with a concrete mixer. In this case, there is no differentiation in the type of concrete mixer, whether 400 l or 600 l, as is the case for concrete production services.

Thus, in order to classify by SINAPI, it is necessary to check the net area of each masonry (Table 3).

Table 3. Masonry quantities

Survey of wall material		
Material	Description	Area (m^2)
Masonry 9 × 19 × 39 cm	Masonry 9 × 19 × 39 cm of the visitors' toilet	8.32
Masonry 9 × 19 × 39 cm	Masonry 9 × 19 × 39 cm of the visitors' toilet	2.65
Masonry 9 × 19 × 39 cm	Masonry 9 × 19 × 39 cm of the visitors' toilet	2.65
Masonry 9 × 19 × 39 cm	Masonry 9 × 19 × 39 cm of the visitors' toilet	0.63
Masonry 9 × 19 × 39 cm	Masonry 9 × 19 × 39 cm of the visitors' toilet	0.81
Masonry 9 × 19 × 39 cm	Masonry 9 × 19 × 39 cm of the visitors' toilet	0.73
Masonry 9 × 19 × 39 cm	Masonry 9 × 19 × 39 cm of the visitors' toilet	0.65
Masonry 9 × 19 × 39 cm	Masonry 9 × 19 × 39 cm of the visitors' toilet	0.81
Masonry 9 × 19 × 39 cm	Masonry 9 × 19 × 39 cm of the visitors' toilet	0.75

According to the table extracted from masonry, only one masonry has a net area greater than 6 m^2. None of the extracted masonry has gaps. Referring to the current unit price table, we see that the difference in the choice, for a masonry of the same area and in the same condition of the existence or not of openings, according to only the type of preparation is less than one real per m^2.

It would also be arbitrary, to assume, what would be a difference in the unit price of service compositions so that there could be a decrease in the possibilities of choice. Or a representative composition could be created, a practice already carried out by SINAPI, which consists of the merger of some similar compositions in order to facilitate the classification process. It occurs in this case, that excluding the need to define the method of preparing the mortar for laying, would already reduce the possibilities of choice and would give greater flexibility to the performer, not tying him in details of the manufacturing method process so that he has a correct remuneration.

In this way, there is an inconsistency in the type 4 classification, because, in this case, there is a lack of information in the composition of the CPOS to elucidate the differences in remuneration.

Rough Cast. Also for rough cast the budget spreadsheet contains the CPOS code 17.02.020 - Chapisco, and its composition includes the use of a concrete mixer for its manufacture. At CPOS there are still four other possibilities for roughcasting that differ by adding material or type of sand, which does not apply.

For SINAPI, there are 22 alternatives for plastering for masonry, the differentiation occurs due to the presence of gaps or not, the means of applying the plaster, the type of mortar and the form of mortar production. Note to the rough-cut, a certain complexity in their choice, because the budgetist must define a series of items, which is often not specified in projects or memorials. This is also reflected in the budget spreadsheet, which at times presents, at random, varieties of types of roughcast, and in execution, there will be virtually no variation after a chosen execution method (Fig. 7).

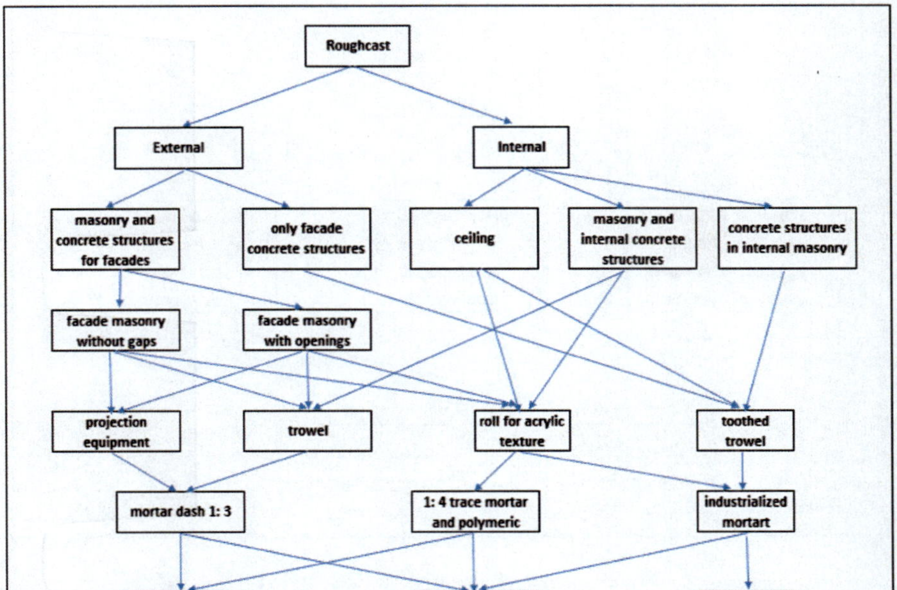

Fig. 7. Types of roughcast from SINAPI

For the use of the classification manager, there was also the possibility of including more than one service in the same element, however, as they are services with different measurement areas (roughcast = 2 × masonry), it is necessary to prepare this multiplication.

Single Plaster/Grease The modeling of the plaster was also a layer of the composite wall. For the bidding spreadsheet, the CPOS code 17.02.140 - Plaster made with polyester foam was chosen. At CPOS there are two more plastering options, one for industrialized mortar and the other for common plaster.

At SINAPI, first, the plaster is divided between two groups, for internal and external masonry. In this division, however, there is a doubt about what should be considered an external wall? Would it be a face that is exposed to the sun and bad weather most of the day? Note the division of the plaster designed, mainly for apartment buildings, with groups facing the plaster of saca-das (internal and external), cloths for facades and internal masonry.

As the masonry is in a covered place, even subject to the weather and exposition to the sun, partially, due to the high ceiling (5 meters), for a SINAPI classification they would all be considered as internal. As there is no application of ceramics, the services of single mass for painting are the most appropriate and in order to define which exact service for the classification, it is necessary to define the type of mortar, the method of its preparation and its form of application, as well as the thickness of the final layer (Fig. 8).

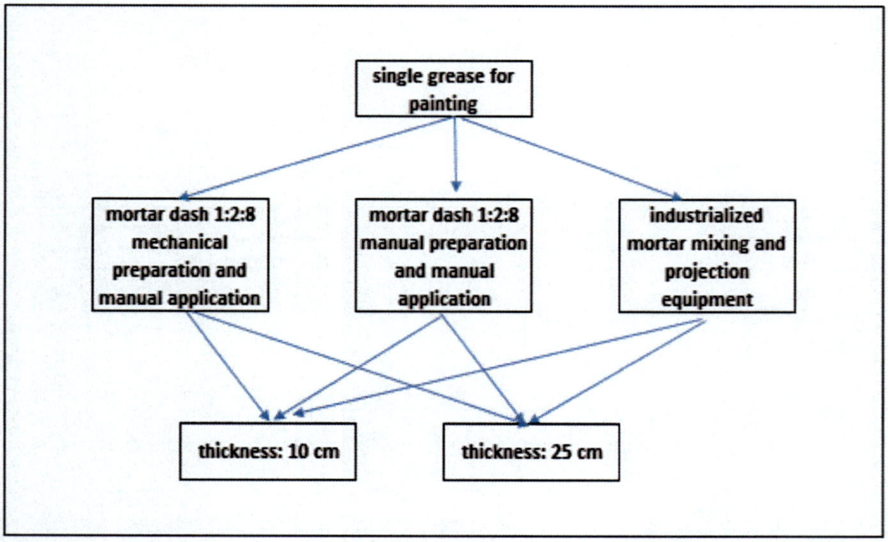

Fig. 8. Types of single mass from SINAPI

Thus, the inconsistency of type 4 is noted, since in the composition of the CPOS there is no specification of the thickness to be applied, thus making comparison difficult. Since the CPOS service is the cheapest, this was possibly the reason for the choice.

6 Conclusion

According to the modeled items, it was possible to point out possible simplifications in the compositions of some services, resulting in less variation for the same type of service. Thus, causing greater flexibility to the constructor, without economically unbalancing the spreadsheet.

The measurement criteria for a modeling process are a guide for adding the necessary information to the elements, that is, they will establish what the **LoI** will be for each one. Therefore, it is understood that the requirement of too much information for each element, so that its classification can be carried out, is a way contrary to the simplification and time gain in the processes that the BIM technology is based on. Thus, a measurement criteria, which requires less information, without causing financial or understanding losses to the constructor, and seem to be more in line with the BIM technology proposal.

This research also serves as an incentive for other countries to verify possibilities of simplifications in the measurement criteria, with the perspective of optimizing the classification of elements in BIM for budgetary purposes.

The inconsistency classification system can still be improved, according to new examples. Its initial version already demonstrates its ability to assist in a process of revising the bidding spreadsheet, even before its publication or as a complement to the classification of irregularities used by TCU.

The Brazilian work public structure can be not a paradigm but, mutatis mutandis, regard its complexities or idiosyncrasies, we believe that it can be used as a case study, especially with some public works that already have some BIM aspects contemplated.

Of course, quantity extraction best practices all over the world needs to be closely followed, i.e., no national effort in organization of its internal process will have success without taking into account what is happening in the major organizations or proposed by serious researches like Mattern, Scheffer and König [16]. This work starts from these propositions to achieve the more positive results possible from BIM modeling associated with a WBS specific to public works and the harmonization of these with the government budget and pricing surveing process.

References

1. BRASIL. Painel de obras. Ministério da Economia. http://transferenciasabertas.planejamento.gov.br/QvAJAXZfc/opendoc.htm?document=painelcidadao.qvw&lang=en-US&host=QVS%40srvbsaiasprd01&anonymous=true. Accessed 23 Jan 2020
2. MDIC. Estratégia Nacional de Disseminação do BIM (2018b). http://www.mdic.gov.br/images/REPOSITORIO/sdci/CGMO/26-11-2018-estrategia-BIM-BR-2.pdf. Accessed 01 Oct 2019

3. Jernigan, F.: BIG BIM little bim - the practical approach to building information modeling - Integrated practice done the right way! Kindle edition, p. 22. 4Site Press (2008)
4. Eastman, C., Teicholz, P., Sacks, R., Liston, K.: Manual de BIM - Um guia de modelagem da informação da construção para arquitetos, engenheiros, gerentes, construtores e incorporadores. Bookman Editora, Porto Alegre (2014)
5. BIM Building Information Modeling NO BRASIL E NA UNIÃO EUROPEIA. http://sectordialogues.org/sites/default/files/acoes/documentos/bim.pdf. Accessed 01 Feb 2020
6. SINAENCO (Brasil). Normas técnicas e BIM serão tema de encontro sobre a modelagem da informação da construção (2018). http://sinaenco.com.br/noticias/normas-tecnicas-bim/. Accessed 14 Jan 2020
7. BRASIL. Decreto nº 7.983, de 8 de abril de 2013. Estabelece regras e critérios para elaboração do orçamento de referência de obras e serviços de engenharia, contratados e executados com recursos dos orçamentos da União, e dá outras providências. Diário Oficial, Brasília, DF, 8 de abril de 2013
8. BRASIL. Lei N° 13.303, de 30 de junho de 2016. Dispõe sobre o estatuto jurídico da empresa pública, da sociedade de economia mista e de suas subsidiárias, no âmbito da União, dos Estados, do Distrito Federal e dos Municípios. Diário Oficial, Brasília, DF, 30 de junho de 2016
9. SINDUSCON – MT. http://sindusconmt.org.br/imprimir-noticia.php?cat=6108. Accessed 01 Feb 2020
10. Eastman, C.M., Afsari, K.: A Comparison of Construction Classification Systems Used for Classifying Building Product Models (2016). http://ascpro0.ascweb.org/archives/cd/2016/paper/CPRT198002016.pdf. Accessed 21 Jan 2020
11. Nunes, H.M.: Sistemas de Classificação de Informação da Construção: Proposta de metodologia orientada para objetos BIM. 2016. 102 f. Dissertação (Mestrado) - Curso de Faculdade de Ciências e Tecnologia, Universidade Nova de Lisboa, Lisboa (2016). https://run.unl.pt/bitstream/10362/20165/1/Nunes_2016.pdf. Accessed 21 Jan 2020
12. Sakamori, M.M.: Modelagem 5D (BIM): processo de orçamentação com estudo sobre controle de custos e valor agregado para empreendimentos de construção civil/ Marcelo Mino Sakamori. – Curitiba (2015). 178 f
13. Felisberto, A.D.: Contribuições para elaboração de orçamento de referência de obra pública observando a nova árvore de fatores do SINAPI com BIM 5D - LOD 300 (2017). 231 p.
14. TCU, Fiscobras (2019). https://portal.tcu.gov.br/biblioteca-digital/fiscobras-2019.htm. Accesses 01 Feb 2020
15. Lacerda, D.P., et al.: Desing Science Research: método de pesquisa para a engenharia de produção. Gestão & Produção, São Carlos, v. 20, n. 4, pp. 741–761, out. 2013
16. Mattern, H., Scheffer, M., König, M.: BIM-Based Quantity Take-Off: Technology Foundations and Industry Practice (2018). https://doi.org/10.1007/978-3-319-92862-3_23

Time-Cost Trade-off Optimization Incorporating Accident Risks in Project Planning

Moein Sadeghi and Ming Lu(✉)

University of Alberta, Edmonton T6G 2R3, Canada
{moein,mlu6}@ualberta.ca

Abstract. Time, cost and safety are essential elements in planning a construction project. A cost-effective project planning method is desired to reduce project cost and duration, while maintaining an acceptable safety level. Although there are many studies addressing safety management and time-cost trade-off optimization separately, less attention has been given to the effect of accident risks in tackling the time-cost trade-off problem. This research has produced a model that factors accident risks in the classic time-cost trade-off problem definition and results in an optimum solution for the time-cost-safety trade-off (TCST) problem. Critical path method is utilized to find the longest duration and the total cost of the project. Then a risk index is developed to add safety assessment to the decision-making process. Finally, the TCST problem is formulated as a nonlinear programming model. A case of a one-story building project with five what-if scenarios is presented to demonstrate and validate the proposed decision-making model using the Excel Solver. The model results show with a proper risk assessment scheme and a proper risk penalty cost setting, accident risks can be incorporated into the TCT optimization while implications of various risk management strategies can be analyzed quantitatively.

Keywords: Time-cost trade-off · Risk index · Nonlinear programming · Construction management · Project scheduling

1 Introduction

Project management aims at achieving on-budget and on-time project delivery subject to multiple constraints. It is a challenge to complete the project with limited resources on time and make clients satisfied due to the uncertain nature of the construction industry [1]. When there is a practical need to complete the project even before the original planned date, the duration of some activities must be reduced by changing execution methods on activities along the project critical path. In general, increasing labor or equipment resources can decrease the duration, but it adds extra direct cost to the project [2]. Therefore, project planners desire to find a trade-off between project cost and time by utilizing techniques like Critical Path Method (CPM) based time-cost trade-off (TCT) analysis [3].

E. Toledo Santos and S. Scheer (Eds.): ICCCBE 2020, LNCE 98, pp. 643–653, 2021.
https://doi.org/10.1007/978-3-030-51295-8_45

The objective of TCT is to reduce the original project duration, to meet a specific deadline with the minimum direct and indirect costs of the project [4]. The direct cost refers to crew installation costs, including material, labor, and equipment. In contrast, indirect costs are the necessary costs to facilitate project execution, which are generally not attributable to particular activities but to the whole project. There are many studies about optimizing the TCT problem by utilizing various computing methods, each having its advantages and disadvantages; some are more practical for large-size projects but may not guarantee to reach the optimum solution [5]. Moreover, researchers have attempted to combine heuristic and analytical methods to address the limitation of the existing TCT optimization methods, for instance, integrating the path-float based CPM and integer programming (IP) to optimize the TCT problem [5].

According to the *2017 Workplace Injury, Disease, and Fatality Statistics* in Alberta, Canada, 38.4 percent of occupational fatalities occurred in the construction industry from 2013 to 2017; moreover, the construction industry fatality rate was 167 fatalities per million person-years in 2017 which is higher than the all-industry average [6]. The workplace accidents also exert financial losses on the workers, project and society [7]. However, investment in safety risk prevention programs can be conducive to reducing the number of accidents, so the cost of prevented accidents can be considered as saving [8]. Recent statistics and related studies show that the likelihood of accident occurrence depends on the accident type; moreover, the cost of an accident is associated with the severity and type of accident [6, 9–11]. Therefore, in this study, the risk index is defined to represent the likelihood and the severity of an accident in the model. These facts justify the need for a decision-making model that can optimize the time-cost tradeoff of the project while considering accident risks.

Numerous methods have been proposed for TCT optimization in literature. Many computational optimization techniques have been used for reducing project duration recently, like Evolutionary Algorithms (EA), including genetic algorithms (GA) and particle swarm optimization (PSO) [12–14]. In this study, the Excel Solver software is utilized as the optimization tool to solve the newly formulated time-cost-safety trade-off (TCST) problem in a case study.

Different sources of information and studies related to this research ranging from project scheduling techniques (including CPM, TCT) and safety science to computational optimization techniques are consulted. Subsequently, the TCST problem is formulated, the optimization model is developed, verified by a case study (i.e. a conceptual one-story building project) consisting of five what-if scenarios to investigate the implications of various risk ratings and the impact of assigned coefficient to factor safety ratings in the cost function in TCT optimization. Finally, the results from the model are analyzed and interpreted before drawing conclusions.

2 Methodology

2.1 Overview of TCST Optimization

The Time-Cost-Safety Trade-off (TCST) optimization problem extends the established TCT optimization problem in project scheduling. To factor in accident cost, a risk index is developed based on assessment of the likelihood of accident occurrence and the severity of the accident in connection with a certain method being planned for a particular activity. The risk index components are rated from zero to ten, with zero and ten denoting the lowest and highest weight, respectively. Next, CPM is applied to find out the critical path and critical activities, resulting in the normal project duration. Then a desired (crash) project duration is set for the project model as a target. Setting the desired project duration helps project managers to identify the optimum solution for the project duration, which is reduced one day in each cycle of the proposed solution by consuming the maximum available crash duration on all the activities. In the next stage, the cost slope and the risk slope of activities are calculated. Then a heuristic method is utilized to reduce duration on selected activities with the focus set on shortening the critical path: critical activities with lower cost slope and risk slope are selected for crashing in order to reach the desired project duration. A penalty cost is set based on the risk index definition in the model, which represents the potential increase on cost due to accident risks and is added to the total cost of the project.

In this research, the TCST problem is formulated as a nonlinear programming model. The model is analyzed by using the Solver software to minimize the total cost factoring in accident risks of the project. It is noteworthy for illustrating TCST optimization; duration and cost of activities remain the same, while the likelihood, severity and penalty cost of accidents related to activities are varied in specification of postulated what-if scenarios in the case study.

2.2 Problem Formulation

A project can be executed faster by crashing the duration of critical activities, but it adds cost and safety risks to the project. The success of the project depends on the ability of the planner to select and crash activities to meet the project completion date in a way that result in the lowest additional cost factoring in accident risks. This study aims to investigate the impact of accident risks associated with duration reduction on the TCT problem by developing a nonlinear model.

Based on the above stated characteristics of the TCST problem, the mathematical model is described as follows. For each activity ($i = A, B, C, \ldots$) in each cycle ($j = 1, 2, 3, \ldots$), the *risk index* (*RI*) is calculated by multiplying the likelihood (*L*) and the severity of the activity (*S*).

$$RI = L \times S \quad (1)$$

The available crash duration (ACD) of an activity is the difference between the normal duration (ND) and the crash duration of an activity (CD).

$$ACD_i = ND_i - CD_i \quad (2)$$

The *cost slope* (*CS*) of an activity is calculated by using the Eq. (3) where (*NDC*) and (*CDC*) are normal and crash direct cost, respectively. In a similar way, the normal risk index (*NRI*) and crash risk index (*CRI*) are utilized to calculate the *risk index slope* (*RIS*) by using Eq. (4).

$$CS_i = \frac{CDC_i - NDC_i}{ACD_i} \tag{3}$$

$$RISi = \frac{CRI_i - NRI_i}{ACDi} \tag{4}$$

The direct cost of an activity (*DC*) and the indirect cost (*IC*) of the project are calculated by following equations. Where (*R*) is the amount of the reduction in the duration of an activity; (*DPD*) is the desired project duration for a cycle and (*ICD*) is the indirect cost per day.

$$DCi = NDCi + (Ri \times CSi) \tag{5}$$

$$IC = DPD_j \times ICD \tag{6}$$

Equation (7) is used to calculate the risk index (*RI*) of an activity. Then, the risk index of activity is multiplied by the *penalty cost (PC) per unit of risk* rating in order to add the accident risks to the objective function of the model called *risk penalty cost* (RPC).

$$RI_i = NRI_i + (R_i \times RIS_i) \tag{7}$$

$$RPC_i = RI_i \times PC \tag{8}$$

The Objective Function. The objective function is minimizing the total cost of the project (T).

$$T = MIN. \sum DC_i + IC + \sum RPC_i \tag{9}$$

Decision Variable. R_i = Amount of the reduction in the duration of an activity.

Constraints. The constraints of the model are the following:

The Maximum reduction constraint.

$$R_i \leq ACD_i \tag{10}$$

The project duration constraint.

$$DPD_j = Desired\, project\, duration\, for\, each\, cycle \tag{11}$$

Integer constraints.

$$R_i = Integer \tag{12}$$

The non-negativity constraint.

$$R_i \geq 0 \tag{13}$$

Critical Path Activities Constraints. For reducing the duration of the project, the selected activities for crashing must be on the critical path (*CP*). This specific constraint is formulated in the Excel spreadsheet in such a way that after each cycle, for each path (*K* = *1, 2, ...*) the path duration (*D*) is updated based on crash duration in the previous cycle, then the path with the longest duration is identified as the *CP*. Next, the *ACD* of the activities on the non-critical paths for the current cycle is set as zero. So, the solver program does not go for the activities with a zero *ACD* according to the maximum reduction and non-negativity constraints. This constraint is formulated as follows:

$$D_{k,j+1} = \sum (ND_{i,k} - R_{i,k,j});\ for\ cycle\ j+1 \tag{14}$$

When the duration of two paths is equal (or where lengths for multiple paths are same and equal to current project duration) the ACD is not changed, but for activities that are specific to a path, the following conditions are applied:

$$If\ D_{k,j+1} < D_{k+1,j+1};\ then\ the\ path\ k+1\ is\ critical;\ so,\ ACD_{i,k} = 0 \tag{15}$$

$$If\ D_{k,j+1} > D_{k+1,j+1};\ then\ the\ path\ k\ is\ critical;\ so,\ ACD_{i,k+1} = 0 \tag{16}$$

This process is repeated for each cycle. Therefore, the model can update the *CP* and the *ACD* of the critical activities while keeping the non-critical activities out of the set of decision variables.

3 Case Study

The proposed model has been applied to a fictitious construction project for concept proof and practicality assessment. The AON diagram of the case study is shown in Fig. 1. Table 1 presents the activities' duration and cost data, which remain the same across all the scenarios. For this case study, based on normal activity settings, the total cost is $543,594 without considering accident risk penalty cost, and the project duration is 100 days. Furthermore, the indirect cost per day is assumed $2000. First, the RPC is set $25 for assessing different risk index settings to compare the impact of different likelihood and severity of accident risks. Next Scenario one with constant ratings for likelihood and severity of accident risks is further investigated by adjusting RPC (namely: $25, $50 and $250) in order to examine the impact of RPC on model's optimization results.

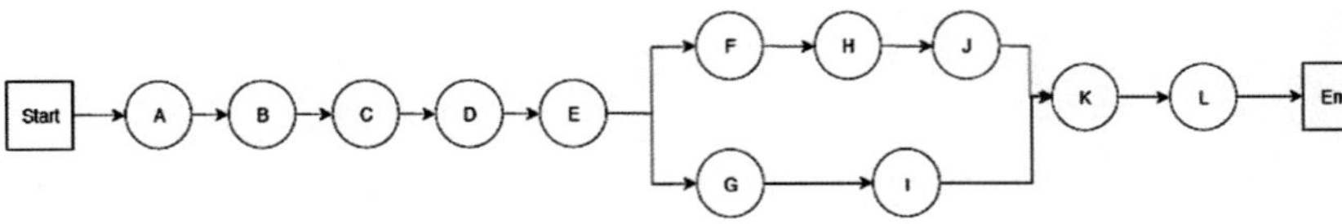

Fig. 1. Activity-On-Node (AON) network diagram: case study

Table 1. Case study project data

Activity	Predecessors	Normal		Crash		Cost slope ($/day)
		Duration (days)	Cost ($)	Duration	Cost ($)	
A		7	13,502	6	14,852	1,350
B	A	1	6,315	1	6,315	0
C	B	12	70,402	11	73,922	3,520
D	C	8	15,338	8	15,338	0
E	D	20	29,279	19	30,743	1,464
F	E	13	65,469	12	68,742	3,273
G	E	26	27,176	24	29,894	1,359
H	F	13	34,881	12	36,625	1,744
I	G	6	42,776	6	42,776	0
J	H	7	18,118	6	19,930	1,812
K	I, J	9	2,438	9	2,438	0
L	K	10	17,900	9	19,690	1,790

Besides, three distinct sets of rating data related to the risk index of project activities under normal duration are provided in Table 2, and data associated with crash duration are given in Table 3. Obviously, in a certain scenario, the risk index on one particular activity is higher in the crash case than in the normal case. In order to investigate model sensitivity to risk index settings, index values associated with activities with similar cost slopes are set to be different in designing the three what-if scenarios in the case study, in order to observe the effect of RPC upon the priority of activities for crashing. For instance, the cost slopes of activities A ($1,350) and E ($1,464) are close, so the risk indexes of Activity A is assigned higher in Scenario two and lower in Scenarios one and three than Activity E.

Table 2. Activities' risk index for each scenario (S_i) in connection with normal duration

Activity	Accident likelihood (1–10)			Accident severity (1–10)			Risk index		
	S1	S2	S3	S1	S2	S3	S1	S2	S3
A	2	7	4	1	7	2	2	49	8
B	6	5	2	4	4	1	24	20	2
C	4	4	7	6	6	10	24	24	70
D	4	8	3	5	10	4	20	80	12
E	5	2	6	4	1	6	20	2	36
F	3	1	3	2	3	5	6	3	15
G	7	2	4	8	4	4	56	8	16
H	3	8	5	1	6	9	3	48	45
I	2	7	3	1	2	5	2	14	15
J	4	6	8	4	9	7	16	54	56
K	3	5	5	1	1	3	3	5	15
L	2	1	2	1	2	1	2	2	2

Table 3. Activities' risk index for each scenario (S_i) in connection with crash duration

Activity	Accident likelihood (1–10)			Accident severity (1–10)			Risk index		
	S1	S2	S3	S1	S2	S3	S1	S2	S3
A	3	8	5	1	7	2	3	56	10
B	6	5	2	4	4	1	24	20	2
C	5	5	8	6	6	10	30	30	80
D	4	9	3	5	10	4	20	90	12
E	6	3	7	4	1	6	24	3	42
F	5	2	4	2	3	5	10	6	20
G	9	3	5	8	4	4	72	12	20
H	5	9	6	1	6	9	5	54	54
I	2	7	3	1	2	5	2	14	15
J	6	7	9	4	9	7	24	63	63
K	3	5	5	1	1	3	3	5	15
L	4	2	3	1	2	1	4	4	3

3.1 Results and Discussion

In this case study, the project manager wishes to gain insight in possible options for reducing project duration and corresponding consequences on costs and accident risks. The model is run with fixed activity cost and duration definitions across all the five what-if scenarios, each having various risk indexes and RPC to investigate the effects of accident risks on the TCST optimization problem. The results of the model for three scenarios with $25 RPC are shown in Tables 4, 5 and 6; Tables 7 and 8 show the model's results given Scenario one with $50 and $250 RPC, respectively.

Table 4. Results from Scenario one (RPC $25)

Cycles	Total safety index	Project duration (days)	Accident risk penalty cost ($)	Total cost ($)	Crashed activities
Normal	178	100	4,450	548,044	
1	179	99	4,475	547,419	A-1
2	183	98	4,575	546,983	E-1
3	185	97	4,625	546,777	H-1
4	**187**	**96**	**4,675**	**546,617**	L-1
5	199	95	4,975	548,087	G-1, J-1
6	209	94	5,225	549,858	C-1
7	221	93	5,525	552,790	G-1, F-1

Table 5. Results from Scenario two (RPC $25)

Cycles	Total safety index	Project duration (days)	Safety penalty cost ($)	Total cost ($)	Crashed activities
Normal	309	100	7,725	551,319	
1	310	99	7,750	550,807	E-1
2	317	98	7,925	550,333	A-1
3	319	97	7,975	550,173	L-1
4	**325**	**96**	**8,125**	**550,067**	H-1
5	336	95	8,400	551,512	G-1, J-1
6	342	94	8,550	553,182	C-1
7	347	93	8,675	555,940	G-1, F-1

Table 6. Results from Scenario three (RPC $25)

Cycles	Total safety index	Project duration (days)	Safety penalty Cost ($)	Total cost ($)	Crashed activities
Normal	292	100	7,300	550,894	
1	294	99	7,350	550,294	A-1
2	300	98	7,500	549,908	E-1
3	301	97	7,525	549,723	L-1
4	**310**	**96**	**7,750**	**549,692**	H-1
5	319	95	7,975	551,087	G-1, J-1
6	329	94	8,225	552,857	C-1
7	336	93	8,400	555,665	G-1, F-1

Table 7. Results from Scenario one with RPC being $50

Cycles	Total safety index	Project duration (days)	Safety penalty cost ($)	Total cost ($)	Crashed activities
Normal	178	100	8,900	552,494	
1	179	99	8,950	551,894	A-1
2	183	98	9,150	551,558	E-1
3	185	97	9,250	551,402	H-1
4	**187**	**96**	**9,350**	**551,292**	L-1
5	193	95	9,650	553,112	C-1
6	209	94	10,450	555,083	G-1, J-1
7	221	93	11,050	558,315	G-1, F-1

The optimum solutions for various scenarios (bolded in above tables) are different in terms of the total cost, the sequence of crashed activities, and the penalty cost due to the various index values on accident risks. Furthermore, the optimum solution has occurred after four days reduction on normal project time (i.e. 100 days), except in

Table 8. Results from Scenario one with RPC being $250

Cycles	Total safety index	Project duration (days)	Safety penalty cost ($)	Total cost ($)	Crashed activities
Normal	178	100	44,500	588,094	
1	**179**	**99**	**44,750**	**587,694**	A-1
2	181	98	45,250	587,938	H-1
3	183	97	45,750	588,228	L-1
4	187	96	46,750	588,692	E-1
5	193	95	48,250	591,712	C-1
6	209	94	52,250	596,883	G-1, J-1
7	221	93	55,250	602,515	G-1, F-1

Scenario one with $250 RPC. In this scenario, after one day reduction on project time, the optimum solution is reached due to the high weight of RPC.

By comparing the results from all the scenarios, it is concluded that although the direct and indirect costs of activities have a major impact on total project cost, the safety penalty costs affect the selection of the crashed activities and the final optimum solution. For instance, in Scenarios one and three the first crashed activity is Activity A, but in Scenario two the first one is Activity E. Table 9 shows the priority of the crashed activities for all scenarios and RPCs compared to basic scenario (S0) which is performing TCT analysis without considering accident risk indexes. The observed change in crashing priority and sequencing (bolded in the below table) is notable in five out of the six scenarios being investigated (except for “S1, $25”).

Table 9. Sequence of crashed activities

Scenario (S), Risk penalty cost ($)	Cycles	1	2	3	4	5	6	7
	Project duration (days)	99	98	97	96	95	94	93
S0		A-1	E-1	H-1	L-1	G-1, J-1	C-1	G-1, F-1
S1, $25		A-1	E-1	H-1	L-1	G-1, J-1	C-1	G-1, F-1
S1, $50		A-1	E-1	H-1	L-1	**C-1**	**G-1, J-1**	G-1, F-1
S1, $250		A-1	**H-1**	**L-1**	**E-1**	**C-1**	**G-1, J-1**	G-1, F-1
S2, $25		**E-1**	**A-1**	**L-1**	**H-1**	G-1, J-1	C-1	G-1, F-1
S3, $25		A-1	E-1	**L-1**	**H-1**	G-1, J-1	C-1	G-1, F-1

Moreover, Fig. 2 illustrates the effect of RPC on the model. The higher RPC makes the optimization model more conservative, while lower RPC lets the model take more risks. For instance, in Scenario one for desired project duration 95 days, where the RPC is $25 the model crashes Activities G and J which have higher risks (total RI = 203); in contrast, when RPC is set as $250, the model crashes C instead which has lower risk (total RI = 193).

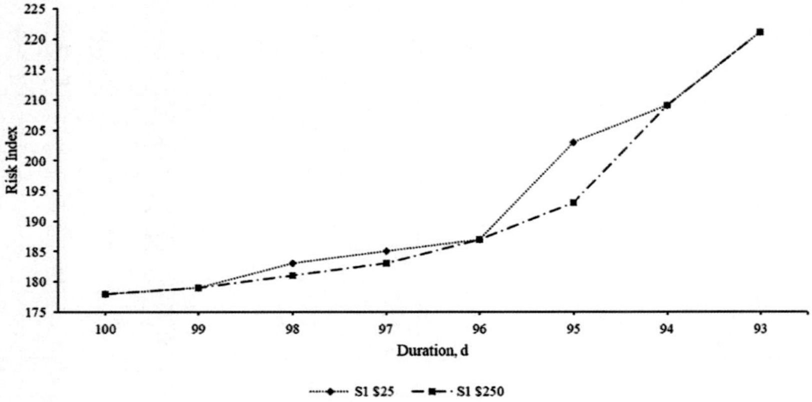

Fig. 2. Risk Index vs. Project Duration during TCST optimization: contrasting Scenario one with $25 and $250 RPC

Consequently, the results of the model have proved that adding accident risks to the TCT decision-making process would impact the optimization outcome and change action plan recommendations to project planners, through preventing accidents by crashing activities with lower safety risks instead of high-risk activities. Moreover, analyzing the results from Scenario one with various RPC settings has proved that with a proper risk assessment scheme and a proper risk penalty cost setting (RPC), accident risks can be incorporated into the TCT optimization by considering implications of various risk management strategies.

4 Conclusion

This research has investigated the impact of the safety related risks on the project time-cost trade-off (TCT) problem in scheduling. CPM method is used to fix critical path and project duration; then, TCT optimization is utilized to expedite project completion. To add accident risk assessment to TCT analysis, a risk index consisting of likelihood and severity of accidents for each activity associated with normal and crash plans is defined. A nonlinear programming approach is utilized to develop a model to optimize the new time-cost-safety trade-off (TCST) problem. The results of the model based on evaluating five scenarios in a case study show that the risk index can change the optimum decision in terms of project schedule and crashed activities. The novelty of the model lies in adding accident risks to the TCT optimization process and facilitating project management to set a target for project duration and investigate time, cost and safety consequences of each alterative.

Nonetheless, the lack of proper and practical information about the consequence of accidents would make it difficult to objectively evaluate safety risks in reality. Therefore, it is essential to conduct broad studies to provide a comprehensive database holding critical information related to construction accidents, including direct and indirect costs of accidents, likelihood, severity and types of accidents in connection with certain construction methods or activities. Then project managers will design simulation experiments based on activity planning and make less subjective decisions to improve risk assessment and risk cost evaluation through trial and error by taking advantage of the proposed methodology. As demonstrated in the case study, with a proper risk assessment scheme and a proper risk penalty cost setting (RPC), accident risks can be incorporated into the TCT optimization by considering implications of various risk management strategies.

Acknowledgements. The presented research is partially funded by Occupational Health & Safety (OHS) Futures Research Funding Program (Grant #095245033) of the Government of Alberta, Canada.

References

1. Chan, A., Scott, D., Chan, A.: Factors affecting the success of a construction project. J. Constr. Eng. Manag. **130**(1), 153–155 (2004). https://doi.org/10.1061/(asce)0733-9364(2004)130:1(153)
2. Modarres, M., Tavakkoli-Moghaddam, R., Pour, N.S.: Time-cost-quality trade-off in project scheduling with linguistic variables. World Appl. Sci. J., 404–413 (2012). https://doi.org/10.5829/idosi.wasj.2012.18.03.874
3. Kelley, J.: Critical-path planning and scheduling: mathematical basis. Oper. Res. **9**(3), 296–320 (1961). https://doi.org/10.1287/opre.9.3.296
4. Alavipour, S., Arditi, D.: Time-cost tradeoff analysis with minimized project financing cost. Autom. Constr. **98**, 110–121 (2019). https://doi.org/10.1016/j.autcon.2018.09.009
5. Nasiri, S., Lu, M.: Path-float based approach to optimizing time-cost tradeoff in project planning and scheduling. In: Computing in Civil Engineering, pp. 585–592 (2019)
6. 2017 Workplace Injury, Disease, and Fatality Statistics Provincial Summary. Alberta Labour, Government of Alberta (2017)
7. López-Alonso, M., Ibarrondo-Dávila, M., Rubio, M.: Safety cost management in construction companies: a proposal classification. Work **54**(3), 617–630 (2016). https://doi.org/10.3233/wor-162319
8. Hallowell, M.: Cost-effectiveness of construction safety programme elements. Constr. Manag. Econ. **28**(1), 25–34 (2010). https://doi.org/10.1080/01446190903460706
9. The Cost of Injury in Canada. Parachute, Toronto, ON (2015)
10. National Work Injury, Disease and Fatality Statistics 2015–2017. Association of Workers' Compensation Boards of Canada
11. Sun, A.Z.P.: Understanding the True Costs of Construction Accidents in Australian Construction Industry (2011)
12. Farmakis, P., Chassiakos, A.: Genetic algorithm optimization for dynamic construction site layout planning. Organ. Technol. Manag. Constr.: Int. J. **10**(1), 1655–1664 (2018). https://doi.org/10.1515/otmcj-2016-0026
13. Chitra, K., Halder, P.: Scheduling project crashing time using linear programming approach: case study. Int. J. Res. Ind. Eng. (2017). https://doi.org/10.22105/riej.2017.96572.1010
14. El-Rayes, K., Khalafallah, A.: Trade-off between safety and cost in planning construction site layouts. J. Constr. Eng. Manag. **131**(11), 1186–1195 (2005). https://doi.org/10.1061/(asce)0733-9364(2005)131:11(1186)

BIM on the Operation and Maintenance Phase/Facilities Managemen

A Methodology for Non-programmers to Automatically Establish Facility Management System with Ontology in Building Information Modeling

Chang-Yuan Liu(✉) and Chien-Cheng Chou

National Central University, Taoyuan, Taiwan
leoliu8268@g.ncu.edu.tw, ccchou@ncu.edu.tw

Abstract. The application of building information modeling to facility management system has proven to help reduce management costs during operation and maintenance phase. However, whenever this type of facility management system is built and changed, it is often necessary to communicate with 1) programmers and facility experts, 2) facility experts, and 3) facility experts and systems. The study established a framework for abstracting the information of BIM to the architecture and attributes of ontology model based on interviews with several facility experts. Also, write a transition function that transforms UML to SWRL. Finally, the execution result of ontology model is automatically obtained through the ontology platform. Therefore, the experts only need to systemize the domain knowledge to form the UML, and then the result of the logic flow chart can be obtained. In the result section, the facility management system is developed to assist the emergency response through different facility managers and firefighters. The results of the implementation found that in the case of no need for programmers and personally establish and alter the system, the facilities related personnel have higher grasp and trust in the system, and save more time in addition to the convenience of personnel communication.

Keywords: Building information modeling · Emergency response system · Facility management system · Ontology · Semantic Web Rule Language · Unified Modeling Language

1 Introduction

The results of the application of ontology in information system development are obvious to all. With interoperability, the domain is fully described to develop the system. From 2003 to 2014, there were a total of 694 articles in related fields, of which 43.52% were related to system and algorithm development and reached a peak in 2012 [1]. In the field of construction engineering, from 1991 to 2015, there were 151 related literatures [2]. However, it is a complex process. The ontology model is highly customized. The similar systems can even reason contradictory results for the identical situation, because different people describe the domain. building and combining different areas of knowledge more efficiently will be key.

E. Toledo Santos and S. Scheer (Eds.): ICCCBE 2020, LNCE 98, pp. 657–671, 2021.
https://doi.org/10.1007/978-3-030-51295-8_46

1.1 Ontological Building Information Modeling

Over the years, in the field of construction engineering, the research on ontology has mainly been in "integration", "building" and "system" [2]. Interoperability is a huge advantage of ontology, making it good at communicating different fields, so it has excellent results in system integration and development. However, the process of "building" has aroused people's attention. The quality of the ontology model depends on the user's experience in the field. In addition, users cannot understand how to describe the field in detail, so they have to learn what ontology is first. This makes everyone focus more on how to build the ontology model correctly [3]. Given this characteristic, ontology is often associated with management issues [4]. In the past research, most of the system types were text-based decision support systems. This makes it a good application in the agent system [5] and the emergency response system [6]. By considering all possible situations and describing them in advance, people can respond immediately when any event occurs. Through a series of research results, it was also found that the ontology lacks a mathematical operation mechanism, so the flexibility of the model is limited. From a practical point of view, the establishment of the model is so rigorous that it lacks versatility [7]. However, since the development of SWRL, the model has begun to change according to circumstances. This situation makes ontology better at application [8]. People have also begun to study how to describe spatial relationships with ontology [9]. And because the building information model contains sufficient geometric information, the ontology has been developed especially in the application of building information models [10]. Among them are more applications in navigation and evacuation [11, 12]. In recent years, with the rise of the Internet of Things, building information modeling (BIM) can also provide more information and accelerate the development of ontological BIM [13–15].

1.2 Building Information Modeling for Facility Management System

BIM is used as a life-cycle management tool to achieve the built environment through software engineering. The goal is to reduce errors and build time. However, the cost of building a model is much higher than traditional methods. In the past few years, when BIM is actually applied in the Facility Management (FM) phase, it is often questioned. Is it really helpful in the operation phase? Autodesk notes that improper data use coupled with highly fragmented teams cost the US capital facilities industry $ 15.8 billion annually. The owner's burden during ongoing operations is approximately two-thirds of these costs. And 55% of owners will face operation and maintenance (O&M) issues. Throughout the supply chain, the adoption of BIM/VDC in Australia can increase industrial productivity by up to 9%. It is reported that the return on investment of BIM implementation is as high as 500%. Based in the United States, Holder Construction Group calculates that based on direct collision detection savings, their BIM returns are three to five times the cost of direct BIM. During the renovation of the US GSA Federal Building in Portland, Oregon, the integrated project delivery team estimated that the use of BIM for coordination helped generate an approximately 300% return on investment [16]. From the above, it can be shown that the benefit of introducing BIM into FM is greater than expected. Nowadays, according to literature statistics, when recent publications apply BIM to FM, they need to face (1) the acquisition and conversion of highly labor-intensive BIM information, (2) the update of BIM information, and (3) the relationship between objects in buildings [17]. This phenomenon indicates that the common difficulty of both the FM system and the ontology model is the high degree of customization. The past research has used the Model-Driven Architecture (MDA) method to automate FM management, so that users who are not familiar with programming can operate building information models and programs only through models and architectures, and promote the improvement of FM issues [18]. However, how to accurately represent the spatial relationship of objects in three-dimensional building? In the O&M phase, BIM always lacks spatial representation and spatial computing capabilities. Therefore, to analyze BIM, laborious conversion and modeling are required. Specifically, BIM has challenges to be overcome in terms of interoperability, data calculation and practical application [19]. Ontology engineering is to provide a clear network for establishing a knowledge system in a certain field to solve the vagueness problem of the system composed of traditional rules of thumb [20]. And because it represents the characteristics of the knowledge system, it is considered a suitable solution to solve the problem of interoperability [21]. Therefore, in order to solve interoperability difficulties in various fields of BIM, scholars use IFC to transform BIM model and use ontology to solve the problem of communication among experts [22].

1.3 Unified Modeling Language to Ontology

For non-programmers, imperative programming is tricky. Some literature also pointed out that to deal with problems that are prone to errors and more careful and subtler, declarative programming methods are effective [23]. And many cases show that declarative programming is obviously better at modeling and solving complex problems [24]. However, with the development of BIM, past research has also found it difficult to query in text form. And using Visual Programming Language (VPL) as an example, it can reduce the difficulty. However, the article states that the situation that SPAQL cannot infer Query3 is practically feasible in SWRL, but the literature also shows that current programming is difficult to express in architectural spaces. It needs to be executed in a higher-than-text form [25]. Similarly, in other studies, knowledge systems have also been used to implement the FM architecture design of BIM. Among them, due to the difficulty of experts in programming, VPL has been added, and a declarative programming system can be used by experts to understand [26]. In general, facility management systems are mostly imperative programming. Compared to SWRL, which is declarative programming, it is more difficult for non-programmers. UML is easier to understand than SWRL's architecture, especially when describing spatial relationships. Based on past experience, research on transforming UML to SWRL is not rare [27]. The author believes that allowing users to build systems in UML will be the key to solving the highly customized ontology and FM systems.

1.4 High Customization

Based on the above, we know that High Customization is a factor that makes building work difficult [2]. The author found that the objects of the ontology model are objective. In contrast, reasoning part is a subjective judgment. This means that the two parts of the ontology model should be operated separately, as shown in Figure 1. For example, the height of the building are specific and absolute, but the performance of the equipment varies from person to person. Therefore, the methodology divides the ontology model into the ontological BIM and the UML to SWRL.

Ontological BIM. The content contains equipment data and building geometric information. Objectively describe all data that will not differ from user to user.

UML to SWRL. The content contains information about the flow and logic of the operation. Users take the ontological BIM as a reference to describe their ideas. The process of the work describes all the reactions the user will make in the event of a situation. Users logically judge the information implied by the equipment's data or the building's geometric information.

Consequently, it can be known that the ontological BIM will be reusable due to its invariance, and the UML to SWRL allows non-programmers to build it themselves. This allows users to build their own systems without having to communicate with the system designer, and solves the problem of High Customization.

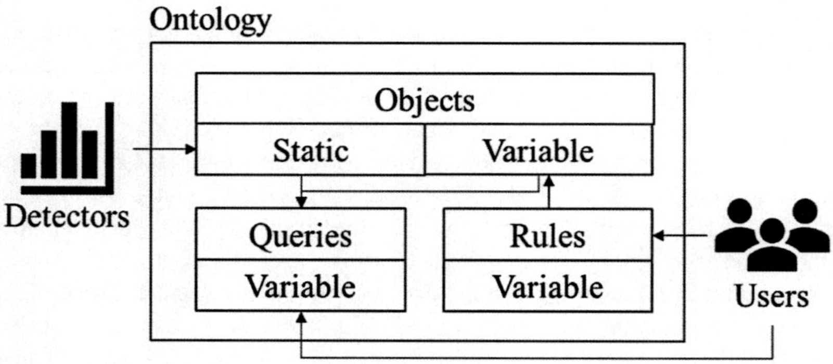

Fig. 1. Explain the variables and invariants in the ontology model.

2 Methodology

This chapter details the process of setting up this system. First, the overall architecture of the system is described in the "System Architecture" section. Secondly, in the "Formalized Building Information Modeling" section, define what information is retrieved from the BIM model. Next, the section "Workflow Diagram to SWRL Converter" describes how to convert UML to SWRL. Finally, the section "Information Transmission Building Information Modeling and Ontology" explains how to convert the BIM model to the ontology model.

2.1 System Architecture

The system architecture, as shown in Figure 2, can be divided into three parts. The BIM model in Autodesk Revit can be different according to building. UML will change in StarUML based on the mangers reference to the BIM Model. The last part is the main part of the system. Through the Revit API and UML API, the Plug-in of Autodesk Revit can read and process UML and BIM information. Then, save the information to XML files. The plug-in will call the encapsulated executable Jar file from the command line and return the result. Jar file will automatically read the information of the XML file and use the OWL API and SWRL API to convert it into a ontology model and reason.

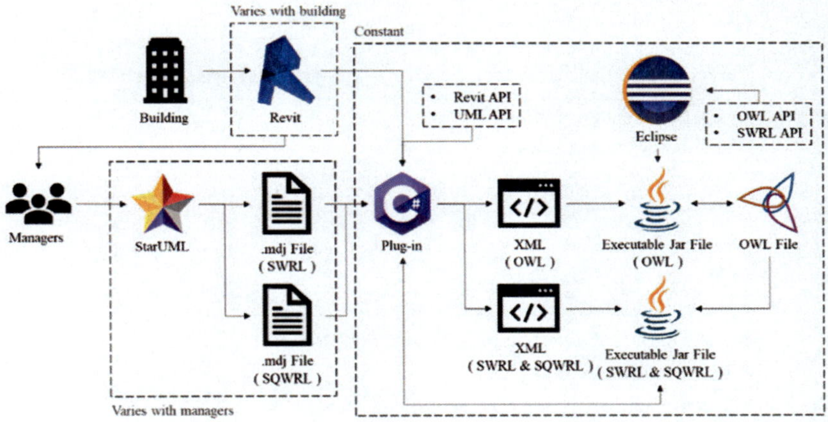

Fig. 2. System architecture diagram

It is worth noting that the UML API is written by ourselves. Since StarUML is open source software, its file format has not encoded. We can easily understand the rules of the file format. In addition, we recommend to separate the SWRL and SQWRL of the ontology model. After the ontology model is created and stored, the .mdj file and the ontology model are read from the plug-in and added to the SWRL and SQWRL. Because the ontology model is changed according to the BIM Model, usually the building will not be changed in a project. Therefore, separate execution will effectively reduce the time required for the reasoning of the ontology model.

2.2 Formalized Building Information Modeling

In the process of building the system, this part is the most complicated. Everyone's habits are different when building a BIM Model. Moreover, the information in the BIM Model is often lacking, especially MEP-related objects, as shown in Table 1. Window thicknesses and materials are relevant for many applications during the O&M phase, such as sunlight analysis, time to failure, and field of view analysis. Power systems and pipelines are often lacking in the BIM Model. The reason is that whenever a BIM Model is established, it is in the design or O&M phases. The design phase is ambiguous for MEP objects and Family Instances, so detailed information will not be established. The O&M phase is because the related objects have been created. If we want to create a BIM Model at this time, people will have to reconfirm all the MEP and Family Instances information, and even some pipelines are located in walls or partitions, which is difficult to check. The information needed during the O&M phase is not taken into consideration when building a BIM Model. Modeling is expensive and time-consuming. For convenience and time saving, every detail of the building is not built up in detail. This situation leads to the need to spend more time on enriching information when building a BIM model to establish a FM system.

Table 1. List the disadvantages when employing FM system in BIM.

Often Lack of Information
• Thickness of Windows
• Material of Objects
• Geometry Information of Pipes
• Power System
Important in O&M Phase
• Walls should divide by stories
• Walls should divide by space boundaries
• Stairs should divide by stories
• Space boundaries are proper to be convex shapes
• Doors and Windows connect to outside should be defined

2.3 Workflow Diagram to SWRL Converter

The author mainly uses the UML Workflow Diagram. Its function is to formulate the judgment of the managers on the performance of the facility. Therefore, there will be great differences due to different users. Diagram can be divided into three types. One is SWRL. The other is SQWRL and another is the process. In simple terms, SWRL which is the logical reasoning of users will add to the ontology model. Then, the information of the reasoning result is presented via SQWRL. The process integrates all the SWRLs and SQWRLs to be used and lists them in the Plug-in. Since StarUML is an open-source software, its file format is not encoded, so after analyzing the rules of its file format, the author wrote the UML API to facilitate reading and convert into SWRL, as shown in Table 2.

Table 2. Explain how to convert diagrams into ontology models.

Name	Property	Content	SWRL
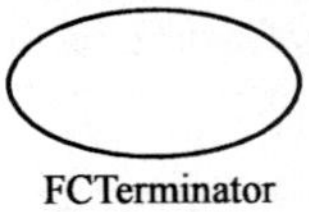 FCTerminator	Text	Start:#SWRL Start:#SQWRL Start:#Process End	None
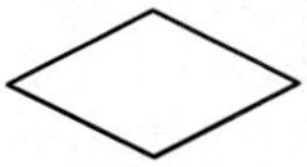 FCDecision	Text	If #Class.#ObjectProperty is #A If #Class.DataProperty is A	Yes: sameAs No: differentFrom Yes: swrlb:equal No: swrlb:notEqual
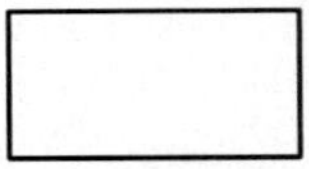 FCProcess	Text	Declaration:#A Declaration:#A.#ObjectProperty(B) Declaration:#A.DataProperty(B) Reason:Result is A Reason:Result is #A Print:Table(#A,#B,#C) Print:"Sentence"	A(?A) B(?A,?B) B(?A,?B) Result(?Result,A) Result(?Result,#A) sqwrl:select(?A,?B,?C)
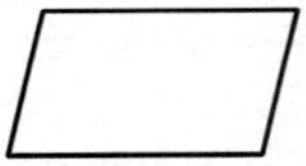 FCData	Text	Data:#Select Data:#Text Data:#SWRL:A.mdj Data:#SQWRL:A.mdj	None
 FCFlow	Text Head Tail	Yes/No	None
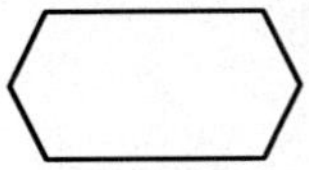 FCPreparation	Text	Use n From a To b	*Change ?A to ?An

There are three points worth noting here. SWRL will execute all branches and cooperate with "FCDecision" to add the results of different judgments to the ontology model. SQWRL is mainly based on One-line operation and "FCPreparation" to increase the flexibility of declarative programming. Process, as an integrated process, adds the required SWRL to the process, and the part that needs to display the ontology model information, adds SQWRL to display.

2.4 Information Transmission Building Information Modeling and Ontology

Through the above-mentioned transformation of UML to SWRL, and after the plug-in is converted into XML files, passing and returning information to the ontology model as a whole can be divided into five stages. The first stage is "Abstract Information from the BIM Model"; the second stage is "Design Mechanism for Transmitting

Information"; the third stage is "Constructing the Format of Transmission File"; the fourth stage is "Mechanism for Automatic Reasoning" The fifth stage is "Constructing the Format of the Ontology Model". The overall structure from the BIM model, plug-in program, XML file, JAVA program to OWL file makes the logical inference of the building information model and returns the results to the plug-in program.

Abstract Information from BIM Model. This study abstracts the BIM information from the model to Add-In using Revit API via Autodesk Revit 2018. Regarding captured object information, its Category (Revit API category) contains Rooms, Walls, Doors, Windows, and Stairs. The attributes of these objects are inconsistent depending on the classification. Rooms and Stairs are passage type objects that do not contain Exterior and Thickness properties, but the passage length is defined by the connected objects. The judgment of the connected objects is mainly divided into two categories, Walls and FamilyInstances. The judgment methods of the two are not the same in the Revit API and if the connected object is found by the coordinate method, there will be an error on the practical level. Therefore, Walls needs to use the GetBoundarySegments method to extract the Wall object from the Room object, and must exclude the Column object. Because the same wall may be associated with the two boundaries of the same room, duplicate information must be deleted by the Distinct method at the end. The FamilyInstances are objects that use the BoundingBoxXYZ class in conjunction with the BoundingBoxIntersectsFilter class, and then select the FamilyInstance object (including the Room Separation Line) through the judgment. If there are the same connected objects in the two passage objects passing through the route, it is determined that the passage objects are connected to each other. Then, the GetBoundingBox method is used to extract the spatial attributes like MinX, MinY, MaxX, and MaxY of the object, thereby obtaining the center point and calculating the distance from the adjacent object. The sum length is the passage length. Thus, there are cases where the same passage object has different passage lengths on different routes. In the stair part, the actual passage length of the stair is calculated by the GetParameters method, in which STAIRS_ACTUAL_RISER_HEIGHT, STAIRS_ACTUAL_-TREAD_DEPTH, and STAIRS_ACTUAL_NUM_RISERS indicate the level, depth, and number, respectively. These three spatial attributes then allow the length of the stairs to be calculated. Other objects belong to the obstacle objects and do not contain the Length property. For the Exterior property, if an object is open to the outdoors, the number of rooms connected is only one, so that the object connected to the outside is captured through the Room attribute. Another attribute of the door thickness is that the GetParameters method takes the thickness of the door with DOOR_THICKNESS. At this stage, the information of the BIM is completely established through information retrieval and spatial attribute judgment.

Design Mechanism for Transmitting Information. Add-In in this study was written using the programming language C#. This differs from Protégé's ontology, which utilized Java. In the past, IKVM.net was used to convert Ontology API and SWRL API into Microsoft .NET Framework. However, it not only requires that the Ontology API itself link to related APIs, but also has compatibility problems between API versions.

Also, IKVM.net stopped updating on April 21, 2017 (37). Finally, this study can only call cmd.exe to execute the Java program because it needs to transfer the information of the BIM to the ontology model and pass the logical reasoning results back to Add-in. A mechanism for transmitting information is required. The XML file for storing information and ProcessStartInfo class for executing the Java program from C# program pass the information in two parts and pass it back to Add-in.

Constructing the Format of Transmission File. The XML file format for transferring information needs to be easy for Add-in to store and read information from Java. Java programs not only need to read information, but they also need to be combined with Ontology API and SWRL API to construct OWL files. Consequently, this study uses LINQ to XML, which construct XML with nodes, as an API for information transfer. The characteristic is that in the case of many objects of the same type, adding a node as a mark to increase the XML node will make the establishment of XML more concise. The main purpose of creating an XML file format is to establish a bridge between the information of the BIM and the other two APIs (Ontology API and SWRL API), where the XML elements include New, Result, Url, Base, Classes, DataProperties, ObjectProperties, Individuals, SWRLs, and SQWRLs. New is the path of the initial ontology model, Result is the path of the final ontology model, Url is the network path of the ontology model, and Base is the ontology model to which the example belongs. Classes, DataProperties, and ObjectProperties are declared categories and attributes of the ontology model. Individuals have child elements including Name, OfClass, ID, Family, Length, Level, and Exterior then there are SWRLs and SQWRLs that include the rules that the system wants to reason. The information of the BIM is completely transmitted to the ontology model through the above XML file format. In the part that returns the reasoning result, the encapsulated Java program is executed via cmd.exe using the ProcessStartInfo class, and the result is read through the StreamReader class. The format of the return result must be defined for the program to read.

Mechanism for Automatic Reasoning. Because of the need to automatically return to Add-In, the Java program will have full control over the contents of the program based on the XML file. Therefore, when deciding the XML file format architecture, it is necessary to consider the information required for automatic reasoning. Ontology API is not fully compatible with the SWRL API version. The newer version of Ontology API is basically incompatible. This study used Eclipse and Apache Maven to complete the setup and writing of the Java programs. Maven can be viewed as project management and as a tool that manages and builds, in particular, Java program projects through the Project Object Model (POM). The study used Maven to complete the version setting and adjustment of the Ontology API, SWRL API, and LINQ to XML. Because the Ontology API is different from the SWRL API team, it has a certain version difference. After the study was tested, owlapi-distribution 4.2.8, swrlapi 2.0.6, swrlapi-drools-engine 2.0.6, and slf4j-simple 1.6.6 were used. In regards to JAVA, XML DOM is used to parse the XML file from the BIM. The whole is for automatic purposes. Therefore, the XML information is read by getElementsByTagName and declared by OWLDEclarationAxiom. The addAxiom is used to add Individuals, and finally the SWRLRuleEngine and SQWRLRuleEngine are used for reasoning.

Constructing the Format of the Ontology Model. The ontology editing software Protégé mainly uses OWL file as the format file. This study uses Java to construct and reason ontology in an automatic way. Therefore, the format structure of the ontology model needs to be determined and designed in advance. It must be convenient for the system to return information to Add-In for easy reading. Therefore, there needs to be more caution when writing.

After passing the information back to Add-In through five parts, the user does not need to obtain the reasoning information from other interfaces and can directly compare the BIM to know whether there is room for improvement in decision making and to promote efficiency.

3 Result

The application of the system was completed through discussions between seven professional firefighters and two facility managers. When generating a workflow, three problems are often encountered: 1) Because normal operations rely on rules of thumb, past experience needs to be integrated. 2) It is vague as to how detailed the work flow is needed. 3) It is necessary to understand the rules for making flowcharts. We will present the process of firefighters going to the fire scene. The judgment of the flowchart is mostly for the text analysis of the string. It should be noted that the user joins the facility management system with his own experience. The study uses a process by which firefighters work with administrators to deal with fire incidents. Among them, some firefighters suggested that what tools should be used to destroy which material.

3.1 Tool Experiment

This study examined the time required for the destruction of various materials by seven career firefighters. The number of seconds it takes for each material of unit thickness to be destroyed by each forcible entry tool is shown in Table 3. As you can see from the table, when the Rock Cutter destroys the Steel Panel, its seconds are judged as "-". This situation means that it cannot be destroyed. In the case of Glass, it was mostly found to be destroyed within 3 seconds because ordinary glass is destroyed instantaneously under normal conditions, so the data obtained by this experiment is recorded in the shortest number of seconds.

Table 3. Unit Thickness Material Destroy Time Experiment

Destructor	Material													
	Aluminum Frame	Copper Wire	Iron Column	Steel Panel	Artificial Timber	Laminated Timber	Green Timber	Concrete	Rock	Tile	Cement	Tempered Glass	Laminated Glass	Glass
Saber saw	76	71	95	123	43	55	81	218	292	187	249	13	10	3
Rock cutter	168	117	192	–	171	198	213	29	49	23	31	29	23	3
Disk cutter	38	36	43	76	39	45	78	94	214	75	89	20	12	3
Chain saw	158	149	218	–	38	42	51	–	0	–	–	81	54	3
Hydraulic spreader	18	15	24	37	43	62	80	52	0	71	57	14	11	3
Hydraulic cutter	11	7	18	31	41	58	83	49	0	65	57	14	11	3
Hydraulic pusher	19	14	42	–	153	219	226	51	198	39	47	55	49	3
Lifting air bag	–	–	–	–	–	–	–	92	212	159	128	–	–	–
Electricity-proof	–	19	–	–	–	–	–	–	–	–	–	–	89	3
Manual cutter	21	17	33	–	–	–	–	–	–	–	–	81	64	3
Crowbar	271	213	284	–	–	–	–	381	–	191	–	139	135	3
Fire axe	259	157	–	–	224	245	379	–	–	165	–	120	112	3

3.2 Workflow Diagram for SWRL

The workflow diagram uses SWRL as an example to reason appropriate tools. Based on the experiment described above, the study currently uses the material, thickness, and position as a reference to generate a workflow diagram, as shown in Figure 3. At the beginning, it will be announced that this is the SWRL process. Also, add a DataProperty named Tool to Space. Then we announced Space. Space mainly refers to rooms, doors, windows and walls. Use the IF process to determine the material of the objects and distinguish whether it is open to the outside. Finally, refer to Table 3 to set the appropriate tools. Note that the SWRL process executes all branches. It means that the appropriate destruction tools for all objects will be identified through 28 SWRL rules. The SWRL rules in ontology model would be like Equation1. The time it takes to destroy an object is the time multiplied by the thickness, similar to Fig. 3. The important part of this section is the need to pay attention to declared variables.

$$\begin{gathered} Space(?\mathrm{s})^\wedge \mathrm{Material}(?\mathrm{s}, ?\mathrm{m})^\wedge \mathrm{swrlb} : \mathrm{equal}(?\mathrm{m}, \mathrm{Laminated\ Glass})^\wedge \\ \mathrm{Outside}(?\mathrm{s}, ?\mathrm{o})^\wedge \mathrm{swrlb} : \mathrm{equal}(?\mathrm{o}, \mathrm{Yes}) \rightarrow \mathrm{Tool}(?\mathrm{t}, \mathrm{Saber\ Saw}) \end{gathered} \tag{1}$$

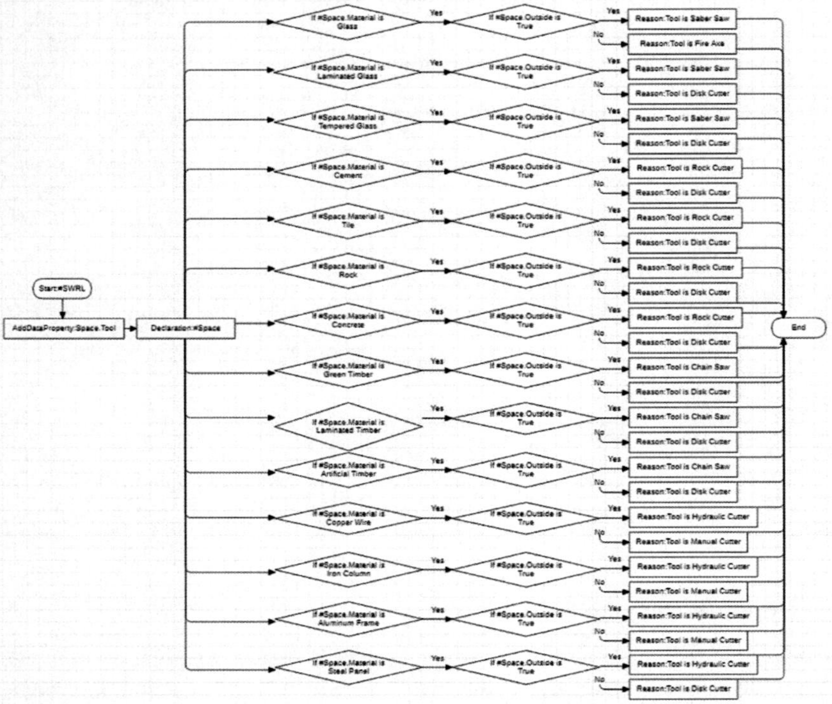

Fig. 3. Workflow for SWRL diagram

3.3 Workflow Diagram for SWRL

The execution result of the system, as shown in Figure 4, presents the required steps in a list. Steps 1 to 5 are after analyzing the Text in UML and displaying it in the form. In listView, SQWRL is mainly used to retrieve the required information. After declaration, use sqwrl: select to display the desired information. Although the establishment of the overall system took a considerable amount of time for the author, firefighters and administrators were very quick to draw within the rules. Most of the time consumption lies in establishing standards for transmitting data and models. Thankfully, according to the methodology, when building a new system, there is no need to consume time again.

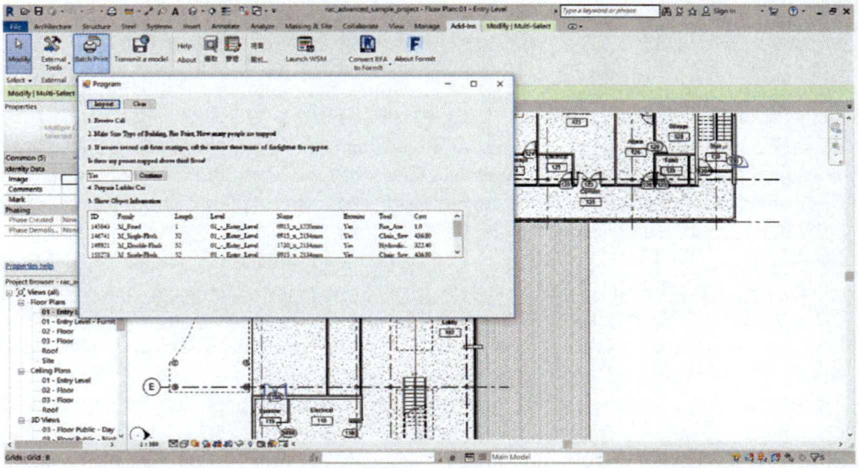

Fig. 4. The result of the fire rescue system

4 Conclusion

The feedback received by the firefighters and managers at the end is positive. When they are using the system, there is no need to explain professional knowledge to system designers, which greatly reduces the problems caused by communication. Although the current program is biased towards a simple mechanism, the problem has been effectively solved. In addition, the study found that when the author wanted to provide the shortest path method as a reference, the managers was reluctant to trust the results of the system. They tend to believe in the data they get from experiments. And, reluctant to bear the risk of inference results which is not intuitive to distinguish the good and the bad. Moreover, they believe that relying on their own rules of thumb may achieve better results in actual situations. This is also the crux of artificial intelligence that can only be used to assist decision-making. To sum up, this methodology can 1) let managers build their own systems without explaining the domain knowledge to system designers. 2) exclude the situation that system designers convince managers to trust the system. 3) save time teaching managers. 4) make the system be reused in different environment. However, here comes a problem that need to teach managers how to draw the diagrams, but it turns out that people are more willing to study from pictures rather than words.

In the future, the author expects to add more applications to this system. The purpose is 1) to understand what information should be abstracted from the BIM model through different applications, so as to improve the ontological BIM. 2) that there are still functions in SWRL and SQWRL that have not yet been applied. Expect complete rules. 3) to use a simpler way to achieve the transformation and transmission between models. Make the methodology as a whole more powerful.

References

1. Otero-Cerdeira, L., Rodríguez-Martínez, F.J., Gómez-Rodríguez, A.: Ontology matching: a literature review. Expert Syst. Appl. **42**(2), 949–971 (2015)
2. Zhou, Z., Goh, Y.M., Shen, L.: Overview and analysis of ontology studies supporting development of the construction industry. J. Comput. Civ. Eng. **30**(6), 4016026 (2016)
3. Deshpande, N., Kumbhar, R.: Construction and applications of ontology: recent trends. DESIDOC J. Libr. Inf. Technol. **31**(2), 84–89 (2011)
4. Charlesraj, V., Paul, C.: Knowledge-based building information modeling (K-BIM) for facilities management. In: 31st International Symposium on Automation and Robotics in Construction and Mining, Oulu, Finland (2014)
5. Ren, Z., Anumba, C.: Multi-agent systems in construction–state of the art and prospects. Autom. Constr. **13**(3), 421–434 (2004)
6. Amailef, K., Lu, J.: Ontology-supported case-based reasoning approach for intelligent m-Government emergency response services. Decis. Support Syst. **55**(1), 79–97 (2013)
7. El-Diraby, T., Lima, C., Feis, B.: Domain taxonomy for construction concepts: toward a formal ontology for construction knowledge. J. Comput. Civ. Eng. **19**(4), 394–406 (2005)
8. Luo, H., Peng, X., Zhong, B.: Application of ontology in emergency plan management of metro operation. Procedia Eng. **164**, 158–165 (2016)
9. Kim, T.W., Fischer, M.: Ontology for representing building users' activities in space-use analysis. J. Constr. Eng. Manag. **140**(8), 04014035 (2014)
10. Leite, F., Akinci, B.: Formalized representation for supporting automated identification of critical assets in facilities during emergencies triggered by failures in building systems. J. Comput. Civ. Eng. **26**(4), 519–529 (2012)
11. Tan, L., Hu, M., Lin, H.: Agent-based simulation of building evacuation: combining human behavior with predictable spatial accessibility in a fire emergency. Inf. Sci. **295**, 53–66 (2015)
12. Tsetsos, V., Anagnostopoulos, C., Kikiras, P., Hadjiefthymiades, S.: Semantically enriched navigation for indoor environments. Int. J. Web Grid Serv. **2**(4), 453 (2006)
13. Lim, Y., Lim, S., Choi, J., Cho, S., Kim, C., Lee, Y.: A fire detection and rescue support framework with wireless sensor networks. In: 2007 International Conference on Convergence Information Technology, Gyeongju, South Korea (2007)
14. Sha, K., Shi, W., Watkins, O.: Using wireless sensor networks for fire rescue applications: requirements and challenges. In: 2006 IEEE International Conference on Electro/Information Technology. East Lansing, East Lansing, Michigan (2006)
15. Nunavath, V., Prinz, A.: Visualization of exchanged information with dynamic networks: a case study of fire emergency search and rescue operation. In: 2017 IEEE 7th International Advance Computing Conference, Hyderabad, India (2017).
16. Autodesk. The Value of BIM for Owners: Save Time and Money During the Building Lifecycle. https://damassets.autodesk.net/content/dam/autodesk/www/solutions/bim/images/stories/The%20business%20benefits%20of%20BIM%20to%20Owners%20(Owners%20PPT)%202016%20update.pdf. Accessed 31 Jan 2020
17. Volk, R., Stengel, J., Schultmann, F.: Building Information Modeling (BIM) for existing buildings—literature review and future needs. Autom. Constr. **38**, 109–127 (2014)
18. Hsieh, C.C., Liu, C.Y., Wu, P.Y., Jeng, A.P., Wang, R.G., Chou, C.C.: Building information modeling services reuse for facility management for semiconductor fabrication plants. Autom. Constr. **102**, 270–287 (2019)
19. Azhar, S.: Building Information Modeling (BIM): trends, benefits, risks, and challenges for the AEC industry. Leadersh. Manag. Eng. **11**(3), 241–252 (2011)

20. De Nicola, A., Missikoff, M., Navigli, R.: A software engineering approach to ontology building. Inf. Syst. **34**(2), 258–275 (2009)
21. Pouchard, L., Ivezic, N., Schlenoff, C.: Ontology engineering for distributed collaboration in manufacturing. In: Artificial Intelligence and Simulation Conference, Tucson, Arizona (2000)
22. Beetz, J.: Facilitating distributed collaboration in the AEC/FM sector using Semantic Web Technologies. Technische Universiteit Eindhoven, Eindhoven, Netherlands (2009)
23. Sivaramakrishnan, K.C., Kaki, G., Jagannathan, S.: Declarative programming over eventually consistent data stores. In: 36th ACM SIGPLAN Conference on Programming Language Design and Implementation, Portland, Oregon (2015)
24. Seipel, D., Hanus, M., Abreu, S.: Declarative programming and knowledge management: Conference on Declarative Programming, DECLARE 2017, Unifying INAP, WFLP, and WLP, Würzburg, Germany, 19–22 September 2017. Revised Selected Papers. Springer, New York (2017)
25. Preidel, C., Daum, S., Borrmann, A.: Data retrieval from building information models based on visual programming. Vis. Eng. **5**(1), 1–14 (2017)
26. Singer, D., Borrmann, A.: A novel knowledge-based engineering approach for infrastructure design. In: The Fourth International Conference on Soft Computing Technology in Civil, Structural and Environmental Engineering, Prague, Czech Republic (2015)
27. Parreiras, F.S., Staab, S.: Using ontologies with UML class-based modeling: the TwoUse approach. Data Knowl. Eng. **69**(11), 1194–1207 (2010)

Impact of COBie on Design Activities

Daibee Bose[1], E. William East[2(✉)], and Raja R. A. Issa[1]

[1] Rinker School of Construction Management, University of Florida, Gainesville, FL, USA
[2] Prairie Sky Consulting, Mahomet, IL, USA
bill.east@prairieskyconsulting.com

Abstract. Delivery of Construction Operations Building information exchange (COBie) data helps facility maintenance, operations, and asset managers to more quickly operate their facilities at lower life-cycle cost. Since its first publication, COBie has been widely adopted in projects around the globe, in part due to the precise definition of what is and is not a COBie deliverables. Metrics for COBie quality control, established in the standard, require the verification of data format and validation of content. While the technical hurdles to correct COBie design data have been largely overcome, the question of what design activities must change if data quality standards are to be met has not been answered. This paper presents the results of a case study created specifically to address this question.

Keywords: COBie · BIM · Process · Quality control

1 Background

Our buildings spend most of their life in the operation and maintenance phase. A poorly managed facility increases ownership cost of the facility and assets. A well-maintained building reduces energy consumption and enables fast and easy replacement of faulty units thus increasing productivity and efficiency. A significant roadblock to the management of building portfolios is the need to manually enter construction handover data into multiple owner/operator computer systems for functions such as maintenance management, operations management, tenant management, inventory control, and fiscal control [12, 13]. In recent years the use of BIM for maintenance of information purposes has been one of the leading topics of research [16]. BIM can potentially be used in several areas of Facility Management (FM) services such as locating building components facilitating real-time data access, checking maintainability, creating and updating digital assets, visualization and marketing and space management [3].

Liu and Issa [17] investigated the use of integrating Geographical Information Systems (GIS) and Building Information Models (BIM) to get information for facility operation and maintenance of sub-surface pipelines. ArcGIS10, Revit and AutoCAD Civil 3D were used in this research. ArcGIS10 was used to view the building location, the surrounding roads, and the subsurface pipeline networks. Revit was used to model the building and its Mechanical, Electrical, and Plumbing (MEP) systems and

E. Toledo Santos and S. Scheer (Eds.): ICCCBE 2020, LNCE 98, pp. 672–682, 2021.
https://doi.org/10.1007/978-3-030-51295-8_47

AutoCAD Civil 3D was used to integrate the BIM model in Revit and the GIS dataset from ArcMap. The model provided precise information related to location, elevation, and properties of each pipe which can be used for maintenance purposes like digging, excavation, and replacement of faulty pipeline.

Gheisari et al. [14] assessed the benefits of using BIM for Facility Management (FM) practitioners and how using Mobile Augmented Reality can enhance facility management practices. A survey conducted to analyze existing FM Technology use suggested that a majority (77%) of the respondents used laptops, 71% used finger touch smartphones 50% used desktop computers for assessing the required data at their job site.

As for the information type, 92% of the respondents accessed the information required to perform their task through electronic format files, e.g. PDF, MS Excel, MS Word, MS Access followed by paper format files (71%) such as built plans, brochures, followed by 2D/3D CAD/visualization files like AutoCAD, Navisworks, Revit. As for using BIM in Facility Management Practices, facility managers do have some insight into the applications of BIM technology in facility management practices but most of the FM professionals either do not use or are not aware of their companies use of any BIM technology for FM purposes.

Facility Managers often use computer software tools such as Computerized Maintenance Management System (CMMS), Computer Aided Facilities Management (CAFM), Building Automation System (BAS), Energy Management System (EMS) to store, track, and control various FM activities. Ideally, CMMS, CAFM, BAS and EMS systems should be able to exchange data among one another with BIM authoring tools, however, this often is not the case as each program is developed by different software developer due to which there is a constant hindrance in information exchange among the programs [15].

To define the subset of construction handover data need by facility managers, and to identify which of the information should be provided by designers and contractors, the US Army created and tested the Construction Operations Building information exchange (COBie) specification [6, 7]. Today, COBie is an Industry Foundation Class Model View Definition [8] published as part of the National Building Information Model Standard - United States®, Version 3.0. [5]. COBie information exchanges from design [9, 11] software are updated during construction and then imported into downstream facility management information systems [18].

Key to successful use of COBie is the two-stage process of COBie Quality Control (QC) described directly in the standard [10]. Step one verifies that the COBie file format conforms with the standard format and addresses all necessary data cardinality constraints. File format and cardinality verification is accomplished using updated, open-source software, COBieQCReporter, used during the later stages of testing by the US Army [4]. Step two checks the content of the information to ensure that the information provided reflects the content of the building.

Despite progress on demonstrating these exchanges, the process of capturing the asset information in COBie format has proven to be challenging [1]. These challenges include incorrect COBie specifications, confusion about the difference between format and content, and user understanding and practice.

2 Objective and Methodology

The objective of this research was to investigate if changes to current design software use practices are required to correctly deliver standardized building information. To accomplish this, a set of design models were created for the building housing the M.E. Rinker School of Construction Management, a 500,000 sqft. academic building on the University of Florida, Gainesville campus. Separate BIM models in Revit software were created for Architectural, Plumbing, Mechanical, Electrical, and Fire protection systems from the existing set of drawings and plans by Construction Management graduate students. Discipline-specific COBie data were exported from these models according to established best-practices [11]. Each discipline-specific set of COBie data were tested using the required two-step COBie QC Process. Where needed, corrections were made to resolve format errors (i.e. verification errors) and content errors (i.e. validation errors). These corrections were evaluated for the relevance for designers and design technicians and documented [2]. The case study methodology is summarized in Fig. 1.

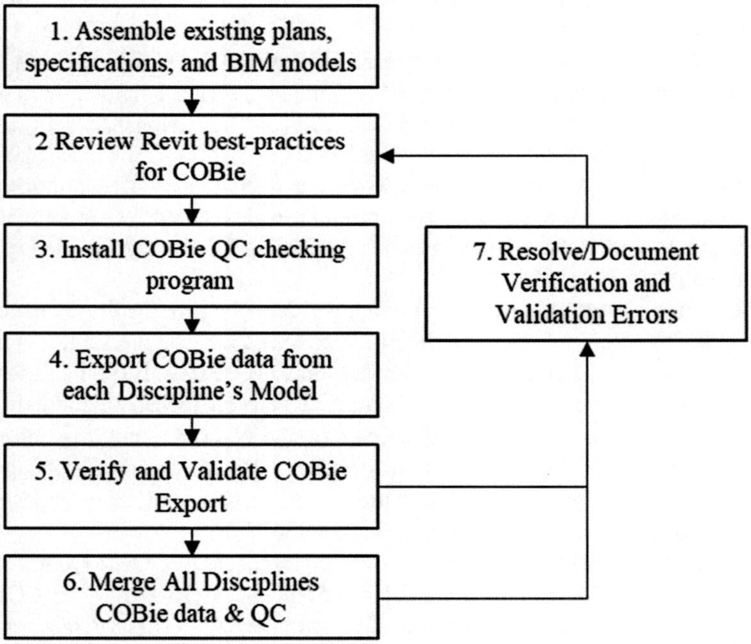

Fig. 1. Case study methodology [2]

3 Discussion

During the resolution of errors found in the individual discipline files and the merged files (Fig. 1, Step 7) several BIM software usage patterns were identified relevant to those designers/design technicians responsible to create BIM models where non-geometric data is required for export.

3.1 Verification Errors

Verification errors are technical errors in the delivery of the COBie file. These issues primarily relate to the way that COBie data must be delivered to ensure proper use. Primarily these issues relate to the relationships among COBie data elements. For example, if one were to assert in a COBie file that "Pump-05 is in room 300," then there must be information about room "300" in the COBie file. We would also expect that for proper FM activities to be accomplished there can only be one "Pump-05" and room "300". There can be no controversy about these data cardinality rules as these constraints are a published part of the COBie standard.

Key to the COBie standard is the idea that FM technicians will need to physically go to the room in which a component (e.g. equipment or fixture) is located. This concept of "Spatial Containment" differs from the "Coordination Position" used in geometric models. To provide the FM technician the precise coordinates for a component will not be helpful, the room number is what is most important for the FM use case. As a result, each COBie Component must have an associated COBie Space Name. In our testing we found that spatial containment was missing from many pieces of equipment. While the design software used for this case study can automatically identify spatial containment, that functionality was not known to the original BIM modeling team. As a result, adjustments to "tails" in Revit "Room Calculation Points" were required, Fig. 2. In addition, differences between architectural rooms and mechanical spaces, in the software system used, required a detailed evaluation of rooms whose ceilings were not the same level across the entire space, Fig. 3.

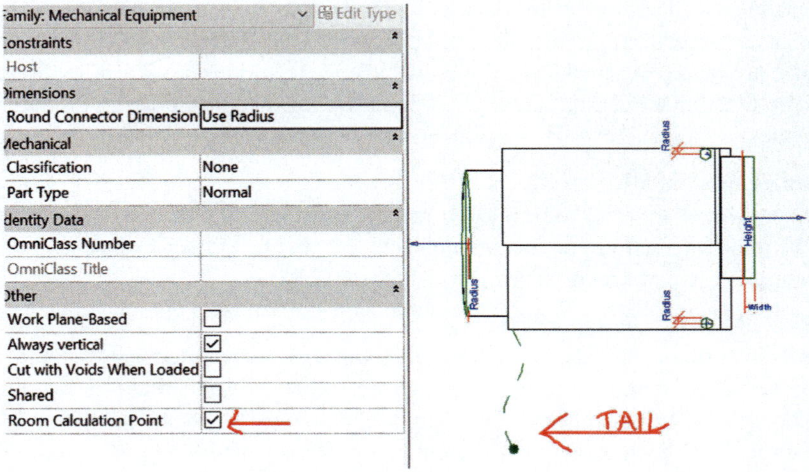

Fig. 2. Room calculation point

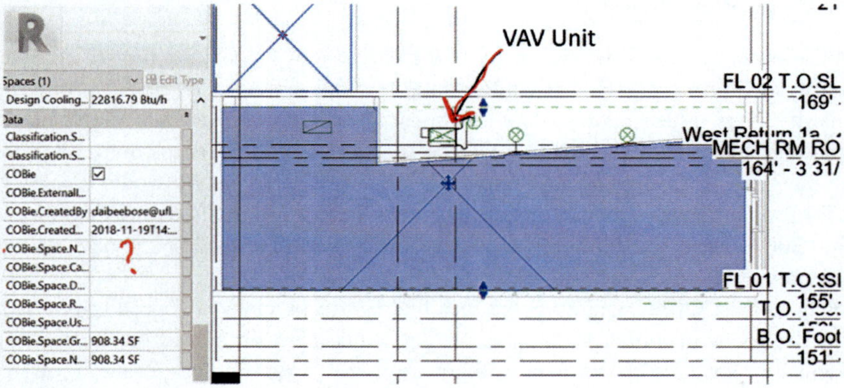

Fig. 3. Variation in room ceiling elevations

3.2 Validation Errors

While a COBie file may be technical correct, it may not contain the necessary information for the building under consideration. As noted in the COBie standard, objects delivered in a COBie file are scheduled products and equipment not found on the "Exclusion List." Since no standard can identify the products and equipment to be provided on every type of building, anywhere in the world, the COBie standard essentially requires the delivery of all manufactured products and equipment that can be identified by manufacturer and model number. Validation testing checks that information on the contract drawings and schedules is the same as information provided in the associated COBie file. Spot-checking typically identifies common issues consistently incorrect across an entire model. The validation issues identified were not software functionality issues, these are differences between the content of models and the content of the contract design drawings.

Missing Components
A comparison of the student-developed models to the original design drawings found that the plumbing and fire protection models, for example, did not include fixture-type elements. These items were added to the models.

Incorrect/Missing Component Types
Several instances were also found where the "type" of mechanical equipment or electrical fixture identified on the drawings was incorrectly identified in the student-developed models. Product types found on design drawings typically identify a "mark" or "call-out", as shown next to each fan in Fig. 4. Resolving differences between Revit models and design drawings to find missing component was a time consuming and cumbersome process.

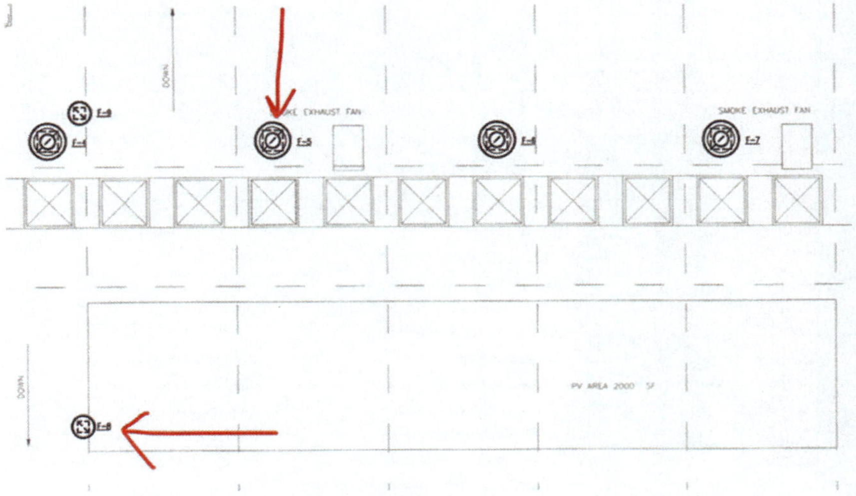

Fig. 4. Mechanical equipment drawing of roof

Incorrect Object Types

Every element in BIM model is typically developed from a template or "Type" object. COBie requires that exported types data match that identified on drawing schedules. Numerous instances were found where the object types on the drawings did not match those found in the model.

While COBie Type data must be based on the products identified in the contractually binding drawing schedules, the information in the BIM model does not. In the student-developed model doors, diffusers, and HVAC components were found to be not "typed" correctly. To resolve this, each model type (i.e. Revit "families") was added to reflect the list of scheduled types and assigned to the correct component.

Another deviation between model, drawing, and COBie data also occurred when Revit "Curtain Walls" were used for modeling but "windows" were called out on the drawings. The software does not transform these curtain walls into windows during the export to COBie and as a result the "windows" on the drawings do not appear in the exported data as is required by the COBie standard. Rather than remodel these curtain walls, manual changes were made directly to COBie data, including the spatial containment of each window (Fig. 5). Further care was taken to ensure that general updates to other COBie export functions did not re-set these manual data entries.

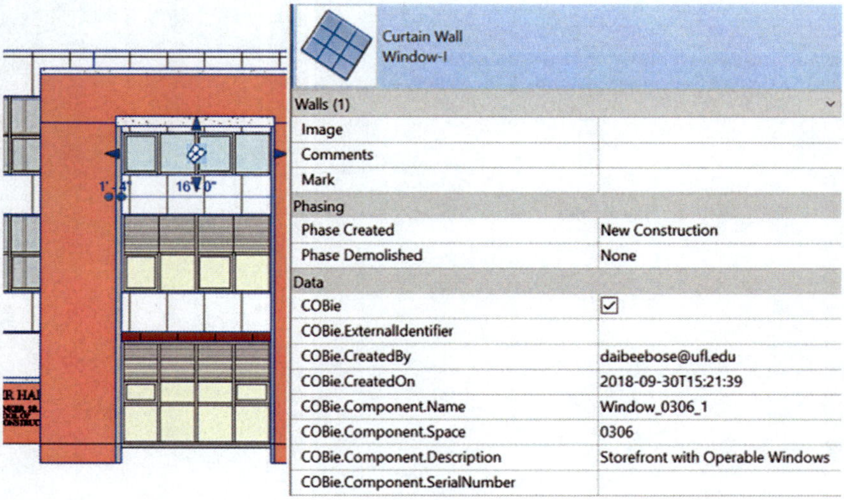

Fig. 5. Identifying COBie window data

Scheduled Attributes

In the original architectural and mechanical models, drawing schedule information was captured in the BIM models and therefore could not be exported to COBie format. A significant amount of time was required to add scheduled information found on the drawings to room finishes, windows, doors, plumbing fixtures, mechanical equipment (and terminal units), fire protection equipment (and sprinklers), and electrical equipment (and terminal units). In many cases, additional parameters were added to native model elements to capture scheduled properties.

Implied Schedule Attributes

Of specialized concern when developing scheduled attributes was to parse the conventions often used when preparing traditional drawing schedules. This parsing is required since human conventions that simplify the presentation of scheduled products must be made explicit if the information content of those schedules is to be delivered in a way that does not require human interpretation.

Incorrect Object Types

Every element in BIM model is typically developed from a template or "Type" object. COBie requires that exported types data match that identified on drawing schedules. Numerous instances were found where the object types on the drawings did not match those found in the model.

While COBie Type data must be based on the products identified in the contractually binding drawing schedules, the information in the BIM model does not. In the student-developed model doors, diffusers, and HVAC components were found to be not "typed" correctly. To resolve this, each model type (i.e. Revit "families") was added to reflect the list of scheduled types and assigned to the correct component.

A representative example is provided in Fig. 6. The first issue to resolve was the identification of names and units for all numerical values. The next issue was to separate multi-valued numeric attributes into individual attributes. For example, the "CFM- Range" column was changed "Flow Rate-Min" and "Flow Rate-Max" with a measured unit of "CFM."

DIFFUSER SCHEDULE

DESCRIPTION	NECK SIZE (IN)	FRAME SIZE	CFM-RANGE	MAX. NC	MANUFACTURER/ MODEL	REMARKS
CEILING SUPPLY DIFFUSER	6	24X24	0 - 125	19	TITUS/OMNI	SEE NOTE BELLOW
CEILING SUPPLY DIFFUSER	8	24X24	100 - 250	19	TITUS/OMNI	SEE NOTE BELLOW
CEILING SUPPLY DIFFUSER	10	24X24	250 - 350	12	TITUS/OMNI	SEE NOTE BELLOW
CEILING SUPPLY DIFFUSER	12	24X24	351 - 500	17	TITUS/OMNI	SEE NOTE BELLOW
CEILING SUPPLY DIFFUSER	14	24X24	501 - 700	20	TITUS/OMNI	SEE NOTE BELLOW
CEILING RETURN REGISTER	14	24X24	0 -1125	20	TITUS/23R	SEE NOTE BELLOW
CEILING RETURN REGISTER	SEE FLOOR PLANS				TITUS/350ZFL	SEE NOTE BELLOW

NOTES:

1. FINISHES SHALL BE SPECIFIED BY ARCHITECT
2. SEE SPECIFICATION FOR ADDITIONAL INFORMATION
3. ALL DIFFUSERS AND REGISTERS SHALL BE SUITABLE FOR THE TYPE OF CEILING CONSTRUCTION THAT THEY ARE BEING INSTALLED IN AND HAVE A UNIFORM FACE APPEARANCE.
4. INSTALL FOUR (4) WAY DIFFUSER ONLY WITH BLANK OFF FOR 2-WAY OR 3-WAY BLOW

Fig. 6. Diffuser and register schedule

Some information is not shown on the schedule at all. For example, the collection of a "ceiling return registers" list required extraction of data from the callout symbols found on the "floor plans." Depending on the specific schedule definitive "Note" information should be created as an actual schedule column. In Fig. 6, none of the notes listed would be relevant to FM. Finally, the schedule in Fig. 6, as with many MEP non-fixture schedules, does not explicitly identify the product "type" to be provided. This information is hidden in the "Manufacturer/Model" column. From the information provided, three product "types" can be inferred from the model numbers listed.

The issue of notes was most pronounced in architectural schedules where room finish, door, and window schedules all referred to abbreviations listed below the schedule. To create the equivalent COBie data, the models required the addition of both the scheduled code values, but also the meaning of the code. In some cases, schedule columns also included multiple codes referring to different attributes that, while easily interpreted by humans, cannot be explicitly represented in electronic exchanges. For example, a door schedule value of "HM./PTD" was separated into a "Door Material" attribute with value "Hollow Metal" and a "Door Finish" attribute with a value of "Painted."

4 Limitations

Before concluding it is also necessary to identify the limitations of this case study. First, this case study was conducted using a single building reflecting the design choices and conventions made in a single set of design drawings. We expect that additional case studies on projects conducted elsewhere will identify additional conventions that must be explicitly described. For example, the second author has participated on projects where fixtures and MEP terminal units are not modeled, and the corresponding drawing schedules are also missing. The identification of fixtures requires the traditional process of manually counting drawing marks.

The second concern of this case study is that the original model and updated COBie compliant models were created by graduate students. This concern is not as important as one might expect given that there are many BIM technicians who are required to use a BIM authoring tool with less instruction than provided graduate students at the University of Florida. As a result, we expect many of the conclusions reached are directly relevant to new modelers. Our emphasis on having "data match the drawings" has indirectly pointed out those issues that the Rinker Building's designers, and by extension all other designers will have to consider when creating their models.

5 Conclusions

An underlying conclusion that for many is now itself an assumption, is that by using software that has undergone increasingly rigorous COBie compliance testing over the last decade, teams can export design files that will meet COBie QC standards. For those new to the COBie standard, this will be conclusion enough. For those who have begun to implement COBie additional, broader, conclusions can also be reached.

When information is exchanged on design drawings it is largely based on the reader's ability to interpret the content. Thus, the content of the drawings is meaningful only when it is understood and correctly interpreted by experienced professionals and practitioners. Directly exchanging the written tokens of convention is insufficient to delivery information outside the context of design drawings. COBie represents the first step at replacing better paper documents by forcing an explicit explanation behind these conventions.

Without taking this first step for the relatively simple set of scheduled rooms, products, and equipment, it is our considered conclusion that solving complex sets of information flows required for higher-level problems will only be solvable in the limited cases where pilot testers manually re-key required properties.

6 Recommendations

We provide the following recommendations for design teams working to deliver COBie data deliverables. While some of the specific resolutions of issues were particular to the software used in this case study, overall recommendations regardless of the software used, include the following:

1. BIM models must be tested to determine if product and equipment data typically found on drawing schedules is in the model or simply pasted as an image on design or construction drawings.
2. BIM operators must be trained on specific software functions allowing correct use and identification of object types, data export mappings, and spatial containment algorithms.
3. To export usable building information useful beyond BIM and drawing conventions, schedule data must not contain multi-valued numeric attributes. Information implied in schedule notes must be explicitly assigned.
4. Since BIM models are developed over time, modelers should expect to revisit and update models to export currently accurate equipment containment and schedule data.

References

1. Anderson, A., Marsters, A., Dossick, C.S., Neff, G.: Construction to operations exchange: challenges of implementing COBie and BIM in a large owner organization. In: Construction Research Congress 2012, pp. 688–697 (2012)
2. Bose, D.: Case study on exporting construction operations building information in COBie format from a BIM. M.S. thesis, University of Florida, Gainsville (2019)
3. Becerik-Gerber, B., Jazizadeh, F., Li, N., Calis, G.: Application areas and data requirements for BIM-enabled facilities management. J. Constr. Eng. Manag. **138**(3), 431–442 (2011)
4. Bogen, A.C.: COBieQCReporter. Prairie Sky Consulting, Version 1.1 (2019). https://github.com/OhmSweetOhm/CobieQcReporter/releases. Cited 20 May 2019
5. bSa. The National BIM Standard - United States®, National Institute of Building Sciences, buildingSMART alliance, National Institute of Building Sciences, Version 3.0, March 2014. (2015). https://www.nationalbimstandard.org/. Cited 20 May 2019
6. East, E.W.: Construction Operations Building information exchange (COBie) Requirements Definition and Pilot Implementation Standard, U.S. Army, Corps of Engineers, Engineer Research and Development Center, TR 07-30, June 2007 (2007)
7. East, E.W.: Pilot Test of the Construction Operations Building information exchange (COBie) Data Format for Army Installations, U.S. Army, Corps of Engineers, Engineer Research and Development Center, TR 08-18, September 2008 (2008)
8. East, E.W., Chimpan, T.: Model View Definition, Industry Foundation Classes: Construction Operations Building information exchange, National Institute of Building Sciences (2013). http://docs.buildingsmartalliance.org/MVD_COBIE/. Cited 20 May 2019
9. East, E.W., Jackson, R.: Delivering Construction-Operations Building information exchange (COBie) in GRAPHISOFT ARCHICAD. Prairie Sky Consulting (2016). ISBN 978-136-526840-3
10. East, E.W., Bogen, A.C.: Construction-Operations Building information exchange (COBie) Quality Control. Prairie Sky Consulting (2016). ISBN 978-1-365-41018-5
11. East, E.W., Keeffe, S.O., Kenna, R., Hooper, E.: Delivering Construction-Operations Building information exchange (COBie) Using Autodesk Revit. Prairie Sky Consulting (2017). ISBN 978-1-387-20091-7
12. Farghaly, K., Abanda, F.H., Vidalakis, C., Wood, G.: Taxonomy for BIM and asset management semantic interoperability. J. Manag. Eng. **34**(4), 04018012 (2018)

13. Foster, B.: Using real-time CMMS asset data capture during construction to improve facility management. J. Build. Inf. Model., 17 (2012)
14. Gheisari, M., Goodman, S., Schmidt, J., Williams, G.: Exploring BIM and mobile augmented reality use in facilities management. In: Construction Research Congress 2014, pp. 1941–1950 (2014)
15. Ghosh, A., Chasey, A.D., Mergenschroer, M.: Building information modelling for facilities management: current practices and future prospects. In: Issa, R.R.A., Olbina, S. (eds.) Building Information Modeling: Applications and Practices, pp. 223–253. American Society of Civil Engineers, Reston (2015)
16. Hwang, S., Liu, L.Y.: BIM for integration of automated real time project control systems. In: Construction Research Congress 2010: Innovation for Reshaping Construction Practice, pp. 509–517 (2010)
17. Liu, R., Issa, R.R.A.: 3D visualization of sub-surface pipelines in connection with the building utilities: integrating GIS and BIM for facility management. In: Computing in Civil Engineering 2012, pp. 341–348 (2012)
18. WBDG. buildingSMART alliance information exchanges: Means and Methods, Whole Building Design Guide, National Institute of Building Sciences (2014). https://www.nibs.org/page/bsa_cobiemm. Cited 20 May 2019

Automating BIM Objects Quantity Take-Off for Lifecycle Costing of Cleaning Operations

Adam Piaskowski(✉) and Kjeld Svidt

Aalborg University, Thomas Manns Vej 23, 9000 Aalborg, Denmark
akptech@outlook.com

Abstract. By 2022, the Danish government will require mandatory evaluation of lifecycle costs (LCC) for new buildings, as opposed to only considering procurement costs. The maturing building industry sector looks towards LCC as a source of competitive advantage. LCCbyg is a tool released by the Danish authorities to calculate lifecycle costs. As the quantity take-off is mostly manual, we investigate ways of automating it. Multiple methods exist for calculating procurement costs at an object level. The same cannot be said about operational costs, where methods mainly rely on building areas. The generic approximations fail to account for object quantity, type, and materials. Therefore, the objective of this study was twofold; firstly, to investigate the potential for basing operational costs on objects, and secondly, to develop and test a data transfer method, integrating independent software packages for design and costing. The industry stakeholders were invited to test the data transfer prototype, showcased on two Building Information Models (BIMs); an operational, and a conceptual model of two office buildings. The proposed approach transferred Revit object quantities to the LCCbyg XML schema, and the operational cost data has been visualized using PowerBI. To conclude, the stakeholders saw the benefits of basing operational data on objects, while also pointed at obstacles concerning a lack of model detail in the early design phase. Despite a successful data transfer, practical implementation will require further integration with cost databases and other subsets of LCC calculations to be considered a complete solution.

Keywords: BIM · VPL · Data exchange · Lifecycle Cost · Quantity take-off

1 Introduction

When making design decisions, it is no longer valid to solely consider initial purchase costs. To assess factors outside of the initial price, lifecycle assessment (LCA) endeavours a holistic outlook of the buildings' operation, disposal and reuse [1]. The direct economic impact can be evaluated using the Lifecycle Cost (LCC) calculation for the individual building components, as well as the building as a whole [2].

The lifecycle cost calculations envelop merging costs incurred during construction, operations, maintenance, renovation and demolition phases. Due to the high cost of human labour in Denmark, a vast proportion of the LCC of a building is accredited to the operational phase [3]. Interior cleaning must be performed daily to maintain a pleasant indoor environment. This is especially true in mass workplaces.

E. Toledo Santos and S. Scheer (Eds.): ICCCBE 2020, LNCE 98, pp. 683–696, 2021.
https://doi.org/10.1007/978-3-030-51295-8_48

LCC of cleaning operations is often generalized and under-costed, resulting in worse material performance and faster building degradation. According to [4] Guillen et al. (2016), up to 60% of the total lifecycle cost belongs to the occupancy phase. Accounting for operational costs in the early design offers the most significant design optimization potential at the lowest cost of changes in design [5].

Sustainability ratings are a commonly agreed procedure for ensuring the buildings are built sustainably and to a high standard. DGNB is one of the most popular certifications, entailing LCC as one of its key economic indicators. LCC is a requirement forming 13.5% of total DGNB score [6].

Following the DGNB TEC 1.5, the cleaning costs are based on areas and functions of rooms [7, pp. 385–387]. When calculating the procurement costs, the square meter price is usually regarded as the guiding metric, whereas detailed estimations rely on individual objects. There is a possibility the area-based calculations conceal the design optimizations only distinguishable at an object level.

Evaluating the costs of cleaning on objects may present interesting information relative to the ease of cleanliness and durability directly concerning the economic lifespan of materials and components. Thanks to a parametric and hierarchical data structure of Building Information Models (BIMs), object-based quantity take-off (QTO) is computable, given correct modelling process.

Although attempts of costing based on BIM models are a popular research domain [8], the uptake is slow, due to lack of software interoperability [9]. According to Hannan [10], mapping objects using BIM-based QTO is not a common skill amongst auditors. Another challenge to QTO automation is the accuracy of BIM models and varying modelling methods [2]. Hence, the current method of calculating lifecycle costs in Denmark involves manual QTO or uses spreadsheet-based calculation schemes. These approaches are prone to human errors [2].

There are successful attempts of automating QTO using Industry Foundation Classes (IFC). An approach proposed by Lawrence [8] underlines a need for integrating costing information in the design model. He also points out that standards can be impractical when guiding complicated workflows involving multiple stakeholders.

There also exist references to the failure of IFC, mainly regarding redundancy and ambiguity of application. Furthermore, IFC interoperability is not bulletproof and often results in transfer data loss [11, 12].

There is a need to align existing popular BIM tools with current work practices, as the accustomed stakeholders are more likely to use the tools adequately [13]. Hence this project entails developing a prototype for the integration of Revit model data with an LCCbyg XML schema.

LCCbyg is a tool released by the Danish Building Research Institute (SBi), at Aalborg University (AAU) on behalf of the Danish Transport, Construction and Housing Agency [14], with the purpose of enabling an accurate and uniform lifecycle cost calculation for all future public developments in Denmark.

Given the popularity of Autodesk Revit in the Danish architecture, engineering and construction (AEC) sector, the authors attempt to bridge interoperability directly, using this proprietary software environment.

Visual Programming Language (VPL) Dynamo for Revit is used to structure data from a BIM model, while Python is used to transfer the data to analysing and visualizing software.

2 Methodology

The study is based on a combination of methods. Existing literature on the subject was reviewed, followed by extensive stakeholder interviews. The Plan-Do-Check-Act (PDCA) methodology [15] (see Fig. 1) was combined with the Contextual Design (CD) methodology [16] to structure the investigation. The prototype transfer was tested on two case studies, and the results were presented in two different tools; LCCbyg for LCC analysis, and PowerBI for data visualization.

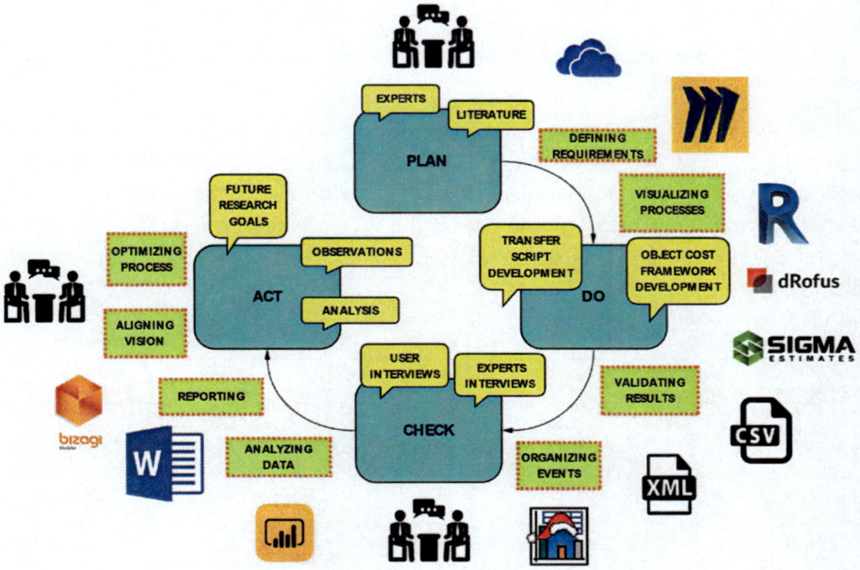

Fig. 1. The development process diagram illustrating software used and aligned with the PDCA milestones.

3 Development of the Object-Oriented Costing Prototype

The building interior cleaning can be subdivided to categories concerning object cleaning, waste removal and replacement of utilities, see Fig. 2. To structure the cleaning information on objects, the existing tools and approaches are first investigated. The frequent and labour-intensive interior cleaning subset generates a significant proportion of the LCC and pose an interesting potential for optimising costs. Following the proposed approach for cleaning, it will likely be possible to cost remaining lifecycle phases and optimize costs holistically.

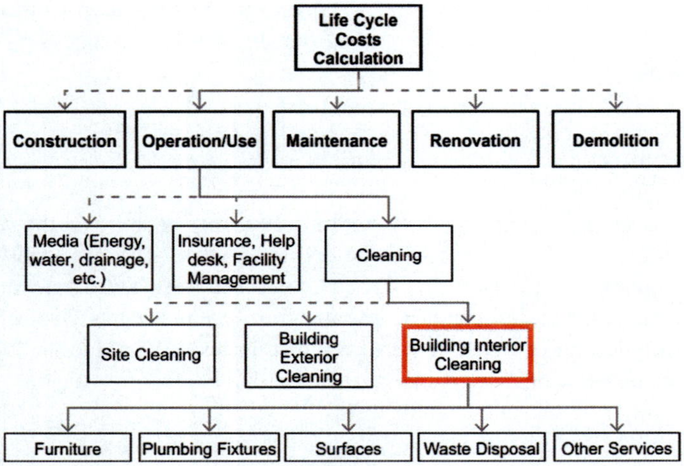

Fig. 2. The building interior cleaning was chosen as the scoped subset of lifecycle costing

The existing costing approach is taken from the DGNB LCCbyg template, which stems from the TEC 1.5 of the DGNB costing manual [7]. The costing approach is based on the sum of the square meter areas. The surfaces are then grouped by finish material and cleaning intensity to organize the QTO data. Although the DGNB template relies on areas only, by default, the LCCbyg XML schema accounts for other units.

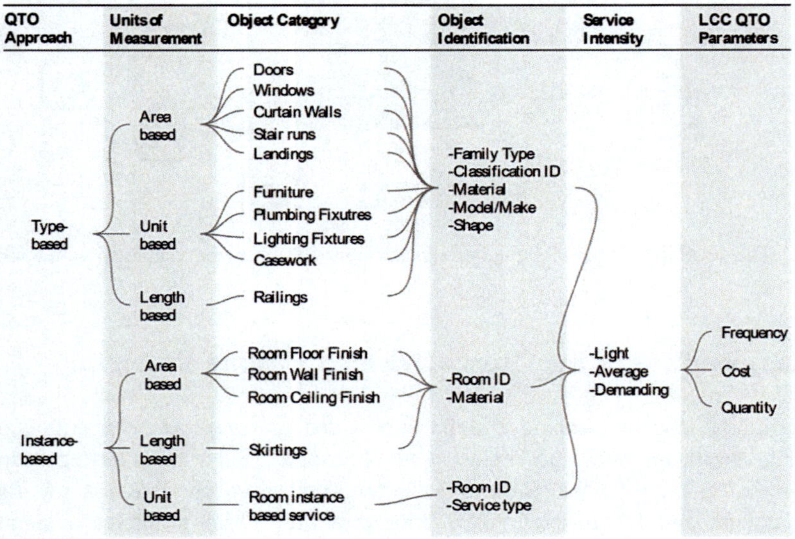

Fig. 3. The sorting and grouping schema converting Revit object parameters to the LCCbyg

This enables the proposed prototype to group information based on 1; object count, for type-based objects such as furniture; 2; length-based measurement (LBS) for, i.e. railings, and 3; using areas on surfaces such as floors. The use of both object types and instances is investigated, and the approach is visualized in Fig. 3.

The price variation depends on the service intensity of cleaning, i.e., light, average, demanding, with corresponding explanation supplemented in the TEC 1.5. In the guideline, the intensity is mainly concerned with the accessibility to the object or surface. If a window requires a ladder to be cleaned, it is classified as demanding.

The parameter can be tailored to the specific use case of the building, expanding on the accessibility difficulty level with issues concerning the thoroughness of cleaning, and machinery required. The cleaning intensity forms the parent relative to the cost, frequency and count of an object, as to allow grouping per intensity level. As the QTO relies on a BIM model, the 3D object instances can intersect with room volumes allowing grouping per room instance. This can be used to visualize cost data in PowerBI.

3.1 Approach to Mapping BIM Model QTO with LCCbyg

The sequence diagram in Fig. 4 describes the overview process of the QTO from a BIM model. Firstly, the information is stored in a database. Using objects classification such as the SfB classification [17], the Cuneco Classification System (CSS) [18] or the BIM7AA classification system [19] enables human-readable, standardized linking of database objects to a use case BIM model.

The objects can be linked to the placeholders in proprietary databases to ensure a single point of reference. The Revit model frequency, cost, count, and unit type parameters must be reflected in the database. dRofus is a tool tailored for Revit which enables storing room and object placeholders, enabling a bi-directional mapping between the model and the database.

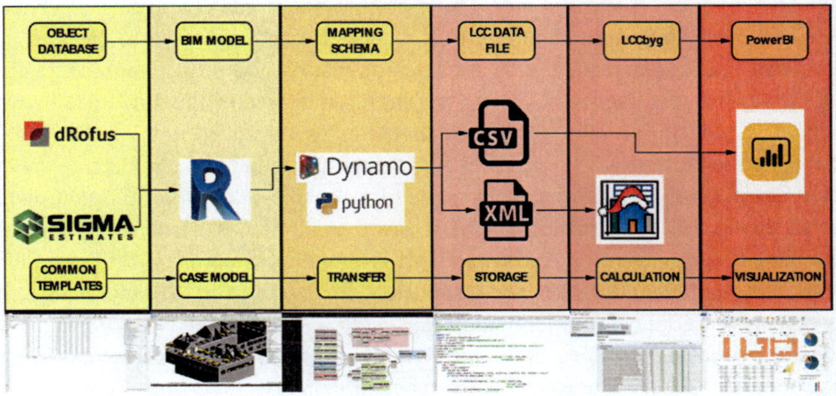

Fig. 4. An overview of the software sequence from the database to data visualization.

This way, the Revit model acts as an environment composed of predefined components which are otherwise stored in a repository. Prior to sending the transfer data, model validation is required, to ensure the objects contain the required information. Using an up-to-date database ensures the data is pre-validated, and the objects are modelled uniformly. If the model was not to rely on a database, or the repository was to contain known errors, geometry and information checks must be in place.

Once the validated information model is ready, the Dynamo data transfer script is run, and the Python code sends the data to the LCCbyg software (see Fig. 5). Following model adjustments, multiple data transfers enable an effortless LCC optimization cycle. The chosen design is then saved as a final, and the reports from LCCbyg can be generated. The resulting Dynamo exports .CSV file, which is then converted to an XML file, by traversing and parsing the schema using Python ElementTree module.

Name	Unit	Quantity	Unit price (DKK)	Frequency (per year)
Buildings, internal				
Version2				
312001 Family Type: 2090 x 1400/ Family: Windows_Fixed/Average	m2	322	1.00	2.00
311001 Family Type: 810 x 2030/ Family: Int.Single-Flush/Average	m2	143	3.00	26.00
311002 Family Type: 912 x 2113/ Family: Int.Single-Flush/Average	m2	61	3.00	26.00
311003 Family Type: 1812 x 2113/ Family: Int. Double-Flush/Average	m2	24	3.00	26.00
586539/Toilet/Tile antiskid/Average	m2	3	1.30	100.00
586542/Toilet/Tile antiskid/Average	m2	3	1.30	100.00

Fig. 5. A sample of quantity take-off on types and instances. LCCbyg 2.2.52, Screenshot.

Dynamo is a Visual Programming Language which directly plugs to the Revit API. It uses the Dynamo Textual Language (DTL), formerly DesignScript. It enables a direct connection with Python 2.7. It is critical to express design intentions and ease the wiring of data and geometry, beyond the out of the box (OOTB) capabilities of the Revit model [20]. The Dynamo version 2.0.3 is used.

A range of Dynamo packages can be used as predefined modules, improving the use of Dynamo OOTB applicability. Using custom nodes enables scaling of Dynamo scripts and controlling the granularity of the code. It is possible first to create a draft code transfer, and later replace it by modules of custom nodes or Python scripts.

If the reports generated by LCCbyg are hard to interpret, or the difference between options is difficult to observe, the information can be visualized in PowerBI (Fig. 6). The results are then made presentable for clients and other stakeholders. A Simple Vector Graphic (SVG) conversion is required to enable interactive 2D plans which correlate to the .CSV data tables. PowerBI was then be used to visualize optimization efforts and resulting savings. In PowerBI, it is possible to compare different design options, visualize which spaces are responsible for aggregating the most costs, and showcase how certain design solutions may reduce operational costs. What is still needed is overlaying cleaning costs with energy consumption, purchase and lifespan costs as the parallel indicators will fully enable a data-driven design.

The design choices can also inform the database once new solutions had proven more cost-effective, thereby improving existing templates and default solutions. The resulting workflow can inform best practices, which can then be applied to a variety of projects.

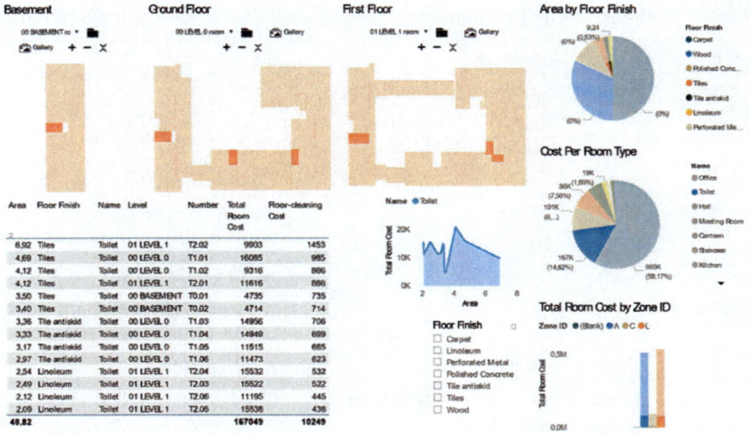

Fig. 6. Structure of the PowerBI interface – Case study 1 – Yearly cleaning costs per room. The plot diagram in the middle is not linear, due to discrepancies in furniture composition.

3.2 Case Studies

Our approach is tested on two case studies located in Aalborg, Denmark. The first case study is a 1,700 m^2 office building which accommodates two IT companies. The latter is an office building of 700 m^2 area, currently in the design phase (Fig. 7).

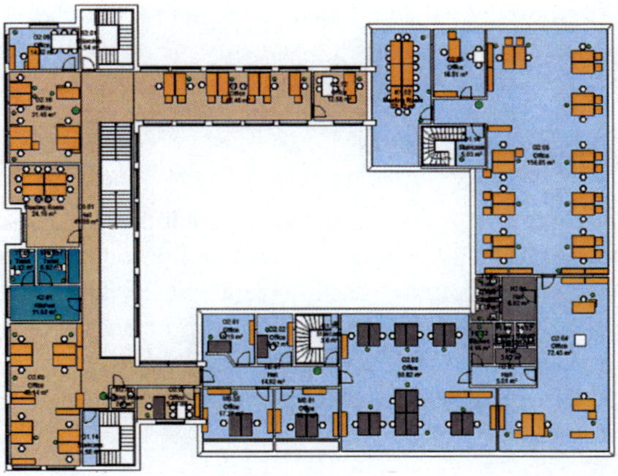

Fig. 7. Case study #1 office building – First-floor plan. Black desks are covered with antiskid surfaces, whereas orange objects are made of laminated wood. The colours of floor surfaces correspond to different floor finish materials.

The list of components used in the Revit model can be seen in Table 1. All components have been successfully transferred to the LCCbyg XML file, automatically. The computation time needed to transfer around 1,000 components was achieved under 15 s, using a regular laptop.

Table 1. Components and component materials included in the Case study model.

Furniture families	*Plumbing fixtures*	*Casework*
Desks, Tables, Chairs, Bins	Toilets, Washbasins, Hand dryers, Towel dispensers, Soap dispensers, Sanitary bag mounts, Mirrors	–
Doors	*Windows*	*Curtain walls*
Single & double leaf	Openable, Fixed	Openable, Fixed
Stair runs	*Lighting fixtures*	*Railings/skirtings*
Steel staircase runs	–	Stair railings, Balusters
Ceilings	*Object materials*	*Floor materials*
–	Smooth laminated wood, Gripped rubber plastic, Metal, Porcelain/Ceramic, Glass, Paint, Steel, Glass	Mosaic Wood, Tile antiskid, Tile smooth, Carpet, Perforated Steel, Linoleum, Entry Mats

The second case study model was obtained from a partnering architectural firm. The validating goal was to check if the Dynamo data transfer can support elements from various models. Some issues were encountered, related to inadequate modelling techniques, rather than the proposed Dynamo prototype. The issues concerned objects with no 3D geometry, incorrectly categorized or generic components, and object groups.

The model has been corrected and validated manually. Following the adjustments, the data transfer was achieved with the transfer time of under 10 s.

4 Chapter 4: Discussion

Calculating lifecycle costs can be beneficial to multiple stakeholders. Generating accurate estimations can inform material choices, and design solutions during design development, as well as help optimize cleaning strategies during the operational phase. BIM models can accommodate QTO, and the data can be automatically translated to a lifecycle costing software. Although the data transfer using the proposed workflow was deemed successful, there exist issues concerning the prerequisite actions needed to accommodate the transfer.

4.1 Could the Object-Oriented Cleaning QTO Inform Other LCC Phases?

The object-oriented costing mechanism has been developed for internal cleaning purposes only, and until the category is merged with the remaining LCC categories, the cost estimation will not be complete. To estimate the costs holistically, information regarding energy consumption, lifespan, maintenance and purchase cost will be required. As the mapping relies on input parameters located on objects, it will be possible to include the remaining information in parallel parameters using flexible mapping. To manage multiple phases a classification system must be developed to map parameters to a specific lifecycle phase, as proposed by Saridaki [21].

In its current form, internal cleaning mapping can be converted to external cleaning such as gardening or washing facades. The same structure can be applied by using quantity, intensity, frequency, and object cost per unit of measure.

Specifying the level of intensity may be undesirably ambiguous, as every object and circumstance may require different cleaning methods and equipment. One approach could be to internally develop object costs based on a longitudinal time measurement taken by a facility management team. Another approach could be to introduce international guidelines for manufacturers to estimate cleaning time for objects in a similar way to the energy rating, applying a grade to the ease of cleanliness, and estimated cleaning time, similar to watts needed to power the objects.

4.2 Developing Object-Oriented Costing Databases

The lack of object-oriented costing database prevents referencing costs to standardized pricing repository. This is an issue, which prevents the proposed approach from generating reliable LCC conclusions, hence hinders the applicability of the tool itself.

The proposed object-oriented costing mechanism will rely on gathering data on time it takes to clean specific objects over a significant period. The consolidated time averages will then inform the costs specific to each object. The variables will further include a specified intensity rate, the overhead costs and other factors affecting the final cost.

Gathering such data for architects responsible for calculating LCC is outside of the scope, hence the data must be either obtained from partnerships or available statistics. Currently, no such object-based cleaning repositories exist in Denmark. The FM teams will likely gather their own proprietary costing data, given the cleaning department is insourced, yet until the data is shared with the designers, the existing, questionably inaccurate methods for evaluating LCC will only be viable.

The lack of cost data is further hindering the proposed workflow, as the costs cannot be integrated with the object database. This delays the introduction of proper use of the database and gives the resistors a solid foundation to oppose the use of the system, thus preventing the integration of LCC analysis in the early design phase.

4.3 The Industry is not Ready

According to the interviewed BIM managers, it is challenging to prescribe correct modelling techniques, especially to the senior architects. It is equally challenging to convince people to use new software on top of the *existing new* software in use.

This further hinders the modelling accuracy as people tend to follow familiar logic, as opposed to the guidelines. Failure to do so hinders consistency which is the key benefit of using a standardized object library across multiple projects. If consistency cannot be achieved, effort must be undertaken by the auditor to validate the model prior to QTO.

The architects question the object-oriented approach by stating that the object placement and count are often unknown during the early design phase, as seen in Table 2. Although the information may exist in brief specification, it may not be converted to Revit. Some practices include using software such as Sketchup or Rhinoceros for early design development while using Revit only for the detail design phase.

Table 2. A summary of stakeholder engagement with regards to data automation

Stakeholder	Main concern	Main benefit	Interest
Architects	Not every project requires LCC, troubleshooting, 3D models in Revit	Increased efficiency, reduced auditing cost. Increased utility, dRofus integration	Interested
BIM managers	Dynamo interaction, coordinating modelling techniques, updating the database, challenging existing methods, issues with grouped objects	Simplified auditing process, standardized workflow, dRofus integration	Interested

That said, the current method of evaluating LCC, in practice, skips the optioneering part during early development and generates cost information only once the design is set. It is almost as if the LCC evaluation was to affirm the operational costs and receive a certificate, rather than to optimize the design in the first place.

Having interviewed the stakeholders from the AEC industry, there seems to be a gap in understanding the importance of LCC information between those assessing the initial cost, and those using this information to base tender contracts and cleaning strategies upon. The consensus (see Table 3) across stakeholders involved in the operational phase is that the costs are essential towards estimating contract values, as well as the Service Level Agreements (SLAs), while the evaluations during design are mainly made to accommodate the requirements specified by the sustainability certifications.

Table 3. A summary of stakeholder engagement with the proposed object-based cleaning.

Stakeholder	Main concern	Main benefit	Interest
Build to sell owners	Extra costs of design analysis	DGNB certificate	Not interested
Build to use owners	Extra costs of design analysis	Long-term savings, FM cost tracking, detailed Service Agreement, Material choice help, DGNB certificate	Interested
Architects	Complexity, Lack of early design info, extra workload	Long-lasting design, DGNB certificate, Extra fees	Partially interested
FM software developers	Relying on geometry, updating models, proprietary software	Sensor integration, relatively easy implementation, value creation	Interested
LCCbyg software developers	Lack of early design info, too specific, rigid transfer, deviation from DGNB, Lack of cost databases	KPIs visualization, FM integration, increased reliability	Partially interested
Hospital FM team	Must be integrated into the same software already used. Older hospitals lack BIM models	Integrated issue reporting, design option support, location-based object ID, IoT potential, Quality Control	Interested
Cleaning management	Expensive software and hardware	Work scheduling, route optimisation, pay per object, tacit knowledge transfer, performance tracking. Accurate tendering. Quality control	Interested
Operational cleaners	Undercosting of objects, reliance on technology, IT competences	Work schedules, more accurate pay, relaxed working time, reduced surveillance	Interested

Although the proposed approach was intentioned for the early design phase, there seems to be a need for an automated QTO system for existing buildings, so that the operational FM teams can better optimize their cleaning expenses. The key value for FM teams is information. Having the right information and the right tools to view the information effectively can furthermore assist the staff training, quality control and wayfinding.

The FM software developers agree that there is a need stemming from public clients, when it comes to an accurate evaluation of operational tenders, especially for buildings requiring outsourcing. Some argued that public spending could be better managed, even for insourced cleaning teams, using the proposed tool.

In Denmark, there are multiple computer-aided facility management (CAFM) software actively used [22]. The two most popular are Dalux FM and MDoc FM, which both use 3D geometry stemming from Revit or IFC models. Integrating the proposed object-based costing approach for operations was deemed useful and applicable, given the parametric approach is similar to costing object-based procurement costs. The future will show if the object-based cleaning sees implementation in the existing tools.

5 Conclusion

To conclude, a prototype for the integration of Revit model data with LCCbyg XML schema was developed successfully. Due to the frequency of operational costs, there seem to be good reasons to calculate cleaning costs at an object level. There are tools available to quantify objects, and there are tools available to calculate the lifecycle cost.

What is still missing is a method to quickly assess the overall building cost of operational maintenance and visualize it effectively to the stakeholders. This gap is bridged in this study with the approach to quantifying, sorting, and mapping data between a BIM model and an estimating tool and visualizing the results.

Using the proposed object-based approach enables making design decisions on fixtures and materials that influence operations costs during early design. Moreover, the visual display of data will likely convey the significance of operational costs to the stakeholders outside of the AEC industry. While the prototype workflow can be deemed successful, and the stakeholders agree on the value it brings, certain observations regarding the modelling of objects and information accessibility may hinder its quick adoption.

To implement the proposed approach effectively, object-based costing databases must be first established, thus enabling a validated reference point for pricing and specifications. Those must be then linked with BIM models, using a centralized object library. A standardized workflow must be adopted by the modellers to encourage uniform modelling techniques and adhere to the established process schemas. Integrating the proposed object-based cleaning with other subsets of LCC will only then fully unlock the potential for data-driven design envisioned in this study.

5.1 Future Research

To enable data transfers between the Revit model and LCCbyg, standard file formats are used; .CSV and .XML to transfer data. Due to early development, the integration of the proposed LCC object data with the IFC schema is yet to be proposed. A connection with IFC transfer is deemed an essential subject for future research, as it is hard to expect FM teams to have access to the Revit models and update them accordingly. IFC will enable interoperability with the FM platforms given a standardized mapping schema will be developed. Approaches of LCC IFC integration had been attempted before [23], though the subject requires further research.

An object-oriented costing and an object-specific cleaning intensity specification will need to be developed, either internally or nationally as a standard, prior to the implementation of the data in the architects' standard object databases. Pilot longitudinal research in collaboration with existing facilities will likely inform object costs, thus forming a reference point much needed for integrating the cost databases with the case-specific BIM models.

Acknowledgements. We would like to thank all stakeholders who took part in the development and discussion about the proposed methods including the C.F Møller Architects, Dalux FM software team, LCCbyg software team, NAU hospital FM team, and Thrane cleaning.

References

1. Vigovskaya, A., Aleksandrova, O., Bulgakov, B.: Life Cycle Assessment (LCA) in building materials industry. In: MATEC Web of Conferences, vol. 106 (2017)
2. Saridaki, M., Psarra, M., Haugbølle, K.: Implementing lifecycle costing: data integration between design models and cost calculations (2019)
3. Jensen, P.A.: Håndbog i Facilities Management. Dansk Facilities management netværk (2011)
4. Guillen, A.J., Crespo, A., Gómez, J., González-Prida, V., Kobbacy, K., Shariff, S.: Building information modeling as assest management tool. IFAC-PapersOnLine **49**(28), 191–196 (2016)
5. Conradie, D.C.U., Roux, E.: Construction project management; design, quality management; simulation. Gazette, pp. 16–18 (2008)
6. Green Building Council: An introduction to DGNB : Ensure the quality of your sustainable buildings in planning, construction, and operation. The DGNB system helps you get there (2013)
7. Green Building Council Denmark: DGNB System Denmark manual for kontorbygninger 2016, vol. 1.1 (2017)
8. Lawrence, M., Pottinger, R., Staub-French, S., Nepal, M.P.: Creating flexible mappings between Building Information Models and cost information (2014)
9. McCuen, T.L.: Underdeveloped and underutilized: cost estimating in BIM. AACE Int. Trans. **1**, 13–21 (2010)
10. Hannon, J.J.: Estimators' functional role change with BIM. AACE Int. Trans. (2007). https://search-proquest-com.zorac.aub.aau.dk/docview/208175596?OpenUrlRefId=info:xri/sid:primo&accountid=8144. Accessed 16 Jan 2020
11. Venugopal, M., Eastman, C.M., Sacks, R., Teizer, J.: Semantics of model views for information exchanges using the industry foundation class schema. Adv. Eng. Inform. **26**(2), 411–428 (2012)
12. Eastman, C.M., Jeong, Y.S., Sacks, R., Kaner, I.: Exchange model and exchange object concepts for implementation of national BIM standards. J. Comput. Civ. Eng. **24**(1), 25–34 (2010)
13. Hartmann, T., Van Meerveld, H., Vossebeld, N., Adriaanse, A.: Aligning building information model tools and construction management methods. Autom. Constr. **22**, 605–613 (2012)
14. Haugbølle, K., Scheutz, P., Saridaki, M., Sørensen, N.L.: Installation Guide and User's Guide LCCbyg version 2.2 (2017)
15. Law, J.: A Dictionary of Business and Management. Oxford University Press, Oxford (2009)
16. Beyer, H., Holtzblatt, K.: Contextual Design: Defining Customer-Centered Systems. Morgan Kaufmann, Burlington (1998)
17. Samarbetskomitén för Byggnadsfrågor: SfB-systemet, pp. 316–320 (2012)
18. BIPS: CCS (2015). https://bips.dk/værktøjsemne/ccs#0
19. BIM7AA: "Bim7Aa Typekodning," Europaplads 2, 11. sal, 8000 Aarhus C (2018)
20. Mccrone, C.: Dynamo Language Manual, pp. 1–57 (2010)
21. Saridaki, M., Psarra, M.: Integration of LCC into BIM concept, no. DTU Byg, shared internally, p. 137 (2017)

22. Basico: An overview of FM systems in Denmark. DFM (2017). https://docplayer.dk/48025613-Fm-systemer-i-danmark-et-overblik.html. Accessed 17 Jan 2020
23. Santos, R., Aguiar Costa, A., Silvestre, J.D., Pyl, L.: Integration of LCA and LCC analysis within a BIM-based environment (2019)

BIM and AM to Manage Critical and Relevant Water and Wastewater Utilities Assets

Wagner Oliveira Carvalho[1,2,3](✉)

[1] Aegea Sanitation, São Paulo, Brazil
wagnerc2011@gmail.com
[2] Zigurat Global Institute of Technology, Barcelona, Spain
[3] Universitat de Barcelona, Barcelona, Spain

Abstract. Global Water and Sewage Utilities must manage a wide variety of essential infrastructure assets while meeting complex social and environmental challenges, and this is especially true in Brazil, where according to SNIS Institute only 53% of the population has sewerage coverage and 17% still does not have access to treated water [16].

The Brazilian government's new regulatory sanitation framework [3] was approved in December 2019, establishing goals to accelerate water and sewerage coverage, stimulating private investors either to build new or to operate existing plants.

However, the asset inventory handover process from the public to the private sector is unstructured, causing numerous operational and maintenance challenges to new operators and investors.

This paper presents the digital transformation framework applied in Aegea, a private group of water and sewage utilities in Brazil, aiming to meet in a sustainable way a social, environmental and financial value (SLO - Social License to Operate) [15] by combining Building Information Modelling (BIM) [17] and Asset Management (AM) [8] disciplines. Aegea's integrated approach, "Infra Inteligente" (Infra), introduces intelligent 3D models of Water and Sanitation business.

The 3D business representations trigger the Digital Twin in unrivaled data and analytics visualization within a virtual reality environment supporting the organization to take strategic and tactical decisions to do the maintenance and operations of the inherited plants. By switching from reactive to proactive maintenance it will save millions of Brazilian Reais, extending the life of critical system assets and prevent the operation team from dealing with frequent and unpredictable equipment breakdowns.

Keywords: BIM · AM · SLO · Digital Twin

1 Introduction

According to the United Nations World Urbanization Prospects [14], the global perspectives on population living in urban areas indicate a steady increase from 55% to 68% by 2050, driving more than 2.5 billion people to the cities. Based on these statics,

E. Toledo Santos and S. Scheer (Eds.): ICCCBE 2020, LNCE 98, pp. 697–720, 2021.
https://doi.org/10.1007/978-3-030-51295-8_49

urban areas are expected to be more resilient and this will be even more challenging in developing countries. Thus, in order to take a step forward toward smart cities, asset managers need to have structured processes in place to build and manage water and sewerage infrastructure.

Taking into account that sanitation is essential for human health, the digital asset transformation can bring practical benefits for citizens' life by increasing the availability and quality of services.

The Brazilian sanitation coverage gap added to the environmental resource's limitation turn imperative the improvement of water infrastructure projects to meet the citizens' demand in a more effective way, balancing cost, performance and risks over asset's lifecycle.

Today, more and more organizations are challenged to share structured and unstructured data and information collected during the entire life of physical assets (revenue generators), including: construction projects, geospatial information (G.I.S), operation, maintenance, condition monitoring among others, according to BuildingSMART International (2019) [13].

Likewise, the BuildingSMART indicates that there is a long run to evolve in BIM standards for the different Infrastructure types. The first Industry Foundation Class (IFC = "Operating system" for open digital data exchanges) specific for infrastructure was launched in 2014, and the steering committee was established to combine, enhance and develop open standards for intelligent data.

Building Information Modeling (BIM) methodology is spreading rapidly in infrastructure management providing a simple and friendly integration of 3D objects (digital twin) [21].

Although many companies are adopting BIM when contracting their projects, and even digitizing some operational structures, there is still a lack of comprehensive studies proving the effectiveness of digital management over the entire life cycle of sanitation assets.

A great success example in BIM implementation for the architecture, engineering, and construction (AEC) industry is the CCC (Consolidated Contractors Company), which is today the major construction company in the Middle East. The Zigurat's master program is a source of knowledge about the CCC BIM story since the first project in Doha in 1997, passing through the CCC BIM Centers until the mega infrastructure projects such as the Midfield Terminal Abu Dhabi Airport in the United Arab Emirates [22].

Responsible for providing the commodity to life, Water and Sewage Utilities play a vital role on cities' development. Operators thus need to effectively reduce risk and associated costs going beyond the business-oriented perspectives to obtain the social license to operate (SLO).

In this context, BIM can serve as a shared knowledge resource for information if unified with AM (Asset management) over the asset life cycle, helping drive compliance with the international management standard ISO 26000 (guideline for social responsibility) [6].

The construction industry can better support sanitation from understanding its needs in a more structured way covering not only the constructed facilities but also the natural world including our interaction with the environment we inhabit.

In addition, the water asset manager along with the design and construction phase has a vital interest in receiving complete, accurate and validated infrastructure information, and with the BIM project completion, the handover will include the facility's digital representation including physical and functional characteristics to the operation team in a structured manner.

This global water context is even more challenging in some Brazilian regions, where the private owner-operators are assuming long term contracts, 30 years on average, with regulatory commitment to increase water and sewage coverage while enhancing operation performance, increasing the availability and quality of services.

Brazil, having 50% sewage coverage and estimated demand for investments around USD 100 billion by 2033 (according to the Brazilian governmental agency - PLANSAB) [1] sets the scene to the challenges ahead. That's why, there are countless places in Brazil to be recovered and with today's technology and a sustainable approach, there are smart ways to do this (see Fig. 1).

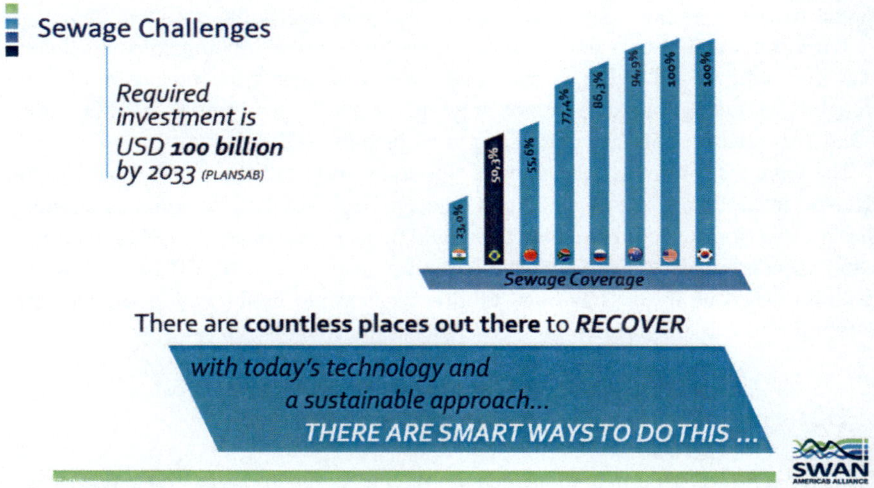

Fig. 1. Sewage Cover Index (Source: Aegea SWAN presentation, Miami 2019) [18]

In face of global sanitation challenge, BIM and AM could be the game-changer having the potential to deploy optimized infrastructure faster and cheaper than any other method.

The private operators are taking considerable risks, as they usually start the bid with inconsistent asset inventory from public sector. The technical information collection is spread in different files formats, split into thousands of 2D drawings, specs, printed documents, and normally without standard for asset nomenclature.

Consequently, assuming the operation of these systems in Brazil means to deal with a lot of water outages, often caused by unknown assets, having to face all the impacts of their failures on safety, environment, reputation, legal, contractual and quality of services, as well as the financial consequences of the unpredictable events.

The transition from reactive to proactive approach is a complex issue and requires the organization to raise asset management maturity index, which includes the implementation of an AM methodology and a computerized maintenance management system (CMMS).

The integrated approach to face the challenges was named "Infra Inteligente" (Infra), and applying this concept to water utilities produces intelligent 3D models based on physical and functional assets characteristics of its digital representation. It means a considerable shift in the traditional way to deal with the information until today, from electronic documents and 2D drawings to 3D model asset data-oriented over the entire concession lifecycle.

To water owner-operators, the Infra benefits is even more significant, considering that they have the decision power through conception, design, construction, operation, and maintenance to turn the business steadier and more profitable.

Making use of the intelligent and friendly 3D BIM models integrating hydraulic models and AM across the business-life, enables water managers to create faster and reliable what-if scenarios simulation to support better decisions.

Therefore, the investors must pursue a joined-up view of funding streams related to Total Expenditure (TOTEX), which can leverage better decisions over the asset through time, taking into account not only the Capital Expenditure (CAPEX) but also as the asset continues its lifespan during the operation (OPEX).

The three main research questions of this study were to first, assess if the Building Information Modeling (BIM) and Asset Management (AM) methodologies applied to other sectors (gas, telecom and energy) would be as effective as at the Brazilian water sector? Secondly, what would be the transformative impact of BIM and AM on a privately-owned utility in Brazil and thirdly, what would be the key lessons for other water utilities to learn from?

2 Objectives

This real case study aims to demonstrate the BIM tools integration to AM approach can deliver reliable water asset lifecycle information into friendly 3D models, incorporating specs, documents, attributes and operational real time data of virtual assets (digital twin).

The general objectives are the evaluation of applying a powerful set of technologies and processes in the sanitation sector, in order to lead the organization to make better, faster and more accurate decisions over the whole phases of the concession contract in compliance with the SLO principles.

In order to achieve and demonstrate this case study objectives the analysis was divided into following steps:

- Brief history report of how the need emerged for structuring asset management processes within the organization;
- Technology selection of a computerized management maintenance system (CMMS);
- Present the planning, benchmarking, classification methodology and implementation of an AM system;
- Advanced data acquisition technologies for digitizing assets in 3D;
- Introduce the BIM execution plan;
- Practical demonstration of a decision process on a critical and relevant asset, balancing cost, risk and performance using reliable data and analytics in a virtual reality environment.

3 Water and Wastewater Utility Study Case

The innovative initiative known as "Infra" was introduced at Aegea corporate environment and it has been revolutionizing the sanitation sector in Brazil. It was a long history of successful experiences in some Units (cities with Aegea sanitation concessions), diagnosis of problems in other Units and, the vision to improve overall business management.

Aegea inspiration relates to 2014 during the development of the Water Master Plan 2041 study for the Unit of Prolagos, which is one of the major Aegea concessions located in the state of Rio de Janeiro. The challenge was to deliver a long-term contract study in order to improve the overall system performance, and the final result presented was remarkable: increase revenue by 30%, save 55% energy consumption, reduce maintenance costs and enhance system security.

At that time Aegea realized the need to possess reliable asset information in order to combine the hydraulic powerful modeling tool used with friendly visualization for each "what-if scenarios" and alternatives proposed.

The application of Bentley WaterGEMS hydraulic modeling tool took the project to be presented at Bentley's 2015 Symposium in London, ranked among the world's best winning the BIM Advancements in Water Networks category by Bentley Systems [2].

The Prolagos master plan a robust and multidisciplinary study taking into account the scope and contract to detail, identifying the improvements and investments envisioned along the entire concession life cycle. Aegea's award-winning master plans have delivered significant ROI benefits by optimizing water and sewage infrastructure through scenario selection and alternatives generated within hydraulic modeling tools (WaterGEMS) (see Fig. 2).

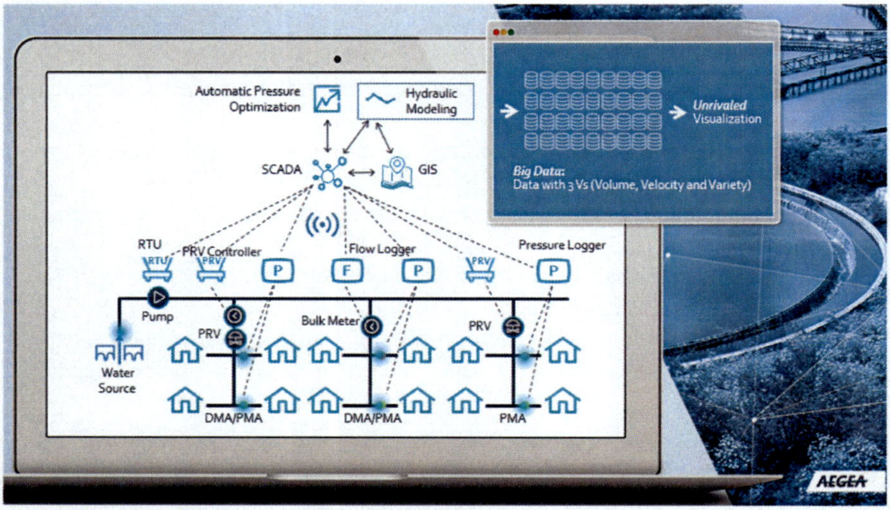

Fig. 2. Aegea Water Architecture – Prolagos Master Plan 2041 (Source: SWAN Forum) [20]

Aegea board, realizing the opportunities to improve asset (revenue generators) management, held several strategic meetings during 2015 reporting the main problems that were impacting the business:

- Difficulty in standardizing operation and maintenance of Water and Wastewater Systems;
- 48 Water and Sewage Units inherited from the public sector in different regions of the country, different organizational cultures, old infrastructures and little information available;
- Difficulty in meeting Capex and Opex planned budget;
- Excessive water supply and sewage system interruptions. Plenty of overtime in maintenance work;
- Difficulty in measuring inherent and residual risks;
- Difficulty in finding better engineering solutions (design) to meet the coverage and efficiency contractual targets;

Aegea board's decision was to evaluate the asset management maturity level of one of its Units.

By March 2015, work was carried out to identify the maturity stage of processes related to asset management in one of Aegea's most significant Water and Wastewater Utility, Águas Guariroba, located at Mato Grosso do Sul state. This Unit serves more than one million customers and it is a representative sample of the other units.

In 2017, a survey and mapping of operating costs were published in the - Costs OpEx Curve ABC - 2017 (Águas Guariroba) document, identifying opportunities for cost optimization. In this survey, a structured evaluation methodology was used in line to ABNT NBR ISO 55001: 2014 requirements and the 39 principles of GFMAM

(Global Forum Maintenance and Asset Management), capable of providing an accurate maturity diagnosis in asset management system [9–11].

The methodology applied compares the Águas Guariroba Unit routines and processes to a global Asset Management best practices generating an array of considerations:

- The absence of a structured policy, strategy, and CMMS (Computerized Maintenance Management System) system. Activities related to assets occurred reactively or in isolation, without any organizational or operational context;
- Physical assets life cycle assessment was still a topic little known by managers, but intuitively used to justify some specific investment projects;
- Absence of warehouses with spare parts and defined processes;
- No specific criteria for defining an inventory item, nor its maximum and minimum levels required;
- Tagging (functional identification) of assets was inconsistent. Their movement, or any type of important information, wasn't managed/administered;
- Absence of a formal system to manage requests for tactical and non-tactical maintenance;
- Documents inconsistency allocated to technical library;
- Absence of a formal process for identifying, resolving and measuring gains from operational improvements in physical assets (revenue generators).

In this scenario of uncertainty of uncertainty there was a need to start looking for better decision-making and prioritize investments in order to improve performance, availability and quality of services?

Then in 2018, given the business needs and operational opportunities, the company launched the "Infra Inteligente" (Infra). The program started with an intense planning phase, mobilizing a multidisciplinary team, asset management specialists' partners and Aegea's key areas representatives.

The program's core team in search of AM benchmarking looked into a successful case at the energy operation of Epcor Utility in Canada. In 2019, a technical visit was made to Edmonton City to validate in practice how they were able to digitize assets, enabling them to quantify asset health and realize a 15% improvement in reliability. In the same year, this Canadian Utility company won Bentley's global infrastructure award competition in Singapore [4].

Aegea Infra Inteligente program aims to:

- Implement BIM (Building Information Modeling) methodology, supported by an asset management system using interoperable digital tools covering the stages of planning, design, analysis, documentation, construction, monitoring, commissioning, operation, maintenance and decommissioning physical assets under the responsibility of Aegea's existing and future concessions.
- Establish process and standards for contracting design, construction and operation;
- Establish a process to manage the life cycle of physical assets;
- Initially establish control and subsequently, optimize maintenance and operation physical assets costs with an estimated reduction in Aegea's OpEx:

  - Energy (Efficiency): reduction up to 18%;
  - Maintenance (OpEx): reduction up to 24%;
  - People involved in maintenance activities (Productivity): increase up to 20%;
- Establish control and mitigation of risks associated to assets (physical, human and informational), considering the areas:
  - Water Intake, Treatment, Storage and Distribution (Water System);
  - Sewage Collection and Treatment (Wastewater System);
  - Various services including partners, third parties, designs, engineering (Services and Support);

4 Methods

As Aegea manages a large number of Units and a variety of sizes, populations ranging from 13 thousand to 1 million inhabitants, and thus the implementation strategy was based on a selection of 4 pilot Units set for proof of concept, classified according to their demographic population as well as one small, two medium and one large pilot Unit.

The application of AM and BIM methodologies at pilot Units guided the entire planning and development of Aegea's program.

Concluded the experimentation (diagnostic phase) the activities followed the Aegea guideline to consolidation which involved: survey and asset analyses; criticality, relevance and risk identification; CMMS tool parameterization and practical application of new decision-making process on a critical and relevant asset (see Fig. 3).

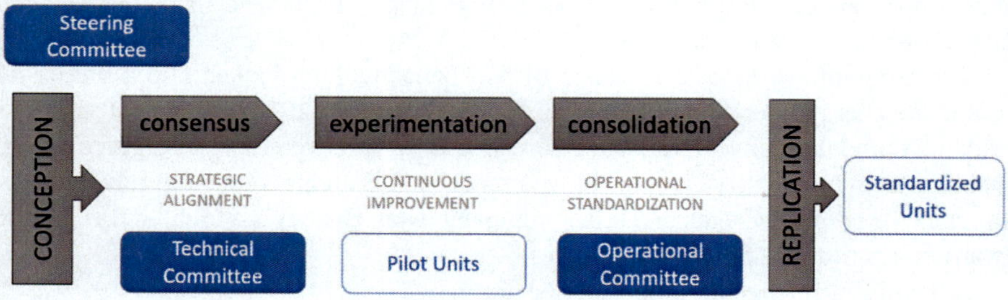

Fig. 3. Methodology flowchart

4.1 Asset Management Guidelines

The first focus was to development Aegea's AM Manual [5], aimed to establish the requirements, standards, norms and technical references to asset managers, through standards and procedures including, but not limited to:

- Definition and Application of Aegea Asset Management recommendation based on ABNT NBR ISO 55.001: 2014;

- Definition and Application of Structural Standards and Procedures to the correct implementation of Aegea Asset Management;
- Definition and Application of Structural Standards and Procedures to the correct parameterization of the CMMS Asset Management system.

To begin the implementation phase, it was necessary to answer three questions about physical assets: What are they? Where are they? What conditions are they?

Considering Aegea size in Brazil managing more than 49 large, medium and small water and sewage Units spread from north to south and estimated 30 thousand equipment (revenue generators) the team had a tremendous challenge ahead.

To answer the basic information was necessary to develop a strategic plan including a business mindset change from a water and sewage operator in to an intensive asset management organization (EAM - Enterprise Asset Management) [12].

An effective physical asset management is the new core of the business, in other words, the reliance on machinery, equipment and infrastructure would deliver real value for all the stakeholders – business health and depends on physical assets health (see Fig. 4).

Fig. 4. Asset Management Mandala, Infra Inteligente Program – Adapted from PAS 55

Thus, at the edge of the mandala seats the business strategy translating the shareholders' expectation, at the center seats the physical assets responsible to materialize the shareholders expectation, and rotating the mandala clockwise is possible to understand the complexity and responsibility of a modern asset management philosophy:

- Human assets: interacts directly or indirectly with physical assets, demanding defined roles and responsibilities, among other things like motivation, knowledge, experience, leadership;
- Information assets: a single source of truth (data + information) about condition, performance, activities, costs and opportunities;
- Intangible assets: reputation, image, moral, compliance, restrictions and social impact;
- Financial assets: consequence of good pillars management, translating in the life cycle costing, capital investment criteria and operation cost;

All the AM dimensions and BIM standards (ISO 19650) [7] inspired Aegea to create the streamline business vision also presented at the SWAN World Conference in Miami on May 2019, showing the water utility lifecycle representation integrating: main business phases and operation phases usually lasting 30 years in average from assumption to end of concession's contract (see Fig. 5).

Fig. 5. Water Concession Lifecycle (Source: Aegea SWAN presentation, Miami 2019)

The divisions created and represented at the Fig. 5, was named: "the Virtual Phase" and "Digital Phase", corresponding to the investment stage (Capex) and operational stage (Opex).

The virtual stage involves the challenge of designing and implementing new infrastructures or renovating existing ones. The macro phases were split in conception and design studies proceeding to execution ending with the new infrastructure handover and marking the beginning of operational cycle.

Succeeding that, we embraced the physical asset digital phase encompassing cycles of operation and maintenance. This last phase is the longest and where the business results coming through revenues of water supply and sewage collection, creating a great potential for improvements by digitalizing the critical and relevant assets and creating the respective digital twins.

4.2 Asset Identification

As asset management comprises decision making throughout its life cycle, standards and record structure are a fundamental part of any management system that aims to manage relevant information.

The Logical Aegea Assets Tree (how the assets are linked together) allows to centralize the entire management process in a single functionality, giving an overview of organizational structure, the interaction and locations of related assets enabling a holistic view of the entire system or process.

For the successful implantation of a structured asset tree is necessary to carry out the first phase, tagging (TAG) and the definition of organizational structure and processes that will be understood/represented by different levels, such as "Installation Sites".

The definition of a structural equipment tree model depends on the exposure form selected meaning the assembly option would be the best way to visualize the process at managers mind, maintainers and those involved in the asset management process.

The Logical asset Tree, through its database and relationships, expands the management potential in the maintenance area, concentrates the execution of operational routines, such as corrective and preventive work orders, reports and indicators, maximizing productivity without burdening the day to day users.

4.3 Tag Representation by Aegea's Asset Tree Model

The Infra Inteligente labels characteristics for each level in Aegea asset tree hierarchy was developed in order to establish a corporate standard (see Fig. 6).

Description Level	Level 1 Organization	Level 2 Region	Level 3 Unit (Utility)	Level 4 Contract	Level 5 Aggregator System	Level 6 Plant	Level 7 Phase	Level 8 Area	Level 9 Asset	Level 10 Place holder	Level 11 Component
Photo		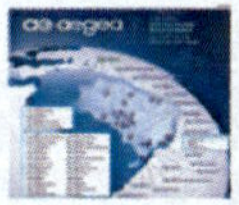			Not applicable		Not applicable	Not applicable		Not applicable	
Tag Layout		Not applicable		Not applicable	Not applicable		Not applicable	Not applicable		Not applicable	
Tag Specification	According to Layout 1 details	Not applicable	According to Layout 1 details	Not applicable	Not applicable	According to Layout 1 details	Not applicable	Not applicable	According to Layout 2 details	Not applicable	According to Layout 3 details

Fig. 6. Aegea asset tree (Source: Infra Inteligente Program)

To implement a sustainable asset management program or system is necessary the knowledge/recognition of existing Aegea's assets and for this purpose a procedure was created to raise the following data: technical, characteristics, attributes, conditions of existing assets in the field.

4.4 Criticality, Relevance and Risk Classification

Criticality is the attribute that expresses the importance of a machine or equipment within a production process, classifying how important and indispensable a certain piece of equipment is within the operational context.

Relevance is the attribute related to replacement cost/value, which can cause downtime and therefore high maintenance costs.

Aegea's Infra program pilot implementation was conducted at Holambra, Matão and Piracicaba Units, cities located in São Paulo state. The first consolidated information of critical and relevant assets was reached at Holambra Unit (see Fig. 7).

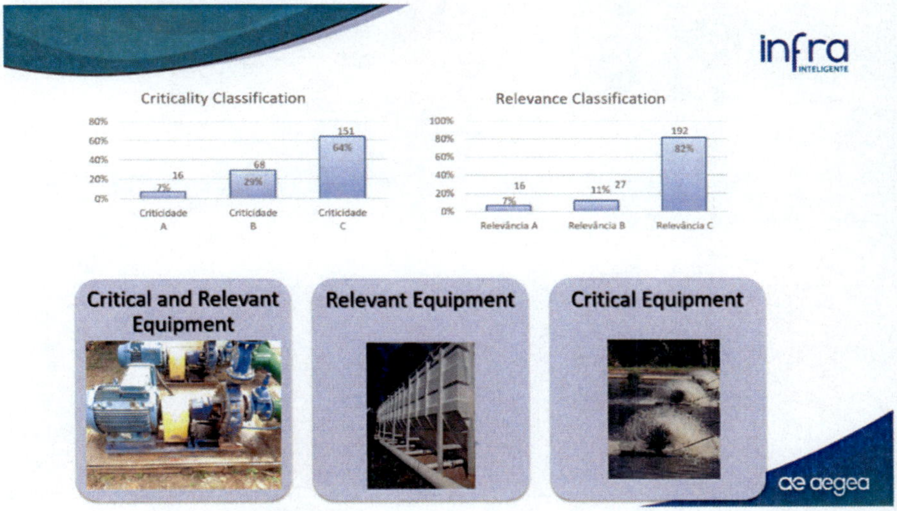

Fig. 7. Criticality and Relevance Classification – Holambra Water and Wastewater Utility

The attributes are stratified according to Aegea technical criteria level A, B and C. The "A" represents the most critical assets which will be subject to risk analysis.

The graphic above shows 7% of the total physical assets in Holambra were classified as the most critical and relevant "A", meaning that sixteen assets are really significant and important to the operation system, and according to the proposal of Infra Inteligente Program they will be digitized (Digital Twin).

The assets risk analysis is carried out through the procedures: "Aegea Risk Analysis", taking into account the weigh criteria of the Aegea Risk Matrix spreadsheet which covers the impact in: maintenance; operation; environment, health and safety; legal and regulatory compliance; reputation and cost of substitution.

Once the risks are identified, they must be classified according to the probability of occurrence, and the severity of the consequences (see Fig. 8).

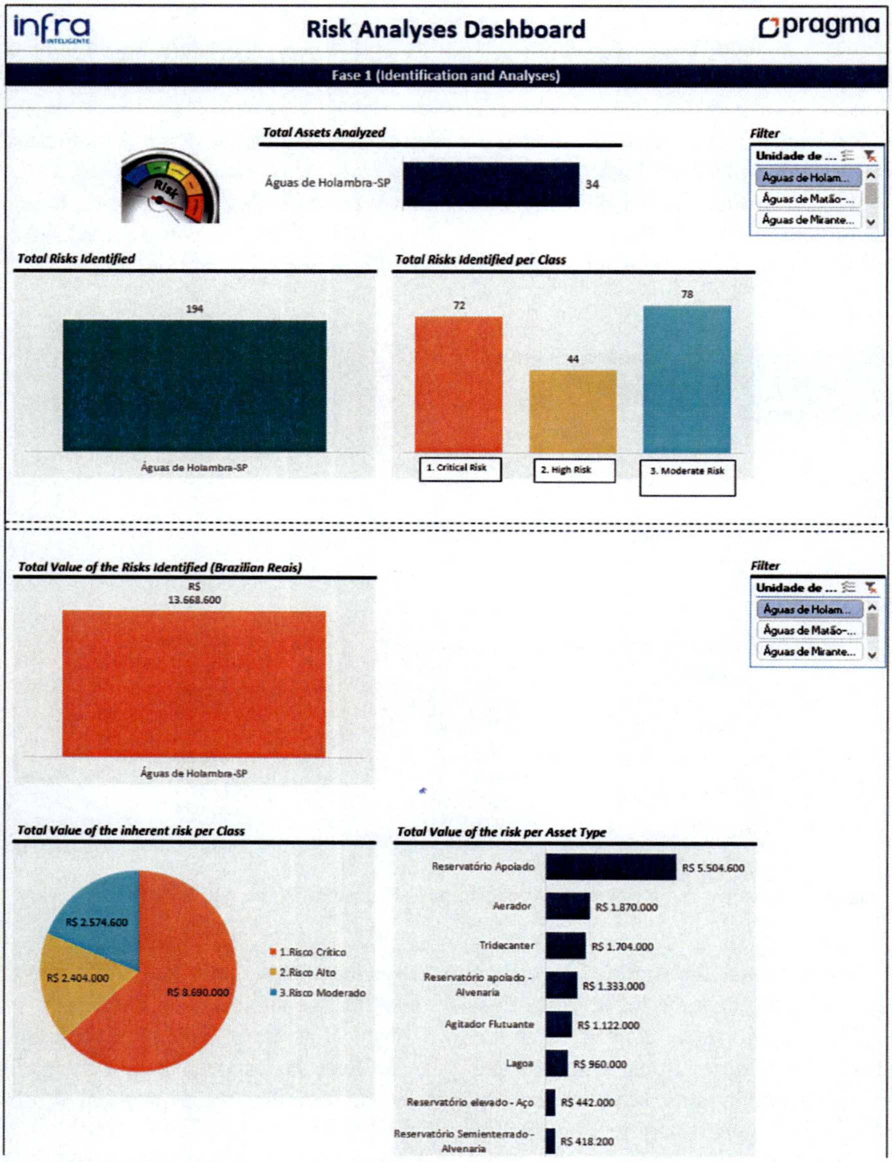

Fig. 8. Asset Risk Analyses – Holambra Water and Wastewater Utility

4.5 CMMS Tool Implementation

The Infra Program required the adoption of a powerful CMMS toll to integrate de information created by BIM activities as well as to capture and recognize the existing physical assets reality at the filed.

An CMMS shall provide few basic important features that serves as basis to control the different water and sewage units. The Infra established the following conditions to be met by the CMMS in order to integrate and support the field activities:

Provide structure and process to optimize maintenance work planning and control (effective utilization of trades people).

Provide information to enable intelligent management decisions.

A tool that can help to implement the optimum mix of maintenance work (breakdown versus scheduled).

A system to record and tracks maintenance cost (material, labor and contractors) over the asset life cycle.

The computerized maintenance management system (Pragma On Key) was implemented to manage the physical assets, bringing valued asset information at the hand of the decision makers allowing them to take responsibility with accountability for their units (see Fig. 9):

- Failures per location;
- Failures per asset;
- Failures per asset type;
- Failures per component;
- Failures per root cause;
- Failures per failure type;
- Failures per repair type;

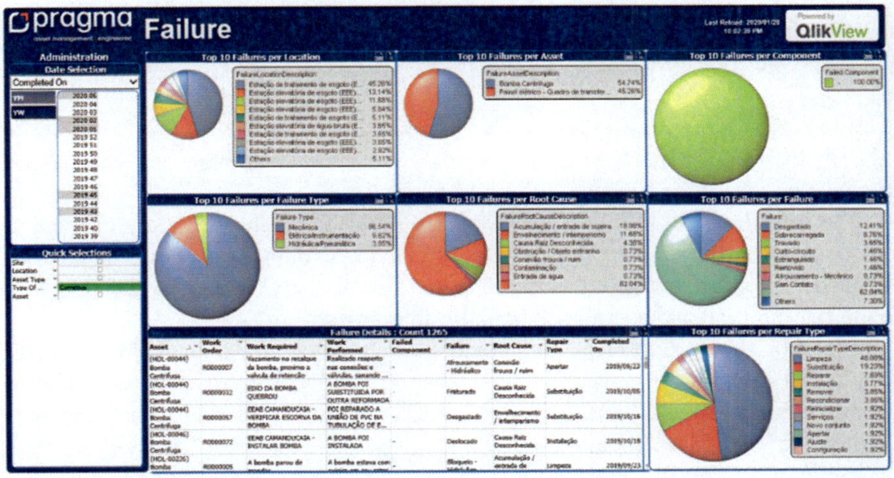

Fig. 9. Failure Dashboard – Holambra, Piracicaba and Matão Water and Wastewater Utilities

The valuable and dynamic information CMMS system is promoting a cultural transformation in the organization and its processes, reaching the initial Infra Intelligent Program targets.

4.6 Critical and Relevant Asset Digitalization

Once the critical and relevant asset were identified the field surveying to digitization process began, involving the combination of advanced acquisition technologies such as drones, GPS antenna, laser scanner, high-resolution cameras, and other equipment (see Fig. 10).

Fig. 10. Example of critical and relevant asset digitalization in a Wastewater Treatment Plant

4.7 BIM Execution Plan Implementation

Initially, the Aegea BEP (BIM Execution Plan) was set to orient new projects and also to guide the use of the digital twin in retrofit cases.

The Infra Inteligente program also created a virtual reality laboratory to promote collaborative discussions with main designers and operational stakeholders using the 3D model, in other words, this space integrates technology, processes and people to figure out the best engineering solutions for the projects.

The lab also aimed to break the current silos along the project evolution, applying Aegea BEP over design review sections every 15 days. At each stage of evolution geometric information is shared, structured data and documentation linked following the BIM Levels of Development (LoD).

The Aegea BEP is underway by the Infra Intelligent Program and the key pillars are displayed on a mental map (see Fig. 11).

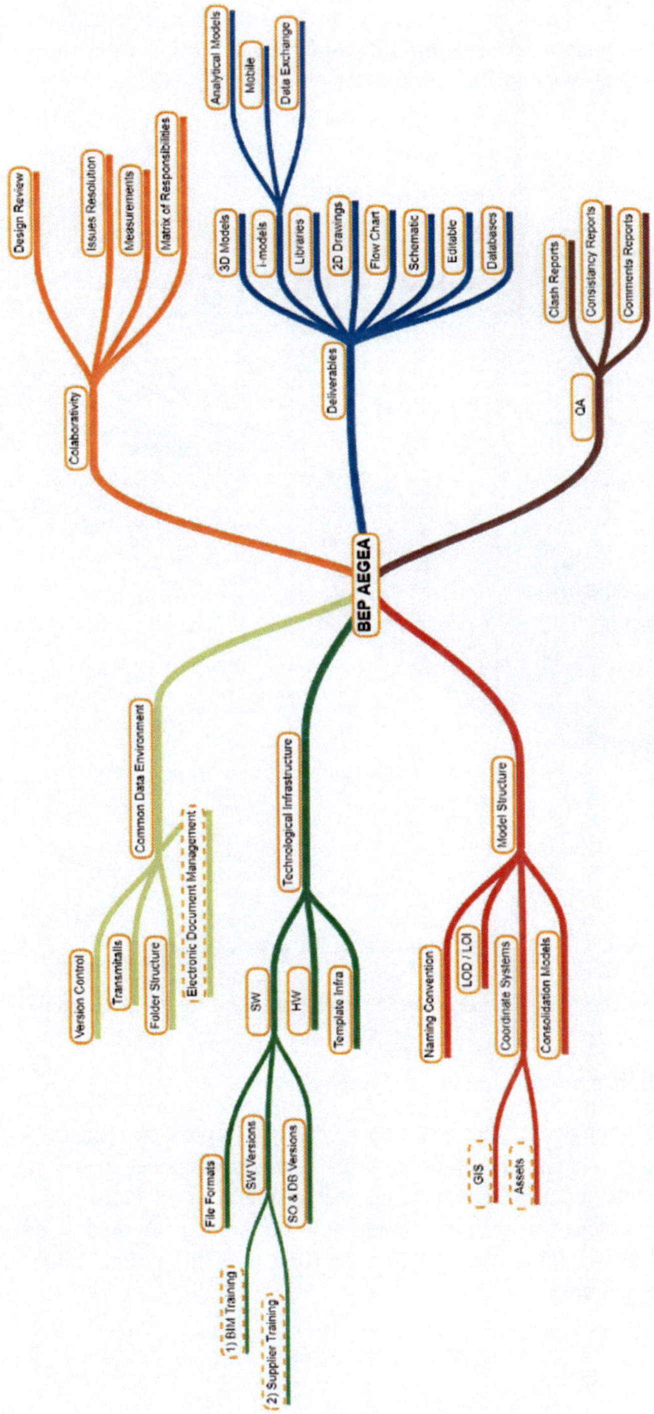

Fig. 11. Mental Map – BEP Aegea

The asset registration catalog was set in BIM platform (OpenPlant Bentley). The outsourced designers worked within the configurated work set to follow classification rules and attributes required for each asset type (see Fig. 12).

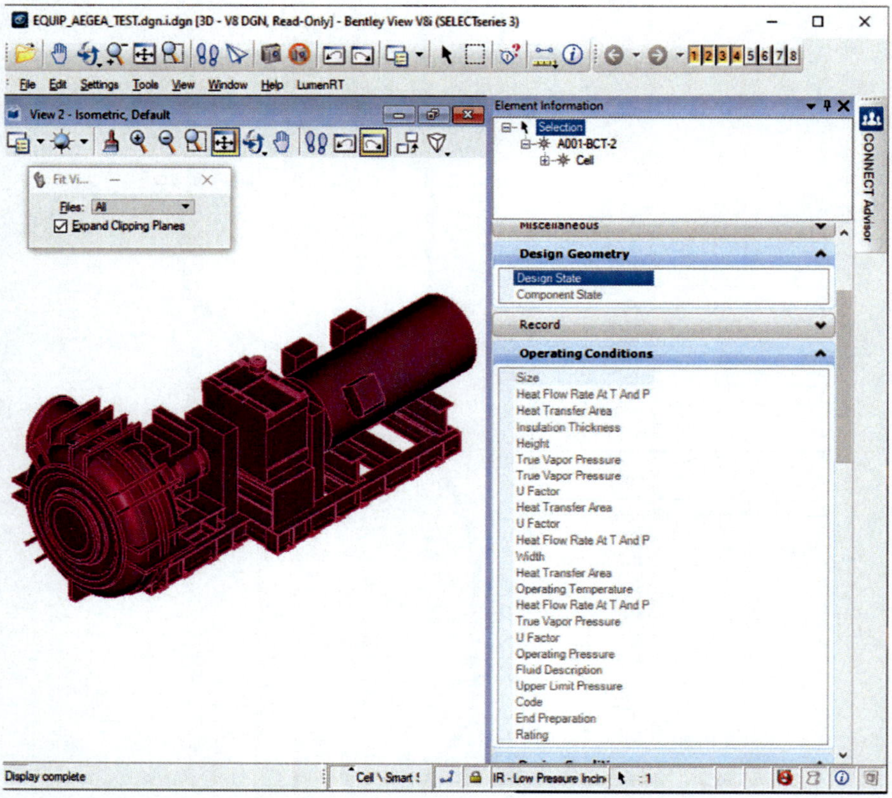

Fig. 12. Asset registration catalog inside Open Plant BIM tool – Infra Inteligente Program

4.8 Cost, Performance and Risk Analyses

The last Infra Inteligente stage methodology aimed to provide accurate information to different strategic, tactical and operational decision-making teams providing Key Performance Indicators (KPI) balancing cost, risk and asset performance.

Making the cross-platform data integration combining physical assets, digital twin and a virtual environment the complex decision over the critical asset proved to be much more effective.

The main difference perceived between the traditional method and the new model, besides the availability of reliable and accurate information, was the new way of discussing and presenting technical alternatives to all stakeholders.

The new dialogue experience increases understanding, even for those not having technical background, thus strengthening the relationship with local communities in accordance with the principles of the SLO. Everything is attractive and friendly as there is the integration of non-graphic information and graphic information (3D model) within a virtual environment (see Fig. 13).

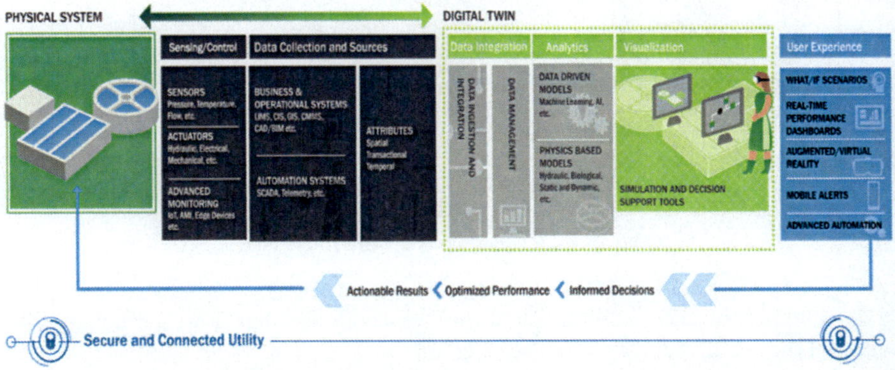

Fig. 13. SWAN Digital Twin Architecture Model [19]

The following practical application demonstration and analysis relates to a critical and relevant asset - decanter centrifuge - at Piracicaba Sewage Utility.

This equipment is considered by many specialists to be the heart of the sewage treatment plant, as it is responsible for the final processing of solid phase contained in the sewage for dewatering in preparation to the final disposal at sanitary landfill.

Hence, the equipment operational failure can mean substantial impact on the environment and consequently to the community, affecting the financial asset with a high cost in corrective repair and sometimes generating substantial operator's liabilities.

Following this chapter, the preventive actions Business Case generated by the AM and BIM methodologies is presented, which will increase the useful life of this equipment avoiding/preventing operational damages (Fig. 14).

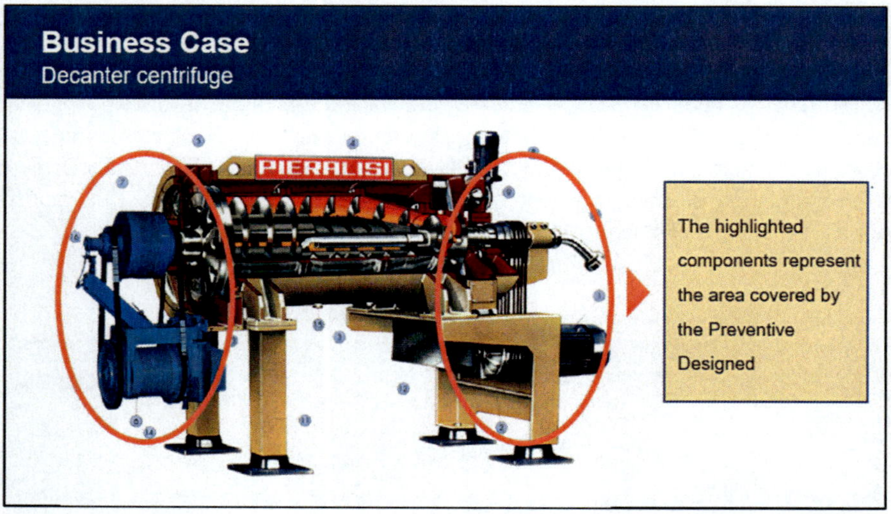

Fig. 14. Critical and relevant asset – Infra Inteligente Program

As highlighted at the CMMS platform, this equipment has a total estimated risk over 740 thousand Brazilian reais (see Fig. 15).

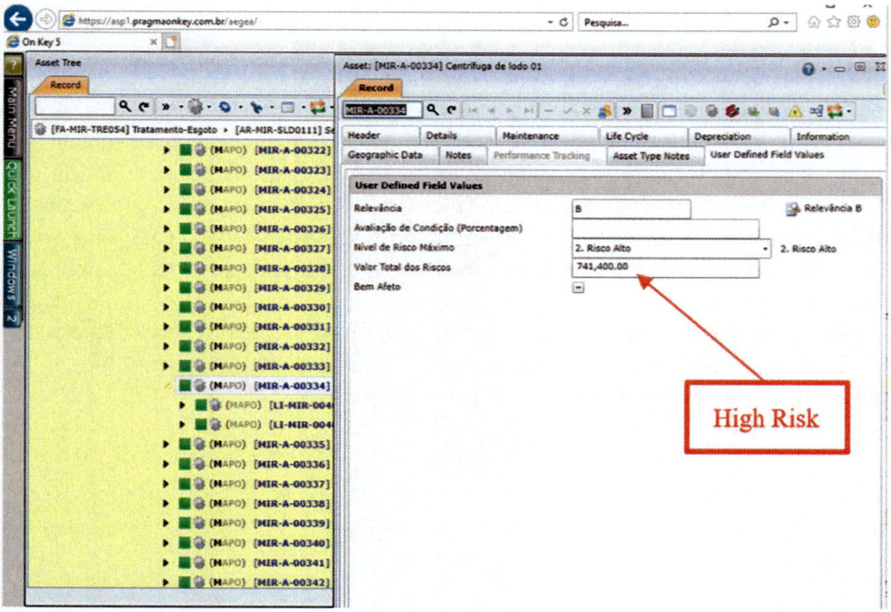

Fig. 15. The On Key CMMS tool highlighting the critical and relevant asset – Infra Inteligente Program

The following Fig. 13, shows the conflicting balancing factors of cost, risk and performance in relation to investments in preventive actions (see Fig. 16).

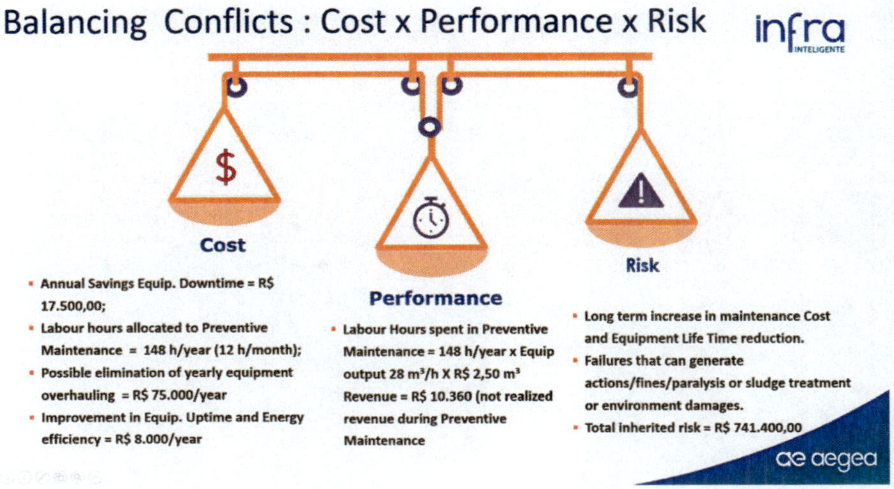

Fig. 16. Balancing conflicts analyses – Infra Inteligente Program

Bringing this case to the triple bottom line (SLO), it means lower environmental risk, higher quality of services for society, and economic benefit for all stakeholders, represented by a fairer tariff to the end user.

By looking and interpreting the available information generated by Infra, the impacts avoidances in addition to all intangible benefits for society is quite remarkable (see Fig. 17).

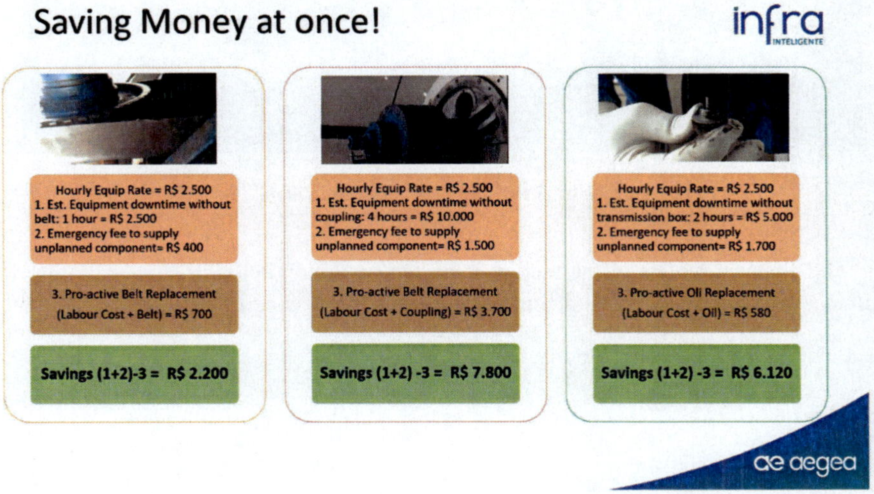

Fig. 17. Preventive tasks on decanter centrifuge – Infra Inteligente Program

The BIM collaboration analyses and decision happens at the virtual reality lab which is equipped with advanced hardware, specific application, and VR glasses.

Inside this immersive environment, the specialists have a better comprehension of the critical and relevant assets represented in detail by its digital twin (Fig. 18).

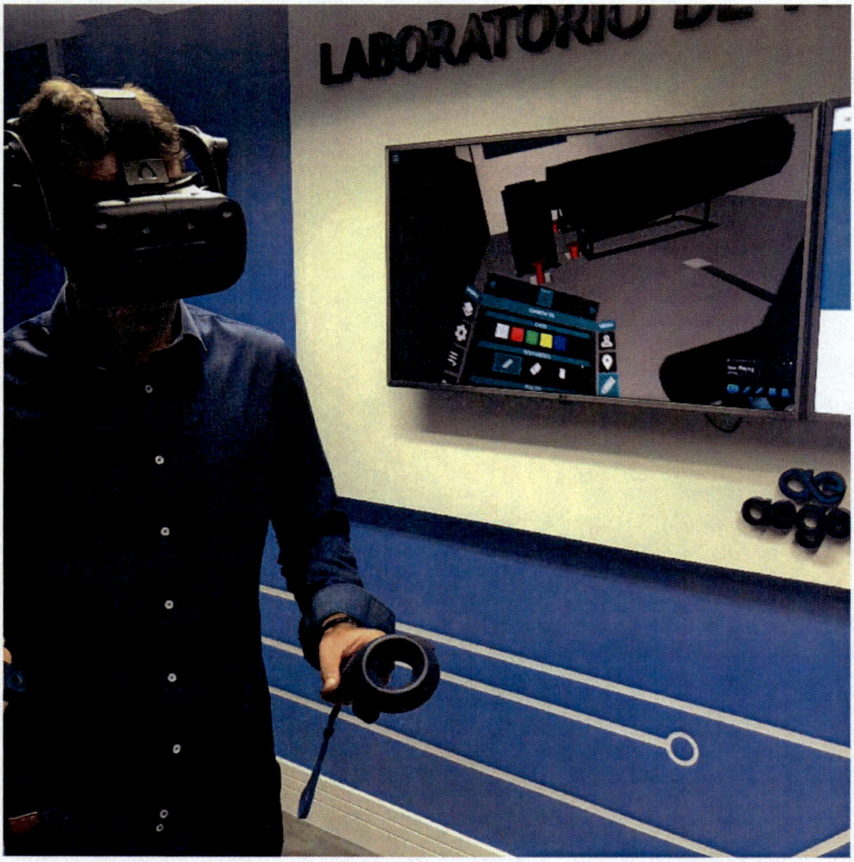

Fig. 18. Infra virtual reality lab - "Imersão Visual" technology

The physical replica of the facility is hosted on the digital twin cloud platform for collaborative analysis and comment among the various experts (Fig. 19).

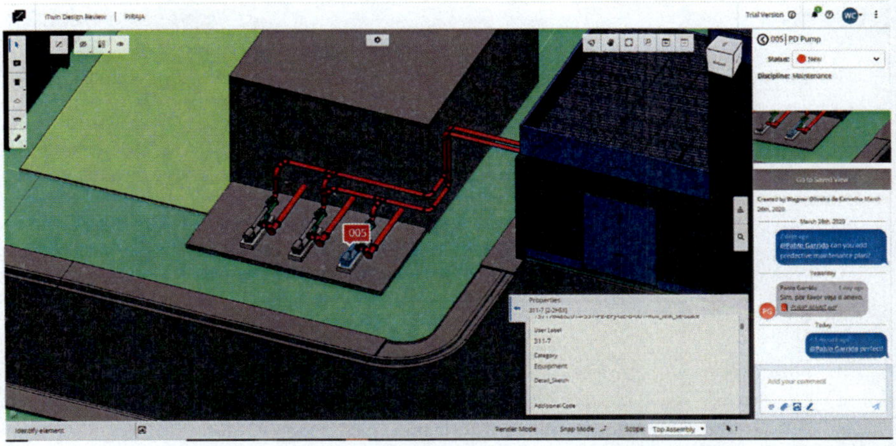

Fig. 19. 3D BIM model inside the Bentley iTwin

5 Main Findings and Contributions

The first pilot case study at Aegea Utilities proved the applicability and effectiveness of Infra Inteligente methodology in sanitation to ensure compliance and adherence to the global good practices of the Social License to Operate (SLO).

Concerning the BIM and AM transforming impact for private sanitation utility in Brazil, the methodologies combination proved to be very effective, mainly by bringing reliable information about assets in friendly 3D graphic models that facilitates decision-making at all levels of the organization.

This experience, developed and applied in practice, can be taken to other water utilities with some key lessons for attention:

- It is worth demanding the application of BIM for winning design contractors;
- Establish partnership with experienced consulting companies in AM and BIM;
- Selecting good technological partners that provide advanced, interoperable and open BIM platforms, as well as provide support for integration;
- Studying and engagement with BIM, AM and digital twins research groups to be updated on the latest trends and protocols.

However, it is important to highlight that as the sanitation sector in the country lacks structured processes, the gradual evolution in good AM and BIM practices will have as main challenge the need to change the organizational culture.

Other possibilities for future works are the Infra Inteligente to support the project's execution phase (4D and 5D) including physical asset handover in a continues update of the digital twin models.

References

1. ABRHIDRO. https://www.abrhidro.org.br/SGCv3/publicacao.php?PUB=2&ID=162&SUMARIO=5250. Accessed 31 Mar 2020
2. Bentley: Case Study. https://www.bentley.com/en/project-profiles/prolagos_aegea_water_master_planc
3. Camara. https://www.camara.leg.br/proposicoesWeb/fichadetramitacao?idProposicao=2207613. Accessed 31 Mar 2020
4. Canadian Manufacturing Homepage, Canadian Utility company won the award from Bentley in global infrastructure competition in Singapore. Accessed 31 Mar 2020
5. Infra Inteligente, Aegea & Pragma: Aegea Manual of Asset Management, São Paulo – Brazil (2020)
6. ISO. https://www.iso.org/iso-26000-social-responsibility.html. Accessed 31 Mar 2020
7. ISO 19650-1:2018 Organization and digitalization of information about buildings and civil engineering works, including building information modelling (BIM)
8. ISO. https://committee.iso.org/home/tc251. Accessed 31 Mar 2020
9. ISO 55000:2014 Asset management – Overview principles and technology
10. ISO 55001:2014 Asset management – Management systems – Requirements
11. ISO 55002:2014 Asset management – Management systems – Guidelines for the application of ISO 55001
12. McKeon, P.: CGI, Darren Ramshaw, EDF Energy: Implementing Enterprise Asset Management. Limited Edition. CGI, West Sussex – England (2013)
13. Phil, J.: Infrastructure Asset Managers BIM Requirements. Technical Report No. TR 1010, BuildingSMART International Infrastructure Room, Version 1 (2018)
14. Rowman. https://rowman.com/ISBN/9789211483192/World-Urbanization-Prospects-The-2018-Revision. Accessed 31 Mar 2020
15. Social License. https://socialicense.com/definition.html. Accessed 31 Mar 2020
16. SNIS, p. 62, Data table 6.1. http://www.snis.gov.br/downloads/diagnosticos/ae/2018/Diagnostico_AE2018.pdf. Accessed 31 Mar 2020
17. Stefan, M., Paul, S., David, P.: Building Information Modeling, 1st edn. Wiley, West Sussex (2016)
18. SWAN Conference. https://www.swan-forum.com/annual-conferences/. Accessed 31 Mar 2020
19. SWAN Digital Twin Work Group. https://www.swan-forum.com/digital-twin-h2o-work-group/. Accessed 31 Mar 2020
20. SWAN Forum. https://www.swan-forum.com/. Accessed 31 Mar 2020
21. Webopedia. https://www.webopedia.com/TERM/D/digital-twin.html. Accessed 31 Mar 2020
22. Zigurat. https://www.e-zigurat.com/en/masters-bim-infrastructure-projects/. Accessed 10 Apr 2020

Extracting Bridge Components from a Laser Scanning Point Cloud

Linh Truong-Hong(✉) and Roderik Lindenbergh

Delft University of Technology, 2628 CN Delft, The Netherlands
{l.truong, r.c.lindenbergh}@tudelft.nl

Abstract. A three-dimensional (3D) geometric model of a bridge plays an important role in inspection, assessment and management of the bridge. As most bridges were built after the second world war, 3D bridge models are rarely available. A recent development in laser scanning offers a cost-efficient method to capture dense, accurate 3D topographic data of the bridge. However, given the typical complexity of the bridge, a current workflow based commercial software to construct the bridge model still requires intensive labour work. This paper introduces a new approach to extract the point cloud of each surface of structural components of a slab/box beam bridge automatically in a sequential order from a superstructure to a substructure. The proposed method first employs a quadtree to decompose the point cloud of the bridge into two dimensional (2D) cells. Second, a kernel density estimation is used to separate a point cloud describing patches of surfaces within the cells. Subsequently, the cell- and voxel-based region growing are developed to segment patches within the cells/voxels for the superstructure and substructure, respectively. Moreover, knowledge of the bridge's components (e.g. position, orientation, or shape) is introduced to allow the proposed method to identify criteria for filtering irrelevant objects, and to establish criteria for extracting the components. An experimental test shows the proposed method successfully extracts all surfaces of the bridge components.

Keywords: Point cloud · Segmentation · Cell-based segmentation · Bridge deficiencies · Bridge inspection · Bridge modelling · Bridge components · Slab/box bridge

1 Introduction

Most bridges around the world were built after the second world war (WWII) and are now in the second half of their service life, are subjected to deterioration due to excessive usage, overloading, aging, and environmental impacts. In a recent survey on bridges from 6 European nations, it showed most bridges were built in the period 1945–1965 and have undergone deterioration [1]. In the US, the American Society of Civil Engineers reported about 9.1% of 614,387 national bridges is deficiencies and 40% of the bridges is more than 50-year-old [2]. Deficiencies cause negative impact on bridge operation that triggers increasing cost of delivered goods. As such, understanding changes of the bridge structures are an important key for decision making over various activities of the bridge, for example operation, inspection, assessment,

E. Toledo Santos and S. Scheer (Eds.): ICCCBE 2020, LNCE 98, pp. 721–739, 2021.
https://doi.org/10.1007/978-3-030-51295-8_50

maintenance and management, which is done based an as-service 3D bridge model. Indeed, there is large demand on 3D bridge models to support to the bridge activities efficiently.

A traditional surveying method, e.g. a total station, is time-consuming and expensive in bridge geometry acquisition because of the complexity of the bridge. Development of laser scanning allows to capture 3D topographic information of 3D scene in high details, which makes it recently adopted for surveying existing bridges [3]. A point cloud is subsequently processed for various applications in a bridge engineering, for example, bridge inspection and monitoring or assessment [4]. However, raw point clouds are insufficient to comprehend the changes of the components and/or to represent the bridge. As such, in a pipeline to convert the raw data points to meaningful information, the point cloud of each component must often be extracted. This procedure can reduce the complexity of the processing and avoids intensive computation on a massive point cloud of an entire bridge in further steps.

In practice, various standalone commercial products and CAD plug-ins, for example Leica cyclone software V9.1.3, Trimble RealWork and Revit, can assist in segmenting a point cloud of a surface and fitting the surface or an object by built-in simple geometric primitives (e.g. planes, cylinders, and spheres or even irregular shapes). However, these packages require to manually select a sub-point cloud and identify input parameters for a segmentation step. Indeed, only experienced practitioners can have ability to handle and process a massive point cloud of a bridge efficiently. Therefore, this paper proposes a new method to automatically extract the point cloud corresponding surfaces of a bridge' components for slab/box beam bridges. The proposed method deploys both spatial information of the point cloud and context knowledge of the components (e.g. locations, orientation, shape and size) derived from a bridge design specification.

2 Related Works

Given a complexity of a bridge structure, recognizing point clouds of surfaces of structural components is a key step toward to automatically inspecting surface deficiencies and generating a 3D geometric bridge model. Many methods have been developed to extract a point cloud objects' surfaces [5, 6] and then to generate 3D models [7]. However, a majority of these methods are applied for building structures instead of bridge structures. This section is restricted to investigate the methods applying for bridge structures excluding historic bridges, while extensive surveying on segmentation and geometric modelling can be found in [8, 9].

Walsh et al. [10] used a region-growing technique to extract a point cloud belonging to individual surfaces of a concrete pier cap. Next, a planar surface was fitted to the point cloud of each individual surface and the complete 3D model can be obtained by modifying these fitting surfaces to their intersections. However, this work also requires user interaction to extract edge points or groups of points. Moreover, Zhang et al. [11] detect a point cloud of planar patches of the bridge's elements involving three main steps: (1) determine a linear dependence relationship between points, (2) cluster the points based on the linear subspace and (3) extract the planar points through singular value decomposition. This method was heavy computation because features of all point were computed. Lu et al. [12] use a slicing algorithm to divide a bridge point cloud into two groups of points representing the deck and pier components based the slice heights. For the pier points, the surfaces of the pier cap or the bottom of the deck are separated based normal vectors of surfaces. For the desk points, a histogram is applied to extract the girder points after aligning the bounding box, whose edges are parallel to the axes of the coordinate system. The proposed method can be applied for slab/girder concrete bridges, but relies on heavy engineering parameters varying in a wide range.

Yan et al. [13] employed a voxel grid to subdivide the point cloud into uniform voxels, and the real boundary of the structure were then modified by using a grid-based mesh generation method proposed by Schneiders [14]. Moreover, nodes, edges and faces between two adjacent elements were adjusted to ensure a continuous mesh. Additionally, Stull and Earls [24] use the software Geomagic to create a parasolid model of a three-span, continuous steel I-girder bridge from a TLS point cloud for finite element analysis.

It shows that although generation of as-service bridge models from a point cloud are high demand for bridge engineering applications, few research studies develop methods to automatically extract structural components of a bridge and construct a 3D model. A restriction may be due to a massive data, a complexity of the data sets representing complex components of the bridge, and data problems like occlusion or uneven data point distribution [15]. Goal of this paper is to resolve those issues by developing a new method for extracting structural components of the bridge in automated manner.

3 Scope of Work

According to the US National Bridge Inventory [16], slab and box beam or girder bridges are account for 23.5% of the total bridge, which is the second largest bridge type on the road network. However, this bridge type is account for about 17.6% of the total bridge deficiencies slightly larger than 13.2% of the largest bridge type, a stringer, multi-beam or girder bridge. Moreover, in UK, 43% of the highway bridges are a slab bridge and 86% of future bridges will be either a slab or slab-beam bridges [17]. Therefore, the slab, slab-beam or beam-box bridges can be high attraction. This study is going to focus on extracting components of this bridge type from the laser scanning.

Components of a bridge can be classified as superstructure and substructure (Fig. 1). The superstructure is the components (girders, slabs, parapets) above the bearings, while the substructure consists of the components below the bearings. The superstructure components are often distributed in a horizontal plane along the longitudinal direction of the bridge while the substructure components are distributed in lateral and vertical directions. Moreover, when using a laser scanning to acquire the bridge geometry, only visible exterior surfaces of the bridge component can be captured. As such, following main context information of the components are used in this study.

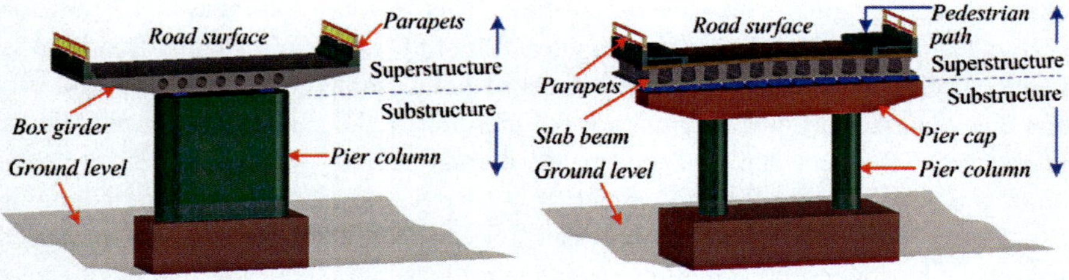

Fig. 1. Typical cross-section of the stab-beam or box bridges

Feature 1: The superstructure is bounded by 3 sets of surfaces in a horizontal direction: (i) the top of the bridge desk including road and pedestrian surfaces ($S_{supstr.top}$), (ii) the bottom of the superstructure ($S_{supstr.bot}$), and (iii) intermediate surfaces ($S_{supstr.int}$) between $S_{supstr.top}$ and $S_{supstr.bot}$, as shown in Fig. 2. Other horizontal surfaces between $S_{supstr.top}$ and $S_{supstr.bot}$ may not be available because they can be invisible. Based on bridge design principles, the road surface is the largest surface in terms of area (or size) and the number of data points. The pedestrian surfaces are smaller surfaces along the road and the elevation of the pedestrian surfaces is either equal or higher than that of the road surface. Moreover, parapets are parts connected to atop $S_{supstr.top}$.

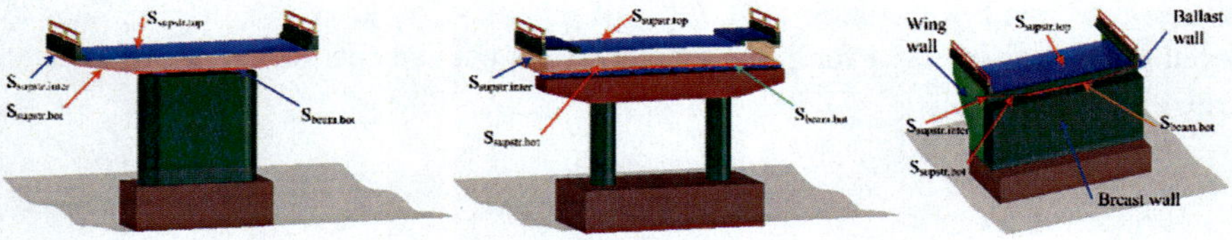

Fig. 2. Surfaces components of a superstructure

Feature 2: The substructure including the piers and abutments is distributed in a transverse and vertical direction, for example piers' columns are in the vertical direction, while the pier cap is in the transverse direction (Fig. 2). In the abutment, the breast wall is a vertical surface and its length is often equal to the bridge width, while the wing walls are usually perpendicular to the breast wall. In the pier, heights of the pier column(s) are generally larger than that of the pier cap. Moreover, the length of the pier, which is the dimension of the pier in the bridge transverse direction, is mostly larger than a half of the bridge width. The superstructure connects to the substructure via bearings or directly. In the former case, as bearing stones are often small and cannot be captured, a gap between the superstructure and substructure is available. In the latter case, which often occurs an integrated structure, the bottom horizontal surface of the beams/girders connects to the vertical surface of the pier cap or column.

Feature 3: The parapets may consist of vehicle and pedestrian railings. The parapets are on the edges of the road and pedestrian surfaces and atop of $S_{supstr.top}$ (Fig. 1). Heights of the parapets often vary from 1000 mm to 1500 mm. Notably, the parapets are not an objective in this study.

4 Proposed Method

The proposed method for extracting point clouds of surfaces of bridge components has 4 main consecutive steps (Fig. 3). First, the point cloud is decomposed in 2D cells in a xy plane and then patches within the cells are extracted. Second, a cell-based region growing (CRG) and a connected surface components (CSC) algorithm are used to extract $S_{supstr.top}$ and $S_{supstr.bot}$ from patches within the 2D cells. Subsequently, intermediate surfaces ($S_{supstr.inter}$) are extracted based on $S_{supstr.top}$ and $S_{supstr.bot}$. Once the superstructure is extracted, remaining points below the superstructure are used as input data for extracting substructures. Next, a cell-based filtering is proposed to extract cells containing candidate points of the abutments and piers through features derived from the z-coordinate of the points within a cell. Moreover, a voxel-based region growing (VRG) is employed to segment the candidate points, while the CSC algorithm is employed to determine the segmented surfaces as parts of the structural components.

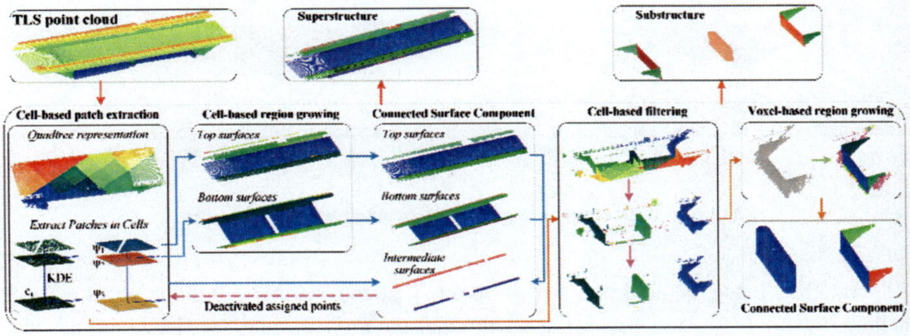

Fig. 3. A workflow of the proposed method

Step 1: Cell-based patch extraction

First, a quadtree is employed to recursively subdivide an initial minimum 2D bounding box of a bridge data points ($\mathbf{P}_{br} \in R^3$) into smaller 2D cells ($\mathbf{C} = \{c_1, \ldots, c_i, c_N\}$, $i = [1, N]$), in which the subdivision is carried out along the x- and y-directions in a Cartesian coordinate system. In this study, a cell size is used as a terminated subdivision condition, which a cell on a leaf node must have size smaller than the predefined cell size. Next, each cell occupies a property as "empty", if the cell contains the number of the points less than a predefined threshold called *min_ptc*; otherwise it is classified as "full". Each cell c_i stores information including x- and y-coordinates of two opposite corners, a list of point indices and the cell's property.

As 2D cells in a xy plane represent 3D data points of the bridge, the cell (c_i) contains a point cloud $\mathbf{P}_{ci} = \{p_{ci} = (x_i, y_i, z_i) \mid \mathbf{P}_{ci} \in \mathbf{P}_{br}\}$ that may belong to many patches $\boldsymbol{\Psi} = (\psi_{i1}, \psi_{ij}, \ldots, \psi_{in})$ of different component surfaces (Fig. 4a). As such, a kernel density estimation (KDE) generated from the z-coordinate of the points are used to extract points corresponding to the patch [18]. In this implementation, a predefined bandwidth can be selected as the minimum distance between two consecutive surfaces of the structural elements in the elevation. Given a probability density shape (PDS) of the KDE, the z-coordinates of the points ($p_{\psi ij}$) of the patch ψ_{ij} are located between two consecutive valleys of the PDS (Fig. 4b), which is expressed in Eq. 1. Notably, the patches in the cell are ordered from the highest to the lowest elevation.

$$\psi_{ij} = \left(p_{\psi ij} \mid z_{valley,k} \leq p_{\psi ij}.z \leq z_{valley,k+1} \right) \tag{1}$$

where $z_{valley,k}$ and $z_{valley,k+1}$ are coordinates of two consecutive valleys of the PDS.

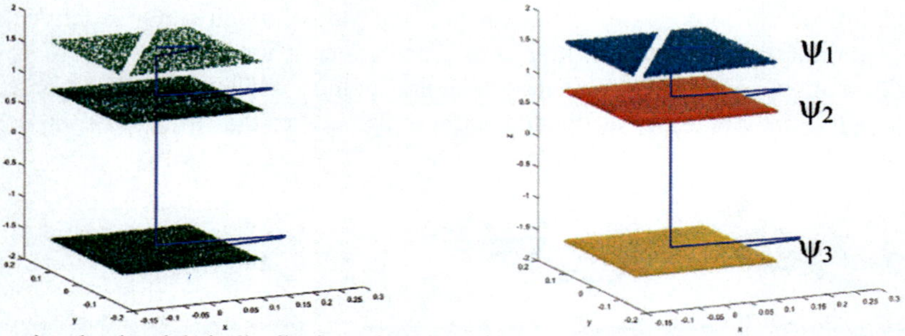

a) A point cloud and their distribution a long elevation

b) Extracting points presenting patches by using KDE

Fig. 4. Extracting data points of patches within a 2D cell

Next, the patch is assumed to represent a planar surface, and its fitting planar surface is computed from $\mathbf{p}_{\psi ij}$ using a principal component analysis (PCA) based on a covariance matrix (Eq. 2 and 3) [19]. The normal vector of the fitting surface is the eigenvector corresponding to the smallest eigenvalue.

$$cov_{\psi ij} = \sum\nolimits_{i=1}^{N} \left(p_{\psi ij} - p_{\psi ij,0} \right) \left(p_{\psi ij} - p_{\psi ij,0} \right)^{T} \tag{2}$$

$$p_{\psi ij,0} = \frac{1}{|P_{\psi ij}|} \sum\nolimits_{i=1}^{N} p_{\psi ij} \tag{3}$$

where $p_{\psi ij,0}$ is the centroid of the points $\mathbf{p}_{\psi ij} \in \psi_{ij}$ computed in Eq. 3, | | is the cardinal of the $p_{\psi ij}$.

Moreover, the residual is defined as the root mean square distances from the points to a fitting plane defined as $\psi_{ij}(p_{\psi ij0}, n_{ij})$, given in Eq. 4 and 5.

$$r_{\psi ij} = \sqrt{\frac{1}{|d_{p\psi ij}|} \sum\nolimits_{i=1}^{N} d_{p\psi ij}^{2}} \tag{4}$$

where $d_{p\psi ij}$ is the distance from a point to $\psi_{ij}(p_{\underline{s}ij,0}, n_{ij})$, is given in Eq. 5.

$$d_{p\psi ij} = \frac{(x_i - x_0)n_{ij}.x + (y_i - y_0)n_{ij}.y + (z_i - z_0)n_{ij}.z}{\sqrt{n_{ij}^2.x + n_{ij}^2.y + n_{ij}^2.z}} \quad (5)$$

where x_0, y_0, z_0 are coordinates of $p_{\psi ij,0}$ while $n_{ij} = (n_{ij}.x, n_{ij}.y, n_{ij}.z)$ is the normal vector

Step 2: Cell-based region growing

This step segments patches within 2D cells to extract $S_{supstr.top}$ and $S_{supstr.bot}$. However, as patches may contain points of multiple surfaces, the CRG consists of 4 sub-steps: cell-cell segmentation (*Step 2.1*), forward and backward filtering (*Step 2.2*), cell-point region growing (*Step 2.3*) and merging co-planar regions (*Step 2.4*), as specified below.

Step 2.1: Cell-cell segmentation

The cell-cell segmentation is similar to the region growing segmentation proposed by Rabbani et al. [5]. It starts with an initial seeding patch ($\psi_{ij} \in c_i$) having the smallest residual value (*min*_$r_{\psi ij}$). The neighboring patches ($\psi_{kj} \in c_k$) of ψ_{ij} are retrieved. ψ_{kj} is added to a region (R_i) if the patch satisfies Eq. 6. Next, ψ_{kj} is considered as the seeding patch in the next iteration if its residual ($r_{\psi kj}$) is smaller than a predefined residual threshold (r_0). The region R_i is incrementally growing until no more seeding patch is available. Moreover, the segmentation iteratively groups patches until all input patches are to be examined. Thus, the segmentation is grouping patches into a set of the regions $\mathbf{R} = (R_1, \ldots, R_i, R_N)$ where $R_i = (\psi_{ij})$.

$$\begin{cases} \angle n_{ij}, n_{kj} \leq \alpha_0 \\ d\left(p_{\psi kj,0}, \psi_{ij}\right) \leq d_0 \end{cases} \quad (6)$$

where $\angle n_{ij}, n_{kj}$ is a deviation of the normal vectors, $d\left(p_{\psi kj,0}, \psi_{ij}\right)$ is the distances from the center of ψ_{kj} to ψ_{ij}, and α_0 and d_0 are respectively a predefined angle and distance threshold.

Step 2.2: Forward and backward filtering algorithm

Patches ψ_{ij} on the boundary of the region R_k may contain data points of adjacent surfaces (Fig. 5a). A forward and backward filtering is developed to adjust incorrect segmented points on the regions' boundaries [20]. First, the patches ψ_{ij} in the region are classified into boundary and interior patches ($\psi_{ij.bound}$ and $\psi_{ij.interior}$), in which ψ_{ij} is $\psi_{ij,bound}$ if its cell connects to less than 8 adjacent cells, otherwise it is $\psi_{ij,interior}$. Next, the filtering algorithm starts with $\psi_{ij.bound} \in R_k$, neighbor patches of $\psi_{ij.bound}$, are extracted. The neighbor patches consist of $\psi_{kj.interior}$ and $\psi_{kj.bound} \in R_k$, and $\psi_{lm.bound} \in R_l$ (Fig. 5b). $\psi_{kj.interior}$ is assumed to represent the local surface $\psi_R \in R_k$, which can be determined using Eq. 2 and 3. $p_{\psi ij} \in \psi_{ij,bound}$ are classified as inlier and outlier points ($p_{\psi ij,inlier}$, $p_{\psi ij,outlier}$), as given in Eq. 7. $p_{\psi ij,inlier}$ and $p_{\psi ij,outlier}$ are remained and discarded out of R_k, respectively. In addition, $\psi_{lm.interior} \in R_l$, neighbor patched of $\psi_{lm.bound} \in R_l$, are extracted. $p_{\psi ij,outlier}$ can be merged to R_l if they satisfies Eq. 7, in which ψ_R is computed from the points within $\psi_{lm.interior}$.

$$p_{\psi ij} \rightarrow p_{\psi ij.inlier} \; if \; d\left(p_{\psi ij}, \psi_R\right) \leq d_0 \tag{7}$$

where $d\left(p_{\psi ij}, \psi_R\right)$ is the distance from the points $p_{\psi ij}$ to the surface ψ_R and d_0 is the distance threshold.

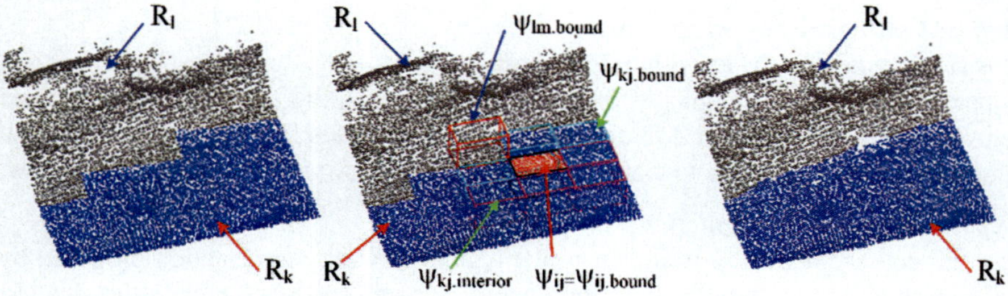

a) Point clouds of adjacent regions b) Extracted neighbouring patches and patch classification c) Point clouds of adjacent regions after filtering

Fig. 5. Illustration of a forward and backward filtering algorithm

Step 2.3: *Cell-point region growing*
In *Steps 2.1* and *2.2*, the process is done based on the specific patches $\psi_{ij} \in c_i$. However, in practice, because of the complexity of the structure and occlusions, $\psi_{ik} \in c_i(k \neq j)$ may be part of the surface. Goal of this step is to introduce a cell-point region growing process to overcome this issue. The process starts with $\psi_{ij,bound} \in c_i$ R_k to search for unsegmented, adjoined cell c_j. $p_{\psi jk} \in \psi_{jk} \in c_j$ are considered to belong to R_k if the distance from the points $p_{\psi jk}$ to ψ_{ij} satisfies Eq. 7. Additionally, the patch ψ_{jk} is considered as a new boundary patch of R_k if the percentage of points $p_{\psi jk}$ to be added to R_k is empirically more than 75% of $|\psi_{jk}|$. The cell-point region growing is iteratively searching points of the region R_k until all $\psi_{ij,bound} \in R_k$ are checked and no more point is added.

Step 2.4: *Merging co-planar regions*
In practice, due to missing data, noisy data or surface defects, a surface of a component can be segmented into multiple regions. The merging step is to overcome this issue. Regions R_k and R_l are considered to belong to the same surface of a component, if their relationships satisfy 3 conditions (Eq. 8–10). The first two conditions check a co-planar of R_k and R_l.

$$\angle n_{Rk} n_{Rl} \leq \alpha_1 \tag{8}$$

$$d\left(p_{Rl,0}, \psi_{Rk}\right) \leq d_1 \tag{9}$$

where n_{Rk} and n_{Rl} are normal vectors of the fitting planes of R_k and R_l, and $p_{Rl,0}$ is the center of the fitting plane of R_l.

The third condition checks for the surface continuity. This starts from $\psi_{ij.bound} \in R_k$ and $\psi_{lm.bound} \in R_l$ to search their neighbors, $\psi'_{lm,bound} \subseteq \psi_{lm.bound}$ and $\psi'_{ij,bound} \subseteq \psi_{ij.bound}$, respectively. The local plane ψ_{Rij} is estimated from the points $\mathbf{p}_{boundary} = \left(p_{\psi'ij} \in \psi'_{ij,bound} \wedge p_{\psi'lm} \in \psi'_{lm,bound}\right)$, as expressed in Eq. 2 and 3. There is a continuity between R_k and R_l if the points $p_{\psi'ij}$ and $p_{\psi'lm}$ satisfy Eq. 10.

$$\begin{cases} \left|p_{\psi'ij,inlier}\right| \geq 0.75\left|p_{\psi'ij}\right| \; where \; d\left(p_{\psi'ij,inlier}, \psi_{Rij}\right) \leq d_0 \\ \left|p_{\psi'lm,inlier}\right| \geq 0.75\left|p_{\psi'lm}\right| \; where \; d\left(p_{\psi'lm,inlier}, \psi_{Rij}\right) \leq d_0 \end{cases} \quad (10)$$

Step 3: Extract a point cloud of a superstructure

In slab beam or box bridges, the superstructure often consists of $S_{supstr.top}$, $S_{supstr.inter}$ and $S_{supstr.bot}$ (Fig. 2). There surfaces have the following main features: $S_{supstr.top}$ and $S_{supstr.bot}$ are nearly horizontal, planar surfaces, while $S_{supstr.inter}$ are mostly inclined or vertical planar surfaces. Moreover, the patches of the surfaces belonging to $S_{supstr.top}$ and $S_{supstr.bot}$ are mostly the first and second patches of a cell. Therefore, the process to extract the superstructure is in an order $S_{supstr.top}$, $S_{supstr.bot}$ and $S_{supstr.inter}$ as following.

Step 3.1: $S_{supstr.top}$

Based on the observation above, only the first patches $\psi_{i1} \in c_i$ are used as an input data in *Step 2.1 and 2.2* instead of all patches. Moreover, other patches in the cell c_i are used as input data for *Step 2.3*. After merging the co-planar regions in *Step 2.4*, the resulting segmentation consist of a set of regions ($\mathbf{R} = \{R_1, \ldots, R_i, \ldots, R_N\}$). Next, relationships between these regions and regions' features are explored to label these regions as well as eliminate incorrect regions. From *Feature 1*, the largest region in a size (length, width and area) is assigned as the road surface ($R_i = R_{road} \equiv S_{road}$). Moreover, the remaining regions ($R' = \mathbf{R} \setminus R_i$) are assigned as the pedestrian (footpath) surfaces if they satisfy Eq. 10.

$$\begin{cases} R'_i.length \geq 0.75 R_{road}.length \\ R'_i.width \geq B_{pedestrian} \\ R_{road}.elevation \leq R'_i.elevation \end{cases} \quad (11)$$

where length and width of a region are computed from a 2D minimum bounding box (mBB) determined from x- and y-coordinates of the points, and $B_{pedestrian}$ is the minimum width of the pedestrian path, which is no less than 1.0 m.

After assigning semantic information to the region $\mathbf{R} = \{R_1, \ldots, R_i, \ldots, R_N\}$ where $R_i = \{\psi_{ij}\}$, $\psi_{ij} \in c_i$ are then deactivated, which means they are no longer consideration in further processing. Moreover, the points associated with the patch $\psi_{kl} \in c_k$ above $S_{supstr.top}$ are also deactivated because those points may represent parapets which are out of the scope of this study.

Step 3.2: $S_{supstr.bot}$

In addition to $S_{supstr.top}$, $S_{supstr.bot}$ is extracted similarly, where only the second patches $\psi_{i2} \in c_i$ are initially used in *Steps 2.1 and 2.2* and the final segments are $\mathbf{R} = \{R_1, \ldots, R_i, \ldots, R_N\}$. Next, a CSC algorithm is introduced to group regions that are parts of the structural component. This is based on the hypothesis is that a component of a set of individual surfaces connected together, and two adjacent surfaces are sharing an edge. However, due to edge loss during data acquisition or segmentation and the complexity of the scene itself, a gap between two adjacent surfaces are often exist. The CSC algorithm starts with boundary patches $\psi_{ij.bound} \in R_k$ and $\psi_{lm.bound} \in R_l$ to search their neighbors, $\psi'_{lm,bound} \subseteq \psi_{lm.bound}$ and $\psi'_{ij,bound} \subseteq \psi_{ij.bound}$, respectively (Fig. 6a). R_k is possibly adjoined R_l if Eq. 11 is satisfied.

$$\begin{cases} \psi'_{ij,bound} \neq \emptyset \\ \psi'_{kl,bound} \neq \emptyset \end{cases} \tag{12}$$

Second, from points within $p_{\psi ij} \in \psi'_{ij.bound} \in R_k$ and $p_{\psi kl} \in \psi'_{kl,bound} \in R_l$, fitting local planes ($S_{k,local}$ and $S_{l,local}$) are respectively estimated by Eq. 2 and 3. Then R_l is considered to connect to R_k, if they are intersection (Eq. 13) and the projection of $\psi'_{ij.bound}$ and $\psi'_{kl,bound}$ onto the intersection line satisfies Eq. 14.

$$\angle n_k n_l \geq \alpha_3 \tag{13}$$

$$\begin{cases} L'_k \geq L_0 \\ L'_l \geq L_0 \\ L'_{kl} \geq 0.5\min(L'_k, L'_l) \end{cases} \tag{14}$$

where n_k and n_l are normal vectors of $S_{k,local}$ and $S_{l,local}$, $L'_k = |P_{k1}\ P_{k2}|$ and $L'_l = |P_{l1}\ P_{l2}|$ are the lengths of the projection of $p'_{\psi ij} \in \psi'_{ij,bound}$ and $p'_{\psi lm} \in \psi'_{lm,bound}$ onto the intersection line $L_{kl} = |P_{k1}\ P_{l2}|$, the length $L'_{kl} = |P_{l1}\ P_{l2}|$ is the overlap length between L'_k and L'_l, and L_0 is a predefined length threshold (Fig. 6b).

The CSC algorithm is iteratively checked if the region R_l connects to region R_k and a cluster describing the surfaces of the component is incrementally growing until no more region can be added. As the implementation does not check all regions, the initial region must be one of surfaces of the components. As such, to extract $S_{supstr.bot}$, the bottom surfaces of the slab beam or box ($S_{beam.bot}$) shown in Fig. 2 are selected as the initial region.

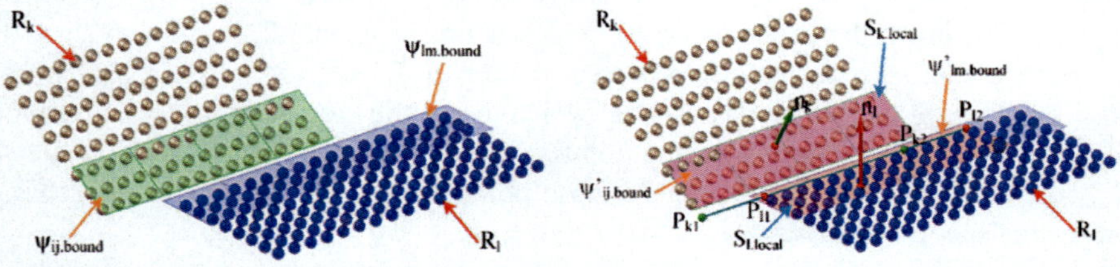

a) Adjacent regions and their boundary patches b) An intersection between two regions

Fig. 6. Determine a connection between two adjacent regions based on cells

In the bridge structure, cross-sections of the slab beams or boxes of short and medium span bridges are mostly constant. This implies the distance between $S_{beam.bot}$ to S_{road} is a constant. A voting score is used to determine $S_{beam.bot}$ from $\mathbf{R} = \{R_1, \ldots, R_i, \ldots, R_N\}$. First, the distance from $\psi_{ij} \in R_k$ to S_{road} is computed using Eq. 5, which is expressed as $d(p_{\psi ij.0}, \psi_{ik})$, where the center $(p_{\psi ij,0})$ of ψ_{ij}, $\psi_{ik} \in S_{road}$, and $\psi_{ij} \wedge \psi_{il} \in c_i$. Using d $(p_{\psi ij.0}, \psi_{ik})$ can give an intuitive distance between $S_{beam.bot}$ and S_{road}, where a curve in the elevation or a lateral slope of the bridge superstructure is available [21]. Continuously, the DBScan algorithm [22] is employed to cluster $d(p_{\psi ij.0}, \psi_{ik})$, and the cluster having the largest distance is extracted. Subsequently, the region containing the maximum number of the patches in this cluster is assigned as $S_{beam.bot}$, which is the initial region for seeding regions of the component in the CSC algorithm to extract $S_{sup,bottom}$.

Step 3.3: $S_{supstr.inter}$

As $S_{supstr.inter}$ locates between the $S_{supstr,top}$ and $S_{supstr,bot}$, the point cloud $p_{supstr.inter}$ belonging to $S_{supstr,inter}$ can be extracted as following. From each cell $c_i \in S_{supstr,top} \cap S_{supstr,bot}$, the patches $\psi_{ij} \in S_{supstr,top}$ and $\psi_{ik} \in S_{supstr,bot}$ are respectively retrieved (Fig. 7a). The $p_{ci} \in c_i$ are considered as $p_{supstr.inter}$ if they satisfies Eq. 15.

$$p_{ci} \epsilon S_{supstr.inter} \; if \begin{cases} sign(d(p_{ci}, \psi_{ij}))sign(d(p_{ci}, \psi_{ik})) < 0 \\ cosin(n_{ij}, n_{ik}) > 0 \end{cases} \tag{15}$$

where $\psi_{ij}(p_{\psi ij.0}, n_{ij})$ and $\psi_{ik}(p_{\psi ij.0}, n_{ik})$ are respectively the patches of $S_{sup,top}$ and $S_{sup,bottom}$, and n_{ij} and n_{ik} are the normal vectors of ψ_{ij}, ψ_{ik}.

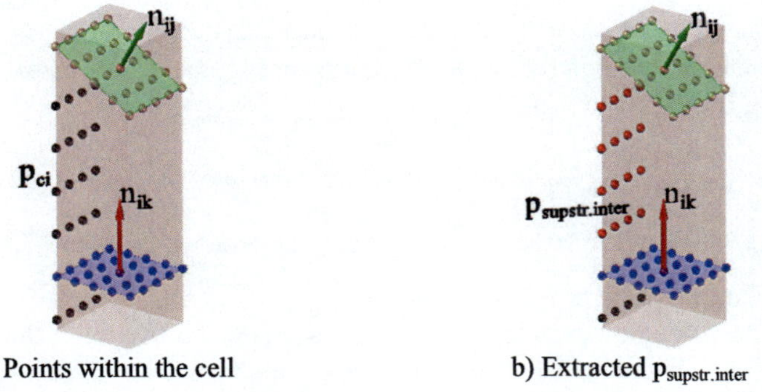

Fig. 7. Extract points of $S_{supstr.inter}$ based on patches of $S_{supstr,top}$ and $S_{supstr,bot}$

Step 4: Extract the substructures

The substructure consists of abutments (S_{abut}) and piers (S_{pier}), which has different geometric configurations (Fig. 1). After extracting point clouds of the superstructure, remaining points are used as input data to extract the substructure. However, the data set may contain the points of irrelevant objects (e.g. ground, plants). The cell-based filtering is proposed to extract candidate points of the substructure. Subsequently, a voxel-based region growing (VRG) and the CSC algorithm (*Step 3.2*) are used to obtain the points of the abutments and piers.

Step 4.1: Cell-based filtering
Points of the cells representing the abutments and piers distribute from the ground to $S_{beam.bot}$. As such, the height of the cell and the maximum gap expressed in Eq. 16 and 17 are used to describe a cell features.

$$H_{ci} = max(p_{ci}.z_j) - min(p_{ci}.z_j) \tag{16}$$

$$\Delta H_{ci} = max(p_{ci}.z_j - p_{ci}.z_{j+1}) \tag{17}$$

where $p_{ci}.z_j$ is the z-coordination of the points $p_{ci} \in c_i$.

As the points cloud of the substructure mainly distributes in the vertical direction, c_i is considered to represent the substructure if the cell features satisfy Eq. 18.

$$c_i \rightarrow c_{supstr.candidate} \in S_{sub} \; if \; H_{ci} \geq H_{c0} \; and \; \Delta H_{ci} \leq \Delta H_{c0} \tag{18}$$

where H_{c0} and ΔH_{c0} are respectively the predefined thresholds for the minimum height of the cell and the maximum gap. H_{c0} can selected based on the dimensions of the substructure in a vertical direction and ΔH_{c0} is empirically selected equal to a half of H_{c0}.

Moreover, in 3D scenes, the cells of irrelevant objects may also satisfied Eq. 18, but they are sparse distribution. The cells $c_{supstr.candidate}$ are clustered based on their connectivity, and then cluster features involve width and length of a 2D mBB computed from x- and y-coordinates of the points. As shown in *Feature 2*, the length of the abutments and piers is no smaller than a half of the bridge width. As such, the clusters are considered to represent the abutments and piers if they are satisfied Eq. 19. Moreover, the cluster representing to the abutment is located in outmost clusters while other are of the piers.

$$c_{supstr.candidate}.length \geq 0.5 S_{road}.width \tag{19}$$

where $S_{road}.width$ is a road width of the bridge determined in *Step 3.1*.

Step 4.2: Extract surfaces of an abutment
The voxel-based region growing (VRG) [23] is employed to extract the points of the surfaces of the abutment. The main difference is the forward and backward algorithm similar to *Step 2.2* is used to refine over- and under-segmentation instead of using a point-based region growing in [23]. Subsequently, the CSC algorithm is employed to determine whether the segmented surfaces are parts of the abutment. In this work, the initial region for the CSC algorithm is a breast wall (Fig. 2), which is (i) a vertical surface and perpendicular to the longitudinal direction of the bridge, (ii) has its length approximately equal to the bridge, and (iii) connects to the ground level. Therefore, the breast wall is the segmented region satisfying the sequential conditions expressed in Eq. 20. By applying the CSC algorithm, the surfaces representing the abutment surfaces can be extracted.

$$\begin{cases} \angle n_{Ri}, t_{road} \leq \alpha_0 \\ R_i.length \geq S_{road}.width \\ R_i.z \rightarrow min \end{cases} \tag{20}$$

where t_{road} and n_{Ri} are the tangent and normal vectors of R_{road} and R_i, $R_i.length$ and $R_i.z$ are the length and the z coordinate of the centre of R_i.

Step 4.3: *Extract surfaces of a pier*
Similar to the *Step 4.2*, the VRG is applied to segment the candidate points of the pier. The pier columns involves the vertical surfaces connected to the ground while the pier cap locates between the top of the pier column and the bottom surfaces of the superstructure. Moreover, heights of the pier columns are also larger than that of the pier cap. Based on these features, the resulted regions ($\mathbf{R}$ = (R_i, i = 1 − N) are divided into two groups based on the region centers, as expressed Eq. 21 and 22.

$$R_{pier,column} = R_i \in R \,|\, R_i.z \leq C_{pier}.z + 0.25C_{pier}.H \tag{21}$$

$$R_{pier,cap} = R_i \in R \,|\, R_i.z \geq C_{pier}.z + 0.25C_{pier}.H \tag{22}$$

where $R_i.z$ is the z coordinate of the region center computed from its points, $C_{pier}.z$ and $C_{pier}.H$ are the z coordinate of the center and the height of the 3D bounding box of the candidate points of the pier.

Subsequently, the CSC algorithm is separately applied to the regions $R_{pier,column}$ and $R_{pier,cap}$. In this work, the largest region in terms of area is used as the initial surface for seeding other surfaces of the components. Once the regions are determined as parts of the components of the pier, other regions are then discarded because they may be of irrelevant objects.

5 Experimental Test, Results and Discussions

To demonstrate performance of the proposed method, the concrete box bridge on CO16 road, Seßlach in Germany is selected. It is a two span bridge with the span length about 10.5 m, and its cross-section consists of 2 traffic lanes about 7.0 m and 2 pedestrian lanes about 2.0 m (Fig. 8a). The bridge was scanned by a Leica ScanStation P20 with a maximum scanning range at 120 m and an angular accuracy of 8 s of arc in both vertical and horizontal [24]. During data acquisition, a sampling step of 6.3 mm at a range of 10 m was used, which translates to an average sampling step of 10 mm on the surface for an average scanning distance of 15 m from the ground level. A total of 5 scanning stations was established to maximize data coverage (Fig. 8b). The point clouds of scanning stations were registered by the Leica Cyclone software [24] and the registration error was about 5 mm. Finally, 28.5 million points of the bridge was exported with x-, y- and z-coordinates for each point as input data for the proposed method.

a) Photo of the bridge b) Point cloud of the bridge

Fig. 8. A case study

The quadtree is employed to decompose the point cloud into 2D cells in a xy plane, in which the cell size of 0.5 m and *min_ptc* of 10 are used as a termining condition and a cell classification (Fig. 9a). The selected cell size is to ensure at least one cell can represent the smallest surface of a superstructure, for example the pedestrian width is often larger than 1.0 m. Moreover, the bandwidth of 0.2 m is used in the KDE to separate data points of an individual patch within the cell (Fig. 9b). This value is chosen smaller than the distance between two surfaces of the slab.

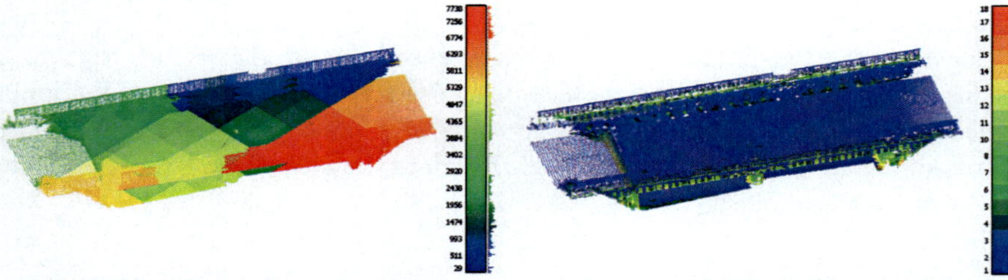

a) Point clouds rendering by cell indices b) Point clouds rendering by patch indices

Fig. 9. Extract the points of patches within the cells by using KDE

Next, the CRG is used to extract $S_{supstr.top}$ and $S_{supstr.bot}$ of the superstructure. This method requires the angle threshold α_0, the distance threshold d_0 and the residual threshold r_0. As the bridge has transverse slopes, the angle threshold must select to ensure to extract patches of the surface (Fig. 10), and $\alpha_0 = 5°$ is selected to cover most of cases. Moreover, d_0 is selected equal to a half of the curb height ($\Delta_{curb} = 0.1$ m), which is the smallest curb height in the bridge. Finally, r_0 by 10 mm is approximate 2 times of a registration error, which can compensate a data error and imperfect smooth of the component surface. Moreover, in *Step 2.4,* $\alpha_1 = 2\alpha_0$ and $d_1 = 2d_0$ are selected. Results of the extracted $S_{supstr.top}$ is illustrated in Fig. 11.

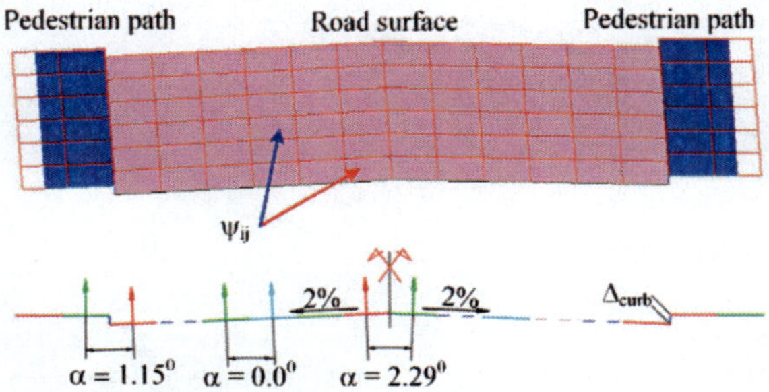

Fig. 10. Identifying criteria for the cell-cell segmentation (*Step 2.1*)

Next, the CSC algorithm are respectively used to filter final surfaces of $S_{supstr.top}$ and $S_{supstr.bot}$ from results of the CRG. For $S_{supstr.bot}$, the predefined length threshold, L_0 is set equal to 2.5 m equivalent to 5 times of a cell size. Finally, resulted surfaces of the superstructure are shown in Fig. 12.

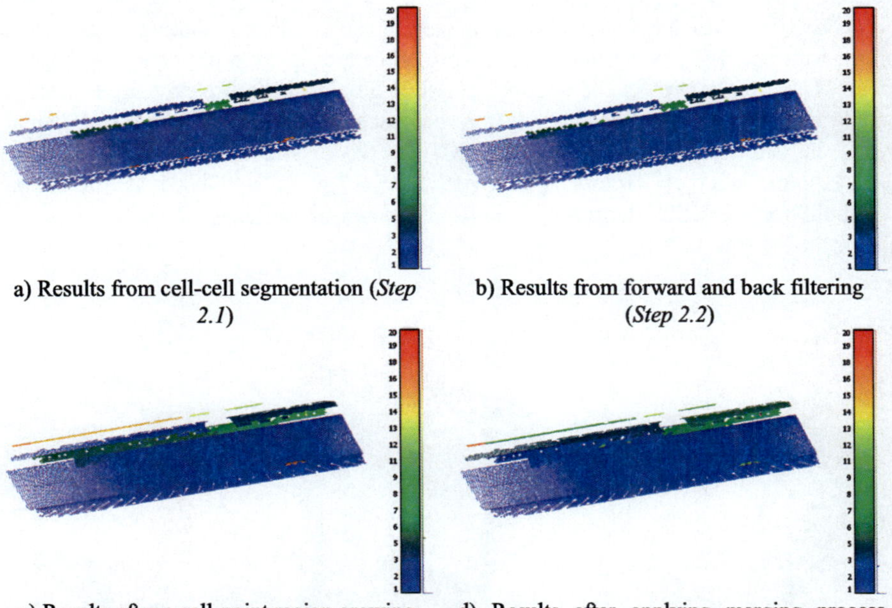

a) Results from cell-cell segmentation (*Step 2.1*)

b) Results from forward and back filtering (*Step 2.2*)

c) Results after a cell-point region growing (*Step 2.3)*

d) Results after applying merging process (*Step 2.4)*

Fig. 11. Illustration of the CRG of $S_{supstr.top}$

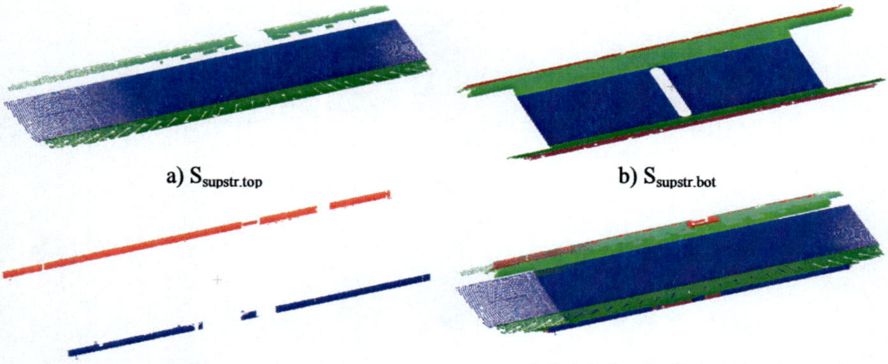

a) $S_{supstr.top}$ b) $S_{supstr.bot}$

c) $S_{supstr.inter}$ a) Point cloud of a supstructure

Fig. 12. Resulted point cloud of the superstructure

Additionally, in the cell-based filtering, $H_{c0} = 0.5$ m and $\Delta H_{c0} = 0.25$ m are chosen to extract candidate points of a substructure. As for the abutment and the pier because they own different types of shapes, when the VRG is applied on candidate points of these structures, different parameters are used. In the abutment, parts of a breast wall and wing walls may not be visible because a slope protection in front of them. However, to protect the bearing, a visible height of the breast wall is often no smaller than 0.5 m, which is arguably the smallest size of the surface in the abutment. As such, the parameters for the VRG are voxel_size = 0.2 m, and $\alpha_0 = 5°$, $d_0 = 10$ mm, and $r_0 = 10$ mm. Moreover, $L_0 = 0.5$ m is also used as the length threshold for the CSC algorithm. Results of extracting surfaces of the abutment is shown in Fig. 13.

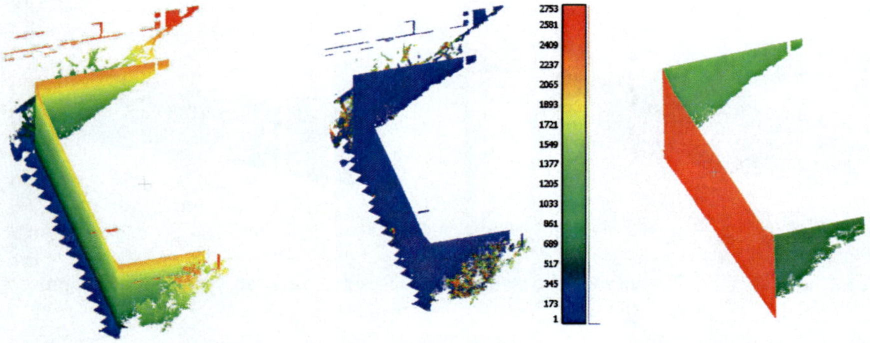

a) A point cloud rendering based on z-coordinate b) Resulted segmentation c) Final surfaces of the abutement

Fig. 13. Resulted point clouds of the abutment

Moreover, for the pier, parameters for the VRG should accommodate to extract both planar and curve surfaces. However, the smallest radius of the curve is no smaller than 0.6 m, which can be in the pier column or a head of the pier cap. Moreover, a surface is mostly perpendicular to its adjoin surface. Therefore, the parameters selected in this study are voxel_size = 0.1 m, and $\alpha_0 = 15°$, $d_0 = 25$ mm, and $r_0 = 25$ mm, while $L_0 = 0.6$ m is used in the CSC algorithm. Resulted extraction of the points of the pier is shown in Fig. 14.

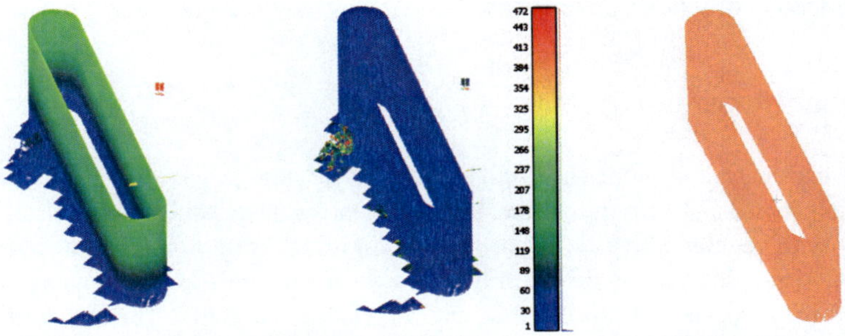

a) A point cloud rendering based on z-coordinate

b) Resulted segmentation

c) Final surfaces of the abutement

Fig. 14. Resulted extraction of point clouds of the pier

Finally, Fig. 15 shows the point cloud of the substructure surfaces and of entire bridge. As the ultimate goal of the study is to generate 3D bridge model, using quality indicators (e.g. completeness, correctness and quality) based on extracted points and ground truth may not fully reflect a performance of the proposed method. That is because both locations and the number of the incorrect/missing points of a surface impact to quality and quantity of the generated surface. As such, in this study, the visualization evaluation is used as a preliminary evaluation, which is showed that 100% surfaces of the structural components are extracted. Moreover, in future work, the quantitative elevation including the level of locational deviation, shape similarity and positional accuracy of the surface are implementing to give details of the performance.

a) Point clouds of substructure surfaces

b) Point clouds of component surfaces of the bridge

Fig. 15. Resulted point cloud of surfaces the structural components of the bridge

Finally, the proposed method is implemented in MATLAB 2019 [25], and the bridge data points with 28.5 million points is processed Dell Precision Workstation with a main system configuration as follows: Intel(R) Xeon(R) W-2123 CPU @ 3.6 GHz with 32 GB RAM. The processing time is 654.9 s including executing time for extracting patches within the cells about 68.2 s while for extracting points of the superstructure and substructure about 300.2 s and 286.5 s, respectively. By comparing the work of Lu et al. [12] when the extract the point cloud of the bridge components for 10 bridges, an average running time is about 481.2 ± 181.2 s for an average data set of 35.7 million points. Notably, the proposed method in [12] was implemented in C# and using functions in a point cloud library.

6 Conclusions

A climate change and increasing freight have cause negative impact to bridges on a transport network. That requires the bridges must be inspected and monitored periodically with regular intervals. Moreover, as digitized bridge models are mostly not available, which lead to inefficient in bridge asset management. An advantage of laser scanning can capture 3D topographic data of visible surfaces of bridge components accurately and efficiently, which subsequently use to detect surface defects and to create a 3D bridge model. This paper introduces a new method to automatically extract point cloud of component surfaces of slab/box beam bridges. The proposed method deploys both spatial information of a point cloud and context information of the bridge structures to extract the bridge components in order from a superstructure to substructure. An experimental test shows the proposed method can extract all surfaces of the structural components of a box bridge, and an executing time is about 23 s/1 million points, which is comparable to the existing method. Moreover, in future work, additional experimental testes could be carried to show a robustness of the method and a detailed evaluation associated designated applications should be reported. Finally, although the proposed method aims to apply for slab/box beam bridges, this work can extent to other common types of the bridges.

Acknowledgement. This work was funded by the generous support of the European Commission through H2020 MSCA-IF, "*BridgeScan: Laser Scanning for Automatic Bridge Assessment*", Grant 799149. The first author is graceful for this support.

References

1. Žnidarič, A., Pakrashi, V., O'Brien, E., O'Connor, A.: A review of road structure data in six European countries. Proc. Inst. Civil Eng. - Urban Des. Plan. **164**, 225–232 (2011)
2. ASCE: 2017 Infrastructure Report Card: Bridge (2017)
3. Truong-Hong, L., Laefer, D.F.: Laser scanning for bridge inspection. Laser Scan. Emerg. Technol. Struct. Eng. **14**, 189 (2019)
4. Riveiro, B., DeJong, M.J., Conde, B.: Automated processing of large point clouds for structural health monitoring of masonry arch bridges. Autom. Constr. **72**, 258–268 (2016)

5. Rabbani, T., van den Heuvel, F., Vosselmann, G.: Segmentation of point clouds using smoothness constraint. Int. Arch. Photogram. Remote Sens. Spat. Inf. Sci. **36**, 248–253 (2006)
6. Tarsha-Kurdi, F., Landes, T., Grussenmeyer, P.: Hough-transform and extended RANSAC algorithms for automatic detection of 3D building roof planes from lidar data. In: ISPRS Workshop on Laser Scanning 2007 and SilviLaser 2007, pp. 407–412 (2007)
7. Truong-Hong, L., Laefer, D.F., Hinks, T., Carr, H.: Combining an angle criterion with voxelization and the flying voxel method in reconstructing building models from LiDAR data. Comput.-Aided Civil Infrastruct. Eng. **28**, 112–129 (2013)
8. Son, H., Bosché, F., Kim, C.: As-built data acquisition and its use in production monitoring and automated layout of civil infrastructure: a survey. Adv. Eng. Inform. **29**, 172–183 (2015)
9. Costin, A., Adibfar, A., Hu, H., Chen, S.S.: Building Information Modeling (BIM) for transportation infrastructure – literature review, applications, challenges, and recommendations. Autom. Constr. **94**, 257–281 (2018)
10. Walsh, S.B., Borello, D.J., Guldur, B., Hajjar, J.F.: Data processing of point clouds for object detection for structural engineering applications. Comput.-Aided Civil Infrastruct. Eng. **28**, 495–508 (2013)
11. Zhang, G., Vela, P.A., Karasev, P., Brilakis, I.: A sparsity-inducing optimization-based algorithm for planar patches extraction from noisy point-cloud data. Comput.-Aided Civil Infrastruct. Eng. **30**, 85–102 (2015)
12. Lu, R., Brilakis, I., Middleton, C.R.: Detection of structural components in point clouds of existing RC bridges. Comput.-Aided Civil Infrastruct. Eng. **34**, 191–212 (2019)
13. Yan, Y., Guldur, B., Hajjar, J.F.: Automated structural modelling of bridges from laser scanning. In: Structures Congress 2017, pp. 457–468 (2015)
14. Schneiders, R.: A grid-based algorithm for the generation of hexahedral element meshes. Eng. Comput. **12**, 168–177 (1996)
15. Khaloo, A., Lattanzi, D.: Robust normal estimation and region growing segmentation of infrastructure 3D point cloud models. Adv. Eng. Inform. **34**, 1–16 (2017)
16. Farhey, D.N.: Structural performances of bridge types in the US national bridge inventory. Infrastructures **3**, 6 (2018)
17. Kim, M.-K., McGovern, S., Belsky, M., Middleton, C., Brilakis, I.: A suitability analysis of precast components for standardized bridge construction in the United Kingdom. Procedia Eng. **164**, 188–195 (2016)
18. Laefer, D.F., Truong-Hong, L.: Toward automatic generation of 3D steel structures for building information modelling. Autom. Constr. **74**, 66–77 (2017)
19. Hoppe, H., DeRose, T., Duchamp, T., McDonald, J., Stuetzle, W.: Surface reconstruction from unorganized points. In: ACM SIGGRAPH 1992, pp. 71–78 (1992)
20. Truong-Hong, L., Chen, S., Cao, V.L., Laefer, D.F.: Automatic Bridge Deck Damage Using Low Cost UAV-based Images (2018)
21. Truong-Hong, L., Lindenbergh, R.: Measuring deformation of bridge structures using laser scanning data. In: 4th Joint International Symposium on Deformation Monitoring (JISDM), p. 7 (2019)
22. Ester, M., Kriegel, H.-P., Sander, J., Xu, X.: A density-based algorithm for discovering clusters in large spatial databases with noise. In: KDD, pp. 226–231 (1996)
23. Vo, A.-V., Truong-Hong, L., Laefer, D.F., Bertolotto, M.: Octree-based region growing for point cloud segmentation. ISPRS J. Photogram. Remote Sens. **104**, 88–100 (2015)
24. Leica Geosystems. https://w3.leica-geosystems.com/downloads123/hds/hds/scanstation_p20/brochures-datasheet/leica_scanstation_p20_dat_en.pdf. Accessed 15 Jan 2020
25. MathWorks: MATLAB Function Reference (2019)

A Framework for Utilization of Occupants' Trajectory Data to Enhance Building Management

S. H. Hsu, W. Han, Y. T. Chang, Y. C. Chan, and S. H. Hsieh(✉)

Department of Civil Engineering, National Taiwan University, No. 1, Sec. 4, Roosevelt Road, Taipei 10617, Taiwan
shhsieh@ntu.edu.tw

Abstract. Modern people spend approximately up to 90% of their time indoors. Moreover, building operational phase contributes to more environmental impact (e.g. energy use and carbon emission) than other phases. Hence, proper building management is critical to the welfare of human beings and mother nature. And occupant behavior is crucial information for building managers. However, in the past, such information was based on statistics and assumptions, which may lead to inefficient and even false operations. As the monitoring techniques developed rapidly, occupant behavior can be tracked more easily. Although research related to building management and occupant behaviors has grown steadily in the past few decades, more direct discussions between occupant trajectory data and building management are still limited. Therefore, the goal of this paper is to propose a procedural framework to collect and utilize occupants' trajectory data for building management. This framework is developed by reviewing existing applications to identify the key factors to be considered during the process and the relationships between these factors are also analyzed. Application scenarios (includes safety, space, energy management) and monitoring areas (public and private) are identified as important inputs to decide what kind of data to collect. Data processing, granularity, and privacy are recognized as essential attributes of trajectory data to consider. Furthermore, the strength and weaknesses of different monitoring techniques to provide different occupant trajectory data are also presented as an additional reference for building managers.

Keywords: Building management · Occupant behavior · Trajectory data

1 Introduction

Occupant-related data are useful for improving overall building management. From a building management's perspective, knowing occupants' presence, movement, and behavior can facilitate demand-driven controls, optimize occupants' schedules, enhance flow management, pre-arrange escaping routes, and so on. As a result, there is a major increase in the number of papers on developing approaches to analyze occupant presence, movement and behavior with collected data [1]. Moreover, this trend fosters corresponding frameworks. For example, Hong et al. [2, 3] proposed an

E. Toledo Santos and S. Scheer (Eds.): ICCCBE 2020, LNCE 98, pp. 740–754, 2021.
https://doi.org/10.1007/978-3-030-51295-8_51

ontology, called "Drivers - Needs - Actions - Systems" (DNAs), to represent energy-related occupant behaviors by characterizing the interaction between occupants and indoor environment.

Among all the available occupants-building interaction data, the trajectory data are considered paramount because it is essential to continuously position the occupants before any further exploration of occupants' behavior [4]. The series of their location points form the trajectory data provide a strong foundation for extended analysis and applications [5]. Arslan et al. [6] designed a framework called 'Occupant Behaviors in Dynamic Environments' (OBiDE) by extending the DNAs framework. The existing ontology was combined with their trajectory enrichment model to improve the part of actions with more information including occupancy and movements. In addition, building data that represent the dynamicity of a building environment were adopted to better determine occupants' activities. As a result, the study presented the fusion of occupants' movements and the latest building information to address a particular issue, safety management. Additionally, the study showed how the researchers utilize the data is strongly correlated to if they can reach their respective goals.

In recent years, the advancement in monitoring techniques makes occupant trajectory data more available and various studies were conducted to maximize the capacity of monitoring campaigns [7]. However, the utilization of trajectory data and its targeted goals are rarely systematically discussed. In most cases, raw data should be collected and processed to obtain occupants-related information. Considering that occupant-related data are stochastic and complex, it is sometimes costly and usually takes a long period of time to collect enough data to achieve applications' target goals. It is important to decide the level of details of data and how to utilize the data before data collecting and processing [8]. Only a few studies have emphasized data processing or reviewed papers from the perspective of data utilization.

Because the requirements and limitations vary in different building management scenarios, and most of the existing frameworks are focusing on behavior rather than occupants' presence and movement (trajectory). In that way, a more explicit framework focusing on the utilization of occupants' trajectory is needed to properly interpret relationships between application scenarios and the characteristics of required data. To fill the gap in integrating research on data utilization, this paper introduces a preliminary framework by analyzing the attributes of trajectory data. The framework is currently composed of five aspects and is capable of initially classifying reviewed papers. And the way how they processed trajectory data is organized. This should encourage more applications based on occupants' trajectory data in building management. More details will be discussed in the following section. After that, some guidelines on collecting and utilizing the trajectory data are developed for building management.

2 Application Review

2.1 Review Scope

This paper adopted the well-known academic database, Scopus, to search for related work. The searching scope is not limited to any year ranges or publication types. Three sets of keywords were applied: (1) building occupant trajectory; (2) tracking building occupant; and (3) building occupant movement, resulting in 52, 128, 161 papers as search results respectively (in total 341 papers). Excluding papers in which trajectory data are not linked to occupant behaviors, research which focus mainly on monitoring techniques than application scenario for building management, and last but not the least, the overlapping ones collected by three times of searching and those written in non-English languages, the final selection of literature comes to 22 papers. Among these selected publications, most of them (19) are journal papers and the rest (3) are in conference proceedings. Table 1 organizes primary classification and notation of our review work. The classification only includes the applications mentioned in the papers reviewed, but other issues such as acoustics management and human social interaction may be included in the future when they are needed in building management. This section mainly discusses three major aspects which are safety, space management and energy management.

Table 1. Selected literature and their related application scenarios

Application Scenarios	Number of papers	Literature Reference
Safety	8	Ma et al. (2017) [9], Peacock et al. (2016) [10], Rahman et al. (2012) [11], Ma et al. (2015) [12], Pinto et al. (2012) [14], Lotfi et al. (2012) [13], Ronchi et al. (2014a) [15], Ronchi et al. (2014b) [16],
Space Management	5	Dzeng et al. (2013) [18], Dzeng et al. (2014) [19], Dzeng et al. (2015) [20], Nassar (2010) [21], Vanclooster et al. (2012) [22]
Energy Management	9	Chandrasiri & Geekiyanage (2018) [23], Liu et al. (2017) [24], Li et al. (2012) [25], Li et al. (2019) [26], Diraco et al. (2015) [27], Tekler et al. (2019) [28], Dziedzic et al. (2017) [29], Rafsanjani et al. (2018) [30], Chen et al. (2019) [31]

2.2 Safety

In our reviewed literature [9–16], applications based on trajectory data in building safety were mainly related to evacuation planning which can be achieved by analysis of occupant trajectory patterns and location of occupants. This paper further explores the above purposes by the proposed framework including application scenario, monitoring area, processing, privacy and granularity to clarify their requirements.

First of all, analysis of occupant trajectory patterns is defined as an interpretation of occupants' movements which are integration of locating points along with time. Thus, trajectory data are usually processed in batches and require higher granularity to deduce complex target patterns. Additionally, complex target patterns can be extended to evacuation route planning [9–12], analysis of interaction between occupants [9] and occupant behavior [13]. They are all valuable information for building design and occupant behavior models to simulate the flow of human. Furthermore, real-time analysis is hard to be realized because of heavy computation. As a result, it is paramount for analysis of occupant trajectory pattern to collect data at a high level of granularity and input in batches instead of real-time data. Next, in contrast, the location of occupants is mostly applied to real-time decision making; therefore, it merely requires low granularity of trajectory data and less computational cost. Studies of decision making including rescue planning [11] and dynamic adjustment to occupants' evacuation routes [12] were generally based on existing rules and subsequently required information about occupants' positions. Last, to summarize, trajectory data not only play a pivotal role in different stages of evacuation planning but also provide real-time location of occupants for decision-making models.

However, few research focuses on other directions for maintaining building safety because it is considered privacy intrusion to collect individual data in a building. This circumstance should hinder the number of related papers and it is a dilemma about how to make a choice between privacy concerns and safety issues. Without the problem, the utilization of trajectory data for maintaining safety should become more vigorous.

2.3 Space Management

Since space management might involve a wide range of research, this paper merely focuses on constructed spaces in a building. In reviewed papers, the objectives of space management can be defined as the optimization of physical utilization and improvement of occupants' experience [17]. To achieve that, it is essential to have a better understanding on space-use because the data are the direct observation of occupants' actions. In this context, trajectory data is crucial to the performance of space management. According to the reviewed papers, trajectory data is primarily applied to analysis of movement conflict and efficiency.

Dzeng et al. [18–20] collected occupants' movement in the daytime of weekdays through RFID set in multiple classrooms and the data were further utilized to a building renovation. The renovation was to increase the usability performance by optimizing the assignment of functions to spaces. As a result, the monitoring area should be public and the batch trajectory data were input into a function–space assignment optimization model to minimize the distance for movement. Subsequently, non-individual information is sufficient while for data granularity, the movements were needed to interpret occupants' daily activities. Additionally, Nassar [21] aimed to predict the level of service (LOS) of building spaces during the design phase. Since the LOS is determined by four indexes including flow rate and speed of occupants, the author proposed the model to estimate the parameters. The real-world data of three different corridors on a campus which belong to public areas and batch analysis within the framework were collected to validate the performance of the model. Furthermore, only points without identities are needed to have occupants' travel times, flow densities and capacities for the spaces. Vanclooster et al. [22] similarly adopted the flow model to calculate exitability of different rooms in a building which represents the quality of access to exit. In this case, possible movements from each room to the exit were pre-defined by the spatial information while occupancy distribution is crucial to exitability values. As a result, both public and private areas should be monitored to measure room occupancy through the real-time collection of occupants' location points.

In summary, reviewed papers usually deployed their methods to a certain scale of spaces such as the whole floor of a building. For a building manager, it is unusual to merely enhance the environment of a single room. Considering the additional cost of sensors, the benefits of collecting trajectory data should be significant enough; therefore, a wide range of spaces is adopted to achieve the purpose. In addition, it is arduous for past studies to reach reliable trajectory data in such monitoring scale through traditional approaches. With the advancement of deep learning in monitoring techniques, the collected data should be more reliable so that applications for space management based on the data will increase in the near future.

2.4 Energy Management

In addition to safety and space management as two main application scopes of occupant trajectory data, energy management is another major focus in the field. Energy consumption in buildings may be largely related to: heating, ventilation, and air conditioning (HVAC) system, lighting and appliance. The relevant occupant behaviors to these three categories can be quite different. Therefore, to identify the relevant attributes of occupant trajectory data for efficient data collection and analysis, a building managers need to understand the application scenario clearly first.

Because the HVAC systems usually account for more energy usage than the other two in most places around the world [23], more research has been conducted on the interaction between occupant behaviors and HVAC systems than others. Among the collected papers on energy management, most of them studied about public areas (e.g. office, exhibition hall) while only few studied private areas (e.g. housing unit). Many of them had utilized real-time occupant trajectory data for automatic control of HVAC and lighting system [23–25], while others collected batch data to understand occupant behaviors for better energy simulation in the future [26–28]. Since the number of occupants in the thermal zone is one critical factor for HVAC system control, many studies merely utilized occupant trajectory data at the "point" level in regard to data granularity [23–25]. Nevertheless, some studies have pointed out that thermal preference of individuals is another influential factor to be considered. Therefore, the phenomenon where occupant identifies matching with their preferences is also important information to be utilized [25]. Furthermore, with the fast developing vision-based tracking system, occupant activity can also be identified and utilized for better HVAC operation [27]. To be noticed, many studies which collected batch data of occupant trajectory not only at the "point" but also "movement" level to understand patterns of occupant behaviors. Beside the level of "point" and "movement" of data granularity, few studies have identified "action" of occupants from collected trajectory data (e.g. entering and exiting room) to match with their energy consumption behaviors (e.g. usage of appliances) to have a clearer picture of occupant behavior patterns for better energy simulation [28].

3 Analysis-Identifying Key Factors

3.1 Key Factors

This paper aims to determine the key factors of trajectory data to analyze data utilization among different applications. Zheng [32] gave a panoramic survey of trajectory data mining including people, vehicles, animals and environment. Furthermore, the author discussed trajectory data's sources, preprocessing, indexing and retrieval, mining task and format. Based on the result, this paper reviews the papers in the building management aspect of the previous chapter, targeting extended research and analyzes them from the perspective of the utilization of the trajectory data. In that way, according to attributes and monitored cases, five key factors are proposed to determine which way of trajectory data utilization should be adopted.

To analyze the reviewed papers, Fig. 1 demonstrates the map of five key factors: application scenario, monitoring area, data processing, data privacy and data granularity. Firstly, the application scenario focuses on issues of building management including safety, space, and energy while monitoring area corresponds to where the applied scene is or the experiment is conducted. Then, the monitoring area for data collection which involves its location and indicates possible limitations is considered the basis of the experiment in each paper. After that, both data processing, data privacy and data granularity are related to the scale of trajectory data. Data processing shows different time scales of data for analysis methods while data granularity indicates if the time information is included. Subsequently, data privacy is decided by that if the collected data contain any individual information which could offend occupants' privacy. By determining factors of data, it ensures that the data utilization becomes more concrete. To sum up, this paper focuses on a deeper interpretation of the trajectory data and aims to provide a better foundation for researchers to quickly obtain related studies. The following will introduce each factor and elaborate on its function.

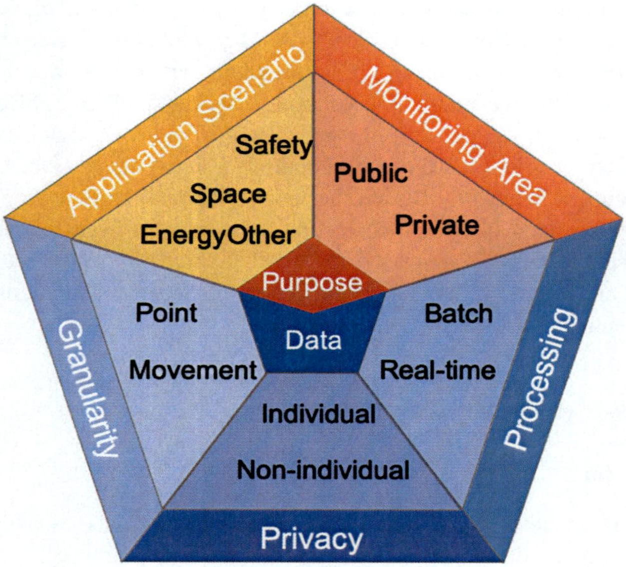

Fig. 1. The overview of the proposed framework for defining trajectory data

Application Scenario.
Since building management aims to efficiently keep the occupants' comfort, it is essential for any decision making to know their moves. Considering the main purpose of the building management, public safety and space assignments are directly related to the quality of environment in a building. Apart from that, because buildings roughly consume 40% of total global energy, the circumstance leads to great interest in the energy issue. Above all, the application scenario in the proposed framework focuses on the three issues, namely energy saving, safety, and space management, and they will be used to classify the papers reviewed.

Monitoring Area.
In different applications, monitoring area can be generally divided into two classes which are public area and private area. The former including open workspaces and corridors should have fewer privacy concerns than the latter; furthermore, collecting data from private areas might not even be allowable in some cases. For a private area, there will only be a specific group of occupants, so it is easier to manage the composition of them. Therefore, occupancy in public areas should be much larger than that in private areas and rising number of identities is too many to be handled by managers. In consequence, utilization of trajectory data in each kind of monitoring area should be different and it also influences the factor of privacy. For example, for an individual office, it is not appropriate to collect occupants' movements and at most, only the presence of occupants can be attained. This situation limits the number of studies and a building manager can only refer to solutions based on the public area. In that way, possible risks are ambiguous and the uncertainty hinders the utilization of trajectory data in practice. Since the difference between monitoring areas basically affects the

applicability of methods, it should be first decided to clarify limitations in the collection of data.

Data Processing.
As the collected trajectory data initially belong to raw data, they will usually be processed to retrieve the usable information such as occupants' presence or analysis of space-use. The proposed factor map emphasizes the required time scale of raw data and poses a binary classification for the way of processing which should be either real-time or batch analysis. To immediately respond to changes in the environmental state, access to the real-time data is crucial to building systems. For example, automatic lighting control depends on occupancy changes and lights should be turned off when the value equals to zero. As a result, real-time processing will be applied to research on making adequate responses according to critical moments. By contrast, studies usually aim to explore latent patterns through batch analysis and then the result will help make strategies for enhancement of the building management. Above all, requirements for each type of processing are close to the efficiency of collecting the trajectory data and the way to store them. This should be determined before discussing the other attributes of data because research, where the data had even the same attributes, encounter unique difficulties with different processing.

Data Privacy.
Among all data-based applications, the privacy concern was commonly mentioned since analysis might inadvertently capture exceeding individual information. The most critical one, the privacy, should be carefully handled because it is the key to associate the result with different targets. If any individual information is included for the research purpose, an evaluation should be made to measure the extent of privacy infringement. Once the measurement shows excessive harm, the beforehand request might be required before collecting the data. In consequence, it is crucial to figure out that if the collected data involves any individual information in case the analysis causes unexpected privacy offense.

Data Granularity.
To specify the utilization of trajectory data, the factor of granularity is essential to define what extent should the information be. In this context, the factor map presents two levels of trajectory data which are point and movement. Movements are considered the basic unit of trajectory data while points can be extracted from them by separating data at different times. Although movement contains more information, its size should be bigger and more computational cost is needed. As a result, it is vital for applications to determine which level should be achieved and then corresponding methods can be developed to resolve their objectives. To start with the location points in trajectory data, they represent the positions of occupants inside a building and without the association, points at different times are separate. As the presence of occupants can be inferred by the point data, only a boolean value has to be recorded and spaces for storage can be reduced. After that, movements are composed of a series of points which are obtained by various approaches like tracking techniques. The movement data indicates how an occupant walks from one area to another and by analyzing movements in a space, it is

possible to identify any route conflict which can help validate the building design. In consequence, the factor specifies which scale of the data is needed for the research goal.

Summary.
To summarize, it is vital to understand if the monitoring area is public in advance, because the difficulties of these two areas are distinct; for example, the amount of collected data should differ significantly. After that, both processing and privacy should be determined before clarifying the requirement of granularity of trajectory data. It is because real-time processing and the involvement of individual information could limit the choice of methods. Finally, the granularity of trajectory data depends on their ability to achieve the objective and meanwhile, maximize the utilization of trajectory data. To better demonstrate the factor map, an example is presented below.

Assuming that there is a teaching building and the school wants to make the building operation fit the curriculum more so that they can better control the energy consumption which is one of the application scenarios. To this end, they decided to monitor the situation of students to make suitable strategies. In that way, trajectory data are collected in each classroom which belongs to a public area while real-time processing is much more suitable since there should be rapid occupancy changes between classes. Furthermore, since it is unnecessary to know who a student is, non-individual information is capable to reach the purpose. Lastly, considering that most students sit on their seats during class, point data for granularity is sufficient to reach the satisfying information about the presence of occupants. According to the information, lightings and air conditioners can be adjusted to save redundant energy costs. By comparing the number of course enrollments with actual occupancy, the rate of ditching the class can be calculated to help measure the efficiency of courses. The above simply presents an example where trajectory data are utilized to enhance the building management and how the proposed factor map is exploited to classify the application.

3.2 Relationships Between Key Factors

After reviewing and analyzing relevant literature of utilizing occupants' trajectory data for building management under the five key factors mentioned above, tendency of data attributes for different application scenarios is revealed. As shown on the left-hand side of Fig. 2 below, more research utilized batch data than real-time one and occupant movement data than presence (point). This indicates that the processing real-time data is still challenging and occupant movement data is of great value for collection. However, while mapping the data processing and granularity of review papers, some clear tendency for attribute combination is shown. For safety and space management related papers, more studies utilized batch occupant movement data than real-time occupant presence, while for energy management, the numbers of papers with these two types of combinations are similar. Therefore, the application scenario is highly relevant to decide the granularity of data collection.

As shown on the right-hand side, most of the research is on public space, and more of them utilized data with identity than the one without. This not only shows that utilizing data to monitor private space is still limitedly explored, but also data with individual information has great value for utilization. However, while mapping the

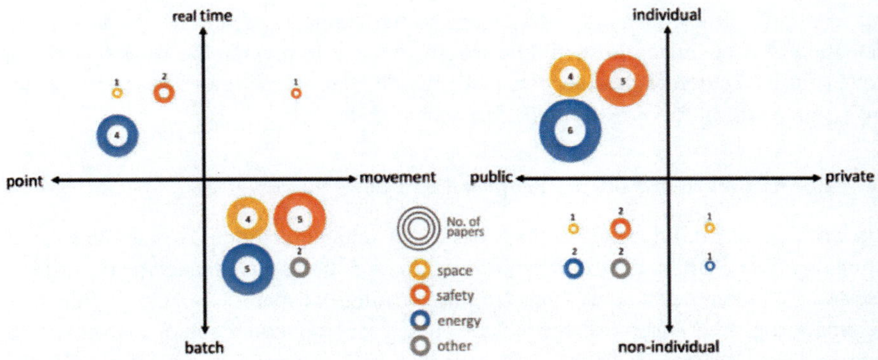

Fig. 2. Analysis of reviewed papers by data attributes for different application scenarios

monitoring area and data privacy of review papers, no absolute tendency for attribute combination is shown, since quite a few research utilized non-identity data for public area. This indicates that the deciding factor for collecting data with or without identity is not merely determined by types of monitoring areas but also the application scenario.

4 Result - A Procedural Framework

As can be seen in Fig. 3, this paper proposes a framework on choosing the most suitable technology for different purposes. Step 1 is to identify the key factors in designing building management systems through where and what the observed objects are. According to the design of building management system, step 2 is to identify the key attributes for collecting trajectory data to clarify which technology is the best or most beneficial technology for the building management system. At last, step 3 is selecting the corresponding technologies.

Based on the framework mentioned above, some guidelines are developed to point out the matters of collecting and utilizing data in each step. Since collecting trajectory

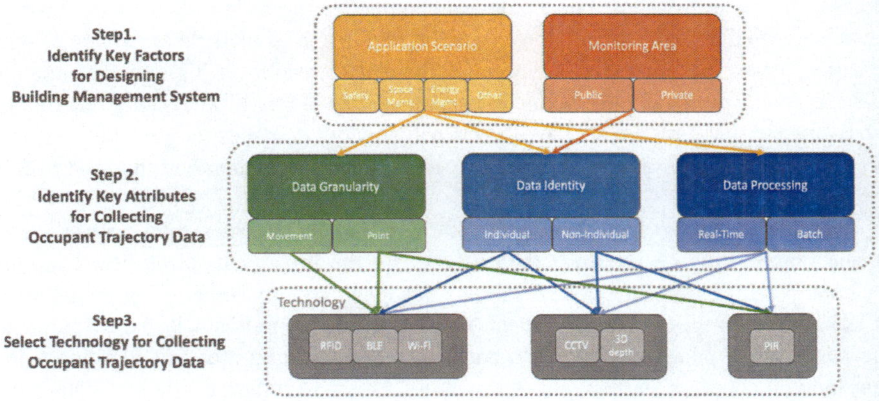

Fig. 3. The process of data collection.

data is usually time-consuming and crucial to the applications, the design and implementation of data collection should be specific enough to prevent the work from being proved futile. Therefore, this section first provides the precautions of each key factor and a procedure framework of collecting technologies.

4.1 Instruction Based on the Proposed Framework

The work of collecting trajectory data has many notable matters. To start with application scenario, after identifying the design for building management systems, researchers should consider if there are any existing occupants' trajectory patterns in the monitoring area; otherwise, this is likely to cause the re-collection of trajectory data. It is because researchers usually discover totally different patterns after analyzing collection of data. As a suggestion, researchers can start with a few tests to check there are no more unexpected patterns. Next, all the existing patterns can be composed by diverse granularity to narrow down into specific two terms, movement and point. Second, the requirements of each level of granularity can be categorized into time interval, batch size and extra information, and this leads to the next branch, processing. Third, the storing problem is the most critical one for both real-time and batch processing because the computing time and storage size heavily influence the performance and traceability. Last, researchers usually face the privacy issue in both two kinds of monitoring areas, private and public, and this should be the highest priority issue to be considered. To sum up, the key to success in collecting data is to carefully decompose occupants' patterns and be cautious when dealing with the above matters.

Every application scenario contains several occupants' trajectory patterns, and these patterns can be classified into different granularity, in order to figure out the requirements of time interval, batch size and extra information. For example, different time intervals may present the same condition differently. Assuming that there is a research on exploring customers' flow in a mall and aiming to determine which store location should be priced higher. In this scenario, the trajectory data collected within the big-time interval cannot show the pattern of stop-and-go. Thus, it needs a more comprehensive design of collecting data to gather all useful patterns.

Data size is a critical problem which influences the performance, traceability and storage size of the system. For example, when collecting trajectory data from video sequences, the tracking technology usually requires a considerable computational resource. Also, the enormous memory of these data results in high cost of storage management. Specifically, the frame per second (FPS) of real-time tracking determines the computational requirement, and the traceable record demands the storage size. As a suggestion, the video should be saved as a result after processing, and the least FPS is recommended.

Public monitoring areas have more complicated legal issues than private areas, because monitoring private areas only need to get the agreements of observed people. By contrast, although there are different laws in every country, monitoring public areas at least need every person's agreement, but the people are not usually the same. Next, the factor privacy in step 2 needs to consider the monitoring area to determine using non-individualized or individualized trajectory data, and different kinds of data bring different privacy issues. Briefly, invasion of privacy comes behind individualized trajectory data. As a suggestion, the collection of data should not only be de-identified but also legal to protect the observed people and researchers themselves. In conclusion, the work of collecting data is time-consuming, so the above presents several suggestions to avoid mistakes.

4.2 Technology for Collecting Occupant Trajectory Data

Trajectory data can be classified into two categories, individual and non-individual, based on whether the observed object can be tracked and identified. Since individualized one faces serious privacy issues, non-individual trajectory data is the first choice if it meets the requirements. According to our review literature, the most popular technology of collecting non-Individualized trajectory data is infrared sensor, but it suffers from two limitations [33]: the data only contain whether the room is occupied or not instead of the number of people; also, infrared sensor cannot distinguish unoccupied between unmoved people. Dedesko et al. [34] proposed the method combining infrared sensors with environmental sensors such as CO2 sensors to identify the presence of occupants and their activities in a hospital. Also, the door sensor detects the actions of going in and out; in this way, it is able to accumulate the number of people inside the observed room. To sum up, non-individualized trajectory data cannot gather the actual coordinate of occupants.

The technology collecting individual trajectory data includes RFID, Wi-Fi, BLE and Ultra-wide band (UWB), and RFID has been reported to be used the most [35]. It is because RFID takes advantage of low cost, low energy consumption, decent error range and flexibility. Moreover, based on the previous research which organizes positioning technology [35], this paper reorganizes it by adding the information of review paper, showing the frequency of adopting these positioning technologies in building management. As shown in Fig. 4, UWB has the highest accuracy of positioning, while the difficulty of its implementation repels users. Besides, because the positioning accuracy of application usually allows moderate errors, Wi-Fi and BLE are better choices with easier implementation, providing 1–3 m positioning error range.

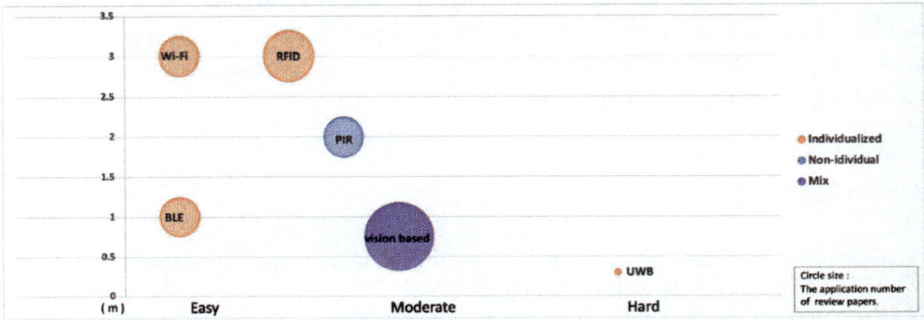

Fig. 4. Comparison of major RTLS technologies (area of circle represents how widely the technology has been adopted in the work reviewed by this paper)

The weakness of technologies collecting trajectory data mentioned above is that its observed object must hold a tag to be tracked which is impossible in some scenarios such as shopping malls, train stations and hospitals. To solve this problem, vision-based tracking is recommended, because CCTV systems are already matured nowadays. Although the vision-based tracking technology is immature, more and more research of computer vision has been published in recent years. Additionally, the merit of vision-based technology is that it can provide both individual and non-individual trajectory data which only uses different processing programs for videos. In comparison to RFID, the error range of vision-based technology is relatively higher. To improve the accuracy of vision-based technology, depth camera is able to detect the observed object's depth to reduce the error of positioning. Nevertheless, the limitation of vision-based tracking technology is the great demand for computational resources. As a result, the applications are usually processed in batch instead of in real-time. Otherwise, RFID, Wi-Fi, BLE and PIR are the better choices for real-time processing applications.

5 Conclusion

Even though the importance of occupant behavior for proper building management and the progress of monitoring techniques for occupant trajectory data have been recognized by many researchers in the field of building study and data science, limited study has drawn close discussion on possibility between these two topics, i.e. occupant trajectory data for building management. Therefore, this study has proposed a framework to collect and utilize occupant trajectory for building management to fill this research gap. By reviewing 24 relevant publications within the research scope, this study has identified some application scenarios of building management (including safety, space and energy management) and targeted monitoring area (public and private) as important inputs to decide what kind of trajectory data is needed. And data progressing (real-time and batch), granularity (occupant presence and movement) and privacy (individual and non-individual) are recognized as key attributes of occupant trajectory data. Several application tendencies of data attributes for different scenarios are also found, providing good references and indication for building managers to design ideal systems and for academic researchers to decide future research directions. Last but not the least, this study also recommends ideal monitoring techniques to building managers and system designers. It is hoped that, with the proposed framework, there will be more academic and empirical work on occupant trajectory data application for building management.

References

1. Balvedi, B.F., Ghisi, E., Lamberts, R.: A review of occupant behaviour in residential buildings. Energy Build. **174**, 495–505 (2018)
2. Hong, T., D'Oca, S., Turner, W.J., Taylor-Lange, S.C.: An ontology to represent energy-related occupant behavior in buildings. Part I: introduction to the DNAs framework. Build. Environ. **92**, 764–777 (2015)

3. Hong, T., D'Oca, S., Taylor-Lange, S.C., Turner, W.J., Chen, Y., Corgnati, S.P.: An ontology to represent energy-related occupant behavior in buildings. Part II: implementation of the DNAS framework using an XML schema. Build. Environ. **94**, 196–205 (2015)
4. Tomé, A., Kuipers, M., Pinheiro, T., Nunes, M., Heitor, T.: Space–use analysis through computer vision. Autom. Constr. **57**, 80–97 (2015)
5. Arslan, M., Cruz, C., Ginhac, D.: Spatio-temporal dataset of building occupants. Data Brief **27**, 104598 (2019)
6. Arslan, M., Cruz, C., Ginhac, D.: Understanding occupant behaviors in dynamic environments using OBiDE framework. Build. Environ. **166**, 106412 (2019)
7. Bolichini, C., Geronazzo, A., Quintarelli, E.: Smart buildings: a monitoring and data analysis methodological framework. Build. Environ. **121**, 93–105 (2017)
8. Hsu, S.H., Hsieh, S.H., Lin, T.H., Dai, T.Y.: A re-identification system for multi-camera tracking of building occupants. In: The 32nd KKHTCNN Symposium on Civil Engineering, Daejeon, Korea, 24–26 October 2019
9. Ma, Y., Zhang, H., Chen, T., Yang, R.: Experimental study on small group behavior and crowd dynamics in a tall office building evacuation. Phys. A **473**, 488–500 (2017)
10. Peacock, R.D., Reneke, P.A., Kuligowski, E.D., Hagwood, C.R.: Movement on stairs during building evacuations. Fire Technol. **53**(2), 845–871 (2016). https://doi.org/10.1007/s10694-016-0603-5
11. Rahman, A., Devi, R., Partiwi, S.G., Sudiarno, A.: Occupant movement tracking in high-rise building evacuation planning using RFID technology and SMS gateway. In: Southeast Asian Network of Ergonomics Societies Conference (2012)
12. Ma, Y., Li, L., Zhang, H., Chen, T.: Decentralized evacuation system based on occupants distribution and building information. In: Proceedings of the ISCRAM (2015)
13. Pinto, N., Coelho, A.L., Gonçalves, M.C., Cordeiro, E.: Characterization of movement on vertical circulations. Build. Simul. **5**, 383–392 (2012)
14. Lotfi, A., Langensiepen, C., Mahmoud, S.M., Akhlaghinia, M.J.: Smart homes for the elderly dementia sufferers: identification and prediction of abnormal behaviour. Ambient Intell. Hum. Comput. **3**, 205–218 (2012)
15. Ronchi, E., Kuligowski, E.D., Peacock, R.D., Reneke, P.A.: A probabilistic approach for the analysis of evacuation movement data. Fire Saf. J. **63**, 69–78 (2014)
16. Ronchi, E.: An analysis of evacuation travel paths on stair landings by means of conditional probabilities. Fire Saf. J. **65**, 30–40 (2014)
17. Becerik-Gerber, B., Jazizadeh, F., Li, N., Calis, G.: Application areas and data requirements for BIM-enabled facilities management. J. Constr. Eng. Manag. **138**(3), 431–442 (2011)
18. Dzeng, R.J., Hsiao, F.Y., Xie, Y.Q.: A simulation and visualization model for occupants' movements in instructional buildings. Comput.-Aided Des. Appl. **9**(1), 95–110 (2012)
19. Dzeng, R.J., Lin, C.W., Hsiao, F.Y.: Application of RFID tracking to the optimization of function-space assignment in buildings. Autom. Constr. **40**, 68–83 (2014)
20. Dzeng, R.J., Wang, W.C., Hsiao, F.Y.: Function-space assignment and movement simulation model for building renovation. J. Civ. Eng. Manag. **21**(5), 578–590 (2015)
21. Nassar, K.: A model for assessing occupant flow in building spaces. Autom. Constr. **19**(8), 1027–1036 (2010)
22. Vanclooster, A., Neutens, T., Fack, V., Van de Weghe, N., De Maeyer, P.: Measuring the exitability of buildings: a new perspective on indoor accessibility. Appl. Geogr. **34**, 507–518 (2012)
23. Chandrasiri, A.P., Geekiyanage, D.: Real-time object detection system for building energy conservation: an IP camera based system. In: Gorse, C., Neilson, C.J. (eds.) Proceeding of the 34th Annual ARCOM Conference, Belfast, UK, 3–5 September 2018, pp. 577–586, Association of Researchers in Construction Management (2018)

24. Liu, S., Yin, L., Ho, W.K., Ling, K.V., Schiavon, S.: A tracking cooling fan using geofence and camera-based indoor localization. Build. Environ. **114**, 36–44 (2017)
25. Li, N., Calis, G., Becerik-Gerber, B.: Measuring and monitoring occupancy with an RFID based system for demand-driven HVAC operations. Autom. Constr. **24**, 89–99 (2012)
26. Li, L., Li, X., Yang, Y., Dong, J.: Indoor tracking trajectory data similarity analysis with a deep convolutional autoencoder. Sustain. Cities Soc. **45**, 588–595 (2019)
27. Diraco, G., Leone, A., Siciliano, P.: People occupancy detection and profiling with 3D depth sensors for building energy management. Energy Build. **92**, 246–266 (2015)
28. Tekler, Z.D., Low, R., Blessing, L.: Using smart technologies to identify occupancy and plug-in appliance interaction patterns in an office environment. In: IOP Conference Series: Materials Science and Engineering, vol. 609, No. 6, p. 062010. IOP Publishing, September 2019
29. Dziedzic, J., Yan, D., Novakovic, V.: Occupant migration monitoring in residential buildings with the use of a depth registration camera. Procedia Eng. **205**, 1193–1200 (2017)
30. Rafsanjani, H.N., Ahn, C.R., Eskridge, K.M.: Understanding the recurring patterns of occupants' energy-use behaviors at entry and departure events in office buildings. Build. Environ. **136**, 77–87 (2018)
31. Chen, J., Chen, H., Luo, X.: Collecting building occupancy data of high resolution based on WiFi and BLE network. Autom. Constr. **102**, 183–194 (2019)
32. Zheng, Y.: Trajectory data mining: an overview. ACM Trans. Intell. Syst. Technol. **6**(3), Article 29, 41 (2015). https://doi.org/10.1145/2743025
33. Li, N., Calis, G., Becerik-Gerber, B.: Measuring and monitoring occupancy with an RFID based system for demand-driven HVAC operations. Autom. Constr. **24**(2012), 89–99 (2012)
34. Dedesko, S., Stephens, B., Gilbert, J.A., Siegel, J.A.: Methods to assess human occupancy and occupant activity in hospital patient rooms. Build. Environ. **90**, 136–145 (2015)
35. Zhang, J., Guo, J., Xiong, H., Liu, X., Zhang, D.: A framework for an intelligent and personalized fire evacuation management system. Sensors **19**, 3128 (2019)

Rule-Based Systems Applications

Virtual Permitting Framework for Off-site Construction Case Study: A Case Study of the State of Florida

Mouloud Messaoudi(✉) and Nawari O. Nawari

University of Florida, Gainesville, FL 32611, USA
{mmessaoudi,nawari}@ufl.edu

Abstract. The rising calls to supply more housing units in the United States real estate market have hit the record in the post-recession era. This pressure has led the investor in the AEC industry to think about alternative non-conventional ways to improve the quality, productivity and labor shortage. The growing business of adopting off-site construction has currently been rising. This new delivery method has helped to control the quality, quantity, and time. However, the off-site construction schedule can be shortened, and the building can be delivered faster by improving the permitting approval process. The current construction permitting process in the U.S is tending to be bureaucratic and not productive. Nevertheless, a growing number of Building Authorities are looking for a better way to boost the productivity of issuing the construction permit. Some of the Building Authorities have adopted online permitting and less paper policy. This paper reviews the barriers and challenges facing off-site construction permitting in the US. The study discusses the incorporation of Building Information Modeling (BIM) in the workflow of the manufactured building permitting. Moreover, the paper presents a Virtual Permitting concept to optimize the current permitting process for off-site construction projects.

Keywords: BIM · Virtual Permitting · Construction permit · Off-site construction · Modular Construction · Manufactured building

1 Introduction

The recent data in the United States is showing an increase in demand to build more housing units across the nation. Moreover, the data gap report prepared by the National Low-Income Housing Coalition (NLIHC) indicated that many cities and Metropolitan areas are struggling to provide affordable housing to low-income renters. The housing stock is facing a shortage of 7.2 Million rental units that need to be constructed to meet the need of low-income households [1]. This shortage has a direct impact on the families' social and economic wellbeing. According to the National Association of Realtors(NAR), the housing supply is decreasing by 5.7% from the last year, and the data shows that the number of units on the market is 1.64 million as of November 2019 compared to 1.74 million as of the same month on 2018 [2].

The problem of reducing the shortage of housing units supply can be addressed by adopting creative ways to expedite construction products such as off-site and Modular

E. Toledo Santos and S. Scheer (Eds.): ICCCBE 2020, LNCE 98, pp. 757–771, 2021.
https://doi.org/10.1007/978-3-030-51295-8_52

Construction (MC). However, the AEC stakeholders in North America are still reluctant to go in this direction compared to their peers in countries such as Japan, Sweden, Australia, UK, China, and Singapore [3]. The data provided by the modular Building institute show that the off-site constructions share in the North American market is not aligned with the overall construction growth. The modular construction share has risen in 2017 by 3.27% up from 3.18% in 2016 [4]. According to the Manufactured Housing Survey (MHS), the shipment of new manufactured homes is slowly increasing since the last recession. The buildings that have been manufactured and shipped were increased by only 15000 units from 2008 to 2018 [5] (see Fig. 1).

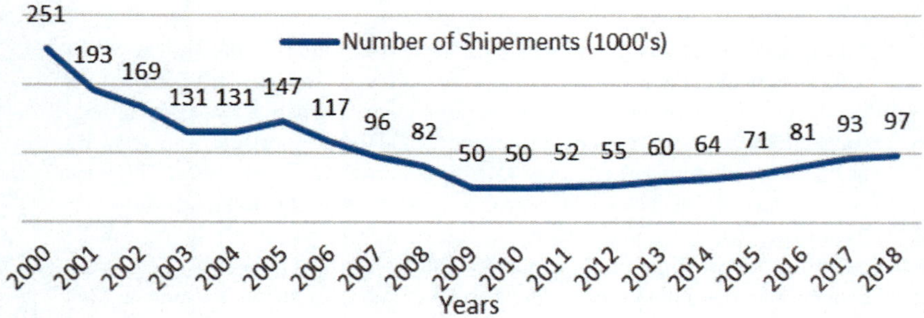

Fig. 1. Shipments statistic (1000's) of new manufactured homes from 2000 to 2018 in the US (modified from [5])

The modular construction is part of a delivery method that is controlled form the design, assembly to the shipping. The production of these structures in manufacture plants goes under thorough testing and inspection to ensure the integrity and the strength of the modular assembly to withstand the shipment and on-site installation process [6]. The modular construction is considered a productive delivery method due to the ability to control construction workflow inside the plant. The elimination of weather delays and the reduction of work environment disruption are among the reasons for increased productivity. This production set can also improve the safety of the workers compared to on-site construction [7]. According to Modular Building Institute, the off-site construction can save the owners and contractors about 30% to 50% of schedule time if it was adopted as a delivery method instead of the traditional on-site construction [6]. The off-site construction management is done on the phased process.

A study has been conducted by the Modular Building Institute analyzed various types of projects showed that the construction schedule was never exceeded one year. The data collected from 41 projects showed that the hospitality projects completion took an average of 207 days (The highest) and the retail project's completion took an average of 95 days (The lowest) (see Fig. 2). This shorter schedule of construction has allowed the owner to occupy the space faster and generate more revenues [4].

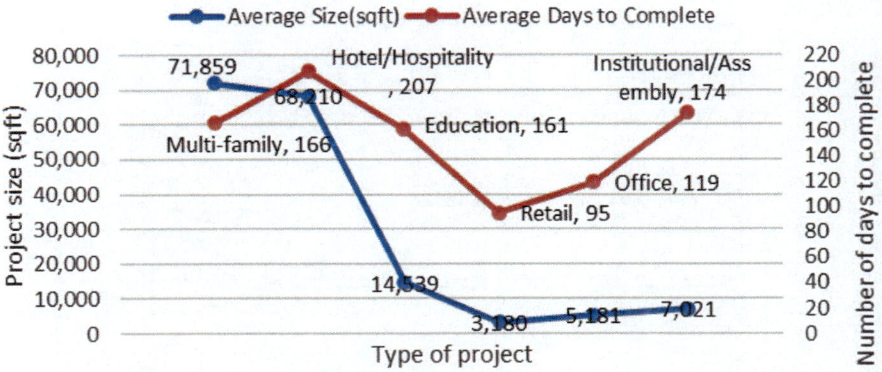

Fig. 2. The relation between the modular projects in type, size and the construction schedule (modified from [4])

2 Statement of the Problem

According to the National Institute of Building Sciences (NIBS) industry 2018 survey, the most significant challenges and barriers that are facing off-site construction include design alteration at the time of finalizing the permitting drawing. This change can result in pushing the construction schedule than cause unnecessary expenses to the owners.

The lack of design flexibility and better esthetic building are the most noted challenges facing modular construction today. Also, the modular construction permitting process is one of the problems that need attention in the modular construction industry. This particular phase on the construction schedule has potential benefits of saving time and cost if improved and streamlined. In the United States, regulating the modular building manufacturing, shipping, and installing involves many government bodies and agencies at the federal, state, and local levels. Modular home or as known mobile home that can be shipped and installed across the United States is regulated by U.S. Department of Housing and Urban Development (HUD) code and standards, while a manufactured building that can be shipped and installed within the same state jurisdiction is subject to the state codes and standards. These types of buildings must go under two different permitting and plan review processes. The first one is for the structure build on the plant by federal or state agency and the second one is for the foundation and site work by local jurisdiction [8] (see Fig. 3).

This paper focuses on the state of Florida permitting process for manufactured building. In Florida, the modular building or mobile homes built under HUD regulations are administered by the Florida Department of Highway Safety and Motor Vehicles (DMV) and the modular buildings built under Florida Building code and standards are governed by Department of Community Affairs (DCA).

This research study centers on the permitting approval process for manufactured building that is administered by DCA and has to comply with Florida Manufactured Building Act of 1979, The Florida Building Code and the Florida Fire Prevention and Life Safety Codes in addition to the local codes that govern on-site construction work such as foundation and site restoration.

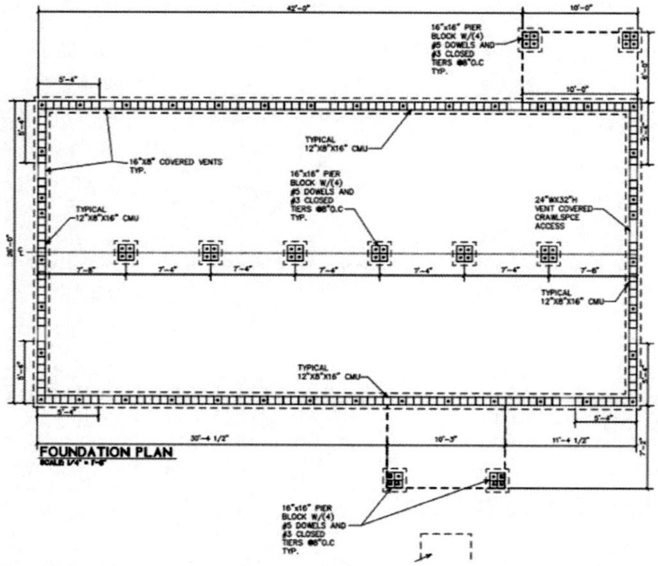

Fig. 3. Example of foundation plan for two modules of a residential building in Florida. Source: [9].

As the pressure on permitting authorities skyrocket due to the boom on construction and need for faster permit approval process, several State Agency and local Building Authorities have learned that the change of their traditional paper-based process to electronic-based process at the time of submitting and reviewing construction permits is the best way to cope with the growth of the construction industry statewide.

According to the Associated General Contractors of America (AGC), the State of Florida has added 4,600 construction jobs from September to October 2019 and is one of the top growing economies in the U.S. Moreover, the requests to obtain construction permits in Florida are the highest since the recession. The U.S. Census Bureau data show that the state has issued approximately 144,427 residential permits in 2018 [10] (Fig. 4).

In the state of Florida, the permit approval process is undergoing improvement and streamlining by introducing new legislation and methods that encourage the technology incorporation for permit submission and review. A legislation of one-stop application and approval has been adopted by the state in 1996. The legislative intent is to limit the review process to 90 days to approve or disapprove the permit. The project submitted for review must meet specific requirements [11]. The state also allowed the owners and contractors to use a private provider to perform the permit review and inspection within a particular scope explained by Florida statues 553.791.

The private provider, or as known by Duty Authorized Representative (DAR) reviews the compliance of the project with building code and provides compliance affidavit to the building department. The building officials have to issue the permit within 30 business days if no revisions are required. Furthermore, some of the

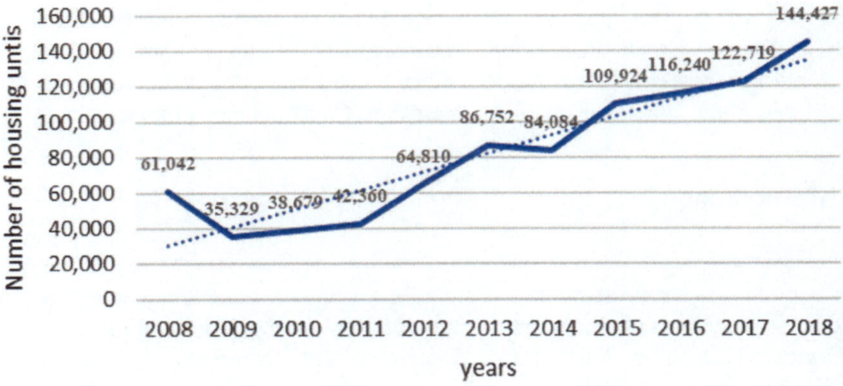

Fig. 4. The number of new residential housing units by state (total) (modified from [10]).

permitting authorities have developed in-house software application such as Professionals Electronic Data Delivery System (PEDDS) created by the Florida Department of Transportation (FDOT) [12] or acquired commercial web-based systems such as ProjectDox developed by Avolve Software. These software and methods have been used to facilitate transmitting the project documents between the stakeholders and to conduct plan review and permitting by multiple departments simultaneously. Yet, more challenges and barriers need to be eliminated or simplified.

As of emerge of BIM as a powerful tool for collaboration across different design professionals, various research teams and organizations are exploring the use of BIM to expedite the permit review process [13–15]. Nawari study [14, 15] explained the limitation of the existing commercial systems that aimed to automate the building design review such as Solibri and Singapore-CORENET. These applications categorized by researchers as a black box and gray box and some were directed toward the fire egress code checking and or the accessibility code checking. The research study has presented the Generalized Adaptive Framework (GAF) as a practical way to implement the semi-automated or automated code checking [15].

This paper approach is to fill the gap and to enhance current state knowledge on this topic by developing a Virtual Permitting Framework (VPF) to assist in expediting the permitting process for Off-Site Construction in Florida based on the current code and standards. This approach uses the Generalized Adaptive Framework (GAF) concept to construct the proposed changes to the current workflow off-site construction permitting in Florida.

3 Goal and Objectives

The main goal of this paper is to present a framework for computable semi-automatic code compliance verifying method that can help the off-site construction permitting expedition in the state of Florida. The existing permitting workflow from requesting the permit to receiving the approval or disapproval of permit goes on two steps, the 1st step is applying for the structure that will be assembled in the plant and the 2nd step is applying for permit site installation. The objectives of this research study are to (i) review the current state of the permitting process (ii) present the components of Virtual Permitting Framework (VPF) for Off-Site Construction in Florida.

4 Methodology

This research approach is to (i) explore the exiting literature and the efforts that have been dedicated to streamlining the construction permitting process in general, (ii) discuss the shortcomings of the current permit workflow for modular building in the State of Florida, (iii) present the proposed framework VPF for off-site construction and (iv) discuss the role of BIM-based GAF as one of the key component of Off-site-VPF.

5 Current State of Construction Permitting Process

According to the 2019 business report [15], the simplicity or complexity of the permitting process plays a significant role in classifying countries' favorability or unaffordability to start a new business. Measuring and assessing the construction permits efficiency are based on four (4) indicators: number of procedures, time, cost, and building quality control index. The report has used a typical building of the warehouse to demonstrate the applicability of the indicators [15]. Many countries are working very hard to enhance their permitting process system to attract more business (see Fig. 5). Some of the process reforms that have been recorded include streamlining the procedure, reducing time for processing permit application, introducing a one-stop-shop, reducing fees, enhancing electronic submission, and introducing risk-based approvals. As an example, in 2012, Taiwan has reformed its rules to obtain construction permits and approve new building inspections in a single stop [16]. This has improved the coordination between multiple departments involved in the process. This fundamental change of consolidating the building permit operation in a one stop shop has helped the city of Taipei to save 31 days and cut 14 bureaucratic procedures according to the Doing Business database [17]

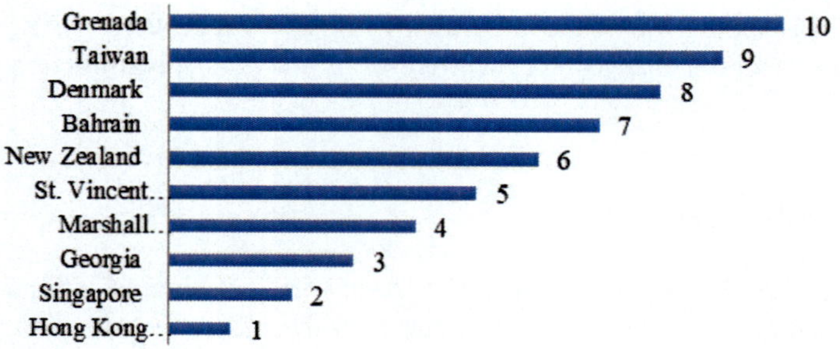

Fig. 5. The top 10 countries where dealing with construction permits is the easiest. (Modified from [16])

Some countries such as Portugal, India, Norway, Malaysia, Greece, Burundi, and Benin have restricted time limits to process construction permit applications. The Netherlands, Panama, Peru, Costa Rica, and the Russian Federation have worked to streamline the process by eliminating some of the unnecessary requirements to obtain a permit [17].

In the U.S., the efficiency of the construction permitting process is based on the number of procedures; numbers of days and the cost varies significantly across different jurisdictions. The majority of these variations originated from the building, zoning, and environmental permit approvals requirements, according to the U.S. Chamber of Commerce Foundation's 2014 Regulatory Climate Index. For a typical small commercial project, the data shows that the average cost of complying with code and regulations are estimated at around 1% of total construction costs, and it takes three months to complete the process of obtaining various approvals. According to the regulatory climate index, a comparison between 10 cities across the U.S. in term of dealing with construction permits shows that the construction permitting fees are the lowest in some cities such as St. Louis, Raleigh, and Dallas, and highest in some cities such as San Francisco and Los Angeles. These fees can go from a low of 0.3% to a high of 3.0% of the total cost of construction. The additional procedures imposed by some cities as part of permitting requirements can cause significant increases in time and cost. As an example, the environmental quality review required by the city of San Francisco costs the building owners an additional $50,000 in fees and 66 additional days expected time to review by the city staff [18], The permit delay has put the US real estate market under pressure due to the growing gap between the number of permits approved and the demand to build more housing units [19, 20] (see Fig. 6).

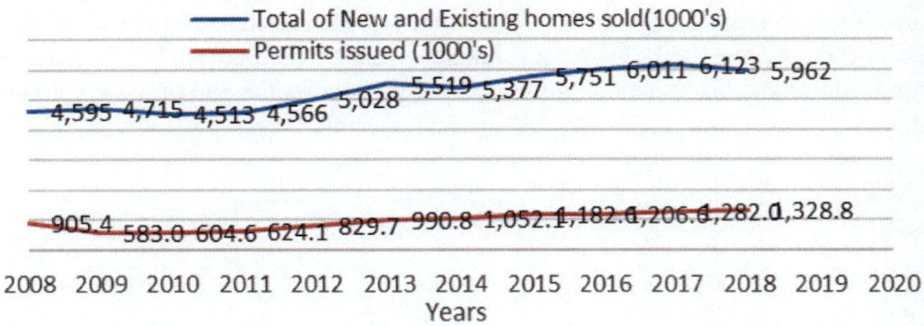

Fig. 6. The current gap between the residential units have been sold and the number of the approved permits (modified from [18, 19])

6 Current State of Modular and Off-site Construction Permitting Process in Florida

The permitting approval process for modular construction is similar to site-built traditional construction. The 1st phase of the process is monitored by the Department of Community Affairs (DCA). In this phase, the permit review process starts when a manufacturer or owner sends a request for modular building permit approval. The DCA or any of its certified plan review agency as stated in chapter 553, Part I, F.S instruct the manufacturer or owner to submit permit plans for approval through the Building Code Information System (BCIS). The plans of the modular building need to be prepared, signed and sealed by Florida licensed architects or engineers. The submittal includes architectural drawings, structural drawings, plumbing drawings, electrical drawings, and any necessary documents needed for the review.

The plan review agency checks and approves or disapproves of the plan's compliance to Florida building code and Florida Fire Prevention Code, latest editions within 30 days. If any revisions are needed, the agency notifies the manufacturer to correct and resubmit the plans according to the Florida Administrative Code. The plans reviewed in this phase are not subject to the local building department review.

The 2nd phase of the permitting approval process is conducted by the local building authority where the manufactured building is installed. In this phase, the permit applicant submits drawings and documents related to foundations, site work, and utility connection to the local permitting authority. The submittal includes (i) modular building plans approved by DCA; (ii) site plan showing utilities and connection; (iii) plans showing parking area, accessibly building accessory such as ramp, stairs, etc.; (iv) design professional signed and sealed document showing that foundation meet High-Velocity Hurricane Zone (HVHZ) requirements in the FBC; (v) signed and sealed foundation plan showing tie-down information; (vi) manufacture certificate document issued by DCA; and (vii) HVAC installation details if needed [21]. The local permitting authority starts reviewing the permit submittal after the end of the 1st phase by receiving approval from DCA certified agency (see Fig. 7).

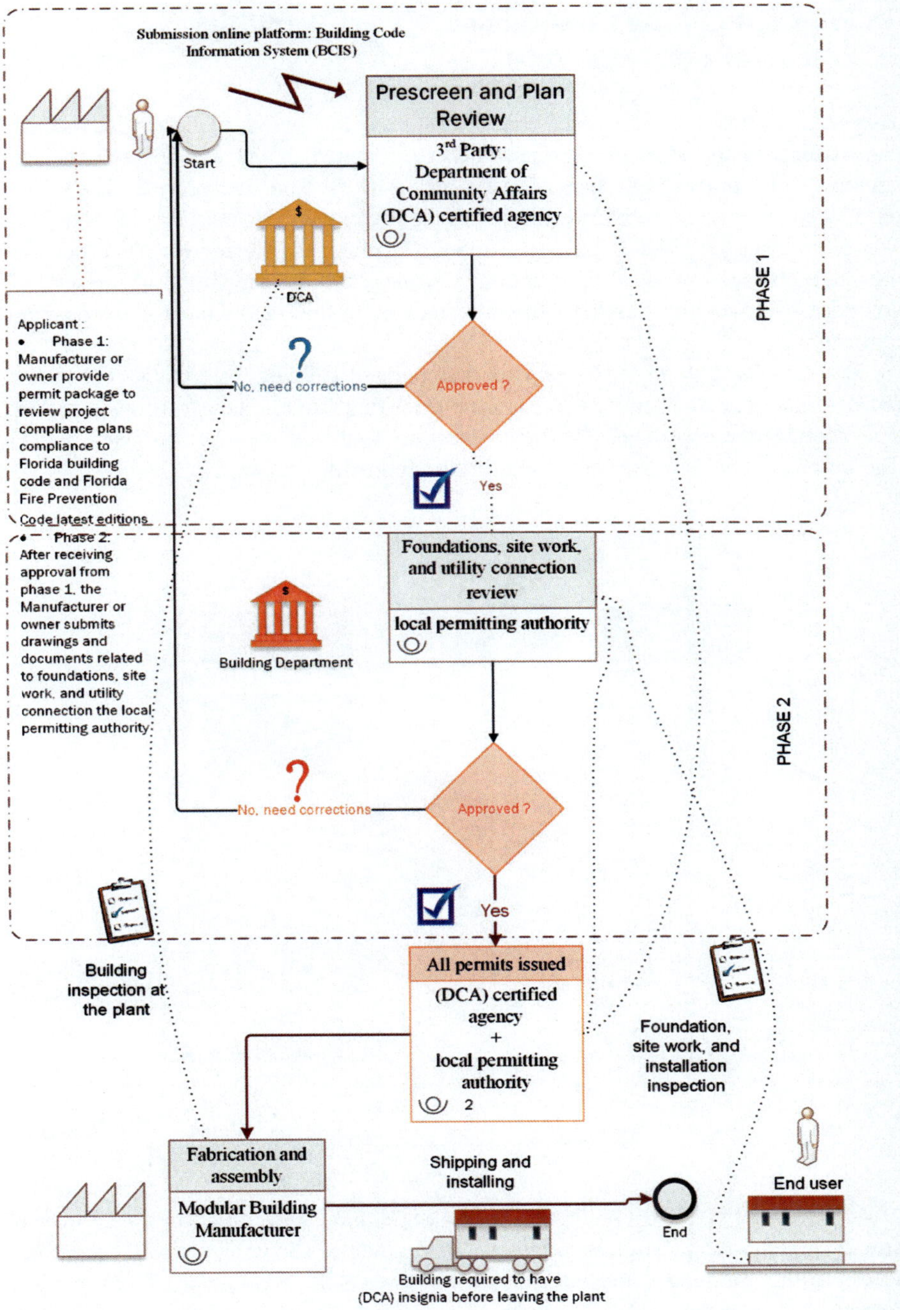

Fig. 7. The current permit workflow for modular building in the State of Florida

7 Proposed Off-site Construction Virtual Permitting Framework (Off-site-VPF)

This section presents the proposed change to the existing permit workflow for modular construction in the state of Florida which is designed to cut the review time and streamline the process. As part of state efforts to expedite the process, lawmakers introduced and passed various rules such as the Electronic Signature Act of 1996 [22]. The Florida statues 668.001 have clarified the advantages of using the electronic signatures such as the boost of the electronic business and the expedition of the services delivered by government entities, and the reduction of the fraud related to the business transactions.

The proposed workflow seeks to incorporate Building Information Modeling (BIM)-based Virtual Permitting (VP) into the existing modular construction permitting and eliminate the dependency of local permitting to the state permitting (see Fig. 11). The proposed permitting process includes the following phases:

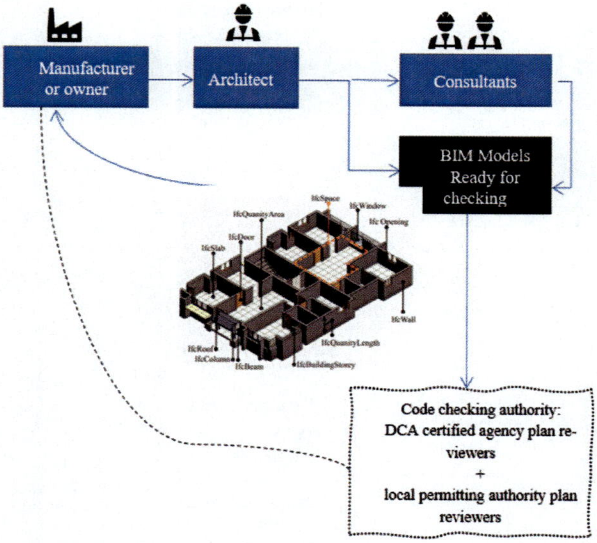

Fig. 8. Information flow and exchange models in phase 1.

Prescreen and Plan Review Phase: In this phase, the manufacturer or owner need to provide a permit application package and BIM models. The applicant will be granted access to the Building Code Information System (BCIS) to upload all the required drawings, documents, and BIM model. The applicant is required to upload a BIM-friendly file format that will be checked later by the automated code checking engine. The BIM-based Exchange Model (EM) needs to have relevant data to perform efficient code compliance verification. At the beginning of the project, the applicant will inform the project architect to provide and prepare a model for code checking purposes. Examples of the architects and their consultants Exchange Models (EMs) are described as the followings:

BIM Models for code checking for DCA certified agency
BIM Models for code checking for local permitting authority review

The plan reviewers from DCA certified agency and local permitting authority will be receiving the request to start reviewing the compliance of BIM exchange models according to the review scope for each party (see Fig. 8). The local permitting review process will start earlier and without waiting for the approval of DCA certified agency. This change can save time and cost.

Semi-automated and Manual Code Checking Phase: In this phase, the Virtual Permitting Framework (VPF) is incorporated as a code checking engine. The assigned reviewers from phase 2 will verify the code compliance of the BIM-based model using the automated checking engine and manual checking as well. The result will be a report showing the list of code provisions violated or passed the code checking (see Fig. 9). At this stage, the engine will check the model to verify the compliance with provisions R 303, R 304, R 305, R 310, and R 311 from Florida Building Code, Residential, 6th Edition (2017) [23]. The list of provisions violated will be communicated to the permit applicant. Resubmission rounds or additional information need to be submitted to address any information deficiency in the plans/BIM model prior to the permit issuance.

7.1 Virtual Permitting Framework (VPF)

The VPF will be integrated in phase 2. It is an essential part of this proposed approach. The goal of this framework is to assist designers using the code checking engine through the Planning, Design and Execution, and Validation stages.

Nawari study [15] proposed the generalized adaptive framework (GAF) as an alternative Automated Design Review (ADR) and innovative concept that can assist in

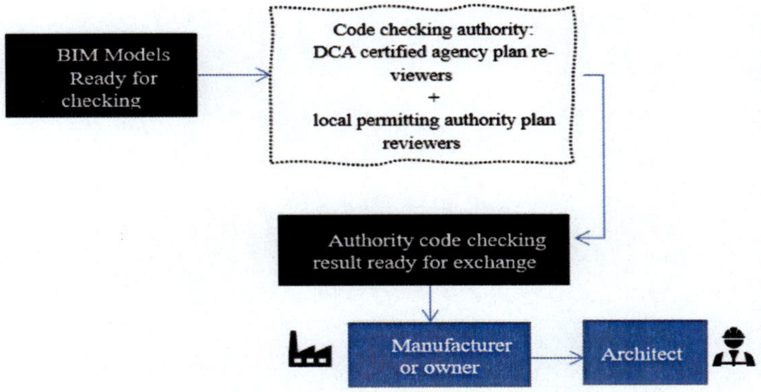

Fig. 9. Information flow and exchange models in phase 2

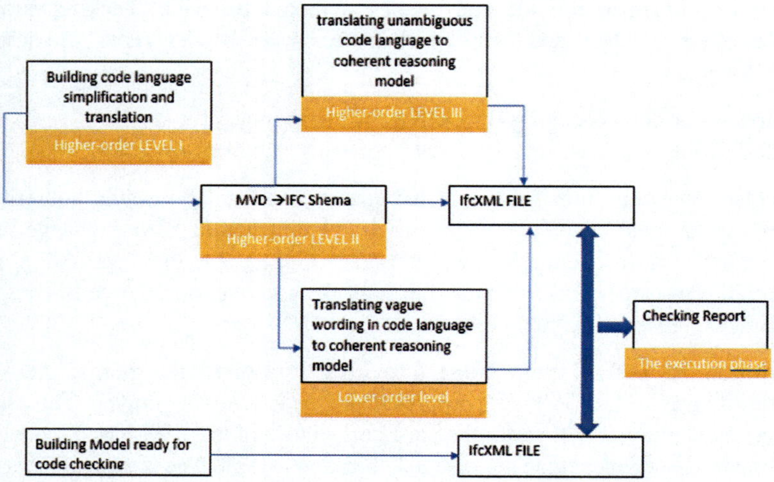

Fig. 10. Generalized adaptive framework for semiautomated buildings design review (modified from [15])

translating different code regulation texts into simplified language that will be used by the machine. The GAF framework components are explained in higher-order level I, higher-order level II, higher-order level III, lower-order level, and the execution phase (see Fig. 10). At the level I, the code texts will be categorized into conditional, content, ambiguous, and dependent; at level II, the data from level I will be used to develop Model View Definition (MVD) and Industry Foundation Classes (IFC) object model. At level III, an extraction algorithm will be developed to translate the ambiguous data. At lower order, fuzzy logic will be used for the partial checking. At the exclusion phase, a report of compliance will be generated based on the previous levels.

This framework adds great value to the domain of integrating computer applications to the building design review. An additional advantage of this system is its ability to use neutral, open-standard data schemas such as Industry Foundation Classes (IFC), which is an international data standard for data sharing and exchanges in the construction industry (ISO 16739:2018) [24].

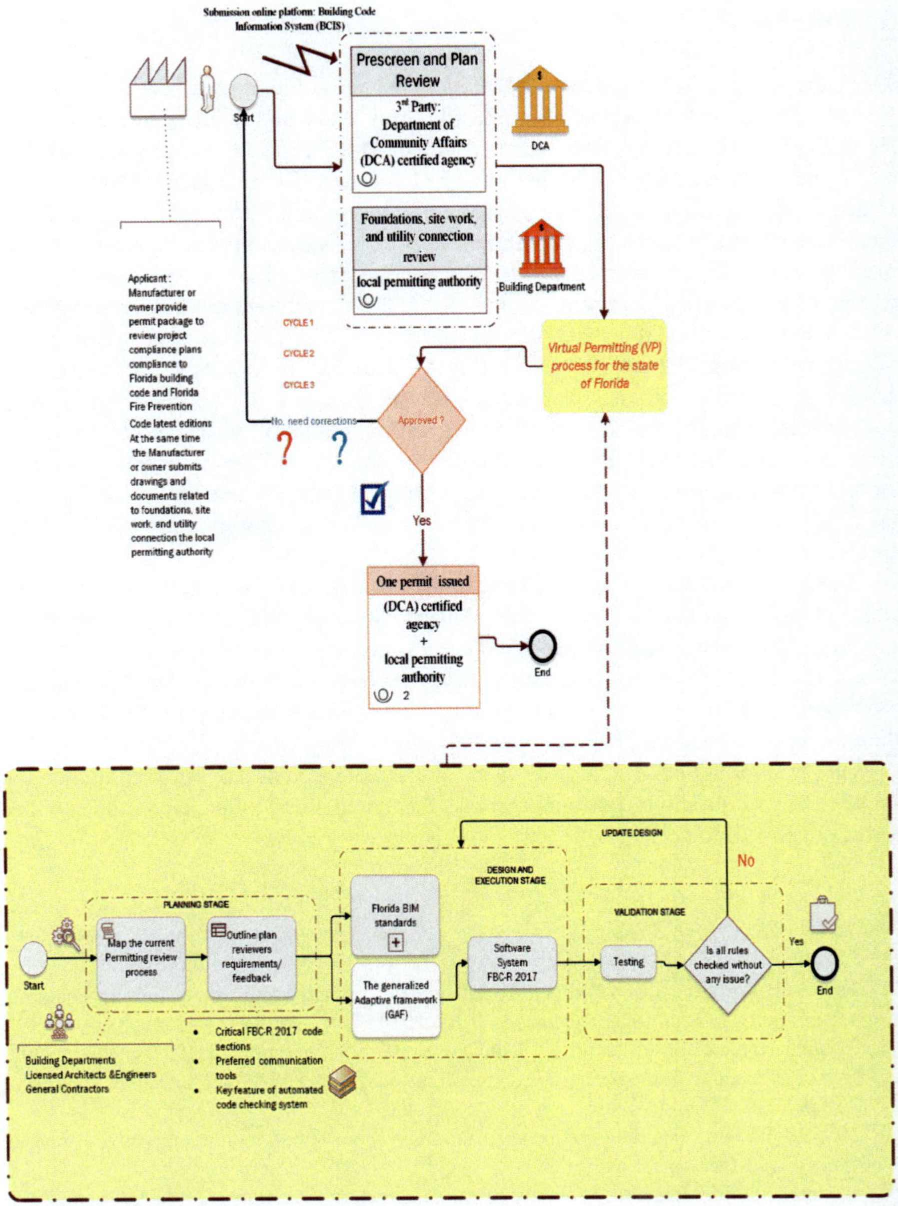

Fig. 11. Proposed off-site construction virtual permitting framework (off-site-VPF)

8 Conclusion

This paper explored the shortcoming that faces the permitting review process in the off-site construction approval process in Florida. This study highlighted the gap between the number of permits issued and the real estate market expectation. This situation has led to increased pressure on the permitting authorities. The modular building is presented as an alternative to solve the shortage of housing units. The study concluded that the design restriction and the complexity of the permit submittal process are among the major problems facing the modular construction industry in the US. The study focused on the off-site building permitting monitored by the Department of Community Affairs (DCA) and local permitting authority in the state of Florida. This paper presents the Virtual Permitting Framework for Off-Site Construction in the state of Florida. This proposed framework has two phases that cover all the permitting processes of modular construction. The first phase is prescreening and plan review, and the second phase is semi-automated and manual code compliance checking. This proposed process is recommending eliminating the dependency on local permitting authority to issue the DCA approval certificate. This framework is based on incorporating BIM and the Generalized Adaptive Framework (GAF).

These changes to the current permitting process will be a win-win for every stakeholder in the modular construction industry. Some of the advantages in addition to the time and cost-saving include increasing tax revenues, empowering citizens, attracting more businesses, managing labor shortage, improving coordination and transparency, increasing the supply of affordable housing, minimizing the impact on the environment, and facilitating post-disaster rebuilding efforts.

Furthermore, future research will focus on the extension of the proposed framework in different permits and inspection processes that are required by the cities and counties in the state of Florida.

References

1. The Gap: A shortage of affordable rental homes, national low-income housing coalition. https://reports.nlihc.org/gap. Accessed 30 Jan 2020
2. Existing-Home Sales Descend 1.7% in November, 19 December 2019. www.nar.realtor, https://www.nar.realtor/newsroom/existing-home-sales-descend-1-7-in-november. Accessed 30 Jan 2020
3. Modular Building Institute: Frequently Asked Question. http://modular.org/HtmlPage.aspx?name=faq. Accessed 30 Jan 2020
4. Modular Building Institute: Summary of permanent modular construction for year-end 2017. http://modular.org/HtmlPage.aspx?name=MA-feature-pmc17-sum. Accessed 30 Jan 2020
5. US Census Bureau: Shipments of New Manufactured Homes, The United States Census Bureau, 10 June 2019. https://www.census.gov/data/tables/time-series/econ/mhs/shipments.html. Accessed 30 Jan 2020
6. Modular Building Institute: What Is Modular Construction?—Modular Building Institute.. http://modular.org/HtmlPage.aspx?name=why_modular. Accessed 20 Jan 2020
7. Wilson, J.: Design for modular construction: an introduction for architects (2019)

8. Stein, A.: Disruptive Development: Modular Manufacturing in Multifamily Housing. University of California, Berkeley (2016)
9. RCA Inc.: Example of foundation plan.JPG (2018)
10. M. C. D. US Census Bureau: New Residential Construction (2019). https://www.census.gov/construction/nrc/historical_data/index.html. Accessed 30 Jan 2020
11. Raepple, C.: Florida's expedited permit review process: streamlining the development of Florida's economy, p. 15 (1998)
12. Florida Department of Transportation: TillmanQuinton-DanaBruce-DigitalDelivery.pdf (2013)
13. Wible, R.C.: Keeping building departments ahead of the curve. Build. Saf. J. 12–16 (2008)
14. Nawari, N.O.: Automating codes conformance. J. Archit. Eng. **18**(4), 315–323 (2012). ASCE
15. Nawari, N.O.: Generalized adaptive framework for computerizing the building design review process. J. Archit. Eng. **26**(1), 04019026 (2020). https://doi.org/10.1061/(ASCE)AE.1943-5568.0000382
16. World Bank, Doing Business 2019: Training for Reform. The World Bank (2018)
17. World Bank: Starting a business, in doing business 2013, 0 vols., pp. 56–97, The World Bank (2012)
18. The U.S. Chamber of Commerce Foundation: Regulatory Climate Index 2014, April 2014
19. U.S. Department of Housing and Urban Development, USHMC-Housing-Demand, National Housing Market Summary and Data (2019). https://www.huduser.gov/portal/ushmc/hd_home_sales.html. Accessed 30 Jan 2020
20. M. C. D. US Census Bureau: US Census Bureau Building Permits Survey (2019). https://www.census.gov/construction/bps/. Accessed 30 Jan 2020
21. Florida International University: Building Permit Requirements for Manufactured (Modular) Buildings, p. 1 (2012). https://facilities.fiu.edu/projects/BT_904/Permit_Information/d._Building_Permit_Requirements_for_Modular_Buildings.pdf
22. Earles, M.D.: Clicking on the dotted line: florida's enactment of the uniform electronic transactions act as a boost to e-commerce. Nova Law Rev. **25**, 27 (2000)
23. Florida Building Code (2017). https://www.floridabuilding.org. Accessed 30 Jan 2020
24. ISO: Industry Foundation Classes (IFC) for data sharing in the construction and facility management industries – Part 1: Data schema, ISO (2018). http://www.iso.org/cms/render/live/en/sites/isoorg/contents/data/standard/07/03/70303.html. Accessed 10 Sept 2019

Rule-Based Semantic Validation for Standardized Linked Building Models

Philipp Hagedorn(✉) and Markus König

Computing in Engineering, Ruhr-University Bochum, Universitätsstr. 150, 44801 Bochum, Germany
philipp.hagedorn-n6v@rub.de

Abstract. Planning data of construction projects consist of distributed data in the manner of building models and accompanying heterogeneous information models. For data drops, these models can be stored and linked within an information container using link sets. When using open standards and file formats, link sets are often created and maintained manually, because an automated traceable generation of links is not yet reasonably provided. In this research, semi-automatic validation is preferred to ensure the syntactic, semantic and technical correctness of link sets for the practical application of so-called linked building models. Therefore, rule validation approaches in the AEC industry are examined regarding their feasibility for linked building models. A method for rule checking is presented and evaluated in a case study, which utilizes a building model of a fictive tunnel construction project and a linked tunnel construction schedule as a linked building model inside an Information Container for Data Drop.

Keywords: BIM · Linked building models · Information Container for Data Drop · Rule validation

1 Introduction

Design and construction data usually consist of distributed heterogeneous information in proprietary serialized files leading to a separation of data. Building Information Modeling (BIM) was established as a method for centralized object-oriented information modeling to connect the existing data islands. The current approaches in the Architecture, Engineering, and Construction (AEC) industry focus on this centralized data repository in a building model [1], while Werbrouck et al. [2] recently showed, that the industry needs a more data-centric approach to connect distributed building data. Focusing on the data, these building models and the additional file-based planning data can be stored and linked within the Information Container for Data Drop (ICDD). This information container is based on recent Semantic Web technologies and works as a common environment for most file-based or model-based information, for example building models, images, spreadsheets or schedules. Within the container environment, the use of generic link sets adds a semantic layer so that information from the different sources can be linked together on an entity level. The semantically interlinked

E. Toledo Santos and S. Scheer (Eds.): ICCCBE 2020, LNCE 98, pp. 772–787, 2021.
https://doi.org/10.1007/978-3-030-51295-8_53

heterogeneous resources generate the so-called linked building model, also referred to as nD-Model in industry.

However, planners often create and evaluate linked building models manually, because an automated traceable generation of links is not yet reasonably provided due to domain-specific file formats and cross-domain data in the unique construction processes [1]. Since construction projects require a large amount of information in several information models, the link sets between these models can also contain a large number of links, so that in practice it is not possible to maintain them manually. Semi-automatic validation is preferred in this paper to ensure the syntactic, semantic and technical correctness of link sets for the practical application of linked building models. The validation relies on a set of parametric rules on the one hand and on the building model, an information model, and the link set connecting both inside the information container on the other hand.

This paper provides a short overview of rule checking and validation approaches in the AEC industry and an overview of different rule languages for automated validation of linked building models using Semantic Web technologies. In the related work, Semantic Web-specific languages like the Semantic Web Rule Language (SWRL), the SPARQL Inferencing Notation (SPIN), and the Shapes Constraint Language (SHACL) are within the scope. Based on the literature, the present study introduces a methodology for applying rule-based validation in particular for linked building models inside the ICDD. For the evaluation, a case study shows a prototypic implementation of the methodology using an exemplary linked building model from the mechanized tunnel construction domain inside an ICDD. This building model is linked to a tunnel construction schedule utilizing a binary directed link set as linked construction schedule model (4D).

2 Related Work

The research on the use of rules for validation and code compliance of building models significantly advanced with the development of BIM and the Industry Foundation Classes (IFC). Mathematical logic defines rules as one paradigm to model and validate knowledge according to a certain domain of logic that can be, for instance, the first-order logic [3]. Rule languages enable the expression, exchange, and interpretation of rules [4]. Regardless of the rule language used, rule checks require formalization of the technical rule, a corresponding interpreter, a prepared building model, a processor that executes rules on a model, and an output of the results [5]. According to Schwabe et al. [6], the rule-based model checking is still expandable and the need for a specific rule language for the checking of building models was identified in recent research. The following chapter gives an overview of the literature on the general use of rule languages in the AEC industry, on the use of rule languages according to Semantic Web approaches and on the use of standardized linked building models referencing the multi-model approach and the ISO-standardized ICDD.

2.1 Use of Rule Languages in the AEC Environment

In terms of computer science, research and practice defines the information model or the amount of information to be checked as the fact base or knowledge base. Depending on the assumption of the completeness of the knowledge base, for the application of the rules to the facts, we distinguish between the validation of the knowledge base in the Closed World on the one hand and reasoning in the Open World on the other hand [7]. Reasoning accordingly describes the information gained from the application of rules. It generates additional information added to the knowledge base. This method is called rule-based reasoning or inferencing [8]. Solihin and Eastman published an overview of seven different application types of rule checking in the AEC environment and a classification into four different classes according to the complexity of the information needed within the facts to process the inference [9]. The classification of Solihin and Eastman distinguishes between a low requirement for explicit information needs, the need for derived information through model preparation, the need for extended data structures for testing, and the need for concept-related and solution-oriented checking modifying the facts to achieve a certain task.

Regarding the implementation of rule checking, three general approaches are considered: the use of hard-coded rules, the use of parametric approaches and the language-driven approach [5]. All three approaches have been investigated in research, while the third approach is most often used for building models. The suitability of the language-driven approach regarding BIM relies on its ability to express complex rules derived from natural language in vendor-neutral exchangeable rules. Hard-coded rules cannot be exchanged properly and parametric approaches do not always offer the ability to formulate complex scenarios. In the following, applications of rule languages in the construction industry and the respective rule languages in their context are presented.

In computer science, information modeling and software engineering are often performed using standardized diagrams from the Unified Modeling Language (UML). The Object Constraint Language (OCL) was developed to define boundary conditions in this information modeling with UML [10]. Boundary conditions in OCL are represented by invariants, which must be valid when the model is instantiated. For validation according to the semantic concept of BIM, Kraft [11] investigated how the use of the standardized constraint language OCL can support the compliance of building models with technical rules. Kraft concludes that the use of OCL as a formalized validation constraint is contradictory to the extendable concept of the Industry Foundation Classes (IFC).

Lee et al. [12] proposed a more BIM-specific language referred to as the Building Environment Rule and Analysis (BERA). This language is developed to deal with building information models from IFC data mapped into a specific object data model. BERA rules and the object data model are transferred into Java code and checked in a Java-based executor. Dimyadi et al. [13] proposed another example for BIM-specific use of rule languages as the BIM Rule Language (BIMRL), which focuses on IFC data and addresses the gap between the required information and the availability of information in the building information model.

Schwabe et al. [6] present another approach of rule-based model checking using the open-source rule engine Drools. Conclusively, they stated that even in current research on rule checking, there is a lack of satisfactory solutions that are open to interpretation and are not based on hard-coded rules. A promising approach for the non-proprietary definition of rules is the use of Semantic Web technologies discussed in the next section.

2.2 Rule Modeling and Checking Using Semantic Web Technologies

Semantic Web describes a technology in which the conventional unstructured data on the Web are enriched with structured data to make the meaning of the information machine-readable. The structuring of the data can be described by ontologies, which in turn are expressed in standardized language constructs based on the Resource Description Framework (RDF) and the Web Ontology Language (OWL) [3].

The RDF is a framework for information representation in the web recommended by the World Wide Web Consortium (W3C) [14]. The RDF is a modeling format for ontologies that creates interoperable data for both human perception and machine processing. OWL is a descriptive language for publishing and exchanging ontologies within the syntax of RDF and mainly extends the expressivity of RDF, e.g. with vocabularies for describing classes. At the abstract level, an ontology in the Web Ontology Language contains a collection of axiomatic facts in the subject-predicate-object pattern [3].

Rattanasawad et al. [4] published a general review of rule languages for the Semantic Web referring to the SWRL, which is a subtype of the declarative logic programming language Rule Modeling Language (RuleML) combined with the description logic of OWL. SWRL can be used to express rules as well as logic and can be written in either a human-readable syntax or RuleML-XML, respectively RDF-XML syntax [15]. The SWRL bases on the predicate logic declaring a body consisting of conditions and a head of consequent implications assuming an open world. This method is on the one hand compatible with the declaration of facts in OWL, but on the other hand, it can be ambiguous if a relation is not restricted by conditions [3]. Furthermore, Pauwels et al. [7] conducted a detailed investigation of the performance of rule languages in the Semantic Web area for use with IFC data. The research focuses on the rule checking using rules in the SWRL and SPARQL Inference Notation (SPIN). As a W3C Member Submission, SPIN is a syntax to represent rules and constraints using the SPARQL Protocol And RDF Query Language (SPARQL) that is a language to query or manipulate RDF-based data [16]. SPIN is used for both the inference of knowledge and the validation of constraints for existing facts. In their research, Pauwels et al. [7] showed that all of the examined languages and implementations have certain characteristics and cannot be compared directly, but can lead to certain conclusions for certain application cases. Zhang et al. [17] presented another approach of rule checking using SPARQL for real-world scale building models. It is based on a conversion of the building model into OWL and the transformation of natural language formalized rules into SPARQL queries on the model. The authors prove that rule compliance checking for large building models is possible in a reasonable time. Werbrouck et al. [2] presented the Shapes Constraint Language (SHACL), which is a successor of SPIN, in a validation approach on distributed linked building data. They proposed to use SHACL to validate a distributed linked building model in a decentralized common data environment.

The SHACL is the most recent W3C Recommendation for the validation of RDF graphs, where one of the graphs in this context is the so-called data graph that is validated against a shapes graph including a set of conditions [18]. According to the W3C Recommendation, a shapes graph consists of target shapes that determine the resources to be validated and of node shapes and property shapes that specify conditions for either the target node or its properties. As SHACL is the successor of SPIN, it also supports the selection of SPARQL-based targets. The output of the validation process is a specific Validation Report specified in the SHACL Vocabulary.

2.3 Standardized Linked Building Models

Linked building models in a container-based environment have been researched especially by Fuchs et al. [1] as the multi-model approach and by Hoeber et al. [19] as the COINS container. The aim of the multi-model approach has been the cross-domain applicability using original models in a standardized, serialized format with persistent and restorable links and metadata. A multi-model is associated with one or more elementary models, which can be any instances of self-contained original data models, e.g. IFC models or schedules.

Furthermore, the multi-model can be associated with any number of linked models, each of which is represented by a non-empty set of links. These links consist of at least two or more linked elements from different elementary models. Each linked element belongs only to a single elementary model and contains the information to identify the corresponding element from the elementary model. The XML-serialized and ZIP-archived expression of the schema and the inherent original data are referred to as the multi-model container.

However, the COINS container is a more industry-driven development targeting the standardization of a flexible information container for linking a certain amount of building data from different domain models using linked data approaches. Like the multi-model container, the COINS container links models using generic extendable link sets. The metadata of the container and the link sets are serialized in RDF and can be exchanged in its entirety as a zip-archived file. The COINS container is the basis for the international standardization of the Information Container for Data Drop in the ISO 21597 [20].

3 Proposed Method

For the effective implementation of the rule-based validation of linked building models, several issues need to be addressed. Initially, to work with linked building models, the container format must be implemented to process the container into facts for validation. The processing of the container requires a library to deserialize and parse the RDF files from the container to get the required metadata information about the domain models and link sets. The second issue is the conversion of the documents from the original data format into an OWL graph-based format to enable the validation using SHACL. After the conversion, the data need to be merged into a single data graph that can be validated by a shapes graph accordingly. Referring to the related work, these two issues can be summarized as the data preparation step. The third issue is the definition of the rules itself based on SHACL and with the inclusion of SPARQL. The rules have to be derived from natural language rules regarding the logical connection of the linked domain models. Corresponding examples are presented in the evaluation through the use case. The last issue to be solved is the actual validation with a SHACL engine.

According to the issues defined before, Fig. 1 shows the general process of the validation of linked building models in this approach. We assume that a linked building model in an ICDD has been properly created in the first process step. The resulting file is imported into the validation framework that was developed in previous research and presented in [21]. This validation framework is based on the latest version of the open-source library dotnetRDF[1] that provides an RDF parser, a SPARQL query processor, and a SHACL validation engine.

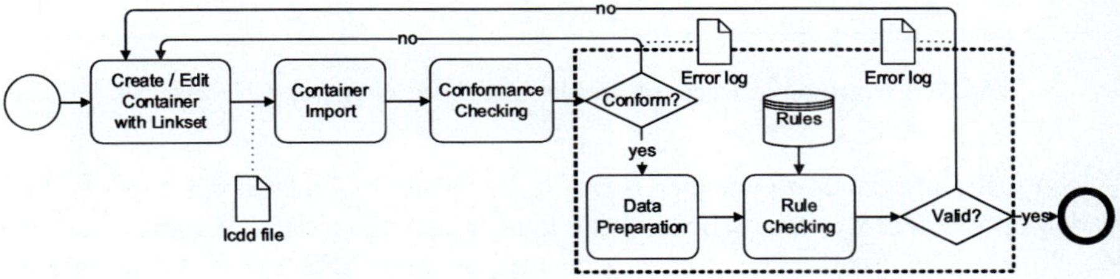

Fig. 1. Process of the validation of linked building models and scope (dotted) of this approach

After the import of the container file, the framework checks the conformance of the container according to the criteria of the underlying ISO standard. The whole process fails if the conformance cannot be checked and the container must be edited to fulfill the conformity until further validation is possible. The focus in this contribution lies on the consequent steps of the data preparation, the rule definition and, finally, the implementation of the rule checking process.

3.1 Data Preparation

Technically, domain models are defined and exchanged according to their underlying schemas. For instance, domain models can be geometry-based building models for design, structural engineering or MEP planning, but also alphanumeric models for cost planning or scheduling. To be able to check the individual models at the entity level, a conversion to a more generic format is necessary. In this approach, we choose RDF and OWL as a common data basis. The data preparation relies on the conversion from the

[1] https://github.com/dotnetrdf/dotnetrdf.

originally defined data schemes into RDF-based data, as depicted in Fig. 2. Therefore, the IFCtoRDF[2] converter is utilized for transferring building models from the IFC EXPRESS schema into the ifcOWL schema.

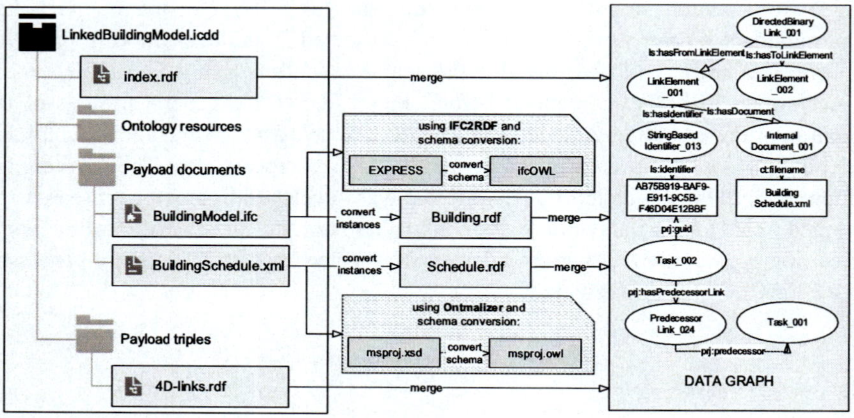

Fig. 2. Converting instances of domain models according to their schemas

Since the information container is not restricted to IFC-related file types, information from other text-based schemas need to be transferred to RDF. Examples are the MS Project-XML schema for project schedules or the GAEB DA XML (a specific format for the bill of quantities in Germany). Based on the respective schema file and the instance file that can be serialized with the schema, an OWL ontology and the corresponding instances can be converted to RDF instances. In this research, the open-source software Ontmalizer[3] is used – a tool for transforming XSD and XML to OWL and RDF data. Ontmalizer relies on an XML schema object model library and the Apache Jena[4] framework. The tool bases on a set of transformation rules that map the types defined in the XSD to respective types in OWL/RDF, for instance, the *xs:complexType* to *owl:Class* or an *xs:element* to either an *owl:DatatypeProperty* or an *owl:ObjectProperty* depending on the type of the element. Both, the IFCtoRDF converter and the Ontmalizer are utilized for the validation framework as compiled jar files. After the data is converted, the generated individual graphs of the models, the parent container graph and the link set graph are merged into a single data graph, which is exemplarily depicted in Fig. 2 indicating an extract from the RDF graph.

[2] https://github.com/pipauwel/IFCtoRDF.

[3] https://github.com/srdc/ontmalizer.

[4] http://jena.apache.org/.

3.2 Rule Definition

Before technical rules can be formalized between the instances of the models, the respective identifiers must first be determined and the structure of the link validated. Therefore, the RDF graph of a directed binary link inside a link set is considered in Fig. 3. The link consists of two linked elements, whereby one is the "from" element and the other one is the "to" element. Both elements have the mandatory relation to a document that is registered in the container and the optional relation to the identifier of the entity inside the respective models.

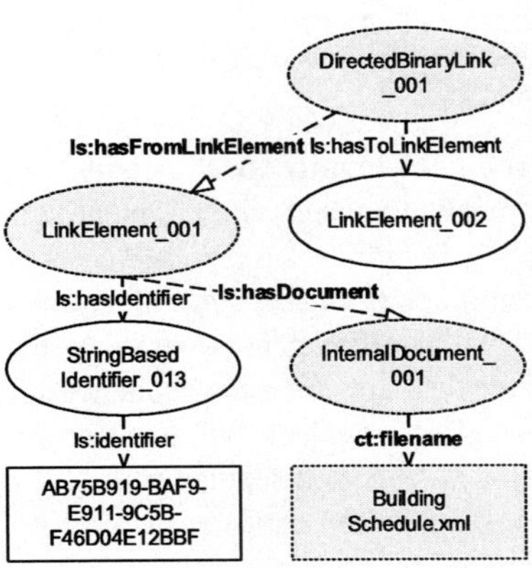


```
icddrule:DirectedBinaryLinkHasFromElementShape
  rdf:type sh:NodeShape ;
  sh:property [
     sh:path ls:hasFromLinkElement ;
     sh:maxCount 1 ;
     sh:minCount 1 ;
     sh:node ls:LinkElement ;
     sh:qualifiedMaxCount 1 ;
     sh:qualifiedMinCount 1 ;
     sh:qualifiedValueShape [
         sh:path ls:hasDocument ;
         sh:qualifiedMaxCount 1 ;
         sh:qualifiedMinCount 1 ;
         sh:qualifiedValueShape [
             sh:path ct:filename ;
             sh:hasValue "BuildingSchedule.xml" ;
           ] ;
       ] ;
   ] ;
 sh:targetClass ls:DirectedBinaryLink ;
.
```

Fig. 3. RDF graph for a directed binary link in the ICDD and corresponding SHACL validation

In order to perform a validation based on the parameters of the entities inside the container, we need to prove whether the link elements have an identifier and whether the identifiers point to the entities of a model. If this is the case, the corresponding entities can be queried from the data graph so that access to the respective parameters is given.

The rule is defined in SHACL, therefore the vocabulary from the SHACL namespace is considered. The core vocabulary of the language has built-in constraint components on nodes like cardinality, value type, value range, and allowed values. Moreover, it is possible to define types of constraints for more complex validation conditions, for example with using SPARQL queries in SHACL shapes. A SHACL shape can be defined as an instance of a *sh:NodeShape*. A *sh:targetClass* can be applied to a sh:NodeShape to determine for which instances from the specified classes in the RDF graph the condition must be fulfilled.

On the right, Fig. 3 shows a rule for the validation which corresponds with the RDF graph on the left. The considered predicates in the rule and in the graph are highlighted with bold font and dotted nodes. This rule validates whether each node of the type *ls: DirectedBinaryLink* is related via an *ls:hasFromLinkElement* predicate to an *ls: LinkElement* node. This node must refer to a document with a defined file name or alternatively the URI of the document node. This concatenated relation over several predicates can be modeled with the help of a *sh:qualifiedValueShape*. This type of shape restricts the path by using the object node of the previous validation as subject node for a subsequent validation. The occurrences of the subsequent path can also be restricted using *sh:qualifiedMinCount* and *sh:qualifiedMinCount*. The identifiers of the link elements can be validated through the path-related validation using the *ls:hasIdentifier* predicate in the same way.

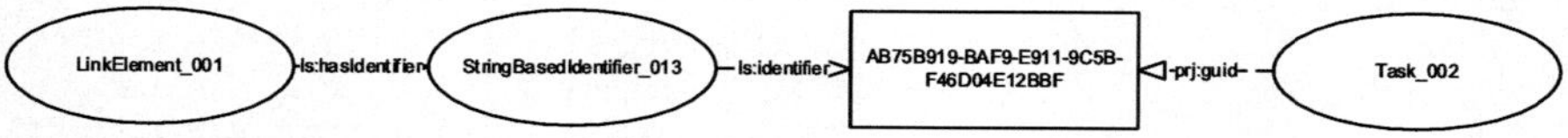

Fig. 4. The path from the identifier of a link to the identifier in the original model

In order to access the original model data for each linked entity for the validation, the respective identifiers from the link element must first be matched with the identifiers in the model data. As shown in Fig. 4, both predicates *ls:identifier* and *prj:guid* point to the same literal node containing the identifier. The modifier *sh:inversePath* is required to adjust the path-based validation accordingly. This modifier can be used in the *sh:path* property of a shape combined with a regular predicate. As a consequence for the validation, the focused node is not anymore treated as a subject, but as an object that must have exactly one incoming triple consisting of inverted regular path and a subject node. According to Fig. 5, this subject node is the linked entity in the original model. The parameters of this entity can be incorporated for semantic validation.

Comprehensive possibilities for validation can be developed on this basis. Considering a linked construction schedule model (4D), the predecessor of a task can be determined by developing a sh:qualifiedValueShape with the vocabulary from the project schedule namespace (*proj*) defined during the conversion of the schedule XML schema. The predicates *proj:hasPredecessorLink* and *proj:predecessor* can be used with the qualified cardinalities to find a preceding task. This shape can be used, for example, to check for a particular validation case whether the time between the end time of the predecessor and the start time of the focused task is less than a determined value.

Besides the core constraint components, the W3C SHACL recommendation offers the possibility of defining SPARQL constraint components to express even richer constraints. These constraint components can be defined once and be reused for several SHACL shapes to ensure the usability of the rules. The components define a constraint for a focus node that is validated based on a *SELECT* query delivering a result set with entries that do not fulfill the constraint.

3.3 Rule Checking

The precondition for the rule check is the existence of both a well-formed data graph and a valid shapes graph. Both are generated in the previous steps of data preparation and rule definition, whereby the well-formedness at the time of implementation is checked only in the validation engine. The validation engine used in this approach is implemented in the dotnetRDF framework (see chapter 3). The framework supports the SHACL core implementation as well as the SHACL-SPARQL implementation and is successfully tested within the W3C SHACL Test Suite[5] as a fully compliant engine.

The validation engine is implemented in a class *IcddRuleValidator* that is instantiated from an existing *IcddContainer* object and the path to the Turtle serialized shapes file. During the creation of the validator, the data graph is generated from the container as described in Sect. 3.1. The validation results are transformed from the single validation report provided by the dotnetRDF framework into an *IcddValidationResult* for each violation of the shapes to be displayed separately for each focus node.

4 Case Study and Evaluation

The validation of linked building models according to the proposed method is performed through a case study presented in the following. Figure 5 presents an exemplary tunnel model in the IFC 4 version and a corresponding construction schedule in an MS Project XML. The tunnel model, the construction schedule and an Information Container for Data Drop comprising these two models were created for demonstration purposes. A link set has been added to the container, which contains a directed binary link for each task, linking the Globally Unique Identifiers (GUID) of the tunnel rings and the tasks. This setup serves as a minimal example and considers only the order in which the tunnel rings are built.

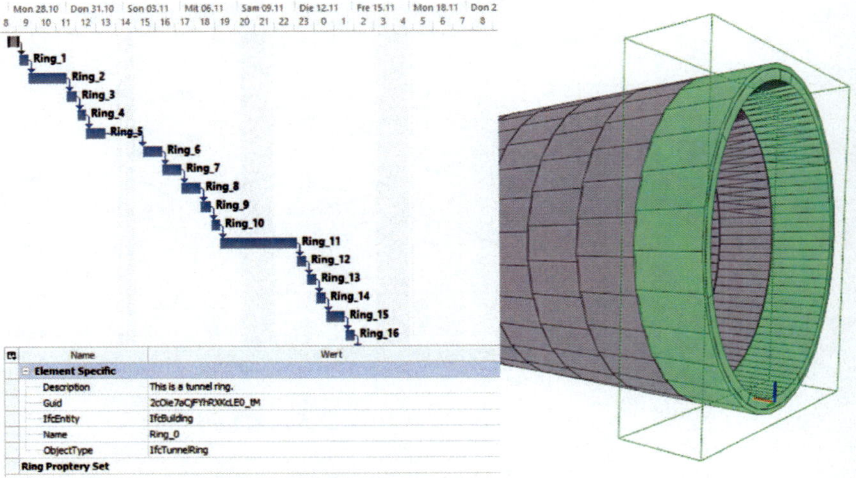

Fig. 5. Use case data: tunnel schedule (left) and tunnel model (right)

5 https://w3c.github.io/data-shapes/data-shapes-test-suite.

It is assumed, that each element in the schedule is assigned at least to one element in the building model, but not every element in the building model must have an assigned element in the schedule. This may be the case, for example, for proxy elements groping a set of elements. Therefore, the usage of the *DirectedBinaryLink* type for the relation between schedule and building model is established. In this case study, the following boundary conditions should be checked in the linked building model:

1. **Conformity level:** For each link element that points to a specific GUID, an associated entity from the two original models must exist in the RDF graph.
2. **Semantic level:** Tunnel rings that are built directly in succession according to the schedule must be adjacent in the building model.

Since the conformity condition has already been defined as a general rule for each linked building model in Sect. 3.2, the focus of the rule design in the next section is on semantic level condition.

4.1 Rule Design

For the rule design, we use an extract from the aggregated RDF data graph shown in Fig. 6 and consider the *DirectedBinaryLink_001* as the focus node. The rule is developed as a SPARQL-based constraint and targets all directed binary links in the link set. In the *SELECT* query of the constraint, the focus node can be used with the pre-bound variable *$this* when integrating the query into the shape.

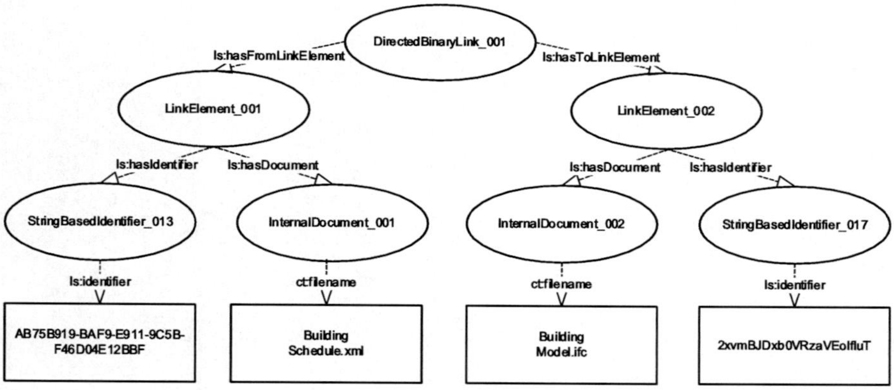

Fig. 6. Extract of a link instance from the data graph

First, the rule defined in natural language must be transferred and formalized. Therefore, all entities required for the examination of the rule must be consulted, which are the two preceding tasks from the schedule and the two adjacently expected building elements from the building model.

Beginning with the access of the tasks, we provide a query via the path from the *LinkElement_001* querying the respective string-based identifier as a literal node. The associated task can be found inverting the query and considering the path *prj:guid* pointing to the same identifier as the *StringBasedIdentifier_013* (see Fig. 7). We get the individual of the task and can retrieve the corresponding predecessor for it, by querying the predecessor link as depicted in Fig. 7. Using the predecessor task (*Task_001*) and its GUID, the corresponding link (*DirectedBinaryLink_002*) in the link set can be determined. We have, then, determined both task individuals as well as the link elements and links that refer to these tasks.

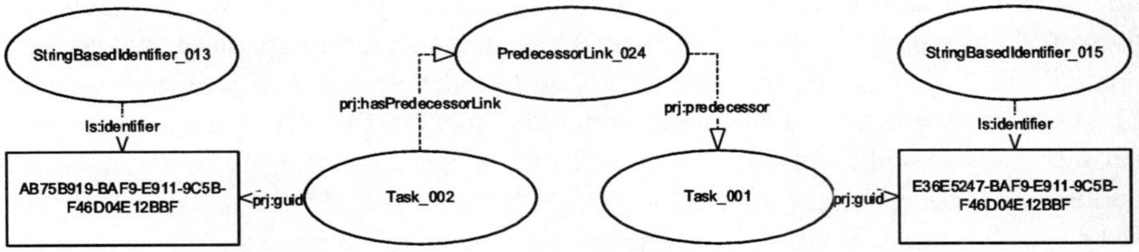

Fig. 7. Extract of the relation between two tasks in the data graph

The directed binary links are instrumentalized for the identification of the associated building elements in the building model. The linked building entity can be determined from the link by querying the path of the *hasToLinkElement* predicate. Following the path of the identifier, the respective global identifier from the IFC schema lead to the two individuals of the building elements *IfcBuildingElement Proxy_2447* and *IfcBuildingElementProxy_2475*. Using the hierarchy from the IFC, the Cartesian coordinates of the two considered tunnel rings can be evaluated using the predicates *ifcowl:objectPlacement_IfcProduct*, *ifcowl:relativePlacement_IfcLocal*

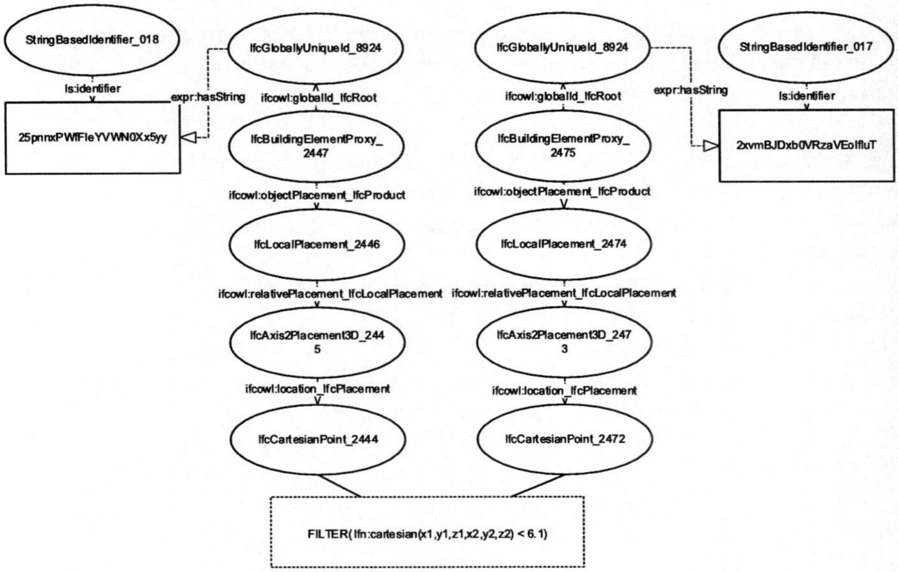

Fig. 8. Extract of the relation between links and building elements

Placement, *ifcowl:location_IfcPlacement* and *ifcowl:coordinates_IfcCartesianPoint* (see Fig. 8). In a last step, the x, y and z coordinates of each element need to be queried.

With the two Cartesian points and the access to the respective coordinates, the distance between these can be determined in a SPARQL expression. Unfortunately, neither the mathematical square root function nor the Cartesian distance is defined in the vocabulary of SPARQL. For this reason, an external vocabulary providing this function must be included. To extend the basic query vocabulary, dotnetRDF has implemented a so-called Leviathan Function Library[6], which contains a variety of numerical extension functions for SPARQL and can be used in queries executed with dotnetRDF. Furthermore, the framework allows developers to implement custom SPARQL expressions for the integration into the query engine. The Cartesian distance between the two tunnel rings in a 3D coordinate system is determined by means of the numerical function *lfn:cartesian(x1, y1, z1, x2, y2, z2)* utilized from this library. In the last step, the distance needs to be compared to the fixed maximum value of the average width of a tunnel ring in the model (here 6.1 m).

Task	LinkedBuildingElement1	hasPredecessor	LinkedBuildingElement2	distanceBetween
◆ prj:Task_002	◆ inst:IfcBuilding_2437	◆ prj:Task_001	◆ inst:IfcBuilding_2369	5.99999999999999e0

Fig. 9. Evaluation of the SPARQL query on the data graph for a specified link

The query defined in SPARQL is tested for correctness before defining the rule on the aggregated dataset and returns a result set including the distance of two consecutive elements as shown in Fig. 9. For the integration into the SHACL constraint, the query is modified using a *FILTER* expression to return only results if the distance between tunnel rings is larger than 6.1 m. This means that the result set will be empty for all positive results and will pass the positive result to the SHACL constraint. The query is nested in the shape defined with an anonymous instance of *sh:sparql* constraint and put into the constraint with the *sh:select* property (see List. 1).

```
@prefix rule: <http://hive.vm.rub.de/semweb/icdd/rules#> .
@prefix sh: <http://www.w3.org/ns/shacl#>.
rule:DirectedBinaryLinkDistanceBetweenPredecessors
  a sh:NodeShape;
  sh:targetClass ls:DirectedBinaryLink;
  sh:sparql [
     a sh:SPARQLConstraint;
     sh:message “ The examined BuildingElements cannot be adjacent”;
     sh:select “““ #SELECT QUERY INSERTED HERE “““;
  ] .
```

Listing 1: SHACL shape

[6] https://github.com/dotnetrdf/dotnetrdf/wiki/DeveloperGuide-SPARQL-Leviathan-Functions.

4.2 Rule Checking and Results

The previously defined rule shape is transferred into the rule engine together with the aggregated data graph. For the validation to take effect, the first link in the link set was connected to a tunnel ring that is not adjacent to the tunnel ring of the predecessor. The result in Listing 2 shows, that the respective link failed the validation as it caused a violation of the defined shape. The validation report with the predicate *sh:conforms* (line 2) states that during the validation violations of the shapes have occurred. These violations are displayed in *sh:result* specifically for each checked node and described according to the defined rule (line 3–10).

```
1 _:autos1 a sh:ValidationReport;
2 sh:conforms false;
3 sh:result [a sh:ValidationResult;
4  sh:focusNode links:DirectedBinaryLink_001;
5  sh:value links:DirectedBinaryLink_001;
6  sh:resultMessage "The examined BuildingElements cannot be adjacent";
7  sh:sourceConstraintComponent sh:SPARQLConstraintComponent;
8  sh:sourceShape rules:DirectedBinaryLinkDistanceBetweenPredecessors;
9  sh:resultSeverity sh:Violation;
10  sh:sourceConstraint []].
```

Listing 2: SHACL validation result set

5 Discussion

The case study presented in this paper has confirmed, that available technologies make it feasible to query and validate linked building models. Although the objective of this research was achieved, the simplified case study has some limitations, as it covers only an exemplary technical rule for the validation of linked building models.

Due to the highly complex relationship between the connected models, the reusability of rules cannot be assumed, and a user-supported adaptation for each application is required. Therefore, a fully automated checking of linked building models for the correctness of the links is not possible at this state of research so that there is only a human-assisted semi-automated checking realized.

This research has disclosed that the use of Semantic Web technologies, especially RDF, SPARQL, and SHACL, provide an advantage over hard-coded rule checking, as rules can be exchanged easily and can be included directly within the information container. The combination of SPARQL and SHACL offer a wide range of possibilities for specifying rules for validation. However, this can be problematic when formulating rules manually, as there is a high potential for errors. Furthermore, the exact knowledge of the RDF graphs of the respective data models is required to formulate appropriate rules.

6 Conclusion and Outlook

Although linked building models are widely used in practical applications, especially for 4D or 5D use cases, the creation and the exchange of those models, e.g. in an information container, is hardly considered either by theory or praxis. This research focused on an easier creation and exchange of linked building models providing a method to validate the correctness of the defined links. As the related work shows, there is a significant number of research approaches regarding rule checking for the construction industry and especially approaches based on building models. This focus should be extended to include rule checking for linked models and thus, for all information within the planning process.

This paper demonstrates the feasibility of rule-based validations in the information container by processing all contained models into RDF and validate the data in a rule engine according to SHACL shapes. This paper demonstrates how natural language rules can be formalized as SHACL shapes for link sets based on the parameters of linked entities of the original models.

Generally, a benchmark with real-world models and different use cases should be carried out to make a dedicated statement about the usability of the method. With regard to the conversion of data into an RDF graph, a long process time can occur with large models and models with many documents. Therefore, it is suitable to implement simplifications for further development, e.g. to consider existing approaches to streamline IFC models in ifcOWL or to remove unnecessary information like the geometry representation from the models.

However, the proposed method shows that the Semantic Web is a valuable basis for performing comprehensive validation. Nevertheless, this approach is not yet practicable for use in construction practice, since the rule definition requires knowledge of Semantic Web technologies as well as the respectively used data schemas to convert natural language rules into formalized rules related to the information container. In order to bridge these knowledge gaps, an extension of the vocabulary with ICDD-specific predefined constructs and simplifications, such as the BIMSPARQL extension, will be considered. The aim for further research is the increase of the usability of the ICDD for practical applications. A further important point is the definition of a simple language for the representation of rules in natural language that can be translated into a conform SHACL-SPARQL constraint component.

References

1. Fuchs, S., Scherer, R.J.: Multimodels — instant nD-modeling using original data. Autom. Constr. **75**, 22–32 (2017). https://doi.org/10.1016/j.autcon.2016.11.013
2. Werbrouck, J., Pauwels, P., Beetz, J., van Berlo, L.: Towards a decentralised common data environment using linked building data and the solid ecosystem. In: Proceedings of the 36th CIB W78 2019 Conference, Northumbria University, Newcastle (2019)
3. Hitzler, P., Krötzsch, M., Rudolph, S.: Foundations of Semantic Web Technologies. CRC Press/Chapman/Hall/CRC textbooks in computing, Boca Raton/London/New York (2010)

4. Rattanasawad, T., Saikaew, K.R., Buranarach, M., Supnithi, T.: A review and comparison of rule languages and rule-based inference engines for the semantic web. In: Proceedings of the 2013 International Computer Science and Engineering Conference (ICSEC), pp. 1–6 (2013). https://doi.org/10.1109/icsec.2013.6694743
5. Eastman, C., Lee, J.-m., Jeong, Y.-s., Lee, J.-K.: Automatic rule-based checking of building designs. Autom. Constr. **18**(8), 1011–1033 (2009). https://doi.org/10.1016/j.autcon.2009.07.002
6. Schwabe, K., Teizer, J., König, M.: Applying rule-based model-checking to construction site layout planning tasks. Autom. Constr. **97**, 205–219 (2019). https://doi.org/10.1016/j.autcon.2018.10.012
7. Pauwels, P., de Farias, T.M., Zhang, C., Roxin, A., Beetz, J., de Roo, J., Nicolle, C.: A performance benchmark over semantic rule checking approaches in construction industry. In: Advanced Engineering Informatics, vol. 33, pp. 68–88 (2017). https://doi.org/10.1016/j.aei.2017.05.001
8. De Meester, B., Heyvaert, P., Arndt, D., Dimou, A., Verborgh, R.: RDF graph validation using rule-based reasoning. Semant. Web **1**, 1–5 (2020)
9. Solihin, W., Eastman, C.: Classification of rules for automated BIM rule checking development. Autom. Constr. **53**, 69–82 (2015). https://doi.org/10.1016/j.autcon.2015.03.003
10. Cabot, J., Gogolla, M.: Object Constraint Language (OCL): a definitive guide. LNCS, vol. 7320 (2012). https://doi.org/10.1007/978-3-642-30982-3_3
11. Kraft, B.: Using object constraint language for the validation of BIM platform object graphs. In: Proceedings of the Lake Constance 5D-Conference 2013, vol. 220, pp. 90–98 (2013)
12. Lee, J.-K., Eastman, C.M., Lee, Y.C.: Implementation of a BIM domain-specific language for the building environment rule and analysis. J. Intell. Robot. Syst. **79**(3–4), 507–522 (2015). https://doi.org/10.1007/s10846-014-0117-7
13. Dimyadi, J., Solihin, W., Eastman, C., Amor, R.: Integrating the BIM rule language into compliant design audit processes. In: Proceedings of the 33rd CIB W78 Conference 2016, Brisbane (2016)
14. Cyganiak, R., Wood, D., Lanthaler, M.: RDF 1.1 Concepts and abstract syntax, W3C Recommendation 25 February 2014
15. Horrocks, I., Patel-Schneider, P.F., Boley, H., Tabet, S., Grosof, B., Dean, M.: SWRL: a semantic web rule language combining OWL and RuleML. W3C Memb. Submiss. **21**(79), 1–31 (2004)
16. Knublauch, H., Hendler, J.A., Idehen, K.: SPIN - overview and motivation. W3C Member Submission 22 February 2011 (2011)
17. Zhang, H., Zhao, W., Gu, J., Liu, H., Gu, M.: Semantic web based rule checking of real-world scale BIM models: a pragmatic method. In: Proceedings of the International Congress and Conferences on Computational Design and Engineering (i3CDE'19), Penang, Malaysia, pp. 130–137 (2019)
18. Knublauch, H., Kontokostas, D.: Shapes Constraint Language (SHACL). W3C Recommendation 20 July 2017 (2017)
19. Hoeber, H., Alsem, D.: Life-cycle information management using open-standard BIM. Eng. Constr. Archit. Manag. **23**(6), 696–708 (2016). https://doi.org/10.1108/ECAM-01-2016-0023
20. ISO 21597-1, 2018: Information Container for Data Drop - Exchange specification - Part 1: Container (ISO/DIS 21597-1:2018), German and English version prEN ISO 21597-1 (2018)
21. Hagedorn, P.: Implementation of a validation framework for the information container for data drop. In: Proceedings 30, Forum Bauinformatik, At Bauhaus Universität Weimar (2018)

Analysis of Urban Legislation of Engineering Projects Using Building Information Modeling (BIM) with the Aid of Graphic Programming

Victor Farias, Bruna Roque, Ingryd Tavares, and Davi Pinheiro(✉)

Instituto Federal de Educação Ciência e Tecnologia do Ceará,
Fortaleza, Ceará, Brazil
victorigor.f@hotmail.com, brunavitalr@gmail.com,
{ingryd,davi}@ifce.edu.br

Abstract. The analysis of urban viability consists of a study about the potential of implementation of a construction being carried out in the preliminary design phase and with the objective of ensuring the best use of the land. The analysis is based on the selection of the possible land and the review of the urbanistic laws that determine the legal parameters for the study of the implantation and the volume of the enterprise on the selected land. Therefore, it aims to use Building Information Modeling (BIM) as a tool to automate the process of feasibility analysis of legal parameters, such as the verification of land use and occupation law. Thereby, a volumetric model of the building in BIM will be built, followed by the modeling of the chosen land and finally, the normative and legal parameters will be assigned through the graphical programming tool, using Dynamo, a programming software to automate work on Revit. The project will be deployed to each selected land with its respective parameters. Lastly the result of each analysis will be compared in order to obtain the best benefit in relation to land and enterprise. greater agility and accuracy in the technical feasibility analysis of the project is expected to be achieved by reducing rework, inaccuracy, errors and time. Therewith greater agility and accuracy in the technical feasibility analysis of the project is expected to be achieved by reducing rework, inaccuracy, errors and time.

Keywords: Feasibility study · Urban legislation · BIM · Dynamo

1 Introduction

BIM (Building Information Modeling) is configured as a modeling technology and a group of processes that allow an integrated practice of production and analysis of the building model (Eastman 2008). The softwares that operate through BIM technology use parametric models of the building elements in order to allow the development of dynamic changes in the graphic representation of the building (Coelho and Novaes 2011).

One of the preliminary studies required in the execution of civil works is the feasibility of implementing the construction, according to its characteristics. It is necessary to evaluate the premises exposed in the law of use and occupation of the land

E. Toledo Santos and S. Scheer (Eds.): ICCCBE 2020, LNCE 98, pp. 788–800, 2021.
https://doi.org/10.1007/978-3-030-51295-8_54

regarding the attributes of the construction to be carried out, analyzing if the legal parameters allow the work to be carried out in the chosen location, according to the height, recessions, utilization rate, land occupation rate, among others.

The objective of this study is to propose a method to evaluate the urban suitability of civil constructions in terms of the land guidelines through BIM technology and using the Dynamo tool, in order to make this phase of the project more agile and efficient.

2 Urban Suitability

As exposed by SEDUH (2020), the Land Use and Occupation Law is classified as an ordering standard in which are defined the uses and indexes that guide constructions, in order to obtain the urban plan control. It fits as a tool for regularization of buildings, as it defines the uses and measures of land occupation according to the capacity of urban infrastructure present or planned for a specific urban area.

Urban viability is characterized by the execution of a study in a specific location, with the objective of analyzing the performance determined as passive or not, interpreted by law, of development, according to the impacts on the road, housing and productive system, as well as changes reduced in construction.

Therefore, this study is carried out by collecting data from the region, evaluating the adequacy of the project, keeping its characteristics harmonious, controlling the real estate construction and maintaining the designations of the master plan according to the region, whether residential, wrong, industrial, commercial, rural, among others.

Every master plan referring to a municipality must be outlined by its development, controlling what and how it should be built, as well as the feasibility study and the impact of each enterprise that will be built in the city, in order to obtain security and follow the specific criteria for each enterprise.

The standardization of buildings aims to establish detailed parameters on aspects of buildings in different areas, including both the relationship of the building with its surroundings (recessions, number of floors, maximum height) and its interior configuration (sunshine, ventilation, dimension of rooms).

According to Law No. 7987 of December 23, 1996, the Municipality's division into Density Microzones and Special Zones, has the function of regulating the use and occupation of the soil taking into account the characteristics of the areas in question, taking into account main objectives:

- Preservation and protection of the natural and cultural environment;
- Ordering the city's functions through the rational use of the territory, natural resources and land parceling;
- Rationalization of the use of installed infrastructure, including road and transport systems, avoiding overload or idleness;
- Compatibility of the density of urban activities with natural conditions;
- Intensification of the land occupation process, as the capacity of the infrastructure expands, preserving the quality of life of the community;
- Compatibility of land use with the function of the road ensuring safety, fluidity, circulation, comfort and physical-operational restrictions;

- Assistance to the social function of urban real estate, as advocated in the Federal Constitution;
- Incentive for areas with concentration and with a tendency to concentrate activities, enabling the development of alternative nuclei to the existing ones, through the application of urban and fiscal instruments.
- The law No. 7987 of December 23, 1996 also points out in its definitions the maximum height of the building as the vertical distance taken in the middle of the facade, and the highest point of the roof, including auxiliary buildings, located above the roof of the last floor (water tank, houses machines, stair halls) and the elements of composition of the referred facade, considering the following factors:
- In relation to the distance between the buildings and the alignment with the public street, the height will be counted from the elevation level of the sidewalk, in the plane of the facade, coinciding with the center of the same;
- In relation to the spacing of the buildings, as for the lateral and bottom boundaries, the height will be counted from the elevation level of the land that coincides with the center of the corresponding facade.

Another important definition presented by the land occupation law of the municipality of Fortaleza, is the occupancy rate, which can be defined as the percentage of land that can be occupied by the projection of the building. The purpose of this indicator is to limit excessive urban growth. The formula for calculating the occupancy rate is as follows: (total constructed area of the first floor + surplus area of the other floors)/total area of the lot.

When calculating the occupancy rate, the number of floors of the work should not be taken into account, but only the horizontal plane. This is because this rate is used strictly in cases where it is intended to know the total sum of the built area of the first floor, regardless of the height of the work or the number of floors.

In comparison to the occupancy rate, another relevant indicator that is also defined in the law referred to above, is the utilization rate, which represents the relationship between the built area and the land area. Corresponds to the total area that can be built at the limit of that land according to parameters, such as the setback for example.

Unlike the occupancy rate, calculating the utilization rate includes the floors. Balconies, eaves and open areas, such as the pool and garden, are not usually counted in the utilization coefficient.

The utilization coefficient is a number that multiplied by the area of the land indicates the maximum number of square meters that can be built, adding up the areas of all floors. Just divide the total built up area (including all floors) by the land area or lot.

The constructed area of the pavement is the floor construction area of the pavement, including those occupied by walls and pillars, including common areas and excluding voids from ventilation and lighting shafts.

Below are some definitions presented in the law No. 7987 of December 23, 1996 of areas referring to the project for a better understanding of this work:

- Covered area - is the measurement of the projection surface, in a horizontal plane, of any building's covering, including surfaces of the projections of walls, pillars, marquees, eaves and other components of the facades;

- Constructed floor area - is the floor construction area of the floor, including those occupied by walls and pillars, including common areas and excluding voids from ventilation and lighting shafts;
- Total constructed area: is the sum of the floor areas of all the main buildings and buildings, including those occupied by common areas;
- Occupied area - it is the lot's surface occupied by the projection of the building in a horizontal plane, not including components for the facades for the calculation of this area, such as: "brise-soleil", planters, marquees, pergolas and eaves;
- Partial building area: is the sum of the partial areas of all floors in a building;
- Partial area of the pavement: it is the built area of the pavement, including those occupied by walls and pillars, excluding the common areas, the voids of ventilation and lighting wells and planters and balconies up to 0.90 m (ninety centimeters) wide;
- Setback area - is the area of non-buildable land, comprised between the boundaries of the land and the alignments of the setbacks;
- Total building area - is the sum of the floor areas of all floors in a building;
- Common use area - is the built-up area or not, which is intended for the common use of the owners or occupants of a land or building, consisting of autonomous units;
- Useful area - is the usable surface of the built area of a part or a building, excluding the parts corresponding to the walls, pillars, planters and balconies up to 0.90 m (ninety centimeters) wide;
- Green area - is the percentage of the area object of installment destined exclusively to squares, parks, gardens for the benefit of the population;
- Area for institutional use - is the percentage of the area object of installment destined exclusively for the implantation of community equipment for the benefit of the population;
- Public areas - these are allotment areas for circulation, for the implantation of urban and community equipment as well as free spaces for public use;
- Area "NON AEDIFICANDI" - is the area located along running and dormant waters, along the tracks of railways, highways and pipelines as well as along urban equipment, defined in federal, state or municipal laws where no building is allowed.

3 BIM Technology

Building Information Modeling (BIM) is a technology that allows the development of a virtual building prototype with three-dimensional visualization. Parametric modeling is used, so that all stages of the project are included since its conception, generating feasibility and monitoring reports (Merlo 2015).

The BIM systems are based on a database made up of attributes and information from the design models, and, as their elements are parametric, it is possible to modify them and obtain instant updates in the construction process, allowing and stimulating experiments and revisions, increases productivity, and streamlines the process (Florio 2007).

According to Tavares Junior (2001), these characteristics of the BIM model allow a systemic view of the projects, that is, an analysis of the reality in variable applications. In this way, it becomes possible to understand the very interdependent parts, being able to evaluate the results and make assertive decisions regarding the needs.

Thus, this technology, now commonly known as Construction Information Modeling or Parametric Model of Virtual Construction, is registered not only as a model of responses from the design space, but also as a digital model composed of data that aggregates all phases and components of the project (TSE and Wong 2005).

Among the BIM tools, there are Dynamo, a visual programming language, defined by Autodesk (2020), as an autonomous environment that allows you to create visual logic in order to explore parametric and automated concepts as activities. In this way, workflows are designed that direct the geometry and behavior of the design models, allowing to improve several processes in the model. These flows are created from connections between "virtual nodes", which have functions and input and output fields, making it possible to connect with each other.

4 BIM Model in Urban Feasibility Study

4.1 Initial Considerations

To exemplify the proposed studies, we used hypothetical data to show the proposed algorithm and present results corroborating its efficiency. The parameters were also simplified in favor of the study. Only the occupancy rate, the index of operation and the recessions were used, these being the most expressive requirements as limiting an enterprise. In recessions, values of frontal and back recessions were considered equal, in order to facilitate the calculation.

Two terrains were used, one with dimensions of 90 m × 60 m and the other with dimensions of 40 m × 35 m, on top of these terrains, a 3D conceptual model with their respective floors was modeled using a visual programming code in Dynamo. Through a control panel in which it was possible to modify three characteristics of a building, 2 models were generated for each terrain, and the result of each model tested was presented (Fig. 1).

The image above shows a macro view of all the visual programming used to generate the isometric models of the construction. In summary what this schedule does is:

1. To read the dimensions of the terrain through the "room" function in Revit, which is capable of delimiting areas between walls, for this purpose such elements were modeled representing the limits of the terrain proposed in each case.
2. With the limits of the legislative parameters inserted in the programming, and with the dimensions of the land extracted, these values are then treated mathematically in order to generate the maximum possible area of occupation and the area built by pavements.
3. These values obtained from mathematical calculations are then reproduced in the project graphically with the creation of a translucent conceptual mass, and the representation of each floor within it.

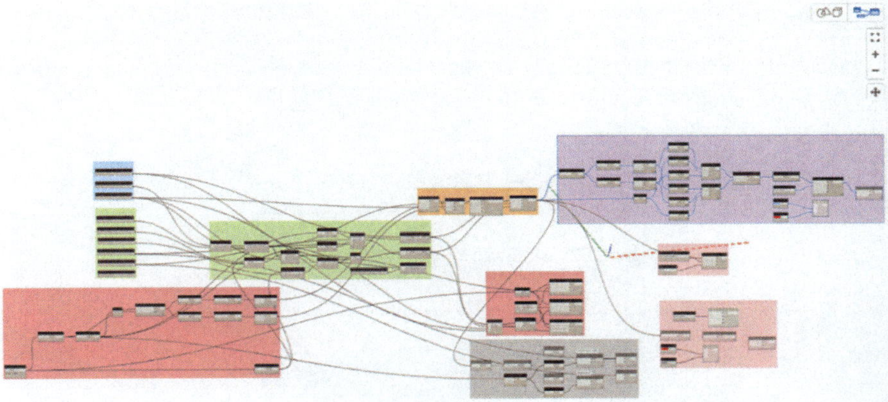

Fig. 1. Macro view of the code in Dynamo.

4. Finally, these mathematical values are also entered in parameters created within the "room" element, so that they are shown in a software schedule and thus the results can be compared to each model created.

In each of the two terrains, different hypothetical parameters were simulated, in order to represent different zonings of a city. And in each terrain, two models were generated, modifying only two out of three adjustable command parameters, namely: width, height and ceiling height.

The program was proposed to generate the building at the limit of the legislative parameters, therefore, when changing the proposed width of a model, the length automatically adjusts to meet the maximum use of the land, without having to have it as an alterable parameter. The recessions limits are always respected, so that regardless of the proposed value for the model, dimensions larger than allowed will never be generated.

The height directly influences the area per floor of the building, so that, at higher heights, there was less area per floor, since the total allowed area would be divided for each floor. This parameter also respects the proposed limits, generating no greater heights than the imposed legislative limit.

The ceiling height influences the number of floors, because with higher heights, the building would reach its limit and result in a smaller number of floors. However, although possible, this parameter has not been modified in the models presented, settled a value of 3 m as a fixed height, a value that is common practice in Brazilian territory.

5 Experiment Arrangement

It will be divided into case 1, model 1.1 and 1.2; and case 2, model 2.1 and 2.2.

5.1 Case 1

It was used a hypothetical terrain of 90 m width and 60 m length, the proposed legislative parameters are shown in the image below (Fig. 2).

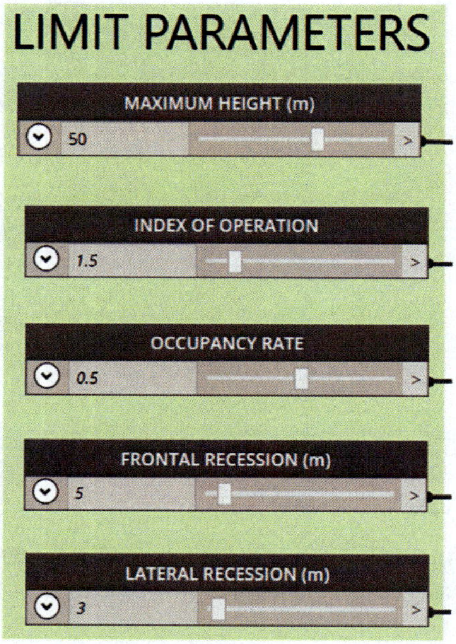

Fig. 2. Limit parameters for case 1.

With that, the following two models were generated (Figs. 3, 4, 5 and 6):

Model 1.1

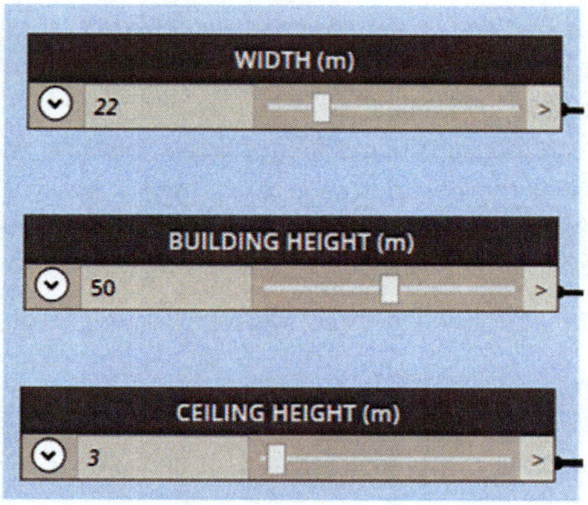

Fig. 3. Commands for model 1.1.

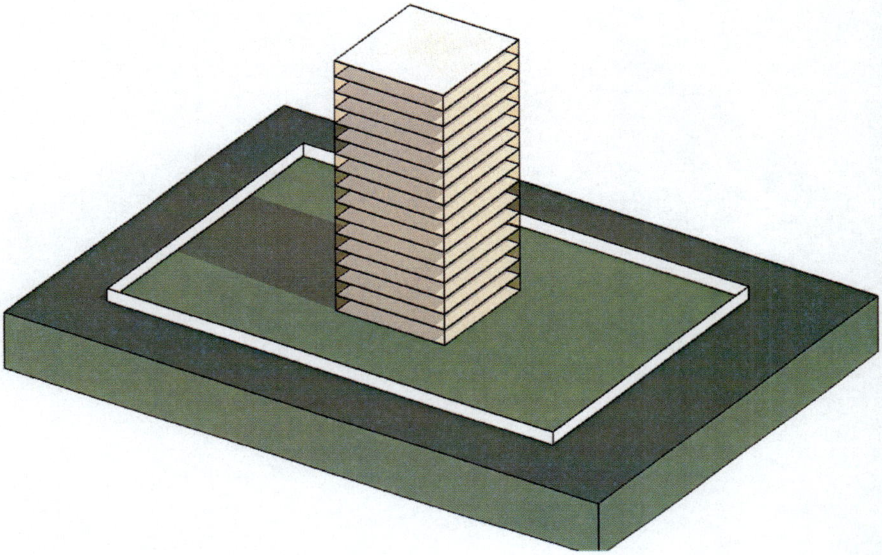

Fig. 4. Graphical model 1.1.

Model 1.2.

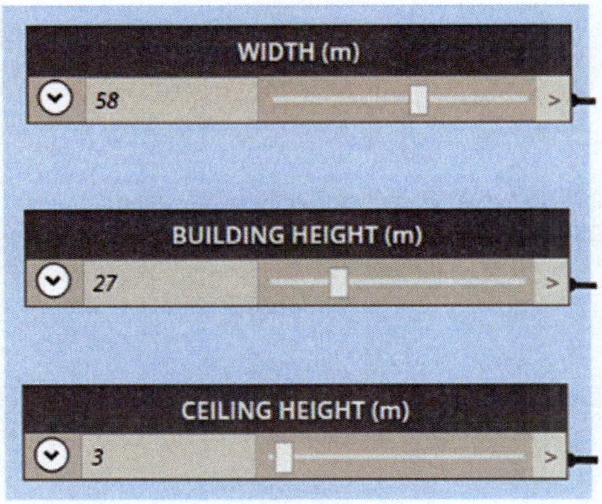

Fig. 5. Commands **for model 1.2**

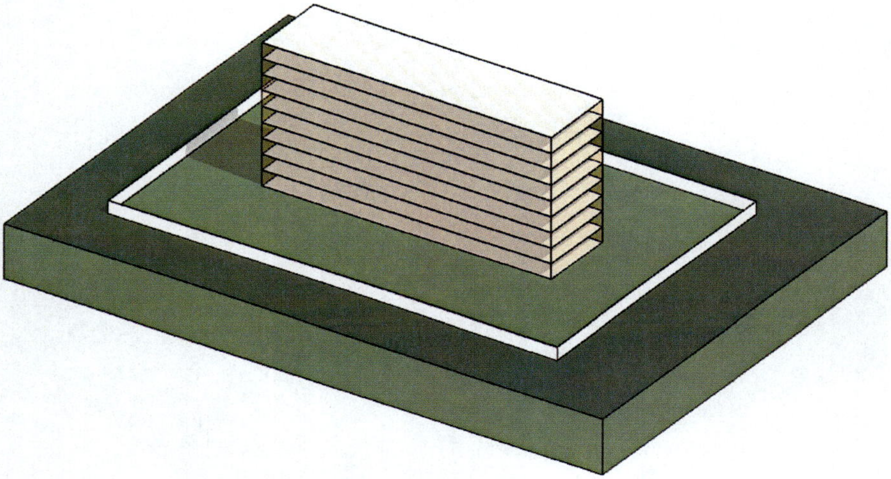

Fig. 6. Graphical model 1.2

5.2 Case 2

In the case of the second land, one 40 m wide by 35 m long was used, generating an area smaller than that of the previous case, on the other hand, the legal parameters were milder, enabling a larger construction within the land proposed as presented in the following models (Figs. 7, 8, 9 and 10):

Model 2.1.

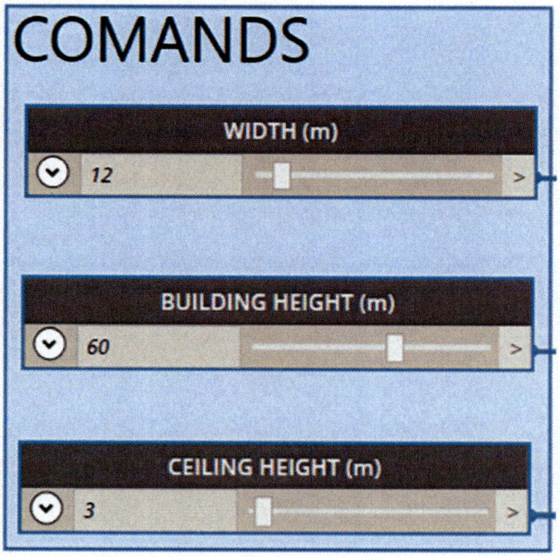

Fig. 7. Commands for model 2.1.

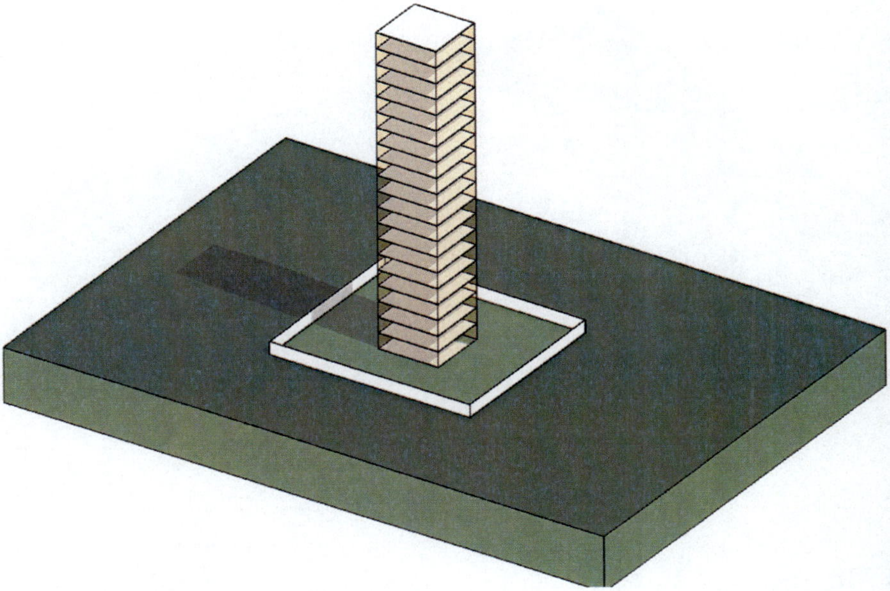

Fig. 8. Graphical model 2.1

Model 2.2.

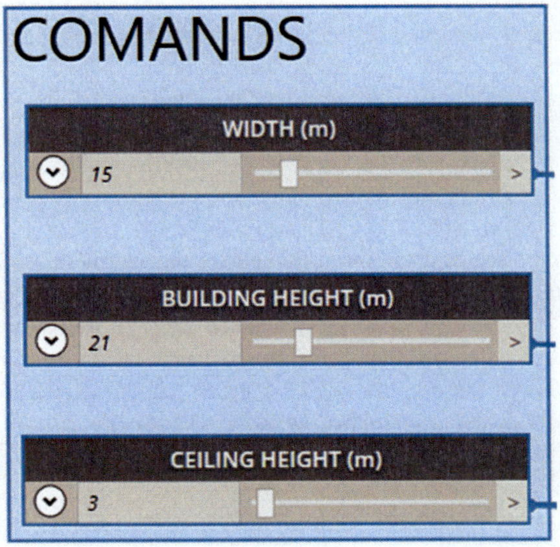

Fig. 9. Commands for model 2.2

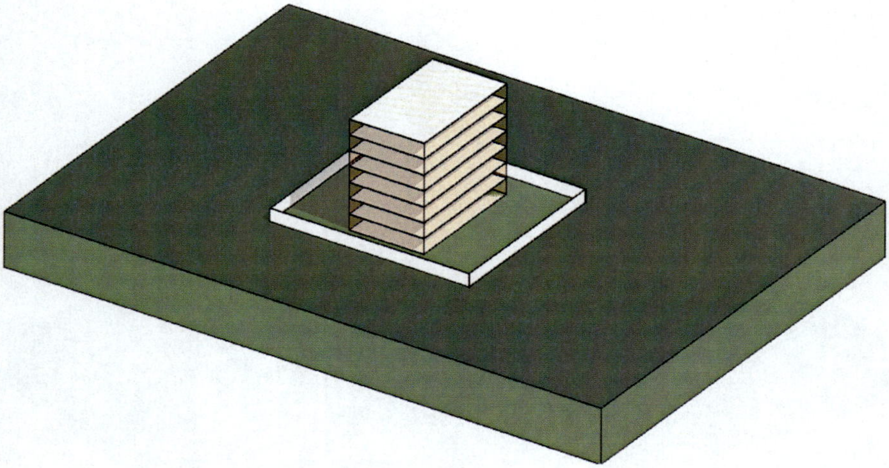

Fig. 10. Graphical model 2.2

6 Results

Among the different types of buildings, the residential model was chosen, resulting in a multifamily building. In an enterprise of this type, the thing that matters to business is the number of apartments, so, as part of the results, a parameters to evaluate the number of apartments per floor has been incorporated, the calculation has the only intention of show the code possibilities and a small perspective of the information that can be analyzed and extracted to compare between different enterprise formats.

In the four models, the need for an 80 m^2 apartment was simulated, and we considered an estimate of 25% of the common area on each floor, with that, the 75% remaining area was then divided by the size of the apartment chosen, and its value was rounded down to find an integer, all the calculus was done within a in Revit schedule, being necessary to insert only the area of the apartment desired, inserted by the operator, and the result of the paving area, as well as the number of pavements, parameters entered in the schedule by Dynamo code.

In the models of the first terrain, although the legislative parameters were more limited in relation to the second one, it was still possible to obtain a great amount of apartments, since the dimensions of the terrain are larger. It is also possible to observe that in both cases, a taller building, generate no better results in the number of total apartments, because as the number of floors grows, the total area allowed by law is divided more times, generating smaller areas for each floor.

6.1 Models Results

(Tables 1, 2, 3, and 4).

Table 1. Model 1.1.

Name	Area	Total area built	Area per floor
Terrain 1	5400 m^2	8100 m^2	506,25 m^2
Numbers of floors	Apartment area	Apartment per floor	Total of apartments
16	80 m^2	4	64

Table 2. Model 1.2.

Name	Area	Total area built	Area per floor
Terrain 1	5400 m^2	8100 m^2	900 m^2
Numbers of floors	Apartment area	Apartment per floor	Total of apartments
9	80 m^2	8	72

Table 3. Model 2.2

Name	Area	Total area built	Area per floor
Terrain 2	1400 m^2	2800 m^2	140 m^2
Numbers of floors	Apartment area	Apartment per floor	Total of apartments
20	80 m^2	1	20

Table 4. Model 2.2.

Name	Area	Total area built	Area per floor
Terrain 2	1400 m^2	2625 m^2	400 m^2
Numbers of floors	Apartment area	Apartment per floor	Total of apartments
7	80 m^2	3	21

7 Conclusions

The results should be seen only as an example of the possibility of improvement that automation using Dynamo can add to projects made in Revit. Tests like the previous models can be generated in a matter of seconds and several possibilities can be tested. The number of apartments shown in the results tables has no scientific corroboration regarding the calculation used, since this was not the focus of the paper, but a deeper study, incorporating methods used by architects, could be implemented, making the code more usable commercially.

The need for more assessment parameters can also be incorporated so that better decisions can be done during feasibility analysis. Some possible needs that could have been attended would be financial analysis, adding the cost per square meter and a sale value; building weight per square meter, being influenced by height and an average weight of materials that would be used, serving to analyze structural methods or solutions for foundations.

References

Autodesk: Computational BIM design software (2020). https://www.autodesk.com/products/dynamo-studio/overview. Accessed Jan 2020

Coelho, S., Novaes, C.: Modelagem de Informações para Construção (BIM) e ambientes colaborativos para gestão de projetos na construção civil. Grupo de Pesquisa Aplicada em Construção Civil (2011). http://www2.pelotas.ifsul.edu.br/gpacc/BIM/referencias/COELHO_2008.pdf. Accessed Jan 2020

Eastman, C., Teicholz, P., Sacks, R., Liston, K.: BIM Handbook: A Guide to Building Information Modeling for Owners, Managers, Designers, Engineers, and Contractors. Wiley, Hoboken (2008)

Florio, W.: Contribuições do Building Information Modeling no processo de projeto em arquitetura, In: SEMINÁRIO TIC 2007 – TECNOLOGIA DA INFORMAÇÃO E COMUNICAÇÃO NA CONSTRUÇÃO CIVIL, 2007, Porto Alegre. Anais, Porto Alegre, TIC 2007 (2007)

Merlo, T.: Gestão De Projetos Modulares Com O Auxílio Da Tecnologia Bim. Revista Científica Tecnológica, vol. 2, pp. 302–321 (2015)

SEDUH: Secretaria De Estado De Desenvolvimento Urbano E Habitação. LUOS (2020). http://www.seduh.df.gov.br/luos-2/#at. Accessed Jan 2020

Tavares Junior, W.: Desenvolvimento de um modelo para compatibilização das interfaces entre especialidades do projeto de edificações em empresas construtoras de pequeno porte. Florianópolis, 2001. 124 p. Dissertação (Mestrado) – Faculdade de Engenharia de Produção da Universidade Federal de Santa Catarina (2001)

Tse, T.C.K., Wong, K.D.A., Wong, K.W.F.: The utilization of building information models in nD modelling: a study of data interfacing and adoption barriers. Electron. J. Inf. Technol. Constr. **10**, 85–110 (2005)

The Relationship Between Requirements Subjectivity and Semantics for Healthcare Design Support Systems

Joao Soliman-Junior[1(✉)], Barbara Pedo[1], Patricia Tzortzopoulos[1], and Mike Kagioglou[2]

[1] Innovative Design Lab, School of Art, Design and Architecture, University of Huddersfield, Huddersfield HD1 3DH, UK
Joao.SolimanJunior@hud.ac.uk
[2] School of Engineering, Western Sydney University, Penrith, NSW 2751, Australia

Abstract. Subjectivity exists in requirements described in the healthcare regulatory framework. This is mainly due to the nature of regulatory requirements and the uniqueness of the design process. Past research identified that subjectivity in regulations is a key issue for automated code and rule checking. The aim of this paper is to discuss how requirements subjectivity could be addressed within building models through semantic enrichment, within the context of automated rule and code compliance checking. The paper presents preliminary findings of a research that follows the Design Science Research approach, framed within the UK healthcare design context. Findings suggest that part of the requirements subjectivity exists due to the implicit relationships between the elements of the healthcare built environment, which also include healthcare services. In order to enable automation, implicit relationships from the regulatory framework should be represented in building models – which could potentially be done through semantic enrichment. The paper discusses some complementarity between relationships identified in regulatory requirements and semantic enrichment operators. Moreover, findings indicate that incorporating semantic relationships in building models can be a promising way to deal with requirements subjectivity, rather than eliminating subjective expressions from regulations.

Keywords: Subjectivity · Semantics · Healthcare regulations · Automated code/Rule checking

1 Introduction

The healthcare regulatory framework includes requirements originating from specific medical needs and technologies, which should represent the needs of all stakeholders involved or influenced by healthcare projects [1]. It is known that healthcare building requirements evolve over time [2], which creates challenges for designers. With the use of digital tools in healthcare design, the iteration between designing solutions and verifying their compliance to the associated regulatory framework can be automated

E. Toledo Santos and S. Scheer (Eds.): ICCCBE 2020, LNCE 98, pp. 801–809, 2021.
https://doi.org/10.1007/978-3-030-51295-8_55

and as such creating opportunities for time and cost savings, in contrast to manual checking approaches. This is influenced by the different typologies and characteristics of regulatory requirements [3], represented by either abstract or concrete information [4], which might be implicit or explicit within requirements definition. Thus, regulatory requirements can, in some cases, be subjective and open to different interpretations [5]. This is because this type of requirement is traditionally created and included into documents and standards to be read, interpreted and used by people, through the use of natural language expressions [5]. This introduces challenges to incorporate information from these requirements into object-oriented and automated interfaces, such as Building Information Modelling (BIM).

The use of automated rule checking for the purpose of addressing building compliance has been extensively explored in research [6, 7]. Despite these advancements, there is still a gap regarding its use in practice, related to the subjectivity in regulatory requirements and how these are used across the design process [5, 8]. Thus, the aim of this paper is to discuss how healthcare regulatory requirements subjectivity could be addressed within building models through the use of semantic enrichment. The discussion provides insight on how implicit information, mostly emerging from subjective requirements, could be verified in the design process through automated approaches.

2 Information in Regulations and Subjectivity

Regulations are created, interpreted and used by people [5]. The way regulations are developed is pointed as one of the reasons for the difficulties observed while using the documents later in practice [9]. This is because creating regulations can be a cumbersome process, due to many iterations performed by multiple people during long periods of time, as well as changes in requirements [9]. As a result, regulations include an extensive and generally complex set of requirements, consisting fundamentally of interdependent elements [9, 10].

Furthermore, there are other factors that contribute to requirements complexity, such as: (i) language structure from requirements; (ii) domain knowledge embedded in the regulations; (iii) logic elements from requirements; and (iv) human knowledge associated to interpreting regulations [9]. In fact, these factors are responsible for an important characteristic of such requirements: they tend to be indeterminate by nature, due to the open-textural elements that are used to define the sentences [11]. This typically implies in vagueness and ambiguity, stemming from the diverse senses information might assume within this context [5]. Indeterminacy also might be increased due to implicit local understandings, relationships and unwritten knowledge by experts involved in the development of regulatory documents [9, 10]. Therefore, subjectivity is often embedded within requirements through open-texts, which hardly can be automated because they are context-dependent and require human judgment to determine compliance [11, 12]. This process demands a significant level of human abstraction so requirements can be properly identified and considered in design [13].

Challenges related to requirements subjectivity are currently discussed, even though these have been identified long ago by e.g. Fenves et al. (1995) [11]. Recent research has explored structuring and translating regulatory documents for use in computational interpretation [6]. Despite these efforts, the importance of understanding and interpreting regulatory information is still under researched [9]. According to Macit İlal and Günaydın [10], identifying the nature of regulations and the hierarchical information associated to its requirements is fundamental to enable a degree of automation in the design process. Solihin and Eastman [9] also point that the analysis of regulatory requirements still needs be done by humans because at the moment, only they are able to understand the implicit content of regulations. This process could be facilitated by using mechanisms of classification, which enable grouping elements according to different classes, i.e. distinguishing them according to their characteristics, such as a requirements taxonomy [3, 14]. Additionally, by classifying the content of regulations it might be also possible to identify different relationships between requirements [10], which can support a better understanding of the degree of embedded subjectivity [3].

3 Semantic Enrichment of Building Models

Semantic enrichment is defined as an automated or semi-automated process which allows incorporating meaningful information to building models, related to its objects and their relationships, through the use of domain-specific rulesets [15, 16]. This information, in turn, can potentially enhance the building model by facilitating its use for specific issues, such as code compliance checking [17, 18]. Relationships among objects in the semantic enrichment process are incorporated according to three steps, defined as tiers [15]. Tier 1 relates to the definition of inference rule sets by the users; tier 2 relates to libraries of concepts and operators used to compile rules; while tier 3 relates to how operators are implemented by programmers [15].

Enhancing building models through semantic enrichment to facilitate code compliance checking emerged due to the lack of practical solutions which are applicable to different types of requirements [18]. This was mostly due to issues with existing modelling approaches, stemming from the need to include additional information which was originally missing or inaccurate, so building models could be used for multiple types of analysis and iterations [17]. The need for semantic enrichment is also explained due to the need to turn implicit semantics and relationships into explicit information by enhancing data from Industry Foundation Classes (IFC) files [18].

Bloch and Sacks [17] reviewed existing applications focused on building code compliance and identified that one of the main difficulties involved in this process emerges due to the lack of sufficient information embedded within building models. In this context, explicitly defined parameters, topological structures and connections among objects [18] are often missing or inaccurate, which is identified when building models are analysed. Another factor that increases the importance of semantic enrichment is the limitations of current checking tools.

The use of automated approaches for code compliance is often limited to quantitative and objective requirements, i.e. numerical constraints or mathematical equations [3, 6, 18–20]. The use of semantic enrichment is an opportunity to enhance building models by including other types of relevant information, so they can be used for different purposes and analysis – acknowledging that automated code checking can be one of the main uses of such approach.

4 Research Method

This paper reports preliminary findings of an ongoing research project investigating opportunities for improving the healthcare design by using design support systems. The research adopts the Design Science Research approach (DSR), which is relevant for solving problems that have both practical and theoretical relevance [21]. The outputs of DSR are artefacts aimed to solve practical problems, while their development, implementation and evaluation enable theoretical insights [21, 22]. By using DSR, the research problem is typically understood at the same time the artefact is designed [23]. This process is fundamentally done through iterative cycles of analysis, understanding, development and refinement. It is important to highlight that the findings presented in this paper do not include final artefact, but are part of the initial stages of the research, and hence involve mainly a better understanding the research problem.

A literature review was developed to understand subjectivity embedded in regulatory requirements and how it could be addressed in the healthcare design process, focusing on automated code checking. The discussions presented in this paper are also partially informed by empirical data from two studies: (i) one in collaboration with an institution responsible for Primary Healthcare buildings across the UK; and (ii) a retrospective study developed with a University Hospital in Porto Alegre, Brazil. The sources of evidence are: (i) analysis of regulations, which supported understanding and identifying requirements characteristics and subjectivity; and (ii) meetings and interviews with designers and other stakeholders, which provided inputs related to how the regulatory framework is used in practice and how subjectivity affects the development of healthcare projects.

5 Key Findings

From the analysis of the information within healthcare regulations, different types of requirements were identified. One key element responsible for such outcome was identified i.e. the degree of abstraction involved in the requirement. Some requirements are objective while others remain subjective. Within this context, much of the identified subjectivity is due to implicit relationships between elements of the healthcare built environment, which also include health services. Examples of these requirements are presented below, through extracts from both Brazilian (RDC 50) and British (HBN 11-01 and HBN 00-03) healthcare regulations.

In the below, the terms in bold are examples where implicit relationships were identified, e.g. arranged with a direct relationship to. These relationships might often manifest across different sets of regulations within the healthcare design context, since requirements tend to repeat their structure and typology [3]. These relationships cannot be included in building models using traditional modelling approaches.

1. *Specific installation systems (e.g. hot water, medical gases and alarming)* ***are needed within*** *the spaces of healthcare facilities, depending on the use of each individual space* (followed by a set of tables). (RDC 50)
2. *The bed-area in the physiotherapy diagnostics and therapy rooms* ***must be equal or superior to*** *2.4* m^2. (RDC 50)
3. *Consulting/examination suites* ***should be arranged possibly with an adjacent*** *suite to enable patients to be referred on from their initial consultation to a specialist consulting/examination suite or treatment suite.* (HBN 11-01)
4. *Generally, community records* ***should be stored within or adjacent to*** *the open-plan admin area. GP records and records on loan from the acute sector* ***should be stored close to*** *the main reception desk or appropriate control point.* (HBN 11-01)
5. *Consulting/examination suites should be* ***arranged with a direct relationship to*** *the main waiting area, and possibly an* ***adjacent*** *suite to enable patients to be referred on from their initial consultation to a specialist consulting/examination suite or treatment suite.* (HBN 11-01)
6. *The dental treatment room* ***will contain*** *specialist built-in cabinetry, a reclining chair, ceiling mounted lamp, wall-mounted inter-oral periapical X-ray machine and a console* ***adjacent to*** *the chair supplying dental gases.* (HBN 11-01)
7. *Seated recovery area* ***should contain*** *a separate zone for clinical hand-washing.* (HBN 00-03)
8. *The space required for the en-suite shower room* ***includes*** *not only the enclosed area but also the temporary manoeuvring space for assisting a patient on both sides of the WC* ***which overlaps*** *the bed space.* (HBN 00-03)

One of the promising possibilities of incorporating implicit information from healthcare regulations to building models is associated to semantic enrichment. Implicit relationships observed in this context occur due to the way information is embedded into regulatory requirements. In order to enhance building models semantically, operators related to concepts, properties, relationships, geometry, spatial orientation, spatial topology and auxiliary information are used [15]. This is a very important step of the semantic enrichment process because by doing so, implicit information can be incorporated as explicit parameters and data within building models, so it can be further utilized for different purposes.

The use of semantic enrichment appears to be a fruitful opportunity to support addressing the issue of automated rule checking for code compliance. Within the scope of healthcare regulations, particular types of information are observed due to the specific characteristics of this context [3]. The *Spatial Topology operators*, which are one of the specific tier-2 elements defined by Belsky et al. [15], appear to be especially relevant in relation to the healthcare regulatory framework. Table 1 is based on the original definition of the Spatial Topology operators [15] and how each of the examples of regulatory requirements described above might be related to them.

Table 1 suggests that by better exploring the spatial topology operators it might be possible to enable an easier automated rule checking approach. This is due to the correlation identified between these operators and specific types of requirements observed in the healthcare regulatory framework. This fact indicates an existing

Table 1. Relationship between spatial topology operators and requirements, focusing on semantic enrichment

Spatial topology operators	Description according to Belsky et al. [15]	Example from Regulatory Requirements (highlighted element)
Adjacency	At least two objects which have a common face or are inserted within a given tolerance distant from each other.	[3] *[space]* ***should be arranged possibly with an adjacent*** *[space]* [4] *[space]* ***should be stored within or adjacent to*** *[space]* [5] *[space]* ***adjacent*** *[space]* [6] *[object]* ***adjacent to*** *[object]*
Contact	Adjacent objects that are in contact, meaning their tolerance distance equals to zero	[4] *[space]* ***should be stored within or adjacent to*** *[space]* [4] *[space]* ***should be stored close to*** *[space]* [5] *[space]* ***arranged with a direct relationship*** *[space]*
Containment	One object is utterly contained within another spatial object	[1] *[objects]* ***are needed within*** *[space]* [4] *[space]* ***should be stored within or adjacent to*** *[space]* [6] *[space]* ***will contain*** *[object]* [7] *[space]* ***should contain*** *[space] [to service]* [8] *[space]* ***includes*** *[space]*
Overlapping	At least two objects have an overlapping volume	[8] *[space]* ***which overlaps*** *[space]*
Volumetric operators	A relative volume of space that is occupied by one or multiple objects	[2] *[space]* ***must be equal or superior to*** *[dimension]*

complementarity between the relationships identified in regulatory requirements and semantic enrichment operators.

- Requirements related to the operators **adjacency** and **contact** are often identified in the healthcare regulations. These are associated to the spatial configuration of buildings and represent information considered at early design stages, e.g. schedule of accommodation, briefing and basic layout and flow definition.
- The operator **containment** can be associated to a type of requirement frequently observed within the healthcare design context. It is associated to the presence of specific systems and objects such as furniture and medical equipment within rooms, according to their use.
- The **volumetric** operator is related to requirements that describe needed object spatial boundaries and/or projection spaces, e.g. minimum spaces in front and on

the sides of beds. The **overlapping** operator is observed in spatial requirements and can be associated to volumetric operators e.g. bed space.

Typically, dealing with requirements subjectivity within current approaches for automated code checking requires human involvement. Building codes and regulations are typically translated into a coded fashion after arduous efforts from human experts, what might involve individual biases and potential misunderstandings. This generally results in partial interpretations due to the fact that subjective information is converted into objective and quantifiable sentences before rule creation. The issue involved in this inductive reasoning is related to a potential harmful approach to the automated rule checking context. By doing so, individual understandings might be generalised, hindering the more creative and idiosyncratic character of the architectural design process. Such issue becomes especially relevant while considering the context of healthcare design and the importance these facilities have on health outcomes.

6 Final Remarks

This paper explored how semantic enrichment can support addressing subjectivity from regulatory requirements for automated rule and code compliance checking. Key findings are related to identifying that implicit information within the healthcare regulatory framework is one of the main sources of subjectivity.

By analysing eight examples extracted from Brazilian and British healthcare design regulations, important characteristics of requirements have been identified. Moreover, some requirements, in fact, match the definition of some semantic enrichment operators, such as adjacency, contact, containment and volume. Therefore, semantic enrichment could potentially be a suitable way of addressing these implicit relationships in BIM. This could support automated design compliance checking.

Additionally, by exploring the relationship between regulatory requirements and semantic enrichment operators, it was possible to discuss how this approach might support the use of subjective information across healthcare design. By turning implicit relationships into explicit data in building models through semantic enrichment, subjectivity is not completely eliminated from the process, whereas no individual biases are introduced in the checking process. This approach might prove to be beneficial to the healthcare design context and it consists of an important shift. While the beneficial aspect of subjectivity would remain, allowing the creative and idiosyncratic reasoning from human designers, the negative aspect of subjectivity would be eliminated. By using semantic enrichment, subjective requirements would not have to be transformed into objective sentences, what could occur through potential biased processes.

The findings presented in this paper are limited as it is based on an exploratory literature analysis and partially informed by empirical data. There is a need to further test the relationships identified and verify in practice if the assumption presented on Table 1 is valid. Further testing should also explore whether semantic enrichment is, in fact, capable of dealing with subjectivity in a more efficient way, by taking into consideration human inputs through hybrid solutions, instead of eliminating subjective expressions from the regulations. Future research should also explore different types of semantic enrichment operators aiming to identify new relationships between those and different types of requirements. Finally, although semantic enrichment is not novel, it could potentially support the future adoption of automated rule checking in healthcare design by also considering subjective requirements.

Acknowledgements. The authors would like to thank the Innovative Design Lab (IDL), University of Huddersfield. Moreover, we would like to thank Community Health Partnerships (CHP) collaborators, project and design team members, for their time, support and the opportunity to help in this investigation.

References

1. Chellappa, J.R.: BIM + healthcare: utilization of BIM in the design of a primary healthcare project (University of Hawai'i) (2009)
2. Kiviniemi, A., Fischer, M.: Requirements management interface to building product models, pp. 1–12 (2004)
3. Soliman-Junior, J., Formoso, C.T., Tzortzopoulos, P.: A semantic-based framework for automated rule checking in healthcare construction projects. Can. J. Civ. Eng. **13**, 1–13 (2019)
4. Gutman, J.: A means-end chain model based on consumer categorization processes. J. Mark. **46**, 60–72 (1982)
5. Nawari, N.O.: The challenge of computerizing building codes in a BIM environment. Comput. Civ. Eng. **1**, 285–292 (2012)
6. Eastman, C., Lee, J., Jeong, Y., Lee, J.: Automatic rule-based checking of building designs. Autom. Constr. **18**, 1011–1033 (2009)
7. Nawari, N.O.: A generalized adaptive framework (GAF) for automating code compliance checking. Buildings **9**, 86 (2019)
8. Dimyadi, J., Amor, R.: Automated building code compliance checking – Where is it at? In: Proceedings of CIB WBC, pp. 172–185 (2013)
9. Solihin, W., Eastman, C.M.: A knowledge representation approach in BIM rule requirement analysis using the conceptual graph. J. Inf. Technol. Constr. **21**, 370–401 (2016)
10. Macit İlal, S., Günaydın, H.M.: Computer representation of building codes for automated compliance checking. Autom. Constr. **82**, 43–58 (2017)
11. Fenves, S.J., Garrett, J.H., Kiliccote, H., Law, K.H., Reed, K.A.: Computer representations of design standards and building codes: US perspective. Int J. Constr. Inf. Technol. **3**, 13–34 (1995)
12. Marchant, D.: The design brief: requirements and compliance. J. Inf. Technol. Constr. **21**, 337–353 (2016)
13. Lee, J.-K., Eastman, C.M., Lee, Y.C.: Implementation of a BIM domain-specific language for the building environment rule and analysis. J. Intell. Robot. Syst. **79**, 507–522 (2014)
14. Hjelseth, E.: Classification of BIM-based model checking concepts. J. Inf. Technol. Constr. **21**, 354–370 (2016)
15. Belsky, M., Sacks, R., Brilakis, I.: Semantic enrichment for building information modeling. Comput. Civ. Infrastruct. Eng. **31**, 261–274 (2016)
16. Sacks, R., Ma, L., Yosef, R., Borrmann, A., Daum, S., Kattel, U.: Semantic enrichment for building information modeling: procedure for compiling inference rules and operators for complex geometry. J. Comput. Civ. Eng. **31**, 1–12 (2017)

17. Bloch, T., Sacks, R.: Comparing machine learning and rule-based inferencing for semantic enrichment of BIM. Autom. Constr. **91**, 256–272 (2018)
18. Sacks, R., Bloch, T., Katz, M., Yosef, R.: Automating design review with artificial intelligence and BIM: state of the art and research framework computing in civil engineering 2019, pp 353–360 American Society of Civil Engineers, Reston, VA (2019)
19. Dimyadi, J., Pauwels, P., Amor, R.: Modelling and accessing regulatory knowledge for computer-assisted compliance audit. J. Inf. Technol. Constr. **21**, 317–336 (2016)
20. Lee, P.-C., Lo, T.-P., Tian, M.-Y., Long, D.: An efficient design support system based on automatic rule checking and case-based reasoning. KSCE. J. Civ. Eng. **23**, 1952–1962 (2019)
21. Lukka, K.: The constructive research approach. Case Study Res. Logist. Publ. Turku Sch. Econ. Bus. Adm. Ser. B **1**, 83–101 (2003)
22. Hevner, A.R., March, S.T., Park, J., Ram, S.: Design science in information systems research. MIS Q. **28**, 75 (2004)
23. Holmström, J., Ketokivi, M., Hameri, A.-P.: Bridging practice and theory: a design science approach. Decis. Sci. **40**, 65–87 (2009)

Blockchain Applications for AEC

Blockchain Technologies: Hyperledger Fabric in BIM Work Processes

Nawari O. Nawari(✉)

College of Design, Construction, and Planning, The University of Florida, Gainesville, FL 32611, USA
nnawari@ufl.edu

Abstract. Blockchain Technology (BCT) is defined as a decentralized data ledger that registers each transaction processed in an encrypted data format. The implementation of such a concept brings about several benefits to the built environment, specifically, Building Information Modeling (BIM) workflow. These include, for instance, security, reliability, accountability, and could potentially accelerate a shift in the work environment dynamics from the current centralized data processing to a decentralized, cooperative chain of transactions by encouraging trust and transparency. Hyperledger Fabric (HLF) is a BCT implementation that generates distributed ledger systems, supported by a modular design, offering a flexible and extensible digital framework, that delivers high levels of confidentiality, and scalability. It has a modular subsystem designed so that different employments can be plugged in and implemented over time. This paper reviews the BCT platforms and describes the details of the various components/modules and their interactions and functions. Furthermore, the study proposes a new work process based on HLF that aims to address some of the current BIM workflow shortcomings. The paper discusses how the HLF system enhances the BIM work process and presents a use case of automated code compliance checking mechanisms and future research directions.

Keywords: Blockchain · BIM · Distributed ledger · Hyperledger fabric

1 Introduction

The blockchain network is a distributed, decentralized ledger of data, resources and all data exchange operations that have been invoked and shared among participants in the network. Such a Distributed Ledger (DL) system was first used in Bitcoin, the first cryptocurrency, which has a public circulation and worldwide appeal. Bitcoin was first abstracted in a whitepaper published in 2008 [1], defining details essential for a procedure to create a decentralized digital currency that runs on a secure and pseudonymous network of users. One 'bitcoin' is basically a tradeable resource generated by the DL network as a way of payment for its operation and maintenance by miners. Subsequent to the achievement of Bitcoin, the nomenclature 'blockchain' was no longer synonymous with digital cryptocurrencies as its broad area of prospective applications was apparent [2].

E. Toledo Santos and S. Scheer (Eds.): ICCCBE 2020, LNCE 98, pp. 813–823, 2021.
https://doi.org/10.1007/978-3-030-51295-8_56

Although the term blockchain and DL Technologies (DLT) are used reciprocally in the literature, there is a distinction between them which is worth mentioning. A blockchain is just an exemplar of a specific kind of ledger where information can be saved in a specific format. There are other kinds of ledgers with various data formats. Recently, DLT has received extensive interest from the industry, the Government and academia. It is viewed as one of the essential technologies to transform the landscapes of numerous industries.

A DL is a ledger comprising of sequential blocks connected together following a stringent set of policies. The ledger is disseminated and saved by the nodes of a Peer-to-Peer (P2P) network, where each block is formed at a predefined period in a decentralized fashion employing a consensus algorithm. The consensus protocol ensures data integrity and reliability [3].

The implementation of DL technologies in the AEC is limited in contrast to finance, e-business, and healthcare areas. Lately, more studies are emerging tackled the subject at a higher level focusing on the capabilities of DL technologies in driving efficiencies in the AEC and how this can transform industry culture and advance innovation [4, 5 and 6].

BIM represents an important element of the digital transformation of the AEC industry, promoting collaboration and trust, and streamlining data exchanges between different stakeholders in the industry. BIM models present a comprehensive design model of the building that can include all aspects of the facility such as architecture, structural, and MEP components. Moreover, several tools in BIM platforms support the simulation of external site conditions, geography, weather, as well as conducting building energy analysis, structural analysis, etc. BIM aims at essentially unifying all aspects of design, construction, and facility management in one platform. Currently, the BIM process still exposes several deficiencies. Some of these shortcomings are attributed to certain limitations in the schema, technological toolset, and management aspects. Examples of these include obstacles in allocating responsibilities and liabilities due to the overlap of roles and obligations, safeguarding intellectual property protection, risk distribution, privacy, third-party dependence and software agents [5, 7], and insufficient cyber-resilience of the software platforms [3].

This research examines the potential of blockchain technology to minimize the shortcomings of the current BIM process. The paper investigates different blockchain platforms to advance the BIM workflow. Hyperledger Fabric is proposed as the essential DL platform to address the major issues with BIM process. Moreover, the paper provides a use case of the automated code compliance checking mechanisms to demonstrate the Proof of Concept (PoC).

2 Statement of Purpose

Certain of the foremost limitations of the current BIM process include cybersecurity, data integrity, tracking and enforcing the accountability of all stakeholders who make changes to the model during a project. In particular, the issue of handling permissions, e.g., who can publish which content, which changes are accepted, and so on is still not resolved in the current workflow. Thus, this study aims to address these issues by incorporating a blockchain framework.

3 Objectives

The chief objectives of this study are: i) to review the BIM work process and BCT platforms; and ii) to analyze the potential integration of DLT in BIM process with regard to cybersecurity, management of permissions, and data ownership.

4 Methodology

The research method is based on scope identification, assembling evidence through literature review, and develop a framework for integrating BIM workflow and BCT. Figure 1 exhibits an overview of the research approach. The first phase of the study is scope identification and assembling evidence. The second phase is the analysis and development of a new framework based on BCT and BIM work environment.

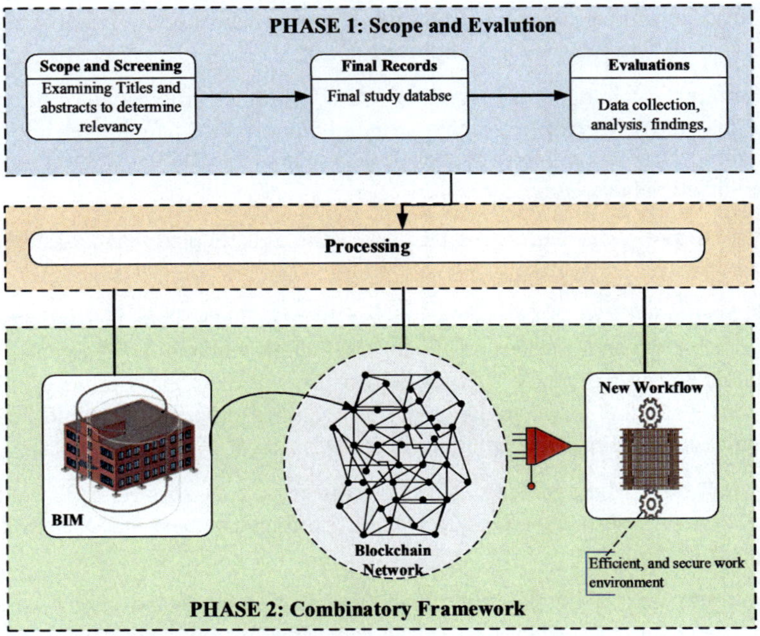

Fig. 1. Overview of the research method.

5 Distributed Ledger Systems

5.1 Basic Principles

The development of the DLT, with robust data integrity, accuracy, and provenance, has opened the door for opportunities in different domains. One example of DL design is blockchain which registers consensus, contracts and transactions in a continually updated list. Even though it was originally developed to support crypto-currency, this peer-to-peer system has come into eminence and earned recognition across a wide range of application domains, as for example, supply chain, healthcare, energy, education, finance, transportation, and governance.

The DLT is based on a distributed ledger that is stored by the nodes of a P2P network where each data block is created at a predefined interval in a decentralized manner by way of a consensus algorithm. This consensus algorithm assures data integrity-related properties and presently, there are many consensus algorithms introduced by deferent blockchain platforms that are intended to meet specific goals. The essential properties of DLT are described next [8]:

Distributed Consensus: One of the fundamental features of any DL is its ability to attain a distributed consensus on the state of the ledger without depending on any third party. This allows one to construct and use a system where every possible state and transactions are certifiable by any approved nodes.

Immutability: Attaining a distributed consensus with the participation of many peers certifies that the ledger state becomes virtually immutable and irreversible after a specific time. Also, this property applies to Smart Contracts (SC), which authorizes the invoking of immutable computer codes.

Data Persistence: Information in a DL is saved in a distributed manner safeguarding its insistence.

Data Provenance: The data storing process in any DL system is facilitated by a mechanism known as a transaction. Every transaction requires to be digitally signed using public-key cryptography (PKI), which confirms the legitimacy of the data source. Fusing this with the immutability and irreversibility properties provides a robust nonrepudiation instrument for any data in the DL system.

Distributed Data Management: ADL guarantees that data saved in the ledger or accessed from the ledger can be conducted in a distributed manner that demonstrates no single point of failure.

Accountability and Transparency: Since the state of the ledger, along with every single exchange among network nodes can be verified by any authorized entity in the networks, DL supports accountability and transparency.

In summary, the inherent properties of DLT include resiliency, integrity, anonymity, decentralization and autonomous control. Depending on the application domains, different DL deployment strategies can be pursued. Based on these strategies, there are two predominant ledger types, namely, Public and Private ledgers.

A public ledger, also known as the non-permissioned ledger, permits a node to create and authenticate blocks as well as to update the ledger state by saving and modifying information. This implies that the ledger status and its transactions along with the saved data is transparent and available to everyone. This raises confidentiality matters for many industries [9, 10]. On the other hand, Private Ledger, which is also known as permissoned ledger, is restricted in the sense that only authorized and trusted entities can participate in the activities within the network. Thus, a private ledger ensures the privacy of ledger data, which can be desirable for many industries include the AEC industry.

5.2 DL Platforms

This section briefly describes selected DL platforms that can be employed in AEC applications, highlighting their key features and functionalities. Table 1 delineates a summary of these platforms.

Table 1. Comparison between Distributed Ledger Platforms.

Platform	Privacy	Description	Scalability	Suitability
Bitcoin	Public	It is based on a public DL allowing anyone in the world to join the network, participate in the process of updating the ledger state and to interact with other entities in the network	Block size limit 1 MB; time 10 m; high energy cost	Banking sector. Being predominantly a payment system, it enables anyone to transfer digital currencies from one entity to another
Ethereum	Public	Ethereum is a DLT system, driven by Smart Contract (SC), and featuring a virtual machine, called Ethereum virtual machine (EVM) whose state, along with the additional ledger data, is stored in the ledger. Like Bitcoin, it is a permissionless DL, and provides similar functionalities for a cryptocurrency payment network	Block size is limited; time 0.5 s; high energy cost	Finance sector

(continued)

Table 1. (*continued*)

Platform	Privacy	Description	Scalability	Suitability
Corda	Private	Corda is a DLT that aims to provide support for the finance domain. It is a permissioned network. One of the primary differences between Corda and other mainstream platforms such as Bitcoin and Ethereum is the use of a blockchain. Corda is a distributed ledger but does not use a blockchain to record transactions [11]	Block size is configurable; time 0.5–2 s; low energy cost	It acts as a financial network as well as a platform enabling it to have the best of both Ethereum and Bitcoin domain
Multichain	Private	It is an open source platform, originating from Bitcoin blockchain that allow users to create and deploy private DL applications either within or among organizations [12]	Block size and time are configurable; low energy cost	It has been successfully implemented in healthcare, insurance security, and food supply chain sectors
IOTA	Private	IOTA is a DL devised for the Internet of Things (IoT). It offers secure communications and payments between IoT devices [13]	Block size and time are configurable; low energy cost	IOTA can be used for application in AEC industries that have the use-cases working with IoT devices
HLF	Private	It is the first major private DL system originated from the Hyperledger systems. It has been devised with strong privacy in mind to ensure that different governmental agencies and industries can take advantage of a DL system in different use-cases. A distinct feature of HLF is that it can maintain multiple ledgers within its ecosystem. This is a useful aspect, which separates HLF from other DL systems	Block size is configurable; time 0.5–2 s; low energy cost	HLF can be useful in many AEC applications. In particular, it can be employed to improve the BIM process

5.3 Hyperledger Fabric (HLF)

Hyperledger is the first major private DLT that originated from the Hyperledger systems. It is a collaborative effort founded by the Linux Foundation in 2016 to advance cross-industry DLT. This study yields that HLF is one of the most suitable platforms for BIM work environment. Because it has been devised with robust privacy in mind to certify that different governmental agencies and industries can take advantage of a DL system in various areas of applications. A distinct feature of HLF is that it can maintain multiple ledgers within its ecosystem. This is a useful feature, which separates HLF from other DL systems.

HLF aims at creating distributed applications devised in standard general-purpose programming languages [14, 15]. It is a cross-industry open standard platform for blockchain that seeks to transform the technique business transactions are conducted universally.

A major strength of the HLF in contrast to the other DL platforms described in the previous section is that it is based on modular and pluggable components [15, 16]. Moreover, HLF can generate channels, granting a subgroup of nodes in the network to create a separate ledger of transactions. This can be particularly valuable characteristic for BIM workflow where subcontractors can exchange data within only a subdivision of the network. For example, the structural engineer of record of a project can exchange information with steel connection subcontractors only while still being part of the HLF network and sharing those transactions with remainder of the network participants [2].

Possible components that can be plugged in and executed in the HLF system include (see Fig. 2):

(a) Membership services: This component focuses on permissioning and serves in managing the identity of members and generate a root of trust during network formation.
(b) Chaincode: A chaincode, or a smart contract, is business logic code saved on the ledger as a part of a transaction. It is written often in Go or Java languages.
(c) Consensus algorithms: These services are vital for any DL network. They facilitate a trust system. The consensus module authorizes digitally signed transactions to be recommended and endorsed by network nodes. The ordering services in HLF represent the trust system.

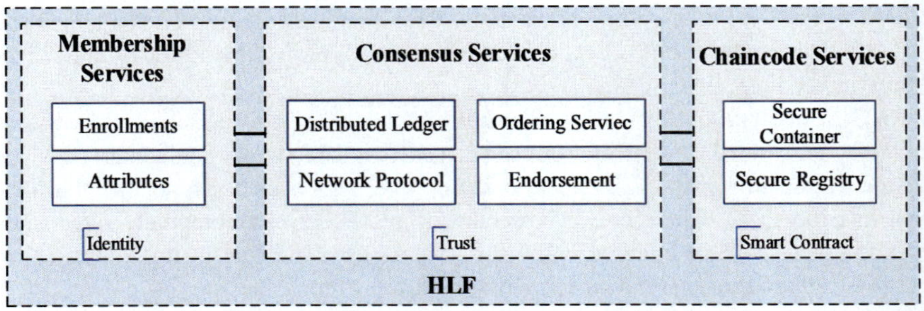

Fig. 2. Hyperledger Fabric (HLF) components.

6 BIM Work Process

BIM is at the leading edge of the digital transformation in the AEC sectors, encouraging collaboration and trust, and streamlining data transactions. However, the BIM process still bears several shortcomings. These include, for instance, limitations in the schema, technical toolset, and management and organizational aspects of the existing process. Some of these major deficiencies comprise:

- Cybersecurity, and no history of BIM model changes,
- the unsuitability of the integrated model for co-designing in real-time [17].
- inadequate exchange standards and BIM model sharing that can lead to interoperability issues between different BIM tools [18, 19]
- impediments in delegating tasks and accountabilities due to the overlap of roles and responsibilities, safeguarding intellectual property, risk mitigations, and privacy [5, 7].
- actory cyber-resilience of the current software platform and subsequent risks of data theft, and other tampering threats,
- absence of legal framework describing model data ownership and lawful contractual concerns [20].

Several of these limitations can be tackled by the application of the DLT in BIM processes. One such example is bcBIM that investigates expanding the combination of these emergent technologies aiming to chiefly merge mobile cloud data with BIM architecture [21, 22]. Another example is the blockchain-based systems for BIM launched by a French tech start-up company, concentrating on immutability, and they seek to link these technologies into a cohesive system and move on from the need for a third-party authorization [23].

A further example is depicted by the concept of virtual currency denoted by #AECoin to provide a reward mechanism in the process. #AECoin is a recently developed cryptocurrency coin created explicitly for design and construction transactions. This effort can contribute to addressing the difficulties in assigning responsibilities and liabilities due to the overlap of roles and responsibilities, ensuring intellectual property protection, risk allocation, privacy in current BIM workflow [5, 7 and 20].

Since BIM workflow encompasses various information exchange between actors, technology, and processes and inter-relationships, it is crucial to consider cybersecurity implications, assess levels of reliability, address current drawbacks, and reinforce security. In the present practice, all BIM data is electronically shared across a common data environment. Thus, it is essential for all project members to understand cybersecurity limitations [24].

In BIM processes, the cyber environment includes the Internet, telecommunication networks, embedded processors, controllers, sensors, data storage and control devices, as well as the information, services, collaborative, and business functions that only exist in cyberspace. Therefore, cybersecurity threats that can potentially affect BIM workflow need to be addressed. DLT can play an essential role in improving cybersecurity in the current BIM work environment.

7 Proposed Framework

This study seeks to expand the existing BIM processes by incorporating the DLT. HLF platform is proposed as the most suitable DLT for BIM work environment. It offers secure identification and authentication of participants and establishes a ubiquitous and reliable infrastructure that serves as a repository for data storage, as well as a consistent platform that facilitates data exchanges. Essential aspects of a blockchain, such as utilizing secure cryptography, asset sharing, auditing trails of data access, immutability, and a resilient peer-to-peer network, presents an innovative and encouraging method of addressing the current shortcomings of BIM workflow.

The framework is illustrated by employing the automated code compliance checking system as a use-case. The automated code checking system utilized in this study is based on the Generalized Adaptive Framework (GAF) [25]. Figure 3 delineates an overview of the unified framework to automate building design compliance checking process. The four main modules of the new framework are the GAF, Smart Contracts (Chaincode), membership services, and ordering services.

The framework proposes that building codes rules and BIM model data are stored in the blockchain network and the chaincode to function as the model checker service and building permit issuer. The details of this algorithm involve (see Fig. 3):

a) The building regulations upon which the BIM model data are to be assessed must be handled using computable expressions. A Smart Contract (chaincode) can be programmed to process the rules from a natural language using the GAF (Nawari, 2020). This chaincode needs to be implemented to incorporate all clauses, terms, and variables used in the building codes and regulations. Following the transformation of the building regulations, the chaincode generates a second appended smart contract that can now be used by the model checker service.
b) The BIM model data must be in IFC format (ifcXML) and retrieved using the scripting language, such as Java or Go, utilized by HLF.

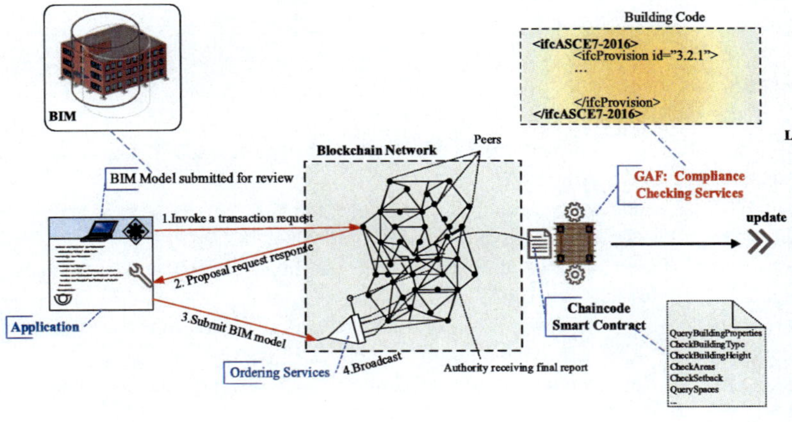

Fig. 3. Overview of the proposed framework.

c) A model compliance checking service is programmed in the form of another chaincode that can extract data from the BIM model and, upon invoking, can verify the extracted data against the translated rules created in step (a).
d) The model compliance checking service executes the code and generates output reports to be sent to an authority to review and confirm the final permit status.

8 Conclusion

The DLT is characterized by a distributed, decentralized ledger of data, resources, and data exchange operations that have been invoked and shared among participants in a network. It offers immutable, transparency, emphasizes reliability and trust between all nodes on the network. Such characteristics can be beneficial in minimizing current limitations with BIM workflow.

Different DL platforms are presently available for developing blockchain applications. However, the HLF is the most appropriate one for enhancing the BIM work process. For the reasons that it has been devised with secure privacy in mind to ensure that different agencies and industries can take advantage of a DL system in different use-cases. A distinct feature of HLF is that it can maintain multiple ledgers within its ecosystem. This is a vital aspect, which separates HLF from other DL systems.

The proposed framework aims to incorporate HLF with the BIM process to strengthen collaboration and trust, cybersecurity, responsibility, and data transaction integrity. The multi-signature (multisig) protection and the transparency of the HLF coupled with the database characteristics of BIM can introduce trust and accountability, which would generate a new viable BIM process.

The application of the suggested framework has been demonstrated in a use case of automated building code compliance checking. The four main modules of this implementation include the Generalized Adaptive Framework (GAF), Smart Contracts (Chaincode), membership services, and ordering services. The example illustrates the potential application of the new framework in a typical BIM workflow as well as several enhancements over the current BIM process.

References

1. Nakamoto, S.: Bitcoin: a peer-to-peer electronic cash system (2008). https://bitcoin.org/bitcoin.pdf. Accessed 1 Aug 2018
2. Nawari, N.O., Ravindran, S.: Blockchain technology and BIM process: review and potential applications. J. Inf. Technol. Constr. (ITcon) **24**(12), 209–238 (2019)
3. Nawari, N.O., Ravindran, S.: Blockchain and the built environment: potentials and limitations. J. Build. Eng. **25**, 100832 (2019)
4. Mathews, M., Robles, D., Bowe, B.: BIM + blockchain: a solution to the trust problem in collaboration? In: CITA BIM Gathering, November 2017 (2017)
5. Turk, Ž., Klinc, R.: Potentials of blockchain technology for construction management. Procedia Eng. **2017**(196), 638–645 (2017)

6. Nawari, N.O., Ravindran, S.: Blockchain technologies in BIM workflow environment. In: Computing in Civil Engineering 2019: Visualization, Information Modeling, and Simulation, pp. 343–352. ASCE (2019)
7. Eastman, C., Teicholz, P., Sacks, R., Liston, K.: BIM Handbook: A Guide to Building Information Modeling for Owners, Managers, Designers. Engineers and Contractors. Wiley, Hoboken (2011)
8. Dinh, T.T.A., Wang, J., Chen, G., Liu, R., Ooi, B.C., Tan, K.-L.: Blockbench: a framework for analyzing private blockchains. In: Proceedings of the 2017 ACM International Conference on Management of Data, pp. 1085–1100. ACM (2017)
9. Buterin, V.: Ethereum white paper: a next generation smart contract & decentralized application platform (2014)
10. Buterin, V.: Chain Interoperability. https://static1.squarespace.com/static/55f73743e4b051cfcc0b02cf/t/5886800ecd0f68de303349b1/1485209617040/Chain+Interoperability.pdf. Accessed 10 Aug 2018
11. Corda, September 2019. https://www.r3.com/platform/. Accessed 12 Sept 2019
12. Ferdous, M.S., Biswas, K., Chowdhury, M.J.M., Chowdhury, N., Muthukkumarasamy, V.: Integrated platforms for blockchain-enablement (2019)
13. Iota, October 2019. https://www.iota.org/. Accessed 15 Oct 2019
14. Macdonald, M., Liu-Thorrold, L., Julien, R.: The blockchain: a comparison of platforms and their uses beyond bitcoin. COMS4507-Advanced Computer and Network Security (2017)
15. Androulaki, E., Barger, A., Bortnikov, V., Cachin, C., Christidis, K., De Caro, A., Enyeart, D., Ferris, C., Laventman, G., Manevich, Y., Muralidharan, S.: Hyperledger fabric: a distributed operating system for permissioned blockchains (2018). https://doi.org/10.1145/3190508.3190538
16. Bartoletti, M., Pompianu, L.: An empirical analysis of smart contracts: platforms, applications, and design patterns, 18 March 2017. Universitia degli Studi di Cagliari, Cagliari, Italy (2017). arXiv:1703.06322v1 [cs.CR]
17. Miettinen, R., Paavola, S.: Beyond the BIM utopia: approaches to the development and implementation of building information modeling. Autom. Constr. **43**, 84–91 (2014)
18. Nawari, N.O.: BIM standard in off-site construction. J. Archit. Eng. **18**, 107–113 (2012)
19. Nawari, N.O., Alsaffar, A.: Advancing BIM standardization: floating structures. In: Proceedings of the International Conference on Civil and Building Engineering Informatics (ICCBEI 2015), Tokyo, Japan, 22–24 April 2015, p. 78 (2015)
20. Ahn, Y.H., Kwak, Y.H., Suk, S.J.: Contractors' transformation strategies for adopting building information modeling. J. Manag. Eng. **32**, 05015005 (2015)
21. Kuo, T.-T., Zavaleta Rojas, H., Ohno-Machado, L.: Comparison of blockchain platforms: a systematic review and healthcare examples. J. Am. Med. Inform. Assoc. **26**(5), 462–478 (2019)
22. Zheng, R., Jiang, J., Hao, X., Ren, W., Xiong, F., Ren, Y.: bcBIM: a blockchain-based big data model for BIM modification audit and provenance in mobile cloud. Math. Prob. Eng. **2019**, 13 (2019). https://doi.org/10.1155/2019/5349538. Article ID: 5349538
23. Cousins, S.: French start-up develops blockchain solution for BIM (2018). http://www.bimplus.co.uk/news/french-start-develops-blockchain-solution-bim/. Accessed Mar 2019
24. Hammi, A., Bouras, A.: Towards safe-BIM curricula based on the integration of cybersecurity and blockchains features. In: Proceedings of INTED 2018 Conference, Valencia, Spain, 5–7 March 2018, pp. 2380–2388 (2018)
25. Nawari, N.O.: Generalized adaptive framework for computerizing the building design review process. J. Archit. Eng. **26**(1), 04019026 (2020). ASCE

Framework for Automated Billing in the Construction Industry Using BIM and Smart Contracts

Xuling Ye(✉) and Markus König

Ruhr-University Bochum, Universitätsstraße 150, 44801 Bochum, Germany
xuling.ye@rub.de

Abstract. Building Information Modeling (BIM) represents the application of digital building models to improve project-wide collaboration. With the help of building information models, elements with tendering information, construction work to be carried out, quantities to be processed, planned times and subcontractors commissioned can be described in an integrated way. However, digital models are currently still little used for the automated billing of construction work. In this paper, an approach is presented to make contract management and payment within construction projects simpler, more transparent and more automated. The implementation takes place with the help of Smart Contracts, which are linked to digital building information models. For this purpose, a billing model and processes for checking the contract are defined on the basis of the building elements, determined quantities and defined construction works. If the contracted construction has been checked and accepted, the payment procedure is started automatically. The Smart Contract, as well as the individual transactions, are managed using Blockchain technology. Detailed contract data and execution information is stored in a common data environment (CDE) and linked to the Smart Contract. The concept for the realization of "BIM contracts" as well as aspects of the implementation are presented in this paper. Finally, we discuss several aspects for improvement and further development in the future.

Keywords: Building Information Modeling · Blockchain · Smart Contract · Common data environment · Information container for data drop

1 Introduction

An increasing need for digitization in the coming decades is expected in the building industry for various reasons. Among the most important reasons are the increasing complexity of construction projects, low productivity, and the growing need for real-time information sharing [1].

Building Information Modeling (BIM) has become a widely accepted solution for making the whole lifecycle of construction project digital. Agarwal et al. [1] shows that low margins and relatively low productivity are the issues of the construction industry for decades. The main reason for this lies in the low digitization degree in the construction sector despite the increasing scale and complexity of construction projects. Relying mainly on drawings or paper-based documentation does no longer satisfy the

E. Toledo Santos and S. Scheer (Eds.): ICCCBE 2020, LNCE 98, pp. 824–838, 2021.
https://doi.org/10.1007/978-3-030-51295-8_57

increasing needs of construction projects. Digitization can meet these needs by facilitating information sharing, data capturing and analysis. In this way, it also enables time-efficiency for construction projects. With the benefits of digitization, Agarwal et al. believe that 5D BIM, which integrates time and cost into a 3D BIM model, is a coming up trend in the construction industry.

By integrating "cost" to the BIM model, a 5D BIM model can instantly generate a cost budget and the corresponding financial representations over time. Advances are reducing the time of quantity take-off (QTO) and generating bill of quantities (BoQ) from weeks to minutes, improving the accuracy of estimation, minimizing disputes caused by ambiguity in CAD data, and allowing cost consultants to spend more time in improving value [2]. QTO specifies the quantity of different materials, while BoQ structures and lists the costs of the quantities.

In this paper it is assumed, that a contract is concluded between the client and the contractor for the construction assignment of the works. The individual construction work items, accounting units and payments are included in the contract. The required information can be taken from the 5D BIM models. The contract specifies additional regulations for the release of deliverables or procedures and retentions in case of defects. The performance report, inspection of the construction work as well as the invoicing can also be partially supported by 5D BIM models. However, the communication and invoicing is usually still conventional without a direct link to the construction contract.

Using current practice has limitations: (1) The construction accounting is currently a manual process and therefore very time-consuming and error-prone. (2) The time from invoicing to receipt of payment is disproportionately long. This fact leads to many bankruptcies in the building industry. (3) The contract structures are very complex and non-transparent. (4) Payment claims are difficult to verify, since detailed documentation of the as-is construction process and the complex interactions of the individual contracting parties are missing (5) Because of the complexity, single construction tasks are combined into large payment blocks.

Due to the limitations of current practice, a new approach for combining BIM with Smart Contracts using Blockchain technology is proposed in this paper. Smart Contracts are self-executing contracts, where the terms of the agreement between the client and contractor are written directly in lines of computer code. The code and the agreements are stored via a distributed, decentralized Blockchain network. Smart Contracts enable trusted transactions and agreements to be carried out between different anonymous parties without the need for a central authority, legal system or external enforcement mechanism. Thanks to the Blockchain technology, Smart Contracts make all transactions traceable, transparent and irreversible.

Smart Contracts contain certain details that are normally described in a textual form in conventional contracts. The contract information, however, is defined by programming code. The principle is: Code is law. For example, a Smart Contract can define that a contractor receives a certain amount of money if he has delivered a certain service. Consequently, Smart Contracts contain if-then rules. If a condition specified in the contract is fulfilled, this automatically results in a consequence. This means that defined activities to implement the intent expression of the contract are automatically carried out when a required event occurs. At the same time, all contractual partners are informed of any status changes in real time.

Smart Contracts based on a Blockchain have some advantages over conventional contract forms: 1) Reliability: If a Smart Contract has been programmed correctly, difficulties in interpreting the terms of the contract are almost impossible. The loss of documents is also excluded. 2) Security: If Smart Contracts are programmed based on a Blockchain, they are protected from hackers by cryptographic encryption procedures. Nobody can change the negotiated contractual conditions afterwards. 3) Efficiency: Programming a Smart Contract takes less time than corresponding bureaucratic processing. This saves contract partners time and money. 4) Independence: Smart Contracts reduce the need for third parties, such as lawyers, notaries and bankers. The unchangeable Blockchain is used to verify a contract. For this reason, the principle "code is law" applies. The program code of a Smart Contract decides whether the contract conditions have been fulfilled correctly or not.

2 Related Work

2.1 5D BIM Model

Agarwal et al. [1] define a 5D BIM model as "a five-dimensional representation of the physical and functional characteristics of any project. It considers a project's time schedule and cost in addition to the standard spatial design parameters in 3-D." However, this definition is not detailed enough and leaves room for interpretation. The paper assumes a 5D BIM model to be a combination of different information. The basis is a digital building information model which contains geometric and semantic information about the building. The Industry Foundation Classes (IFC) are used for data exchange, especially in building construction. The bill of quantities (BoQ) are created for the tendering and awarding of construction works [21]. An international standard for the description of BoQ information does not yet exist. Nevertheless, there is a standardized format for data exchange in Germany called GAEB (in German: Gemeinsamer Ausschuss Elektronik im Bauwesen) [20] provided by the Joint Committee for Electronics in the Construction Industry. The GAEB XML format is supposed to serve the agreement of a uniform standard for the exchange of building information and thus to support all requirements to electronic processes concerning public notice, contracting and accounting in the execution of construction works [20], and is linkable with other information models [24, 25]. The building model can be used to determine quantities that are assigned to the individual BoQ elements. For an in-situ concrete wall, for example, these are formwork area, reinforcement in kilograms or the concrete quantity in cubic meters. The calculation of the individual quantities based on the components of the BIM model is called quantity take-off (QTO). For the data exchange of the calculated quantities simple lists or the mentioned GAEB XML format can be used [20].

Subsequently, based on the individual construction works and calculated quantities, a schedule and time planning can be made, taking into account the design and the building site. For this purpose, individual activities are defined, connected and saved as a network plan [22]. The durations can be determined, for example, considering the selected construction methods and calculated quantities. The components of the BIM model are then linked to the individual activities using the calculated quantities. In the current approach, the schedule is not taken into account, as it can be stored and exchanged separately. The combination of building model, BoQ and QTO is referred to in the paper as 5D BIM model. In practice, these three data sources or files are linked together [23].

2.2 Blockchain and Smart Contracts

Blockchain is a distributed ledger technology [3–5], where "distributed" is managed by a decentralized peer-to-peer (P2P) network [3] and "ledger" is realized as an electronic database [5] comprising a chain of blocks [4]. In this network, the transactions are verified, validated and combined into blocks. The blocks are then chained together. Thus, the procedure is based on a chained block structure that grows linearly. The data is stored multiple times in this network to be able to cope with a failure of one or more nodes. The data in the Blockchain is unchangeable and cannot be manipulated or deleted. Using a P2P network in the Blockchain was first proposed in 2008 as a Bitcoin cash system to solve the double-spending problem [6], which is a problem since digital currency in the internet could be used more than once since it can easily be copied and broadcasted by hackers. This problem was solved 1983 [7] by using a client/server (C/S) network, where a C/S network provides a central server as a trusted third party between two nodes. However, compared to the C/S mode, the P2P network can provide users with a higher degree of autonomy and direct-interaction capability, can further distribute data and control, load-balance requests across the network and eliminate the risk of a single point of failure [8].

The Blockchain concept was developed from the Bitcoin technology firstly proposed by the pseudonymous Nakamoto in 2008 [6]. Brito and Castillo first mention the term "block chain" based on the Bitcoin policy in 2013 [9]. Since then, the term Blockchain has been widely accepted and used. However, the Bitcoin (i.e. the origin of Blockchain) was a non-programmable system suitable only for cash transactions [10]. The situation starts to change after Smart Contracts come into the stage of Blockchain. The Smart Contract, first proposed by Szabo in 1996 [11], is a digital agreement which self-executes itself. Since Buterin et al. [12] have embedded Smart Contracts in their Ethereum Blockchain system in 2014, Blockchain has become a platform with programmable applications. This way, more possibilities and more applications in more fields than just finance have come to the Blockchain stage. A Blockchain can be set up in three simplified ways: as a public, private or consortium Blockchain [13].

The public Blockchain is a completely decentralized network without a higher-level authority. Each transaction is verified and synchronized by each node of the Blockchain before it is written to the system. This makes this variant relatively slow and very resource-intensive, but transparent and secure. With the private Blockchain, a company operates the system and has control over the transactions. Only one participant verifies and writes each transaction into the system. In addition, the read rights for the transactions per user can be restricted, which allows higher data protection than the public variant. Since the size of the private Blockchain is limited by the company limits, the resource requirements and computing time per transaction are also kept within limits. However, when the completely decentralized approach is abandoned, part of the reliability is also lost. The consortium Blockchain is a kind of hybrid of private and public variants. Here, a group of participants shares the decision-making power over the verification and distribution of read rights for transactions. This offers many advantages of the private Blockchain such as efficiency and data protection of transactions without putting the entire sovereignty in one hand. For implementation within a construction project, it is intended to use a consortium Blockchain. The contracting parties as well as banks, insurance companies and, if necessary, associations, chambers or the public sector can be involved.

In the construction industry, a great deal of research has been done on using Blockchain due to its enormous potential. For example, Turk and Klinc [14] point out that the organizational and legal issues, which cannot be solved by BIM technology, can be addressed by using Blockchain to manage BIM files. Wang et al. [15] explain three categories of potential Blockchain applications that can solve three challenges in construction engineering management: notarization-related applications for trust issue in contract management, transaction-related applications for asset management (e.g. equipment leasing), and provenance-related applications for supply chain management. In 2019, Li et al. [16] reviewed 73 papers from 2014 to 2018 about the applications of Blockchain in the construction industry. Those applications lie in smart energy, smart cities and the sharing economy, smart government, smart homes, intelligent transport, building information modelling (BIM) and construction management, as well as business models and organizational structures.

However, at this moment, due to limited Blockchain storage and relatively slow transaction speed, it is worthless, time-consuming and overloaded to store all the data of a construction project in the Blockchain. Many researchers suggest the notion of off-chain data storage, which is a conventional or a distributed database to store the data outside of the Blockchain. For example, Xu et al. [17] propose an architecture using Blockchain for connecting application layer within off-chain data and off-chain control.

2.3 CDE and ICDD

In the construction sector, the common data environment (CDE) could be used as an off-chain data storage to connect with the Blockchain. A CDE stores all the data (i.e. geometrical and semantical information as well as the documentation) related to the lifecycle of a construction project [18]. Due to its functions of bringing all information together and serving as central data management tool, the CDE is widely recognized as a solution for implementing the BIM method [19]. Meanwhile, Preidel et al. [19] list several existing approaches which could be an application of CDE: BIMserver.org, A360 & Forge, BIMcloud, BIMcollab, bim+ and Trimble connect.

ISO 21497-1 was developed for the standardized data exchange of linked data sources using a generic information container format. According to ISO 21497-1, “a container is a file that shall have an extension ‘.icdd’ and shall comply with ISO 21320-1 also known as ZIP64”. In this sense, the ICDD can store the 5D BIM model data and be uploaded in a CDE.

A container includes a header file in the top-level folder. A container shall have at least three folders. The 'Ontology resources' folder may be used to store ontologies providing the object classes and properties that shall be used to specify the contents of and links between the documents within the container. The 'Payload documents' folder and 'Payload triples' folder may contain nested folders to allow groups of associated digital resources to be held together and referenced as a group (e.g. a building information model with its associated reference files or a set of linked spreadsheet documents) (cf. Fig. 1). A container according to ISO 21497-1 can also be used to exchange data from 5D BIM models and forms the information basis for the approach presented here. Building models, BoQs and QTOs are linked with the help of a container and used in the creation of a building contract.

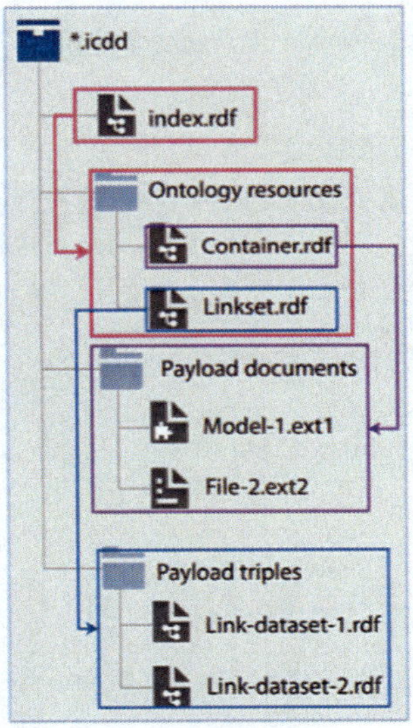

Fig. 1. The structure of an information container for a data drop (according to ISO 21497-1)

3 Methodology

For the definition of a Smart Contract in the construction industry and its implementation based on distributed ledger technology, it is necessary to specify exactly which contractual regulations are to be implemented and which information has to be stored. The presented approach assumes that not all contractual regulations will be transferred into Smart Contracts. Certain aspects such as information on insurance, cancellation policies or regulations on the limitation of claims for defects are still described in conventional ways. Therefore, a paragraph is included in the paper-based contract that refers to the Smart Contract, i.e. if the conventional contract is signed, the Smart Contract is also accepted and set up (cf. Fig. 2). To enable an evaluation, it must be ensured that the Smart

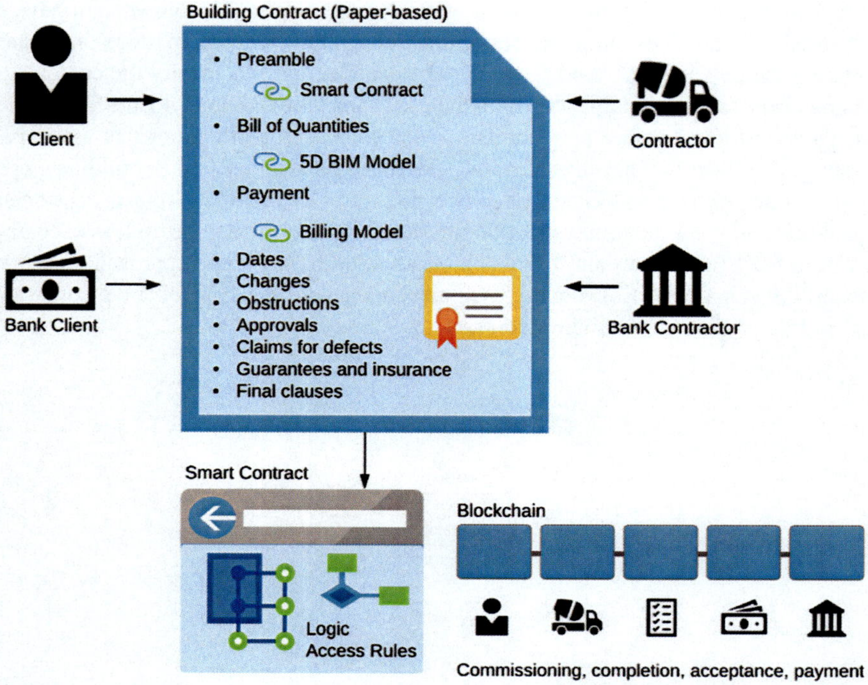

Fig. 2. Integration of a Smart Contract into a paper-based building contract

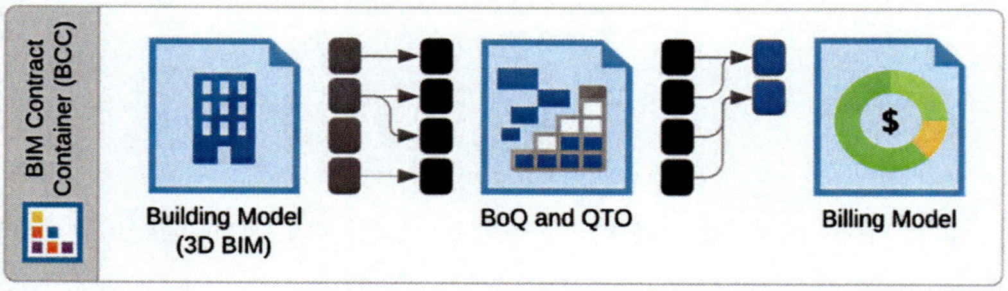

Fig. 3. Linking of 3D BIM model, BoQ, QTO and billing model

Contract can be viewed and understood by the parties involved. The textual or visual representation of a Smart Contract will not be discussed in this paper.

The 5D BIM model describes the construction work to be commissioned. For the payment, a billing model proposed in this paper must be developed based on the 5D BIM model. Depending on the type of contract used, the structure and the content of a billing model can be different. In consequence, a billing model is the digital equivalent of a payment plan and contains single billing units (e.g. in terms of area and cubic volumes or individual components). Additional conditions, interactions and processes are defined so that an automatic payment can be started. For example, a payment can only be started once the corresponding billing unit has been confirmed by the ordering party. The Smart Contract is implemented on a Blockchain or distributed ledger technology. This means that the transactions for automated payment are checked and stored via the Blockchain.

3.1 BIM Contract Container

The 5D BIM models already provide information for the description of construction works. Within the concept, it is assumed that a digital building model in IFC format is linked to the Bill of Quantities in GAEB XML format (more specific, in X86 format of GAEB XML). Single items of the BoQ are usually grouped together into larger billing units. The items can be spatially or logically grouped together using the building model. The billing units form an independent digital billing model that is stored in addition to 5D BIM information. Each unit has a globally unique ID (GUID).

A simple XML-based data format has been defined to store the billing units (Fig. 4). Beside the GUID information about costs, the delivery date and descriptions are also stored. The corresponding links to single quantity splits of the BoQ elements are specified by using a special payload triples document in the information container (cf. Fig. 5).

```xml
<?xml version="1.0" encoding="utf-8"?>
<!--Billing Model Example-->
<BillingModel ID="1YIsmNMZj5bQrUEeiAMoKE">

  <BillingUnit ID="12SBg5R7DEOPAG2AUZkifj">

     <Description>Shell Construction Building Part A</Description>
     <PaymentAmount Unit="EURO">500,000.00</PaymentAmount>
     <PaymentDueDate>2019-01-27</PaymentDueDate>

  </BillingUnit>

  <BillingUnit ID="0JZF8RWkzBoQ3pK$81Mctt">
    ...
  </BillingUnit>
  ...
</BillingModel>
```

Fig. 4. XML-based data format of the billing model

In the current approach, the schedule is not taken into account. To store the essential 5D BIM information (the 3D BIM, BoQ, QTO and billing model) for the construction project, a BIM Contract Container (BCC, cf. Fig. 3) is needed. The information in this BCC is linked and stored as an information container for data drop (ICDD) according to ISO 21497-1 (Fig. 5).

3.2 Acceptance Procedure

After specifying the data to be stored in the Smart Contract and consequently in the Blockchain, the contractual regulations have to be defined. Depending on the

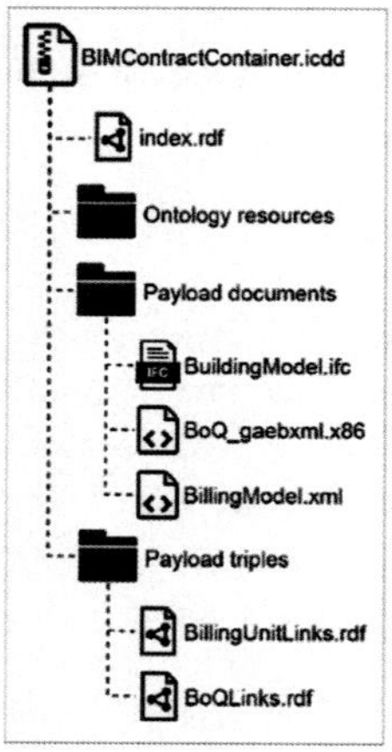

Fig. 5. BIM Contract Container according to ISO 21497-1

contractual situation, the individual rules can become extremely complex. Therefore, the approach is that individual general rule components exist which can be combined to form a complex set of rules. Each rule component consists of single transactions, interactions and if-then statements.

The standard rule for each billing unit with a simple procedure for the removal of construction defects can be described as follows (see Fig. 6). The contractor documents that he will start work on a billing unit (e.g. BU1). The contractor later informs that he has completed the work. The client checks the construction work performed and documents the results. There are now various options to proceed. If no defects are found, the agreed payment is initiated. If defects are found, the client has a claim for subsequent performance. The billing unit is divided into a correct (BU1.C) and a defect part (BU1.D). For the partial work performed without defects, the client defines an appropriate payment and initiates the payment. The initiated payments (complete or partial) are confirmed by the bank. The contractor is then requested to correct the documented defects. The client will set a reasonable deadline for this. The open work with the associated payment is defined as an open billing unit. Before the correction of defects is activated in the Smart Contract, a confirmation from the contractor must be obtained. If the contractor accepts the correction of defects, the fulfilment, acceptance, and payment procedure is started for the open billing unit. If the correction of defects was not accepted, an arrangement outside of the Smart Contract must be made based on legal framework conditions.

The individual information and transitions are stored in the underlying Blockchain. A block of the Blockchain includes a link to the BCC and the corresponding billing unit. A block can only be added if the conditions defined in the Smart Contract are fulfilled (cf. Fig. 7). It is very important that only the necessary information is stored in the Blockchain. The combination of using the billing model and the Smart Contract is the key for automated payment. Aspects of personal data protection must be taken into account. Feedback from the participating banks is also stored in the Blockchain. In principle, it would also be possible to use a digital currency.

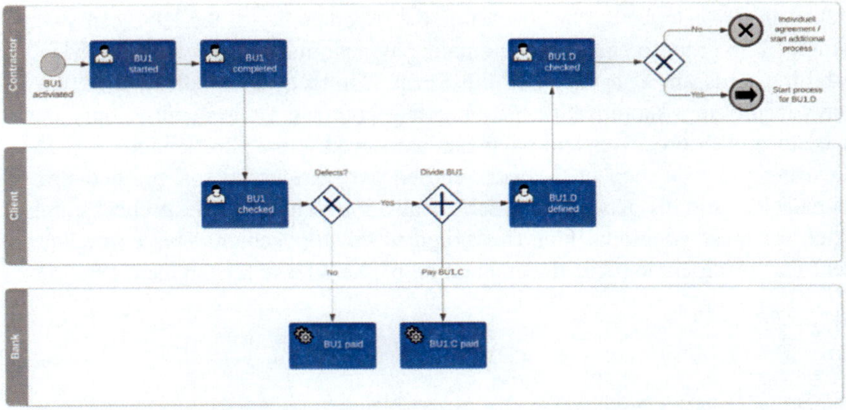

Fig. 6. Process for automated reporting, checking and payment of a billing unit.

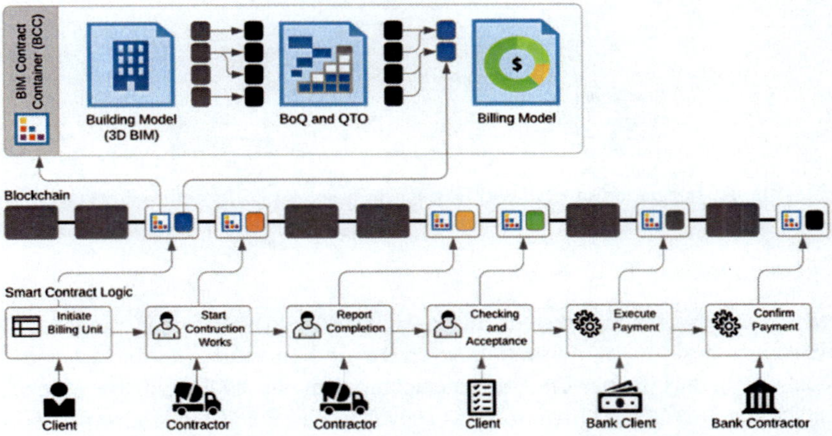

Fig. 7. Linking the BIM Contract Container via Smart Contract using Blockchain technology

3.3 System Architecture

During the contract negotiations, a BIM Contract Container (BCC) was agreed upon and subsequently also contractually agreed upon. However, due to the size of the data, it does not make sense to store a complete BCC in the Blockchain. Usually a common data environment (CDE) is provided to collect, manage and share all the project information. In the presented approach, the BCC is stored in the CDE and the contractual partners are given access to the CDE and the BCC. Using CDE as an off-chain (i.e. outside of Blockchain) data storage can lighten the load on the Blockchain. In this way, only the important data (e.g. payments and processes) is recorded in the Blockchain, and the rest data is stored in the CDE. To ensure that the BCC was not modified, a checksum is subsequently created for the container using Secure Hash Algorithm 3 (SHA-3). The checksum is included in the Smart Contract and stored in the

Blockchain. By calculating the checksum, the immutability of the BBC can always be guaranteed. In order to enable an automatic payment, the contractually agreed costs of the billing units are also stored in the Smart Contract. A modification of the BCC always results in a modification of the actual contract. Consequently, only the data related to project processes and payments are stored in the Blockchain.

During the inspection of the work carried out, various photos, protocols or other documents are usually generated, which are also stored in the CDE. A checksum is also created for these additional files and stored in the Blockchain. This ensures a transparent and manipulation-free documentation of the release for payment (Fig. 8).

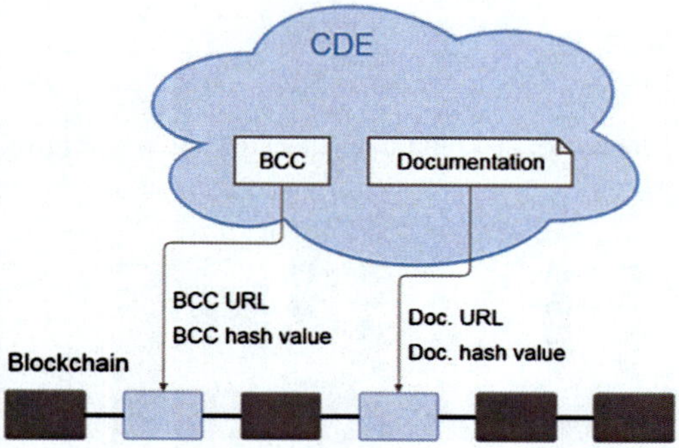

Fig. 8. Integration of CDE and Blockchain based on URLs and hash values

3.4 User Interface Framework

Based on the system architecture, the user interface framework is proposed. The framework is designed to illustrate how clients or contractors can interact with this application. In this framework, the interaction with the CDE and the Blockchain technology are divided into two tools as shown in Fig. 9 and Fig. 10, respectively.

The smart contract setup tool (Fig. 9) loads the 5D BIM data from the CDE. The 3D BIM model and the BoQ information can be used to define the billing model and the Smart Contract. For each billing unit, the acceptance and paying procedure can be defined. The user can select out of predefined Smart Contract templates relevant components for the current project to specify an individual Smart Contract. Furthermore, individual Smart Contract rules can be created by using the Smart Contract setup tool. After the client and the contractor have agreed on a billing model and digital contract modules, the Smart Contract is digitally signed. Consequently, a BCC with a 3D BIM model, a BoQ data and a billing model is generated and uploaded to the CDE. The CDE returns the URL and the hash value of the BCC, which will be stored together with the Smart Contract rules in the Blockchain and become the initial data of the Blockchain.

Once the paper contract has been signed, work can begin on the construction site. The Smart Contract defines processes and interactions. All interactions with the Smart Contract are stored in the Blockchain. In the following, only the acceptance by the client is highlighted. For checking, documenting and starting the payment, the client

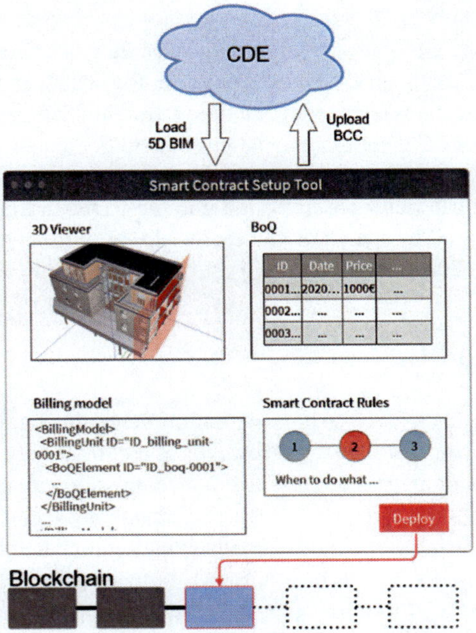

Fig. 9. The user interface framework of the Smart Contract setup tool

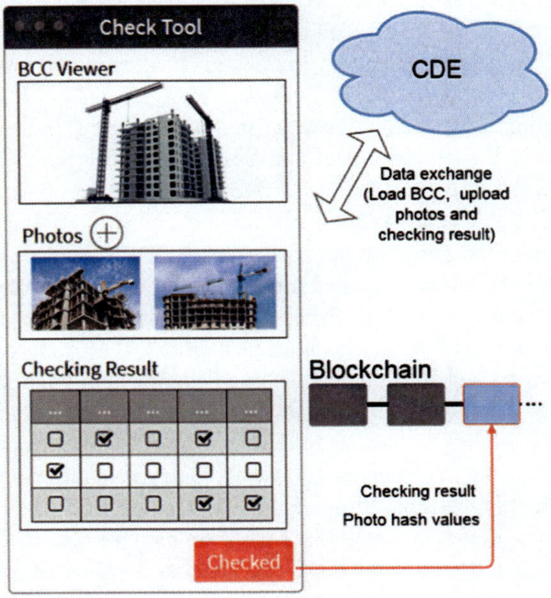

Fig. 10. The user interface framework of the check tool

should use a user-friendly application (e.g. an app for a mobile device, Fig. 10). During the assessment, photos are often taken, various checklists are filled out and the entire assessment process is documented. For this purpose, the client can navigate through the 3D BIM model, view individual contracted construction works and also check the billing units. Furthermore, it should also be possible to document any shortcomings. If everything is correct, the construction work can then be released. During the release, the information and documents are uploaded into the CDE. In the next step the URLs and hash values returned by the CDE and the original results of the checking process are stored in the Blockchain.

4 Conclusion and Outlook

Digitization is widely considered as the solution for the optimization and automation of processes in the construction industry. However, the current procedure of a construction project still relies on paper contracts and conventional communications, which are time-consuming and nontransparent. Using a 3D modeling technology linked to cost information (i.e. 5D BIM) can simplify the payment and construction process. However, the usage of 5D BIM currently is at a very early stage and still cannot automate the construction project procedure. This paper presents a framework for automated billing by combining the BIM Contract Container (BCC) with Smart Contracts. All the project-required data are stored and linked in the BCC. Compared to the current procedure of a construction project, the whole construction process proposed in this paper could be simplified and automated. All the linked 5D BIM data and the billing model can be stored in the CDE using BCC, and the sensitive and important information (i.e. acceptance procedures and payments) is safely stored and recorded in the Blockchain.

For implementing automated payment of this framework, it is vital to have a more specific definition of the billing model and Smart Contract rules, which will be discussed further in the future. To select a suitable CDE for the proposed framework, the feasibility tests of the existing CDE applications (e.g. bim+, Trimble connect, etc.) should be provided in the future research. A suitable CDE should be able to return the URL and the hash value of the uploaded files. The following five factors should be first considered for selecting a suitable Blockchain platform: 1) Does it support Smart Contract applications? 2) How much data can it store? 3) How fast is its transaction speed? 4) What kind of chain (e.g. public chain, consortium chain or private chain) does it support? 5) Could it interact with the user interfaces we proposed? For the development of the Smart Contract setup tool and the check tool, it is essential to take the following four points into account: 1) Are there any existing open source 3D BIM viewers that can be plugged into the tools? 2) How can a BCC file be generated and hashed in the Smart Contract setup tool? 3) How can the Smart Contract rules be defined? 4) What data stored in the CDE can generate a checklist and show it in the check tool?

As an initial attempt, this proposed framework is simplified and does not cover all the factors of the construction process which should be considered in the future. For example, in this framework, only the client and the contractor are taken into account.

However, in a real construction project, the subcontractors also play an important role and should be considered. The different requirements for different participants also need to be more specific. This framework supposes that the client can define an appropriate payment for the construction project, which is not practical currently. Moreover, as the Smart Contract rules still need to be set manually in the Smart Contract setup tool, this framework is not fully automated yet.

References

1. Agarwal, R., Chandrasekaran, S., Sridhar, M.: Imagining construction's digital future. McKinsey Productivity Sciences Center (2016)
2. Kamardeen, I.: 8D BIM modelling tool for accident prevention through design. In: 26th Annual ARCOM Conference, vol. 1, pp. 281–289. Association of Researchers in Construction Management, Leeds (2010)
3. Dorri, A., Steger, M., Kanhere, S.S., Jurdak, R.: Blockchain: a distributed solution to automotive security and privacy. IEEE Commun. Mag. **55**(12), 119–125 (2017)
4. Bahga, A., Madisetti, V.K.: Blockchain platform for industrial Internet of Things. J. Softw. Eng. Appl. **9**(10), 533 (2016)
5. Sikorski, J.J., Haughton, J., Kraft, M.: Blockchain technology in the chemical industry: machine-to-machine electricity market. Appl. Energy **195**, 234–246 (2017)
6. Nakamoto, S.: Bitcoin: a peer-to-peer electronic cash system (2008). http://bitcoin.org/bitcoin.pdf
7. Chaum, D.: Blind signatures for untraceable payments. In: Advances in Cryptology, pp. 199–203. Springer, Boston (1983)
8. Barkai, D.: An introduction to peer-to-peer computing. Intel Developer Update Magazine, pp. 1–7 (2000). https://www.researchgate.net/publication/265873606_An_Introduction_to_Peer-to-Peer_Computing
9. Brito, J., Castillo, A.: Bitcoin: A Primer for Policymakers. Mercatus Center at George Mason University, Arlington (2013)
10. Swan, M.: Blockchain: Blueprint for a New Economy. O'Reilly Media Inc, Beijing (2015)
11. Szabo, N.: Smart contracts: building blocks for digital markets. EXTROPY: J. Transhumanist Thought (16), 18, 2 (1996)
12. Buterin, V.: A next-generation smart contract and decentralized application platform. Ethereum White Paper **3**(37) (2014)
13. Buterin, V.: On public and private blockchains. Ethereum blog 7 (2015)
14. Turk, Ž., Klinc, R.: Potentials of blockchain technology for construction management. Procedia Eng. **196**, 638–645 (2017)
15. Wang, J., Wu, P., Wang, X., Shou, W.: The outlook of blockchain technology for construction engineering management. Front. Eng. Manag. **4**(1), 67–75 (2017)
16. Li, J., Greenwood, D., Kassem, M.: Blockchain in the built environment and construction industry: a systematic review, conceptual models and practical use cases. Autom. Constr. **102**, 288–307 (2019)
17. Xu, X., Pautasso, C., Zhu, L., Gramoli, V., Ponomarev, A., Tran, A.B., Chen, S.: The blockchain as a software connector. In: 2016 13th Working IEEE/IFIP Conference on Software Architecture (WICSA), pp. 182–191. IEEE (2016)
18. British Standards Institution: PAS 1192-2:2013 - specification for information management for the capital/delivery phase of construction projects using building information modelling (2013)

19. Preidel, C., Borrmann, A., Oberender, C., Tretheway, M.: Seamless integration of common data environment access into BIM authoring applications: the BIM integration framework (2015)
20. GAEB. https://www.gaeb.de. Accessed 27 Jan 2020
21. Brook, M.: Estimating and Tendering for Construction Work. Butterworth-Heinemann, Oxford (2008)
22. Penttinen A.: Chapter 10 – Network planning and dimensioning. Lecture Notes: S-38.145 - Introduction to Teletraffic Theory. Helsinki University of Technology (1999)
23. Lu, Q., Won, J., Cheng, J.C.: A financial decision making framework for construction projects based on 5D Building Information Modeling (BIM). Int. J. Proj. Manag. **34**(1), 3–21 (2016)
24. Fuchs, S., Scherer, R.J.: MMQL—a language for multi-model linking and filtering. In: eWork and eBusiness in Architecture, Engineering and Construction: ECPPM 2014, p. 273 (2014)
25. König, M., Habenicht, I., Spieckermann, S.: On-site logistics simulation in early planning phases. In: Proceedings of the International Conference on Construction Applications of Virtual Reality, pp. 728–737 (2011)

Using Blockchain Technology to Implement Peer-to-Peer Network in Construction Industry

Meiling Shi(✉), André Hoffmann, Anna Wagner, Tim Huyeng, Christian-Dominik Thiele, and Uwe Rüppel

Institut für numerische Methode und Informatik im Bauwesen, Technische Universität Darmstadt, Darmstadt, Germany
shi@iib.tu-darmstadt.de

Abstract. Collaboration is the intrinsic element of the construction industry since construction projects involve a dynamic grouping of several companies. [1] studied the relationships among the companies and noted that some communication patterns among participants show a Peer-to-Peer nature. Digital technology which implements this Peer-to-Peer communication pattern is the Blockchain technology (BCT). BCT has gained significant attention across different industry sectors in the past years, such as finance, health care, supply chain management, etc. BTC can solve the trust and security problem of the Peer-to-Peer network. It enhances effectiveness, transparency, productivity, and cybersecurity in collaborative works. While BCT has been intensively researched in fields like the financial industry, healthcare, and supply chain management, the research in the construction industry is still in its early stages. Besides that, the majority of articles related to BCT in construction are still concepts. This paper gives an architecture of using BCT in collaboration with P2P nature and presents the proof of concept by implementing the BCT in the distributed product data platform.

Keywords: Collaboration · Peer-to-Peer network · Blockchain technology · Product data service

1 Introduction

The construction industry has a multi-disciplinary participants circle: architects, engineers, construction companies, manufacturers of building material and building components, etc. They work concurrently or serially on a construction project. Collaboration exists in several scenarios in the construction industry. Such as the design phase, where several disciplines are working parallel, and information exchange among small disciplinary groups is necessary. Or in construction logistics and supply chain management, where several suppliers, distributors, and construction companies arrange and keep track of goods information at a different time. Collaboration in those fields can reduce waste and improve productivity significantly [2]. Within the collaboration, organizations usually need to share and exchange their operating data and sensitive business information. [1] studied the degree to which the relationships concerning data sharing among the construction participants are hierarchical or Peer-to-Peer. They noted that

E. Toledo Santos and S. Scheer (Eds.): ICCCBE 2020, LNCE 98, pp. 839–849, 2021.
https://doi.org/10.1007/978-3-030-51295-8_58

particularly in information-intensive phases, centralization of information was necessary because of technology restriction. When dealing with communication patterns, which shows a Peer-to-Peer (P2P) relationship among participants, a un-constraining technology such as the P2P network is more suitable.

A P2P network is one of the architectures of distributed systems. In a P2P network, the participants share a part of their hardware resources, e.g., processing power or storage. These shared resources are necessary to provide the service and content offered by the network, such as file-sharing or shared workspaces for collaboration. They are accessible by other peers directly, without passing intermediary entities. The participants of such a network are resource providers as well as resource requestors. Any single, arbitrary chosen peer can be removed from the network without having the network suffering any loss of network service [3]. The P2P network is resilient to single server failure and more cost-efficient compares to a central server. However, the P2P network also has several drawbacks regarding "Trust": Firstly, since there is no central authority management and peers are free to join or leave, the network is not trustable and the membership is unmanaged; Secondly, since there is no central server that manages the peers' memberships, it is difficult to connect to a P2P network without knowing the IP-address of one peer in advance. Thirdly, there is no central address book that registers which peer has which resources. The absence of an address book is also a big problem for users who are searching for specific resources in this network. To implement a trustful P2P network for collaboration in the construction industry, the authors propose a concept of using a permissioned BCT as an address book in the P2P collaboration and gives a proof of concept by applying BCT in distributed product data services.

This paper is structured as follows: Sect. 2 explains the background knowledge of the centralized system, the P2P network, the BCT, as well as the consensus mechanism. Section 3 lists research in the last years that applies BCT in the construction industry. Section 4 describes the suggested concept of using blockchain in P2P systems with the example of implementing BCT in the manufacturers' product data service. Sections 5 and 6 give a discussion of results and the conclusion of the authors, respectively.

2 Background

2.1 Centralized System

A centralized system is one that uses client/server architecture. In this system, a major server communicates with its client nodes, while nodes cannot connect directly without through the central server. The central server organizes the authorities and memberships of participants and acts as a central authority. Centralized systems are being used very commonly in the construction industry because of authority, trust, and manageable nodes in the network. For example, the online shopping platform, company intern SAP system, or the BIMserver. A centralized system also has some limitations: It is vulnerable to cyberattacks and system failure. Vertically it can't scale up without big expense after reaching a limit. Since only one central server responds to the requests from clients, communication bottlenecks can appear when too many clients try to connect with it. Also, the centralized system is less transparent since the central server only belongs to one specific company. Some resources in the system or some clients can be preferred than others, which can lead to unfairness.

2.2 Peer to Peer System

A peer-to-peer (P2P) system is one in which participant peers or nodes rely on each other for service, rather than solely relying on dedicated and often centralized infrastructure [4]. Each peer connects to a relatively small set of neighbors, which in turn also connects to other peers. Each peer in this network is a server, as well as a client at the same time. In a P2P system, users can submit queries to any random peer. The receiving peer first evaluates the query in its resource base. If the peer can serve resources to the query, it returns the user the answer. Otherwise, the peer forwards the query to its neighbors randomly. To enable peers forward queries more specifically, destination indices can be used [5]. BCT is a suited method for redirecting queries.

2.3 Blockchain

By design, blockchain is well suited for systems that require integrity, transparency, and immutability [6]. A blockchain is essentially a chain of chronological blocks which is maintained by nodes in a distributed network who do not trust each other. A node is a participant in the blockchain network, which has a similar meaning as a peer. Every node in the blockchain network holds the replicated data in the form of blockchains. Each block consists of a set of transactions and the digital signature of its generator. Transactions in BCT are executions done by participants in the blockchain network, which arises status changes, such as change of the asset's ownership. In a blockchain network, nodes conduct transactions with each other. Every once in a while, transactions are collected into a block by a miner. Miners are nodes with transaction collection and verification rights. Within the mining process, miner packs transactions that are executed in the nearest timespan in one block. This block consists of the hash value of its previous block, the collected transactions, a timestamp by which the current block is created, and a hash value of the current block. The current hash value and the hash value of the previous block links block chronologically together, which makes a chain of blocks. Subsequently, all nodes in the network give their votes to this block, which is called consensus. If the consensus rule is satisfied, for example, more than 51% of the nodes have given their endorsement, the miner broadcast this block to all other nodes in the network. These nodes in the network then append the new block to their local blockchain.

Blockchain is considered to be very secure. Every node in a blockchain network is equal-valued and can respond to clients' requests individually, which ensures scalability. Replication of the same data on distributed nodes and consensus ensures fault tolerance. Moreover, since blocks are linked together chronologically through hash values, the manipulation of data is even more challenging. Because as long as one node edit transactions, the hash value of the block changes. Once the nodes append a block into the blockchain, it is almost impossible to manipulate or erase this entry [7], since one needs to manipulate data in all nodes. By signing a digital signature in each transaction, the identity of the transaction executor can be traced later. Thus, by using cryptography, distributed database, consensus algorithms, and digital signature, blockchain provides "Trust" in a distributed network without the need for a certified central authority.

According to its publicness, there are three kinds of blockchains: public, permissioned, and private blockchain. In a public blockchain, anyone can read and write blocks in it or participate in the consensus process to determine the "valid" state. A permissioned blockchain, also called consortium blockchain, only grants the write permission to a consortium of known nodes in a given business context. Permissioned blockchains have methods to identify the nodes that can access and update the shared states. A private blockchain only allows a single node to operate blocks [8]. In the collaboration process in the construction industry, usually, multiple partners who are connected by contracts need to occasionally share operation data and information which should not be seen by nodes out of the consortium. Thus the permissioned blockchain, which provides transaction privacy for consortium from the public, is most suitable.

2.4 Consensus

In a p2p network, individual nodes can crash, behave maliciously, or act against the common goal. To make the P2P network dependable and resilient, BCT runs a fault-tolerant consensus mechanism. The consensus mechanism ensures that all nodes agree on the same blockchain status [8]. There are several consensus mechanisms designed for different blockchain types. PBFT (Practical Byzantine Fault-Tolerance) is one of them designed for permissioned blockchain [8]. PBFT requires that every node is known to the network [9]. It can tolerate Byzantine faults and handle up to 1/3 malicious nodes [10]. In the mining round, where a new block is created, the P2P network will elect a miner node according to some predefined rules. This miner node is responsible for collecting and validating the transactions, as well as creating a new block. Afterward, the miner node broadcasts the new block to rest nodes in the network. The rest nodes verify the block's correctness and give their votes to it. As long as the miner node receives 2/3 of all nodes in the network, the new block will be appended into the blockchain, and the next mining round starts.

3 BCT in Construction Industry: Related Work

The emergence and widespread use of electronic collaborative technologies for distributed project management have opened up opportunities for sharing data and collaborative development, such as digital cloud platforms. Because of the complexity and short of professional personals, these platforms are usually outsourced and provided by software development companies. Besides, most tools and techniques for project management focus on static, long term relationships when inter-organizational relationships are becoming more dynamic and geographically separate [11]. [1] studied the relationships concerning data sharing among the construction participants are hierarchical or peer-to-peer. They noted when using un-constraining technology, communication patterns among participants show a peer-to-peer relationship. Other than the p2p relationship that exists in the construction industry intrinsically, there are concerns that a single third-party provider can threaten the security and privacy of shared data. Thus the demand for distributed systems, which are independent of central third party control, increases.

Nowadays, BCT is one of the leading technology to implement a p2p system. It has the benefit of a traditional p2p network and solves the trust problem in the absence of central administration. BCT has been successfully applied in several fields, such as data management, finance, supply chain, and health. However, there are limited research and applications of BCT in the AEC industry [12]. [13] looks into the innovation that has been invented to solve the "Trust "problem in the absence of administrative overhead and suggests that the combination of BIM+blockchain has potential as a platform for true collaboration. This collaboration platform brings simulated design, lean process, and energy efficiency design together so that the intrinsic intangible value of the engineers' collaborative professional service over the buildings' lifecycle can get rewarded. [14] investigated the potential of applying BC to three sectors in construction management: Notarization-related applications to eliminate the verification time of construction documents' authenticity, Transaction-related applications to facilitate automated procurement and payment, and Provenance-related applications to improve transparency and traceability of construction supply chains. Applying BCT in those three sectors can avoid disputes, reduce time and cost. (This paper also arises the technical, construction business-related and human-related challenges.) [1] discusses the suitability of BCT for construction management and gives four scenarios to use BCT for managing building information. According to [1], integrating BCT with a BIM Server is the most promising one of the four scenarios since the data size of the BC can be expected only one order of magnitude larger than the BIM Server database. [12] contributes a thorough review of 420 scientific publications in BCT and its current application in the AEC industry and provides approaches for integrating BC with BIM workflow. [15] describes a concept of combining RFID technology with BCT to reduce losses from human factors and intentionally false information and to strengthen trust between the participants in the supply chain management of construction materials. [16] present a proof of concept implementation of an integration strategy for connecting CAD with blockchain. The authors use Rhino/Grasshoper and Ethereum (an open-source blockchain platform) to record the differences between different versions of CAD plans.

4 Example of a P2P Network in the Construction Industry: Manufacturers' Product Data Service

With buildings and infrastructures are getting more sophisticated, the requirement of direct embedding digital product data from manufacturers into building information models in design phases increases. Through integrating product data into digital models, building analyses, such as energy simulation or construction simulation, can be done more precisely. To find the most suitable product for a building, planners spend much time searching for and comparing products. There are several platforms to compare products from third-party, such as BIMobject, buildup, However, Heinzer-BIM, and Plan.One [17]. The main problems of these platform providers are the uncertainty of a central authority, concerns of data security, and lack of transparency. Manufacturers have a legitimate interest in retaining control over the disclosure of their product data and to make sure that their products are not neglected intentionally by platform providers while being searched from the internet. Thus, to satisfy the demand for data autonomy, data security, and transparency, a P2P network is needed.

The basic idea of the approach presented in this paper is a network of manufacturer data services. Each of the data services provides product descriptions that contain information created in the design and development phase. Each manufacturer can integrate a data service into its IT infrastructure, e.g., via an API gateway that orchestrates requests and forwards them to the data service. Thus the data services are in the responsibility of manufacturers. The data service can be requested by users as well as by other manufacturers to answer the user queries they received. To redirect queries more efficiently across the P2P network, the manufacturer data services are registered in a destination index [5], which is called a distributed product data catalog (PDC) in this paper. The PDC stores the IP addresses of the manufacturer data services and the specific product data types the manufacturers offering.

This PDC allows users and search applications to find services relevant to them. Instead of using a centralized authority for PDC, which is vulnerable to malicious attacks, the authors suggest making the replication of PDC a part of the manufacture's data service. The PDC must stay synchronous in the P2P network via a process that is accepted by all parties involved. However, the synchronization of transactions in a P2P is very demanding, especially if the system has enormous peers. In the synchronization process, the following requirements must be satisfied: The data in the distributed nodes shall remain consistent; malicious data manipulation by unauthorized persons must be avoided; All updates in local PDC of each manufacturer need to be traceable, who has done which updates to the sharing data catalog; The entering of new nodes in the system shall be flexible and controllable without a central administration. Because of the natural properties of distributed systems, say concurrency, geographical separation, and lack of a global clock, it is not straightforward to synchronize the replicated data across the distributed network without problems. The traditional methods for synchronization, such as Master-Slave and 2-Phase-Commit, are not suitable because the synchronization process is managed by one central peer, which is vulnerable to attack and cannot scale well.

This article looks into using BCT to realize the synchronization of PDC in a P2P network. With blockchains, the P2P can continuously update the executed transactions and automatically synchronize them over the network to all peers. Through the chaining property of blockchain, all updates and transactions taken in distributed systems are traceable. With the consensus procedure and the cryptographic encryption, the data in distributed systems is tamper-proof, since it takes substantial computing power to tamper with the data in all peers of a blockchain network. Moreover, a blockchain network can be publicly, privately, or permssioned accessible according to projects' requirements. Considering all the requests, the authors decide to use permissioned BCT to synchronize the PDC. With the help of the approach presented, changes in shared PDC in the P2P collaboration are traceable, transparent, tamper-proof, and trustful.

4.1 Concept and Implementation

This paragraph presents the architecture of a system named "Blockchain Service" (see Fig. 1), which is a part of the product data service to synchronize distributed PDC. The architecture of the system includes:

1. A local database stores "Product Data Catalog" (PDC), which will be synchronized.
2. A Data Catalog API (DC-API), which collects and edit transactions (insert, update and delete queries in a database) from the local PDC;
3. An Encrypt Service (EC), which signs the edited transactions with the manufacture's digital signature and sends the signed transactions to the miner node according to Membership Service;
4. A Membership Service (MeS), which manages the digital certificates of manufactures in this blockchain consortium;
5. A Mining Service (MiS), which receives transactions from other Blockchain Service nodes and sends the generated block back to the nodes;
6. A Block Service (BS), which packs the identified transactions in a block and updates the local Blockchain.

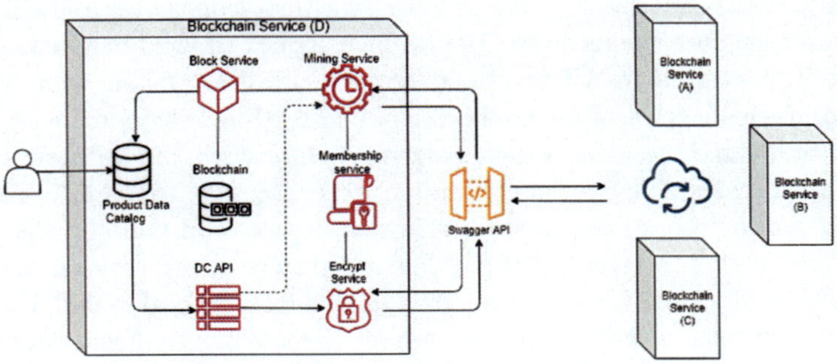

Fig. 1. Architecture of Blockchain Service

Before starting the blockchain network, manufacturers need to get a digital certificate for the future BS from a trusted institution (e.g., buildingSMART). The digital certificates will then be stored in the Membership Service (MeS) in each Blockchain Service instance, which forms the blockchain consortium. Every manufacturer in this consortium needs to install the Blockchain Service instance in their product data service. Each manufacturer that implemented the product data service will be called as a "node" in the P2P network. The consortium elects an initial node that acts as a miner to start the first mining round. The duration of a mining round will be set before starting the blockchain network by the consortium. Within a mining round, the network communication works as follows: A user (people who are responsible for the manufacture's PDC) edit the local PDC. Most of the database system possesses a logging function which saves the users' operation steps into a log file. Whenever users edit, insert or delete the product data catalog, the transactions will be automatically saved with their corresponding timestamps in the log file. The DC-API catches the local transactions during a mining round. To identify which transactions are being collected in a mining-round, the DC-API stores the timestamp of the start of the round and the duration of a mining round. If a mining round is reached since the previous mining process, an event will be triggered, and the DC-API looks up in the log file and get all transactions since the last timestamp. To make the updating of transactions more efficiently later in the miner node, the local logs from the DC-API will be firstly edited before sending them to the mining node. The relevant transactions will be combined if they concern the same row. For example, if one row is inserted and then updated afterward in the same round, these two transactions will be combined so that there will be only one insert transaction with the latest information in it. Or suppose a row is inserted and then deleted later in the same round, the DC API will not collect this row.

After that all the local transactions during the previous mining round are collected, the finishing timestamp of collecting transactions will replace the old timestamp. The DC-API then hand the edited transactions over to ES. ES will sign the edited transactions with this node's digital signature and send it according to the MiS to the miner node in this blockchain network.

The miner node receives the signed transactions from nodes in the consortium and verifies its identity with MiS. The MiS references the MeS for digital signatures of the senders' IP addresses. If it is true, the MiS hand the transactions over to the BS; if not, the MiS turns the transactions down. The BS then applies all valid transactions into its own PDC. The transactions will be used to create a new Block, along with its creation timestamp, the hash value of the previous block in the Blockchain, the hash value of the actual block, and the miner's digital signature. Afterward, the MiS sends the new block to other nodes' MiS in the consortium for consensus. This article uses PBFT consensus, which requests a positive verification of two-third nodes in the network. The MiSs of other nodes verifies if the digital signature of the miner is correct and whether their transactions are stored correctly in the block. If so, they add this block to their local blockchain and apply the transactions in the block to their local databases. Otherwise, they will send a "Block is faulty" notification to other nodes in the network. Nodes that receive this notification and verified its correctness would turn to refuse the block. As long as one-third node complains the accuracy of the block, the blockchain network will break down and the consortium needs to take a closer look into the blockchain. Otherwise, the next miner will be selected and start mining.

If the blockchain reaches a size limit, it can also be exported from the Blockchain Service and stored elsewhere. Before exporting the old blockchain, the consortium should make sure of the data consistency in all nodes. A new blockchain will start with the last hash value of the last block in the old blockchain.

5 Results and Discussion

The Program BS was implemented with MySql Server and python programming language. The product data service forwards quires getting from users to other data services who have the being requested product according to PDC. The PDC, which is a MySQL database, has a logging function that writes files for documenting what activity is taking place. The BS filters out transactions of the current mining round based on timestamps, which will be synchronized periodically with BCT. The services described in Sect. 4 were programmed in the python programming language. Existing python packages (such as flask for web development, json for exchanging data, hashlib and crypto for encryption and decryption, etc.) were used not to reinvent the wheel.

To validate the proposed concept, the authors have set up 4 product data services in a windows machine with Intel(R) Core(TM) i7 CPU and 16 GB RAM. All 4 data services have the same PDC at the beginning, which is being updated separately afterward. A mining round takes 2 min. The data services communicate through the local network with each other.

The results confirm the proposed concept. Miners are selected randomly by the services. Transactions of a mining round can be filtered out from logging files and sent to miner nodes. The miner node verifies the authorities of senders' IP-addresses by decrypting the message with senders' public keys. After collecting transactions from all nodes, the miner generates a block and sending it back to other nodes, who then append the new block into the local blockchains and apply the transactions in their PDC.

The implementation also shows two shortcomings:

1. Scalability: since only one miner node receives transactions from all other nodes and usually one server has only limited ports, the miner node can easily reach his computing limitation. One possible solution would be the partition of the P2P network, along with a two-phase mining process. There will be two-level of miners in each mining round: the global miner and the local miners in each subgroup network. The local miner collects all transactions from its subgroup and verifies the validity of the transactions. Afterward, the local miner sends the verified transactions to the global miner, who then creates a new block from it. The two-level-miner process can be extended to an n-level-miner process, which reduces the web traffic logarithmically.
2. Performance: While testing, the implemented product data service and blockchain service usually takes seconds or even minutes, depends on the queries' complexity, to return responses and synchronize the BC with a new block. One possible reason may be the nature of blockchain having relatively low throughput [18]. One option to increase throughput is to use off-chain blockchain [18]. Another reason may be that the authors didn't pay too much attention to code performance during development. To find whether it is a result of the system architecture or because of the programming, new implementation with marketable BC platforms like Hyperledger or Ethereum is necessary. If the performance problem is owing to system architecture, off-chain blockchain [19, 20] could be one solution.

6 Conclusion

This article describes an idea and the proof-of-concept of applying BCT in P2P Systems in the Construction industry. The proof-of-concept presents the implementation of blockchain technology in a distributed manufacturer product data service. The implementation conforms advantages that the BCT provides: immutability, traceability, and transparency of cryptically secured transactions. Those properties build up trust in a P2P network without a central authority, which is the weak point for cyberattacks. One of the assumptions to use blockchain successfully is that the majority of participants in the network works honestly. Otherwise, the system can fail on the majority attacks [21] or the byzantine fault.

The concept we proposed can be further investigated and implemented in other P2P relationships for collaboration in the construction industry, such as supply chain management for construction materials, RFID for tracing the building components during their life cycle, and paralleled design, etc. The smart contract would also be an interesting extension to the architecture to automatic rule-based execution—for example, user and access rights management in multi-disciplined construction project platform with smart contracts.

Acknowledgment. This work is part of the research project EnOB: Semantic Construction Project Engineering, founded by the German federal ministry for Economic Affairs and Energy (BMWi).

References

1. Turk, Ž., Klinc, R.: Potentials of blockchain technology for construction management. Procedia Eng. **196**, 638–645 (2017). https://doi.org/10.1016/j.proeng.2017.08.052
2. Fulford, R., Standing, C.: Construction industry productivity and the potential for collaborative practice. Int. J. Proj. Manag. **32**(2), 315–326 (2014). https://doi.org/10.1016/j.ijproman.2013.05.007
3. Schollmeier, R.: A definition of peer-to-peer networking for the classification of peer-to-peer architectures and applications. In: Proceedings of the First International Conference on Peer-to-Peer Computing, Linkoping, Sweden, pp. 101–102 (2002). https://doi.org/10.1109/p2p.2001.990434
4. Gribble, S., Halevy, A., Ives, Z., Rodrig, M., Suciu, D.: What can peer-to-peer do for databases, and vice versa? In: Proceedings of the WebDB (2001)
5. Crespo, A., Garcia-Molina, H.: Routing indices for peer-to-peer systems. In: Proceedings of the 22nd International Conference on Distributed Computing Systems, Vienna, Austria, pp. 23–32 (2002). https://doi.org/10.1109/icdcs.2002.1022239
6. Wüst, K., Gervais, A.: Do you need a blockchain? In: 2018 Crypto Valley Conference on Blockchain Technology (CVCBT), pp. 45–54 (2018). https://doi.org/10.1109/CVCBT.2018.00011
7. Crosby, M.: Blockchain technology: beyond bitcoin, no. 2, p. 16 (2016)
8. Cachin, C., Vukolić, M.: Blockchain consensus protocols in the wild. arXiv:1707.01873 Cs, July 2017

9. Zheng, Z., Xie, S., Dai, H., Chen, X., Wang, H.: An overview of blockchain technology: architecture, consensus, and future trends. In: 2017 IEEE International Congress on Big Data (BigData Congress), pp. 557–564 (2017). https://doi.org/10.1109/BigDataCongress.2017.85
10. Castro, M., Liskov, B.: Practical byzantine fault tolerance, p. 14 (1999)
11. Qureshi, S., Liu, M., Vogel, D.: The effects of electronic collaboration in distributed project management. Group Decis. Negot. **15**(1), 55–75 (2006). https://doi.org/10.1007/s10726-005-9006-6
12. Nawari, N.O., Ravindran, S.: Blockchain technology and bim process: review and potential applications, p. 31 (2019)
13. Mathews, M., Robles, D., Bowe, B.: BIM + blockchain: a solution to the trust problem in collaboration?, p. 11 (2017)
14. Wang, J., Wu, P., Wang, X., Shou, W.: The outlook of blockchain technology for construction engineering management. Front. Eng. Manag. **4**(1), 67 (2017). https://doi.org/10.15302/J-FEM-2017006
15. Lanko, A., Vatin, N., Kaklauskas, A.: Application of RFID combined with blockchain technology in logistics of construction materials. In: MATEC Web Conference, vol. 170 (2018). Article Number: 03032. https://doi.org/10.1051/matecconf/201817003032
16. Dounas, T., Lombardi, D.: A CAD-blockchain integration strategy for distributed validated digital design, p. 8 (2018)
17. Der Vergleich von Produktvergleichsplattformen - Deutsche BauZeitschrift. https://www.dbz.de/artikel/dbz_Der_Vergleich_von_Produktvergleichsplattformen_3381292.html. Accessed 22 July 2019
18. Croman, K., et al.: On scaling decentralized blockchains: (a position paper). In: Clark, J., Meiklejohn, S., Ryan, P.Y.A., Wallach, D., Brenner, M., Rohloff, K. (eds.) Financial Cryptography and Data Security, vol. 9604, pp. 106–125. Springer, Heidelberg (2016)
19. Eberhardt, J., Tai, S.: On or off the blockchain? Insights on off-chaining computation and data. In: De Paoli, F., Schulte, S., Broch Johnsen, E. (eds.) Service-Oriented and Cloud Computing, vol. 10465, pp. 3–15. Springer, Cham (2017)
20. Poon, J., Dryja, T.: The bitcoin lightning network, p. 59 (2015)
21. Lin, I.-C., Liao, T.-C.: A survey of blockchain security issues and challenges. J. Netw. Secur. **19**, 653–659 (2017). https://doi.org/10.6633/IJNS.201709.19(5).01

A Secure and Distributed Construction Document Management System Using Blockchain

Moumita Das, Xingyu Tao, and Jack C. P. Cheng(✉)

The Hong Kong University of Science and Technology, Clear Water Bay, Hong Kong
cejcheng@ust.hk

Abstract. Construction Document Management Systems (CDMS) track, manage, and store a variety of documents such as 3D model files, schedules, specifications, and images that are large in size and have different ownerships. The primary objectives of CDMSs are to facilitate document approval workflows, document version management, and security properties such as integrity of data, audit-trail, and approvals. A popular approach for CDMS is to consolidate all the project documents into an on premise or cloud-based centralized location that is owned and managed by a project participant or a trusted third party. Such centralized approaches for electronic documents pose risks such as single points of failures causing loss or corruption of documents and deliberate blocking of access during disputes between the owner of CDMS and other project participants. This situation is particularly challenging in construction projects where project participants cannot fully trust each other due to its inherently fragmented project organizational structure. Therefore, in this paper, a distributed construction document management system using Blockchain and distributed content-addressable storage technologies is presented. Blockchain is a peer-to-peer technology that facilitates distributed computation and irreversible data recording through smart contract and blockchain ledger technologies respectively. Immutability of records and computational logic is facilitated through the unique cryptographic data structure of blockchain ledgers and probabilistic consensus algorithms. In this paper, smart contracts are deployed to facilitate document approval workflows to support processes such as design review and request for information in construction projects. A blockchain ledger data model for tracking workflows and document version management is proposed. Public-key cryptography is deployed to facilitate data confidentiality and integrity in endorsements. A cryptographic indexing structure to support blockchain ledger in document versioning and to validate the authenticity of document search results for CDMS is proposed. The proposed framework also deploys peer-to-peer content-addressable storage for preventing single points of failure and data integrity of documents through data partitioning, data replication, and cryptography. The proposed CDMS is a distributed yet unified platform for managing construction documents. A demonstration of the proposed CDMS is presented with a case-based scenario on request for information (RFI) management.

E. Toledo Santos and S. Scheer (Eds.): ICCCBE 2020, LNCE 98, pp. 850–862, 2021.
https://doi.org/10.1007/978-3-030-51295-8_59

Keywords: Blockchain · Construction Document Management System · Document workflow · Versioning · IPFS · Security

1 Introduction

Construction document management systems (CDMS) are systems that coordinate and control tracking, storage, retrieval, processing, and distribution of electronic documents in construction projects [1, 2]. The primary objectives of construction document management systems (CDMS) are to facilitate document approval workflows and support versioning and searching of documents [3, 4]. Construction documents contain sensitive information such as financial information, patented drawings, and participant's personal information. Therefore, security is also one of the primary requirements of construction document management systems [1, 5, 6]. Existing document management systems include frameworks that manage documents and workflows through a centralized approach [2]. Caldas and Soibelman [7] presented an SVM-based document management system that classifies construction documents according to standard construction information classification systems such as CSI MasterFormat [8]. Park et al. [9] deployed ontologies representing domain-specific knowledge of construction documents to improve document searchability, by inducing consistency in the vocabulary of search keywords among multi-disciplinary participants of construction projects. Commercial software such as Aconex [10] and Autodesk 360 [11] deploy a centralized cloud-based platform for the management of construction documents and workflows (shown in Fig. 1). However, cloud platforms are vulnerable to security risks such as data loss, denial of data access, and partial control over sensitive data which is well-documented in the existing literature [12, 13]. Moreover, due to the inherent contractual nature of construction projects, project participants do not fully trust each other and therefore are not confident about entrusting a central entity with the right to store and manage documents. Therefore, existing centralized systems for document management that require consolidation of project documents and workflows on a physical or cloud-based centralized platform that is owned by a central project participant or a trusted third party is inapt.

Therefore, this paper proposes a decentralized but unified construction document management system using blockchain and distributed content-addressable storage. Blockchain [14] is a peer-to-peer technology that facilitates recording and computation of data in a distributed and trustless manner via distributed ledger technology (DLT) and smart contracts respectively. In the existing literature, smart contracts and blockchain ledger in combination with other decentralized technologies such as IPFS storage systems and IoT networks have been explored for checking the authenticity of videos [15], securing AI applications [16], automatic monetization and security of IoT networks [17, 18]. For construction projects, blockchain has been deployed for several applications such as measuring construction productivity [19], contract management [20], pre-cast supply chains [21]. In this paper, smart contracts technology of blockchain that facilitates modeling of executable code for automating processes in a distributed manner is deployed to facilitate document approval workflows in construction projects. For developing the smart contract framework, standard components and relations in document approval workflows in construction projects are identified from existing literature. Based on the finding, smart contracts capturing the logic (associating roles to their corresponding tasks) of various approval processes such as design review, are developed. As workflows may be unique to organizations and projects, smart contract logic to orchestrate custom workflows by re-using standard sub-processes in document approval workflows is proposed. A blockchain ledger data model for tracking workflows and document versioning is proposed. A cryptographic indexing structure to facilitate searchability and authenticity (in terms of completeness and consistency) of query results is proposed. The proposed indexing structure maintains the references to older versions of a document and requires no reconstruction when newer versions are added. Public-key cryptography is used to facilitate data confidentiality and integrity of endorsements. A distributed peer-to-peer content-addressable storage such as Interplanetary File System (IPFS) is deployed to facilitate data integrity and prevent single points of failure through cryptographic hashing and data replication. The proposed framework takes a distributed approach for document management to ensure that workflows and documents are not controlled by a single entity while at the same time deploys measures to counter security risks to data confidentiality, integrity, and authorization due to decentralization. The proposed framework is supported by a case-based scenario on request for information management that demonstrates its key features and advantages such as automatic workflow execution and tracking and security properties such as data authenticity, data confidentiality, and user integrity.

2 The Proposed Framework

In this section, a framework for decentralized construction document management using blockchain and distributed content-addressable storage is proposed (as shown in Fig. 1). As shown in Fig. 1, the proposed framework takes a distributed approach towards workflow deployment, document versioning, audit-trail recording, and document storage using smart contracts, distributed ledger technology, cryptographic data structures, and content-addressable storage as discussed in Sects. 2.1–2.5.

2.1 The General Structure of Document Approval Workflows

Figure 2(a) shows the general components of workflows for document management in construction projects [22–24] – (1) processes such as design review, request for information, and change order management, (2) roles such as architect, main-contractor, and owner, and (3) data such as instructions to initiate or documents to support processes. Processes may be further sub-divided into sub-processes like 'creating a new document', 'providing endorsement', and 'feedback' as shown in Fig. 2(a). Figure 2(b) shows the structure of a typical document management workflow which consists of a series of processes and sub-processes facilitated by roles through supporting data.

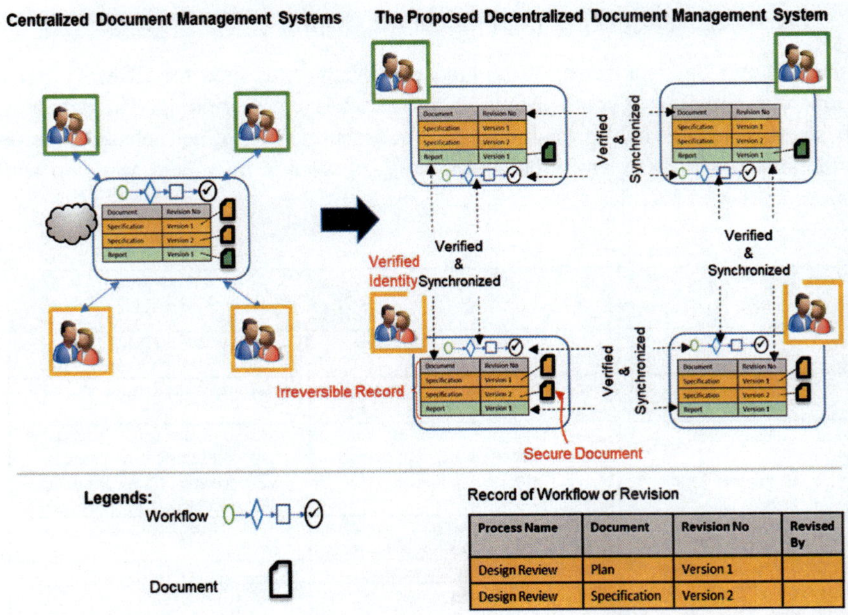

Fig. 1. The proposed framework for construction document management

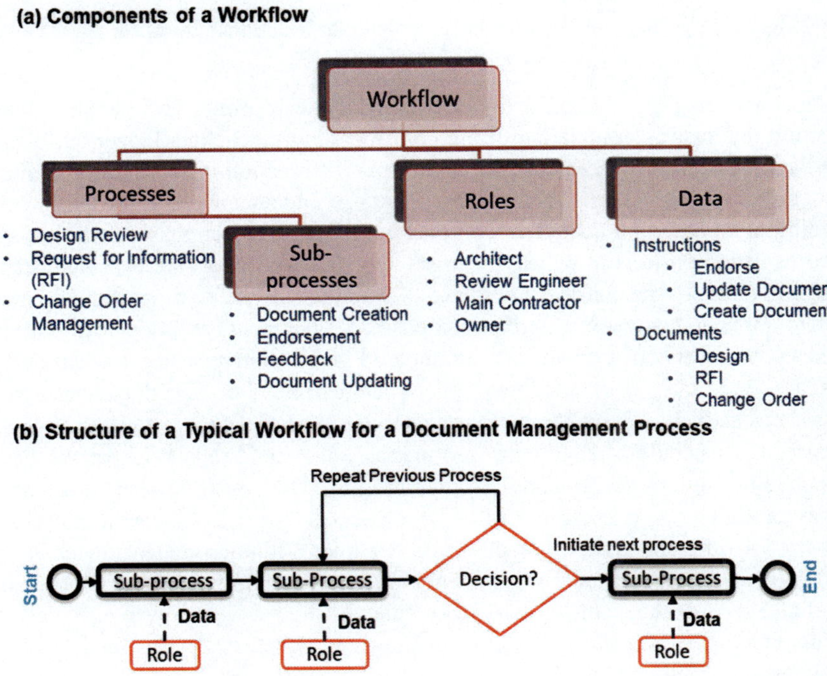

Fig. 2. The general structure of workflows for document management in construction projects

2.2 The Blockchain Ledger Data Model for CDMS

Figure 3 shows the data model of the blockchain ledger required for CDMS to record document approval workflows (discussed in Sect. 2.3) and versioning. The data model consists of metadata such as transaction ID, transaction type, project phase, document ID, the status of a document, time, and identity of roles in document approval workflows as shown in Fig. 3.

Attributes of the data model		Description	Example
Transaction ID		ID of a Blockchain Transaction	f4184fc596403b9d638783cf57adfe4c75c60 5f6356fbc91338530e9831e9e16
From Account		User or smart contract that sends a request to invoke a smart contract function	Process orchestrator Smart Contract/ End-User
To Account		A smart contract that receives a request from an end-user or another smart contract	Process orchestrator Smart Contract/ Sub-process Executor Smart Contract
Transaction Type		Type of sub-process	Add Document , Update Document, Endorse Document, Send Feedback
Payload	Project Phase	The phase of the construction project in which a process workflow is executed	Design/ Construction
	Document ID	Unique identifier of a document	17-111-12
	Status	The current status of a workflow process	RFI submitted, RFI endorsed
Time		Time at which a workflow process is executed	20191219 1220
Sender's Signature		Digital signature of the end-user who has invoked a sub-process smart contract	Role's Signature

Fig. 3. Data model of ledger in the proposed blockchain framework for CDMS

As shown in Fig. 3, 'Transaction ID' is a unique identifier for a smart contract execution that results in updating of the corresponding blockchain ledger and is generated automatically by the blockchain platform. 'Transaction type' in this data model refers to sub-process such as document creating, endorsement, or feedback corresponding to a phase of a document approval workflow. 'Project phase' is the phase of the construction projects in which document approval workflows are executed such as the design phase or the construction phase. 'Document ID' in the proposed data model refers to a unique document identifier composed of document type, document number, and document version number. For example, 17-111-12 represents a document ID where 17, 111, and 12 are document type, document number, and document version number respectively. 'Document type' refers to the type of a document such as 'notice to proceed', 'schedule', 'request for information', and 'field report' as per the Omniclass classification for construction information [25]. 'Document number' is an integer unique for the type of document. Unlike 'Document ID', the 'Document number' does not vary for different versions of the same document. Similarly, 'Document version number' is an integer unique for document number. Along with this, the proposed data model also records the status of workflows, time at which a workflow is executed, and the identity of the workflow executor as shown by 'Status', 'Time', and 'Sender's Signature', as shown in Fig. 3.

2.3 A Smart Contract Framework for Distributed Workflows for Document Management

Figure 4 shows the smart contract framework for facilitating document approval workflows for construction projects. The framework is designed by considering custom workflows to suit the needs of different organizations and project types. Therefore, two types of smart contract logics namely – (1) sub-process executor and (2) process orchestrator smart contracts are developed to facilitate reuse of sub-process for orchestrating different workflows.

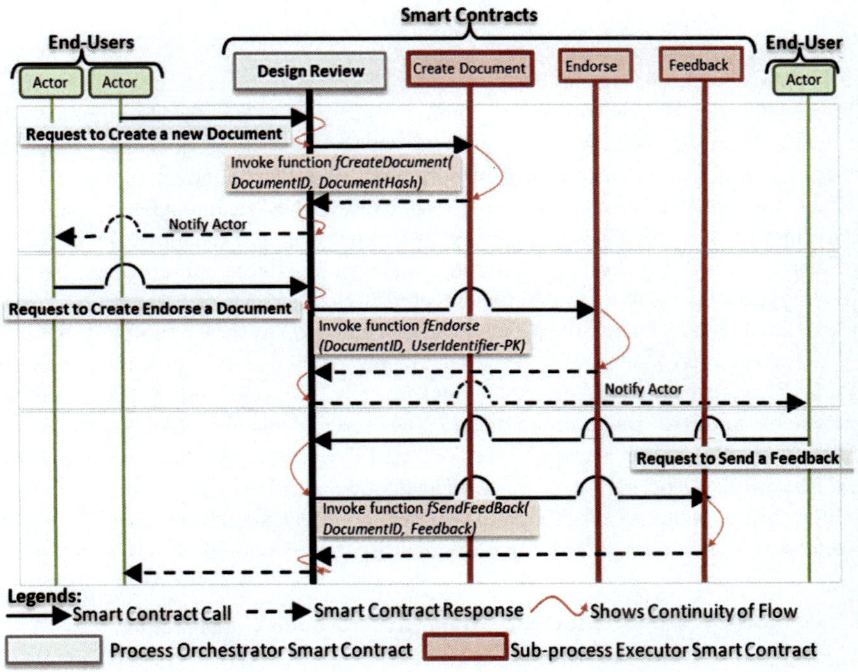

Fig. 4. Interactions between actors and smart contract for document approval workflows

2.3.1 Sub-process Executor Smart Contracts

As shown in Fig. 2, sub-processes such as document creation, document updating, document endorsement, and feedback are individual tasks that make a document approval workflow process. In this section, the logic to deploy sub-processes through smart contract functions is presented. Solidity which is the smart contract language of Ethereum [14] blockchain platform is used.

As shown in Fig. 4, *fCreateDocument* is a function in the '*Create Document'* smart contract that records the creation of a new document. The inputs to *fCreateDocument* function are Document ID (as discussed in Sect. 2.2) and a Document Hash (which is a unique identifier of the content of a document and will be discussed in Sect. 2.4). Upon execution, this function creates a record in the blockchain ledger (data model shown in

Sect. 2.2) and sends a notification to the project participant (as shown by 'actor' in Fig. 4) who is responsible for executing the subsequent sub-process of the document approval workflow. Sub-process executor smart contracts are created at the beginning of a construction project and deployed to the distributed blockchain network so that they may be verified for authenticity by all project stakeholders. Each sub-process is deployed as a separate smart contract to facilitate reuse and variation in orchestrations as required by different document approval workflows (discussed in Sect. 2.3.2). Furthermore, this approach imparts scalability to the proposed framework by permitting deployment of new smart contracts for special sub-tasks if required at a later stage or a project with special requirements.

2.3.2 Process Orchestrator Smart Contract

The objective of a process orchestrator smart contract is to orchestrate the calls to different sub-process executor smart contract functions as per the requirement of a document approval workflow. A process orchestrator smart contract facilitates orchestration of a document approval workflows via smart contract functionality to facilitate interactions between two smart contacts [26] by using ABIs (Application Binary Interface) of sub-process executor smart contracts. For example, the document approval workflow for the design review process consists of sub-processes such as uploading a new plan by the project architect, followed by endorsement from a review engineer, and feedback from the project owner. Therefore, as shown in Fig. 4, the logic of the process orchestrator smart contract for the design review process consists of sequential calls to functions *fCreateDocument* of '*Create Document*' smart contract, *fEndorse* of '*Endorse*' smart contract and *fSendFeedback* of '*Feedback*' smart contract facilitated by the architect, review engineer, and the project owner respectively. Figure 4 shows the logic of the process orchestrator smart contract for design review. Similarly, smart contracts for different document approval workflows such as a request for information and change order management may be deployed.

2.4 A Blockchain-Based Indexing Data Structure for Construction Document Management

In this paper, an indexing structure (as shown in Fig. 5) based on Radix tree [27] and Merkle tree [28] cryptographic data structures is presented for facilitating document querying according to document version and other attributes and enabling the security of consistency and completeness in the query results. A Radix tree is an ordered data structure for storing, indexing, and retrieving in-memory data that goes through many updates [27]. In a Radix tree, data is stored in a key-value format, where a "key" is a path (arranged in lexicographical order) in the tree data structure that leads to a "value" at a leaf node. As shown in Fig. 5, in the proposed indexing data structure, the 'Document ID' composed of "document type – document number – version number" (such as "14-13-1") is deployed as a branch (path) that leads to a leaf node. As shown in Fig. 5 the value is a document hash which is a unique identifier based on the content of a document generated by content-addressable storages like IPFS (Interplanetary File System) [29]. IPFS is a peer-to-peer distributed file system that stores documents in a key-value format such that the key is a cryptographic hash generated based on the content of the document and the value is the document itself. Cryptographic hashes [30, 31] are deterministic encryption functions that convert a document or a text into a unique 256-bit hexadecimal number. This means that the hash of a document varies with very small changes in the document content and hence may be used as proof of authenticity.

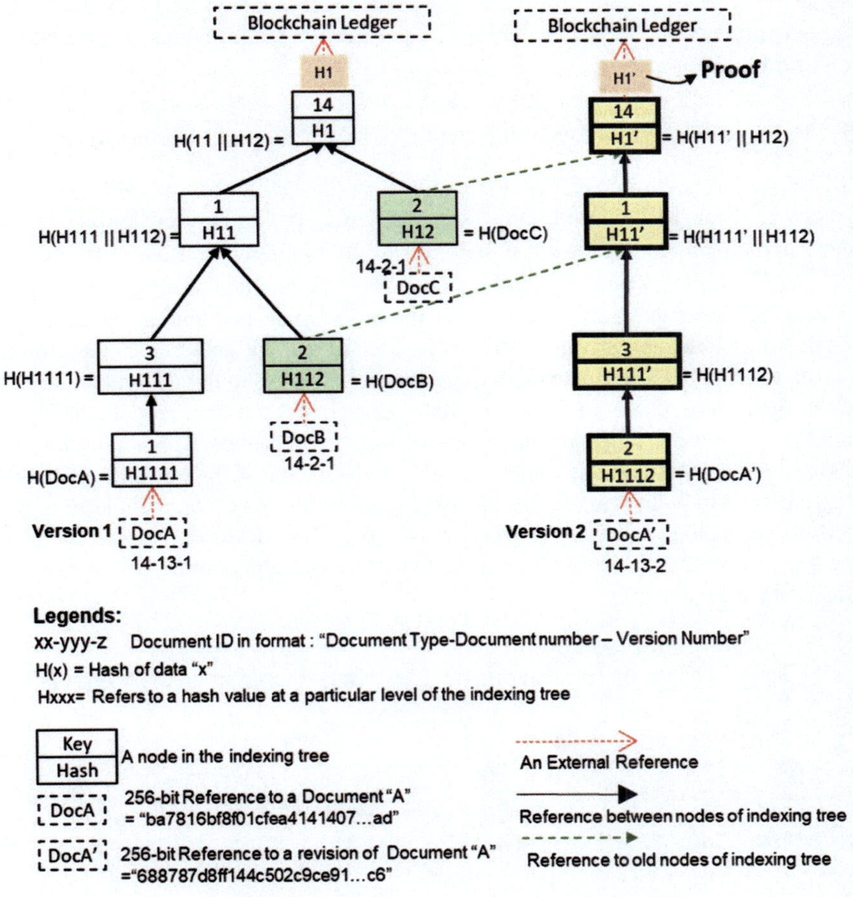

Fig. 5. A blockchain-based indexing data structure for document versioning

The second type of tree data structure used in the proposed framework is Merkle tree [28] to facilitate the integrity of datasets in distributed environments. Merkle trees are tree-like data structures in which the leaf nodes are deterministic cryptographic encryptions (Hashes) of the data stored in them and the non-leaf nodes are hashes of their child nodes. As shown in Fig. 5 the leaf node *DocA* of the proposed indexing data structure represents a 256-bit hexadecimal document hash. Its parent node contains the hash of this unique identifier as shown by *H1111*. Similarly, other parent nodes contain hashes generated from their corresponding child nodes. As shown in Fig. 5, when a new version of *DocA* is added, it is included in the indexing tree through 4 new nodes and references to old nodes which prevents the need to reconstruct the whole tree as necessary in other indexing data structures such as B+ trees [27]. The authenticity of a document version may be verified using its corresponding root hash (proof) as shown in Fig. 5. For example, $H1'$ is the proof of completeness and consistency of the new document version, $DocA'$. It is difficult to modify $H1'$ as it depends on unique document hash and references to previous document hashes and is stored on blockchain ledger which is irreversible in nature. In Sect. 3, a case-based scenario is presented to demonstrate this property.

2.5 Public Key Cryptography for Data Confidentiality and Endorsement

In this paper, data confidentiality and endorsement is facilitated via public-key cryptography [32] for the proposed framework. To deploy public-key cryptography, every project participant should have a public-private cryptographic key pair (as shown as "$PK^i - SK^i$" in Fig. 6 each. The public key, "PK^i" is the public identity of a user "i" and may be shared publicly while SK^i is the private key that should be kept confidential. Both public and private keys may be used to encrypt data to facilitate different types of security. Figure 6 shows two approaches where public and private keys are used in opposite manners to deploy digital endorsement and data confidentiality. For endorsement, a user A (as shown in Fig. 6) may encrypt data with his private key. Anyone may authenticate the validity of this endorsement if he can decrypt the data using the corresponding public key of user A. Similarly, data confidentiality is facilitated by encrypting data with a user's public key. Therefore, only those users who have the corresponding private key are able to access the encrypted data hence facilitating data confidentiality.

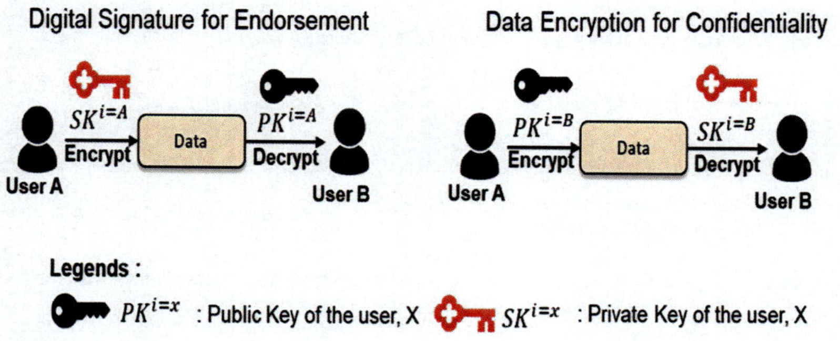

Fig. 6. Data confidentiality and endorsement security using public-key cryptography

3 Validation

In this section, the key properties of the proposed framework which are workflow automation, document searching, consistency and completeness (data integrity), and integrity of endorsement are validated using a case-based scenario of request for information. Figure 7 shows a standard document approval workflow for a request for information [11, 33]. This workflow primarily consists of four sub-processes that are – (1) creation of an RFI document, (2) endorsement of the RFI document, (3) addition of comments thereby updating the RFI document, and (4) sending a notification to the project owner.

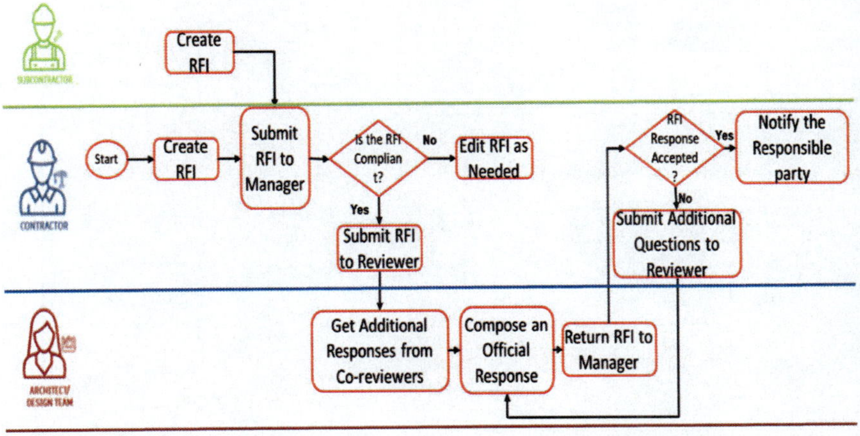

Fig. 7. A standard 'request for information' (RFI) workflow

Figure 8(a) shows the pictorial depiction of the RFI workflow that is deployed using the proposed blockchain-based framework. Figure 8(c) shows how the RFI workflow is tracked on the blockchain ledger by recording information such as document ID, name of the sub-process, and a digital endorsement of the authorized end-user. Endorsements are provided through digital signatures using the private keys of project participants. This facilitates integrity in endorsements as a digital endorsement once recording on an irreversible blockchain ledger cannot be refuted at a later stage.

Figure 9 shows the simplified blockchain ledger and the indexing data structure deployed by the proposed framework to facilitate the searching and integrity of search results. A particular version of a document may be searched by first parsing the blockchain ledger indexed with a local indexing structure such as Skip List [34] to find the document ID followed by traversing the indexing data structure shown in Fig. 9. In order to validate the authenticity of a document say, $DocA'$ (as shown in Fig. 9), the nodes (Shown in green in Fig. 9) of the indexing data structure should be requested from the network. The receiver may be assured of the correctness of a document if he can recreate the root node corresponding to a document. As shown in Fig. 5, the new state root, $H1'$ can be generated by an end-user if he has $DocA'$ and the elements (as shown in green) in Fig. 5. To verify the integrity, he can compare it with the state root stored in the blockchain ledger. If he has the right version of $DocA'$, he should be successful.

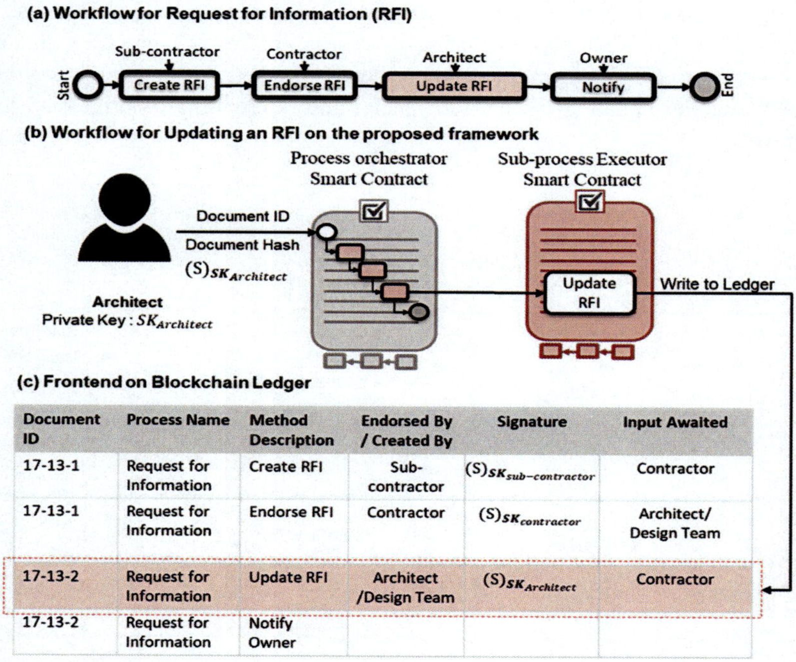

Fig. 8. Document approval workflow for RFI on the proposed framework

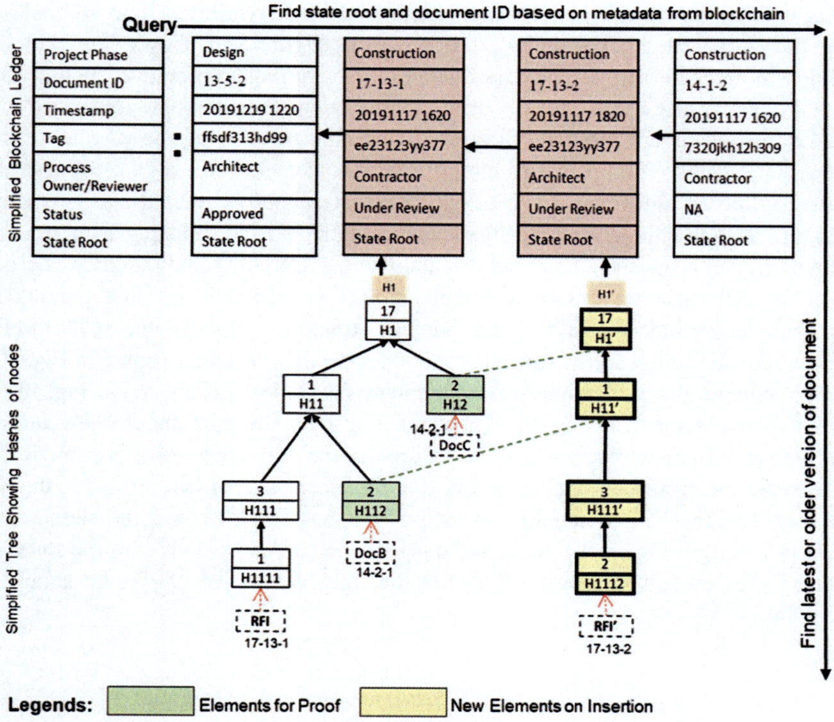

Fig. 9. The proposed blockchain ledger and indexing data structure for RFI management

4 Conclusion

Construction Document Management Systems (CDMS) facilitate coordination among project participants through document approval workflows and facilitating versioning, and searching for documents. Effective document management facilitates a smooth flow of information for various processes among different project participants thereby improving the quality of construction projects. Existing document management systems in construction projects are centralized which are unsuitable due to the fragmented contract-based project organization structure of construction projects. Furthermore, centralized approaches pose several risks such as loss and corruption of data. Therefore in this paper, a distributed construction document management system using blockchain and distributed content-addressable storage is proposed. This framework deploys blockchain smart contracts to facilitate distributed but synchronized or unified document approval workflows and document indexing. A blockchain ledger is deployed to track workflows and versioning of documents to create a credible and irreversible audit-trail. Future work will be done to extend this framework to include complex document approval workflows. The data model of the blockchain ledger and the indexing data structure will be extended to include document metadata to facilitate querying based on multiple criteria.

References

1. Fernando, H., Hewavitharana, T., Perera, A.: Evaluation of electronic document management (EDM) systems for construction organizations, pp. 273–278 (2019)
2. Asılı, H., Tanrıover, O.O.: Comparison of document management systems by meta modeling and workforce centric tuning measures. J. Comput. Sci. Eng. Inf. Technol. **4**(1), 57–67 (2014)
3. Opitz, F., Windisch, R., Scherer, R.J.: Integration of document-and model-based building information for project management support. Procedia Eng. **85**, 403–411 (2014)
4. Cha, H., Lee, D.: Framework based on building information modelling for information management by linking construction documents to design objects. J. Asian Archit. Build. Eng. **17**(2), 329–336 (2018)
5. Xie, L.: Evaluation of electronic document and record management program in Canadian municipality. The University of British Columbia (UBC) (2006)
6. Bazlamit, S.M., Ahmad, H., Ayoush, M.: Document management systems in small and medium size construction companies in Jordan. In: Proceedings of the 6th International Conference on Engineering, Project, and Production Management (EPPM 2015). Griffith University, Gold Coast (2015)
7. Caldas, C.H., Soibelman, L.: Automating hierarchical document classification for construction management information systems. Autom. Constr. **12**(4), 395–406 (2003)
8. CS Institute: MasterFormat (1995)
9. Park, M., Lee, K.-W., Lee, H.-S., Jiayi, P., Yu, J.: Ontology-based construction knowledge retrieval system. KSCE J. Civ. Eng. **17**(7), 1654–1663 (2013)
10. ACONEX: Cloud-based information management: a guide for IT. WhitePaper, Oracle (2018)
11. Autodesk: Autodesk collaboration for Revit security overview. Whitepaper (2016)

12. https://blogs.cisco.com/smallbusiness/the-top-5-security-risks-of-cloud-computing. Accessed 15 Jan 2019
13. Studnia, I., Alata, E., Deswarte, Y., Kaâniche, M., Vincent, N.: Survey of security problems in cloud computing virtual machines, pp. 61–74 (2012)
14. Wood, G.: Ethereum: a secure decentralised generalised transaction ledger. Ethereum Project Yellow Paper (2014)
15. Hasan, H.R., Salah, K.: Combating deepfake videos using blockchain and smart contracts. IEEE Access **7**, 41596–41606 (2019)
16. Khaled, S., Habib, M., Nishara, N., Ala, F.: Blockchain for AI: review and open research challenges. IEEE Access (2019)
17. Mistry, I., Tanwar, S., Tyagi, S., Kumar, N.: Blockchain for 5G-enabled IoT for industrial automation: a systematic review, solutions, and challenges. Mech. Syst. Sig. Process. **135**, 106382 (2020)
18. Chaer, A., Salah, K., Claudio, L., Ray, P.P., Sheltami, T.R.: Blockchain for 5G: opportunities and challenges. In: Proceedings of IEEE GLOBECOM 2019, Waikoloa, USA (2019)
19. Heiskanen, A.: The technology of trust: how the Internet of Things and blockchain could usher in a new era of construction productivity. Constr. Res. Innov. **8**(2), 66–70 (2017)
20. Wang, J.: The outlook of blockchain technology for construction engineering management. Front. Eng. Manag. **4**(1), 67 (2017)
21. Wang, Z., Wang, T., Hu, H., Gong, J., Ren, X., Xiao, Q.: Blockchain-based framework for improving supply chain traceability and information sharing in precast construction. Autom. Constr. **111**, 103063 (2020)
22. Froese, T.M.: The impact of emerging information technology on project management for construction. Autom. Constr. **19**(5), 531–538 (2010)
23. Craig, N., Sommerville, J.: Information management systems on construction projects: case reviews. Rec. Manag. J. **16**(3), 131–148 (2006)
24. Borrmann, A., König, M., Koch, C., Beetz, J.: Building information modeling (2018)
25. CSI: OmniClass (2012)
26. https://medium.com/coinmonks/creating-smart-contracts-with-smart-contract-d54e21d26e00. Accessed 27 Jan 2020
27. Leis, V., Kemper, A., Neumann, T.: The adaptive radix tree: ARTful indexing for main-memory databases, pp. 38–49 (2013)
28. Szydlo, M.: Merkle tree traversal in log space and time, pp. 541–554. Springer, Heidelberg (2004)
29. https://arxiv.org/pdf/1407.3561.pdf. Accessed 19 Mar 2019
30. Gilbert, H., Handschuh, H.: Security analysis of SHA-256 and sisters, pp. 175–193. Springer, Heidelberg (2004)
31. NSA: SHA256 hash generator (2012)
32. Rivest, R.L., Shamir, A., Adleman, L.: A method for obtaining digital signatures and public-key cryptosystems. Commun. ACM **21**(2), 120–126 (1978)
33. http://www.hka-tech.com.au/. Accessed 26 Jan 2020
34. Xu, C., Zheng, C., Xu, J.: vChain: enabling verifiable boolean range queries over blockchain databases, pp. 141–158

BIM and Internet of Things (IoT) Frameworks and Applications

Conceptual Framework for Tracking Metallic Formworks on Construction Sites Using IoT, RFID and BIM Technologies

Caroline Silva Araújo(✉), Leandro Cândido de Siqueira, Emerson de Andrade Marques Ferreira, and Dayana Bastos Costa

Federal University of Bahia, Salvador, BA 40210-630, Brazil
araujo.caroline92@gmail.com

Abstract. The difficulty in tracking physical resources at the construction site causes problems in the information flow required for better management and quality control. The integrated use of Building Information Modeling (BIM), Internet of Things (IoT) and Radio Frequency Identification (RFID) technologies can enable tracking and improve construction site-related processes. Specifically, there is a lack of studies regarding the tracking of metallic formworks during the execution phase of the cast in place concrete wall constructive system, in which quality issues often imply rework and delays. The main objective of this paper is to design a conceptual framework for tracking physical resources at construction sites using BIM, IoT, and RFID, in which the metallic formworks are the object of study. This study is based on a Systematic Literature Review, which was conducted to identify requirements to connect the concepts and technologies in the conceptual framework and to develop an integrated system that uses the aforementioned technologies. Moreover, a preliminary proposal of an integrated system associating the RFID tags with the ID of equivalent virtual elements in BIM and storing information from the formworks in a database, which communicates with the virtual model, was presented. The conceptual framework and the preliminary proposal will be used for the design of the system prototype for future implementation in real construction environments.

Keywords: Tracking · Metallic formwork · IoT · BIM

1 Introduction

The construction site is a dynamic and complex environment that involves a large amount of physical resources such as personnel, equipment, and materials in its production processes, implying the need for increasingly efficient and transparent management procedures (Park et al. 2016a; Park et al. 2016b; Oliveira and Serra 2017). The improvement in the information flow regarding those resources may contribute to better decision-making capacity of the managers concerning production planning and control (Cordova and Brilakis 2008; Park et al. 2016b).

E. Toledo Santos and S. Scheer (Eds.): ICCCBE 2020, LNCE 98, pp. 865–878, 2021.
https://doi.org/10.1007/978-3-030-51295-8_60

Recent developments in information technology and Industry 4.0 have changed different types of construction projects. Thus, several tasks related to project design, planning, execution, quality control, and site management now require a wide variety of identification and location data to ensure the proper flow of information (Fang et al. 2016).

The activities related to the execution and management of construction depend heavily on the visibility and traceability of the information required and the coordination of different elements that are dispersed on-site or off-site (Xue et al. 2018). Tracking of elements on the construction site is the ability to determine the current state of this element at any time, which can help increase productivity (Zhong et al. 2017). The integrated use of technologies such as Building Information Modeling (BIM), Internet of Things (IoT), and Radio Frequency Identification (RFID) to track physical resources on the construction site can improve different processes.

Regarding physical resources, metallic formworks are an object of interest to several companies that deal with the cast in place reinforced concrete wall constructive system, especially in the Brazilian context for housing construction. This is a monolithic system, in which the structure and seals are composed of self-supporting elements of cast-in-place reinforced concrete (ABNT NBR 16055 2012). The removable and modulated metallic formworks are assembled preceding the concreting, according to the architectural design and respecting the necessary openings, such as windows and doors. The electrical and hydraulic installations of the walls and slabs are also fixed before concreting, as well as reinforcements and details of the facade such as frieze and recesses (ABNT NBR 16055 2012).

In the last decade, the use of that constructive system has gained strength in the construction industry in Brazil due to its large-scale use for low-income housing program "My House My Life", in Portuguese *"Minha Casa Minha Vida"*, created by the Brazilian government to reduce the housing deficit. Some of the main advantages of that constructive system are the reduction of construction steps, cleaning improvement of the construction site, waste reduction, productivity, and execution speed increasing (Nunes 2011).

However, after the removal of the metallic formworks, there is a persistent stage of defect corrections, causing rework and delay of the project due to wall quality issues. Some causes of the great number of defects can be attributed to the difficulty in controlling the quality of the construction process in the concrete wall system, as well as the integrity and service life of each metallic formwork. Therefore, this study assumes that digital technologies, such as BIM, IoT, and RFID have the potential to improve the processes related to the cast in place reinforced concrete wall constructive system and support the aforementioned problems.

Therefore, the main objective of this article is to design a conceptual framework for tracking physical resources in construction sites using BIM, IoT and RFID, in which the metallic forms manipulated in the concrete wall construction system are the object of study.

This study was based on a Systematic Literature Review to identify requirements such as software, hardware, protocols, communication networks and methodologies required to develop an integrated system that uses such technologies.

2 Overview of IoT, Clouding Computing, RFID and BIM Technologies

This section presents an overview of the technologies adopted in this research for tracking physical resources at construction sites.

IoT is one of the key components of Industry 4.0 (Roblek et al. 2016) and can be defined as a set of physical artifacts that can be associated with embedded systems of electrical, mechanical, computational, and communication mechanisms that allow for exchange of data via the Internet (Thames and Schaefer 2016). The IoT can allow communication between all real or virtual devices that exist in human daily life and to detect status changes of those devices in real-time (Isikdag 2015). IoT's reach has grown in recent years because of the offer of low-cost devices in the market and due to the evolution of the Internet (Thames and Schaefer 2016).

Cloud Computing can promote a virtual infrastructure focused on IoT that integrates monitoring devices, analytical tools, visualization platforms and delivery of interfaces suitable for end-users (Gubbi et al. 2013). The IoT model based on Cloud Computing will allow the provision of end-to-end services for companies and private users to access information wherever they are, at any time (Gubbi et al. 2013). In turn, Big Data is an emerging paradigm that has an abstract concept, but it can be understood as large data sets that cannot be perceived, acquired, managed and processed by traditional Information Technology (IT) tools in a timely manner (Chen et al. 2014). Big data tools have strict requirements regarding storage capacity and processing capacity, as well as transmission capacity for the network (Chen et al. 2014). Databases used in IoT projects are often associated with Cloud Computing platforms.

RFID is a radio frequency identification technology that supports a large number of unique IDs (Want 2006). Its development served as the basis for building IoT as it is known today (Sundmaeker et al. 2010). Nowadays, RFID systems are beginning to conquer new mass markets due to their advantages compared to other identification systems (Finkenzeller 2010). Although there are widely adopted technologies for external tracking of elements in the construction site, such as the Global Positioning System (GPS), most construction activities take place indoors and the implementation of robust indoor location systems for physical resource management remains a major challenge for the construction industry (Fang et al. 2016). Compared to other internal location technologies such as Inertial Browsing (INS), Ultra-Broadband (UWB), Wireless Local Area Network (WLAN), and Bluetooth, RFID stands out among researchers and construction professionals because of its level of technological maturity and its relatively economical infrastructure (Fang et al. 2016). In general, a typical structural RFID system consists of a reader (data collector or interrogator), electronically programmed tags located on the object of interest and the control software (Finkenzeller 2010; Oliveira and Serra 2017). After obtaining the data from the tags, the reader sends them to the system, which is responsible for re-knowing the communication protocols and giving destination to the data (Oliveira and Serra 2017). An RFID system can use low frequency (LF), high frequency (HF), ultra-high frequency (UHF) or microwave (MW) and the RFID tags can be of two types: passive (without batteries) or active (with batteries) (Xue et al. 2018).

BIM has benefited construction because of its ability to produce increasingly accurate virtual models to represent the reality that facilitate communication and collaboration between owners, architects, consultants, contractors, manufacturers, and operators, promoting the development of more efficient and integrated projects (Eastman 2011). In the construction phase, BIM can provide essential information for construction progress monitoring processes and also for physical building resource tracking (Fang et al. 2016). BIM models offer the potential to become "digital twins" of the represented built environment, providing a refined spatial context for IoT systems (Gao et al. 2018). There is a tendency to use BIM in the Cloud to receive information collected by RFID more efficiently (Xue et al. 2018). RFID-enabled BIM systems allow traceability and visibility of information related to physical objects of the building in real-time (Xue et al. 2018).

3 Research Method

The Systematic Literature Review (SLR) aims to map previous studies, analyze known empirical solutions, as well as identify in an integrated way the relationships between the concepts and variables involved in the study. Considering that no study was found in the literature regarding the specific tracking of metallic formworks at the construction site, this gap was converted into the following research question to guide the literature review in this study: "What are the main requirements for the development of an RFID/IoT/BIM system for tracking metallic formworks in the cast in place reinforce concrete wall constructive system?"

The review was conducted through the following steps: (a) choice of databases, (b) definition of review inclusion criteria, (c) definition of the exclusion criteria from the review, and (d) analysis of the resulting articles.

Scopus, Compendex and Web of Science were the three academic databases selected for the research due to their wide coverage of relevant academic articles. Among the inclusion criteria adopted, the selected search terms are into the following groups, which also included terms considered equivalent: a) IoT and RFID, b) BIM, c) Tracking, and d) Construction and Metallic Formwork. Such terms should be included in the title, abstract or keywords of the articles. All available period in the databases was included in the search. The review was limited to selecting articles published in peer-reviewed scientific journals.

The exclusion criteria adopted to reach the final sample of articles were the removal of duplicates and the removal of articles that did not address BIM, IoT, and RFID as the main theme. The final sample resulted in 16 articles, all written in English due to the universality of this language. The specification of the search terms used in each database to establish the conceptual boundaries of the review are indicated below:

- ("Internet of Things" OR "IoT" OR "RFID" OR "Radio Frequency" OR "Radio-Frequency" OR "Radio Frequency Identification" OR "Pervasive Computing" OR "Ubiquitous Computing") **AND** ("Building Information Modeling" OR "BIM" OR "Building Information Model") **AND** ("Track" OR "Tracking" OR "Traceability" OR "Addressing") **AND** ("Construction" OR "Metallic Formwork").

The content analysis of the articles in the resulting sample aimed to identify requirements such as technologies, software, hardware, protocols, communication networks and methodologies necessary to develop this study. Based on the requirements identified after the sample analysis, a conceptual framework was developed to connect the concepts and technologies involved. A proposed integrated system, considering the requirements, was also designed aiming its future implementation in real construction environments. Figure 1 illustrates the procedures adopted in the review.

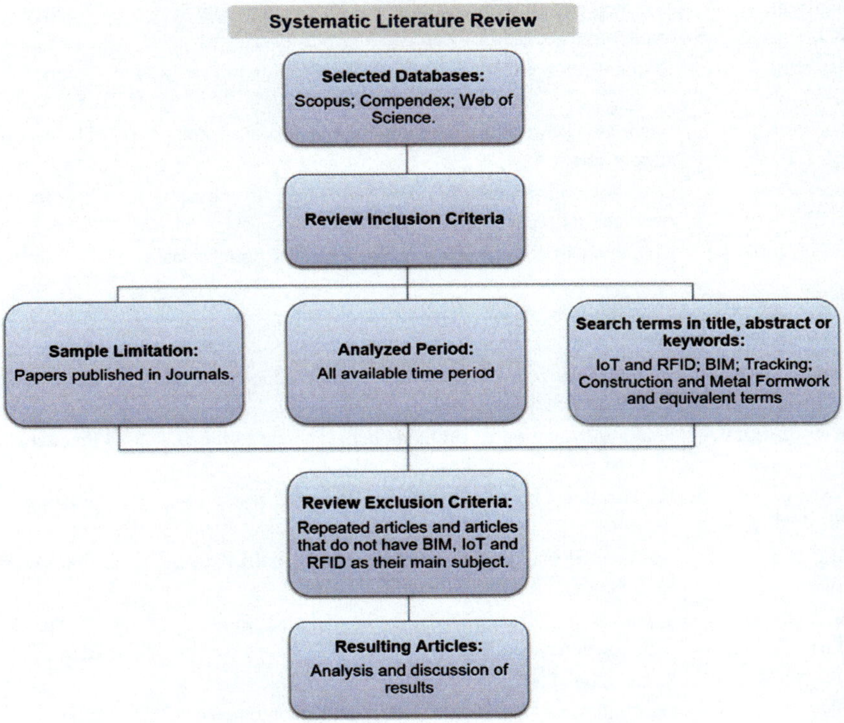

Fig. 1. Procedure adopted for conducting SLR.

4 Sample Analysis

The Table 1 below shows the articles that are part of the sample obtained with SLR, with an indication of the main topic addressed and whether the study considered any practical application in the field, if it presented a prototype in the laboratory or if it provided theoretical contributions.

Several articles in the sample address the theme of the construction elements prefabricated/precast and mainly deal with prefabrication production, logistics and on-site assembly, such as Feng et al. (2015), Li et al. (2018a), Li et al. (2018b), Li et al. (2017), Xue et al. (2018), Zhong et al. (2017), Zhong et al. (2015), Ikonen et al. (2013), and Ma et al. (2019).

Table 1. List of 16 papers in the final sample.

Author	Main theme	Application
Akanmu et al. (2012)	Location tracking and status of general construction components	Prototype
Akanmu et al. (2016)	Data capture in real-time to auto-generated site layout	Practice case
Costin et al. (2015)	Location tracking and status for building protocol control focused in safety	Practice case
Fang et al. (2016)	Location tracking and monitoring for management application focused on safety	Practice case
Feng et al. (2015)	Status of components and materials logistics information	Practice case
Iacovidou et al. (2018)	Facilitate construction components reuse back to the supply chain	Theoretical
Ikonen et al. (2013)	Logistics and production decision-making for precast concrete elements	Practice case
Li et al. (2017)	Management and schedule performance of prefabricated house construction	Theoretical
Li et al. (2018a)	On-site assembly of prefabricated construction	Practice case
Li et al. (2018b)	Manufacturing and logistics to prefabrication housing production	Prototype
Ma et al. (2019)	Status Information of prefabricated concrete building focused on reducing waste	Practice case
Motamedi et al. (2013)	Localization of assets during the operation phase of facilities	practice case
Wang et al. (2013)	Augmented reality (AR) to enable the visualization of each construction activity or task	Theoretical
Xue et al. (2018)	Status of prefabricated facades for manufacturing, logistics and assembly	Practice case
Zhong et al. (2015)	Status of prefabricated housing construction for manufacturing, logistic, assembly and maintenance	Practice case
Zhong et al. (2017)	Status of prefabricated construction for manufacturing, logistic and assembly	Practice case

Considering the studies with a practical application in the sample, all authors used at least, as minimum physical infrastructure, an RFID reader and tags in the elements of interest. In most studies, a portable reader was used, which is portable. However, the authors Akanmu et al. (2016), Costin et al. (2015), Fang et al. (2016), Feng et al. (2015), Ma et al. (2019) used fixed readers.

According to Xue et al. (2018), UHF RFID systems provide a satisfactory solution for most applications in practice in the construction phase, such as in the status tracking of construction elements or for maintenance purposes. If the application requires a communication distance above 1-meter, those authors recommend using active tags, whereas, for distances less than 1-meter, passive tags are more recommended.

Some authors of the sample use additional resources to promote the location of elements on the construction site. Fang et al. (2016) and Ma et al. (2019) reported the use of auxiliary antennas, while Ikonen et al. (2013), Li et al. (2017), Li et al. (2018), Zhong et al. (2015), and Zhong et al. (2017) reported the use of GPS technologies associated with RFID.

Regarding the infrastructure of systems enabled for RFID, IoT, and BIM, the authors Li et al. (2017), Li et al. (2018), Xue et al. (2018), Zhong et al. (2017) used it in their studies or recommended the use of a smart gateway. A smart gateway is a device configured to collect data in real-time that can communicate and interact with decision-making systems to support RFID systems (Li et al. 2018). The gateway can be enabled to communicate with RFID readers via Bluetooth and then send data to cloud servers via communication networks such as 4G or Wi-Fi (Li et al. 2018). A gateway is software middleware that can convert the information collected through a message exchange protocol such as XML, which enables communication with BIM servers (Xue et al. 2018).

Concerning databases, Akanmu et al. (2012), Akanmu et al. (2016), Ikonen et al. (2013), Li et al. (2017), Ma et al. (2019) reported the use of SQL (structured) databases. According to Xue et al. (2018), an important trend in RFID systems is the use of BIM associated with Cloud Computing to establish platforms to receive real-time information collected (Xue et al. 2018). The authors Costin et al. (2015), Fang et al. (2016), Li et al. (2017), Li et al. (2018), Xue et al. (2018), Zhong et al. (2017), and Zhong et al. (2015) mention cloud servers in their studies.

Regarding BIM, cloud BIM can be deployed through a commercial platform such as Autodesk BIM 360 or Graphisoft BIM, however, these platforms maybe not provide the necessary protocols or the necessary flexibility to receive the data captured by the RFID system (Xue et al. 2018). A second approach is to develop the BIM model in commercial software such as Autodesk Revit, export the BIM data to an open format, such as Industrial Foundation Classes (IFC) and convert it to the cloud through rendering using WebGL (Web Graphics Library), for example (Xue et al. 2018). In this second approach, the network protocols for systems to communicate can include HTTP (or HTTPS), SFTP (Secure File Transfer Protocol) and XML/JSON (JavaScript Object Notation) data management services (Li et al. 2018).

Considering the data visualization interface, at least one of the following devices is used in all sample studies as physical display devices: laptop, tablet, mobile phone, smartphone, personal digital assistant (PDA) or desktop computer. The communication networks Bluetooth, 3G/4G, Ethernet, Wi-Fi and ZigBee were mentioned in the studies.

5 Results

This section presents a conceptual framework and a preliminary proposal of an Integrated System for tracking metallic formwork, including its requirements, based on the literature review.

5.1 Conceptual Framework

Based on previous studies about BIM, IoT, and RFID, and considering the concepts and definitions presented throughout the literature review, a conceptual framework was built for tracking physical resources at construction sites using BIM, IoT, and RFID, also taking into account the metallic formwork handled in the cast in place concrete wall system which is the object of study. Figure 2 presents the main concepts addressed and their interconnections, the operationalization of those concepts and the products generated.

At the construction site, tracking physical resources can support the production planning and control process during construction. Tracking Physical Resources can be done through Digital Technologies associated with improvements in management models. Among the digital technologies, BIM, Cloud Computing and IoT are the ones used in conceptual framework. To operationalize those concepts into practice, Virtual Models, a Database/Cloud Server, and a RFID system can be used, respectively, as shown in Fig. 2. As a product of the integration of those technologies and aligned with the aims of this research, it is expected to develop an Integrated System and the User Interface to extract information to track physical recourses such as the metallic formworks handled in the cast in place concrete wall system. To develop this system, it is necessary to formalizing practical requirements for implementing the proposed integration of technologies.

5.2 Requirements for the Integrated System for Tracking Metallic Formwork

The proposed system consists of associating the RFID tags with the ID of equivalent virtual elements in BIM and storing the information from these formworks in a database, which communicates with the virtual model.

Regarding the physical requirements of the system, it will be used a UHF RFID system according to the recommendations of Xue et al. (2018). This system consists of a UHF RFID reader and passive UHF tags. According to the studies of most authors and considering that metallic formworks are resources that move around the construction site, the reader will be a hand-held reader. As indicated by Li et al. (2017), Li et al. (2018), Xue et al. (2018), Zhong et al. (2017), a smart gateway will act as the system's middleware. The device used for this purpose can be a smart device such as the Raspberry Pi. Initially, interactions of the RFID system with other technologies such as GPS and auxiliary antennas will not be required in this study. According to Li's et al. (2018) recommendations, the smart Gateway will communicate with the reader at the construction site via Bluetooth.

Based on the studies developed by Akanmu et al. (2012), Akanmu et al. (2016), Ikonen et al. (2013), Li et al. (2017), Ma et al. (2019), the adopted database will probably be of the SQL type, although it is necessary to evaluate the advantages that a NoSQL database can promote in terms of data structuring and scalability. Considering the observations presented by Xue et al. (2018) and other studies previously mentioned, the database should be based on Cloud Computing. Regarding the BIM model, the

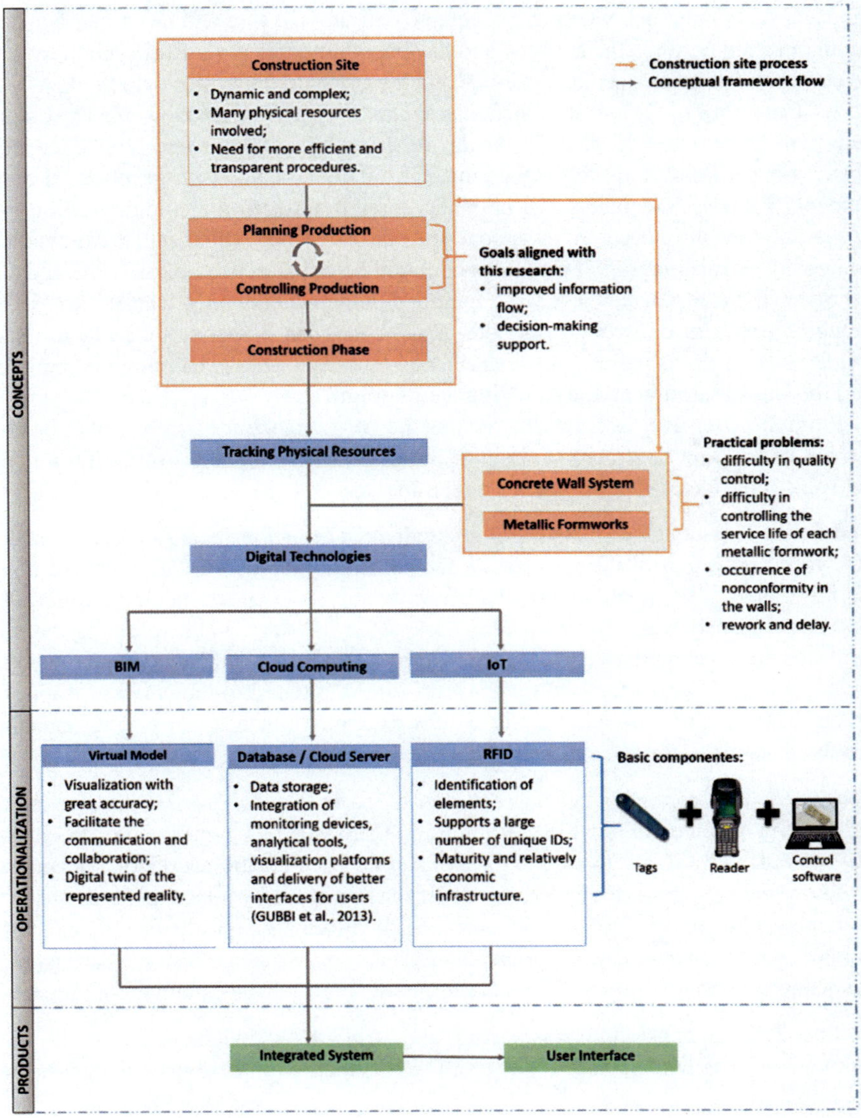

Fig. 2. Conceptual framework.

second approach suggested by Xue et al. (2018) will be used. First, the model will be generated in commercial software and then it will be translated for use in the system.

As shown in Fig. 3, in a Real Construction Site, unique identification UHF tags will be applied to each of the formwork studied (or parts associated with the formworks). Then, the process will involve the use of a mobile UHF reader to capture the data contained in the tags by reading whenever necessary. The data is sent to a smart Gateway installed locally that is responsible for establishing communication with the Cloud

Database via the internet, where the structured data storage step will be carried out. The communication between the gateway and the reader takes place via Bluetooth, however, the communication between the gateway and the database takes place via Wi-Fi.

As shown in Fig. 3, in a Virtual Construction Site, each of the elements used in the study will be represented virtually in the BIM model with its respective unique ID, which will be associated with the equivalent tag ID contained in the physical environment. Such an association will be made possible through the communication that occurs between the virtual environment and the Database, which will be previously configured for this purpose. The BIM model will be translated to be easily viewed and manipulated by managers in a User Interface, which could be a website accessible through computers or through portable devices that can be easily taken to the construction site, such as tablets and smartphones. The data flow is bidirectional between the Physical Environment and the Virtual Environment.

From the User Interface, which contains the BIM project, the manager will be able to track in real-time information about the characteristics and status of the formworks and walls, considering the following three modules.

Module 1. This consists of registering essential information regarding each metallic formwork that is part of the set used on the constructive process. The user will insert this information. To enable this module functionality, one tag containing a unique ID must be previously placed on each formwork. After this step, when accessing the user interface and selecting this module, there will be a request to read the tag with the RFID reader in order to direct the information that will be inserted to the correct location in the database. Such information could be, for example, dimensions, weight, manufacture date, acquisition date, and numerical identification code of the formwork.

Module 2. This refers to the maintenance of each metallic formworks previously registered in Module 1. From the user interface, there will be a request in this module to read the tag with the RFID reader. This enables the automatic access of information already registered about the maintenance of the formwork or else enable the user to insert new information into the database. Such information could be, for example, changes in part dimensions, last maintenance date, maintenance history records, next maintenance forecast, service life forecast, reports, and related photographic records.

Module 3. This consists of associating the tracking of metallic formworks with information related to concrete walls. From the beginning of the project, the unique tag ID of each formwork registered in Module 1 must be associated with the unique ID of a corresponding virtual object in the BIM model. This module has different functions. The first function aims to provide support during the installation phase of the formworks to verify and correct assembly problems. From the virtual construction site, it is possible to track and locate a formwork on the real construction site. For that, it is necessary to click on the virtual element of interest and read the environment with the RFID reader. In this case, the identification is through an audible proximity alert. Once the formwork is located, it is possible to check if the position is correct according to the project. Considering the bidirectional data flow, it is also possible to read the tag in a metallic formwork from a real construction site to identify which is the corresponding virtual element in the virtual construction site. In this case, the identification is through

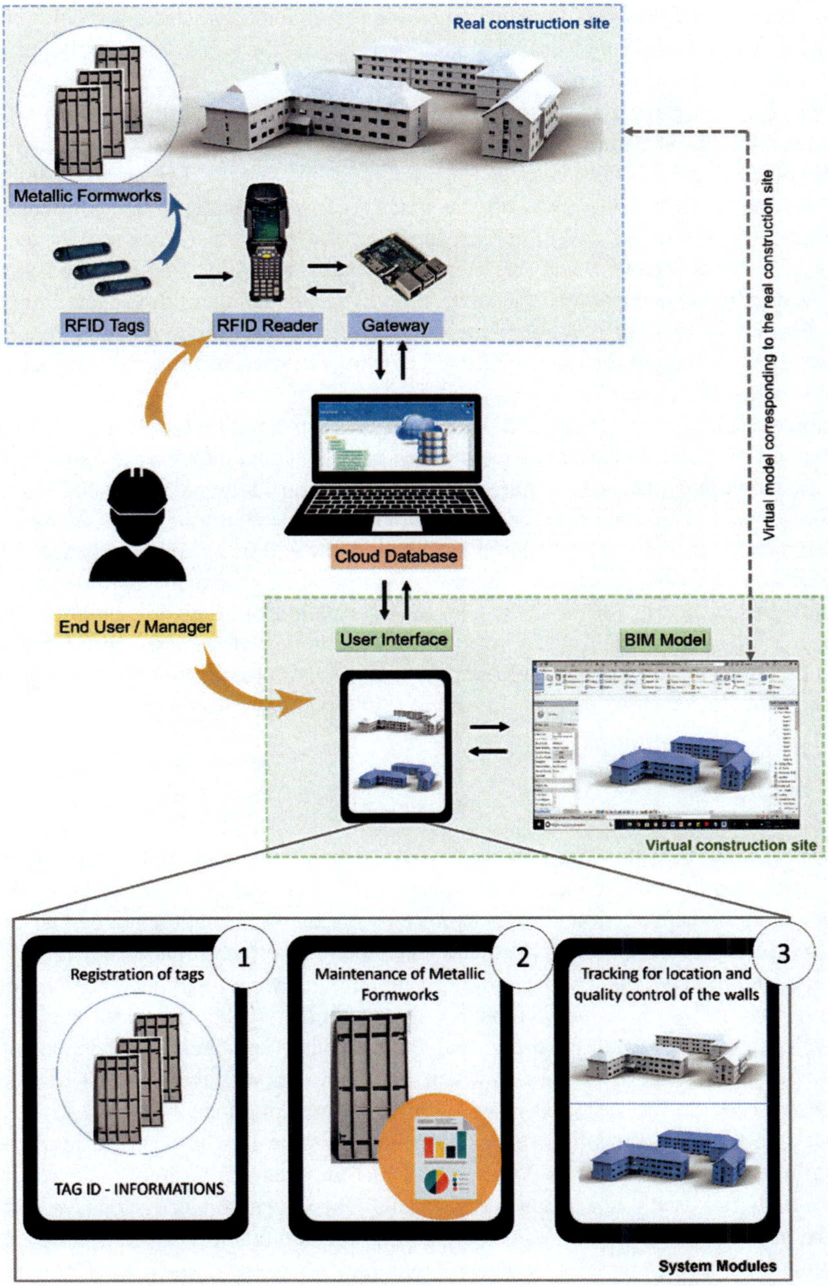

Fig. 3. Design of the integrated system for tracking metallic formworks.

a color change of the virtual element, according to a previously established color code. After correcting problems identified in the first function, the second function requires a final sequential reading of the tags that must be made for automatic comparison with the sequence provided for in the project, aims to ensure the correct location of the formworks used. The third function is for the quality control phase of the concrete walls and assumes that the assembly is according to the project. The main objective is to compare the data collected with the quality of the concrete walls verified after removing the formworks. Within this function, the user must click on the corresponding virtual element and register information regarding the non-conformity observed in that respective wall. Subsequently, by triangulating all data, including data from Module 2, it will be possible to identify whether problems with the integrity of the formwork are affecting the quality of the formworks to prevent future problems in the quality of the constructive process.

Some challenges for practical implementation of the system were already identified: (a) the potential and limitations of the devices, such as signal interference between tags and the minimum distances required for readings, considering the metallic surface; (b) the best way to structure and configure the database used; (c) the characteristics and the availability of the communication established in the process, preferably via Wi-Fi or Bluetooth; (d) the necessary adjustments to the BIM model to support the other technologies involved; (e) the most relevant information to be made available on the User Interface, as well as the way of visual representation of the interactions between equivalent physical and virtual elements (for example, through color codes), and (f) the requirements to manage this data according to the company needs.

6 Conclusion

This study introduces the development of a conceptual framework and a proposal of an integrated system using digital technologies BIM, IoT and RFID to track physical resources at the construction site, focusing on the metallic forms for the cast in place wall reinforce concrete system. Requirements such as software, hardware, protocols, communication networks, and necessary methodologies were identified. The proposed system consists of three main modules that associate RFID tags with the ID of equivalent virtual elements of the virtual BIM model. The status and location information are then stored in a database, which communicates with the virtual BIM model. End-user access to the system is done through the user interface.

Despite the challenges that still need to be overcome for the practical implementation in real construction sites, the potential that an integrated system that associates the technologies used in this study is remarkable. The system to be developed would be able to involve the process of collecting, storing, managing and visualizing data on an integrated platform. As one of the general requirements adopted, the system must have its low cost as a differential when compared to commercial solutions and flexibility to adapt to other types of physical resources at the construction site.

The next steps in this study involve the development of a prototype of the integrated system proposed on a conceptual level and its respective user interface. Further, a pilot implementation of this system in real construction sites will be developed, following by the assessment of the system to which extent this application improves the information flow, has a positive impact on Production Planning and Control, is useful and easy to adopt by the construction projects. This system is expected to contribute to improving the information flow on construction sites that use a cast in place concrete wall system, assisting to managers' decision making and promoting improvements in terms of management and quality control.

References

ABNT NBR 16005: Associação Brasileira De Normas Técnicas: Parede de concreto moldada no local para a construção de edificações - Requisitos e procedimentos. ABNT, Rio de Janeiro (2012). (in Portuguese)

Akanmu, A.A., Anumba, C.J., Messner, J.I.: An RTLS-based approach to cyber-physical systems integration in design and construction. Int. J. Distrib. Sens. Netw. (2012). https://doi.org/10.1155/2012/596845

Akanmu, A., et al.: Auto-generated site layout: an integrated approach to real-time sensing of temporary facilities in infrastructure projects. Struct. Infrastruct. Eng. **12**(10), 1243–1255 (2016). https://doi.org/10.1080/15732479.2015.1110601

Chen, M., Mao, S., Liu, Y.: Big data: a survey. Mob. Netw. Appl. **19**(2), 171–209 (2014). https://doi.org/10.1007/s11036-013-0489-0. Accessed 20 Jan 2020

Cordova, F., Brilakis, I.: On site 3D vision tracking of construction personnel. In: 16th Annual Conference of the International Group for Lean Construction, Manchester, United Kingdon, pp. 809–820 (2008)

Costin, A.M., Teizer, J., Schoner, B.: RFID and BIM-enabled worker location tracking to support real-time building protocol control and data visualization. J. Inf. Technol. Constr. (ITcon) **20**, 495–517 (2015). http://www.itcon.org/2015/29. Accessed 17 Jan 2020

Eastman, C., et al.: BIM Handbook: A Guide to Building Information Modeling for Owners, Managers, Designers, Engineers and Contractors, 2nd edn. Wiley, Hoboken (2011)

Fang, Y., et al.: Case study of BIM and cloud-enabled real-time RFID indoor localization for construction management applications. J. Constr. Eng. Manag. **142**(7) (2016). American Society of Civil Engineers (ASCE). https://doi.org/10.1061/(asce)co.1943-7862.0001125

Feng, C., et al.: An intelligent logistics management model in prefabricated construction. Front. Eng. Manag. **2**(2), 178–181 (2015). https://doi.org/10.15302/j-fem-2015038. Editorial Department of Engineering Sciences

Finkenzeller, K.: RFID Handbook: Fundamentals and Applications in Contactless Smart Cards, Radio Frequency Identification and Nearfield, 3rd edn. Wiley, London (2010)

Gao, X., et al.: Foundational research in integrated building Internet of Things (IoT) data standards – Research report. Center for the Development and Application of Internet of Things Technologies, Georgia Institute of Technology (2018). https://cdait.gatech.edu/sites/default/files/georgia_tech_cdait_research_report_on_integrated_building_-_iot_data_standards_september_2018_final.pdf. Accessed 14 Mar 2020

Gubbi, J., et al.: Internet of Things (IoT): a vision, architectural elements, and future directions. Future Gen. Comput. Syst. **29**(7), 1645–1660 (2013). https://doi.org/10.1016/j.future.2013.01.010. Accessed 20 Jan 2020

Iacovidou, E., Purnell, P., Lim, M.K.: The use of smart technologies in enabling construction components reuse: a viable method or a problem creating solution? J. Environ. Manag. **216**, 214–223 (2018). https://doi.org/10.1016/j.jenvman.2017.04.093

Ikonen, J., et al.: Use of embedded RFID tags in concrete element supply chains. J. Inf. Technol. Constr. (ITcon) **18**, 119–147 (2013). http://www.itcon.org/2013/7. Accessed 01 Jan 2020

Isikdag, U.: Enhanced Building Information Models, 1st edn. Springer, Istanbul (2015)

Li, C.Z., et al.: An Internet of Things-enabled BIM platform for on-site assembly services in prefabricated construction. Autom. Constr. **89**, 146–161 (2018a). https://doi.org/10.1016/j.autcon.2018.01.001

Li, C.Z., et al.: Integrating RFID and BIM technologies for mitigating risks and improving schedule performance of prefabricated house construction. J. Clean. Prod. **165**, 1048–1062 (2017). https://doi.org/10.1016/j.jclepro.2017.07.156

Li, X., et al.: RBL-PHP: simulation of lean construction and information technologies for prefabrication housing production. J. Manag. Eng. **34**(2) (2018b). https://doi.org/10.1061/(asce)me.1943-5479.0000577

Ma, G., Jiang, J., Shang, S.: Visualization of component status information of prefabricated concrete building based on building information modeling and radio frequency identification: a case study in China. Adv. Civ. Eng. 1–13 (2019). https://doi.org/10.1155/2019/6870507

Motamedi, A., Soltani, M.M., Hammad, A.: Localization of RFID-equipped assets during the operation phase of facilities. Adv. Eng. Inform. **27**(4), 566–579 (2013). https://doi.org/10.1016/j.aei.2013.07.001

Nunes, V.Q.G.: Análise estrutural de edifícios de paredes e concreto armado. Dissertação (Mestrado - Programa de Pós-Graduação e Área de Concentração em Engenharia de Estruturas). Escola de Engenharia de São Carlos da Universidade de São Paulo (2011). (In Portuguese)

Oliveira, V.H.M., Serra, S.M.B.: Controle de obras por RFID: sistema de monitoramento e controle para equipamentos de segurança no canteiro de obras. Ambiente Construído **17**(4), 61–77 (2017). https://doi.org/10.1590/s1678-86212017000400185. (In Portuguese)

Park, J., Kim, K., Cho, Y.: Using BIM geometric properties for BLE-based indoor location tracking. In: Seoul International Conference on Applied Science and Engineering, Seoul (2016a)

Park, J., Cho, Y., Ahn, C.: A wireless tracking system integrated with BIM for indoor construction applications. In: Construction Research Congress, San Juan, pp. 2660–2668 (2016b). https://doi.org/10.1061/9780784479827.265

Roblek, V., Meško, M., Krapež, A.: A complex view of industry 4.0. SAGE Open **6**(2), 1–11 (2016). https://doi.org/10.1177/2158244016653987

Sundmaeker, H., et al.: Vision and Challenges for Realising the Internet of Things, 1st edn. Publications Office of the European Union, Luxembourg (2010)

Thames, L., Schaefer, D.: Software-defined cloud manufacturing for industry 4.0. In: Changeable, Agile, Reconfigurable & Virtual Production, vol. 52, pp. 12–17 (2016)

Wang, X., et al.: A conceptual framework for integrating building information modeling with augmented reality. Autom. Constr. **34**, 37–44 (2013). https://doi.org/10.1016/j.autcon.2012.10.012

Want, R.: An introduction to RDIF technology. IEEE Perv. Comput. **5**(1), 25–33 (2006). https://doi.org/10.1109/MPRV.2006.2. Accessed 28 Jan 2020

Xue, F., et al.: Linking radio-frequency identification to building information modeling: status quo, development trajectory and guidelines for practitioners. Autom. Constr. **93**, 241–251 (2018). https://doi.org/10.1016/j.autcon.2018.05.023

Zhong, R.Y., et al.: Towards physical internet-enabled prefabricated housing construction in Hong Kong. IFAC-PapersOnLine **48**(3), 1079–1086 (2015). https://doi.org/10.1016/j.ifacol.2015.06.227

Zhong, R.Y., et al.: Prefabricated construction enabled by the Internet-of-Things. Autom. Constr. **76**, 59–70 (2017). https://doi.org/10.1016/j.autcon.2017.01.006

BIM and Automation of Building Operations in Japan: Observations on the State-of-the-Art in Research and Its Orientation

Jeferson Shin-Iti Shigaki(✉) and Tomonari Yashiro

Institute of Industrial Science, The University of Tokyo, Tokyo, Japan
jshigaki@iis.u-tokyo.ac.jp

Abstract. Japan has long-term automation experience in several manufacturing industries. Recently, players of the real estate and building sector have turned strategic attention and research investments to the automation of building operations, aware of the business relevance of the occupancy phase of facilities. This investigation analyzed R&D programs and experimental projects in Japan to foresee the role of BIM in the automation of building operations. A survey of disclosed information helped to understand the trends in automation research with the adoption of digital technologies for immediate and future use. Appliances from electronics manufacturers have been incorporated into complex cyber-physical systems to communicate with the built facility through a BIM-IoT framework. The evolution of ICT has influenced building management systems attributes which require new definitions of the set of data to be delivered and maintained by project players. There is recent evidence of efforts to turn BIM as the platform that creates, manages and also stores information for future BIM-FM uses, even though the integration with BAS/BEMS is not fully solved. Finally, this study identified connections between prior construction automation achievements by Japanese contractors and future orientation related to the technological challenge of context-awareness and servitization.

Keywords: BIM · Automation · Building operation · BAS · BEMS · IoT

1 Introduction

1.1 Context of Building Automation in Japan

Japan has long-term automation experience in several manufacturing industries. Over the last four decades, construction companies have invested in the development of construction robots and automated building construction systems [1]. Recently, players of the real estate and building sectors including owners, developers, general contractors, building service subcontractors, energy service providers, electronics manufacturers, home appliance retailers, and ICT vendors have turned attention to the automation of building operations, aware of the business relevance of the building's occupancy phase. Therefore, this investigation aimed to observe State-of-the-art research on BIM and Automation of Building Operations in Japan and its orientation.

E. Toledo Santos and S. Scheer (Eds.): ICCCBE 2020, LNCE 98, pp. 879–894, 2021.
https://doi.org/10.1007/978-3-030-51295-8_61

Building operation expenses vary greatly depending on how managers operate, repair and renew facilities. In addition to energy savings, building automation has promised to increase safety, comfort and quality of life of occupiers. Users empowered by technologies can easily access and control devices and systems through the internet. Thence, construction companies have strived to provide product-service systems; and the automation of buildings operation became part of this strategic shift.

In Japan, the idea of Building Automation Systems (BAS) often links with Building Energy Management Systems (BEMS). Despite covering specific functions, residence-oriented HEMS (Home Energy Management System) has become attainable for the general public both as a concept and as a commercial solution. According to the Yano Research Institute [2], the estimated worth for domestic BEMS/BAS market in Japan for Fiscal Year 2019 was 140bi JPY (1.28bi USD). Drivers for sales were the increased demand for construction of new buildings following the redevelopment of major cities such as Tokyo, Nagoya and Osaka and the context of Olympic Games Tokyo 2020. Accordingly, despite drastic cost reduction with the adoption of international standard protocols (i.e. BACnet® and LonWorks®) and stronger competition with new vendors, the market tends to keep rising due to an increase of new buildings. In the future, the market will shift from new buildings to renovation projects of existing facilities as new functions and services with cutting-edge technologies are improved for these projects.

Recent automation research orientation also relates to a strategic policy known as Society 5.0 [3]. The national authority aims at creating a society where they can resolve various social challenges by incorporating the innovations of the fourth industrial revolution into every industry and social life. The deployment of IoT, Big Data, AI, Robot and Sharing Economy would lead Japan to realize a "super-smart society", continuously creating new values and services, making people's life more comfortable and sustainable. The abundant accumulation of real data and technology cultivated from "monozukuri" are advantages that make Society 5.0 possible. The Japanese Government has supported Zero Energy Houses (ZEH) and smart communities' initiatives as part of the carbon emission reduction commitment, and it is implementing subsidies for the installation of HEMS in households by 2030 [4].

1.2 Context of Building Automation in Japan

There are many direct and indirect players involved with R&D in Building Automation and Control Systems in Japan. It ranges from Governmental agencies to technology institutes or R&D departments of construction, manufacturing and services companies. It also covers academic societies and research institutes which provide consultancy and manage research partnerships and consortiums.

The Building Automation Association (BAA) was established in 1997, first as a point of contact for the Ministry of Land, Infrastructure, Transport and Tourism (MLIT) with the building automation industry [5]. In 2011, it became a general incorporated association with electronics manufacturing and control system services companies aiming the further development of the industry. Besides conducting research and educational activities on relevant topics either guided by the Government or identified by the sector, it organizes industry-wide meetings and disseminates information.

The Japan Smart Community Alliance (JSCA) was established in 2010 to support the future growth of Japan's leading energy-saving and new energy technologies, as well as to promote collaboration between the public and private sectors [6]. Four working groups conduct research and collaboration activities: International Strategy, International Standardization, Roadmap, and Smart House and Building. Activities encompass the identification of technology, policy and market trends and the development of strategies to support Japanese companies. In cooperation with the Ministry of Economy, Trade and Industry (METI), the alliance works on studying global trends in standardization, future preparedness through roadmaps developing scenarios for next-generation societies and further dissemination of smart buildings.

The challenge of researching building automation is admittedly tough. So, despite the competition, industry players have ever more worked together. One finds in their partners the market understanding in specific business branches, background experience within the area of application or similar type of technological solution, technical information to solve engineering integration problems, and enhanced capital investment capacity.

In some cases, the provider of HEMS and the house manufacturer belong to the same business group such as PanaHome and Panasonic [7]. Similarly, NTT Facilities took advantage of its real estate, telecommunication and computer systems background in developing its BAS/BEMS solution [8]. The "NTT Data I System" is open network-compatible enabling multi-vendor support for lower-cost realization through BACnet® and LonWorks®. The BAS Partners Portal is a channel providing further technical and commercial information [9]. In other cases, partnerships already exist from other product sectors and were built out again for developing solutions for the automation of building operations as Denso Co. with Toyota Housing Co. and Misawa Homes [10].

Moreover, companies collaborate for specific objectives in demonstration experiments which can involve a Governmental agency and or universities. One example is the Smart House HEMS prototype developed by Fujitsu with Tokyo Electric Power Company Holdings (TECPO) also involving Waseda University Advanced Collaborative Research Organization for Smart Society within a project by the METI's Agency for Natural Resources and Energy [11].

For a long time, Japanese manufacturing companies have adopted an in-house technology development strategy. Nevertheless, the formation of the associations above suggests that such a complex problem should be faced by cooperative organizations to speed up the accomplishment of business objectives. Together, they may not only share the technological infrastructure but the background knowledge to engineer integrated solutions which depend on multiple players to perform adequately.

2 Methodology

2.1 Sources of Information and General Procedure

The approach consisted of a literature exploration conducted through surveying disclosed information from relevant players of the Japanese construction, electronics manufacturing and operation service industries. The reason for using disclosed information is that some development programs still undergo without a commercial solution or are kept in industrial secrecy. Nevertheless, Governmental agencies have promoted general concepts; while industry associations have diffused BAS/BEMS technology information aiming to expand the market. Some companies released their 'visions' and have received valuable feedback to improve their R&D programs. Consultancy firms and research institutes have published insightful articles giving a hint on the direction of current R&D efforts. Findings of this paper are primarily based on the observation of publicized information and focus on the strategic bias of on-going research rather than technical output details.

As a secondary source of information, a survey on the J-Stage Platform [12] intended to overview the debate in Japan from an academic perspective. The objective was to find specific integration solutions between BIM and BAS/BEMS or interoperability research findings in this area. Therefore, the keywords used were "BIM" AND "BAS" OR "BEMS" OR "HEMS" all in alphabetic characters because even in Japan, it is common to refer to these terms in Romanized writing. All publication types were admitted, both in Japanese and English. The subject areas were limited to "Engineering and Technology", including Architecture and Civil Engineering, and Electrical and Electronic Engineering. However, this search resulted in nine documents from which just one was considered relevant after excluding unavailable documents and papers from non-related engineering fields. That paper described the situation of BACS development and briefly pointed out the challenge of creating data models linking BACS and BIM [13].

A search with "BIM" AND "IoT" resulted in another ten documents related to its application for site management or structure performance monitoring; thus, not related to the automation of building operations. A search excluding the term "BIM" resulted in a higher amount of publications, especially in the electronics field, but their scope did not meet the objective of this exploration. These keywords were also found in the titles of technical papers in magazines of construction companies. Although only a few were available online, it suggested on-going developments and their research orientation.

2.2 Identification of Major Research Streams

Despite the initial purpose of this investigation in identifying R&D programs concerned with BIM and BAS/BEMS systems integration, most findings related to the advancement of BAS/BEMS technologies and its potential applicability as commercially deployable solutions regardless of a BIM orientation. Building Automation research in Japan is closer to IoT-related societies rather than the BIM community.

Therefore, the main differences between Japan and overseas possibly rely on the context and structure of research initiatives rather than the technical outputs. In a sense, Japan is striving to adopt international standards based on the commonly agreed problem of heterogeneity [5, 6]. Yet, many initiatives are the result of a collaboration between domestic partners toward the accomplishment of national policies.

There are two main streams for R&D efforts.

New technology development research: Intend to consolidate and expand markets with the creation of cutting-edge technologies (technology push); and reduce acquisition, installation, operation and maintenance costs (response to market pull). Service provision and sustainability tenets have focused on improving usability and user interface of systems as well as designing marketable solutions for inexperienced potential users. Technology development may be held internally or in partnership with providers holding specific expertise in a field of application.

Technology management for solving threats raised with emerging technologies: problems raising with the deployment of new technologies may represent serious hazards for users, society, and the building automation business itself. Despite risk mitigation is managed during technology development, some problems not expected in early conceptual stages express real threats for its sustainability and endurance.

These streams, however, do not disconnect from each other. For instance, 'Technological domain' research for advancing hardware, software, network systems of BA are closer to the technology development stream but implies strategic make-or-buy decisions. Then, despite the technical nuance of security problems, R&D on 'Cyber-security issues' have emphasized their management implications along with the efforts to advance technology. Besides, 'Response to lifestyle change' may first sound closer to the management stream, it deals with solving information exchange pitfalls between devices of multiple vendors to expand the market.

Such type of dynamic interaction is an intrinsic characteristic of Japanese industrial technology research in line with its unique approach to managing the creation of new knowledge [14]. Another peculiarity observed in this investigation is the propensity and ability to cross the industry boundaries; so, companies are open to using diverse technologies to create new products that transform markets [15]. The cases presented below demonstrate how the Japanese industry has managed to advance in this field.

3 Research in Automation of Building Operation in Japan

3.1 Technological Domain and In-house Technology Development

Japanese automobile companies tended to adopt an in-house automation technology development strategy, which was not only a low-cost but had several advantages. These included using internal know-how to design machinery "with just enough functions" compatible with the Gemba, increased control over maintenance, and easier functional expansion [16]. The building sector first conducted automation research turned to improve production processes. Construction companies strived to automate construction operations by adapting robotic technologies to satisfy specific operation conditions. Site automation efforts mostly resulted in single-task types of robots [17].

Obayashi Corporation started to develop the Automated Building Construction System (ABCS) in the 1980s to increase productivity and improve the working environment resulting in climbing mechanisms, welding robots and other site technologies [18]. Meanwhile, Shimizu Corporation worked on fireproofing spray robots, decommissioning robots for nuclear power plants and even robots for space programs [19]. These activities took place at R&D departments, referred to as Technology Research Institutes, which can reutilize prior automation experiences for new research aiming at the operation phase.

However, the characteristics of research programs have changed. The former generation was mechanic-oriented, with most devices installed in a fixed location. Old versions of HEMS were arranged in relatively closed architecture networks and struggled with interoperability issues. In turn, the current generation is ICT-oriented and strives to be movable and far-reaching. They are more flexible and easier to communicate through Open Systems. Smart objects, such as self-operating windows and housekeeping buddies, incorporate robotics, IoT and AI features.

Construction companies owning a real estate investment division have merged background knowledge to set visions of the Next Generation building operations. Facilities managers have worked closer to R&D to explore the potential use of BIM data in the operation phase. They not necessarily adopt off-the-shelf CAFM platforms but develop custom solutions "with just enough functions" which are exchangeable with their existing databases.

One change is that technology development, predominantly held in-house or with companies of the same group, is gradually conducted with external strategic partners holding specific know-how and infrastructure. As an example, recently Kajima Corporation began to provide Building Management services utilizing a "Kajima Smart BM" platform to predict energy consumption and equipment malfunctioning, improving operations and reducing running costs [20].

Traditionally, building operation data had been stored for a short period due to the limited capacity of monitoring equipment. In this new system, data collected from BAS, BEMS and multiple IoT sensors are automatically stored in the cloud to be then analyzed using AI. For the development of this platform, the two companies of the Kajima Group (one for construction and other for integrated facilities management) combined their knowledge and technology to build data infrastructure and implement Microsoft Azure receiving full technical support of Microsoft Japan.

3.2 Response to Lifestyle Change at Various Scales

Society 5.0 has raised public attention and has fostered R&D investments to the realization of Smart Cities and Smart Buildings. Japan has also strived to contribute to the UN's Sustainable Development Goals (SDG). The popularization of the concept and potential benefits of ZEH by families has accelerated the improvement of HEMS both in technological and commercial ways. Such energy and environmental policies have oriented research efforts in various levels, from building parts, building units and urban scales. Home appliance manufacturers, builders and real estate companies have used the flag of ZEH and Automation as a way of promoting healthy and convenient lifestyles, making their customer feel they are continuously contributing to the environment and saving energy costs.

LIXIL, a building materials and component maker, has developed the "IoT-Home Link". Various IoT devices connect with the "Life-Assist", from where one can check the status of the house and remotely control it from a single smartphone application [21]. The assist rule function combines a life scene and a trigger to take actions with sensors and actuators connected to building parts or equipment configured to behave according to a desired lifestyle or automatically respond to critical situations. House manufacturers, such as Sekisui Heim [22] and Daiwa House [23], have prepared advertisement materials highlighting the benefits of HEMS, provide easy explanation during visiting sessions, and distribute user-friendly manuals for using their products.

Applications of performance-oriented solutions have affected architectural planning. Therefore, the physical and spatial characteristics of the building are made ready for the installation of devices, movement of equipment and adequate operation of network systems. Building automation demands flexible designs for maintenance, the occasional substitution of controlling devices and eventual transit of surveying machines for system updates. Yet, research on smart services robots might change the image of automation of building operations.

In the urban scale, Toyota just revealed the plan to build the "Woven City", a fully connected ecosystem prototype at the base of Mt. Fuji [24]. The vision of making the city a "living laboratory" is audacious even for Japanese standards. Toyota sees the project as an opportunity to develop future technologies for the city's infrastructure. The digital operating system presumes advancing interdisciplinary technologies, so they plan to collaborate with commercial and academic partners and attract scientist and researchers to work together. The breadth of technologies and industries covered by the group enables the exploration of new forms of urbanity changing the way of inhabiting and navigating cities.

3.3 Cybersecurity Issues and the Involvement of Multiple Players

Previous generations of Building Control Systems were not easily targeted by cyber-attacks because they have been separated from the internet or used system-specific protocols. However, it is no longer possible to ignore cybersecurity issues since system designs increasingly assume external connections for new functionalities.

Aware of its potential threat to the consolidation and expansion of systems for building operations, the Ministry of Economy, Trade and Industry (METI) organized a sub-working group to specifically discuss cybersecurity measures for building systems under the Industrial Cybersecurity Study Group WG1 [25]. In March 2019, this group released a draft entitled "Guidelines for cyber-physical security measures in building systems", providing comprehensive guidelines to building users and services providers, as well as countermeasures against identified threats at each stage of the building lifecycle. Members of this group included academics, local authorities, general contractors, and companies and associations of the real estate, insurance, telecommunication, electronics and automation service sectors. These measures, however, covered BACS systems and do not encompass BIM or BIM-BAS yet.

4 Cases of Research and Development Initiatives

4.1 Electronics Manufacturers Consortium: ECHONET

ECHONET (Energy Conservation & Homecare Network) Consortium was established in 1997 to promote the standardization of home network infrastructure technology. It strives to interconnect household appliance, residential equipment, sensors, controllers and other devices from multiple vendors. Take part of the consortium seven managing members (Hitachi, Mitsubishi Electric, NTT, Panasonic, Sharp, TEPCO, and Toshiba), 166 general members and 32 academic members. It has strengthened partnerships with other organizations to disseminate HEMS and IoT Smart House.

In 2011, the Consortium released the ECHONET Lite Specifications (Fig. 1). which are compatible with standard protocols in Japan and overseas. After that, they launched ECHONET Lite AIF (Application Interface) specifications aiming to improve interoperability between vendors. The international standard for ECHONET Lite AIF is approved under ISO/IEC 14543-4-3 and IEC 62934 [26]. Between 2012 and 2019, 705 devices acquired the ECHONET Lite certifications. Meanwhile, 478 AIF certifications were conceded (217 for devices and 261 for controllers). The consortium estimates the number of ECHONET Lite specification compliant products exceed 48 mi units by 2017 [27]. The Government understands this plan encourages people to introduce HEMS and strengthen competitive advantages of Japanese products in the international market.

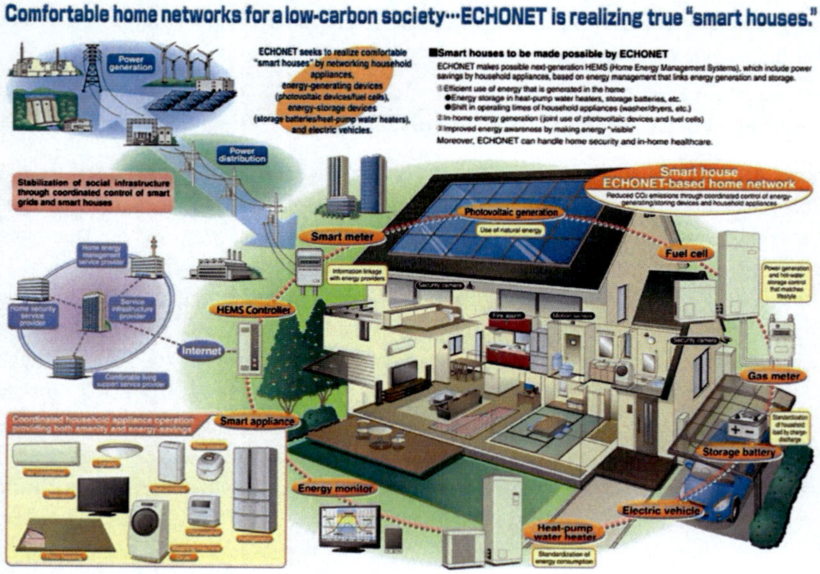

Fig. 1. ECHONET-based smart house concept (Source: ECHONET).

4.2 Joint Demonstration Experimental Project: COMMA HOUSE

The COMMA HOUSE (COMfort MAnagement House) initiated in August 2011 as a joint demonstration experimental house project between LIXIL and the Institute of Industrial Science of The University of Tokyo with several industry research partners (i.e. Toshiba Solutions Co., NEC Co., Family Net Japan Co., Tokyo Electric Power, Nomura Research Institute) and contributors [28]. It aimed to establish and demonstrate concepts and technologies for smart houses that would become popular by 2020. The intent of wisely increasing energy-saving performance incorporates HEMS to optimally control domestic machines and equipment to make it a Zero Energy Building (ZEB). The prototype was built at The University of Tokyo (Fig. 2).

Fig. 2. COMMA HOUSE prototype (photo: author).

This project works on the specificities of "Domestic IoT" in comparison with Industrial IoT with a prototyping orientation. It attempts to deploy atypical combinations of things to meet the needs of everyday life through the concept of an "Open System". Such a term alludes to the successful Japanese experience on open systems for building production. Nevertheless, here, the idea is to safely and quickly connect home appliances of diverse manufacturers flexibly responding to the various needs of users. In other words, it re-signifies the concept for the building operation phase while considering the current technology challenges and with end-user consideration.

The technical approach for ubiquitous connectivity in the COMMA HOUSE was the adoption of a Web-API method. It is different from "local integrators" since it is not an application but an interface for applications to operate things through multiple communication protocols. The IoT Special Study Group actively disseminated information about the central idea of Web-API to the IETF (Internet Engineering Task Force), which influenced de-facto standards [29]. An internet-draft entitled "Problems in and among industries for the prompt realization of IoT and safety considerations" report findings including some obtained through the COMMA HOUSE project [30].

4.3 Development by a Large Contractor: Obayashi's WellnessBOX®

Obayashi has developed the WellnessBOX®, a Smart Building Management System utilizing IoT-AI [31]. Demonstration operation started in office buildings in December 2017. The system aims to realize optimal building management while improving the wellness of workers. The strategy of providing wellness for building occupiers follows the need of work-style reforms in a society struggling with the decline of the working-age population.

WellnessBOX® uses IoT technology to transmit individual comfort information obtained through building equipment sensors and location information of building users using beacons through the internet. By using AI technology, this Building Management System provides an optimal environment for individual users and realizes energy saving due to the fine-grained building control using information stored in the cloud. Moreover, it helps to reduce the load on facilities managers since building equipment settings can are adjustable from remote locations according to the user's needs (Fig. 3).

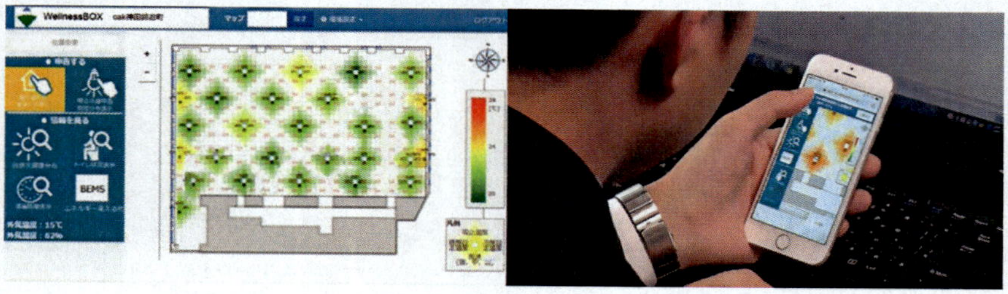

Fig. 3. WellnessBOX® operation screen of AC control (source: Obayashi).

The system has three main features. The first is a "user menu" to improve wellness in terms of comfort, health, convenience and security, evaluating and controlling the building and indoor environment. The second feature is "opening environment information of the building in use". Management information in thousands of points is obtained from standardized sensors to make possible the control of lighting and air conditioning to provide a comfortable environment for each user. And the third is the "provision of one-stop service". Although BAS directly controls building equipment but retrieves operation and control information from the cloud. Therefore, users can enjoy a variety of services through their smartphones, even use from remote locations.

Information on operation conditions obtained from the WellnessBOX® can exchange with a BIM platform also developed by Obayashi. BIMWill® utilizes the BIM model at the time of construction to improve the prediction accuracy of equipment deterioration and update timing of preventive maintenance before failures occur. Thus, by combining both systems, the organization have the potential to develop building management services utilizing a Digital Twin.

5 Research Orientation and the Future Role of BIM

5.1 BIM Preparedness to Expand Automation Functions

Despite the expectations of identifying technical solutions or hints for BIM and BAS/BEMS integration, this is still an open problem. There is evidence on the awareness of the potential of BIM for FM and Automation of Building Operations. Nonetheless, the availability of deployable solutions is still scarce. Prior research initiatives conducted mainly by electronics-related organizations dedicated to connecting physical devices to BAS/BEMS from an IoT perspective regardless of BIM.

Those researches tending to an 'electronics-orientation' focused on standardization for improving interoperability are not in a much different direction compared to the latest works carried out overseas, as in [32]. About BIM and IoT integration, Japanese R&D held by industrial players tends to be less comprehensive in scope and more function-specific compared with Western academic research such as [33, 34]. Although technical details remain concealed, Obayashi's WellnessBOX® and BIMWill® solutions seem to be unique with the merit of testing in real facilities. Nevertheless, in Japan, it could not be found any report covering BIM and BAS exchange explicitly using BACnet® and IFC as in [35].

Lately, Japanese companies and researchers have also considered BIM-IoT frameworks. Discussions often touch future scenarios with Digital Twins, exploring possibilities for automation with BIM and IoT and Robotics, encompassing Common Data Environment, Web Services, and API. However, there are some fundamental questions not yet solved. These are 'How to collect and use BIM data for building automation purposes', 'How to integrate BIM workflow with physical equipment?' and 'How to build and collect digital data in such a way it can be accessed by different stakeholders?'. These questions imply in further research within the realm of the two streams mentioned before: new technology development research and technology management for solving threats raised with emerging technologies.

5.2 The Challenge of 'Context Awareness' and BIM-Linked Robots

R&D in construction automation has gradually advanced in BIM and Robotic integration. Shimizu's Robo-Buddy Multipurpose Robot (one of the three developments of "Shmz Smart Site" system along with Robo-Carrier and Robo-Welder) detects its location using lasers by referencing BIM information to move to the designated work location and install ceiling and floors grid frames [36]. Real-time spatial data obtained from sensors is compared with other spatial data acquired from BIM models.

Moving to the operation phases, connecting position information with BIM models would expand new possibilities to improve FM systems. However, context awareness is still a tough challenge that not embedded in service robots. The accuracy of position information required will demand the further development of technologies as well as real-time communication mechanisms with these robots in indoor spaces.

Research might also explore educating robots about spatial context to make them able to safely perform certain operations within the building during the occupancy phase. In addition to networked devices fixed in building objects (including those with moving capabilities), the Next-Generation R&D pursues giving them the ability to freely move around the building, such as autonomous vehicles, to increase the breadth and convenience of services and improve operational management.

Part of these efforts has related to indoor positing technology. By improving navigation capabilities, facilities managers can think and develop a new set of service applications. NTT DATA Corporation and Narita International Airport partnered to release "NariNAVI", an airport navigation app using high-precision indoor location information technology (BLE beacons and geomagnetic signals) to real-time navigation in indoor spaces [37].

In turn, larges-scale infrastructure projects developed within a BIM workflow may benefit from context-awareness capabilities to monitor and manage operations with a lifecycle management perspective. In Japan, BIM for Civil Engineering projects, such as road structures, is commonly named CIM (Construction Information Modelling, different to CIM as City Information Modeling as used in other countries). The CIM Implementation Guidelines (for infrastructure projects) by the MLIT suggests that CIM models are suitable for acquiring 3D terrain data obtained through Mobile mapping systems to support decision-making on road improvement and repair works and prevent the omission of inspection work in the future [38].

Context-awareness also fits the urban scale. BIM is a component of the 'digital city' in the cyberspace which needs to connect with things and people in the physical space. Geospatial data collected from sensors are expected to connect with GIS datasets and BIM data, expanding the scope of management possibilities. Smart communities supported with a Digital Twin will enable the automation of urban equipment operations communicating with buildings and objects; and thus, expand the impact in people's daily life and contribute to achieving environmental goals. The interaction of automation in building scale and city-scale will require communication mechanisms that are not clear yet but are under consideration.

6 Discussion and Conclusion

This paper introduced a prospect of research in the automation of building operations by observations on Japanese R&D programs and experimental projects. The outcome was finding that automation research takes part in servitization strategies in a broad sense. Technology advancement programs come along with technology management efforts for solving threats raised with emerging technologies and focus on value creation from a lifecycle perspective. The structure and prior knowledge of Japanese companies in related areas influence the scope and orientation of R&D programs as well as the configuration of collaboration schemes.

Guidelines provided by the Government and industry associations, or joint efforts have exerted influence over the characteristics of R&D programs. On this basis, companies and research institutes have built frameworks to set an effective strategy to support future society and ensure business competitiveness. However, despite the intention of integrating BIM data with automation systems, such as BAS/BEMS and, especially, HEMS which has attracted attention after the plans encouraging ZEH, there is scarce evidence of research progress in this area from the disclosed information. The compendium of research initiatives extracted from non-academic media helped to illustrate the situation, especially for the non-Japanese speaking community.

Instead of adopting off-the-shelf automation tools, there is an apparent trend in the development of custom solutions, either held in-house or with strategic partners. After companies become more experienced with BIM in planning, design and construction phases and built more robust databases, BIM applications for BA and FM purposes are likely to become more common.

Recent achievements of mobility research in technologies to recognize the space and provide intelligence to machines have been a source of inspiration for the building sector. In the recent past, Japanese automobile production systems inspired a new model of managing the construction site in terms of both machine technology and management philosophy. Now, the breadth of AI-powered services and the technological solutions developed for communicating the cyberspace with the physical space became inputs in the creation of intelligent operating building components as well as indoor services robots for performing complex tasks.

As a tentative conclusion, current R&D efforts in Japan are under change and reflect background experience and the organizational structure for automation research in related areas. There are new drivers, such as those policies and goals related to Society 5.0, modifying the orientation of upcoming programs. Besides, key players such as real estate and facilities managers are more attentive to the potential value of automation in the operation phase of the building lifecycle. A provisional representation of the Japanese R&D orientation is drawn below and expresses the probable new wave of automation development investment (Fig. 4).

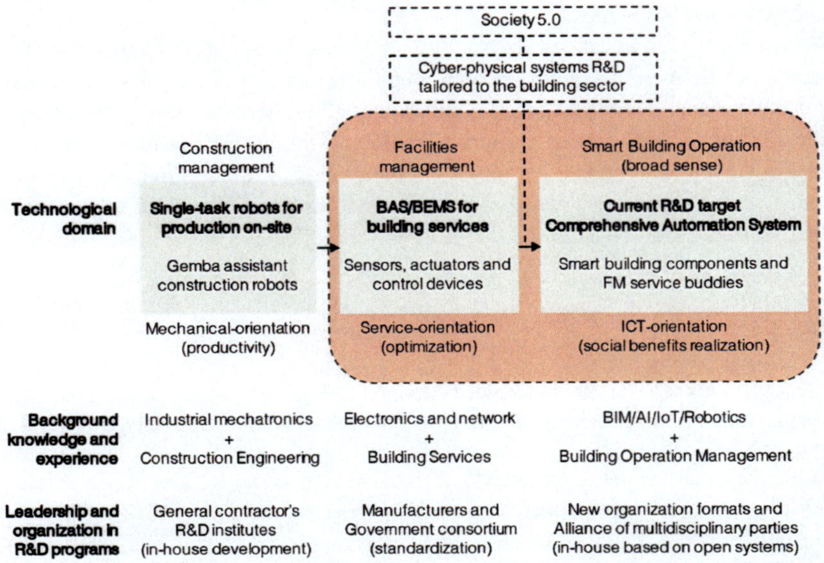

Fig. 4. Background and orientation for Japanese R&D in smart building operation.

In the first stage, mechanical-oriented construction robotics research learned from the successful Japanese Monozukuri to tackle productivity and quality issues. Construction companies began to engineer specific robots to perform construction operations and automate assembly works in the complicated conditions of the Gemba out of controlled factories. Such mechanical orientation moved to an electronics orientation as the focuses moved from improving construction management to facilities management employing automation gadgets. However, physical devices and networks systems involved in BAS/BEMS required solutions from several players to communicate with each other. Current R&D have strived automation solutions to the realization of Smart Buildings. Its ICT-orientation encompasses a set of technologies to support building lifecycle management, from smart operating components to service assistant robots.

Despite the historical and ongoing achievements of Japanese R&D programs, audacious projects at the scale of Toyota's Woven City are an exception today compared with the construction automation research storm in Japan 1980s. However, Japanese construction companies have great potential to develop breakthrough solutions based on consolidated R&D structure, investment and experience. Despite the meaningful outcomes of Kaizen culture, the magnitude of the challenges calls for a more disruptive R&D strategy attitude, such a fostering Kaikaku, to maintain Japanese research in a global leadership position. Collaboration arrangements have built to overcome such a complex challenge.

One example is the partnership between Kajima and Microsoft to offer a new platform for building management services with AI. Another indicative of change is the just-announced technical cooperation between construction giants Kajima and Takenaka for advancing construction robotics and IoT. The "Construction RX (Robotics Transformation) Project" is the first comprehensive collaboration after some joint research on specific themes [39].

Based on the observations above, future developments shall keep exploring the integration of BIM and IoT with upcoming building automation systems. An advantage of Japanese general contractors is their potential to benefit from the expertise of business units under the same corporate group and the possibility to use robust databases. However, the realization of BIM integration with automation systems for building operations is yet not widely diffused as commercial service at the moment. Further research shall re-examine the progress and effects of the current initiatives.

References

1. Bock, T., Linner, T.: Robot-Oriented Design: Design and Management Tools for the Deployment of Automation and Robotics in Construction. Cambridge University Press, New York (2015)
2. Yano Research Institute Ltd.: BEMS & BAS Market in Japan: Key Research Findings (2019). https://www.yanoresearch.com/en/press-release/show/press_id/2204. Accessed 13 Jan 2020
3. Government of Japan: Realizing Society 5.0. https://www.japan.go.jp/abenomics/_userdata/abenomics/pdf/society_5.0.pdf. Accessed 13 Jan 2020
4. National Institute for Environmental Studies: Environmental Technology Commentary on HEMS. http://tenbou.nies.go.jp/science/description/detail.php?id=17. Accessed 31 Jan 2020. (in Japanese)
5. Building Automation Association: Business Plan. http://ba-system.org/business_plan/. Accessed 31 Jan 2020. (in Japanese)

6. Japan Smart Community Alliance. https://www.smart-japan.org/english/about/index.html. Accessed 31 Jan 2020
7. Panasonic: About HEMS. https://sumai.panasonic.jp/aiseg/hems/index.html. Accessed 31 Jan 2020. (in Japanese)
8. NTT Facilities: Building Integrated Management Systems. https://www.ntt-f.co.jp/service/fm/mon_basbems/. Accessed 31 Jan 2020. (in Japanese)
9. NTT East Japan: BAS Partners Portal. https://www.bas-partners.com. Accessed 31 Jan 2020. (in Japanese)
10. Denso: Denso Develops Home Energy Management Systems. http://www.globaldenso.com/en/newsreleases/111207-01.html. Accessed 31 Jan 2020
11. Fujitsu Journal: Developing Smart Houses with HEMS Using Electricity to Suit People's Lifestyles. https://journal.jp.fujitsu.com/en/2016/02/24/01/. Accessed 31 Jan 2020
12. J-STAGE: Advanced Search. https://www.jstage.jst.go.jp/search/global/_search/-char/en. Accessed 31 Jan 2020
13. Nakahara, N.: The state-of-the-art BACS development at the ISO TC205/WG3. J. Inst. Electr. Install. Eng. Japan **33**(2), 87–91 (2013). (in Japanese)
14. Nonaka, I., Takeuchi, H.: The Knowledge-Creating Company: How Japanese Companies Create the Dynamics of Innovation. Oxford University Press, New York (1995)
15. Kodama, F.: Technology Fusion and the New R&D. Harvard Business Review, Brighton (1992)
16. Fujimoto, T.: Introduction to Production Management II – Production Resources and Technology Management. Nihon Keizai Shimbun Inc., Tokyo (2001). (in Japanese)
17. Bock, T.: The future of construction automation: technological disruption and the upcoming ubiquity of robotics. Autom. Constr. **59**, 113–121 (2015)
18. Kudoh, R.: Implementation of an automated building construction system. In: Proceedings of the 13th International CIB World Building Congress Research and Technology, pp. 17–28. CIB, Amsterdam (1995)
19. Yoshida, T.: A short history of construction robots research & development in a Japanese company. In: Proceedings of the 23rd International Symposium on Automation and Robotics in Construction ISARC 2006, pp. 188–193. IAARC, Amsterdam (2006)
20. Kajima Corporation: Provision of building management services utilizing a new platform: Kajima Smart BM. https://www.kajima.co.jp/news/press/201912/pdf/4a1-j.pdf. Accessed 31 Jan 2020. (in Japanese)
21. LIXIL: IoT Home Link. https://www.lixil.co.jp/lineup/zehiot/iot/. Accessed 31 Jan 2020. (in Japanese)
22. Heim, S.: Ohisama HEMS. https://sekisuiheim.com/ohisama/hems/. Accessed 31 Jan 2020. (in Japanese)
23. Daiwa House: Smart Eco possibilities link D-HEMS. https://www.daiwahouse.co.jp/jutaku/smarthouse/dhems.html. Accessed 31 Jan 2020. (in Japanese)
24. TOYOTA: Toyota to Building Prototype City of the Future. https://global.toyota/en/newsroom/corporate/31171023.html. Accessed 31 Jan 2020
25. METI: Guidelines for cyber-physical security measures in building systems - First Draft. https://www.meti.go.jp/press/2018/03/20190311001/20190311001-2.pdf. Accessed 31 Jan 2020
26. METI: ECHONET Lite Application Communication Interface (AIF) Specifications. https://www.meti.go.jp/english/press/2019/0206_002.html. Accessed 31 Jan 2020
27. ECHONET Consortium. https://echonet.jp/about_en/. Accessed 31 Jan 2020
28. COMMA HOUSE: COMfort MAnagement House. http://www.commahouse.iis.u-tokyo.ac.jp/english/index.html. Accessed 31 Jan 2020

29. Yashiro, T., Baba, H.: A Book to Understand Domestic IoT: Innovation that Connects Things of Life with the Internet and its Challenges. Impress R&D Next Publishing, Tokyo (2017). (in Japanese)
30. IETE: Problems in and among industries for the prompt realization of IoT and safety considerations. https://tools.ietf.org/id/draft-baba-iot-problems-06.html. Accessed 31 Jan 2020
31. Obayashi Corporation: Applying WellnessBOX, a smart building management system utilizing IoT-AI to office buildings in Tokyo for the first time. https://www.obayashi.co.jp/news/detail/iotaiwellnessbox.html. Accessed 31 Jan 2020. (in Japanese)
32. Domingues, P., Carreira, P., Vieira, R., Kastner, W.: Building automation systems: concepts and technology review. Comput. Stand. Interf. **45**, 1–12 (2016)
33. Dave, B., Buda, A., Nurminen, A., Främling, K.: A framework for integrating BIM and IoT through open standards. Autom. Constr. **95**, 35–45 (2018)
34. Gao, X., Pishdad-Bozorgi, P.: BIM-enabled facilities operation and maintenance: a review. Adv. Eng. Inform. **39**, 227–247 (2019)
35. Tang, S., Shelden, D.R., Eastman, C.M., Pishdad-Bozorgi, P., Gao, X.: BIM assisted Building Automation System information exchange using BACnet and IFC. Autom. Constr. **110**, 1–14 (2020)
36. Shimizu Corporation: Robots under Autonomous Control at Robot Lab. https://www.shimz.co.jp/en/company/about/news-release/2018/2018006.htm. Accessed 31 Jan 2020
37. NTT DATA: Japan's First High-Precision Indoor Positioning Driven by Geomagnetic Technology Used at Narita Airport. https://www.nttdata.com/global/en/media/press-release/2018/october/japans-first-high-precision-indoor-positioning-driven. Accessed 31 Jan 2020
38. MLIT: CIM Implementation Guideline (Draft) Part 1 – Common. https://www.mlit.go.jp/tec/it/pdf/guide01.pdf. Accessed 31 Jan 2020. (in Japanese)
39. Nihon Keizai Shimbun, Kajima and Takenaka collaborate on technology development - labour shortage solved by labour saving. https://www.nikkei.com/news/print-article/?R_FLG=0&bf=0&ng=DGXMZO55028750Q0A130C2X12000. Accessed 31 Jan 2020. (in Japanese)

BIM, IoT and MR Integration Applied on Risk Maps for Construction

Lorena C. de S. Moreira[1(✉)], Paula Pontes Mota[2], and Fernanda Almeida Machado[3]

[1] Federal University of Bahia, Rua Caetano Moura, 121, Federação, Salvador, BA, Brazil
lorenasm@ufba.br

[2] Sippro Engenharia, Av. Padre Antônio Tomas, 2420 - cj. 605, Fortaleza, CE, Brazil

[3] Rua James Joule, 65 - Cidade Monções, São Paulo, SP, Brazil

Abstract. In Brazil, the Architecture, Engineering, and Construction (AEC) industry is a high-risk segment that accumulates workplace injuries. National standards concerning Risk Management in construction require the production of Risk Maps, aiming at increasing users' awareness about risks and hazards on site. However, the traditional process of implementing Risk Maps has limitations that reduce their efficiency in a real application. The adoption of innovative technologies such as Building Information Modeling (BIM), Internet of Things (IoT), and Mixed Reality (MR) may benefit communication resources of Risk Maps and improve Risk Management. This paper presents a conceptual framework of Risk Management for the construction phase to achieve a Dynamic Risk Map using BIM, IoT, and MR technologies. Also, this paper presents a proof of concept based on the framework proposed. The methods include (i) identifying Risk Maps limitations; (ii) mapping their traditional process to integrate innovative technologies; (iii) creating a framework for supporting Risk Maps improvement; and (iv) simulating Dynamic Risk Maps concerning tracking, sensing, and exhibition resources. Research outcomes highlight that the dynamic status of a Risk Map increases feedback capabilities regarding predicted and actual risks on site and context awareness. That increase is due to communication enrichment and the assurance of inspection activities based on prevention through design and real-time monitoring.

Keywords: Risk management · Risk maps · BIM · Internet of Things · Mixed reality

1 Introduction

Risk Management in Architecture, Engineering, and Construction (AEC) industry is a global issue. According to [1], failure to handle risks appropriately may not only lead to challenges in meeting project purposes but also reflects on the future growth of cities. In Brazil, the Ministry of Labor and Employment [2] considers the AEC industry as a high-risk segment, due to the distribution of labor accidents by economic sectors.

E. Toledo Santos and S. Scheer (Eds.): ICCCBE 2020, LNCE 98, pp. 895–906, 2021.
https://doi.org/10.1007/978-3-030-51295-8_62

Brazilian Federal Government's historical database of accidents, when restricted to Building Construction, reinforces those statistics, as exhibited in Fig. 1.

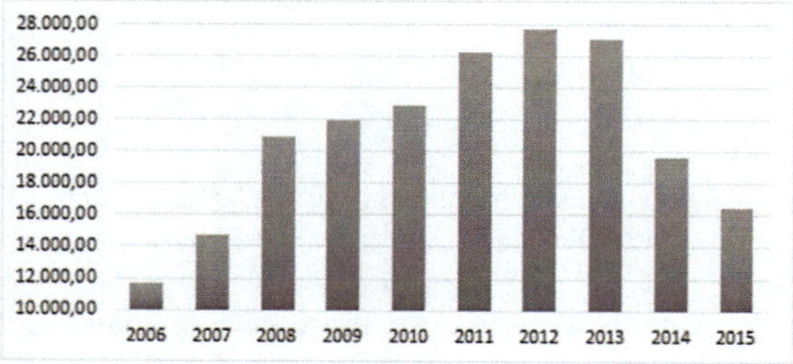

Fig. 1. National quantity of civil construction work accidents.

Considering such a scenario, Brazil has regulated the NBR ISO 31000 as the national standard for Risk Management. That standard defines Risk Management as a set of coordinated activities that directs and controls an organization regarding risks [3]. Risks comprise time overruns, economic risks, industrial conflicts, and workplace injuries [4]. One Brazilian Regulatory Standard concerning the prevention of workplace injuries is the NR5: Internal Commission of Accident Prevention (CIPA). The NR5 prescribes the Risk Map as an essential tool to prevent workplace injuries since its graphical resources represent the risks and hazards of a particular environment.

Despite the prescription of Risk Map as an essential tool, there are several limitations in its production and delivery, ranging from planning, the graphical representation itself, until the applied methods and context-based updates. In short, CIPA has challenges in conceptualizing a Risk Map, which reflects on the information exposure, visualization, and understanding concerning the potential user [5]. Besides, its implementation is still a manual undertaking as well as its decision making is frequently based on knowledge and intuition - decreasing efficiency in the real environment [6].

Otherwise, changing in technological, technical, or standard resources may reduce the number of workplace injuries [7]. In that sense, according to [1], the interest and adoption of Information and Communication Technologies (ICTs) for Risk Management have increased. Recognizing that trend, this paper considers that the use of ICTs by CIPA in Brazilian circumstances may enhance the traditional process of Risk Map, lowering current issues and assuring prevention to risk and hazard in the AEC industry. We assume that ICTs such as Building Information Modeling (BIM), Internet of Things (IoT), and Mixed Reality (MR) have resources that can directly support communication, comprehension, and access to Risk Maps.

Previous studies such as [1] identified gaps concerning the integration of BIM and related digital technologies with traditional methods, processes, and techniques of Risk Management. Moreover, there is a need for combining BIM-based and traditional Risk Management to improve practical applicability. Hence, this paper proposes a conceptual framework of Risk Management for the construction phase to achieve a Dynamic Risk Map using BIM, IoT, and MR technologies. Also, we present the proof of concept based on the framework proposed. The authors expect that ICTs adoption and the dynamic status of a Risk Map may increase the effectiveness of its planning and implementation, and contribute to the practical reduction of accidents at the construction site.

2 Context

As previously highlighted, risks and hazards in the construction phase are recurrent [8], which transforms the AEC industry in one of the economic sectors that registers the most accidents in Brazil. The traditional Brazilian process of Risk Management for construction requires an Internal Commission of Accident Prevention regulated by NR5. CIPA also responds to requirements of ABNT NBR ISO 31000 and other Regulatory Standards such as NR18, for assuring safety in the work environment. Furthermore, the commission has the support of the Specialized Services in Occupational Health and Safety (SESMT). CIPA composition usually involves building representative stakeholders such as consultants, clients, designers, the construction team, and others. The Commission's main role is to define practical tools for achieving effective Risk Management. One of those tools is the Risk Map, which supports education, mainly among the construction team, comprising risks and hazards on site. That support is based on controlling, reducing, and eliminating workplace injuries [9]. Figure 2 synthesizes that traditional process.

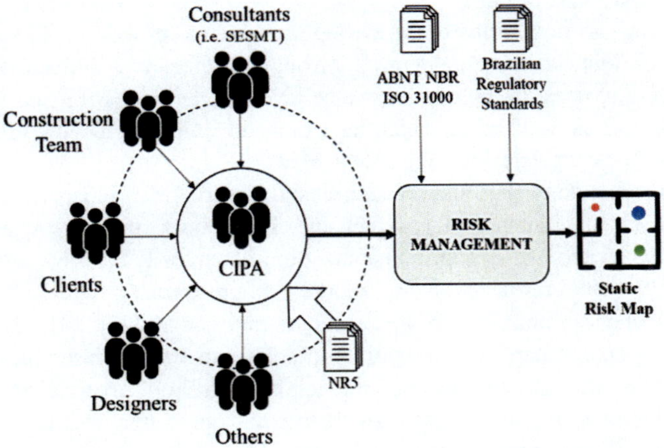

Fig. 2. Brazilian risk management in construction phase.

Risk Map is a graphical representation of the work environment (i.e., construction site's floorplan) with its risks and hazards. That conjunction usually follows as standardized annotations: (i) the risk groups by room, considering their agents (e.g., physical, chemical, biological, ergonomic and accident) and related causes; (ii) the degree of intensity for each group, according to worker's perception of risk; and (iii) the number of workers exposed to the predicted risks [10]. Graphically, the risk groups comprehend mark-ups such as circles combined with color patterns (e.g., green, red, brown, yellow, and blue), sizes, and quantities regarding risk magnitude (Table 1).

Table 1. Brazilian risk map standard.

Agents	Degree of intensity			Causes	Examples
	Little	Medium	Big		
Physical				Forms of energy	Noise, vibration, abnormal pressures, extreme temperatures
Chemical				Substances, compounds or products	Gases, vapors and mists
Biological				Microorganisms	Bacilli, bacteria, fungi, protozoa, parasites, viruses
Ergonomic				Psychophysiological characteristics	Heavy physical work, incorrect postures, uncomfortable positions, repetition, monotony, excessive rhythm
Accident				Inappropriate physical arrangements	Defective, inadequate or non-existent machinery, equipment and tools, danger of fire or explosion

Considering the concurrent changes related to the construction environment, Risk Maps, despite their static resources and 2D format, should attend to more than one construction stage. That attendance in the traditional process includes the removal or decrease of circles or their features (i.e., size or quantity). Furthermore, [10] recommends the location of Risk Maps at accessible areas to facilitate the construction team's awareness of risks and hazards.

Recognizing the limitations of Risk Map's traditional process, [11] states that for ensuring a safe and unsurprising work environment, there is a demand for adopting innovative technologies. In that regard, we believe that the increasing adoption of BIM in AEC industry, as well as its potential extension and link to MR and IoT technologies, should be explored to assist Risk Maps.

BIM is a technology that plays a significant role in Risk Management during the Construction and Maintenance of a project since it enhances efficiency regarding safety through the exploration of dynamic visualization resources [12]. BIM benefits in that context involve the use of BIM 3D Model's main features, which facilitate risk identification and communication for accident prevention [13, 14]. Besides, BIM Model uses to comprehend the possibility of risk and hazard assessment [15] supporting the prevention through design approach [16]. In short, potential applications of BIM for Risk Management concerning construction phase may include: (i) the association between BIM Model's objects and risk and hazard's types; (ii) the visualization of solutions that help in environment's safety; and (iii) the risk control proposal through BIM Model's revisions [17].

IoT is a paradigm defined as a dynamic global network infrastructure [18] that links real objects and beings with the virtual world and environment, enabling connectivity in any place, anytime, not only for anyone but also for anything [19]. In IoT, "things" exchange sensing data and information about the environment, while reacting to the real-world events and influencing it by running processes that trigger actions, create and deliver services [18]. [15] indicates a BIM/IoT Interfacing in which the BIM Model is connected to that IoT network and benefited from real-time data provided by monitoring and control (e.g., Wireless Sensor Network - WSN) as well as tracking and tagging (e.g., Radio Frequency Identification – RFID, Near Field Communication – NFC, Bluetooth Low Energy - BLE) technologies. Concerning Risk Management and BIM/IoT Interfacing, [20] present solutions of real-time environmental monitoring for confined spaces and hazardous zones. Moreover, [21] and [22] demonstrate the monitoring and control of security violations on construction sites. Finally, [23] explores BIM Model's for identifying risks and displaying actual data.

Mixed Reality involves the merging of real and virtual worlds. It is considered the area between the two extremes, where both the real and the virtual are mixed. Virtual reality and augmented reality are part of Mixed Reality and works together in *reality-virtuality continuum* [24]. Virtual Reality (VR) is a computational interface that involves simulation in real-time [25]. VR has the characteristic of virtual environments visualization and interaction with objects by the user, in addition to stimulating the other senses such as touch and hearing [26]. Augmented Reality (AR) connects computer-generated objects in a real environment, in real-time. AR can apply to all senses, such as vision, hearing, and smell [27]. Concerning Risk Management and AR, previous studies present, for instance, applications involving emergency response regarding AR-based mobile escape guidelines applied to self-evacuation [28], as well as segment displacement inspection during tunneling construction, using AR-based system superimposed with BIM Models into the real structure [29]. Nevertheless, regarding Virtual Reality (VR) and its relation to safety learning in construction and engineering, [30] highlights research gaps in Risk Management stages such as risk evaluation, risk response planning, and risk monitoring and controlling.

Therefore, Risk Management is a potential area to integrate Mixed Reality with BIM Models and IoT technologies. The integration benefits may embrace the visualization of the Risk Map superimposed in the real work environment with not only predicted risks by CIPA but also actual risks provided by real-time monitoring. After identifying Risk Maps limitations, mapping the traditional process, and evaluate the potential of integrating innovative technologies, it was possible to propose a framework of Risk Management for producing Risk Maps based on the benefits of BIM, IoT, and MR (applying augmented reality and virtual reality technologies). Those resources may enhance the communication of Risk Maps aiming at reducing risks and hazards in construction sites. As following, it was presented two conceptual simulations of the framework considering tracking, sensing, and exhibition resources.

3 A Framework Integrating BIM, IoT and Mixed Reality

Regarding the gap of integrating BIM and related digital technologies with the traditional framework of Risk Management, we identified the potentiality of inputting resources of such technologies in the product itself: the Risk Map. As a result, we suggest transforming the Risk Map that has static features in a Dynamic Risk Map aiming at providing an interactive and accessible visualization (Fig. 3).

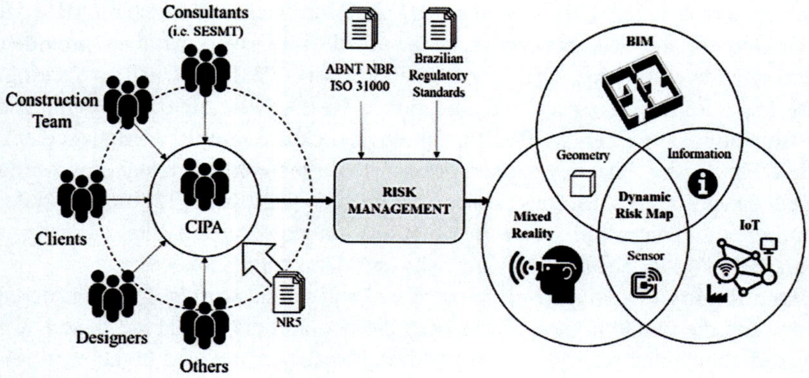

Fig. 3. BIM-based risk management integrated with IoT and MR in construction phase.

The interface between these technologies has different approaches: BIM and MR may share 3D Models and building-specific information that will be organized and exhibited using an established device. MR and IoT may interface with each other using sensor-based tracking for activating the visualization and/or through the exhibition of real-time data collected by IoT solutions.

Finally, BIM and IoT may be integrated through the association of virtual building data and actual data, respectively, and/or inputting that actual data in 3D Models for geometric and semantic contextualization. The simulation of these integrated technologies is highlighted in the next framework presented in Fig. 4.

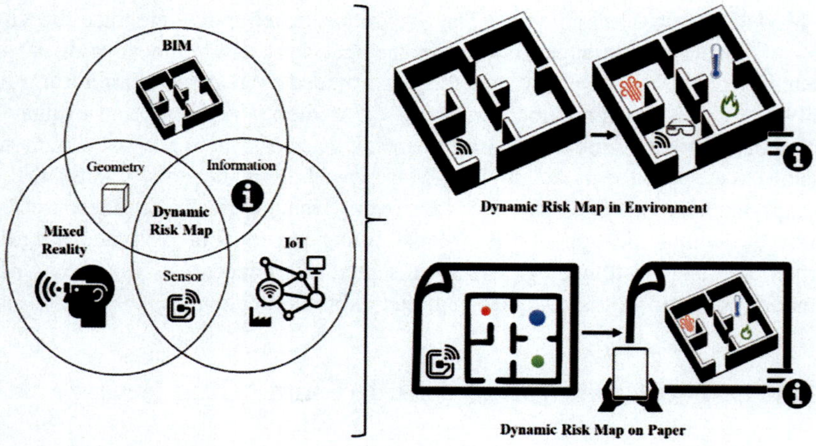

Fig. 4. Dynamic risk map: visualization forms.

The Dynamic Risk Map in Environment is a condition in which users can access each construction zone with a display device (e.g., smart glasses, tablet, smartphone) to visualize within a system application the sorts of risks predicted by CIPA. In this case, the risk map is visualized with Augmented Reality technology, and it is localized on each floor, as recommended by CIPA, and also individualized in each room. Concerning the risks themselves, their visualization is through 3D symbols related to each predicted risk, colored according to the group risk pattern defined by NR5. In turn, the BIM Model is displayed in the natural environment and scale (1:1), becoming essential for contextualizing the risks through geometric visualization of the several construction phases identified in the 4D dimension. Construction phases can be previously inputted into the system for later access and AR visualization. Thus, it is possible to comprehend the construction status of geometrically and predicted risks by CIPA associated with that specific phase.

Besides, users can also visualize actual risks due to sensor-based information feedback provided by an IoT solution. Sensors in that scenario are defined from CIPA predicted risks and installed in each related local. In Table 2, we present an association between CIPA Risk Factors and Sensor Types regarding that situation. For instance, in a zone that has risks of high temperatures (e.g., confined spaces), CIPA may require an IoT solution that collects temperature data in real-time and delivers warnings based on indicators of risk and hazard. It is a situation of context-awareness.

Table 2. CIPA risk factors and sensor types correlation.

Risk Group	Agents	Causes	BIM Objects	IoT Sensors
Group I	Physical	Forms of Energy		Temperature, Humidity, Pressure, Microphone, Water Detection
Group II	Chemical	Substances, Compounds or Products		NO2, SH2, CO, CO2, Hydrocarbons, CH4, O3, Ultraviolet (UVA, UVB)
Group III	Biological	Microorganisms		Not identified
Group IV	Ergonomic	Psychophysiological Characteristics		Not identified
Group V	Accident	Inappropriate Physical Arrangements		O2, H2, CH4, Ethanol, Isobutane, Accelerometer

MR tracking for visualization in that context can be practiced through the use of fiducial markers or sensors. Considering fiducial markers, images previously registered in the system application will activate a model visualization by the user's activity of scanning. Regarding the second possibility, sensors will activate the visualization during the detection of the user's and/or display device's presence by technology such as RFID, NFC, or BLE. Regarding both types of markers for MR tracking, any display device mentioned above can be utilized.

The Dynamic Risk Map on Paper is a condition in which the 2D Risk Map is already localized in a strategic place of the construction environment, and the users can utilize a display device to visualize within a system application the predicted risks. That risks are contextualized in a 4D BIM Model following its scale and exhibited in Virtual Reality. In this scenario, users can also utilize a tracking system through a fiducial marker or sensor to activate the application. The risk representation is the same as in the case of the Dynamic Risk Map on Environment, that is, symbolic objects representing risks with colors related to their group risk considering NR5 standards. That application enhances the user's awareness of risks and hazards and gives an overview of the construction environment. In item 4, we will present the proof of concept for the Dynamic Risk Map on paper.

4 Application

The application was developed using a BIM Model (Fig. 5) of the Faculty of Architecture at Federal University of Bahia (FAUFBA) annex building (Fig. 6), located in the Federação district in Salvador, Bahia. The building has four floors, a mixed structure (concrete and metal), is under construction and is expected to be completed by the end of 2020.

Fig. 5. FAUFBA annex building.

The BIM Model (Fig. 6A) was developed in the authoring application Autodesk Revit 2018, the BIM/IoT interfacing was designed in the BIM middleware solution Dynamo and the MR application was created with Autodesk Revit Live. The Revit Live opened a connection between the authoring environment and an engagement viewer for users that can be distributed in desktop and mobile devices.

The authors developed a proof of concept using the Risk Group I (Table 2) and a desktop device. The goal was the Dynamic Risk Map simulation considering the resources of tracking, sensing, and exhibition through the verifying of sensor data and its visualization in virtual reality.

For this prototype, an IoT platform database was used to register environmental data (e.g., temperature and humidity) collected by a sensor network. A script (Fig. 6B) was developed through Dynamo to retrieve sensor data, get the last value of each

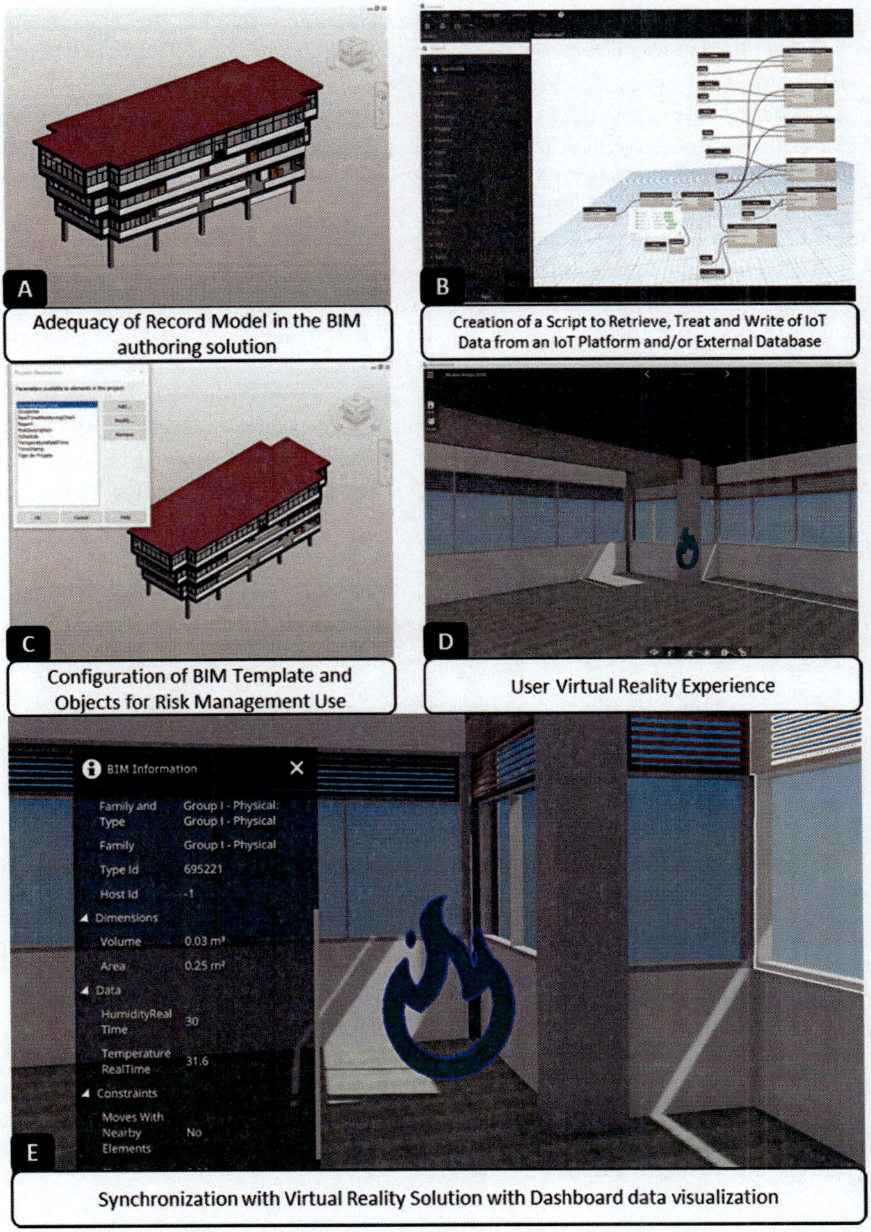

Fig. 6. Dynamic risk map on paper developing process.

reading by room, and through a conditional statement, write these values with an alert to specific object parameters.

The adequacy of the Record Model involved the settings to build shared parameters (Fig. 6C) with this aim. The script also considered inserting the symbolic risk objects modeled as generic types by room based on the values.

Considering the validation of these first steps, with the BIM/IoT interfacing working for Risk Management purposes with a periodic script running, the next goal was to synchronize the updated 3D view sets from the BIM authoring environment to the MR solution (Fig. 6D). A MR solution is essential to share information among non-BIM users, especially those with a mobile device on-site. The synchronizing depends on the software and can be user-based or through automation programming to achieve live status. The actual 3D view set with IoT and risk data was synchronized with Autodesk Revit Live to be accessed by any desktop device or mobile application (Figure E). According to the risk results, the risk symbols were viewed in different scale following the standard and their IoT data consulted through the properties browser. As mentioned before, regarding the degree of intensity in Table 1, the bigger the symbol, the bigger the risk of the user. With real-time presentation data (MR synchronization), users can access quantitative risk information. Besides, one of the limitations of this prototype was the no success to insert textual parameters in the 3D symbol (Fig. 7).

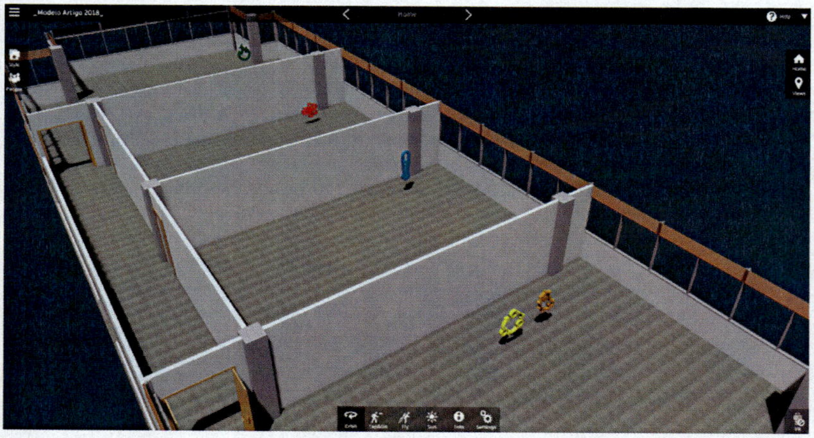

Fig. 7. Risk symbols on the environment.

5 Conclusion and Future Work

Concerning the gaps identified in previous studies comprising the creation, understanding, and visualization of Risk Maps as well as techniques for improving Risk Management, this paper presented a conceptual framework integrating BIM, IoT, and MR with a proof of concept. That integration considered Risk Management in the Construction phase. We presented the framework involving the relations between BIM Model and MR technology through geometry, IoT, and MR technologies through sensor-based information and/or sensor-based tracking; and BIM Model and IoT through the contextualization of actual data. These relations were highlighted in two simulations derived from the framework: a Dynamic Risk Map in Environment and a Dynamic Risk Map on Paper. In the first one, users can visualize, in Augmented Reality, both risks predicted by CIPA and actual risks provided by IoT solutions by room in a construction zone. Risks are contextualized by BIM Models on natural scale. In the second one, users can visualize, in Virtual Reality, only predicted risks in an overview of the construction zone. Risks are contextualized by BIM Models superimposed on paper following the scale of the traditional 2D Risk Map. In both simulations, the visualization is in Augmented Reality to increase awareness about the surroundings, which should make the system application adaptive and intelligent. Concerning actual risks for the Dynamic Risk Map in Environment, we also presented an association between risks grouped by CIPA and types of sensors. Finally, a proof of concept was carried out to validate the Dynamic Risk Map on Paper framework created. Besides, the IoT information was linked with success to the Record Model by an authoring tool. From that link, new risk symbols were shown by risk map group concepts. Concluding, the frameworks developed in this paper, with a proof of concept, will help the CIPA team in the planning and implementation of Risk Maps for Construction and Commissioning phase. The benefits include the rising awareness of risk and hazard on-site as well as the reduction of workplace injuries. The framework brings a potential for future work aiming at implementing these proposed systems for AR sensors solutions and measuring their performance through interviews with users and safety indicators.

References

1. Zou, Y., Kiviniemi, A., Jones, W.: A review of risk management through BIM and BIM-related technologies. Saf. Sci. **97**, 88–98 (2017)
2. MTE - Ministério do Trabalho e Emprego: Estratégia Nacional para Redução dos Acidentes do Trabalho 2015-2016. Ministry of Labor and Employment Report (2015)
3. Associação Brasileira de Normas Técnicas: Risk management – Principles and guidelines. ABNT NBR 3100 (1996)
4. The BIMe Dictionary. https://bimdictionary.com. Accessed 2 Dec 2019
5. Mattos, O., Freitas, M.: Brazilian risk map: limited applicability of a worker model. Cad. Saúde Pública **10**(2), 251–258 (1994)
6. Shim, C., Lee, K., Kang, L.S., Hwang, J., Kim, Y.: Three-dimensional information model-based bridge engineering in Korea. Struct. Eng. Int. **22**, 8–13 (2012)
7. Oddone, I., Marri, G., Gloria, S., Briante, G., Chiattella, M., Re, A.: Ambiente de Trabalho: A Luta dos Trabalhadores pela Saúde. Hucitec, São Paulo, Brazil (1986)
8. Pandaggis, L.R.: Uma Leitura da Árvore de Causas no Atendimento de Demanda do Poder Judiciário: Um Fluxograma de Antecedentes. University of São Paulo, São Paulo (2003)
9. NR-5 - Regulatory Standard: Comissão Interna de Prevenção de Acidentes. Ministério do Trabalho e Emprego (2011)
10. Manual de Elaboração Mapa de Risco. http://www.sgc.goias.gov.br/upload/arquivos/2012-11/manual-de-elaboracao-de-mapa-risco.pdf. Accessed 10 Dec 2019
11. Malekitabar, H., Ardeshir, A., Sebt, M.H., Stouffs, R.: Construction safety risk drivers: a BIM approach. Saf. Sci. **82**, 445–455 (2016)

12. Zou, Y., Kiviniemi, A., Jones, S.W.: Developing a tailored RBS linking to BIM for risk management of bridge projects. Eng. Constr. Archit. Manag. **23**, 727–750 (2016)
13. Liu, W.P., Guo, H.L., Li, H., Li, Y.: Using BIM to improve the design and construction of bridge projects: a case study of a long-span steel-box arch bridge project. Int. J. Adv. Rob. Syst. **11**(8), 125–135 (2014)
14. Hardin, B., Dave, M.: BIM and construction management: proven tools, methods, and workflows, 2nd edn. Wiley, New York (2011)
15. Succar, B., Saleeb, N., Sher, W.: Model uses: foundations for modular requirements clarification language. In: Australasian Universities Building Education (AUBEA 2016), Cairns, Australia, pp. 1–12 (2016)
16. Zhang, S., Teizer, J., Lee, J., Eastman, C.M., Venugopal, M.: Building Information Modeling (BIM) and safety: automatic safety checking of construction models and schedules. Autom. Constr. **29**, 183–195 (2013)
17. Kamardeen, I.: 8D BIM modelling tool for accident prevention through design. In: 26th Annual ARCOM Conference Proceedings, Leeds, UK, pp. 281–289 (2010)
18. Vermesan, O., Friess, P., Guillemin, P., Gusmeroli, S., Sundmaeker, H., Bassi, A., Jubert, I. S., Mazura, M., Harrison, M., Eisenhauer, M., Doody, P.: Internet of things strategic research roadmap. Internet Things-Glob. Technol. Soc. Trends **1**, 9–52 (2011)
19. Sundmaeker, H., Guillemin, P., Friess, P., Woelfflé, S.: Vision and challenges for realising the Internet of Things. In: Cluster of European Research Projects on the Internet of Things 3, European Commission (2010)
20. Arslan, M., Riaz, Z., Kiani, A.K., Azhar, S.: Real-time environmental monitoring, visualization and notification system for construction H&S management. ITcon **19**, 72–91 (2014)
21. Costin, A.M., Teizer, J., Schoner, B.: RFID and BIM-enabled worker location tracking to support real-time building protocol and data visualization. ITcon **20**, 495–517 (2015)
22. Fang, Y., Cho, Y.K., Zhang, S., Perez, E.: Case study of BIM and cloud-enabled real-time RFID indoor localization for construction management applications. J. Constr. Eng. Manag. **142**(7), 05016003-1–05016003-12 (2016)
23. Park, J., Kim, K., Cho, Y.: Framework of automated construction-safety monitoring using cloud-enabled BIM and BLE mobile tracking sensors. ASCE J. Constr. Eng. Manag. **143**(2), 05016019 (2016)
24. Milgram, P., Kishino, F.: A taxonomy of mixed reality visual displays. IEICE Trans. Inf. Syst. **77**(12), 1321–1329 (1994)
25. Burdea, G., Coiffet, P.: Virtual Reality Technology. Wiley, New York (1994)
26. Tori, R., Kirner, C., Siscoutto, R. (org.): Fundamentos e Tecnologia de Realidade Virtual e Aumentada. SBC, Porto Alegre (2006)
27. Azuma, R.: A survey of augmented reality. Presence: Tele-oper. Virtual Environ. **6**(4), 355–385 (1997)
28. Tsai, M., Yau, N.: Enhancing usability of augmented-reality-based mobile escape guidelines for radioactive accidents. J. Environ. Radioact. **118**, 15–20 (2013)
29. Zhou, Y., Luo, H., Yang, Y.: Implementation of augmented reality for segment displacement inspection during tunneling construction. Autom. Constr. **82**, 112–121 (2017)
30. Kassem, M., Benomran, L., Teizer, J.: Virtual environments for safety learning in construction and engineering: seeking evidence and identifying gaps for future research. Visual. Eng. **5**(1), 1–15 (2017). https://doi.org/10.1186/s40327-017-0054-1

An Ontology-Based Mediation Framework for Integrating Federated Sources of BIM and IoT Data

Mehrzad Shahinmoghadam(✉) and Ali Motamedi(✉)

Department of Construction Engineering, École de technologie supérieure, Montréal, Canada
mehrzad.shahinmoghadam.1@ens.etsmtl.ca, ali.motamedi@etsmtl.ca

Abstract. Building Information Modeling (BIM) and Internet of Things (IoT) are two emergent technologies that promise substantial improvements in the lifecycle management of facilities. Recently, research on BIM-centered IoT applications has been gaining an increasing momentum. Currently, there exist some major challenges that impede the realization of the synergistic potentials of BIM and IoT. Among them is to maintain an effective semantic interoperability between the BIM and IoT data ecosystems. Indeed, an accurate integration between IoT-enabled raw sensory data and existing BIM-based digital models of facilities requires a profound understanding of contextual clues (time, location, occupants' profiles, etc.), which are embedded within actual observations derived from IoT sensors. In this paper, a framework is presented for integrating disparate sources of BIM and IoT data through an ontology-based mediation mechanism. This framework enables an integrated access to local sources of BIM and IoT data thorough query-rewriting processes. In order to demonstrate the applicability of the proposed framework, query rewriting examples have been provided in the context of indoor comfort analysis for facility occupants. In the case demonstrations, the geospatial data acts as the main clue for the derivation of semantic dependencies between local instances of BIM and IoT data.

Keywords: Building Information Modeling · Internet of Things · Ontologies · Semantic interoperability

1 Introduction

Since its emergence, Building Information Modeling (BIM) has kept the vision for delivering an integrated and real-time access to the numerous and heterogeneous sources of data that are continuously generated across the whole lifecycle of facilities. The lifecycle data can be captured and maintained by various systems (e.g., a wireless sensor network linked to a computer-aided facility management program) and then be integrated within a BIM environment in the cloud. Due to the rapid advancements in the domains of remote sensing/control technologies and cloud computing, Internet of Things (IoT)-enabled applications have provided the opportunity for collecting huge

E. Toledo Santos and S. Scheer (Eds.): ICCCBE 2020, LNCE 98, pp. 907–923, 2021.
https://doi.org/10.1007/978-3-030-51295-8_63

amounts of facility lifecycle data in a real-time manner. Recently, researchers have shown an increasing interest in a new research paradigm that can be referred to as "connected BIM". In a connected BIM setting, live observations over various aspects of a given facility can be readily made available within the mainstream of BIM processes. Ideally, if IoT architectures can be effectively merged within cloud-enabled BIM environments, physical components of facilities can be interconnected with the digital entities representing various aspects of them. As a result, digital twins of facilities can be created. Moreover, by applying advanced data mining methods over integrated sources of live (IoT-enabled) and recorded (BIM-enabled) data, and developing various decision support systems, digital twins of facilities can be equipped with the power of learning from past experiences.

In practice, however, providing an integrated access to disparate sources of BIM and IoT data will be substantially challenging as both domains are abundant in proprietary formats and deployments that are tightly coupled with closed data ecosystems. In addition to syntactical and schematic heterogeneities among BIM and IoT data representations, establishing semantic relationships between fragmented sources of such data, which remains essential for maintaining the same interpretations of data by different systems, hampers the realization of a connected BIM environment. This issue, which is known as enabling semantic interoperability between software components, has been an ongoing research effort within both BIM and IoT research domains. Moreover, for enabling derivation of meaningful information from low-level data (e.g., raw sensory data), a particular attention must be paid to establishing proper semantic connections between context-independent data and underlying contextual clues (e.g., the time and location data that corresponds to a particular observation). The importance of connecting IoT data with contextual data about its origins (mainly spatial data) has been most recently highlighted for BIM-IoT integrated frameworks in [1].

Within the last few decades, ontology-based approaches and semantic web technologies have been at the center of attention for addressing syntactic and semantic interoperability issues in a wide range of research domains. In the context of BIM, researchers have been attempting to semantically enrich the data represented in BIM models by providing formal definitions of the relevant concepts and the relationships among them within ontology-based schemas. Examples of the main outputs relevant to this line of research are "ifcOWL", which is an ontology-based representation of the IFC schema supported by the buildingSMART community, and the "Linked Building Data" initiative, which is supported by the W3C[1] community. On the other hand, a combination of ontology-based formalisms and semantic web tools have shown considerable potentials for creating context-aware frameworks which consists of formal definitions of the underlying semantics between sensor data, events, and contextual dependencies [2, 3]. Accordingly, researchers have been investigating such potentials for creating smart IoT solutions for a wide range of domain applications. Recent examples of using ontologies and semantic web tools for developing context-aware IoT middleware can be found in [4–7].

[1] World Wide Web Consortium (https://www.w3.org/).

Despite the potentials mentioned above, a most recently published literature review over the publications that made contributions with respect to the BIM-IoT integration subject [8], shows that the investigation of ontology and semantic web-based approaches has been considerably limited within the existing body of knowledge. In light of this knowledge gap, this paper presents an ontology-based framework for integrating disparate sources of BIM and IoT data through the application of semantic web tools, namely RDF, OWL, and SPARQL. The proposed framework first requires local sources of BIM and IoT data to be available in the form of federated RDF-formatted instances of local data. In order to create such RDF-based representations of BIM and IoT data, existing ontological formalisms will be considered. Thereafter, an ontology-based mediation mechanism included in the framework provides an integrated access to the federated data instances by reformulating queries posed according to a common (shared) ontology. The context-awareness power of the framework is gained by defining semantic correspondences between contextual clues, such as location and time data. In order to light upon the capabilities of the proposed framework for addressing the issues around BIM-IoT data integration problem, case demonstrations of two main query-rewriting scenarios are elaborated in this paper. The query rewriting examples are given in the context of thermal comfort assessment for facility occupants.

2 Background

The construction sector is known for being one of the most fragmented industries. For a given facility to be designed, built and operated, huge amounts of data/information from numerous fragmented disciplines (e.g., architecture, structural, mechanical and electrical engineering) will be generated. This fragmentation is, to a large extent, accompanied by the use of various data representation languages and data model schema definitions for the same sources of data, thereby resulting in a significant syntactic heterogeneity for BIM data. This fragmentation and heterogeneity inevitably leads to serious challenges for data/information exchange among various tools/databases used by different parties involved at different stages of the facility's lifecycle. Although the shift from closed data ecosystems towards openly interoperable BIM environments has been gaining momentum, application of proprietary formats is still commonplace in the BIM data ecosystem. Nonetheless, application of standardized serialization formats and data representation schemas such as ifcXML for web-based BIM data exchange, will address the interoperability problem only at the syntactic level. Yet, a more profound problem exists for BIM interoperability, which is semantic interoperability. At the semantic level, the meaning of the data and its combination with other pieces of data/information in a meaningful way is of particular interest [9]. Up until now, a major portion of previous research attempts have been dedicated to solving BIM interoperability at the syntactic level. This existing disproportionate balance between syntactic and semantic-oriented research efforts has been explicitly mentioned in [10] for the case of BIM-GIS integration. A recent review over1244 peer-reviewed journal articles about BIM [11], highlights "interoperability" as the second top keyword with the highest count. This means that delivering effective solutions to address BIM interoperability issues still remains as an open research problem.

Addressing semantic interoperability issues is an open research problem for the IoT domain as well. A number of key factors contribute to a lack of sufficient interoperability, which is hindering IoT solutions to be widely adopted in different domains of application. One major factor among them is that sensory data, as the principle category of the data that is generated in a typical IoT deployment, are heterogeneous and distributed in nature [12]. Insufficient existing standards for representing sensor data [13], and deploying resources that are usually tightly coupled with closed data ecosystems [14], are among other major factors.

Developments in the fields of ontology engineering and the semantic web, have brought about new lines of research for addressing semantic interoperability for both BIM and IoT domains. Ontologies arise out of the field of formal knowledge representation, which is one of the key branches in the artificial intelligence discipline. An ontology uses a logical language to formally represent the concepts that are the subject of interest in a particular domain (ontology classes) and the semantic relationships between those concepts (ontology axioms). The semantic web [15], on the other hand, holds the vision of providing machines and humans with a shared understanding of the all resources available in the Internet and formal ontologies play an essential role in this respect. To date, many different tools/technologies have been used/devised to meet the vision of the semantic web, such as OWL[2] (a logic-based language for ontology authoring), RDF[3] (a standard representation model for metadata), SPARQL[4] (a query language for RDF-formatted data), and SWRL[5] (a logical language for expressing rules that are compatible with OWL constructs).

A combination of formal ontologies and semantic web technologies offers a considerable potential for resolving syntactic and semantic heterogeneities existing in almost any data/information integration scenario. By providing a shared vocabulary for the conceptualization of a subject domain in a machine-understandable manner, ontologies offer significant potentials for addressing syntactic and semantic heterogeneities among distributed sources of data/information, at the same time. Once the intended data source is described through ontology classes and the logical relationships among those classes (semantics), queries (e.g., SPARQL queries over RDF-formatted data) can be posed to retrieve the relevant data. By recognizing the potentials of formal ontologies and semantic web technologies for integrating distributed sources of data/information, researchers have been investigating such approaches in the context of the built environment research. Examples of such efforts can be found in [16–18]. However, investigating such approaches for addressing BIM-IoT data integration in particular, is limited within the existing body of knowledge.

Based on the guidelines provided in [19], generating a single ontology for each of the BIM and IoT domains through *ontology merging* approach, and then integrating

[2] https://www.w3.org/TR/owl-ref/.

[3] https://www.w3.org/TR/rdf-concepts/.

[4] https://www.w3.org/TR/rdf-sparql-query/.

[5] https://www.w3.org/Submission/SWRL/.

those ontologies into a single global ontology will be neither sensible nor feasible as BIM and IoT environments share the characteristics of being highly dynamic and distributed. Hence, *ontology alignment* can be a key to an effective ontology-based integration of BIM and IoT data. Ontology alignment is the process/technique of establishing connections between two or more ontologies (usually from complementary domains), while keeping the local ontologies separate from each other. This will be done based on the semantic correspondences between the local ontologies to be aligned [19]. In other words, in an ontology-based data integration system that works on an ontology alignment basis, ontologies serve as the mediated schemas required for answering queries in terms of the vocabularies used in the local schemas. Such ontology-based data integration systems can be developed by taking two main approaches, namely: Global-as-View (GaV) and Local-as-View (LaV) [20]. While in the former approach, entities of the global schema will be associated with views over the local schemas, in the latter, constructs of the local schemas will be defined with reference to views over a global schema. The main difference between these approaches lies at the complexity of query processing for retrieving the data instances from the local data sources stored in the local databases. Particularly, in the GaV-based approach, query processing involves less complexity as it is done by expanding the atoms of the query according to their relative definitions with the local schema constructs. In the LaV-based approach, on the other hand, inference mechanisms will be required for re-expressing the atoms of the global schema in accordance with atoms of the local schemas [21]. However, in the LaV approach, local schemas can be changed independently of the global schema, while evolution of the local schemas in the GaV approach can be problematic for maintaining the functionality of the data integration system [20].

In the context of BIM-IoT data integration, the key question is that which strategy will be a more effective one? In light of the aforementioned points, it can be said that choosing between these two main approaches depends on two main factors: (i) the inherent complexity of query processing, and (ii) the maturity level of the existing ontologies (in terms of standardization and stability) for the particular domains of interest for a given data integration scenario. Hence, due to the complex nature of the built environment data and lack of currently existing global ontologies that encompass BIM and IoT relevant entities for a wide range of use-case applications (e.g., construction progress monitoring, safety management and energy analysis), the choice of a GaV-based approach will be evidently a more efficient decision for integrating BIM and IoT data sources.

The application of a common (shared) ontology as a means for integrating distributed sources of data about buildings have been investigated previously. In [16], for instance, researchers developed a new ontology for integrating the data coming from three main sources, namely BIM-based, sensor-based, simulation-based data (represented in three local ontologies). The main theme of the mentioned study can be viewed as a BIM-IoT integrated scenario as the research objective has been to evaluate the gap between the design of the buildings and their actual performance by linking real-world observations (derived from sensors) and the simulated data about the building performance. However, the data integration method proposed in that study can be viewed as an example of merging semantically related ontologies. In other words, the common ontology is a merged ontology. On the contrary, the method proposed in the present study investigates an ontology-based mediation approach which is based on the principles of ontology alignment. Accordingly, the focus of the present study remains on rewriting queries for accessing disparate sources of data, on the basis of semantic correspondences between local schemas used for representing BIM and IoT data.

3 Overview of the Proposed Framework

In this paper, a conceptual generic framework for creating an integrated and context-aware access to the disparate sources of BIM and IoT data is presented as shown in Fig. 1. The proposed framework contributes to the middleware layer of the three-layered architecture of generic BIM-IoT integrated deployments proposed in [22]. The main components of the framework and the main steps required for delivering the integration between local sources of BIM and IoT data are described as follows:

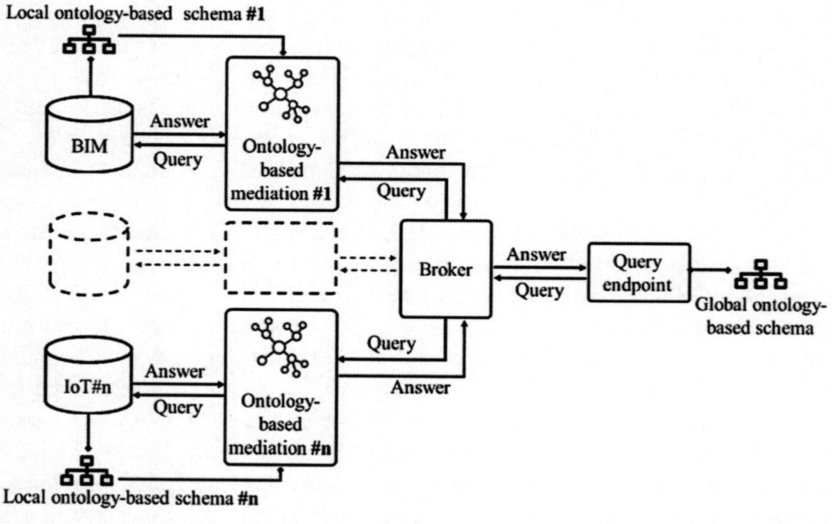

Fig. 1. Conceptual framework for BIM-IoT data integration through ontology-based mediation

As shown in the figure, local ontology-based schemas are one of the main components of the proposed framework. The local schemas will be responsible for representing the disparate sources of data coming from IoT sensors/devices and the BIM data provided within digital models of facilities. An example of local ontology-based schema for BIM data will be the ifcOWL ontology. As it can be inferred from the figure, the proposed design of the framework allows different BIM or IoT based sources of data to come in different local formats (e.g., encoded in XML or JSON serializations). Additionally, various ontologies can be used for delivering RDF-based representations of BIM and IoT data. At the first step of the data integration process, instances of local BIM and IoT data represented in terms of their local schemas need to be made available with reference to the formalisms of their intended local ontology-based schema. As a result, disparate sources of BIM and IoT data will be provided as federated RDF data sources about the facility. The second main component of the framework is the global ontology-based schema, which can be viewed as a virtual common ontology that is semantically interrelated with all local schemas. The next main component of the framework are the ontology-based mediation modules. These modules are primarily responsible for preserving the semantic understandings required for query answering by providing semantic correspondences (ontology mappings) between the entities of the global ontology-based schema and the relevant entities of the local sources of BIM and IoT data. Hence, an ontology-based mediation module is established for each local ontology-based schem. Similar to the approaches taken in [16, 18], semantic correspondences between local ontology schemas and the common (shared) ontology can be formally defined. Finally, with these semantic correspondences in place, the integrated access to the federated RDF data sources can be provided using the query-rewriting patterns presented in [23]. In particular, queries expressed in SPARQL will be rewritten with reference to the semantic correspondences (mappings) between the OWL ontologies used for representing local sources of BIM and IoT data and an OWL ontology standing as the common ontology (global schema). In this paper, two main SPARQL query-rewriting scenarios are elaborated for using the proposed framework. Examples of using each of these two scenarios are given in the next section.

As an integrated access to the local sources of BIM and IoT data becomes available through the above-mentioned procedure, a context-aware interpretation of BIM and low-level IoT data can be facilitated by adding a context model ontology into the proposed framework. Contextual information, such as current time of the day/year or weather data, can be included as another local source of the data. In case a context model ontology is added, a hybrid inference architecture similar to those proposed in [24, 25], which are built upon the constructs of OWL and SWRL reasoning, will facilitate derivation of context-aware interpretations for linking BIM and IoT data. Although, inclusion of the context model within the proposed framework is out of the scope of the present study, the role of contextual clues for establishing connections between disparate sources of BIM and IoT data is highlighted within the examples provided in the next section.

4 Case Demonstration

In this section two examples of using the framework proposed in this study are described to demonstrate the role of formal ontologies and semantic web tools for BIM-IoT data integration. The examples in both case studies are given in the context of indoor comfort analysis for building occupants based on the PMV (Predicted Mean Vote) as mentioned in the latest edition of ASHRAE Standard 55. Among the main variables that need to be defined for calculating the level of thermal comfort for facility occupants, operative temperature and relative humidity have been considered for the given examples. As mentioned in [26], to calculate the operative temperature, Mean Radiant Temperature (MRT) needs to be defined in addition to the level of air temperature captured by sensors. Moreover, for calculating the MRT value, in addition to the average surface temperature (captured by a thermal camera), the angle factor between the occupant and the main surrounding surfaces (i.e., walls, ceiling, floor) need to be defined. Calculations of the angle factor will be performed based on the width and height of the surface, and the distance between the occupant and the target surface. Hence, the rich geospatial data available in the BIM models, along with an IoT deployment similar to the one presented in [27] for capturing sensor readings about air temperature and humidity, can facilitate the live calculation of thermal comfort level. Accordingly, the following scenario has been considered for the case demonstrations.

Figure 2 shows a partial 3D view of the area and the main experiment room, located at École de technologie supérieure, Montréal, Canada. On the other hand, an IoT-enabled monitoring system has been implemented for detecting the levels of temperature and humidity through DHT11 sensors installed at 9 spots as depicted in Fig. 2 (c).

For the case demonstrations, ontology-based representations of BIM data has been provided with reference to ifcOWL[6] and GeoSPARQL[7] ontologies. On the other hand, a combination of SSN[8], DUL[9], and GeoSPARQL vocabularies has been used for the IoT data. As mentioned in Sect. 2, ontological representations of BIM data based on ifcOWL has been previously explored by the researchers. However, combining ifcOWL and GeoSPARQL semantics is an emerging research theme. In particular, this decision is motivated by the research work most recently reported in [28]. In that study, a method has been presented for converting the ifcOWL-based representations of geometric data into GeoSPARQL-based representations with Well Known Text (WKT) serializations. The main benefit of providing WKT serializations within an ontology-based representation of BIM data is that existing vocabularies for representing geospatial data can be re-used for linking data instances from federated sources that are semantically interrelated through geospatial connections. In the context of the case examples presented here, the WKT serialization of the IoT data representing the geometry of the room in which sensors have been deployed, can act as a semantic clue for writing queries to retrieve the dimensions of the walls contained in the same room from the BIM database to perform angle factor calculations. Hence, for the sake of semantic representation of the geospatial data regarding the IoT system (e.g., locations of the DHT11 sensors), the vocabularies from GeoSPARQL and the DUL have been considered as a complementary meta-data schema to the SSN ontology. The

[6] https://technical.buildingsmart.org/standards/ifc/ifc-formats/ifcowl/.

[7] https://www.opengeospatial.org/standards/geosparql.

[8] https://www.w3.org/TR/vocab-ssn/.

[9] http://www.ontologydesignpatterns.org/ont/dul/DUL.owl#.

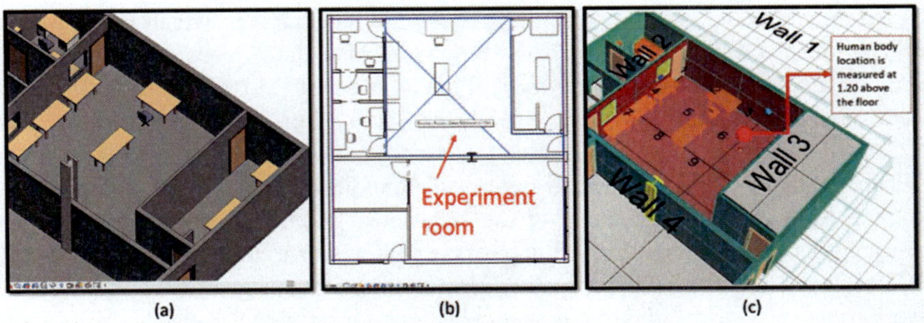

Fig. 2. (a) BIM model used in the experiment, (b) main experiment room space, (c) IoT sensor installation spots

justification for using GeoSPARQL and DUL vocabularies along with the SSN ontology can be found under the "Common Modeling Questions"[10] section provided at the W3C's official documentation.

Some key assumptions have been taken for conducting each case demonstration scenario. The first assumption is that, instances of data sources can be made readily available in the RDF format with reference to the specific ontologies considered for each local data source. The next important assumption is that for a particular use-case application, in order to access the local data sources through ontology-compatible queries (i.e., SPARQL queries in this paper), there exists a finite number of ontology mappings between the global schema and the local schemas that can be derived. Finally, although the query processing (rewriting) has been performed manually, it is assumed that SPARQL queries made with reference to the constructs of the global ontology (the virtual common ontology) can be *broken* by a software component and reformulated into a number of queries over the relevant local sources of data.

Example 1: rewriting queries posed over the BIM local schema in terms of the IoT schema The first example gives an indication of how formal semantic mappings between the ontologies used for delivering RDF-based representations of BIM and IoT data can be used for rewriting SPARQL queries from one source to another and vice versa. This example can be viewed as an instance of data integration in semantic peer-to-peer systems in which, a query Q1 posed over Peer-1 can be rewritten to a query Q2, which involves the vocabulary used in the local schema of the Peer-2 system. For this experiment, the query example shown in Listing 1 will be the source query. As it can be seen within the prefix namespaces in the query's header, the query has been defined in terms of the vocabularies used for representing BIM-based data according to ifcOWL (line 2) and GeoSPARQL (lines 3-4). The query presented in Listing 1 can be simply translated as: "*Return all sensors contained in the experiment room along with their corresponding location represented in the WKT format*". In Listing 1, lines 8-12 refer to the containment of the sensors in the experiment room. According to the IFC4.1 standard schema, placement of an `ifcElement`, which in our case will be an

[10] https://www.w3.org/TR/vocab-ssn/#location.

`ifcSensor`, will be relative to the local placement of its container, which in our case is the experiment room represented as an `ifcSpace`. Lines 13-14 refer to the location of sensor placement in the experiment room. Finally, the first `FILTER` condition (lines 15-16) restrict results according to the experiment room geometry data, and the second `FILTER` condition (line 17) ensures that sensor locations are contained within the experiment room. It is important to note that although the exact WKT values corresponding to the polyline geometry of the experiment room in line 16 have not been indicated, this query and the following ones are presented based on the assumption that WKT representations of the geospatial data of interest, i.e., geometric shape representation and placement data for ifc products, can be provided in the ifcOWL-based RDF representation of BIM data through the method presented in [28].

Listing 1. Source query posed over the BIM local

```
1.  PREFIX rdf: <http://www.w3.org/1999/02/22-rdf-syntax-ns#>
2.  PREFIX ifcowl: <http://www.buildingsmart-tech.org/ifcOWL/IFC4_ADD1#>
3.  PREFIX geo: <http://www.opengis.net/ont/geosparql#>
4.  PREFIX geof: <http://www.opengis.net/def/function/geosparql/>
5.  SELECT     ?sensor    ?sensorNodeWKT
6.  WHERE {
7.  ?sensor rdf:type ifcowl:IfcSensor .
8.  ?a  ifcowl:relatedElements_IfcRelContainedInSpatialStructure     ?sensor .
9.  ?b  ifcowl:containsElements_IfcSpatialElement       ?a .
10. ?b  geo:hasGeometry         ?c .
11. ?c  geo:asWKT      ?roomWKT .
12. ?d  ifcowl:relatedElements_IfcRelContainedInSpatialStructure     ?sensor .
13. ?d  geo:hasGeometry         ?e .
14. ?e  geo:asWKT          ?sensorNodeWKT .
15. FILTER (geof:sfWithin(?roomWKT,
16. "LINESTRING (. . , . . . , . . . , . . . , . . .)"^^geo:wktLiteral))
17. FILTER (geof:sfContains(?roomWKT, ?sensorNodeWKT) && !sameTerm(?c, ?e)) }
```

The query in Listing 2 shows the target query which is rewritten in terms of the IoT local schema (lines 2-6). For rewriting the source query, first the initial patterns were rewritten with reference to the semantic correspondences between the predicate parts of ifcOWL, SSN, and DUL. For instance, a mapping pattern has been defined to relate `ifcowl:containsElements_IfcSpatialElement` with `sosa:hosts`. Next, the rewriting process continues according to the semantic mappings between the object parts. For instance, the object `ifcowl:IfcSensor` in the source query is mapped to the `sosa:Sensor` in the target query.

Listing 2. Target query posed over the IoT local schema

```
PREFIX rdf: <http://www.w3.org/1999/02/22-rdf-syntax-ns#>
PREFIX ssn: <http://www.w3.org/ns/ssn/>
PREFIX sosa: <http://www.w3.org/ns/sosa/>
PREFIX dul: <http://www.ontologydesignpatterns.org/ont/dul/DUL.owl#>
PREFIX geo: <http://www.opengis.net/ont/geosparql#>
PREFIX geof: <http://www.opengis.net/def/function/geosparql/>
SELECT  ?sensor   ?sensorNodeWKT
WHERE {
?sensor rdf:type sosa:Sensor .
?a  rdf:type  sosa:Platform .
?a  rdf:type  dul:SpaceRegion .
?a  sosa:hosts     ?sensor .
?b  rdf:type  dul:SpaceRegion .
?b  dul:hasPart      ?a .
?b  geo:hasGeometry      ?c .
?c  geo:asWKT     ?roomWKT .
?d  sosa:hosts     ?sensor .
?d  geo:hasGeometry      ?e .
?e  geo:asWKT        ?sensorNodeWKT .
FILTER (geof:sfWithin(?roomWKT,
"LINESTRING (. . , . . , . . , . . , . .)"^^geo:wktLiteral))
FILTER (geof:sfContains(?roomWKT, ?sensorNodeWKT) &&
!sameTerm(?c, ?e))  }
```

Example 2: rewriting queries posed over the global schema in terms of the local BIM and IoT schemas

The second example is given in the context of retrieving sensory data that is required for the calculation of thermal comfort level at the experiment room. This case demonstration can be viewed as the data integration scenario, in which a high-level query can be posed with reference to the global ontology vocabulary, then, the query can be reformulated into sub-queries over disparate sources of BIM and IoT data. In this example, the query shown in Listing 3 is posed in terms of the global (mediated) ontology, which in addition to its native vocabulary (defined with `my: namespace`) uses GeoSPARQL vocabularies. The source query can be translated as: "*Return temperature and humidity values from the sensors installed at the experiment room, at a particular time, as well as the geometric data representing the shape of the walls contained in the experiment room*".

Listing 3. High-level SPARQL query in terms of the global ontology

```
PREFIX rdf: <http://www.w3.org/1999/02/22-rdf-syntax-ns#>
PREFIX my:  <http://www.ontologies.com/SourceOntology.owl#>
PREFIX geo: <http://www.opengis.net/ont/geosparql#>
SELECT    ?tempValue   ?humValue   ?wallGeo
WHERE {
?a  rdf:type  my:assessmentInstance .
?a  my:hasAssessmentTime  ?time
?a  my:hasAssesmentSpot  ?b .
?b  geo:hasGeometry      ?c .
?c  geo:asWKT  ?spotWKT .
?a  my:hasAssesmentIndoorSpace  ?d .
?d  geo:hasGeometry      ?e .
?e  geo:asWKT  ?spaceWKT .
?d  my:hasSurrondingWall ?f .
?f  geo:hasGeometry      ?g .
?g  geo:asWKT   ?wallGeo .
?a  my:hasTempObservationValue  ?tempValue .
?a  my:hasHumObservationValue  ?humValue .
FILTER (geof:sfWithin(?spotWKT,
"Point(... ...)"^^geo:wktLiteral))
FILTER (geof:sfWithin(?spaceWKT,
"LINESTRING (. . , . . , . . , . . , . .)"^^geo:wktLiteral))
FILTER ( ?time = "2020-01-01T14:30:00-05:00"^^xsd:dateTime )   }
```

Queries depicted in Listing 4 and Listing 5, show the sub-queries posed over the BIM and IoT local sources of data, respectively. As it can be seen, the high-level query which involves the concepts `my:hasAssesmentIndoorSpace` and `my:hasSurrondingWall`, is translated to the query involving `ifcowl:IfcWall` and `ifcowl:IfcRel ContainedInSpatialStructure`. Moreover, using the correspondences between the common ontology and the IoT system ontology, such as between `my:assessmentInstance` and `ssn:Observation`, or between `my:hasTempObservationValue` and `sosa:hasSimpleResult`, the initial query is translated to a query that is aligned with the local schema of the IoT system.

Listing 4. Sub-query posed over the BIM local schema

```
PREFIX rdf: <http://www.w3.org/1999/02/22-rdf-syntax-ns#>
PREFIX ifcowl: <http://www.buildingsmart-tech.org/ifcOWL/IFC4_ADD1#>
PREFIX geo: <http://www.opengis.net/ont/geosparql#>
PREFIX geof: <http://www.opengis.net/def/function/geosparql/>
SELECT  ?wall     ?wallGeo
WHERE {
?wall rdf:type ifcowl:IfcWall .
?a  rdf:type    ifcowl:IfcRelContainedInSpatialStructure .
?a  ifcowl:relatedElements_IfcRelContainedInSpatialStructure    ?wall .
?b  rdf:type    ifcowl:IfcSpace .
?b  geo:hasGeometry     ?c .
?c  geo:asWKT       ?roomGeo .
?wall  geo:hasGeometry      ?d .
?d  geo:asWKT       ?wallGeo .
FILTER (geof:sfWithin(?roomGeo,
"LINESTRING (. . , . . , . . , . . , . .)"^^geo:wktLiteral))   }
```

The above examples demonstrate how an ontology mediation approach, with a common ontology developed as a global view over local sources of BIM and IoT data, can provide an integrated access to the data being separately generated and stored within two distinct systems. The first example showed that how a set of semantic correspondences, known as "ontology alignment", between local ontologies used to describe disparate sources of BIM and IoT can provide an integrated access to BIM and IoT data by translating queries posed over each of the systems to the ones that is understandable by the other system, without the need of translating data instances in terms of the common ontology schema. The second example showed that how the process of querying the local sources of BIM and IoT data one by one, can be replaced by retrieving the required data from them only by writing queries over a common ontology. In particular, this example illustrates that how high-level queries posed over a global ontology can eliminate the need for providing an acute awareness over the vocabularies and semantic relationships within local schemas for obtaining data from them. In the given examples, geospatial data played a key role in the process of BIM-IoT data integration. More precisely, in both examples WKT-based serializations of geometric data can be viewed as the main semantic clue for linking the disparate sources of BIM and IoT data. An alternative scenario for linking BIM and IoT data could be as follows: making use of sensors GUID available in the IFC-based BIM data, as the main reference for making sensors uniquely identifiable in disparate sources of data. In this case, the IoT deployment in our experiment design would have no clues about the sensors in the BIM database unless an understanding of the GUID values are somehow included within the local schema considered for the IoT system. On the contrary, by relying on geospatial data as one of the main references for identifying sensor instances and establishing the mappings between geospatial data representations in local schemas, the BIM and IoT systems would have a shared understanding of the sensor that is located in a unique spatial entity. However, relying on a single semantic clue, such as geospatial data as discussed in this paper, will not be sufficient. For example, if sensors are attached to mobile objects, or in case that more than one sensor is placed at a specific location, relying solely on geospatial data will be problematic. Hence, by considering various semantic clues and designing effective ontology and rule-based inference mechanisms, the precision of retrieving data form federated sources of BIM and IoT data can be improved.

Listing 5. Sub-query posed over the IoT local schema

```
PREFIX rdf: <http://www.w3.org/1999/02/22-rdf-syntax-ns#>
PREFIX ssn: <http://www.w3.org/ns/ssn/>
PREFIX sosa: <http://www.w3.org/ns/sosa/>
PREFIX dul: <http://www.ontologydesignpatterns.org/ont/dul/DUL.owl#>
PREFIX geo: <http://www.opengis.net/ont/geosparql#>
PREFIX geof: <http://www.opengis.net/def/function/geosparql/>
SELECT  ?sensor  ?property  ?value
WHERE {
?a  rdf:type  sosa:Platform .
?a  rdf:type  dul:SpaceRegion .
?a  sosa:hosts  ?sensor .
?a  geo:hasGeometry   ?b .
?b  geo:asWKT  ?spotWKT .
?c  rdf:type  ssn:Observation .
?c  ssn:observedBy ?sensor .
?c  ssn:observedProperty  ?property .
?c  sosa:hasSimpleResult  ?value .
?c  sosa:resultTime ?time.
FILTER (geof:sfWithin(?spotWKT,
"Point(... ...)"^^geo:wktLiteral))
FILTER ( ?time = "2020-01-01T14:30:00-05:00"^^xsd:dateTime )    }
```

5 Conclusion

This paper presented a conceptual framework for integrating disparate sources of BIM and IoT data through the application of ontological formalisms and semantic web tools. The core rationale behind the proposed framework is to establish semantic connections between disparate sources of BIM and IoT data with the aid of ontologies as mediators between local data representation schemas and a common virtual ontology. OWL ontologies, along with other semantic web tools, namely RDF, SPARQL are the building blocks of the proposed framework. Integration of data sources is achieved by reformulating SPARQL queries with reference to the semantic correspondences (i.e., ontology mappings) between the entities of the local ontology-based schemas and the global ontology-based schema. This framework adds to the body of knowledge by investigating the potentials of formal ontologies and semantic web tools for addressing syntactic and semantic interoperability issues that impede an efficient integration of live IoT data within BIM environments.

The focus of the current study remained on presenting a conceptual framework and for the case demonstrations, finding the relevant ontology mappings and rewriting queries according to those mappings were done manually. However, an effective use of the proposed framework for allowing IoT services to openly interoperate with BIM environments requires that recognition of the mappings between ontologies, and query-rewriting processes be done in an automated manner. The query broker component (i.e., software components as query mediators) that has been considered within the design of the proposed framework can automatically perform the task of transforming queries into new queries to be posed over the local data sources and translating the answers into a form that is understandable in terms of the global schema (common ontology). Hence, our current research investigates the design of the automated procedures for finding semantic mappings and developing query brokers to be included within the implementation of the proposed framework. Moreover, the purpose of the examples provided in this study was to support only a conceptual implementation of the proposed framework. For the future works, the applicability and effectiveness of the proposed framework need to be validated by conducting multiple complementary case studies through a detailed implementation of the framework. Moreover, the presented framework will be expanded by adding elements and techniques that compensate the known shortcomings of the SPARQL language (e.g., verbosity and complexity). One potential scenario in this respect is to add a GraphQL-based layer to the current design of the proposed framework. Such an additional query layer can ease the process of working with complex SPARQL queries by transforming more developer-friendly GraphQL-based queries into the equivalent SPARQL queries over RDF triples.

References

1. Dave, B., Buda, A., Nurminen, A., Främling, K.: A framework for integrating BIM and IoT through open standards. Autom. Constr. **95**, 35–45 (2018). https://doi.org/10.1016/j.autcon.2018.07.022
2. Bettini, C., et al.: A survey of context modelling and reasoning techniques. Pervasive Mobile Comput. **6**(2), 161–180 (2010). https://doi.org/10.1016/j.pmcj.2009.06.002
3. Ye, J., et al.: Semantic web technologies in pervasive computing: a survey and research roadmap. Pervasive Mobile Comput. **23**, 1–25 (2015). https://doi.org/10.1016/j.pmcj.2014.12.009
4. Alirezaie, M., et al.: An ontology-based context-aware system for smart homes: E-care@home. Sensors **17**(7), 1586 (2017). https://doi.org/10.3390/s17071586
5. Goel, D., Chaudhury, S., Ghosh, H.: Smart water management: an ontology-driven context-aware iot application. In: Pattern Recognition and Machine Intelligence, Cham, pp. 639–646 (2017). https://doi.org/10.1007/978-3-319-69900-4_81
6. Khowaja, S.A., Prabono, A.G., Setiawan, F., Yahya, B.N., Lee, S.-L.: Contextual activity based healthcare internet of things, services, and people (HIoTSP): an architectural framework for healthcare monitoring using wearable sensors. Comput. Netw. **145**, 190–206 (2018). https://doi.org/10.1016/j.comnet.2018.09.003

7. Song, R., Vanthienen, J., Cui, W., Wang, Y., Huang, L.: Context-aware BPM using IoT-integrated context ontologies and IoT-enhanced decision models. In: 2019 IEEE 21st Conference on Business Informatics (CBI), vol. 01, pp. 541–550 (2019). https://doi.org/10.1109/cbi.2019.00069
8. Tang, S., Shelden, D.R., Eastman, C.M., Pishdad-Bozorgi, P., Gao, X.: A review of building information modeling (BIM) and the internet of things (IoT) devices integration: present status and future trends. Autom. Constr. **101**, 127–139 (2019). https://doi.org/10.1016/j.autcon.2019.01.020
9. Panetto, H.: Towards a classification framework for interoperability of enterprise applications. Int. J. Comput. Integr. Manuf. **20**(8), 727–740 (2007). https://doi.org/10.1080/09511920600996419
10. Karan, E.P., Irizarry, J.: Extending BIM interoperability to preconstruction operations using geospatial analyses and semantic web services. Autom. Constr. **53**, 1–12 (2015). https://doi.org/10.1016/j.autcon.2015.02.012
11. Lemaire, C., Rivest, L., Boton, C., Danjou, C., Braesch, C., Nyffenegger, F.: Analyzing BIM topics and clusters through ten years of scientific publications. J. Inf. Technol. Constr. (ITcon) **24**(15), 273–298 (2019)
12. Sheth, A., Henson, C., Sahoo, S.S.: Semantic sensor web. IEEE Internet Comput. **12**(4), 78–83 (2008). https://doi.org/10.1109/MIC.2008.87
13. Compton, M., et al.: The SSN ontology of the W3C semantic sensor network incubator group. J. Web Semant. **17**, 25–32 (2012). https://doi.org/10.1016/j.websem.2012.05.003
14. Cirani, S., Ferrari, G., Picone, M., Veltri, L.: Internet of Things: Architectures. Protocols and Standards. Wiley, Hoboken (2018)
15. Berners-Lee, T., Hendler, J., Lassila, O.: The semantic web. Sci. Am. **284**(5), 28–37 (2001)
16. Corry, E., Pauwels, P., Hu, S., Keane, M., O'Donnell, J.: A performance assessment ontology for the environmental and energy management of buildings. Autom. Constr. **57**, 249–259 (2015). https://doi.org/10.1016/j.autcon.2015.05.002
17. Jiang S., Wang N., Wu J.: Combining BIM and ontology to facilitate intelligent green building evaluation. J. Comput. Civ. Eng. **32**(5), 04018039 (2018). https://doi.org/10.1061/(ASCE)CP.1943-5487.0000786
18. Niknam, M., Karshenas, S.: Integrating distributed sources of information for construction cost estimating using semantic web and semantic web service technologies. Autom. Constr. **57**, 222–238 (2015). https://doi.org/10.1016/j.autcon.2015.04.003
19. Choi, N., Song, I.-Y., Han, H.: A survey on ontology mapping. SIGMOD Rec. **35**(3), 34–41 (2006). https://doi.org/10.1145/1168092.1168097
20. Cruz, I.F., Xiao, H.: The role of ontologies in data integration. Eng. Intell. Syst. Electr. Eng. Commun. **13**(4), 245 (2005)
21. Euzenat, J., Shvaiko, P.: Ontology Matching, vol. 18. Springer (2007)
22. Shahinmoghadam, M., Motamedi, A.: Review of BIM-centered IoT deployment: state of the art, opportunities, and challenges. In: 2019 Proceedings of the 36th ISARC, Banff, Alberta, Canada, pp. 1268–1275 (2019). https://doi.org/10.22260/ISARC2019/0170
23. Makris, K., Gioldasis, N., Bikakis, N., Christodoulakis, S.: Ontology mapping and SPARQL rewriting for querying federated RDF data sources. In: Meersman, R., Dillon, T., Herrero, P. (eds.) On the Move to Meaningful Internet Systems, OTM 2010, vol. 6427, pp. 1108–1117. Springer, Heidelberg (2010). https://doi.org/10.1007/978-3-642-16949-6_32
24. Kim, K., Kim, H., Kim, S.-K., Jung, J.-Y.: i-RM: an intelligent risk management framework for context-aware ubiquitous cold chain logistics. Expert Syst. Appl. **46**, 463–473 (2016). https://doi.org/10.1016/j.eswa.2015.11.005

25. Shahinmoghaddam, M., Nazari, A., Zandieh, M.: CA-FCM: towards a formal representation of expert's causal judgements over construction project changes. Adv. Eng. Inform. **38**, 620–638 (2018). https://doi.org/10.1016/j.aei.2018.09.006
26. Natephra, W., Motamedi, A., Yabuki, N., Fukuda, T.: Integrating 4D thermal information with BIM for building envelope thermal performance analysis and thermal comfort evaluation in naturally ventilated environments. Build. Environ. **124**, 194–208 (2017). https://doi.org/10.1016/j.buildenv.2017.08.004
27. Natephra, W., Motamedi, A.: Live data visualization of IoT sensors using Augmented Reality (AR) and BIM. In: 2019 Proceedings of the 36th ISARC, Banff, Alberta, Canada, pp. 632–638 (2019). https://doi.org/10.22260/ISARC2019/0084
28. O'Donovan, J., O'Sullivan, D., McGlinn, K.: A method for converting IFC geometric data into GeoSPARQL. In: Proceedings of the 7th Linked Data in Architecture and Construction Workshop - LDAC2019 (2019)

Digital Twins in Architecture, Engineering, Construction and Operations. A Brief Review and Analysis

Ramy Al-Sehrawy(✉) and Bimal Kumar

Department of Architecture and Built Environment, Northumbria University, Newcastle, UK
ramy.alsehrawy@northumbria.ac.uk

Abstract. Advances in and pervasiveness of technologies like Internet of Things (IoT), Big Data (BD) and Big Data Analytics (BDA), Artificial Intelligence (AI), Cloud Computing (CC) and Cyber Physical Systems (CPS) have allowed for the dissemination of intelligence and innovative solutions across different industries including Architecture, Engineering, Construction and Operation (AECO). Recently, the ideas behind "Digital Twin" (DT) have been gaining ground in the AECO sector. This paper briefly reviews and analyzes relevant literature to discover the origins of the DT idea; crystallizes a clear definition; identifies DT's key features; highlights some case studies carried out so far and outlines some of the key applications and the challenges for wider adoption.

Keywords: Built environment · AECO · Digital Twin (DT) · Life-cycle · System of Systems (SoS)

1 Introduction: Origin of Digital Twin

In 1991, David Gelernter; a specialist in the field known as massively parallel computation at Yale introduced the notion of "Mirror Worlds" as a city-scale model of reality that is continuously being fed by massive amount of data enabling the user to zoom in or out for realizing desired levels of details on a computer screen "in a single dense, live, pulsing, swarming, moving, changing picture." (Gelernter 1991 p. 30). Years later; in 2002, Grieves in one of his presentations depicted a conceptual model named "conceptual ideal for PLM" comprising real space, virtual space, virtual sub-spaces and the link of data flowing between both virtual and real spaces for the purpose of Product Lifecycle Management (PLM). Similar concept was embedded in the first PLM Course in 2003 at the University of Michigan and referred to as "Mirrored spaces model". The same term was published in a journal article in 2005 (Grieves 2005), yet a year later, Grieves in his book (Grieves 2006) referred to this concept as "information Mirroring Model".

While the same concept kept expanding and maturing, Piascik et al. (2010) coined the term Digital Twin in one of NASA's strategic technological roadmaps as an equivalent term to the "long-term vision" named "Virtual Digital Fleet Leader". Since

E. Toledo Santos and S. Scheer (Eds.): ICCCBE 2020, LNCE 98, pp. 924–939, 2021.
https://doi.org/10.1007/978-3-030-51295-8_64

then, the term "Digital Twin" has been widely adopted; whereas it has embraced other precedent terms (i.e. Information Mirroring Model) in later researches (Grieves 2011).

2 Definition of Digital Twin

There has been an obvious increase of interest over time in the term "Digital Twin" during the past few years (Fig. 1). Gartner expected that by 2022 two-thirds of companies already implementing IoT will be deploying at least one DT (Costello and Omale 2019). While this exponential growth in interest has significantly built research momentum towards DT advancements, yet it has also facilitated for the term to slowly enter into mainstream (Costello and Omale 2019) while still in infancy and with no established common definition across the literature (Kritzinger et al. 2018) to underpin the emerging area. Urban innovation labs [uil] (2019) noted that many have used the term DT without having an understanding of what it really means, or in some cases even using misleading definitions (Tomko and Winter 2019). Batty's (2018) use of the term "cliché" when referring to DT seems justifiable considering widespread lack of understanding that has led to difficulties in meeting expectations and unrealized benefits (Hicks 2019).

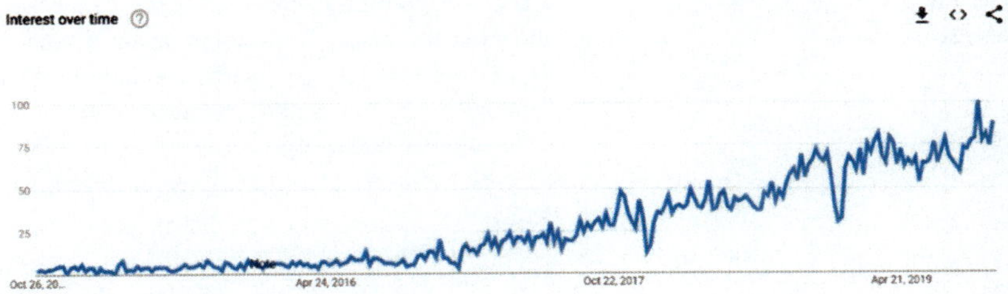

Fig. 1. Interest in "Digital Twin" over last 5 years. *Source:* https://trends.google.com/

Definitions of DT as given by different multidisciplinary researchers, when reviewed and analysed were found to comprise three different aspects although some less comprehensive definitions include a combination of two of these aspects or only a single aspect in few cases. First aspect is focused on the identity, nature or the structure per se of the DT (1). The second aspect is aim, purpose or function oriented (2). While the third highlighted the main constituents, elements, components or aspects of a DT (3). Table 1 includes a breakdown of all reviewed definitions from different sectors including manufacturing and aerospace and Architecture, Engineering, Construction and Operation [AECO] based on the aforementioned three aspects. The main focus was only on explicit definitions of the term DT given by the cited authors without interpretations or analyses of any of their further discussions.

Table 1. DT definitions in reviewed literature

Author/Aspect	Identity, nature or structure of a DT	Aim, purpose or function of a DT	Main constituents of a DT
Manufacturing & Aerospace			
Glaessgen and Stargel (2012)	"An integrated multi-physics, multi-scale, probabilistic simulation of a complex product"	"to mirror the life of its corresponding twin."	"uses the best available physical models, sensor updates, etc." "digital twin consists of three parts: physical product, virtual product, and connected data that tie the physical and virtual product"
Hochhalter et al. (2014)	"Digital twin is an emerging concept"	"To enable high-fidelity modeling of individual vehicles throughout their service lives"	"which employs modeling and simulation of the as-built vehicle state, as-experienced loads and environments, and other vehicle-specific history"
Boschert and Rosen (2016)	N.A.	The simulation models making up the DT are specific for their intended use and apply the suitable fidelity for the problem to be solved The DT evolves along with the real system along the whole life cycle The DT is not only used to describe the behaviour but also to derive solutions relevant for the real system	the linked collection of the relevant digital artefacts including engineering data, operation data and behaviour descriptions via several simulation models "integrates the currently available knowledge about it [whole life cycle]"
Grieves and Vickers (2017)	"a digital informational construct about a physical system could be created as an entity on its own"	"This digital information would be a "twin" of the information that was embedded within the physical system itself and be linked with that physical system through the entire lifecycle of the system"	N.A.

(*continued*)

Table 1. (*continued*)

Author/Aspect	Identity, nature or structure of a DT	Aim, purpose or function of a DT	Main constituents of a DT
Söderberg et al. (2017)	"The concept of using a digital copy of the physical system"	"to perform real-time Optimization"	N.A.
Tao et al. (2017)	N.A.	"… that [connected data] tie the physical and virtual product"	"it [DT] consists of three parts: physical product, virtual product, and connected data…"
Vachálek et al. (2017)	"A DT is essentially a functional system"	"…continuous process optimization"	"…physical production lines [physical systems in manufacturing context] with a digital "copy"
Negri et al. (2017)	"a virtual representation of a production system that is able to run on different simulation disciplines that is characterized by the synchronization between the virtual and real system"	"to forecast and optimize the behaviour of the production system at each life cycle phase in real time"	"sensed data and connected smart devices, mathematical models and real time data elaboration"
Parrott and Warshaw (2017)	"a near-real-time digital image of a physical object or process"	"that helps optimize […] performance"	N.A.
Boschert et al. (2018)	"A comprehensive physical and functional description together with	N.A.	"all available operational data of a component, product or system, which includes more or less all information which could be useful in all - the current and subsequent - lifecycle phases"
Kritzinger et al. (2018)	"It is the virtual and computerized counterpart of a physical system"	"A Digital Twin can be used to simulate it for various purposes, … and is able to decide between a set of actions with the focus to orchestrate and execute the whole production	"exploiting a real time synchronization of the sensed data originating from the field-level" Fully integrated Bidirectional communication and flow of data

(*continued*)

Table 1. (*continued*)

Author/Aspect	Identity, nature or structure of a DT	Aim, purpose or function of a DT	Main constituents of a DT
		system in an optimal way. This Results in a higher efficiency, accuracy and gains economic benefits in the production"	
Tao et al. (2019)	"The models serve as a communication and recording mechanism" "The DT is another concept associated with cyber–physical integration"	"to help interpret the behaviors of machines or systems and to predict their future state"	"based on real-time data, historical data, experience, and knowledge, as well as on data from models"
Architecture, Engineering, Construction & Operation [AECO]			
National Infrastructure Commission [NIC] (2017)	"A digital model: a dynamic representation of a system…" "A virtual representation of a physical object or system across its lifecycle"	"mimics its [digital model] real-world Behavior (and, in some cases, the surrounding environment)" "to enable understanding, learning and reasoning"	"real-time, updated collection of data, models, algorithms or analysis"
Bolton et al. (2018)	"a realistic digital representation of assets, processes or systems in the built or natural environment"	"They unlock value by enabling improved insights that support better decisions, leading to better outcomes in the physical world"	N.A.
Dawkins et al. (2018)	"The concept of 'Digital Twin' refers to the coupling of a physical system with its digital representation in a computer…"	"…such that any relevant change of state in the physical system is detected and triggers a flow of data that causes a corresponding change in the state of its digital counterpart"	N.A.
Batty (2018)	"A mirror image of a physical process that is articulated alongside the process in question…"	"…usually matching exactly the operation of the physical process…"	"which takes place in real time"
Tomko and Winter (2019)		"capable not only of reaction and prediction,	"properties of an organism: (a) it can

(*continued*)

Table 1. (*continued*)

Author/Aspect	Identity, nature or structure of a DT	Aim, purpose or function of a DT	Main constituents of a DT
	"a cyber-physical-social system with coupled properties"	but increasingly also of action (rather than being the passive reflection of a mirror)"	sense the (physical or digital) environment and updates its counterpart accordingly, (b) it has agency and can change the (physical or digital) environment based on instructions, possibly from the counterpart environment and including engagement with people, and (c) it has a moderation system that attempts to preserve its operational inner state, here called an immune system"
Farsi et al. (2019)	a concept or a workflow	to monitor its past and present operational state, detect issues, accurately forecast and predict its future state, maintenance as well as its prediction of remaining useful life (RUL) DT determines the best course of action	Uses collected real-time using sensors and IoT, along with data-driven analytics and intelligence (e.g.: Machine Learning algorithms) to create a continuously adaptable model representing the structure, behaviour and context of a physical asset Physical entity in real space, Sensors & IoT for real-time data acquisition

A deeper review of the second aspect in literature gives a better feel for the potential of DTs and the possible emerging set of aims and uses. The core aim of DT as mentioned by most of the researchers is the ability to predict the future state of a physical object or system. This creates several opportunities including detection of issues or anomalies sooner and giving early warnings (Parrott and Warshaw 2017), which helps in higher performance during operational phase of the physical system through proactive maintenance (Farsi et al. 2019), detection of inefficient processes (Vachálek et al. 2017) and operational optimization (Tao et al. 2017).

It is quite clear when considering the third aspect of DT that the models and data are its core constituents. Moreover, if the major aim of a DT is to predict future state and subsequently assist in taking strategic decisions, give feedback and recommend course of action, then the need to simulate the behaviour and functions of the physical system with exceptionally high levels of accuracy is inevitable (Schleich et al. 2017; DebRoy et al. 2017). Attaining this requires what is defined as the "input" to a DT: an unprecedented huge amounts of data as well as models representing all aspects of the physical system's nature and context. However, to leverage this input to create the "output" (i.e.: High-fidelity Virtual Model; Simulation capabilities and feedback), it is easily realized that "New IT" (Tao et al. 2019) including Big Data Analytics (BDA) and Artificial Intelligence (AI) are fundamental components and facilitators for a DT to deliver its basic outputs and as a result achieve its main aim.

Based on the reviewed definitions, a DT can be further defined as:

An approach for connecting a physical system to its virtual representation via bidirectional communication (with or without human in the loop) using temporally updated Big Data (primary data collected from this physical system and supplementary data from the surrounding environment interacting with it for the purpose of contextual awareness) to allow for exploitation of Artificial Intelligence and Big Data Analytics by harnessing this data to unlock value through optimization and prediction of future state.

3 Key Features of Digital Twin

3.1 Data Communication (C)

The Digital Model (DM) in AECO industry has been around for decades since the emergence of CAD which comprised geometric information of the relevant asset. Later the idea of a DM evolved further through the rise of BIM, which is mainly concerned about combining asset information with the geometric information along with providing a common environment as the only source of truth to enhance collaboration and data integration (Garber 2014), thus attaining a higher level of data fusion (Klein 2004). DMs in the paradigm of BIM have witnessed considerable evolution from static 3D up to dynamic 6D simulation models (Al-Sehrawy et al. 2019) with possibly more dimensions to embed. However, a DM (CAD or BIM) is typically modelled manually since it has no links or bridges in terms of communication and connectivity with the physical asset, thus has no minimum capabilities to maintain a flow of data with the physical asset, neither in a unidirectional nor in a bidirectional way.

Many researchers within the construction industry headed towards the idea of a digital Shadow (DS) (Kritzinger et al. 2018) through building a uni-directional route for data to flow from the physical to the virtual system using different technologies (e.g.: laser scanners, RFID) (Chin et al. 2008; Golparvar-Fard et al. 2009; Motamedi and Hammad 2009; Sorensen et al. 2010; Akanmu et al. 2013). Recently a DT approach has evolved from being merely descriptive (i.e.: DS) to a more of an actionable and proactive one (Grieves and Vickers 2017). One can easily draw from the recent definitions of a DT that a DT is no more than just a combination of pure data (i.e. DS) but also an intelligent system that leverages data and models to establish a bi-directional communication with the physical world by providing predictions of the future state, recommendation for best courses of actions to support decisions to optimize performance. This intelligence-based unique capability keeps the DT at a higher level in terms of data fusion and communication compared to DM or DS.

In other words, in a DT not only the physical world changes the virtual one, but the virtual world is also able to change the physical world. This bi-directional communication enabled by a DT takes place in either a passive or an active form. Conventionally, a "Passive DT" establishes a passive bi-directional communication through an open-ended feedback loop where humans have to interfere to execute data-driven-decisions into the physical world based on the predictions and recommendations given. Although an "Active DT" is not necessarily more intelligent than a "Passive DT", it is capable of changing the physical world without human intervention through a closed-ended feedback loop using actuators and Cyber-Physical-System (CPS) technologies that are meant to apply the data-drivendecisions. Tao et al. (2019) pointed out that combining DTs with CPS technologies can achieve precise and efficient management and control.

3.2 Vertical Integration (V)

Unlike other systems (i.e.: DS) a DT can "mirror the life of its corresponding twin" (Glaessgen and Stargel 2012) and not merely at the very moment when data from its physical counterpart is acquired. The future insights given by the powerful analytic capabilities of the DT enables predicting the remaining useful life (RUL) of the physical system, blends its life cycle phases and provides a complete digital footprint for it (Tao et al. 2017) offering a comprehensive description and extensive information about the whole life cycle phases (Boschert and Rosen 2016) from the beginning to end (Schroeder et al. 2016) and even beyond by reaching out to early life cycle phases of newly developed products (Boschert et al. 2018). Combining all life-cycle phases of a one system as well as the merging of successive life-cycles of domain-similar systems is recognized as "Vertical Integration".

3.3 Horizontal Integration (H)

The collaboration between different separate entities (i.e.: cyber or physical) is known as "Horizontal Integration" and can be categorized into a Unit level, System Level or System of Systems (SoS) Level. The Unit Level is a single, standalone component at the most basic level that can form a part of a system. For example, a unit level in a conventional building can be a door, structural beam, lighting fixture…etc. Multiple unit levels combined in an interoperable way enabling a wider coordination between the different unit levels sharing a common environment to deliver mutual aim(s) constitutes a System Level. At a system level, interdependencies between different units can be easily comprehended and their outcomes are quite predictable. Following the same example, a system for controlling Indoor Environmental Quality (IEQ) in a building may consist of AC units, lights, Windows and blinds. Furthermore, an integration of different system forms a System of Systems (SoS). The SoS Level is more concerned about complexity than count of units or systems involved albeit being directly proportional in many cases. Perrow (1984) defined complex interdependencies as those whose outcomes cannot be expected or planned for, leading to unpredicted emergent behaviors (Grieves and Vickers 2017). This complexity results from the integration of independent systems that normally provide useful services independently. Extending the above example, a SoS for the building may take into account its interdependencies with other independent systems and possibly from different domains such as power grid, water, wastewater, transport, rail networks, meteorology, demographics…etc. (Rinaldi et al. 2001; NIC 2017).

Figure 2 below depicts the three key features (C, H & V) of a DT in a 3-dimensional space of virtual world. It shows that the journey of development of a DT can go from manual- to active-bidirectional in terms of data communication level (C); from unit to SoS level of horizontal integration (H) and from a single phase to multi life-cycles in terms of vertical integration (V).

3.4 Fidelity

In terms of "Fidelity" of a DT, previous research raised questions like "how well these virtual systems mirror their physical counterparts?" (Grieves and Vickers 2017), or wondered about the level of "faithfulness" of a DT's representation of the physical environment (Tomko and Winter 2019). All of this shows how all were mainly concerned with the highest levels of fidelity that can be reached – atomic level (Grieves and Vickers 2017) – in the attempt of chasing the idealization of an exact mirror. Nonetheless, Batty (2018) pointed out that models or systems are not supposed to be an exact mirror despite how rational it may be to try to bring the virtual system closer to the real one.

In an attempt to measure the level of fidelity of a DT, Grieves (2006) introduced "Grieves Tests of Virtuality" (GTV) which are three tests carried out to measure three different aspects of fidelity, viz. Sensory Visual Test; Performance Test and Reflectivity Test. Despite such attempts to develop methods for measuring DTs' levels of fidelity, none – to the best of the authors' knowledge – have been proposed or applied yet within the AECO industry. However, inspired by the GTV, it is proposed that the following could be a possible method of measuring "fidelity" within AECO industry:

Visual-Fidelity: Measures the level of geometric detail and granularity of graphical information in terms of reflecting the appearance of the real physical system. (i.e.: Level of Detail (LOD) (British Standards Institution [BSI], 2019)).

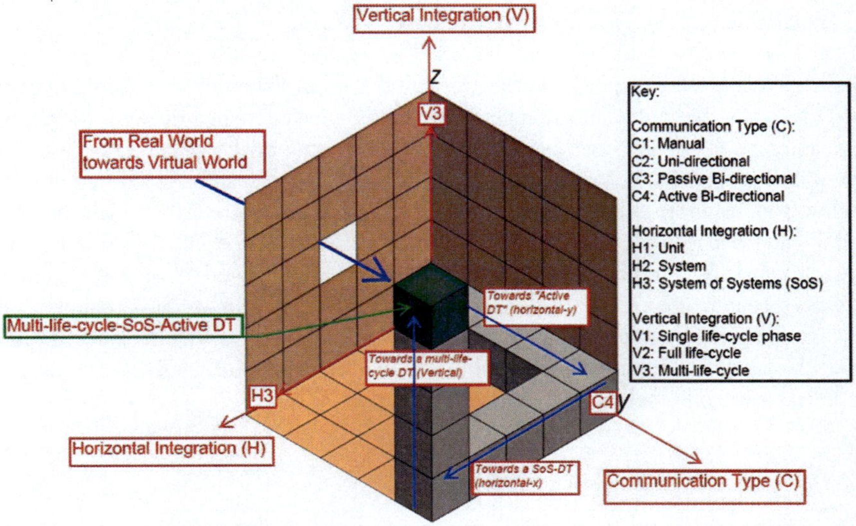

Fig. 2. 3 dimensions of digital models (C-H-V)

Reflectivity-Fidelity: Measures the level of detail and granularity of non-graphical information. (i.e. Level of Information (LOI) (BSI 2019)). In principle, all GTV including the reflectivity test are passed when the ultimate level is achieved with zero tolerance, however, BSI (2019) highlights that "one purpose of defining the level of information need is to prevent delivery of too much information". The levels of reflectivity in AECO have witnessed a breakthrough by the virtue of ubiquitous sensors and IoT which have paved the way for the level of reflectivity not only to be higher but also to shift from a static to a real-time dynamic paradigm.

Performance-Fidelity: Measures the level of accuracy in predicting the true and real behavior of the physical system. While visual and reflectivity fidelities are fairly well represented in aforementioned ISO standards, the representation of the performance-fidelity in AECO is quite nascent. Figure 3 shows level of fidelity may increase over time when following a "Vertical Integration" approach resulting in a decrease in the percentage of unpredicted behaviors of a SoS by a DT compared to real behavior. Height of point (A) corresponds to the percentage of predicted behavior on the "Performance-fidelity" vertical axis. As the construction phase of the physical asset reaches completion, still the same percentage of predicted behavior during design simulations is unchanged, yet DT now starts to rely on real sensed data instead of design assumptions and hypothetical models. Increase in percentage of "Performance-fidelity" from point (B) to point (C) happens due to learning some of the unpredicted behavior during design by the virtue of DT and real-time data collected throughout the operation phase. By the end of the life-cycle and approaching disposal or refurbishment, transfer of knowledge and legacy takes place to inform the new-born DT, thus starting at a much higher level of "Performance-fidelity" at point (D) compared to the predecessor DT.

4 Applications and Case Studies in AECO

Within the built environment, DTs are showing a great potential in multiple applications like asset operation and maintenance; faults and anomalies detection in equipment; interactions of systems; performance optimization; cost reduction; security and many other areas (NIC 2017; Johnson Controls 2019), which can be at a different scale, ranging from a simple unit-scale to a complex city-scale. Wildfire (2018) identified four possible approaches for a DT to tackle such applications. These are: Reactive for real-time interventions; Predictive for long-term planning and answering what-if questions; Iterative/feedback loops and responses to enable informed-pre-emptive actions; and Sandbox scenario testing for strategic planning for future assets or extensions. From a life-cycle perspective, applications of DT may differ from one life-cycle phase to another as explained below.

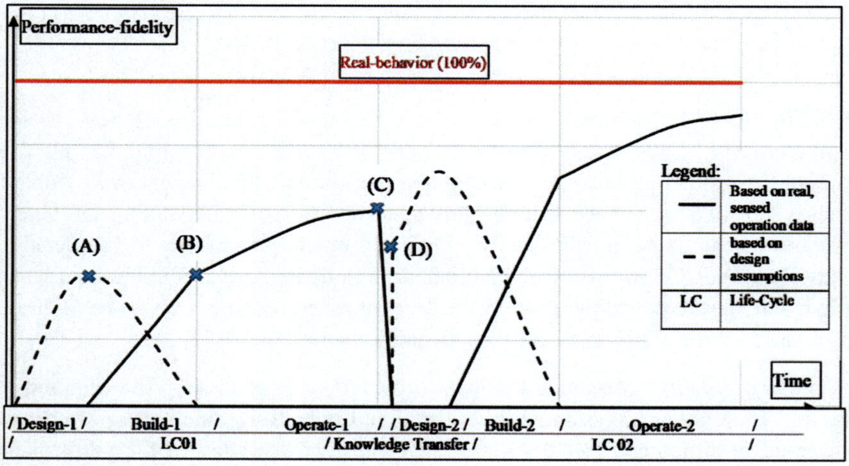

Fig. 3. Performance-fidelity VS Time with vertical integration

4.1 Design

Informed-Design. Establishing informed-design method using user-related information to better understand users' behaviors, usage patterns and their interactions with the asset in question. Having a single DT is not the final step but the first. The greater value and potential are unlocked by federating multiple DTs (Costello and Omale 2019) through "horizontal integration"; forming a SoS capable of analyzing the interdependencies between the elements of these systems. Information facilitating informed-design can also be acquired from the operational stages of previous similar DTs (Vertical Integration). Exploiting the gathered knowledge and information using horizonal and vertical integrations of DTs will help in running simulations of higher performance-fidelity levels, thus reducing uncertainties and unpredicted emergent behaviors for the remaining life leading to the delivery of more user-centered and human interactive buildings and subsequently more value to users.

Impact-Aware Design. The approach of a SoS-DT can change the way designers think today by dealing with any newly built system as an intervention to existing systems and not only an addition to them. Having access to DTs of the existing built and natural environments interacting with the asset being designed will help in running high performance-fidelity scenarios to better understand and manage the impact of a newly introduced asset throughout its lifecycle on other tangible (i.e. built assets) or intangible (economic, environmental and social) systems and perhaps enable trade-offs between these systems.

4.2 Construction

Strike-Less Construction. DT can assist in visualization and awareness of all buried service locations, interactions, current state and predicted future state resulting in no clashes or conflicts at site with buried services or systems (NIC 2017).

Impact-Aware Construction. DT can help in monitoring, controlling and predicting the bidirectional impacts of the construction activities on the other systems and environments (i.e. built and natural) and vice versa. This endows construction management support in decision making and reaching optimum solutions to achieve project goals (i.e.: prevent late delivery or cost overrun before any happen; predict, avoid and mitigate health, safety and environmental issues, risks and hazards…etc.) and follow a more sustainable construction approach for public good.

4.3 Operation and Services

Servitisation. DT-based approach will help leverage big data to shift into a world of servitisation that offers Everything-as-a-Service, through delivering innovative services and implying the move from paradigms of ownership and siloed services to another of an integrated, user-centered services, such as Energy-as-a-Service or Mobility-as-a-Service (uil 2019).

Resilient Built-Environment. The move from reactive asset management thinking to a proactive one can be achieved by gaining insights from a DT about future state of infrastructures in order to intervene early enough during construction phase to prevent late delivery or cost overrun; or to predict, avoid and mitigate health, safety and environmental issues, risks and hazards for a more sustainable construction. Finally, during operational phase this approach could prevent a failure due to disasters or degrading or a drop in quality of service provided or improve resilience.

4.4 End-of-Life

Linear to Circular Life-Cycle. DT can help the built environment to transform the pattern of the life-cycle from a linear to a circular one through obtaining value and transferring knowledge from the DT end of life to inform the new life of the next DT.

Waste Management. Whether it is a disposal, decommissioning, refurbishment or upgrade, DT can help in reducing impact on environment and the considerable amounts of waste usually produced during this phase.

5 Challenges

5.1 Open Data

While DT holds a great potential, it also presents some challenges that must be addressed to realize its several promised benefits. The main challenge is related to the DT's main pillar, which is open data; how data can be shared, integrated and made accessible across multiple sectors and entities. This challenge must be dealt with from different aspects including:

Technical. A problem inherited from the BIM paradigm and augmented in the DT context is how to integrate incompatible big data from disparate, diverse and heterogeneous sources (Qiuchen et al. 2019) and collected in varying quality levels and temporal granularity where some may not be fit for purpose (Bolton et al. 2018), or perhaps lost the quality later while going through knowledge abstraction processes (e.g. during vertical integration between successive lifecycle phases). Lack of interoperability and absence of a common language is a considerable barrier.

Contractual. Including governance, data ownership and liability;

Commercial. How to reach a perfect balance between sharing enough data for public good without harming the competitive advantage of profitable bodies.

Cultural. How to apply change management techniques with old-school stakeholders and closed mentalities resisting the shift and failing to realize the value of sharing data.

5.2 Other Challenges

Connectivity. Another challenge is DT Connectivity. During power loss or DT upgrade or maintenance; keeping the DT online and in-service, and the stream of data flowing with no disruption or termination (Tomko and Winter 2019) to avoid loss in value or inaccurate and misleading predictions.

Security. Security of the source as the data provider and security of the DT per se comprising a huge pool of data and federated models through proper data protection arrangements. Securing data is directly related to the security of the DT users who might be heavily relying on the services offered by the DT.

Privacy. How to maintain appropriate levels of Privacy for all users and data sources to reserve personal and commercial rights (NIC 2017).

Inclusiveness. How to have the built environment legacy, existing pre-digital buildings and infrastructures as well as all range of users on board, and ensure no assets are left behind and no users are kept "offline".

6 Summary

This paper briefly reviewed and analyzed literature relating to the origins and definitions of the term "Digital Twin" (DT). While no consensus over a one crystallized definition has been developed, it is clear that a DT is a model that captures an abstraction of reality and not an absolute duplication of real life. The seamless bidirectional connection with physical world and intelligence of a DT, however, are its core features compared to preceding generations of models (e.g.: BIM). These unique capabilities have burgeoned by the virtue of 'New IT' allowing for unprecedented amounts of data to be collected and subsequently analyzed using state-of-the-art Big Data Analytics and AI techniques. DT is still in its infancy within the AECO industry, nonetheless, it shows a great potential across multiple applications and uses.

References

Akanmu, A., Anumba, C., Messner, J.: Scenarios for cyber-physical systems integration in construction. J. Inf. Technol. Constr. (ITcon) **18**(12), 240–260 (2013)

Al-Sehrawy, R., Kumar, B., Amoudi, O.: Exploitation of BIM in planning and controlling the construction phase on-site carbon emissions: a 6D BIM case study. Innov. Prod. Constr.: Transf. Constr. Emerg. Technol. **279** (2019)

Batty, M.: Digital twins. Environ. planning B: Urban Anal. City Sci. **45**(5), 817–820 (2018). https://doi.org/10.1177/2399808318796416. http://journals.sagepub.com/doi/abs/10.1177/2399808318796416

Bolton, A., Butler, L., Dabson, I., Enzer, M., Evans, M., Fenemore, T., Harradence, F., Keaney, E., Kemp, A., Luck, A., Pawsey, N., Saville, S., Schooling, J., Sharp, M., Smith, T., Tennison, J., Whyte, J., Wilson, A., Makri, C.: Gemini principles. Report (CDBB_REP_006) (2018). https://doi.org/10.17863/CAM.32260. Accessed 9 July 2019

Boschert, S., Heinrich, C., Rosen, R.: Next generation digital twin. In: Horvath, I., Suarez Rivero, J.P., Hernandez Castellano, P.M. (eds.) Proceedings of TMCE, Las Palmas de Gran Canaria, Spain (2018)

Boschert, S., Rosen, R.: Digital twin—the simulation aspect. In: Mechatronic Futures, pp. 59–74. Springer, Cham (2016)

British Standards Institution: Organization and digitization of information about buildings and civil engineering works, including building information modelling (BIM). Information management using building information modelling. Delivery phase of the assets (BSENISO19650) (2019). https://shop.bsigroup.com/ProductDetail?pid=000000000030389655

Chin, S., Yoon, S., Choi, C., Cho, C.: RFID + 4D CAD for progress management of structural steel works in high-rise buildings. J. Comput. Civ. Eng. **22**(2), 74–89 (2008)

Costello, K., Omale, G.: Gartner survey reveals digital twins are entering mainstream use (2019). https://www.gartner.com/en/newsroom/press-releases/2019-02-20-gartner-survey-reveals-digital-twins-are-entering-mai. Accessed 4 Nov 2019

Dawkins, O., Dennett, A., Hudson-Smith, A.: Living with a digital twin: operational management and engagement using IoT and mixed realities at UCL's Here East Campus on the Queen Elizabeth Olympic Park. GIScience Research UK (2018)

DebRoy, T., Zhang, W., Turner, J., Babu, S.S.: Building digital twins of 3D printing machines. Scr. Mater. **135**, 119–124 (2017)

Farsi, M., Daneshkhah, A., Hosseinian-Far, A., Jahankhani, H.: Digital twin technologies and smart cities (2019). https://doi.org/10.1007/978-3-030-18732-3

Garber, R.: BIM Design: Realising the Creative Potential of Building Information Modelling. Wiley, Hoboken (2014)

Gelernter, D.: Mirror Worlds: or the Day Software Puts the Universe in a Shoebox…How It Will Happen and What It Will Mean. Oxford University Press, Oxford (1991)

Glaessgen, E., Stargel, D.: The digital twin paradigm for future NASA and US Air Force vehicles. In: 53rd AIAA/ASME/ASCE/ AHS/ASC Structures, Structural Dynamics and Materials Conference 20th AIAA/ASME/AHS Adaptive Structures Conference 14th AIAA 1818 (2012)

Golparvar-Fard, M., Sarvarese, S., Pena-Mora, F.: Interactive visual construction progress monitoring with D4AR—4D augmented reality—models, building a sustainable future. In: Proceedings of the 2009 Construction Research Congress, pp. 41–50 (2009)

Grieves, M.: Product lifecycle management: the new paradigm for enterprises. Int. J. Prod. Dev. **2**(1/2), 71–84 (2005)

Grieves, M.: Product Lifecycle Management: Driving the Next Generation of Lean Thinking. McGraw-Hill, New York (2006)

Grieves, M.: Virtually Perfect: Driving Innovative and Lean Products Through Product Lifecycle Management. Space Coast Press, Cocoa Beach (2011)

Grieves, M., Vickers, J.: Digital twin: mitigating unpredictable, undesirable emergent behavior in complex systems (2017)

Hicks, B.: Industry 4.0 and digital twins: key lessons from NASA (2019). https://www.thefuturefactory.com/blog/24. Accessed 4 Nov 2019

Hochhalter, J.D., Leser, W.P., Newman, J.A., Glaessgen, E.H., Gupta, V.K., Yamakov, V., et al.: Coupling damage-sensing particles to the digital twin concept. NASA Center for AeroSpace Information, Hanover (2014)

Johnson Controls: Applying digital twins to the built environment. Johnson Controls (2019). https://www.johnsoncontrols.com/-/media/jci/insights/2019/bts/jci-661_dv_digital_twin_white_paper_020819_6p_f3.pdf

Klein, L.A.: Sensor and Data Fusion: A Tool for Information Assessment and Decision Making, vol. 324. SPIE Press, Bellingham (2004)

Kritzinger, W., Karner, M., Traar, G., Henjes, J., Sihn, W.: Digital twin in manufacturing: a categorical literature review and classification. IFAC-PapersOnLine **51**(11), 1016–1022 (2018). https://doi.org/10.1016/j.ifacol.2018.08.474

Motamedi, A., Hammad, A.: Lifecycle management of facilities components using radio frequency identification and building information model. In: ITcon, Special Issue Next Generation Construction IT: Technology Foresight, Future Studies, Roadmapping, and Scenario Planning, vol. 14, pp. 238–262 (2009)

National Infrastructure Commission: Data for the public good, National Infrastructure Commission, London, UK (2017)

Negri, E., Fumagalli, L., Macchi, M.: A review of the roles of digital twin in CPS-based production systems. Procedia Manuf. **11**, 939–948 (2017)

Parrott, A., Warshaw, L.: Industry 4.0 and The Digital Twin: Manufacturing Meets its Match. Deloitte University Press, Dallas (2017)

Perrow, C.: Normal Accidents: Living with High-Risk Technologies. Basic Books, New York (1984)

Piascik, R., Vickers, J., Lowry, D., Scotti, S., Stewart, J., Calomino, A.: Technology area 12: Materials, structures, mechanical systems, and manufacturing road map. NASA Office of Chief Technologist (2010)

Qiuchen Lu, V., Parlikad, A.K., Woodall, P., Ranasinghe, G.D., Heaton, J.: Developing a dynamic digital twin at a building level: using Cambridge campus as case study. In: International Conference on Smart Infrastructure and Construction 2019 (ICSIC) Driving Data-Informed Decision-Making, pp. 67–75. ICE Publishing (2019)

Rinaldi, S.M., Peerenboom, J.P., Kelly, T.K.: Identifying, understanding, and analyzing critical infrastructure interdependencies. IEEE Control Syst. Mag. **21**(6), 11–25 (2001)

Schleich, B., Anwer, N., Mathieu, L., Wartzack, S.: Shaping the digital twin for design and production engineering. CIRP Ann. **66**(1), 141–144 (2017)

Schroeder, G.N., Steinmetz, C., Pereira, C.E., Espindola, D.B.: Digital twin data modeling with AutomationML and a communication methodology for data exchange. IFAC-PapersOnLine **49**(30), 12–17 (2016)

Söderberg, R., Wärmefjord, K., Carlson, J.S., Lindkvist, L.: Toward a digital twin for real-time geometry assurance in individualized production. CIRP Ann. **66**(1), 137–140 (2017)

Sørensen, K.B., Christiansson, P., Svidt, K., Jacobsen, K., Simoni, T.: Ontologies to support RFID based link between virtual models and construction components. Comput.-Aided Civ. Infrastruct. Eng. **25**(4), 285–302 (2010)

Tao, F., Cheng, J., Qi, Q., Zhang, M., Zhang, H., Sui, F.: Digital twin-driven product design, manufacturing and service with big data. Int. J. Adv. Manuf. Technol. **94**, 3563–3576 (2017). https://doi.org/10.1007/s00170-017-0233-1

Tao, F., Qi, Q., Wang, L., Nee, A.Y.C.: Digital twins and cyber–physical systems toward smart manufacturing and Industry 4.0: correlation and comparison. Engineering **5**(4), 653–661 (2019). https://doi.org/10.1016/j.eng.2019.01.014

Tomko, M., Winter, S.: Beyond digital twins–A commentary. Environ. Plan. B: Urban Anal. City Sci. **46**(2), 395–399 (2019)

Urban innovation labs: Integration and optimisation of services embedded in the built environment (2019). CDBB_REP_21. https://doi.org/10.17863/CAM.40459. Accessed 4 Nov 2019

Vachálek, J., Bartalsky, L., Rovny, O., Šišmišová, D., Morháč, M., Lokšík, M.: The digital twin of an industrial production line within the Industry 4.0 concept. In: Proceedings of the 2017 21st International Conference on Process Control (PC), Štrbské Pleso, Slovakia, 6–9 June 2017, pp. 258–262. IEEE, Washington, DC (2017)

Wan, L., Yang, T., Parlikad, A.: City-level digital twin experiment for exploring the impacts of digital transformation on journeys to work in the Cambridge sub-region. (CDBB_REP_34) (2019). https://doi.org/10.17863/CAM.43317

Wildfire, C.: How can we spearhead city-scale digital twins? Infrastructure Intelligence (2018). www.infrastructure-intelligence.com/article/may-2018/how-can-we-spearhead-city-scale-digital-twins

Whyte, J., Chen, L., Gamble, C., Genes, C., Pierce, K., Fitzgerald, J., Coca, D., Mayfield, M., Babovic, F., Pedro, A., Shah, N.: Analysing systems interdependencies using a digital twin, 2018/19 General Project funded by CDBB, Final Report (2019)

Information Technologies Applications on the Construction Site

Digital Situation Picture in Construction – Case of Prefabricated Structural Elements

Rita Lavikka[1](✉), Pertti Lahdenperä[2], Markku Kiviniemi[2], and Antti Peltokorpi[3]

[1] VTT Technical Research Centre of Finland, Vuorimiehentie 3, 02150 Espoo, Finland
rita.lavikka@vtt.fi
[2] VTT Technical Research Centre of Finland, Visiokatu 4, 33720 Tampere, Finland
{pertti.lahdenpera,markku.kiviniemi}@vtt.fi
[3] Aalto University, Rakentajanaukio 4, 02150 Espoo, Finland
antti.peltokorpi@aalto.fi

Abstract. Efficient construction supply chain management necessitates that all parties possess an adequate situation picture (SP) of construction projects. Currently, an SP of project operations is formed by combining information from phone calls, emails and site visits. As a result, the scope, quality and accessibility of the information is weak and does not support supply chain management. Although several studies have been conducted on technologies supporting data capture, data exchange and analysis, a digital SP is still missing in the construction sector. This study aims to advance understanding of how to create a digital SP of construction operations. A digital SP in construction refers to an up-to-date context-specific machine-to-machine readable information, which can be used by different project parties to manage construction operations. The empirical study is conducted in the context of developing supply chain management of prefabricated structural elements by way of interviews and participatory workshops. The findings indicate that specific data, such as – the identification number of the prefabricated structural element, installation blocks, installation order and its schedule, transport loads and their orders and element statuses – should be shared in a digital format. The study describes inter-company data flows that are needed to create a digital SP employing seven use cases. These findings contribute to the SP literature in construction by mapping opportunities of a digital SP of construction for supporting efficient construction supply chain management.

Keywords: Digital situation picture · Prefabricated structural elements · Data integration

1 Introduction

Efficient construction supply chain management necessitates that all parties possess an adequate situation picture (SP) of the construction project and its operations. An SP is formed by combining data on the design, production plans and status of construction.

E. Toledo Santos and S. Scheer (Eds.): ICCCBE 2020, LNCE 98, pp. 943–958, 2021.
https://doi.org/10.1007/978-3-030-51295-8_65

Currently, an SP of construction operations is formed by combining information from meeting memos, phone calls, emails and site visits. As a result, the scope, quality and accessibility of the information is weak and does not support supply chain management [1].

A digital SP refers to an automatic, machine-to-machine, formulation of an SP. The data for forming the SP could be retrieved from several sources, such as building information modelling (BIM), companies' enterprise resource planning (ERP) systems, 3D scanning and photogrammetry and sensors that collect data for understanding, e.g. the location of materials and humidity conditions. A digital SP would allow a reliable and up-to-date SP that is accessible to all project parties at any time via their chosen medium, such as a mobile phone or a tablet. A digital SP would allow all parties to understand the current situation and forecast into the future to better plan their future activities [2]. Typical questions that could be answered through a digital SP are the following: Are we in schedule and budget? Is there a need to adjust plans due to actualised deviations? Have the preceding tasks been successfully executed? Are the needed tools and materials available to execute a task in the right location?

Several parties are interested in these questions during the project, especially on-site activities because the questions relate to the three cornerstones of project management - cost, time and quality. Unfortunately, technical challenges still prevent the formulation of a digital SP. For example, the lack of digital interpretable data and data in the wrong format are usually problems in creating a digital SP [3]. A digital SP is still missing in the construction sector.

Although several studies have been conducted on technological trials and solutions, they have not been implemented into the construction sector on a broader scale. One explanation might be that the previous studies have mainly focused on technological issues and somewhat ignored the contextual process, business model and people issues, even though these issues are crucial for technology implementation. For example, studies on BIM implementation have shown that the implementation of technology is a change management task that necessitates that process and people issues are taken into account [4, 5]. In practice, practitioners should accept the new technology by implementing it to construction practices and receive benefits from using the technology [6].

For this reason, this study will adopt a socio-technical approach acknowledging that the meaning and use of technology are negotiated by the practitioners [4, 7, 8]. In practice, the researchers will not try to implement a technology for data capture, data sharing and analysis but instead, focus on capturing the current practices for managing the construction supply chain. The understanding on the current practices will allow the researchers to reveal further development areas. After these research actions, the researchers will apply participatory methods to reveal use cases for efficient construction supply chain management based on a digital SP. The underlying logic is to make companies and people commit to the needed changes by involving them in planning and determining the future. The commitment of companies and people is vital because resistance to change can prevent the digitalisation of construction processes [9].

The study aims to advance understanding of how to create a digital SP of construction operations. The empirical study is conducted in the context of developing the supply chain management of prefabricated structural elements. These elements are engineered-to-order (ETO) which means that their production environment is full of uncertainty as these elements are only detailed after the customer has placed the order and delivery dates are set [10, 11]. The management of ETO products is intensive because their specification is demanding, their cost is high, and they are often delivered to the site in small batches [12]. The uncertainty could be reduced by digitally managing the information concerning the supply chain of ETO products [13]. This study aims to contribute to the SP literature in construction by describing the inter-company data flows through use cases that are needed for creating a digital SP supporting efficient construction supply chain management of prefabricated structural elements.

2 Theoretical Background

2.1 Studies on Creating a Digital SP of Construction Operations

A digital SP of construction operations is still a vaguely defined term. Some researchers state that an SP in construction defines the as-is conditions which help in the daily planning and controlling activities [2, 3]. In this study, a digital SP in construction refers to context-specific information that is compiled by algorithms from different data sources. In practice, all data needs to be machine-to-machine readable. All project parties can use the compiled information to manage construction operations. A digital SP allows a comprehensive understanding of the past, current and future construction activities: What has happened? What is happening? What will happen?

Two main arguments for a digital SP seem to be the optimisation of productivity in the form of reduced costs and duration and improved construction worker safety on-site [3, 10, 14, 15]. Other reasons for a digital SP are the need for a low level of inventories and the elimination of waste in the form of rework and unnecessary transport operations [10, 16].

Data needed to create a digital SP of construction operations can be retrieved by tracking and sharing the status of plans, prefabricated products and assembly between the project parties. Also, the location and status of construction resources, such as workers, materials, vehicles and tools, can be tracked and shared [15, 17]. The resources are naturally part of a construction activity that takes place at a specific location. Thus, one of the purposes of technology is to make resources location-aware [18].

Several research papers have aimed to create a digital SP in construction by testing reality capture technologies, such as 3D imaging [19], in-situ sensing [20, 21], electronic tagging [22–24] and global positioning system (GPS) with wireless web technologies [14]. BIM has also been integrated with a geographic information system (GIS) to provide a visual representation of the construction supply chain process for monitoring resources [25]. In conclusion, the purely technological studies seem to dominate the discussion on the creation of a digital SP of construction operations.

Only a few studies have acknowledged the need for understanding the socio-technical viewpoint where the interaction between technologies and humans' ways of working is considered essential. As an example, one study focused on logistics management, and combined BIM 4D modelling and lean principles, and as a result, they found out that synergies are created when the technology and process approaches are combined [16]. The researchers were able to support both collaborative planning meetings and control meetings with the 4D BIM technology [16]. Another study, focusing on the implementation of 3D modelling into the Danish construction sector, demonstrated that the critical enablers for the successful adaption of information technology were, among others, client demand and shared understanding on the implementation, such as how to exchange 3D models [8]. The research also revealed barriers to implementing information technologies: they were related to technologies, process and people. Technological barriers concerned the shortcomings of software or non-user-friendly interfaces. Process and people barriers concerned the different ambitions and aims of people, different working cultures, individual attitudes and lack of skills and experience. [8]

2.2 Forming a Digital SP for the Supply Chain Management of Prefabricated Structural Elements

The trajectory of structural elements consists of the phases of design, detailing, fabrication, transportation, installation and inspection [26]. An architect initiates the elements by designing the overall structure of a building, after which a structural engineer designs the structural elements. Then, the general contractor selects the element fabricator who, then, selects the factory and details the elements to fit the factory settings, and fabricates the elements, according to the schedule agreed with the contractor. After the transportation and possible short-term storage, the sub-contractor of the general contractor (or the fabricator) installs the elements, according to the schedule that is updated throughout the project following the real situation on the site. Finally, assembly contractor inspects the elements. In the best case, supply chain management related data from all of these phases would be available to all project parties. In reality, the supply chain management of prefabricated structural elements suffers from an incomplete SP; data access problems, duplicate information processing, information delays and inadequacy to react to quality issues are usual [27].

The structural elements are given several statuses during their lifecycle. The information related to the status of the element is currently mainly handled by email, phone and fabricator's ERP system [13]. Thus, the creation of the SP is a manual task, which is prone to human errors and takes time. As a result, the current information processing practice means that a digital SP for the supply chain management of prefabricated structural elements is missing, and cannot be used by the construction parties to plan and control their operations.

All construction project parties – owners, site workers, designers, engineers, fabricators and managers – would benefit from a digital SP concerning the status and quality of structural elements [26, 28]. The project parties should be provided with information that is relevant to them at that particular time, in that location and in that situation. Thus, up-to-date data should be provided in the form of context-specific information to allow data-based decision-making. For example, the structural designer might want to know whether design changes are still possible or whether the elements are already in production. General contractor, on the other hand, might want to know whether the elements are just about produced and ready for transportation in order to start the detail-level planning of the assembly activities.

The creation of a digital SP of prefabricated structural elements has been the topic of several research papers. These studies have approached the topic mainly from the technological viewpoint, creating and testing data retrieval technics and data analysis methods. For example, tracking construction resources has been a topic of several studies. Prefabricated structural elements have to be individually tracked since they are customised products. One study proposed an automated system that applies radio frequency identification (RFID) technology together with GPS technology to track prefabricated concrete components in a storage yard of a manufacturing plant [24]. Another study created an automated system to identify and localise engineered components on industrial sites and estimated that the steel erection productivity was improved a bit [29]. A third study followed the suggestion of a preceding study [13] and tracked prefabricated components to produce load plans quickly [10]. They included data from the 4D model, such as dimension and location, the date of assembly and production factories' status reports. A literature study, on the other hand, concluded that laser scanning, RFID, camera, GPS, GIS, and augmented reality (AR) were the technologies used for capturing and analysing real-time information from physical project processes [30]. These technology-oriented studies provide a good starting point for understanding data flows and use cases for efficient supply chain management of ETO products, but the studies seem to lack a trajectory perspective. Thus, no study has revealed all useful use cases for efficient supply chain management of ETO products.

3 Data Collection and Analysis

The empirical study was conducted in the context of developing the supply chain management of prefabricated structural elements in Finland. The current supply chain management is largely based on manual data collection and analysis, while the supply chain parties have expressed their need for a digital SP.

The starting point for the empirical study was the theoretical understanding that currently available technologies, construction processes and people need to be aligned to succeed in enabling a digital SP. For this reason, the research adopted an interpretive approach because it helps to understand and describe social processes and complex factors [31, 32].

Researchers interviewed Finnish companies representing different disciplines: construction, structural element design and fabrication to understand the needs and expectations of different supply chain parties for a digital SP. Each interview lasted between one to two hours, and they were audio-recorded. Researchers also took detailed notes during the interviews. The semi-structured interviews were designed to reveal the process of managing the design, production and assembly of prefabricated structural elements. Snowball sampling was applied in the selection of the interviewees [33].

In addition to interviews, participatory action workshops were organized to create a shared understanding of what information should be shared between the parties to create a digital SP. Another aim of the workshops was to understand whether the parties' already possessed information that could be integrated to form a digital SP for managing the supply chain of prefabricated structural elements. Also, it was discussed when and how real-time status data of the elements should be gathered for the efficient supply chain management. Finally, also, the obstacles and enablers for information sharing were discussed. The research efforts also supported understanding the needed developments for BIM-based supply chain management of ETO structures. These findings will be reported in another paper due to word limits.

The selection of participatory action workshops as a data collection method was based on the idea that humans tend to better commit to those decisions they have been making [34]. Table 1 provides details on the practitioners involved in the various types of activities identified in the leftmost column: interviews (I), group interviews (GI) and workshops (W). The table also presents the focus of each interview and workshop carried out during 2019. As a whole, the interviews and workshops allowed the researchers to model and validate the current supply chain management process of prefabricated structural elements, and the needed digital data flows to enable a digital SP of these elements. Two or three researchers attended the interviews and workshops to increase the validity of data collection.

The study followed the recommendations of Miles and Huberman [35] in the analysis of the interview data. First, the researchers read the detailed notes from the interviews and workshops and listened to the audio recordings when details needed to be checked. Then, the researchers read and underlined sentences that seemed relevant, mainly relating to the supply chain management of prefabricated structural elements. These sentences concerned current challenges, data flows, intra-organizational and inter-organizational tasks and development ideas. Four researchers conducted the data analysis, and the findings of the researchers were compared to receive a consensus on the meaning and relevance of data. Data collected from the workshops was mainly in the form of notes and post-it notes. Finally, the researchers visualised the findings from interviews and workshops into use cases and a process model that describes how supply chain management related data flows in the current situation and how it should flow in a digital format.

Table 1. Details of interviewees and participants of the workshops.

Activity	Focus	Organization(s)	Title(s)	Date
Interview	Parties' needs and expectations for a digital SP	Structural designer	Technology manager	15.2
Interview		Structural steel fabricator	Project engineer	12.3
Interview		Structural steel fabricator	Project manager	12.3
Interview		Structural steel fabricator	Production manager	12.3
Interview		Structural steel fabricator	Design manager	12.3
Interview		Precast concrete fabricator	Sales manager	19.3
Interview		Precast concrete fabricator	Project manager	19.3
Interview		Precast concrete fabricator	Production manager	19.3
Group interview	Use cases for efficient construction supply chain management	Structural steel fabricator Precast concrete fabricator Structural designer	Director of project management IT solution specialist Development manager	6.6
Group interview	Obstacles and enablers for information sharing	Structural steel fabricator Precast concrete fabricator	Director of project management IT solution specialist	28.6
Group interview		IT technology provider IT technology provider IT technology provider	Product manager Digital construction manager Software architect	28.8
Workshop	Integration of existing information for status and schedule sharing	General contractor General contractor General contractor IT technology provider IT technology provider	Technology manager Business development manager Digital director Digital construction manager Software architect	3.10
Group interview	Perspective of on-site workers and managers	General contractor General contractor General contractor	Business development manager Site manager Responsible site manager	7.10

(*continued*)

Table 1. (*continued*)

Activity	Focus	Organization(s)	Title(s)	Date
Group interview	Business perspective	Structural steel fabricator Structural steel fabricator	R&D manager Technology director	23.10
Workshop	Application program interfaces	IT technology provider IT technology provider General contractor General contractor	Digital construction manager Software architect Business development manager Digital director	28.10
Workshop	Process model and digital data flow validation	IT technology provider General contractor Precast concrete fabricator Structural steel fabricator	Digital construction manager Business development manager IT solution specialist Director of project management	8.11
Workshop	Use of status data for the efficient supply chain management	IT technology provider General contractor General contractor Precast concrete fabricator Precast concrete fabricator	Digital construction manager Business development manager Digital director Project manager Development manager	2.12

4 Findings

4.1 A Process Model of the Data Flows in the Supply Chain Management of Prefabricated Structural Elements

The interviews and workshops confirmed that data on the status, schedule and quality of prefabricated structural elements are not always digitally collected, let alone shared in a digital format between companies. The informants mentioned that no standardised process of managing these data exists. However, the informants were willing to start sharing and utilising more digital data. They had realised that the supply chain management process is not that smooth as it could be if data on status, schedule, and quality of prefabricated structural elements were available in a digital format and guiding the operations of the parties. The informants hoped that the construction sector would develop and agree on a data model for managing these data. The data model should define the formal naming of data and relationships between different data points to enable machine-to-machine data flows. The informants considered it essential that instead of point-to-point data interfaces, open and commonly agreed data models would be used for information exchange in a domain where stakeholders are changing project by project.

The informants, together with researchers, envisioned methods to collect and share data on the status, schedule and quality of prefabricated structural elements. Sensor technology was considered playing a crucial part in providing data on the locations of materials, tools, and site workers. Image data, on the other hand, was envisioned to be used for checking the quality and installation location of elements. As an example, drones could be used to get accurate point clouds of installations, which could then be compared to BIM models to detect any differences between them. Similar results were envisioned to be achieved with crane-mounted cameras, although with lower accuracy. Data during transportation and installation could be collected through GPS. The informants agreed that supply chain management data should be collected and shared as automatically as possible to avoid time lag and human mistakes. The transfer of data necessitates open application programming interfaces (API) between the parties' information management systems.

Figure 1 depicts a process model presenting, firstly, the current data flow in the supply chain management of prefabricated structural elements. The figure also points to the needed data flows (numbered from one to seven in the process model) that should be in a digital format in the future.

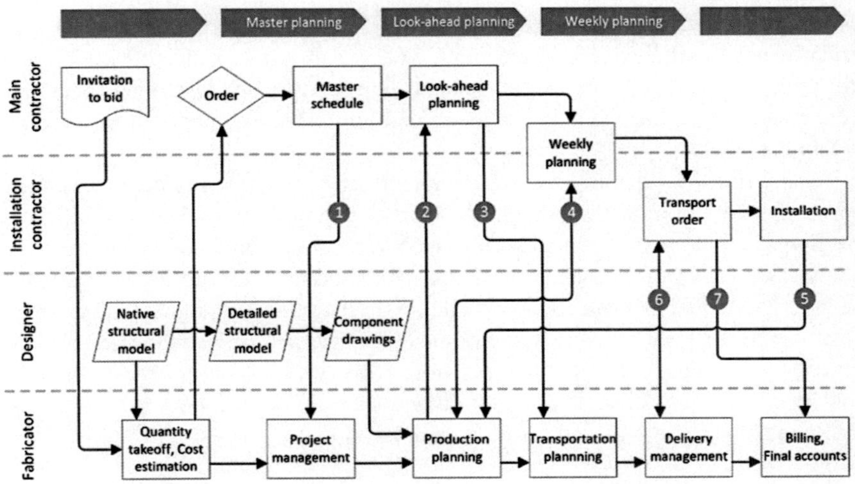

Fig. 1. The data flow in the supply chain management of prefabricated structural elements. Numbers point to the needed data flows in a digital format.

The process model includes four parties: main contractor, installation contractor, designer and fabricator. The process is divided into three indicative phases: The first phase is the master planning, which ends up with a master schedule defined by the main contractor. Look-ahead planning forms the second phase that forecasts work to be made in the near future, six to eight weeks ahead. The third phase, weekly planning, focuses on day-to-day tasks over one week.

The process model is built on the presumption that the fabricator can apply a native structural model, which is used to calculate the quantities of the elements according to their type and according to the installation blocks defined by the fabricator. As the process is currently tied to a native structural model, data transfer is not entirely open. The change to an IFC-based process would require significant development efforts. IFC stands for Industry Foundation Classes, which is a data model that describes architectural, building and construction sector data in an open file format.

4.2 Use Cases Supporting Efficient Supply Chain Management of Prefabricated Structural Elements

Based on the interviews, seven use cases supporting efficient supply chain management of prefabricated structural elements were identified. The use cases aim to cover the whole trajectory of the elements. The use cases are based on the idea that needed data would flow between the project parties in machine-to-machine format. Efficient supply chain management brings project benefits. For example, the general contractor could improve its planning accuracy if it had accurate schedule data on the delivery of the elements from the fabricator. To enable just-in-time delivery from the site's perspective, the fabricator needs up-to-date assembly schedule data from the site manager. Based on the assembly data, the fabricator can optimise the schedule of the element production process.

Another example of the benefit of sharing inter-organizational data in a digital format relates to site delivery and control. The site manager could provide the fabricator with data on the quality of the elements before assembly. Up-to-date data on the element's quality at the site is an essential input to the quality management process in the fabricator's factory.

Table 2 describes the use cases (numbered from one to seven in the process model in Fig. 1) and the data that should flow in a digital format. The interviews and workshops indicated that specific data, such as – the identification number of the prefabricated structural element, installation order and times of blocks, transport loads and their orders and element statuses – should be shared in a digital format to create a digital SP that allows efficient supply chain management of prefabricated structural elements. These data could be collected from BIM (Building information modelling), simulations, ERP systems, Internet of Things (IoT) sensors, indoor positioning systems, photos, point clouds and onsite communication.

Table 2. Use cases and their needed digital data. The number refers to the number in the process model.

Use case • Type of data	Detailed description
1. Optimisation of the schedule of designing and producing elements • Installation block breakdown structure • Installation order of blocks • Installation times of blocks	The element fabricator manages production at granular planning level by dividing the full delivery by expected installation blocks at the site and collecting the installation timing data from the master schedule. Based on these data, the fabricator can optimise the schedule of designing and producing the elements in the multi-project fabrication environment Currently, the master schedule does not usually specify different blocks within the entirety. The fabricator and the general contractor may define the order of block realisation differently. The start and end dates of elements' assembly are not shared from the master schedule to the fabricator. The lack of these dates complicates the just-in-time production planning at the factory
2. Optimisation of the look-ahead planning • Status of an element • The identification number of an element	The contractor needs from the fabricator the data concerning the status – in design, in production, delivered – of the elements in order to optimise the look-ahead planning. These data are critical for the site managers as they plan their site production schedule. The fabricator could share this information through its enterprise resource planning (ERP) system Also, the identification number of elements needs to be digitally shared throughout the supply chain. These data should be shared in a generic format agreed by all fabricators, which would make it easier for the contractors to apply the data in their information systems Currently, the identification number of the elements is not shared in a digital format. However, some fabricators include a Quick Response (QR) code to each element. The QR code includes the identification number of the element. The QR code remains hidden after the assembly of the element, and thus cannot be read Some fabricators share the status of the elements through their ERP system, but the general contractor needs to log into a proprietary system to get the data, and the data format is not standard

(*continued*)

Table 2. (*continued*)

Use case • Type of data	Detailed description
3. Planning of the shipment of elements in transportation • Installation order of elements in planned blocks	The fabricator needs to receive data concerning the installation order of elements in planned blocks from the contractor. Based on these data, the fabricator can better plan the shipment of the elements for transportation Currently, the planning of the shipment is made in the same order the elements were ordered from the fabricator, which does not necessarily support the installation order of the elements on site. In some cases, the information in installation order is available, but even then, ad hoc practices are followed instead of sharing digital interpretable data
4. Delivery capacity confirmation • Element status data • Installation plan of the next 1–2 week	The contractor needs to receive element status data from the fabricator to confirm delivery capacity on time to keep the project schedule. On the other hand, the fabricator needs look-ahead planning data from the contractor to forecast when transportation capacity is needed Currently, the contractors do not usually share the detailed installation plan with the fabricator in a digital format
5. Planning of subsequent shipments • Systematic installation status recordings	The fabricator needs to receive data on realised installations from the assembly contractor for synchronizing production and planning of subsequent shipments. The method for recording the used statuses needs to be standardised and system-independent because elements can come from multiple fabricators. Standardisation would allow the supply chain data to be analysed and distributed automatically, in a transparent way Currently, the fabricator does not receive data concerning the realised installations. Data needed could cover whether the elements were of good quality and whether they have successfully been installed
6. Standardised transportation management • Transport order and receipt	The parties need a standardised transportation order method, and transport orders and receipts should be handled in a digital format to allow a better digital SP of the elements' statuses during transportation Currently, transportation orders are made using email, and only the sender and receiver know about the communication. Usually, one worker receives the materials, and other onsite workers do not necessarily receive the information that materials are on site

(*continued*)

Table 2. (*continued*)

Use case • Type of data	Detailed description
7. Issue management • Issue	Contractors' issues and reclamations concerning the elements should be handled in a digital format and be shared with other companies when necessary. The digitalisation of issue management would enable learning from mistakes and avoiding them in the future as well as facilitating the commercial procedures between the parties. When enough issue and reclamation data would be gathered, a more detailed analysis could be performed by algorithms which could lead to learning in the long run Currently, issues are handled manually, through emails, and their processing takes time, which means that only certain people know about the issues and their root causes are hard to track

5 Discussion and Conclusions

The study aimed to advance understanding of how to create a digital SP of construction operations. The empirical study was conducted in the context of developing the supply chain management of prefabricated structural elements. The SP of a construction project must also contain data of upstream operations of the supply chains. The findings reveal that the sharing of digital data concerning the status, schedule and quality of prefabricated structural elements is not yet common in the Finnish construction projects although the construction stakeholders, especially interviewed fabricators manage the data with ERP systems and BIM-based planning tools.

The primary method for SP in fabricators' supply management is maintaining a status value of individual components. Specific tracking technologies are not yet used to collect the needed data. However, the companies participating in the study have identified several tracking technologies, such as sensor technologies and GPS, which could be used in the future.

Several studies have focused on creating the technological prerequisites for an SP that would allow efficient supply chain management of structural elements [13, 16, 24, 26–28]. Despite the numerous theoretical and empirical studies, the developed technological solutions of up-to-date data sharing are not yet part of the supply chain management of prefabricated structural elements.

This study adopted a socio-technical approach [8, 16] that allowed focusing on process and people issues, in addition to technological issues. As a result, this study adds new knowledge to SP literature in construction by reporting inter-company data flows that are needed to create a digital SP employing seven use cases that support efficient supply chain management of prefabricated structural elements. What is different from previous study efforts is that the use cases cover the whole trajectory of the elements, which hopefully will lead to a more integrated approach in implementing the digital SP for managing the supply chain, which has been called for [12].

The study provides implications to practice. An open data model is needed to enable data transfer, concerning the status, production schedule and quality of structural elements, between several project parties. The data do not have to reside in a common data environment [7], but it should be retrievable from several linked data sources. The major issue is how to develop and widely disseminate the needed data model and processes in the project-based domain at national and international levels. From the research perspective, new approaches should be applied like results of ecosystem development and innovation management research.

Interestingly, the informants did not express any significant financial concerns related to supply chain management. One explanation is that the informants had not thought about them, or they considered that the expenses would not be huge because these organizations represented the largest ones in Finland. For example, small and medium-sized companies might not have the funds to bear the costs associated with data collection, analysis and sharing services [7]. Another explanation is that the study did not explicitly focus on financial issues. However, it may be possible that the current conventional project contracts need to be changed to enable flexible changes in the delivery schedules of elements. At the moment, the schedules of deliveries are fixed in the contracts, so no party wants to deliver late and pay penalties.

Commonly, the construction schedules are relatively broad and detailed planning is done when the work proceeds. This practice is natural due to the challenges related to a one-time nature of a project, and when this is the situation, the contractor expects the fabricator work flexibly. Delays and deviations are common, but the contractor is not willing to assume the further risk of delay by reporting the fabricator the precise situation of the site assembly. The risk is that the fabricator postpones the fabrication while any hardship would further delay the assembly. Nevertheless, the fabricator may be in trouble due to uneven workload while serving numerous projects, and the unnecessary rush may cause it to use outsourcing. This vicious cycle may cause extra costs, and when the delay becomes evident, temporary storing causes further troubles. This, again, is an additional risk from the viewpoint of quality and late changes.

This study has only surfaced the topic of creating a digital SP that would allow more efficient supply chain management of prefabricated structural elements. The study mapped opportunities for enabling a digital SP, but it has not empirically tested any use cases in a real project environment, which can be considered as a limitation of the study. The authors of this study will focus their further research endeavours on conducting a pilot study for collecting the needed digital data for efficient supply chain management of prefabricated elements.

The empirical data is collected in the context of managing the prefabricated element supply chain, which poses another limitation to the study. Research on creating a digital SP in construction operations should also be conducted in other similar contexts where the use of digital data is still limited. The study did not focus on capturing the digital situation picture of the design phase of ETO products. This topic also deserves to be further scrutinized.

References

1. Vrijhoef, R., Koskela, L.: The four roles of supply chain management in construction. Eur. J. Purch. Supply Manag. **6**, 169–178 (2000)
2. Kärkkäinen, R., Lavikka, R., Seppänen, O., Peltokorpi, A.: Situation picture through construction information management. In: Lill, I., Witt, E. (eds.) 10th Nordic Conference on Construction Economics and Organization (Emerald Reach Proceedings Series), vol. 2, pp. 155–161. Emerald Publishing Ltd. (2019)
3. Akinci, B.: Situational awareness in construction and facility management. Front. Eng. Manag. **1**, 283–289 (2014)
4. Schweber, L., Harty, C.: Actors and objects: a socio-technical networks approach to technology uptake in the construction sector. Constr. Manag. Econ. **28**, 657–674 (2010)
5. Khosrowshahi, F., Arayici, Y.: Roadmap for implementation of BIM in the UK construction industry. Eng. Constr. Archit. Manag. **19**, 610–635 (2012)
6. Voordijk, H., Adriaanse, A.: Engaged scholarship in construction management research: the adoption of information and communications technology in construction projects. Constr. Manag. Econ. **34**, 536–551 (2016)
7. Adamu, Z.A., Emmitt, S., Soetanto, R.: Social BIM: co-creation with shared situational awareness. J. Inf. Technol. Constr. **20**, 230–252 (2015)
8. Moum, A., Koch, C., Haugen, T.I.: What did you learn from practice today? Exploring experiences from a Danish R&D effort in digital construction. Adv. Eng. Inform. **23**, 229–242 (2009)
9. Ratajczak, J., Riedl, M., Matt, D.T.: BIM-based and AR application combined with location-based management system for the improvement of the construction performance. Buildings **9**, 1–17 (2019)
10. Bataglin, F.S., et al.: Model for planning and controlling the delivery and assembly of engineer-to-order prefabricated building systems: exploring synergies between Lean and BIM. Can. J. Civ. Eng. **47**(2), 165–177 (2020). https://doi.org/10.1139/cjce-2018-0462
11. Wesz, J.G.B., Formoso, C.T., Tzortzopoulos, P.: Planning and controlling design in engineered-to-order prefabricated building systems. Eng. Constr. Archit. Manag. **25**, 134–152 (2018)
12. Li, N., Becerik-Gerber, B.: Life-cycle approach for implementing RFID technology in construction: learning from academic and industry use cases. J. Constr. Eng. Manag. **137**, 1089–1098 (2011)
13. Ergen, E., Akinci, B.: Formalization of the flow of component-related information in precast concrete supply chains. J. Constr. Eng. Manag. **134**, 112–121 (2008)
14. Oloufa, A.A., Ikeda, M., Oda, H.: Situational awareness of construction equipment using GPS, wireless and web technologies. Autom. Constr. **12**, 737–748 (2003)
15. Cheng, T., Teizer, J.: Real-time resource location data collection and visualization technology for construction safety and activity monitoring applications. Autom. Constr. **34**, 3–15 (2013)
16. Bortolini, R., Formoso, C.T., Viana, D.D.: Site logistics planning and control for engineer-to-order prefabricated building systems using BIM 4D modeling. Autom. Constr. **98**, 248–264 (2019)
17. Zhong, R.Y., Peng, Y., Xue, F., Fang, J., Zou, W., Luo, H., Thomas Ng, S., Lu, W., Shen, G.Q.P., Huang, G.Q.: Prefabricated construction enabled by the Internet-of-Things. Autom. Constr. **76**, 59–70 (2017)
18. Edirisinghe, R.: Digital skin of the construction site - smart sensor technologies towards the future smart construction site. Eng. Constr. Archit. Manag. **26**, 184–223 (2019)

19. Turkan, Y., Bosché, F., Haas, C.T., Haas, R.: Toward automated earned value tracking using 3D imaging tools. J. Constr. Eng. Manag. **139**, 423–433 (2013)
20. Turkan, Y., Bosche, F., Haas, C.T., Haas, R.: Automated progress tracking using 4D schedule and 3D sensing technologies. Autom. Constr. **22**, 414–421 (2012)
21. Zhao, J., Seppänen, O., Peltokorpi, A., Badihi, B., Olivieri, H.: Real-time resource tracking for analyzing value-adding time in construction. Autom. Constr. **104**, 52–65 (2019)
22. Ko, H.S., Azambuja, M., Felix Lee, H.: Cloud-based materials tracking system prototype integrated with radio frequency identification tagging technology. Autom. Constr. **63**, 144–154 (2016)
23. Ghimire, S., Luis-Ferreira, F., Nodehi, T., Jardim-Goncalves, R.: IoT based situational awareness framework for real-time project management. Int. J. Comput. Integr. Manuf. **30**, 74–83 (2017)
24. Ergen, E., Akinci, B., Sacks, R.: Tracking and locating components in a precast storage yard utilizing radio frequency identification technology and GPS. Autom. Constr. **16**, 354–367 (2007)
25. Irizarry, J., Karan, E.P., Jalaei, F.: Integrating BIM and GIS to improve the visual monitoring of construction supply chain management. Autom. Constr. **31**, 241–254 (2013)
26. Ikonen, J., Knutas, A., Hämäläinen, H., Ihonen, M., Porras, J., Kallonen, T.: Use of embedded RFID tags in concrete element supply chains (2013)
27. Törmä, S., Backman, J., Kiviniemi, M., Aro, J., Nissilä, J.: Three methods for exchanging status information of building elements. In: Martens, S., Mahdavi, Scherer (eds.): eWork and eBusiness in Architecture, Engineering and Construction, pp. 173–180. Taylor & Francis Group, London (2014)
28. Jussila, A., Kiviniemi, M., Talvitie, U.: Piloting a new information sharing method in a construction supply chain. In: eWork and eBusiness in Architecture, Engineering and Construction - Proceedings of the European Conferences of Product and Process Modelling, ECPPM 2012, pp. 707–712 (2012)
29. Grau, D., Caldas, C.H., Haas, C.T., Goodrum, P.M., Gong, J.: Assessing the impact of materials tracking technologies on construction craft productivity. Autom. Constr. **18**, 903–911 (2009)
30. Chen, K., Lu, W., Peng, Y., Rowlinson, S., Huang, G.Q.: Bridging BIM and building: from a literature review to an integrated conceptual framework. Int. J. Proj. Manag. **33**, 1405–1416 (2015)
31. Schwandt, T.A.: Constructivist, interpretivist approaches to human inquiry. In: Denzin, N. K., Lincoln, Y.S. (eds.) Handbook of Qualitative Research, pp. 118–136. Sage Publications, Thousand Oaks (1994)
32. Merriam, S.B.: Qualitative Research: A Guide to Design and Implementation. Jossey-Bass, San Francisco (2009)
33. Biernacki, P., Waldorf, D.: Snowball sampling: problems and techniques of chain referral sampling. Sociol. Methods Res. **10**, 141–163 (1981)
34. Argyris, C., Schön, D.A.: Participatory action research and action science compared. Am. Behav. Sci. **32**, 612–623 (1989)
35. Miles, M.B., Huberman, A.M.: An Expanded Sourcebook – Qualitative Data Analysis. Sage Publications, California (1994)

Dynamic Crane Workspace Update for Collision Avoidance During Blind Lift Operations

Leon C. Price[1], Jingdao Chen[1], and Yong K. Cho[2(✉)]

[1] Institute for Robotics and Intelligent Machines, Georgia Institute of Technology, 777 Atlanta Dr. N.W, Atlanta, GA 30332-0355, USA
{lprice8, jchen490}@gatech.edu

[2] School of Civil and Environmental Engineering, Georgia Institute of Technology, 790 Atlanta Dr. N.W, Atlanta, GA 30332-0355, USA
yong.cho@ce.gatech.edu

Abstract. Cranes are associated with high levels of accidents and struck-by fatalities in many industrial settings. This is due to the congested nature of crane operation sites and the high cognitive load for the crane operator to track the load position throughout the lifting operation, especially during blind lifts. Current research on crane safety assistance systems focuses on preventing collisions with static entities such as existing building structures but do not adequately handle dynamic entities such as workers and movable containers. This research proposes a dynamic crane workspace updating method for collision avoidance during blind lift operations. The position and orientation of the crane load, as well as the position and orientation of surrounding obstacles, are automatically tracked and updated in a 3D crane workspace model during lifting operations. The load base position is first estimated using forward kinematics obtained from an encoder system. Then the load swing and load rotation are corrected for using a vision-based load detection algorithm. A static map of surrounding obstacles is initially obtained using 3D laser scanning. The map is then updated over the course of a lifting operation through semi-automated obstacle placement. In addition, a vision-based worker detection algorithm is used to track the position of workers in proximity to the load. Finally, the crane workspace is updated in real-time for the crane operator through a 3D user interface and warnings are issued whenever a potential collision scenario is detected. The proposed method was evaluated at a crane yard where multiple blind lift scenarios were carried out. Experimental results indicate that the proposed method can potentially improve the situational awareness of the crane operator for blind lift operations.

Keywords: Crane · Blind lift · Collision avoidance · Computer vision

1 Introduction

Cranes are some of the most used pieces of equipment for handling large loads and materials in industrial settings. They are used for lifting and horizontal movements, and they are often a key component of many operations. Unfortunately, cranes are also

E. Toledo Santos and S. Scheer (Eds.): ICCCBE 2020, LNCE 98, pp. 959–970, 2021.
https://doi.org/10.1007/978-3-030-51295-8_66

associated with high levels of accidents and fatalities. Depending on the environment, operators may be subject to circumstances where they have no line of sight to the load that they are moving, which is classified as a blind lift. Blind lifts have been pegged as one of the most dangerous by the industry and pose the most risk for accidents with personnel and for damage to the load. For example, offshore crane operators are often required to perform blind lifts on a regular basis. Loads are delivered to the offshore facilities by boats which are down at sea level while the cranes are on the deck. When lifting the loads off the delivery boats, the first part of the lift is completely blind, and the crane operators must rely on instructions from a spotter via radio communication.

In order to minimize the risks during blind lifts, the following paper proposes a methodology which provides the operator with a 3D visualization that is updated in real-time to help guide the lift and provide warnings for potential collision with objects in the environment. There are three main components of the dynamic update to this system: 1) load swing and rotation estimation 2) worker detection and tracking and 3) the updating of dynamic assets in the workspace. This paper follows with a review of previous related work. Then it outlines the methodology which includes a 3D reconstruction of the static workspace, load position monitoring as well as novel approaches for vision-based load sway monitoring and a method for dynamically updating the workspace. The results of the proposed methodology are then presented based on a field test at a manufacturing yard for offshore cranes. Finally, the novel additions to this branch of research are discussed as well as its shortcomings and potential future work.

2 Literature Review

Crane safety assistance systems can come in many forms including lift planning [1], collision avoidance [2], and tracking of moving entities such as workers [3]. One simple and cost-effective approach is to install cameras, whether in the form of Closed-Circuit Television (CCTV) systems [4] or camera-equipped Unmanned Aerial Vehicles (UAV) [5] to monitor the crane surroundings. However, the downside of this method is that cameras lack depth perception and cannot accurately estimate the distance between the crane and a potential obstacle. Another line of research uses a 3D point cloud to represent the crane workspace for the purposes of collision avoidance [6–8]. These methods focused on collision avoidance in simple cases where the surrounding environment is assumed to be static. The load is also assumed to be in a fixed orientation. However, the crane workspace during an actual lift scenario is usually highly dynamic due to load swing, load rotation as well as the movement of workers and surrounding equipment. [9] tracked dynamic assets on a crane workspace in a real-time setting but only worked with very specific objects such as workers and cars. It also did not account for the movement of the crane load. [10] performed vision-based load sway estimation but did not account for the movement of surrounding objects.

Tracking and detection of workers as well as other equipment at construction sites have been done largely using wireless technology such as Radio Frequency Identification (RFID), Bluetooth Low Energy (BLE)-based location technology and Global Positioning System (GPS) [11, 12]. However, computer vision-based methods have also been utilized. [13, 14] used computer vision feature detection methods and background suppression to track and detect construction workers from a static camera overlooking the construction site.

Another method for collision avoidance is to outfit the load with sensors. [15] proposes utilizing Cyber Physical Systems for both planning and monitoring with an Inertial Measurement Unit (IMU) attached to the load. However, the framework was not tested on a physical system. Additionally, [8] used an IMU for load position and sway monitoring. [16] demonstrated a successful blind lift using a variety of sensors including a CCTV camera, ultrasonic sensors, a gyroscope and laser distance sensor mounted to the load for a high capacity blind lift. While a viable method for certain lifts, often, attaching sensors to the load is not an option.

3 Methodology

This research proposes a dynamic crane workspace updating method for collision avoidance during blind lift operations. The position and orientation of the crane load, as well as the position and orientation of surrounding obstacles, are automatically tracked and updated in a 3D crane workspace model during lifting operations. The crane workspace is displayed to the crane operator through a Unity 3D (a gaming engine developed by Unity Technologies) interface and alerts are automatically given whenever a potential collision is detected.

Figure 1 below outlines the overall approach; the crane encoder, which measure the different crane poses, data streams combined with the feed from a camera mounted at the boom tip are utilized to monitor the position of the load. The camera also detects workers in the environment. The workers and the load position are then updated real-time in the aforementioned Unity interface. The operator also has the ability to manually update the interface with new obstacles that may have entered the workspace. Then the system dynamically monitors collisions between the obstacles, workers and crane during the lift and gives a warning to the crane operator when the entities come within a certain proximity threshold.

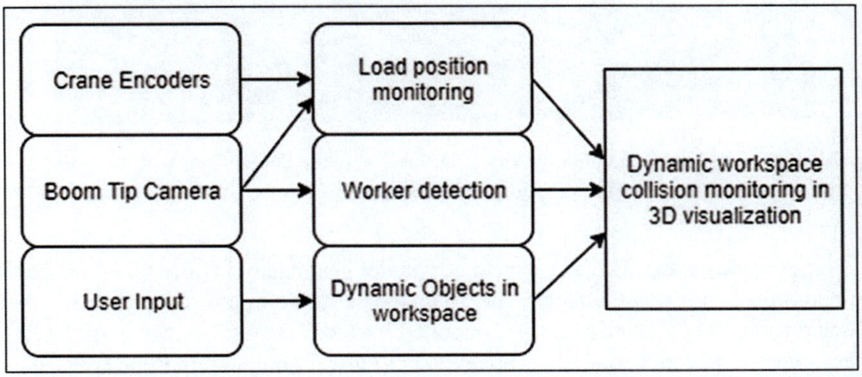

Fig. 1. Overall framework for dynamically updating the visualization workspace for collision monitoring.

3.1 3D Reconstruction of Static Crane Workspace

3D point clouds of the workspace were captured to model and monitor static objects in the visualization. Figure 2 below shows the 3D point clouds generated by two different methods of acquiring 3D point clouds. The first method uses a terrestrial 3D laser scanner along with registration key point objects. Scans were taken at 8 different locations. Then the individual scans were registered in post processing. In this study, the scans were registered manually but automated registration methods [17, 18] can be used as well. In the second method, an unmanned aerial vehicle (UAV) captured 89 photo images during 8 min of flight time. Using Structure-from-Motion [19], these photos were used to form a sparse but larger 3D point cloud. Both point clouds were used for reference and to reconstruct the 3D Unity visualization.

Fig. 2. 3D point cloud generated by terrestrial laser scanner (TLS) (top) and 3D point cloud generated by Unmanned Aerial Vehicle (UAV) (bottom).

Figure 3 shows the 3D crane workspace model generated in Unity based on the 3D geometry from the point clouds. The static objects, the crane and the loads were modeled using 3D CAD software and imported into the Unity scene. Some of the static objects were modeled as bounding boxes to help lower computation time when doing collision detection in Unity 3D.

Fig. 3. 3D crane workspace model with bounding boxes around static obstacles

3.2 Crane Load Pose Estimation

The position and orientation of the crane load is updated dynamically throughout the lift operation using a combination of multiple encoders and a camera (Fig. 4). The crane position estimation utilized the onboard encoders installed by the crane manufacuturer. The slew encoder was mounted on the top of the kingpost and monitored the angular rotation about the kingpost. The boom angle was measured by an angle sensor that was mounted up near the tip of the boom. The rope payout, which is the cable length from the boom tip to the load, was monitored by the encoders on the large cable spools.

Fig. 4. Sensor setup used to estimate pose of the crane load: camera mounted on the boom tip and encoders that measure the boom angle, slew angle, and hoist cable length

Key coordinate transforms were selected in the Unity scene to specify each axis of rotations for the crane model. After reading the sensor values, the boom angle and slew angle were updated in the visualization, and the Unity backend software handled the forward kinematics to give the position of the boom tip. There were some scaling factors applied to convert the raw encoder values into radians, which are required as the input for the visualization. Additionally, calibration and set-up were required to initially align the visualization with the real-world.

Then the load position is approximated using a vision system to calculate the swing and rotation of the load relative to the boom tip. Given the boom tip location from the encoder angles, we are then able to specify the global position of the load in the visualization. The vision system for monitoring load swing and rotation uses images retrieved from the camera mounted on the boom tip. Initially, a Hue-Saturation-Value (HSV) color threshold is applied to the image; then corners and the center of the load are found. In Fig. 5, the corners are marked in green while the center of the load is marked in red. The offset in pixels of the load center from the neutral hang position (which is set while the crane and load are static) is calculated. Then, using the load payout value from the rotary encoders on the cable spool, the pixel value is converted into meters and updates the load position in the Unity visualization.

Fig. 5. Visualization of workers (blue dots) detected in proximity to the crane load (green dots)

3.3 Vision-Based Worker Detection

Computer vision methods are used to detect workers in proximity to the crane load which may constitute collision hazards. Using the live video feed from the boom-mounted camera, the crane workspace monitoring system situated inside the crane cabin can perform worker detection and tracking and convey the results to the crane operator. The proposed method takes advantage of the fact that workers usually wear bright-colored hard hats on the jobsite. First, a circle detection algorithm is applied to the input image to obtain the outline of the hard hats. Circles that are too big or too small are filtered out. Next, the mean-shift algorithm is used to merge overlapping circles and prevent duplicate detections. Finally, an intensity-based filter is applied to only keep bright circles which correspond to the workers' hard hats. The center of each remaining circle is projected into 3D space according to the camera parameters and stored as the worker position.

The worker positions are also tracked from frame to frame by direct association. Each worker detected in the current image frame is associated with the nearest neighbor in the previous image frame. Any new detections that are not associated with the previous frame are added to the list of workers (e.g. when a worker enters the crane workspace for the first time). Similarly, any previous detections that are not associated with the current frame are removed from the list of workers (e.g. when a worker leaves the crane workspace).

Figure 5 shows an example of the worker detection results where the detected workers are indicated by blue dots. The position of the crane load is also indicated by red and green dots for reference. In this example, all three workers in proximity to the crane load are successfully detected by the proposed algorithm.

3.4 Dynamic Crane Workspace Update

Whenever a new container or large equipment is moved within the crane workspace, the workspace can be updated in a semi-automated manner. First, the overhead camera view is displayed on the bottom right corner of the user interface. From this view, the crane operator is able to visualize any new container or large equipment that moved into the crane workspace. Through the user interface, the crane operator clicks and drags around the corresponding object in the overhead camera view to create a rotated 2D bounding box (Fig. 6, bottom right). Next, the object location in the 3D workspace can be obtained by projecting using the camera matrix [9] since the camera height and camera calibration parameters are known. Finally, the crane operator uses a slider bar to specify the height of the newly-added object. This creates a 3D bounding box in the crane workspace (Fig. 6, left) centered around the object of interest with the appropriate size. This 3D bounding box is added to the list of static 3D bounding boxes (Sect. 3.1) that will create an alert to the operator whenever the crane load moves too close to the bounding box. The overall process is semi-automated since the 2D bounding box is manually drawn while the corresponding 3D bounding box is automatically populated. Due to the user interaction required, this process should be carried out before or in between lifts and not during a lift operation for safety purposes.

4 Results

To evaluate the effectiveness of the proposed crane collision avoidance system, multiple blind lift tests were carried out at a crane yard. The test environment within the crane yard encompasses an area of 73 m × 79 m. Figure 7 shows the experimental setup for the blind lift tests. There are many potential obstacles that have been placed in the crane yard such as shipping containers, equipment, and moving workers. The crane

Fig. 6. Adding a new container to the crane workspace through the user interface

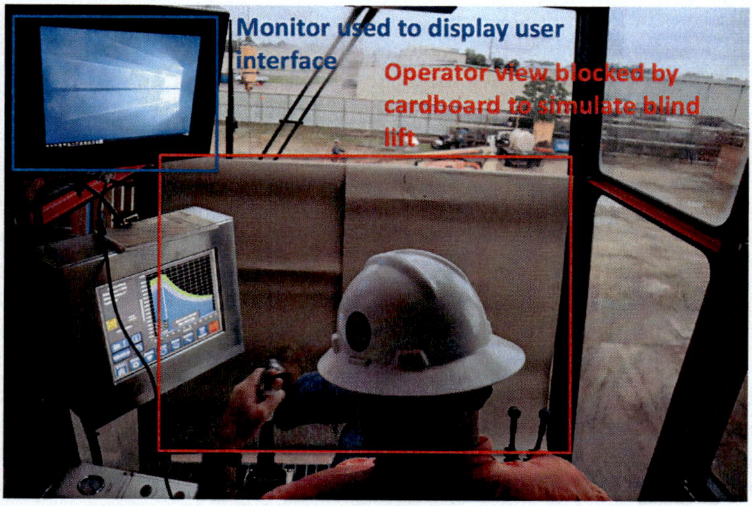

Fig. 7. Experimental setup for blind lift crane operation

operator is tasked with moving the load from one point to another while avoiding collisions. During the lift operation, the crane operator view is artificially blocked by cardboard to simulate blind lift conditions. Thus, the crane operator must rely on the 3D visualization displayed on a monitor inside the cabin to perceive the surrounding environment of the load in order to avoid collisions. The blind lift tests were carried out with two different crane operators, with each test taking around 15 min.

Figure 8 shows the collision detection results displayed to the crane operator through the user interface. The left panel, as well as the top right panel, shows the area around the load highlighted in red due to a worker that is standing too close to the load. Whereas, the bottom right panel shows the actual overhead scene observed from the boom-mounted camera. The crane operator is alerted to the potential collision through these visual warnings as well as a series of audio warnings. Thus, the crane operator can take necessary preventive actions.

Fig. 8. Collision detection results displayed on the user interface: (left) 3D workspace model, (top right) close-up 3D workspace model, (bottom right) overhead camera view.

During the blind lift, the operator could see the workers that had entered the workspace as seen in Fig. 5. There were occasional false positives due to reflective surfaces that appeared to be bright circles like hardhats; however, there were no false negatives, which would have created a potentially dangerous collision scenario.

There were several shipping containers added to the workspace after the 3D point clouds were collected and were considered dynamic objects. These objects were added to the visualization using the method described in Sect. 3.4. This update helped to guide the operator during the lift by cueing the operator on when to maintain adequate clearance.

5 Discussion

The proposed crane workspace model is able to assist the crane operator in carrying out blind lifts by continuously monitoring the load position and tracking both static and dynamic assets in the workspace. While not being fully automatic, this framework did provide a comprehensive method for monitoring all possible collisions that could occur. Table 1 shows a comparison of the achieved safety criteria between our proposed method and other crane safety assistance systems. [7] previously achieved a crane safety assistance system that could handle static assets and load sway but did not take into account dynamic assets or load rotation. On the other hand, [10] mostly focused on load sway and load rotation monitoring without tracking the possibility of collisions with surrounding assets. Thus, the main advantages of our proposed method are (i) the ability to handle dynamic assets (ii) the ability to track nearby workers and (iii) a practical implementation that is demonstrated to work with blind lifts in field tests.

Table 1. Comparison of achieved safety criteria between different methods

Safety criteria	Static assets	Dynamic assets	Worker tracking	Load Sway/Rotation	3D Visualization	Tested with blind lifts
[7]	Yes	No	No	Only sway	Yes	No
[10]	No	No	No	Yes	Yes	No
Our method	Yes	Yes	Yes	Yes	Yes	Yes

6 Conclusion

In conclusion, this paper presents a method to dynamically update the crane workspace for collision avoidance purposes. The proposed method is expected to enhance the crane operator's situational awareness during blind lift operations by assisting in monitoring the load position and possible collisions with both workers, static and dynamic objects in the workspace.

The proposed method automatically detects any workers that enter the field of view without requiring manual identification. Similarly, for the load sway, previous methods were merely used for tracking the load, while the proposed method can identify new loads as they are attached and detached. The semi-automated workspace update improves upon the existing workspace object detection methods by offering greater flexibility. While the method used in [20] could automatically identify specific objects, it neglected many others that could cause potential collisions.

This work is limited in its ability to scale worker detection and load position monitoring to different settings as it relies upon detection schemes utilizing hardcoded color information. Additionally, the lack of full autonomous function for detecting dynamic objects in the workspace is a downside and forces the operator to manually intervene. In future work, methods for upgrading the semi-autonomous workspace to a more autonomous approach will be explored. Additionally, strategies for more robust worker detection that allow for a variety of hard-hats will be investigated. It is expected that improvements in both areas will be largely driven using deep learning with artificial neural networks as well as other machine learning techniques.

References

1. Haas, C.T., Hornaday, W.C.: Computer-aided planning for heavy lifts, **119**, 498–515 (1994)
2. Hwang, S.: Ultra-wide band technology experiments for real-time prevention of tower crane collisions. Autom. Constr. **22**, 545–553 (2012). https://doi.org/10.1016/j.autcon.2011.11.015
3. Li, H., Chan, G., Skitmore, M.: Integrating real time positioning systems to improve blind lifting and loading crane operations. Constr. Manag. Econ. **31**, 1–10 (2013). https://doi.org/10.1080/01446193.2012.756144
4. Chi, S., Caldas, C.H.: Automated object identification using optical video cameras on construction sites. Comput.-Aided Civ. Infrastruct. Eng. **26**, 368–380 (2011). https://doi.org/10.1111/j.1467-8667.2010.00690.x
5. Roberts, D., Bretl, T., Golparvar-Fard, M.: Detecting and classifying cranes using camera-equipped UAVs for monitoring crane-related safety hazards. In: Computing in Civil Engineering (2017). https://doi.org/10.1061/9780784409374
6. Fang, Y., Cho, Y.K.: Effectiveness analysis from a cognitive perspective for a real-time safety assistance system for mobile crane lifting operations. J. Constr. Eng. Manag. **143**, 05016025 (2016)
7. Fang, Y., Cho, Y.K., Chen, J.: A framework for real-time pro-active safety assistance for mobile crane lifting operations. Autom. Constr. (2016). https://doi.org/10.1016/j.autcon.2016.08.025
8. Fang, Y., Cho, Y.K.: Crane load positioning and sway monitoring using an inertial measurement unit. Comput. Civ. Eng. **2015**, 700–707 (2015)
9. Chen, J., Fang, Y., Cho, Y.K.: Real-time 3D crane workspace update using a hybrid visualization approach. J. Comput. Civ. Eng. **31**, 04017049 (2017). https://doi.org/10.1061/(ASCE)CP.1943-5487.0000698
10. Fang, Y., Chen, J., Cho, Y.K., Kim, K., Zhang, S., Perez, E.: Vision-based load sway monitoring to improve crane safety in blind lifts. J. Struct. Integrity Maint. **3**, 233–242 (2018). https://doi.org/10.1080/24705314.2018.1531348
11. Park, J., Kim, K., Cho, Y.K.: Framework of automated construction-safety monitoring using cloud-enabled BIM and BLE mobile tracking sensors. J. Constr. Eng. Manag. 05016019 (2016). https://doi.org/10.1061/(asce)co.1943-7862.0001223
12. Fang, Y., Cho, Y.K., Zhang, S., Perez, E.: Case study of BIM and cloud–enabled real-Time RFID indoor localization for construction management applications. J. Constr. Eng. Manag. **142** (2016). https://doi.org/10.1061/(asce)co.1943-7862.0001125
13. Park, M.W., Brilakis, I.: Construction worker detection in video frames for initializing vision trackers. Autom. Constr. **28**, 15–25 (2012). https://doi.org/10.1016/j.autcon.2012.06.001
14. Park, M.W., Brilakis, I.: Continuous localization of construction workers via integration of detection and tracking. Autom. Constr. **72**, 129–142 (2016). https://doi.org/10.1016/j.autcon.2016.08.039
15. Kan, C., Fang, Y., Anumba, C.J., Messner, J.I.: A cyber-physical system (CPS) for planning and monitoring mobile cranes on construction sites. Proc. Inst. Civ. Eng.: Manag. Procure. Law **171**, 240–250 (2018). https://doi.org/10.1680/jmapl.17.00042
16. Zhou, C., Luo, H., Fang, W., Wei, R., Ding, L.: Cyber-physical-system-based safety monitoring for blind hoisting with the internet of things: a case study. Autom. Constr. **97**, 138–150 (2019). https://doi.org/10.1016/j.autcon.2018.10.017
17. Kim, P., Chen, J., Cho, Y.K.: Automated point cloud registration using visual and planar features for construction environments. ASCE J. Comput. Civ. Eng. **32**, 1–13 (2018). https://doi.org/10.1061/(ASCE)CP.1943-5487.0000720

18. Kim, P., Chen, J., Cho, Y.: Automatic registration of laser scanned color point clouds based on common feature extraction. In: International Symposium on Automation and Robotics in Construction (ISARC), Auburn, AL (2016)
19. Westoby, M.J., Brasington, J., Glasser, N.F., Hambrey, M.J., Reynolds, J.M.: 'Structure-from-Motion' photogrammetry: a low-cost, effective tool for geoscience applications. Geomorphology **179**, 300–314 (2012). https://doi.org/10.1016/J.GEOMORPH.2012.08.021
20. Chen, J., Fang, Y., Cho, Y.K.: Mobile asset tracking for dynamic 3D crane workspace generation in real time. In: Congress on Computing in Civil Engineering, Proceedings, American Society of Civil Engineers (ASCE), pp. 122–129 (2017). https://doi.org/10.1061/9780784480830.016

Construction Field Management Using a Popular Text Messenger

Ghang Lee[1(✉)], Jehyun Cho[1], Taeseok Song[1], Hyunsung Roh[1], Jeaeun Jung[1], Jihoon Chung[1], Gunwoo Yong[1], and David Jeong[2]

[1] Yonsei University, Yonsei-ro 50, Seodaemun-gu, Seoul 03722, Korea
glee@yonsei.ac.kr
[2] Texas A&M, 334 Francis Hall, 3137 TAMU, College Station, TX 77843-3137, USA

Abstract. This paper introduces a method for managing daily construction tasks using a text messenger and a tool to support the method. Many construction field management tools have been developed and used in practice. In addition, each company has developed its own construction task management tools. Nevertheless, due to the number of various tools available, contractors and subcontractors suffer from installing at least three or four applications on mobile devices to support different projects. To resolve this issue, the authors developed ProjectHub-CL (Construction Listener), a method and tool to manage daily construction tasks using a popular text messenger that doesn't require the installation of an additional mobile app. It helps field engineers and subcontractors to exchange and manage field tasks using text messages and additional features on a text messenger and can automatically produce a construction daily report at the end of the day as a result. The system was implemented on top of KakaoTalk, the most popular text messenger in South Korea. Its applicability and usability in the field were validated with contractors and subcontractors in four different projects (conducted by different general contractors).

Keywords: Construction daily task management · Daily work report · Field management · App fatigue · Text messenger · Mobile

1 Introduction

Daily construction field management is a fundamental part of construction management and the basis of short-term and long-term schedule management (such as weekly and master schedule management). A construction daily report is an outcome of the daily construction field management tasks and is a crucial process of converting various data of the project status into a specific form. Nevertheless, the report generation process often becomes an intrusive burden on contractors and subcontractors, because it requires approximately thirty minutes to a few hours of non-value-added activity for them each day simply to generate the report [1]. To make the reporting process more efficient, many construction field management tools have been developed. These tools are enhanced by building information modeling (BIM), and therefore, are often referred

E. Toledo Santos and S. Scheer (Eds.): ICCCBE 2020, LNCE 98, pp. 971–979, 2021.
https://doi.org/10.1007/978-3-030-51295-8_67

to as field BIM tools. Examples of well-known commercial tools include Procore, Aconex, Autodesk BIM 360 Field, and FINALCAD. In addition, many companies have their own custom-developed construction task management tools to meet their own needs.

The abundance of field BIM tools allows field engineers to manage a construction field more efficiently than before but created another problem. Each project often requires project participants to install a new set of multiple construction management tools. A contractor needs to install a client's system, a headquarters system, and a field management system. A subcontractor's case is worse. A subcontractor works on multiple construction projects, and each project uses different applications (apps). This causes "app fatigue"—users becoming tired of installing, deploying, and learning new software applications—and, needless to say, multiple data entry and software cost issues.

To alleviate these app-fatigue issues, the authors developed the ProjectHub-CL (Construction Listener), a method and tool to manage daily construction tasks using a popular text messenger that doesn't require the installation of an additional mobile app. The development of the system was motivated by two facts. First, more than 80% of field engineers exchange and share information related to daily construction management through a commonly used text messenger [1], although it is often recommended not to exchange project management information through a text messenger. Second, text messages carry more information than one might expect.

ProjectHub-CL helps field engineers and subcontractors to exchange and manage field tasks using text messages and additional features of the text messenger. It also helps to automatically produce a construction daily report at the end of a day as a result. The system was implemented on top of KakaoTalk, the most popular text messenger in South Korea. Its applicability and usability in the field were validated with contractors and subcontractors in four projects (conducted by different general contractors).

2 Background

Quite a few studies have been conducted to make the daily construction report (DCR) generation process more efficient. Russel (1993) [2] proposed and developed a daily construction project management system, which is similar to the current project management information systems (PMIS) and field BIM tools. The goal was to rapidly report and share site information and the progress of the project among participants. Kim et al. (1999) [3] developed a similar PMIS that could manage labor, equipment, daily tasks, and generate a DCR. Shiau et al. (2003) [4] proposed a construction management information system (CMIS) that had a DCR module, a cost management module, and a design-change management module based on daily tasks. Na et al. (2003) [5] focused on the DCR as a basis for managing construction sites. Kim et al. (2003) [6] proposed a work process that could improve the DCR generation process. Seo et al. (2004) [7] emphasized the importance of subcontractor-focused daily construction site management and proposed a web-based system to support such a subcontractor-centric process.

After the advent of BIM and cloud technologies, many BIM- and cloud-based field BIM tools, including Aconex, Procore, BIM 360 Field, FINALCAD, and others became available. Bowden et al. (2005) [8], Lee et al. (2010) [9], Kim et al. (2011) [10], Lee et al. (2011) [11], Tserng et al. (2005) [12], Irizarry et al. (2009) [13], Park et al. (2013) [14], Chun et al. (2013) [15], and Shin et al. (2016) [16] have studied the use of mobile technologies on construction sites. These studies reported that mobile devices, such as a mobile phone and a digital tablet, were used in managing defects, safety, soil conditions, materials, daily task management, personnel employed, and other issues on site. Nevertheless, a study conducted by Sattineni et al. (2015) [17] regarding the use of mobile devices on construction sites revealed that the use of mobile devices in practice still focused on basic functions, such as emailing, photo sharing, document viewing, and file sharing, rather than complex or advanced task management functions identified by previous studies. Their study also revealed that there were concerns about safety issues due to distraction, privacy invasion, adoption cost, subcontractor buy-in, older worker reluctance, data synchronization, and software bugs.

Some studies proposed the use of other systems to improve field management. Ahn et al. (2012) [18] claimed that site access control systems were limited in acquiring actual information about personnel employed on the day and proposed a labor management system based on a global positioning system (GPS). Hegazy et al. (2012) [19] proposed an email-based construction project management system to acquire site information. Kim et al. (2013) [20] proposed a mobile system that could acquire information from CCTV (closed circuit television) images. Choi et al. (2011) [21] and Moon et al. (2016) [22] proposed a field management system based on radio-frequency-identification (RFID) and quick response (QR) codes. Ferrada et al. (2016) [23] proposed a mobile-based expert system where experience and lessons learned from experienced engineers could be shared. Ahn et al. (2014) [24] and Kim et al. (2015) [25] proposed the use of social media to collect and manage construction site and defect information. Chin et al. (2004) [26] proposed a location-based field management system by combining a location tracker and a mobile-based PMIS.

As reviewed thus far, many studies have been conducted to improve field management, some by improving the DCR generation process and some by adopting information and communication technologies, including mobile apps, PMISs, GPS, RFIDs, QR codes, and emails. Nonetheless, no study was found to use IMs as a medium to collect information for the generation of a DCR.

A preliminary study conducted by the authors with 62 field engineers on the data collection and exchange methods for daily construction work in South Korea showed that 54% of the respondents used text messengers [1]. That is, there was a high probability to use an instant message (IM) project management information system as a solution to collect information for generating a DCR and resolve "app fatigue" issues if the system was developed successfully.

Still, there were several challenges to overcome. The first challenge was to identify what information could be collected and should be collected. Second, because of the nature of an IM, its add-on functions would be restricted compared to those of a stand-alone mobile app. Despite the restriction, a simple user authentication process for different projects would be critical. Another challenge was how to make the overall system design easy enough for the field engineers to use every day on site.

3 Research Method

This study was conducted as follows. First, the information requirements for DCRs were identified by collecting and analyzing DCR forms used in various projects. Second, the sources of the information were analyzed to understand what information could be collected from an IM. Third, the system was implemented and tested for field applicability.

4 Information Requirements for Daily Construction Reports

To develop a system that could automatically produce a DCR, the first step was to identify the information requirements for DCRs. This section describes those requirements, which were identified by analyzing DCR forms collected from 16 construction projects. The forms were used in various building, civil infrastructure, and plant construction projects. No new information was identified after analyzing 10 DCR forms, but the analysis was continued for 16 forms to assure that there was no information overlooked. A total of 66 information items in six categories (general information, personnel employed, material management, equipment management, task management, and workload) were identified. The workload category was found only in the civil infrastructure project (Table 1).

Table 1. Information items required by DCRs

Information category	Number of information items
General information	14
Personnel employed	7
Material management	9
Equipment management	5
Task management	19
Workload (Civil project only)	12

오류! 참조 원본을 찾을 수 없습니다summarizes the six categories and the number of items in each category. The specific number was dependent on how each item was counted. For example, today's number of personnel employed and the accumulated number of personnel employed could be counted as one item or two. In such cases, each item was counted as one for the purposes of this study.

The general information included project name, contractor name, date, weather, project director's name, etc. The personnel employed included job type, company name, number of personnel employed, etc. The material management included material name, material specifications, material unit, planned vs. actual amounts, etc. The equipment management included equipment name, equipment specifications, amount of equipment, etc. The task management included key task list, task name, person in charge, planned vs. actual progress, safety issues, civil claims, environmental requirements, site cleaning, etc. The workload included work location, work type, work unit, work amount, designed vs. actual workload, work progress, etc.

5 Information Source Analysis

The information requirements were recategorized by possible information source. The weather information could be collected from a local weather report or from weather-related sensors installed on site. The personnel employed and the equipment information could be gathered through a construction site access control system. Some of the material information could be checked through a RFID system, but not all.

To analyze the information items obtainable from IMs, a total of 53,134 chat bubbles (i.e., single messages from one individual to another) were collected from seven different projects, ranging in duration from 43 days to 720 days; six projects were in South Korea, and one was located in Singapore. The duration of the data collection did not affect the number of collectable information types. For example, the chat bubbles were collected for only 43 days from the sewer project, and its number of collected chat bubbles was not the highest. The coverage of information provided, however, was very high.

The analysis results showed that the material management information could not be collected from the text messages, but a large portion of the personnel, equipment, and task management information as well as general information, such as dates, could be obtained from text messages. The context analysis of the chat bubbles also verified that the exchange of daily workload information was limited to civil projects, and it was rare to exchange workload information in building projects (Table 2).

Table 2. Information items obtainable through text messages

DCR information category	Project						
	A	B	C	D	E	F	G
General information	√	√	√	√	√	√	√
Personnel employed	√	√	√	√	√		√
Material management							
Equipment management	√	√	√	√	√		
Task management	√	√	√	√	√		√
Workload	√	√					

A: Sewer, B: Highway, C: Factory, D: Office, E: OfficeTel, F: Apartment Complex, G: Office

6 Implementation

The ProjectHub-CL system (focusing on personnel, equipment, and task management information) was implemented on top of KakaoTalk, the most dominant text messenger that was used by over 68.5% of South Koreans. This was done by using the KakaoTalk and PlusFriend Application Programming Interface (API). Amazon Web Services (AWS) was used for cloud and web hosting services. For database management, the AWS Relational Data Services (RDS) and MySQL were used. A DCR was automatically generated in Microsoft Excel format from the databases when requested by a general contractor.

Figure 1 provides screenshots from ProjectHub-CL. ProjectHub-CL began with the registration of project team members and a user authentication process. A general contractor could request personnel, equipment, or task information from subcontractors as a text message through KakaoTalk. Subcontractors could report the status of personnel, equipment, or tasks in one or more text messages. Then, the general contractor could copy and paste the report message to the user interface.

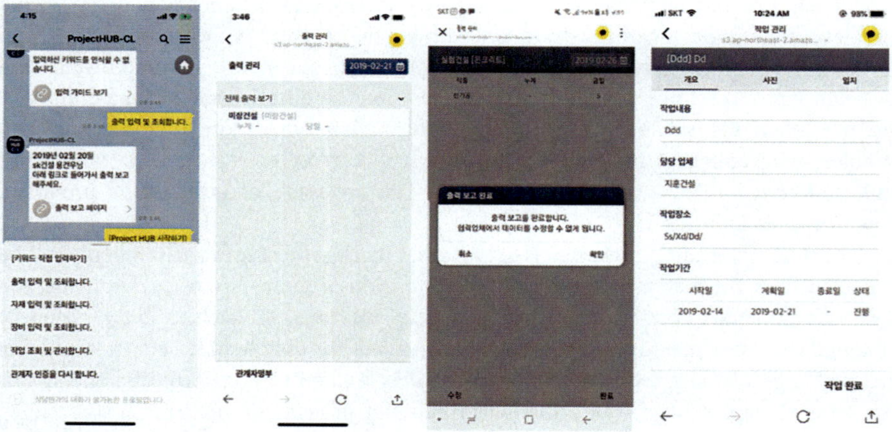

Fig. 1. Screenshots of ProjectHub-CL

The implemented system was tested through a structured field applicability test with 18 field engineers (10 of them were subcontractors and eight were general contractors). Figure 2 illustrates the DCRs. The implemented system could successfully produce DCRs from the data collected through the text messages.

공 사 일 보				결재	담당	공사	공무	관리	소장
현 장 명	JH연습	허가일	2018-10-08	날씨	기온		날 짜	2019년 07월 29일 월요일	
시공업체	연습	착공일	2018-10-08	구름	25℃	31℃	작 성 자	조재현	

공종	업체	전일누계	오늘	누계	비고
골조	한솔건설	30	15	45	일반 5, 철근 10
설비	DH건설	15	16	31	일반 6, 타설 10
마감	재온건설	35	15	50	철골 10, 일반 5
총계		80	46	126	

특기사항

전일 자재 투입 현황

자재명	규격	단위	전일누계	오늘	누계
목 창호	230 *180	개	3	5	8
철근	D35	Ton	5	10	15

전일 장비 투입 현황

장비명	규격	전일누계	오늘	누계
펌프카	A	5	5	10

금일 작업 현황

공종	작업명	작업내용	담당업체	작업장소	상태
안전	라바콘 철거	도로변 라바콘 철거	재온건설	1/1/0/상세위치	완료
안전	라바콘 철거	도로변 라바콘 철거	강승우	1/1/1/도로변 라바콘 철거	완료
마감불량	할석 및 미장	계단참과 계단 경사 접합부 마감불량	한솔건설	1/7/3/1	완료

2019-07-29

Fig. 2. Automatically generated DCRs

7 Conclusion

A method and system to automatically generate a DCR from text messages exchanged through a commonly used text messenger was proposed and developed. The advantage of this approach over other mobile-app-based approaches is that it does not require installation of additional apps, which will relieve field engineers from "app fatigue" in installing multiple field management apps.

By analyzing 16 DCR forms, a total of 66 information items required to generate DCRs were identified. The content analysis of 53,134 chat bubbles collected from seven different projects, ranging from 43 days to 720 days, showed that the personnel, equipment, and task management information could be collected from text messages exchanged on site via IMs. The system was developed as an add-on to KakaoTalk, the most popular text messenger in South Korea. A field-applicability test with 18 field engineers confirmed that the proposed approach had a high potential to be used as a method that could help field engineers efficiently collect information and generate DCRs.

To our knowledge, this is the first system of this kind and as such, has much room for improvement. With a tight connection to the PMIS of a construction site, the system would be able to acquire a full set of information required to generate a DCR with less manual input. Moreover, if a chatbot system was deployed, the manual copying-and-pasting of text messages to the input module could be greatly reduced.

Acknowledgements. This work was supported by an Institute for Information & Communications Technology Promotion (IITP) Grant funded by the Korea Government (MSIT) (No. 2019-0-01559-001, Digitalizing Construction Project Requirements Using Artificial Intelligence and Natural Language Processing).

References

1. Song, T.: Extracting daily report information from text messages between the field managers. Master, Architecture & Architectural Engineering, Yonsei University (2018)
2. Russell, A.D.: Computerized daily site reporting. J. Constr. Eng. Manag. **119**(2), 385–402 (1993)
3. Kim, H., Choi, J., Lim, H., Song, Y.: A study on the improvement of work management through the rationalization of daily construction reports. Daelim Tech. Rev. **14** (1999)
4. Shiau, Y.-C., Wang, W.-C.: Daily report module for construction management information system. In: Proceedings of the 20th International Symposium on Automation and Robotics in Construction (2003)
5. Na, K.-s., Kim, T.-h., Kim, S.-k.: A study on development of project management system with daily report, pp. 459–462. Architectural Institute of Korea (2003)
6. Kim, B.-n., Kim, C.-D., Ee, H.-S., Kim, S.-K., Seo, S.-W.: The analysis of current daily-report system and the proposal of improvement framework. Kor. J. Constr. Eng. Manag. **4** (2), 66–73 (2003)
7. Seo, H.-J., Koo, K.-J., Um, I.-J.: Web-based coordination system for daily work report between contractor and subcontractor. In: Conference of Korea Institute of Construction Engineering and Management, pp. 370–375 (2004)
8. Bowden, S., Dorr, A., Thorpe, A., Anumba, C., Gooding, P.: Making the case for mobile IT in construction. Comput. Civ. Eng., 1–12 (2005)
9. Lee W.-J., Choe W.-U.: Case of application of Smartphone construction industry - GS EIP + TPMS using smart phone. Kor. Inst. Constr. Eng. Manag. **11**(5), 30–34 (2010)
10. Kim, B.-S., Kang, Y.-m., Kim, M.-J., Park, J.-M., Kim, O.-K., Choi, B.-J.: Labor management process of a construction site using Smart Phone. In: Conference of the Regional Association of Architectural Institute of Korea, pp. 357–360 (2011)
11. Lee, K.-P., Lee, H.-S., Park, M., Kim, E.: Construction material management using smart mobile computing. In: Conference of Korea Institute of Construction Engineering and Management, pp. 59–69 (2011)
12. Tserng, H.P., Dzeng, R.J., Lin, Y.C., Lin, S.T.: Mobile construction supply chain management using PDA and bar codes. Comput.-Aided Civ. Infrastruct. Eng. **20**(4), 242–264 (2005)
13. Irizarry, J., Gill, T.: Mobile applications for information access on construction jobsites. In: Computing in Civil Engineering, pp. 176–185 (2009)
14. Park, M., Kim, E., Lee, H.-S., Lee, K., Suh, S.-W.: Real time safety management framework at construction site based on smart mobile. Kor. Inst. Constr. Eng. Manag. **14**(4) (2013)
15. Chun, S.B., Lee, B.B., Yoon, J.G., Kim, H.D.: Development of the BIM-based mobile field management system. In: Conference of Korean Society of Civil Engineers, pp. 4–7 (2013)
16. Shin, Y., Kim, D., Kim, T.-Y., Kim, G.-H.: Daily reporting system using digital pen at construction site. J. Kor. Inst. Build. Constr. **16**(2), 177–183 (2016)
17. Sattineni, A., Schmidt, T.: Implementation of mobile devices on jobsites in the construction industry. Procedia Eng. **123**, 488–495 (2015)
18. Ahn, C.-S., Yoon, S.-W., Chin, S.-Y.: Diligence and indolence management system for specialty contractor on construction site -using GPS of smart phone. Kor. J. Constr. Eng. Manag. **13**(3), 56–66 (2012)
19. Hegazy, T., Abdel-Monem, M.: Email-based system for documenting construction as-built details. Autom. Constr. **24**, 130–137 (2012)
20. Kim, C., Park, T., Lim, H., Kim, H.: On-site construction management using mobile computing technology. Autom. Constr. **35**, 415–423 (2013)

21. Choi, Y.-g., Yoon, S.-W., Chin, S.: Integrated manpower information systems on construction site - using RFID and QR-code. Kor. Inst. Constr. Eng. Manag. **12**(4) (2011)
22. Moon, J.-B., Seo, C.-Y., Kim, C.-D.: Analyzing the improvement of the RFID system and the proposal of project manpower management system using QR-code. In: Conference of Korea Institute of Construction Engineering and Management, p. 4 (2016)
23. Ferrada, X., Sepúlveda, M., Serpell, A., Núñez, D., Neyem, A.: A lessons-learned mobile system for construction companies: motivation and design. Procedia Eng. **85**, 157–165 (2014)
24. Ahn, J., Kim, N., Chin, S.: An SNS based construction information communication and sharing system. Kor. J. Constr. Eng. Manag. **15**(5), 115–126 (2014)
25. Kim, Y.-H., Kim, G.-H., Kim, C.-D.: The improvement of the apartment housing defect management using group type SNS application. In: Conference of Korea Institute of Construction Engineering and Management, pp. 70–73 (2015)
26. Chin, S., Kang, W.-Y., Kim, Y.-S.: A development of task generator through an analysis of task location information structure. J. Kor. Inst. Constr. Eng. Manag. **5**(2), 64–71 (2004)

Study of IMU Installation Position for Posture Estimation of Excavators

Jingyuan Tang, Han Luo, Peter Kok-Yiu Wong, and Jack C.P. Cheng(✉)

Hong Kong University of Science and Technology, Hong Kong SAR, China
{jtangap, han.luo, kywongaz}@connect.ust.hk, cejcheng@ust.hk

Abstract. Posture estimation provides basic mechanical information for remote control of construction equipment, safety analysis of construction sites and productivity analysis of construction projects. The rapid development of micro-machined electromechanical systems (MEMS) inertial sensors enables wide application of posture estimation in the construction industry. Previous studies have put efforts to orientation estimation, trajectory tracking, and activity recognition of construction equipment based on Inertia Measurement Units (IMU). As excavators are common and important heavy construction equipment on site, posture estimation of excavators is closely related to the safety of construction sites and automation of construction processes. Although there have been studies on IMU-based posture estimation of excavators, there is still a lack of studies that rigorously discuss the impact of IMU installation positions on posture estimation of construction equipment, especially excavators. Placing IMU sensors in appropriate positions on excavators would affect the accuracy of posture estimation. More practically, the installability issue also governs the installation positions. Hence, this paper investigates the influence of different IMU installation positions on posture estimation of excavators. Firstly, a systematic framework is proposed to estimate posture of excavators based on data collected by IMUs. Our experiments verify the feasibility of our proposed framework. Furthermore, the in-fluences of different installation positions of IMU are investigated, on the basis of accuracies represented by Euler angle. Considering accuracy and applicability, a scheme of installing IMU on excavators is discussed. This paper will provide a theoretical basis and a reliable operation scheme for IMU-based posture estimation of excavators in the future.

Keywords: Inertial Measurement Unit (IMU) · Excavator · Posture estimation · Installation position

1 Introduction

Posture estimation can provide basic posture information for many industrial applications of construction equipment, such as remote control of construction equipment, dynamic monitoring of equipment operation safety, and productivity analysis. Excavators are important and common heavy construction machines on construction sites. The research on posture estimation of excavator is tightly related to construction safety,

E. Toledo Santos and S. Scheer (Eds.): ICCCBE 2020, LNCE 98, pp. 980–991, 2021.
https://doi.org/10.1007/978-3-030-51295-8_68

automation construction and smart construction site. With many surveillance cameras installed, previous studies have attempted to estimate partial or full body postures of excavators using computer vision techniques from videos and images captured from construction sites [1, 2]. However, such techniques may not maintain satisfactory performance when environmental conditions of construction sites become worse. For example, low light conditions will damage the quality of surveillance videos and images, and thereby the computer vision techniques will have lower accuracies when estimating postures of excavators.

With the rapid development of micro-machined electromechanical systems (MEMS) inertial sensors, the sharp decline in the price of Inertial Measurement Unit (IMU) device allows widespread use of such devices in the construction industry. An IMU sensor usually have three parts, including an accelerometer, a gyroscope and a magnetometer. It can provide the rotation information of an object in 3D space with good accuracy. There are many advantages of adopting IMU sensors, including light weights, small volumes. In addition, IMU sensors are more insensitive to environmental conditions, compared to common RGB cameras. Therefore, IMU sensors have been utilized to collect data by many scholars in different research directions such as trajectory tracking and activity classification of construction equipment [3]. However, the application of IMU sensors for estimating postures of excavators are still in its infancy.

The IMU-based posture estimation of excavators makes up with two purposes. Firstly, it is to address the inadequacy of computer vision techniques in challenging scenes such as night and rainy day. Secondly, the IMU-based posture estimation results could potentially complement that obtained by computer vision techniques. To adopt IMU sensors, a crucial issue is to select appropriate installation positions of IMU sensors. In industrial applications related to the IMU-based posture estimation of excavators or even other construction equipment, installation positions should be carefully chosen for accurate posture estimation as well as their installability. However, there is still a lack of studies that rigorously discuss the impact of IMU installation positions on posture estimation of construction equipment, especially excavators.

In this context, this paper aims to analyze the effect of different installation positions of IMU sensors on the accuracy of posture estimation. Moreover, this paper considers the installability of IMU sensors when deciding the installation positions. Based on the experimental findings, this paper will provide a theoretical basis and a reliable operation scheme for IMU-based posture estimation of excavators in the future.

The rest of paper is organized as follows. Section 2 introduces existing work related to posture estimation of construction equipment and IMU performance evaluation. Section 3 describes the methodology and related technologies of this study in detail. In Sect. 4, an illustrative experiment is designed and conducted to validate the proposed methodology. Besides, the effect of different installation positions of IMU sensors on the posture estimation of excavator is evaluated and discussed through the results. Conclusion are drawn in Sect. 5.

2 Related Works

2.1 Posture Estimation of Construction Equipment

In previous studies, computer vision techniques are commonly used to estimate postures of construction equipment. In 2017, Soltani et al. [1] estimated partial posture of excavators (i.e. arm and boom) and track the location of excavators by applying background subtraction and k-means algorithms to surveillance video. Recently, Luo et al. [2] identified full body postures of construction equipment automatically from images captured from surveillance camera on construction sites by structuring a deep learning model. Previous computer vision-based studies on posture estimation of construction equipment have achieved satisfactory performances, which means such methods can provides reliable posture information for construction safety analysis and other fields. However, due to the intrinsic characteristics of RGB cameras, vision-based excavator posture estimation approaches may lose accuracy when environmental conditions (e.g. lighting and occlusion) of a construction site are poor.

To address such limitations, IMU-based posture estimation of construction equipment can be investigated, which is less sensitive to insufficient lighting and visual occlusion. Recent studies tend to rely on multi-sensor fusion. Kim et al. [4] presented a method for estimating poses and position of dozer blade by a sensor fusion system, including 2 IMU and 2 RTK GPS. Im et al. [5] proposed a remote-control station for excavator autonomous driving, including: attribute-sensing module, path-tracking and posture control algorithm. In their study, IMU was applied in attribute-sensing module as data collection sensor of posture information of excavators. However, the study related to IMU-based posture estimation of construction equipment, especially excavator, is still lacking in construction industry.

2.2 Performance Evaluation of Inertial Measurement Unit

Previous studies have provided methods for evaluating performances of IMU sensors, considering the installation position, speed and other factors. Sessa et al. [6] presented a reliable methodology for comparing IMU sensors or other attitude estimation devices in a Vicon environment as ground-truth. In their study, dynamic time warping (DTW) is proposed to align misalignments of time series due to different receiving terminals or different reference frames. Mourcou et al. [7] evaluated the performance of embedded IMU sensors in the 3 best-selling smart phones under different algorithms in trials. Under the condition of static protocol and dynamic protocol, the effect of position and speed on the embedded IMU sensors in smart phones is analyzed in term of root mean square and variance, in order to discuss whether the orientation estimation result of embedded IMU sensor in smart phone can be reliably used in the clinic applications.

Although the performance evaluation methods of IMU sensor in previous studies have provided theoretical ideas, there still lacks the related discussion for construction industry such as equipment posture, particularly for excavators. Apart from performance evaluation, there also lacks discussion about the influence of installing IMU sensors in different positions of excavators.

3 Methodology

In this study, the proposed approach is divided into four steps: (1) Data Collection, (2) Data Preprocessing, (3) Posture Estimation and (4) Accuracy Evaluation & Result Analysis, which is as shown in Fig. 1. In this study, for each excavator, the bucket, arm, boom and cabin can be regarded as independent motion segments. In the step of Data Collection, raw data for estimating postures of excavators or orientations of segments, is collected by IMU sensors installed in different positions. Afterwards, in the step of Data Preprocessing, the raw data collected in Data Collection are preprocessed through calibration and noise reduction. Next, the calibrated data are transformed into rotation represented by quaternion using orientation filters in the step of Posture Estimation. Then, the rotation related to posture of excavators is conversed to Euler Angle from quaternion representation for better understanding and expression. In the step of Accuracy Evaluation & Result Analysis, the rotation, which is obtained from different IMU sensors in previous step, is aligned. In the end, the results are compared and discussed by Root Mean Square (RMS) and Standard Deviation (Std).

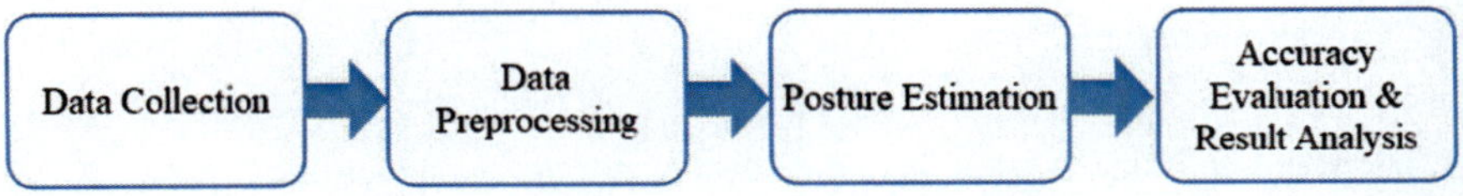

Fig. 1. Workflow of the proposed approach

3.1 Data Collection and Data Preprocessing

In the step of Data Collection, several IMU sensors are installed in different positions of the same segment of an excavator. Then, raw data of IMUs, which describe the orientation of the given segment, are collected from different IMU sensors. After obtaining the raw data, data preprocessing is necessary before estimating postures of the excavator. Data preprocessing includes two parts: (1) noise reduction and (2) magnetometer calibration. In this study, moving average filter and DBSCAN are adopted to denoise and smooth the data collected by accelerometer, gyroscope and magnetometer in IMU sensors. Through the ellipsoid fitting of the magnetic field data collected by IMU sensors in the ferromagnetic environment, which is same as the experiment, and obtaining the corresponding parameters, the flux related to the posture of excavators or orientation of segments is calibrated effectively. After data preprocessing is done, the calibrated data are obtained, which contain angular rate, acceleration and flux.

3.2 Posture Estimation

Posture of Excavators and Orientation of Segments. For an excavator that moves continuously, its posture records the rotation state of the excavator at a certain time without semantic information. Figure 2 shows the common technical framework for estimating the posture estimation of an excavator using IMUs. For each excavator, the bucket, arm, boom and cabin can be regarded as independent motion segments. As shown in Fig. 2., the posture of the excavator is obtained by combining the orientation estimation results of each independent motion segment of the excavator at each time and rotating based on the given initial posture. Hence, the orientation estimation of the segment is an independent unit of the posture estimation, and the estimated posture is an ordered combination of the estimated orientation of the segment.

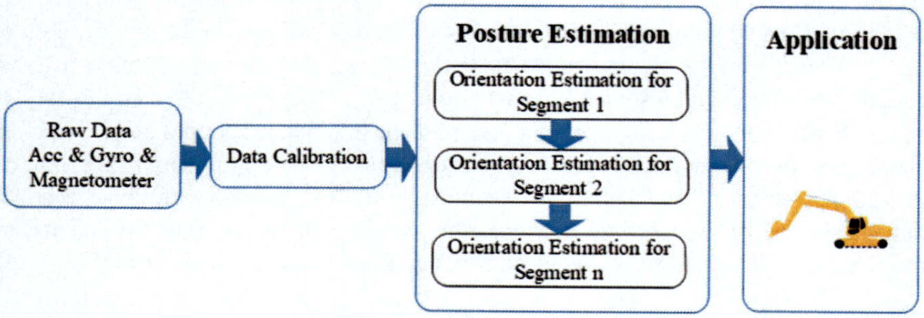

Fig. 2. Common technical framework for estimating the posture of an excavator using IMUs

Therefore, in order to investigate the effect of different installation positions of IMU sensors on posture estimation of excavators, this study selects an independent motion segment of the excavator (i.e. boom) as the research object.

According to the same approach as the Fig. 2., in this study, data are collected from IMU sensors which are installed in different installation positions of the same excavator motion segment. Then, the raw data are preprocessed by denoising and calibration. The preprocessed data from different IMU sensors are calculated into the rotation state represented by quaternion through orientation filter. The orientation filter selected in this study will be introduced in detail in the next subsection.

Madgwick Filter. Madgwick filter [8] is a classical orientation algorithm for inertial sensors. It is widely used in human motion detection, inertial navigation system etc. The advantages of Madgwick Filter are shown as follows: 1) Low computational load. 2) Shown stable and efficient performance, especially in IMU with low sampling frequency. 3) Fewer parameters need to be defined in advance. Therefore, in this study, Madgwick is selected as the orientation filter to calculate the original data collected from accelerometers, gyroscopes and magnetometers into rotation represented by quaternions.

The basic principle of Madgwick, shown as Eq. (1), is to combine the orientation estimation result from accelerometer and magnetometer at time t and the orientation estimation result from gyroscope at time t by weight γ_t.

$$ {}^S_E q_{est,t} = \gamma_t {}^S_E q_{\nabla,t} + (1-\gamma_t) {}^S_E q_{w,t}, \; 0 \le \gamma_t \le 1 \quad (1) $$

where E denotes the stationary object toward the north horizontally in the earth frame. S denotes the state of the object after a motion in the sensor frame. ${}_E^S q$ denotes the orientation change of state in the sensor frame relative to its state in the earth frame.

3.3 Accuracy Evaluation and Result Analysis

According to the study from Sessa [6], the IMU sensor will cause misalignment to some extent because different receiving PC terminals result in different receiving data timestamps through wireless transmission. Therefore, an enlarged error will occur if the calculated results are compared directly. Therefore, before final evaluation, the results need to be firstly aligned through dynamic time warping (DTW). The basic principle of DTW will be introduced in detail as follows. Afterwards, this section will introduce evaluation metrics and analytical methods of the proposed method on the basis of dynamic time warping.

Dynamic Time Warping (DTW). DTW [9] is a representative method on solving misalignment issue with elastic distance measures. The target of DTW is to align two time series with different length.

Basic principle of DTW is shown as follows. Suppose there are two time series S and T, of length n and m respectively, and a warping path W, which records the distance between the closest points in series S and T.

$$S = s_1, s_2, \ldots, s_i, \ldots s_n \tag{2}$$

$$T = t_1, t_2, \ldots, t_i, \ldots t_m \tag{3}$$

$$W = w_1, w_2 \ldots w_k \tag{4}$$

where s_i is the ith point in S, and t_i denotes the ith point in T, w_i represents the possible node on warping path, corresponding to s_i and t_i.

The minimum cumulative distance between start point and end point should be found over potential warping paths, as shown in Eq. (5).

$$DTW(S,T) = minw[\sum\nolimits_{k=1}^{p} \delta(w_k)] \tag{5}$$

where is a distance measure between two time series elements, minw denotes the minimum cumulative distance, and p is the number of points included in the warping paths.

According to the warping paths, the correspondence between S and T is found and the two sequences are aligned.

Evaluation Methods and Measurement Metrics. In this study, Root Mean Square (RMS) and Standard Deviation (Std) are served as measurement metrics. RMS of the pairwise differences can be used to measure distance between the average error and zero. Std is a common indicator that can be used to measure the discrete degree of data.

$$X_{RMS} = \sqrt{\frac{1}{N}\sum\nolimits_{n=1}^{N} |X_n|^2} \quad (6)$$

$$X_{Std} = \sqrt{\frac{1}{N}\sum\nolimits_{n=1}^{N} |X_n - X|^2} \quad (7)$$

where X_{RMS} and X_{Std} mean the value of RMS and the value of Std respectively, X_n is the value of sample n, X means the average value, and N is sample size.

Figure 3 shows the workflow of alignment and comparison between IMUa and IMUb on Roll. Firstly, the orientation estimation results represented by quaternions obtained by the IMU sensors (i.e. IMUa and IMUb) installed in different positions of the same segment of the excavator in the step of posture estimation are conversed into Euler angles for more intuitive comparison and interpretation. Next, the Euler angle calculated and conversed from data collected from different IMU sensors are aligned by DTW on pitch and roll. The difference between warped Euler angle related to different IMU sensors are represented as error series on each axis respectively. In the end, the RMS and Std of error series are calculated, and the results are analyzed and discussed.

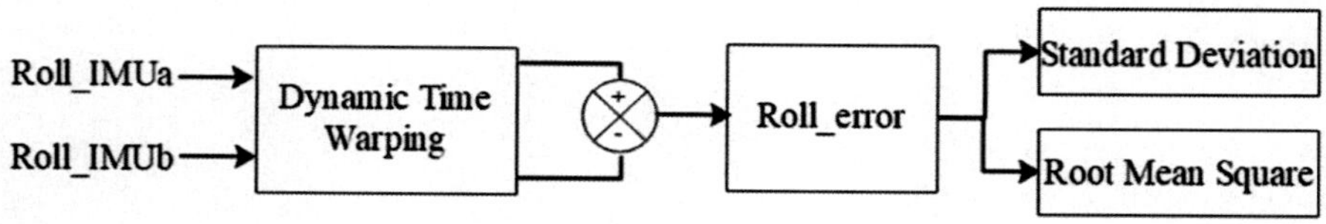

Fig. 3. Example of the alignment and accuracy evaluation of Roll between IMUa and IMUb

4 Experiment

4.1 Experiment Setup

Experiment Environment and Materials. The experiment aims to investigate whether varying the installation positions of IMU sensors would affect the accuracy of the posture estimation of excavators in practice. In our study, this is related to estimating the orientation of each segment, with IMU sensors installed in different positions on the mechanical arm of an excavator. Based on the objective, the experiment is designed for comparing the estimation results obtained from an independent motion segment of an excavator, with respect to rotation in 2 directions from the perspective of Euler Angles. Specifically, the rotation in 2 directions are the Roll, corresponding to the rotation of the X-axis, and the Pitch, corresponding to the rotation of the Y-axis.

Due to the high operative difficulty as well as the complexity to use actual excavator in the real environment of construction sites, this study was conducted in indoor micro-experimental environment in order to improve the operability of repeated tests and data collection. In order to ensure the accuracy and reliability of the experimental results, the real complex conditions and available specifications of the construction site were simulated as much as possible. Factors including ferromagnetic environment simulation, equipment selection and measurement device selection were carefully considered.

Regarding the complex ferromagnetic environment that may exist in a real construction site, the experiment was conducted with various electronic devices that caused notable interference, instead of an open environment. In terms of equipment selection, all test sensors were installed on a scaled-down, standard-size engineering model of 1:14 relative to a real excavator, as shown Fig. 4(a). In terms of measurement device selection, the LPMS-B2 IMU sensor was used as a data acquisition device in the experiment, as shown in Fig. 4(b). The sensors were equipped with a Bluetooth transmitting and receiving module, with the highest sampling frequency up to 400 Hz. Consequently, information within the range of 20 m could be collected and stored by mobile phone or PC terminal.

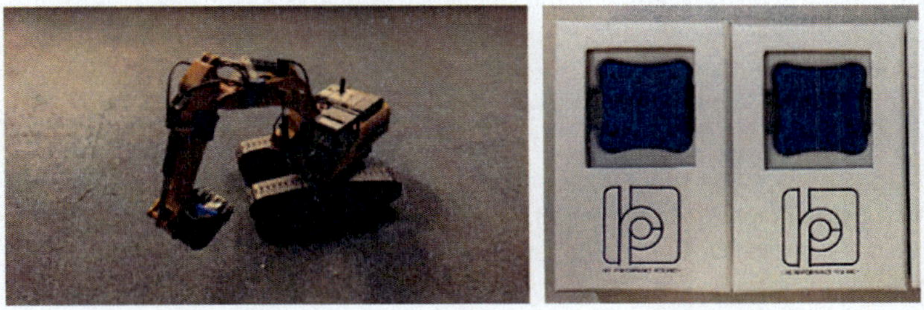

(a) (b)

Fig. 4. (a) Excavator Model (b) LPMS-B2 IMU sensor

Experiment Design. In this experiment, the IMU sensors were installed in two different positions of the excavator boom (i.e. IMU1 and IMU2 in Fig. 5.). One is at the gravity center of boom and the other is near the junction of boom and arm. The two IMU sensors. collected raw data for the subsequent estimation of excavator posture. As for the operation of the excavator model, our experiment was designed to better explain the results. Specifically, the continuous movements were simplified, according to the actual operation of the excavator in construction sites. Each segment rotated and moved separately in each operation. In this experiment, the excavator model simulated the actual operation of digging and dumping, and then recorded two continuous working cycles in each trial. Each working cycle starts with a given initial posture, and goes through cabin left turn, boom lift, arm lift, cabin right turn, boom drop, arm drop, and finally cycle back and stop at the initial posture. In order to justify the repeatability of our experimental results, the experiment was repeated four times in two different ferromagnetic environments.

In addition, as the complex ferromagnetic environment may have a direct influence on experimental results, the magnetometer data in experimental ferromagnetic environment and the corresponding device ID of two IMU sensors were recorded for the magnetometer calibration of posture estimation.

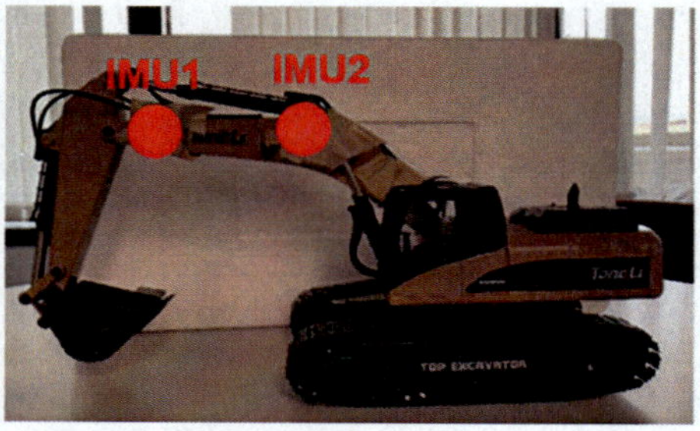

Fig. 5. Different installation positions of IMU sensors and the given initial posture in the experiment

In the step of Data Preprocessing, the magnetometer was calibrated by ellipsoid fitting through the collected calibration data. The pre-processed original data were used for orientation estimation of the independent segment of the excavator through Madgwick algorithm. Finally, a quaternion representing the excavator segment rotation state was obtained. By converting the quaternion into Euler Angle, a more understandable representation of the experimental results was obtained as time series at Roll and Pitch, which were then compared by two measurement metrics namely Root Mean Square and Standard Deviation. Experimental results are presented in the next section, which would imply the performance of our proposed framework and the effect of different installation positions.

4.2 Experiment Result and Discussion

According to the raw data collected by two IMU sensors installed in different positions on the same segment of the excavator boom, the Table 1. shows the results at the Roll and Pitch under experimental conditions. By the data of the four experimental trials in the Table 1, estimation of excavator segment orientation in two different installation positions can be obtained. Average Roll RMS is 5.42°, Std is 5.23°, while average Pitch RMS is 3.87°, Std is 3.48°. And by observing the trend of data in Fig. 6., the Roll generally has a larger deviation and dispersion degree than the Pitch.

Table 1. Effect of position, Root Mean Square (RMS), Standard Deviation (Std) for Roll and Pitch in degree

Trial	RMS (Roll)	Std (Roll)	RMS (Pitch)	Std (Pitch)
1	6.11	5.45	5.48	4.46
2	4.43	4.37	4.23	4.02
3	6.08	6.07	3.81	3.60
4	5.04	5.03	1.95	1.84
Average	5.42	5.23	3.87	3.48

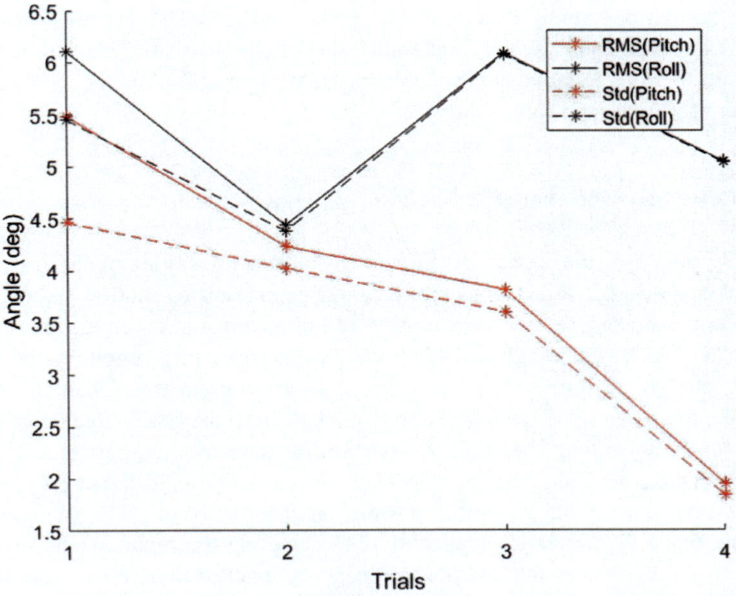

Fig. 6. Effect of position, Root Mean Square (RMS), Standard Deviation (Std) for Roll and Pitch in degree

Based on the average RMS, the deviation of our experimental result is in a reasonable range of the actual measurement, compared with IMU sensors in smart phones under the same algorithm in dynamic protocol [7]. The induced error in our study may be due to the high sensitivity of Madgwick algorithm to the changing magnetic field environment. In particular, the excavator operation is relatively extensive. In the actual operation of the excavator, the average deviation and dispersion degree shown in the Table 1. will not have a significant impact on the actual application of the excavator posture estimation in construction sites.

The reasons for the results can be inferred as follows: firstly, as a rigid body, the rotation is the same everywhere. Therefore, whether the IMU sensor is positioned near the junction and the gravity center or any other points, it should ideally be able to record the same set of data of rotation states. Secondly, the unavoidable estimation error due to the influence of magnetic field environment and IMU sensor hardware overweighs the error due to the inertial load applied to the accelerometer at different positions.

Therefore, from the perspective of the accuracy, IMU sensors installed at different positions have no significant impact on the excavator posture estimation. The induced error is within the acceptable range for industrial applications.

However, considering the application conditions and practical operation of excavators in construction sites, the installability is also an important factor. For buckets in particular, any installation position not near the junction may be subject to falling off due to collision with earth and rock during normal operation or other conditions affecting normal operation. Therefore, the installation plan of IMU sensor should be discussed in combination with installability. It is believed that the position near the junction is a more reliable and suitable installation position in the application of excavator posture estimation.

5 Conclusion and Future Work

This study discusses the impact of different installation positions of IMU sensors on posture estimation of excavators or orientation estimation of motion segments. The experimental results show that the average RMS of the error between the gravity center and near the junction is less than 5.42° under normal operating conditions in complex ferromagnetic environments. Such an error value belongs to the reasonable range of error in the actual measurement, and also can result from the reasonable range that can be accepted by an actual operation of excavator in industrial applications. Therefore, from the perspective of the accuracy, IMU sensors installed in different positions have no significant impact on excavator posture estimation, or the error is within the acceptable range for industrial applications. Considering the requirements of installability, this study proposes that the position near the junction should be a more reliable and suitable installation position in the application of excavator posture estimation.

Two aspects could be considered for potential improvement of this study. Firstly, only Madgwick algorithm was used in posture estimation of excavator. As the algorithm is sensitive to the magnetic field environment, the error may be enlarged due to such an influence. In the future, multiple algorithms could be applied for comprehensive evaluation of the effect of installation position on posture estimation results. Secondly, statistical method was not utilized to evaluate the results. This study intuitively explains the influence degree of error, based on considerations of industrial application. In future studies, statistical methods could be adopted to provide measurable impact evaluation criteria in order to enhance the reliability of the result analysis.

References

1. Soltani, M.M., Zhu, Z., Hammad, A.: Skeleton estimation of excavator by detecting its parts. Autom. Constr. **82**, 1–15 (2017)
2. Luo, H., Wang, M., Wong, P.K., Cheng, J.C.: Full body pose estimation of construction equipment using computer vision and deep learning techniques. Autom. Constr. **110**, 103016 (2020)
3. Rashid, K.M., Louis, J.: Times-series data augmentation and deep learning for construction equipment activity recognition. Adv. Eng. Inform. **42**, 100944 (2019)
4. Kim, S., Lee, Y., Sun, D., Lee, S., Yu, B., Jang, S., Kim, W., Han, C.: Development of bulldozer sensor system for estimating the position of blade cutting edge. Autom. Constr. **106**, 102890 (2019)
5. Im, S., Choi, S., Lee, J.: Autonomous traction control for the intelligent excavator system. In: 28th International Symposium on Automation and Robotics in Construction, ISARC 2011 (2011)

6. Sessa, S., Zecca, M., Lin, Z., Bartolomeo, L., Ishii, H., Takanishi, A.: A methodology for the performance evaluation of inertial measurement units. J. Intell. Robot. Syst. **71**, 143–157 (2013)
7. Mourcou, Q., Fleury, A., Franco, C., Klopcic, F., Vuillerme, N.: Performance evaluation of smartphone inertial sensors measurement for range of motion. Sensors **15**, 23168–23187 (2015)
8. Madgwick, S.: An efficient orientation filter for inertial and inertial/magnetic sensor arrays. Rep. x-io Univ. Bristol (UK) **25**, 113–118 (2010)
9. Berndt, D.J., Clifford, J.: Using dynamic time warping to find patterns in time series, vol. 10, pp. 359–370 (1994)

BIM-Based Concrete Printing

Kay Smarsly[1(✉)], Patricia Peralta[1], Daniel Luckey[1], Sebastian Heine[2], and Horst-Michael Ludwig[2]

[1] Bauhaus University Weimar, Chair of Computing in Civil Engineering, Weimar, Germany
kay.smarsly@uni-weimar.de
[2] Bauhaus University Weimar, F.A. Finger Institute for Building Material Engineering, Weimar, Germany

Abstract. Extrusion-based additive manufacturing (AM), or three-dimensional (3D) printing, has matured into a set of advanced methods to automate the construction of large-scale concrete structures, while minimizing cost and material waste. However, current AM data models are inadequate for 3D concrete printing due to insufficient incorporation of information on the relationships between process, material, and geometry, which may cause redundancy, information loss, and inconsistencies. Aiming at improving AM data modeling for concrete printing, this paper proposes a metamodeling approach for AM of concrete structures, referred to as "printing information modeling", which takes advantage of building information modeling (BIM). As will be shown in this paper, the BIM-based printing information model, serving as a metamodel, incorporates the digital data triplet of process, material, and geometry parameters to generate computer numerical control (CNC) commands that may readily be used for concrete printing. A validation test is performed, which instantiates the printing information model, using a BIM model, for generating CNC commands, enabling optimal digital data exchange from BIM models to concrete printers. As a result of this study, it is demonstrated that printing information modeling adequately defines the information required for AM of concrete structures using a BIM-based approach, showing promising potential to improve current AM data modeling efforts.

Keywords: Additive manufacturing · Concrete printing · Printing information modeling (PIM) · Metamodeling · Building information modeling (BIM)

1 Introduction

Driven by recent Industry 4.0 developments, the architecture, engineering, and construction (AEC) industry is experiencing an era of transformation and ongoing automation (Tay et al. 2018). Recent approaches towards automating construction methods are centered on extrusion-based additive manufacturing (AM), or three-dimensional (3D) printing, of large-scale concrete structures (Bos et al. 2016; Buswell et al. 2007). ASTM International, formerly known as American Society for Testing and Materials, defines AM as a process of generating objects from 3D model data by joining materials, usually layer upon layer (ASTM 2012). Research to date has focused

E. Toledo Santos and S. Scheer (Eds.): ICCCBE 2020, LNCE 98, pp. 992–1002, 2021.
https://doi.org/10.1007/978-3-030-51295-8_69

on developing viable AM hardware and materials to achieve concrete structures with certain degrees of geometrical precision, while reducing cost and waste (Buswell et al. 2018). In an attempt to improve geometrical precision and process control, sensor technologies in AM of concrete structures have been presented by Wolfs et al. (2017), illuminating the advantages of real-time monitoring to adjust process parameters during printing. However, the full potential of deploying sensing technologies, which has become common in other field within the AEC industry (Dragos and Smarsly 2016; Hartmann et al. 2011), has not yet been fully exploited in AM.

In recent years, efforts have been undertaken towards AM data modeling for concrete printing, where converting 3D model data into printed components is derived in the same way as in conventional AM (Peralta et al. 2019). In conventional AM, data modeling is mainly based on Standard Tessellation Language (STL) and G-code, i.e. ISO 6983-1 (Bonnard et al. 2018). However, current data models are not suitable for concrete printing of components of high load-bearing capacity and geometric precision. Shortcomings regarding AM data modeling for concrete printing include a lack of knowledge on how processes, materials, and geometries interrelate in AM data modeling, causing redundancy, information loss, and inconsistencies. Thus, current data modeling approaches limit concrete printing to an inconsistent and unreliable process, where expertise is required to determine ideal printing process settings, toolpaths, and printing strategies, taking into account variations in material properties of fresh concrete over time. The main challenge in AM data modeling, therefore, is to understand the relationships between AM processes, materials, and geometries, i.e. the digital data triplet.

To compensate for the shortcomings and to optimize AM data modeling for concrete printing, a metamodel for AM of concrete structures that takes advantage of building information modeling (BIM) is proposed in this paper. The BIM-based approach towards concrete printing enables AM data modeling and digital data exchange without breaks in the digital thread, incorporating real-time monitoring of material properties and adjustments of process parameters, based on previous studies (Theiler and Smarsly 2018). The metamodel represents a step necessary for standardizing the digital data triplet into a unified printing information model for concrete printing in civil engineering.

The remainder of this paper is organized as follows. First, the metamodeling approach, referred as printing information modeling (PIM), is shown. The PIM model defines input parameters and relationships within the AM digital data triplet to generate computer numerical control (CNC) code for providing hardware control of concrete printers. Second, using the PIM model as a formal basis for concrete printing, a validation test is devised using BIM models that are defined according to the Industry Foundation Classes (IFC) standard as input sources to generate CNC code, enabling an optimal digital thread from BIM models to concrete printers. Finally, the paper concludes with a summary and an outlook on potential future research efforts.

2 Printing Information Modeling for Concrete Printing

The printing information modeling approach outlines a metamodel, which comprises the main input parameters of the AM digital data triplet, describing the relationship between the parameters to semantically represent specifications for concrete printing jobs. Following a unified modeling language (UML) modeling approach, classes are defined to coherently group the input parameters of the digital data triplet. As a result, breaks in the digital thread are reduced, while optimizing digital data exchange for concrete printing. As a result, the PIM model represents an understandable and instantiable metamodel that may be implemented into technology-independent semantic descriptions of buildings, such as the IFC standard. In the following subsections, the main input parameters of the digital data triplet, the relationships, and the PIM model for concrete printing are introduced.

2.1 Main Parameters and Relationships

According to the digital data triplet, the main parameters relevant to the PIM model presented in the following subsection are subdivided into process parameters, material parameters, and geometry parameters. When describing the parameters, elements directly reflected in the PIM model are printed in italics.

Process Parameters. Most concrete printers have adopted an extrusion-based AM method, referred to as "fused deposition modeling (FDM)". Main elements of printing hardware related to data modeling in FDM printing are pump system, printhead, and motion-controlled printing system. The *pump system* is devised to transport concrete from the pump system to the printhead without segregation and bleeding. Relationships regarding pump system parameters have been identified, e.g., in studies of Bos et al. (2016) and Paul et al. (2018a). Pump system parameters include pump speed and pump pressure, which is associated with pump speed. The *printhead* allows the concrete to be printed at target locations with target speed under target angles, and it is composed of several parts, including a nozzle at its end. Printhead parameters include printhead speed, printhead position and orientation, extruder speed, and nozzle shape. Relationships regarding printhead parameters have been reported, e.g., by Salet, et al. (2017), Paul, et al. (2018a), Lim, et al. (2012), and Wolfs, et al. (2017) for printhead speed, printhead orientation, nozzle shape, and printhead height above the printing surface. The geometry of the concrete *filament* extruded from the printhead influences the *toolpath* definition and the buildability of structural components to be printed. Monitoring the geometry of the filament provides *control* of the geometric precision during the printing process. The motion-controlled printing system processes the CNC code, which defines toolpaths, controls the multiple axes, and monitors and adjusts the printhead performance based on sensing technologies. Influences of toolpaths definitions and sensing technologies on the properties of the concrete layers have been described, e.g., in studies of Salet, et al. (2017) and Wolfs et al. (2017).

Material Parameters. Concrete printing involves *concrete* as main printing material and *reinforcement*, which may be incorporated before, during, or after the printing process (Asperone et al. 2018; Buswell et al. 2018). The composition of concrete, defined in the *mix design*, determines whether a concrete mix satisfies printing process requirements and *material specifications*. Setting reactions of concrete show strong interactions with the process parameters and with the printing strategy applied for manufacturing. Relationships between *material properties* of fresh state concrete (e.g. green strength and layer interval time) and process parameters, together with the component size, are relevant to the way the printing process should be conducted to ensure extrudability, buildability, and layer adhesion, as identified in recent studies (Bos et al. 2016; Paul et al. 2018b; Salet et al. 2017; Wolfs et al. 2017).

Geometry Parameters. Definitions of *dimensions*, *points*, and *geometric objects* are necessary to describe 3D geometries of the structural components to be printed. The definitions may advantageously be inherited from digital 3D models, such as BIM models. Dimensional parameters are defined to ensure geometric precision. Point parameters are used to define vectors and geometries. Geometric object parameters, which define component geometries, depend on the geometric representation used in the digital 3D models. For example, Correa (2016) has demonstrated that IFC-compliant geometric representations, such as boundary representation and tessellated representation, are adaptable to 3D printing. In addition, geometric parameters related to the AM processes should be considered together with the geometrical descriptions, such as *contour lines* resulting from slicing.

2.2 Printing Information Model

As described earlier, several studies have documented relationships between the parameters of the data triplet. Herein, the studies are taken as basis to develop the PIM model, shown in the UML class diagram in Fig. 1. The PIM model describes the printing information necessary for concrete printing according to the digital data triplet with three main components materialized in the UML class diagram in terms of the abstract classes,

- *ProcessInformation*,
- *MaterialInformation*, and
- *GeometryInformation*.

Printing process data and monitoring data are included in the *ProcessInformation* abstract class, categorized into the *PumpSystemData*, *PrintheadData*, *FilamentData*, *ToolpathData*, and *ControlData* subclasses. The data of the material employed in the printing process, such as concrete and reinforcement, is described in the *MateriaInformation* abstract class with the subclasses *ReinforcementData* and *ConcreteData*, the latter including the subclasses *MaterialSpecification*, *MixDesign*, and *MaterialProperty*: Material specifications define a concrete mix design, which, in turn, has a set of material properties that may be monitored. Finally, geometry data that may be inherited from BIM models is defined in the *GeometryInformation* abstract class, categorized into the *Dimension*, *Point*, *GeometricObject* and *ContourLine* subclasses, to achieve a complete geometrical description.

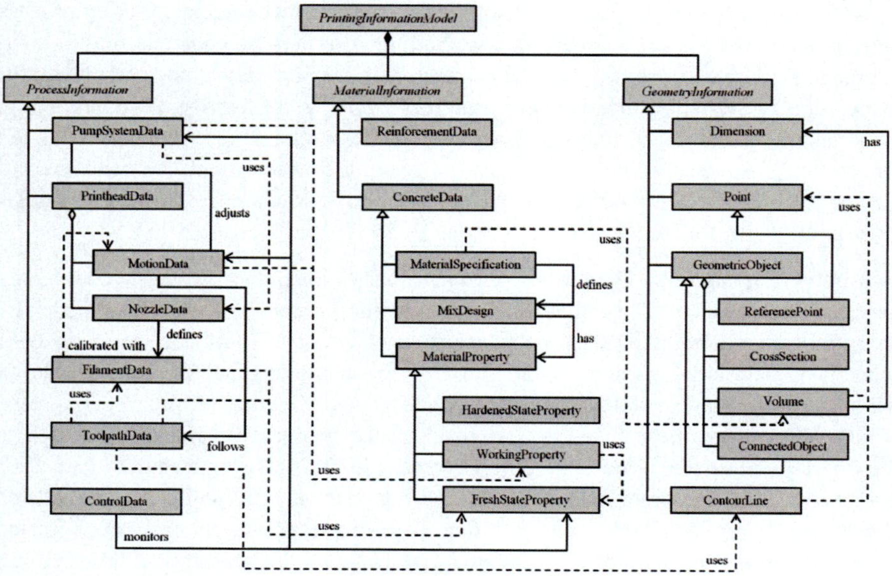

Fig. 1. Extract of the PIM model.

3 A Building Information Modeling Approach Towards Concrete Printing

Building information modeling is a promising means to provide complete information relevant to AM of concrete structures, incorporating the data triplet in a single data model. As shown in previous studies, with the IFC schema being extended to describe sensor systems (Theiler and Smarsly 2018) or cyber-physical systems (Fitz et al. 2019), incorporating control data for monitoring and adjustments of concrete printing jobs becomes possible on a standardized BIM basis. A BIM approach that implements the PIM model as a formal basis for concrete printing is presented in this section. The geometry parameters and several (but not all) material parameters required for AM are already specified in common BIM models compliant to the IFC schema. In this study, by incorporating all parameters and parameter relationships defined in the PIM model into a BIM model, printing strategies and toolpaths are defined and, then, commands that provide hardware control for concrete printing are generated as CNC code. Clearly, IFC schema extensions are needed to incorporate the process parameters and the material parameters that are not yet defined in the IFC schema, which is, however, not within the scope of this paper. For validating the BIM-based approach towards concrete printing, a software tool, the PIM test application, is designed and implemented that extracts information related to the geometry and material parameters from IFC-compliant BIM models. The PIM application generates CNC commands from the geometry and material parameters, while process parameters are added manually as user-defined parameters. The PIM application and the validation test are described in the following subsections, complemented by a summary and a discussion of the validation test results.

3.1 Design and Implementation of a PIM Test Application

To generate CNC code, the PIM application, developed in Java language, instantiates PIM models for concrete printing jobs based on process parameters entered manually as well as material and geometry parameters extracted from BIM models, respectively. In other words, the input data for the PIM application is the data stemming (i) from BIM models and (ii) from user-defined inputs (e.g. printhead and pump parameters), while the output is CNC code for hardware control of concrete printers. Furthermore, an algorithm for calculating slicing (to obtain contour line data) and an algorithm for calculating filling (to obtain toolpaths) is implemented into the PIM application. The slicing algorithm solves the plane intersection problem originating from slicing geometric objects into layers. The filling algorithm implements a raster-scanning approach to determine the infill pattern for each layer of the geometric object and defines toolpaths, using a user-defined printing strategy (i.e. unidirectional or bidirectional) and deploying filament data to increase geometric precision. The CNC code is generated in terms of G-code files, incorporating specifications of printhead, pump, material, and toolpaths. So called "PIM factors" are applied to modify the printhead and pump parameters, aiming to compensate for variations in the material properties. The software architecture of the PIM application, with main packages and classes, is shown in Fig. 2.

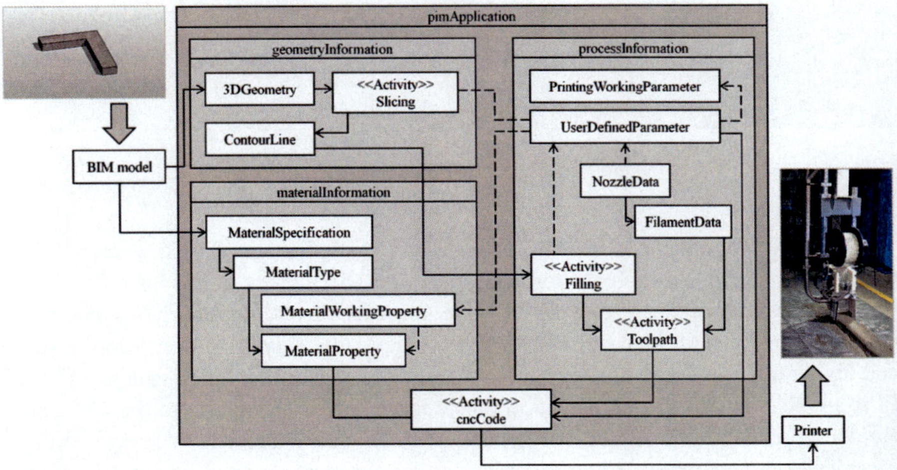

Fig. 2. PIM test application for validating the BIM-based approach.

3.2 Validation Test

The validation test is performed for an L-shaped concrete wall of 80 mm height, 150 mm thickness and a base of 800 mm × 600 mm (length × width), designed as an IFC-compliant BIM model. The validation test is conducted on a gantry concrete printer, shown in Fig. 3, and a custom-designed concrete mix with a design strength of 30 MPa. The gantry concrete printer has a capacity to print 6000 cm^3 of concrete, a nozzle of 150 mm × 40 mm (length × width), and a printing area of 2000 mm 1000 mm (length × width). The geometry and material parameters are extracted from the IFC file of the BIM model and used as input parameters to the PIM application. Regarding the process parameters, a printing speed of 100 mm/s, a pumping speed of 22 mm/s for a pump pressure of 1 MPa, a layer height of 40 mm and a unidirectional printing strategy are specified manually.

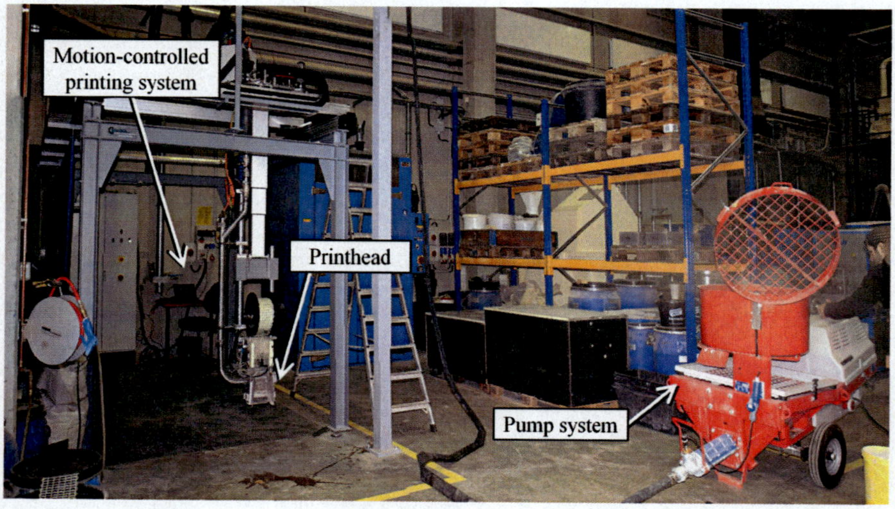

Fig. 3. Gantry concrete printer used for the validation test.

The PIM application slices the L-shaped wall, defined in the BIM model, into layers and generates a toolpath for each layer (i.e., two layers here). Specifically, the PIM application generates the CNC code for each layer, taking into account the material properties of the custom-designed concrete mix, such as slower printing speed and higher pump pressure when printing the first layer. The CNC code generated by the PIM application and the visualization of the toolpath corresponding to the CNC code are shown in Fig. 4. The CNC code has a start command block to set initial process parameters, a positioning command block to set the start position of the printhead, a printing command block that controls the extrusion of the concrete printer, and a homing command block to finalize printing. The extrusion of concrete by the pump system is controlled with commands to turn the pump on and off. Finally, the CNC code generated by the PIM application is transferred to the motion-controlled printing system of the gantry concrete printer, and the wall is printed.

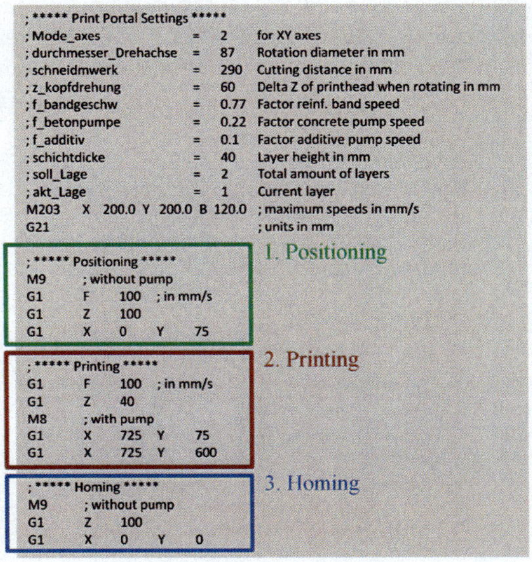

```
; ***** Print Portal Settings *****
; Mode_axes                  =  2     for XY axes
; durchmesser_Drehachse      =  87    Rotation diameter in mm
; schneidmwerk               =  290   Cutting distance in mm
; z_kopfdrehung              =  60    Delta Z of printhead when rotating in mm
; f_bandgeschw               =  0.77  Factor reinf. band speed
; f_betonpumpe               =  0.22  Factor concrete pump speed
; f_additiv                  =  0.1   Factor additive pump speed
; schichtdicke               =  40    Layer height in mm
; soll_Lage                  =  2     Total amount of layers
; akt_Lage                   =  1     Current layer
M203   X  200.0 Y  200.0 B  120.0     ; maximum speeds in mm/s
G21                                   ; units in mm

; ***** Positioning *****             1. Positioning
M9        ; without pump
G1        F     100   ; in mm/s
G1        Z     100
G1        X     0     Y     75

; ***** Printing *****                2. Printing
G1        F     100   ; in mm/s
G1        Z     40
M8        ; with pump
G1        X     725   Y     75
G1        X     725   Y     600

; ***** Homing *****                  3. Homing
M9        ; without pump
G1        Z     100
G1        X     0     Y     0
```

a) CNC code generated by the PIM application.

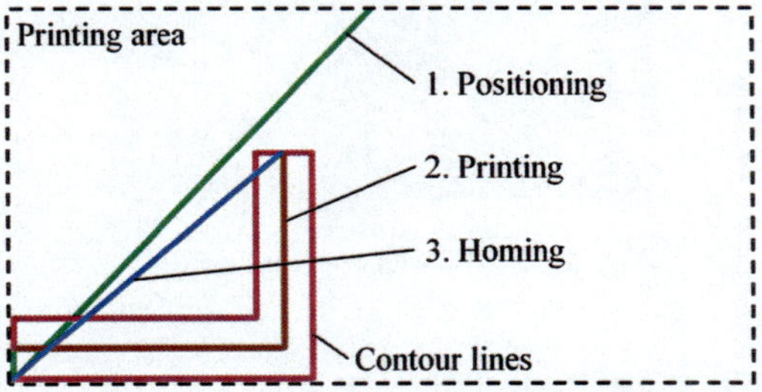

b) Visualization of the toolpath provided by the PIM application.

Fig. 4. Output of the PIM application.

3.3 Results and Discussion

The validation test demonstrates the feasibility of the BIM-based approach towards concrete printing, using the PIM model as formal basis. From the IFC file of the BIM model, geometry and material parameters have successfully been processed. Additional material parameters have manually been specified within the PIM application for the custom-designed concrete mix, such as an open time of 90 min, a setting time of one minute, and a maximum layer interval time of 5 min. While printing, which is shown in Fig. 5, the pump speed and the printhead speed are adjusted from layer to layer to compensate for the changes in the material properties, caused by a loss of moisture of the concrete over time. However, it must be noted that the algorithms of the PIM application are not suitable for AM data modelling of structural components with complex geometries. The slicing algorithm implemented in the BIM application is limited to the IFC entity *IfcSolidModel* defined as *IfcSweptAreaSolid* for generating contour lines. Furthermore, the scanning distance of the filling algorithm is defined by a filament width of 150 mm, resulting in a single infill line for the L-shaped wall. Also, it has been observed that the scanning-raster approach may generate faulty toolpaths for components with more complex shapes.

Fig. 5. BIM-based concrete printing.

4 Summary and Conclusions

A BIM-based metamodeling approach, referred to as "printing information modeling", has been developed as a first step towards standardizing the digital data triplet for additive manufacturing of concrete structures. Validating the PIM model in a test application, a BIM model has been used as input source to generate CNC code for a gantry concrete printer. It has been demonstrated that the PIM model, as a metamodel, semantically describes the AM digital data triplet, facilitating instantiable models for BIM-based concrete printing. Although a successful validation test can be reported, there exists room for improvements. For example, further studies are required to optimize the filling algorithm for calculating toolpaths and to further advance digital models that implement the PIM model to generate CNC code for BIM-based concrete printing able to incorporate sensor data for process monitoring and real-time printing adjustments. Another direction for future work, which deserves being explored, is the extension of the IFC schema with process parameters and (further) material parameters, enabling additive manufacturing of concrete structures with IFC-compliant BIM models as a sole formal basis.

Acknowledgments. Financial support of the German Research Foundation (DFG) through grant SM 281/7-1 is gratefully acknowledged. Major parts of this work have been conducted in the "Structural Health Monitoring Laboratory", sponsored by the European Union through the European Fund for Regional Development (EFRD) and the Thuringian Ministry for Economic Affairs, Science and Digital Society (TMWWDG) under grant 2016 FGI 0009. Any opinions, findings, conclusions, or recommendations expressed in this paper are those of the authors and do not necessarily reflect the views of DFG, EFRD, or TMWWDG.

References

Asperone, D., Menna, C., Bos, F., Salet, T.A., Mata-Falcón, J., Kaufmann, W.: Rethinking reinforcement for digital fabrication with concrete. Cem. Concr. Res. **112**(2018), 111–121 (2018)

ASTM: F2792. 2012 Standard terminology for additive manufacturing technologies. ASTM International, West Conshohocken, PA, USA (2012)

Bonnard, R., Hascoët, J.-Y., Mognol, P., Stroud, I.: STEP-NC digital thread for additive manufacturing: data model, implementation and validation. Int. J. Comput. Integr. Manuf. **31** (11), 1141–1160 (2018)

Bos, F., Wolfs, R., Ahmed, Z., Salet, T.: Additive manufacturing of concrete in construction: Potentials and challenges of 3D concrete printing. Virt. Phys. Prot. **11**(3), 209–225 (2016)

Buswell, R.A., Leal de Silva, W.R., Jones, S.Z., Dirrenberger, J.: 3D printing using concrete extrusion: A roadmap for research. Cem. Concr. Res. **112**(2018), 37–49 (2018)

Buswell, R.A., Soar, R.C., Gibb, A.G., Thorpe, A.: Freeform construction: mega-scale rapid manufacturing for construction. Autom. Const. **16**(2), 224–231 (2007)

Correa, F.R.: Robot-Oriented Design for Production in the context of Building Information Modeling. In: Proceedings of the 33rd International Symposium on Automation and Robotics in Construction, Auburn, AL, USA, 21 July 2016 (2016)

Dragos, K., Smarsly, K.: Distributed adaptive diagnosis of sensor faults using structural response data. Smart Mater. Struct. **25**(10), 105019 (2016)

Fitz, T., Theiler, M., Smarsly, K.: A metamodel for cyber-physical systems. Adv. Eng. Inform. **41**(2019), 100930 (2019)

Hartmann, D., Smarsly, K., Law, K.H.: Coupling sensor-based structural health monitoring with finite element model updating for probabilistic lifetime estimation of wind energy converter structures. In: Proceedings of the 8th International Workshop on Structural Health Monitoring. Stanford, CA, USA, September 13, 2011 (2011)

Lim, S., Buswell, R., Le, T., Austin, S., Gibb, A., Thorpe, T.: Developments in construction-scale additive manufacturing processes. Autom. Const. **21**(2012), 262–268 (2012)

Paul, S., Tay, Y.D., Panda, B., Tan, M.: Fresh and hardened properties of 3D printable cementitious materials for buildings and construction. Arch. Civ. Mech. Eng. **18**(1), 311–319 (2018a)

Paul, S., van Zijl, G., Tan, M., Gibson, I.: A review of 3D concrete printing systems and material properties: current status and future research prospects. Rapid Prototyping J. **24**(4), 784–798 (2018b)

Peralta, P., Mthunzi, E. M., Heine, S., Ludwig, H.-M., Smarsly, K.: A metamodel for 3D concrete printing. In: Proceedings of the Sixteenth International Conference on Civil,

Structural & Environmental Engineering Computing, Riva del Garda, Italy, 16 September 2019 (2019)
Salet, T.A., Bos, F.P., Wolfs, R.J., Ahmed, Z.Y.: 3D concrete printing – a structural engineering perspective. In: Proceedings of the 2017 fib Symposium – High Tech Concrete: Where Technology and Engineering Meet, Maastricht, NL, 12 June 2017 (2017)
Tay, S.I., Lee, T.C., Hamid, N.Z.A., Ahmad, A.N.A.: An overview of industry 4.0: definition, components, and government initiatives. J. Adv. Res. Dyn. Control Syst. **10**(14), 1379–1387 (2018)
Theiler, M., Smarsly, K.: IFC monitor – an IFC extension for modeling structural health monitoring systems. Adv. Eng. Inform. **37**(2018), 54–65 (2018)
Wolfs, R.M., Bos, F.P., van Strien, E.F., Salet, T.M.: A real-time height measurement and feedback system for 3D concrete printing. In: Proceedings of the 2017 fib Symposium – High Tech Concrete: Where Technology and Engineering Meet, Maastricht, NL, 12 June 2017 (2017)

Field BIM: Establishing a Requirements Framework for Mobile BIM Technologies

Benjamin Jowett and Mohamad Kassem(✉)

Northumbria University Newcastle, Newcastle NE1 8ST, UK
{benjamin.jowett,Mohamad.Kassem}@northumbria.ac.uk

Abstract. Field BIM and its enabling Mobile BIM Technologies (MBT) are increasingly recognised for their role in improving collaboration and integration between project teams on construction sites. Very limited research about the development and applications of Field BIM and MBT is available in the literature. The few available studies have generally focused on specific types of MBT and their application on a single project or individual use cases. A framework to clarify the technical and business requirements of MBT is still missing. This paper is part of a wider research aiming to develop a holistic framework for MBT and Field BIM consisting of a requirements taxonomy, a benefits evaluation approach linking use cases to relevant performance metrics, and a project and supply chain performance analytics dashboard. This paper presents the findings about the technical and business requirements of MBT, which represents the first step towards the development of the requirements taxonomy.

Keywords: BIM · Field BIM · Mobile BIM technology

1 Introduction

Inefficient practices and low levels of productivity are consistently diagnosed as key symptoms affecting the construction sector (e.g. Latham 1994; Egan 1998; Wolstenholme 2009; Hackitt 2018; Farmer 2016; KPMG 2014; CITB 2018). One of the frequently cited issue contributing to such symptoms is the low level of collaboration between project participants. There is a consistent proposition across the literature and the digital transformation strategies of many countries that the adoption of BIM is the one key measures for a more productive and efficient Construction Industry. Digital transformation of the construction sector through BIM has gained significant attention in many countries (Kassem and Succar 2017). For example, the UK BIM Framework – formerly called BIM Level 2 – has been implemented to improve the productivity within the UK construction sector and requires all centrally funded public sector projects to implement information management processes according to the standard and guidance of the UK BIM Framework.

Many model uses and their enabling BIM technologies and processes have been widely discussed in the literature. Field BIM, despite its potential impact on the productivity of construction projects, has received little attention as evidenced in this research. According to Koseoglu and Nurtan-Gunes (2018) the BIM literature shows

E. Toledo Santos and S. Scheer (Eds.): ICCCBE 2020, LNCE 98, pp. 1003–1013, 2021.
https://doi.org/10.1007/978-3-030-51295-8_70

that where the benefits of BIM are discussed, it is generally focused on the preconstruction and design stages. Research into different aspects of construction stage, such as defect management (Lin et al. 2016) and its enhancement through BIM have not been considered from field BIM and MBT perspectives. There is also a limited understanding of the terms used and the technological and business requirements of MBT for field BIM. There are also no established ways for evaluating the benefits of BIM (Bockstael and Issa 2016; Leite et al. 2011) in general and of field BIM in particular. This gap in literature is significant when considering that pre-construction and design- based roles are predominantly desk-based occupations whereas field- based occupation account for 73.5% of the total construction workforce (Lepanjuuri and Humphrey 2016). Hence, the application of field BIM and MBT is also understood to be a way of exposing a significant proportion and a wider range of occupations to BIM-based processes. This paper presents the first step towards the establishment of requirements taxonomy for MBT. Section 2 reviews related studies and confirm the knowledge gap in this field. Section 3 described the methodology. Section 4 presents the preliminary findings. Section 5 discusses the findings and concludes the paper.

2 Literature Review

The concepts of both Mobile BIM Technologies (MBT) and Field BIM are not widely discussed, and established definitions of both concepts are also lacking. A few terms including Mobile BIM, Field BIM, Site BIM, Cloud BIM, and Mobile Apps are used interchangeably in existing studies (such as in Abanda et al. 2018; Ma et al. 2017; Rolfsen et al. 2019). A cloud- based BIM technology may not have field functionalities and vice versa. To avoid this misconception between the terms, this paper separates between two concepts: Field BIM as one of the applications or uses of BIM on construction sites; and MBT as the enabling technology of Field BIM. Field BIM is defined as a BIM application (or BIM use or model use) where project information (Building Information Model, documents, and data) is utilised on construction sites to assist in the management of workface operation. Mobile BIM Technologies (MBT) enables Field BIM by providing the necessary information management capabilities (e.g. accessing relevant project information, viewing and editing project information, submitting requests, etc.).

Some existing studies have investigated the benefits of these solutions (Chong et al. 2014; Wong et al. 2014; Wang and Chong 2015; Abanda et al. 2018). Chuang et al. (2011)suggested that the utilisation of cloud- based technologies is a critical component to enhance the control and monitoring of a project between participating organisations. Abanda et al. (2018) noted that cloud- based technology that support the BIM process can lead to higher levels of cooperation and collaboration amongst project organisations. Cloud- based BIM technology is also able to provide a remote platform through which real- time communication amongst project team members is possible (Abanda et al. 2018).

A review of cloud- based technologies within the construction sector identified that 21 of the 31 (67.7%) applications reviewed facilitated mobile access whereas only 4 of the 31 applications (12.9%) facilitated BIM model integration (Chong et al. 2014).

Applications that were both mobile and had model access equated to 3 of the 31 (9.7%) applications reviewed. This point is of significance when distinguishing MBT from broader cloud- based technology as current research tends to not distinguish between these two types.

Chu et al. (2018) found that there are efficiencies to be gained through digital processes facilitated via MBT but highlighted the risk of over burdening site teams with too much data. Appearing to support this supposition, Kerosuo et al. (2015) also noted that site teams tended to prefer 2D paper drawings due to the perceived increase in effort of accessing information via a 3D model.

Mahamadu et al. (2013) highlighted the need to consider information security measures in mobile BIM technologies which should consider the influencing mechanisms of security risks across the technological, organisational and environmental context. In the UK construction sector, the standard PAS 1192-5 (BSi 2015) sets out the processes to be undertaken regarding security measures to be taken within a BIM process.

3 Research Methodology and Methods

To develop the theoretical constructs and models for Field BIM and MBT, a grounded theory approach has been adopted. The grounded theory is reflected by (1) the inductive nature of the research where data is gathered, synthesized, analysed, and conceptualised for the purpose of building the theoretical constructs; (2) adequate sampling techniques of participants and case study projects; and (3) the close and longitudinal interaction of the researchers with the data in the actual field.

Several research methods were combined as part of this methodology. These included: (1) *a case study project* which comprised of (i) a launch workshop at the start of the case study project to understand the initial requirements of a MBT to support Field BIM, (ii) *a survey* at the end of the case study project (after one year from the start) to understand the usage of MBT; (2) *independent expert interview*, external to the case study project, to corroborate and complement the identified requirements within the case study project; (3) a *focus group* with experts to identify MBT business requirements; and (4) and *survey of industry* to understand general usage of MBT and Field BIM and potential risks. A synthesis of findings from across all the aforementioned research items is used to develop the MBT requirements taxonomy. This paper reports on the outcomes from the first two research methods (i.e. the case study project, and the independent expert interview).

The following sections respectively describe the research design and the findings from the Case Study project (i.e. launch workshop; a survey at the end of the case study project), and the independent expert interview.

4 Case Study Project

The criteria for selecting a suitable case study project included: (1) it should be a project that relied upon the use of federated 3D geometrical models for design and construction; (2) it should be implementing Field BIM using a relevant MBT and its associated workflows; (3) the adoption of BIM should be governed by some established framework for information management (e.g. the UK BIM Framework); and (4) the adoption of non- proprietary data formats to reduce the software- bias in the research outcomes. The selected case study consisted of a £40 m Design and Build Higher Education project situated on a city centre site in Northern England. The project was delivered to 'Level 2 BIM' standards (BSi 2013). It comprised of six linked blocks which were stepped in height that ranged from five storeys to eight storeys high creating a floor area of 15,500 m2. The projects delivery was also aligned with the stages of the RIBA Plan of Works (2013).

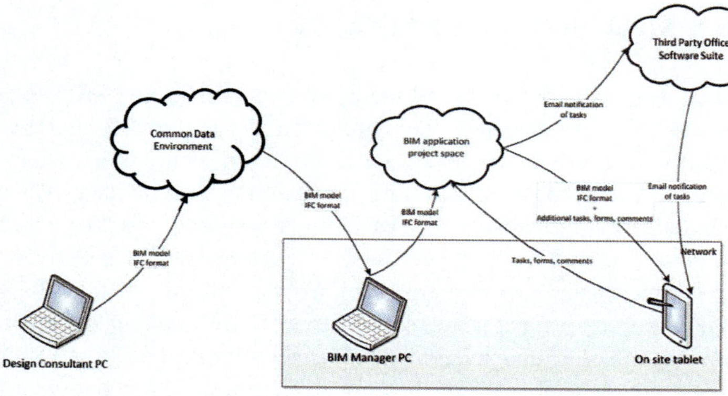

Fig. 1. Field BIM workflows within the case study project.

The Field BIM workflows involved in the case study project are described in Fig. 1. The MBT application was utilised over a period of one year and gathered qualitative and quantitative data about Field BIM or on- site uses of MBT. 3D models were transferred utilising the IFC file format, whilst 2D based information, including drawings and specifications, utilised the PDF format.

4.1 Launch Workshop

A launch workshop was undertaken with the project team before the start of the case study project. The workshops aimed to capture high level requirements about Field BIM and the capabilities of enabling MBT. The workshop agenda was developed and issued one week prior to the workshop and consisted of; (1) Explanation of BIM deliverables on the case study project, (2) introduction to the MBT application, and (3) open sharing of views on MBT requirements. The workshop involved 10 site managers who were active participants in the case study project. They had no prior experience with any proprietary MBT was deemed necessary to reduce potential bias in requirements identification. The workshop was held during the middle of RIBA Stage 4 in the construction site offices and facilitated by the project BIM Manager who also captured the MBT requirements of each manager. The high- level categories of requirements that were captured are summarised as follows:

- View 3D Model Files;
- View 2D drawings, specifications and similar associated construction documents;
- Produce digital tasks that are linked to both the 3D and 2D environment;
- Produce digital forms that are linked to both the 3D and 2D environment;
- Perform analysis on the produced digital tasks and forms; and
- Ability to work on both IOS and Android tablet devices.

These high- level requirements are fulfilled by the selected MBT which was introduced to the project supply chain and the necessary training was provided. The project supply chain has accepted the introduction of the solution despite it was not part of the specified technology requirements at the project outsets. The table within Appendix A sets out the identified intersections between the identified MBT requirements and use cases by management discipline. Subsection 4.2 describes the perceived performance of the project's supply about Field BIM and MBT.

4.2 Survey of Participants at End of Case Study

A year after the workshop launch, the 10 site managers in the case study project team were asked to complete a questionnaire on the overall performance of the MBT. The questionnaire was executed via an online platform, sent via email to each individual manager. Responses were obtained over a period of 1 week and comprised of 9 Likert-Scale questions. One free text response was also executed to obtain further information that may not have been previously considered in the other questions. The results were extracted from the online platform and reported in Table 1. The result clearly shows a high level of satisfaction of the project supply chain with the implementation of Field BIM and the MBT. It is important to note that additional requirements, representing MBT functionality that emerged during the case study's implementation, are also rated in Table 1.

The participants were given the opportunity to provide qualitative comments about their experience in implementing Field BIM. One manager (M6) commented that the MBT was brought late into the project which led to a combination of two workflows on site, one involving information coming to site via MBT and another workflow occurring outside the MBT; this feedback is attributed to the outlier result evident in the final question where M6 also noted that the team had *"too many ways of accessing information"*. M7 commented upon both the strengths and the weaknesses of the MBT and suggested some further development: *"the mobile application is more intuitive than the desktop version which has a complicated GUI. The model viewing excellent, but issue of forms from the mobile application needs improvement. Use of the location service on mobile devices to automatically pin- point current location within the building and adding augmented reality to view areas with the model overlaid are*

Table 1. Perceived performance of MBT by the project's supply chain.

	Poor	Fair	Good	Very Good	Excellent
Viewing of 2D drawings	0%	0%	0%	60%	40%
Viewing of 3D models	0%	0%	0%	40%	60%
Ease of finding correct room/space in a 3D model	0%	0%	0%	40%	60%
Ease of finding correct equipment in a 3D model	0%	0%	0%	40%	60%
Download of 2D drawings	0%	0%	0%	90%	10%
Download of 3D models	0%	0%	10%	80%	10%
Effectiveness of communication with project team and subs	0%	0%	20%	70%	10%
Understanding of design intent	0%	0%	0%	80%	20%
How good is the application at creating/viewing issues and forms?	10%	0%	0%	90%	0%

functionalities that requires improvements". M5 appreciated the ability to "*look at a 3D model to see how the area we are working on is supposed to look like"*. M8 summed up some of the efficiencies enabled by the MBT: *"I have not needed to download any drawings from the mobile application, just to view them. The capability to send an email alert with a summary snagging items - new, open, updated - is useful"*.

5 Independent Expert Interview

The 10 site managers on the case study project had no prior experience of Field BIM and MBT. To substantiate and augment the identified requirements through the workshop with the 10 site managers, an additional interview was performed with a project manager who is experienced in Field BIM and MBT. The interviewed expert had to be external to the case study project team but working within the same organisation so as not to inhibit the findings and offer new insights. The selected Project Manager was not involved in the case study project but had delivered a £70 m commercial project in London utilising MBT. The interview was undertaken whilst the case study project was in construction. The interview took place in isolation from all case study participants over a period of 2 weeks of part- time questioning. Open ended probing questions covering the functionalities and capabilities of MBT that are required for Field BIM were posed. The themes that guided the initial sets of question emerged from the project case study and covered the use of MBT for:

- Viewing Models, Drawings and Documents;
- Creation and Management of Digital Tasks and Forms;
- Storing and Accessing of Documents; and
- Quality Control and Assurance.

The content of the interview was transcribed and key highlights are included in this section. The Project Manager noted that the MBT application was able to manage *"all of the forms, tasks, processes, meeting minutes, diaries and general day to day documentation that managers require to run a modern day construction site"*. This complements the findings of from the case study project which showed that site managers were predominantly utilising the MBT application for the generation of forms and tasks.

When asked about the utilisation of the Building Information Model in tandem with associated documentation, the Project Manager highlighted *"the system links into a building information model or current up- to- date drawings which all are kept and accessed on the one system. This gives us the ability to add and upload further information needed by the UK BIM framework such as British Standards, manufacturers recommendations, test and inspection sheets, functionality inspections as well as Mechanical & Electrical completion certifications like NIC EIC certificates, among others"*. This also supports the findings from the Case Study project which showed uses of MBT for the uploading and viewing of project documentation.

The Project Manager further highlighted that a full usage of MBT can "support the Golden Thread[1] of information as mentioned within the current Hackett Report." At the end of the interview the project manager highlighted that a compulsory and thorough training for both site and subcontractors teams is critical for the successful implementation of MBT.

6 Discussion and Conclusions

The paper identified some important gaps in literature about Field BIM and MBT including. There is no clarity around the terms used; limited understanding of requirements at both project and organisation level; and lack of a framework linking together MBT requirements, use cases, and benefits evaluation. This paper started addressing some of these gaps. It separated between the concepts of Field BIM and its enabling MBT to avoid the on- going interchangeable use of terms (e.g. Mobile BIM, Field BIM, Site BIM, Cloud BIM, and Mobile Apps).

Although not investigated further within the scope of this research, there remains further challenges to overcome prior to the broader understanding of the application of MBT on construction projects; (I) Data Security; both physical and virtual, needs to be more widely understood, particularly with the advent of the remote accessibility of potentially sensitive data facilitated via MBT (II) Excess of data; as on site data generation may experience significant growth, the creation of unusable data may also increase, compromising the ease of access to useable data (III) Skills gap and ease of use; consideration of usability of the MBT and the training and capability of the workforce present clear barriers to the successful use of MBT and need to be understood as part of wider research in this area.The longitudinal case study with a launch

[1] A Digital Record of the building capturing key as planned and as built information such products used and their characteristics. It is intended to underpin an effective understanding of the constructed building across the life cycle.

workshop at the beginning and a survey after one year, with the involvement of 10 site managers, showed very positive perceptions of Field BIM and MBT. It also contributed to identify preliminary categories of high- level MBT requirements that are necessary to enable Field BIM. The positive experience of the project's supply chain positive with the implementation of the MBT experience and the high- level requirements captured from the project were corroborated and complemented through an independent interview with an expert in MBT. The results from the interview with the expert shows that in addition to the capability of handling all different types of site information, Field BIM and MBT can play a significant role in quality control and assurance.

It is clear from the results that Field BIM, enabled by adequate MBT, has a significant impact on the performance of site operation. It was accepted on a significant construction project despite its late introduction to a supply chain with no prior experience in Field BIM and use of MBT. A major project, that has implemented Field BIM using MBT, confirmed the feasibility of managing all types of site information and revealed further potential for MBT.

These results will be extended in future in two directions, each with a specific purpose: (I) *confirmation through further qualitative and quantitative data*: this will implement a larger survey of the construction sector to confirm perception about usage, identify potential risks of MBT adoption, and perform a quantitative analysis of MBT usage on real projects; and (II) *extension of requirements to cover both project and organisation requirements:* this will aim to extend the high- level requirements for project level (identified in this paper) and identify organisation- level requirements. Together these two strands should provide adequate knowledge to develop a requirements taxonomy for MBT. The resultant taxonomy is expected to play a key role in the subsequent development of the framework for Field BIM and MBT. Both the requirements taxonomy and the Field BIM framework will have a number of practical uses (e.g. use of the requirements taxonomy to develop new MBT and/or perform an evaluation of exiting MBT; development of a standard operating procedures for Field BIM; automated approaches to site performance analysis) and theoretical implications (e.g. benefits evaluation approach; enablement of new findings through integration between BIM, Lean Construction theory and methods; new methods for lean supply chain management; data- driven supply chain performance measurement systems).

Appendix A – MBT Requirements and Use Case Interaction

MBT Requirements	Core Management Disciplines						
	Health & Safety	Quality	Environmental	Site Records & Programme	Design	Cost	BIM
View 3D Model Files in the field	- Understand potential construction risks in design – e.g. working at height i - Access to relevant H&S data within the model	- Access to relevant Quality data within the model – e.g. specification references - Understand design intent	- Access to relevant Environmental data within the model – e.g. BREEAM, EPC, Life Span of product	- Compare site progress versus model in the field - Record progress in the field	- Compare design intent versus installation - Understand coordination issues	- Compare site progress versus model in the field	- Access to model data in the field - Compare model data to installed assets - Share model issues from the field with real life context
View 2D Drawings & Documentation in the field	- Understand detail that may not be recorded within the model – e.g. CDM risks - Access to relevant specifications, standards and guidance	- Understand detail that may not be recorded within the model – e.g. interface details - Access to relevant specifications, standards and guidance	- Understand detail that may not be recorded within the model – e.g. BREEAM information - Access to relevant specifications, standards and guidance	- Understand detail that may not be recorded within the model – e.g. objects that may not be modelled - Access to relevant specifications, standards and guidance	- Understand detail that may not be recorded within the model – e.g. objects that may not be modelled - Access to relevant specifications, standards and guidance	- Ability to visualise elements that are costed but not modelled - Access to relevant specifications, standards and guidance	- Monitor compliance with BEP and associated standards remotely - Access to wider BIM documentation for reference in real life environment for compliance
Produce Digital Tasks linked to either 2D or 3D environment in the field	- Action on site hazard observations - Action H&S Good Practice - Address non-compliance issues - Accessible Site Audit Trail	- Action on site quality issues; e.g. snagging, defective materials - Action incomplete works - Accessible site audit trail	- Address on site environmental issues; e.g. water discharge, dust, noise, vibration - Accessible site audit trail	- Action incomplete works - Accessible site audit trail - Record status of project at regular intervals for future reference; e.g. condition of scaffolding, fire stopping installation	- Technical Queries from the field - Requests for Information from the field - Address non-compliance issues – e.g. incorrect materials - Accessible site audit trail	- Access to all site activity to understand which organisations are performing which activities - Monitor progress against costs	- Address BIM Compliance issues - Utilisation of BCF to communicate back to design team - Establish connections between site data and model data
Produce Digital Forms linked to either 2D or 3D environment in the field	- Production of regular forms – e.g. permits, inspections, drills, - Accessible evidence of controlling the works - Access to historical data over duration of project	- Production of regular quality forms – e.g. Inspection Checklist, Pre- Start Checklist, Quality Plans - Access to historical and live site data	- Production of regular quality forms – e.g. Environmental Inspection Checklist, Waste Records, Quality Plans - Access to historical and live site data	- Record site progress; planned vs actual - Access to live site data to understand progress over duration of project	- Manage change control and understand impact on design - Detailed technical queries and requests for information - Approve design solutions on site - Access to site data audit trail	- Manage change control and understand impact on design - Understand flow of information and impact on progress and costs - Manage sign off of work packages for payments	- Record Model Compliance against agreed deliverables - Establish connections between site data and model data

(*continued*)

(*continued*)

MBT Requirements	Core Management Disciplines						
	Health & Safety	Quality	Environmental	Site Records & Programme	Design	Cost	BIM
Perform Analysis on Digital Data created in the field	- Understand where current issues are quickly, and actions required - Establish trends in data against variables – e.g. amount of cleanliness issues by contractor discipline or building area	- Understand where current issues are quickly, and actions required - Establish trends in data against variables – e.g. amount of fire issues by contractor discipline	- Understand where current issues are quickly, and actions required - Establish trends in data against variables – e.g. amount of missing BREEAM data by design discipline	- Understand where current issues are quickly, and actions required - Establish trends in data against variables – e.g. amount of delays in planned vs actual by contractor discipline	- Understand where current issues are quickly, and actions required - Establish trends in data against variables – e.g. amount of coordination issues by design discipline	- Understand where current issues are quickly, and actions required - Establish trends in data against variables – e.g. amount of outstanding work by contractor discipline to substantiate payments	- Understand where current issues are quickly, and actions required - Establish trends in data against variables – e.g. amount of outstanding model issues by model author organisation

References

Abanda, F.H., Mzyece, D., Oti, A.H., Manjia, M.B.: A study of the potential of cloud/mobile BIM for the management of construction projects. Appl. Syst. Innov. **1**, 9 (2018)

Bockstael, D., Issa, M.H.: A methodology for contractor clash detection using building information modelling on commercial construction projects. J. Inf. Technol. Constr. **21**, 233–249 (2016)

BSi: PAS 1192-5: 2015 Specification for security- minded building information modelling, digital built environments and smart asset management, 1st edn. BSi, London (2015)

Chong, H.Y., Wong, J.S., Wang, X.: An explanatory case study on cloud computing applications in the built environment. Autom. Constr. **44**, 152–162 (2014)

Chu, M., Matthews, J., Love, P.E.D.: Integrating mobile building information modelling and augmented reality systems: an experimental study. Autom. Constr. **85**, 305–316 (2018)

Chuang, T.H., Lee, B.C., Wu, I.C.: Applying cloud computing technology to BIM visualization and manipulation. In: 28th International Symposium on Automation and Robotics in Construction, pp. 144– 149. ISARC, Seoul (2011)

CITB: UK Industry Performance Report, 1st edn. CITB, Kings Lynn (2018)

Egan, J.: Rethinking Construction, 1st edn. HMSO, London (1998)

Farmer, M.: The farmer review of the UK construction labour model, 1st edn. CLC, London (2016)

Hackitt, J.: Building a Safer Future, 1st edn. HMSO, London (2018)

Kassem, M., Succar, B.: Macro BIM adoption: comparative market analysis. Autom. Constr. **81**, 286–299 (2017)

Kerosuo, H., Paavola, S., Miettinen, R., Maki, T.: Challenges of the expansive use of Building Information Modeling (BIM) in construction projects. Produção **25**(2), 289–297 (2015)

Koseoglu, O., Nurtan-Gunes, E.T.: Mobile BIM implementation and lean on construction site: a case study of a complex airport project. Eng. Constr. Architect. Manag. **25**(10), 1298–1321 (2018)

KPMG: UK Construction Barometer: Margins Under Pressure, 1st edn. KPMG, London (2014)

Latham, M.: Constructing the Team, 1st edn. HMSO, London (1994)

Leite, F., Akcamete, A., Akinci, B., Atasoy, G., Kiziltas, S.: Analysis of modeling effort and impact of different levels of detail in building information models. Autom. Constr. **20**(5), 601–609 (2011)

Lepanjuuri, K., Humphrey, A.: Survey of Employment by Occupation in the Construction Industry, 1st edn. NatCen, London (2016)

Lin, Y.C., Chang, J.X., Su, Y.C.: Developing construction defect management system using BIM technology in quality inspection. J. Civ. Eng. Manag. **22**(7), 903–914 (2016)

Ma, Y.P., Hsu, C.C., Lin, M.C.: Combine BIM- based and mobile technologies to design on- site support system for the communication and management of architectural heritage conservation works. In: 2017 International Conference on Applied System Innovation, pp. 307–310. ICASI, Sapporo (2017)

Mahamadu, A.M., Mahdjoubi, L., Booth, C.: Challenges to BIM-cloud integration: Implication of security issues on secure collaboration. In: 2013 IEEE 5th International Conference on Cloud Computing Technology and Science, pp. 209–214. IEEE, Bristol (2013)

Rolfsen, C.N., Lassen, A.K., Mohamed, M., Shakari, A., Yosefi, H.: On- site quality assurance: moving from forms to digital capture. In: Creative Construction Conference 2019, pp. 886–891. Diamond Congress, Budapest (2019)

Wang, X., Chong, H.Y.: Setting new trends of integrated Building Information Modelling (BIM) for construction industry. Constr. Innov. **15**(1), 2–6 (2015)

Wolstenholme, A.: Never Waste a Good Crisis, 1st edn. Constructing Excellence, London (2009)

Wong, J., Wang, X., Li, H., Chan, G., Li, H.: A review of cloud- based BIM technology in the construction sector. J. Inf. Technol. Constr. **19**, 281–291 (2014)

Improving Construction Job Site Safety with Building Information Models: Opportunities and Barriers

Muhammad Tariq Shafiq and Muneeb Afzal(✉)

College of Engineering, Department of Architectural Engineering, United Arab Emirates University, Al-Ain 155551, Abu Dhabi, UAE
{muhammad.tariq,muneeb.afzal}@uaeu.ac.ae

Abstract. Effective management of construction job site safety is a critical aspect of successful management of construction projects. Research has shown that safety planning at construction jobs sites relies on 2D documents, manual observations, which is a labor-intensive, time-consuming and thus highly ineffective and inefficient process. In addition, the link between safety planning and work task execution is often weak due to communication barriers, ambiguous interpretation of information and subjective judgment of Jobsite safety workers. Safety knowledge transfer is a difficult process when using 2D documents, instruction manuals and verbal instructions, especially in projects with multicultural and multilingual construction workforce, such as in the case of the UAE construction industry. This paper explores the potential of using Building Information Models (BIM) to overcome the problems of safety management & knowledge transfer in the construction jobsite in the UAE. BIM-enabled 3D models and visualization tools can effectively simulate actual job site conditions and can educate the job site crew in an interactive way to enhance their understanding of the job site conditions and safety requirements. The research has used a survey questionnaire to understand the safety knowledge transfer practices in the UAE and has investigated various aspects of safety management and where BIM can bring radical improvements. This study is useful for construction stakeholders, and policymakers in the UAE, to leverage BIM potential for improving construction Jobsite safety. This study also invites future research to devise tools and strategies towards BIM-enabled visualization centric approaches for safety management, training and knowledge transfer in the UAE construction industry.

Keywords: Building Information Modeling (BIM) · United Arab Emirates (UAE) · Safety · Construction · Barriers · 3D models · Visualization

1 Introduction

The construction industry is well known for its complex and adversarial nature of on-site work patterns, which impose several safety hazards for all the people working at a construction job site, as well as for the project itself. Poor safety planning and execution can lead to serious worker injuries, loss of production time & resources and

E. Toledo Santos and S. Scheer (Eds.): ICCCBE 2020, LNCE 98, pp. 1014–1036, 2021.
https://doi.org/10.1007/978-3-030-51295-8_71

severe social implications (Dawood et al. 2014). Construction Jobsite safety is a tedious and challenging process as (1) temporary structures in construction projects are fundamentally important and requires installation and dismantling as the work progress, which are high-risk activities on a construction job site; (2) Jobsite conditions, temporary structures, movement of material & equipment and machinery operations are typically not addressed in design & contract documents; and (3) existing safety planning tools are heavily based on 2D paper-based information which is often complex and difficult to understand for a Jobsite worker. In addition, safety knowledge transfer is difficult to transfer by paper-based safety regulations, which is a key contributor to lack of safety understanding and misinterpretation of safety instructions on construction sites (Zhang et al. 2015a, b). Due to the temporary nature of construction projects, construction workers often move among projects, companies and countries, which may have different safety cultures & requirements. Though many job sites require safety orientation and training, it is demanding for workers to acquire the knowledge in a short time and stick to the rules accordingly when performing design, planning, and work tasks. This situation is worse in multicultural construction environments, such as in the UAE, where the construction workforce belongs to several different countries and speaks multiple languages. For example, the top tier of the construction workforce in the UAE belongs to GCC & western countries, including the USA, UK, European, Australian, Canadian expatriates. The middle management tire consistent of engineering professionals from Pakistan, India, Philippines, Egypt, Jorden etc. The lower tire of construction workforce belongs to trade workers and labours from several south Asian countries (including Pakistan, India, Bangladesh, Nepal etc.). Jobsite safety planning is performed at the top & middle tire of the construction workforce, which is generally in English & Arabic. However, the lower tire who performs the construction tasks cannot understand Arabic or English and do not have any common language which can effectively translate the safety knowledge. This situation invites translation of safety instructions in several languages (e.g. Urdu, Hindi, Punjabi, Bengali, Tillage, Tamil, Pashto etc.), which is a tedious, error-prone and ineffective process. In addition, the global construction industry accounts for a high number of workplace-related incidents with a primary reason for human errors in hazard recognition (Mo et al. 2018). Several other industries, such as manufacturing, have used technology to transfer their safety management practices, which is lacking in construction worksites.

This research position itself strongly to believe that safety planning process tools must be simple and convenient enough for the construction managers to effectively implement them on construction job sites. This could be achieved using the latest digital technologies, such as Building Information Models and visualization tools, by creating information-rich 3D models with embodied safety provisions to communicate, execute and management construction job site safety requirements. BIM is being used in the UAE, mainly due to legislation imposed by regulating authorities (such as Dubai Municipality), but its applications for safety provision modelling has not been explored yet. The objectives of this study are to explore the potential of safety improvements with the use of BIM and highlight the barriers which hinder the BIM implementation in the region, especially in the context of construction job site safety management. The findings of this quantitative based study will help local stakeholders and researchers understand the potential of adopting BIM to improve construction safety and introduce necessary policies or prototype tools which can benefit UAE's construction industry minimizing job site safety threats. Therefore, this research contributes to the advancement of knowledge, as well as benefits the local construction industry.

2 Related Work

Safety management is a key concern in construction industry, due to complex and adversarial nature of on-site work patterns. As compared to other industries, construction is more prone to injuries and accidents (Waehrer et al. 2007; Zou et al. 2017). These construction related incidents are mostly recurring (Kim et al. 2013) and have a detrimental effect on society (Zhou et al. 2013), as the indirect costs of these incidents are six times more than that of the direct costs (Fisher 1995). Despite the efforts of governments, law enforcement agencies and safety professionals, there has not been a significant decline in fatalities, injuries associated with construction (Azhar 2017). For example, in USA alone although the number of fatal injuries decreased from 5,190 in year 2016 to 5,147 in year 2017, this number is still considered very high (Bureau of Labour Statistics 2017).

Many researchers have addressed the issues to improve the safety planning, analysis, control and execution process at construction job sites. For example, Yi and Langford (2006) analyzed historical safety records and presented a theory on safe planning by estimating the risk distribution of a project. They proposed a method of safety planning by estimating situations of concentrated risk and then adjusting the schedule to avoid the risk peaks. Saurin et al. (2005) integrated a Safety Planning and Control Model (SPC) into the production planning and control process. A specific Construction Job Safety Analysis (CJSA) tool was developed by Rozenfeld et al. (2010). Before that Tam et al. (2002) proposed to apply Non-Structural Fuzzy Decision Support System (NSFDSS) to evaluate safety management systems and prioritize the measures with the consideration of various decision criteria, and further to facilitate more realistic decision making. The underpinning assumptions in all the proposed safety planning methods and tools are based on using historical data to identify potential causation of job site accidents. Analysis of historical data provide valuable insight into safety measures but this, however, is not sufficient to predict potential hazards and causation of accidents at unique construction projects. This has led researchers to the advent of information technology enabled approaches, such as Building Information Modelling (BIM) to support the construction job site safety analysis, planning and execution processes.

Building Information Model (BIM) provides a 3D representation of all the physical and functional characteristics of a project in an object-oriented data repository (Eastman et al. 2018). Effective communication and quick decision-making are one of the pillars of success within the construction industry. BIM has emerged as a revolutionary resource that has enhanced the stream of communication, collaboration and decision-making throughout the life cycle of a construction project. Recently, the applications of BIM are further enhanced using virtual reality (VR) wearable devices such as Google Glass®, and Oculus Rift® (Popov et al. 2010). The purpose of these tools is to create simulated immersive environment that can help decision-makers in better understanding the situation and make more effective and informed decisions. The information-rich intelligent BIM models can be used to embed safety requirements at the design stage (i.e. modelling clearances etc.) and can provide simulations of construction operations, using VR tools, to aid the safety planning, training and execution at the construction job sites. The utilization of BIM technology can result in improved occupational safety by connecting the safety issues more closely to the construction planning, providing more illustrative site layout and safety plans, providing methods for managing and visualizing up-to-date plans and site status information, as well as by supporting safety communication in various situations, such as informing site staff about making safety arrangements or warning about risks.

This study is focused on UAE construction industry which is the 5th largest contributor to the UAE's economy and accounts for several mega projects of national and international importance. At present, several mega infrastructure projects worth over 55 billion USD are underway in the UAE, which includes the construction of metro lines, Expo 2020, Airport extensions etc. Effective management of construction jobsite safety is a waxing issue for the UAE construction industry due to the large number of international focused mega projects, and an international construction workforce. According to a report published by IOSH for Middle East region (Witt and Birt 2014), there has been a decline in fatalities in UAE since the implementation of 2009, "Code of Construction Safety Practice". As per this report from 2014, 69% of construction companies lacked health and safety (H&S) implementation policies; as most of the small and medium scale companies do not have well established written H&S protocols. According to a study related to safety concerns of UAE (Zekri 2013) 38% of onsite deaths are due to lack of supervision, where as 25% die due to lack of training. Majority of respondents claimed that poor supervision along with lack of training are main reasons for deaths. The study also found that 'falls' is the main reason for injury followed by 'slips and trips' followed by 'stuck by object'.

Table 1. Examples of BIM usage for improving construction safety.

Sr. No	BIM functionality	Potential for construction safety improvement	Authors
1	Safety model to identify hazards and develop strategies	Safety management	(Alizadehsalehi et al. 2018)
2	Retrieval System to reduce the no. of safety queries	Precise information about safety incidents from past, to take measures according	(Kim et al. 2013)
3	Safety risk identification system (SRIS) and early warning system (SREWS)	Risk analysis and monitor safety status of site	(Li et al. 2018)
4	Modularized electronic scaffolding and formwork objects for BIM models	Design smarter and safer buildings and infrastructure	(Chi et al. 2012)

(*continued*)

Table 1. (*continued*)

Sr. No	BIM functionality	Potential for construction safety improvement	Authors
5	Cloud based safety information and communication system	Improve Safety performance of infrastructure projects	(Zou et al. 2017)
6	4-D based safety model based on rule based checking	Safety management of metro construction	(Zhou et al. 2013)
7	Workforce location tracking, visualize and analyze workspaces	Safety planning	(Zhang et al. 2015b)
8	Automated safety rule checking	Detect safety hazards and suggest preventive measures	(Zhang et al. 2013a,b)
9	Automated rule-checking framework for incorporating safety into BIM models	Safety planning	(Zhang et al., 2015a, b)
10	Risk analysis with RFID and BIM models	Safety management	(Zhang et al. 2014)
11	Construction Safety Audit Scoring System (ConSASS)	Improve safety performance	(Teo et al. 2016)
12	Integration of automated safety rule checking with BIM models	Eliminate hazards, regulate safety provisions, implement safety rule checking throughout different phases	(Takim et al. 2016)
13	Visualization features of BIM to help students understand potential threats	Safety training	(Soemardi 2017)
14	Near miss database with BIM	Analyze and visualize near misses areas during construction	(Shen and Marks, 2016)
15	Risk evaluation on assigned spaces of BIM models	Safety risk analysis	(Setayeshgar et al. 2013)
16	BIM and RFID tags to track workers, equipment and materials	Safety monitoring	(Sattineni and Azhar 2010)
17	BIM based serious game	Build realistic game to quantify human behavior on fire safety	(Rüppel and Schatz 2011)
18	BIM and sensor based data management system	Safety monitoring	(Riaz et al. 2017)
19	CoSMoS-integration of BIM and wireless sensor	Safety monitoring of confined spaces	(Riaz et al. 2014)
20	SMVS (safety management and visualization system)	Improve communication and risk recognition capability and reduce risks	(Park and Kim 2012)

The problem of jobsite knowledge transfer due to a multi-tired and multilingual construction workforce is already highlighted in the introduction of this paper. The UAE government has taken several initiatives towards improving construction safety, and has introduced digital strategies, including BIM to delivery and manage construction projects in the region. Despite the huge potential of BIM to radically improve the construction jobsite safety (Table 1), its implementation in the construction industry, including UAE, is slower than anticipated. With this background, the aim of this study are to explore the potential of safety improvements with the use of BIM and highlight the barriers which hinder the BIM implementation in the region, especially in the context of construction job site safety management.

3 Methodology

The targeted population for this research were AEC professionals from UAE industry. The targeted participants were selected on a basis of their exposure with BIM and construction safety, handling key organizational roles in construction industry. Convenient sampling technique was used in order to meet suitable sample size for the study. BIM experts from industry, researchers from local universities and construction experts from ongoing project, participated in the conducted research. The three sample groups were chosen as per their relevancy and adoption of BIM and latest construction safety trends, resulting in their likelihood of providing valuable and accurate feedback. Since the nature of the study conducted was descriptive, the research aimed to evaluate respondent's opinion pertaining to awareness, potential and barriers of BIM for construction safety. In this case the suitable technique for data collection was to conduct an analytical survey (Forza 2002), carried out in the form of a research questionnaire.

The research methodology for this study can be summarized into the following stages:

Stage 1: In the first step comprehensive literature review was conducted in order to understand the construction dynamics of UAE industry. Research work pertaining to BIM and construction safety conditions were analyzed and trends were deconstructed so that a questionnaire could be devised which incorporates maximum data and knowledge of the respondents.

Stage 2: The prepared questionnaire was distributed among the study sample. Once the required response rate was achieved the data was then analyzed statistically.

Stage 3: Statistical analysis was performed for mainly two kinds of variables derived from the literature; 1) practical applications, provided BIM is implemented for safety, 2) barriers impeding the implementation of BIM for safety. These were then ranked as per their relative importance index (RII).

Stage 4: Based on this analysis, conclusions were drawn and suggestions were made, for the better implementation of BIM to help improve safety conditions of region.

4 Data Analysis

4.1 Analysis of Data

Last parts of the questionnaires 4 and 5 adopted a five and four point Likert scale questions respectively. Standard mechanism was used in order to establish the safety factors which BIM can improve and barriers in adopting BIM for construction safety in UAE. The relative importance index (RII), was used in this regard to tabulate and rank all these factors and barriers on the frequency of responses. The relative importance index formula is defined as:

$$\text{Relative Importance Index (RII)} = \frac{\sum w}{AN} = \frac{5n_5 + 4n_4 + 3n_3 + 2n_2 + 1n_1}{5N} \quad (1)$$

where:

w = is the weighting given by each respondent. This ranges from 1 to 5 for part 4 and 1 to 4 for the part 5 as shown in the Table 2 below:

Table 2. Likert scale values.

n value	Part 4	Part 5
1	Not at all important	Not a barrier
2	Not so important	Somewhat of a barrier
3	Somewhat important	Moderate barrier
4	Very important	Extreme barrier
5	Extremely important	–

A = highest weight given, which is 5 and 4 for parts 4 and 5 respectively.
N = total no. of responses.

4.2 Reliability of Questionnaire

In order to check if the responses provided by the respondents are reliable and follow internal consistency, Cronbach's Coefficient Alpha method is used, which is most common marker of reliability of the questionnaire. The value for this coefficient should be greater than 0.7, in order for data to be reliable. Table 3, shows the calculated values for the Cronbach's Alpha Coefficient which was 0.935 for the construction safety factors that could be improved using BIM and 0.733 for the barriers in implementing

Table 3. Cronbach's Alpha values for parts 4 and 5.

Section	No. of variables	Cronbach's alpha
Safety factors that could be improved using BIM	17	0.935
Barriers in BIM to improve safety	11	0.733
Total	28	0.912

BIM for safety. The general reliability of the questionnaire is 0.912 which is greater than 0.7 hence the reliability of the survey is mandated.

4.3 Normality of Questionnaire

Kolmogorov-Smirnov (K-S) test was performed in order to find out if the data of the questionnaire follows a normal distribution or not. The test results show if the nature of the questionnaire is parametric or non-parametric in nature. Significance values for both parts 4 and 5 were 0.000 which are less than 0.05 and it can be concluded that data is not normally distributed and hence non parametric tests were conducted for further analysis. Below Table 4 and 5 shows the results for K-S test conducted for safety parameters that may improve using BIM and barriers in BIM to improve safety:

Table 4. K-S test for Safety Factors that could be improved using BIM.

Sr. No.	Variable	Statistic	df	Sig.
1	Hazard identification and recognition	0.333	26	0.000
2	Site layout	0.305	26	0.000
3	Analyzing confined and congested spaces	0.251	26	0.000
4	Safety training and capacity building of workers	0.288	26	0.000
5	Safety induction of workers	0.305	26	0.000
6	Safety reports, internal and external audits	0.283	26	0.000
7	Tool box talk meetings	0.235	26	0.001
8	Safety inspections	0.258	26	0.000
9	Compliance of OSHA rules and regulations	0.196	26	0.011
10	Safety designing and planning of temporary structures (scaffolds, stairs and frameworks)	0.31	26	0.000
11	Fall hazard prevention strategies	0.289	26	0.000
12	Active site safety monitoring for critical activities	0.282	26	0.000
13	Risk assessment and mitigation	0.235	26	0.001
14	Better communication between stakeholders	0.266	26	0.000
15	Designing of emergency and evacuation plans	0.332	26	0.000
16	Emergency drills	0.347	26	0.000
17	Inspection of heavy equipment and machinery	0.218	26	0.003

4.4 Non Parametric Analysis

Since it is established above that the nature of data is non parametric, Kruskal-Wallis test was performed in order to further analyze the data. Construction stakeholders as per their organization role were grouped together and the significant values against the variables for parts 4 and 5 were quantified as shown below in Tables 6 and 7. Significant values were found to be greater than 0.05, hence it can be established that respondents shared similar perception on the variables asked.

Table 5. K-S test results for barriers in adopting construction safety.

Sr. No.	Variable	Statistic	df	Sig.
1	Lack of knowledge about VDC technologies	0.262	46	0.000
2	High setup, implementation and maintenance costs	0.241	46	0.000
3	Lack of effective training	0.241	46	0.000
4	Lack of knowledge about return on investment for VDC technologies	0.287	46	0.000
5	Temporary nature of construction projects	0.209	46	0.000
6	Lack of guidance and SOP for proper implementation of VDC technology for safety	0.286	46	0.000
7	Not the demand of client	0.209	46	0.000
8	Resistance to technological adaptation	0.259	46	0.000
9	Overall poor safety culture in construction industry	0.233	46	0.000
10	Lack of standard practices in construction industry	0.239	46	0.000
11	Contractual limitations	0.213	46	0.000

Table 6. Kruskal-Wallis test statistics [a, b].

Sr. No.	Variable	df	Asymp. Sig.
1	Hazard identification and recognition	5	0.523
2	Site layout	5	0.105
3	Analyzing confined and congested spaces	5	0.378
4	Safety training and capacity building of workers	5	0.846
5	Safety induction of workers	5	0.854
6	Safety reports, internal and external audits	5	0.918
7	Tool box talk meetings	5	0.794
8	Safety inspections	5	0.844
9	Compliance of OSHA rules and regulations	5	0.938
10	Safety designing and planning of temporary structures (scaffolds, stairs and frameworks)	5	0.865
11	Fall hazard prevention strategies	5	0.466
12	Active site safety monitoring for critical activities	5	0.457
13	Risk assessment and mitigation	5	0.551
14	Better communication between stakeholders	5	0.304
15	Designing of emergency and evacuation plans	5	0.630
16	Emergency drills	5	0.620
17	Inspection of heavy equipment and machinery	5	0.583

a. Kruskal Wallis Test for safety factors that could be improved by BIM
b. Grouping variable: Organization Role

Table 7. Kruskal-Wallis test statistics [a, b].

Sr. No.	Variable	df	Asymp. Sig.
1	Lack of knowledge about VDC technologies	5	0.182
2	High setup, implementation and maintenance costs	5	0.435
3	Lack of effective training	5	0.266
4	Lack of knowledge about return on investment for VDC technologies	5	0.146
5	Temporary nature of construction projects	5	0.957
6	Lack of guidance and SOP for proper implementation of VDC technology for safety	5	0.937
7	Not the demand of client	5	0.267
8	Resistance to technological adaptation	5	0.034*
9	Overall poor safety culture in construction industry	5	0.300
10	Lack of standard practices in construction industry	5	0.540
11	Contractual limitations	5	0.435

a. Kruskal Wallis Test for barriers in adopting BIM for construction safety
b. Grouping variable: Organization Role
*. Resistance to technological adaption Sig. Value = 0.034 < 0.05

Table 8. Basic demographic information of respondents.

	Client		Consultant		Project Management Consultant (PMC)		Prime contractor		Sub-contractor		Academia		All respondents	
Item	Freq.	%	Freq.	%	Freq.	%	Freq.	%	Freq.	%	Freq.	%	Freq.	%
Highest Academic Qualification														
Diploma	0	0	1	17	0	0	1	10	4	24	1	20	7	15
Bachelors	1	33	4	66	4	80	5	50	8	47	4	80	26	56
Masters	2	67	1	17	1	20	4	40	5	29	0	0	13	29
Doctorate	0	0	0	0	0	0	0	0	0	0	0	0	0	0
Professional Experience in Years														
0-5 years	1	33	2	33	1	20	5	50	11	64	2	40	22	48
6-10 years	0	0	2	33	2	40	3	30	3	18	3	60	13	28
11-15 years	1	33	1	17	1	20	2	20	1	6	0	0	6	13
16-20 years	0	0	0	0	1	20	0	0	0	0	0	0	1	2
21 and above	1	33	1	17	0	0	0	0	2	12	0	0	4	9
Organization Type														
Government	1	33	0	0	0	0	0	0	1	6	4	80	6	13
Semi-Government	2	67	0	0	1	20	0	0	1	6	1	20	5	11
Private	0	0	6	100	4	80	10	100	15	88	0	0	35	76

5 Results and Discussions

5.1 Part 1: Basic Demographic Information

The first part of the questionnaire deals with respondents providing their basic information. The findings concluded that 56 of the respondents were equipped with

Bachelors education, whereas 29 had masters. It was found that respondents between 0–5 years' experience bracket were 48, between 6–10 years 28 and between 11–15 years 13. Majority of the respondents belonged from the private sector i.e. 76, and then the government sector with 13. Table 8, provides all the details regarding every stakeholder, pertaining to their education, professional experience and organization type.

5.2 Part 2: Current Safety Conditions in UAE Industry

Current Jobsite Safety Conditions in UAE. Figure 1 shows the current satisfaction levels of the participants working in UAE's construction industry. 9 of the participants are very satisfied whereas, majority of the people (59) are satisfied with current safety conditions of the industry. This high percentage is an indication that strict rules regarding safety precautions are implemented and regulated in UAE but leaves room for improvement incorporating BIM into the traditional workflows. 17 of the participants were neither satisfied nor dissatisfied whereas 11 of the respondents were dissatisfied. A small portion of the respondents (4), were very dissatisfied by the current safety conditions of industry.

Please try to avoid rasterized images for line-art diagrams and schemas. Whenever possible, use vector graphics instead (see Fig. 1).

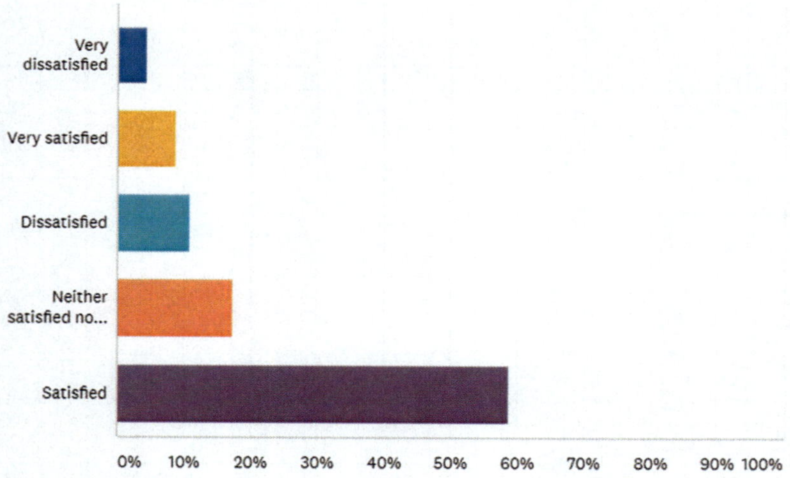

Fig. 1. Satisfaction levels of current state of safety in UAE's construction industry.

Mode of Communication for Safety Instruction in UAE. Figure 2 shows that majority of the participants i.e. 48, claimed that verbal means were used in order to communicate safety related information during construction. Slightly less than that 35 preferred digital means in order to transfer safety related information. This information could be regarding training or particular set of safety instructions to be laid out at site.

This percentage indicates that there is good potential of BIM technology to be used as a means to monitor and execute safety related decisions.

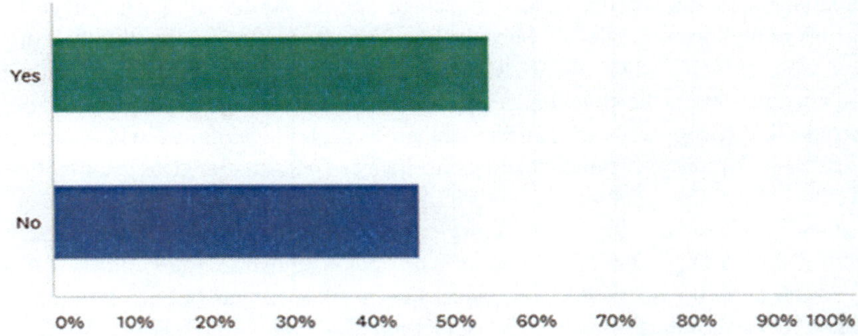

Fig. 2. Means for conducting safety communication.

Language Difference Between Upper Hierarchy and Labor of Site. Figure 3 shows the stats to the yes or no answered question, where participants were asked to answer if there was a difference between the language spoken for imparting instructions and the language spoken by on-site workers. A significant majority (54) answered this question to be yes, showing that there is clear difference in language spoken by managers and labors. This is currently a major problem of the multi-lingual construction industry of UAE. This is where BIM can play a very important role by helping the workers understand the safety instructions with the help of 4-D based simulations, training and visualizations.

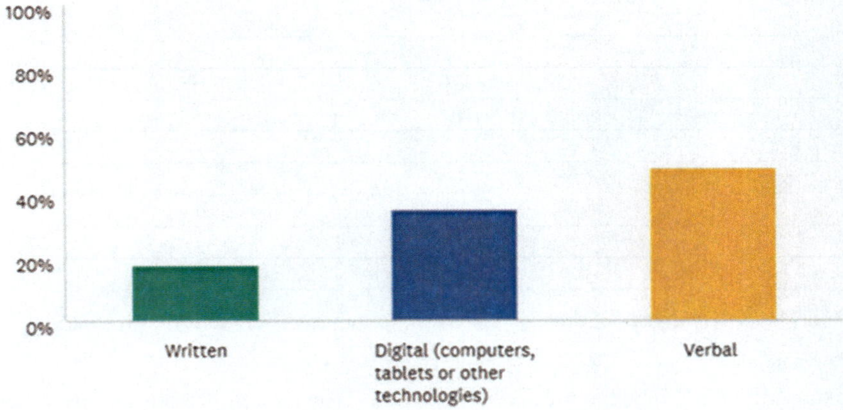

Fig. 3. Language difference between workers and managers.

Languages Spoken for Construction Safety Related Communication. Figure 4 shows a pie chart illustrating all the languages spoken to communicate safety related information. The chart illustrates that the majority (70) of the communication is been transcended forward in English, whereas 26 use Urdu/Hindi as their primary language to communicate. A relatively small percentage (4) use Arabic as the language to be used for communicating safety related instructions. If we further the analysis, all the clients who participated in the survey use English as their primary language for communication. 80 of the consultants chose to speak English, whereas the rest of them preferred Hindi/Urdu as their primary means of communicating. Almost same trend was followed by project management consultants. The situation was a bit different, from the contractor end of the population. Primary contractors used English, Urdu/Hindi and Arabic as the language for communication and same was the case with subcontractors. The percentage of people who spoke English, Urdu/Hindi and Arabic were 40, 50 and 10 respectively. 75 contractors mentioned that English was the main language for communication whereas Urdu/Hindi and Arabic were spoken by 20 and 5 of the subcontractors respectively. Majority of the participants who filled this survey were on the executive or had upper hierarchical positions. The workers in construction industry associated with lower stratum are mainly from South Asian countries such as India, Pakistan and Nepal. So a number of languages are spoken at the lower strata of the construction hierarchy. These languages include Malayalam, Telugu, Hindi, Pushto, Punjabi, Urdu, Nepali etc. This difference in the languages spoken between both workers and supervisors and senior decision makers tends to cause major lapse in construction safety.

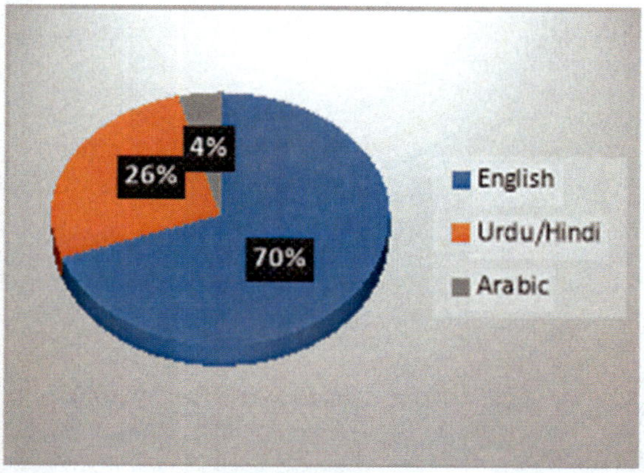

Fig. 4. Percentage of different languages spoken to deliver safety instructions.

Satisfaction of BIM Usage for Safety in UAE. When asked if the respondents were satisfied with the BIM usage for the purposes of ensuring construction safety on job-site, majority of the people (39), responded that they were satisfied whereas, only 6 claimed that they were very satisfied with the current usage of BIM to be used for construction safety purposes. A significant 30 responded this question as neither satisfied nor dissatisfied. Percentages of respondents who were dissatisfied and very dissatisfied from the potential of BIM to be used in conventional safety purposes in UAE, were 22 and 2 respectively.

5.3 Part 3: Safety Related Applications of BIM

Table 9 enlists different aspects of construction safety, which might be improved provided BIM is used for construction safety in UAE. The participants marked on the importance scale, if the particular factor of safety could be enhanced using BIM. It must be noted that the only participants who had prior awareness of BIM were made to rank these factors.

"Hazard identification and recognition" was ranked 1st, having potential of improving safety protocols provided BIM is incorporated in workflows. The relative importance index for this factor was 88. Significant studies such as (Zhang et al. 2015, Zhang et al. 2013) have already been done to see the potential of BIM in order to help improve hazard identification to mitigate safety threats.

Both "Safety designing and planning of temporary structures (scaffolds, stairs and frameworks)" and "designing of emergency and evacuation plans" ranked at 2nd position with RII of 87. According to OSHA as per year 2018 report, scaffolding was the third most significant reason for the deaths caused during construction in USA (Commonly Used Statistics 2018). Considering the current scenario of deaths caused by this factor, BIM can be extremely useful to assuage the safety threats. Kim and Cho (2015), devised an algorithmic approach towards incorporating BIM to analyze and improve safety risk pertaining to temporary structures. Similarly in UAE, the construction industry has to go through rigorous approvals for their safety and fire evacuation plans for during and after construction. The very process has a huge potential to be improved if taken through BIM. Since the models and 2-D drawings are already present, n no. of possible plans and scenarios can be extracted through visualizing the models and then extracting related plans.

"Safety training and capacity building of workers", "fall hazard prevention stages" and "better communication between stakeholders" are three aspects of safety that had equal potential of improvement according to participants with equal relative importance index of 85. Three of the safety related aspects see huge potential of improvement due to the fact that BIM enhances communication between stakeholders. As per question 5 already answered in part 2, we already know that there are different languages spoken in multi-lingual, multi-cultural construction industry of UAE. It was also pointed out by a safety manager that he faces lots of problems and huge amount of effort gets wasted in order to have the labor understand the basic safety instructions. BIM and VR simulations can be of great help in abridging this communication barrier between different multi-lingual hierarchies operating in construction industry of UAE.

Safety induction of workers had a 7th rank, whereas "active site safety monitoring for critical activities" and "tool box talk meeting" both shared 8th position with 83 relative importance index. BIM models can be connected with drones which are already being used in various construction projects for monitoring purposes. Alizadehsalehi et al. (2018) integrated 4-D BIM model and Unmanned Aerial Vehicle

(UAV) to design, analyze and then implement accident mitigation strategies onto the construction site. Similarly 10th rank with relative importance index of 82, was shared by four aspects of safety that have potential to be improved using BIM. "Safety inspections", "emergency drills", "site layout" and "risk assessment and mitigation" are still held in a traditional manner on construction site. The very culture of safety is now improving in Middle East and hence, research needs to be conducted in the domain of safety and BIM to make BIM useful enough to improve safety of the region as well.

"Analyzing confined and congested spaces", "inspection of heavy equipment and machinery", "compliance of OSHA rules and regulations" and "safety reports, internal and external audits" were 14th, 15th, 16th and 17th ranked with relative importance indexes of 81, 79, 78 and 73 respectively.

The low ranking of these factors suggest that although BIM has been taking its market place in construction industry of UAE, still lots of measures are needed in order to leverage BIM usage to further construction safety practices.

Table 9. Safety factors potential of improvement using BIM.

Sr. No.	Factors that could be improved using BIM	Frequency of responses					Total respondents (N)	Weighted total	RII	Rank
		"5"	"4"	"3"	"2"	"1"				
1	Hazard identification and recognition	14	9	3	0	0	26	115	88	1
2	Safety designing and planning of temporary structures (scaffolds, stairs and frameworks)	13	9	4	0	0	26	113	87	2
3	Designing of emergency and evacuation plans	14	7	5	0	0	26	113	87	2
4	Safety training and capacity building of workers	12	9	5	0	0	26	111	85	4
5	Fall hazard prevention strategies	12	8	6	0	0	26	110	85	4
6	Better communication between stakeholders	11	10	5	0	0	26	110	85	4
7	Safety induction of workers	13	6	6	1	0	26	109	84	7
8	Active site safety monitoring for critical activities	8	14	4	0	0	26	108	83	8
9	Tool box talk meetings	10	7	7	0	0	24	99	83	8
10	Safety inspections	11	8	6	1	0	26	107	82	10
11	Emergency drills	6	17	3	0	0	26	107	82	10
12	Site layout	7	15	3	1	0	26	106	82	10
13	Risk assessment and mitigation	10	9	6	1	0	26	106	82	10
14	Analyzing confined and congested spaces	8	12	5	1	0	26	105	81	14
15	Inspection of heavy equipment and machinery	9	8	8	1	0	26	103	79	15
16	Compliance of OSHA rules and regulations	8	9	8	1	0	26	102	78	16
17	Safety reports, internal and external audits	3	13	8	2	0	26	95	73	17

5.4 Part 4: Barriers in Adopting BIM for Construction Safety in UAE

Table 10 enlists 11 barriers according to their rank devised by their relative importance index (RII). This final ranking of BIM barriers for safety has been a cumulative analysis on the individual ranking by each stake holder. The categorization has been made in order to understand which barrier impedes the BIM process for individual category of stakeholder.

"Lack of knowledge about ROI on BIM" was the predominant barrier ranked highest with an overall RII of 86. Client, both PMC and consultants marked this barrier to be major hurdle in implementing BIM to improve construction safety, whereas for contractors this was one of the top barriers observed. Although the concept of BIM is not novel to the UAE construction industry, BIM for safety has not been under consideration; since the very idea in itself is under experimentation or research phase and has yet to be formulated to become part of standard operating procedures for the industry to reap benefits out of it.

"Lack of effective training" and "Lack of knowledge about BIM" with the RII of 81 and 80 respectively, were the second and third significant barriers associated with training and education. Capacity building is one of the major reasons (Arayici, et al. 2011; Enshassi and Ayyash 2016) across the globe for successful implementation of BIM, let alone leveraging technology in order to improve onsite safety. The subcontractors rated this reason to be their top barrier with RII of 76. In countries where BIM has been mandated, trickledown effect is observed when it comes to training. Although client mandates BIM, in order to fulfill level of detail requirements for shop drawings and model submissions, sub-contractors aren't able to comply along the technical and man power requirements. Usually the client conducts workshops and meeting to discuss these issues, with the concerned consultant, this training is then transferred to the contractor and eventually to the prime contractor. By the time, a subcontractor has access to this knowledge, on-site construction is altogether in different phase.

"Not the demand of client" is now not considered as one of the top barriers with an RII of 76, according to research conducted by Mehran (2016), 52 of the participants predicted widespread BIM usage across UAE in less than 5 years time frame. The same trend was observed in this study also where most of the stakeholders were either associated with BIM inclined projects or had definitive knowledge about it.

"Resistance to technological adaptation" also had a 4th rank along with the demand for BIM set by client. For clients this was one of the leading barriers whereas oppositely the contractors ranked it to be insignificant. This opposite trend means that the clients face a great deal of objection from contractors in order to implement BIM, whereas contractors have to comply by the instructions of clients, for successful handover of the project.

Barriers such as "high setup, implementation and maintenance costs", "lack of guidance and SOP for proper implementation of BIM for safety" and "contractual limitations" all shared the 6th rank. High financial constraints have been pointed out by many studies (Migilinskasa et al. 2013; Hosseini 2015; Newton and Chileshe 2012) conducted in different regions of the world for successful implementation of BIM. These barriers indicate that a lot needs to be done across all stakeholders for widespread usage and acceptability of BIM. The legislative state of UAE is such that it

Table 10. Relative Importance Index (RII) and ranking of barriers.

		Client		Consultant		Project Management Consultant		Prime Contractor		Sub-Contractor		Academia		Overall rank	
Sr. No.	Barriers	RII (%)	Rank	RII (%)	Rank	RII (%)	Rank	RII (%)	Rank	RII (%)	Rank	RII (%)	Rank	RII (%)	Rank
1	Lack of knowledge about return on investment for BIM	92	1	83	1	95	1	85	2	74	2	90	2	86	1
2	Lack of effective training	83	3	71	5	75	4	85	2	76	1	95	1	81	2
3	Lack of knowledge about BIM	75	7	71	5	85	2	93	1	72	3	85	4	80	3
4	Not the demand of client	83	3	83	1	85	2	73	7	63	5	70	7	76	4
5	Resistance to technological adaptation	92	1	79	3	75	4	70	10	57	11	80	5	76	4
6	High setup, implementation and maintenance costs	83	3	67	9	75	4	73	8	62	7	65	9	71	6
7	Lack of guidance and SOP for proper implementation of BIM for safety	75	7	71	5	60	9	73	9	71	4	75	6	71	6
8		83	3	75	4	60	9	75	5	60	8	70	7	71	6

(*continued*)

Table 10. *(continued)*

		Client		Consultant		Project Management Consultant		Prime Contractor		Sub-Contractor		Academia		Overall rank	
Sr. No.	Barriers	RII (%)	Rank	RII (%)	Rank	RII (%)	Rank	RII (%)	Rank	RII (%)	Rank	RII (%)	Rank	RII (%)	Rank
	Contractual limitations														
9	Overall poor safety culture in construction industry	67	9	67	9	75	4	78	4	59	10	50	10	66	9
10	Temporary nature of construction projects	58	11	63	11	65	8	55	11	60	8	90	2	65	10
11	Lack of standard practices in construction industry	67	9	71	5	60	9	75	5	63	5	50	10	64	11

doesn't have any viable policy pertaining to safety, off the available provisions most are generic and do not address to the technical delicacies related to construction safety (Turley 2010).

"Overall poor safety culture" was ranked 9th amongst the barriers, it must be taken into consideration that although the current state of BIM is in implementation stages, it was observed from visiting the sites and interacting with respondents that the authorities are taking stringent measures to implement safety measures and heavy fines were imposed in case of not abiding by the safety standards set by the client. But since industry isn't much aware of the benefits that BIM could reap in order to improve construction safety, the traditional safety practices are common.

"Temporary nature of construction projects" was ranked at 10th place with RII of 65. This low RII indicates that although construction projects are of temporary nature, as long as there is a demand of client in order to induce certain technology or a safety policy into the workflows, they will be mandated by contractors and other stakeholders related with the project.

Lastly "lack of standard practices in construction industry" was marked at 11th the last place of all ranks. This result in an indication that UAE construction industry is one of the sophisticated construction industries of the world where standard operating procedures have made their way into the conventional workflows and are now not just implemented but regulated as well.

6 Conclusions and Recommendations

Building Information Modeling has changed the overall landscape for construction industry. It has improvised every aspect of construction from initiation to closure. Over the recent years, great amount of research has been conducted on BIM and its associated technologies and tools. This paper evaluated the current state and potential of improving safety conditions of multi-cultural, dynamic construction industry of UAE using BIM. UAE's construction industry can be termed as one of the finest across the globe, with large number of vertical constructions, each with their own unique innovative architecture designs. With such advancements, UAE has incorporated strict safety regulations, which have resulted in overall improved safety conditions. Although UAE has adopted BIM since 2014, construction industry has not been able to extract maximum benefits out of it.

UAE's construction industry has a multi-lingual environment due to on-site workers from different regions of the world speaking different languages. This makes difficult to pass on and understand construction safety related information between different stratums of hierarchies. Respondents also believed that construction industry is not taking advantage from BIM to improve traditional safety workflows.

Since BIM concept isn't novel to UAE industry, majority respondents claimed that BIM has potential for improving safety, for different project types. When asked about the aspects of construction safety, that could be improved incorporating BIM, respondents, claimed that "hazard identification and recognition", "safety designing and planning of temporary structures"and "designing of emergency and evacuation plans" have most potential to be improved provided BIM is incorporated. When respondents were asked about major barriers that could impede BIM implementation for improving construction safety, they claimed that "lack of knowledge about return on investment" and training related barriers to be most significant for BIM to improve safety.

This research paper covers construction safety, BIM and current market scenario of UAE in all aspects, hence it is a stepping stone for the researchers and stakeholders to be aware of the current condition across these dimensions. UAE government has already enforced BIM and safety related policies, but great work needs to be done in order to infuse the concept of BIM for safety. Researchers need to develop tools and plugins incorporating local safety clauses to enhance safety using BIM. Necessary steps should also be taken across all the levels of stakeholders to take initiatives to improve construction safety aspects and in this regard capacity building measures should be taken. Industry and academia should work together hand in hand in order to improve construction safety. Joint ventures between government municipalities, universities and private entities can drive forward the current path of BIM to improve construction safety.

References

Abreu Saurin, T., Torres Formoso, C., Borges Cambraia, F.: Analysis of a safety planning and control model from the human error perspective. Eng. Constr. Architectural Manage. **12**(3), 283–298 (2005)

Alizadehsalehi, S., Yitmen, I., Celik, T., Arditi, D.: The effectiveness of an integrated BIM/UAV model in managing safety on construction sites. Int. J. Occup. Saf. Ergon. 1–36 (2018)

Arayici, Y., Coates, P., Koskela, L., Kagioglou, M., Usher, C., Reilly, K.O.: Technology adoption in the BIM implementation for lean architectural practice. Autom. Constr. 189–195 (2011)

Azhar, S.: Role of visualization technologies in safety planning and management at construction jobsites. Procedia Eng. **171**, 215–226 (2017)

Barlish, K., Sullivan, K.: Automation in construction how to measure the benefits of BIM—a case study approach **24**, 149–159 (2012). https://doi.org/10.1016/j.autcon.2012.02.008

Bureau of Labour Statistics (2017). United States Deparment of Labor. https://www.bls.gov/news.release/pdf/cfoi.pdf

Chen, A., Golparvar-Fard, M., Kleiner, B.: SAVES: a safety training augmented virtuality environment for construction hazard recognition and severity identification. In: International Conference on Construction Applications of Virtual Reality (CONVR), London, UK, pp. 373–383 (2013)

Chi, S., Hampson, K.D., Biggs, H.C.: Using BIM for smarter and safer scaffolding and formwork construction: a preliminary methodology. In: Tjandra, K.I., Ofori, G., Teo, E.A.L. (eds.) Proceedings of the CIB W099 International Conference on Modelling and Building Health and Safety, 10–11 September 2012, Singapore (2012)

Choe, S., Leite, F.: Automation in construction construction safety planning: site-specific temporal and spatial information integration. Autom. Constr. **84**(September), 335–344 (2017). https://doi.org/10.1016/j.autcon.2017.09.007

Commonly Used Statistics, 30 Sept 2018. OSHA. https://www.osha.gov/oshstats/commonstats.html

Croft, A.: BIM implementation in the UAE on the rise, 18 Sept 2017. LEXOLOGY. https://www.lexology.com/library/detail.aspx?g=75c26dc9-e653-4ea7-ae38-7e9efd983ef5
Dawood, N., Miller, G., Patacas, J., Kassem, M.: Construction health and safety training: the utilisation of 4D enabled serious games. J. Inf. Technol. Constr. **19**, 326–335 (2014)
Eastman, C.: Bim Handbook: a Guide to Building Information Modeling for Owners, Managers, Designers, Engineers and Contractors. Wiley, Hoboken (2018)
Enshassi, A., Ayyash, A.: BIM for construction safety improvement in Gaza. Int. J. Constr. Manag. 249–265 (2016)
Enshassi, A., Ayyash, A., Choudhry, R.M.: BIM for construction safety improvement in Gaza strip : awareness, applications and barriers. **3599**(April) (2016). https://doi.org/10.1080/15623599.2016.1167367
Fisher, D.J.: Development of a computerized system for **28**(3), 457–466 (1995)
Forza, C.: Survey research in operations management: a process-based. Int. J. Oper. Prod. Manag. **22**, 152–194 (2002)
Goucher, D., Thurairajah, N.: Advantages and challenges of using BIM: a cost consultant's perspective. In: 49th ASC Annual International Conference, California Polytechnic State University (Cal Poly), San Luis Obispo, California (2012)
Grilo, A., Jardim-Goncalves, R.: Automation in construction value proposition on interoperability of BIM and collaborative working environments. Autom. Constr. **19**(5), 522–530 (2010). https://doi.org/10.1016/j.autcon.2009.11.003
Kim, H., Lee, H.-S., Park, M., Chung, B., Hwang, S.: Information retrieval framework for hazard identification in construction. J. Comput. Civil. Eng. **29**(3), 04014052 (2013). https://doi.org/10.1061/(ASCE)CP.1943-5487.0000340
Hosseini, M.R., Azari, E., Tivendale, L., Chileshe, N.: Barriers to adoption of building information modeling (BIM) in Iran: preliminary results. In: The 6th International Conference on Engineering, Project, and Production Management (2015)
Kim, K., Cho, Y.: BIM-based planning of temporary structures for construction safety. Comput. Civil Eng. (2015)
Latiffi, A.A., Brahim, J., Fathi, M.S.: The development of building information modeling (BIM) definition. Appl. Mech. Mater. **567**, 625–630 (2014)
Li, M., Yu, H., Jin, H., Liu, P.: Methodologies of safety risk control for China' s metro construction based on BIM. Saf. Sci. **110**(February), 418–426 (2018). https://doi.org/10.1016/j.ssci.2018.03.026
Malekitabar, H., Ardeshir, A., Sebt, M.H., Stouffs, R.: Construction safety risk drivers: a BIM approach. Saf. Sci. **82**, 445–455 (2016)
Marefat, A., Toosi, H., Hasankhanlo, R.M.: A BIM approach for construction safety: applications, barriers and solutions. Eng. Constr. Archit. Manag. **26**(9), 1855–1877 (2019). https://doi.org/10.1108/ecam-01-2017-0011
Mehran, D.: Exploring the adoption of BIM in the UAE construction industry for AEC firms. Procedia Eng. **145**, 1110–1118 (2016). https://doi.org/10.1016/j.proeng.2016.04.144
Migilinskasa, D., Popovb, V., Juoceviciusc, V., Ustinovichius, L.: The benefits, obstacles and problems of practical bim implementation. In: 11th International Conference on Modern Building Materials, Structures and Techniques, pp. 767–774 (2013). Procedia Engineering
Mo, Y., Zhao, D., Du, J., Liu, W., Dhara, A.: Data-driven approach to scenario determination for VR-based construction safety training. Constr. Res. Congr. **2018**, 116–125 (2018). https://doi.org/10.1061/9780784481288.012
Newton, L.K., Chileshe, N.: Awareness, usage and benefits of building information modelling (BIM) adoption-the case of South Australian construction organizations. In: 28th Annual Association of Researchers in Construction Management (ARCOM) Conference. Edinburgh (2012)

Park, C., Kim, H.: Automation in construction a framework for construction safety management and visualization system. Autom. Constr. (2012). https://doi.org/10.1016/j.autcon.2012.09.012

Popov, V., Juocevicius, V., Migilinskas, D., Ustinovichius, L., Mikalauskas, S.: The use of 1132 a virtual building design and construction model for developing an effective project 1133 concept in 5D environment. Autom. Constr. **19**, 357–367 (2010). 1134. https://doi.org/10.1016/j.autcon.2009.12.005

Puerto, C.L., Clevenger, C.M.: Enhancing safety throughout construction using BIM/VDC. In: EcoBuild Conference Proceedings, ECOBuild 2010 BIM Academic Forum. Washington DC (2010)

Riaz, Z., Arslan, M., Kiani, A.K., Azhar, S.: Automation in construction CoSMoS: a BIM and wireless sensor based integrated solution for worker safety in confined spaces. Autom. Constr. **45**, 96–106 (2014). https://doi.org/10.1016/j.autcon.2014.05.010

Riaz, Z., Parn, E.A., Edwards, D.J., Arslan, M., Shen, C., Pena-Mora, F.: BIM and sensor-based data management system for construction safety monitoring. J. Eng. Des. Technol. **15**(6), 738–753 (2017). https://doi.org/10.1108/jedt-03-2017-0017

Rozenfeld, O., Sacks, R., Rosenfeld, Y., Baum, H.: Construction Job Safety Analysis. Saf. Sci. **48**, 491–498 (2010)

Rüppel, U., Schatz, K.: Designing a BIM-based serious game for fire safety evacuation simulations. Adv. Eng. Inform. **25**, 600–611 (2011). https://doi.org/10.1016/j.aei.2011.08.001

Sattineni, A., Azhar, S.: Techniques for tracking RFID tags in a BIM model. In: Proceedings – The 27th International Symposium on Automation and Robotics in Construction (2010). https://doi.org/10.22260/isarc2010/0037

Setayeshgar, S., Hammad, A., Vahdatikhaki, F.: Real-time Safety Risk Analysis of Construction Projects Using BIM and RTLS, November 2017 (2013). https://doi.org/10.22260/ISARC2013/0080

Shen, X., Marks, E.: Near-miss information visualization tool in BIM for construction safety. J. Constr. Eng. Manag. **142**(4), 04015100 (2016). https://doi.org/10.1061/(asce)co.1943-7862.0001100

Soemardi, B.W., Erwin, R.G.: Using BIM as a tool to teach construction safety, vol. 05007, pp. 1–8 (2017). https://doi.org/10.1051/matecconf/201713805007

Sulankivi, K., Makela, T., Kiviniemi, M.: BIM-based site layout and safety planning. In: 1st International Conference on Improving Construction and Use through Integrated Design Solutions CIB IDS 2009, Technical Research Center of Finland, pp. 125–140 (2009)

Takim, R., Zulkifli, M.H., Nawawi, A.H.: Integration of automated safety rule checking (ASRC) system for safety planning BIM-based projects in Malaysia. Procedia Soc. Behav. Sci. **222**, 103–110 (2016). https://doi.org/10.1016/j.sbspro.2016.05.195

Tam, C., Tong, T., Chiu, G., Fung, I.: Non-structural fuzzy decision support system for evaluation of construction safety management system. Int. J. Proj. Manag. **20**, 303–313 (2002)

Teo, E., Lin, A., Ofori, G., Tjandra, I., Kim, H.: Framework for productivity and safety system using BIM in Singapore (2016)

The Business Value of BIM in the Middle East (2017). https://www.construction.com/toolkit/reports/business-value-bim-middle-east

Turley, D.: United Arab Emirates: Health & Safety in the Construction Sector in the UAE, 31 March 2010. http://www.mondaq.com/x/97312/Building+Construction/

Waehrer, G.M., Dong, X.S., Miller, T., Haile, E., Men, Y.: Costs of occupational injuries in construction in the United States. Accid. Anal. Prev. **39**, 1258–1266 (2007). https://doi.org/10.1016/j.aap.2007.03.012

Witt, J., Birt, A.: Managing liability for worksite accidents. In: Middle East Annual Conference 2014 ((2014))

Yi, K.J., Langford, D.: Scheduling-based risk estimation and safety planning for construction projects. J. Constr. Eng. Manag. **132**, 626–635 (2006)

Zekri, M.K.S.: Construction safety and health performance in Dubai. (Unpublished thesis). Heriot Watt University, Dubai (2013)

Zhang, C., Chen, J., Yang, Y.: CONVR2013-72-Cheng experimental investigation of using RFID integrated bim model for safety and facility management 1 March (2014)

Zhang, S., Sulankivi, K., Kiviniemi, M., Romo, I., Eastman, C.M., Teizer, J.: BIM-based fall hazard identification and prevention in construction safety planning. Saf. Sci. **72**, 31–45 (2015a). https://doi.org/10.1016/j.ssci.2014.08.001

Zhang, S., Teizer, J., Lee, J., Eastman, C.M., Venugopal, M.: Automation in construction building information modeling (BIM) and safety: automatic safety checking of construction models and schedules. Autom. Constr. **29**, 183–195 (2013b). https://doi.org/10.1016/j.autcon.2012.05.006

Zhang, S., Teizer, J., Pradhananga, N., Eastman, C.M.: Automation in Construction Workforce location tracking to model, visualize and analyze workspace requirements in building information models for construction safety planning. Autom. Constr. **60**, 74–86 (2015b). https://doi.org/10.1016/j.autcon.2015.09.009

Zhou, Y., Ding, L.Y., Chen, L.J.: Automation in construction application of 4D visualization technology for safety management in metro construction. Autom. Constr. **34**, 25–36 (2013). https://doi.org/10.1016/j.autcon.2012.10.011

Zou, P.X.W., Lun, P., Cipolla, D., Mohamed, S.: Cloud-based safety information and communication system in infrastructure construction. Saf. Sci. **98**, 50–69 (2017). https://doi.org/10.1016/j.ssci.2017.05.006

Trends of Research and Development on Construction Robotics Considering the Supporting Technologies and Successful Applications

Shiyao Cai, Zhiliang Ma(✉), and Xinglei Xiang

Tsinghua University, Beijing, China
mazl@tsinghua.edu.cn

Abstract. Although construction robotics has been researched and developed over the past decades and the practical needs keep rising, its application remains limited. This study is aimed at exploring the trends of research and development (R&D) on construction robotics based on a thorough investigation of the numerous existing publications and case studies of successful applications in this field. First, a scientometric review is conducted to identify and discuss the supporting technologies with significant contributions in the field during the past decades and their trends, as well as the hottest ones at present. Then case studies on two Chinese construction companies that developed and applied robotic products successfully are presented to reveal how the supporting technologies were implemented in practices and how the R&D and application were followed up. The hottest supporting technologies at present were found to be computer vision, 3D printing, Building Information Modeling (BIM), and Artificial Intelligence (AI), as well as their emerging branches including Simultaneous Localization and Mapping (SLAM), machine learning, and deep learning. Three critical factors were identified for the follow-up of the R&D of construction robotics, i.e., mutual promotion between the R&D and application of construction robot products and the development of supporting technologies, asynchronous development in different application areas, and multidisciplinary cooperation among construction companies and professional technological companies. The findings are expected to be valuable for the related stakeholders including researchers, robot developers, and construction companies to conduct further R&D and application of robotics in the construction industry.

Keywords: Construction robotics · Scientometric literature review · Case study

1 Introduction

Construction robotics have been researched and developed over the past decades. Although the practical needs keep rising, the application rate remains limited. To promote the future development in the field, many efforts have been devoted to the analysis of the development trends. For example, Bock [1] argued that the conventional construction methods have reached their limits to the growing needs of the construction

E. Toledo Santos and S. Scheer (Eds.): ICCCBE 2020, LNCE 98, pp. 1037–1051, 2021.
https://doi.org/10.1007/978-3-030-51295-8_72

industry and predicted a future trend of the ubiquity of robotics. Cai et al. [2] explored the development of construction automation and robotics for high-rise buildings and identified three key research areas, i.e., steel beam assembly, automated construction and logistic systems, and façade works. Son et al. [3] analyzed the subjects of papers in the International Symposium on Automation and Robotics in Construction (ISARC) during the period 1990–2008 to identify the research trends of automation and robotics in the construction industry.

These works have presented a general picture of the field, pointed out the key research areas and analyzed the current and ongoing research and development (R&D) directions. However, in recent years, the latest development of some supporting technologies has made unprecedented progress for construction robotics [2], which have rarely been analyzed systematically. Besides, although the application rate is low as mentioned above, there are many application cases with successful experience, while few research efforts have been made to learn from these cases.

Therefore, this paper aims to explore the trends of R&D on construction robotics by identifying the supporting technologies that played a key role in the development of construction robotics and understanding the way to successful R&D and application. First, a scientometric literature review was conducted to identify and discuss the supporting technologies with significant contributions in the field and their trends during the past decades, as well as the hottest ones at present. Then two successful cases of R&D and application of construction robotics were presented to reveal how the supporting technologies were implemented in practices and how the R&D and application were followed up. Finally, based on the results of the literature review and the case studies, research findings were concluded. The findings are expected to be valuable for the related stakeholders including researchers, robot developers, and construction companies to conduct further R&D and application of robotics in the construction industry.

2 Method

A mixed-method was employed in this paper, combining a scientometric literature review and case studies. In the literature review, a scientometric analysis tool named CiteSpace was used to obtain an overview of the technologies implemented in the field of construction robotics. Then, case studies were conducted to present successful applications as well as the supporting technologies implemented.

2.1 Scientometric Review of Literature

The literature retrieval was conducted in December 2019. Among the major databases, Scopus was selected because it covered a large range of literature in the field of construction robotics [2]. The following retrieval code was TITLE-ABS-KEY (construction AND robot*), in which, "TITLE-ABS-KEY" means title or abstract or keywords, while the asterisk (*) denotes a fuzzy search. A screening process was conducted successively to exclude the papers from irrelevant areas (e.g., biology, medicine, agriculture, etc.). The time span of the remaining records was 1974–2020. Because only four papers were published in 2020, they were also excluded because of the little statistical significance. Finally, a total of 5,522 bibliographic records were obtained.

To extract information from the large number of papers, a scientometric method is used. It is a statistical analysis technique, which processes enormous bibliographic data and links literature concepts to help researchers make systematic literature-related discoveries that may be overlooked in manual reviews [4]. Many studies have summarized the tools that are available for scientometric analysis [5, 6]. In this paper, CiteSpace was selected. It is a free Java application that is strong in identifying the emerging trends and pivotal points in a specific domain [7] by providing various accessible graphs systematically [8].

2.2 Case Studies of Successful Applications

The cases in this paper were selected from Chinese construction companies and the reason is as follows. Although a large number of robots have been developed, many of them stopped at the end of pilot tests without further steps for the follow-up R&D and on-site applications. Nevertheless, we know that a number of Chinese construction companies have successful follow-ups. They implemented the latest supporting technologies to develop robotic products and improve the products iteratively based on the applications in their own construction projects. It is expected that their experiences are valuable references for both robot providers and adopters.

Based on a thorough investigation of the well-known Chinese construction companies and their achievements in the field of construction robotics, two case companies were selected. They are referred to as Company A and B respectively, and the overview of their profiles is shown in Table 1. Both companies are state-owned. In China, state-owned companies constitute about one-fourth of the contractors and most of them are stronger technologically than the non-state-owned ones.

Table 1. Overview of the case companies.

Case company	Main business	Annual turnover (CNY)	No. of employees
A	Specialty contractor of steel structures & general contractor	7.7 billion	2,376
B	Specialty contractor of steel structures & general contractor	20.1 billion	About 5,000

Then interviews were conducted with engineers, managers, and technical directors who play an important role in the R&D and application of the robotic products. The profiles of the interviewees are shown in Table 2. The interviews are about the processes from R&D to applications and the supporting technologies that have been implemented. A list of questions was prepared and sent to the interviewees for them to prepare themselves in advance. Each interview lasted about 2 h, being conducted by two members of the research team and tape-recorded.

Table 2. Profiles of the interviewees.

Case company	No.	Position	Years of experience
A	1	Chief engineer	18
	2	Vice president of the engineering research institute	27
	3	Technical director	11
	4	Director of BIM department in the engineering research institute	11
B	1	General manager of informatics department	18
	2	Manager of the research institute	10
	3	Manager of informatics department of Southern China Branch	17
	4	Vice general manager of Southern China Branch	20
	5	Director of R&D center of Southern China Branch	19
	6	Vice director of the R&D center of Southern China Branch	8

3 Analysis of Supporting Technologies Based on a Scientometric Review

3.1 Overview of Retrieved Papers

The distribution of the 5,522 papers from 1974 to 2019 is shown in Fig. 1. Two turning points can be identified in 1982 and 2003. The period before 1982 is an "infancy" stage, during which the R&D of construction robotics started to appear. The two decades from 1983 to 2003 is a "stable development" stage. In each year, the number of papers was almost between 50 to 100. From 2004, the research in this field entered a "bursting" stage. The number of papers soared with fluctuations.

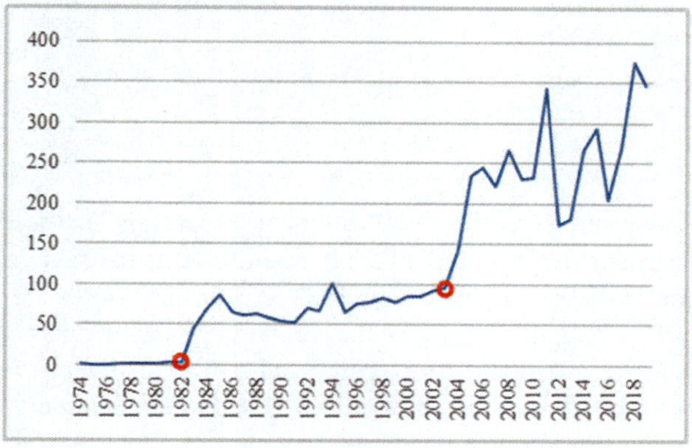

Fig. 1. The number of papers from 1974 to 2019.

3.2 Identification of the Supporting Technologies

To identify the supporting technologies in the field, the bibliographic data of the 5,522 papers was fed into CiteSpace for visualization. First, a keyword co-occurrence network is generated, as shown in Fig. 2. Because keywords are representative and concise descriptions of the research contents, the keyword co-occurrence network is expected to help researchers identify the hot topics in the field [7]. However, the result in this study provides little information because all keywords seem to gather together and no specific clusters were formed in the figure to indicate corresponding research areas. Although CiteSpace provides functions of cluster generation and analysis, the result was also unsatisfactory because some clusters were labeled with robotic and mechanic theories, some with construction tasks, and some with research methods. This problem might be due to the complexity of the field. Construction robotics involves multiple disciplines, multiple stakeholders, and multiple construction phases, which needs to be analyzed from multiple dimensions and is difficult to be simply separated into several clusters without specifying a certain dimension.

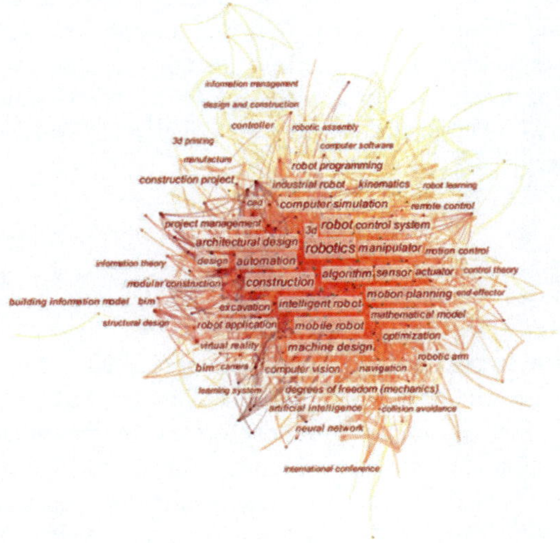

Fig. 2. Keyword co-occurrence network.

Therefore, a document co-citation analysis (DCA) was then tried out. DCA studies a network of co-cited references. Generally, the references cited in the papers provide the knowledge base of these papers [9]. Since the supporting technologies provide the knowledge base for R&D in construction robotics, DCA is expected to be useful in identifying the supporting technologies. The result is shown in Fig. 3. Seven significant co-citation clusters were identified automatically based on the keywords of the papers cited in each cluster in CiteSpace. CiteSpace provides several algorithms for clustering and the Log-Likelihood Ratio (LLR) algorithm was applied in this paper. This is because the LLR method can select the best cluster labels in terms of uniqueness and coverage [10]. Among the seven clusters, three of them refer to specific technologies, i.e., computer vision (#0), Radio Frequency Identification (RFID) (#5), and additive manufacturing (#6); two of them refer to the construction tasks in specific construction patterns, i.e., construction assemblies (#1), and off-site construction (#4); and two of them refer to the functions or characteristics of the robotic system, i.e., automated system (#2) and mobile system (#3), while the "automated system" is a wide and general concept.

Then the top 10 terms of each cluster are listed in Table 3, and the terms of specific technologies are marked in bold font. The terms were from the titles, abstracts, and keywords of the papers, and CiteSpace can extract the top 100 terms for each cluster. It needs to be explained that the terms in Table 3 have been processed manually based on the results generated by CiteSpace. In the process, similar terms were combined, such as "deep learning technique" and "using deep learning", and some terms providing little information of the cluster were removed, such as "critical review" and "construction site". Table 3 not only presents the supporting technologies for specific tasks and functions (#1, #2, #3, and #4) but also indicated the close relations between the supporting technologies (#0, #5, and #6). For example, computer vision is a supporting technology for many kinds of robots, including automated systems and mobile systems. Besides, the Building Information Model (BIM) and Artificial Intelligence (AI) technologies including deep learning and machine learning are also important supporting technologies. BIM promotes the development of construction assemblies and off-site construction, while AI technologies promote the development of computer vision and additive manufacturing.

3.3 Discussion

For further understandings, the citation histories of the terms related to the above technologies are also presented in Fig. 4 to Fig. 7. In Fig. 4, it is indicated that computer vision has been popular throughout the past decades and is still hot at present. Meanwhile, Simultaneous Localization and Mapping (SLAM) is a new branch of computer vision, facilitating the mobile robots to move in the unknown environment and establish the map at the same time. Figure 5 shows that RFID is hot during the second half-decade of the 2000s. Additive manufacturing (or 3D printing) and BIM are new technologies that became popular in the 2010s, as shown in Fig. 6 and Fig. 7. Figure 8 suggests that the development of the AI technologies guide the way of their implementations in construction robotics. The expert system appeared and developed in the 1970s and the 1980s to simulate the knowledge and experience of human experts to solve problems in specific fields, which realized a major breakthrough in AI from theoretical research to practical application, from general reasoning strategy to the use of specialized knowledge. This is the beginning of AI implementation in construction robotics. Then it increased significantly in the 2000s, which might be due to the development of the Internet, promoting the innovations in AI. In recent years, with the rapid development of information technologies, new directions such as machine learning and deep learning furtherly promote the AI technologies to play a more important role in construction robotics.

From Figs. 4, 5, 6, 7 and 8, it is observed that the recent hot supporting technologies include computer vision, 3D printing, BIM, and AI, as well as their emerging branches, including SLAM, machine learning, and deep learning. In other words, it can be expected that these supporting technologies are likely to be used in the R&D of construction robotics more than the other supporting technologies.

In addition, it is obvious that some supporting technologies have been studied and implemented continuously since the stable development stage (1983-2003), such as computer visioning and AI, and new branches of these technologies were developed in the bursting stage (2004-present). Besides, many new supporting technologies appeared in the bursting stage (2004-present), such as RFID and BIM, promoting the realization of new construction patterns.

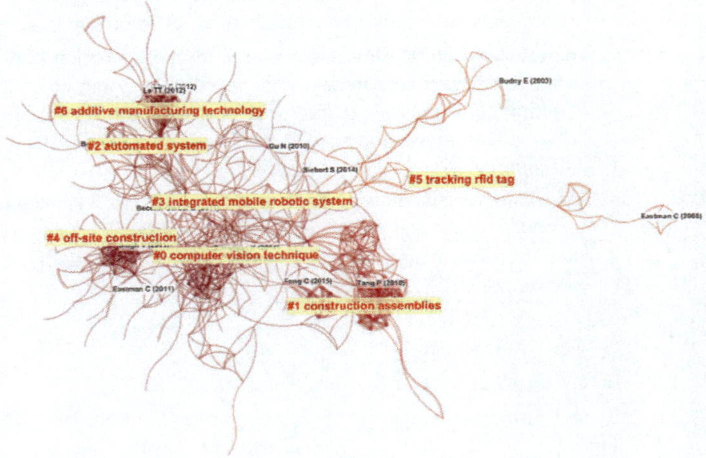

Fig. 3. Document co-citation network.

Table 3. Top 10 terms in each cluster.

Cluster	Top 10 terms
#0 computer vision	Maintenance phases; **deep learning**; automated concrete delamination assessment; **machine learning**; **SLAM-driven robotic mapping**; predicting safety hazard; point cloud; robust control approach; teaching robot; facility operation
#1 construction assemblies	Automated assembly discrepancy feedback; **building information model**; automatic realignment; robotic kinematics analogy; automatic as-built modeling; structural frames realignment planning; **point cloud**; automated progress monitoring; plant construction; quality-assured fabrication

(*continued*)

Table 3. (*continued*)

Cluster	Top 10 terms
#2 automated system	Modular integrated construction; digital fabrication; automated mobile marking robot system; **computer vision**; high-rise building construction; **monocular video camera**; 3d human body reconstruction; ergonomic posture analysis; **mobile robot**; hand tracking
#3 mobile system	Robotic fabrication; **visual SLAM**; ground-level structure; **active vision-based mapping**; motion planning strategy; vocational education; semi-dense visual-inertial odometry; **mixed reality approach**; building maintenance; capture position recognition
#4 off-site construction	**Building information model**; digital information management; industrialized house-building; model-based approach; earthwork project; construction document phase; **IoT application**; **5G technology**; in-situ construction method; Chinese super mason
#5 RFID	**Augmented navigation**; construction resource location tracking; **ultra-wideband technology**; high-speed railway lot; color tracking; multiple robot control; earthmoving operation; tracking error; effective construction logistics management; integrated advance data storage technology
#6 additive manufacturing	**3D printing**; materials properties; **augmented reality**; **deep learning**; geopolymer concrete; mechanical properties; feedback control system; mobile robot; maintenance management; **artificial intelligence technique**

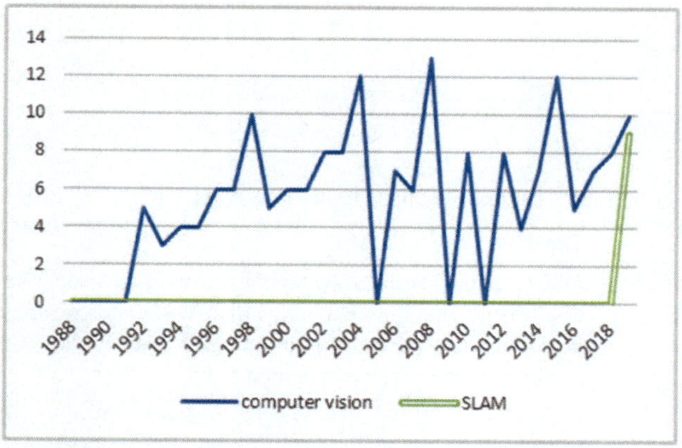

Fig. 4. Citation histories of the terms related to computer vision.

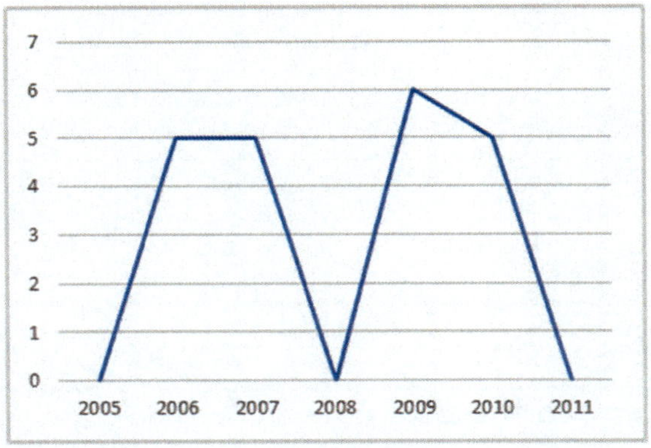

Fig. 5. Citation history of the term "RFID".

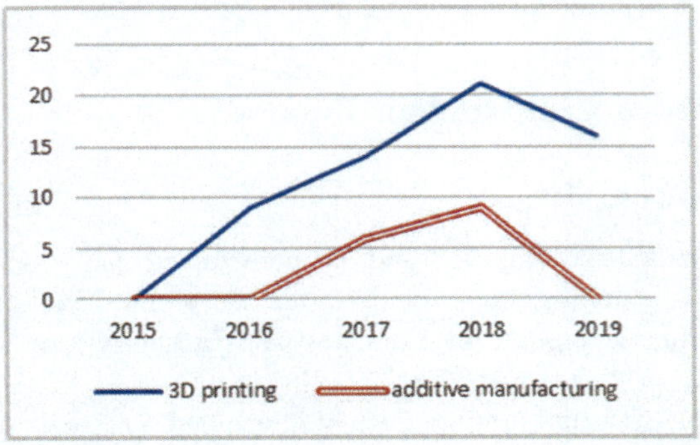

Fig. 6. Citation histories of the terms related to additive manufacturing.

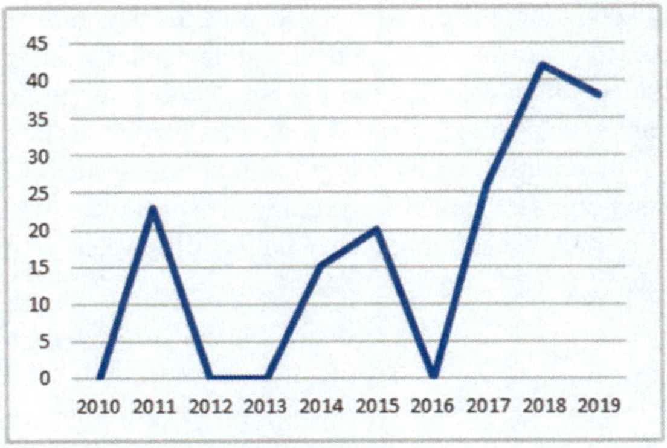

Fig. 7. Citation history of the term "building information model".

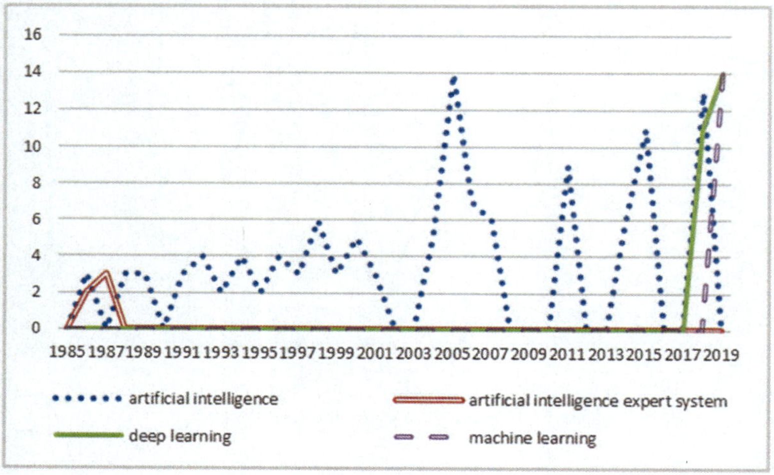

Fig. 8. Citation histories of the terms related to AI technologies.

4 Case Studies of Construction Robotics

4.1 Case 1: Company A

Company A was established in 1958 and has constructed many landmark projects in China. It is both a general contractor and a specialty contractor of steel structures. A representative robotic product of Company A is a gantry-type 3D printing robot system.

The gantry-type 3D printing robot system developed by Company A consists of a 3-axis gantry machine tool, a high-speed 3D printing nozzle, and a high-precision numerical control system, as shown in Fig. 9. With an effective printing size of 2500 mm × 3600 mm × 2000 mm, one of the most important features of this robot system is the large volume of its product, which leads to changes in the printing techniques and materials compared with those used in the manufacturing industry. Therefore, Company A developed a compaction technology to improve interlayer adhesion and conducted hundreds of experiments to optimize the printing parameters for large components such as printing speed and extruding speed. Besides, Company A also cooperated with a professional company for 3D printing materials to research and improve a new high polymer material for printing. This material has high strength and toughness, as well as high weather resistance ability, which can satisfy the need in the construction industry.

Fig. 9. 3D printing robot system.

The R&D work of the 3D printing robot system was initiated by a construction project of a 3D printed landscape bridge. In this project, the length of the bridge is 15 m and the weight is 30 t. In 2018, the bridge was printed at the construction site successfully in 35 days. Later on, Company A undertook more projects of 3D printing bridges. In 2019, the second bridge was printed in components and then installed at the site. In this time, the printing efficiency was improved to four to five times compared to the first one, and the quality is also improved. The robot is going to be used for the third bridge in 2020.

With the successful 3D printing robot system and its application experience, Company A is expanding and cultivating its market to promote follow-up applications in new areas. It has been used for making landscape decorations such as artificial hills, making molds for wind turbine blades, and repairing ancient architectures.

In addition, Company A is also conducting R&D on the implementation of BIM for the welding robots on the steel production line. Based on in-depth observation and analysis of the production techniques in the factory, the production processes can be expressed digitally with BIM data. Then the BIM model can be used to guide the steel production processes.

According to the interviewees from Company A, based on the current technologies, the construction robots in the factories have more opportunities to be widely used because the working environment is more standard and stable in the factories rather than the construction sites.

4.2 Case 2: Company B

Company B was established in 1995. It is both a general contractor and a specialty contractor of steel structures. It owns five regional companies in China (Eastern China, Southern China, Western China, Northern China, and Central China) with modern steel factories, reaching a total annual production capacity of more than 1.2 million tons, ranking the first in China's structural steel industry. Considering the implementation of robotics as a future direction, Company B built an intelligent production line in the Southern China factory from 2015 to 2017, which is the first intelligent steel production line in China. It integrated a number of advanced technologies, including robotic welding stations, as shown in Fig. 10.

Fig. 10. Robotic welding station.

The welding robots were imported from Japan and Company B improved them with supporting technologies and adapt them to the Chinese market. For example, based on cooperation with professional companies, computer vision technology was integrated into the robots for tracking the trajectory of the end-effector and rectifying the deviation. Besides, because most steel components for super high-rise buildings are nonstandard in China, robot operators need to write a large number of welding programs for different components. Therefore, Company B developed a parametric programming platform for the welding robot to simplify the process of programming.

According to the interviewees of Company B, the development of robotics and the supporting technologies have been changing the market environment. With the improvement of technology maturity in recent years, the application of robotics has become a popular direction in many industries in China. Many startup companies providing robots, robotic accessories, and technological services emerged in this field, and the government also enacted many encouragement policies for these companies. Therefore, the costs have decreased sharply, especially for the hardware. This market change promotes not only the R&D and application of construction robotics but also the integration of the supporting technologies.

4.3 Discussion

Among the key supporting technologies identified in Sect. 3, 3D printing (or additive manufacturing), computer vision, and BIM are implemented in the two cases. Besides, other technologies such as materials also played an important role. It is indicated that construction companies are active in implementing the supporting technologies in their robotic products. Based on the analysis of the implementation processes, the following factors are identified to be critical for the follow-up of the R&D of construction robotics.

Mutual Promotion Between Construction Robotics and Supporting Technologies. The development and maturity of the supporting technologies provide new opportunities for construction robotics and improve the market demand, while the market demand promotes the R&D of construction robotics. Meanwhile, during the R&D of construction robotics, the supporting technologies are implemented and also improved to adapt to the construction practices. In Case 1, the project owners appointed to build the bridge with 3D printing technology because it is popular and successful in many other fields. Then the need of the owner drove Company A to conduct the R&D of the 3D printing robot system. During the R&D, 3D printing technology has also been improved in several aspects. In Case 2, the technological development lowered the costs, encouraged the market, and promoted technology implementation.

Asynchronous Development in Different Application Areas. As we know that in the construction industry, robotics is promising in both the factories and the construction sites. In the factories, it is easier to learn from the experience of industrial robots and promote large-scale applications because of a similar working environment. At the construction sites, the environment is more complicated and nonstandard. Therefore, the future R&D of on-site robotics needs to study the approaches that are adaptable to construction sites. In addition, the application of construction robotic products is not necessarily limited to the construction industry. It is possible for the developers to explore new areas of application with the existing products.

Multidisciplinary Cooperation. Since construction robotics is a multidisciplinary field and the supporting technologies come from various fields, cooperative efforts are necessary for the R&D and application. In both cases, the construction companies implemented the supporting technologies with professional companies because the successful applications need to combine their construction experience with the knowledge and products of professional companies.

5 Conclusions

Construction robotics is expected to improve the construction industry in many aspects. Although many efforts have been made in this field over the past decades, the application rate is still limited. Therefore, this study explored the R&D trends in this field based on a scientometric literature review and case studies of two Chinese construction companies that developed and applied robotic products. In the literature review, the key supporting technologies in the past decades are identified, and their trends are discussed. The hottest ones at present include computer vision, 3D printing, BIM, and AI, as well as their emerging branches including SLAM, machine learning, and deep learning. From the case studies, it is indicated that the supporting technologies are implemented actively by the construction companies that develop and apply their own robots. Three factors are identified to be critical for the follow-up of the R&D of construction robotics, i.e., mutual promotion between the R&D and application of construction robotics and the development of supporting technologies, asynchronous development in different application areas, and multidisciplinary cooperation among construction companies and professional technological companies.

The findings are expected to be valuable for the related stakeholders including researchers, robot developers, and construction companies to conduct further R&D and application of robotics in the construction industry. Although the cases in this paper were based on Chinese companies, it is valid for the construction industry in other countries because the discussion and analysis were not confined to China.

Acknowledgments. The study has been supported by the Tsinghua - Glodon Joint Research Center for Building Information Modeling. We also acknowledge all the interviewees for their help in the case study: Xiaoming Chen, Linfeng Sheng, Jiqing Li, and Liang Ma, from Shanghai Mechanized Construction Group Co. LTD.; Jiantao Wang, Qingchuan Feng, Chunlin Wang, Jianfang Tian, Zhiyong Zuo, and Dongrong Xie, from China Construction Science and Industry Corporation LTD.

References

1. Bock, T.: The future of construction automation: technological disruption and the upcoming ubiquity of robotics. Autom. Constr. **59**, 113–121 (2015). https://doi.org/10.1016/j.autcon.2015.07.022
2. Cai, S., Ma, Z., Skibniewski, M.J., Bao, S.: Construction automation and robotics for high-rise buildings over the past decades: a comprehensive review. Adv. Eng. Inform. **42**, 100989 (2019). https://doi.org/10.1016/j.aei.2019.100989
3. Son, H., Kim, C.H., Kim, H., Han, S.H., Kim, M.K.: Trend analysis of research and development on automation and robotics technology in the construction industry. KSCE J. Civil Eng. **14**(2), 131–139 (2010). https://doi.org/10.1007/s12205-010-0131-7
4. Su, H.-N., Lee, P.-C.: Mapping knowledge structure by keyword co-occurrence: a first look at journal papers in Technology Foresight. Scientometrics **85**(1), 65–79 (2010). https://doi.org/10.1007/s11192-010-0259-8
5. Börner, K., Huang, W., Linnemeier, M., Duhon, R., Phillips, P., Ma, N., Zoss, A., Guo, H., Price, M.: Rete-netzwerk-red: analyzing and visualizing scholarly networks using the network workbench tool. Scientometrics **83**(3), 863–876 (2010). https://doi.org/10.1007/s11192-009-0149-0
6. Cobo, M.J., López-Herrera, A.G., Herrera-Viedma, E., Herrera, F.: Science mapping software tools: review, analysis, and cooperative study among tools. J. Am. Soc. Inf. Sci. Technol. **62**(7), 1382–1402 (2011). https://doi.org/10.1002/asi.21525

7. Zhong, B., Wu, H., Li, H., Sepasgozar, S., Luo, H., He, L.: A scientometric analysis and critical review of construction related ontology research. Autom. Constr. **101**, 17–31 (2019). https://doi.org/10.1016/j.autcon.2018.12.013
8. Chen, C.M.: CiteSpace: A Practical Guide for Mapping Scientific Literature. Nova Science Publishers (2016). ISBN 978-1-53610-280-2
9. Li, X., Ma, E., Qu, H.: Knowledge mapping of hospitality research−a visual analysis using CiteSpace. Int. J. Hospital. Manag. **60**, 77–93 (2017). https://doi.org/10.1016/j.ijhm.2016.10.006
10. Zhao, X.: A scientometric review of global BIM research: analysis and visualization. Autom. Constr. **80**, 37–47 (2017). https://doi.org/10.1016/j.autcon.2017.04.002

Unmanned Aerial Systems (UAS) Applications in Construction

Using UAS for Roofs Structure Inspections at Post-occupational Residential Buildings

Bruno Silveira(✉), Roseneia Melo, and Dayana Bastos Costa

Federal University of Bahia, Aristides Novis 02, Salvador, Bahia 40210-630, Brazil
brunofalcon.s@hotmail.com

Abstract. The traditional roof inspection tends to be a slow and challenging task, mainly due to accessibility limitations and the risk of falls from height. In recent years, the adoption of new technologies, such as the Unmanned Aerial System (UAS) to support the inspection process has proven to be an effective option to overcome such problems. Therefore, this study aims to evaluate the use of UAS for roof inspection on post-occupation residential buildings, seeking to assist in the decision-making process. The research method adopted was the case study, structured in the following steps: (a) evaluation of items that can be viewed with UAS, (b) protocol for flight planning and data collection with UAS, (c) analysis of non-compliance checking of roof conditions, and (d) evaluation of users' perceptions concerning the usability of UAS to support roof inspection. A total of 10 residential projects of a large construction company were inspected, totalizing 167 roofs. The analysis of the visualization potential shows that 72% of the requirements contained in the construction company's checklist can be inspected with the UAS. Besides that, 50,5% of the non-conformities identified are related to poor execution during the construction phase, highlighting the need for improvements. According to the maintenance team's perceptions, the UAS can contribute to increase worker safety, reduce time and effort during the data collection process, and assist in the identification of non-conformities.

Keywords: Roof inspection · Unmanned aerial system · Building maintenance and operation

1 Introduction

Inspection is an essential task in the building maintenance process, since it helps the previous identification of anomalies, reducing the cost with repairs (Garcez et al. 2012). And also, it has a function of ensuring that customers continue to get the value of their products obtained.

During the building operating, the roof has a complex and useful role due to its function of isolating, which aims to create a sealed and protected interior environment (Conceição et al. 2017). The barrier acts against the exterior agents, such as variable temperatures, rain, and biological agents (Conceição et al. 2017; Garcez et al. 2012). In addition to environmental conditions, other factors that can influence the roof performance are the incorrect design and poor execution (Conceição et al. 2017).

E. Toledo Santos and S. Scheer (Eds.): ICCCBE 2020, LNCE 98, pp. 1055–1068, 2021.
https://doi.org/10.1007/978-3-030-51295-8_73

Besides the importance, roofs are not regularly inspected, making it difficult the previous identification of anomalies that may cause problems (Garcez et al. 2012). The small defects, when not treat adequately, can lead to problems in the roof structures and other parts of the buildings, increasing the cost to repair and upkeep (Garcez et al. 2012). Thus, determine the severity of damage previously is, therefore, essential to keep costs down, while at the same time, making sure that those roofs that do need to be repaired are done so on time (Hezaveh et al. 2017).

The main shortcomings faced by the roof inspection process are related to the inefficiency of manual visual inspection, the difficulty of accessing roofs, especially the sloped, and the time consuming to inspect all elements (Bown and Miller 2018). Moore et al. (2018) discuss that depending on the size of the building and the type of roof (flat or slope), the inspections can take a couple of days or a few weeks. In some cases, lifting equipment is required to carry out the service. Another concerning is related to worker safety during the inspection processes, once falls from height represent one of the leading causes of fatal accidents in the construction industry (Bown and Miller, 2018; Moore et al. 2018).

Those drawbacks have driven the construction sector to search for new technologies that can facilitate the execution of inspection processes, reducing time, the risk of accidents and errors during collection, and processing data. One of the technologies that have been gaining attention in engineering to perform monitoring and inspection is the use of UAS (Unmanned Aerial Systems). The use of UAS to perform inspections is promising since it eliminates the need for lifting equipment and specialized operators. In addition to ensuring increased safety during the inspection process and performing the activity more efficiently and economically (Eschmann et al. 2012; Morgenthal and Hallermann, 2014).

Therefore, researches have been developing in several fields in order to enhance the inspection process, such as task-related to visual inspection and damage detection (Kumar et al. 2013; Morgenthal and Hallermann 2014; Torok et al. 2014); maintenance of facades and mapping of historical monuments (Eschmann et al. 2012; Emelianov et al. 2014; Roca et al. 2014); monitoring of bridges (Metni and Hamel, 2007; Ellenberg et al. 2015); safety conditions at construction sites (Costa et al. 2016, Melo et al. 2017), and construction progress monitoring (Álvares and Costa 2019).

Regarding roof inspection, most of the applications of the use of UAS are focused on thermal imaging as the primary tool to identify damages (Zhang et al. 2015; Krawczyk et al. 2015; Moore et al. 2018; Rakha et al. 2018). Only a few studies have been using photos of high resolution to assess the roof conditions (Bown and Miller 2018). This shows the high potential of use, once the visual inspection is an essential step in the diagnosis phase and, based on the first results, other complementary diagnosis methods can be applied (Garcez et al. 2012).

Moore et al. (2018) pointed out that UAS is a suitable option for a performance roof inspection in a faster and safe way. After the data collection using the UAS, the inspector can analyze the conditions, identify anomalies, and create a plan to repair it. The UAS makes it possible to increase transparency, simplify steps, reduce inspection time and reduce activities that do not add value by allowing the focus on preparing more accurate reports and corrections in a timely manner (Melo et al. 2017).

The use of visual technologies is mainly targeted to reduce the amount of waste, time, and effort and generate value to the process. Therefore, despite the literature pointed out the potential of UAS to support in identifying damage and anomalies on a roof inspection, its use and method to apply should be better investigated in order to understand how this technology can be incorporated into facilities management operations (roof structures). Therefore, the main objective of this study is to evaluate the use of UAS for roof inspection on post-occupational buildings and propose a protocol for flight planning and data collection of this kind of structure. This study is based on a set of data of 167 building roofs.

2 Use of UAS for Structure Inspections

As discussed previously, the UAS has been showing as a promising tool to perform inspections in several fields in engineering. In civil engineering, some studies and their contributions are better presented in this chapter.

Eschmann et al. (2012) propose the use of UAS to scan the facades of buildings through a high-resolution digital camera. The method used was divided into two processes. In the first process, data acquisition takes place, in which a set of photos are captured in order to ensure the cover-up of the entire façade of the building. In the second process, digital processing of the images is performed, the high resolution allows to identify damage and cracks in the precision of mm. The results obtained showed that the use of UAS equipped with high-resolution cameras generates a database for visual inspection of buildings (Eschmann et al. 2012). Kumar et al. (2013) carried out mapping a tower in India 300 m high, in which the structure was examined from different angles. To this end, we opted for a circular and vertical flight. Through visual assets, it was possible to visualize cracks, internal structural damage, exposure of steel armor, and rust, among other pathological manifestations.

Another increasing application is the use of UAS to monitor large structures, such as bridges and viaducts. Methi and Hamel (2007) assess the feasibility of the UAS to inspect viaducts. During the experiment, a sequence of videos and photos was obtained through the onboard camera, and later, the visual assets were shown to experts in this type of inspection so that they could be evaluated. The analysis of the images showed the enormous potential of using UAS to perform visual inspections, where after an additional treatment and due to the excellent resolution of the photos, it was possible to identify cracks up to 0.1 mm. Ellenberg et al. (2015) implemented the use of UAS in practical applications for monitoring structural health monitoring in bridges. The use of the UAS showed promising results in the detection of defects and damage, with errors that can lead to a much more quantitative evaluation when compared to the traditional visual inspection method. Gillins et al. (2016) proposed a methodology to inspect bridges. According to the authors, the UAS can be used remotely to produce high-resolution close-up images from various viewing angles. Such image resources provide resolution comparable to what can be visually observed by an inspector at an arm-length distance.

Regarding roofs, more studies are concerning the use of thermal images to inspect roofs. Zhang et al. (2015) proposed an algorithm to detect thermal anomaly automatically. The results show that the use of algorithms can improve the contrast between warm and cold areas on the roof surface. As well as produces more constant thermal signatures of different roof insulations or surfaces, facilitating both visual interpretation and computer-based thermal anomaly detection. Krawczyk et al. (2015) showed that aerial thermal images associated with UAS could be advantageous for the buildings sector. Moore et al. (2018) selected five buildings to be analyzed for potential damages on the roofs. A UAS equipped with a thermal camera and a visual camera was selected. As a result, they found several types of damages. The authors highlighted that through the visual assets collected, the maintenance teams can determine whether the roofs need further maintenance.

Bown and Miller (2018) performed tests to evaluate the use of high-resolution images to roof inspection. According to the results, the optimal flight elevation should be between 3–4, 6 m above the roof; below this elevation, the quality of the image was excellent, but more images were needed for the inspection, which makes it slower to analyze all data. Above 4, 6 m elevation, the image quality was not sufficient detail to support the decision-making. These were the first parameters found in the literature about the use of UAS to make roof inspection. The authors also analyze that after approximately 10 h of flight time, new pilots start to develop some basic skills that allow them to focus on the inspection project first of all, and flying the UAS as a secondary concern. The authors recommend the use of individual images using a zooming approach.

As the main limitation regarding the use of UAS, some researchers pointed out the difficulty of operating the aircraft in different weather conditions such as rain, snow, and high wind (Morgenthal and Hallermann 2014; Álvares and Costa 2019).

3 Research Method

The research strategy adopted was the case study. A case study allows the researcher to study phenomena in real-life, considering several contexts and interactions (Yin 2014; Taylor et al. 2011). As benefits, it also shed light on new insights, lines of inquiry, and theoretical models that can enhance and expand construction engineering management research (Taylor et al. 2011). This research was a demand of one construction company that due to the high number of roofs in the post-occupational building to inspect, seeks alternatives to make the process faster and reduce cost. In addition, to explore the UAS for roof inspections, this research also involved training (theory and practical lessons) for the maintenance team about the use of UAS for increase their capabilities in using UAS technology.

The present study was divided into the following stages, detailed in the next sections: (a) identification of a set of items on the roof structure that can be inspected through images collected with UAS; (b) development of case studies (UAS data collection protocol, field data acquisition, and preliminary data analysis); (c) training of maintenance teams to use of UAS for roofs inspections; (d) statistic data analysis and assessment of results.

The first stage involved the analysis of the construction company checklist used to perform the roof inspection on post-occupational buildings. After a previous analysis, a checklist containing 28 elements related to the roof was used for this study, according to Table 1.

Table 1. List of pathologies for analysis in the checklist

Checklist Items (Pathologies)			
1	General aspect	*15*	Caulking of small parapet wall openings (when applicable)
2	Antenna box cover	*16*	Valley flashing integrity and caulking
3	Installation and integrity of the Lightning protection systems (when applicable)	*17*	Integrity of the parapet wall top (when applicable)
4	Installation of the hatch and its caulking	*18*	Integrity and waterproofing of the parapet wall frame (when applicable)
5	Aircraft warning lights installation (when applicable)	*19*	Waterproofing/texture on the inside of the parapet wall
6	Stair integrity (where applicable)	*20*	Base of tubes that pass through the duly waterproofed tiles
7	Coverage in the elevator shaft	*21*	Railing integrity
8	Tiles Shift	22	Gutter integrity and cleanliness
9	Deformations in the tiles	23	Roof ridge integrity
10	Presence of leaks	24	Ceiling Trapdoor integrity
11	Ventilation duct painting	25	Ventilation duct integrity
12	Cracks in the tiles/broken tiles	26	Rusted Metals
13	Presence of residues on the roof	27	Integrity of photovoltaic plates (when applicable)
14	Accumulation of algae, lichens and mosses	28	Tiles painting

The study was carried out at ten residential post-occupational condominiums located in Brazil. The projects studied were selected based on the regulation's requirements established by the National Agency for Civil Aviation (ANAC), which allows flights with a minimum radius distance of 5 km from airports and heliports.

For each project, a field visit was carried out for data collection with UAS. The data flights for each project are presented in Table 2. For the inspections, a DJI Phantom 4 model with 2 batteries (autonomy 22 min each) and a DJI Phantom 3 with 3 batteries (15 min each) were used. The flights were performed in manual mode to ensure safety due to the presence of physical obstacles and to maintain the quality of images. In all projects, the date to perform the inspection was scheduled previously with the administration of the residential condominium. Thus, the residents were communicated through a letter about the roof inspections with UAS beforehand.

Table 2. Visual assets data collection and flight data

Project	Date	Nº blocks/roofs	Nº units	Year of construction	Nº photos collected	Nº of flights	Total flight duration
A	13/ago/18	46	920	2018	359	3	00:32:15
B	21/nov/18	20	400	2018	230	2	00:19:00
C	21/nov/18	4	170	2014	119	1	00:14:00
D	19/dez/18	12	600	2015	353	2	00:27:58
E	19/dez/18	10	460	2015	245	2	00:22:01
F	02/abr/19	31	620	2017	489	2	00:32:01
G	18/abr/19	13	520	2015	331	2	00:24:23
H	18/abr/19	10	640	2017	396	2	00:25:02
I	30/jul/19	18	780	2014	448	2	00:34:02
J	22/ago/19	3	360	2017	180	2	00:24:04
	–	167	5470	–	3150	20	04:14:46

A protocol with a sequence of steps and activities to perform the inspection with UAS safely was established before the flights. A set of forms were used to support the application of UAS based on Irizarry et al. (2015) and Melo et al. (Melo et al. 2017), as described below:

- Planning meeting form: it determines the points of take-off and landing; how many flights need to be performed to collect all roofs in order to save time; and also identifying possible barriers to the UAS flight mission.
- UAS flight mission checklist: it records the flight's technical data (such as location, date, weather conditions, and others) and also presents a checklist of all necessary procedures for a UAS flight mission according to safety requirements.
- Flight log worksheet: it lists all the flight's data recorded in the DJI Go app and UAS flight mission checklist, including information about unwanted situations that could happen during the flight, as loss signal.

Figure 1a shows an example of UAS flight planning developed to perform roof inspections. In the example presented was performed two flights to inspect 20 roofs. Figure 1b presents the details of the UAS flight path carried out in each roof.

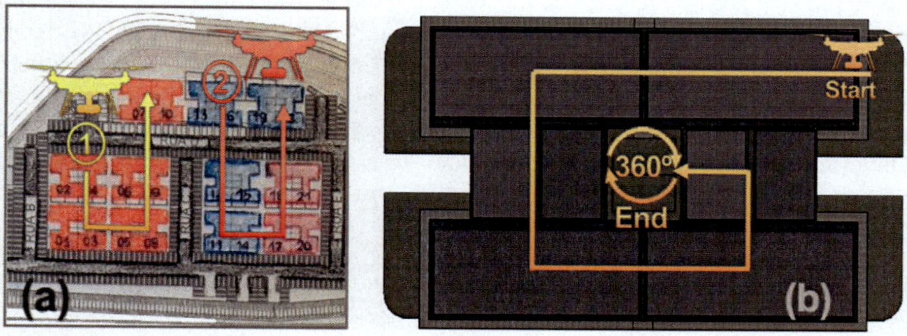

Fig. 1. Flight path (a) UAS Flight planning; (b) Example of flight planning for each building

To ensure that the pathology analysis could be identified in each roof individually, the procedure was to collect images from each roof separately. In each roof, a 360º flight was performed to record the visual data assets (Fig. 1b). The flight altitude above the building was approximately 5 meters, certifying the high resolution in the images. The maintenance team had access to all visual assets collected.

Three analyses were carried out based on images collected. The first one consisted of verifying whether the roof and structure checklist could be visualized based on the photos collected with UAS. The items were analyzed according to the following criteria, (1) visualized, (2) non-visualized due to technology limitations or image without enough details to decision-making and, (3) non-applied. The technology limitations are concerning the image quality in cases that not provide enough details to decision-making or items inside the buildings that could not be inspected by UAS.

The second analysis aimed to evaluate the compliance of the inspected items, being classified as (C) compliant items, (NC) non-compliant items, and (NA) non-applied items to requirements that were not observed in the project. The third analysis consisted of identifying the most recurring pathologies and make a comparison of the amount of non-compliance found between projects and their construction age.

In parallel to roofs inspections, it was made training with the maintenance team to transfer the use of technology to them. The training was composed of theoretical lessons about the types of UAS, sensors, legislation requirements, and guidelines to use the UAS safely. The researchers used the DJI simulator to provide a first experience with the flight piloting. After that, a practical lesson was carried out in an open field, and all participants had the opportunity to pilot for a short time. A total of six-person (two civil engineers, two trainees, one quality analyst and one installations engineer) participated in theoretical and practical lessons; each lesson had a duration of 4 h.

In the end, a questionnaire was applied to all participants to collect the perceptions about the usability of the UAS. The data were analyzed through the Relative Importance Index (RII), as presented in Eq. (1):

$$IIR = \frac{\sum W}{A \times N} \tag{1}$$

W is the weight given by the participant for each element using Scale in 10 levels (1 - Very Low up to 10 - Very High); A is the highest level in this study it is 10; N is the sample size (6 participants).

Also, interviews to gather the maintenance team's perception concerning the barriers and benefits were conducted to identify the points of improvement to incorporate the technology into facilities tasks.

4 Results and Discussions

4.1 Roof Inspections Analysis

The first analysis shows that 72% of the roof checklist (20 items) could be verified through visual assets collected with UAS. The others 28% could not be inspected due to technology limitation (21%) and the image without enough details to decision-making (7%). The technology limitations were related to items inside the buildings, which makes it impossible to be viewed using UAS. The image without enough details corresponds to the items associated with roof tiles conditions, especially those that had an accumulation of lichens and mosses or became worn.

In the second analysis, 167 roofs were inspected based on the items that can be visualized using UAS. Table 3 presents the number of pathologies identified in each project. Based on the results, the Project I recorded a more amount of pathologies, totaling 34, followed by Project D (30 pathologies) and Project A (27 pathologies). However, analyzing the average of pathologies/roof, Project C recorded 4,25 non-conformity/roof, followed by Project D (2,5 non-conformity/roof) and Project E (2,20 non-conformity/roof). Project C and I are the oldest, delivered in 2014.

Table 3. Results of conformity data analysis

Project	Nº blocks/roofs	Amount of non-conformities	% non-conformity	Rate of non-conformity/roof	Year of construction
A	46	27	13,4	0,59	2018
B	20	20	9,9	1,00	2018
C	4	17	8,4	4,25	2014
D	12	30	14,9	2,50	2015
E	10	22	10,9	2,20	2015
F	31	11	5,4	0,35	2017
G	13	18	8,9	1,38	2015
H	10	17	8,4	1,70	2017
I	18	34	16,8	1,89	2014
J	3	6	3,0	2,00	2017
Total	167	202	–	–	–

Figure 2 shows the number of pathologies according to the delivered age. It can be seen that the new buildings presented fewer pathologies than the oldest when designed and executed adequately. However, one of the reasons for the increase of pathologies throughout the years is the use and operation carried out by residents, such as irregular TV antennas installations, lack of cleaning and maintenance and others.

According to the results, two projects delivered in 2017 presented a high rate of non-conformity/roof (Project J and H) comparing to the buildings delivered in 2015. These projects are towers with 12 floors (Project J) and eight floors (Project H), while the others are buildings with five floors each. Therefore, the number of residents may be a factor that influences the number of pathologies. Once a higher number of residents may have requested services involving access to the roof, causing pathologies such as displacement or breaking tiles and installations damages.

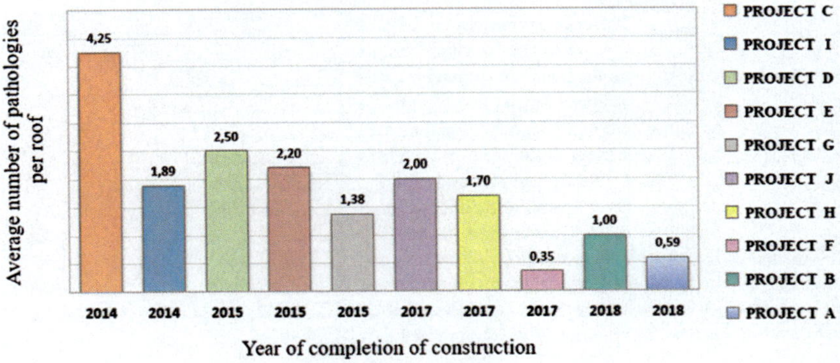

Fig. 2. Rate of non-conformity per roofs throughout the years

The four most recurring pathologies represent 50,5% of the pathologies (102 occurrences) identified as presented below.

- Presence of waste in the roofs (materials, plastic, and others);
- Structure and installations metal covered with rust, due to the lack of adequate painting and maintenance;
- Failures in waterproofing of the plateau, which has the function of preventing undesirable water seepage, or directing it to drainage points outside the area that needs to be protected;
- Accumulation of lichen and moss in the rain gutters, due to the lack of regular cleanliness

Figure 3 shows the top 10 most recurring pathologies considering all projects. The causes of the pathologies were classified as related to use and maintenance or related to construction problems. Figure 4 presents an example of failure in the integrity of the flashing.

The roof flashings have the function to protect the encounters of the walls with tiles, preventing infiltration of rainwater besides preventing seepages by capillarity. Thus, to ensure the roof integrity, it is vital to carry out the adequate correction of these small pathologies to avoid problems regarding seepage in the future. Figure 4 also shows problems regarding damage in the tiles, pointing out the need for replacement, as preventive actions. According to Fig. 3, the pathologies related to maintenance correspond to 49,5% of the total pathologies. This indicates that there is room for improvement in the construction stage (highlighting the points that the company needs to improve), in addition to making it possible to alert the condominiums about the need for maintenance.

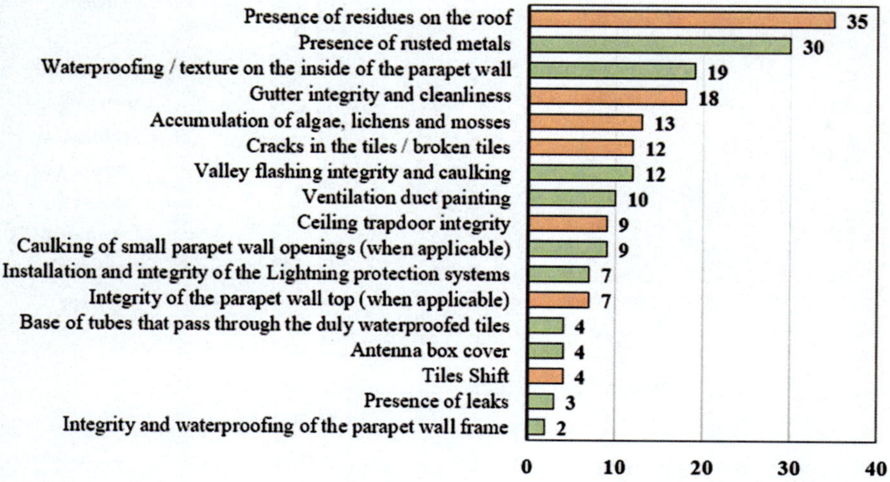

Fig. 3. Most recurring pathologies identified during all roof's inspections

Fig. 4. Example of pathologies identified, such as roof flashing and tiles damages

4.2 Analysis of the Maintenance Team' Perceptions

This analysis was based on the maintenance team's perception concerning the usability of the UAS to roof inspections and the utility of the training carried out. The data analysis was performed using the Relative Importance Index (RII) presenting in Table 4.

Table 4. Maintenance team' perception concerning the usability of the UAS (n = 06)

Category/construct	Item	IIR
Training/utility	Quality of theoretical lessons (Basic knowledge about UAS, rules, recommendations, and others)	1,00
Inspection/usability	Reduction of physical efforts to perform the inspection	1,00
Inspection/usability	Reduction of the time spent to perform the inspection in the field	0,98
Inspection/usability	Reduction of time spends to elaborate on the report	0,98
Inspection/usability	Number of photos collected during the inspection to be analyzed	0,98
Inspection/usability	Increasing of inspector safety during the process of inspection	0,97
Inspection/usability	Use of visual assets to assist in the decision-making process	0,93
Training/utility	Time spent to carry the practical training	0,92
Training/utility	Confidence to pilot the UAS after the training	0,90

On the whole, the result shows that both inspections using UAS and the training carried out had high relative importance perceptions, evidencing the technology potential to assist roof inspections. The lower items are related to the time spent to carry the training and the confidence to pilot the UAS. The practical training was dimensioned based on the batteries' autonomy, which can be improved thought the increase of equipment and hours to practice. Regarding the theoretical lessons, 4 h sounds to be enough to present the basic knowledge.

In the items related to the usability of the technology, the result presents that 100% of the participants agree that the UAS can reduce the physical efforts during the roof inspection. The time spends to perform the inspection, and the report was also evaluated positively. Despite that, the report continues demanding more time due to the number of photos to be analyzed, as discussed in the interviews. Through the interviews were possible to identify the benefits and barriers regarding the use of UAS for technical assistance purposes (Table 5).

The interviewers pointed out improvements in the roof's inspection performance, due to the previous identification of the non-conformities and failures caused by the inadequate use and operation by the residents. The most regular cases were the inadequate antennas' installations and the lack of preventive maintenance. Besides, at the beginning of the inspections, the maintenance team was concern about the residents' acceptance; however, none reclaims was registered, and some residents praised the attitude developed by the company to seek other ways to perform the inspection.

Table 5. Benefits and barriers regarding the use of UAS to perform a roof inspection

Benefits	Barriers
✓ Greater agility in the inspection process, contributing to increasing quality of the data used in the reports, enabling the zoom and the search for specific details. ✓ Good acceptance by residents. ✓ Allow easy and fast access to roofs. ✓ Increasing inspector safety during the inspection process. ✓ Facilitate the identification of the roof during the elaboration of the report due to the aerial perspective. ✓ Make faster the inspection of a high number of roofs before the raining season, facilitating the planning of preventive actions. ✓ The time save during the inspection, can be used to improve the report analysis.	ℵ Batteries Autonomy, since due to the high number of roofs per project, it was necessary to recharge the batteries to continue the inspections. Besides that, with more batteries, it will be possible to record more details about the roofs. ℵ Cost of the equipment. ℵ Time spends during the manual elaboration of the report, which makes it impossible to have a result in real-time. ℵ Difficulty in identifying items related to gutters rust, flashing caulking and small cracks in the tiles. ℵ Impossibility to perform flight in roofs near to the airport, as established by ANAC. ℵ Possibility of accidents due to inadequate flight planning and mechanic issues. ℵ Curiosity awakened in the residents can cause distractions to the pilot during the flight performance

5 Conclusions

The objective of this study was to evaluate the use of UAS for roof inspection on post-occupational buildings. A protocol for flight planning and data collection is presented, and a set of data of 167 roofs were analyzed.

The application of the UAS provided the visualization of 72% of the items present in the checklist, the rest of the items not being observed due to technology limitation (21%), or lack of detail in the collected images (7%). Based on the visualization of the items, it was possible to identify non-conformities related to the use and maintenance of roofs, such as the presence of residues, presence of rusted metals and the lack/damage of the waterproofing. 50,5% of the non-conformities are related to poor execution during the construction phase, highlighting the need for improvement. The other 49,5% are related to poor use and lack of maintenance, which could be reduced thought regular cleanness and preventive maintenance.

In addition, through the interviews conducted by the researchers with the technical assistance team, it was realized that: (a) there is a need to increase the hours of practical classes with the UAS, seeking to increase the pilots' confidence in handling the equipment; (b) the UAS provides greater agility and ease in obtaining the data; (c) increasing the inspector's safety during the inspection process; (d) difficulty in viewing certain items, such as small cracks; and (e) the residents' curiosity can distract the pilot during the inspection, being recommended the presence of one or more employees to work around this problem.

In general, the findings point out the potential of UAS application for roof inspections, providing a great amount of data in a short time and allowing the visualization of pathologies in difficult areas. Due to the higher visualization and process transparency, more reliable data is delivered to the customers and the professionals responsible for the repair.

Such visual technology can be used to improve the information flow of communication, comprehension, evaluation and application of corrective actions of roof pathologies. Therefore, as further research, the interviewer suggested the report automatization aiming associated the photo with the non-conforms items during the inspections process, allowing elaborating the report in almost real-time.

References

Álvares, J.S., Costa, D.B.: Construction progress monitoring using unmanned aerial system and 4D BIM. In: Pasquire, C., Hamzeh, F.R. (eds.) Proceedings of the 27th Annual Conference of the International. Grupo para Construção Enxuta (IGLC), Dublin, Irlanda, pp. 1445–1456 (2019)

Alves, T.D.C.L., et al.: Understanding inspection challenges in the EPC industry: a simulation approach. In: IGLC 2016 - 24th Annual Conference of the International Group for Lean Construction, pp. 23–32 (2016)

ANAC Agência Nacional de Aviação Civil: Regulamento Brasileiro da aviação Civil Especial (RBAC-E nº94). ANAC, Brasília (2015)

Aziz, R.M., Hafez, S.M.: Applying lean thinking in construction and performance improvement. Alexandria Eng. J. **52**(4), 679–695 (2013)

Bown, M., Miller, K.: The use of unmanned aerial vehicles for sloped roof inspections – considerations and constraints. J. Fac. Manag. Educ. Res. **2**(1), 12–18 (2018)

Conceição, J., et al.: Inspection, diagnosis, and rehabilitation system for flat roofs. J. Perform. Constr. Facil. **31**(6), 1–15 (2017)

Costa, D.B., Melo, R.S., Álvares, J.S., Bello, A.A.: Evaluating the performance of unmanned aerial vehicles for safety monitoring. In: Proceedings of 24th Annual Conference of the International. Group for Lean Construction, Boston, MA, USA, Sect. 11, pp. 23–32 (2016)

Ellenberg, A., Branco, L., Krick, A., Bartoli, I., Kontsos, A.: Use of unmanned aerial vehicle for quantitative infrastructure evaluation. J. Infrastruct. Syst. **21**(3), 04014054 (2015)

Emelianov, S., Bulgakow, A., Sayfeddine, D.: Aerial laser inspection of buildings facades using quadrator. Procedia Eng. **85**, 140–146 (2014)

Eschmann, C., Kuo, C.M., Kuo, C.H., Boller, C.: Unmanned aircraft systems for remote building inspection and monitoring. In: 6th European Workshop on Structural Health Monitoring, Dresden, Germany, pp. 1–8 (2012)

Garcez, N., et al.: System of inspection, diagnosis and repair of external claddings of pitched roofs. Constr. Build. Mater. **35**, 1034–1044 (2012)

Gillins, M.N., Gillins, D.T., Parrish, C.: Cost-effective bridge safety inspections using unmanned aircraft systems (UAS). In: Proceedings of the Geotechnical and Structural Engineering Congress, pp. 1931–1940 (2016)

Hezaveh, M.M., Kanan, C., Salvaggio, C.: Roof damage assessment using deep learning. In: Proceedings - Applied Imagery Pattern Recognition Workshop, vol. 2017, pp. 6403–6408 (2018)

Irizarry, J., Costa, D.B., Kim, S.: Potential applications of unmanned aerial systems for construction management tasks. In: Construct15 – Global Knowledge, Global Solutions, Perth. International Conference on Innovative Production and Construction (IPC 2015). Perth (2015)

Krawczyk, J., et al.: Infrared building inspection with unmanned aerial vehicles. Trans. Inst. Aviat. **240**(3), 32–48 (2015)

Kumar, S.K., Rasheed, A.M., Kumar, R.K., Giridharan, M., Ganesh: "Dhaksha, the Unmanned Aircraft System in its New Avatar Automated Aerial Inspection of India's Tallest Tower", UAV-g 2013. Rostock, Germany (2013)

Melo, R., Costa, D.B., Alvares, J.S.: Applicability of unmanned aerial system (UAS) for safety inspection on construction sites. Saf. Sci. **98**, 174–185 (2017)

Metni, N., Hamel, T.: UAV for bridge inspection: visual servoing control law with orientation limits. Autom. Constr. **17**, 3–10 (2007)

Moore, J., et al.: Facility inspection using UAVs: a case study in the University of Georgia campus. Int. J. Remote Sens. **39**(21), 7189–7200 (2018)

Morgenthal, G., Hallermann, N.: Quality assessment of unmanned aerial vehicle (UAV) based visual inspection of structures. Adv. Struct. Eng. **17**, 289–302 (2014)

Rakha, T., et al.: Heat mapping drones: an autonomous computer-vision-based procedure for building envelope inspection using unmanned aerial systems (UAS). Technol. Archit. Des. **2** (1), 30–44 (2018)

Roca, D., Armesto, J., Laguela, S., Dıaz-Vilarino, L.: Lidar-equipped UAV for building information modelling. Int. Arch. Photogram. Remote Sens. Spatial Inf. Sci. **40**(5), 523 (2014)

Taylor, J.E., Dossick, C.S., Garvin, M.: Meeting the burden of proof with case-study research. J. Constr. Eng. Manag. **137**(4), 303–311 (2011)

Tezel, A., Aziz, Z.: From conventional to IT based visual management: a conceptual discussion for lean construction. J. Inf. Technol. Constr. (ITcon) **22**, 220–246 (2017)

Torok, M.M., Golparvar-Fard, M., Kochersberger, K.B.: Image-based automated 3D crack detection for post-disaster building assessment. J. Comput. Civ. Eng. **28**, A4014004 (2013)

Yin, R.K.: Case Study Research Design and Methods, 5th edn., 282 p. Sage, Thousand Oaks (2014)

Zhang, J., et al.: Thermal infrared inspection of roof insulation using unmanned aerial vehicles. In: International Archives of the Photogrammetry, Remote Sensing and Spatial Information Sciences - ISPRS Archives, vol. 40, no. 1W4, pp. 381–386 (2015)

Multi-scale Flight Path Planning for UAS Building Inspection

Paul Debus(✉) and Volker Rodehorst

Bauhaus-Universität Weimar, Weimar, Germany
paul.debus@uni-weimar.de

Abstract. Unmanned aircraft systems (UAS) show large potential for the construction industry. Their use in condition assessment has increased significantly, due to technological and computational progress. UAS play a crucial role in developing a digital maintenance strategy for infrastructure, saving cost and effort, while increasing safety and reliability. Part of that strategy are automated visual UAS inspections of the building's condition. The resulting images can automatically be analyzed to identify and localize damages to the structure that have to be monitored. Further interest in parts of a structure can arise from events like accidents or collisions. Areas of low interest exist, where low resolution monitoring is sufficient.

From different requirements for resolution, different levels of detail can be derived. They require special image acquisition parameters that differ mainly in the distance between camera and structure. Areas with a higher level of detail require a smaller distance to the object, producing more images.

This work proposes a multi-scale flight path planning procedure, enabling higher resolution requirements for areas of special interest, while reducing the number of required images to a minimum. Careful selection of the camera positions maintains the complete coverage of the structure, while achieving the required resolution in all areas. The result is an efficient UAS inspection, reducing effort for the maintenance of infrastructure.

Keywords: UAS · Flight path planning · Infrastructure assessment · Image-based SHM · Drones · Multi-scale monitoring

1 Introduction

The technological progress in the field of unmanned aircraft systems (UAS) in recent years has enabled many important applications of those mobile sensor platforms in all areas of life [1], especially for remote sensing [2]. UAS facilitate the fast and easy access to places that are not without problems reachable or dangerous for humans. Equipped with modern camera hardware, they allow for a fast and automated acquisition of image data, that can be used in many applications [3].

The maintenance of infrastructure is important for the functioning of a modern society. Many bridges and similar structures have been built in the last century and are nearing the end of their design life. In Germany, more than 60% of all road bridges are older than 30 years [4], while normal design life of a bridge is around 50 years. To

E. Toledo Santos and S. Scheer (Eds.): ICCCBE 2020, LNCE 98, pp. 1069–1085, 2021.
https://doi.org/10.1007/978-3-030-51295-8_74

allow for an efficient management of the necessary repair work, a digital maintenance strategy is required, enabling a high utilization of available resources. Part of this strategy has to be the support of human actors with digital technologies, modern data acquisition and processing tools and transparent and understandable decision aids.

The use of UAS for the condition assessment of large infrastructure has the potential to greatly reduce the effort, cost and time required for reliable monitoring [5], while increasing reliability and reproducibility of the results. The availability and mobility of UAS allow them to capture the surface of a structure far more efficiently and rapidly, than human inspectors could. In combination with modern data processing pipelines like Structure from Motion (SfM) [6] for three dimensional (3D) reconstruction or automated image analyses, they allow for a fast initial assessment of the condition of the structure. Especially modern machine learning tools enable the automated detection of anomalies like cracks and other damages to the structure [7]. Together with high-resolution deformations, computed from the 3D models, simulations and predictions of the current and future condition of the structure are possible and enable adequate maintenance [8].

One of the biggest advantages of this technology is to allow human bridge inspectors, the experts that are finally responsible for the correct maintenance, to focus their attention on relevant parts of the structure, after the UAS based inspection identified regions of high and low interest. Not only does this speed up the inspection process, it also allows to pay more attention to critical parts, where potential damages were detected in the images.

One challenge of the application of UAS-based visual inspection is the amount of data it generates. To cover the entire structure in high resolution, many images are required, that need to be processed for the analysis. To reduce the number of these images to a manageable amount, careful selection of the camera positions, from which the images are taken, is required. The task of UAS flight path planning has been part of active research in recent years, creating flight routes that cover the entire structure with the required resolution. The number of necessary images can still be very high, if a high resolution of the inspection is required.

One approach to aid with this is to not capture the entire structure with the same resolution, but to consider the relevance and interest of different parts of the structure for the flight path planning. If previous inspections detected areas where damages exist or are developing, those areas have to be monitored more closely than other parts where no damage is suspected. On the other hand, some areas may not need high resolution at all, for example parts that are not part of the supporting structure, like railings or noise barriers.

1.1 State of the Art

The application of UAS in the monitoring and inspection of infrastructure buildings has increased significantly in recent years [2, 3], especially in the field of visual inspection and remote sensing. In [9], the authors combine the application of UAS with bridge information modelling (BrIM) tools, to collect and aggregate valuable lifecycle information. The combination of UAS and installed monitoring systems is evaluated in [10], which shows that both technologies can successfully combined.

In a related publication [8], the authors of this work have proposed a framework for an integral digital maintenance strategy based on visual UAS inspections that also includes numerical simulations and virtual reality (VR) technologies, to increase the lifespan of bridges.

In a study for the State Department of Transportation in the United States [11], the authors compare different existing solutions with regard to their applicability and find that the principal applicability of UAS is high, while practical limitations hinder a wide adoption.

One important requirement for the application of UAS for inspection tasks is a suitable flight path. The full potential can only be used, if the mission is planned in advance to achieve all the goals of the inspection. Several approaches have been proposed, for example in [12], a bridge is partitioned into planar surfaces, for each of which a local route is computed that are later combined.

In [13], the authors introduce a method for flight planning that focuses on the image based 3D reconstruction, while also considering environmental limits, like no-fly zones. They show that their proposed method is able to compute flight paths that cover the structure of interest well for object resolutions of few centimeters.

The authors of [14] describe the application of UAS for the inspection of pipelines and canals, where the drone follows the long structures and conducts more detailed inspections of areas with detected anomalies. While this method works on a much larger scale, the findings support the approach proposed here, that different levels of inspection resolution can provide value, while saving time and effort.

2 Multi-scale Building Inspection

One challenge of bridge inspection is the scale of the inspected structures combined with the necessity to inspect even small phenomena like cracks in concrete. One approach to handling this challenge is to not process the entire structure with the same detail, but to identify parts, that require higher resolution and inspect the rest with lower resolution. This reduces the overall effort for the inspection and allows focusing on parts that deserve attention.

This approach also values previous knowledge about a structure that would not be exploited, if the entire structure was inspected equally. If damaged areas have been detected in previous inspections, those require closer attention and need to be inspected with higher resolution. In addition, the expert knowledge and experience of human inspectors can be applied here. Inspectors often have a good knowledge, which parts of a structure are susceptible to damages and which parts generally do not suffer from damages visible in UAS images. Special parts of a structure, like bearings, hinges and joints suffer from strong wear over time and have to be inspected closely in most cases. On the other hand, non-bearing elements, like railings, do not require close inspection and can be covered with low-resolution inspection. This application of human expert knowledge can improve the automated inspection process.

In the case of an accident, for example the impact of a vehicle in a bridge pier, this specific area might need fast and precise inspection, if the stability of the structure is compromised. In this case, not only the directly impacted part is of interest, but also the entire structure, if deformations appear that are symptoms of deeper damage to the bridge.

2.1 Definition of Levels of Interest

We propose the definition of three interest levels, to not only mark regions of high interest that have to be inspected very closely, but also regions of explicitly low interest that can be covered with low resolution, as shown in Table 1. The object resolution is the size of one pixel in an image on the surface of the structure, equivalent to the smallest reliably measurable size in the resulting data. The object distance is computed from that target resolution under the assumption of the following properties of the used camera hardware:

- Model: Sony α 7R II
- Resolution: 7360 px × 4912 px
- Focal length: 50 mm

While the defined levels of interest are very useful for this applications, they can be varied in other cases, according to the specific requirements of the inspection task. The resulting parameters are only inputs into the computation and can be changed for other values if required.

Table 1. Defined interest levels for the inspection and example purposes

Interest level	Object resolution	Object distance	Purpose
1 (low)	2.0 mm/px	22.2 m	Rough geometry
2 (medium)	1.0 mm/px	11.1 m	Detailed geometry
3 (high)	0.1 mm/px	1.1 m	Crack detection

In most UAS-based inspection applications, the normal resolution level (Level 2) is realized, with 10 m being a distance to the object that common path planning procedures can generate [15]. An inspection of this level can be used to make measurements in the resulting 3D model, look for bigger damages that are well visible in the images and extract the 3D geometry of the structure for further analyses and simulations.

Another very common inspection level is the low interest inspection (Level 1), in many cases based on aerial images from above and some images from the sides, from a far distance away. These inspections allow for a visual overview for human inspectors that for example check if all required railings or traffic signs exist on a bridge. The low resolution only allows for rough measurements that are not sufficiently precise for simulation models.

The highest interest level (Level 3) is the most valuable for precise analysis of a buildings condition. It enables accurate image analysis for the detection and localization of damages, such as cracks or spallings, where even small shapes are well visible. The specific resolution of 0.1 mm was chosen based on the requirements of automated crack detection in the resulting images. For an object to be detectable in an image, it has to be contained in at least 2px according to the Nyquist-Shannon sampling theorem [16]. With one pixel corresponding to 0.1 mm, a crack of 0.2 mm can reliably be discovered in images, creating a sensible trade-off between detecting all cracks and limiting the number of required images.

Each level on its own can be used for different inspection tasks, but the combination of those levels allows for a complete inspection of the entire structure, while providing high resolution for parts, where it is required.

2.2 Derivation of Acquisition Parameters

While the target resolution was defined in Sect. 2.1 for the different levels, the required distance between camera and object was computed using the properties of the used camera and the following equation

$$\frac{d_{obj}}{dist} = \frac{d_{pix}}{f}$$

Where d_{obj} is the smallest measurable distance on the object, $dist$ is the distance between the camera and the object, d_{pix} is the size of one pixel on the sensor in the camera and f is the focal length of the lens. With the assumption that one can measure with the precision of 1px in the image, the required distance between camera and object can be computed as

$$dist = f\frac{d_{obj}}{d_{pix}}$$

depending on the required object resolution d_{obj} in mm/px. For the camera used in this work with $d_{pix} = 0.00451\,\mathrm{mm/px}$ and $f = 50\,\mathrm{mm}$, this results in

$$dist = d_{obj} * 11086px$$

for the computation of the values in Table 1.

One important result of the UAS-based structure inspection is the generation of 3D models. For this, the SfM pipeline, as shown in Fig. 1, is commonly used. This method exploits changing views of a scene while moving a camera, where the perspective distortion of objects in the images is different depending on their distance to the camera, to simultaneously compute the camera poses and the 3D geometry of the scene, from which the images where taken. It is based on finding corresponding points in multiple images, from which the relative orientation of the cameras to each other can be computed. By incremental combination of all image pairs, an image bundle can be constructed, which contains all camera positions and the identified object points in the same 3D space.

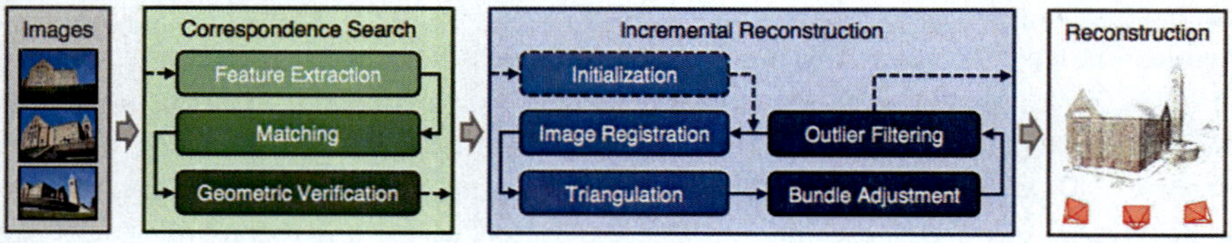

Fig. 1. The structure from motion pipeline [6]

For this process to work reliably and produce good results, the acquired images have to fulfill certain criteria. To find the same points in different images, those images have to overlap to a certain degree. In practice, an overlap of 70% in flight direction and 20% in the orthogonal direction have proven to be a good compromise between a stable reconstruction and a reduced number of images. Further, it is beneficial if points are visible in more than two images, to strengthen the image bundle and reduce errors.

3 Flight Path Planning

The goal of the flight path computation is to define an ordered set of camera positions and viewing directions, from which the structure that is to be inspected is completely covered with overlapping images in the defined resolution. Those viewpoints are then transferred as waypoints to the control of the UAS for the automated image acquisition.

The computation of the viewpoints is based on an existing 3D model of the structure. This model only needs to provide the approximate shape of the structure, precise to ca. 10 cm, to be able to compute the paths with full coverage and no collisions. Those models can come from previous UAS inspections with low resolution, created using the same SfM pipelines as used for the high-resolution inspection. Another option is to use digital models created during the planning of the structure, if such models are available. Using for example a CAD, FEM or BIM model, the geometry of the structure can be extracted and transformed into a world coordinate system. This is required, as the computed camera positions have to be approached by the UAS using a GNSS, which operates in this coordinate system.

The proposed approach to computing the viewpoints is based on selecting only the minimal set of required camera positions from a larger set of possible camera positions.

3.1 Computation of Camera Positions

The 3D model, on which the computation is based, is represented as a 3D polygon mesh of the structure. The mesh consists of a set of n triangles F called faces that form the surface of the structure. They are defined via a set of m vertices V, that define the locations of the triangles, such that $V_{i;1}$, $V_{i;2}$, $V_{i;3}$ are the three vertices of face F_i.

To obtain a large set of possible camera positions that guarantee complete coverage of the model, a corresponding viewpoint C_i is computed for each of the triangles F_i of the mesh. Each face belongs to one of the levels of interest defined in Sect. 2.1 and therefore has an associated distance f_i, from which it has to be captured. For each face F_i, its center c_i and outward pointing unit normal vector n_i can be computed via the average of the triangles vertices and the cross product of two its clockwise edge vectors as shown in the following equations.

$$c_i = \frac{1}{3}\sum_{j=1}^{3} V_{i,j}$$

$$\mathrm{n_i} = (V_{i,1} - V_{i,2}) \times (V_{i,3} - V_{i,2})$$

The position C_i^p and the viewing direction C_i^d corresponding to face F_i are computed as follows

$$C_i^p = c_i + n_i * f_i$$

$$C_i^d = -n_i * f_i$$

Each of the computed viewpoints is the defined distance away from the surface of the structure and its image contains a part of the structure. To compute the visible part of the surface, the vertices of the mesh are used as object points to represent the structure. The viewpoints are defined such that at least the three vertices of the corresponding face are visible, while generally more of the surrounding is visible in an image. To ensure the correct coverage of the surface, each vertex is only counted as visible in viewpoints corresponding to the highest interest level of its adjacent faces.

For the selection of the minimal subset of viewpoints that covers the entire structure, the complete visibility relations between the viewpoints and the object points has to be available. These relations are represented as a $m \times n$ visibility matrix M, with one column per viewpoint and one row per object point, where

$$M_{ij} = \begin{cases} 1 & \text{if point i is visible from viewpoint j} \\ 0 & \text{otherwise} \end{cases}$$

This matrix is computed column wise, for each viewpoint individually, in three steps:

1. Selection of only those object points, whose interest level is equal the interest level of the considered viewpoint.
2. Projection of all remaining object points into the image, to select only those points, whose projections lie in the image.
3. Computation of intersections of the corresponding viewing rays with the structure. If the first intersection is closer than the original point, something is occluding the point from the camera and the point is discarded. If the first intersection is the original point, the point is visible in the image and the corresponding entry in the M matrix is set to 1.

S_i is the set of object points, that are visible from viewpoint C_i and is equivalent to the 1 entries in $M_{:i}$, the i-th column of M.

3.2 Selection of the Required Camera Positions

With the computed visibility matrix M, the selection of the required camera positions can be described as a set multi-cover problem as stated in [17]: "Given a universe V of *m* vertices and a collection of sets $S = \{S_1, ..., S_n\}$, where each S_i is a subset of *V*, find a minimum cardinality sub-collection $R \subseteq S$, such that *R* covers *V*; in other words, the union of the sets in *R* is *V*. [...] In [the set multi-cover] version, each element $e \in V$ has an integer demand or requirement *d(e)* and a multi-cover is a sub-collection $R \subseteq S$ such that for each $e \in V$ there are *d(e)* **distinct** sets in R that contain *e*."[1]

The demand *d(e)* defines the number of images, each object point has to be visible in. This value can be derived from the requirements of the photogrammetrical 3D reconstruction using SfM. As described in Sect. 2.2, a point has to be visible in at least two images for the 3D reconstruction to work. The process can be improved and the result stabilized by finding correspondences in more than two images, which also allows for deviations in the image acquisition and other inaccuracies in the process. For these reasons, a constant demand of $d(e) = \forall e \in V$ is defined for all points.

The set multi-cover problem is **NP-HARD** [17], no polynomial time exact algorithms are available, so it is generally solved using approximation algorithms. This topic has been part of active research in recent years, and many approaches have been proposed [17–20]. In this work, a simple greedy algorithm, shown in Algorithm 1, is used that takes the visibility matrix *M* as input and produces a set of viewpoints that cover the entire structure.

The algorithm starts with fixing those viewpoints, that need to be included, because some of the object points they contain are only visible in *d(e)* images. Then the next viewpoint with the highest value is chosen iteratively, until complete coverage is

Algorithm 1. Greedy set multi-cover algorithm

```
Data: Visibility matrix M, vertex demand d(e)
Result: Smallest set of columns for complete coverage
m, n = number of rows and columns of M;
Initialize result vector R to a vector of n zeros;
foreach row i of M do
  /* count columns that cover row i */
  visibilityCount = Mi:;
  if visibilityCount < d(i) then
    Abort the algorithm, row i not sufficiently covered;
  end
  if visibilityCount == d(i) then
  /* all columns covering row i have to be used */
  set all columns that cover row i in R to 1;
  end
end
Compute the initial value of each column;
while not every row is covered d(e)-times do
  find column j where Rj== 0 with the highest value;
  set Rj to 1;
  compute the updated value for each column;
end
```

[1] Symbols changed to match those used in this work.

reached. The value of a column is the number of rows it covers, that have not been sufficiently covered by the already selected columns.

After computing the set of viewpoints, a flight path connecting them has to be computed. This is equivalent to the order, in which the UAS will approach the viewpoints during the automated image acquisition. The computation has to not only find an efficient connection between all positions, but also consider some practical aspects. The route may not come close to, or even touch, objects, either of the structure or the surrounding. These collisions have to be avoided under all circumstances. Further, the flight duration of the UAS is limited, forcing multiple sub-routes instead of one continuous route. With these considerations, the path-planning problem can be modeled as a distance constrained vehicle routing problem, a special case of the traveling salesperson problem. This sort of problems is well described in literature and many solution approaches exist. In this case, the route planning is solved using the Lin-Kernighan-Helsgaun solver [21].

4 Application of the Method

The proposed method aims at being a general approach, to computing flight paths for visual UAS inspections of structures. Especially since no implicit assumptions about the structure, the geometry or the mesh used for the computation are made, the method can be applied to a variety of structures. For practical validation of the approach, a reference object has been selected, for which flight routes have been computed that include defined regions of interest based on real observations. The results have been evaluated to show the applicability of the proposed method.

Fig. 2. The Scherkonde bridge, reference for the validation of the method

Fig. 3. Image acquired during a previous UAS inspection

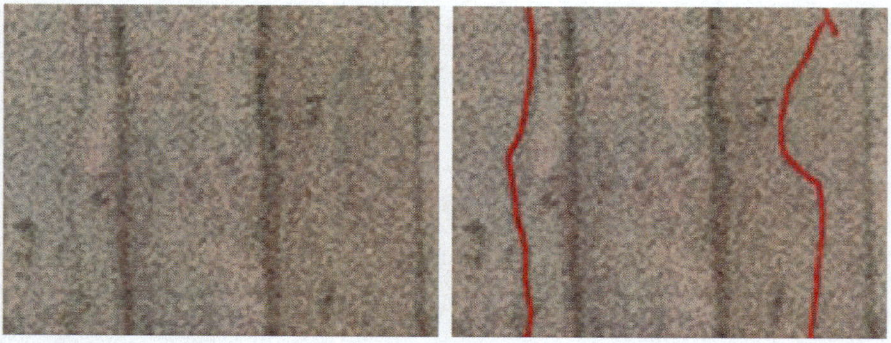

Fig. 4. Detail section and hypothesis of possible cracks on the surface

4.1 The Reference Object

For the validation of the proposed method, the Scherkonde bridge shown in Fig. 2 that is located on the railway high-speed track between Erfurt and Berlin, Germany, has been selected. The bridge was completed in 2011 after three years of construction and is built in semi-integral pre-stressed concrete with monolithic conjunctions between the superstructure and the supports. This leads to large horizontal forces in the piers, which

Fig. 5. Reconstructed 3D model of a part of the Scherkonde bridge, created from previous UAS inspections

require close monitoring. The bridge has a length of 576 m and 13 piers, the highest of which measures 34 m.

The bridge has been object to previous UAS inspection missions, so image data, simulations and 3D models are available. Figure 3 shows an image from one of those inspections in 2019, containing the center part of a pier. Figure 4 shows a section of that image in full resolution, approximately 1 mm/px on the surface. The vertical planking patterns have a width of around 10 cm or 100px. Using the method described in [22], a preliminary crack detection on the images using Convolutional Neural Networks (CNN) has been performed. The CNN produced a response for two regions of the image, where cracks could be located. Similar effects occur in other regions of the image. Due to the low resolution of the image, no precise detection is possible, so the result is vague and could well be a false positive detection.

In addition to the analysis of the images, a 3D model has been computed using the SfM pipeline. A section of that model is shown in Fig. 5. That part consists of 500000 triangles and 250500 vertices, contains the pier that is shown in Fig. 3 and is used in the following as the reference model.

4.2 Assignment of Interest

Even though the bridge under consideration is relatively new, it is subject to reoccurring inspections as a maintenance measure. In a previous UAS inspection, images of the bridge have been captured and analyzed. In images of one pier, the automated crack detection showed signs of possible damages to the structure. To generate a new data set, that provides higher resolution coverage of the possibly damaged area, an adaptive multi-scale flight path is to be computed.

The multiple scales for the flight path are defined equivalent to the levels defined in Sect. 2.1, as they each provide the required amount of information and resolution for

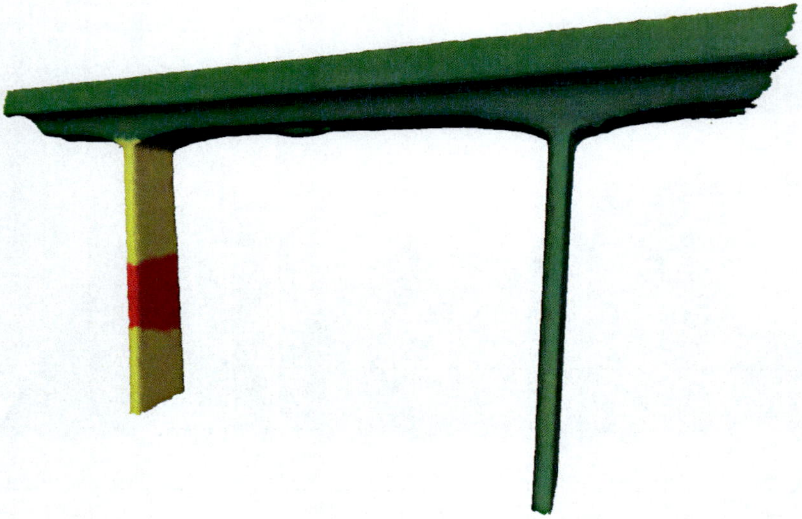

Fig. 6. Model of the Scherkonde bridge with faces colored according the level of interest

the different parts of the structure. The area of the pier, that is visible in Fig. 3, shows a hypothesis for possible cracks on the surface and has to be monitored closely, so it is assigned to the highest level of interest (Level 3). The rest of that pier could also be in a condition that requires more attention, so it is assigned the medium level of interest (Level 2), while the rest of the model is assigned to Level 1, as it is only required for completeness and context. This context is especially important for the georegistration of the 3D model computed from the images, as the points with precisely measured locations are distributed over the entire structure and are generally not all contained in images of the higher interest levels.

Figure 6 shows the model of the bridge, colored according to the defined interest levels. As the 3D model was computed using SfM, the orientation of the images is also known. With this knowledge, the exact visible part of the pier from the image of the damaged area can be computed and is highlighted in red.

4.3 Resulting Flight Path

After assigning the interest levels to the different parts of the structure, the flight path was computed according to the method described in Sect. 6. Table 2 shows the resulting number of viewpoints for each interest level, together with the corresponding area that needs to be covered.

The table shows that reducing the interest level for large parts of the structure, which are indeed of lower interest, greatly reduces the number of required images. High interest inspections produce around 120 times more viewpoints for the same area, than low interest inspections (0.04vp/m^2 vs. 4.84vp/m^2). This corresponds well with the fact that the area visible in an image grows with the square of the distance to the surface. Had the entire structure been captured in high resolution, more than 15000

Table 2. Interest levels and corresponding viewpoints

Level	Area	Viewpoints	Density
1	2796 m^2	122	0.04vp/m^2
2	372 m^2	62	0.17vp/m^2
3	58 m^2	281	4.84vp/m^2
Total	3225 m^2	465	0.14vp/m^2

images would have been required, compared to the 465 that are the result of the adaptive multi-scale flight path.

4.4 Validation of the Damage Hypothesis

After computing a flight path for the detailed and targeted inspection of the possibly damaged area, an automated UAS inspection was performed on the bridge. Using the computed waypoints and the GNSS navigation system, the second UAS inspection was conducted shortly after the initial mission.

The evaluation of the newly created images allowed a closer inspection of the surface. Figure 7 shows one of the resulting images with a full resolution detail of the area, where a possible damage had been suspected previously, shown in Fig. 8. Running the automated damage detection on those images showed that one of the detections was a false positive that was caused by a mud stain on the surface. The other

Fig. 7. Image from the resulting high-resolution flight path acquired from ca. 1 m

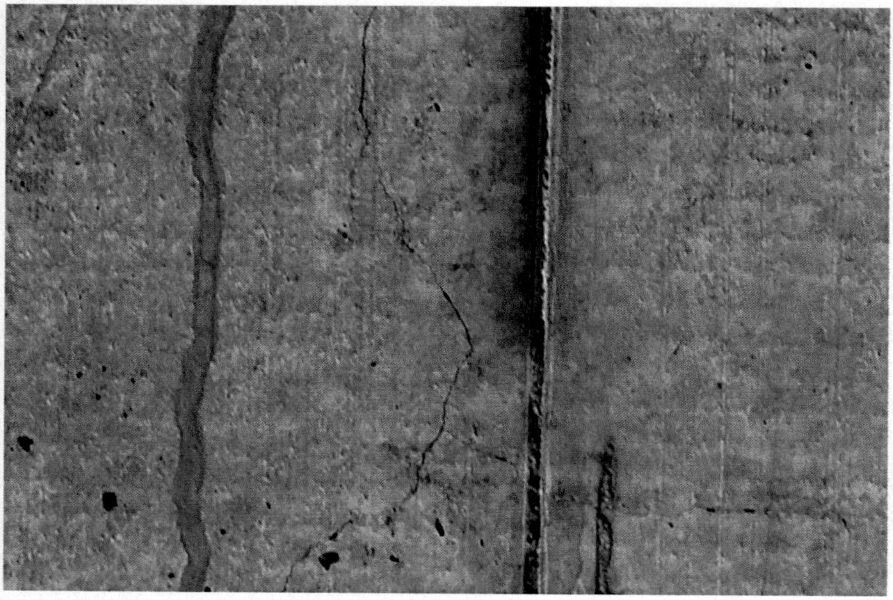

Fig. 8. Detail of the high-resolution image in the original size containing a small crack

hypothesis, which is visible in Fig. 8, was correct and corresponded to a crack in the concrete. With the high resolution however, it was determined by an engineer that the crack is not relevant for the safety and structural health of the bridge, the width corresponding to roughly 0.15 mm. It is only a small damage to the surface that does not have any further implications for the structure. This shows the possibilities this new technology provides, enabling the human experts to quickly obtain high-resolution information about regions of interest. In this case, the suspected damages were either false alarms or not relevant, but in other cases a disaster could be prevented by having an early warning system for damages, before they happen.

5 Conclusion

In this paper, a method for the computation of flight paths for visual UAS building inspection is proposed that takes into account different levels of interest defined for parts of a structure. In Sect. 2.1, the levels of interest are defined as high interest for areas that need very close attention, medium interest for areas that have to be well visible in the resulting images, and low interest for areas that are not important for the inspection itself, but for context information and completeness, especially for the georeferencing of the reconstructed 3D model. The assignment of a certain interest to parts of the structure can be based on many factors, like expert knowledge of the structure or an impact event. One important reason to mark an area for high interest can be the suspicion of cracks or other damages that can be detected in images.

For each of the defined levels of interest, viewpoints have to be computed that cover the corresponding area well, while reducing the number of images to a minimum. For this, a method is proposed that is based on selecting a minimum covering set from a larger set of possible camera positions. This large set is computed by assigning one viewpoint to each triangle of the model of the structure. Using a simple greedy solver for the set multi-cover problem, good solutions can be computed quickly.

A case study has been conducted, in which information from a previous inspection was used to compute a flight path that provides special attention to areas that could be damaged. Automated image analysis of the low detail images indicated the possibility of cracks in the concrete surface. After conducting the high resolution and detailed inspection of the identified area, it showed that the suspected damages were either false alarms or small surface cracks without any implications for the structural health and safety. The method itself proved to work well for the generation of flight paths that cover specific areas of a structure in high resolution for details, while also covering the other parts of the structure in lower resolution for context.

This shows that visual UAS monitoring can help protect important infrastructure by quickly inspecting large structures, allowing for automated analysis of the results and the dispatch of more detailed inspections, should possible damages be detected. As part of a digital maintenance strategy, it helps human experts to focus on the important aspects of the structures, while making sure that nothing is missed.

6 Further Research Possibilities

While the presented approach already produces good results for the studied test cases, the wide variety of building shapes can introduce further challenges that need to be overcome. The current implementation does not try to incorporate the surrounding of the structure while placing the possible viewpoints, but only during the computation of the visibility relations. Changing this would allow placing closer viewpoints in regions where space is limited and the computed larger distance between camera and object is not possible.

Currently, a very simple greedy solver for the set multi-cover problem is used. A more advanced solver could provide solutions that are closer to the mathematical optimum. The solution quality can also be improved by incorporating special knowledge concerning UAS flights and photogrammetric reconstructions as well as automated image analysis. While this may not influence the mathematical quality of the solutions, the real world applicability can greatly benefit from adapting the solution algorithm to additional constraints.

At the moment, the selection of a viewpoint does not affect the value of surrounding viewpoints, while in an improved version, viewpoints very close to already selected points should be discouraged, to produce a sufficiently large baseline (distance) between adjacent viewpoints. Further attention should be paid to special geometric features of the structure, for example corners or very fine details. For a stable SfM reconstruction, adjacent images may not be too different from each other, especially with regard to rotation, as otherwise the matching of corresponding feature points becomes very difficult.

While the 3D reconstruction could be computed successfully in the studied cases, the reconstruction at the borders between the interest levels showed some artifacts, especially between levels 2 and 3, where the distances increases by factor 10. In these areas, additional viewpoints with intermediate distances could help improving the quality of the reconstruction, while increasing the number of images only slightly.

Acknowledgement and Funding. This work was supported by the German Federal Ministry of Education and Research as part of the AISTEC project under grant numbers 13N14657 to 13N14660. We would also like to thank our project partners, especially the Deutsche Bahn, for making this work possible.

References

1. Shakhatreh, H., Sawalmeh, A.H., Al-Fuqaha, A., Dou, Z., Almaita, E., Khalil, I., Othman, N. S., Khreishah, A., Guizani, M.: Unmanned aerial vehicles (UAVs): a survey on civil applications and key research challenges. IEEE Access **7**, 48572–48634 (2019)
2. Xiang, T.-Z., Xia, G.-S., Zhang, L.: Mini-unmanned aerial vehicle-based remote sensing: techniques, applications, and prospects. IEEE Geosci. Remote Sens. Mag. **7**, 29–63 (2019)
3. Nex, F., Remondino, F.: Preface: latest developments, methodologies, and applications based on UAV platforms. Drones **3**, 26 (2019)
4. Bundesanstalt für Straßenwesen, "Zustandsnoten der Brücken" (2019). https://www.bast.de/BASt_2017/DE/Statistik/Bruecken/Zustandsnoten-excel.html
5. Reagan, D., Sabato, A., Niezrecki, C.: Feasibility of using digital image correlation for unmanned aerial vehicle structural health monitoring of bridges. Struct. Health Monit. **17**, 1056–1072 (2017)
6. Schonberger, J.L., Frahm, J.-M.: Structure-from-motion revisited. In: Proceedings of the IEEE Conference on Computer Vision and Pattern Recognition (2016)
7. Mohan, A., Poobal, S.: Crack detection using image processing: a critical review and analysis. Alexandria Eng. J. **57**, 787–798 (2018)
8. Morgenthal, G., Hallermann, N., Kersten, J., Taraben, J., Debus, P., Helmrich, M., Rodehorst, V.: Framework for automated UAS-based structural condition assessment of bridges. Autom. Constr. **97**, 77–95 (2019)
9. Barrile, V., Candela, G., Fotia, A., Bernardo, E.: UAV survey of bridges and viaduct: workflow and application. In: Computational Science and Its Applications – ICCSA 2019, vol. 11622, pp. 269–284. Springer, Cham (2019)
10. Agdas, D., Rice, J.A., Martinez, J.R., Lasa, I.R.: Comparison of visual inspection and structural-health monitoring as bridge condition assessment methods. J. Perform. Constr. Facilit. **30**(3) (2016)
11. Dorafshan, S., Maguire, M., Hoffer, N.V., Coopmans, C.: Challenges in bridge inspection using small unmanned aerial systems: results and lessons learned, pp. 1722–1730. IEEE
12. Yu, K., Shanthakumar, P., Orevillo, J., Bianchi, E., Hebdon, M., Tokekar, P.: View Planning and Navigation Algorithms for Autonomous Bridge Inspection with UAVs, p. 9
13. Koch, T., Körner, M., Fraundorfer, F.: Automatic and semantically-aware 3D UAV flight planning for image-based 3D reconstruction. Remote Sens. **11**(13), 1550 (2019)
14. Martin, R., Blackburn, L., Pulsipher, J., Franke, K., Hedengren, J.: Potential benefits of combining anomaly detection with view planning for UAV infrastructure modeling. Remote Sens. **9**(5), 434 (2017)

15. Peng, C., Isler, V.: Adaptive view planning for aerial 3D reconstruction. In: 2019 International Conference on Robotics and Automation (ICRA) (2019)
16. Shannon, C.E.: Communication in the presence of noise. In: Proceedings of the IRE, vol. 37, pp. 10–21, January 1949
17. Chekuri, C., Quanrud, K.: On approximating (sparse) covering integer programs. In: Proceedings of the Thirtieth Annual ACM-SIAM Symposium on Discrete Algorithms (2019)
18. Chekuri, C., Clarkson, K.L., Har-Peled, S.: On the set multi-cover problem in geometric settings. In: Proceedings of the 25th Annual Symposium on Computational Geometry - SCG 2009 (2009)
19. Chen, A., Harris, D.G., Srinivasan, A.: Partial resampling to approximate covering integer programs. In: Proceedings of the Twenty-Seventh Annual ACM-SIAM Symposium on Discrete Algorithms (2016)
20. Dinur, I., Guruswami, V., Khot, S., Regev, O.: A new multilayered PCP and the hardness of hypergraph vertex cover. SIAM J. Comput. **34**, 1129–1146 (2005)
21. Helsgaun, K.: An extension of the Lin-Kernighan-Helsgaun TSP solver for constrained traveling salesman and vehicle routing problems. Roskilde: Roskilde University (2017)
22. Benz, C., Debus, P., Ha, H.K., Rodehorst, V.: Crack segmentation on UAS-based imagery using transfer learning. In: 2019 International Conference on Image and Vision Computing New Zealand (IVCNZ) (2019)

Integrating UAV Photogrammetry and Terrestrial Laser Scanning for Three-Dimensional Geometrical Modeling of Post-earthquake County of Beichuan

Xiaoxi Chen, Dongfeng Jia, and Weiping Zhang(✉)

Key Laboratory of Performance Evolution and Control for Engineering Structures of Ministry of Education, Tongji University, Shanghai 200092, China
weiping_zh@tongji.edu.cn

Abstract. UAV photogrammetry and terrestrial laser scanning are non-contact measurement technologies, and have become increasingly popular for three-dimensional geometrical modeling of a single building or a city in recent years. By integrating UAV photogrammetry and terrestrial laser scanning, we are able to obtain the buildings' present status and their reconstruction models, which plays a great role in protection of post-earthquake sites. This paper presents the 3D modeling and application process of one of the most important post-earthquake sites in the county of Beichuan in China. First, texture characteristics were extracted from buildings through site investigation and the database of buildings' earthquake damage characteristics was established. Then, in order to realize the complementary advantages of UAV photogrammetry and terrestrial laser scanning technologies, effectively solve the deficiencies of the single technology, this paper combined tilt photogrammetry and terrestrial laser scanning technology for 3D modeling of post-earthquake county of Beichuan. The results indicate that the 3D model established by integrating UAV photogrammetry and terrestrial laser scanning is tested to be of high quality and high precision. Meanwhile, the integration of two sources of point cloud data effectively reduced the workload of field control survey work. Moreover, the resulting 3D models can be used for further analysis such as earthquake disaster assessment, feature fracture extractions and 3D measurement.

Keywords: UAV photogrammetry · Terrestrial laser scanning · 3D geometrical modeling

1 Introduction

Earthquakes are considered as major disasters because they can destroy buildings and greatly affect human lives and environments. After the earthquake, it is important to assess the scale of the damage in order to rescue and recovery effectively. Therefore, researchers have been carrying out a lot of research with the respect of earthquake damage assessment. Conventional method employed to earthquake damage information acquisition is field survey [1]. However, when it is carried out in large-scale area,

E. Toledo Santos and S. Scheer (Eds.): ICCCBE 2020, LNCE 98, pp. 1086–1098, 2021.
https://doi.org/10.1007/978-3-030-51295-8_75

substantial amounts of manpower and time is required. Since satellite imagery can assess earthquake damage in a wide area using a relatively short time, it has become very important data for post-earthquake damage assessment in recent years. Saito et al. performed visual damage inspection to buildings after 2001 Gujarat, India earthquake by comparing the post-event IKONOS image and pre-event other satellite images [2]. QuickBird images with the maximum spatial resolution of 0.6 m were also employed by Yamazaki et al. to detect collapsed buildings after the 2003 Bam, Iran, earthquake [3]. The previous work on post-earthquake damage assessment has indicated that satellite images can be helpful to detect collapsed buildings due to an earthquake. However, satellite images only provide the central projection of the buildings from the top view, thus it is difficult to detect the damage of building facades from the satellite images [4]. Hence, the accuracy of building damage information acquisition based on satellite imagery is relatively low. Compared with manually on-site assessment or satellite imagery, terrestrial laser scanner (TLS) and unmanned aerial vehicle (UAV) photogrammetry can provide faster, cheaper, and more flexible post-earthquake information acquisition and damage assessment. In recent years, significant interest has emerged in using TLS and UAV for regional seismic damage assessment of buildings [5, 6].

TLS is an approach which emits a high-speed laser that performs noncontact scanning measurements, and can provide high-quality 3D data for structural damage detection, such as deformation or cracks. The feasibility of TLS was demonstrated by Anil by presenting that the measurable width, depth, and orientation of cracks according to the sampling interval of the laser scanner [7]. Giri and Kharkovsky also proved that a laser displacement sensor can measure a cracks' width up to 0.7 mm on a concrete specimen [8]. Laefer et al. configured the fundamental mathematics to determine the detectable minimum crack width with a terrestrial laser scanner in unit-based masonry by a parametric study using orthogonal offset, interval scan angle, crack orientation, and crack depth [9]. Besides, TLS is able to obtain the position of a huge amount object points in high precision and relatively short time. The dense 3D points cloud with geometry information can be reconstructed into 3D geometrical models, and have been widely used for cultural or historical heritage 3D documentation and visualization. For example, Lyn Wilson created a 3D digital documentation of Mackintosh building in the aftermath of fire by means of TLS. The 3D point clouds and derived models have also proven beneficial in visualizing proposals for restoration [10]. Another example of 3D documentation project is the work of Yahya Alshawabkeh et al. In their research, an approach combining in situ surveying and laboratory analysis was applied to provide sufficient and comprehensive data regarding the documentation and evaluation of the status of the Al-Deir monument in Petra. And an accurate 3D realistic model was established by laser scanning and digital photogrammetry [11]. However, TLS are relatively expensive and need a flat base as well as clear field of view, which may not be available in most of post-disaster situation. Some researchers have employed TLS sin field survey work after earthquakes and have acquired point cloud data for damaged buildings and secondary disaster areas. For example, Qiang Li used the Riegl VZ1000 TLS to collect data of Beichuan post-earthquake site. After data processing, a three-dimensional model of the seismic site was established for subsequent seismic damage information acquisition and damage assessment [12]. Enrico Quagliarini proposed a fast, low cost and safe methodology by using TLS. And the methodology was applied to a case study, the Santa Maria in Portonovo church after an earthquake. Quite evident mechanisms were found in vestibule, building façade and original apse, demonstrating that it is possible to accurately evaluate the building safety after an earthquake by TLS [13]. However, studies on the quantitative extraction and analysis of building damage information from point cloud data are insufficient.

UAV photogrammetry is considered as a cost-effective and efficient way for reconstructing high-quality 3D digital models in terms of hardware requirements, knowledge background, and man-hours. Lately, UAV has been exploited for capturing images of a building of interest from multiple different perspectives, which results in a 3D model of the building [14]. Hoskere et al. established a 3D mesh model of an individual building through multi-view images obtained by a multi-rotor UAV [15]. In hard-to-reach or dangerous areas, UAV-based photogrammetry is less restricted and more safety. Moreover, compared with conventional remote-sensing platforms, UAV photogrammetry is capable of cruising at a relatively low altitude and providing more detailed sensing data of building facades, which means higher accuracy [16, 17]. Thus, this technique is increasingly being used in regional seismic damage assessment. Galarreta proposed a method for damage assessment of individual buildings based on the multi-view images and 3D point clouds acquired by a multi-rotor UAV [5].

However, there is no single sensor can acquire the complete information. Valero [18] have investigated different sensors to assess which provide a more objective approach for analysis of historic masonry, noting that a combination of TLS and digital photogrammetry may prove to be a good idea. However, few of the previous studies have taken full account of limitations of a single data source. For the purpose of sustainable restoration of the late Gothic St. Petri cathedral of Bautzen, Henze et al. combined photogrammetry methods and TLS methods, to create a form-correct and deformation-true documentation of the facades [19].

In order to analyze the earthquake damage of the structure and combine the advantages of the above two methods, effectively solve the deficiencies of the single method, this paper presents an approach to integrate UAV photogrammetry and TLS point cloud for 3D modeling of damaged buildings. Post-earthquake county of Beichuan is of high importance because it has witnessed the Wenchuan Ms8.0 earthquake and should be properly maintained. Results show that the 3D model established by integrating two datasets is tested to be of high quality and high precision and can effectively reduce the workload of field control survey work. Moreover, the resulting 3D models can be used for further analysis such as earthquake disaster assessment, feature fracture extractions and 3D measurement.

2 Reverse Geometrical Modeling of a Post-earthquake County

2.1 Brief Introduction to Post-earthquake County of Beichuan

On May 12, 2008, the Wenchuan Ms8.0 earthquake occurred in Sichuan Province, which was one of the most destructive earthquakes since the founding of the People's Republic of China. The epicenter was located at 31.01°N, 103.42°E. Most of the buildings of Beichuan County (Fig. 1a), which were located in close to the epicenter, were seriously damaged. And other secondary disasters caused by the Wenchuan earthquake severely exacerbated the losses attributable to the disaster, which destroyed a large number of basic and public facilities and caused enormous social and economic losses [20]. Post-earthquake county of Beichuan have retained both diversity type of seismic damage of buildings (Fig. 1b) and secondary disaster scenes such as mudslides and landslides. After necessary preventive strengthening measures were carried out, it is reserved as an earthquake site for the purpose of mourning the dead and natural disaster study. Therefore, post-earthquake county of Beichuan were selected as the study area, and a Faro laser scanning system as well as a DJI Phantom 4 UAV were used to acquire and process the building 3D data in order to monitor and protect the important post-earthquake site.

(a) (b)

Fig. 1. (a) Beichuan post-earthquake site. (b) Buildings with different degrees of earthquake-induced damage.

2.2 Data Acquisition and Post-processing of UAV Photogrammetry

As an information acquisition approach, UAV photogrammetry is becoming attractive for many surveying applications in civil engineering. To achieve a large-scale and reliable 3D geometrical modeling, a post-earthquake site inspection based on UAV photogrammetry framework is needed. Three primary stages are proposed and described in Fig. 2, i.e. data acquisition, data post-processing and 3D reconstruction.

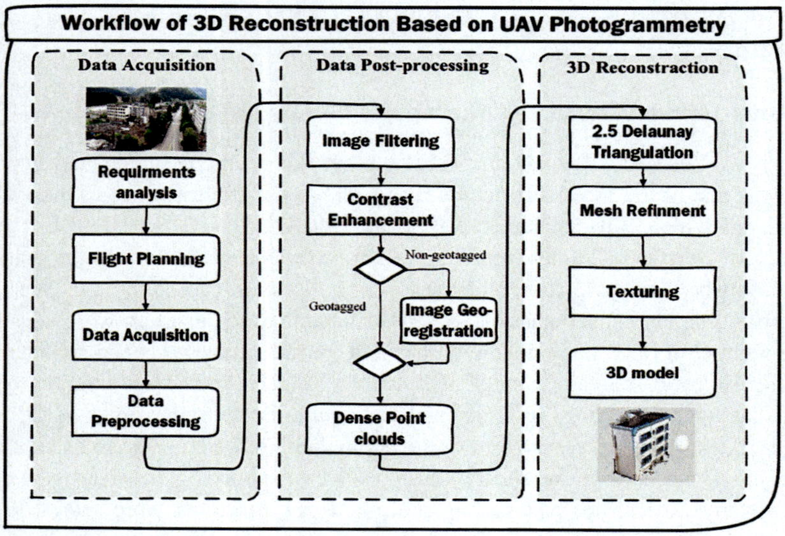

Fig. 2. Workflow of 3D reconstruction based on UAV photogrammetry.

Flight Planning and Data Acquisition

Flight planning is unarguably the most important step of data acquisition since it has the strongest impact on data quality. Before flight planning, the requirements of the mission need to be analyzed in order to cover the areas of interest with the appropriate level of precision. Light conditions, camera angle, offset distances, flight pattern, and degree of overlap between images should be carefully considered. Then, a flight plan is created in the form of waypoint files. In data acquisition step, it is necessary to take enough number of images to ensure the required overlap between different images, allowing high levels of detail in both meshes and textures. Once all the data has been collected, it is organized and prepared for post-processing.

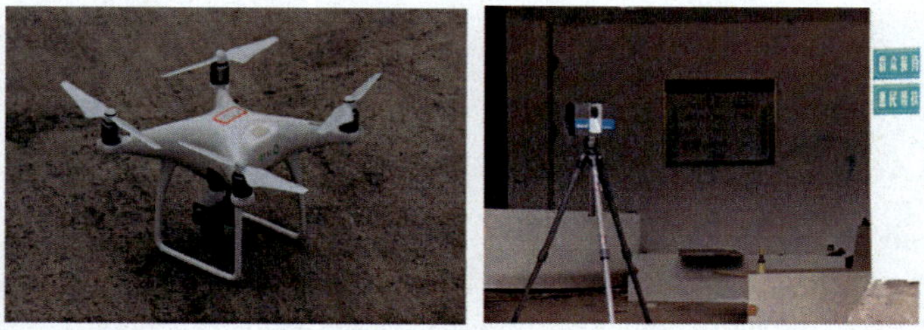

(a) (b)

Fig. 3. (a) DJI Phantom 4 UAV. (b) Faro Focus S350 laser scanning system.

In this research, an area of approximately 218,000 m^2 was covered with a cross-section of 250 m width and 872 m long. The data acquisition was based on 3D methodologies by using DJI Phantom 4 UAV (Fig. 3a), whose sensor size is 6.17 × 4.55 mm, focal length is fixed at 3.55 mm, and pixel numbers are 4,000 × 3,000. The flight plan was designed to cover the whole area of the settlement. The UAV flight took place at 45 m above the ground in the area of interest, with 75% forward overlap and 45% sidelap. Flight path specially designed for 3D modeling was executed to collect imagery. 1368 images of object area were taken and used as an input for further processing.

Data Post-processing

The UAV images collected from the flight planning and data acquisition stage need to be processed in order to generate a georeferenced 3D point cloud from the unordered, overlapping images of the object area. In this step, Structure from Motion (SfM) algorithms, as first introduced by Ullman (1979) is applied [21]. This method can recover scene geometry by processing a number of unordered images from perspective viewpoints with high overlaps. By detecting key points in each image, the geometric relationship between images can be calculated and used for triangulation, from which the depth information of key points is derived and placed into a unique coordinate system.

In this research, data was processed by a commercially-existing software solution by AgiSoft PhotoScan [22], called PhotoScan, which has been recently optimized for the use of UAV photogrammetry. After photographs were imported, a first orientation of the set of images was made. Next, for the purpose of geo-referencing, an overall number of 24 Ground Control Points (GCPs) measured using a TLS, were introduced to geo-reference and provide geometric accuracy to the scene, this allowed accelerated surveying operations and improved accuracy of the final geo-referencing. Then, the images were readjusted, and finally the definitive point cloud and triangular mesh were processed.

2.3 Data Acquisition and Post-processing of TLS

In this research, TLS was used to make different scanning from different perspective of the post-earthquake site and these scanning have been aligned using the software Autodesk Recap. The result was a 3D model faithful to reality, from which it has been obtained the deformation of the damaged building.

Preparation and Scanning

In the step of preparation, the placements of the scanning station sites should be carefully planned considering the buildings' locations, surroundings, and shapes, thus all the target data can be obtained. In addition, to meet the requirements of data registration, each station must have a certain amount of overlap. In this research, data collection was performed in the post-earthquake area using a Faro Focus S350 phase-shift scanner (Fig. 3b) [23]. In order to gain as much damage information as possible, 126 scans have been performed for the whole post-earthquake area, including scans for both exterior and interior part of buildings. Consequently, we acquired high-density point cloud data of the ground, collapsed buildings, damaged buildings, and other ancillary facilities in Beichuan County.

Registration

In order to express the entire object area with a point cloud obtained based on each local coordinate system, a coordinate conversion and combination process called registration is required. In this research, the processing and registration of each scan were performed with the Autodesk Recap software, which processes the data obtained directly from the TLS. After registration, cleaning and filtering, the final point clouds were formed by about 29 billion of points (Fig. 4a). Meanwhile, 24 TLS points with a high precision were used as ground control points of the UAV dataset.

(a) (b)

Fig. 4. (a) Registration of each scans. (b) 3D point cloud collected by TLS.

2.4 Integration of Two Datasets

As is known, both UAV photogrammetry and TLS methods have their limitations. How to integrate the two methods to make up for their own shortcomings has become a key issue. For example, when using UAV photogrammetry to acquire an image, a control network needs to be deployed, which will undoubtedly increase the labor intensity. The TLS can improve the speed of data collection, and reduce the labor intensity of measurement. Moreover, as the scanner's location is fixed during scanning, the line of sight nature of the technology may potentially result in occlusions. However, UAV photogrammetry can flexibly collect data points, which is beneficial to collect inaccessible information of roof structures and other missing details which TLS are unable to collect. Meanwhile, TLS can capture richer details of building facades than UAV. Hence, the integration of two datasets is helpful to enhance the collection of comprehensive information as well as improve measurement accuracy [24].

In this research, imagery and 3D points were collected simultaneously from a DJI Phantom 4 UAV and Faro Focus X330 phase-shift scanner, respectively. The 3D UAV point clouds were generated from downwards-looking camera using a SfM approach, thus the UAV 3D data were in a local 3D photographic coordinate system. Since there was no prior information available, both data sets were in different coordinate systems. In order to realize the complementary advantages of UAV photogrammetry and TLS techniques, and effectively overcome the defect of the single way, in this paper a coarse-to-fine integration method is proposed for these two techniques (Fig. 5).

The method begins by manually extracting feature points, for example, buildings corners and road boundary, from TLS point clouds and back-projecting these points onto the point clouds generated from UAV imagery around the whole area. To improve this initial alignment, dense 3D point clouds are generated from the UAV images. Then, the dense point cloud of the whole area is segmented into certain buildings with the requirement of high precision, fast speed and easily programming. After that, the refined transformation parameters of each building are acquired by applying the Iterative Closest Point (ICP) algorithm on the 3D UAV points and the TLS point clouds. Finally, 3D reconstruction of objective building was accomplished, and the resulting 3D model can be used for further analysis such as earthquake damage assessment, crack detection and 3D measurement.

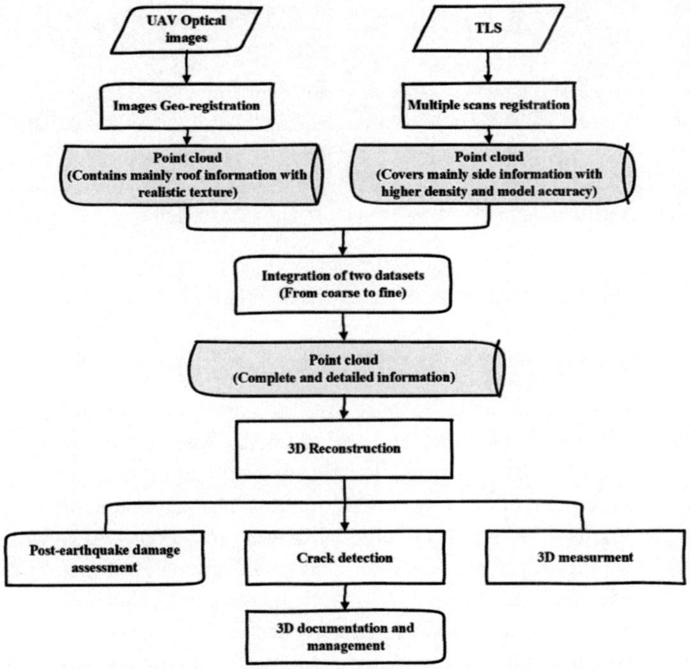

Fig. 5. Workflow of 3D documentation and management of post-earthquake site with multi-source point clouds.

2.5 3D Geometrical Modeling

The "Three Centers" site was one of the most characteristic sites of Beichuan Qiang Autonomous County. It was composed of: the Yu and Qiang Cultural Center (the place to display books and relics about Yu and Qiang culture), the People's Center (the place to help the poor and vulnerable), and the Social Affairs Center (the platform to improve administrative effectiveness and provide services). Unfortunately, the "Three Centers" site was destroyed by Wenchuan Ms8.0 earthquake, and all the facilities, cultural relics and historical records have become ruins. In this paper, the 3D reconstruction of the "Three Centers" site was conducted based on the integration of UAV photogrammetry and TLS point cloud (Fig. 6a). The results indicate that the 3D model established by integrating UAV photogrammetry and terrestrial laser scanning is tested to be of high quality and high precision.

(a) (b)

Fig. 6. (a) 3D reconstruction of the "Three Centers" site. (b) VR tour of the post-earthquake site.

3 Applications of Post-earthquake 3D Model

3.1 Post-earthquake Damage Assessment

Earthquakes often induce collapse or cause extreme damage to large areas of buildings, which is the main cause of casualties and property losses. The purpose of post-earthquake damage assessment is to analyze the losses of earthquake quickly in order to provide emergency rescue effectively. Therefore, it has to guarantee time-savings, high precision and security during the operational stages. In recent years, with the rapid development of measuring instruments and data processing algorithms, different types of remote sensing data such as UAV optical images, synthetic aperture radar (SAR) imagery, and airborne LiDAR data have been used in damage assessments [25]. However, due to the limitation of the spatial resolution of the image by the aerospace or aerial remote sensing, the damage level of the buildings visually judged is limited, and generally only three categories such as uncollapsed, partially collapsed and collapsed can be identified. To date, because of the low cost, high speed and security of UAV systems for collecting images, UAV photogrammetry is already a reliable replacement over satellite and manned vehicle. Moreover, it has overcome the disadvantage of low flexibility and expensive of aerial imagery. Therefore, in terms of seismic damage assessment, this paper uses high resolution UAV images combined with TLS point cloud to judge the damage level of buildings. In further study, we attempt to establish evaluation criteria of different levels of building damage and classify the building seismic damage into 5 grades from intact to destroyed according to cracks and inclines of façades.

3.2 Crack Detection and 3D Measurement

After an earthquake, it is important to identify and record seismic damage directly on the structures in earthquake-stricken areas for earthquake emergency operations. Compared to costly and time-consuming manual detection, TLS technology can directly obtain the coordinates of mass points while maintaining a high measurement accuracy, thereby providing the means to directly extract quantitative information from surface cracks on damaged buildings. The building structures in Beichuan County primarily include brick masonry, bottom frame masonry, and frame-filled wall structures. During the earthquake, the walls of buildings throughout Beichuan County were commonly subjected to shear damage. This occurred because the masonry material is relatively brittle and thus boasts low tensile and shear strengths, leading to the generation of oblique cracks. Since cracks are critical to judge the state of buildings, in this research, we present a framework for extracting wall cracks from high-density TLS point clouds. The interior wall of a damaged but not collapsed building in the "Three Centers" site was selected as primary example to extract wall cracks (Fig. 7a). Meanwhile, a bottom frame masonry building was 3D measured by calculating the normal vectors for each plane as well as measuring edge line inclination (Fig. 7b). The real length and width of the cracks as well as deformation of the selected building were measured, clearly reproduced the seismic damage observed in the field and provided the basic data for safety evaluation of post-earthquake structure.

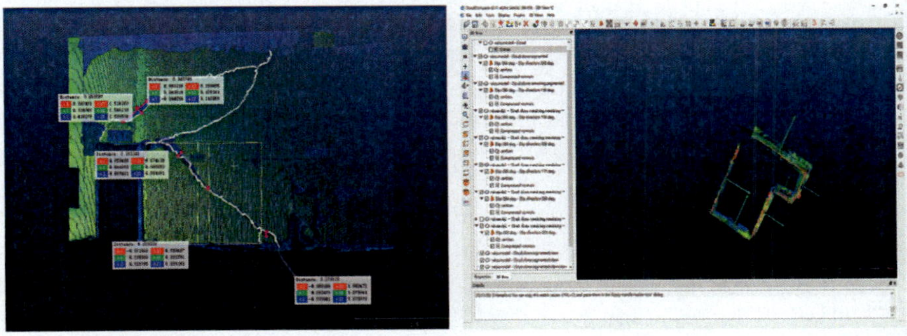

(a) (b)

Fig. 7. (a) Detection and measurement of cracks. (b) 3D Measurement of exterior walls.

3.3 Digital Conservation and Management

Beichuan earthquake site is of great social and research value. However, structures of these buildings are constantly exposed to natural threats such as flood and rain. It is necessary to act to ensure its preservation over time. 3D data, when captured at sufficiently high resolution and processed effectively, can be a rich digital resource with immediate applications in disaster management situations. In recent years, the use of 3D digital modeling by means of UAV photogrammetry and TLS has increased because of their efficiency in regard to time, cost, accuracy and usefulness, and a large number of research papers have been published on the use of 3D modeling technology to archive historical buildings [26].

In this paper, by using UAV photogrammetry and TLS, we are able obtain the building's present status and its reconstruction modeling, which plays a great role in protection of damaged earthquake sites. Therefore, this paper focused on developing a possible detection and monitoring approach, which used non-contact techniques for documentation of post-earthquake buildings. The primary data, after a series of data processing, expecting to quickly and efficiently get high precision 3D geometrical model, so as to carry out a series of further analysis on the damaged buildings.

After data processing, we are able to establish a building information model (BIM) of the Beichuan earthquake site, documenting locations, surroundings, building deformation, building material and state of conservation to ensure the stability and safety of buildings. Moreover, by taking the advantages of geometry and texture information, we can produce a photorealistic 3D visualization model (Fig. 6b) of the earthquake site that helps people understand the process of the "5.12" Wenchuan earthquake more easily.

4 Conclusions

In this paper, a post-earthquake site at Beichuan Qiang Autonomous County after Wenchuan Ms8.0 earthquake was selected as an area of interest. And a method for the integration of rotor UAV and TLS platforms for 3D geometrical modeling was proposed in order to provide a rapid and accurate means of assessment within a short period of time following the earthquake.

In this research, 126 individual laser scans of the whole post-earthquake area were performed and registered together. Meanwhile, 1368 images collected by UAV at a low altitude of the study area were processed. Then, the suggested method was used to integrate two datasets, generating a complete and precise 3D geometrical post-earthquake building model. Above all, the study demonstrated that UAV-TLS cooperative method can offer significant advantages in equipment cost, surveying time, point distribution, and ultimate coverage, which allows the problem of surveying buildings in post-earthquake sites or dangerous areas to be solved effectively. And the combination of the two sources of point clouds enhances the possibility to extract cracks and deformation which are helpful for damage assessment for post-earthquake buildings. Further work will be conducted including analyzing the distribution of geometric features, 3D documentation as well as deformation monitoring.

Although the methodology discussed above demonstrates the high potential of 3D models in the display, detection and monitoring of post-earthquake structures. Problems remain including long post-processing time and high noise levels. To solve these problems, future research should focus on how to optimize 3D reconstruction algorithms and develop better noise removal methods.

References

1. Miyamoto, H.K., Gilani, A.S., Wada, A.: Reconnaissance report of the 2008 Sichuan earthquake, damage survey of buildings and retrofit options. In: Proceedings of 14th World Conference on Earthquake Engineering, Beijing, China, pp. 1–10 (2008)
2. Saito, K., Spence, R.J., Going, C., Markus, M.: Using high-resolution satellite images for post-earthquake building damage assessment: a study following the 26 January 2001 Gujarat earthquake. Earthq. Spectra **20**(1), 145–170 (2004)
3. Yamazaki, F., Yano, Y., Matsuoka, M.: Visual damage interpretation of buildings in Bam City using QuickBird images following the 2003 Bam, Iran, earthquake. Earthq. Spectra **21** (S1), 329–336 (2005)
4. Lu, X., Zeng, X., Xu, Z., Guan, H.: Improving the accuracy of near-real-time seismic loss estimation using post-earthquake remote sensing images. Earthq. Spectra **34**(3), 1219–1245 (2018)
5. Galarreta, J., Fernandez, K., et al.: UAV-based urban structural damage assessment using object-based image analysis and semantic reasoning. Nat. Hazards Earth Syst. Sci. **15**(6), 1087–1101 (2015)
6. Olsen, M.J., Cheung, K.F., Yamazaki, Y., et al.: Damage assessment of the 2010 Chile earthquake and tsunami using terrestrial laser scanning. Earthquake Spectra **28**(1), 179–197 (2012)
7. Anil, E.B., Akinci, B., Garrett, J.H., Kurc, O.: Characterization of laser scanners for detecting cracks for post-earthquake damage inspection. In: Proceedings of the International Symposium on Automation and Robotics in Construction and Mining (ISARC), Montreal, QC, Canada, 11–15 August 2013
8. Giri, P., Kharkovsky, S.: Detection of surface crack in concrete using measurement technique with laser displacement sensor. IEEE Trans. Instrum. Meas. **65**, 1951–1953 (2016)
9. Laefer, D.F., Truong-Hong, L., Carr, H., Singh, M.: Crack detection limits in unit based masonry with terrestrial laser scanning. NDT and E Int. **62**, 66–76 (2014)
10. Wilson, L., Rawlinson, A., Frost, A., et al.: 3D digital documentation for disaster management in historic buildings: applications following fire damage at the Mackintosh building, The Glasgow School of Art. J. Cul. Herit. **31**, 24–32 (2018)
11. Alshawabkeh, Y., Bal'aw, F.: 3D digital documentation, assessment, and damage quantification of the Al-Deir monument in the ancient city of Petra, Jordan. Conserv. Manag. Archaeol. Sites **12**(2), 124–145 (2010)
12. Qiang, L., Jingfa, Z., Qisong, J.: Seismic damage characteristics and digital preservation of remote sensing image of Beichuan earthquake site. Sci. Technol. Eng. **17**(18), 159–167 (2017)
13. Quagliarini, E., Clini, P., Ripanti, M.: Fast, low cost and safe methodology for the assessment of the state of conservation of historical buildings from 3D laser scanning: the case study of Santa Maria in Portonovo (Italy). J. Cult. Herit. **24**, 175–183 (2017)
14. Ordoukhanian, E., Madni, A.M.: Introducing resilience into multi-UAV system-of-systems network. In: Madni, A., Boehm, B., Ghanem, R., Erwin, D., Wheaton, M. (eds.) Disciplinary Convergence in Systems Engineering Research, pp. 27–40. Springer, Cham (2018)
15. Hoskere, V., Narazaki, Y., Hoang, T.A., Spencer Jr., B.F.: Towards automated post-earthquake inspections with deep learning-based condition-aware models. In: Proceedings of 7th World Conference on Structural Control and Monitoring, Qingdao, China, pp. 1–10 (2018)
16. Dominici, D., Alicandro, M., Massimi, V.: UAV photogrammetry in the post-earthquake scenario: case studies in L'Aquila Geomatics. Nat. Hazards Risk **8**(1), 87–103 (2017)

17. Mavroulis, S., et al.: UAV and GIS based rapid earthquake-induced building damage assessment and methodology for EMS-98 isoseismal map drawing: The June 12, 2017 Mw 6.3 Lesvos (Northeastern Aegean, Greece) earthquake. Int. J. Disaster Risk Reduct. **37**, 101–169 (2019)
18. Valero, E., Bosché, F., Forster, A., Hyslop, E.: Historic digital survey: reality capture and automatic data processing for the interpretation and analysis of historic architectural rubble masonry. In: Aguilar, R., Torrealva, D., Moreira, S., Pando, M.A., Ramos, L.F. (eds.) Structural Analysis of Historical Constructions. RILEM Bookseries, vol. 18, pp. 388–396. Springer, Cham (2019)
19. Henze, F., Wulf-Rheidt, U., Schneider, D., Bienert, A.: Photogrammetric and geodetic documentation methods at St. Petri Cathedral, Bautzen. In: Proceedings of the CIPA 2005 XX International Symposium, Torino, Italy, pp. 366–371 (2005)
20. Li, S., Tang, H.: Building damage extraction triggered by earthquake using the UAV imagery. In: ISPRS TC III Mid-term Symposium: Developments, Technologies and Applications in Remote Sensing, Beijing, pp. 929–936 (2018)
21. Ullman, S.: The interpretation of structure from motion. Proc. R. Soc. Lond. B Biol. Sci. **203** (1153), 405–426 (1979)
22. AgiSoft PhotoScan: Website http://www.agisoft.ru2012. Accessed 25 June 2012
23. FARO: FARO® Laser Scanner Focus. https://www.faro.com/en-sg/products/construction-bim/faro-laser-scanner-focus/. Accessed 18 June 2019
24. Wójcik, A., Klapa, P., Mitka, B., Piech, I.: The use of TLS and UAV methods for measurement of the repose angle of granular materials in terrain conditions. Measurement **14** (6), 780–791 (2019)
25. Calantropio, A., Chiabrando, F., Sammartano, G., Spanò, A., Losè, L.T.: UAV strategies validation and remote sensing data for damage assessment in post-disaster scenarios. In: International Archives of the Photogrammetry, Remote Sensing and Spatial Information Sciences XLII-3/W4, Istanbul, Turkey, pp. 121–128 (2018)
26. Wilson, L., Rawlinson, A., Frost, A., Hepher, J.: 3D digital documentation for disaster management in historic buildings: Applications following fire damage at the Mackintosh building, The Glasgow School of Art. J. Cult. Herit. **31**, 24–32 (2018)

Unmanned Aerial Vehicles and Digital Image Processing with Deep Learning for the Detection of Pathological Manifestations on Facades

Ramiro Daniel Ballesteros Ruiz(✉),
Alberto Casado Lordsleem Júnior, Bruno José Torres Fernandes,
and Sérgio Campello Oliveira

Universidade de Pernambuco, Recife, PE, Brazil
ramiro.dbr@hotmail.com

Abstract. In the diagnosis phase of pathological manifestations in facades, the visual inspection stage deserves special attention due to its inherent complexity (height, size, access difficulties and exposure conditions). In recent years, the use of deep learning techniques to detect and classify specific features in images and videos has been increasing, which when combined with the use of Unmanned Aerial Vehicles (UAVs) for capturing images, is a potential useful tool that can assist and automate the visual inspection procedure of facades. This paper aimed to perform the analysis of Digital Image Processing (DIP) for automatic detection of cracks in building ceramic tiles, associated with UAV, which would potentially result in benefits (time, cost and safety) with respect to diagnosis. Thus, the research results showed the technical feasibility of detection of cracks by DIP techniques, however, the main limitation for this purpose is the lack of a dataset of images of pathological manifestations in facades publicly available, restricting the computer learning process and consequently compromising the recognition capacity. Still, the project was able to develop a simple and efficient methodology to what was proposed.

1 Introduction

In civil construction, the performance requirements of the constructions need to be established in the design stage (design life) and, after occupation, the building must be able to withstand the action of degradation agents for a predetermined or predicted time [1]. Whether or not the service life is fulfilled, maintenance, recovery or repair activities are necessary to reestablish the minimum performance requirements previously established.

For this purpose, the visual inspection process of the building's facades is an activity of great importance, since the facade is the main responsible for the building's performance, having the function of protecting it against external degradation agents and, still, serving to provide functional and aesthetic comfort to its users [2].

The visual inspection of facades aims to assess the state of conservation of the vertical fence and the different elements that comprise it, providing guidelines for

E. Toledo Santos and S. Scheer (Eds.): ICCCBE 2020, LNCE 98, pp. 1099–1112, 2021.
https://doi.org/10.1007/978-3-030-51295-8_76

effective and cost-effective maintenance. With this practice, it is possible to guarantee the smooth functioning of the vertical sealing subsystem, allowing it to fulfill the functions for which it was built and, possibly, extend its useful life.

To that end, the most effective and safest method that reduces both the time and the cost of inspection is visual inspection using an Unmanned Aerial Vehicle (UAV) or also known as a drone [3]. With the help of this technology it is possible to carry out a complete filming and capture photos in high resolution HD (High Definition). Inspection through the UAV can register problems not seen with the naked eye seen from the ground floor, such as: infiltrations, detachment or holes in the coating and/or cracks and cracks in the facade.

Additionally, the approach of visual inspection of different elements using images or photographs has proven to be the easiest access and most effective resource since the last century, and it is no different for the inspection of pathological manifestations on building facades [4]. However, in recent years, the use of deep learning techniques (field of computational sciences that seeks to simulate computationally the brain as a learning machine), to detect and classify specific characteristics in images and videos, it has been growing more and more, bringing more accurate results compared to human perception in some cases [5].

Deep learning also has a more robust inspection performance when compared to traditional techniques such as edge detection using regression (intense image smoothing model eliminating irrelevant information and selectively maintaining features of interest in the selected area) [6].

In order to carry out the inspection, it is necessary to fully understand an image, which according to Zhao *et al.* [7], one must not only classify several different images, as well as to accurately estimate the locations of the objects contained in the scenarios.

Digital image processing, within the context presented, represents a promising alternative for the diagnosis of pathological manifestations in buildings, in particular, assisting in the detection of cracks in ceramic wall cladding.

The pathological manifestations of façades in the form of cracks are especially important due to the concern about the risk to safety and loss of performance of the building, since they also account for aspects related to comfort, hygiene and health [8, 9].

In the diagnostic phase of the pathological manifestations in facades, the visual inspection step (survey) deserves a prominent attention because of the inherent complexity (height, size, difficulties of access and exposure conditions).

Currently, in Brazil the visual inspection of building facades is mostly performed by the traditional method, in which the professional of this activity (industrial mountaineering) performs the visual verification and photographic record of the main points to be analyzed.

On the other hand, as a result of the development and improvement of new technological tools in the construction industry, it is clear that the visual inspection procedure, and particularly the photographic record, can be aided and optimized using the Unmanned Aerial Vehicle (UAV), commonly known as drone, as a tool for capturing aerial images [10–13].

The UAV associated with digital image processing for automatic detection of cracks in ceramic tiles in buildings, would result in benefits (term, cost and safety) with regard to diagnosis. Consequently, assisting in the process of preventive or corrective maintenance of the facade identified with pathological manifestations.

However, there is no publicly available database (dataset) of pathologies images in civil construction, with cracks in elements of the concrete structure or mortar and ceramic coverings, especially with images labeled at the pixel level for conducting the segmentation training [10].

Thus, this work had as main objective to carry out a study of detection and segmentation of pathologies in civil construction through deep learning, focusing on cracks in ceramics.

Considering the limitation found, a brief preliminary dataset was developed to carry out the inspection of cracks study in ceramics using deep learning, which will be presented in the following topics.

2 Theoretical Reference

2.1 Unmanned Aerial Vehicle (UAV)

The Unmanned Aerial Vehicle, in a succinct form, refers to a class of aircraft piloted remotely or automatically through the use of pre-defined coordinates, being an emerging technology known for its role in military applications [13].

More recently, both in Brazil and in other countries, the potential use of these aerial vehicles in civilian environments has become increasingly common in areas such as agriculture, forestry, archeology, architecture and civil construction.

What was exclusive for military use, is now also present in several other sectors, highlighting that the types of UAVs and uses of this technology by the military are totally different from current and common civilian applications, bringing an innovative way to use this technological tool.

According to the Brazilian Ministry of Defense the Unmanned Aerial Vehicle (UAV), is all types of aircraft that can be controlled on the 3 axes and that do not need pilots on board to be guided. This type of aircraft is controlled remotely by electronic and computational means, under the supervision of humans, or even without their intervention, through Programmable Logic Controllers. It can be classified according to aerodynamic characteristics, such as fixed wings, rotating wings, airships and beater wings [14], as shown in Fig. 1.

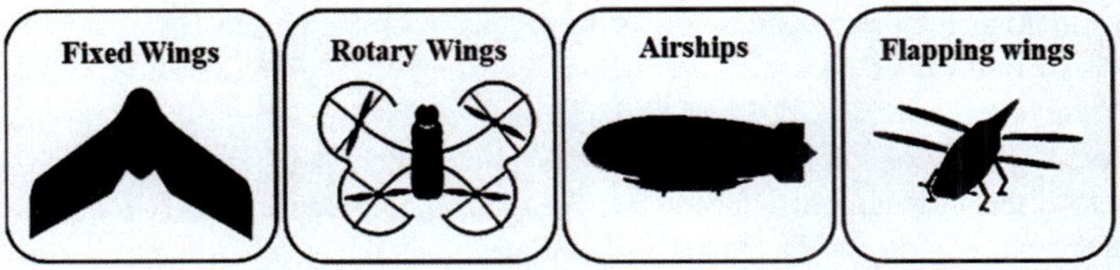

Fig. 1. UAV types by aerodynamic characteristics.

In recent years, different scientific studies have been developed using the UAV as a visual inspection tool, essentially for various elements of infrastructure. The results analyzed allow us to infer that the UAV of rotating wings has more suitable characteristics for vertical inspections. These aircraft perform vertical takeoff and landing, using a support system that allows high maneuverability and stability in the air [3, 11, 13, 15].

Such characteristics have driven the growth of aircraft applications within this classification, particularly in inspection activities, this technological tool allows the collection of images efficiently [16].

Likewise, researchers have been testing the use of different types of cameras and sensors embedded in the UAV for the detection of pathological manifestations in infrastructure, a fraction of these studies has an emphasis on building facades, in most cases in mortar coverings and medium-sized buildings height [15, 17, 18, 19]. Where the most popular embedded collection devices are digital cameras, infrared thermographic cameras and mobile scanners [20].

However, under current conditions, the UAV and digital camera set has been presented as the most affordable alternative in terms of cost - benefit for the purpose of facade inspection [15, 21]. Since most of these aircraft are already commercialized with this type of cameras incorporated in their structure, providing greater reliability to the performance of both integrated technologies, both in hardware and software.

Recently, authors such as Serrat *et al.* [18] and Walter *et al.* [19] studied the combined use of UAVs with an embedded digital camera for inspection of mortar facades. These authors found that the process of collecting digital images with this aircraft has wide potential for use, allowing the identification of pathological manifestations such as dirt stains, expansion in the coating and/or painting, execution failures and cracks, both horizontal and vertical. With the main limitation of operational range and flight duration [17].

In short, from studies conducted by Melo Jr. [15] and Patriota [22], it is possible to infer great benefits of incorporating digital photogrammetry techniques in collecting facade of images, especially in terms of quality and standardization of photographs. This technique, accompanied by a protocol, will structure the entire procedure for collecting and documenting photographs, even when computational processing to manufacture photogrammetric products is not intended.

The main characteristics of this technique associated with UAV include the positioning of the camera, overlay of the photographs and finally, the sequence of capturing the images. In which the parallelism between the photographic shots is essential, always looking for the perpendicularity to the plane of the facade [15].

It is recommended to ensure an overlap between photographs of 60%–80%, both longitudinal and lateral [23], in addition to maintaining a relationship of the distance between the position of the stations in the photo shoot and the distance from the stations to the photographed facade between 0.1–0.5; because if it is too small, the automatic correlation of the points is facilitated, but the geometric precision in the digital reconstruction of the facade is impaired [24].

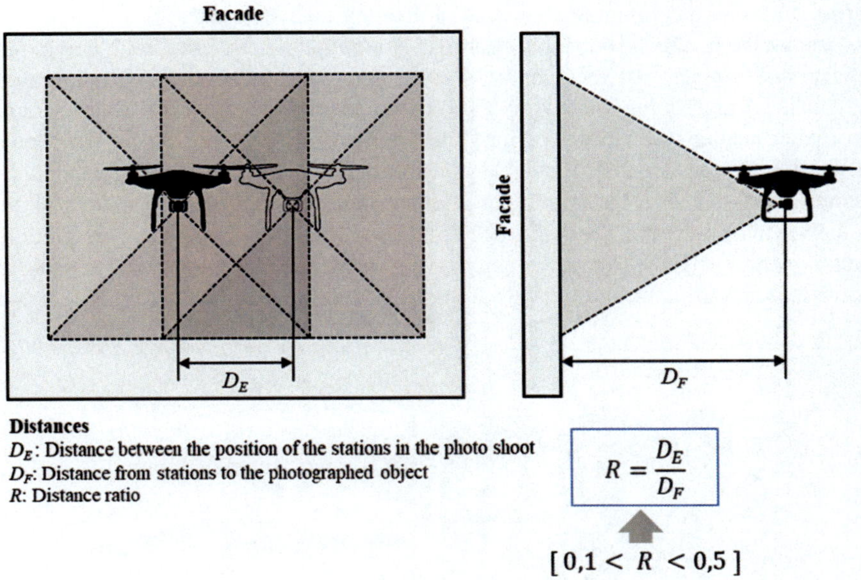

In summary, this technique provides the standardization of photographs in terms of the collection sequence, parallelism between capture points, perpendicularity to the facade plane, and especially, overlapping photographs. The latter allows to record each point of the facade in at least 3 different angles, minimizing and even eliminating the occurrence of omissions in the facade inspection [21].

In short, it appears that all these characteristics are potentially exploitable to automate the process of detecting pathological manifestations. For this, the use of Digital Image Processing (DIP) aimed at the extraction of attributes, that is, recognition and classification of characteristics or patterns of a photograph, would allow to widely optimize the facade inspection service.

2.2 Digital Image Processing (DIP)

Computer vision is an area of study that deals with methods used by computers to obtain a high level of understanding of digital images [25]. Thus, the application of concepts such as Deep Learning allows to train computers to perform tasks like human beings, which includes the identification of different patterns in image and predictions.

In this context, digital image processing for the purpose of detecting pathological manifestations, can be largely benefiting from the standardized qualities of photographs collected with UAV and digital photogrammetry techniques, facilitating the recognition of patterns in structurally similar images. In this sense, the research must investigate the performance of this technique for the recognition of damages, selecting the cracks, due to the difficulty inherent in the size and visual characteristics that hinder the detection of this pathology.

Thus, with the proposal to detect and segment the cracks, several studies were carried out, with the common challenge of creating their own datasets.

Among the methods adopted, Coudray [26] highlights four: Threshold, being quite popular and simple, but inefficient; Morphology, which combines Threshold with mathematical morphology, showing a strong dependence on the initial static parameters; Multi-Scale, which even having a better performance than the previous ones, obtains more false negatives; and Neural Network, which is more efficient, but can become expensive with the training or learning stage.

Following the neural network method, Cha *et al.* [25] presented a Convolutional Neural Network (CNN) model architecture focused on the detection of cracks in concrete, demonstrating good results from 10.000 images in its dataset, as shown in Fig. 2.

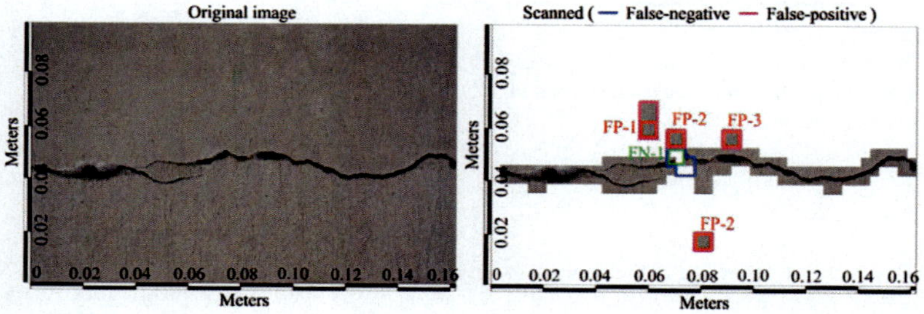

Fig. 2. Example image tested on a CNN [25].

With a slightly different methodology, Silva and Lucena [27] introduced the implementation of a Machine Learning model to classify images in an infrastructure inspection. For this, a pre-trained model (VGG16) and the transfer learning technique were used, aiming to enable the classification of images with and without fissures, as shown in Fig. 3.

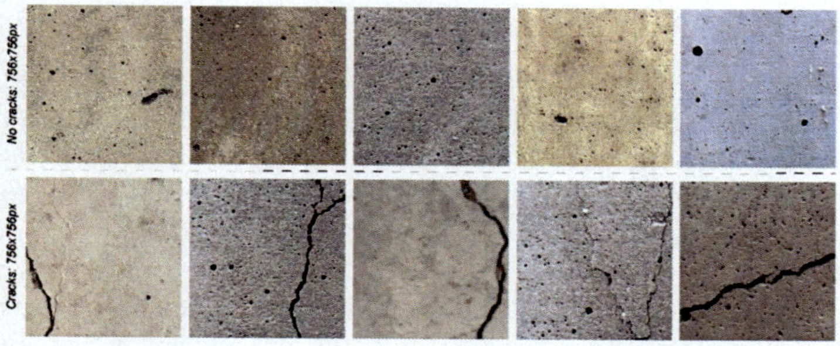

Fig. 3. Example of images with and without cracks or cracks that make up the dataset [27].

The study by Yang *et al.* [10], presented the InspectionNet model, inspired by the HED [28] and U-net [29] models. These three models are intended for pixel level detection and segmentation. Thus, the study focused on the detection of concrete fragmentation, shown in Fig. 4.

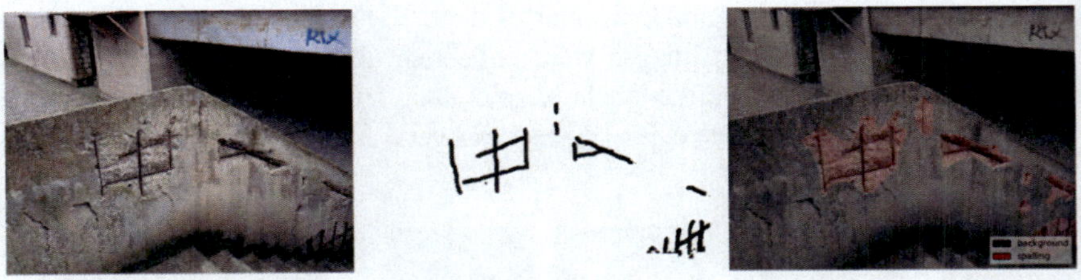

Fig. 4. Original image, followed by its label and the detection of fragmentation [10].

Unlike the aforementioned studies, the present study was carried out using the neural network method, whose deep learning model used was the U-net [29], since it requires few images for good performance.

Next, the methodology will be detailed, as well as the stages of the work, such as the dataset developed, the pre-processing of the images and the chosen model.

3 Method

For the project construction, it was necessary to separate the entire process into modules, as shown in Fig. 5. Starting with the creation of the dataset (database) t, the data is pre-processed with digital image processing (DIP) techniques and finally, the deep learning training and testing stage.

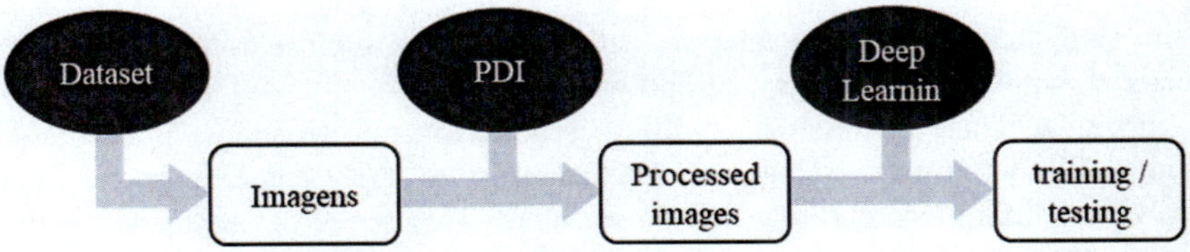

Fig. 5. Process steps, separated by modules.

3.1 Building the Dataset

With the absence of a public dataset to the given problem, was necessary to create a small data base, seeking to meet the minimum requirements of deep learning, even knowing that for good results thousands of images would be needed for training.

The U-net adopted model [29] brought great versatility in this limitation, since it managed to cope well with a small amount of data, compared to other models.

In this way, 167 different images were collected, all showing cracks in ceramic tiles. Consecutively the respective labels were created so that the deep learning interprets the pathological occurrences as "correct answers", as shown in Fig. 6.

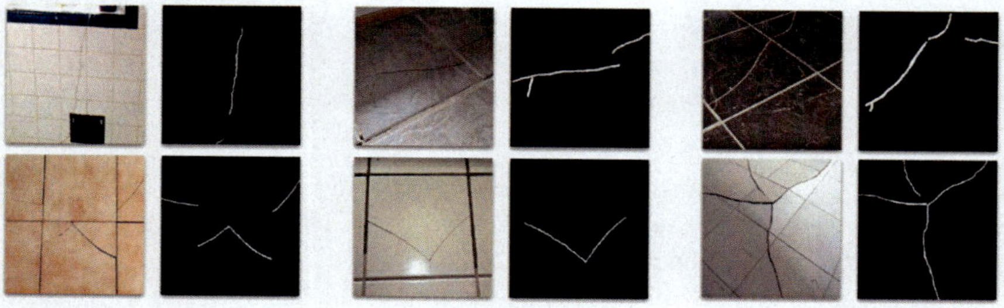

Fig. 6. Example of the images collected and their respective labels.

From the collected data set, 67 images were separated for the test stage and 100 for training. Among the training images, the data augmentation process was carried out, in which DIP techniques are applied to each image, varying some of its aspects, such as angulation, zoom and displacement. With this technique it was possible to create 10,000 new images and increase the training base, resulting in 10,100 images.

3.2 Pre-processing

In this step, the images were treated so that the result was as close to the labels, that is, binarized, with a black background and white strokes.

In digital image processing, a widely used binarization technique is Otsuthresholding [30]. Before performing it, it was necessary to apply some other techniques, involving lighting, morphology and noise filters. This series of techniques made it possible to visualize the object of study (the cracks) in most of the images that previously presented problems, as shown in Fig. 7.

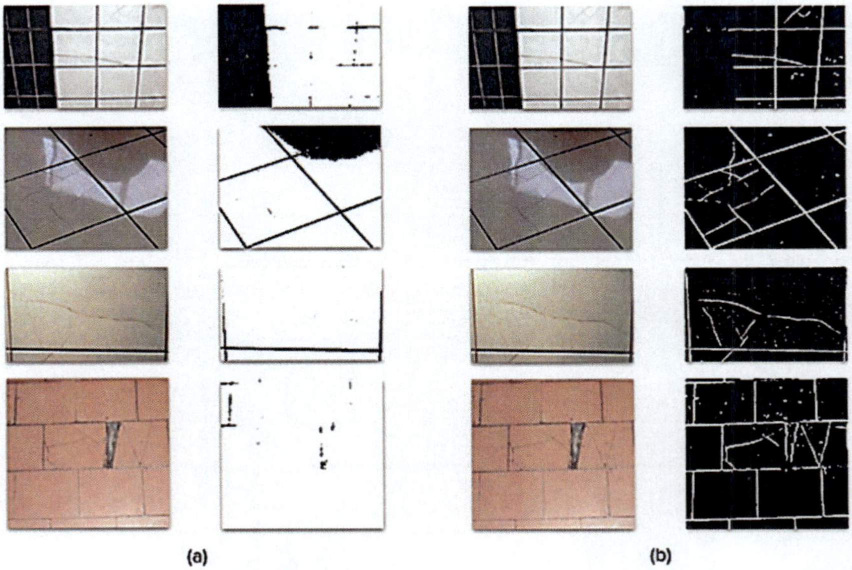

Fig. 7. (a) Example of images only with Otsuthresholding and (b) with the series of applied techniques.

However, the resulting images presented a wide variety of scenarios, lighting, angulation and distance. This meant that the pre-processing did not meet all the images, as shown in Fig. 8.

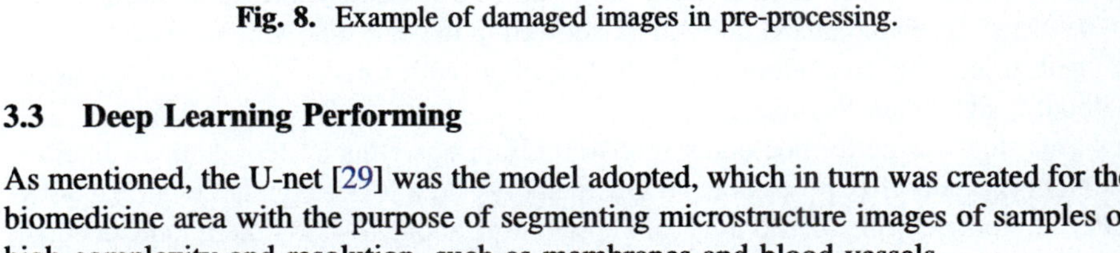

Fig. 8. Example of damaged images in pre-processing.

3.3 Deep Learning Performing

As mentioned, the U-net [29] was the model adopted, which in turn was created for the biomedicine area with the purpose of segmenting microstructure images of samples of high complexity and resolution, such as membranes and blood vessels.

In this way, the architecture was planned to offer good performance even with a small set of images (compared to the other deeplearning models).

The U-net architecture, shown in Fig. 9, consists of two activities. The first activity corresponds to the contraction, applying repeated convolutions, in which each is followed by a Rectified Linear Units (ReLU - Rectifier of neural networks) and a grouping operation. The second activity corresponds to expansion, through convolutions and concatenations with high resolution resources [29].

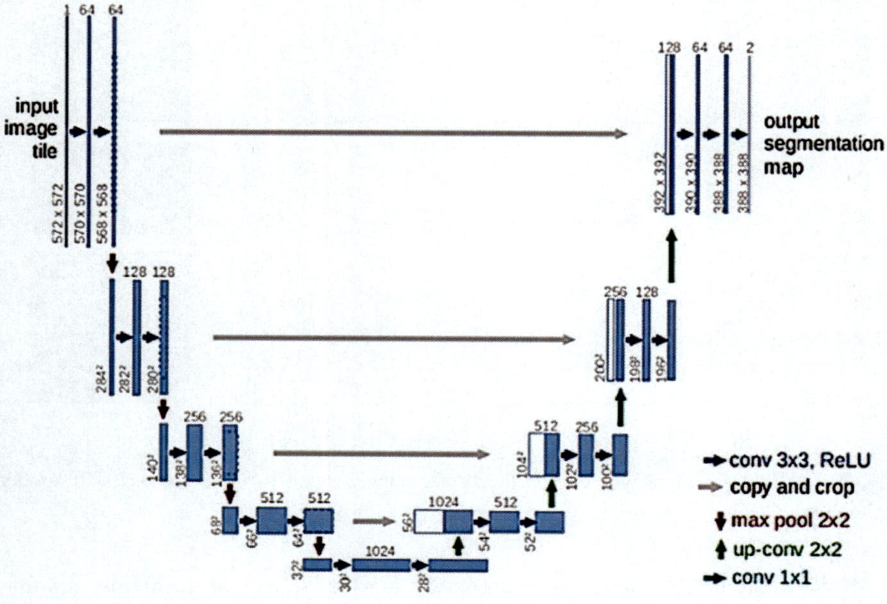

Fig. 9. U-net architecture, adaptation of [29].

In other words, u-net is a convolutional network architecture for fast and accurate segmentation of images. The strategy used by u-net, succinctly, is the classification of the pixel considering the context of the entire image. As for convolution, it is the filtering operation in the intensity domain (as the image is presented in practice, with the values of each pixel) [29].

For training, model learning was monitored, avoiding overfitting situations (overfitting or overfitting - when a statistical model fits very well into the previously observed data set, but is ineffective to predict new results) and underfitting (sub-adjustment - when the model does not adapt well to the data with which it was trained), as well as the implementation of Early Stopping, with tolerance of up to 5 seasons without a significant learning rate.

This step was performed under the Arch Linux operating system, with an Intel i7-7500U (4) 3,500 GHz processor, 16 GB memory and a 4 GB Nvidia GeForce 940 M GPU. Performed with 10,100 images, the step took about 10 h to obtain more accurate results in the 67 test images. In a way, the model's result tends to improve as the data volume is increased, making it possible to spend more time in the training stage and suffer less overfitting problems.

4 Results

The result of the study provided, even under the limitations, a good analysis of the scenario and good segmentations in the images as shown in Fig. 10. Both images with a lot of noise, as well as with discrete fissures, the inspection through deep learning proved to be adaptable to complex situations and flexible to obstacles.

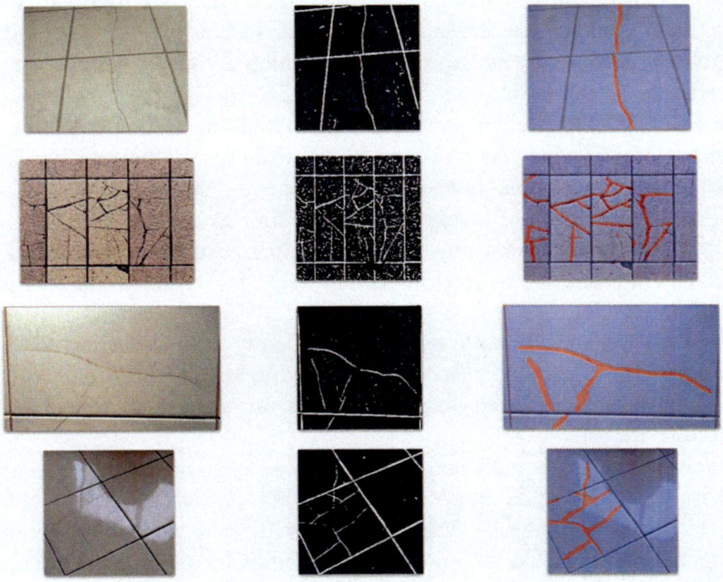

Fig. 10. Example of the original, pre-processed images and the result of the segmentation superimposed on the original image.

However, the lack of a public dataset available for this type of problem, directly influenced the performance of deep learning, since it was necessary to search for images that represented the object of study to be segmented (cracks in ceramics).

It is important to emphasize the importance of a wide dataset to work with deep learning, with good quality and labeled images, in order to obtain better results.

Even after selecting the most relevant data to compose the dataset, even so, in many cases there was great variation in the images, such as angulation, distance and quality.

In addition, the pre-processing of the data also impacted some of the results, as it did not meet all the images in the dataset, mainly due to the variation in lighting.

5 Conclusion

The present study proposed an exploratory analysis on the use of the DIP system, with an emphasis on deep learning, in the inspection process for pathological manifestations on facades, specifically, in the detection and segmentation of cracks and cracks in ceramic tile systems.

The detection of cracks by DIP techniques is feasible from a technical point of view, however, it is considered a complex job when there is a high variation in the study images, and in the face of a limiting scenario, such as the lack of public datasets for the problem, the project still managed to develop a simple and efficient methodology to what was proposed.

Despite the difficulties encountered, it was possible to develop a simple and efficient methodology to what was proposed. Being even more efficient, in training time and precision in segmentation, to the most common problems (with a well-defined data set), such as the inspection of cracks in concrete, for example.

Finally, there is a promising way to improve the project, seeking to further simplify the approach (minimizing the pre-processing step, for example) and improve the accuracy of the results.

The absence of public datasets referring to different pathological manifestations of facades, in general, opens the space for new studies that evaluate and reinforce the creation of these data sets according to the need for automated inspection for each problem in the area.

In conclusion, there is sufficient evidence to emphasize the potential of Digital Image Processing for facade inspection. This method, combined with photographs collected by Unmanned Aerial Vehicles, especially by collection techniques that standardize the photographic record, denotes viability for the joint inspection procedure. Which, in general, would provide a service with greater security, greater speed and lower costs, requiring less resources compared to traditional inspection methods, such as industrial mountaineering.

References

1. Possan, E., Demoliner, C.: Desempenho, durabilidade e vida útil das edificações: abordagem geral. Revista técnico científica do CREA-PR **1**(9) (2013)
2. Bauer, E., Milhomem, P., Aidar, L.: Evaluating the damage degree of cracking in facades using infrared thermography. J. Civil Struct. Health Monitor. **8**(3), 517–528 (2018)
3. Melo, R., Costa, D.: Uso de veículo aéreo não tripulado (VANT) para inspeção de logística em canteiros de obra. In: SIBRAGEC-ELAGEC, 1, 2015. São Carlos (2015)
4. Adhikari, R., Moselhi, O., Bagchi, A.: Image-based retrieval of concrete crack properties for bridge inspection. Automat. Construct. **39**, 180–194 (2014)
5. Liu, Y., Kohlberger, T., Norouzi, M., Dahl, G., Smith, J., Mohtashamian, A., Olson, N., Peng, L., Hipp, J., Stumpe, M.: Artificial intelligence–based breast cancer nodal metastasis detection. Archives Pathol. Lab. Med. **143**(7), 859–868 (2018)

6. Jahanshahi, M., Masri, S.: Adaptive vision-based crack detection Using 3D scene reconstruction for condition assessment of structures. Autom. Construct. **22**, 567–576 (2012)
7. Zhao, Z., Zheng, P., Xu, S., Wu, X.: Object detection with deep learning: a review. IEEE Trans. Neural Netw. Learn. Syst. **30**, 1–21 (2019)
8. Ribeiro, D., Sales, A., Souza, C., Almeida, F., Cunha, M., Lourenço, M., Helene, P.: Corrosão em estruturas de concreto armado: Teoria, controle e métodos de análise, 1st edn. Elsevier, Rio de Janeiro (2013)
9. Costa e Silva, A.: Descolamento nos revestimentos cerâmicos de fachada na cidade do Recife. Escola Politécnica, Universidade de São Paulo (Thesis - Master's degree) São Paulo (2008)
10. Yang, L., Li, B., Li, W., Jiang, B., Xiao, J.: Semantic metric 3D reconstruction for concrete inspection. In: IEEE/CVF Conference on Computer Vision and Pattern Recognition Workshops, Boston (2018)
11. Hung, M., Sampaio, T., Schultz, G., Siefert, C., Lange, D., Marangon, F., Santos, I.: Levantamento com veículo aéreo não tripulado para geração de modelo digital do terreno em bacia experimental com vegetação florestal esparsa. In: Revista Ra'e Ga Espaço Geográfico em Análise, vol. 43, pp. 215–231. Curitiba (2018)
12. Reagan, D., Sabato, A., Niezrecki, C.: Feasibility of using digital image correlation for unmanned aerial vehicle structural health monitoring of bridges Structural Health Monitoring. SAGE Publishing, vol. 17, no. 3, pp. 289–302. Massachusetts (2017)
13. Irizarry, J., Costa, D.: Exploratory study of potential applications of unmanned aerial systems for construction management tasks. J. Manage. Eng. **32**(3), 05016001 (2016). Atlanta
14. Angelov, A.: Sense and Avoid in UAS: Research and Applications, 1st edn. Wiley, United Kingdom (2012)
15. Melo, C.: Metodologia para geração de mapas de danos de fachadas a partir de fotografias obtidas por veículo aéreo não tripulado e processamento digital de imagens. Universidade de Brasilia (Thesis – Doctoral degree), Brasília (2016)
16. Amorim, A., Silva, L., Christakou, E.: Uso de VANT em documentação arquitetônica. In: Seminário Nacional de Documentação do Patrimônio Arquitetônico como uso de Tecnologias Digitais. Belém (2012)
17. Kerle, N., Nex, F., Gerke, M., Duarte, D., Vetrivel, A.: UAV-based structural damage mapping: a review. Int. J. Geo-Inf. **9**(1), 14 (2019). Noruega
18. Serrat, C., Cellmer, A., Banaszek, A., Gibert, V.: Exploring conditions and usefulness of UAVs in the BRAIN massive inspections protocol. Open Eng. J. **1**(9) (2019). Berlin
19. Walter, B., Schaeffer, G., Christ, R., Pacheco, F., Tutikian, B.: Uso do VANT para identificação das manifestações patológicas de fachadas: estudo de caso em edificações. In: CONPAT 2019. Chiapas (2019)
20. Acosta, L., Torres, A., Krejza, Z., Gibert, V., Serrat, C.: Exploring technological alternatives to the visual inspection method in the built environment. IOP Conference Series: Earth and Environmental Science, vol. 222 (012027). Barcelona (2019)
21. Ballesteros, R.: Inspeção de manifestações patológicas de fachadas utilizando Veículo Aéreo Não Tripulado (VANT): estudo exploratório. Escola Politécnica, Universidade de Pernambuco (Thesis - Master's degree). Recife (2020)
22. Patriota, R.: Mapeamento de fachadas usando aeronave remotamente pilotada. Pontifícia Universidade Católica do Paraná (Monograph – Specialization). Curitiba (2017)
23. Groetelaars, N., Amorim, A.: Dense Stereo Matching (DSM): conceitos, processos e ferramentas para criação de nuvens de pontos por fotografias. In: SIGRADI 2012. Fortaleza (2012)

24. Pierrot-Deseilligny, M., De Luca, L.: Remondino, F.: Automated image-based procedures for accurate artifacts 3D modeling and orthoimage generation. Geoinformatics CTU FCE (2011)
25. Cha, Y., Choi, W., Büyüköztürk, O.: Deep learning-based crack damage detection using convolutional neural networks. Comput. Aided Civil Infrastructure Eng. **32**, 361–378 (2017)
26. Coudray, N., Karathanou, A., Chambon, S.: Multi-resolution approach for fine structure extraction: application and validation on road images. In: International Conference on Computer Vision Theory and Applications (VISAPP), pp. 17–21. Angers (2010)
27. Silva, W., Lucena, D.: Concrete cracks detection based on deep learning image classification. In: Proceedings of the 18th International Conference on Experimental Mechanics (ICEM 2018). Brussels (2018)
28. Xie, S., Tu, Z.: Holistically-nested edge detection. In: IEEE International Conference on Computer Vision, pp. 1395–1403. Washington (2015)
29. Ronneberger, O., Fischer, P., Brox, T.: U-Net: Convolutional networks for biomedical image segmentation. In: Medical Image Computing and Computer-Assisted Intervention – MICCAI 2015, pp. 234–241. Munich (2015)
30. Otsu, N.: A threshold selection method from gray-level histograms. IEEE Trans. Syst. Man Cybernetics **9**, 62–66 (1979)

Vision and Image-Based Applications in Construction

A Real-Time Automated Approach for Ensuring Proper Use of Personal Protective Equipment (PPE) in Construction Site

Shi Chen[1](✉), Kazuyuki Demachi[1], and Manabu Tsunokai[2]

[1] The University of Tokyo, Tokyo, Japan
shichen@g.ecc.u-tokyo.ac.jp
[2] IIU Corporation, Tokyo, Japan

Abstract. Construction sites are one of the most perilous environments where many potential hazards may occur. Even though workers are trained to stay away from potential dangers, there are still many types of risks that can occur within only a few minutes of carelessness. Personal Protective Equipment (PPE) is an important safety measure used to protect construction workers from accidents. However, PPE usage is not strictly enforced among workers due to all kinds of reasons. This paper proposes the combination of deep learning-based object detection and individual detection using geometry relationships analysis to automatically identify non-PPE-use (NPU); i.e., if a worker is wearing hardhat, eye protection visors, dust masks, or both, to help to facilitate the safety monitoring work of construction workers to ensure PPE are appropriately used. The experimental results demonstrate that the approach was capable of detecting NPU workers with high precision (84.13%) and recall rate (93.10%) while ensuring real-time performance (7.95 FPS on average).

Keywords: Construction safety · Personal Protective Equipment (PPE) · Deep learning · Object detection

1 Introduction

Construction work is much more dangerous than most other occupations, where many potential hazards may occur. According to the United States' Bureau of Labor Statistics, the number of construction fatalities in the US has gradually increased from 933 to 1013 between 2014 and 2017 [1]. Similarly, in Japan, there were 926 construction fatal accidents during 2016–2018 and the Ministry of Health, Labor and Welfare of Japan is aiming to reduce construction fatalities by at least 15% (relative to the 2017 level) by 2022 [2].

The construction fatalities are always caused by the combination of different factors that involve occupational risk exposures (e.g., electricity), tools (e.g., grinder), equipment (e.g., crane), and environment (e.g., dust and noise) in the construction sites. The consequences of head injuries caused by falling from height or being stuck by vehicles and other moving plants and equipment are one of the most serious of all construction accidents. A total of 2,210 construction fatalities occurred because of traumatic brain injury (TBI) which represented 25% of all construction fatalities during

E. Toledo Santos and S. Scheer (Eds.): ICCCBE 2020, LNCE 98, pp. 1115–1126, 2021.
https://doi.org/10.1007/978-3-030-51295-8_77

2003 and 2010 [3]. The Occupational Safety and Health Administration (OSHA) in the United States stipulated that workers working in areas where there is a possible danger of head injury from impact, or from falling or flying objects, or from electrical shock and burns shall be protected by hardhats [4]. Furthermore, construction-related occupational eye injuries are an important cause of vision loss. According to the National Institute for Occupational Safety and Health (NIOSH), an average of 2,000 U.S. workers require medical treatment for job-related eye injuries every day [5]. The majority of construction-related ocular injuries are preventable. The reasons cited for the majority of ocular injuries include the non-wearing of available eye protection or wearing of inappropriate eye protection for the current task [6]. On the other hand, fine dust and particles, gases and vapors can be produced when using machine tools and silica dust from bricks can cause lung and airway diseases such as emphysema, bronchitis and silicosis, and may increase cancer risks. Personal Protective Equipment (PPE) such as respirators or dust masks are used to controls these hazards [7]. OSHA indicates the workers shall be ensured to wear eye or face protection when exposed to eye or face hazards from flying particles, molten metal, liquid chemicals, acids or caustic liquids, chemical gases or vapors, or potentially injurious light radiation [8]. Nonetheless, the workers do not precisely follow the construction site's safety regulations due to all kinds of reasons, even if they have been previously educated and trained. Thus, an automated monitoring approach is necessary to be conducted to help to facilitate the safety monitoring work of construction workers to ensure PPE are appropriately used.

In this paper, we present a real-time approach to automatically identify non-PPE-use (NPU) in response to the limitations of monitoring systems in construction sites. The goal is to detect hardhats, eye protection visors and dust masks in each observed frame, and to identify whether individuals on construction sites are wearing PPE. Firstly, we detect entities in each observed frame using deep learning-based models: PPE(s) are recognized and localized using Yolov3 [9] and individual(s) are characterized by extracting their joint positions using OpenPose [10]. Subsequently, we associate detected PPE(s) with the detected individual(s). Finally, we identify whether PPE(s) are appropriately used by analyzing the geometric relationships of detected PPE(s) and individual(s). To summarize, this work contains the following contributions: (1) In contrast to commonly used object detection-based NPU identification approaches, this study provides a novel solution to automatically identify any failure to wear PPE by the combination of object detection and individual detection using geometry relationships analysis which is more effective and robust with viewpoint changes and different individual postures. (2) To the authors' knowledge, almost no research studies have been conducted concerning multi-class NPU identification other than NHU. This paper explores the possibility of multi-class NPU identification for non-hardhat-use (NHU), non-mask-use (NMU), and non-visor-use (NVU). (3) This paper demonstrates that the proposed approach can rapidly handle NPU identification to meet the industrial requirements of real-time processing.

2 Related Works

At present, a number of methods have been studied for automatic NPU detection [11–16], which can be divided into two categories: sensor-based detection and vision-based detection.

Sensor-based detection primarily relies on remote locating and tracking techniques, such as radio frequency identification (RFID) and wireless local area networks (WLANs). Kelm et al. [11] developed a mobile RFID portal for checking PPE compliance of personnel. The RFID readers were located at the construction site entrance, and therefore only the individuals who enter the construction site can be checked. Dong et al. [12] developed the real-time location system (RTLS) and virtual construction for worker location tracking to identify whether the worker should wear a hardhat by placing a pressure sensor in the hardhat to identify whether a hardhat was being worn, and if not, to transmit a warning. Generally, existing sensor-based methods relying on physical tags or sensors employed in PPE have difficulty in identifying whether any individuals on the sites are wearing PPE or not. Moreover, the practical use of the tags or sensors will lead to high costs due to the large volume of devices required.

Vision-based methods are nonintrusive and less device-intensive because of the wide application of on-site surveillance cameras. Shrestha, et al. [13] use edge detection algorithms to recognize the edge of objects inside the upper head region i.e. hardhats. However, this method relies on the recognition of facial features, therefore workers who turn their face away from the cameras cannot be recognized. Park et al. [14] proposed a vision-based NHU detection method that detects both a human body and a hardhat simultaneously in each obtained frame. The detected human body region and hardhat region are then matched for the detection of NHU. In general, these methods rely heavily on hand-crafted features to analyze the individuals. Consequently, they may fail in the case of complicated scenes with weather variability, different viewpoints and/or occlusions.

Recently, deep learning-based object detection methods have shown remarkable performance on most visual tasks in the construction industry. Fang et al. [15] proposed a Faster R-CNN based method to detect construction workers' NHU automatically. A total of 81,000 image frames were collected from various construction sites as a training dataset to train the Faster R-CNN model. In the training phase, the worker-of-interest (WOI) in the image was annotated as the ground truth for training. In the test phase, the NHU workers were detected and the rest were considered the background. Wu et al. [16] deployed a Single Shot Multibox Detector (SSD) based model combined with the presented reverse progressive attention (RPA) to propagate contextual information back to bottom layers discriminately. A benchmark dataset GDUT-HWD was generated by downloading Internet images retrieved by search engines to train the SSD-RPA model. However, existing deep learning-based detection methods are mainly focused on learning to localize only PPE(s) or NPU individual(s) in the obtained images, which may fail in cases of uncommon human gestures or appearance. Also, almost no research studies have been conducted concerning multi-class NPU identification other than NHU and are limited in practical application to real scenarios. In response to these limitations, the overall objective of this paper is to develop a new approach for monitoring workers and evaluate whether the proposed approach could be used to detect failure use of hardhats, visors and masks in the construction site.

3 Methodology

The associated generic pipeline is illustrated in Fig. 1, and it follows the subsequent stages:

(1) For each observed frame, individual(s) are detected, together with their keypoints coordinates using OpenPose [10]. PPE(s) are recognized and localized using YOLOv3 [9].
(2) Detected PPE(s) are associated with the detected individual(s).
(3) NPU identification is performed by analyzing the geometry relationships of the individual's keypoints and the detected PPE(s).

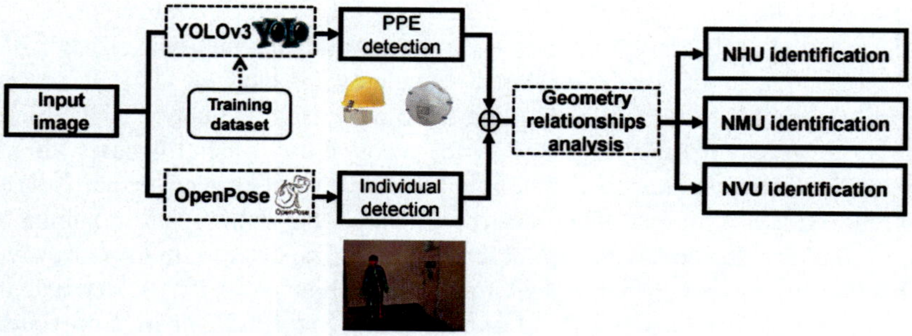

Fig. 1. Generic pipeline of the proposed approach.

3.1 Entity Detection

PPE Detection. We deploy YOLOv3 [9] to detect the PPE(s). YOLOv3 uses Darknet-53, a CNN with 53 layers, to extract image features. Subsequently, YOLOv3 makes predictions at three scales, which are precisely given by downsampling the dimensions of the input image by 32, 16 and 8 respectively. This method allows YOLOv3 to get more meaningful semantic information from the upsampled features and finer-grained information from the earlier feature map. The prediction result of the network is a 3-d tensor that encodes bounding box, objectness score and prediction over classes:

$$\mathrm{N} \times \mathrm{N} \times (3 * (4 + 1 + \mathrm{C})) \quad (1)$$

where $N \times N$ is the number of the grid cells of the system, and C is the number of the classes to train the network on. Besides, YOLOv3 predicts a confidence score for each bounding box using logistic regression, which is overwhelmingly beneficial especially considering that one image might enjoy multiple labels, and not all the labels are guaranteed to be mutually exclusive.

In a word, predictions of YOLOv3 are carried out from one single network, which can be trained end-to-end to improve accuracy. Higher efficiency and better performance on small object detection make YOLOv3 a reasonable option for real-time processing for industrial purposes.

Individual Detection. We characterize the workers' postures by extracting the joint positions of the individual in the images using OpenPose [10], which processes images through a two-branch multi-stage Convolutional Neural Network (CNN) and uses Part Affinity Fields (PAFs) to learn to associate body parts with individuals in the image to output the 2D keypoints for all people in the image. OpenPose provides the positions of 18 body joints (pre-trained using COCO 2016 keypoints challenge dataset [17], see Fig. 2). The choice of OpenPose is motivated for its functionality on RGB images or videos taken by on-site surveillance cameras in real-time. This provides a huge benefit in comparison to the skeletal tracking capability of RGB-D devices (e.g., Microsoft Kinect) which depend on depth information. To further improve the estimation speed, we deploy a light-weight architecture, Mobilenetv2 [18] as the feather extractor instead of VGG-19 [19] in the original paper.

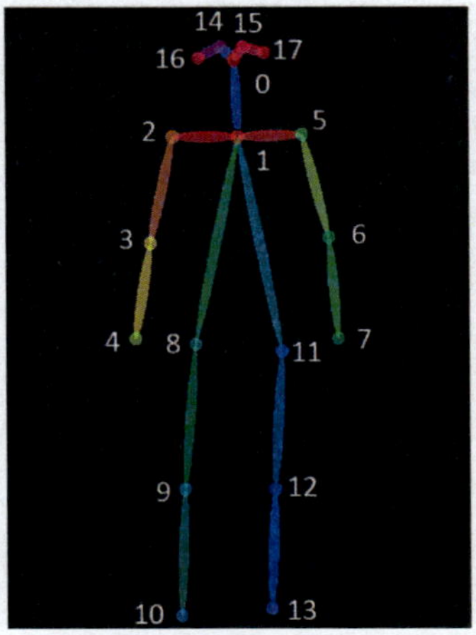

Fig. 2. OpenPose Output Format.

However, Openpose may fails in detecting individuals without their full body visible in the near field. In this case, we characterize individuals by localizing their faces using YOLOv3.

3.2 NPU Identification

We formulate the output of YOLOv3 as a set of objects bounding boxes $B = \{B_1, B_2, \ldots, B_I\}$, where I is the number of detected objects bounding boxes in the obtained frame. Each bounding box $B_i = (x_i, y_i, w_i, h_i, c_i)$, $I \in \{1, 2, \ldots, I\}$ contains five elements, where (x_i, y_i) and (w_i, h_i) are respectively the bounding boxes position and size and c_i represents the class of objects in the bounding box. Let $H = \{H_1, H_2, \ldots, H_J\}$ be the set of the detected individual(s) via OpenPose, where J is the number of detected individual(s) in the obtained frame and $H_j = \left\{\left(x_j^{(0)}, y_j^{(0)}\right), \left(x_j^{(1)}, y_j^{(1)}\right), \ldots, \left(x_j^{(17)}, y_j^{(17)}\right)\right\}$ represents the detected bodyparts of the j_{th} individual. We associate a detected PPE i^* to a specific individual j^* by searching the minimum Euclidean distance between bounding boxes B and detected neck keypoints $H^{(1)} = \left\{\left(x_0^{(1)}, y_0^{(1)}\right), \left(x_1^{(1)}, y_1^{(1)}\right), \ldots, \left(x_J^{(1)}, y_J^{(1)}\right)\right\}$ (bodypart 1 in Fig. 2) that satisfy the geometric constraints to make sure each PPE is in the upper area of the candidate associated individual:

$$i^*, j^* = \underset{i \in \{1,2,\ldots,I\}, j \in \{1,2,\ldots,J\}}{\operatorname{argmin}} \sqrt{\left(x_i - x_j^{(1)}\right)^2 + \left(y_i - y_j^{(1)}\right)^2} \quad s.t.\ y_{i^*} < y_{j^*}^{(1)} \tag{2}$$

Subsequently, for each associated PPE and individual, we identify whether the PPE is appropriately used by analyzing key lengths. We take advantage of Euclidean distance among detected neck keypoints and hip keypoints (bodyparts 8 and 11 in Fig. 2) as reference threshold when the distance between the individual and the camera changes, the reference threshold change synchronously:

$$\beta_{i^* \leftrightarrow j^*} = max\left(\sqrt{\left(x_{j^*}^{(1)} - x_{j^*}^{(8)}\right)^2 + \left(y_{j^*}^{(1)} - y_{j^*}^{(8)}\right)^2}, \sqrt{\left(x_{j^*}^{(1)} - x_{j^*}^{(11)}\right)^2 + \left(y_{j^*}^{(1)} - y_{j^*}^{(11)}\right)^2}\right) \cdot \gamma \tag{3}$$

where γ is the scaling coefficient to strike different NPU identification. If the Euclidean distance between the detected PPE and detected neck keypoint of the associated individual is smaller than the reference threshold $\beta_{i^* \leftrightarrow j^*}$, then the detected PPE is identified as being appropriately used by the individual (Fig. 3 (a)); otherwise the condition is identified as NPU (Fig. 3 (b)).

In the case of OpenPose failure, PPE-individual association is performed using detected face bounding boxes (Fig. 3 (c), (d)). Let $B' = \{B'_1, B'_2, \ldots, B'_K\}$ be the detected face bounding boxes, where K is the number of detected face bounding boxes in the obtained frame. Similarly, we associate a detected PPE i^* to a specific individual

k^* by searching the minimum Euclidean distance between PPE and face bounding boxes:

$$i^*, k^* = \underset{i \in \{1,2,\ldots,I\}, k \in \{1,2,\ldots,K\}}{\operatorname{argmin}} \sqrt{\left(x_i - x'_k\right)^2 + \left(y_i - y'_k\right)^2} \tag{4}$$

where (x_i, y_i) is the position of the i_{th} PPE bounding box in B and (x_k, y_k) is the position of the k_{th} face bounding box in B'. The reference threshold $\beta_{i^* \leftrightarrow k^*}$ is given as follow:

$$\beta_{i^* \leftrightarrow k^*} = max\left(w'_{k^*}, h'_{k^*}\right) \tag{5}$$

where $\left(w'_{k^*}, h'_{k^*}\right)$ is the size of the associated face bounding box in B'.

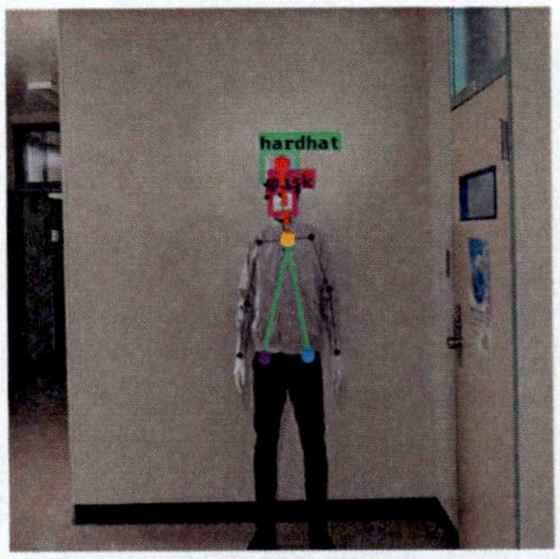

(a) PPE proper use identified by OpenPose keypoints.

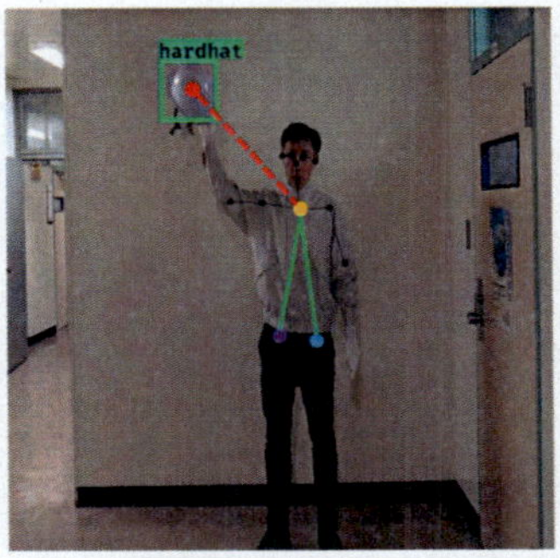

(b) NPU identified by OpenPose keypoints.

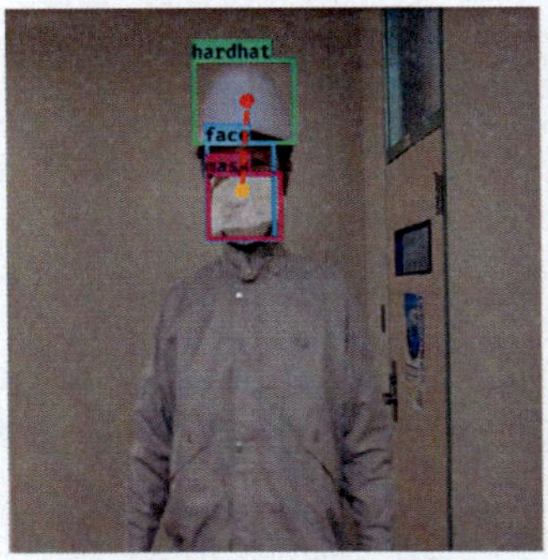

(c) PPE proper use identified by detected face bounding boxes.

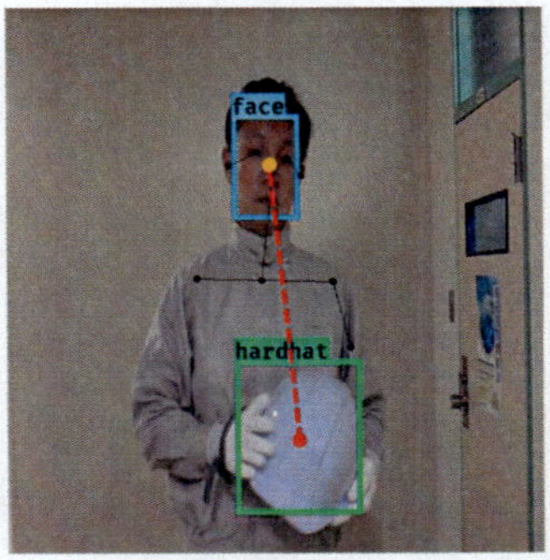

(d) NPU identified by detected face bounding boxes.

Fig. 3. NPU identification strategies.

4 Experiments and Results

4.1 Experimental Dataset

To create the training dataset of PPE detection, we collected hardhat, eye protection visor, and dust mask images from two sources: downloading Internet images retrieved by search engines using keywords, and capturing real-world images using the webcam as listed in Table 1. Also, we randomly selected 2,000 images from the WIDER FACE dataset [20] to learn the features of face. A total of 7,929 images were collected and annotated to train a YOLOv3 model.

Table 1. Information of collected training dataset.

	Internet images	Real-world images	Total
Hardhat	933	1,209	2,142
Mask	983	1,222	2,205
Visor	110	736	846
Face	2,000	736	2,736
Overall	4,026	3,903	7,929

Furthermore, to create the testing dataset to validate the performance of the trained model, six volunteers were instructed to perform normal working behaviors while wearing PPEs at different distances to the camera. As surveillance cameras are placed in different locations on construction sites and the trajectory of workers is stochastic, construction workers were captured in different resolutions in the surveillance videos. Thus, different distance conditions (1 m, 3 m, 5 m) were considered in our experiments to validate the robustness of our proposed approach. Finally, we randomly selected 1,500 images (500 images for each distance condition) from the collected image sequences and created the testing dataset. The details are provided in Table 2 where positive samples refer to the NPU (including NHU, NMU, and NVU) individuals and negative samples referred to the individuals who are wearing PPE properly.

Table 2. Information of collected testing dataset.

Distance (m)	NPU Categories	Images	Positive samples	Negative samples
1	NHU	500	318	182
	NMU		275	225
	NVU		334	166
3	NHU	500	263	479
	NMU		278	464
	NVU		452	290
5	NHU	500	274	457
	NMU		288	443
	NVU		444	287
Overall		1,500	2,926	2,993

4.2 Evaluation Metrics

We adopted precision and recall to evaluate the performance of our approach:

$$Precision = \frac{TP}{TP + FP} \tag{6}$$

$$Recall = \frac{TP}{TP + FN} \tag{7}$$

Where *TP* is defined as the number of correct detections of NPU individuals. *FP* is the number of wrong detections of NPU individuals, while *FN* is the number of the ground truth not detected as defined in Table 3.

Table 3. Definitions of TP, FP, and FN.

	Predicted	Ground truth
TP	NPU	NPU
FP	NPU	Proper use
FN	Proper use	NPU

4.3 Implementation Details

We initialized the YOLOv3 model based on pre-trained weights on the ImageNet dataset [21]. Training of YOLOv3 was performed in two stages. We first froze all convolutional layers up to the last convolutional block in Darknet-53 and train with frozen layers to get a stable loss. Subsequently, we proceed to unfreeze all convolutional layers of Darknet-53 to perform fine-tuning. The learning rates for the first and the second stage are 1e−3 and 1e−4 respectively. Adaptive Moment Estimation (Adam) optimizer was adopted to adjust the learning rate during optimization automatically. α (initial learning rate), β_1 (exponential decay rate for the first moment estimates), β_2 (exponential decay rate for the second-moment estimates) and ξ (a very small number to prevent any division by zero in the implementation) were set to 1e−3, 0.9, 0.99, 1e−8, respectively.

4.4 Results

We report the identification results on the testing dataset in Table 4. As the distance between the camera and individuals increases, the resolution of the individual's critical regions in the image gradually becomes smaller. Although the precision and recall rate of NHU identification gradually decreased, the precision and recall rate remained higher than 80% and 81%, respectively. For NMU identification, the precision and recall rate declined only slightly from 1 m to 3 m, and even the masks were quite small in far-field (5 m) images, our approach still achieved precision and recall rates of approximately 80%. For NVU identification, the precision and recall rate between 1 m to 3 m remained higher than 85% and 97%, respectively. However, the performance is decreased in the 5 m case since it is challenging to detect transparent visors with relatively low resolution. The overall precision and recall rates are 84.13% and 93.10%, respectively, which demonstrates the robustness of the proposed approach in PPE wearing identification at different distances. Figure 4 illustrates the identification examples on the testing dataset.

Table 4. Identification results under different distance.

NPU Categories	Distance (m)	TP	FP	FN	Precision (%)	Recall (%)
NHU	1	318	13	0	96.07	100
	3	241	31	22	88.60	91.64
	5	222	54	52	80.43	81.02
NMU	1	247	28	28	89.82	89.82
	3	250	40	28	86.21	89.93
	5	225	58	63	79.51	78.13
NVU	1	325	47	9	87.37	97.31
	3	452	75	0	85.77	100
	5	444	168	0	72.55	100
Overall		2724	514	202	84.13	93.10

(a) Distance: 1m (b) Distance: 3m (c) Distance: 5m

Fig. 4. Identification examples on the testing dataset.

4.5 Computational Efficiency Analysis

To meet the industrial requirements of real-time processing, we also conducted computational efficiency analysis experiments. Computational efficiency analysis results are presented in Table 5, the inference time of our approach outperforms other state-of-the-art methods while preserving high-quality results. It is able to run at about 7.95 FPS in a machine with a GeForce GTX 1080 Max-Q with 8 GB of GDDR5X memory and 2560 CUDA cores and it indicates that our approach is more effective compared to the Faster R-CNN approach adopted by Fang et al. [15] and SSD-RPA method proposed by Wu et al. [16].

Table 5. Computational efficiency analysis results.

Approach	Input size	FPS
Faster R-CNN [13]	300 × 500	4.88
SSD-RPA [14]	304 × 304	3.22
Ours	416 × 416	7.95

5 Conclusion

This paper has presented a new vision-based approach to address the difficulties of PPE proper use management in the construction sites. Firstly, we created a dataset using Internet images and real-world images to train the YOLOv3 model to recognize hardhats, eye protection visors, and dust masks. Subsequently, we conducted NPU identification using geometric relationships of the outputs of OpenPose and YOLOv3. The performance of the proposed approach was experimentally evaluated under various distance conditions. The test results indicate that the approach was capable of detecting NPU workers with high precision (84.13%) and recall rate (93.10%) while ensuring real-time performance (7.95 FPS on average). Further extensions of this work follow the consideration of on-site system implementation and performance improvement based on time-series analysis.

Acknowledgements. This work was supported by Japan Society for the Promotion of Science (JSPS) Grants-in-Aid for Scientific Research Grant No. 19K05324.

References

1. U.S. Bureau of Labor Statistics, Construction: NAICS 23. https://www.bls.gov/iag/tgs/iag23.htm. Accessed 13 Jan 2020
2. The Japanese Ministry of Health, Welfare, The 13th occupational safety health program. https://www.mhlw.go.jp/content/11200000/000341159.pdf. Accessed 13 Jan 2020
3. Konda, S., Tiesman, H.M., Reichard, A.A.: Fatal traumatic brain injuries in the construction industry, 2003–2010. Am. J. Ind. Med. **59**(3), 212–220 (2016)

4. U.S. Department of Labor, Occupational Safety and Health Administration, Head protection. https://www.osha.gov/laws-regs/regulations/standardnumber/1926/1926.100. Accessed 26 Jan 2020
5. The National Institute for Occupational Safety and Health, Eye safety. https://www.cdc.gov/niosh/topics/eye/. Accessed 13 Jan 2020
6. Dannenberg, A.L., Parver, L.M., Brechner, R.J., Khoo, L.: Penetrating eye injuries in the workplace: the national eye trauma system registry. Arch. Ophthalmol. **110**(6), 843–848 (1992)
7. Government of Western Australia, Department of Commerce, Guide to using dust masks in construction work (2019). https://www.commerce.wa.gov.au/sites/default/files/atoms/files/guide_to_using_dust_mask.pdf
8. U.S. Department of Labor, Occupational Safety and Health Administration, Eye and face protection. https://www.osha.gov/laws-regs/regulations/standardnumber/1910/1910.133. Accessed 13 Jan 2020
9. Redmon, J., Farhadi, A.: YOLOv3: an incremental improvement, arXiv preprint arXiv:1804.02767
10. Cao, Z., Hidalgo, G., Simon, T., Wei, S.E., Sheikh, Y.: Openpose: realtime multi-person 2D pose estimation using part affinity fields, arXiv preprint arXiv:1812.08008
11. Kelm, A., Laußat, L., Meins-Becker, A., Platz, D., Khazaee, M.J., Costin, A.M., Helmus, M., Teizer, J.: Mobile passive Radio Frequency Identification (RFID) portal for automated and rapid control of Personal Protective Equipment (PPE) on construction sites. Autom. Construct. **36**, 38–52 (2013)
12. Dong, S., He, Q., Li, H., Yin, Q.: Automated PPE misuse identification and assessment for safety performance enhancement. In ICCREM 2015, pp. 204–214 (2015)
13. Shrestha, K., Shrestha, P.P., Bajracharya, D., Yfantis, E.A.: Hard-hat detection for construction safety visualization. Journal of Construction Engineering **2015**, 1–8 (2015)
14. Park, M.W., Elsafty, N., Zhu, Z.: Hardhat-wearing detection for enhancing on-site safety of construction workers. J. Construct. Eng. Manage. **141**(9), 04015024 (2015)
15. Fang, Q., Li, H., Luo, X., Ding, L., Luo, H., Rose, T.M., An, W.: Detecting non-hardhat-use by a deep learning method from far-field surveillance videos. Autom. Construction **85**, 1–9 (2018)
16. Wu, J., Cai, N., Chen, W., Wang, H., Wang, G.: Automatic detection of hardhats worn by construction personnel: a deep learning approach and benchmark dataset. Autom. Construction **106**, 102894 (2019)
17. Lin, T.Y., Maire, M., Belongie, S., Hays, J., Perona, P., Ramanan, D., Dollár, P., Zitnick, C.L.: Microsoft coco: common objects in context. In: European Conference on Computer Vision, pp. 740–755. Springer, Cham (2014)
18. Sandler, M., Howard, A., Zhu, M., Zhmoginov, A., Chen, L.C.: MobileNetV2: inverted residuals and linear bottlenecks. In: Proceedings of the IEEE Conference on Computer Vision and Pattern Recognition, pp. 4510–4520 (2018)
19. Simonyan, K., Zisserman, A.: Very deep convolutional networks for large-scale image recognition (2014). arXiv preprint arXiv:1409.1556
20. Yang, S., Luo, P., Loy, C.C., Tang, X.: Wider face: a face detection benchmark. In: Proceedings of the IEEE Conference on Computer Vision and Pattern Recognition, pp. 5525–5533 (2016)
21. Russakovsky, O., Deng, J., Su, H., Krause, J., Satheesh, S., Ma, S., Huang, Z., Karpathy, A., Khosla, A., Bernstein, M., Berg, A.C.: Imagenet large scale visual recognition challenge. Int. J. Comput. Vis. **115**(3), 211–252 (2015)

Vision-Based Pose Forecasting of Construction Equipment for Monitoring Construction Site Safety

Han Luo[1], Mingzhu Wang[2], Peter Kok-Yiu Wong[1], Jingyuan Tang[1], and Jack C. P. Cheng[1(✉)]

[1] The Hong Kong University of Science and Technology, Hong Kong SAR, China
{han.luo,kywongaz,jtangap}@connect.ust.hk, cejcheng@ust.hk

[2] Imperial College London, South Kensington, London, UK
m.wang@imperial.ac.uk

Abstract. Construction sites suffer from higher hazard rates compared to other occupational workplaces, which can be attributed to dynamic activities of construction equipment. Hence, it is important to track locations, poses, and movements of construction equipment for on-site safety monitoring. With the wide installations of surveillance cameras, previous studies focused on tracking locations of construction equipment from videos using computer vision techniques. However, the location of many construction equipment remains unchanged during operation while the equipment poses are varying constantly. The variation of equipment poses can cause severe hazards, such as the collision with surrounding workers and other equipment. To avoid such potential hazards, it is important to monitor and forecast the poses of equipment. So far, there is limited research on automatically forecasting poses of construction equipment, which is necessary to provide hazard alerts and prevent spatial conflicts. Therefore, a vision-based pipeline is proposed in this paper to automatically forecast dynamic poses of construction equipment based on historical on-site surveillance videos. The proposed pipeline firstly utilizes a deep learning -based equipment pose estimation model to estimate poses of construction equipment so as to obtain historical poses of construction equipment. Then, one type of recurrent neural network (RNN), namely Gated Recurrent Unit (GRU), is adopted to learn the temporal features of the generated sequential poses and forecast potential poses of construction equipment. To validate the proposed method, a dataset containing equipment keypoint-based poses is created and annotated for training the model. The experiment results based on our created dataset demonstrate the capability of the proposed pipeline.

Keywords: Convolutional neural network (CNN) · Computer vision · Pose prediction · Pose forecasting · Construction equipment · Construction safety · Deep learning

E. Toledo Santos and S. Scheer (Eds.): ICCCBE 2020, LNCE 98, pp. 1127–1138, 2021.
https://doi.org/10.1007/978-3-030-51295-8_78

1 Background

The construction industry is one of the most labor-intensive and equipment-intensive industry. Consequently, it is extremely significant to ensure safety of construction projects. However, construction sites have very high hazard rates among all workplaces, and thereby improving safety of construction sites is a major concern [1]. Statistics suggests that unsafe operation of construction equipment is an essential reason of fatal hazards occurred on construction sites. In the U.S., over 38% of construction accidents are caused by interactions between construction-related objects (e.g. workers and equipment). Hence, it is important to monitor movements of construction equipment on sites.

Up till now, most site managers monitor the working states of construction equipment by watching surveillance videos and evaluating potential risks manually. Such manual observations of onsite safety condition are error-prone and time-consuming because they are very dependent on physical status and expertise of the inspector. With wide installations of surveillance cameras, researchers have attempted to apply computer vision techniques for processing onsite videos and monitor construction sites automatically. For example, background subtraction, edge detection and convolutional neural networks (CNN) have been adopted for detecting deformable parts of machines and tracking locations of construction equipment [2–4]. Most of those studies only attempt to locate and identify construction equipment, while research on tracking equipment poses or more precise movements is still lacking. As heavy machines (e.g. excavators and cranes) seldom change their location frequently in the working process, it is vital to monitor their poses which varying constantly when deformable components of construction equipment are operated.

To track the equipment poses, some papers have proposed methods for estimating poses of construction equipment from videos, recently. Soltani et al. [2] extracted two-dimensional poses of construction equipment based on a part-based pattern matching method, and then they extended equipment pose estimation into three dimensions through stereo visions constructed by two RGB cameras. Besides, Liang et al. [5] and Luo et al. [6] have utilized deep learning techniques to estimate partial or full-body poses of excavators working on construction sites under complex working conditions. It should be noted that those efforts only focused on estimating the captured poses of construction equipment which has occurred. With the aim to avoid potential hazards, practitioners are more concerned about the variation and future poses of equipment on construction sites, which can help evaluate the safety conditions in a future time frame. For the sake of safety monitoring and hazard prevention, understanding the current poses of construction machines is not sufficient. Effective approaches should be developed for forecasting future poses of construction equipment.

To address the limitations of previous studies, the objective of this paper is to forecast the poses of construction equipment such as to facilitate the on-site safety monitoring. Therefore, this paper proposed a pipeline for automatically forecasting future poses of construction equipment in consecutive video frames based on historical keypoint-based poses using computer vision and deep learning techniques. In this study, a time-series keypoint-based equipment pose dataset is built with 8,000 consecutive frames of one on-site video. In the video, there are five different actions performed by an excavator, and keypoints have different visibility conditions in those video frames. With the dataset available, the pipeline is capable of processing a number of preceding frames to estimate poses of equipment, after which it can forecast the poses in the following frames based on the historical time-series information.

The rest of this paper is outlined as follows. Section 2 explains the methodology of this paper in detail, which includes descriptions of the proposed pipeline for forecasting construction equipment poses, model training strategies and evaluation metrics. Section 3 presents experiments conducted using an excavator as the example and discussion about the experiment results. In the end, conclusion is stated in Sect. 4.

2 Methodology

Aiming at addressing previous limitations, a pipeline is developed in this paper to forecast poses of construction equipment based on the keypoint definition by [6] using computer vision and deep learning techniques. The following sections introduce the proposed pose forecasting method, the model training details and the evaluation metrics respectively.

2.1 The Proposed Pipeline for Forecasting Construction Equipment Poses

The proposed pipeline of forecasting poses of construction equipment is shown in Fig. 1, which mainly contains two modules. The first module is a pose estimation model which estimates equipment poses from the input video frames using computer vision and deep learning techniques and outputs keypoint feature vectors. Then, the generated keypoint feature vectors are used as the input of the second module, which extracts the temporal features of historical keypoint-based equipment poses and finally forecasts the future poses of equipment. More details are illustrated below.

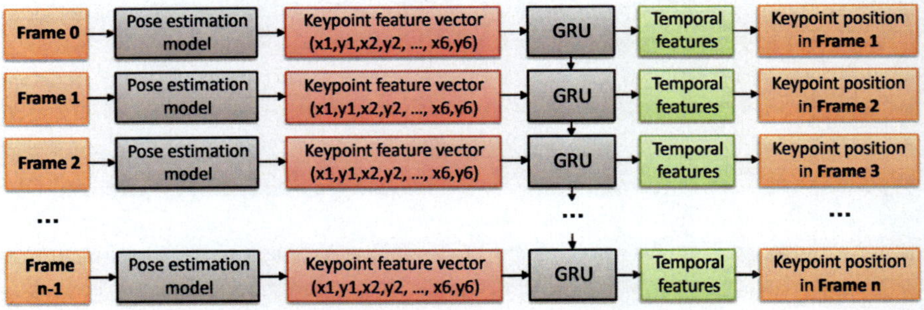

Fig. 1. The framework of the proposed pose forecasting model.

Pose Estimation Model. Apose estimation model is designed to automatically estimate the poses of construction equipment from surveillance videos or images based on keypoints. For estimating equipment poses, the method based on computer vision and deep learning techniques has achieved great performances in our previous work [6]. Among the three algorithms compared in that work, Cascaded Pyramid Network (CPN) is chosen as the basis of this paper considering the trade-off of three model performance evaluation metrics, i.e. accuracy, training time and detection speed. CPN is a state-of-the art deep learning architecture for 2D human pose estimation and also obtains ideal performances when estimating equipment poses from videos and images captured from various construction sites. The CPN model firstly utilizes a GlobalNet to extract coarse visual features from surveillance video frames by a U-shape structure. Then, a RefineNet is attached to the GlobalNet so as to improve the localization results of equipment keypoints. In the end, CPN obtains a keypoint feature vector, which contains x and y coordinates of all equipment keypoints in the input video frame.

Gated Recurrent Unit (GRU) Model. Gated recurrent unit model [7] is a type of recurrent neural network (RNN), which is used for processing time-series data and learning complex dynamic temporal features from the input sequences. Compared to traditional RNN, GRU tries to solve the problems of long-term memory and gradients in backpropagation. The main reason to apply GRU in this work is that GRU demonstrates good performances similar to long-short term memory (LSTM), but requires less computational power with higher efficiency. Within GRU, there are two gates to control and update the state of the GRU model, i.e. update gate and reset gate. These two gates build a mechanism to determine which part of the input should be kept or abandoned. Based on this mechanism, GRU performs well in time-series related tasks.

In the proposed pose forecasting pipeline, the sequence of a number of preceding keypoint feature vectors is the input of GRU model. After processing by the GRU mechanism, the forecasting result of equipment poses in the future frames will be generated. In this paper, one layer of bidirectional GRU with hidden size of 128 as the RNN cells is used to more precisely extract temporal features in both forward and backward directions. To output the final forecasting result of the equipment poses, two fully connected layers are utilized following the GRU model. One of the fully connected layers has a hidden size of 128 followed by a ReLU nonlinearity layer. The other has a hidden size of 12, which directly exports the coordinates of the future poses of construction equipment.

2.2 Model Training

When training the pose forecasting model based on equipment keypoints, a sequence of keypoint coordinates of the past poses are fed into the recurrent neural network and the forecasting results of future poses are generated through the forward inference. Then, the L2 – norm loss function (as defined in Eq. (1)) is adopted as the loss function to calculate the training loss, which is fed back into the network through back-propagation for optimizing the model weights.

$$L_2 = \sum_i (Y_i - f(x_i))^2 \tag{1}$$

where Y_i is the ground-truth coordinates, x_i is the input information, f is the functions included in the deep learning model, and $f(x_i)$ means the forecasted keypoint coordinates of construction equipment of the i_{th} sample. The L2 – norm loss computes the error between forecasting coordinates of construction equipment and the actual values.

The objective of training deep learning models is to minimize the L2 – norm loss by calculating the gradient with respect to the model parameters, the process of which is called gradient descent. During the training process, a learning rate controls the speed of the gradient descent, such that the model can converge efficiently. In this study, Adam algorithm [8] is used as the optimizer for optimizing the L2 – norm loss. Adam algorithm is a popular algorithm for deep learning because it requires less computation power and it is convenient to be implemented. Compared to other gradient descent algorithms that update weights of deep learning models using a fixed learning rate, the learning rate adopted by Adam algorithm is more adaptive for different parameters during the training process. For the sake of improving the efficiency of the optimizer, min-max normalization is adopted to normalize the input data into the interval of [0,1].

2.3 Model Evaluation

Two metrics are adopted to evaluate the performance of the pose forecasting model based on equipment keypoints, i.e. normalized error (NE) and percentage of correct keypoints (PCK). When calculating NE and PCK, we require a distance normalization parameter since bounding boxes of the detected equipment will have different sizes in video frames. The distance normalization parameter (s_m) is computed according to Eq. (2).

$$s_m = \max(h, w) \tag{2}$$

where h and w are the height and width of the bounding box of an excavator in the mth frame, respectively.

Normalized Error (NE). Normalized error (NE) is the average normalized distance between the forecasted keypoints and the actual values [9]. Only the keypoints included in the input frame (i.e. $v \geq 0$) are counted. A smaller NE suggests a better model performance. The calculation of NE is shown in Eq (3).

$$NE = \frac{1}{M}\sum_{m=1}^{M}\frac{\sum_{k=1}^{K}\left(\frac{d_{m,k}}{s_m}\cdot\delta\left(v_{m,k}\geq 0\right)\right)}{\sum_{k=1}^{K}\delta\left(v_{m,k}\geq 0\right)} \tag{3}$$

where M is the total number of test images, K is the total number of a piece of equipment keypoints, $d_{m,k}$ is the Euclidean distance between the predicted position and the ground-truth label of the kth keypoint in the mth image, s_m is the distance normalization parameter in the mth image (see Eq. (2)), and $v_{m,k}$ means the visibility of the kth keypoint in the mth image.

Percentage of Correct Keypoints (PCK). PCK evaluates the accuracy by measuring the percentage of keypoints correctly forecasted over the total amount of keypoints in one category or the whole annotation set [10]. As shown in Eq. (4), a candidate keypoint is assumed correctly located if its normalized error falls within a given threshold α, where $\alpha \in [0, 1]$. Being consistent with [6], $\alpha = 0.05$ is adopted as the upper bound value so as to assess the performance of the proposed pipeline more precisely.

$$PCK(\alpha) = \frac{1}{M}\sum_{m=1}^{M} NE_m < \alpha, \alpha \in [0, 0.05] \tag{4}$$

where α is the threshold for calculating and plotting the PCK curves, and the definition of other parameters are illustrated in Eq. (3).

3 Experiment

The objective of the proposed pipeline is to automatically forecast the poses of construction equipment using computer vision techniques and a recurrent neural network (RNN), namely Gated Recurrent Unit (GRU) through predicting keypoints positions. Our previous work already demonstrates the performance of different models for estimating poses of construction equipment through experiments [6], which is the main objective of the first module. Therefore, in this paper, we only focus on the experiment of the second module which is to forecast poses of construction equipment using GRU based on keypoint feature vectors. Experiments are conducted using video frames of working excavators as the illustrative example.

3.1 Experiment Setup

Dataset. To evaluate the performance of the proposed method for equipment pose forecasting based on keypoints, a time-series equipment keypoint dataset is created on the basis of the construction site video published by [11]. Originally, there are bounding box annotations for each equipment shown in the video. To create the dataset for keypoint-based equipment pose forecasting, each excavator is labeled with six keypoints following our previous work [6]. In the time-series equipment keypoint dataset, there are 8,000 consecutive frames labeled with bounding boxes and six keypoints, 80% of which are used as the training set for training the proposed equipment pose forecasting model, 10% are used as the validation set for choosing a model with the highest accuracy, and the rest 10% are used as the test set for evaluating the model performance. In the experiments, during each training iteration, a time window of 50 consecutive frames is used to aggregate historical poses for forecasting. For example, the past 50 consecutive frames are used as the input sequence for the model, after which the equipment poses of the 51st frame will be forecasted (Fig. 2).

Fig. 2. An example of our dataset.

Experiment Environment. The experiment is conducted based on Pytorch, which provides libraries for constructing deep learning models. The experiments environment is Ubuntu 16.04 operating system with an Intel Core i7-6700 CPU @ 3.40 GHz × 8, a NVIDIA GeForce GTX 1080 GPU.

3.2 Experiment Result and Discussion

Loss Curves. Loss curves shows the change of the training loss and validation loss when the model keeps updating, which indicates the convergence and reliability of a trained model. In our experiment, the curves of the training loss and the validation loss are shown in Fig. 3. It is obvious that both the training loss and the validation loss decrease dramatically at the beginning of the training process, after which they slowly achieve a similar value without intensive fluctuation in the end. There is no obvious gap between the training loss and validation loss, which indicates that no over-fitting problem of the model is observed.

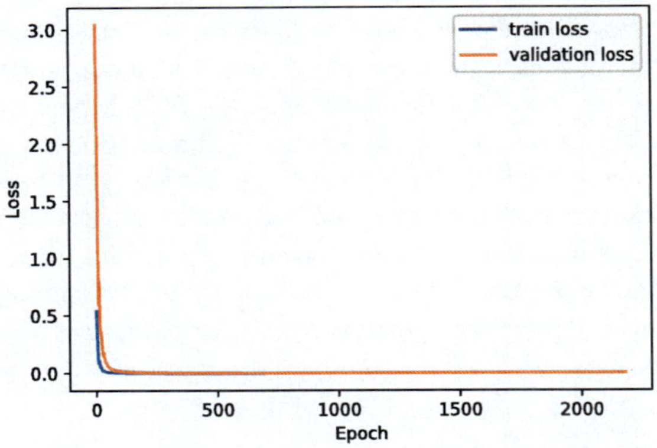

Fig. 3. Curves of training loss and validation loss.

Quantitative Evaluation. The proposed method is preliminarily evaluated by normalized error and percentage of correct keypoints.

Normalized Error (NE). NE is the first metric to evaluate the accuracy of the proposed equipment pose forecasting model. A smaller NE value indicates a higher accuracy of the model. In our experiment, the average NE value of the proposed equipment pose forecasting model is 8.315×10^{-3} when the model is tested in the proposed time-series equipment pose dataset. It is observed that normalized errors have huge differences among different equipment keypoints. As shown in Table 1, most of the NE values are smaller than 0.01. Cab_boom keypoint achieves the lowest NE value of 0.764×10^{-3}, while the worst forecasting accuracy shows in the Right_bucket_end keypoint with a normalized error of 14.204×10^{-3}. In particular, Left_bucket_end keypoint has a NE value of 14.061×10^{-3} which is very close to that of Right_bucket_end keypoint.

Table 1. The normalized error (NE) results of the proposed method

Keypoint	Normalized Error ($*10^{-3}$)
Body_end	4.712
Cab_boom	**0.764**
Boom_arm	7.349
Arm_bucket	8.799
Left_bucket_end	14.061
Right_bucket_end	**14.204**
Average	8.3148

Percentage of Correct Keypoint (PCK). Besides NE, PCK is another metric for evaluating the model accuracy. Higher PCK value means higher model accuracy. The PCK curves of our experiment are plotted in Fig. 4 with an increasing threshold $\alpha(\alpha \in [0, 0.05])$ to demonstrate the percentage of keypoints that are correctly forecasted by the proposed equipment pose forecasting model. Figure 4 illustrates that most of the PCK values increases when the threshold α changes from 0 towards 0.05. PCK values of most keypoints have significant increment when $\alpha \in [0, 0.04]$, and those PCK curves reach a plateau when α exceeds 0.04. Eventually, the PCK curves also prove that the proposed model performs the best when forecasting future positions of the Cab_boom keypoint since its curve is the highest among all the curves. Apart from the PCK curve of the Cab_boom keypoint, the curves of the Body_end is always higher than the PCK curves of other keypoints. Similar to the conclusion based on the normalized errors, Fig. 4 suggests that the positions of two keypoint at a bucket's end could be forecasted with lower accuracies than other keypoints since their curves are significantly lower.

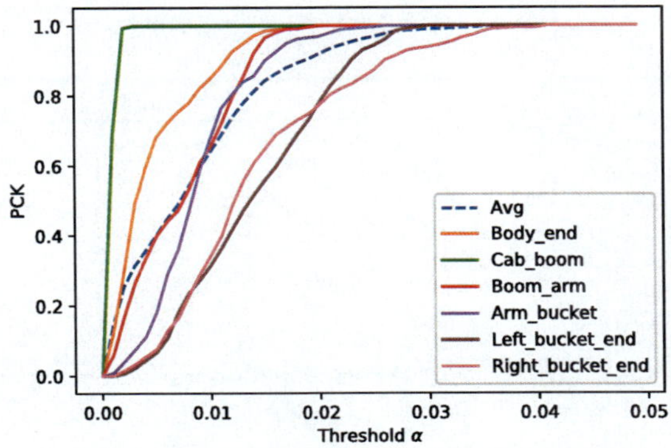

Fig. 4. PCK curves of pose forecasting results.

The evaluation based on both NE and PCK points out that the proposed model has the highest accuracy for forecasting the Cab_boom keypoint, whereas the forecasting accuracies of two keypoints at bucket's end are much worse. After investigating the characteristics of the dataset, it is observed that the Cab_boom keypoint is usually visible with a smaller range of movement. However, the Left_bucket_end and the Right_bucket_end are most likely to be occluded when an excavator is working. The occlusion can arise from their own body and the working environment (e.g. the soil and other on-site objects). To strengthen the robustness of the proposed method, more videos should be collected from various construction sites to enrich the dataset. Particularly, adding more images with occluded keypoints to the training dataset may yield a more robust model upon testing.

Although limitations remain, the quantitative results based on NE and PCK indicate that most keypoints of an excavator can be correctly forecasted with small errors. Therefore, this method has high potential of forecasting poses of construction equipment.

Qualitative Evaluation. Qualitative evaluation is provided by comparing the differences of the actual coordinates (in orange color) of keypoints and the forecasted coordinates (in blue color) of keypoints in 1,600 consecutive frames. As shown in Table 2, the comparison between the real and the forecasted values of the X and Y coordinates is listed respectively for the six keypoints of the excavator. Obviously, the forecasted coordinates and the real values have very similar trends even when the excavator performs various activities during the working process. Such results indicate that the proposed model has the ability to accurately forecast the future poses of construction equipment. There are, however, still some discrepancies between the real and the forecasted values, e.g. the curves of the Cab_boom keypoint. It is observed that larger differences are more likely to occur when a keypoint stays in the same location or only has slight movements for a period of time, but the movement trends of the two curves remain the same. Such phenomenon may result from cumulative error, which need to be further investigated in the future.

Table 2. Keypoint forecasting results.

Keypoint	X coordinate	Y coordinate
Body_end	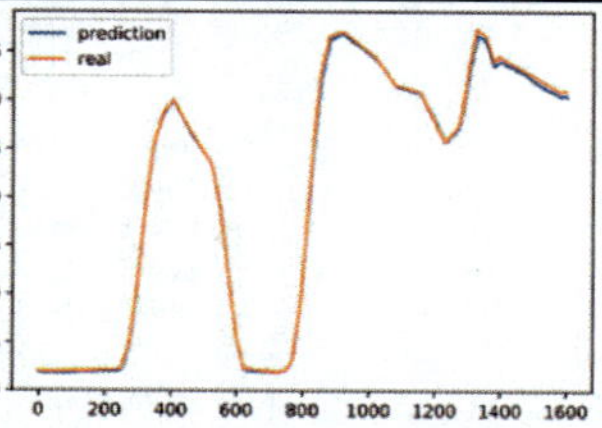	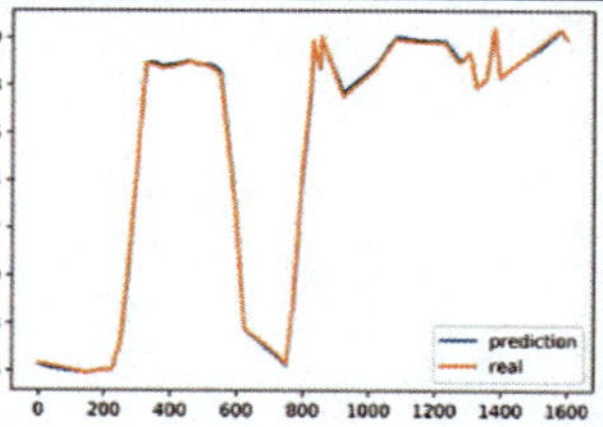
Cab_boom	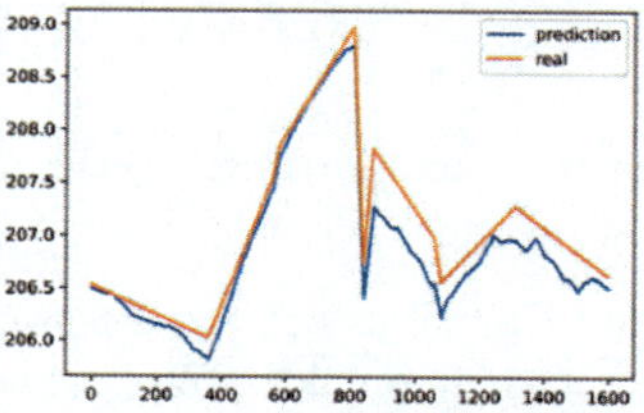	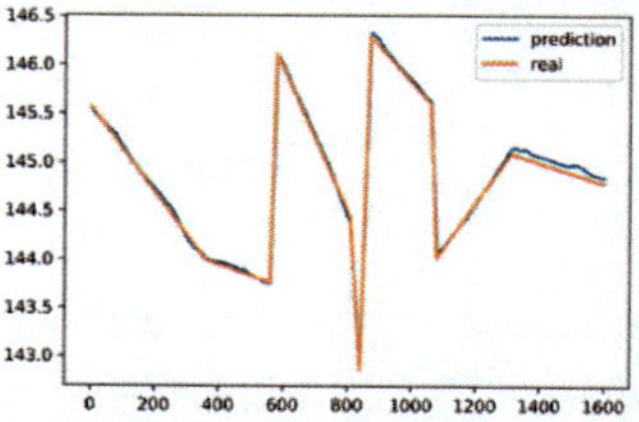
Boom_arm	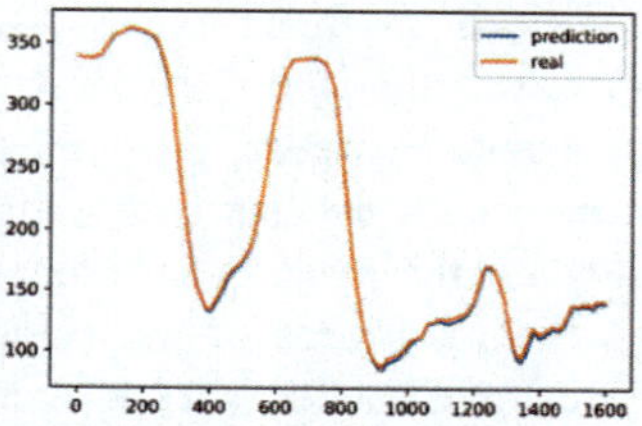	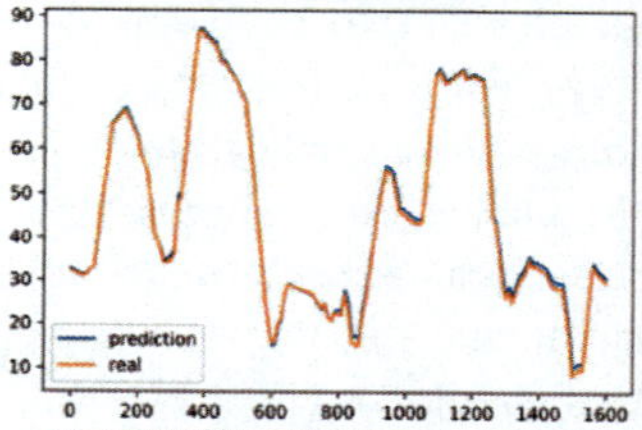
Arm_bucket	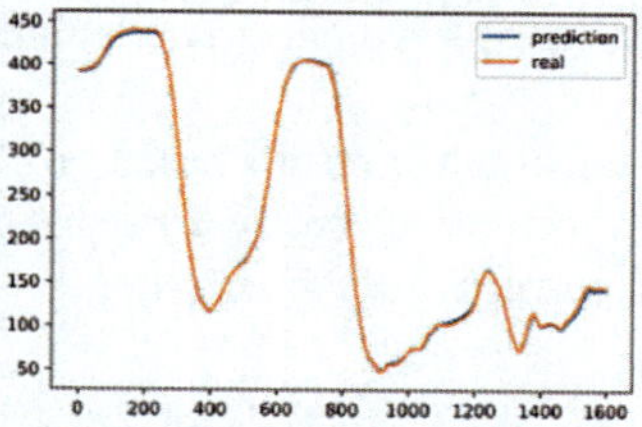	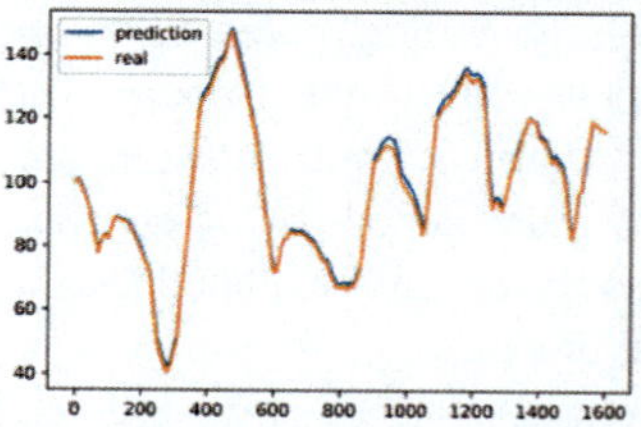
Left_bucket_end	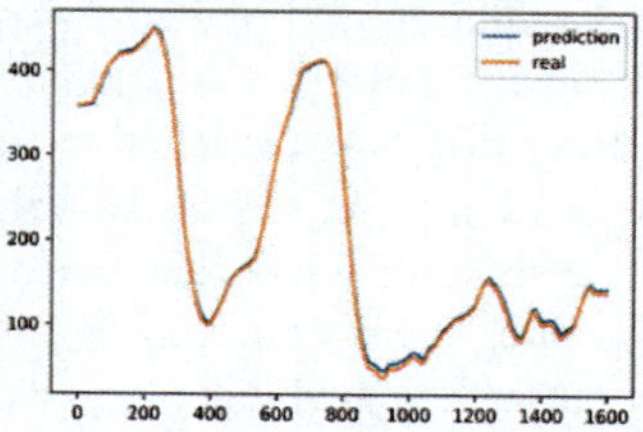	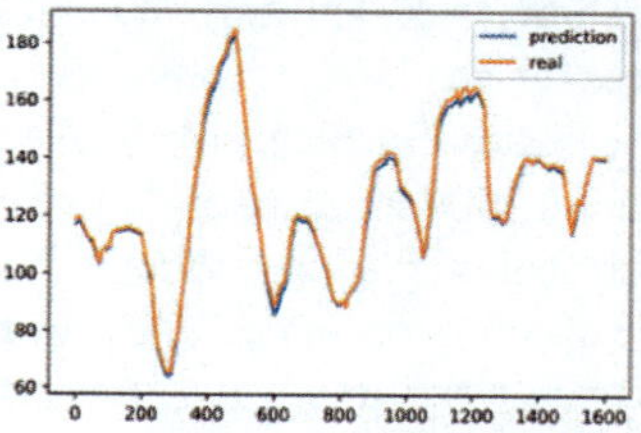
Right_bucket_end	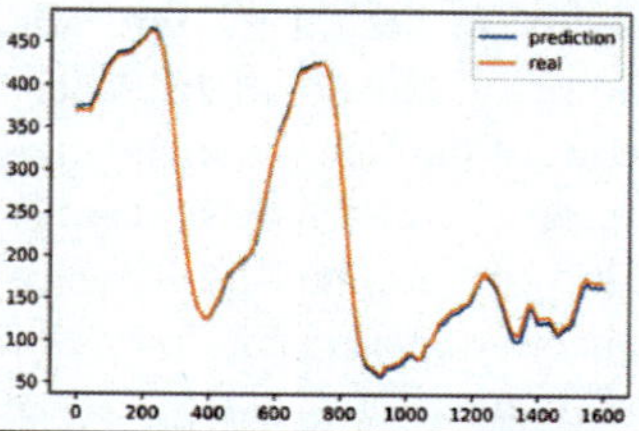	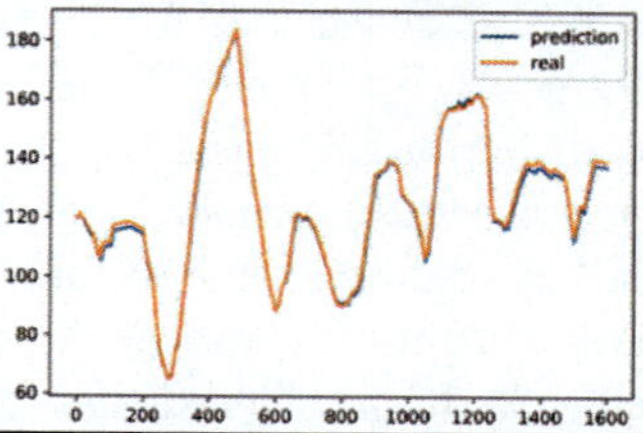

4 Conclusion

Safety is the first priority of all construction projects. Construction equipment is a major source of safety issues on construction sites due to their frequent interactions with workers and other construction-related objects. Therefore, it is necessary to monitor states (i.e. locations, poses and movements) of construction equipment for avoiding potential collisions and other accidents. Besides tracking the past and the current states of construction equipment, evaluating the future states of equipment should also be emphasized such that site managers can have a better understanding of the onsite safety conditions for preventing potential hazards. With many surveillance cameras installed, videos of construction sites are available and accumulated. Although computer vision and deep learning techniques have achieved good performances in multiple tasks (e.g. on-site object detection, object tracking and time-series action classification), few studies focus on analyzing the pose variation and movement of equipment, which can facilitate the inspectors to monitor the past, current and future states of construction equipment.

In this paper, a vision-based pipeline is proposed for automatically forecasting the future poses of construction equipment based on historical video frames captured from surveillance cameras on construction sites. The proposed pipeline consists of two modules. The first module is a pose estimation model which estimates the past poses from historical video frames. The second module is a recurrent neural network, i.e. Gated Recurrent Unit (GRU), which is attached to the first module and forecasts the future poses of construction equipment through spatiotemporal analysis of the past poses estimated by the first module. Based on the RGB videos annotated with bounding boxes and equipment keypoints, the proposed pipeline can forecast the full body poses of a given construction equipment based on the features of past poses. As our previous work already elaborated the performances of pose estimation models, the experiment in this paper only concentrates on the results of the second module, which is to forecast future poses of construction equipment using GRU.

The performance is evaluated by normalized error (NE) and percentage of correct keypoint (PCK). Experiments were conducted using our created time-series equipment pose dataset, which includes 8,000 consecutive frames from on-site videos. The experiment results indicate that most future poses of an excavator can be forecasted with high accuracies. Particularly, the best performance goes to Cab_boom keypoint, with NE value of 0.764×10^{-3} and significantly higher PCK than other keypoints. A possible reason is that in the video Cab_boom keypoint is clearly captured and with a small range of movement. On the other hand, accuracies for the left and right bucket_end keypoints are sensitive to occlusion by the excavator itself or surrounding objects. It is inferred that an enriched dataset including more images with occluded keypoints would further enhance the robustness of our forecasting model. In summary, the proposed method has high potential of forecasting future poses of construction equipment. Most importantly, the proposed method lays the foundations for providing early warnings of potential hazards on construction sites, which allows site managers to prevent hazards with forward thinking.

References

1. U.S. Bureau of Labor Statistics (BLS) National Census of Fatal Occupational Injuries in 2017 (2018)
2. Soltani, M.M., Zhu, Z., Hammad, A.: Skeleton estimation of excavator by detecting its parts. Autom. Construction **82**, 1–15 (2017). https://doi.org/10.1016/j.autcon.2017.06.023
3. Mulligan, I.J., Mackworth, A.K., Lawrence, P.D.: A model-based vision system for manipulator position sensing. In: Proceedings of Workshop on Interpretation of 3D Scenes. IEEE Comput. Soc. Press, pp 186–19 (1989)
4. Rezazadeh Azar, E., McCabe, B.: Part based model and spatial-temporal reasoning to recognize hydraulic excavators in construction images and videos. Autom. Construction **24**, 194–202 (2012). https://doi.org/10.1016/J.AUTCON.2012.03.003
5. Liang, C.-J., Lundeen, K.M., McGee, W., et al.: Stacked hourglass networks for markerless pose estimation of articulated construction robots. In: Proceedings of the 35th International Symposium on Automation and Robotics in Construction (ISARC) (2018)
6. Luo, H., Wang, M., Wong, P.K.Y., Cheng, J.C.P.: Full body pose estimation of construction equipment using computer vision and deep learning techniques. Autom. Construction **110**, 1–41 (2020). https://doi.org/10.1016/j.autcon.2019.103016
7. Cho, K, Van Merriënboer, B, Gulcehre, C, et al.: Learning phrase representations using RNN encoder-decoder for statistical machine translation. In: Proceedings of the Conference on Empirical Methods in Natural Language Processing. Association for Computational Linguistics (ACL), EMNLP 2014. pp 1724–1734 (2014)
8. Kingma, D.P., Ba, J.: Adam: a method for stochastic optimization. In: the 3rd International Conference for Learning Representations. San Diego (2014)
9. The Alibaba Group, the Institute of Textile and Clothing of The Hong Kong Polytechnic University. FashionAI Apparel Keypoints Dataset (2018). http://fashionai.alibaba.com
10. Andriluka, M., Pishchulin, L., Gehler, P., Schiele, B.: 2D human pose estimation: new benchmark and state of the art analysis. In: 2014 IEEE Conference on Computer Vision and Pattern Recognition, pp 3686–3693. IEEE (2014)
11. Roberts, D., Golparvar-Fard, M.: End-to-end vision-based detection, tracking and activity analysis of earthmoving equipment filmed at ground level. Autom. Construction **105**, 102811 (2019). https://doi.org/10.1016/J.AUTCON.2019.04.006

Image Sensing-Based In-Building Human Demand Estimation for Installation of Automated External Defibrillators

Wen-Xin Qiu, Albert Y. Chen(✉), and Tsung-Yin Hsieh

National Taiwan University, Taipei 10617, Taiwan
{r07521505, albertchen, r08521507}@ntu.edu.tw

Abstract. The installation of automated external defibrillators (AEDs) in public places is encouraged, and the determination of the installation locations could be formed into mathematical programming models. In previous studies, data and parameters for those models were generated experimentally or manually. To estimate the parameters in practice for better reliability of the model outcomes, this research aims to design a procedure for the estimation of the real human demand distribution through automatic approaches. Inputs of the proposed work include the image sequences taken inside the building and the network describing the in-building geometry. Images are analyzed via computer vision methods for detection and tracking of people, and also augmented with the extra dimension of image depth estimation through a deep convolutional neural network. The key indicator to project the image back to the real space is identified by the correspondence between the location of humans appear in the image and the location of intersections of camera field of view with the network. Finally, the spatial-temporal human count is recorded to be the output and could serve as input parameter for related AED installation problems. The process is validated in image sequences taken at a corridor in a school building; the performance of each step and the influence of each step is discussed. In conclusion, our research provides a more automatic and robust approach to obtain the real spatial-temporal human distribution in public buildings.

Keywords: Automated external defibrillator (AED) · In-building human distribution · Computer vision (CV) · Human detection and tracking · Depth estimation

1 Introduction

An automated external defibrillator (AED) is an emergency medical device for the treatment of out-of-hospital cardiac arrest. Due to its effectiveness to provide better survival rate, public access defibrillation programs were proposed and the placement of AEDs in public places have been suggested by international guidelines [1–3]. The American Heart Association [4] has concluded a median survival rate of 53.0% of patients treated by public placed AEDs by lay first responders. While the evidence has shown the importance of the response time for defibrillation, the guidelines do not provide the requirements for appropriate locations of AEDs. In previous studies, the

E. Toledo Santos and S. Scheer (Eds.): ICCCBE 2020, LNCE 98, pp. 1139–1151, 2021.
https://doi.org/10.1007/978-3-030-51295-8_79

optimal location of such facilities are formulated into mathematical programming models [5, 6]. The optimal location could either maximize the coverage of demand of AEDs or minimize the response distance. However, these models assume the distribution of AED demand, which are critical parameters in these models, are known or assumed to follow particular spatial distribution. To enable those models to be practical, this work aims to design a process that estimates the real demand. The spatial-temporal human distribution with automatic approaches are to be analyzed, particularly for buildings equipped with cameras.

For in-building AED locations, Lee et al. [7] focused on the generation of the network from a Building Information Model (BIM) and mixed integer programs for optimal decisions, while Chen et al. [8] proposed the collection of data by human detection in videos. Both studies indicated the necessity of an automatic demand estimation.

The objective of this study is to accumulate the spatial-temporal distribution in buildings with camera installed. Though it could be done by manually counting on-site or from video records as well as installation of counting sensors, such as infrared sensors, in each space interested, we hope to establish an approach with minimal infrastructure requirements. Image sequences taken from surveillance cameras are input to our approach which consists of computer vision and data mining methods.

Computer vision methods provide the recognition of features in the image. In this study, detection and tracking [9–11] are used to trace the humans, and to determine the locations humans appear and disappear in the image. Detect-and-track [9] is a state-of-the-art deep learning model for this purpose, whereas deepSORT [11] is a stronger tracking model and it is combined with a real-time detection model – YOLOv3 [10]. As for finding the correspondence between the image space and the real space, the homography (or perspective transformation) [12] was tested but abandoned due to its strong requirement of knowing exact coordinates of 4 spatial locations for each camera. Depth from the camera [12] is considered as a critical piece of information. A deep learning model, rather than range sensors, is adopted in our process for less infrastructure requirement. The deep convolutional neural network (DCNN) model "monodepth" by Godard et al. [13] is based on unsupervised data which takes only stereo pairs as training data. Furthermore, we consider using cluster analysis method [14] as an unsupervised method to spatially group data.

2 Approach

The analysis of the proposed approach is shown in Fig. 1. The input is a sequence of images and the network of the study space. The output is the human count of each location, represented by a node in the network.

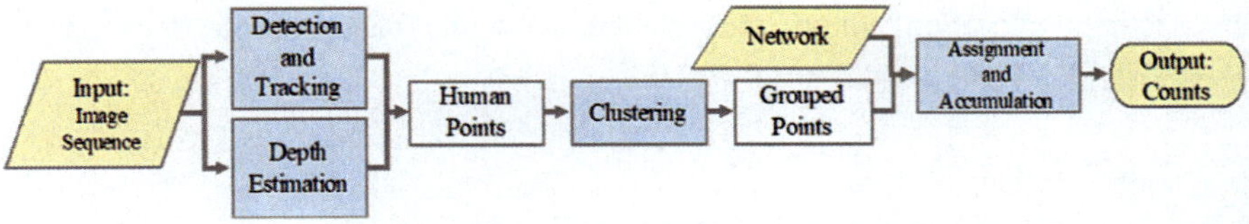

Fig. 1. Procedure of our approach

2.1 Network

The network required in this work not only provides the geometric relationship of the real space (or called object space in some discipline) for pathfinding steps in decision-making problems but also contains the information about the relationship with cameras. The base network could be constructed conventionally, such as manually setting, while Lee et al. [7] also provides an approach to utilize the BIM model. From the former model, each camera is also set as a node and specified their location, rough orientation, and their field of view (FoV) in order to calculate the vector to original nodes.

In this work, "sensing nodes" represents locations where counting is performed, similar to where conventional sensors are installed. There are three kinds of nodes in our network, shown in Fig. 2. The distance from camera nodes to sensing nodes is basically defined as the distance along the camera orientation, which is the depth from the view of the camera.

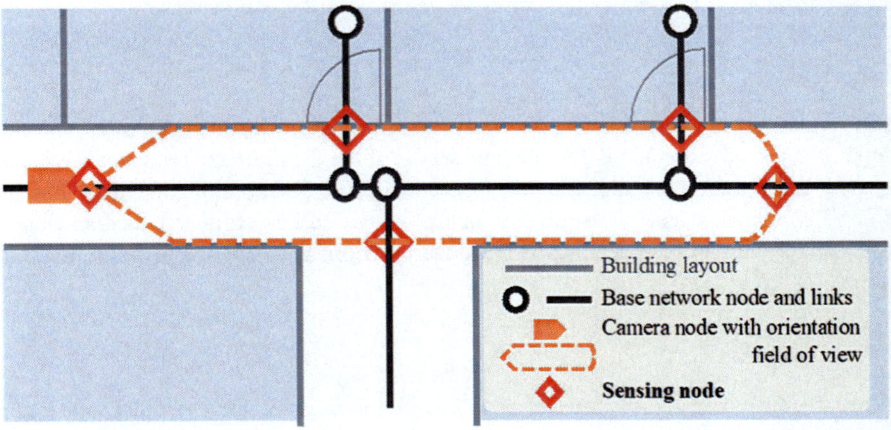

Fig. 2. Types of nodes in our network

2.2 Detection and Tracking

Detection is performed on the image sequence for recognition of humans and their tracking from appearing to disappearing. The appearing and disappearing location of each person represent passing through of a pair of "sensing nodes." The two locations are recorded as points which are set at the bottom-center of the detection bounding box, representing the location on floor where the human stands. These recorded locations are defined as "appearing/disappearing points."

2.3 Depth Estimation

The depth estimation extends the dimension from the 2D image; the depth is defined as the distance from the camera along the optical axis [12].

Since the pre-trained model is used in our test case, the baseline and focus length are not the same as our device. Thus, the depth values do not represent absolute true value but the relative ones. This simplification, illustrated by the following equations, does not impact the final result.

Original form:

$$depth = \frac{baseline \times focal\ length}{disparity} \tag{1}$$

Simplified form:

$$relative\ depth = 1/disparity \tag{2}$$

Moreover, only several frames without detected human are used to estimate the depth due to the reduction of calculation in practice.

The depth has two functions in the following steps: (i) to be a feature of the data for clustering and (ii) to identify the correspondence with the real space.

2.4 Human Points

The outputs from the previous two steps, the appearing/disappearing points and the depth images, are combined: The median depths of the depth image sequence are taken at the image locations of appearing/disappearing points. Each point, defined as "human point," contains: x and y coordinates in the image space, estimated median depth, whether the human is appearing or disappearing from the image, and frame number (time).

2.5 Clustering

When people pass the same sensing node in a period of time, human points appear at a close proximity. Due to the error in detection and the width of real space, the points distribute in a small spatial range. The points are to be grouped to determine the sensing node. In this work, k-means clustering is adopted due to its unsupervised property and simple computation. Values in x, y dimensions are scaled by dividing the length of the longer side, while depth values are scaled by min-max scaling. The conversion is to convert values in the [0, 1] interval in order to eliminate the strong impact from any dimension.

2.6 Assignment and Accumulation

The final step is to assign the clustered points to the sensing nodes. Since the assumption that each node is passed by at least one human, the clusters should have a one-to-one correspondence to the sensing node. Here the correspondence is simply sorting by the depth and assigned by the order.

After the assignment, the human counts are accumulated at each sensing node by the recorded frame number (time) and appearing or disappearing information of the points. The final output is expected as shown in Fig. 3, which each sensing node have their count in each interval of time. In other words, this represents the number of humans with the spatial-temporal relationship.

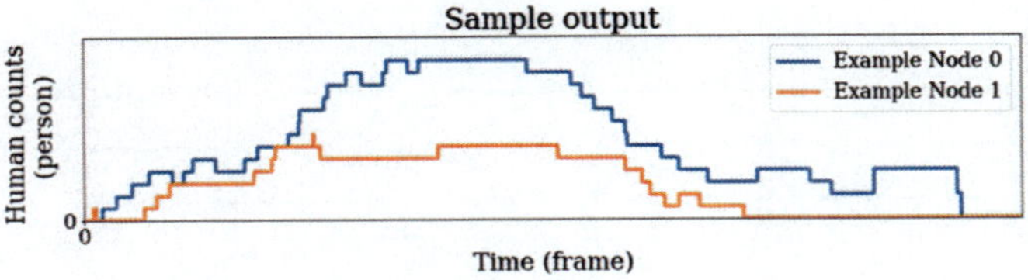

Fig. 3. Illustration of the count-time relationship

3 Validation Result

3.1 Case Description

Our approach is tested in a 3-min video sequence. The scene is at a corridor with one side of four doors at the Civil Engineering Building at National Taiwan University, as shown in Fig. 4. Figure 5 illustrates the network containing five sensing nodes; the distances in Table 1 are measured by a tape ruler. Noted that if people will go into the FoV from the far side or backside of the camera, the sensing nodes should be located. Yet, the distance to the far side (node 5 in our case) is not an exact value in our case, so the distance is recorded as unknown. Additionally, because no person passes through the backside node in this case, the node was not considered in this case. The image sequence contains 19 counts of humans in total: the ground truth of spatial-temporal distribution is labeled manually in Fig. 6.

Fig. 4. The scene of the test case

Table 1. Distance of the network

Start node	End node	Distance (cm)
0	1	250.0
0	2	736.0
0	3	1260.4
0	4	2226.4
0	5	Unknown

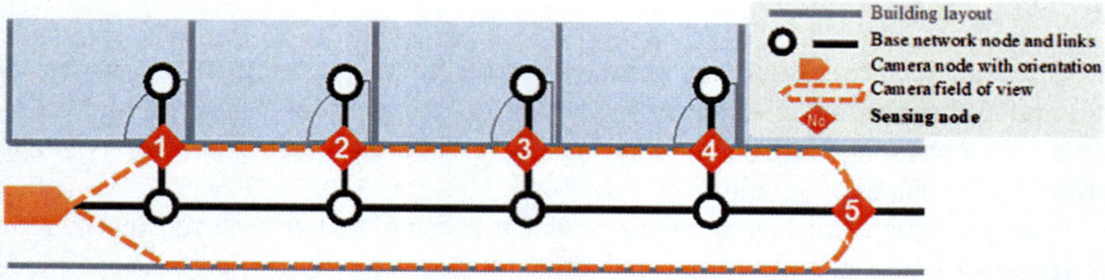

Fig. 5. The network of the test case

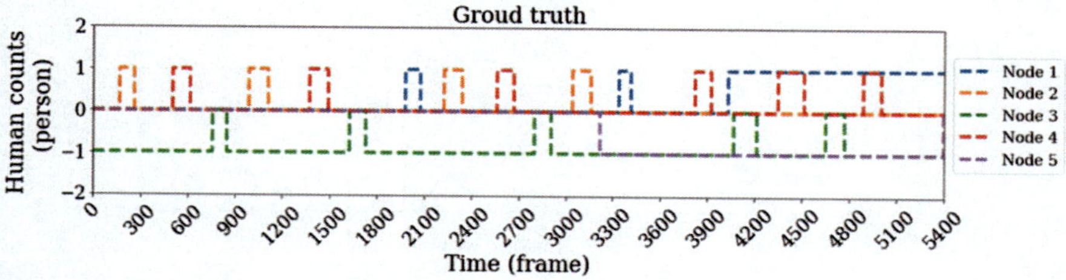

Fig. 6. Ground truth of spatial-temporal distribution (manually labeled) Since the counts only accumulate the period of the image sequence, the negative counts result from a person already in the room at the beginning of the record.

Table 2. Detection and tracking results (number of ground truth tracks = 19)

Case	Model	Frame rate	Number of tracks from model	Error (%)
(a)	deep_sort_yolo_v3 [15]	30	21	10.5
(b)	detect-and-track [9]	30	51	168.4
(c)	deep_sort_yolo_v3	30/8	8	57.9
(d)	detect-and-track	30/8	26	36.8

3.2 Detection and Tracking

Two deep learning models are tested with the original frame rate (30 fps) and the reduced frame rate (30/8 fps). The results are shown in Table 2. Among the four combinations, (a) in Table 2 has the least error of track numbers and is used in the further steps, while (d) is the second and has the potential to be adopted due to the lower frame rate.

By inspecting the output from this step, the error was found highly related to the occlusion of multiple humans in the scene. Additionally, humans who are far from the camera may reach the limitation of the resolution of detection; they are too small in the image to be clearly detected.

3.3 Clustering and Assignment

The number of clusters k is a parameter that should be given when using k-means clustering. In our test case, the k was tested from 2 to 10 and the result is shown in Fig. 7. By plotting the inertia (Fig. 8), the elbow method suggests $k = 3$–5 to be the better choices. The accuracy of the manually labeled ground truth shows that the $k = 5$ case would be the best choice. However, in practice, k should be determined without having the ground truth labels, while the elbow methods may not provide a specific answer. Thus, k is suggested to be equal to the number of sensing nodes.

The depths of cluster centers are compared with the ground truth distances to sensing nodes in Table 3. Although the depths are not exact values, a high correlation between them and ground truth distances indicates the rationality of the matching step as well as the estimation of the depth.

The elbow method is to find the point before the inertia converges.

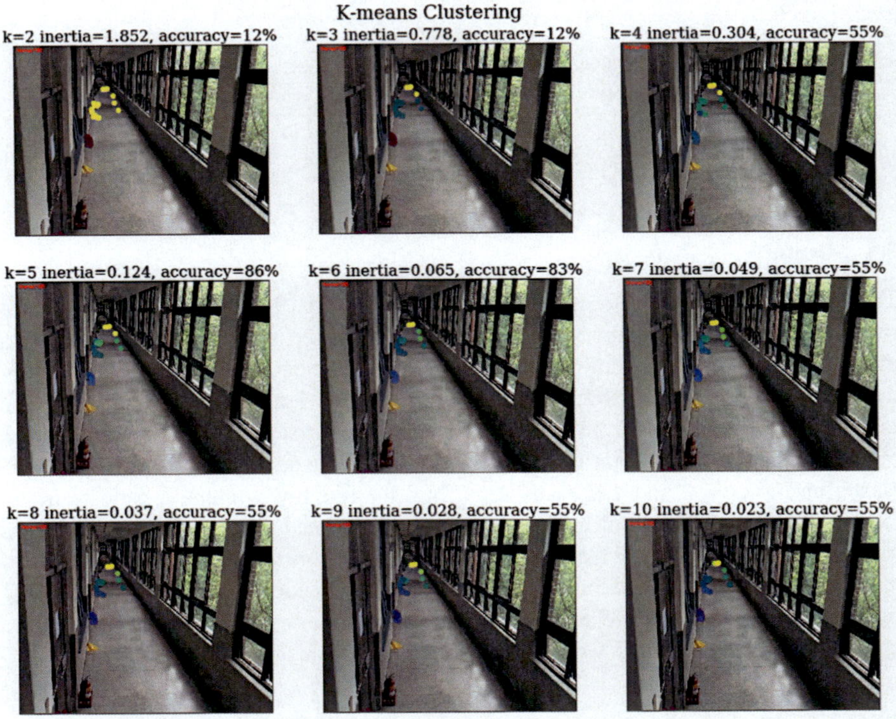

Fig. 7. Clustering results have the best accuracy 86% when $k = 5$.

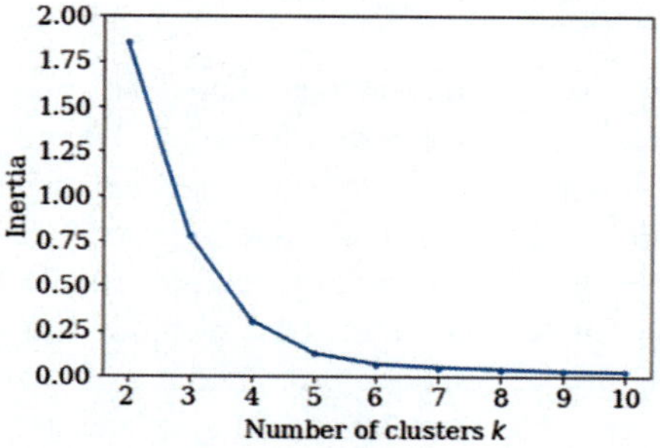

Fig. 8. Inertia of the clustering results

Table 3. Depth of cluster centers and corresponding network distances

Cluster no.	Center depth (relative value)	Start node	End node	Distance (cm)
1	0.00364877	0	1	250.0
2	0.02877908	0	2	736.0
3	0.24428719	0	3	1260.4
4	0.41213060	0	4	2226.4
5	0.87687506	0	5	Unknown

3.4 Output

The final output of each sensing node is shown as Fig. 9 and Fig. 10 and Table 4 summarize the error of each node. In nodes 1 and 2, the error appearing as small peaks mainly occurs at the appearing and disappearing moment because the model could only detect an almost complete human in the images while human eyes could identify a human from a small part of the body. The detection interval makes only several frames of error, which does not greatly impact the result. On the contrary, the error from false clustering in nodes 3 and 4 is easily seen from the roughly symmetric patterns in the latter half in Fig. 10, causing major impact. On the other hand, the error in node 5 is mainly from two kinds of detection and tracking errors due to the occlusion and resolution limit, mentioned in Sect. 3.2. Fortunately, node 5 represents the sensing node of the FoV limit and seldom connects to a room interested but another part of the space. The result of node 5 would not be used in further decision-making steps.

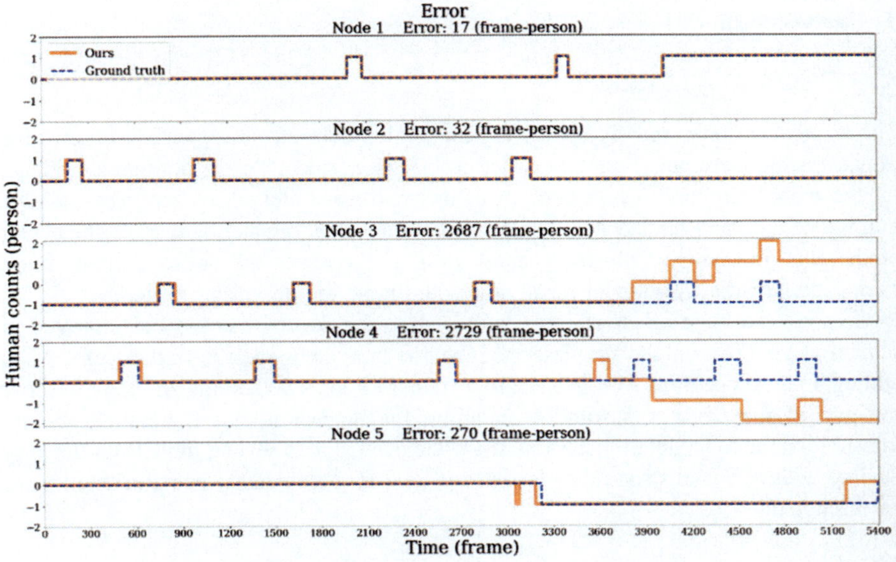

Fig. 9. Person-time output of each sensing node

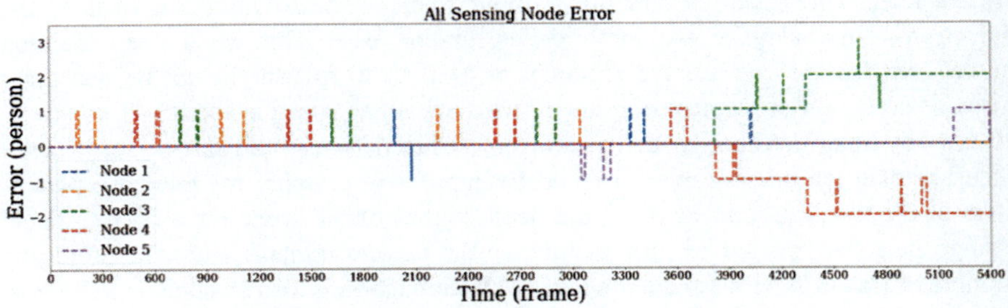

Fig. 10. Errors of each sensing node are estimated by the difference between output from our approach and the ground truth

Table 4. Error values of each sensing node

Node	Total error	
	(frame-person)	(second-person)
1	17	0.567
2	32	1.067
3	2687	89.567
4	2729	90.967
5	270	9.000

4 Discussion

4.1 Detection and Tracking

The error in this step mainly splits the true single track into many pieces. One of the impacts is the increment on the count of each sensing node. This is not that significant for the spatial-temporal results because of its small time-interval. Nevertheless, a huge number of incorrect human points might affect the clustering step and make the center of the clusters diverge from the ground truth. The error of the cluster centers might even influence the assignment result if it is too large and the order switches.

Despite the huge effect of error in this step, the accuracy could be improved in several ways. First and the simplest one is to fine-tune the models to fit the scene. Also, adding some procedures to tolerate a few frames of miss detections and thus to close the gap of the track is a solution handling the occlusion, but it requires another threshold value to be determined. Finally, if the occlusion is severe, models particularly dealing with occlusions should be tested, and adjusting the camera position should also be discussed.

4.2 Frame Reduction

For both the "Detection and Tracking" and "Depth Estimation" steps, the input frames are reduced. The reason is that more input frames lead to more calculation; the reduction could alleviate the intensive calculation especially when deep learning models are used. However, the accuracy of 1/30 s (30 fps) might not be necessary when detecting the movement of humans, whose walking speed is about 1–3 m/s (3.3–10 cm per frame). Although the tracking of reduced frames did not perform better accuracy than the original ones, it could be improved by tuning the tolerance parameters of the tracking models. Also, the depth values of the scene have no significant change in a short period of time as long as the camera remains still. Therefore, the number of frames is suggested to be lower. This reduction is also reasonable in the real scene where the surveillance cameras might not have a high frame rate due to the limit on transmission bandwidth and storage size.

4.3 Clustering and Assignment

From the results, we suggest that the number of clusters could be set as the number of sensing nodes. It is reasonable based on the assumption of the one-to-one relationship between clusters and sensing nodes. Nevertheless, some errors in previous steps are found to generate more points than expected. When the occlusion intensively occurs at a specific location, an improvement in accuracy could be achieved by introducing an additional cluster. However, this action could hardly be done without a former manual investigation of the image sequences, which is in contrast to the spirit of the automatic process.

Besides, normalizing the data before clustering could also be challenging since the true dimensions are not equal. Yet, the same difficulty is that the scales between the dimensions are hardly known. Even the geometry of the FoV in the real space could be measured, the exact correspondence between the real space and the image space should be calculated with other details of the camera such as focus, orientation and distortion parameters.

A distinct alternative that highly increases the accuracy is to manually carry out these two based on the information given by previous steps. Although this solution seems to violate the automation purpose, it not only provides quite different confidence that the grouped human points could correspond to the doors but also eliminates the assumption of at least one passing human for all the nodes which leads to the one-to-one correspondence. Since the result of our test case is not complicated, it could be done automatically; in case of some complicated scenes, such as sensing nodes in different directions or some path are seldom used, manually checking by this alternative could maintain the accuracy.

4.4 Camera Location

Although locations of the camera are usually fixed in the real cases, those which could have better performance through our approach is still worthy of investigation for the evaluation of fitness and recommendation to further installations or adjustments.

First, as the depth is a key feature of the data in our approach, the clusters and sensing points are corresponded by their depths; if the sensing points are at similar depths in the image, they could not be identified well. In contrast, the sensing points located in a row of different depths are relatively easy to be identified. The examples are shown in Fig. 11.

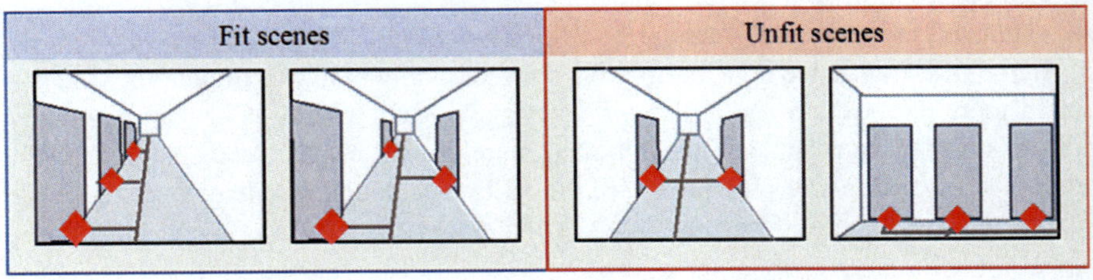

Fig. 11. Example scenes that are fitted to our models and those which are not

Furthermore, if the sensing nodes are in a sequence of the same depth interval and the alternate ones have no human passing, as shown in Fig. 12, this set of nodes might be incorrectly corresponded in the assignment step. Thus, this scene might need an alternative manual correction.

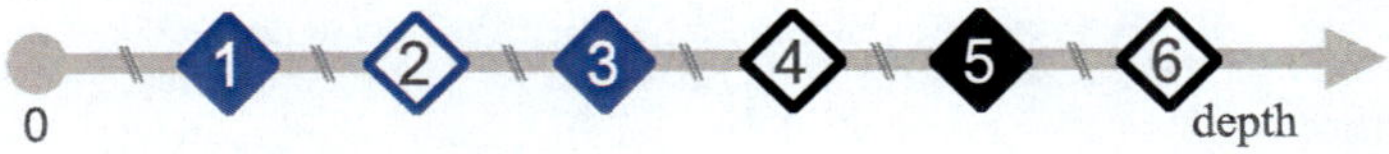

Fig. 12. Illustration of incorrect assignment. The 6 diamond shape represent sensing nodes with the equal intervals, whereas only the hollow ones (2, 4, 6) are passed by humans. The three clusters might correspond to (2, 4, 6), (1, 2, 3) or even (1, 3, 5) because of the relative depth.

Also, in the future of our study, we are going to consider the scene that is observed by multiple cameras. Either stereo matching or sensor fusion methods are to be discussed its fitness and whether it improves the accuracy.

4.5 Summary of Discussion

In brief, the impact of error in each step is discussed in this chapter. The settings in our approach are challenged and described their reasons. From the choice of the methods in the steps, we encountered some trade-offs between automation and accuracy. It is noteworthy that the models of the same functions in each step could be replaced; for example, detect-and-track and deepSORT are tested in our case, while other tracking models could be tested for a better outcome or better efficiency.

5 Conclusion

This study provides a procedure for decision-makers to estimate the spatial-temporal human distribution from relatively simple and fewer requirements through machine learning methods. Because locations of AEDs are suggested to be at noticeable positions, we focus on observing the spaces connecting corridors and hallways.

Installation of AED is usually conducted in existing public places, and cameras are fixed and difficult to adjust and to acquire the precise parameters. Considering these real factors, our work demonstrates an applicable process.

A notable finding is that depth could serve as a valuable feature. Despite a lack of the exact translation between the real and image space, the depths from the network and the estimated ones show a potential correspondence.

In the near future, we hope to improve our methodology in two directions. One is to obtain more test cases including crowd scenes and scenes from other camera orientations in order to investigate applicable scenes and the robustness of the approach. The other is to estimate more parameters for the AED installation problem, such as the age and the walking speed of humans.

References

1. U.S. Department of Health and Human Services and General Services Administration: Guidelines for Public Access Defibrillation Programs in Federal Facilities, pp. 41133–41139 (2009)
2. 公共場所必要緊急救護設備管理辦法 [Public Emergency Medical Service Equipment Control Regulations], Taiwan (2013)
3. Japan Heart Foundation: AEDの設置基準の条件 [Guidelines for the appropriate Installation Criteria Requirements of AED] (2019)
4. Bækgaard, J.S., Viereck, S., Møller, T.P., Ersbøll, A.K., Lippert, F., Folke, F.: The effects of public access defibrillation on survival after out-of-hospital cardiac arrest. Circulation **136** (10), 954–965 (2017). https://doi.org/10.1161/CIRCULATIONAHA.117.029067

5. Dao, T.H.D., Zhou, Y., Thill, J.C., Delmelle, E.: Spatio-temporal location modeling in a 3D indoor environment: the case of AEDs as emergency medical devices. Int. J. Geogr. Inf. Sci. **26**(3), 469–494 (2012)
6. Chan, T.C.Y.: Rise and shock: optimal defibrillator placement in a high-rise building. Prehosp. Emerg. Care **21**(3), 309–314 (2017). https://doi.org/10.1080/10903127.2016.1247202
7. Lee, C.T., Lee, Y.C., Chen, A.Y.: In-building automated external defibrillator location planning and assessment through building information models. Autom. Constr. **106**, 102883 (2019). https://doi.org/10.1016/j.autcon.2019.102883
8. Chen, C.C., Chen, A.Y.: Video-based indoor human detection for decision-making of the installation locations for automated external defibrillators. In: ASCE International Workshop on Computing in Civil Engineering 2017, pp. 444–449, June 2017. https://doi.org/10.1061/9780784480823.053
9. Girdhar, R., Gkioxari, G., Torresani, L., Paluri, M., Tran, D.: Detect-and-track: efficient pose estimation in videos. In: Proceedings of the IEEE Conference on Computer Vision and Pattern Recognition, pp. 350–359 (2018)
10. Redmon, J., Farhadi, A.: YOLOv3: an incremental improvement. Technical report (2018)
11. Wojke, N., Bewley, A., Paulus, D.: Simple online and realtime tracking with a deep association metric. In: 2017 IEEE International Conference on Image Processing (ICIP), pp. 3645–3649 (2017). https://doi.org/10.1109/icip.2017.8296962
12. Szeliski, R.: Computer Vision Algorithms and Applications. Texts in Computer Science. Springer, London (2011)
13. Godard, C., Aodha, O.M., Brostow, G.J.: Unsupervised monocular depth estimation with left-right consistency. In: 2017 IEEE Conference on Computer Vision and Pattern Recognition (CVPR), pp. 6602–6611 (2017). https://doi.org/10.1109/cvpr.2017.699
14. Han, J., Kamber, M., Pei, J.: Data Mining: Concepts and Techniques, 3rd edn. Elsevier Science, Burlington (2012). https://doi.org/10.1016/C2009-0-61819-5
15. Deep_sort_yolov3. https://github.com/Qidian213/deep_sort_yolov3. Accessed 22 July 2019

Construction Scene Parsing (CSP): Structured Annotations of Image Segmentation for Construction Semantic Understanding

Yujie Wei(✉) and Burcu Akinci

Carnegie Mellon University, Pittsburgh, PA 15213, USA
yujiew@andrew.cmu.edu, bakinci@cmu.edu

Abstract. Images ubiquitously support visualization and documentation of as-built status of built infrastructure. Images captured on-site during construction contain rich semantic information, such as object categories, materials and topological relationships, which are useful for many applications, such as progress monitoring, crack detection, quality control, and safety management. Recent advancements in deep learning and convolutional neural network can effectively extract various types of semantic information from images, but they require annotated datasets for model training. Most existing scene parsing datasets contain annotated images captured at project closeout and there lacks a dataset that can be used for construction site scene understanding. In order to support construction scene understanding, we present the Construction Scene Parsing (CSP), an annotated image dataset that contains over 150 construction scenes with image segmentation labelled by experts. The CSP dataset have two primary contributions: 1) It provides a hierarchical semantic structure rather than a unitary label for each image to deal with incomplete and changing components presented on construction images; 2) It provides pixel-wise annotations for every scene and can support various types of scene understanding tasks, such as object recognition, semantic segmentation, instance segmentation and panoptic segmentation. The dataset can be accessed at https://github.com/yugitw/Construction-Scene-Parsing.

Keywords: Semantic segmentation · Deep learning · Convolutional neural network · Construction · Images

1 Introduction

Over the last decade, as cameras become more ubiquitous, the number of images captured in the AEC industry has increased significantly, providing an opportunity to automate semantic understanding of things captured on images [1]. Recently, researchers argued that scene understanding techniques, such as object recognition, object classification and semantic segmentation, are needed to support many construction and facility management activities, such as inspection, progress monitoring, resource tracking, and as-built modeling [2].

Conventional semantic understanding on AEC images heavily rely on hand-crafted features to identify facility components of interest. Previous researchers have proposed

E. Toledo Santos and S. Scheer (Eds.): ICCCBE 2020, LNCE 98, pp. 1152–1161, 2021.
https://doi.org/10.1007/978-3-030-51295-8_80

using lines, patches, and planes on images to detect components, such as beams, columns and walls [1, 3]. When images are reconstructed into a point cloud, geometric features, such as normal and curvature, are also widely used for supporting semantic understanding [4–6]. Some researchers further used known primitives in a building to fit objects shown in visual data to help semantic understanding, such as fitting cuboids to find columns and beams [7] and fitting cylinders to detect pipes [8, 9].

Advancements in deep learning and convolutional neural network in previous years began to leverage features from labelled data to support semantic understanding and have outperformed hand-crafted features in many scene understanding tasks [10]. For example, recognizing construction resources, such as workers, equipment and materials, has been one of the areas that learning-based technologies have been successfully applied. Many studies used deep learning and computer vision approaches to monitor activities on site to support productivity and safety management [11–13]. Besides construction resources, learning-based technologies also play an important role in detecting various types of construction damages, such as cracks on structural components [14].

Recently, many researchers proposed using as-built visual data to create a digital twin for physical facilities, raising the need for a panoptic scene understanding approach with finer details [15, 16]. For example, when using Structure-from-Motion approaches to reconstruct a semantically rich 3D model from images, geometries of components are estimated by evaluating pixel-wise correspondences. In order to assign a semantic label to each point in the reconstructed point cloud, it is necessary to have pixel-wise semantics as well. Though some research studies have started to focus on image segmentation for indoor scenes, such as ADE20k [17] and 3DFacilities [18], there lacks an image segmentation dataset to support semantic understanding on construction site images. Therefore, we present the Construction Scene Parsing (CSP) dataset to address the research gap mentioned above.

In Sect. 2, we will first evaluate existing research on scene parsing and identify semantic segmentation challenges that are specifically associated with images captured at construction sites. Section 3 will introduce our structured annotated dataset and compare it with existing datasets and Sect. 4 will introduce how we can standardize the semantic labeling by following existing building classification standards. We will conclude with discussion on contributions, limitations, and future research directions in Sect. 5.

2 Related Work

As learning-based methods are emerging, many datasets have been published to support semantic understanding of both indoor and outdoor scenes. We will review previous research according to the level of details of their annotations below.

Object Classification/Detection. Early semantic understanding datasets, such as Imagenet [10], Pascal [19], and KITTI [20], often contain rough labels only, such as annotations of an image or bounding boxes of interested objects. Imagenet was one of the earliest image classification datasets that contains over ten million images and 32,326 categories, but it primarily focuses on objects rather than scenes. Similarly, though Pascal and KITTI datasets provided bounding boxes for multiple objects presenting on images, they did not cover that many building scenes. Besides general-purpose datasets mentioned above, many AEC-related image datasets have been proposed to detect specific objects, such as equipment [21] and workers [22]. Figure 1 shows that existing object classification/detection datasets can only provide labels for a small subset of a scene and cannot support comprehensive scene understanding of an image.

Fig. 1. Example object detection results (a) equipment detection [21] (b) worker detection [22]

Semantic Segmentation. Semantic segmentation on both indoor and outdoor scenes is an active research area in the recent years. Many image segmentation datasets have been collected and labeled to help support model training and benchmarking [17, 18, 23, 24]. Many datasets contain pixel-wise annotations for objects in a static scene, such as NYU-V2 [25], CityScapes [26], 2D-3D-semantics [23], Matterport3D [24], and 3DFacilities [18]. However, these datasets cannot be used for construction images directly for two reasons: 1) they are created and labeled for general scene understanding rather than focusing on AEC scenes specifically. 2) they do not consider possible ambiguities of construction images due to scene changes. As shown in Fig. 2, components in a construction site might change as construction goes on and can have several possible labels depending on the end user's requirements. Therefore, it is necessary to provide a contextual hierarchical structure for annotations that describes relationships between pixels for construction images. For example, one feasible way of labeling Fig. 2(a) is to not just marking all metal studs, but also label them as parts of a wall.

Fig. 2. Possible labeling ambiguities on construction images. (a) Should we label each metal stud or mark them as a wall? (b) Should we label the components as columns or formwork?

Structured Semantic Understanding. Instead of providing one label for each pixel, many datasets, such as Core and Pascal, started to assign multiple labels to each pixel and model their relationships. For example, one pixel can be labeled as "head" and "human" at the same time with a hierarchical structure defining that a head is part of a human. ADE20k [17] further advanced this effort by annotating instance of each class and their parts. The cascading way of annotation provides a hierarchical structure to help model the "part-of" relationships among various component instances and can be used for modeling contextual information. Despite that ADE20k has collected high-resolution imagery of a wide range of scenes, it cannot be directly used for scene parsing for applications in AEC industries due to the fact that as a general-purpose scene parsing dataset, ADE20k missed many common construction related categories, such as temporary structures and HVAC components.

In summary, the development of existing datasets has provided a good starting point for training general-purpose scene understanding methods. However, there still lacks a specialized dataset that concentrates on AEC industries, especially for construction. Therefore, it is necessary to develop a dataset to support scene understanding for images captured during construction.

3 CSP: Structured Annotations of Construction Scenes

In this section, we will give a detailed introduction to the Construction Scene Parsing (CSP) dataset and compare it with existing datasets to highlight our contributions.

3.1 Dataset Summary

Since training a semantic segmentation model from scratch would be very time-consuming, the dataset is created to support the fine tuning of a general-purpose semantic segmentation model to work on construction scenes. Therefore, we focused on collecting high-resolution images and providing pixel-wise annotations for each image rather than simply adding more frames. The dataset currently contains 185 annotated images captured during the construction of an academic building in the US. The annotation in the dataset are still growing. Figure 3 shows example images and annotations from the dataset.

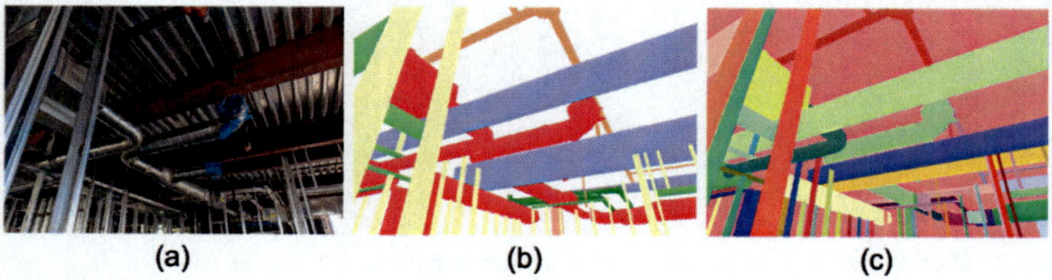

Fig. 3. Example raw image (a), class annotation (b), and instance annotation (c) in the CSP dataset

3.2 Data Collection, Pre-processing and Annotation

All images in the CSP dataset are captured during construction. Since the goal of the dataset is to support scene parsing rather than construction resource tracking, most images are captured when nobody was working. They have the same resolution of 5472 × 3648 and are captured by a digital camera in RAW format to allow for lighting condition calibration. Images in the CSP dataset were annotated by a single expert civil engineer using Adobe Photoshop and PSD-tools[1]. Images captured at construction sites are more likely to have dynamic range issues (overexposure and underexposure regions) due to partial occlusions. Therefore, we calibrated all images to guarantee there is no overexposure or underexposure on the facility component region using Adobe Camera Raw. Figure 4 shows an example image before and after calibration. After calibration, we examined each image manually and provided three types of information for each object shown on an image: object categories, object segments, and object parts. The object category information defines the type of a region, the object segment is a set of pixels that belong to the component, and object parts are subregions that also contain a category name and an image segment. Notice that many datasets predefined a set of object categories based on prior knowledge. However, working with construction site images using such a predefined list of categories is not possible in CSP due to the limited number of images. For example, for images captured from a building that mainly uses steel structures, a predefined category of concrete columns might not contain that many instances. Therefore, we updated the category list during labeling and maintained a JSON list to keep track of the current hierarchical structure and all categories to ensure consistency in classification. Section 4 will provide further discussions on how to convert our labels to existing standards to guarantee interoperability.

[1] https://psd-tools.readthedocs.io/en/latest/index.html.

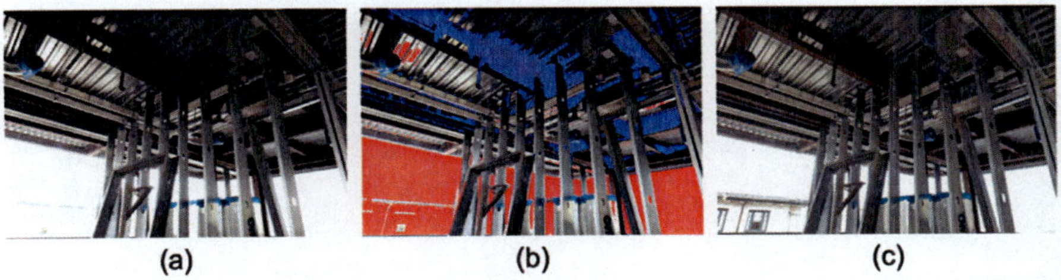

Fig. 4. Example exposure calibration process (a: original, b: visualization of overexposure and underexposure regions, c: image after calibration)

3.3 Comparison with Existing Datasets

Table 1 compares key statistics between the CSP dataset with existing datasets, such as ADE20k, 3DFacilities, Matterport3D, and NYU-V2. From the table, it can be concluded that the CSP dataset is the only dataset that focuses on construction scene parsing in a holistic way. Due to complexities of construction scenes, CSP dataset also has the highest number of instances per image (Datasets without instance labeling have 0 instance per image in the table).

Table 1. Comparison between CSP and existing datasets

Dataset	Scene types	Size	Instance per Img	Structured	Resolution (# of pixel)
CSP (ours)	**Construction site**	**185 images**	**~24.2**	**Y**	**20 M**
ADE20k [17]	General	22,210 images	~9.9	Y	1.3 M
3DFacilities [18]	Building interior	50 scenes	0	N	50 k
Matterport [24]	Building interior	10,800 views	0	N	1.3 M
NYU-V2 [25]	Building interior	1449 images	0	N	51 k

Moreover, the CSP dataset provides structured annotations that can be used for understanding contextual relationships among different facility components. With structured annotations, the CSP dataset not only can be used for training conventional semantic segmentation, but also allows users to select and filter objects and labels based on their requirements. For example, Fig. 5 shows how users can extract different levels of annotations.

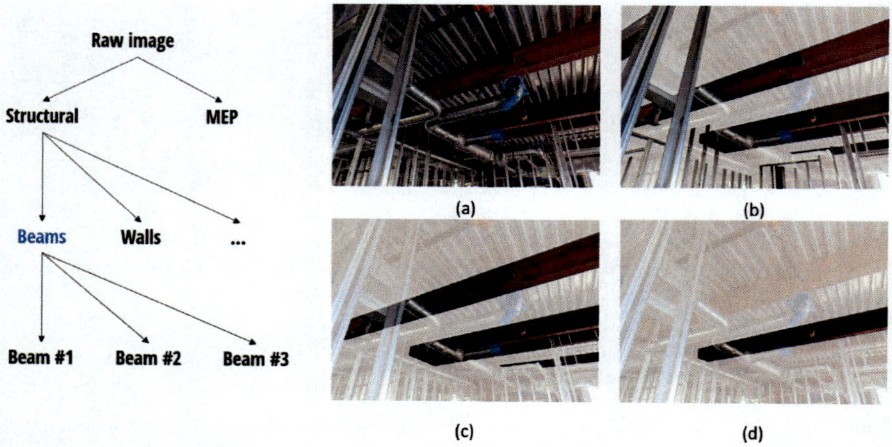

Fig. 5. An example structured annotation (a: raw, b: structural components, c: beam class, d: a beam instance)

The structured annotations can potentially support many new applications. For example, Fig. 6 shows an example digital twin that is organized in a similar hierarchical data format as our structured annotations. In other words, if a model can not only recognize facility components on images but also estimate their relationships, it will help create a semantically rich as-built model much more conveniently.

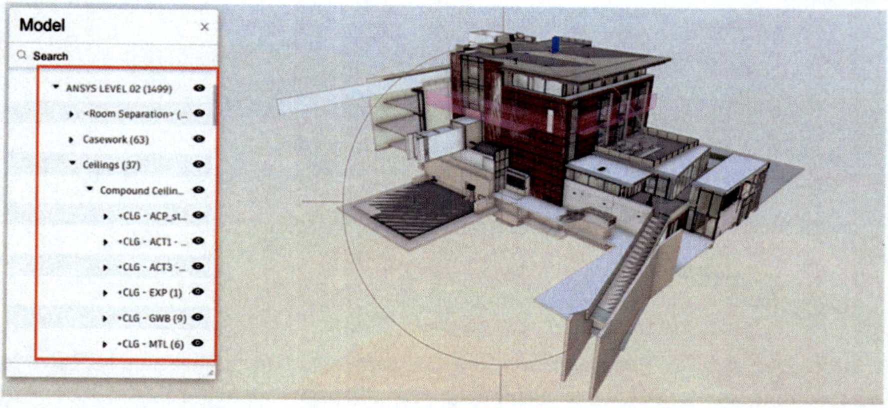

Fig. 6. Tree structure presented in an example digital twin data format.

4 Standardized Structured Labeling

In the previous section, we argued that the structure of labels is subject to objects presented on existing images and therefore did not employ a predefined category list. However, it can be foreseen that as the annotating work continues, a unified and standard structure for labels is needed for comprehensive construction scene understanding.

Though there are many existing building classification standards that are widely used in the industry, such as UniFormat [27], MasterFormat [28], and OmniClass [29], none of them can be directly used for structured annotations on construction site photos. MasterFormat was developed to provide a material-based organization for building content. For example, the level-1 code "03" in MasterFormat refers to concrete structures while the code "04" refers to masonry structures. Such a material-based classification is often used in cost estimates instead of image-based scene parsing. In contrast, UniFormat uses a system-oriented way to classify building components. However, it does not include temporary objects and construction resources, which are often presented on construction images. OmniClass integrates the two standards above and provide a more complicated structure for classifying building components. By using multiple tables, OmniClass can decompose a building in different ways: function, form, elements, work results, products, etc. However, it is difficult to infer such a complex structure from images.

In order to leverage existing building classification standards, we also provided a UniFormat-based category list for our dataset. The list was included in the dataset as well. In short, besides the original labels existing in UniFormat, the list used for image-based semantic understanding also contains temporary objects (such as formwork, scaffolding, etc.) and construction resources (equipment, people, stacked materials, etc.).

5 Conclusions and Future Work

In this paper, we introduced a new structurally annotated construction image dataset with the instances of facility components and their parts to support semantic understanding during construction. The dataset will be maintained and further extended to support training from scratch. Below are several future research directions that the CSP dataset can potentially support:

Instance Segmentation. In the experiments, we only focus on semantic segmentation rather than instance segmentation where individual instances of all classes will also be recognized. Since the CSP dataset contains instance labeling for each type of object, it is possible to extend regular semantic segmentation methods to support instance segmentation.

Structured Segmentation. One important contribution of the CSP dataset is that it annotates the "part-of" relationship among components along with conventional class and segment labels. Therefore, the CSP dataset can be used to develop a structured segmentation model that predicts object semantics and their relationships simultaneously.

As-Built Model Reconstruction. Using the output of a model trained on the CSP dataset, it is possible to reconstruct a semantically rich as-built model from images captured during construction. For example, geometries of components can be obtained from structure-from-motion methods, while the semantic information of each component can be predicted on each image and then reprojected back to the 3D model.

References

1. Golparvar-Fard, M., Peña Mora, F., Silvio, S.: D4 AR – a 4-dimensional augmented reality model for automating construction progress monitoring data collection, processing and communication. Electron. J. Inf. Technol. Constr. **14**, 129–153 (2009). http://www.itcon.org/paper/2009/13
2. Wei, Y., Kasireddy, V., Akinci, B.: 3D imaging in construction and infrastructure management: technological assessment and future research directions. In: Lecture Notes in Computer Science (including subseries Lecture Notes in Artificial Intelligence and Lecture Notes in Bioinformatics), pp. 37–60 (2018). https://doi.org/10.1007/978-3-319-91635-4_3
3. Xiao, J., Furukawa, Y.: Reconstructing the world's museums. Int. J. Comput. Vis. **110**(3), 243–258 (2014). https://doi.org/10.1007/s11263-014-0711-y
4. Xiong, X., Adan, A., Akinci, B., Huber, D.: Automatic creation of semantically rich 3D building models from laser scanner data. Autom. Constr. **31**, 325–337 (2013). https://doi.org/10.1016/j.autcon.2012.10.006
5. Bosché, F.: Plane-based registration of construction laser scans with 3D/4D building models. Adv. Eng. Inform. **26**, 90–102 (2012). https://doi.org/10.1016/j.aei.2011.08.009
6. Furukawa, Y., Curless, B., Seitz, S.M., Szeliski, R.: Manhattan-world stereo. In: 2009 IEEE Computer Society Conference on Computer Vision and Pattern Recognition Workshops, CVPR Workshops 2009, pp. 1422–1429 (2009). https://doi.org/10.1109/CVPRW.2009.5206867
7. Barazzetti, L.: Parametric as-built model generation of complex shapes from point clouds. Adv. Eng. Inform. **30**, 298–311 (2016). https://doi.org/10.1016/j.aei.2016.03.005
8. Bosché, F., Guillemet, A., Turkan, Y., Haas, C.T., Haas, R.: Tracking the built status of MEP works: assessing the value of a scan-vs-BIM system. J. Comput. Civ. Eng. **28**, 05014004 (2014). https://doi.org/10.1061/(ASCE)CP.1943-5487.0000343
9. Bosché, F., Ahmed, M., Turkan, Y., Haas, C.T., Haas, R.: The value of integrating scan-to-BIM and scan-vs-BIM techniques for construction monitoring using laser scanning and BIM: the case of cylindrical MEP components. Autom. Constr. **49**, 201–213 (2015). https://doi.org/10.1016/j.autcon.2014.05.014
10. Jia, D., Wei, D., Socher, R., Li-Jia, L., Kai, L., Li, F.-F.: ImageNet: a large-scale hierarchical image database. In: 2009 IEEE Conference on Computer Vision and Pattern Recognition, pp. 248–255 (2009). https://doi.org/10.1109/CVPRW.2009.5206848
11. Yang, J., Park, M.W., Vela, P.A., Golparvar-Fard, M.: Construction performance monitoring via still images, time-lapse photos, and video streams: Now, tomorrow, and the future. Adv. Eng. Inform. **29**, 211–224 (2015). https://doi.org/10.1016/j.aei.2015.01.011
12. Kim, D., Liu, M., Lee, S.H., Kamat, V.R.: Remote proximity monitoring between mobile construction resources using camera-mounted UAVs. Autom. Constr. **99**, 168–182 (2019). https://doi.org/10.1016/j.autcon.2018.12.014
13. Zhang, B., Zhu, Z., Hammad, A., Aly, W.: Automatic matching of construction onsite resources under camera views. Autom. Constr. **91**, 206–215 (2018). https://doi.org/10.1016/j.autcon.2018.03.011

14. Li, D., Cong, A., Guo, S.: Sewer damage detection from imbalanced CCTV inspection data using deep convolutional neural networks with hierarchical classification. Autom. Constr. **101**, 199–208 (2019). https://doi.org/10.1016/j.autcon.2019.01.017
15. Wei, Y., Akinci, B.: A vision and learning-based indoor localization and semantic mapping framework for facility operations and management. Autom. Constr. **107**, 102915 (2019). https://doi.org/10.1016/j.autcon.2019.102915
16. Czerniawski, T., Leite, F.: Semantic segmentation of images of building facilities. In: CEUR Workshop Proceedings (2019)
17. Zhou, B., Zhao, H., Puig, X., Fidler, S., Barriuso, A., Torralba, A.: Scene parsing through ADE20K dataset. In: Proceedings - 30th IEEE Conference on Computer Vision and Pattern Recognition, CVPR 2017, pp. 5122–5130 (2017). https://doi.org/10.1109/CVPR.2017.544
18. Czerniawski, T., Leite, F.: 3DFacilities: annotated 3D reconstructions of building facilities. In: Workshop of the European Group for Intelligent Computing in Engineering, pp. 186–200. Springer, Cham (2018). https://doi.org/10.1007/978-3-319-91635-4_10
19. Everingham, M., Eslami, S.M.A., Van Gool, L., Williams, C.K.I., Winn, J., Zisserman, A.: The pascal visual object classes challenge: a retrospective. Int. J. Comput. Vis. **111**(1), 98–136 (2014). https://doi.org/10.1007/s11263-014-0733-5
20. Geiger, A., Lenz, P., Urtasun, R.: Are we ready for autonomous driving? The KITTI vision benchmark suite. In: Proceedings of the IEEE Computer Society Conference on Computer Vision and Pattern Recognition, pp. 3354–3361 (2012). https://doi.org/10.1109/CVPR.2012.6248074
21. Arabi, S., Haghighat, A., Sharma, A.: A deep learning based solution for construction equipment detection: from development to deployment (2019)
22. Fang, W., Ding, L., Zhong, B., Love, P.E.D., Luo, H.: Automated detection of workers and heavy equipment on construction sites: a convolutional neural network approach. Adv. Eng. Inform. **37**, 139–149 (2018). https://doi.org/10.1016/j.aei.2018.05.003
23. Armeni, I., Sax, S., Zamir, A.R., Savarese, S.: Joint 2D-3D-semantic data for indoor scene understanding (2017)
24. Chang, A., Dai, A., Funkhouser, T., Halber, M., Niebner, M., Savva, M., Song, S., Zeng, A., Zhang, Y.: Matterport3D: learning from RGB-D data in indoor environments. In: Proceedings - 2017 International Conference on 3D Vision, 3DV 2017, pp. 667–676 (2018). https://doi.org/10.1109/3DV.2017.00081
25. Gupta, S., Arbeláez, P., Malik, J.: Perceptual organization and recognition of indoor scene from RGB-D images semantic segmentation with RGB-D. In: CVPR, pp. 1–9 (2013). https://doi.org/10.1109/ICCVW.2011.6130298
26. Cordts, M., Omran, M., Ramos, S., Rehfeld, T., Enzweiler, M., Benenson, R., Franke, U., Roth, S., Schiele, B.: The cityscapes dataset for semantic urban scene understanding. In: Proceedings of the IEEE Computer Society Conference on Computer Vision and Pattern Recognition, pp. 3213–3223 (2016). https://doi.org/10.1109/CVPR.2016.350
27. Building Systems, Savings-to-Investment Ratio, Cost Risk Analysis: Standard classification for building elements and related sitework—UNIFORMAT II 1. ASTM (2005). https://doi.org/10.1520/E1557-09.2
28. MasterFormat. https://www.csiresources.org/standards/masterformat
29. OmniClass. https://www.csiresources.org/standards/omniclass

Vision-Based Pavement Marking Detection – A Case Study

Shuyuan Xu[1(✉)], Jun Wang[2], Peng Wu[1], Wenchi Shou[3], Tingchen Fang[4], and Xiangyu Wang[5,6]

[1] Curtin University, Perth, WA 6107, Australia
shuyuan.xu@postgrad.curtin.edu.au
[2] Deakin University, Geelong, VIC 3220, Australia
[3] Western Sydney University, Sydney, NSW 2115, Australia
[4] Shanghai Construction Group Co., Ltd., Shanghai 200092, China
[5] College of Civil Engineering, Fuzhou University, Fuzhou, China
[6] Australasian Joint Research Centre for Building Information Modelling, Curtin University, Perth, Australia

Abstract. Pavement markings take responsibility to communicate with road users regarding travel regulations and guidance. Due to their irreplaceable role in ensuring the safety and order on road, it would be beneficial for road agencies to keep an as-is inventory record of the pavement markings on their roads for managerial operations. However, faced with the sheer volume of their responsible assets, manual inspection would be time-consuming and error prone. This study proposes a vision-based method to automatically detect and classify longitudinal markings using videos of road pavement. Not only line markings, audible markings, as a special category, were also identified in the images. The proposed method can achieve an average 0.89 detection accuracy for line markings and 0.82 for audible markings. Limitations and future work are also proposed. This study tests the possibility of utilising visual data to assist road agencies with an informative management of their civil assets.

Keywords: Pavement marking · Detection · Computer vision · Case study

1 Introduction

Pavement markings and signs constitute the most fundamental way to communicate with road users and they are, in most cases, the most effective way to regulate and guide traffic [1]. And this efficient communication on roads can greatly contribute to a safe and ordered traffic environment. This paper focuses on pavement markings, and especially longitudinal line markings among all the other types. Pavement markings have been designed to be highly standardised, in terms of their colour and appearance, aiming to deliver unambiguous instructions to road users and expect immediate responses from them. Colours used for road markings are limited to white, yellow and blue (very rarely). Detailed dimensions of all types of pavement markings, e.g., lines, symbols, numerals and messages, are all defined in national standards and strictly followed by local agencies.

E. Toledo Santos and S. Scheer (Eds.): ICCCBE 2020, LNCE 98, pp. 1162–1171, 2021.
https://doi.org/10.1007/978-3-030-51295-8_81

In addition to line markings, the importance of audible markings has been increasingly raised in recent years. Essentially, audible markings are created by installing a line of small projections in the road surface, on the top of or alongside ordinary line markings. Therefore, they can provide both audio and vibratory warnings to drivers who are running off the lane due to fatigue or bad driving conditions. They are generally associated with edge lines and especially beneficial for roads in rural areas. Figure 1 illustrates audible markings in concepts and real-life images.

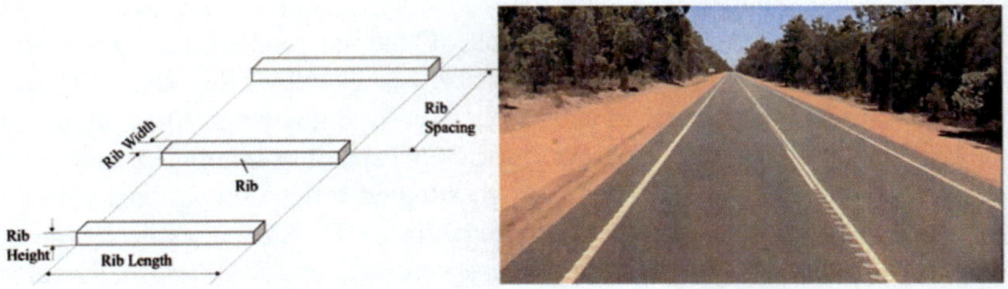

(a) Conceptual (b) Realistic

Fig. 1. Illustrations of audible markings

To maintain road safety and efficiency for citizen's daily commutes, road agencies and traffic organizers have been making extensive efforts to ensure that pavement markings remain in satisfactory condition and keep regulating the traffic. They take responsibility to inspect and accordingly, perform proper maintenance. An as-is inventory data of all pavement markings forms a basis for managerial operations. However, manual collection of such information can be rather time-consuming and due to a lack of human resources, some local agencies and/or government fail to keep an updated record of pavement markings.

Nowadays, computer vision has gained heated attention from both the academia and industry. Its capability to save tedious manual work, automate image processing, and generate reliable detection results has been applied in many disciplines. The AEC/FM industry is, with no doubt, among them. Applications during civil assets' operation and maintenance phases include that defects on structures can be identified on images automatically, and as-is 3D model can be recreated using point clouds or photogrammetry. Such computer vision systems deploy either a combination of image processing techniques and machine learning algorithms, or deep neural networks, to achieve detection, tracking and high-level understanding.

This study collects visual data of road pavement from the Main Roads Western Australia, who governs and manages major routes in Western Australia. A vision-based method was developed and tested to automatically identify different types of pavement markings and classify them according to Australian standards. The rest of the paper is structured as follows: Sect. 2 reviews previous researches of computer vision technology in road pavement management; Sect. 3 illustrates the proposed framework for pavement marking detection and classification; Sect. 4 presents a case study using the video data collected from a road agency in Australia; Sect. 5 indicates limitations and potential future work; and Sect. 6 concludes.

2 Related Work

A major line of studies proposed vision-based lane detection methods to support both advanced driver assistance systems (ADAS) and autonomous driving industry. On the basis of extracted lane line markings on road pavement, various models have been developed to fit the geometric shape of lanes, including, but are not limited to, linear model, parabolic model, hyperbolic model, Clothoid model and spline model. Distinctive features for lane line detection vary. For example, Son et al. [2] utilised colour information while Parajuli et al. [3] relied on local gradients. Alternatively, deep learning algorithms have been employed, for their capability in efficient and accurate object detection. For example, Zang et al. [4] adopted fully convolutional neural network to achieve pixel-wise detection of lane lines. To further assist autonomous driving, other road markings, such as arrows, speed signs and RAIL signs, are targeted for recognition and/or classification. Chen et al. [5] proposed a general framework for intelligent transportation system, by employing Binarized Normed Gradient (BING) features and PCA network to classify road markings.

Another group of studies fall into the area of digital asset management. They focused on using a combination of image processing techniques and machine learning algorithms, to obtain management information for road assets. They analyse road condition by identifying the presence of defects, classifying the damage based on their types and even evaluating their severity. Defects targeted in previous studies included cracks, rutting, potholes and patches. Regarding the first task (i.e., detection), Hoang et al. [6] developed an approach to automatically detect cracks on asphalt pavement, by combining a multi-class SVM classifier and an artificial bee colony optimizer. Chun et al. [7], otherwise, achieved this by integrating image processing techniques with a naïve Bayes-based machine learning algorithm. Similar methods were applied to detect potholes by Azhar et al. [8], which followed by explicit localisation using graph-cut segmentation. Jo and Ryu [9] used a commercial black-box camera and identified potholes based on lane detection results. Tedeschi and Benedetto [10] developed a real-time mobile-based system to recognise several common types of damage: pothole, longitudinal-transverse crack and fatigue crack. But the aforementioned studies can only deal with a particular type of distress at one time. Zalama et al. [11] classified longitudinal and transverse cracks using Gabor filters and an AdaBoost algorithm. Karaköse et al. [12] proposed an approach to further categorise pavement cracks into superficial, crocodile, linear and transverse, and their method was adaptive to different types of road.

Similar to the first application, several studies have successfully employed deep neural networks to detect road damage. By feeding raw images directly into deep learning algorithms, discriminative features of distresses are learned without any image processing techniques. For example, CrackNet, proposed by Zhang et al. [13], employed modified convolutional neural network (CNN) to achieve crack detection on 3D asphalt surfaces, with pixel-level accuracy. Fan et al. [14] modelled pavement crack detection as a multi-label problem and solved it based on structured prediction with CNN. Zhang et al. [15] developed a transfer learning-based framework, to classify cracks and sealed cracks, and extract them at pixel level at the same time. Maeda et al. [16] proposed a smartphone-based application to determine damage status of targeted road, in terms of "no damage", "minimal damage" and "damage needs repair".

Though pavement management has been intensively studied nowadays, few studies focused on pavement marking and their condition assessment. Road agencies need such information as inspection records, and to make informed decision on repainting. An example is the use of end-to-end deep neural networks, which was proposed by Kawano et al. [17], to identify blurry road markings. Another example is done by Maeda et al. [18], who included blur detection of two types of markings, i.e., white line and crosswalk, into their multi-classifier for road damage. Therefore, an efficient framework to inspect pavement markings on images/videos and evaluate their condition with numeric indicators is still lacking in the literature. This research will propose a vision-based framework for pavement marking management, identifying line marking types for inventory purposes and estimating their severity of blur/wear for maintenance decision making.

3 Vision-Based Pavement Marking Detection

The proposed vision-based framework for pavement marking detection can be best understood as a flowchart (as shown in Fig. 2).

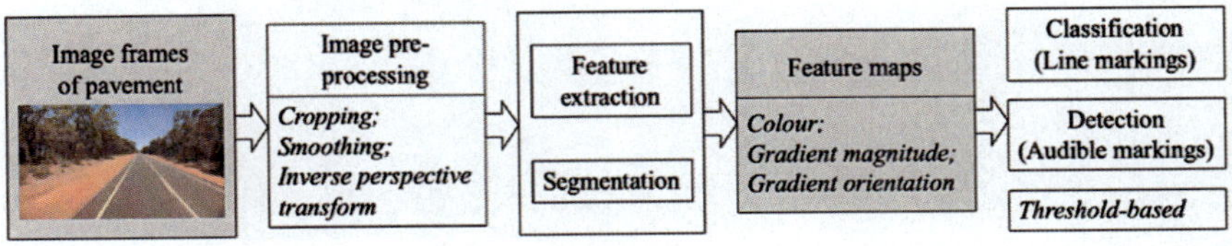

Fig. 2. Flowchart of vision-based line marking detection

3.1 Image Pre-processing

The visual data to be used in this study are videos collected by an in-vehicle camera. The camera was mounted in the front to capture the front-view of road pavement as the vehicle moves forward. During the image pre-processing, image frames are first extracted from the video, cropped and organised in sequence. Smoothing algorithms are then applied to prepare the image for the following processing steps. Examples are Gaussian blur filter and median filter, which has been proven to be efficient in eliminating "salt-and-pepper" noise in images. Another important is image rectification, which removes the perspective effects and generates road-plane images. Inverse perspective transform (IPM) is used and a prerequisite to it is to calculate the transform matrix. One way to compute that is to conduct camera calibration and make sure the stability of the camera system throughout the data collection process. In this way, both intrinsic parameters (i.e., the optical centre and focal length of the camera) and extrinsic parameters (i.e., the position of the camera in the 3D scene) of the entire system can be obtained. Typically, a chessboard is utilised to calibrate and several ad-hoc toolboxes [19] are available. An alternative way to achieve a proper IPM is to utilise at least two pairs of feature points in the image. Such feature pairs are selected if they locate on two parallel straight lines in the bird's eye view. After this step, line markings would be in parallel with each other in the road-plane images, which makes the segmentation of each marking instance easier to achieve. Examples of such top view images are shown in Fig. 3a.

3.2 Feature Extraction and Segmentation

The performance of feature detectors can be adversely affected by variations in lighting conditions, occlusion issues and noises. Potential noises include pavement distresses (e.g., rutting, bleeding and cracking), previous repairs, such as patches and sealed cracks, shadows of road-side facilities and plants, etc. To cope with these environmental factors, a hybrid feature detector, integrating both colour information and gradient features [20], is utilised to generate feature maps. To be specific, colour values of the saturation channel in HSL colour space are extracted to filter out yellow and white pixels in the image, while Sobel operation [21] is used to compute both gradient magnitude and orientation. This hybrid detector was tested to be efficient in eliminating noises like shadows and surface rutting, as shown in Fig. 3.

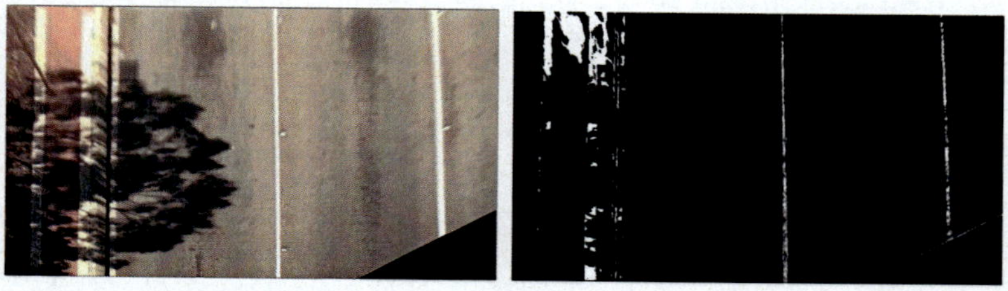

(a) Input (IPM results) (b) Output of feature extraction

Fig. 3. Test of the feature extraction method

On top of the extracted features, individual line markings are segmented. This process is also known as grouping or fitting [22]. To locate each marking instance, Canny [23] is used for edge detection, followed by Hough line Transform [24]. Outliers, such as lines with unreasonable slopes and those locating on non-marking areas, are removed. A grouping method is then applied to extract individual line marking instances.

3.3 Detection and Classification

This step aims to classify different types of pavement markings according to the specifications in Table 1. It lists common types of longitudinal line markings as specified in Standards [25]. A few categories, i.e., single barrier line, edge line, lane line and transition line, convey different instructing messages to road users yet have the same manifestation pattern, thus they are still assembled under a single code (Code 5). The functional differences between them require further reasoning methods to be identified. In addition, a special type of line marking, i.e., audible marking, is targeted for detection.

Table 1. Category of longitudinal line marking in Australian standards

Code	Type	Standard pattern	Illustration
1	Dividing line	Broken, a 3-metre painted stripe and a 9-metre gap	
2	Barrier line (two-way)	Two parallel continuous lines	
3	Barrier line (one-way)	Two parallel lines, one continuous and the other broken	
4	Continuity line	Broken, a 1-metre painted stripe and a 3-metre gap	
5	Single barrier line; Edge line; Lane line; Transition line	Continuous line with 80, 100 or 150-mm width	

To achieve detection and classification, images of extracted and segmented line marking are first analysed and converted into numeric features, that is, width lists representing the width of marking along the vehicle's forwarding direction. Threshold-based methods are then applied on the statistical analysis results of such width lists to detect or classify. Both colour and position information are also utilised to support the process.

4 Case Study Results

To evaluate the proposed vision-based method for pavement marking detection and classification, a case study was carried out. Main Roads Western Australia (MRWA), road agency of the Government of Western Australia, archives pavement video data of their roads. The camera used for recording was mounted on a vehicle which travelled at a uniform speed and in both directions of traffic. Thus, the analysis of each image frame can only focus on the twenty-metre pavement in the near field and save the far-field regions for later. In addition, the two-direction filming is especially necessary for road segments in metro areas, since they are typically too wide to fit into the camera's field of view. The pavement video of a 28-km route in Western Australia was retrieved.

The video data in this case study included five types of longitudinal line markings, i.e., edge line, continuity line, transition line, lane line and dividing line. They were assigned to four classes, as indicated in Table 2. An additional category for classification was introduced to represent non-marking objects, such as road-side railings, kerbs and other vehicles that appear in the region-of-interest. The threshold-based method for detection and classification mainly utilised the distinctive shape information of different line markings. Specifically, dividing lines (code "01") and continuity lines (code "03") are both broken markings with different length of strips and gaps, while marking types in code "02" all appear to be single continuous lines. As for audible markings, they are essentially densely located narrow ribs and thus, have a jagged pattern.

Table 2. Line marking classification results

Code	01	02	03	04	Overall
Type(s)	Dividing line	Edge line, transition line & lane line	Continuity line	Others	
Accuracy	0.92	0.87	0.85	0.92	0.89

Table 2 shows the accuracy of classifying line markings. The detection failures can be attributed to several environmental noises. An example is the pavement in Fig. 4 which suffers from surface distresses and leads to a misleading feature extraction result (as shown in the right image). As a result, both the left edge line and the dividing line were wrongly detected as *Others*. Occlusions by vehicles cause temporary loss of tracking, which can be compensated by referring to the context. Other noises, e.g., kerbs, islands and railings alongside the road, can also confuse the classifier.

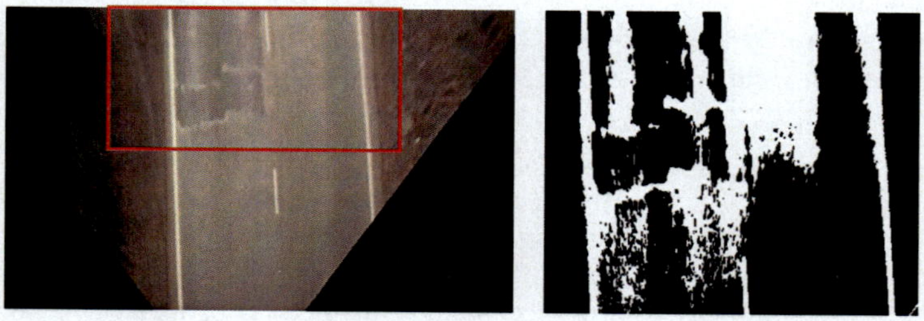

(a) Polished pavement (b) Feature extract result

Fig. 4. Noises on the pavement

The average accuracy of identifying whether they are present reaches 0.82. The performance of detection is highly subject to variations of illuminations and the distance between the targeted markings and camera. Statistics indicate that the detection of markings which delineate the lane our vehicle ran on is more accurate than that of far-away markings (0.52). The detection loss is also partly resulted from the wear of ribs themselves due to intensive traffic.

5 Discussion and Limitation

This section discusses about a few limitations of the proposed method, given the case study results presented in the previous section. One major issue that might negatively affect its applicability in real-life projects lies in the fact that manual selection of features points for the removal of perspective effects in images is required. In addition, such human intervention is not one-off since shooting direction of the camera tends to shift while the vehicle is moving forward. There might also be incremental deviations from the initial camera settings regarding the lane throughout the data collection process. As a result, re-calibration of camera at a certain frequency would be necessary.

Also, from the practical perspective raised by our industrial partners, it would be beneficial if we can use computer vision technology to automatically measure and quantify the condition of detected pavement markings. Such condition assessment is multi-fold. One concern is the physical integrity of these painted markings. Pavement markings suffer from either the loss of a portion/patch at certain location, or general wear and tear on the entire route. Another important property of pavement markings is their reflectorisation [1], which is the capability to be seen under dark driving conditions. The level of their reflectivity can only be tested by professional consultants nowadays, which calls for the development of cost-efficient and measurable indicators. As for the audible markings, the extent of their wearing down by traffic can be informed by measuring the elevation of ribs in 3D laser-scanning profile data, as proposed by Zhang et al. [26]. Potential future agenda can be developed to use vision-based method to automate the condition assessment of various types of pavement markings.

6 Conclusion

This research proposes a vision-based method to automatically collect as-is inventory data of pavement markings. A hybrid feature detector integrating colour and spatial features was employed, individual markings were then segmented, and a threshold-based method was used to classify different types of line markings and detect audible markings as well. The detection can achieve an accuracy of 0.82 and a classification of 0.89 across all types of line markings. Future work will attempt to eliminate the need for human intervention by pre-calibrating the camera system and maintaining its stability throughout the data collection. Vision-based methods for condition evaluation will also be explored.

Acknowledgement. This research was funded by the Australian Government through the Australian Research Council's Linkage Projects (#LP180100222), Discovery Project (#DP180104026). It was also supported by Shanghai Economic and Information Commission Special Fund Programs (Shanghai J-2018-27). We would like to express our cordial thanks to Tom McHugh and Qindong Li (Main Roads Western Australia, MRWA) for their support to this research.

References

1. Austroads: Guide to traffic management part 10: traffic control and communication devices (2019)
2. Son, J., Yoo, H., Kim, S., Sohn, K.: Real-time illumination invariant lane detection for lane departure warning system. Expert Syst. Appl. **42**(4), 1816–1824 (2015)
3. Parajuli, A., Celenk, M., Riley, H.B.: Robust lane detection in shadows and low illumination conditions using local gradient features. Open J. Appl. Sci. **3**(01), 68 (2013)
4. Zang, J., Zhou, W., Zhang, G., Duan, Z.: Traffic lane detection using fully convolutional neural network. In: 2018 Asia-Pacific Signal and Information Processing Association Annual Summit and Conference, APSIPA ASC 2018 - Proceedings, pp. 305–311 (2019)
5. Chen, T., Chen, Z., Shi, Q., Huang, X.: Road marking detection and classification using machine learning algorithms. In: 2015 IEEE Intelligent Vehicles Symposium (IV), pp. 617–621 (2015)
6. Hoang, N.-D., Nguyen, Q.-L., Bui, D.T.: Image processing-based classification of asphalt pavement cracks using support vector machine optimized by artificial bee colony. J. Comput. Civ. Eng. **32**(5), 04018037 (2018)
7. Chun, P.-J., Hashimoto, K., Kataoka, N., Kuramoto, N., Ohga, M.: Asphalt pavement crack detection using image processing and naive Bayes based machine learning approach. J. Jpn. Soc. Civ. Eng. Ser. E1 (Pavement Eng.) **70**(3), 1_1–1_8 (2015)
8. Azhar, K., Murtaza, F., Yousaf, M.H., Habib, H.A.: Computer vision based detection and localization of potholes in asphalt pavement images. In: 2016 IEEE Canadian Conference on Electrical and Computer Engineering (CCECE), pp. 1–5 (2016)
9. Jo, Y., Ryu, S.: Pothole detection system using a black-box camera. Sensors **15**(11), 29316–29331 (2015)
10. Tedeschi, A., Benedetto, F.: A real-time automatic pavement crack and pothole recognition system for mobile android-based devices. Adv. Eng. Inform. **32**, 11–25 (2017)
11. Zalama, E., Gómez-García-Bermejo, J., Medina, R., Llamas, J.: Road crack detection using visual features extracted by Gabor filters. Comput.-Aided Civ. Infrastruct. Eng. **29**(5), 342–358 (2014)
12. Karaköse, M., Akarsu, B., Parlak, K., Sarimaden, A., Erhan, A.: A fast and adaptive road defect detection approach using computer vision with real time implementation. Int. J. Appl. Math. Electron. Comput. **4**(Special Issue-1), 290–295 (2016)
13. Zhang, A., Wang, K.C.P., Li, B., Yang, E., Dai, X., Peng, Y., Fei, Y., Liu, Y., Li, J.Q., Chen, C.: Automated pixel-level pavement crack detection on 3D asphalt surfaces using a deep-learning network. Comput.-Aided Civ. Infrastruct. Eng. **32**(10), 805–819 (2017)
14. Fan, Z., Wu, Y., Lu, J., Li, W.: Automatic pavement crack detection based on structured prediction with the convolutional neural network. arXiv preprint arXiv:1802.02208 (2018)
15. Zhang, K., Cheng, H., Zhang, B.: Unified approach to pavement crack and sealed crack detection using preclassification based on transfer learning. J. Comput. Civ. Eng. **32**(2), 04018001 (2018)
16. Maeda, H., Sekimoto, Y., Seto, T.: Lightweight road manager: smartphone-based automatic determination of road damage status by deep neural network. In: Proceedings of the 5th ACM SIGSPATIAL International Workshop on Mobile Geographic Information Systems, pp. 37–45. ACM (2016)
17. Kawano, M., Mikami, K., Yokoyama, S., Yonezawa, T., Nakazawa, J.: Road marking blur detection with drive recorder. In: 2017 IEEE International Conference on Big Data (Big Data), pp. 4092–4097. IEEE (2017)

18. Maeda, H., Sekimoto, Y., Seto, T., Kashiyama, T., Omata, H.: Road damage detection and classification using deep neural networks with smartphone images. Comput.-Aided Civ. Infrastruct. Eng. **33**(12), 1127–1141 (2018)
19. MathWorks®: Camera calibrator app in: computer vision toolbox™ (2013). https://au.mathworks.com/help/vision/ref/cameracalibrator-app.html
20. Lane-lines-detection-python-opencv. https://github.com/tatsuyah/Lane-Lines-Detection-Python-OpenCV. Accessed 11 Dec 2019
21. Sobel, I., Feldman, G.: A 3×3 isotropic gradient operator for image processing. A Talk at the Stanford Artificial Project in, pp. 271–272 (1968)
22. Forsyth, D.A., Ponce, J.: Computer vision: a modern approach. Prentice Hall Professional Technical Reference (2002)
23. Canny, J.: A Computational Approach to Edge Detection. Morgan Kaufmann, San Francisco (1987)
24. Duda, R.O., Hart, P.E.: Use of the hough transformation to detect lines and curves in pictures. Sri International Menlo Park Ca Artificial Intelligence Center (1971)
25. Australian Standards: Manual of uniform traffic control devices, part 2: traffic control devices for general use (2009)
26. Zhang, D., Xu, X., Lin, H., Gui, R., Cao, M., He, L.: Automatic road-marking detection and measurement from laser-scanning 3D profile data. Autom. Construct. **108**, 102957 (2019)

Virtual Reality for AEC

Design Principles Affecting Motivational and Cognitive Requirements for VR Learning Environments in Engineering Education

Judith Krischler[1(✉)], Andrea Vogt[2], Patrick Albus[2], and Christian Koch[1(✉)]

[1] Faculty of Civil Engineering, Chair of Intelligent Technical Design, Bauhaus-Universität Weimar, Marienstraße 13A, 99423 Weimar, Germany
{judith.krischler,c.koch}@uni-weimar.de

[2] Faculty of Engineering, Computer Science and Psychology, Department Learning and Instruction, Universität Ulm, Albert-Einstein-Allee 47, 89081 Ulm, Germany
{andrea.vogt,patrick.albus}@uni-ulm.de

Abstract. Virtual reality is currently the subject of various studies on its use in teaching as it supports advantageous effects like faster and more effective information processing by human brains. It is also believed to benefit from positive factors such as novelty, embodiment and further motivational effects. These advantages of VR have already been evaluated, so that its use in teaching structural mechanics also promises benefits. A virtual mechanics laboratory provides students with opportunities to consolidate and deepen their knowledge by means of various experiments and to build competences. In this context, the psychological concept of the mental model plays an important role. A mental model is created at the end of each learning process and implies the understanding of the learned knowledge in the overall context. This paper presents a study concept, which investigates the influence on the learning success and the formation of a mental model as well as a qualitative understanding of the students based on various prototypes for virtual reality. Students of engineering science complete experiments of varying degrees of difficulty. For the evaluation, both cognitive and motivational parameters of the test persons are collected, and the results of the different test groups are evaluated by means of success control. The prototypes are designed as applications for VR headsets. Special attention is paid to the necessary degree of attention control and the displayed annotations in order to achieve the best possible learning result without cognitively overloading the user.

Keywords: Virtual reality · Mixed reality · Teaching · Engineering

E. Toledo Santos and S. Scheer (Eds.): ICCCBE 2020, LNCE 98, pp. 1175–1186, 2021.
https://doi.org/10.1007/978-3-030-51295-8_82

1 Background

1.1 Significance of Problem

Franz E. Weinert describes competence as the cognitive abilities and skills available to individuals or learnable by them to solve certain problems, as well as the associated motivational, volitional and social willingness and ability to use problem-solving in variable situations successfully and responsibly [1]. Analytical, systematic thinking and a certain affinity to scientific subjects such as mathematics, physics and chemistry are considered common skills in engineering professions. Spatial imagination and the general ability to mentally transfer abstract technical systems to reality also play a major role.

The traditional engineering education at universities, especially the structural engineering, has always been determined by abstract exercises, which have to be worked out by hand and in a high repetition rate by students. In contrast to the much-commented competence orientation in adult education [2], i.e. the targeted training towards a competence and not towards the mere transfer of knowledge, the university-specific engineering education still represents rather the latter. The above-mentioned competencies for engineering professions are already a prerequisite at the university and are not, or only to a very limited extent, the subject of targeted training. According to Lersch, however, in the competence "knowledge and skills coincide" [3].

The handling of the new working methodology Building Information Modeling (BIM) and the oversized information availability suggests a changed competence profile of future engineers. In a study conducted by the Albert-Ludwigs-University of Freiburg, Germany, on the topic of "Requirements for the engineer of the future", about one third of those questioned named "new hires" as a suitable measure to compensate for competence deficits in the company [4]. In the causal chain, this also places the universities and their teaching under an obligation to educate future engineers accordingly.

1.2 Importance of Cognitive Text and Image Processing for the Choice of Teaching Methods

According to Lersch, the formation of competence requires not only pure knowledge, but also the understanding of what has been learned. In connection with the learning process, the so-called "mental model" is also used. The term, which was coined by P. N. Johnson-Laird, describes the simplified representation of a subject in (human) consciousness [5]. Ideally, this representation should reflect all relevant characteristics of the real facts and make it possible to recognize connections and solve problems based on this deep understanding.

The path to the formation of a mental model can be taken with the help of text and image information. Text information is not only written continuous text in the form of books or lecture notes, but also the spoken word in the form of lectures or audio commentaries. Image information, however, also includes videos, images on slides or in scripts as well as what is seen in (one) reality. For the human brain, pictorial information is easier to process than textual information, and in cognitive image processing the path to the mental model can be shortened [6]. Based on these findings, virtual reality scenarios offer the possibility of contributing to the more efficient formation of a mental model due to their pictorial nature. Figure 1 shows the various intermediate steps in the formation of the mental model. In image processing, the human brain requires fewer processing steps to comprehend what is seen.

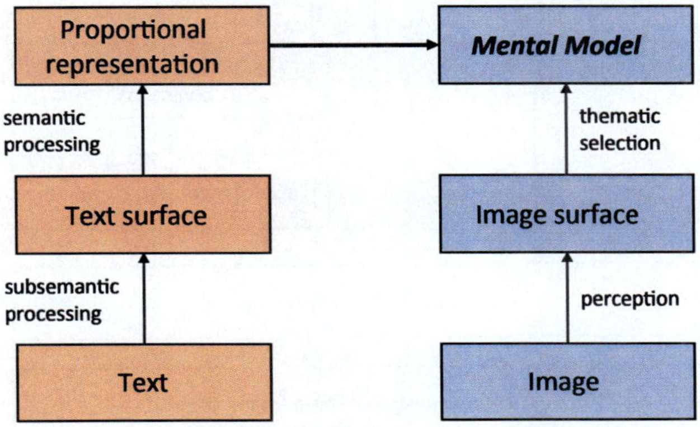

Fig. 1. The path to the mental model via pictorial elements is shorter than via textual elements (figure by Seufert, T., according to Schnotz and Bannert 1999 [6])

Related to this, VR offers further advantages as a means of choice to complement university engineering education. For example, this type of learning allows the exploration of objects and their behavior without destroying them, e.g. in experiments on material failure. Due to the high degree of immersion, the learning process can take place on real objects, under real-looking proportions and conditions. The (perceived) reference to reality thus causes an increase in experience and an integration of what is seen into the mental model via the shortcut of cognitive image processing. Further advantages of VR with regard to the thesis of the embodiment, i.e. positive effects on learning success when body and movements are included in the learning process, can be assumed on the basis of studies [7].

2 Methodology

In the context of this paper, the psychological principles for effective knowledge transfer for the described task, the improvement of engineering theory by means of VR, were first examined. On the basis of the civil engineering bachelor curriculum, suitable examples from the field of mechanics were selected, which are usually taught within the first year of study. The individual learning objectives were derived from the examples and translated into playable VR learning scenarios. After completing the literature research and developing the learning scenarios, the VR prototypes were implemented using Unity and C# scripts. The previously evaluated design principles were incorporated into them. A study design was developed to compare different configurations of the VR prototypes. The most advantageous app design for the learning success is to be found out. The following Fig. 2 shows the methodology.

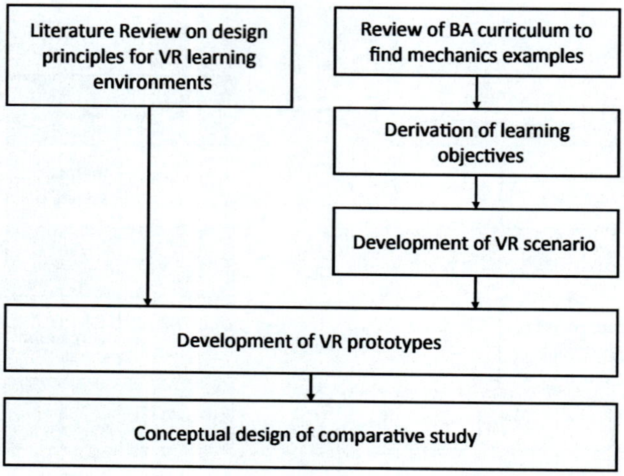

Fig. 2. Methodology of the presented paper

3 Developed Prototypes

3.1 Design Principles

To design appropriate learning materials in conventional learning environments, one can have a look at the multimedia design principles published by Mayer [8]. These design principles include different recommendations to support the learners while learning and to facilitate cognitive learning processes. Some of these principles have already been investigated in virtual reality learning environments and suggest that these multimedia design principles are also applicable for virtual reality learning environments [9]. One possibility to follow the multimedia design principles in virtual reality learning is to include annotations or labels. Placing these annotations close to the corresponding pictorial element follows the recommendation of the spatial contiguity principle. Its positive effect on learning outcomes has already been explored in many studies [10].

Analyzing the effects of annotations more deeply, one could classify annotations as a certain kind of highlighting or signaling, which is also one multimedia design principle [11]. The signaling principle outlines the importance of highlighting relevant information in the learning environment to facilitate the cognitive learning processes by making visual search more efficient and guiding attention [12].

Another possibility to guide attention in a virtual reality learning environment is to use verbal guidance to emphasize the importance of a certain aspect or to explain related element. These auditory prompts can on the one hand include learning content or can on the other hand give additional strategic information how to interact with the learning material [13, 14]. For instance, in conventional learning environments cognitive prompts are successfully used to trigger the use of effective cognitive learning strategies and meta cognitive prompts are used to help monitor and organize the learning process in an efficient way [14]. As in virtual reality learning environments learning content can be displayed by different kinds of representation to outline the relations between these, multiple representations are crucial for learning successfully. Because the aim of learning is to generate one coherent mental model to map corresponding information, one can distinguish these mapping processes into element-to-element and relation-to-relation mapping processes [15, 16]. The term element-to-element mapping refers to supporting the learners on a surficial and syntactical level. In contrast, the term relation-to-relation mapping refers to support the learners to find similarities of processes [16]. Implementing these kinds additional help might reduce the cognitive load while learning.

Baddeley and colleagues assume that only 7 ± 2 units can be stored in working memory [17, 18]. To avoid overloading the working memory, it is important to design the learning environment in such a way that the working memory is used to its most efficient extent [19]. The Cognitive load theory from Chandler and Sweller [20] provides a theoretical framework for this concept and describes that every learning process is associated with cognitive load and that effectively designed teaching and learning materials promote learning, since cognitive capacities can be directed towards the actual learning process [19, 20].

In general, the cognitive load is divided into three dimensions that describe the overall cognitive load and have an additive effect. The intrinsic load, extraneous load and the germane load [19].

The intrinsic load describes the content complexity of the material. Elemental interactivity plays a decisive role in this process. If individual facts can be learned independently of each other, i.e. if one fact can be understood without having to understand other facts at the same time, elemental interactivity is low. If the element interactivity is low, the intrinsic load is also low [19]. Learners reach their capacity limits with complex learning materials because these materials have a high elemental interactivity which also increases the intrinsic load [19].

The extraneous load describes the load that is influenced by the external presentation of the learning material. It can occur if teaching or learning materials are inadequately designed and restrictions of the human working memory are not taken into account [19]. The learner has to deal with irrelevant information generated by, or around the material, which may make it more difficult to concentrate on the actual learning material.

The last form, the learning-related germane load, results from understanding and processing processes during the learning of the material. This part of the load arises when schemata are built up or activated through active engagement with the learning material. It is assumed that the greater the learning-related cognitive load, the better the learning and understanding performance [20]. The use of virtual reality has only an indirect influence on this form of cognitive load [21].

3.2 Conception of Learning Scenarios

For the conception of learning scenarios, individual learning objectives were first defined, which are oriented towards the lecture content for the basics of mechanics for civil engineering students in their first year of study. The focus was on the intention to implement first prototypes under consideration of the design principles described in Sect. 3.1 and to pilot them promptly with first-year students.

However, the thematic preparation of engineering topics as virtual reality applications does not only have potential for first-year students. Seufert [22] shows that, with appropriate complexity, knowledge gains can be expected above all for advanced students. The integration of new knowledge on a known topic and into an already detailed mental model is much easier for learners due to their overview of the topic. Beginners, on the other hand, have to acquire the new knowledge fragments more laboriously, since they still lack this overview [16].

The selected exemplary scenario "beam on two supports/simply supported beam" is intended to introduce the students systematical to the individual components of the equation for deflection, which is the overall learning objective. The following equation is the core objective within this level.

$$f_m = \frac{F * l^3}{48 * E * I_y} [mm]$$

Figure 3 depicts the division of the overall learning objective into several sublevels, each dealing with one component of the superordinate.

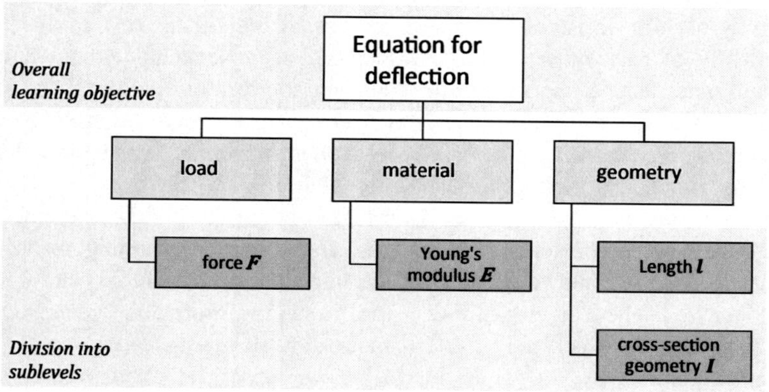

Fig. 3. Splitting the overall learning objective into sublevels

This step-by-step procedure should provide the students a learning scenario with low intrinsic load. Thus, the elements can be understood independently and do not imply high element interactivity. By means of annotations only the information necessary at that moment is conveyed to keep the extrinsic load low. At the end of a level, the relevant learning content is visibly recorded on a virtual blackboard and thus visualizes the ongoing learning progress. Figure 4 shows the implementation of the first sublevel within the VR prototype. The mentioned equation for deflection as well as the discussed part of it is highlighted.

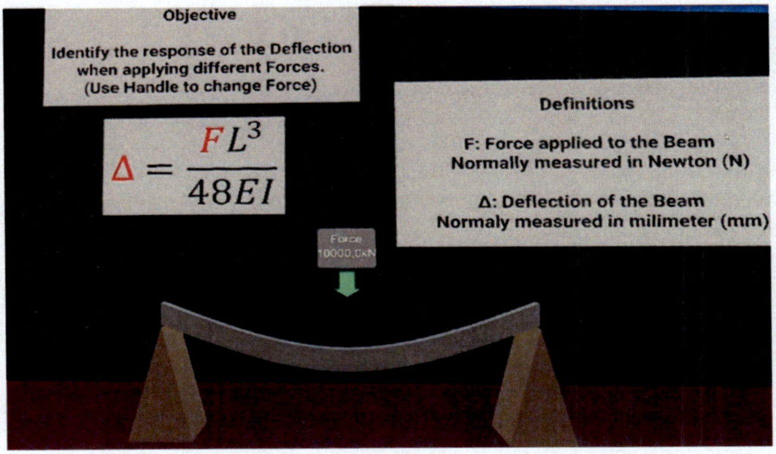

Fig. 4. First level of beam experiment

Each sublevel consists of three phases: the experimentation and observation phase, the conclusion phase and the explanation phase. Figure 5 shows the process that applies for each sublevel.

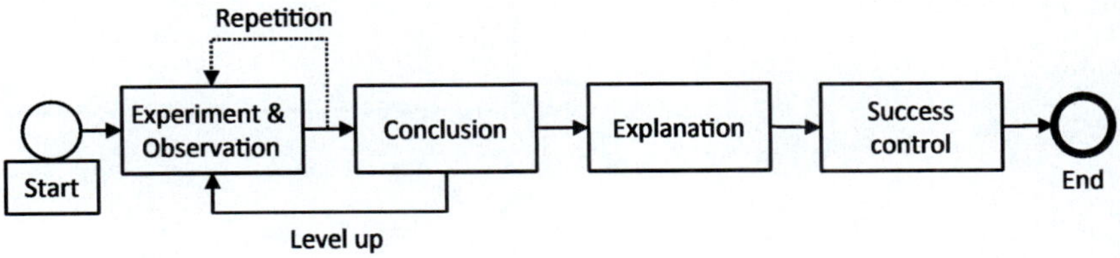

Fig. 5. Phases of level implementation

In the first phase, the participant interacts with the experiment. He/she changes the load application points, the material or the cross-section and observes the influence of these individual parameters on the deflection. After the experiment was performed the participant is asked for his/her conclusion in the form of multiple-choice questions, which is the second phase. The possibility to repeat the experiment is given. Past studies have shown that the success rate of test subjects increases if they are given the opportunity to practice beforehand [22].

If the answer is correct the level moves on to phase 3, the explanation phase, else the participant has the opportunity to repeat the experiment. To avoid infinite loops, the participant has only a limited amount of failures and is automatically moved to the next level after this limit is reached. In phase 3 the participant receives an explanation why his/her observation is correct and which parameter or phenomenon is responsible. The core statement of this explanation is added to the virtual blackboard and available for the next sublevel.

3.3 Technical Framework of the VR Prototypes

For the development of the VR scenarios, the game development environment Unity and Visual Studio for the integration of C# scripts are used. The finished application is exported as.exe from Unity and can be used with commercially available VR glasses and a convenient computer. In this case, HTC Vive Pro head mounted display was used for development. The calculations of the load-bearing behavior are hard-coded depending on the equation which is used for the specific level. For experiments that are more complex a finite element method (FEM) library can be used. The following Fig. 6 depicts the technical setup.

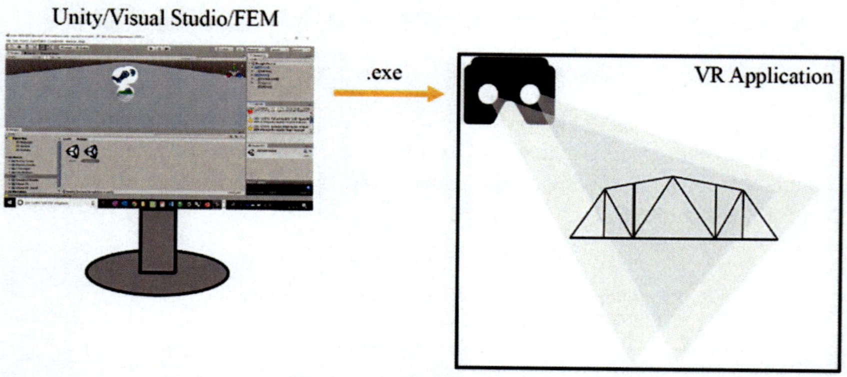

Fig. 6. Technical setup of VR application

4 Study Design

4.1 Execution/Implementation

The main scope of the study is to evaluate the most advantageous design for VR applications in order to impart engineering knowledge and to create best possible conditions to support the formation or enrichment of a mental model, even if the learning content is of varying complexity. In other words, the evaluation which design principles are positively influencing the time required for learning and whether competence development occurs earlier. Thus, the planned study addresses both thematically affine and thematically obverse participants to investigate the effects of the VR design on both laypersons and experts.

A prior knowledge test will be used to determine the thematic affinity of the test subjects and later be deducted as a control variable. Hence, randomization can be maintained even in mixed test and control groups of persons with different knowledge levels.

Studies show that specific prior knowledge has a beneficial effect on image processing and thus on the embedding of new knowledge in existing knowledge structures [22]. Therefore, it is to be expected that well designed VR applications will bring advantages especially for advanced students.

Before the study can be executed, some pretests have to be completed by the participants. Figure 7 depicts the experimental design.

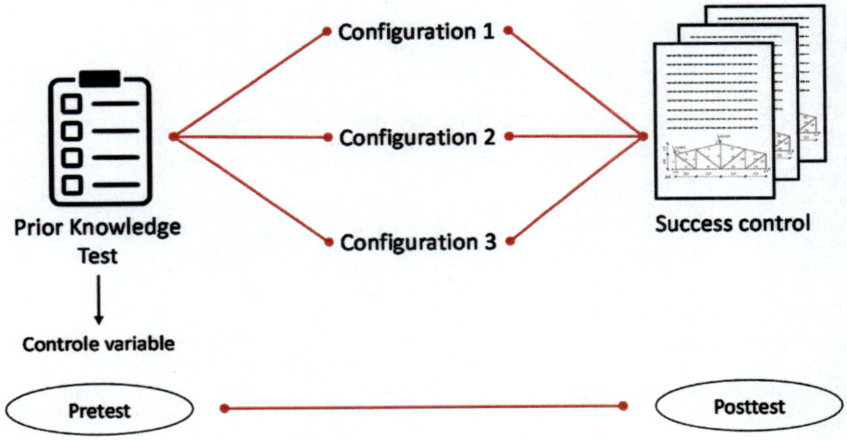

Fig. 7. Experimental design

The experimental design includes first of all a prior knowledge for all participants. The test consists of various questions on mathematical and physical facts, e.g. a question on Newton's second law or related units for describing forces. The questions can be answered in equal parts freely as well as in multiple choice. The answers are scored according to an evaluation matrix and thus enable the classification of the individual prior knowledge of the test person in an appropriate variance. With the help of this knowledge assessment conclusions can be drawn on how well the evaluated design works for beginners as well as for experts.

Before and after the test runs, motivational factors such as self-efficacy, fear of failure, etc. and cognitive factors such as mental effort or cognitive stress (among others, according to Rheinberg et al. [23]) are also collected in the form of pre- and post-tests for further evaluation. In past studies, the novelty effect has shown a positive motivational boost for the test subjects, which could also be effective in the beginning for the aforementioned VR applications. The correlation between motivation and immersion has already been established in previous studies [24], so that a general motivational boost when using VR technique can be expected.

The success control after executing the experiments will depict how well the learning objectives have been internalized and can be applied and combined by the participants. The success of the knowledge transfer is determined on the basis of an unknown problem. To make the different VR design options comparable the same learning content is provided for all test groups, only the design differs.

4.2 Configuration of Design Options

The configuration of the individual VR App variants takes different design elements into account. An overview of the elements resting on the reviewed principles from former studies can be seen in Table 1. The VR options to be compared are composed of these elements.

Table 1. VR key design options matrix

<table>
<tr><td></td><td colspan="6">Elements</td></tr>
<tr><td>Strategical guidance</td><td colspan="3">Text</td><td colspan="3">Audio</td></tr>
<tr><td>Handling assistance</td><td colspan="3">Yes</td><td colspan="3">No</td></tr>
<tr><td>Participation</td><td colspan="3">Active</td><td colspan="3">Passive</td></tr>
<tr><td>Visual annotations, labels, diagrams, equations</td><td colspan="2">Much</td><td colspan="2">Little</td><td colspan="2">No</td></tr>
<tr><td>Signaling & Highlighting</td><td colspan="2">Much</td><td colspan="2">Little</td><td colspan="2">No</td></tr>
</table>

The benefits of auditory or visual strategic guidance have already been assessed in previous studies [13]. All options are first implemented using visual strategic guidance in the form of text instructions.

An assistant in the form of instructions on the controller or equivalent is intended to give users confidence in their interaction in the VR environment. This is intended to reduce additive cognitive stress resulting from action uncertainty as much as possible.

Participation refers to an active or passive role in the VR scenario. Active in this case means that the user not only acts as an observer, but actively influences the experiment and has to take actions him/herself. All VR scenarios developed so far are oriented towards active participation.

In terms of imparting knowledge through visual elements, the degree of use of annotations, labels and diagrams is varied through the different design options.

Signaling and Highlighting of objects, in order to guide attention and considering the multimedia principle as well as the spatial contiguity principle, will be another parameter to evaluate.

The design elements regarding the multimedia principle are expected to help the participants to more efficiently process the learning content and to have a positive influence on learning success as well as to provide a learning environment with low extrinsic cognitive load. To estimate these effects, not only the success control as a posttest is taken into account, also the time the user needed to complete the task as well as the number of failed attempts. Other process data that will be logged during the use of the VR app, e.g. the direction of view, so what the participants look at, how often specific functions are used etc. The final list of process data is currently the subject of preliminary studies.

5 Conclusions and Outlook

This paper presented a literature review as well as a study concept on how to design VR applications in order to optimally support learning of engineering topics. To figure out the most promising configurations of VR apps that are worth to be compared in a study, several didactical and psychological concepts have been taken into account. For this purpose, the game engine Unity and a HTC Vive Pro were used to implement VR prototypes. A basic mechanical concept, the deflection of beams, was selected as a learning scenario and divided into several sublevels that can be applied in VR. The division into sublevels is intended to prevent the cognitive overload of users, especially for the ones not familiar with the topic. Based on the perceptions of various design principles, selected parameters were considered for the configuration. In order to best support the effective creation and enhancement of a mental model, multimedia principles such as visual annotating, labeling, highlighting and signaling were included as elements.

The effect of these elements on learning success will be examined randomly in a first study, considering three different configurations of the same VR prototype. The assessment of success of these effects, pre- and posttests evaluate several cognitive as well as motivational factors of the user. In order to pass the levels within VR also internal success rates can be logged, e.g. the number of failures within a level or the time needed by the user to complete a level successfully. A prior knowledge test allows the maintenance of randomization, as affinity with content can be represented as a control variable. In retrospect, this variable allows effects to be distinguished between beginners and advanced users.

With the help of the study results, recommendations for future design and integration of VR media in university engineering education can be derived.

References

1. Weinert, F.E.: Leistungsmessungen in Schulen. 2, unveränd. Aufl., Dr. nach Typoskript edn. Beltz, Weinheim (2002)
2. Frank, S., Iller, C.: Kompetenzorientierung: mehr als ein didaktisches Prinzip. Report: Zeitschrift für Weiterbildungsforschung **4**, 32–41 (2013)
3. Lersch, R., Schreder, G.: Grundlagen Kompetenzorientierten Unterrichtens. Barbara Budrich-Esser, Leverkusen-Opladen (2013)
4. Albert-Ludwigs-Universität Freiburg: Arbeitswelt Industrie – Anforderungen an den Ingenieur der Zukunft. Weiterbildungsprogramm Intelligente Eingebettete Mikrosysteme.

Hg. v. Albert-Ludwigs-Universität Freiburg (2015). https://www.masteronline-iems.de/fileadmin/user_upload/Downloads/Arbeitswelt_Industrie_IEMS.pdf
5. Johnson-Laird, P.N.: Mental Models. Cambridge University Press, Cambridge (1983)
6. Schnotz, W., Bannert, M.: Einflüsse der Visualisierungsform auf die Konstruktion mentaler Modelle beim Text- und Bildverstehen. Exp. Psychol. **3**, 217–236 (1999)
7. DeSutter, D., Stieff, M.: Teaching students to think spatially through embodied actions: design principles for learning environments in science, technology, engineering, and mathematics. Cogn. Res.: Principles Implications **2**(1), 1–20 (2017)
8. Mayer, R.E.: The Cambridge Handbook of Multimedia Learning. Cambridge University Press, Cambridge (2005)
9. Makransky, G., Borre-Gude, S., Mayer, R.E.: Motivational and cognitive benefits of training in immersive virtual reality based on multiple assessments. J. Comput. Assist. Learn. **6**, 691–707 (2019)
10. Mayer, R.E., Gallini, J.K.: When is an illustration worth ten thousand words? J. Educ. Psychol. **4**, 715–726 (1990)
11. McTigue, E.M.: Does multimedia learning theory extend to middle-school students? Contemp. Educ. Psychol. **2**, 143–153 (2009)
12. Ozcelik, E., Arslan-Ari, I., Cagiltay, K.: Why does signaling enhance multimedia learning? Evidence from eye movements. Comput. Hum. Behav. **1**, 110–117 (2010)
13. Johnson, A.M., Butcher, K.R., Ozogul, G., Reisslein, M.: Learning from abstract and contextualized representations: the effect of verbal guidance. Comput. Hum. Behav. **6**, 2239–2247 (2013)
14. Berthold, K., Nückles, M., Renkl, A.: Do learning protocols support learning strategies and outcomes? The role of cognitive and metacognitive prompts. Learn. Instr. **5**, 564–577 (2007)
15. Gentner, D., Rattermann, M.J., Forbus, K.D.: The roles of similarity in transfer: separating retrievability from inferential soundness. Cogn. Psychol. **4**, 524–575 (1993)
16. Seufert, T., Jänen, I., Brünken, R.: The impact of intrinsic cognitive load on the effectiveness of graphical help for coherence formation. Comput. Hum. Behav. **3**, 1055–1071 (2007)
17. Baddeley, A.: The concept of episodic memory. Philos. Trans. R. Soc. London. Ser. B Biol. Sci. **365**(1413), 1345–1350 (2001)
18. Peterson, L.R., Peterson, M.J.: Short-term retention of individual verbal items. J. Exp. Pschol. **58**(3), 193 (1956)
19. Sweller, J.: Implications of cognitive load theory for multimedia learning. In: The Cambridge Handbook of Multimedia Learning, no. 2, pp. 19–30 (2005)
20. Chandler, P., Sweller, J.: Cognitive load theory and the format of instruction. Cogn. Instr. **4**, 293–332 (1991)
21. Frederiksen, J.G., Sørensen, S.M.D., Konge, L., Svendsen, M.B.S., Nobel-Jørgensen, M., Bjerrum, F., Andersen, S.A.W.: Cognitive load and performance in immersive virtual reality versus conventional virtual reality simulation training of laparoscopic surgery: a randomized trial. Surg. Endosc. **34**(3), 1244–1252 (2019)
22. Seufert, T.: Training for coherence formation when learning from text and picture and the interplay with learners' prior knowledge. Front. Psychol. **10**, 193 (2019)
23. Rheinberg, F., Vollmeyer, R., Burns, B.D.: FAM: Ein Fragebogen zur Erfassung aktuller Motivation in Lern- und Leistungssituationen. Diagnostica **2**, 57–66 (2001)
24. Georgiou, Y., Kyza, E.A.: Relations between student motivation, immersion and learning outcomes in location-based augmented reality settings. Comput. Hum. Behav. **89**, 173–181 (2018)

BIM-GIS Integration in HoloLens

Ralph Tayeh[1(✉)], Fopefoluwa Bademosi[2], and Raja R.A. Issa[1]

[1] Rinker School of Construction Management, University of Florida, Gainesville, FL 32603, USA
{ralph.tayeh,raymond-issa}@ufl.edu
[2] Wentworth Institute of Technology, Boston, MA 02115, USA
bademosif@wit.edu

Abstract. In recent decades, technological advancements have led to the introduction of wearable computing devices allowing visualization using virtual, augmented, and mixed reality. The Architecture, Engineering, and Construction industry has seen an increase in the use of such wearable technology, especially with the introduction of information modeling. Building Information Modeling (BIM) has been used extensively on vertical construction projects to better communicate information among project stakeholders and facilitate the construction process. More recently, infrastructure projects started using information modeling along with Geographic Information Systems (GIS) on local or cloud services. The integration of GIS and BIM has the potential to improve the construction process and aid in decision-making, especially when combined with newer visualization techniques. This paper presents a platform for GIS and BIM integration in Mixed Reality. The proposed application will allow a seamless transfer of data from BIM and GIS software into game engines, and the visualization of the BIM-GIS integration on Microsoft HoloLens 2. The application will allow users not only to visualize models, but to explore information in model elements, and make model changes in the mixed reality environment.

Keywords: BIM · GIS · HoloLens · Visualization

1 Introduction

With the beginning of the Fourth Industrial Revolution (Industry 4.0), global industries are facing major drivers of transformation. Advancements in machine learning, artificial intelligence, holography, 3D printing, robotics, nanotechnology, Internet of Things (IoT), to name a few, are all shaping the future of many industries. At the cusp of this revolution, it is essential to understand how these technologies work and how they can affect the daily workflow processes of different industries.

The Architecture, Engineering, Construction, and Operations (AECO) industry is known to have a slow adoption rate of technologies. While many industries have been using these technologies extensively, the AECO industry is still facing many challenges that can be solved using new technological approaches. One of the latest paradigms shifts in the AECO industry was the use of Building Information Modeling (BIM). BIM allows the big amount of data shared on construction projects to be

E. Toledo Santos and S. Scheer (Eds.): ICCCBE 2020, LNCE 98, pp. 1187–1199, 2021.
https://doi.org/10.1007/978-3-030-51295-8_83

aggregated in one model that can be used throughout the project lifecycle. BIM has been widely used on vertical projects and has started to gain attention on horizontal and infrastructure projects. Traditionally, Geographic Information Systems (GIS) has been used to aid in horizontal construction, create maps, analyze tabular data, and visualize results on two-dimensional (2D) maps or three-dimensional (3D) scenes. The integration of GIS in BIM software combines the benefits of both technologies and allows for better visualization and decision-making.

Among the technological advancements of Industry 4.0, virtual reality (VR), augmented reality (AR), mixed reality (MR) and holography have been used in many industries to visualize and interact with virtual models in real-time. Construction researchers have explored the use of VR and AR in depth; however, the use of MR and holography has not been fully discussed, especially using head-mounted devices such as Microsoft HoloLens 2.

The aim of the study is to evaluate how holographic applications of integrated BIM-GIS can improve the learning experience of construction management students. BIM and GIS have been integrated in the literature through different methods. The integration has been proven effective in urban planning and in better understanding BIM models with respect to the surroundings. However, visualization of the integration was limited to computer monitors or tablets. Newer methods of visualization were not considered. The study will evaluate the benefits of visualizing BIM and GIS data in holographic environments and the impact of this visualization on students' spatial understanding and collaborative problem-solving skills. This paper will present the framework of the study, discussing the gaps in the literature, the methodology, and the expected results

2 Literature Review

Considering the complex nature of the AECO industry, it is of utmost importance to benefit from technological advancements and tailor them to facilitate the evaluation and communication of information on construction projects [1]. Major changes in the construction industry started with the introduction of computer-aided design (CAD) and grew later on with information systems. Today, wearable technologies are starting to gain attention in the construction industry to leverage the vast amount of information on construction projects, better collaborate between project stakeholders, and efficiently deliver projects.

2.1 Information Systems

Since the earliest recorded human history, information systems have been used in the form of maps to preserve and transmit geographic data by means of visual representation. More recently, maps have evolved into more robust platforms with the introduction of GIS, allowing users to better visualize and analyze data. Construction information systems evolved from 2D drawings to CAD drawings, and in the 21st century to BIM models. BIM technology is mainly centered on buildings and indoor environments and provides a rich data source on all building components (architectural, structural, and mechanical) that can be used throughout the project lifecycle. GIS, on the other hand, focuses on outdoor environments and provides insights on urban and logistics planning. Integrating BIM and GIS extends the benefits of building models to outdoor environments [2] and allows for better visualization of the utilities infrastructure.

Geographic Information Systems. GIS is a science and a technology that relies on organizing data into georeferenced layers. GIS is defined as a framework for collecting, managing, and analyzing data [3]. GIS can integrate different types of data as layers of information used for the visualization of maps and 3D scenes. In addition, GIS helps users in the decision-making process based on its ability to structure data into patterns, situations, and relationships. Recently, GIS platforms started shifting from offline desktop applications to web platforms based on cloud computing, thus allowing users to use layers from different providers in the development of their own applications on the web, smart devices, or desktops.

GIS models were originally hosted offline with local files on a single computer. They then evolved into a central database environment hosted on servers. The latest advancements for GIS models are online platforms or Web GIS. Web GIS is an online platform for map creation and geographic information analysis. Layers in Web GIS each have a specific web address to facilitate online data sharing. Also, since layers are georeferenced, users can benefit from millions of data layers from different providers and use them to build their own applications on the web, smart devices, or desktops. Furthermore, dataset layers in GIS are dynamic, cover a given area, and represent fixed or moving objects, boundaries, events, or spatial phenomenon.

In addition to being used in virtually every field, analysis in GIS is very beneficial in the construction industry. In the preconstruction phase, GIS has been successfully used for site layout planning [4], route planning for large vehicles [5], and human and materials resource management [6, 7]. During construction, GIS can be used for scheduling construction activities with respect to the building surroundings [2, 8], managing earthmoving activities [9], and planning for safety [10].

Construction Information Systems. BIM, created in the mid-1970s, is a process developed to integrate complete graphical and non-graphical information to represent building components and their spatial relationships at a single platform [11]. It is a practical and innovative integrated data platform to visualize and exchange building information during the entire lifecycle of a project [12]. BIM is a central database enabling better control of the construction process by minimizing information loss and promoting better internal coordination, cross-disciplinary coordination, decision-making support, productivity improvement, and risk management [13–15]. The construction literature has documented the effectiveness of BIM in reducing errors in design and consequent reworks [16], and in improving the efficiency of the delivery process of projects [17].

The adoption rate of BIM has increased significantly in the past decade in the United States, the United Kingdom, Europe, Japan, and Korea [18–20]. This increase in adoption is mainly caused by the ability of BIM to address many of the construction industry's challenges such as low productivity, rework, poor functionality, and poor waste management [20–22]. In addition to being used mainly in vertical construction,

BIM practice has been extended to horizontal, heavy, and green BIM [23]. This widespread of BIM applications has increased the demand for a competent and skilled workforce [23], and motivated research in technological perspectives [24] and managerial perspectives [25, 26].

Interoperability Between BIM and GIS. Over the last decade, many research studies focused on the integration of BIM and GIS in the construction industry. It was proven that the integration could be useful for collaboration and cooperation [27]. Different methods of integration were discussed in the literature; Ma and Ren [28] categorized them into three types, (1) extracting BIM data to a GIS software, (2) extracting GIS data to a BIM platform, and (3) extraction and deployment of GIS and BIM onto a third party platform. The method developed by de Laat and Van Berlo [29] focusing on extracting BIM data to GIS was the mainstream of the integration methods discussed in the literature [30].

This first method of integration uses the Industrial Foundation Class (IFC) and City Geography Markup Language (CityGML) to exchange semantic information from BIM and 3D geography from GIS [31]. The exchange of data between BIM and GIS led to many research studies. For example, Mignard and Nicolle [32] integrated BIM and GIS using semantic graphs and ontologies to manage urban facilities. Wu and Zhang [33] used a similar integration approach to help in urban spatial planning and shadow analysis. Salimzadeh et al. [34] used the integration for urban energy assessment. Furthermore, GIS and BIM can be integrated using web platforms. Wu et al. [35] utilized a GIS-BIM platform to create a cloud-based Virtual Facility Energy Assessment tool to help decision-makers identify building energy-efficient improvement opportunities and assess real-time energy performance. This method integrated the details of individual buildings with the contextual geospatial information. Niu et al. [36] also developed a web-based GIS-BIM integration platform to support building energy designers and urban planners. Their study used Green Building XML (gbXML) to export BIM data from Autodesk Revit® and integrate it with Keyhole Markup Language (KML) from GIS data.

In addition to being used for energy analysis and urban planning [37, 38], the integration of BIM and GIS can be used in construction management applications. Liu and Issa [39] integrated multiple software (ArcGIS, Autodesk® Civil 3D®, and Revit®) to visualize building mechanical ductwork and pipelines in outdoor networks. Their approach focused on exporting BIM and GIS data into Autodesk® Civil 3D® to visualize the position and elevation of mechanical components in the building site surface area. In a similar context, Cheng and Deng [40] developed a platform to convert 2D CAD files of underground utilities into 3D BIM models and integrate it with GIS for automatic design and clash detection in GIS. Kim et al. [41] developed a BIM-GIS integration system for earthwork calculations in highway construction by using web semantics to retrieve data. Their platform was able to perform cut and fill simulations and generate an optimal equipment plan. More recently, Wang et al. [42] used the integration to visualize the status of construction supply chains, and Deng et al. [43] integrated 4D BIM and GIS to provide sufficient construction information and real-time analysis for the supply chain management.

In short, integrating GIS and BIM offers scalable solutions for preconstruction, construction, and facility management beyond the building perimeter. It also helps decision-makers visualize and analyze the building performance in the context of a broader enterprise.

2.2 Construction Technologies

In many engineering and science fields, good visualization of the subject matter can improve the user's cognition and learning experience. Moreover, visualization helps better communicate information and apply abstract concepts to real-world circumstances. In the construction industry, projects are becoming more complex with a great amount of information shared between project stakeholders. Widespread use of BIM has revolutionized the means construction professionals visualize their projects. BIM models provide rich visualization environments allowing the user to rotate and walk-through models on computer and tablet monitors. More recently, VR has been used to give users a more realistic and immersive walkthrough in the model. The construction literature has explored many use cases of VR headsets for several applications including cost estimating [44], safety training [45], construction equipment operations [46], sustainable design construction [47], and architecture and engineering education [48, 49].

Contrarily to VR where the user is completely immersed in the virtual world, AR blends real-world objects with digital VR content, allowing computer-generated data to be fed into the user's view of a real scene in real-time and interactively. AR helps users efficiently interpret digital and physical content and the relations between them and gives the illusion of holograms that can be interacted with using gestures, gaze, and voice commands. In the medical field, AR has been extensively used to visualize the human body and better communicate medical information [50, 51]. In the AECO industry, the applications of AR are recently emerging, especially with the introduction of head-mounted AR devices such as Microsoft HoloLens.

Augmented Reality. The first applications of AR go back to the 1960s when Sutherland [52] started developing a head-mounted device to visualize 3D graphics. However, the term augmented reality was not used until 1992 at Boeing Corporation during the development of an experimental system aiming at helping workers assemble wiring harnesses [53]. Azuma [54] established three main characteristics of a successful AR/MR environment: (1) the combination of both real and virtual objects should happen in the real world, (2) the combination should run in real-time, and (3) objects should be aligned and registered in 3D. The first applications of AR for the AECO industry was published by Feiner et al. [55] who developed an application to provide information about buildings in a 3D graphical tour. The next milestone in AR applications was the use of 3D markers that could be captured by AR devices to visualize virtual objects [56, 57]. More robust AR/MR applications and games became available with the introduction of Google Glasses in 2014 and Microsoft HoloLens in 2016.

In the AECO industry, the applications of AR can be divided into three main categories, namely mobile computing, coordination, and communication and visualization. Since construction sites are continually changing, mobile computing is necessary to access information at any place, and any time, AR is a technology that can be used to integrate the design model with the actual as-built project. For example, Roberts et al. [58] developed an AR application to visualize underground utilities and help workers with the excavation process. The second category of AR applications is coordination, which is a process during which project stakeholders solve issues arising throughout the construction process. For instance, Georgel et al. [59] developed an application to supervise the construction progress, inspect the final product, and check for differences between design models and as-built facilities. Dai et al. [60] and Golparvar et al. [61] visualized project progress deviation using 4D models and site photographs. For positioning and inspection. Finally, other researchers have focused on the third category, visualization and communication, and developed applications for AR/MR to improve performance of virtual site visits [62] and enhance collaboration opportunities [1], visualize interior design [63], interact with buildings and their thermal environments [64], visualize buildings as interactive holograms [65], and other applications related to construction operations planning, inspection, safety, and education [66].

Holography. In the medical field, holograms have the potential to help medical workers and doctors in their diagnosis of complex medical conditions [67]. In the construction field, Kalarat [68] developed building holograms using the hologram pyramid. The pyramid can display the hologram from three sides, and the user can rotate the building using hand gestures. The application was created using a game engine. Even though the developed holographic experience was proven as an effective visualization tool, Kalarat [68] did not address the issue of information interoperability between BIM modeling platforms and the used game engine. Moreover, the functionalities of the hologram pyramid were only centered on simple visualization of architectural components of buildings.

2.3 Limitations and Contribution

After reviewing the literature, the main limitation that is driving the scope of this research is the lack of a study discussing the benefits of holography in construction management education. The literature discussed several applications for BIM and GIS when used separately or when integrated together. Moreover, many studies evaluated the use of new techniques, including VR, AR, and MR, to visualize BIM models or GIS data. However, none of the studies focused on an integrated visualization of BIM and GIS information in holographic environments. The use of AR in the construction literature was mostly evaluated using phones and tablets, and only a few studies focused on the use of new wearable devices such as Microsoft HoloLens 2 to visualize models. The main contribution of this study is to evaluate the benefits of using holography to visualize integrated BIM-GIS environments. The methodology section of this paper further elaborates on the different steps needed to accomplish this purpose. The steps followed in the creation of the application addressed the limitations discussed in this section. Moreover, the application will be tested in an experimental study to evaluate the effectiveness of holography and its ability to improve construction management students' understanding of construction activities.

3 Methodology

To accomplish the purpose of this study, a three-step process was followed. First, a review of commercially available products that help with the development of the HoloLens application was conducted. After this review, the application was developed using commercial products and C# coding. Once the application was developed, it was tested through an experimental study, and the results were analyzed using descriptive and inferential statistics.

First, the previously developed interactive hologram [65] was further developed to include the functionalities needed in this study. Second, a HoloLens application was developed in Unity. For both applications, the BIM model was exported from Revit® using a previously developed plugin [65]. Similarly, a script was developed to extract GIS data from Infraworks. The exported data from Revit® and Infraworks was combined in Unity, and an application was developed to visualize BIM-GIS environments on an interactive hologram or on the HoloLens.

The Xbox Kinect was added to the interactive hologram to allow for user input using hand gestures and voice. Different plugins were used to achieve the required data interoperability. The developed applications were evaluated to achieve a good user's comfort, holographic frame, and spatial mapping.

3.1 Application

Revit® Plug-in. The author previously developed a plugin to facilitate the export of models from Revit® to Unity. With the introduction of Unity Reflect and the PiXYZ Plugin, the export has become easier. However, these plugins focus on the export of just colors, materials, and textures. Therefore, the previously developed plug-in [65] was revised to improve it, integrate the new Unity plugins, and export more data from Revit® to Unity. The plugin was developed using C# and Revit® 2020 API.

Infraworks Script. With BIM being used more and more on infrastructure projects, Autodesk® Infraworks is the software of choice for export of GIS data. Infraworks allows the user to add multiple GIS layers and visualize them in 3D. Moreover, models can be exported from Revit® to Infraworks and placed in the corresponding scene. This integration between Revit® and Infraworks facilitates the seamless transition of data into Unity. A script will be developed using JavaScript to export GIS data from Infraworks to Unity and integrate it with data from the BIM model exported using the developed plugin.

Unity and HoloLens 2. Once the models were exported from Revit® and Infraworks to Unity, the latter was used to develop the HoloLens application. The Mixed Reality Toolkit (MRTK) was used in the development. The MRTK is an open-source toolkit for mixed-reality applications, and it helps in the development of the input system, hand and eye tracking, voice commanding, gaze and selection, teleportation, UI controls, and spatial understanding.

3.2 Experimental Study

Hypotheses of this Study. The purpose of this study is to evaluate the benefits of using holograms (interactive hologram or HoloLens) to visualize integrated BIM-GIS data and to study how this visualization enhances students' spatial understanding and collaborative problem-solving skills. The holographic intervention will be compared to traditional visualization methods, including two-dimensional drawings or maps and three-dimensional models or scenes. Therefore, the null hypothesis is formulated as follows,

- Holographic representations of integrated BIM-GIS improve students' spatial understanding and collaborative problem-solving involving indoor and outdoor environments in comparison to traditional visualization methods.

This hypothesis aims first at comparing the interactive hologram to traditional visualization methods and then comparing the developed HoloLens application to the interactive hologram and traditional visualization methods.

Scope of the Experiment. The hypothesis will be evaluated considering the three main phases of a construction project lifecycle, preconstruction, construction, and post-construction. Students will be assigned tasks for each phase to solve using the different methods, first individually and then collaboratively.

Examples of tasks used to evaluate the student's spatial understanding of preconstruction tasks will include,

- Evaluating the location of underground utility lines,
- Evaluating the positioning of the construction project vis-à-vis its surroundings,
- Evaluating the best alternatives for pipe connections between the building and the outdoor environment.

Examples of tasks used to evaluate the student's spatial understanding of construction tasks will include,

- Coordination of different building components,
- Coordination of building components with outdoor underground or aboveground utilities,
- Deciding on the best site logistics plan.

Examples of tasks used to evaluate the student's spatial understanding of post-construction tasks will include,

- Locating hidden elements within the building, such as mechanical equipment beyond the ceiling,
- Understanding preexisting connections between the building and outdoor environment,
- Visualizing the impact of expanding an existing building on the surrounding environment.

Each of the above tasks will be completed using first traditional methods, then the interactive hologram, and then the HoloLens application. Following the individual tasks, students will be paired in teams to solve similar problems and come up with collaborative decisions. The group tasks will also be divided per project phase. For example, group tasks during the preconstruction phase will include,

- Deciding on the best route for earthmoving vehicles,
- Taking off building quantities and estimating quantities of outdoor underground utility lines.

Examples of group tasks used to evaluate the students' collaborative problem solving during the construction phase of the project will include,

- Visualizing the effect of change orders and coming up with solutions to minimize rework time and cost,
- Visualizing safety hazards in indoor and outdoor environments and deciding on the best safety plan to follow.

Finally, examples of group tasks used to evaluate the students' collaborative problem solving during the construction phase of the project will include,

- Visualizing the consequences of underground maintenance on the building surrounding and deciding on the best process sequences for earthmoving activities,
- Visualizing building upgrades causing changes to the connections between indoor and outdoor environments.

Similar to individual tasks, each group of students will be asked to evaluate each scenario using first traditional visualization methods, then using the interactive hologram, and finally using the developed application for HoloLens. The data collected from the different parts of the experimental study will be statistically analyzed to accept or reject the null hypothesis answering thus the research question and assessing the benefits of using holography and mixed reality during the lifecycle of the project. The data will also be used to decide on the best method to use during each task execution.

4 Expected Results

The integration between BIM and GIS is expected to be effective and easy to use. It is also expected that this integration will be the basis of other research projects, especially with more infrastructure projects using BIM in Infraworks. Moreover, the scripts developed to extract data from models will be beneficial to develop other applications in game engines that focus not just on visualization, but on investing in the information embedded in the model elements. Finally, the application itself is expected to be beneficial for students and industry professionals during the lifecycle of the project. The building holograms will enhance the users' understanding of the models and allow them to make markups and communicate information efficiently.

The developed experimental study will serve as a use case to validate the effectiveness of the development platform. Further studies on the developed hardware and software can occur through three different use cases. First, a study can be conducted with the same pool of participants but using different building models to affirm the results obtained from the original study. Moreover, participants can be sought to develop their own holographic environments using the developed applications to evaluate the ease of use of such technologies on any construction project. Finally, industry professionals can be sought to evaluate the benefits of holography and mixed reality from a professional standpoint and further study their applications during everyday tasks on the construction site.

References

1. Rankohi, S., Waugh, L.: Review and analysis of augmented reality literature for construction industry. J. Visual. Eng. **1**, 1–9 (2013)
2. Karan, E.P., Irizarry, J., Haymaker, J.: BIM and GIS integration and interoperability based on semantic web technology. J. Comput. Civ. Eng. **30**(30) (2015)
3. Harder, C., Brown, C.: The ArcGIS Book: 10 Big Ideas About Applying the Science of Where. Esri Press, Redlands (2017)
4. Su, X., Andoh, A.R., Cai, H., Pan, J., Kandil, A., Said, H.M.: GIS based dynamic construction site material layout evaluation for building renovation projects. Autom. Constr. **27**, 40–49 (2012)
5. Varghese, K., O'Connor, J.T.: Routing large vehicles on industrial construction site. J. Constr. Eng. Manage **121**(1), 1–12 (1995)
6. Anumba, C., Dainty, A., Ison, S., Sergeant, A.: The application of GIS in the construction labour market planning. Constr. Innov. **5**(4), 219–230 (2005)
7. Li, H., Kong, C.W., Pang, Y.C., Shi, W.Z., Yu, L.: Internet based geographic information systems system for e-commerce application in construction material procurement. J. Constr. Eng. Manag. **129**(6), 689–697 (2003)
8. Bansal, V.K., Pal, M.: Generating, evaluating and visualizing construction schedule with geographic information systems. J. Comput. Civ. Eng. **22**(4), 233–242 (2008)
9. Hassanein, A.: Planning and scheduling highway construction using GIS and dynamic programming. Ph.D. thesis, Concordia University, Canada (2002)
10. Atay, H., Toz, G.: GIS as a planning tool for health and safety management in linear construction projects. J. Earth Sci. Eng. **1**, 132–139 (2011)
11. Wix, J., See, R.: IAI (International Alliance of Interoperability): an introduction to the International Alliance for Interoperability and the industry foundation classes, Oakton, VA (1999)
12. Hwang, B.G., Zhao, X., Yang, K.W.: Effect of BIM on rework in construction projects in Singapore: status quo, magnitude, impact, and strategies. J. Constr. Eng. Manag. **145**(2) (2018)
13. Eastman, C., Teicholz, P., Sacks, R., Liston, K.: BIM Handbook: A Guide to Building Information Modeling for Owners, Managers, Designers, Engineers and Contractors, 2nd edn. Wiley, Hoboken (2011)
14. Lu, Q., Chen, L., Lee, S., Zhao, X.: Activity theory-based analysis of BIM implementation in building O&M and first response. Autom. Constr. **85**, 317–332 (2018)
15. Volk, R., Stengel, J., Schultmann, F.: Building information modeling (BIM) for existing buildings—literature review and future needs. Autom. Constr. **38**, 109–127 (2014)
16. Porwal, A., Hewage, K.N.: Building Information Modeling (BIM) partnering framework for public construction projects. Autom. Constr. **31**, 204–214 (2013)
17. Dimitrov, A., Golparvar-Fard, M.: Vision-based material recognition for automated monitoring of construction progress and generating building information modeling from unordered site image collections. Adv. Eng. Inf. **28**(1), 37–49 (2014)

18. Jung, W., Lee, G.: The status of BIM adoption on six continents. Int. J. Civ. Environ. Struct. Constr. Archit. Eng. **9**(5), 444–448 (2015)
19. McGraw-Hill Construction: The Business Value of BIM in North America: Multi-Year Trend Analysis and User Ratings (2007–2012). Smart Market Report. McGraw-Hill Construction, New York (2012)
20. NBS: NBS international BIM report 2016. Newcastle upon Tyne, UK (2016)
21. Badrinath, A.C., Chang, Y.T., Hsieh, S.H.: An overview of global research trends in BIM from analysis of BIM publications. In: 16th International Conference on Computing in Civil and Building Engineering, Osaka, Japan (2016)
22. Hu, M.: BIM-enabled pedagogy approach: using BIM as an instructional tool in technology courses. J. Prof. Issues Eng. Educ. Pract. **145**(1), 1–12 (2018)
23. Zou, P.X.W., Xu, X., Jin, R., Painting, N., Li, B.: AEC students' perceptions of BIM practice at Swinburne University of Technology. J. Prof. Issues Eng. Educ. Pract. **145**(3), 1–12 (2019)
24. Bademosi, F., Tayeh, R., Issa, R.R.A.: Skills assessment for robotics in construction education. In: Workshop on Intelligent Computing in Engineering, Leuven, Belgium (2019)
25. He, Q., Wang, G., Luo, L., Shi, Q., Xie, J., Meng, X.: Mapping the managerial areas of building information modeling (BIM) using scientometric analysis. Int. J. Proj. Manag. **35**, 670–685 (2017)
26. Oraee, M., Hosseini, M.R., Papadonikolaki, E., Palliyaguru, R., Arashpour, M.: Collaboration in BIM-based construction networks: a bibliometric-qualitative literature review. Int. J. Proj. Manag. **35**(7), 1288–1301 (2017)
27. Kim, M., Chai, S.: The impact of supplier innovativeness, information sharing and strategic sourcing on improving supply chain agility: global supply chain perspective. Int. J. Prod. Econ. **187**, 42–52 (2017)
28. Ma, Z., Ren, Y.: Integrated application of BIM and GIS. An overview. Procedia Eng. **196**, 1072–1079 (2017)
29. de Laat, R., Van Berlo, L.: Integration of BIM and GIS: the development of the CityGML GeoBIM extension. In: Advances in 3D Geo-Information Sciences, pp. 211–225. Springer, New York (2011)
30. Song, Y., Wang, X., Tan, Y., Wu, P., Sutrisna, M., Cheng, J.C., Hampson, K.: Trends and opportunities of BIM-GIS integration in the architecture, engineering and construction industry: a review from a spatio-temporal statistical perspective. ISPRS Int. J. Geo-Inf. **6**(12), 39–48 (2017)
31. Gröger, G., Plümer, L.: CityGML—interoperable semantic 3D city models. ISPRS J. Photogramm. Remote Sens. **71**, 12–33 (2012)
32. Mignard, C., Nicolle, C.: Merging BIM and GIS using ontologies application to urban facility management in ACTIVe3D. Comput. Ind. **65**(9), 1276–1290 (2014)
33. Wu, B., Zhang, S.: Integration of GIS and BIM for indoor geovisual analytics. In: XXIII ISPRS Congress (2016)
34. Salimzadeh, N., Sharif, S.A., Hammad, A.: Visualizing and analyzing urban energy consumption: a critical review and case study. In: Construction Research Congress, Reston, VA, pp. 1323–1331 (2016)
35. Wu, W., Yang, X., Fan, Q.: GIS-BIM based virtual facility energy assessment (VFEA) – framework development and use case of California state university, Fresno. In: Computing in Civil Engineering, pp. 339–346 (2014)
36. Niu, S., Pan, W., Zhao, Y.: A BIM-GIS integrated web-based visualization system for low energy building design. Procedia Eng. **121**, 2184–2192 (2015)

37. Sebastian, R., Bohms, H.M., Bonsma, P., van den Helm, P.W.: Semantic BIM and GIS modelling for energy-efficient buildings integrated in a healthcare district. In: ISPRS 8th 3DGeoInfo Conference & WG II/2 Workshop, Istanbul, Turkey, pp. 255–260 (2013)
38. Zhang, X., Arayici, Y., Wu, S., Abbott, C., Aouad, G.: Integrating BIM and GIS for large scale (building) asset management: a critical review. In: 12th International Conference on Civil, Structural and Environmental Engineering Computing, Funchal, Madeira, Portugal (2009)
39. Liu, R., Issa, R.R.A.: 3D visualization of sub-surface pipelines in connection with the building utilities: integrating GIS and BIM for facility management. In: Computing in Civil Engineering, pp. 341–348 (2012)
40. Cheng, J.C.P., Deng, Y.: An integrated BIM-GIS framework for utility information management and analyses. In: Computing in Civil Engineering, pp. 667–674 (2015)
41. Kim, H., Chen, Z., Cho, C.S., Moon, H., Ju, K., Choi, W.: Integration of BIM and GIS: highway cut and fill earthwork balancing. In: Computing in Civil Engineering, pp. 468–474 (2015)
42. Wang, T.K., Zhang, Q., Chong, H.Y., Wang, X.: Integrated supplier selection framework in a resilient construction supply chain: an approach via analytic hierarchy process (AHP) and grey relational analysis (GRA). Sustainability **9**(2), 289–300 (2017)
43. Deng, Y., Gan, V.J.L., Das, M., Cheng, J.C.P., Anumba C.: Integrating 4D BIM and GIS for construction supply chain management. J. Constr. Eng. Manag. **145**(4) (2019)
44. Bademosi, F., Tayeh, R., Issa, R.R.A.: An immersive approach to construction cost estimating. In: Computing in Civil Engineering, Atlanta, GA, pp. 48–54 (2019)
45. Mo, Y., Zhao, D., Du, J., Liu, W., Dhara, A.: Data-driven approach to scenario determination for VR-based construction safety training. In: Construction Research Congress, New Orleans, LA (2018)
46. Lu, W., Peng, Y., Shen, Q., Li, H.: Generic model for measuring benefits of BIM as a learning tool in construction tasks. J. Constr. Eng. Manag. **139**(2), 195–203 (2013)
47. Ayer, S.K., Messner, J.I., Anumba, C.J.: Augmented reality gaming in sustainable design education. J. Archit. Eng. **22**(1) (2016)
48. Maghool, S., Moeini, S., Arefazar, Y.: An educational application based on virtual reality technology for learning architectural details: challenges and benefits. Int. J. Archit. Res. **12** (3), 246–272 (2018)
49. Wang, P., Wu, P., Wang, J., Chi, H., Wang, X.: A critical review of the use of virtual reality in construction engineering education and training. Int. J. Environ. Res. Public Health **15**, 1204 (2018)
50. Basoglu, N., Goken, M., Dabic, M., Ozdemir Gungor, D., Daim, T.U.: Exploring adoption of augmented reality smart glasses: applications in the medical industry. Front. Eng. Manag. **5**(2), 167–181 (2018)
51. Pratt, P., Ives, M., Lawton, G., Simmons, J., Radev, N., Spyropoulou, L., Amiras, D.: Through the HoloLensTM looking glass: augmented reality for extremity reconstruction surgery using 3D vascular models with perforating vessels. Eur. Radiol. Exp. **2**(1), 6 (2018)
52. Sutherland, I.E.: A head-mounted three-dimensional display. In: Fall Joint Computer Conference, pp. 757–764 (1968)
53. Caudell, T.P., Mizell, D.W.: Augmented reality: an application of heads-up display technology to manual manufacturing processes. In: Twenty-Fifth Hawaii International Conference, pp. 659–669 (1992)
54. Azuma, R.T.: A survey of augmented reality. Teleoper. Virtual Environ. **6**(4), 355–385 (1997)

55. Feiner, S., MacIntyre, B., Hollerer, T., Webster, A.: A touring machine: prototyping 3D mobile augmented reality systems for exploring the urban environment. In: Wearable Computers, pp. 74–81. IEEE (1997)
56. Klein, G., Murray, D.: Parallel tracking and mapping for small AR workspaces. In: 6th IEEE and ACM International Symposium, pp. 225–234 (2007)
57. Reitmayr, G., Drummond, T.: Going out: robust model-based tracking for outdoor augmented reality. In: 5th IEEE and ACM International Symposium on Mixed and Augmented Reality, pp. 109–118 (2006)
58. Roberts, G., Evans, A., Dodson, A.H., Denby, B., Cooper, S., Hollands, R.: The use of augmented reality, GPS and INS for subsurface data visualization. In: FIG XIII International Congress, pp. 1–12 (2002)
59. Georgel, P., Schroeder, P., Benhimane, S., Hinterstoisser, S.: An industrial augmented reality solution for discrepancy check. In: IEEE and ACM International Symposium on Mixed and Augmented Reality (2007)
60. Dai, F., Lu, M., Kamat, V.R.: Analytical approach to augmenting site photos with 3D graphics of underground infrastructure in construction engineering applications. J. Comput. Civ. Eng. **25**(1), 66–74 (2011)
61. Golparvar-Fard, M., Peña-Mora, F., Savarese, S.: D4AR-A 4-dimensional augmented reality model for automating construction progress data collection, processing and communication. J. Inf. Technol. Constr. **14**, 129–153 (2009)
62. Kamat, V., Martinez, J., Fischer, M., Golparvar-Fard, M., Peña-Mora, F., Savarese, S.: Research in visualization techniques for field construction. J. Constr. Eng. Manag. Spec. Iss.: Constr. Eng.: Opportunity Vis. Educ. Pract. Res. **137**(10), 853–862 (2011)
63. Siltanen, S., Oksman, V., Ainasoja, M.: User-centered design of augmented reality interior design service. Int. J. Arts Sci. **6**(1), 547–563 (2013)
64. Malkawi, A.M., Srinivasan, R.S.: A new paradigm for human-building interaction: the use of CFD and augmented reality. J. Autom. Constr. **14**(1), 71–84 (2004)
65. Tayeh, R., Bademosi, F., Issa, R.R.A.: Interactive holograms for better construction information communication. In: International Conference on Computing in Civil Engineering, pp. 112–119 (2019)
66. Behzadan, A.H., Dong, S., Kamat, V.R.: Augmented reality visualization: a review of civil infrastructure system applications. Adv. Eng. Inform. **29**(2), 252–267 (2015)
67. Mishra, S.: Hologram the future of medicine - from star wars to clinical imaging. Indian Heart J. **69**(4), 566–567 (2017)
68. Kalarat, K.: The use of 3D holographic pyramid for the visualization of Sino-Portuguese architecture. J. Inf. Syst. Technol. Manag. **2**(5), 18–24 (2017)

Installing Reinforcement Rebars Using Virtual Reality and 4D Visualization

Martina Mellenthin Filardo(✉), Tino Walther, Sireesha Maddineni, and Hans-Joachim Bargstädt

Bauhaus-Universität Weimar, Weimar, Germany
martina.mellenthin.filardo@uni-weimar.de

Abstract. Improving the visual conception of concrete reinforcement plans is a big step for the digitalization of the construction site. Visualizing reinforcement with virtual reality (VR) technology requires the reinforcement to be modelled, ideally in a building information model. It also builds the ground for better information management between stakeholders, for example for formwork and built-in items.

Visualizing reinforcement using VR allows workers on the construction site to better understand and realize possible obstacles during installation and assembly.

Moreover, the visualization offers the possibility of a last visual collision test by the workers themselves. In addition, the sequencing of the rebar layout is a key factor for the performance during execution. An optimum layout sequence is reached by means of 4D simulation as well as simultaneously including formwork and other installation items.

This paper proposes visualizing concrete reinforcement on site using VR as well as rebar layout sequencing through a 4D model. This model will be applied on a real case study in infrastructure and civil works in Germany.

Keywords: Virtual reality · Building Information Modeling · 4D modeling · Visualization techniques · Communication and collaboration technologies · Decision support systems

1 Introduction

Keeping up with technological developments such as Virtual Reality (VR) or new methods such as Building Information Modeling (BIM) is a big challenge for the rather conservative German construction industry [1]. While it is already possible to model the geometry of a building or structure, the purpose of BIM is rather to aggregate information, attributes and properties to geometrical building components throughout its lifecycle [2].

Advantages of VR technology are widely known [3]. VR uses in the architecture, engineering and construction (AEC) industry are very varied and extend from immersive user experience for marketing purposes, learning environments for heavy machine operating as well as avoiding dangerous on-site situations [4]. The latter was

E. Toledo Santos and S. Scheer (Eds.): ICCCBE 2020, LNCE 98, pp. 1200–1216, 2021.
https://doi.org/10.1007/978-3-030-51295-8_84

primarily considered for this study, not only to avoid said situations but also to allow better understanding of complicated rebar layouts as well as obstacles beforehand.

While it is challenging to implement BIM methods within a project, it is even more challenging to implement them while combining their potential with another technology such as VR simultaneously.

2 Methodology

2.1 BIM to VR Workflow

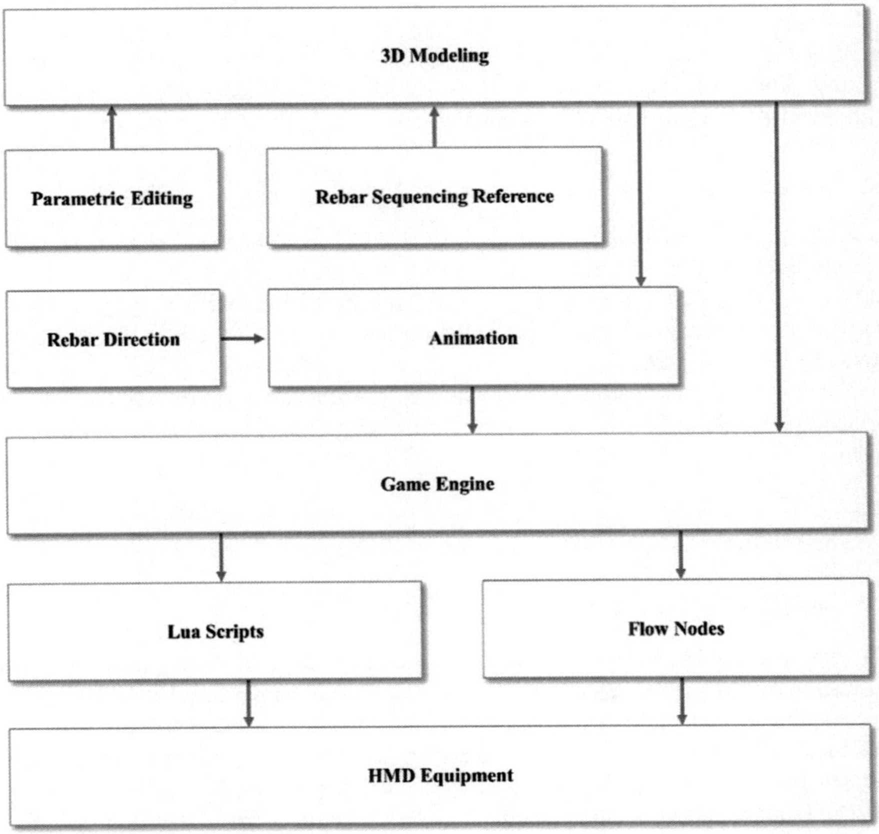

Fig. 1. Scheme of the proposed workflow used to achieve the visualization and interaction of the modelled V-pillar with VR equipment.

To implement the visualization of the reinforcement model in a VR system the workflow depicted in Fig. 1 is proposed and further described in this paper.

To achieve its goal a combination of market ready applications were used, ranging from modelling, over scheduling and animation, to implementation in a game engine and VR equipment in form of a Head Mounted Display (HMD).

In the proposed workflow, the model was generated with Autodesk Revit Structure, supported by Sofistik Reinforcement Detailing tools to extract and alter rebar set information. Subsequently, the model was imported to an imaging application, Autodesk 3ds Max, where it was enriched with information regarding rebar installation sequencing as well as direction. Rebar sequencing information was derived from traditional scheduling methods. Thereafter the model was again imported to Autodesk 3ds Max Interactive (Stingray), a game engine. Regardless of the very similar names, 3ds Max and 3ds Max Interactive are two separate applications, which will later be explained. Next, flow nodes were added for controlling the animations of the rebar sets within the game engine. Furthermore, Lua scripts were employed in the game engine, enabling the communication with the HMD as well as user interaction through the controls. The final step constitutes of visualizing the enriched reinforcement model with the HMD (in this case Oculus Rift).

2.2 Verification

In theory [5], VR solutions promise significant improvements to the construction and planning processes [6]. To assess the impact of the proposed workflow it was implemented in a case study, more specifically for an infrastructure project in Germany, in which the reinforcement layout was planned using conventional methods and tools – hence the 2D PDF drawings.

The V-shaped pillar with a vertical varying thickness has asymmetrical notches. Its top measures 9,60 m in overall width, while its height reaches 10,25 m and the rebar thickness ranges from 12 mm to 25 mm, divided into 24 types.

3 Case Study

3.1 Modeling

Conventional 2D drawings as PDF files were used for the modeling but conventional 2D CAD-based planning was not considered as a part of the proposed workflow, since it would not be compatible with the task of visualizing a 3D object. As expected, the additional step of modeling a finished drawing resulted in an additional visual examination. The assessed planning mistakes were corrected within the modeling process.

The reinforcement was modelled in Autodesk Revit Structure 2019. For modeling the reinforcement, the pillar was divided into five reference levels, based on which adaptive points were placed and adaptive components created from parametric 'structural rebar' families, which were used and combined to rebar sets.

Regarding material properties, the Eurocode2 was considered, leading to concrete strength class C45/55, while B500B steel was used for modeling the reinforcement steel.

Geometrical properties were also assigned. To ensure rebar coverage, cover properties were attached to the rebar objects and parametric distribution was adjusted for adequate spacing. In addition, the previously mentioned rebar sets were selected according to the varying shapes and distribution. These sets assume a pivotal role in the later visualization of the pillar.

Rebar types were divided into freeform shape and sketch rebars according to the assembly direction.

The freeform shape tool was used for vertical rebars, by specifying the reference surface, given the form of the pillar and the resulting irregular number of rebars. For example, rebar type 17 consists of 68 evenly distributed rebars, divided into eight sets, as shown in Fig. 2.

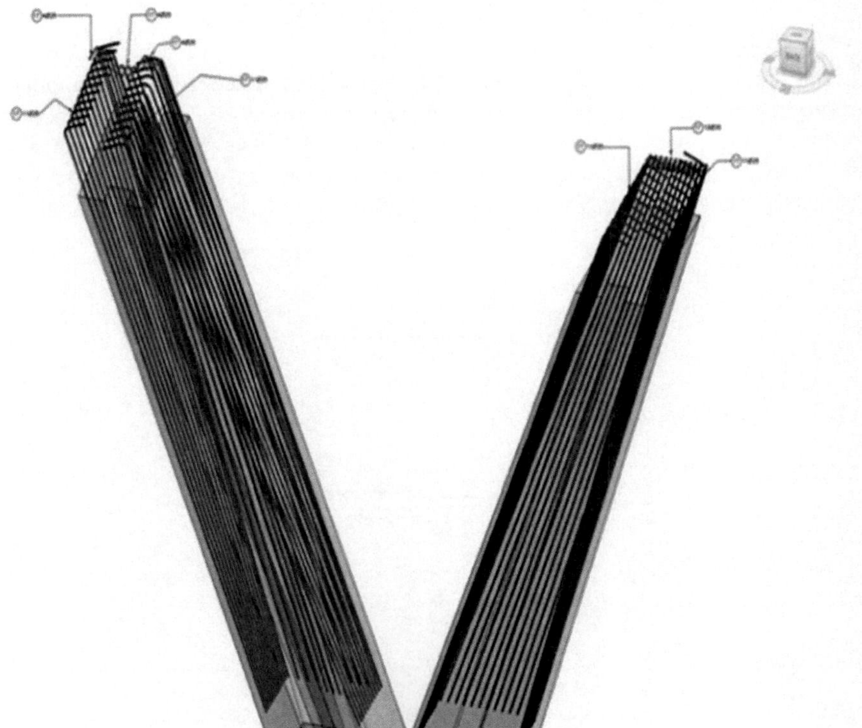

Fig. 2. 3D representation of rebar type 17 showing the rebars divided into eight sets.

The rebar type 17 sets range from one containing 12 rebars, three containing four rebars and four of them consisting of 11 rebars. The size of the sets depend on the position of the rebars within the overall layout, as shown in Table 1. This procedure was repeated on similar rebar types.

Table 1. Overview of the used freeform rebar types (vertical).

Rebar type	Rebar diameter (mm)	Rebar length (m)	No. of bars	Hook length (m)
1	20	5.10	28	Straight bars
2	20	5.15	29	0.35
3	20	5.20	3	0.50
4	20	2.60	10	V-shaped bars
5	20	3.30–4.50	16	Straight bars
6	20	3.80	16	Straight bars
7	20	4.50	4	Straight bars
17	25	8.80	67	0.40
18	25	8.20	37	Straight bars
ME1	14	7.00	8	Straight bars

The sketch tool was used for horizontal rebar shapes, since it is more suitable for the varying dimensions given its parametric properties. Horizontal rebars consist of two overlaying hoops with varying dimensions (X1 and X2, also referred to as A and B below), while the overlapping length (C) stays constant, as depicted in Fig. 3. The parameter C stays constant for all horizontal rebars.

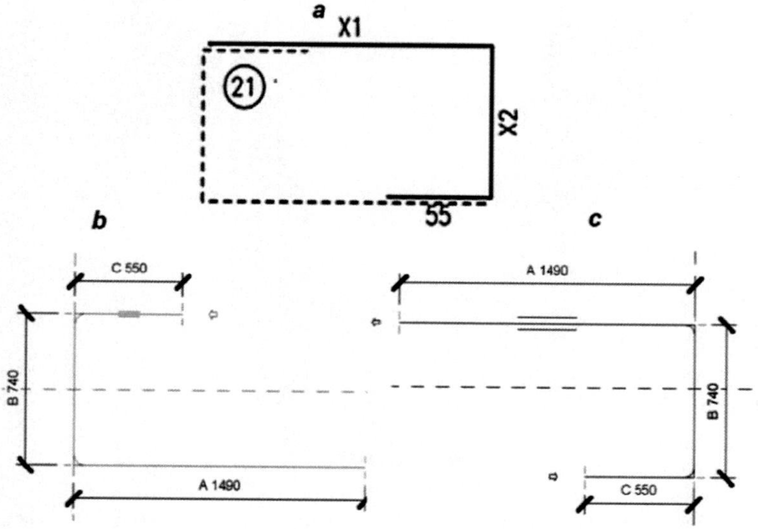

Fig. 3. 2D representation of the parametric dimensions of rebar type 21 (horizontal).

Table 2 shows an overview of the used sketch rebars for the horizontal reinforcement of the V-shaped pillar under the same parameters as shown in Table 1.

Table 2. Overview of the used sketch rebar types (horizontal).

Rebar type	Rebar diameter (mm)	Rebar length (m)	No. of bars	Hook length (m)
8	12	1.93–2.14	16	0.55
9	12	2.90–4.65	42	0.55
10	12	2.08–2.23	4	0.55
11	12	2.12–2.37	156	0.55
12	12	2.28–3.53	16	0.55
13	12	1.69–2.98	28	0.55
14	12	2.22–2.35	92	0.55
15	12	2.35–2.78	92	0.55
16	12	2.64–2.77	23	0.55
19	12	2.40–2.79	160	0.55
20	12	2.37–2.79	176	0.55
21	12	2.78–2.98	176	0.55
22	12	2.63–2.21	176	0.55
ME2	12	3.24–4.10	26	0.55

Furthermore, Fig. 4 shows a 3D representation of rebar type 8 (and 8(1)) exemplarily, to better illustrate the resulting horizontal sketch rebar sets. This rebar type is divided into four sets of four bars each, forming two rebar cages. Type 8 refers to the rebars on the +Y direction, while 8(1) applies to the rebars on the −Y side (front and back). This procedure was also repeated on similar rebar types.

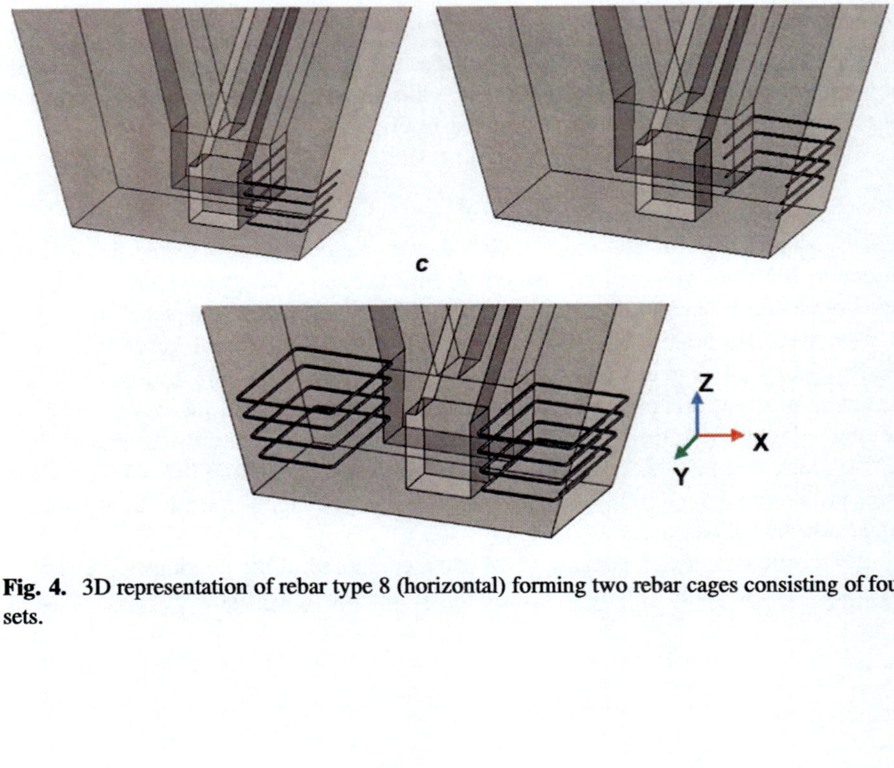

Fig. 4. 3D representation of rebar type 8 (horizontal) forming two rebar cages consisting of four sets.

The Sofistik Reinforcement Plugin was also used in this study as a tool to extract rebars with identical diameters from the applied rebar sets and combine them into a structural rebar container as well as to modify parametric sketch rebar dimensions.

3.2 Rebar Scheduling

The resulting reinforcement design was further used for generating a schedule of the layout using the scheduling tool in Revit Structure 2019.

Rebar tying and placing times depend on the bar diameter as well as assembly direction (vertical, horizontal) and were determined according to [7]. Assuming bar bending is also considered as well as ancillary crane loading time, a skilled worker would need on average 35.2 h per tonne of steel for the assembly and installation on site.

The reinforcement steel for the V-pillar amounts to roughly 7.3 tonnes of steel, which result in 254 h of bar bending and installation, which divided into working hours on site amount to 37 days, as shown in Fig. 5.

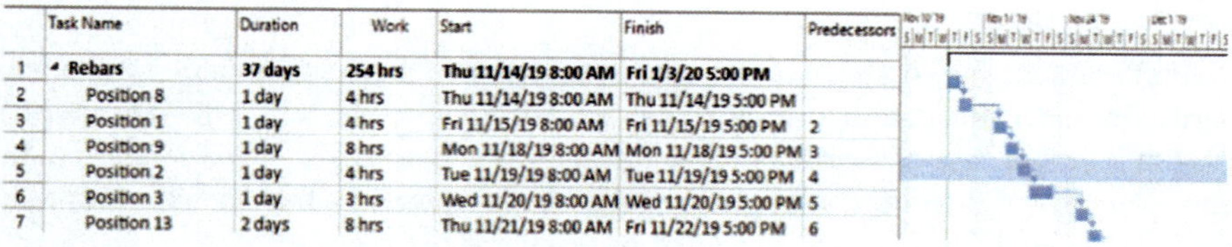

	Task Name	Duration	Work	Start	Finish	Predecessors
1	◂ Rebars	37 days	254 hrs	Thu 11/14/19 8:00 AM	Fri 1/3/20 5:00 PM	
2	Position 8	1 day	4 hrs	Thu 11/14/19 8:00 AM	Thu 11/14/19 5:00 PM	
3	Position 1	1 day	4 hrs	Fri 11/15/19 8:00 AM	Fri 11/15/19 5:00 PM	2
4	Position 9	1 day	8 hrs	Mon 11/18/19 8:00 AM	Mon 11/18/19 5:00 PM	3
5	Position 2	1 day	4 hrs	Tue 11/19/19 8:00 AM	Tue 11/19/19 5:00 PM	4
6	Position 3	1 day	3 hrs	Wed 11/20/19 8:00 AM	Wed 11/20/19 5:00 PM	5
7	Position 13	2 days	8 hrs	Thu 11/21/19 8:00 AM	Fri 11/22/19 5:00 PM	6

Fig. 5. Excerpt of the Gantt diagram depicting the rebar laying sequence schedule for the V-shaped pillar.

Furthermore, the scheduling was indispensable for defining the gradual sequencing of the rebars, starting at the bottom. Based on this schedule, a color-coded visualization of the entire pillar was created, as depicted in Fig. 6.

3.3 4D Animation

For the sequencing animation, the modelled reinforcement layout created using Revit Structure 2019 was imported into Autodesk 3ds Max and combined with the generated schedule based on literature values, as described in the previous chapter.

Autodesk 3ds Max is a digital content creation tool (DCC) used for visual effects, rendering as well as animation [8]. Given that both Revit Structure and 3ds Max are Autodesk products, interoperability was not an issue. Nonetheless, the Revit model file needed to be linked to the 3ds Max project, in order to obtain not only an image of the pillar, which is usually suitable for rendering purposes. By linking the 3D model, the material information (B500B) as well as the family types (now referred to as entities) could also be added.

The created animation consists of 38 parts, composed of the previously mentioned 24 rebar types in addition to half of the divided sets according to the direction of rebar

Fig. 6. Color-coded 3D view of the rebar sets according to sequencing information.

positioning, as explained for rebar types 8 and 8(1). The animation was generated based on the movement between sequencing frames, which resulted in the visualization of the arrival of the rebar sets in their respective order from the corresponding direction, e.g. rebar type 8 comes from the +Y and type 8(1) from the –Y directions, as shown previously in Fig. 4. The transition between frames, also referred to as key-frame [9], was the basis for the animation.

As previously explained for rebar types 8 and 8(1), the movement resulted from changing the value in the direction of movement (in this case, the Y-direction), as illustrated in Fig. 7.

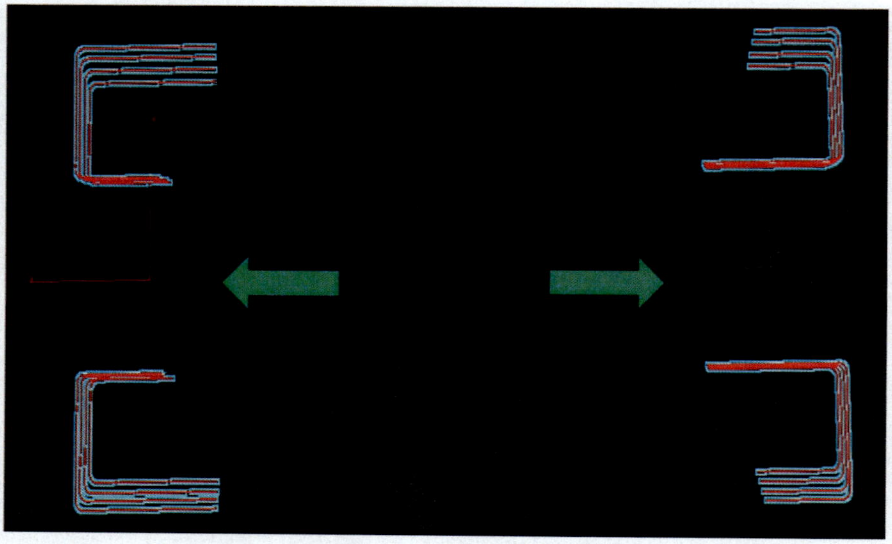

Fig. 7. Representation of rebar types 8 and 8(1), showing the backwards movement in –Y and +Y directions respectively.

By assigning this movement to key-frames, e.g. frames 0 and 20, the rebars 8 and 8 (1) meet after moving in from opposite directions.

As shown in Table 3, the rebar laying sequence from the previously described schedule was assigned to frames and combined with the moving direction (X, Y, and Z).

Table 3. Overview of the used rebar types with assigned directions and sequencing.

Rebar type	Rebar set division	Relation start/end positions X; Y; Z	Start frame		End frame
8	8	0; +500; 0;	0	to	20
	8(1)	0; −500; 0;	0	to	20
1	1	0; 0; +500;	25	to	45
9	9	+500; 0; 0;	50	to	70
	9(1)	−500; 0; 0;	50	to	70
2	2	0; 0; +500;	75	to	95
3	3	0; 0; +500;	100	to	120
13	13	0; +500; 0;	125	to	145
	13(1)	0; −500; 0;	125	to	145
4	4	0; 0; +500;	150	to	170
5	5	0; 0; +500;	175	to	195

(*continued*)

Table 3. (*continued*)

Rebar type	Rebar set division	Relation start/end positions X; Y; Z	Start frame		End frame
6	6	0; 0; +500;	200	to	220
7	7	0; 0; +500;	225	to	245
15	15(1a)	+500; 0; 0;	250	to	270
	15(1b)	−500; 0; 0;	250	to	270
	15(2a)	+500; 0; 0;	275	to	295
	15(2b)	−500; 0; 0;	275	to	295
11	11	0; +500; 0;	300	to	320
	11(1)	0; −500; 0;	300	to	320
14	14	0; +500; 0;	325	to	345
	14(1)	0; −500; 0;	325	to	345
10	10	0; +500; 0;	350	to	370
12	12	+500; 0; 0;	375	to	395
16	16	+500; 0; 0;	400	to	420
17	17	0; 0; +500;	425	to	445
18	18	0; 0; +500;	450	to	470
19	19	0; +500; 0;	475	to	495
	19(1)	0; −500; 0;	475	to	495
20	20	0; +500; 0;	500	to	520
	20(1)	0; −500; 0;	500	to	520
21	21(1a)	+500; 0; 0;	525	to	545
	21(ab)	+500; 0; 0;	525	to	545
	21(2a)	−500; 0; 0;	550	to	570
	21(2b)	−500; 0; 0;	550	to	570
22	22	+500; 0; 0;	575	to	595
ME1	ME1	+500; 0; 0;	600	to	620
ME2	ME1	0; 0; +500;	625	to	645

The frame rate was defined as 30 fps (frames per second), corresponding to the NTSC (National Television System Committee) standard, used in the USA, among other countries [10].

Fig. 8. Still from the animation of the rebar types sequences and directions, based on the start and end positions described in Table 3.

As depicted in Fig. 8, the main added value of the generated animation is that it does not only show the sequence in which each rebar should be installed, but also the direction from which each specific bar should be inserted into the reinforcement layout, mirroring the ideal reinforcement installation.

This information is critical for effective assembly on site – especially regarding big, complex structures with dense reinforcement layouts.

3.4 VR Implementation

For the implementation in a VR environment, Autodesk 3ds Max Interactive was used. It is a game engine used for creating real-time rendered experiences, suitable for VR as well as for augmented reality (AR). At its core is the Autodesk Stingray engine [11]. For this study, it provided a suitable VR environment, since it is compatible with the previously used proprietary applications. Interoperability was neither an issue, nor part of the study.

The virtual model of the rebar layout as well as the created animation clips of the rebar sets with sequencing and direction information were imported as assets into 3ds Max Interactive from a FBX file, while observing material and texture importing options as well.

The project was created using a template provided by the application, where objects were placed for visualization in a scene on a single level.

To achieve the desired behaviors and mechanisms (e.g. updating each frame, triggering flow events), Lua programing language [12] as well as visual node-based flow scripting were used.

Lua scripts offer an interface between the game engine and the immersive technology of the HMD [13]. A set of Lua scripts, referred to as Appkit, is integrated in 3ds Max Interactive, providing the necessary structure for a typical interactive 3D project.

The Lua boot script starts with the gaming engine and contains the packages needed to generate the project in the VR environment (boot packages) as well as a project template. It subsequently loads script files inside the project directory, which is where other custom scripts can be added (e.g. *oculus_vr.lua*), as depicted in Fig. 9. It allows users to write further needed script files required by the boot script.

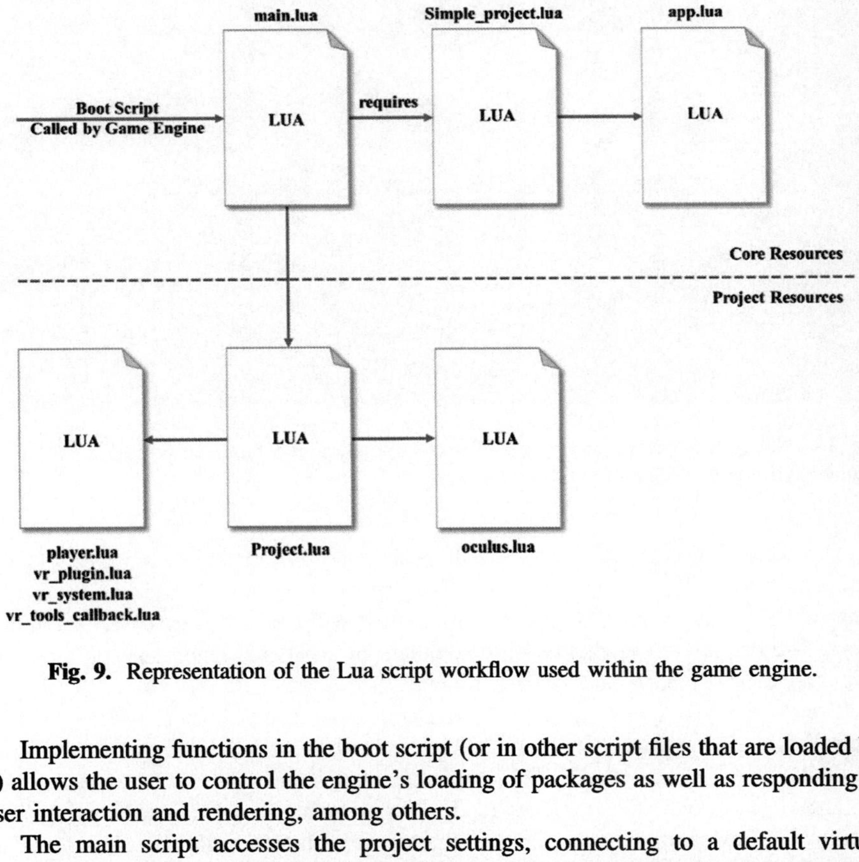

Fig. 9. Representation of the Lua script workflow used within the game engine.

Implementing functions in the boot script (or in other script files that are loaded by it) allows the user to control the engine's loading of packages as well as responding to user interaction and rendering, among others.

The main script accesses the project settings, connecting to a default virtual environment. It requires a simple project object, which is used by template projects by default and handles e.g. camera and level configurations provided by the Appkit, thus allowing users to customize its behavior in the project code. The *app.lua* file, also a default script, defines an object capable of managing engine worlds, levels as well as other utility objects.

Project.lua is the main script within the project. It loads project settings (from *vr_plugin.lua*) as well as camera settings. It imports the player (from *player.lua*) and the user actions (from *oculus.lua*). The *player.lua* script defines the player controls, including eye height. One of *Project.lua*'s most important functions is to enable the flow nodes, which are responsible for triggering the animations for each object (in this case the rebar sets), as depicted in Fig. 10. The actions and reactions the user transmits via the Oculus Rift touch controllers are controlled by flow nodes and must be defined beforehand.

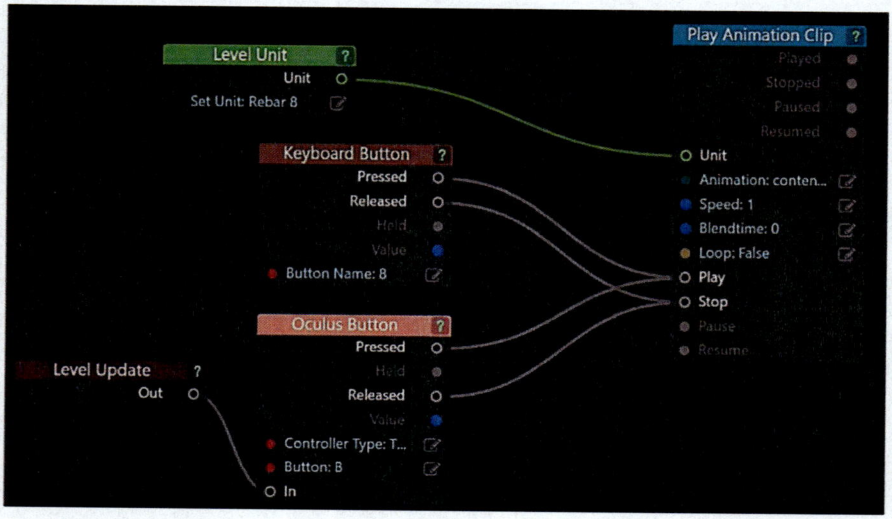

Fig. 10. Node-based visual programming of the animation clip for rebar 8, controlled by the virtual hand (Oculus Button).

The animation clips are connected to their respective rebar set as well as to the Oculus controller button (play and stop). The animation in the VR environment is also controllable via keyboard shortcuts, although these will not be further described in this paper. The display of the changes is transmitted by the Level Update event, ensuring that changes in the scene are visible.

The chosen VR equipment (Oculus Rift, consisting of a headset as well as the two controllers, called virtual hands and its sensors) also needed to be connected. The previously mentioned *oculus_vr.lua* script ensures this connection.

The *vr_tools_callback.lua* script is also imported by *Project.lua*. This script was implemented to control the display of rebar set information within the VR level, as shown in Fig. 11. With this function, users immersed in the animation can view information about the displayed rebar set by pushing a button on the virtual hand, resulting in the display of rebar types and diameters, as depicted in Fig. 11.

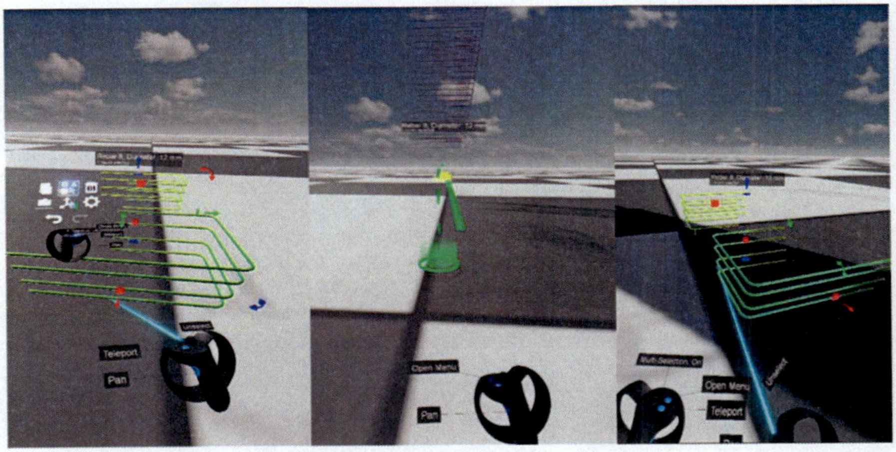

Fig. 11. Rebar set information within the VR environment controlled by and displayed for the user.

3.5 Final Rebar Assessment and Visualization

As described in the previous chapters, the modelled reinforcement layout as well as the animation clips for each set were imported and the functions implemented within the game engine. An excerpt of the resulting animation is depicted in Fig. 12.

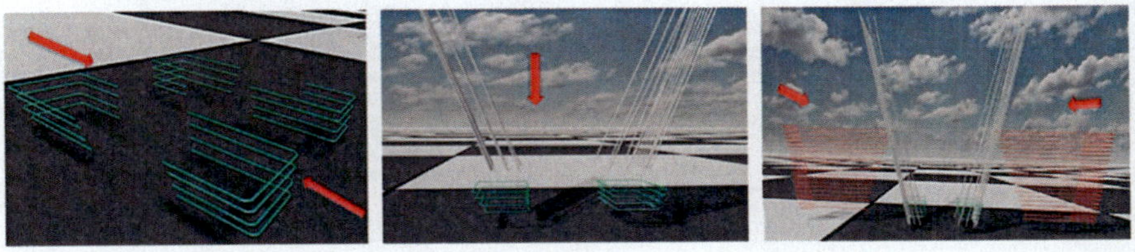

Fig. 12. Stills from the rebar sets flying in from different directions within the VR environment.

The animation clips continue to be displayed gradually, finally resulting in the completed representation of the V-shaped reinforcement. Through the generated interactive animation users are shown not only the final position of the sets, but also the direction from which they should be inserted into the layout as well as the relation with other surrounding objects.

The user can walk around the model and play, pause and rewind within the VR environment. These functions alone should allow for a better understanding of the reinforcement layout for installation and assembly.

In addition, the user can also interact with the animated rebar sets through the controller, thus displaying further information about the rebar set in question (type and diameter), as explained in the previous chapter. Figure 13 displays the user view of the completed reinforcement layout of the pillar within the VR level.

Fig. 13. Representation of the assembled rebar layout in the game engine level.

4 Conclusion and Outlook

4.1 Summary and Conclusion

The proposed workflow started from a conventional setting, shaping the 2D drawings into a 3D geometry, adding parametric as well as material properties, and thus making a BIM-model fit for usage in the field.

Independently, a simple bar bending and installation schedule was created, from which indispensable 4D sequencing information was derived for the visualization. Furthermore, the BIM-model was enriched with the mentioned 4D sequencing information. The resulting model was subsequently processed into animation clips for the rebar types and divided into five directions from which the reinforcement reaches its' final position within the V-shaped pillar. All this information was implemented in a game engine, resulting in a visualization of the model, which allows user immersion as well as interaction options.

This study achieved the desired results by stretching over many different (proprietary) software applications, employing not just new methods and tools such as Building Information Modeling and Virtual Reality, but also by applying simple approaches such as traditional scheduling as well as fps-based animation techniques.

It shows that not all approaches and methods within the construction industry are outdated and obsolete. Moreover, it establishes that old and new can and should work well together.

The delivered visualization is a one-of-a-kind reinforcement model given its information regarding sequencing and most importantly the direction of rebar sets and types.

It serves as a visual aid for workers on or before reaching the construction site. It also offers another chance for a final visual examination, as mentioned in Sect. 3.1. In addition, it eases the understanding of complex, dense reinforcement layouts not only for new, unexperienced workers, but can also assist seasoned personnel in their own understanding of the task as well as in collaborating, instructing and teaching.

Regarding usability by regular workers, the proposed workflow does not address the usual construction worker skill set. It requires several software applications with varying user interfaces, licenses as well as high-performance and sensitive hardware not usually available on construction sites. Nonetheless, the proposed visualization workflow is intended as a substitute for 2D plans, which have a designated on-site sheltered and mobile table or storage, thus serving as laptop and HMD areas for VR visualization purposes. Alternatively, site offices and lodging facilities can be used, although they tend to be further from the actual reinforcement procedure. The required skills are largely covered by the arrival of digital natives in the workforce, which also encourages the aforementioned exchange between seasoned and new workers.

While the resulting visualization is fit for usage, it allows addressing efficiency and comparing the consumed resources to the benefit of the employed methods. Furthermore, it begs the question if a 4D clip with the same user interaction options shown on a tablet (but without the user immersion) would not be equally suitable for the task.

4.2 Outlook

To extend the application of VR for the visualization of reinforcement, next steps should address visualizing not only the direct path of the rebar into its final position but making it flexible by simulating more than one path – e.g. with obstacles on the construction site. In addition, visualizing more complex reinforcement layouts should increase the value of VR uses on the field.

Beyond that, further options of future work regard visualizing the terrain as well as surroundings of the structure and formwork in order to assess more proportional scenarios regarding efficiency.

Regarding applications, next steps include testing the proposed workflow with other software solutions as well as open exchange formats and evaluating the differences.

All of the above-mentioned further applications and extensions would be irrelevant as long as the visualization model in the gaming engine as well as the animation clips are not linked to the original reinforcement model, thus updating geometric and semantic information. Given that the proposed VR visualization is intended as a final visual examination, it is highly likely that all subsequent steps within the workflow would become obsolete after the slightest change, which is neither an efficient nor an overseeable long-term development.

References

1. Borrmann, A., König, M., Koch, C., Beetz, J. (eds.): Building Information Modeling: Technology Foundations and Industry Practice. Springer, Cham (2018). https://doi.org/10.1007/978-3-319-92862-3
2. Sacks, R., Eastman, C., Ghang, L., Teicholz, P.: BIM Handbook, 3rd edn. Wiley, Hoboken (2018). https://doi.org/10.1002/9781119287568
3. Cipresso, P., Chicchi Gigloli, I.A., Maya, M.A., Riva, G.: The past, present, and future of virtual and augmented reality research: a network and cluster analysis of the literature. Front. Psychol. **9**, 2086 (2018). https://doi.org/10.3389/fpsyg.2018.02086
4. Zaker, R., Coloma, E.: Virtual reality-integrated workflow in BIM-enabled projects collaboration and design review: a case study. Vis. Eng. **6**(1), 1–15 (2018). https://doi.org/10.1186/s40327-018-0065-6
5. Dörner, R., Broll, W., Grimm, P., Jung, B.: Virtual und Augmented Reality (VR/AR). Springer, Heidelberg (2013). https://doi.org/10.1007/978-3-642-28903-3
6. Whyte, J.: Innovation and users: virtual reality in the construction sector. Constr. Manag. Econ. **21**(6), 565–572 (2003). https://doi.org/10.1080/0144619032000113690
7. Zentralverband des Deutschen Baugewerbes e.V.: Arbeitszeit-Richtwerte Hochbau: ARH [Tabellen]. Bonn (1982)
8. Autodesk Knowledge Homepage. http://help.autodesk.com/view/3DSMAX/2018/ENU/?guid=__interactive_help_linking_with_dcc_send_assets_dcc_html. Accessed 31 Jan 2020
9. University of Virginia Lecture. http://web.arch.virginia.edu/arch545/handouts/keyframing.html. Accessed 31 Jan 2020
10. Taylor, A.: Design Essentials for the Motion Media Artist. Focal Press, Waltham (2011). https://doi.org/10.4324/9780080953922
11. Geospatial World Homepage. https://www.geospatialworld.net/news/autodesk-3ds-max-2018-1-now-includes-3ds-max-interactive/. Accessed 31 Jan 2020
12. Lua Programming Language Homepage. https://www.lua.org/docs.html. Accessed 31 Jan 2020
13. Autodesk Knowledge Homepage. https://help.autodesk.com/view/3DSMAX/2018/ENU/?guid=__interactive_help_getting_started_get_started_vr_html. Accessed 31 Jan 2020

Scenario Simulation of Indoor Post-earthquake Fire Rescue Based on Building Information Model and Virtual Reality

Xinzheng Lu[1(✉)], Zhebiao Yang[1], Zhen Xu[2], and Chen Xiong[3]

[1] Tsinghua University, Beijing 100084, China
luxz@tsinghua.edu.cn
[2] University of Science and Technology Beijing, Beijing 100083, China
[3] Shenzhen University, Shenzhen 518060, China

Abstract. Post-earthquake fires occur frequently, and the seismic damage to non-structural components results in dangerous fire and evacuation environments. Both firefighters and trapped people will face serious threats to life during the rescue. Consequently, a simulation framework of an indoor post-earthquake fire rescue scenario based on building information model (BIM) and virtual reality (VR) is proposed herein. Information of structural and non-structural components is extracted from the BIM. Subsequently, the seismic damage of non-structural components is predicted according to the time-history results and fragility curves. Moreover, a smoke visualization method combining volume rendering and the particle system is also proposed. A case study of a 19-story hospital is conducted using the proposed method for establishing the fire rescue scenario. The results show that the established VR scenario has high realism and smooth interactivity, and the influence of smoke on the fire rescue is more significant than that of fallen debris.

Keywords: Earthquake · Fire rescue · BIM · Virtual reality · Non-structural components

1 Introduction

Earthquakes can cause structural damage to buildings, and various non-structural components, such as sprinkler systems, suspended ceilings, and infilled walls, can also be damaged. The damage to structural and non-structural components will lead to a higher risk of post-earthquake fire [1–3]. People may be trapped indoors due to injuries or moving difficulties, especially in some special types of buildings (e.g., hospitals, nursing homes). When firefighters enter the building to search for trapped people, they are not familiar with the indoor environment. Both firefighters and trapped people might face serious risks. Owing to the combination of various hazards, the influencing factors on post-earthquake fires are very complex. Research on the construction of post-earthquake fire scenarios is very limited, and almost no system exists for post-earthquake fire rescue training. To establish a post-earthquake fire rescue VR scenario, the following two key problems need to be addressed:

E. Toledo Santos and S. Scheer (Eds.): ICCCBE 2020, LNCE 98, pp. 1217–1226, 2021.
https://doi.org/10.1007/978-3-030-51295-8_85

(1) The combination of multiple hazards. Post-earthquake fire rescue involves several combined hazards, including post-earthquake fire, seismic damage of non-structural components, and fallen debris. The damage of the sprinkler system will affect the spread of a post-earthquake fire. Xu et al. [4] proposed a seismic damage assessment method based on a building information model (BIM) of the sprinkler system. Many experiments on seismic damage to suspended ceiling have been conducted and fragility curves of suspended ceiling have been proposed [5–7]. These studies provide a basis for determining the damage of suspended ceilings and the corresponding fallen debris. Existing VR scenarios do not consider the influence of smoke and fallen debris on pedestrian speed [8–10].

(2) VR scenario construction and smoke visualization. Owing to the complexity of the indoor environment, the modeling workload a post-earthquake fire rescue scenario is quite labor intensive. To avoid the labor-intensive model generation, BIM, which is a three-dimensional model with details [11, 12], is adopted to establish the VR scenario. Existing methods for smoke visualization include particle systems and volume rendering [13]. Particle systems can quickly model smoke and provide a certain degree of realism [14, 15], but the smoke motion generated by a particle system does not match the actual diffusion process [9, 16]. By contrast, volume rendering method can provide a realistic diffusion process [8, 10]. However, this method will consume substantial computer resources.

Regarding the problems mentioned, a simulation framework for a post-earthquake fire rescue scenario is proposed in this study. Based on the Unity platform, the modeling and visualization of the post-earthquake fire scenario of a 19-story hospital building is accomplished, and the fire spread and personnel movement in different scenarios are analyzed. The methodology proposed herein can assist firefighters to conduct specialized drills and provide an effective technical support for fire rescue training.

2 Methodology

As shown in Fig. 1, the proposed framework consists of five modules: (1) BIM; (2) Unity modeling; (3) Seismic damage assessment; (4) Post-earthquake fire simulation; (5) VR scenario construction.

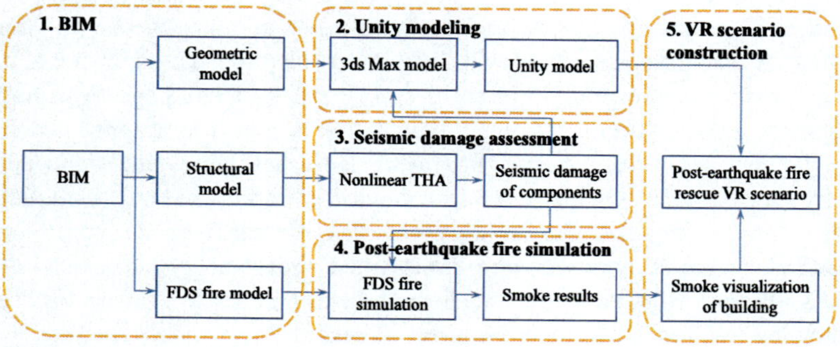

Fig. 1. The proposed framework of post-earthquake fire rescue scenario simulation.

2.1 BIM

BIM is used to establish the geometric, structural, and FDS fire models. These three models form the basis of subsequent analyses and simulations. Specifically, the FBX file is exported from Revit to generate the geometric model needed for subsequent modeling [17]. The structural model of the building can also be constructed using BIM to conduct the seismic response analysis [18, 19]. The pre-processing software for FDS, PyroSim [20], is used to convert the data from BIM to FDS.

2.2 Unity Modeling

The FBX file exported from BIM contains the geometric information of the building. However, some components are not included in the BIM. Therefore, the FBX file of the building obtained in Module 1 is imported into the 3ds Max software [21], and the professional editing tools provided by 3ds Max are used to add texture and illumination in the geometric model to enrich the details. The completed model will be exported to Unity to generate the VR scenario.

2.3 Seismic Damage Assessment

Analysis of Seismic Damage of Non-structural Components. In the post-earthquake fire rescue scenario, the impacts caused by the seismic damage of the sprinkler system and suspended ceilings are the most significant. The fragility curves of non-structural components are provided in FEMA P-58 [5], and the probabilities of various damage states can be determined based on the time history results and fragility curves. Fragility curves are employed to evaluate the seismic damage of each sprinkler component, and the overall seismic damage of the sprinkler system is determined through a tree data structure [4]. The suspended ceiling (C3032) is an acceleration-sensitive component [5], which has three damage states, i.e., DS_1, DS_2, and DS_3. A certain percentage of area will fall off at each damage state. Therefore, the damage state and its probability can be determined by the corresponding fragility curve based on the peak floor accelerations (PFAs). The probability of DS_1, DS_2, and DS_3 can then be obtained from the fragility curves based on the PFAs.

The fallen area of the suspended ceiling can be determined by multiplying the total area by the probability of the corresponding damage state. The total fallen area can be calculated as expressed in Eq. (1),

$$S_{\text{debris},j} = S_{\text{total},j} \times \sum_{i=1}^{3} \left[P_s(DS_i) \times P_j(DS = DS_i) \right], j = 1, 2, 3, \ldots, n \tag{1}$$

where $S_{\text{debris},\,j}$ represents the fallen area of the suspended ceiling on the jth story, $S_{\text{total},j}$ represents the total suspended ceiling area on the jth story, $P_s(\text{DS}_i)$ represents the percentage of fallen area of the suspended ceiling at DS_i, and $P_j(\text{DS} = \text{DS}_i)$ represents the probability that the suspended ceiling on the jth story is at DS_i.

After determining the total fallen area of the suspended ceiling, the model of fallen debris can be established in 3ds Max. The fallen debris is randomly distributed on the ground [22], and this area is equal to the results calculated by Eq. (1).

Influence of Seismic Damage of Non-structural Components on the Fire Scenario. The influence of the seismic damage of non-structural components on the fire scenario includes two parts: (1) the influence of sprinkler system damage on fire spread, and (2) the influence of smoke and fallen debris on pedestrian speed. Because the sprinkler system works as a whole, the damage of water pipes will eventually influence the spray flow of the connected sprinkler drops. The influence of the sprinkler system damage on fire spread can be determined through the FDS simulation in Module 4. Both smoke and fallen debris will reduce pedestrian speed [23–25], and the combined effects are complex. In the present study, the upper limit of the pedestrian speed under different scenarios is first determined according to the soot density, and the impact of fallen debris on the pedestrian speed is further discussed.

2.4 Post-earthquake Fire Simulation

Based on the FDS model established in Module 1, combined with the seismic damage of the sprinkler system obtained in Module 3, the performance of the sprinkler system in the fire analysis model is determined. The FDS software [26] can then be used to perform post-earthquake fire simulations to calculate the movement of smoke particles. Smoke results should be extracted from the FDS results. Specifically, smoke monitoring grids are set inside the building. The soot densities of each grid are output, which provides the input data for establishing the fire VR scenario in Unity.

2.5 VR Scenario Construction

The fire rescue VR scenario is established in Unity, in which smoke visualization is achieved by combining volume rendering and the particle system. Volume rendering is adopted in the initial stage of the fire, and the particle system is adopted in the smoke spread stage. This combination can fully exploit the strengths of the two methods to ensure the visual effects and smooth interactivity for rescue training. Module 4 has already output the soot density D_s of the FDS grids. The opaqueness values α of each grid are calculated, which are adopted to perform smoke rendering [10]. To mimic the process of smoke diffusion, the number and locations of the smoke particle systems should be set, whose parameters (e.g., duration, start speed, start size, emission, and shape) also need to be adjusted.

In the smoke spread stage, the diffusion scenarios of smoke using volume rendering and the particle system are shown in Fig. 2. The particle system can realize a smoke visualization scenario similar to that obtained with the volume rendering method. It should be noted that the volume rendering method occupies nearly 1200 MB of memory. Meanwhile, the particle system method only occupies 400 MB of memory and the modeling process is convenient. Therefore, the particle system method is adopted to realize smoke visualization in the smoke spread stage.

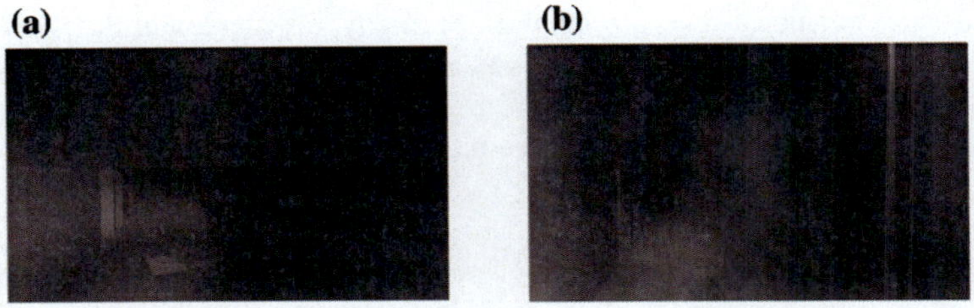

Fig. 2. Scenario comparison of two methods in the smoke spread stage. (a) Volume rendering. (b) Particle system.

3 Case Study

3.1 Case Introduction

Hospitals are densely populated and contain complex facilities, which results in a high risk of fire. A 19-story hospital is chosen as the case study for a post-earthquake fire rescue. The 3D BIM of the hospital is shown in Fig. 3(a). El-Centro ground motion record is selected, whose PGA is scaled to 400 cm/s^2. The seismic response of the building is obtained through nonlinear THA. The maximum PFA occurs on the 10th story. Fire is thus assumed to be ignited in a room on the 10th floor. Two fire scenarios are analyzed: the sprinkler system with seismic damage (namely "damage scenario") and the sprinkler system without seismic damage (namely "non-damage scenario"). As shown in Fig. 3(b), the ignited room is associated with fire initiation. The fire data (e.g., soot density, and FED) are monitored during the fire simulation.

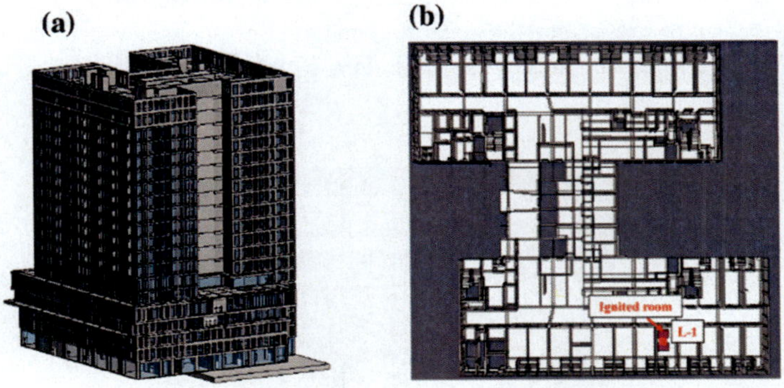

Fig. 3. Schematic diagram of the hospital. (a) BIM model. (b) 10th floor.

3.2 Results of Post-earthquake Fire Simulation

The fire data under both the non-damage and damage scenarios are shown in Fig. 4. When no seismic damage occurs, the sprinkler system works normally. The maximum soot density (Fig. 4(a)) does not exceed 4×10^{-4} mg/m^3 under the non-damage scenario, and it remains at 1×10^{-4} mg/m^3. Similarly, the maximum FED value in the room is 23-times that of the non-damage scenario (Fig. 4(b)). It can be seen that the fire suppression ability of the sprinkler system is significantly weakened owing to the damage, which has an extremely negative influence on the movement of personnel.

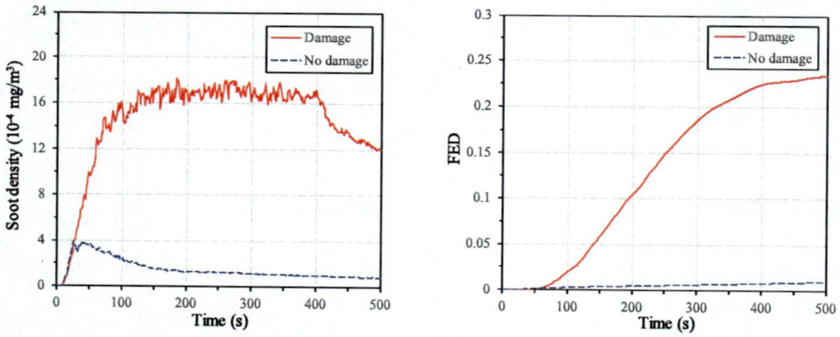

Fig. 4. Quantitative comparisons between two scenarios in the ignited room. (a) Soot density curves. (b) FED curves.

3.3 VR Scenario

To investigate the influence of the seismic damage to the non-structural components on the rescue, it is assumed that some people are trapped in Room A, and the firefighters start from the stairs to perform the rescue (Fig. 5). Three VR scenarios are established herein:

Scenario 1: No fire, and the seismic damage is not considered;
Scenario 2: Fire exists, but the seismic damage is not considered;
Scenario 3: Fire exists, and the seismic damage is considered.

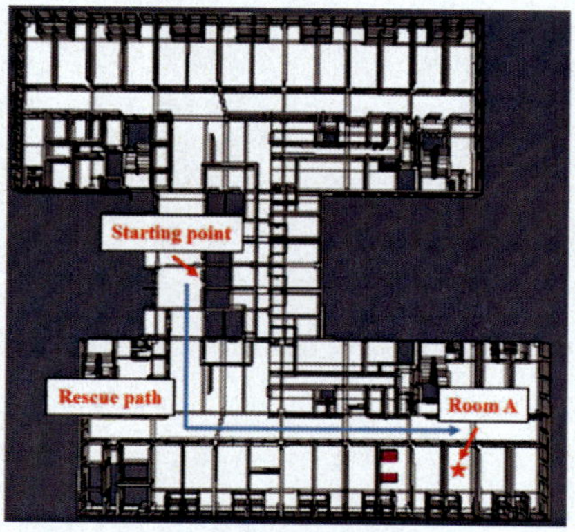

Fig. 5. Schematic diagram of rescue path.

Due to the influence of smoke and fallen debris, the pedestrian speeds in these three rescue scenarios are different. The pedestrian speeds in Scenarios 2 and 3 are 3.73 and 0.42 m/s [23], respectively, which are set as the upper speed limits. When the trainees control their movements in the VR scenario, the speed and route are influenced by smoke and fallen debris conjunctively. The average rescue time of these scenarios is shown in Fig. 6(a). The integrated hazards dose (IHD^{path}) proposed by Xu et al. [10] is selected as the indicator to calculate and compare the risk of the rescue route in these three scenarios (Fig. 6(b)). As Fig. 6 shows, in Scenario 1 the firefighter only needs 12 s to reach the room where the trapped person is located. The firefighter requires 16 s to reach the room in Scenario 2, which is close to the time required in Scenario 1. By contrast, in Scenario 3, owing to the damage of the sprinkler system, the smoke is diffused throughout the floor. The firefighter requires 144 s to reach the room, which is 9-times that required in Scenario 2. Similarly, IHD^{path} of Scenario 3 is 33-times that of Scenario 2. Through the comparison, it can be seen that the soot density has a significant influence on the rescue time and risk.

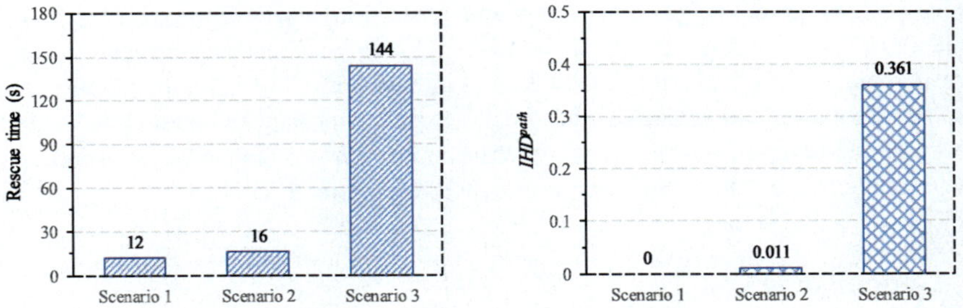

Fig. 6. Comparison of different rescue scenarios. (a) Rescue time. (b) IHD^{path}.

To clarify the combined effect of smoke and fallen debris, Scenarios 4 and 5 are established. Because there is no smoke in Scenario 4, the upper speed limit is set as 4.60 m/s [24], which is equal to the pedestrian speed in a smokeless environment. The keyboard and mouse are used to control the movement in the VR scenario. The comparison of Scenarios 3–5 is presented in Table 1.

Table 1. Comparison of scenarios considering the influence of smoke and fallen debris.

Scenario	Conditions	Average rescue time (s)	Average speed (m/s)	Speed reduction (m/s)
3	Both smoke and fallen debris exist	144	0.38	4.22
4	Only fallen debris exists	19	2.91	1.69
5	Only smoke exists	130	0.42	4.18

Compared with Scenario 1, the average speed in Scenario 4, v_{4r}, is reduced by 1.69 m/s owing to a collision with fallen debris. The average speed in Scenario 5, v_{5r}, is reduced by 4.18 m/s owing to the low visibility caused by smoke. However, when considering the combined effect (i.e., Scenario 3), the average speed, v_{3r}, is reduced by 4.22 m/s. This is not simply equal to the sum of v_{4r} and v_{5r}, where the effects of the two hazards are considered separately. It is therefore necessary to establish a post-earthquake fire scenario in which the combined effect of smoke and fallen debris is considered.

Through an establishment of the post-earthquake fire scenario, the feasibility of the proposed framework is demonstrated. References regarding the equipment configuration of firefighters can be provided. Different respiratory protection equipment is required in different fire environments. In Scenario 2, the fire inside the building is not serious. Firefighters only need to wear gas masks to perform a rescue. However, in Scenario 3, the fire in the building is serious with limited visibility. Therefore, the firefighters need to be equipped with an air breathing apparatus and other professional equipment (e.g., firefighter flashlights and equipment protecting the firefighters from a fall). The weight and price of different types of equipment also differ. For example, a firefighter gas mask weighs only 1 kg and costs 400 RMB (approximately 57 US dollars), whereas a self-contained positive pressure air breathing apparatus weighs approximately 15 kg and costs 4000 RMB (approximately 570 US dollars). Based on a reliable simulation and analysis, it has been deemed important to be equipped with the most suitable equipment when in a fire environment, which not only can improve the rescue efficiency of the firefighters, it can also protect their safety.

4 Conclusions

In this work, a framework for VR scenario modeling of post-earthquake indoor fires is proposed. The following conclusions are drawn:

(1) The generation of a BIM-powered model is clearly convenient for establishing the VR scenario, structural analysis model, and fire analysis model. The combined impacts of earthquake and fire to the rescue scenario can be clearly considered.
(2) The seismic damage of non-structural components and its impact on the personnel movement must be considered in the post-earthquake fire rescue environments.
(3) Both the realism and computer performance can be balanced using the proposed smoke visualization method combining volume rendering and the particle system.
(4) The methodology proposed herein provides a scientific basis for the establishment of a highly realistic post-earthquake fire scenario. As a result, an immersive interactive fire training VR environment can be accomplished. This study is expected to provide a useful reference and technical support for fire rescue training.

Acknowledgements. The authors would like to acknowledge the financial supports of the National Key R&D Program (No. 2017YFC0702902).

References

1. Todd, D.R., Carino, N.J., Chung, R.M., Lew, H.S., Taylor, A.W., Walton, W.D.: 1994 Northridge Earthquake: Performance of Structures, Lifelines and Fire Protection Systems. National Institute of Standards and Technology (NIST), Gaithersburg (1994)
2. Sekizawa, A., Ebihara, M., Notake, H.: Development of seismic-induced fire risk assessment method for a building. Fire Saf. Sci. **7**, 309–320 (2003)
3. Mousavi, S., Bagchi, A., Kodur, V.K.R.: Review of post-earthquake fire hazard to building structures. Can. J. Civ. Eng. **35**(7), 689–698 (2008)
4. Xu, Z., Zhang, Z.C., Lu, X.Z., Zeng, X., Guan, H.: Post-earthquake fire simulation considering overall seismic damage of sprinkler systems based on BIM and FEMA P-58. Autom. Constr. **90**, 9–22 (2018)
5. Federal Emergency Management Agency: Seismic performance assessment of buildings volume 1–methodology, Technical report FEMA-P58. FEMA, Washington DC (2012)
6. Soroushian, S., Rahmanishamsi, E., Ryu, K.P., Maragakis, M., Reinhorn, A.M.: Experimental fragility analysis of suspension ceiling systems. Earthq. Spectra **32**(2), 881–908 (2016)
7. Soroushian, S., Rahmanishamsi, E., Jenkins, C., Maragakis, E.M.: Fragility analysis of suspended ceiling systems in a full-scale experiment. J. Struct. Eng. **145**(4), 04019005 (2019)
8. Cha, M., Han, S., Lee, J., Choi, B.: A virtual reality based fire training simulator integrated with fire dynamics data. Fire Saf. J. **50**, 12–24 (2012)
9. Smith, S.P., Trenholme, D.: Rapid prototyping a virtual fire drill environment using computer game technology. Fire Saf. J. **44**(4), 559–569 (2009)
10. Xu, Z., Lu, X.Z., Guan, H., Chen, C., Ren, A.Z.: A virtual reality based fire training simulator with smoke hazard assessment capacity. Adv. Eng. Softw. **68**, 1–8 (2014)
11. Cerovsek, T.: A review and outlook for a 'building information model' (BIM): a multi-standpoint framework for technological development. Adv. Eng. Inform. **25**(2), 224–244 (2011)
12. Ding, L., Zhou, Y., Akinci, B.: Building information modeling (BIM) application framework: the process of expanding from 3D to computable nD. Autom. Constr. **46**, 82–93 (2014)
13. Williams-Bell, F.M., Kapralos, B., Hogue, A., Murphy, B.M., Weckman, E.J.: Using serious games and virtual simulation for training in the fire service: a review. Fire Technol. **51**(3), 553–584 (2015)
14. Ren, A.Z., Chen, C., Luo, Y.: Simulation of emergency evacuation in virtual reality. Tsinghua Sci. Technol. **13**(5), 674–680 (2008)
15. Sun, Q.Y., Wan, W.G., Yu, X.Q.: The simulation of building escape system based on Unity3D. In: 2016 International Conference on Audio, Language and Image Processing, pp. 156–160. IEEE, Shanghai (2016)
16. Cao, L.J., Lin, J., Li, N.: A virtual reality based study of indoor fire evacuation after active or passive spatial exploration. Comput. Hum. Behav. **90**, 37–45 (2019)
17. Autodesk Revit: Built for BIM. http://www.autodesk.com/products/revit-family/overview. Accessed 21 Oct 2019
18. Hu, Z.Z., Zhang, J.P.: BIM-and 4D-based integrated solution of analysis and management for conflicts and structural safety problems during construction: development and site trials. Autom. Constr. **20**(2), 167–180 (2011)

19. Barazzetti, L., Banfi, F., Brumana, R., Gusmeroli, G., Previtali, M., Schiantarelli, G.: Cloud-to-BIM-to-FEM: structural simulation with accurate historic BIM from laser scans. Simul. Model. Pract. Theory **57**, 71–87 (2015)
20. PyroSim. http://www.thunderheadeng.com/pyrosim/. Accessed 21 Oct 2019
21. Murdock, K.L.: Autodesk 3ds Max 2013 Bible. Wiley, USA (2012)
22. Cimellaro, G.P., Ozzello, F., Vallero, A., Mahin, S., Shao, B.: Simulating earthquake evacuation using human behavior models. Earthq. Eng. Struct. Dyn. **46**(6), 985–1002 (2017)
23. Fridolf, K., Andrée, K., Nilsson, D., Frantzich, H.: The impact of smoke on walking speed. Fire Mater. **38**(7), 744–759 (2014)
24. Seike, M., Kawabata, N., Hasegawa, M.: Experiments of evacuation speed in smoke-filled tunnel. Tunn. Undergr. Space Technol. **53**, 61–67 (2016)
25. Lu, X.Z., Yang, Z.B., Cimellaro, G.P., Xu, Z.: Pedestrian evacuation simulation under the scenario with earthquake-induced falling debris. Saf. Sci. **114**, 61–71 (2019)
26. McGrattan, K., Hostikka, S., McDermott, R.: Fire Dynamics Simulator Technical Reference Guide, Volume 2: Verification. National Institute of Standards and Technology, Gaithersburg (2013)

Integrated Application of BIM and eXtended Reality Technology: A Review, Classification and Outlook

Shaoze Wu, Lei Hou(✉), and Guomin (Kevin) Zhang

School of Engineering, RMIT University,
124 La Trobe Street, Melbourne, VIC 3000, Australia
lei.hou@rmit.edu.au

Abstract. The Architecture, Engineering and Construction (AEC) sector has been benefitted from the continuous development of innovative information and visualisation technologies such as Building Information Modelling (BIM), Virtual Reality (VR), Augmented Reality (AR) and Mixed Reality (MR). These techniques have been widely researched to address traditional challenges such as the lack of capabilities to cope with extensive project information, safety concerns and multitudinous uncertainty. These studies have different research purposes and have led to a variety of outcomes. To understand the state-of-the-art of these research outcomes and identify the gaps and challenges among them, an application-focused literature review is conducted and reported in this paper. Methodically, a classification method is proposed to examine the trend of developing various BIM-XR applications. The gaps, challenges, and future works are summarised and elaborated. This paper will enhance the AEC sector's confidence in addressing the encountering problems associated with conventional project delivery and technological evolvement and make use of the lessons learnt from this study into the futuristic digital engineering practice.

Keywords: Information and Communications Technology (ICT) · Building Information Modelling (BIM) · eXtended Reality (XR) · Virtual Reality (VR) · Augment Reality (AR) · Mixed Reality (MR) · BIM-XR

1 Introduction

Building Information Modelling (BIM) is a technology that has been widely utilised by the Architecture, Engineering and Construction (AEC) sector. The application of BIM can effectively increase design efficiency and the ability to handle complex construction workflows. BIM can also be used to widen the information management areas throughout the project life-cycle, such as facilities management and occupational health and safety. Apart from BIM, Virtual Reality (VR), Augmented Reality (AR) and Mixed Reality (MR) are three visualisation and interaction concepts under the notion of eXtended Reality (XR) [1]. The XR technology merges the real-world and virtual world to allow a user to interact with virtual objects through actions. Based on the different definitions, the visualisation and interaction can be realised in an immersive virtual scenario (VR) or the real-world (AR, MR).

E. Toledo Santos and S. Scheer (Eds.): ICCCBE 2020, LNCE 98, pp. 1227–1236, 2021.
https://doi.org/10.1007/978-3-030-51295-8_86

According to Pour Rahimian [2], XR can be integrated into BIM, and such integration (thereinafter referred to as BIM-XR) has the potential to complement the visualisation performance of BIM without hampering data accessibility. A BIM-XR application can be perceived as a system which constitutes a database for BIM and a human-machine interactive interface that allows context-aware visualisation and interaction [3]. XR can convert and feed the modification and update of building information into the BIM end so that the BIM model can be amended. On the other hand, the output of BIM-XR can proffer intuitive visual representations of the BIM context and an interactive experience within this context. Moreover, BIM-XR is also an excellent technology to achieve the digital twin concept, which refers to a mapping between a real-world entity and its real-time digital replica.

The maturity of BIM-XR technology, as evidenced by various studies, also results in system complexity and diversity. In other words, research and development of BIM-XR techniques in the civil engineering domain demonstrate significant heterogeneity. For the usage of these techniques, it is imperative for the AEC researchers and practitioners to clearly understand application branches, development criticalities and research gaps, upon which possible solutions and reconciled efforts can be made towards a more streamlined practice of using intrinsically complicated BIM-XR methodologies. To this end, this study aims to compile the publications related to a diversity of BIM and XR applications, identify the trendy research areas, and investigate the advanced application in order to conquer emerging civil engineering challenges.

2 Review Methodology

In this study, two common literature review databases *Scopus* and *Web of Science* were used to retrieve articles. The search in the *Scopus* and *Web of Science* engines was set to "Article title, Abstract, Keywords" and "Topic" respectively. The "Topic" term can search four fields including title, abstract, author keywords and keywords plus, and the keywords plus provided by Web of Science augments traditional keyword to broaden the search. The title, abstract and keywords are the concentrated expression of the content of an article. These fields can be used to search related articles effectively. The keywords related to BIM were selected as *building information model* and *BIM*. The keywords related to XR were selected as *virtual reality*, *augmented virtuality*, *augmented reality*, *mixed reality*, and *extended reality*. Asterisk was used to search BIM-related articles but was not used to search XR-related articles, because BIM has several lexical interpretations. Double quotes were used to encompass these keywords to narrow down the scope of the search. Compared to other paper types such as conference papers, journal articles are typically peer-reviewed and more exhaustive to be selected as the review paper type. All these methods had led to 168 results. Among these articles, 118 articles were ruled out due to lack of relevance, and 50 articles [3–52] were got to be reviewed. Based on the review, the articles were classified into four categories for the analysis to summarise the status quo in different application directions. Moreover, an outlook based on the status quo was proposed for the integrated application of BIM and XR.

3 Classification and Analysis

XR is an ICT technology that has been significantly developed and has become the backbone of the graphical context within the entertainment industry. For everyone involved in the design and construction of buildings, XR techniques are increasingly being adopted worldwide to improve productivity while reducing errors, decrease safety issues, and cutting costs. With the development of BIM-XR, the research is becoming comprehensive, and the purpose of articles also tends to be diversified. The current research provides many innovation points but lacks a comprehensive summary to help AEC researchers directly, which results from that searching, screening, and understanding the literature are time-consuming tasks. This study sorted out the published articles and classified them into four application categories to summarise the status quo.

AEC can be divided into four parts, including on-site construction, design, project management, and building the construction team. Correspondingly, the applications of BIM-XR can also be divided into the following four categories: task guidance and information retrieval [3, 5, 8–11, 13, 15, 18, 22, 24, 29, 31, 37, 40, 41, 46, 49]; design review and refinement [4, 7, 16, 23, 27, 30, 33, 35, 36, 42, 52]; process planning and control [6, 19, 21, 26, 28, 32, 34, 45, 48, 50, 51], and upskilling of the AEC workforce [12, 14, 17, 20, 25, 38, 39, 43, 44, 47].

3.1 Task Guidance and Information Retrieval

BIM-XR can provide new solutions to guide tasks and retrieve information. In traditional construction and maintenance processes, the methods are mainly based on voice, text and images to manage and deliver information. For example, a construction manager tells the construction workers the details of a work. The workers note the key points and retrieve the information from handbooks and construction drawings. In these traditional methods, two points can be promoted: retrieving information and explaining tasks. Information retrievals are time-consuming, and oral explaining tasks can lead to misunderstanding. These inefficient and inaccurate methods raise the workload of employees and increase the probability of errors. However, as shown in Fig. 1, BIM-XR provides a better method which can extract information and tasks from BIM and display them by multimedia [41]. The multimedia includes step-by-step instructions and animated demonstrations to assist employees in accomplishing the work. This method based on the information management functions of BIM and the visualisation and interactivity characteristics of XR is more automated and user-friendly [13].

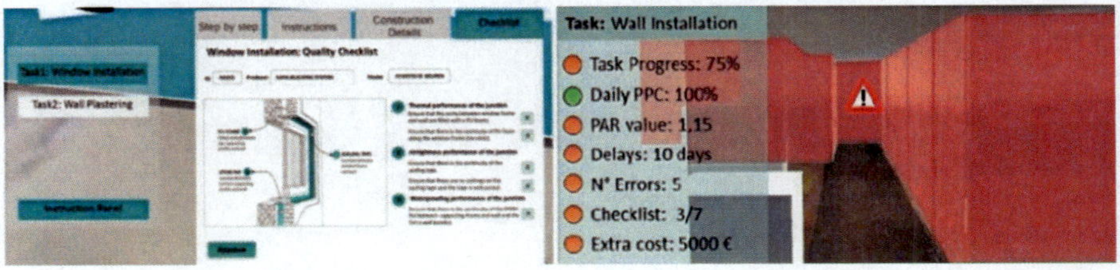

Fig. 1. Context-specific information proffered by the BIM-XR prototype [41].

3.2 Design Review and Refinement

The visualisation and interactivity characteristics of BIM-XR allow workers to review and refine a design effectively. Visualisation on traditional mediums such as computer screens lacks some degrees of immersion which may prevent the full utilisation of a model, but VR as the most immersive medium can solve the problem [52]. In an immersive VR scenario, an immersive experience similar to the real-world can be given to the user without creating real-scale physical mock-ups. In an MR/AR environment, buildings or building components to be reviewed can be superimposed on a real-world scenario [4]. MR can enable a designer without prior maintenance experience to identify the maintainability of a building component [27]. Moreover, designs can also be adjusted through BIM-XR interactive functions, which allows stakeholders to refine the design in an XR environment directly and increase the communication efficiency between clients and designers.

3.3 Process Planning and Control

To implement process planning and control, AEC staff should consider the project information, including the time, workforce and cost. BIM has been developing to adapt four-dimensional (4D) scheduling, five-dimensional (5D) cost planning and n-dimensional (nD) information management, but it lacks a user-friendly interface with management functions, which results in that the AEC staff prefer using non-BIM software to plan and control a project. However, as a visual interaction tool, XR can provide users with a programmable interface. By programming, BIM-XR can realise all the functions needed [51]. Process information can be extracted from BIM files directly to avoid omission in data transmission. Geometric models can be provided to show correspondence between the process information and model components. As shown in Fig. 2, various types of schedule information based on 4D models can be implemented in the real-time field images through AR to reflect the actual field conditions [28].

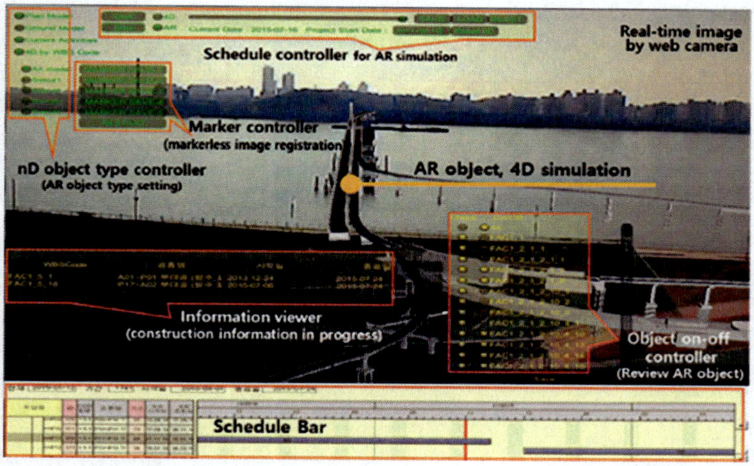

Fig. 2. Remote management using AR to visualise construction site information [28].

3.4 Upskilling of the AEC Workforce

BIM-XR provides an innovative approach for upskilling. Practice is a useful method to improve the skills of the AEC workforce, but it causes the waste of materials, which result in the preciousness of the practice chance. In contrast, BIM-XR can provide virtual practice chances reducing the cost of upskilling. The BIM-XR allows a user to operate virtual objects in immersive virtual scenario or real-world, and the corresponding real-time feedback can be delivered to the user directly, which is similar to the real practice. Users can try different operations and get corresponding results, even if the steps are wrong. Therefore, the users can deepen their understanding through constant experimentation. Moreover, BIM-XR has specific advantages: 1) the integration of BIM and XR has proven effective in improving communicability, passive problem finding and poor spatial cognition [47]; 2) VR can provide comprehensive safety training avoiding physical injuries; 3) all the user's actions can be recorded through sensors for further analysis; and 4) as a database, BIM can provide users with sufficient information support, so when users want to learn more, they can get the in-depth information.

4 Outlook

The current research of BIM-XR technology is fruitful, but there is still a long way to go before the research results can be put into practice. The key point to apply the technology is integrating existing applications in a universal system, which is also the future work of BIM-XR research. The fact can be found through the review that a significant amount of the applications was developed by cross-platform game engines such as Unity. The game engines were proved that they are powerful to develop offline XR application [36] and online XR application [14]. With the help of scripts, the developed applications can exchange data with BIM software, such as Autodesk Revit and Autodesk Navisworks. Both game engines and BIM software have excellent expandability, which is suitable to become a developing platform [41]. Based on the platform, a systemic framework can be proposed. The framework contains two parts, namely the client and the server, and the relationship of them is shown in Fig. 3.

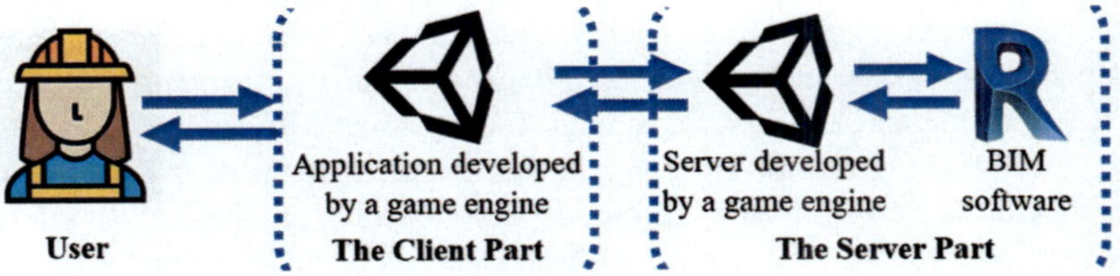

Fig. 3. BIM-XR systemic application framework.

4.1 The Client Part

As a technology that provides the visualisation and interaction for a single person, BIM-XR can give staff a virtuous cycle consisting of the training, assistance, and feedback of work, as shown in Fig. 4. Before an actual work, VR can provide workers with repeatable training to let workers comprehend the purpose and details of the work. The training system can show users recommended operation steps and allow users to try operation steps they prefer. During the actual work, users have an MR device to assist them in their tasks. For different types of work, MR has a customised operation interface. For construction workers, MR can provide task lists, recommended work steps and real-time on-site hazard warnings. For maintenance staff, MR can give damage predictions, maintenance sequences and see-through vision of structures. For construction managers, MR can provide details of the virtual construction site scenario, worker activities, and project schedules. All the staff can connect experts for remote guidance to solve unexpected situations. At the same time, the situations can be recorded in a training library, which is available for related employees to improve their resilience. Finally, workers' daily work status being recorded can be fed back to themselves to avoid work injuries and improve their work efficiency in the future.

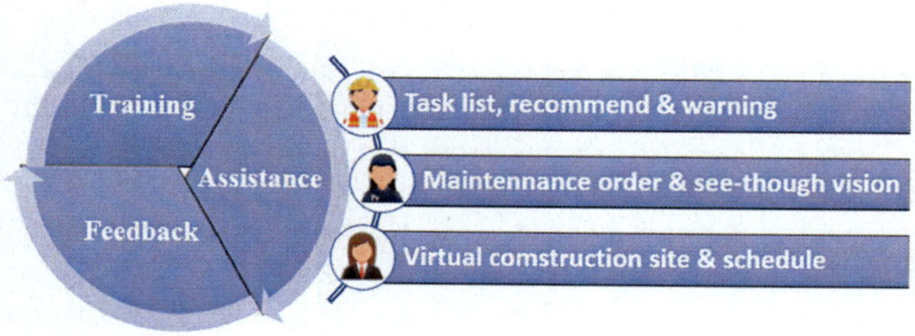

Fig. 4. Virtuous cycle based on the client part.

4.2 The Server Part

The server part can carry a BIM information flow, as shown in Fig. 5. Information flow of a building can be divided into three layers, namely *design & construction*, *maintenance & service*, and *demolish*. Each layer can be divided into four quadrants according to two dimensions which are *physical-digital* and *virtuality-reality*. Information is transmitted in each layer through four quadrants of sequential in *intention*, *VR*, *digital twin (MR)*, and *physical objects*. For the layer of design & construction, the information flow starts with the design and planning of a building. With the help of digital technology, BIM models are established and can be reviewed and refined by VR technology. Then, the BIM geometry model and corresponding information can be imported into the MR to guide the construction. At this stage, BIM will also be modified according to the actual construction situation to realise the digital twins. Finally, when the building is completed, all the building information has been updated and delivered to the next layer. For the layer of maintenance & service, after receiving the information, practitioners can identify inspecting emphases, predict component damages, and make maintenance plans. Then, the VR can be used to determine the feasibility of the plan and familiarise employees with operations. During the implementation phase, MR can provide employees with the corresponding information and a see-through vision of structures to assist in completing maintenance and services. At the same time, relevant task information will be uploaded to the BIM for updates and backups. After the task is completed, the information is retained for the next task or passed on to the demolish layer. The demolish layer is similar to the previous two layers, and it consists of four steps of *planning*, *rehearse*, *implement* and *close-out*. The difference is that the purpose of this layer is to demolish the building.

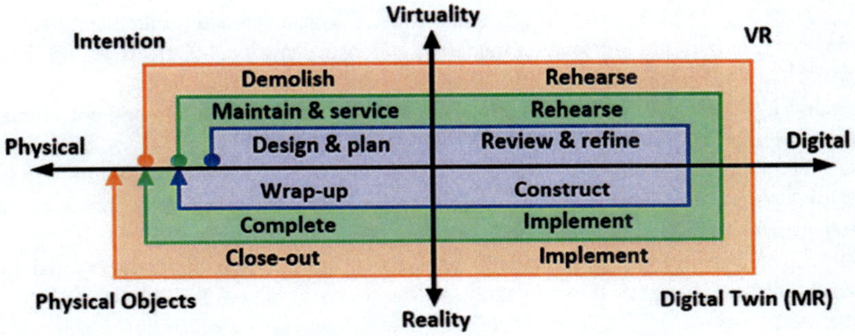

Fig. 5. BIM information flow.

5 Conclusion

The construction industry has witnessed a growth of innovative BIM-XR applications over the last two decades. The combination of BIM and XR technologies has advantages from both the advanced digital engineering area and the advanced visual user interface area, which can effectively cope with extensive project information, safety concerns and multitudinous uncertainty. On the other hand, the integrated application of the two technologies requires a large amount of research, which result in the various research branches. To summarise and elaborate the gaps, challenges, and future works of BIM-XR in AEC practice, this study focused on BIM-XR application research.

BIM-XR research covers various disciplines including civil engineering, ICT, computer vision, human factors and ergonomics. Such multidisciplinary research not only can stimulate the inventiveness of researchers but also brings challenges to improve BIM-XR further. This study demonstrated four current trendy application branches and established a reliable summary for researchers and practitioners who need useful insights into the BIM-XR area. Researchers and practitioners can seek academic cooperation opportunities based on the analysis of contributions to BIM-XR research. Future work about application research of BIM-XR is also anticipated in this paper. The research emphasis of BIM-XR will shift to the integration of existing applications in a universal system. A systemic framework was proposed in this study. The framework demonstrates the potential of BIM-XR in the AEC field and indicate research directions for scholars and practitioners who are interested in BIM-XR.

References

1. Fujiuchi, K., Riggie, J.: Academic library collections in the age of extended reality (XR). Collect. Manag. **44**(2–4), 296–303 (2019)
2. Pour Rahimian, F., Chavdarova, V., Oliver, S., Chamo, F.: OpenBIM-tango integrated virtual showroom for offsite manufactured production of self-build housing. Autom. Constr. **102**, 1–16 (2019)
3. Lee, J., Kim, J., Ahn, J., Woo, W.: Context-aware risk management for architectural heritage using historic building information modeling and virtual reality. J. Cult. Herit. **38**, 242–252 (2019)
4. Alsafouri, S., Ayer, S.K.: Mobile augmented reality to influence design and constructability review sessions. J. Archit. Eng. **25**(3), 04019016 (2019)
5. Bae, H., Golparvar-Fard, M., White, J.: High-precision vision-based mobile augmented reality system for context-aware architectural, engineering, construction and facility management (AEC/FM) applications. Vis. Eng. **1**(1), 3 (2013)
6. Boton, C.: Supporting constructability analysis meetings with immersive virtual reality-based collaborative BIM 4D simulation. Autom. Constr. **96**, 1–15 (2018)
7. Cha, S.H., Koo, C., Kim, T.W., Hong, T.: Spatial perception of ceiling height and type variation in immersive virtual environments. Build. Environ. **163**, 10 (2019)
8. Chalhoub, J., Ayer, S.K.: Using mixed reality for electrical construction design communication. Autom. Constr. **86**, 1–10 (2018)
9. Chen, K.Y., Chen, W.W., Li, C.T., Cheng, J.C.P.: A BIM-based location aware AR collaborative framework for facility maintenance management. J. Inf. Technol. Constr. **24**, 360–380 (2019)
10. Chen, Y.C., Chi, H.L., Kang, S.C., Hsieh, S.H.: Attention-based user interface design for a tele-operated crane. J. Comput. Civ. Eng. **30**(3), 04015030 (2016)
11. Chu, M., Matthews, J., Love, P.E.D.: Integrating mobile building information modelling and augmented reality systems: an experimental study. Autom. Constr. **85**, 305–316 (2018)
12. De Amicis, R., Riggio, M., Badr, A.S., Fick, J., Sanchez, C.A., Prather, E.A.: Cross-reality environments in smart buildings to advance STEM cyberlearning. Int. J. Interact. Des. Manuf. - IJIDEM **13**(1), 331–348 (2019)
13. Diao, P.H., Shih, N.J.: BIM-Based AR maintenance system (BARMS) as an intelligent instruction platform for complex plumbing facilities. Appl. Sci.-Basel **9**(8), 1592 (2019)
14. Du, J., Shi, Y.M., Zou, Z.B., Zhao, D.: CoVR: cloud-based multiuser virtual reality headset system for project communication of remote users. J. Constr. Eng. Manag. **144**(2), 04017109 (2018)
15. El Ammari, K., Hammad, A.: Remote interactive collaboration in facilities management using BIM-based mixed reality. Autom. Constr. **107**, 102940 (2019)
16. Garagnani, S.: Archaeological building information modeling: beyond scalable representation of architecture and archaeology. Archeol. E Calcolatori **28**(2), 141–149 (2017)
17. Getuli, V., Giusti, T., Capone, P., Sorbi, T., Bruttini, A.: A project framework to introduce virtual reality in construction health and safety. Bo-Ricerche E Progetti Per Il Territorio La Citta E L Architettura **9**(13), 166–175 (2018)

18. Gheisari, M., Irizarry, J.: Investigating human and technological requirements for successful implementation of a BIM-based mobile augmented reality environment in facility management practices. Facilities **34**(1–2), 69–84 (2016)
19. Goulding, J.S., Rahimian, F.P., Wang, X.: Virtual reality-based cloud BIM platform for integrated AEC projects. J. Inf. Technol. Constr. **19**, 308–325 (2014)
20. Gurevich, U., Sacks, R.: Examination of the effects of a KanBIM production control system on subcontractors' task selections in interior works. Autom. Constr. **37**, 81–87 (2014)
21. Hallowell, M.R., Hardison, D., Desvignes, M.: Information technology and safety. Constr. Innov. **16**(3), 323–347 (2016)
22. Hernández, J., Martín Lerones, P., Bonsma, P., van Delft, A., Deighton, R., Braun, J.-D.: An IFC interoperability framework for self-inspection process in buildings. Buildings **8**(2), 32 (2018)
23. Imaizumi, J.: Assessment of the process for designing an apartment building through IM & VR. Int. Rev. Spat. Plan. Sustain. Dev. **5**(1), 45–54 (2017)
24. Irizarry, J., Lavy, S., Gheisari, M., Williams, G., Roper, K.: Ambient intelligence environments for accessing building information. Facilities **32**(3/4), 120–138 (2014)
25. Ji, X., Fang, X., Shim, S.H.: Design and development of a maintenance and virtual training system for ancient Chinese architecture. Multimed. Tools Appl. **77**(22), 29367–29382 (2018)
26. Jiao, Y., Zhang, S.H., Li, Y.K., Wang, Y.H., Yang, B.M.: Towards cloud augmented reality for construction application by BIM and SNS integration. Autom. Constr. **33**, 37–47 (2013)
27. Khalek, I.A., Chalhoub, J.M., Ayer, S.K.: Augmented reality for identifying maintainability concerns during design. Adv. Civ. Eng. **2019** (2019)
28. Kim, H.S., Kim, S.K., Borrmann, A., Kang, L.S.: Improvement of realism of 4D objects using augmented reality objects and actual images of a construction site. KSCE J. Civ. Eng. **22**(8), 2735–2746 (2018)
29. Kwon, O.S., Park, C.S., Lim, C.R.: A defect management system for reinforced concrete work utilizing BIM, image-matching and augmented reality. Autom. Constr. **46**, 74–81 (2014)
30. Lai, D.S., Leung, A.K., Chan, D., Ching, S.H.: Cultural heritage preservation using new media methods: Yingxian Wooden Pagoda, China. Virtual Archaeol. Rev. **10**(21), 103–115 (2019)
31. Lee, S., Akin, O.: Augmented reality-based computational fieldwork support for equipment operations and maintenance. Autom. Constr. **20**(4), 338–352 (2011)
32. Lin, T.H., Liu, C.H., Tsai, M.H., Kang, S.C.: Using augmented reality in a multiscreen environment for construction discussion. J. Comput. Civ. Eng. **29**(6), 04014088 (2015)
33. Lin, Y.C., Chen, Y.P., Yien, H.W., Huang, C.Y., Su, Y.C.: Integrated BIM, game engine and VR technologies for healthcare design: a case study in cancer hospital. Adv. Eng. Inform. **36**, 130–145 (2018)
34. Mirshokraei, M., De Gaetani, C.I., Migliaccio, F.: A web-based BIM-AR quality management system for structural elements. Appl. Sci. (Switz.) **9**(19), 3984 (2019)
35. Motamedi, A., Wang, Z., Yabuki, N., Fukuda, T., Michikawa, T.: Signage visibility analysis and optimization system using BIM-enabled virtual reality (VR) environments. Adv. Eng. Inform. **32**, 248–262 (2017)
36. Natephra, W., Motamedi, A., Fukuda, T., Yabuki, N.: Integrating building information modeling and virtual reality development engines for building indoor lighting design. Vis. Eng. **5**(1), 1–21 (2017)
37. Olbrich, M., Graf, H., Kahn, S., Engelke, T., Keil, J., Riess, P., Webel, S., Bockholt, U., Picinbono, G.: Augmented reality supporting user-centric building information management. Vis. Comput. **29**(10), 1093–1105 (2013)

38. Park, C.S., Kim, H.J.: A framework for construction safety management and visualization system. Autom. Constr. **33**, 95–103 (2013)
39. Park, C.S., Le, Q.T., Pedro, A., Lim, C.R.: Interactive building anatomy modeling for experiential building construction education. J. Prof. Issues Eng. Educ. Pract. **142**(3), 04015019 (2016)
40. Park, C.S., Lee, D.Y., Kwon, O.S., Wang, X.: A framework for proactive construction defect management using BIM, augmented reality and ontology-based data collection template. Autom. Constr. **33**, 61–71 (2013)
41. Ratajczak, J., Riedl, M., Matt, D.T.: BIM-based and AR application combined with location-based management system for the improvement of the construction performance. Buildings **9**(5), 118 (2019)
42. Shen, W., Zhang, X., Shen, G.Q., Fernando, T.: The user pre-occupancy evaluation method in designer-client communication in early design stage: a case study. Autom. Constr. **32**, 112–124 (2013)
43. Wang, B., Li, H., Rezgui, Y., Bradley, A., Ong, H.N.: BIM based virtual environment for fire emergency evacuation. Sci. World J. **2014**, 589016 (2014)
44. Wang, C., Li, H., Kho, S.Y.: VR-embedded BIM immersive system for QS engineering education. Comput. Appl. Eng. Educ. **26**(3), 626–641 (2018)
45. Wang, X.Y., Truijens, M., Hou, L., Wang, Y., Zhou, Y.: Integrating augmented reality with building information modeling: onsite construction process controlling for liquefied natural gas industry. Autom. Constr. **40**, 96–105 (2014)
46. Williams, G., Gheisari, M., Chen, P.J., Irizarry, J.: BIM2MAR: an efficient BIM translation to mobile augmented reality applications. J. Manag. Eng. **31**(1), A4014009 (2014)
47. Wu, T.H., Wu, F., Liang, C.J., Li, Y.F., Tseng, C.M., Kang, S.C.: A virtual reality tool for training in global engineering collaboration. Univers. Access Inf. Soc. **18**(2), 243–255 (2019)
48. Xie, H., Shi, W., Issa, R.R.A.: Using RFID and real-time virtual reality simulation for optimization in steel construction. Electron. J. Inf. Technol. Constr. **16**, 291–308 (2011)
49. Yeh, K.C., Tsai, M.H., Kang, S.C.: On-site building information retrieval by using projection-based augmented reality. J. Comput. Civ. Eng. **26**(3), 342–355 (2012)
50. Yu, Z., Peng, H., Zeng, X., Sofi, M., Xing, H., Zhou, Z.: Smarter construction site management using the latest information technology. Proc. Inst. Civ. Eng.: Civ. Eng. **172**(2), 89–95 (2018)
51. Zaher, M., Greenwood, D., Marzouk, M.: Mobile augmented reality applications for construction projects. Constr. Innov.-Engl. **18**(2), 152–166 (2018)
52. Zaker, R., Coloma, E.: Virtual reality-integrated workflow in BIM-enabled projects collaboration and design review: a case study. Vis. Eng. **6**(1), 4 (2018)

Simulation, Parametric Modelling and other Technologies for Innovation in AEC

A Participative Framework Covering Urban Planning Process with A Parametric Approach

Elie Daher and Sylvain Kubicki(✉)

Luxembourg Institute of Science and Technology,
Esch-Sur-Alzette, Luxembourg
{elie.daher, sylvain.kubicki}@list.lu

Abstract. Urban planning design process in general is a manual process based on the expertise of professionals addressing the design, exploitation and use of land spaces. Starting by manual sketching configurations do not allow the evaluation of the design performance. Parametric modelling approach, and algorithmic thinking, are currently helping in controlling and informing architectural and urban design. However, the participation of the citizens in the urban planning remains limited and not fully explored. This paper describes how a parametric modelling approach, coupled with a technological framework, allow a participative design thinking based on parametric capabilities to generate and explore different solutions of urban planning. This research consists of developing a framework and implement it into several technological prototypes that cover different stages of the design process. Two main applications are shown in this paper with the devices that are associated to each of them.

Keywords: Urban planning · Parametric design · Participation

1 Introduction

The change of urban meanings and infrastructure is taking place at different levels and varying time-scales due to the increasing number of urban spaces. This increasing number of population living in urban areas estimated 60% by 2050[1], will create new challenges regarding the environment and human health. In order to anticipate and overcome these challenges, climate and energy targets have been set for the upcoming years.

Urban planning is defined as the theory of space planning focusing on physical forms, association and organization of different urban spaces and functions with the aim to shape the activity of human [1]. The urban planning addresses the development, design, exploitation and use of land spaces for human-activities purposes. Urban planning is also related to social aspects as it is intended to enrich human life at the social, cultural and environmental levels [2]. For this purpose, criteria such as development, resources conservation, people safety, protection, and environment should be taken into consideration. In terms of outcomes, living areas planning should provide

[1] https://www.un.org/en/development/desa/population/publications/pdf/ageing/WPA2015_Report.pdf.

E. Toledo Santos and S. Scheer (Eds.): ICCCBE 2020, LNCE 98, pp. 1239–1251, 2021.
https://doi.org/10.1007/978-3-030-51295-8_87

recommendations and values for the needed urban uses and the allocated lands (for example, public spaces, facilities, housing, road, commercial, etc.). The current practice of urban planning is often based on the expertise of urban planners, developers or investors and other factors [3]. Top-down approach currently dominating the urban planning process is based on the fact that the architects and experts conceive the urban environment with limited contribution of the community and citizens. Some involvement from the community may be implemented but remains limited and insufficient. There is a serious criticism for this approach since it is not grounded to the community level [4, 5] which is creating, in some situation, a failure to urban environments. The process used to conceive urban environment has an importance in providing viable and maintainable communities [6]. Different approaches have been identified with different components and processes. These processes, slightly different, share some common components [7–9] which can be classified as following (Fig. 1):

- Collection of needed information, where the planning team conducts a study of the area by leading a series of surveys and visits to the site and analyzing the activities related,
- General Study, identification of problems and investigation in possible solutions, where, based on the study, the planning team will identify opportunities and possible solutions based on the analysis,
- Development of solutions and comparison between different alternative
- Decision-making and communication.

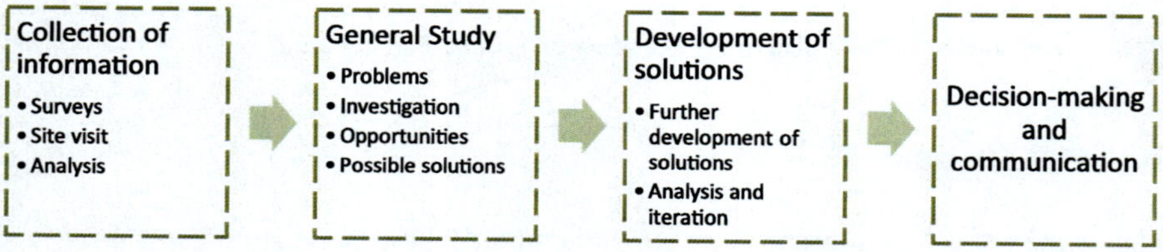

Fig. 1. Main components for urban planning process

2 Waves of Participation and Technological Development in the Urban Planning

The mismanagement of the physical environment is a major factor contributing to social ills [10]. Engaging the user in the design-related process issues will ensure meeting the users' requirements in relation with the physical environment. The user participation has emerged as a key to bridging a gap between the requirements and the solutions proposed by the designers [10].

The community participation has experimented different waves of implementation in the urban design through different processes in relation or not with technological development. Policy makers and academia have recognized the use of modern technologies as an essential feature for smart cities development [11]. Different waves of participation (Table 1) were explored in urban planning in the western countries context during the last 50 years [12–16].

Table 1. Waves of Participation in Urban Planning in the Western Context.

	Milestones	Characteristics
1968–1970s	Emergence of Participation	The redistribution of power that enables the have-not citizens, presently excluded from the political and economic processes, to be deliberately included in the future [12]. Radical experiments with participation
1980s	The participation boom	Neoliberalism taking action. As a reaction, flourishing of non-government organizations in seeking alternatives to top-down outsider driven development. Acknowledgement of the value of local knowledge
1990s	The participation imperative	Legalization and institutionalization of participation. Participation in terms of collecting feedback but no follow up mechanisms present. Participation as a part of the social sustainability in the development field. Participation as a condition for funding. Proliferation of guidebooks and courses
2000s	E-participation and crowdsourcing	Using ICTs to collect citizen feedback e-governance, volunteered geographic information [17], crowdsourcing [15, 18]
2010s	Critical spatial practice	Criticism of the participation, introduction of crossbenching. Transdisciplinary co-creation, networked and performative action [16]

Another important wave of technological capabilities that emerged in the urban planning design as a way to support the decision-making and a ground for participatory design is offered by the capabilities and functions presented in the following methods or systems:

- Computer aided design
- Virtual and augmented environments
- Physical artifacts

Computer Aided Design. The adoption of new technologies and methods in architecture and construction and the accessibility of computers for professional and educational use especially in the 1980's and 1990's, has created a new research wave based on the computer aided design and boosted the innovation in the technology used for architecture and construction to its beyond. Urban planners, architects and experts started exploring in the 2010s the new trends of technology applications. In particular, parametric modelling has changed considerably the way processes are developed. Parametric modelling started to emerge in the urban planning design with a first motivation to automatically generate and optimize urban spaces with a consideration of constraints and different parameters.

As a definition, the parametric modelling is used more and more into the design process, to manage information workflows or to help in the generation of forms in contemporary architecture [19]. It reflects a recent and innovative approach in computing design and enables, based on a visual programming language, an automatically generation of large number of solutions and alternatives [19–22] for the design offering the designer the possibility to explore the design performance [22, 23].

Computer application technologies demonstrate that the use of computing can facilitate the integration of the user into the decision-making process. However, in a research related to tools used for the participative urban planning, Al-Kodmany [24] showed that an important aspect for the tools and methods used in participative urban planning is related to the clear and understandable interface. A confusing interface or unclear one can help in losing the potential of high-tech tools [24]. Urban planning applications are emerging and they are evolving from sharing information on the selected environment, to sharing opinions and discussions between citizen and other stakeholders. These applications, even based on technological development, remain informative and communicative. CAD applications enable as well the simulation of future urban plans. Beirao presented a dynamic urban planning system built with parametric programming tools [25]. Another research presents a methodology to quantify solar energy potential from photovoltaic systems in the urban context [26]. The methodology was based on a parametric approach aiming to optimize the demand supply balance for better energy performances of the cities. A recent research explored the potential of parametric for urban design participation [27, 28]. The outcome of these twos researches are a framework for engaging people in urban design decision-making process. However, in this research urban professionals had to elaborate most parts of the design in their mind, few users were also directly involved in this research. Steino [29, 30] showed also that for an operative participation and collaboration between professionals and users, a digital platform with an interactive interface and real-time simulations is needed. Most of these applications are covering the energy simulation, the participation aspect is not fully covered in parametric current application, and thus, the participatory approaches remain limited to the information or consultation.

Virtual and Augmented Environments. Nowadays, we are also experimenting several new solutions for city and urban spaces visualization in virtual or augmented reality [31–36]. These solutions are providing spatial visualization allowing merging real city with virtual three-dimensional (3D) buildings, which enhance the immersion level of urban planning solution [35]. From this perspective, [37] has tested the community engagement with virtual world in the planning of a neighborhood park in Boston, Massachusetts. This project demonstrates that the virtual reality, when properly used, enables productive and meaningful public participation. However, the participation remains in this case limited to information level of participation or consultation according to the ladder of Arnstein [12].

Other benefits for the application of VR and AR in the urban planning were related to the time saved in the design process and the presentation of realistic and objective results for simulations [38]. Despite the visualization function that can be provided by virtual and augmented reality, the main challenge is to find a way to use augmented reality to allow the participation in the urban process.

Physical Artifacts. The technology now is very advanced and provides multiple interactive environments allowing the visualization of the information. However, adding physical shapes to the displayed information and to the controllers (sliders, inputs…) will enhance the human capacity to perceive the solutions and to more understand the input by controlling and manipulating the related objects [39]. The tangible interfaces aim use the tangible object as input to control and interact with the design [40]. Other physical interfaces combined with digital environment such as tangible tabletop with physical tangible objects are perceived as important in order to enhance a transparent collaboration process. Tangible tabletop systems can be foreseen as a participatory technology that helps to increase the engagement of users and community in urban planning by manipulating physical objects [41, 42].

3 Research Questions and Objectives

The main goals of the participation in the urban planning is to allow the involvement of the community-in-place to contribute in the decision making of their own urban places and to increase the satisfaction of the proposed urban spaces. Solutions and decisions are more likely to meet the real need of the locals. However, this participation is not clearly articulated nor organized with a generic and defined process. As a result, the platforms created to enhance the participation of the community with digital technologies are often below the expectations and lack clarity and easiness of use.

The diversity of actors and designers, in addition to the constraints in the local urban environment and the need of integrating different points of views make the urban planning a complex task. In addition, we noticed, that the parametric process is not adapted to the urban planning process; professionals have to develop their own 'definitions' (i.e. creating a definition in a parametric modelling tool that allow the generation of an urban district according to one design strategy) and algorithm and adapt it to specific situations. In the construction industry, digitalization is still under development. Designers, architects and engineers are working in 3D modelling environment, which is considered as their comfort zone. Working in 3D modelling environment and in particular parametric tools allows designers to explore different alternatives to the design. As well, the participation of communities in the design can be possible thanks to the advent technologies and devices. However, these solutions address mainly issues related to the energy, environment and economy. Even if defining a district remains a debatable investigation topic, we believe that creating a framework and associating this framework to different type of participation will enhance the urban planning process by the ability to gather needed planning information. The research question that we address in the context of our work is the following: *Level of involvement of users can depend on the type of the projects and politics aspects, how can it be implemented into parameters in a multi-objective optimization process of urban planning?*

The purpose is to help in the decision-making by performing spatial usage-driven planning in a fast way to respond to the actual and future needs, while taking into account the technical issues in organizing the usage-driven spaces and in particular enabling the participation of different users in the decision-making process related to the urban planning. The objectives of the proposed research are related to the integration of new capacities of intervention in the urban planning, by creating new systems and methods for non-expert and enabling a smooth collaboration and information exchanges between different parties.

4 Methodology

Our research methodology relies on a case study implementation to validate the conceptual framework outlined with technical features after the state-of-the-art analysis. In a later stage, we developed generic prototypes taking into account the different stage of the conceptual framework which was adapted to the specific situation of the use case. This prototype was developed in two phases. Applied on an industrial site in Luxembourg, the prototype is considering remarkable aspects of the site (e.g. pollution, visual comfort, proximity…). It is tested with its two parts with different categories of users: experts and community.

4.1 Technical and Conceptual Framework

In distinction of what is shown in the section related to the current practice of urban planning, we developed a framework where the engagement of the community was essential for each stage and sub-stage. The framework has an objective to rethink the way the design process is developed enabling the participation process.

The framework is constructed taking into account the capabilities offered by Grasshopper, a plugin for Rhino3D which enables the parametric modelling and the optimization through plugins and that are required to ensure that the algorithm operates properly: (1) Ladybug & Honeybee: These two plug-ins are used to run environmental simulations. They are used for the studies related to the simulation of the view and well-being comfort in each use case. (2) Optimization plugins such as Galapagos for multi-optimization. The objective of the framework is to develop interactions with a set of tools and guidelines to support different scale of users' participation in the different scale of decision-making process in early design phase. The framework is based on three main design interactions Fig. 2:

Interaction 1: Setting Requirements. This part is related to the objective setting selection that should be taken into account in the optimization and design process. The citizens and the participants in the design process will set the requirements with different means and devices. In this prototype, we are using a tangible tabletop that allow with physical objects to set the input as requirements. Other requirements can be set by surveys.

Interaction 2: Dependencies and Requirements Generation. The major lines to follow are set in this part of the framework; this part will differ from one use case to another. Here, we define a methodology based on algorithmic formulas where the users will be exploring the influence of their requirements and exploring different options.

This part is related to the site exploring and site organigram disposal. For this step of optimization of the organigram into the site, we are using space syntax and Kangaroo Physics, two plugins in Grasshopper that allow an optimization of the spaces and functions.

Interaction 3: Design Proposal Geometry Generation. This interaction is dealing more with the 3D generation and simulation details influenced by the input of the citizens. An optimization solver for Grasshopper, based on evolutionary algorithm and adapted for multi-objective optimization is used for this interaction. This plug-in is particularly used in the urban planning use case to generate the 3D geometry of the buildings with an optimization of the objectives required (such as solar and daylight factors), and to the layout of the container shelter in order to optimize the solar and daylight factor.

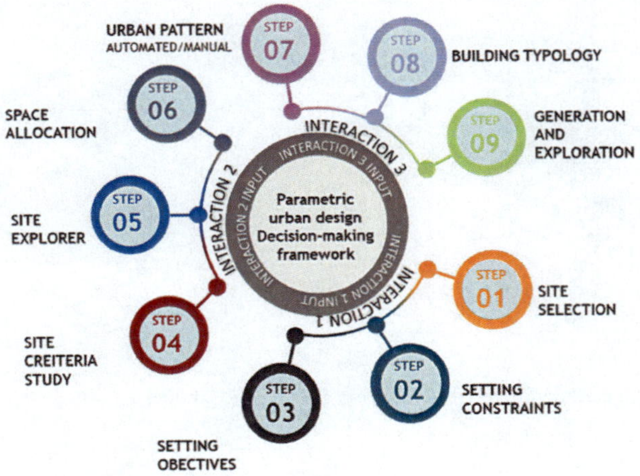

Fig. 2. Illustration of the framework

5 Development and Experimentation

Since urban planning consists of multiple and different tasks to perform. the development considers two prototypes developed simultaneously reflecting different steps from the urban planning process. In the following, we expose each part of the development for both of the prototypes. It should also be mentioned that even if we were exposing these prototypes in separate way, the aspects shown in each of the prototypes would be joined to form one system in the future work.

5.1 First Phase of the Prototype

Development. The first development of the prototype is based on a definition created in Grasshopper, a plugin to Rhino 3D. The objective is to cover different aspect and strategies of the urban planning in a fast way and to show the potential of a tool that can help in the generation of urban spaces. This prototype allows different steps of the design in the framework mainly for the third interaction Fig. 2:

Step 1: Urban pattern. The urban pattern consists on a manual sketching (where user will manually draw urban spaces that are computed directly in the system), or automatic generation of urban spaces based on several geometrical parameters (rectangular, ununiformed), for instance the input parameters for this step are the setback, roads dimensions, area needed and green area ratio.

Step 2: Building topology. The 3D generation Fig. 3, which consist of the generation of 3D shapes of potential urban districts taking into account different design strategies. The parameters responsible of the 3D generation are the maximum height and depth of building, central court, type of the building generation and number of floors (step 8 of the framework).

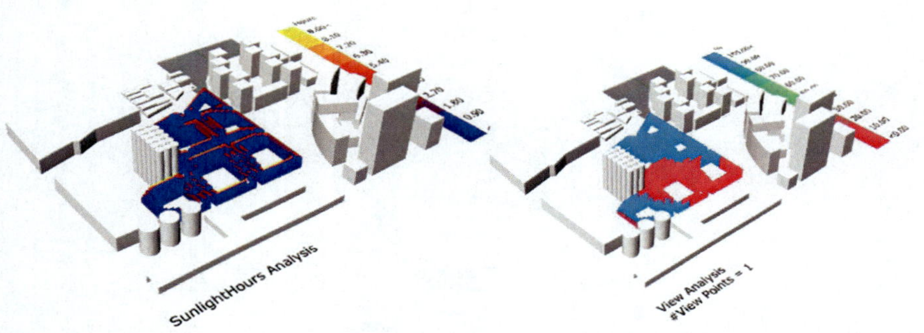

Fig. 3. Different generated geometrical shapes for a potential district with evaluation of different indicators.

Step 3: Generation and exploration. For each iteration Fig. 3, the users are able to analyze the performance indicators and compare between different iterations (step 9 of the framework). It is also possible to perform a multi-objective optimization for the 3D generation based on a plug-in integrated in grasshopper (Octopus) based on evolutionary algorithm adapted for multi-objective optimization. The reason of using this algorithm is that we are tackling different criteria to optimize.

The participation is only limited to communication and information where it was possible for the users to be informed about the decision-making in addition to the value of the criteria. Other possible interaction with the design was that users could interact with the model by modifying the values of the input geometrical parameters and see the modification of the geometry in real-time. The simulation indicators for each solutions is not in real-time and it needs time to be calculated.

Experimentation. To assess this first stage of the prototype, we conducted different interviews where we exposed the prototype to the experts in the field of urban planning. The activities of the 11 structures questioned vary between architecture, urbanism, organization for territory management and public organizations. The interview was held in a form of small workshop, divided into 3 parts:

In the first part, a small introduction about parametric and computational design, this was essential since parametric modelling remains a new approach in Luxembourg and not all offices are aware of this approach. After the presentation of parametric modelling approach, we presented the urban planning framework. The idea from that, is to confront our framework to the experts and practitioners in the field in order to make some adjustments if need. A live demo is exposed presenting different aspects implemented in the prototype.

The workshop ended with an open discussion recorded that allowed us to understand and adapt the prototype according to the need. In the end of the workshop each person was asked to fill in a questionnaire about this application.

5.2 Second Phase of the Prototype

Development. It was clear that the first development was to evaluate the potential of creating a generic system to design and plan urban spaces with parametric approaches. Although this might be a complicated task to perform, parametric approaches showed a great potential in generating the urban spaces. However, for real use cases, adaptations should be considered to answer the exact needs of citizens and the constraints related to the site. This will join the first step of the traditional urban design process mentioned in the introduction corresponding to the site analysis. From this point, the second phase of the prototype is developed to answer this issue. Indeed, we tried in the second prototype to cover the participatory approach in the early stage of the definition of the urban project; this development mainly covers the second interaction of the framework Fig. 2. It is important for the citizens to understand the analysis of the site to more understand the decision made. In this development, we consider some of the steps in the design process and develop a novel digital prototype with different interactions with users. The first step considers the visualization of the site analysis. A tangible user interface with touch screen is used as an auxiliary tool for visualizing the simulations through indicators for the selected site allowing.

Step 4: Site criteria study (or site diagnostic). Experts will perform a site analysis according to needed criteria, users are able to explore and diagnostic the actual situation of the site. For this scenario, the criteria that were simulated for this site are related to the occlusion, pollution, noise, annual daylight and view.

Step 5: Site explorer. For the site exploration, we are generating some of recommendation; we developed an algorithm based on the fuzzy logic. The fuzzy logic is a practical parameter configuration which easily transforms a multi-objective problem to a single objective function. In this step Fig. 4, the recommendation allows users to better understand and explore different solutions by having a real time recommendation about each solutions. The role of the optimization in this case was to recommend solutions and propose better solutions when the chosen one does not answer the requirements. The recommendation system is based on fuzzy logic, which indicates how much percent that each function to place in the site suffices the optimal requirement needed of the corresponding function. Following this, experts and users have to understand the identity of the district to be placed (residential, industrial or infrastructure district).

Fig. 4. Illustration of interface showing the recommendation system of functions

Step 6: Space allocation. After the exploration and recommendation step where the identity of the future urban space is defined Fig. 5, the users with the experts are invited to allocate the urban spaces in the selected site. In this step, users are using 3D printed objects to allocate each space in the site. Real-time recommendations are being displayed on the tabletop whenever the allocated space does not meet with the requirements. Citizens being a part in the design process are also informed about the performance of their propositions. This algorithm is also based on Fuzzy logic.

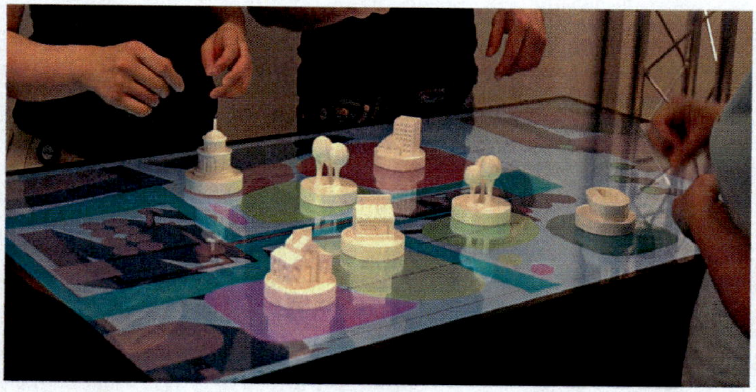

Fig. 5. Illustration for space allocation

Experimentation. To experiment this prototype, we are currently working on one case study in Luxembourg where Schifflange site, an old industrial site, is awaiting its new vocation to be transformed into an urban district. A competition was proposed to develop this site based on participative approach; the participation was limited to verbal communication. The experimentations will include two workshops using the system created to compare the solutions proposed by experts during the workshop and the solutions proposed by the system.

5.3 Findings/On-Going Research

To experiment these prototypes developed for this use case, we started experiments in Luxembourg. We conducted different interviews where we exposed the prototype to the experts in the field of urban planning. This interview was held in a form of small workshop, divided into 3 parts:

In the first part, we had a short introduction about parametric and computational design, this was essential since parametric modelling remains a new approach in Luxembourg and not all architecture and urban planning design firms are aware of this approach. After the presentation of the parametric modelling approach and the field of application of it, we exposed the urban planning process that was presented in previous sections. The idea from that, is to confront our comprehension of the urban planning process to the experts and practitioners in the field in order to adjust the process if need.

A quick presentation was made to explain the platform for the urban planning, before starting a real-time demo. This demo showed all the aspect that have been implemented in the prototype and the parameters that can influence the design. The demo also showed the fast transfer from an automated system to create urban planning spaces to one more manual with the advantage of have in real time the simulation of the different designs created. The workshop ended with an open discussion recorded that will allow us in the future to understand and adapt the prototype. The persons participating in the interviews were also asked to fill questionnaires about the utility of this application.

In later stage, we will conduct a workshop with professional in Luxembourg and Brussels to evaluate the system. Another workshop is going to take place with citizens (current discussion are being held with some municipalities in Luxembourg for a possible evaluation of the system with citizens).

6 Conclusion

The issues raised after presenting the first development of the prototype to the experts and after the questionnaire are related to the need to perform a decision-making in a quick time taking into account multiple criteria and objectives for design optimization. The experts answers showed also the need for a system to compare between different solutions proposed from the early design stages, this system to be used mainly by the public authorities to explore the different possibilities of design. Another issue is related to the transparency of decisions as well as the involvement of users in the urban planning. Therefore, a gap is raised between the design process of urban spaces and the participation in a digital environment. To answer this gap, we developed a system that can interact with the citizens at different design stages. As for the participation approach, the system proposed covers the participation for instance in the exploration and diagnostic of the site, in a later stage, the interaction of the citizens is in the allocation of urban spaces with a recommendation system.

References

1. Carmona, M.: Re-theorising contemporary public space: a new narrative and a new normative. J. Urban. **8**(4), 373–405 (2015)
2. Sandercock, L.: Towards a planning imagination for the 21st century. J. Am. Plan. Assoc. **70** (2), 133–141 (2004)
3. Irvin, R.A., Stansbury, J.: Citizen participation in decision making: is it worth the effort? Public Adm. Rev. **64**(1), 55–65 (2004)
4. Dias, N., Curwell, S., Bichard, E.: The current approach of urban design and its implications for sustainable urban development. Procedia Econ. Financ. **18**(September), 497–504 (2014)
5. Mubita, A., Libati, M., Mulonda, M.: The importance and limitations of participation in development projects and programmes. Eur. Sci. J. ESJ **13**(5), 238 (2017)
6. Katz, P., Scully, V., Bressi, T.W.: The New Urbanism: Toward an Architecture of Community. McGraw-Hill, New York (1994)
7. Greed, C., Marion, R.: Approaching Urban Design: The Design Process, vol. 5. Routledge, Abindon (2014)
8. Moughtin, C., Cuesta, R., Sarris, C., Signoretta, P.: Urban Design: Method and Techniques. Routledge, Abindon (2003)
9. Foroughmand Araabi, H.: A typology of urban design theories and its application to the shared body of knowledge. Urban Des. Int. **21**(1), 11–24 (2016)
10. Sanoff, H.: Community participation methods in design and planning. Landsc. Urban Plan. **50**(4), 270–271 (2000)
11. Marsal-Llacuna, M.L., Colomer-Llinàs, J., Meléndez-Frigola, J.: Lessons in urban monitoring taken from sustainable and livable cities to better address the Smart Cities initiative. Technol. Forecast. Soc. Change **90**(PB), 611–622 (2015)
12. Arnstein, S.R.: A ladder of citizen participation. J. Am. Plan. Assoc. **35**(4), 216–224 (1969)
13. Claridge, T.: Designing social capital sensitive participation, June (2004)
14. Kelly, D.: Community participation in rangeland management, no. 01 (2001)
15. Meeus, B., Pak, B.: Reflections on the counter-mapping of urban 'arrival neighborhoods' through Geoweb 2.0 in Brussels and Ghent. In: Participatory Design Theory. Routledge Research in Planning and Urban Design, pp. 40–55. Routledge, New York (2019)
16. Miessen, M.: Crossbenching: towards a proactive mode of participation as a critical. University of London (2017)
17. Goodchild, M.F.: Citizens as sensors: the world of volunteered geography. GeoJournal **69**, 211–221 (2007)
18. Howe, J.: The rise of crowdsourcing. Wired (2006). http://www.wired.com/wired/archive/14.06/crowds.html
19. Boeykens, S., Quintero, M.S., Neuckermans, H., Leuven, K.U., Arenberg, K., Heverlee, B.: Improving architectural design analysis using 3D modeling and visualization techniques (2008)
20. Stavric, M., Marina, O.: Parametric modeling for advanced architecture. Int. J. Appl. Math. Inform. **5**(1), 9–16 (2011)
21. Elezkurtaj, T., Franck, G.: Genetic algorithms in support of creative architectural design. In: eCAADe Conference Proceedings, pp. 645–651 (1999)
22. Kensek, K.: Visual programming for building information modeling: energy and shading analysis case studies. J. Green Build. **10**, 28–43 (2016)
23. Turrin, M., Von Buelow, P., Stouffs, R.: Design explorations of performance driven geometry in architectural design using parametric modeling and genetic algorithms. Adv. Eng. Inform. **25**(4), 656–675 (2011)

24. Al-Kodmany, K.: Using visualization techniques for enhancing public participation in planning and design: process, implementation, and evaluation. Landsc. Urban Plan. **45**(1), 37–45 (1999)
25. Beirão, J., Arrobas, P.: City information modelling: parametric urban models including design support data, no. 3, pp. 1122–1134 (2011)
26. Amado, M., Poggi, F.: Solar urban planning: a parametric approach. Energy Procedia **48**, 1539–1548 (2014)
27. Chowdhury, S., Schnabel, M.A.: A communication system in urban design process : an algorithmic neighbourhood design approach a communication system in urban design process: an algorithmic neighbourhood design approach, October 2018
28. Chowdhury, S., Schnabel, M.A.: Communication and participation in virtual environments, pp. 727–734 (2018)
29. Steinø, N., Veirum, N.E.: A parametric approach to urban design tentative formulations of a methodology, pp. 679–686 (2005)
30. Steinø, N., Benbih, K., Obeling, E.: Using parametrics to facilitate collaborative urban design. An attempt to overcome some inherent dilemmas, pp. 1–13 (2013)
31. Gün, A., Demir, Y., Pak, B.: Urban design empowerment through ICT-based platforms in Europe. Int. J. Urban Sci. **24**, 189–215 (2019)
32. Ben-Joseph, E., Ishii, H., Underkoffler, J., Piper, B., Yeung, L.: Urban simulation and the luminous planning table: bridging the gap between the digital and the tangible. J. Plan. Educ. Res. **21**(2), 196–203 (2001)
33. Chowdhury, S., Schnabel, M.A.: An algorithmic methodology to predict urban form: an instrument for urban design, May 2018
34. Hanzl, M.: Information technology as a tool for public participation in urban planning: a review of experiments and potentials. Des. Stud. **28**(3), 289–307 (2007)
35. Cirulis, A., Brigis, K.: 3D outdoor augmented reality for architecture and urban planning. Procedia Comput. Sci. **25**, 71–79 (2013)
36. Kaftan, M., Friedrich, E., Urquiza, R., Fihn, J., Galaso, J.L., Gillespie, D.: Informing architecture and urban modeling with real-world data on 3D tangible interfaces and augmented displays, pp. 226–233 (2011)
37. Gordon, E., Koo, G.: Placeworlds: using virtual worlds to foster civic engagement. Space Cult. **11**(3), 204–221 (2008)
38. Shen, B.J., Wu, Y., Liu, H.: Urban planning using augmented reality. Manager **127**(3), 118–125 (2002)
39. Rodrigues, F., Sato, F., Botega, L., Oliveira, A.: Augmented reality and tangible user interfaces integration for enhancing the user experience. In: Proceedings - VRCAI 2012 11th ACM SIGGRAPH International Conference on Virtual-Reality Continnum and Its Applications in Industry, pp. 67–70, December 2012
40. Ishii, H., Ullmer, B.: Tangible bits, pp. 234–241, September 1997
41. Wagner, I., Basile, M., Ehrenstrasser, L., Maquil, V., Terrin, J.J., Wagner, M.: Supporting community engagement in the city: urban planning in the MR tent. In: Proceedings of Fourth International Conference Communities Technologies, pp. 185–194 (2009)
42. Maquil, V., Psik, T., Wagner, I.: The ColorTable: a design story. In: Proceedings of TEI 2008, pp. 97–104 (2008)

Multi-Objective Optimization of a Free-Form Surface Based on Generative Designs

Chankyu Lee, Sangyun Shin, and Raja R. A. Issa(✉)

M.E. Rinker, Sr School of Construction Management, University of Florida, P.O. Box 115703, Gainesville, FL 32611, USA
{chankyu, sang87yun, raymond-issa}@ufl.edu

Abstract. With the improving of technology, many architects adopt curved panels as a type of free-form surfaces. Although these surfaces can be realized by utilizing curved panels including single-curved and double-curved panels, the use of more curved elements result in higher costs. Therefore, it is essential to seek an acceptable balance, in which not only the design intents are met but also the construction of such complex design at a reasonable cost in order to realize free-form surfaces as close to the original design as possible. Therefore, the aim of this study is to develop a Multi-Objective Optimization (MOO) approach which optimizes the process of free-form surface modeling and balances between the design intent and cost efficiency with a clear objective goal by reducing the use of curved panels. A generative and parametric design can facilitate the optimizing process, generate multiple design alternatives and assist architects to decide the optimal solution in the early stages of the design. In this paper, Project Refinery, which is a generative design plug-in for the Autodesk™ Dynamo-Revit environment, is used for this purpose.

Keywords: Generative design · Parametric design · Multi-Objective Optimization · Free-form construction

1 Introduction

With the advances in Computer-Aided Design (CAD), free-form surface modeling has been extensively applied in conceptual design of free-form construction because of its ability to visually express objects with complex shapes, especially those, which cannot be represented accurately by analytical functions. As the use of free-form surface modeling has been further enhanced, more curved elements are adopted for free-form surfaces and the design and construction process of free-form buildings are more complicated, time-consuming, and high-priced. Hence, the process of free-form construction that are labelled representation, parametrization, and rationalization, are becoming one of the main issues in the Architecture, Engineering, Construction, and Operation (AECO) industry. To respond to this trend, the demand for generative and parametric design has grown rapidly, and much attention has been paid to the application of free-form surface modeling.

E. Toledo Santos and S. Scheer (Eds.): ICCCBE 2020, LNCE 98, pp. 1252–1261, 2021.
https://doi.org/10.1007/978-3-030-51295-8_88

The most important factors to realize free-form surfaces are design intent and cost efficiency [2]. A balance between these variables is required during the rationalization process. Due to high cost of curved elements, the best way of rationalization is to cover a maximum area of a free-form surface, which is designed by an architect, with the same sized flat panels, lesser single-curved panels, and a few double-curved panels without giving up the original design and functional requirements. However, it is difficult to do that and almost impossible for a human designer to simultaneously consider all variables, that occur in constructing such a surface. Thus, the goal of this research is to suggest a framework for free-form surface modeling through generative and parametric designs with design intent and cost efficiency taken into consideration by using Multi-Objective Optimization (MOO).

2 Literature Review

2.1 Generative and Parametric Design

Parametric design software has emerged over the last decade, which is thoroughly changing the design process. In a traditional approach, the designer internalized all of design constraints and objectives into a single design solution to solve design problems through their skill and experience and manually created multiple versions of different design solution. In the parametric approach, a parametric model is designed through a clear objective function, which includes the constraints and goals of the design problem. Using parametric design tools, the designer can automatically generate many design solutions by changing the critical parameters that drive different variations without rebuilding design alternatives. It allows the designer to easily adapt to changing conditions.

Generative design is an algorithmic process from which various potential design solutions can be generated automatically by using clear design goals and defining parameters. By establishing clear design objectives, architectural and practical factors can be considered during the design stage without redrawing design alternatives.

Through the parametric approach, the possibilities of design and the boundaries of human-computer interaction in the design process are broadened, however there a few limitations exist because of the abilities of the human designer. Although by changing the parameters set by the constraints, architects could easily create many design options, they are required to investigate these alternatives by manually varying individual parameters and evaluating each individual option using their design intents. This approach is not much different from traditional design methods.

In the concept of generative design, such design alternatives are automatically explored by a computer, and then the promising options for further analysis are generated for review by the architect. This process can be performed faster and more extensive than an architect can do, so that it is allowed to explore and discover a larger variety of design solutions and used in the optimization of a given model to achieve maximum possible performance based on concrete objectives [9]. For the process of generative design, a basic parametric model must be described by clear objective functions, since a computer cannot determine which designs are better than others without the goals. In addition, the input parameters of the model can be controlled and tuned intelligently to find better solutions through the algorithms used for the model. The most promising algorithm is the multi-objective optimization using Genetic Algorithms (GA) [5].

2.2 Multi-Objective Optimization (MOO)

In reality, most problems have multiple objectives, which conflict with each other and prevent concurrent optimization of each objective. These problems are practical and difficult to solve comparing to single scalar objective problems. Multi-objective optimization is used for such realistic complex problems. Two different approaches are applied for the MOO: classical methods and non-traditional methods [4]. Classical methods are to combine multi-objective functions using mathematical principles into a single scalar objective problem by integrating a user-provided weight for each objective (e.g. the weighted sum methods) [10]. In the non-traditional methods, stochastic rules are used to determine a set of Pareto-optimal solutions that are non-dominated with respect to each other. Pareto frontier (also called as Pareto set or Pareto front) is the set of choices, which are Pareto efficient or optimal, and corresponds to the boundary of the feasible objective function space. The final solution is representing a set of good trade-offs between two or more conflicting objectives. A large search space is addressed compared to the classical methods [15].

Recently, generative and parametric modeling platforms are integrated and widely applied in the field of AECO, and the GA, a popular meta-heuristic algorithm, is adapted for such problems as well. In the GA approach, various inputs defined by a designer are plugged into a MOO. If the multi-objective function is not satisfied, the next population of agent points will be generated by a GA, and then using the agent points in the hierarchical agent system could alter the geometry. The simulation should be repeated with the changed geometry until the multi-objective function is satisfied or the maximum iteration is reached.

2.3 Related Work

There has been few studies performed by applying parametric and generative designs. Yi [15] proposed a parametric design process for building optimization to satisfy the user's visual preference and daylighting performance through a case study using an existing building's façade.

Nagy et al. [11] developed a generative design workflow for the design of a new office space for Autodesk™ in Toronto. They applied Multi-Objective Genetic Algorithm (MOGA) for a space planning model by describing six unique objectives, such as adjacency preference, work style preference, buzz, productivity, daylight, views to outside.

Agirbas [1] introduced a conceptual design method for the non-Euclidean façade form production. Multiple design solutions were generated using meatball and Voronoi geometries with a swarm intelligence algorithm based on the analysis of daylight-intake. This process was performed in Grasshopper developed by Robert McNeel & Associates.

3 Research Methodology

3.1 Computer Software

In this research, Autodesk™ Dynamo 2.0.2 is used as a basic platform, which acts as a plug-in to Revit. Project Refinery developed by Autodesk™ and released in Jan, 2019 is used for the basis of the generative design and MOO.

Dynamo is a parametric tool and a visual programming language specifically developed for designers exploring and customizing various Revit™ building information workflows. It allows designers to create variety utilization beyond the traditional design methods with its algorithmic capabilities. In this program, 'program flow' can be developed using 'nodes' without the need for knowledge of computer programming languages. While this program flow is being developed, various add-ons can be easily included in the 'library' as preprogrammed nodes [14]. These add-ons serve different purposes. For example, the Clockwork add-on can be used for list management, mathematical operations, string operations, geometric operations, and paneling. The Lunchbox add-on can be used to manage generative form making, paneling, rationalization, and interoperability.

Project Refinery, which is newly developed application for generative design workflows by Autodesk™, is used along with Dynamo. Designers are allowed to explore, evaluate, and optimize their Dynamo designs. A series of solutions can be generated through multiple design goals, and then decisions can be made by automating the creation and evaluation of designs.

3.2 Objective Functions

To measure the performance of each design generated automatically, a clear objective goal is required, which balance design intents and construction costs. This objective function forms the set of output values that is used to evaluate how well each design option performs and to discover more efficient design.

For a free-form surface, one of the easiest means of construction is utilizing flat panels, such as triangular, Planar Quadrilateral (PQ), and Planar Hexagonal (PH) meshes. Utilizing developable panels, also called single-curved panels, is an elegant way to approximate the original curvedness if the given original surface has one low principal curvature. As described in the definition of single-curved panels, these panels can be unfolded into a plane without any stretching, tearing, or distortion. Therefore, planarity is one of the important characteristics for flat and single-curved panels. Many variables are used to express planarity. For instances, if four angles in a rectangular surface sum up to 2π, this surface is planar. In addition, distance was used for planarity of flat and single-curved panels by using the least-square methods as described Eq. 1. The planarity constraint was proposed for a N-polygon $f = v_0 v_1 \cdots v_{N-1}$; a plane $p = \hat{n_f} \cdot x + d_f = 0$ (n_f is the unit normal of p) is first identified to fit f in the least-square sense. If the distance between every vertex of f and the fitted plane p is zero, the planarity of f could be attained. Planarity is also defined by the distance between two surfaces in space [13].

As mentioned before, the design intent and construction cost are the most important factors in free-form surface modeling. To measure these requirements, closeness and smoothness have been applied in lots of studies [3, 6, 7, 12, 13]. Most studies reviewed used angle differences between surfaces for smoothness. In this study, planarity is applied for closeness to measure the distances between generated and original surfaces (see Eq. 2).

$$\sum_{i=1}^{n} r_i^2 = \sum_{i=1}^{n} (y_i - f(x_i))^2 \quad (1)$$

$$P_{dist} = \sum_i \|s_{i,0} - s_i\|^2 \quad (2)$$

$s_{i,0}$ is the initial surface of s_i and P_{dist} is the distance between s_i and $s_{i,0}$.

P_{area} is the sum of each single and double-curved panel's area. It is minimized to maximize the use of flat panels. P_{cost} is the sum of all panels' cost. Panel costs per square meter ($701 for flat panel, $766 for single-curved panel, and $1,518 for double-curved panel) are assigned based on the research [8].

$$\min_{x \in X} f(x) = \sum_{i=1}^{n} \left[(P_{dist})_n + (P_{area})_n + (P_{cost})_n \right] \quad (3)$$

Thus, the objective function is defined to minimize the paneling cost, area of single and double-curved panels, and distance between the original surface and surface rationalized (see Eq. 3).

3.3 Program Flow Set-Up

Revit, Archicad, and Rhino are the tools mostly used for the design of buildings by architects. Usually, the designs are created in such tools, and then optimized and analyzed through other tools by importing these designs. For this study, a free-form surface was designed in Revit, and then imported into Dynamo. The number of UV grids was assigned to parametric values and inputs. To prevent the creation of a failed design, the UV grid ranged from 5 to 15 and 20 (respectively u grid and v grid). The surface imported was divided by the UV grid and classified into multiple paneling areas through parametric designs. Different design solutions were generated and classified automatically by changing the UV grid. Figure 1 shows the program flow for this research. The area of flat and single-curved panels is colored by green. Red shows the area of double-curved panels.

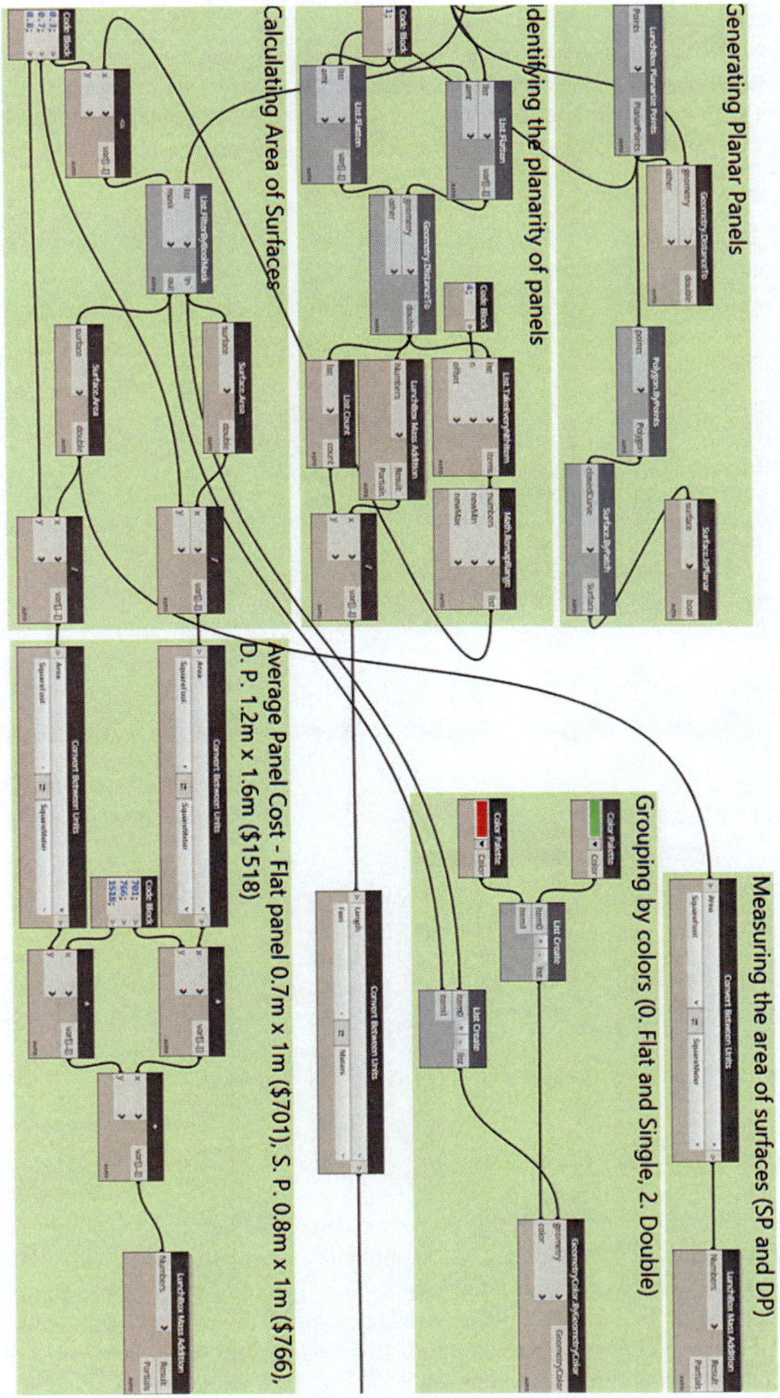

Fig. 1. A program flow

Generative designs and MOO were achieved by using Autodesk Project Refinery, which works as a plug-in to Dynamo (see Fig. 2). Based on the objective function, the optimization process was performed, multiple alternatives were generated and the outcomes were evaluated. If these outcomes did not meet the objective functions, the parametric values were changed and the simulation was restarted.

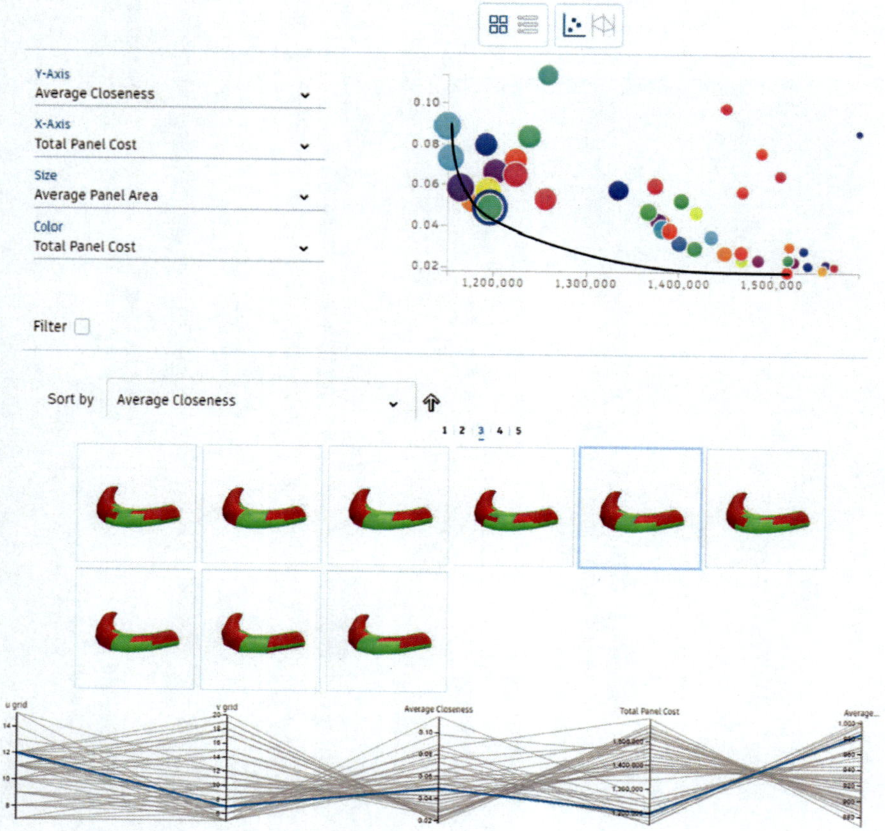

Fig. 2. The result of generative designs in Project Refinery

4 Result and Discussion

As shown in Table 1 and Fig. 2, this process generated 43 design alternatives. In order to choose the final design solution, Pareto optimality was applied using two conflicting objectives, average closeness and total panel cost. The top-right side in Fig. 2 shows six points that formed the Pareto frontier. The point on the top-left side, between the ranges of average closeness from 0.08 to 0.1, shows the best solution in terms of total panel cost. The point on the bottom-right side is the closest design solution to the original design. The other points represent other possible trade-offs. The design option

Table 1. The result of generative designs.

Average closeness (m)	Total panel cost ($)	Total SP and DP (m^2)	U grid	V grid
0.018	1,517,382	899	12	20
0.019	1,554,756	887	12	19
0.020	1,567,280	877	11	19
0.021	1,539,640	887	12	17
0.022	1,558,124	882	11	18
0.023	1,524,858	893	11	17
0.024	1,468,126	904	15	12
0.024	1,485,900	902	12	15
0.024	1,518,281	893	10	18
0.027	1,449,638	918	13	12
0.028	1,467,670	913	11	14
0.028	1,535,174	887	10	15
0.029	1,417,431	923	12	12
0.030	1,519,639	890	10	14
0.032	1,400,639	927	12	11
0.035	1,435,320	915	11	11
0.038	1,390,573	924	11	10
0.039	1,382,154	938	12	9
0.042	1,380,645	944	10	10
0.042	1,378,354	931	11	9
0.047	1,419,224	911	10	9
0.047	1,366,899	933	11	8
0.049	1,196,678	985	12	7
0.052	1,402,617	922	8	10
0.053	1,256,228	964	11	7
0.053	1,179,571	999	15	5
0.057	1,468,680	903	12	6
0.057	1,194,807	989	14	5
0.058	1,334,917	951	9	8
0.059	1,164,691	1003	10	7
0.060	1,374,321	929	7	10
0.065	1,509,455	896	8	8
0.065	1,224,573	979	9	7
0.066	1,202,437	982	12	5
0.072	1,224,800	967	11	5
0.073	1,153,888	995	8	7
0.076	1,489,179	900	9	6
0.080	1,192,062	969	10	5
0.084	1,238,474	964	7	7
0.086	1,593,822	867	8	6
0.089	1,150,032	997	9	5
0.098	1,450,411	897	7	6
0.114	1,257,763	956	7	5

selected was one example of the Pareto frontiers, divided by a 12 × 7 UV grid and had an average closeness (distance) of 0.049 m, with total panels cost of $1.2 M, and total panel area of single and double-curved panels of 985 m^2.

Although one of the designs was chosen by the algorithm, the process does not end. Once human designers select a set of optimal designs, they can find various high performing designs through a deep analysis. New potential design strategies could be suggested by their trade-offs, so that the designer could further explore to achieve a final best design for discussion with the stakeholders.

The algorithm or program flow can be improved by integrating other types of modeling, particularly machine learning. Because aspects of the designs such as comfort, beauty, or novelty, which are crucial to good design, are difficult or impossible to quantify and compute through direct calculation. These difficulties could be the promising research areas in the future.

5 Conclusion

This paper described the development of a generative design workflow for the rationalization of a free-form surface to find out optimal design solutions using the criteria of design intent and construction cost. A free-form surface was divided by a UV grid and panelized using three types of panels, flat, single-curved, and double-curved, to realize the surface as close as the original design in Dynamo-Revit environment. In the optimization process, the multi-objective function was defined by the average planarity, area of single and double-curved panels, and total cost of these panels, and then was minimized to consider such criteria. A variety of design solutions were generated and explored by using Autodesk™ Project Refinery. Human designers can choose the best design solution among the multiple design solutions on the Pareto frontier.

Although the result of this study is encouraging, there are many limitations (for example, classification methods and the design sample). In future research, free-form surface modeling algorithms, which are more advanced, will be applied to rationalize free-form surfaces, and then validate the algorithms through a case study of the actual projects.

References

1. Agirbas, A.: Façade form-finding with swarm intelligence. Autom. Constr. **99**, 140–151 (2019)
2. Bo, P., Pottmann, H., Kilian, M., Wang, W., Wallner, J.: Circular arc structures. ACM Trans. Graph. (TOG) **30**(4), 1–12 (2011)
3. Bouaziz, S., Deuss, M., Schwartzburg, Y., Weise, T., Pauly, M.: Shape-up: shaping discrete geometry with projections. In: Computer Graphics Forum, vol. 31, no. 5, pp. 1657–1667. Blackwell Publishing Ltd, Oxford (2012)
4. Emmerich, M.T., Deutz, A.H.: A tutorial on multiobjective optimization: fundamentals and evolutionary methods. Nat. Comput. **17**(3), 585–609 (2018)

5. Konak, A., Coit, D.W., Smith, A.E.: Multi-objective optimization using genetic algorithms: a tutorial. Reliab. Eng. Syst. Saf. **91**(9), 992–1007 (2006)
6. Li, Y., Liu, Y., Wang, W.: Planar hexagonal meshing for architecture. IEEE Trans. Vis. Comput. Graph. **21**(1), 95–106 (2014)
7. Liu, Y., Xu, W., Wang, J., Zhu, L., Guo, B., Chen, F., Wang, G.: General planar quadrilateral mesh design using conjugate direction field. ACM Trans. Graph. (TOG) **30**(6), 1–10 (2011)
8. Lim, J.S., Ock, J.H.: Development of the preliminary cost estimate method for the free-form building façade trade in conjunction with the panel optimization algorithm process. Korean J. Constr. Eng. Manag. **15**(4), 95–106 (2014)
9. Marler, R.T., Arora, J.S.: Survey of multi-objective optimization methods for engineering. Struct. Multidisc. Optim. **26**(6), 369–395 (2004)
10. Marler, R.T., Arora, J.S.: The weighted sum method for multi-objective optimization: new insights. Struct. Multidisc. Optim. **41**(6), 853–862 (2010)
11. Nagy, D., Lau, D., Locke, J., Stoddart, J., Villaggi, L., Wang, R., Zhao, R., Benjamin, D.: Project discover: an application of generative design for architectural space planning. In: Proceedings of the Symposium on Simulation for Architecture and Urban Design, p. 7. Society for Computer Simulation International (2017)
12. Pottmann, H., Liu, Y., Wallner, J., Bobenko, A., Wang, W.: Geometry of multi-layer freeform structures for architecture. In: ACM SIGGRAPH 2007, pp. 65–es. Association for Computing Machinery, New York (2007)
13. Pottmann, H., Schiftner, A., Bo, P., Schmiedhofer, H., Wang, W., Baldassini, N., Wallner, J.: Freeform surfaces from single curved panels. ACM Trans. Graph. (TOG) **27**(3), 1–10 (2008)
14. The Dynamo Primer Homepage. http://primer.dynamobim.org. Accessed 21 Jan 2020
15. Yi, Y.K.: Building façade multi-objective optimization for daylight and aesthetical perception. Build. Environ. **156**, 178–190 (2019)

The Influence of Wall Boundary Modeling on the Unphysical Frictional Loss Inside Horizontal Main Drain

Lucas Soares Pereira(✉), Rubens Amaro Junior, and Liang-Yee Cheng

Department of Construction Engineering, Escola Politécnica of University of São Paulo, Av. Prof. Almeida Prado, 83, São Paulo, SP, Brazil
lucas_pereira@usp.br

Abstract. Sustainable provisions to reduce the water consumption pose new challenges for an optimal design in building drainage infrastructure. Hence, the accurate prediction of the water flow through building drainage system to assure its self-cleaning performance is an important issue associated with sanitary and construction fields. In order to investigate complex scenarios considering a variety of geometries and loading regimes, as well as the presence of solid wastes and sewages, numerical modeling and simulation are nowadays a promising approach. In the present work, a particle-based method is used to simulate transient free-surface flow inside a horizontal pipe. Since the modeling of the pipe geometry has significant influence on the numerical accuracy, an investigation on the unphysical frictional loss inside the pipe is carried out. Two wall boundary modeling are adopted: in the first one all domain is modeled by particles; in the second one the fluid domain is represented by particles whereas wall boundaries are modeled by triangular polygons. The results demonstrate that the polygon wall representation generates more accurate results than the original discrete wall particle model.

Keywords: Frictional loss · Drainage system · Computational fluid dynamics · Polygon wall boundary · Particle method · MPS

1 Introduction

The continued growth of urban areas and the sustainable provisions to reduce the water consumption pose new challenges for an optimal design in building drainage infrastructure to ensure its self-cleaning performance. The water flow through building drainage system is a highly nonlinear phenomenon that involves multiphase free-surface flows. In case of the partially filled horizontal pipes, it is predominantly a transient free-surface flow that interacts with solid wastes of complex geometries or sewages with a variety of rheological proprieties. Simplified analytical or semi-empirical approaches have been proposed to understand and predict the flow within pipes and to evaluate the self-cleaning performance of the drainage systems [1–5]. In order to investigate in-deep more complex scenarios under a variety of geometries and loading regimes, as well as considering the solid wastes and sewages, numerical

E. Toledo Santos and S. Scheer (Eds.): ICCCBE 2020, LNCE 98, pp. 1262–1275, 2021.
https://doi.org/10.1007/978-3-030-51295-8_89

modeling and simulations are nowadays a promising approach. Among the numerical methods used for this purpose, the mesh-free particle-based methods, such as smoothed particle hydrodynamics (SPH) [6, 7] and moving particle semi-implicit (MPS) [8], have shown a great potential in dealing with such kind of phenomena [9–13].

In the conventional implementation of particle-based methods, wall boundaries are modeled by discrete wall and dummy (ghost) particles. A smoother representation of circular geometries using the particle modeling requires very high resolution, otherwise, it may result in inaccurate geometrical representations that may cause abnormal flow near the wall boundaries or unphysical frictional forces on the near wall fluid particles. In order to address this issue, the adoption of polygonal mesh to represent the wall boundaries is an alternative modeling technique. Aiming to verify the effectiveness of the polygonal mesh modeling, in the present work the MPS is used to simulate a transient free-surface flow inside an horizontal pipe considering two pipe wall modeling: in the first one all domain is modeled by particles; in the second one the fluid domain is represented by particles whereas wall boundaries are modeled by triangular polygons. A numerical investigation on the unphysical frictional loss inside the pipe demonstrates that the polygon wall modeling leads to more accurate results than the original discrete wall particle model.

2 Numerical Method

2.1 Governing Equations

The governing equations of incompressible flow are expressed by the conservation laws of mass and momentum:

$$\frac{D\rho}{Dt} = \rho\nabla \cdot \mathbf{u} = 0, \tag{1}$$

$$\frac{D\mathbf{u}}{Dt} = -\frac{\nabla P}{\rho} + \nu\nabla^2\mathbf{u} + \mathcal{F}, \tag{2}$$

where ρ is the fluid density, $\mathbf{u}$ is the velocity vector, P is the pressure, ν is the kinematic viscosity and $\mathcal{F}$ is the external body force per unit mass vector.

2.2 Moving Particle Semi-implicit (MPS)

In MPS Method, the differential operators of the governing equations of continuum are replaced by discrete operators derived based on a weight function ω_{ij}, which accounts the influence of a particle j in the neighborhood of the particle i, and is given by:

$$\omega_{ij} = \begin{cases} \frac{r_e}{|\mathbf{r}_{ij}|} - 1 & |\mathbf{r}_{ij}| \leq r_e \\ 0 & |\mathbf{r}_{ij}| > r_e \end{cases}. \tag{3}$$

The effective radius r_e, here assigned to $2.1l_0$, limits the range of influence. l_0 is the initial distance between two particles. $|\mathbf{r}_{ij}| = |\mathbf{r}_j - \mathbf{r}_i|$ is the distance between i and j.

The summation of the weight of all the particles in the neighborhood of the particle i is defined as its particle number density pnd_i, proportional to the fluid density:

$$pnd_i = \sum_{j \neq i} \omega_{ij}. \tag{4}$$

For a function ϕ, the gradient and Laplacian operators are, respectively, defined as:

$$\langle \nabla \phi \rangle_i = \frac{d}{pnd^0} \sum_{j \neq i} \frac{\phi_j - \phi_i}{|\mathbf{r}_{ij}|^2} \mathbf{r}_{ij} \omega_{ij}, \tag{5}$$

$$\langle \nabla^2 \phi \rangle_i = \frac{2d}{\lambda^0 pnd^0} \sum_{j \neq i} (\phi_j - \phi_i) \omega_{ij}, \tag{6}$$

where d is the number of spatial dimensions and pnd^0 is the particle number density for a fully filled compact support at the instant $t = 0$. Finally, λ^0 is a correction parameter so that the variance increase is equal to that of the analytical solution, and is calculated at the beginning of the simulation ($t = 0$) as:

$$\lambda^0 = \frac{\sum_{j \neq i} \omega_{ij}^0 |\mathbf{r}_{ij}^0|^2}{\sum_{j \neq i} \omega_{ij}^0}. \tag{7}$$

To solve the incompressible viscous flow, a semi-implicit algorithm is used in the MPS method. At first, predictions of the particle's velocity ($\mathbf{u}_i^*$) and position ($\mathbf{r}_i^*$) are carried out explicitly by using viscosity and external forces terms of the momentum conservation, see Eq. (2). Then the pressure of all particles is calculated by the pressure Poisson equation (PPE) as follows [14]:

$$\langle \nabla^2 P \rangle_i^{t+\Delta t} - \frac{\rho}{\Delta t^2} \alpha_c P_i^{t+\Delta t} = \gamma_1 \frac{\rho}{\Delta t^2} \frac{pnd^0 - pnd_i^t}{pnd^0} + \gamma_2 \frac{\rho}{\Delta t} \nabla \cdot \mathbf{u}_i, \tag{8}$$

where Δt is the time step, pnd_i^t is the particle number density calculated at the instant t, α_c is the coefficient of artificial compressibility and γ_1 and γ_2 are the relaxation coefficients. The coefficients α_c, γ_1 and γ_2 are used to improve the stability of the computation method. In order to prevent instability issues induced by attractive pressure and reduces the effect of nonuniform particle distribution, we adopted the first order pressure gradient [15] for the simulations using particle wall modeling:

$$\langle \nabla P \rangle_i = \left[\sum_{j \neq i} \omega_{ij} \frac{\mathbf{r}_{ij}}{|\mathbf{r}_{ij}|} \otimes \frac{\mathbf{r}_{ij}^T}{|\mathbf{r}_{ij}|} \right]^{-1} \left[\sum_{j \neq i} \frac{P_j - \hat{P}_i}{|\mathbf{r}_{ij}|^2} \mathbf{r}_{ij} \omega_{ij} \right], \tag{9}$$

where $\hat{P}_i$ is the minimum pressure between the neighborhood of the particle i. Lower order pressure gradient, see Eq. (5), was adopted for the simulations using polygons.

Finally, the velocity and new positions of the particles are updated by:

$$\mathbf{u}_i^{t+\Delta t} = \mathbf{u}_i^* - \frac{\Delta t}{\rho} \langle \nabla P \rangle_i^{t+\Delta t}, \tag{10}$$

$$\mathbf{r}_i^{t+\Delta t} = \mathbf{r}_i^* + \Delta t \left(\mathbf{u}_i^{t+\Delta t} - \mathbf{u}_i^* \right). \tag{11}$$

2.3 Solid Wall Boundary Condition

Particle Wall Modeling

Solid wall might be modeled by using three layers of fixed particles, see Fig. 1. The particles that form the layer in contact with the fluid are denominated wall particles, of which the pressure is computed by solving the PPE, see Eq. (8), together with the fluid particles. The particles that form two other layers are denominated dummy particles, which are used to ensure the correct calculation of the particle number density of the wall particles and fluid particles near the walls. The absence of pressure calculation in dummies corresponds to the condition where the pressure of a dummy particle j is equal to that of the particle i. In other words, the Neumann boundary condition, where the pressure gradient is zero, is assigned at the walls. As boundary condition of rigid walls, the zero-velocity condition is considered on the wall.

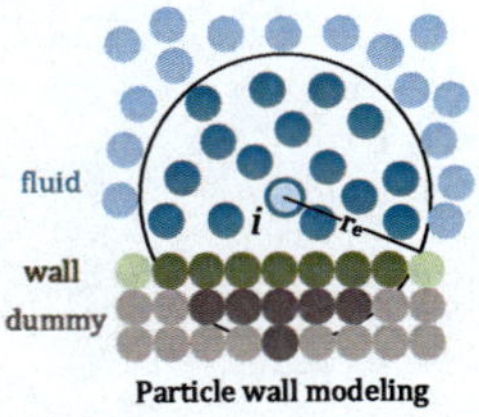

Fig. 1. Representation of the particle wall modeling

Polygon Wall Modeling

The explicitly represented polygon (ERP) wall boundary model [16] might also be used to model the solid wall. The ERP model represents solid walls as triangular polygons in an explicit way without using distance functions. Moreover, the creation of neighbor virtual particles or special adaptations for angled edges are not required. It is worth noticing that in the ERP the Neumann boundary condition for pressure and the free-slip/no-slip condition for velocity on the walls are satisfied.

An axis-aligned bounding box (AABB) hierarchy, implemented using the C+ + library libigl [17], is adopted in the present study to accelerate the calculation of the distances between the particles and polygons.

Since the solid walls are represented by polygons in the ERP model, the compact support of fluid particles near the walls is not fully filled with particles. Hence, the numerical operators of these particles are divided into the contribution due fluid particles $\langle\rangle^{particle}$, see Eqs. (5) and (6), and due polygon walls $\langle\rangle^{wall}$, see Fig. 2.

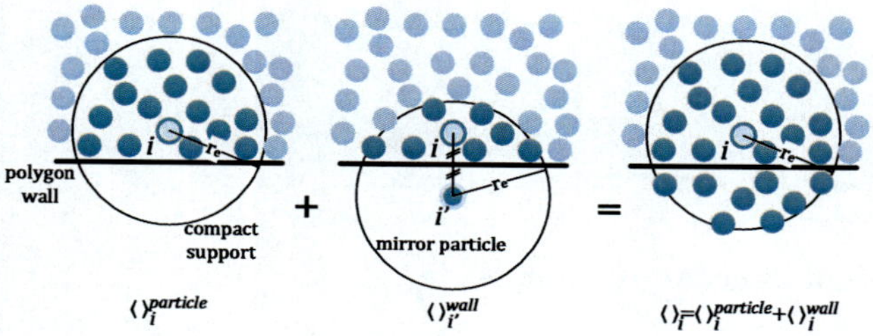

Fig. 2. Polygon wall modeling. Calculation of numerical operators with ERP model

To calculate the numerical operators $\langle\rangle^{wall}$, first, the position of the mirror particle i' corresponding to particle i is computed by:

$$\mathbf{r}_{i'} = 2\mathbf{r}_i^{wall} - \mathbf{r}_i, \tag{12}$$

where $\mathbf{r}_i^{wall}$ is the closest point on the polygon to particle i.

Afterwards, the pressure gradient and Laplacian of the velocity for the no-slip boundary condition of particle i' are computed considering its neighbors, which is $\Omega_{i'}$, and are multiplied by $\mathbf{R}_i^{ref}$ and the identity matrix $\mathbf{I}$ [16], respectively:

$$\langle\nabla P\rangle_i^{wall} = \mathbf{R}_i^{ref} \frac{d}{pnd^0} \sum\nolimits_{j\in\Omega_{i'}} \left(P_j + P_i - 2\hat{P}_i\right) \frac{(\mathbf{r}_j - \mathbf{r}_{i'})}{\left|\mathbf{r}_j - \mathbf{r}_{i'}\right|^2} \omega_{i'j}, \tag{13}$$

$$\left\langle\nabla^2\mathbf{u}\right\rangle_i^{wall} = -\mathbf{I}\frac{2d}{\lambda^0 pnd^0} \sum\nolimits_{j\in\Omega_{i'}} \left(\mathbf{u}_j - \mathbf{u}_{i'}\right)\omega_{i'j}. \tag{14}$$

Finally, the operators are added to the numerical operators $\langle\rangle^{particle}$ of particle i:

$$\langle\nabla P\rangle_i = \langle\nabla P\rangle_i^{particle} + \langle\nabla P\rangle_i^{wall}, \tag{15}$$

$$\left\langle\nabla^2\mathbf{u}\right\rangle_i = \left\langle\nabla^2\mathbf{u}\right\rangle_i^{particle} + \left\langle\nabla^2\mathbf{u}\right\rangle_i^{wall}. \tag{16}$$

The transformation matrix for reflection across the plane $\mathbf{R}_i^{ref}$, whose unit normal vector at the position of particle i is $\mathbf{n}_i^{wall}$, is expressed as:

$$\mathbf{R}_i^{ref} = \mathbf{I} + 2\mathbf{n}_i^{wall} \otimes \mathbf{n}_i^{wall}. \tag{17}$$

The Eq. (14) represents the Laplacian of velocity for the no-slip boundary condition on a wall whose velocity of the mirror particle is:

$$\mathbf{u}_{i'} = -\mathbf{I}\left\{\mathbf{u}_i - 2\left[\mathbf{u}_i^{wall} - \left(\mathbf{n}_i^{wall} \cdot \mathbf{u}_i^{wall}\right)\mathbf{n}_i^{wall}\right]\right\}, \tag{18}$$

where $\mathbf{u}_i^{wall}$ is the velocity of the wall at the closest point on the polygon to particle i.

A repulsive force $\mathbf{f}_i^{rep}$ is added to Eq. (15) to prevent penetrations of the fluid particles at curved edges of the bodies:

$$\mathbf{f}_i^{rep} = \begin{cases} -\alpha_{rep}\left(\frac{0.5l_0}{|\mathbf{r}_{ij}|} - 1\right)\mathbf{n}_i^{wall} & |\mathbf{r}_{ij}| \leq 0.5l_0 \\ 0 & |\mathbf{r}_{ij}| > 0.5l_0 \end{cases}, \tag{19}$$

where α_{rep} is a repulsive coefficient empirically determined.

The particle number density pnd_i also is partitioned into the contribution due the fluid particles $pnd_i^{particle}$ and the polygon walls pnd_i^{wall}. Under the assumption that the wall near the fluid particle is flat, the dummy particles j' are arranged in a uniform particle distribution below the flat wall, and pnd_i^{wall} is evaluated using the dummies:

$$pnd_i^{wall} = f(d_{wall}) = \sum \omega_{ij'}, \tag{20}$$

where f is determined by a linear interpolation of precomputed values at a given discrete distance d_{wall}, which represents the normal distance between the fluid particle i and the nearest polygon wall. It should be emphasized that the values used to obtain f are computed at a few points within the effective radius r_e at the beginning of the simulation and are stored in a lookup table, then saving the processing time.

2.4 Free Surface Boundary Condition

In the present work, the neighborhood particles centroid deviation (NPCD) method [18] is adopted to identify the free-surface particles. The NPCD method improves the stability and accuracy of the pressure computation by eliminating spurious oscillations due to misdetection of free-surface particles inside the fluid domain. The Dirichlet pressure boundary condition is imposed to the free-surface particles and it is considered during the implicit step of the method. By avoiding misdetection of free-surface particles inside fluid domain, NPCD provides relatively stable pressure computation even by using only the particle number density deviation as the source term in Eq. (8).

3 Descriptions of the Case

The simulated model represents a horizontal pipe of diameter $D = 0.05$ m and length $L = 1.0$ m subject to a constant inflow $Q = 0.35$ l/s combined with outflow at the end of the pipe (see Fig. 3). Three inflow heights (h_i) were studied, namely 0.05, 0.025 and

0.0125 m, given the relations $h_i/D = 1.0, 0.5$ and 0.25. The fluid density and viscosity were adopted as 1000 kg/m^3 and 10^{-6} m^2/s, respectively. The Reynolds number for $h_i/D = 1.0$, 0.5, 0.25 are $Re = 9 \times 10^3$, 1.8×10^4, 2.7×10^4, respectively. Reynolds number is given by $Re = 4R_hU/\nu$, where the hydraulic radius is $R_h = A/P_w$, A is the flow cross-sectional area, P_w is the wetted perimeter and U is the inflow velocity. In order to monitor the water level in the pipe, 2 probes were placed at 0.50 m (S1) and 0.75 m (S2) from the upstream inflow section. The duration of simulations was 20 s.

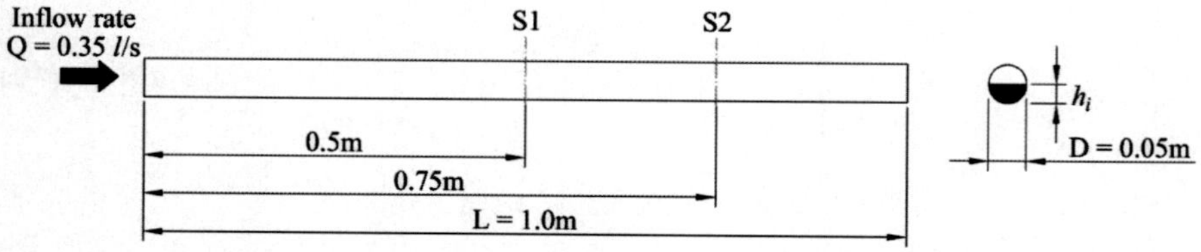

Fig. 3. Main dimensions of the simulated model. S1 and S2 are the sections with wave probes

The numerical parameters adopted in the particle wall and polygon wall modeling are given in Table 1 and Table 2, respectively. Three particle distances (l_0) were simulated, 0.005, 0.002 and 0.001 m, representing the resolutions $D/l_0 = 10$, 25, 50.

The simulations were performed on the same machine which is equipped with Intel© Xeon© Processors E5–2680 v2 (10 cores, 2.8 GHz and 128 GB of memory each).

Table 1. Particle wall modeling: numerical parameters, approximate number of particles and computational cost

Distance between particles (l_0) [mm]	5	2	1
Resolution (D/l_0)	10	25	50
Initial number of particles	31908	164424	643836
Final number of particles	43155	321249	1831708
Time step [ms]	0.2	0.1	0.05
Computational time [hr]	1.6	42.3	1111.1
Relaxation coefficient γ_1	0.0016	0.0025	0.0025
Relaxation coefficient γ_2	0.04	0.05	0.05
Compressibility coefficient α_c [ms^2/kg]	10^{-8}	10^{-8}	10^{-8}

Table 2. Polygon wall modeling: numerical parameters, approximate number of particles and computational cost

Distance between particles (l_0) [mm]	5	2	1
Resolution (D/l_0)	10	25	50
Initial number of particles	552	2268	7368
Final number of particles	6203	105530	834817
Time step [ms]	0.2	0.1	0.05
Computational time [hr]	1.2	35.7	1090.6
Relaxation coefficient γ_1	0.0016	0.0025	0.0025
Relaxation coefficient γ_2	0.04	0.05	0.05
Compressibility coefficient α_c [ms^2/kg]	10^{-8}	10^{-8}	10^{-8}
Repulsive coefficient (α_{rep}) [N/m^2]	10^6	10^6	10^6

4 Results and Discussions

The snapshots of the simulations using both boundary modeling for the inflows $h_i/D =$ 1.00, 0.50 and 0.25, at t = 10 s, are presented in Fig. 4, Fig. 5 and Fig. 6, respectively. The colors are related to fluid velocity magnitude.

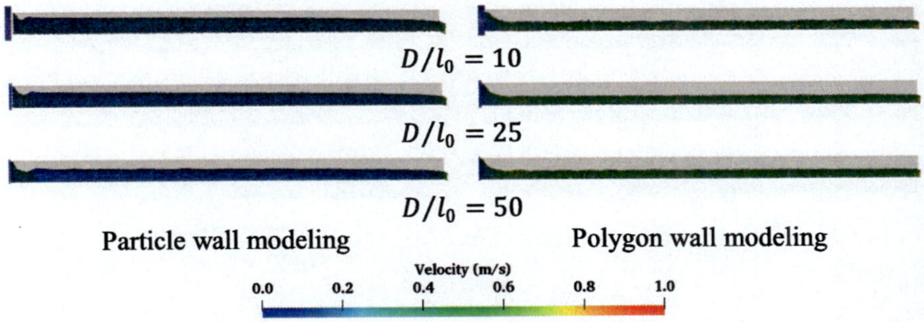

Fig. 4. Snapshots of the simulations with inflow $h_i/D = 1.00$ at t = 10 s. Particle (left) and polygon (right) wall modeling. Resolutions D/l_0 = 10 (top), 25 (middle) and 50 (bottom)

Figure 4 shows that the mean water levels computed by using the particle wall modeling are higher than those obtained using the polygon wall modeling for the inflow relation $h_i/D = 1.00$. Moreover, the flow velocities computed using the particle wall modeling are lower than those calculated by the polygon wall modeling.

For the simulations with inflow $h_i/D = 0.50$, see Fig. 5, results similar to those presented in Fig. 4 were obtained. The snapshots show that, along the pipe, the water levels computed using the particle wall modeling decrease significantly while the water levels computed by using polygon wall remain almost constant.

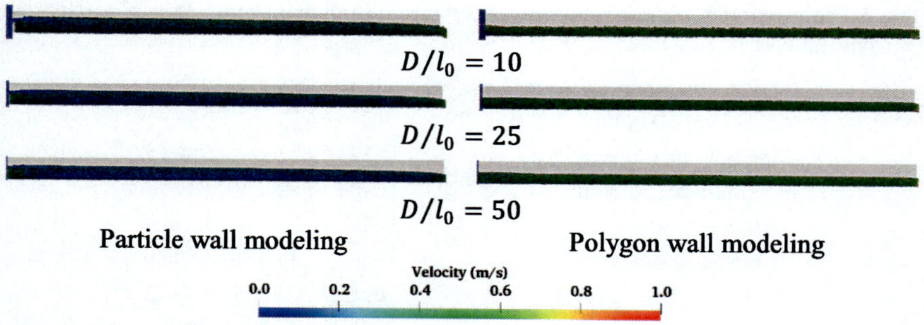

Fig. 5. Snapshots of the simulations with inflow $\boldsymbol{h_i/D} = \mathbf{0.50}$ at t = 10 s. Particle (left) and polygon (right) wall modeling. Resolutions $\boldsymbol{D/l_0}$ = 10 (top), 25 (middle) and 50 (bottom)

Figure 6 shows the snapshots for the inflow $h_i/D = 0.25$. Similar to the previous cases, the water levels computed using the particle wall and polygon wall modeling are different, and much higher flow velocities associated to significantly lower water level occur when polygon wall modeling is adopted. Furthermore, turbulence and hydraulic jump are formed in the simulations with the particle wall modeling, while the simulations using polygon wall modeling present a smooth free-surface profile independent of the model resolution.

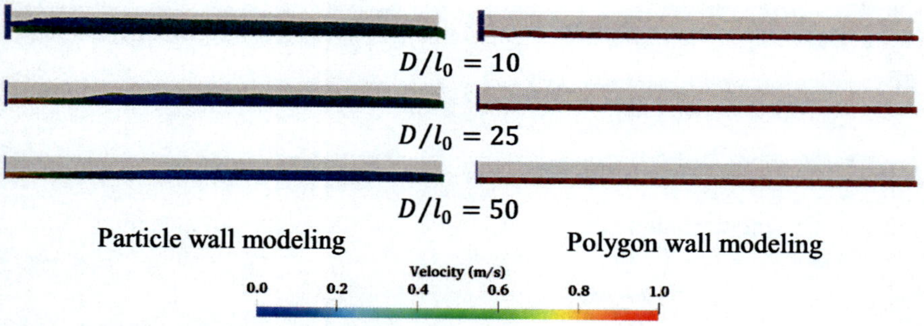

Fig. 6. Snapshots of the simulations with inflow $\boldsymbol{h_i/D} = \mathbf{0.25}$ at t = 10 s. Particle (left) and polygon (right) wall modeling. Resolutions $\boldsymbol{D/l_0}$ = 10 (top), 25 (middle) and 50 (bottom)

Figure 7, Fig. 8 and Fig. 9 give the water level computed at the sections S1 and S2 by using both boundary modeling for the inflow heights $h_i/D = 1.00$, 0.50 and 0.25, respectively.

Figure 7 shows that, regardless of the model resolution, the flow reaches the sections S1 and S2 at the same instant for the simulations with the polygon wall modeling for $h_i/D = 1.00$. On the other hand, when the particle wall modeling is adopted there are small delays depending on the model resolution, and coarse resolution models lead to longer delays. When the numerical solutions reach the steady state, the water levels are almost the same for the cases with polygons, regardless of the model resolution, while lower resolution models result in a higher water level for the particle wall modeling.

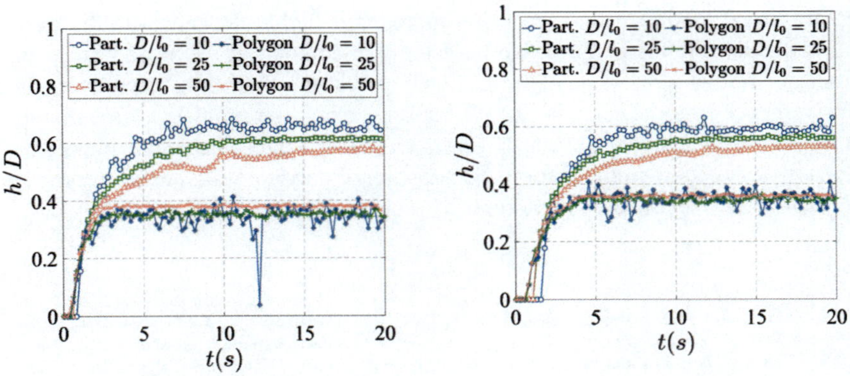

Section S1 Section S2

Fig. 7. Comparison of water levels at section S1 (left) and S2 (right) using both wall modeling and inflow height $\boldsymbol{h_i/D = 1.00}$

For the cases with $h_i/D = 0.50$, see Fig. 8, once again, regardless of the resolution, the flow reaches the sections S1 and S2 at the same instant when the polygon wall modeling is adopted. Larger delays of the lower resolution particle wall modeling are also computed. Although the steady water levels computed using polygons wall modeling with different resolutions are slightly different, good convergence is obtained. For the simulations adopting only particles, the increase of the resolution leads to a decrease of the water level and the convergence is not achieved.

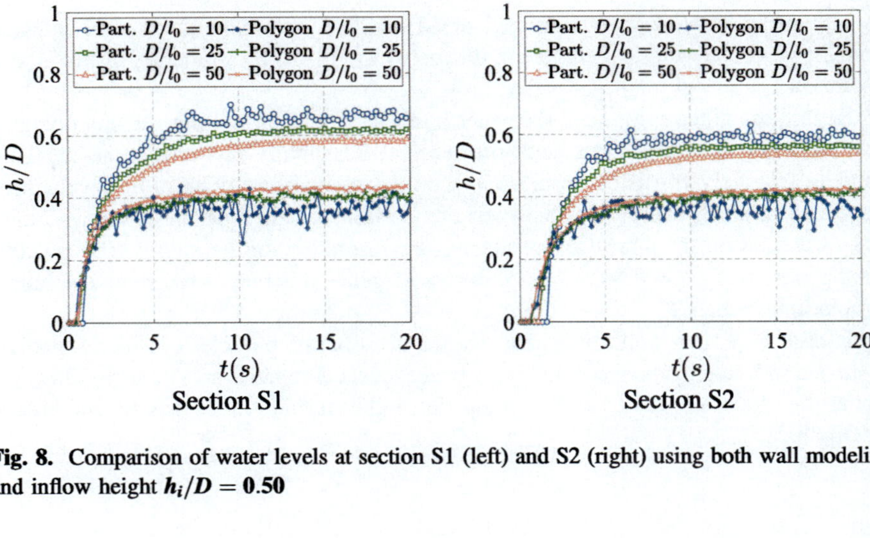

Section S1 Section S2

Fig. 8. Comparison of water levels at section S1 (left) and S2 (right) using both wall modeling and inflow height $\boldsymbol{h_i/D = 0.50}$

Figure 9 illustrates the computed results for $h_i/D = 0.25$. As in the previous results, the polygon wall modeling results in almost the same instants when the flow reaches the sections S1 and S2, while delays occur in the results of the lower resolution simulations of particle wall modeling. The delays computed by using particle wall modeling reveal that the accuracy of this approach is highly dependent to the resolution. This is because the non-uniform particle representation of the smooth inner pipe wall surface results in an unphysical frictional loss of the fluid flow. Moreover, since the convergence is not reached in particle wall modeling, the simulated cases require a finer resolution, which demands much larger computational resources. It is important to point out that these issues are fixed by the adoption of polygon walls, which reproduces more accurately the boundary condition.

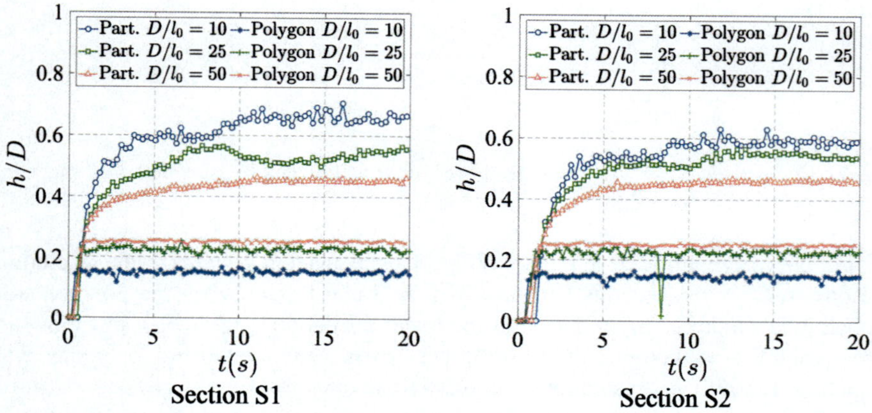

Fig. 9. Comparison of water levels at section S1 (left) and S2 (right) using both wall modeling and inflow height $\boldsymbol{h_i/D = 0.25}$

The mean water levels computed at S1 and S2 after 10 s for both boundary modeling and three inflow heights are illustrated in Fig. 10 as a function of the model resolution.

Regardless of the resolution, the water levels at S1 and S2 computed using polygon wall modeling are very close, with the level at S2 slightly higher than at S1. This provides relatively consistent results and indicates a small reduction of flow velocities associated to small energy loss between two sections of the pipe. In addition to this, mean water levels are almost independent of the resolution for the inflow $h_i/D = 1.00$. For the inflows $h_i/D = 0.50$ and 0.25, the convergence is achieved when the resolution D/l_0 is higher than 25.

In case of particle wall modeling, due the non-smooth particle wall representation of the smooth inner pipe wall, the mean water levels computed at S1 are significantly higher than those computed at S2, except for higher resolution models for the inflow $h_i/D = 0.25$.

The mean water levels at S1 and S2 computed using particle wall modeling are much higher than those obtained using polygon wall modeling. These results show relatively large unnatural energy losses caused by particle wall modeling that lead to much lower flow velocity and higher water levels at same flow rate.

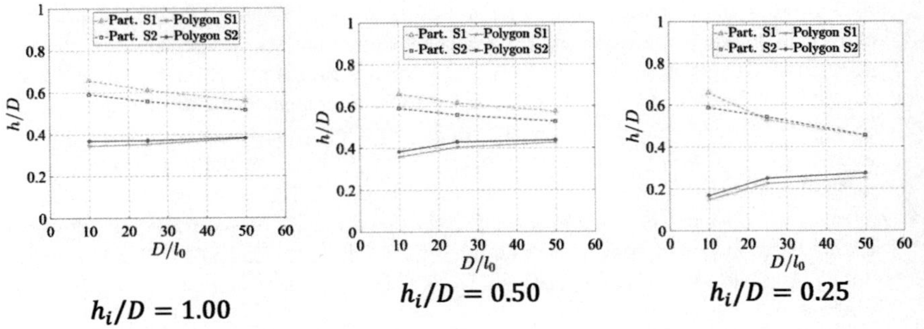

Fig. 10. Comparison of mean water level after 10 s for both wall modeling and three inflow heights: $h_i/D = 1.00$ (left), $h_i/D = 0.50$ (middle), $h_i/D = 0.25$ (right)

5 Final Remarks

In the present work, a particle-based method is used to simulate a free-surface flow inside a horizontal pipe. The unphysical frictional loss inside the pipe is investigated by considering two models: particle and polygon wall modeling. From the outcome results and comparisons conducted here, important conclusions can be drawn as follows:

- The increase of the resolution leads to a decrease of the water level and the convergence is not achieved for the simulations adopting particle wall modeling. On the other hand, convergence is achieved for all simulations with polygon wall modeling.
- Compared to the results of the polygon wall modeling, lower flow velocity and higher water levels are computed in simulations with particle walls due the significant unphysical friction, i.e., a large unnatural energy loss.
- A considerable decrease of the water levels along the pipe occurs for the simulations with particle wall modeling, while they remain almost constant for the simulations with polygon walls.

In summary, more accurate results are provided by the simulations with the polygon wall modeling. Furthermore, the convergence was not reached in the simulations with the discrete particle wall modeling, which requires higher resolution models, and demands much larger computational resources.

Acknowledgements. Authors would like to acknowledge the Conselho Nacional de Desenvolvimento Científico e Tecnológico (CNPq), and Coordenação de Aperfeiçoamento de Pessoal de Nível Superior - Brasil (CAPES) - Finance Code 001 for providing scholarship support for the first and second authors, respectively.

References

1. Gormley, M., Jean, N.: Modelling solid transport in shallow gradient pipe installations: application to simplified sewerage in an international development context. In: The Proceedings of the 38th International Symposium CIB W062 on Water Supply and Drainage for Buildings (2012)
2. Silva-Afonso, A., Bernardo, J., Pimentel-Rodrigues, C.: Implication of reduced flush volumes in building drainage: an experimental study. In: The Proceedings of the 39th International Symposium CIB W062 on Water Supply and Drainage for Buildings (2013)
3. Akiyama, K., Otsuka, M., Shigefuji, H.: Basic study on a method for predicting the waste-carying performance in the horizontal drain pipe of a water-saving toilet. In: The Proceedings of the 40th International Symposium CIB W062 on Water Supply and Drainage for Buildings (2014)
4. Akiyama, K., Otsuka, M., Shigefuji, H.: Further consideration on the prediction method of carrying performance of horizontal drain pipes for water-saving toilets. In: The Proceedings of the 41st International Symposium CIB W062 on Water Supply and Drainage for Buildings (2015)
5. Gormley, M., Williams, L.A., Davidson, J.S., Davidson, A.: Application of building drainage solid transport modelling to village level: a case study of Marikuppam, India. In: The Proceedings of the 42nd International Symposium CIB W062 on Water Supply and Drainage for Buildings (2016)
6. Gingold, R.A., Monagham, J.J.: Smoothed particle hydrodynamics: theory and application to non-spherical stars. Mon. Not. R. Astron. Soc. **181**, 375–389 (1977)
7. Lucy, L.B.: A numerical approach to the testing of the fission hypothesis. Astron. J. **82**, 1013–1024 (1977)
8. Koshizuka, S., Oka, Y.: Moving-particle semi-implicit method for fragmentation of incompressible fluid. Nucl. Sci. Eng. **123**(3), 421–434 (1996)
9. Cheng, L.Y., Oliveira, L.H., Favero, E.H.: Particle-based numerical analysis of drainage flow inside building system. In: The Proceedings of the 38th International Symposium CIB W062 on Water Supply and Drainage for Buildings (2012)
10. Cheng, L.Y., Oliveira, L.H., Favero, E.H., Oliveira, I.B., Goncalves, O.M.: Simulation of drainage system in building using particle-based numerical method. In: The Proceedings of the 39th International Symposium CIB W062 on Water Supply and Drainage for Buildings, pp. 77–91 (2013)
11. Cheng, L.Y., Oliveira, L.H., Ferracini, D.A., Oliveira, I.B., Gonçalves, O.M.: A numerical study on waste transport in main drain. In: The Proceedings of the 40th International Symposium CIB W062 on Water Supply and Drainage for Buildings (2014)
12. Cheng, L.Y., Oliveira, L.H., Osello, P.H.S., Amaro Jr, R.A.: A numerical modeling of solid waste transport in main drain. In: The Proceedings of the 42th International Symposium CIB W062 on Water Supply and Drainage for Buildings (2016)
13. Cheng, L.Y., Oliveira, L.H., Osello, P.H.S., Amaro Jr, R.A.: A numerical investigation on the hydrodynamic impact loads of the solid waste transport inside main drains. In: The Proceedings of the 43th International Symposium CIB W062 on Water Supply and Drainage for Buildings (2017)
14. Zhang, S., Morita, K., Shirakawa, N., Fukuda, K.: Simulation of the Rayleigh-Taylor instability with the MPS method. Mem. Fac. Eng. **64**(4), 215–228 (2004)
15. Wang, L., Jiang, Q., Zhang, C.: Improvement of moving particle semi-implicit method for simulation of progressive water waves. Int. J. Numer. Methods Fluids **85**(2), 69–89 (2017)

16. Mitsume, N., Yoshimura, S., Murotani, K., Yamada, T.: Explicitly represented polygon wall boundary model for the explicit MPS method. Comput. Part. Mech. **2**(1), 73–89 (2015)
17. Jacobson, A., Panozzo, D., et al.: libigl: a simple C++ geometry processing library (2018)
18. Tsukamoto, M.M., Cheng, L.-Y., Motezuki, F.K.: Fluid interface detection technique based on neighborhood particles centroid deviation (NPCD) for particle methods. Int. J. Numer. Methods Fluids **82**(3), 148–168 (2016)

Quantifying Resource Inter-activity Utilization Efficiency Through Simulation-Based Scheduling

Leila Zahedi([✉]) and Ming Lu

University of Alberta, Edmonton T6G 2R3, Canada
{lzahedi,mlu6}@ualberta.ca

Abstract. This research investigates resource inter-activity utilization efficiency arising from resource shared use and transfer observed in planning projects at a structural steel fabrication shop. Aided with discrete event simulation-based resource-constrained scheduling analysis, the necessity of defining the resource inter-activity utilization efficiency factor is justified in project scheduling and budgeting in practice. The Simplified Scheduling Simulation (S3) platform based on the simplified discrete-event simulation approach (SDESA) is utilized to create a resource loaded schedule for a simple project case in order to facilitate analyzing resource inter-activity utilization efficiency. Further, the problem of how to optimize labor inter-activity utilization efficiency is formulated conceptually, which is vitally important to project cost budgeting and earned value management in applications resembling operating a major steel fabrication facility.

Keywords: Resource Inter-activity utilization efficiency factor · Simplified Scheduling Simulation (S3) · Project budgeting · Earned value management (EVM) · Cost estimate

1 Literature Review

The industrial fabrication features made-to-order products, repetitive process-jobs, shared use of limited resources, and resource interactions, challenging the effectiveness of existing methods and tools for scheduling and budgeting in project management [1, 10]. Project Management Institute (PMI) defines project budget as the total amount of authorized financial resources allocated for a particular purpose for a specific period of time [2]. A reliable cost budget encompasses well-defined cost estimate and project schedule, which is indispensably to implement effective cost control throughout project execution leading up to successful project delivery [3]. Steel fabrication projects require the utilization of resources including skilled trades, materials, equipment, and management, to produce fabricated steel products subject to engineering design specification and client requested cost budget and deliver time [4]. Labor cost, which usually accounts in the range of a quarter to a half of the total project cost, is considered as the crucial project element and is determined independently of equipment use in industrial construction [3]. Therefore, labor hours (LH) based budgeting is a common

E. Toledo Santos and S. Scheer (Eds.): ICCCBE 2020, LNCE 98, pp. 1276–1287, 2021.
https://doi.org/10.1007/978-3-030-51295-8_90

practice in making budgets for labor costs on industrial construction projects [5]. In the remainder of this paper, *resource* is specifically referred to as labor resources or trades relevant to industrial construction.

In planning fabrication shop floor operations, resource use efficiency can be distinguished as: *resource inter-activity utilization efficiency* versus *resource intra-activity utilization efficiency*. Resource intra-activity utilization efficiency is the time utilization rate for which resources actively perform a particular job on a given activity by using tools and produce valuable outputs that can be measured. It is defined as the ratio of the effective effort of laborers to the total effort allocated to a particular activity on a project. For example, a journeyman spends 45 min effective time working on a welding activity during 60 min activity duration, resulting in 75% labor inter-activity utilization efficiency; note 15 min is for breaks occurring during the activity duration.

On the other hand, resource inter-activity utilization efficiency is based on how efficient a resource flows between executing consecutive activities. In a resource constrained environment, resource inter-activity utilization efficiency is associated with *Semi-Productive* resource time as a result of (1) shared use of resources such as skilled trades involved in different activities at different locations from different projects, (2) resource transfer between activities in executing concurring jobs and (3) setting up, mobilizing, and getting ready prior to starting the next activity for locating materials, confirming drawings and checking safety hazards etc. Figure 1 shows a graphical representation of semi-productive laborer time based on one resource allocation plan simulated from the case study project consisting of nine activities (A to K). In Fig. 1 (a), the total allocation of journeyman resources in fabrication of the project is denoted with the total area delimited with red frame, while the black boxes represent journeymen's semi-productive time. Note in Fig. 1 (a), those colored areas marked with Activity ID actually symbolize intra-activity utilization of particular resources allocated to a particular activity for its time duration. Essentially, Fig. 1 (a) is a detailed resource work schedule subject to technological constraints on the project. In Fig. 1 (b), the two distinct patterns of journeymen's resource utilization are clearly highlighted.

The resource inter-activity utilization efficiency –which makes a considerable impact on scheduling and cost budgeting– is commonly ignored in current project planning and scheduling practice. This is partly due to the fact that it would not be possible to explicitly analyze resource sharing and transfer when planning made-to-order jobs at a fabrication shop by using popular scheduling tools (such as P6 or MS Project) [6]. As such, the budget derived from commercial software simply overlooks the impact of resource inter-activity utilization efficiency on project budget. At best, seasoned schedulers might apply a factor based on experience to compensate for such impact.

To overcome the limitations and improve the effectiveness of critical path method (CPM) scheduling under limited resource constraints, Lu et al. [7] developed a resource scheduling platform based on the simplified discrete-event simulation approach (SDESA) called Simplified Scheduling Simulation (*S3*), which provides (1) the capability to observe and analyze utilization of associated individual resources on a certain activity in a particular time period and (2) simple user interfaces to schedule a project with resource availability and activity interruption constraints. *S3* is a project planning and resource scheduling tool that enables project modeling, simulation analysis, and resource use optimization in making a project scheduling plan. Particle Swarm

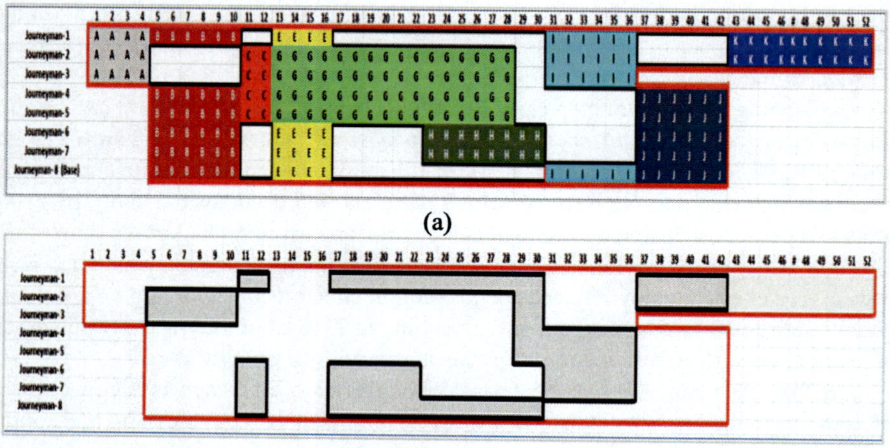

(a)

(b)

Fig. 1. (a) Detailed resource work schedule subject to technological constraints on the project; (b) Semi-productive resource time (black blocks) vs. Total journeymen's efforts are highlighted (red framework)

Optimization (PSO) based optimizer underlies *S3* to identify alternative resource provisions so as to achieve the optimization goal of the shortest project duration or the least project cost. Providing each individual resource working schedule, *S3* enables users to monitor and control individual resources' behavior by precisely generating a sequence of executed activities [7]. Note that PSO is a computational method that optimizes a problem by iteratively attempting to improve a candidate solution with regards to a specifically defined measure of the solution's fitness. Analogous to the Genetic Algorithm, PSO provides a better alternative in complex problems, but not necessarily the best one [8]. In *S3*, the measure of a solution's fitness can be either shorter total project duration or lower total project cost [7].

In this research, a steel fabrication project schedule simulation model is developed by using the *S3* computer platform. The results are conducive to analyzing resource inter-activity utilization efficiency factor and building a realistic labor-hours budget. The effect of applying resource inter-activity utilization efficiency on deriving the Budgeted Cost for Work Scheduled (BCWS) is also investigated. Note BCWS is represented as an S-curve to display cumulative budgeted costs against the project time, which underpins earned value management in evaluating the performance of project schedule control.

2 Problem Statement

In a steel fabrication shop, a series of activities including cutting, fitting, welding, and finishing operations are performed in the fabrication process and entails substantial resource sharing and transfer between activities [6]. Table 1 shows a fictitious fabrication project with time unit being hour, which is adapted from a CPM example in Hegazy [9]. CPM analysis under no resource availability constraints results in project duration of 32 h (Fig. 2).

Table 1. Duration, resource requirement and predecessors of each activity

Activates	Predecessors	Duration (hr)	Resource requirement (Journeyman)
A	–	4	3
B	–	6	6
C	–	2	4
D	A	8	–
E	D	4	4
F	B	10	–
G	B	16	4
H	F	8	2
I	E, H	6	4
J	C	6	5
K	G, J	10	2

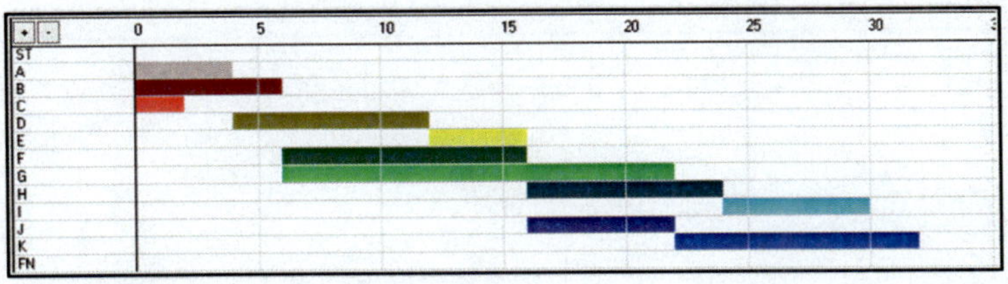

Fig. 2. CPM activity bar-chart schedule under no resource availability constraints

Resulting journeyman resource profile usage is given in Fig. 3. Next, setting the availability limits as 8 journeymen. Resource over allocation for the journeymen is also shown in Fig. 3, which will be resolved by *S3* through resource scheduling simulation as described in the next section.

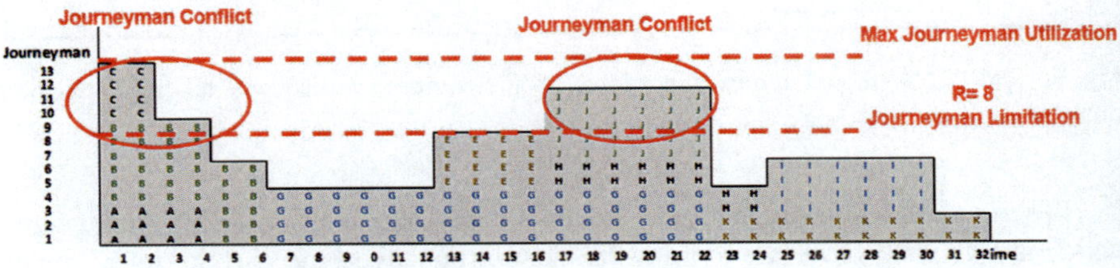

Fig. 3. Resulting journeyman resource use profile revealing resource conflicts

3 Resource Scheduling

Figure 4 is the input interface of the *S3* platform. Project data can be imported directly into the *S3* from databases such as *MS Access*. The S3/SDESA model representing the fabrication activities, the AON network, plus imposed resource constraints is generated automatically by computer, as shown in Fig. 5. With 8 journeymen available, total project duration is extended to 52 h, in contrast with the original 32 h without loading resource limit constraints.

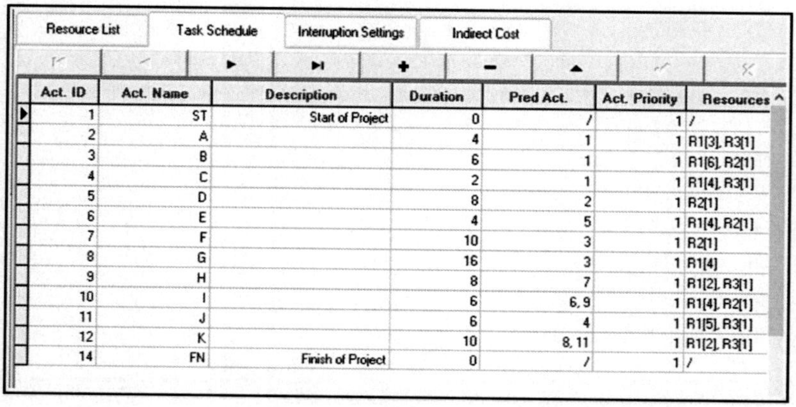

Act. ID	Act. Name	Description	Duration	Pred Act.	Act. Priority	Resources
1	ST	Start of Project	0	/	1	/
2	A		4	1	1	R1[3], R3[1]
3	B		6	1	1	R1[6], R2[1]
4	C		2	1	1	R1[4], R3[1]
5	D		8	2	1	R2[1]
6	E		4	5	1	R1[4], R2[1]
7	F		10	3	1	R2[1]
8	G		16	3	1	R1[4]
9	H		8	7	1	R1[2], R3[1]
10	I		6	6, 9	1	R1[4], R2[1]
11	J		6	4	1	R1[5], R3[1]
12	K		10	8, 11	1	R1[2], R3[1]
14	FN	Finish of Project	0	/	1	/

Fig. 4. Input window of the S3

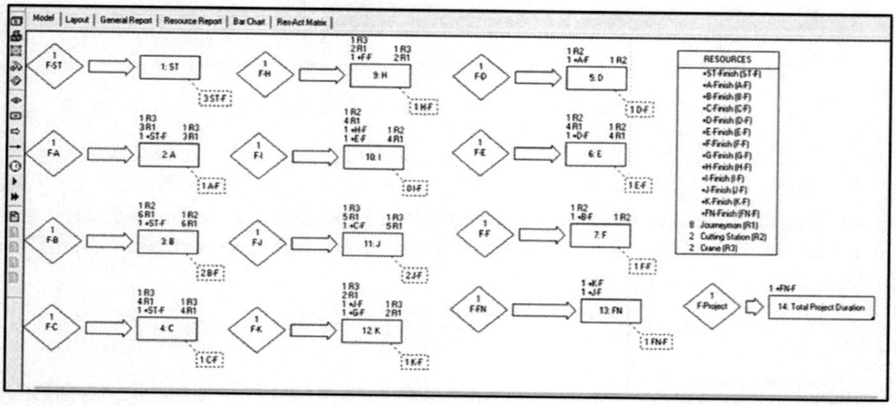

Fig. 5. S3/SDESA model representing fabrication activities, precedence relationships, plus imposed resource constraints

The simulation model generates the time-dependent resource utilization matrix, which allows analysis of the journeyman semi-productive time and the inter-activity efficiency (Fig. 6).

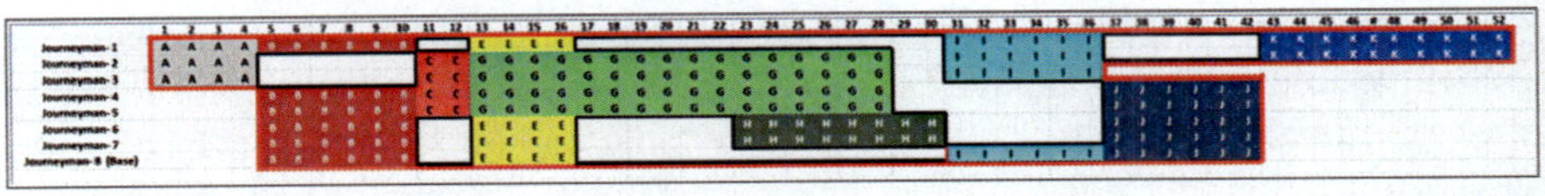

Fig. 6. S3 simulation generated solution in form of time-dependent resource utilization matrix

As per Fig. 6, resource semi-productive resource time blocks are highlighted with black border lines. Out of the total journeymen hours spent (330 h; which is calculated from the start of the first task to the end of the last task for each individual journeyman and denoted by the total area inside the red framework in Fig. 6), 226 h are actively engaged in executing fabrication activities with the remainder of 104 h identified as inter-activity "Semi-Productive". Thus, the rate of *68.5%* is calculated (i.e., 226/330) as the resource inter-activity utilization efficiency for the crew of 8 journeymen based on the simulated project plan.

4 Fabrication Project Budgeting

Given Resource requirement and duration of each activity shown in Table 1, the common procedure for LH budget calculation is given in Table 2. Generally, when applying CPM scheduling under resource constraints, the derived budget denotes resource utilization in activity execution without considering any inter-activity semi-productive time, which is marked in Fig. 6 as the blank cells in the black boxes. It is important to note such resource semi-productive time, which is due precisely to resource use sharing, resource transfer, reassembling, setting up at the next activity in the fabrication shop, is not considered as "unproductive." Reducing "semi-productive" resource time and turning it into "productive" time presents a practical opportunity for optimization in this case.

Table 2. Common procedure for budget calculation based on CPM schedule

Hour	Activity	Direct labor	Duration (Hr)	Direct labor-hour	Cumulative labor-hour
1 to 4	A	3	4	12	12
5 to 10	B + D	6	6	36	48
11 to 12	D + C	4	2	8	56
13 to 16	E + F+G	8	4	32	88
17 to 22	F + G	4	6	24	112
23 to 28	G + H	6	6	36	148
29 to 30	H	2	2	4	152
31 to 36	I	4	6	24	176
37 to 42	J	5	6	30	206
43 to 52	K	2	10	20	226

Planned Value (PV) or Budgeted Cost Work Scheduled (BCWS) is represented as an S-curve to display cumulative budgeted costs against project duration. The PV in terms of cumulative laborer-hour needs to be adjusted by factoring in labor inter-activity utilization efficiency in each time span. Figure 7 illustrates how the resource inter-activity utilization efficiency factor is determined, based on which how to adjust the plotted BCWS S-curve. Note the blue-line represents the baseline where resource inter-activity utilization efficiency is not considered (as obtained from existing commercial software for project scheduling); the orange line is the adjusted S-curve precisely reflecting the budget as per the current resource-loaded project schedule.

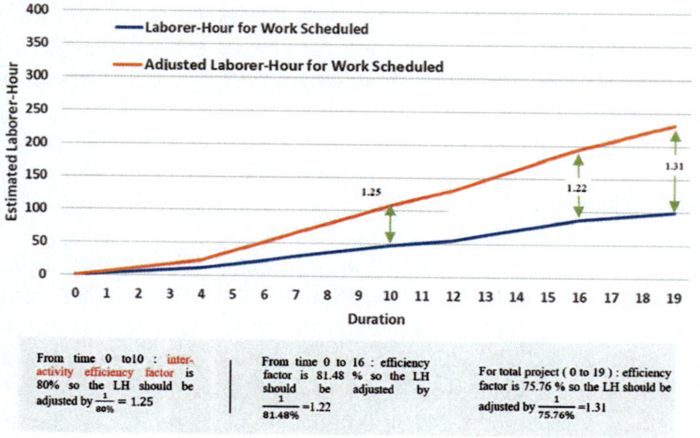

Fig. 7. Determining inter-activity utilization efficiency factor and adjusting BCWS S-curve

For the current resource schedule (Fig. 6), S-curves of the baseline BCWS vs. adjusted BCWS are shown in Fig. 8. Note at project end, 68.5% journeymen inter-activity utilization efficiency translates into a factor of 1.45 applicable to adjust the total labor hours budget.

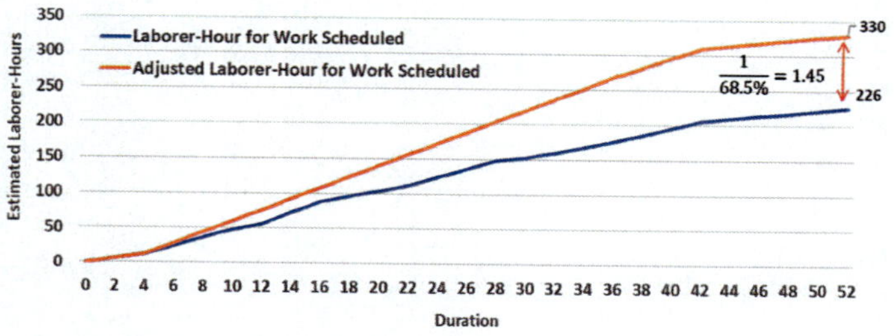

Fig. 8. S-curves of the baseline BCWS vs. adjusted BCWS for the current resource schedule in case study (i.e. Fig. 6).

5 Conceptual Formulation of Optimization Problem

The journeymen's inter-activity Semi-Productive Hours can be used to further illustrate the formulation of the resource inter-activity utilization efficiency optimization problem. The earliest possible start time of an activity according to the journeyman availability is equal to the early finish time of the preceding activity, as per Eq. (1). (for the first scheduled activity it is zero).

$$SL_i^n = Early\ Fininsh\ Time_{(i-1)}^n \tag{1}$$

Where: SL_i^n is earliest start time activity i with respect to journeyman availability and n is the number of journeymen in the shop.

The earliest possible start time according to activity availability with consideration of technology constraint is the earliest date the activity can start if there are no constraints due to journeyman availability (Eq. (2))

$$SA_i^n = Scheduled\ Fininsh\ Time_{(i-1)}^n \tag{2}$$

Where: SA_i^n is earliest start time of activity i with respect to only technology constraints.

Since activity cannot start until its predecessor has finished or until all the required journeymen working on the activity are available, the earliest start time for each activity is set as the latest of SA_i^n and SL_i^n (Eq. (3)).

$$\text{Early Start Time}_i^n = Max(SL_i^n, SA_i^n) \tag{3}$$

For activity i, if the earliest start time according to technology-constrained precedence relationships, SA, is later or equal to the earliest start due to journeyman availability, SL, the activity could start immediately. If the earliest start time, SA, is earlier than the earliest start due to journeyman availability, SL, then the activity would not proceed until the journeyman has finished working on the previous task (Eq. (4)). At any point of time, the number of journeymen engaged in ongoing activities deducted from the total number of journeymen results in the number of journeymen being idle according to Eq. (5).

$$ASD_i^n = \begin{cases} SL_i^n - SA_i^n & if SL_i^n > SA_i^n \\ 0 & if SL_i^n \le SA_i^n \end{cases} \text{journeyman } Idle_i^n \tag{4}$$

$$\text{Journeyman } Idle_i^n = Total \text{ journeymen} - JRAy_i^n \tag{5}$$

Where: ASD_i^n is activity semi-productive duration and $JourneymenIdle_i^n$ is idle number of journeymen for activity i and JRA_i^n is journeyman required for activity i. Therefore, journeymen's inter-activity semi-productive hours (JIH) under above constraints is calculated according to Eq. (6), given the "n journeyman available" scenario.

$$JIH = \sum\nolimits_{1}^{i} \left(\text{Journeyman } Idle_{i-1}^{n}\right) \times \text{ASD}_{i}^{n} \quad (6)$$

The quantity of journeymen which is required for performing a particular fabrication activity depends on its work content. Generally, the location of machines is fixed at particular working zones in the fabrication shop, named *workstation*. In contrast, the journeymen move between the workstations to handle different activities belonging to multiple bridge projects being precessed simultaneously.

6 Simulation Solutions

The S3/SDESA simulation model is conducive to illuminating on above conceptual model formulation, which produces verifiable outputs for postulated scenarios with a different number of journeymen and cross checks the analytical formulation given above. In the remainder of this paper, simulation results from eight scenarios of the case study are examined to evaluate the inter-activity efficiency factor. Results for two selected scenarios are given: Scenario 1 is simulated with 6 journeymen, increasing project duration to 54 h (Fig. 9). Scenario 8 has 12 journeymen and resulting duration is decreased to 42 h (Fig. 10).

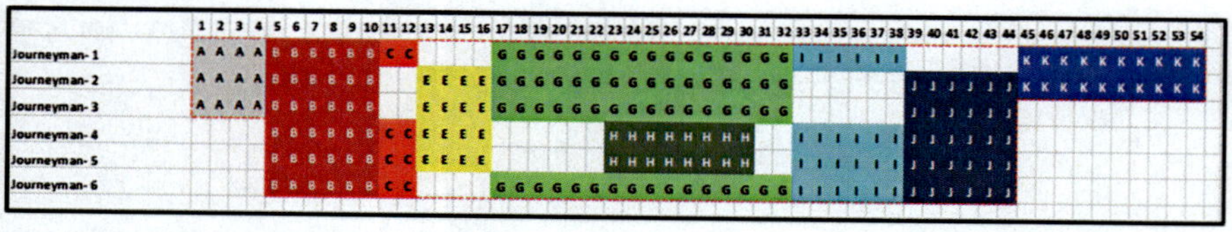

Fig. 9. Scenario 1 with 6 journeymen, increasing project duration to 54 h

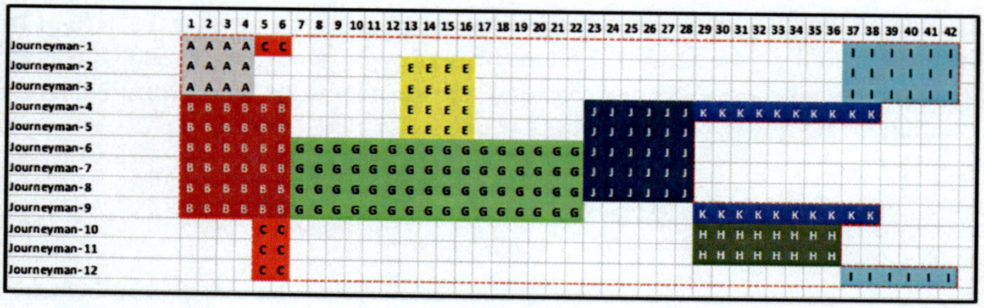

Fig. 10. Scenario 8 with 12 journeymen and resulting duration decreasing to 42 h

The journeyman inter-activity utilization efficiency along with project duration in different scenarios is summarized in Table 3. Scenario 1 has the maximum journeyman inter-activity utilization efficiency (83.1%). In short, the objective function in the conceptual formulation (i.e. minimizing journeymen's inter-activity semi-productive hours) can be realized by adjusting the number of resources and activity sequence for resource allocation under technological constraints, ultimately resulting in the maximum journeyman inter-activity utilization efficiency. Note Scenario 1 does not represent the optimum solution, but the best alternative obtained by running *S3* simulation.

Table 3. Simulation results for different scenarios based on case study project

ID	No. Journeyman	Duration (Hr)	Inter-Act Semi LH	Total LH	Inter-Act Eff %
1	6	54	46	272	83.1%
2	7	54	86	312	72.4%
3	8 (Base Case)	52	104	330	68.5%
4	9	46	116	342	66.1%
5	10	46	130	356	63.5%
6	11	38	130	356	63.5%
8	12	38	154	380	59.5%

As crew size (the number of journeymen) increases, the journeymen' inter-activity utilization efficiency factor decreases, while the total labor cost budget (LH) increases, as shown in Fig. 11.

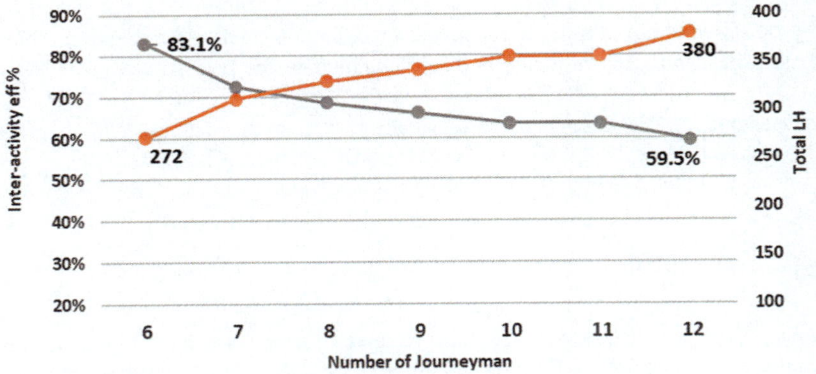

Fig. 11. The effects of number of journeymen on inter activity efficiency and total laborer-hour

7 Conclusion

Labor cost, which usually accounts in the range of a quarter to a half of the total project cost, is considered as the crucial project element and is determined independently of equipment use in industrial construction. Therefore, labor hours (LH) based budgeting is a common practice in making budgets for labor costs on industrial construction projects.

In a resource constrained environment, resource inter-activity utilization efficiency is associated with Semi-Productive resource time as a result of (1) shared use of resources such as skilled trades involved in different activities at different locations from different projects (2) resource transfer between activities in executing concurring jobs and (3) setting up, mobilizing, and getting ready prior to starting the next activity for locating materials, confirming drawings and checking safety hazards etc. Resource inter-activity utilization efficiency is commonly ignored in current project planning and scheduling practice, as the budget derived from commercial software simply overlooks the impact of resource inter-activity utilization efficiency on project budget. At best, seasoned schedulers might apply a factor based on experience to compensate for such impact.

This research has formally defined resource inter-activity utilization efficiency along with how to quantity its impact on scheduling and cost budgeting. A resource scheduling platform based on the simplified discrete-event simulation approach (SDESA) called Simplified Scheduling Simulation (S3) is utilized to conduct case study and develop problem statement. *S3* precisely generates a sequence of executed activities for each individual resource. In this research, a steel fabrication project schedule simulation model is developed for case study. By running *S3* simulation and examining results for different what-if scenarios, the effect of the inter-activity efficiency factor upon project labor cost budget is clarified.

In conclusion, this research is conducive to not only building a realistic budget for labor cost but also formulating the analytical optimization problem. Further, the conceptual formulation of a new optimization problem of minimizing the inter-activity semi-productive hours in order to maximize resource inter-activity utilization efficiency factor will be enhanced, validated and implemented in the near future.

Acknowledgment. The research was substantially funded by the National Science and Engineering Research Council of Canada (NSERC: CRDPJ-501012-16). Kevin Guile, Ken Bird, David Fritz of Supreme Group are acknowledged for support.

References

1. Azimi, R., Lee, S.H., AbouRizk, S.M.: Resource performance indicators in controlling industrial steel projects. In: Construction Research Congress 2010: Innovation for Reshaping Construction Practice, pp. 21–30 (2010)
2. Project Management Institute: A Guide to the Project Management Body of Knowledge (PMBOK guide), 6th edn. Newtown Square, Pennsylvania (2017)
3. Lu, M.: Productivity studies using advanced ANN models, Ph.D. dissertation. University of Alberta (2001)
4. Hu, X., Lu, M., AbouRizk, S.: BIM-based data mining approach to estimating job man-hour requirements in structural steel fabrication. In: Proceedings of the Winter Simulation Conference 2014, Savanah, GA, 2014, pp. 3399–3410 (2014)
5. Mohsenijam, A., Lu, M.: Achieving sustainable structural steel design by estimating fabrication labor cost based on BIM data. J. Procedia Eng. **145**, 654–661 (2016)
6. Siu, M.F., Lu, M., AbouRizk, S.: Optimizing job sequence and resource use in structural steel fabrication scheduling. In: 16th ICCCBE, Osaka, Japan (2016)

7. Lu, M., Lam, H.C., Dai, F.: Resource-constrained critical path analysis based on discrete event simulation and particle swarm optimization. Autom. Const. **17**(6), 670–681 (2008)
8. Clerc, M.: Practical Swarm Optimization, 1st edn. Wiley, Canada (2013)
9. Hegazy, T., Kassab, M.: Resource optimization using combined simulation and genetic algorithms. Const. Eng. Manag. **129**(6), 698–705 (2003)
10. Liu, J., Lu, M.: Constraint programming approach to optimizing project schedules under material logistics and crew availability constraints. J. Const. Eng. Manag. **144**(7), 04018049 (2018)

Life Cycle Assessment for Modular Roof Systems of Large-Span Building

Othman Subhi Alshamrani(✉)

Department of Building Engineering, College of Architecture and Planning, Imam Abdulrahman Bin Faisal University, Dammam, Kingdom of Saudi Arabia
osalshamrani@iau.edu.sa

Abstract. Material extraction, production, transportation, construction, operation, and maintenance consume huge amounts of energy that contribute to environmental pollution and contribute to climate change. In 2017, the building sector was found to be responsible for 39% of energy consumption and associated greenhouse gas (GHG) emissions [1]. In Canada, the industry and building sectors are the highest contributors to total energy consumption. Hence, the embodied and operating energy and their associated GHGs in the building materials, and the operational primary energy over the whole life cycle of buildings, must be assessed [2]. This paper thus presents a Life Cycle Assessment (LCA) approach to evaluating the environmental impact of different modular roof systems, including precast hollow-core concrete, composite roof system, steel roof system, and wood roof system. A case study of a large-span industrial building in Calgary, Alberta, is selected and then analyzed to assess the associated environmental impact of different alternatives of roof systems at each stage in the life-cycle. GHGs and environmental impact parameters and metrics such as embodied and operating energy, fossil fuel consumption, global warming potential, ozone depletion, air, land, and water emissions, and smog potential are analyzed using ATHENA® impact estimator. Wood and steel roof systems are shown to have the lowest energy consumption and lowest GHG emissions.

Keywords: Life cycle assessment · Environmental impact · Roof system · Offsite construction · Climate change · Global warming

1 Introduction

The environmental impacts of the industrial and building sectors have drawn increasing concern in recent decades. The building sector alone consumes about 30% of total energy worldwide and released about 25% of total CO_2 emissions in 2013. In addition, the building sector has been found to be responsible for approximately 75% of the world's overall waste. The construction industry in particular has recorded the largest usage of materials among industrial sectors due to relative growth in recent years and its social import. This industry has also played a major role in pollution and is considered as a target of growing stakeholder scrutiny in North America and Europe [3, 4].

E. Toledo Santos and S. Scheer (Eds.): ICCCBE 2020, LNCE 98, pp. 1288–1303, 2021.
https://doi.org/10.1007/978-3-030-51295-8_91

Therefore, a comprehensive evaluation of environmental impacts and assessment of construction activities is urgently needed. Several quantitative environmental impact evaluation methods and tools, such as LCA, have been developed in recent decades. LCA is a scientifically-based evaluation method that measures the environmental impacts of any service, product, or project over its life cycle span [5]. In practice, LCA is mainly applied when building design is already complete. However, LCA-based quantification is recommended to be applied actively from the early design stages for design optimization and material selection [6].

LCA as a systematic analysis method has been in use in the construction industry since the 1990s. Where it is has been deployed in policymaking in Europe, e.g., for building labeling schemes and for regulating environmental products. It has been also employed in the construction industry in the development of methodologies and decision-making tools [7].

An area of 82 billion square meters, representing about 60% of the world's total building stock, is expected to be constructed in urban areas worldwide by 2030 [8]. Therefore, minimizing the environmental impact of building construction is emerging as an essential component of efforts to achieve the planned target of the Paris Agreement, according to which both embodied carbon (EC) and operational carbon (OC) must be addressed [9]. However, the environmental impact assessment of whole buildings certainly involves many variables and uncertainties, since buildings are complex assemblies with a high quantity of materials, unpredictable future uses, and difficult to predict life spans [10]. LCA can be incorporated into other sustainability metrics such as LEED to evaluate building systems using an integrated sustainability assessment method [11]. It can be also applied to measure the associated environmental impact with applying different structural materials and envelope systems [12].

2 Methodology

The methodology underlying this research consists of five stages, as presented in Fig. 1. The first stage involves identification of the major building parameters, such as building type, gross area, location, height, number of floors, orientation, climate zone, building shape, and envelope system, as shown in Table 1. A case study is considered in order to investigate the selection of the best structural roof systems from the point of view of minimal environmental impact (LCA and GHG emissions). Four different roof systems are involved in this study—composite roof, precast roof, steel roof, and wood roof system—as shown in Table 2. The next stage is preparation of the data for energy simulation purposes, which requires identification of the building and energy parameters in the simulation software. These parameters include the internal energy loads, occupancy schedule, heating and cooling systems, wall system, roofs layers, and U-values.

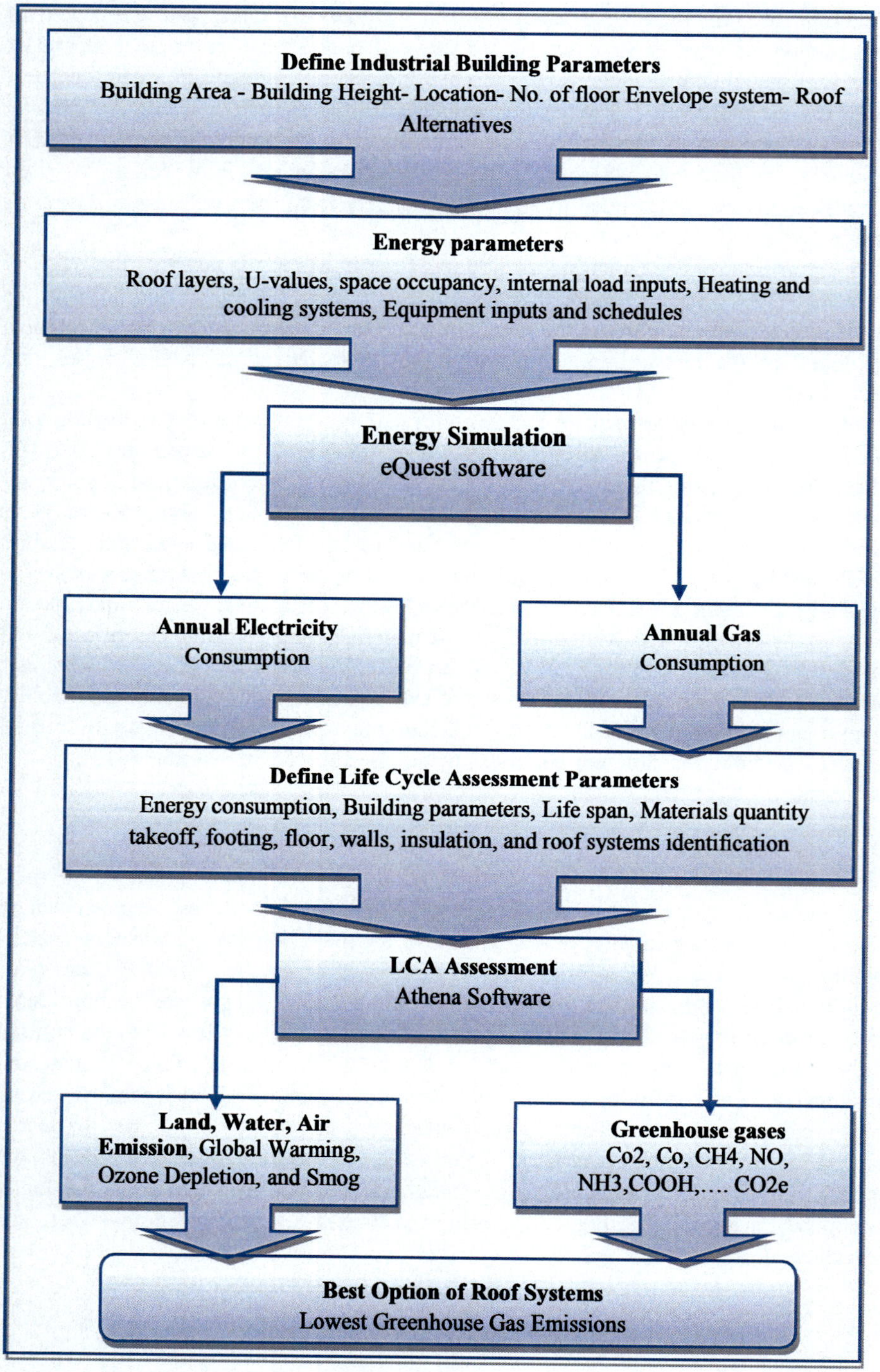

Fig. 1. Flowchart of the methodology.

Table 1. Parameters of the evaluated case study

Item	Description
Building type	Industrial building
Total gross area	65,000 ft^2 (180 ft × 360 ft)
No. of floor	1 floor
Building height	25 ft
Building plan shape	Rectangular 360 ft × 180 ft
Builiding orientation	Long axis-oriented (east-west)
Location	Calgary, Alberta, Canada
Climate zone	4a
Wall type	Insulated concrete block
Roof systems alternative	Composite, precast, steel, and wood roof
Structural system	Steel frame (cloumns and beams)
Life span	30 years

Energy simulation is performed in the next phase in order to assess the annual energy consumption of the building under different roof system scenarios. Annual operating electricity and gas consumption are estimated for the purpose of applying them in the LCA process. The simulation is performed using eQUEST software, which is an analysis tool that provides high quality assessment results by incorporating a building creation wizard and an energy efficiency measure wizard. eQUEST helps user to perform multiple energy simulations and to compare the evaluated alternative results in side-by-side graphics. It also enables users to estimate energy cost and helps them to evaluate a variety of options in order to select the preferable one (Crawley et al. 2005). Figure 2 presents the energy simulated model of the industrial building using eQUEST.

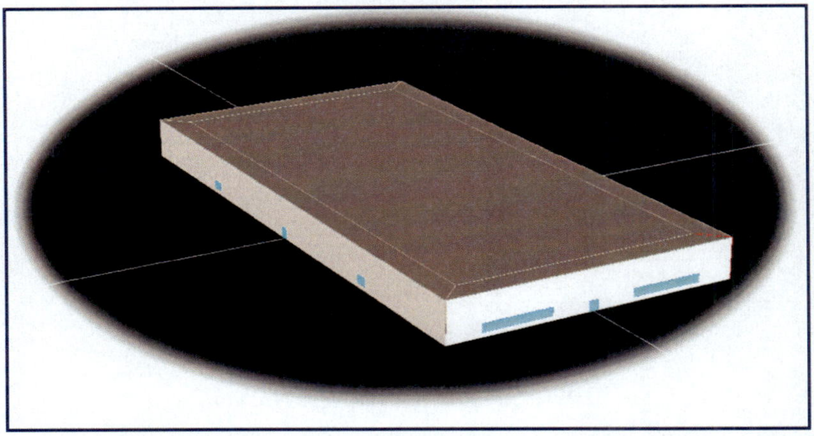

Fig. 2. Energy simulated 3D model.

Table 2. Description of roof system alternatives.

Roof system	Roof description	Total thickness (mm)	Conductivity U- value (m^2C/W)
Composite Roof	Concrete topping – metal deck on steel trusses	200	U(0.591)
Precast roof	Hollow-core slab	300	U(0.822)
Steel roof	Metal deck – steel joist	175	U(0.256)
Wood roof	Wood roof – wood truss	175	U(0.220)

In the next step, the LCA parameters, such as energy consumption, building life span, footings, slabs on grade, columns, beams, walls, roof, insulations, and material quantity take off, are defined. LCA is then conducted using the environmental impact estimator software, ATHENA, which is an impact estimator tool for buildings in North America that assesses whole assemblies, materials, and building components based on an LCA approach. It focuses mainly on the assessment of structural and construction elements, such as footings, floors, walls, roofs, insulations, and finishing materials. The software enables users, architects, engineers, and decision makers to assess and compare alternatives in advance of design, either for new planned buildings or for major repairs and renovations. This tool uses a database that covers a range of building structure and envelope systems used in residential, commercial, and industrial buildings, as shown in Fig. 3.

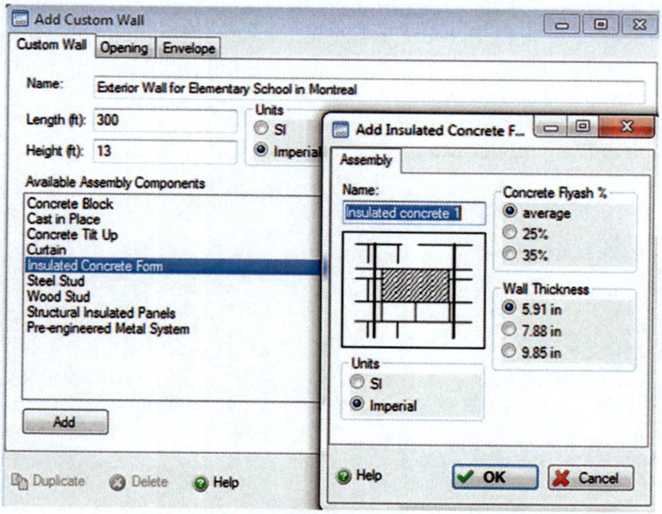

Fig. 3. Wall assembly page in ATHENA software.

LCA is applied in this study to assess at the conceptual design stage the environmental impact and GHG emissions associated with applying different roof systems to industrial building. The environmental impact measures used in this research include ozone depletion, global warming, acidification potential, embodied and operating energy use, emissions to water, solid waste emissions, air emissions, smog potential, and consumption of natural resources. Table 3 shows a sample of the various materials used in the different structural and construction components for the evaluated prototype. Building parameters such as the number of footings, floor area, roof area, number of columns and beams, bay and span sizes, and column and beam dimensions are defined according to the structural framing plan. Four structural roof systems are investigated in this study based on 30 years of building operation across all stages of the life cycle: material extraction, material manufacturing, transportation from and to the site, construction, operation, maintenance, major renovation, end of building life, and beyond building life.

In the final stage of the process, the results of the LCA are presented for all the roof systems tested in order to compare the associated environmental impact of each alternative. The lowest and highest environmental impact are assessed, as well as the percentage of impact reduction achievable by the different alternatives. Finally, the structural roof system that minimizes GHG emissions overall life cycle stages is determined.

Table 3. Sample of materials used in case study.

Meterial	Unit	Foundation	Walls	Columns and beams	Roofs
Aah	kg	5.62E−09	1.58E+00	0.00E+00	1.72E−05
Carbon dioxide, in air	kg	1.02E+02	4.12E+02	1.81E+02	5.44E+03
Clay & Shale	kg	9.27E+03	1.13E+03	0.00E+00	4.45E+03
Coal	kg	2.72E+04	1.98E+04	4.28E+03	1.11E+05
Coarse Aggregate	kg	1.38E+06	3.60E+03	0.00E+00	1.07E+06
Crude Oil	L	4.24E+04	1.29E+04	1.05E+03	6.51E+04
Crude Oil as feedstock	L	5.15E−02	2.48E+03	0.00E+00	1.46E+05
Dolomite	kg	4.03E+02	1.54E+03	9.56E+02	2.15E+04
Ferrous scrap	kg	6.25E+03	2.83E+04	1.25E+04	3.25E+05
Fine Aggregate	kg	9.13E+05	6.94E+04	0.00E+00	8.16E+05
Gypsum (Natural)	kg	3.02E+01	1.90E+04	0.00E+00	1.53E+01
Gypsum (Synthetic)	kg	0.00E+00	4.21E−04	0.00E+00	0.00E+00
Iron Ore	kg	1.56E+04	−2.54E+01	2.8SE+03	4.50E+04
Lignite	kg	7.56E+00	4.33E+02	2.99E+02	8.82E+03
Limestone	kg	S.36E+05	3.17E+04	S.43E+02	2.71E+05

3 Results

This section presents the results of the case study applying both energy simulation and the life cycle assessment.

3.1 Energy Simulation Results

The building performance energy simulation is performed on a case study prototype model for a one-floor industrial building in Calgary, Alberta, Canada. The simulation results show that the peak gas consumption occurs in the months of December and January, followed by February, since these months have extremely cold weather which necessitates high energy consumption for space and water heating, as presented in Fig. 4. The month of July features the highest temperatures, followed by June and August, where these months require high electricity consumption for space cooling. Moderate electricity consumption is recorded from October to Mayas well as in March and November, as presented in Fig. 4 and Table 4.

All of the gas consumption goes to space heating and the water heating, while most of the electricity consumption is directed to area lighting, miscellaneous equipment use, ventilation fans and pumps, and auxiliary equipment, as presented in the pie chart in Fig. 5.

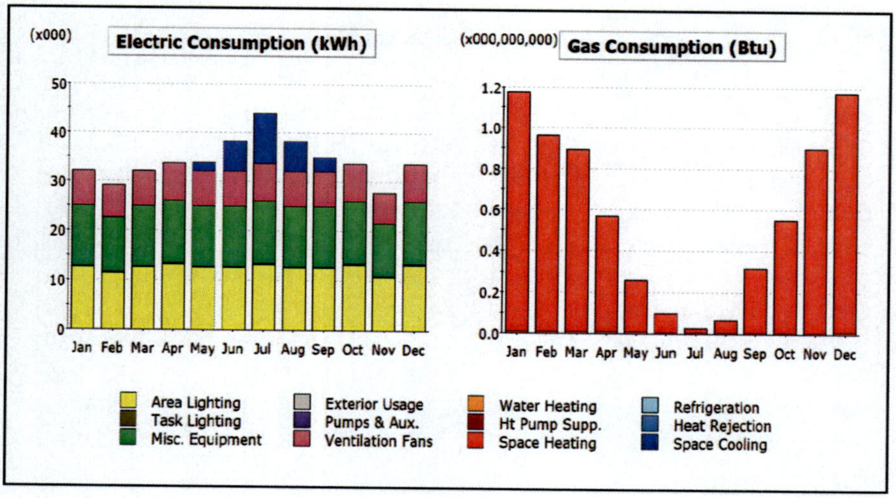

Fig. 4. Monthly electricity and gas consumption.

Table 4. Monthly electricity and gas consumption by end-use.

Electric Consumption (kWh × 000)													
	Jan	Feb	Mar	Apr	May	Jun	Jul	Aug	Sep	Oct	Nov	Dec	Total
Space cool	–	–	–	–	0.59	2.80	5.22	3.22	1.01	–	–	–	13–83
Heat reject.	–	–	–	–	–	–	–	–	–	–	–	–	–
Refrigeration	–	–	–	–	–	–	–	–	–	–	–	–	–
Space heat	–	–	–	–	–	–	–	–	–	–	–	–	–
HP supp.	–	–	–	–	–	–	–	–	–	–	–	–	–
Hot water	–	–	–	–	–	–	–	–	–	–	–	–	–
Vent. pans	3.13	3.82	3.13	3.27	3.13	3.12	3.27	3.12	3.12	3.27	3.67	3.27	37.30
Pumps & auk.	0.04	0.03	0.04	0.03	0.02	0.01	0.00	0.00	0.02	0.03	0.04	0.04	0.30
Ext. usage	–	–	–	–	–	–	–	–	–	–	–	–	–
Misc. equip,	12.01	10.86	12.01	12.48	12.01	11.97	12.52	12.01	11.97	12.52	10.42	12.52	143.30
Task lights	0.22	0.20	0.22	0.23	0.22	0.22	0.23	0.22	0.22	0.23	0.19	0.23	2.63
Area lights	12.72	11.51	12.72	13.32	12.72	12.72	13.32	12.72	12.72	13.32	10.91	13.32	152.01
Total	28.10	25.43	28.10	29.33	28.67	30.83	34.56	31.29	29.05	29.37	24.23	29.38	348.36
Gas consumption (Btu × 000,000)													
	Jan	Feb	Mar	Apr	May	Jun	Jul	Aug	Sep	Oct	Nov	Dec	Total
Space cool	–	–	–	–	–	–	–	–	–	–	–	–	–
Heat reject.	–	–	–	–	–	–	–	–	–	–	–	–	–
Refrigeration	–	–	–	–	–	–	–	–	–	–	–	–	–
Space heat	580.9	462.1	434.2	259.9	114.7	27.3	3.6	7.3	110.6	239.0	425.3	575.8	3,240.6
HP supp.	–	–	–	–	–	–	–	–	–	–	–	–	–
Hot water	9.2	8.5	9.4	9.6	8.7	8.1	8.0	7.4	7.4	8.0	7.1	9.1	100.4
Vent. fans	–	–	–	–	–	–	–	–	–	–	–	–	–
Pumps & aux.	–	–	–	–	–	–	–	–	–	–	–	–	–
Ext. usage	–	–	–	–	–	–	–	–	–	–	–	–	–
Misc. equip.	–	–	–	–	–	–	–	–	–	–	–	–	–
Task lights	–	–	–	–	–	–	–	–	–	–	–	–	–
Area lights	–	–	–	–	–	–	–	–	–	–	–	–	–
Total	590.1	470.7	443.6	269.6	123.4	35.4	11.6	14.6	118.0	247.0	432.3	584.8	3,341.1

Annual Electricity Consumption

The results of analyzing the annual operating energy consumption of different roof systems show that the precast hollow-core concrete records the highest annual electricity consumption at 411,500 kWh. The lowest electricity consumption is recorded for the wood roof system, with an annual energy consumption of 348,400 kWh (i.e., 15% lower compared to the precast roof, as shown in Table 5). The second-lowest consumption is recorded for the steel roof, 14% lower than that of the precast system.

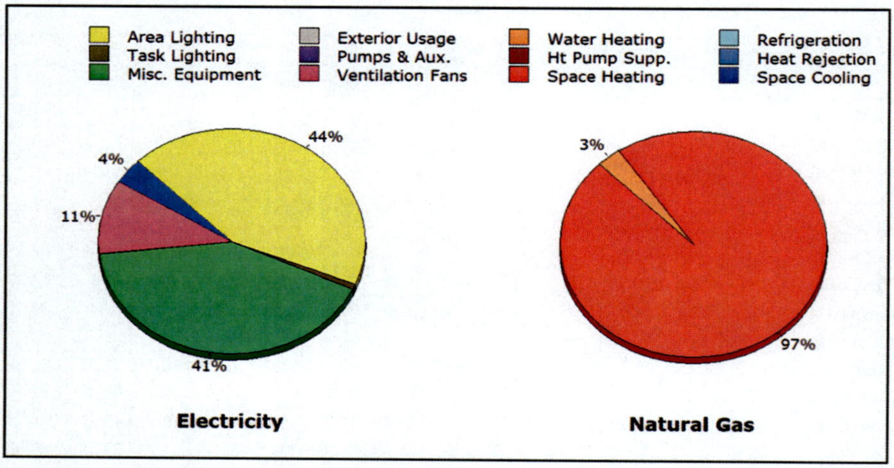

Fig. 5. Annual energy consumption by end-use

Annual Gas Consumption

The results of analyzing the annual operating energy consumption of different roof systems show that the precast hollow core concrete records the highest annual gas consumption at 7,010,000 kBtu. The lowest gas consumption is recorded for the wood roof system with annual gas consumption of 3,240,000 kBtu (i.e., 54% lower than the precast roof, as shown in Table 5). The second-lowest consumption is recorded for the steel roof at 3,650,000 kBtu, 48% lower than that of the precast roof system.

Table 5. Annual electricity and gas consumption by alternatives.

#	Roof system	Annual electricity consumption (kWh)	Annual gas consumption (KBtu)
1	Composite roof	389,500	5,860,000
2	Precast roof	411,500	7,010,000
3	Steel roof	352,600	3,650,000
4	Wood roof	348,400	3,240,000

3.2 Life Cycle Assessment Results

LCA is performed on an industrial building (case study) to measure the environmental impact associated with constructing various structural roof systems overall life cycle stages for 30 years of operation. These life cycle stages include material extraction, material production, transportation, construction, operation, maintenance, end of building life, and beyond building life. The results of the LCA of each structural roof system alternative overall life cycle stages are presented in the following sub-sections

Global Warming Potential (GWP)

GWP is measured in this study as an equivalency relative to CO_2 emissions in kg. The greatest global warming emission by assembly during the production and construction stages is generated by the roof system, as presented in Fig. 6. This finding points to the high impact of the roof system in terms of global warming, underscoring the significance of this research. It is also worth noting that the obvious resulting variation of impact by different roof systems further underscores the importance of conducting comparisons of these systems. For instance, precast and composite roof systems generate a GWP impact more than twice the impact of steel roof and eight times that of the wood roof system. The second-highest GWP for assembly is generated by the foundation followed by the walls.

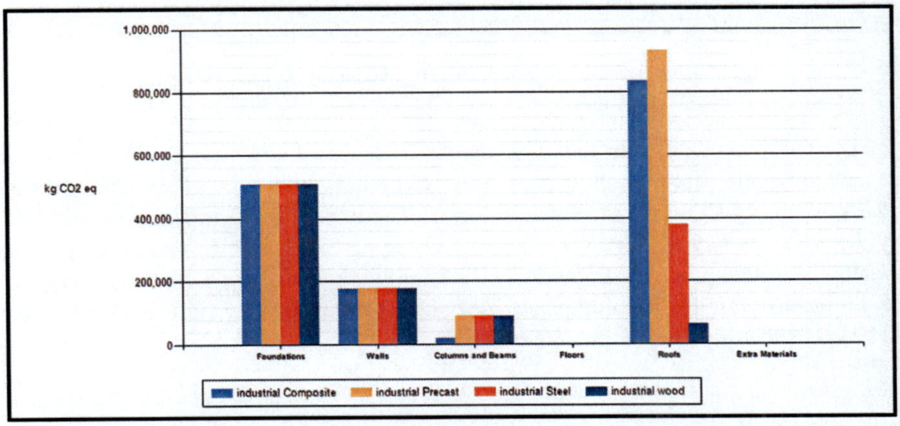

Fig. 6. Comparison of global warming potential by assembly

The greatest global warming emission overall life cycle stages is generated during the operating stage due to the lengthy duration of this stage (30 years) as well as its high energy consumption. GWP emission is highest during the operating stage (90% of the overall effects), followed by the production stage, as presented in Fig. 7. This result indicates the high impact of the operating stage across the life cycle, again underscoring the significance of LCA. It is also worth noting that the obvious resulting variation of impact during these different stages indicates the importance of conducting comparisons of this nature and to take mitigating measures accordingly.

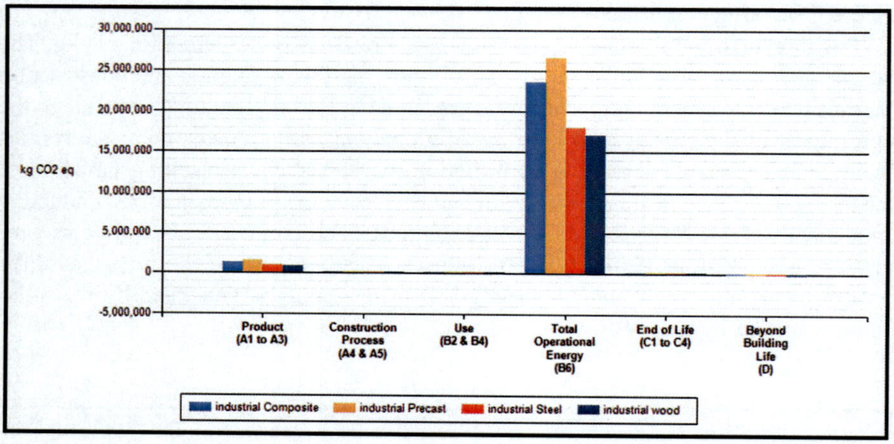

Fig. 7. Comparison of global warming potential by life-cycle stage

The analysis of different roof systems indicates that the highest GWP emission is generated by the precast hollow-core concrete system, with total GWP emissions of 28.3 million (kg CO_2 eq.) as shown in Fig. 8. The lowest GWP emission is produced by the wood roof system at 18 million kg CO_2 eq., 37% lower than that of the precast system. The steel roof system produces 19.2 million kg CO_2 eq., 32% lower than that of the precast system. The composite roof system produces 25.1 million kg CO_2 eq., 11% lower than that of the precast system.

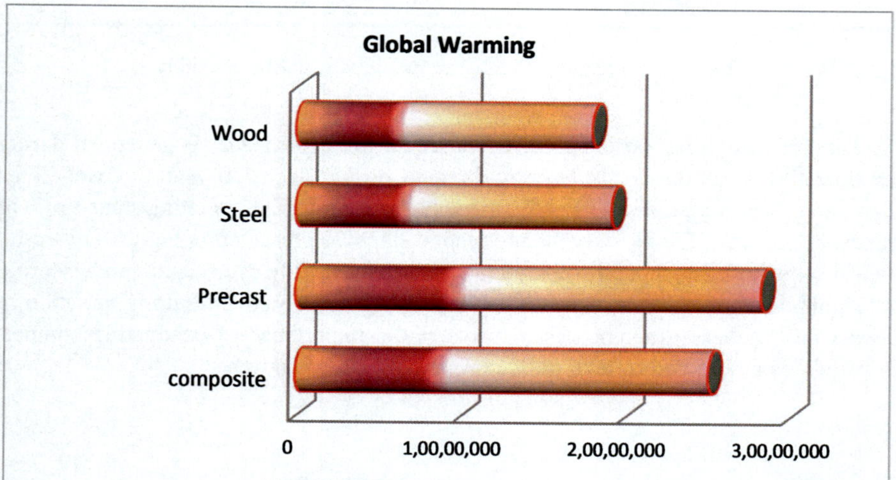

	composite	Precast	Steel	Wood
Global Warming kgCO2 eq.	2,51,00,000	2,83,00,000	1,92,00,000	1,80,00,000
% of Reduction	11%	0%	32%	37%

Fig. 8. Comparison of overall global warming potential by alternative

Emission to Land

The analysis of different roof systems indicates that the highest GHG emissions to land is generated by the composite roof system, with total emission of 3.6 million kg of GHG, as shown in Fig. 9. The lowest GHG emission to the land is produced by the wood roof system at 2.7 million kg of GHG, 23% lower than that of the composite roof system. The steel roof system produces 2.8 million kg of GHG, 22% lower than that of the composite roof system. The precast roof system produces 2.9 million kg of GHG, 18% lower than that of the composite roof system.

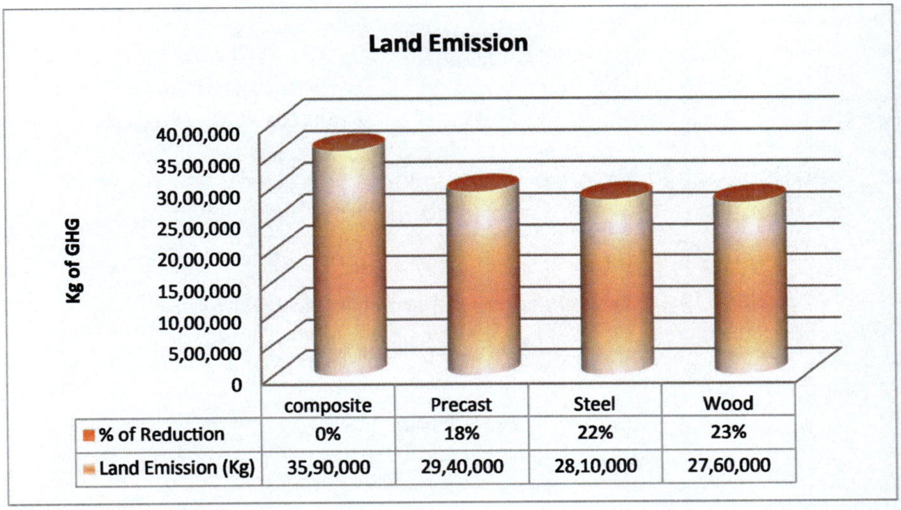

Fig. 9. Comparison of overall emission to land by alternatives

Emission to Water

The analysis also indicates that the highest GHG emission to water is generated by the precast roof system, with total emission of 2.7 million kg of GHG, as shown in Fig. 10.

The lowest GHG emission to the water is produced by the wood roof system at 1.78 million kg of GHG, 34% lower than that of the precast roof system. The steel roof system produces 1.9 million kg of GHG, 28% lower than that of the precast roof system. The composite roof system produces 2.35 million kg of GHG, 13% lower than that of the precast roof system.

Emission to Air

The analysis also indicates that the highest GHG emission to air is generated by the precast roof system, with total emission of 447,000 kg of GHG, as shown in Fig. 11. The lowest GHG emission to the air is produced by the wood roof system at 294,000 kg of GHG, 34% lower than that of the precast roof system. The steel roof system produces 305,000 kg of GHG, 32% lower than that of the precast roof system. The composite roof system produces 397,000 kg of GHG,11% lower than that of the precast roof system.

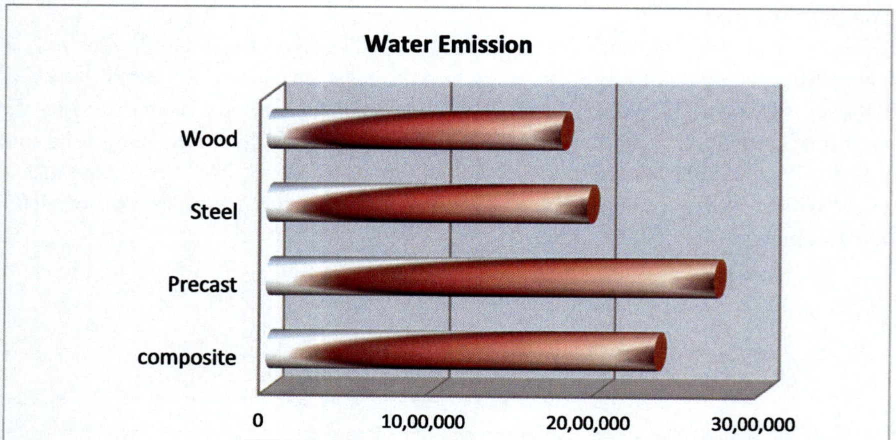

	composite	Precast	Steel	Wood
Water Emission (kg)	23,50,000	27,10,000	19,40,000	17,80,000
% of Reduction	13%	0%	28%	34%

Fig. 10. Comparison of overall water emissions by alternative

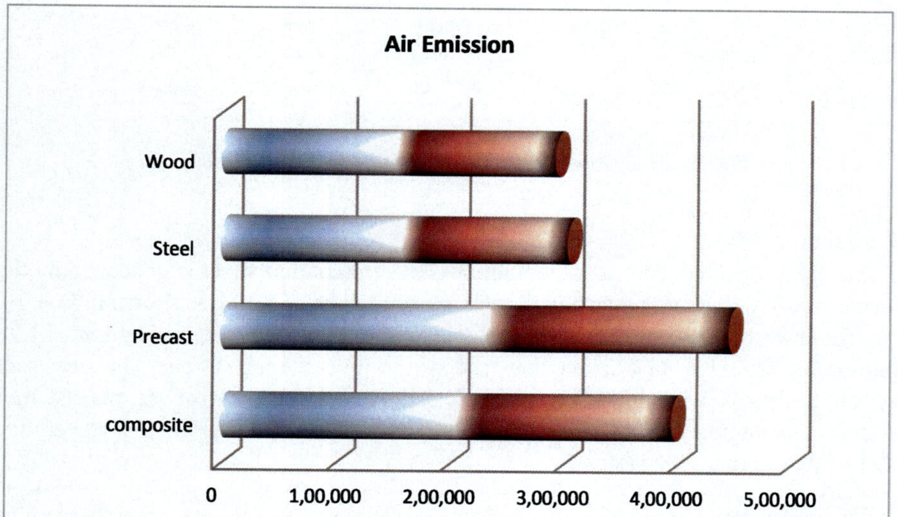

	composite	Precast	Steel	Wood
Air Emission (kg)	3,97,000	4,47,000	3,05,000	2,94,000
% of Reduction	11%	0%	32%	34%

Fig. 11. Comparison of overall emission to air by alternatives

Greenhouse Gas Emissions

Acidification Potential
The highest acidification potential emission is generated by the precast roof system, with total emissions of 199,000 kg CO_2 eq., as shown in Table 6. The lowest acidification potential emission is produced by the wood roof system at 120,000 kg CO_2 eq., 40% lower than that of the precast roof system.

Eutrophication Potential
The highest eutrophication potential emission is generated by the precast roof system, with total emissions of 2,200 kg N eq. as shown in Table 6. The lowest eutrophication potential emission is produced by the wood roof system at 1,200 kg N eq., 42% lower than that of the precast roof system.

Ozone Depletion Potential
The highest ozone depletion potential emission is generated by the precast roof system, with total emissions of 4,400 kg CFC-11 eq., as shown in Table 6. The lowest ozone depletion emissions is produced by the wood roof system at 1,370 kg CFC-11 eq., 69% lower than that of the precast roof system.

Smog Potential
The highest smog potential emission is generated by the precast roof system, with total emissions of 718,000 kg O_3 eq. as shown in Table 6. The lowest smog potential emission is produced by the wood roof system at 418,000 kg O_3 eq., 42% lower than that of the precast roof system.

Fossil Fuel Consumption
The highest fossil fuel consumption is associated with the precast roof system, with total consumption of 420 million (MJ) over 30 years of operation as shown in Table 6. The lowest fossil fuel consumption is associated with the wood roof system at 262 million (MJ) over 30 years of operation, 38% lower than that of the precast roof system.

Table 6. Comparison of overall GHG emissions by alternatives.

LCA measure	Roof system	Precast roof	Composite roof		Steel roof		Wood roof	
	Unit	Emission value	Emission value	% difference	Emission value	% difference	Emission value	% difference
Global warming potential	kg CO2 eq	2.83E+07	2.51E+07	−11.20%	1.92E+07	−32.17%	1.80E+07	−36.48%
Acidification potential	kg SO2 eq	1.99E+05	1.74E+05	−12.69%	1.28E+05	−35.73	1.20E+05	−39.64%
Eutrophication potential	kg N eq	2.22E+03	1.90E+03	−14.60%	1.37E+03	−38.48%	1.29E+03	−41.72%
Ozone depletion potential	kg CFC−11 eq	4.46E−03	1.37E−03	−69.22%	1.39E−03	−68.93%	1.37E−03	−69.23%
Smog potential	kg O3 eq	7.18E+05	6.09E+05	−15.18	4.41E+05	−38.50%	4.18E+05	−41.80%
Fossil fuel consumption	ML	4.20E+08	3.71E+08	−11.74%	2.78E+08	−33.81%	2.62E+08	−37.74%

4 Conclusion

LCA of a case industrial building is performed in the research presented in this paper in order to measure the environmental impact associated with constructing various structural roof systems overall life cycle stages based on 30 years of operation. Various modular roof systems are selected for the evaluation of environmental impact, such as precast hollow-core concrete, composite roof system, steel roof system, and wood roof system.

The wood roof system is found to have the lowest operating and embodied energy consumption, while the precast roof is found to have the highest consumption. The wood roof system is shown to emit 70% less GHGs compared to the precast system.

The results of this study demonstrate the high variability of environmental impact associated with different roof systems, in turn underscoring the significance of the present research. The results also point to the high impact of the building operation stage and the importance of considering the entire life cycle in selecting materials and building systems at the early stages of design. This study can be expected to assist decision makers and industrial building owners in selecting the optimal roof system alternative with respect to minimum energy use and environmental impact.

References

1. UN Environment. IEA, Global Alliance for Buildings and Construction 2018 Globa Status Report, p. 325 (2018). http://www.ren21.net/wp-content/uploads/2018/06/17-8652_GSR2018_FullReport_web_final_pdf
2. Weiler, V., Harter, H., Eicker, U.: Life cycle assessment of buildings and city quarters comparing demolitionand reconstruction with refurbishment. Energy Build. **134**, 319–328 (2017). [CrossRef]
3. Horvath, A.: Construction materials and the environment. Ann. Rev. Environ. Resour. **29**, 181–204 (2004)
4. European Commission-Joint Research Centre-Institute for Environment and Sustainability. International Reference Life Cycle Data System (ILCD) Handbook (2014)
5. ISO 14040-Environmental management-Life CycleAssessment—Framework and principles. International Organization for Standardization (2006). Genève, 31 p
6. Zimmermann, R., Kanafani, K., Rasmussen, F., Birgisdóttir, H.: Early design stage building LCA using the LCAbygtool: comparing cases for early stage and detailed LCA approaches. IOP Conf. Ser.: Earth Environ. Sci. **323**, 12118 (2019)
7. Lasvaux, S., Habert, G., Peuportier, B., Chevalier, J.: Comparison of generic and product-specific life cycle assessment databases: application to construction materials used in building LCA. Int. J. Life Cycle Assess. **20**, 1473–1490 (2015)
8. Architecture 2030, Roadmap to Zero Emissions. Submission to the Ad Hoc Working Group on the Durban Platform for Enhanced Action (2014). https://architecture2030.org/wp-content/uploads/2018/12/Roadmap-to-Zero-Emissions.pdf
9. Saade, M., Guest, G., Amor, B.: Comparative whole building LCAs: how far are our expectations from thedocumented evidence? Build. Environ. **167**, 106449 (2020)
10. Guimarães, G., Saade, M., Zara, O., Silva, V.: Scenario uncertainties assessment within whole building LCA. IOP Conf. Ser. Earth Environ. Sci. **323**, 012059 (2019)

11. Alshamrani, O., Galal, K., Alkass, S.: Integrated LCA-LEED sustainability assessment model for structure and envelope systems of school buildings. Build. Environ. **80**, 61–70 (2014). https://doi.org/10.1016/j.buildenv.2014.05.021
12. Alshamrani, O.S.: Life cycle assessment of low-rise office building with different structure-envelope. Can. J. Civ. Eng. **43**(3), 193–200 (2015)

Constructech Companies: Systematisation of Knowledge and Case Studies

Michelli Tomaz Vasconcelos Fialho(✉)
and Alberto Casado Lordsleem Júnior

Escola Politécnica de Pernambuco – UPE, Recife, Brazil
{mtvf,acasado}@poli.br

Abstract. The construction industry employs approximately 7% of the world's population, representing one of the largest sectors of the global economy, with over US$ 10 trillion invested in goods and services. On the other hand, it is one of the least digitalized sectors, being above only agriculture. In this scenario, constructech companies emerge, whose aim is to seek solutions to the problems of civil construction through technology. The purpose of this work is to systematize the knowledge about construtech companies with the construction industry. The methodology of the work included a descriptive research whose strategy was the case study. Structured interviews were conducted in four construtech companies and four construction companies in the cities of Recife and São Paulo. The results allowed to demonstrate how construtech companies developed, their resources and area of operation, in addition to investigating how construction companies are prepared to adopt available technologies and which processes are of greatest development interest. In summary, the main characteristics perceived regarding the construtech business are the high scalability, the reduced staff, and the agile strategic planning. The construction companies are interested in acquiring solutions that aim at the improvement of the production processes of the works.

Keywords: Constructech · Contech · Technology in civil construction · Innovation · Management · Startup

1 Introduction

The construction industry represents an important share of the world's gross domestic product, resulting in a high level of investment, both for companies in the sector and for customers who purchase it (Azevedo et al. 2011).

It is an extensive and complex activity, but of paramount importance in inducing the growth of the sectors that serve it as input suppliers. According to data from the Federation of Industries of the State of São Paulo (FIESP 2019), the construction production chain brings together a set of activities that total more than 12 million people, about 13% of the workforce employed in the country. Investments in construction, on the other hand, represented around 10% of the national GDP in 2016.

However, according to the report by the McKinsey Global Institute (2017), there are several problems in the construction industry: the low productivity of the workforce, the

E. Toledo Santos and S. Scheer (Eds.): ICCCBE 2020, LNCE 98, pp. 1304–1312, 2021.
https://doi.org/10.1007/978-3-030-51295-8_92

lack of coordination between projects and execution in the field, the lack of inclusion of incentives in contracts for risk sharing and inovation, the precariousness of performance management, as well as the absence of more sophisticated practices in the supply chain.

The same report points out, among other actions, that with the infusion of digital technology, productivity can be increased by 14 to 15% in the construction industry, based on a survey pointing out the implementation of such solutions and the expected difference in terms of impact on productivity.

The study by the McKinsey Institute (2018) shows that the construction industry is among the least digitized sectors, being only above agriculture.

The difficulty in adopting digital technologies in the construction industry can be explained by several factors: the culture of the industry, the low investment in research and development by companies, especially the small ones, and the slowness in changing routines and operations of large companies.

In this sense, in a scenario with great possibilities of expansion, construtech companies, so called the startups of the construction industry, which are technology-based companies, emerged with the aim of solving a problem of this industry, in a scalable and repeatable business model. They stand out for their ability to identify the needs of the industry, and create solutions using innovative technologies for existing traditional models (Loreto 2018).

The article seeks to systematize the knowledge about construtech companies, as well as to identify how construction companies are opportunizing the use of digital technologies, obtaining the benefits of innovation.

2 Theoretical Framework

According to the Oslo Manual (OCDE/EUROSTAT 2018), "innovation is a new or improved product or process (or combination thereof) that differs significantly from previous products or processes of the company". These characteristics must be relevant to the company or to the users/clients.

Around the world, organizations see innovation as an impact factor in the planning and billing of companies, is what the Innovation Survey Report (ABDI 2017) shows, when quoting the Global Innovation Survey, conducted by Price Waterhouse Coopers - PwC (2013): 93% of the 1,757 executives interviewed indicated that "organic growth through innovation would increase the proportion of their companies' revenue growth.

On the other hand, another survey cited in the same report (ABDI 2017), also conducted by the PwC (CEO Pulse Survey of Innovation), indicated that the factors limiting investment to innovation according to the 246 CEOs interviewed in several countries are: financial resources (43%), existing organizational culture not compatible with innovation (41%), lack of talent (30%), regulatory and political factors (21%).

Among the limiting factors presented, the reduction of resources and also the traditional organizational culture are currently more adherent to the Brazilian construction industry.

According to Ozorhon (2013), the decision for innovation in civil construction is impacted mainly by factors related to the project, followed by factors related to management and production, which corroborates the insertion of management mechanisms, with a strong trend towards an innovation model focused on efficiency.

For Barros (1996), part of the introduction of innovation within a company is the involvement of management, favorable organizational conditions, compatibility between innovation and company values, making the environment receptive to innovation.

According to Brockmann et al. (2016), the construction industry is often criticised for its lack of innovation. However, this same industry plans and builds the world's largest projects.

The authors emphasize that, as some innovations occur over the years, they are not easily perceived, besides the fact that civil construction is a conservative industry, with low orientation to the future and to the customer, and still present a low performance in terms of productivity, quality and functionality of the product compared to other industries.

However, it is in an environment where there is still little use of technology that there are ample possibilities for its use, which will meet the demands of the sector, is what Gradvohl et al. (2011) believe.

The constructech companies emerged as new technology-based businesses for civil construction, and this is the result of some factors in addition to those already mentioned: the advance in the use of technology tools and the Internet, access to social networks and their connections, the search for people to undertake in non-conventional businesses with more flexible schedules, and the reduction of fixed jobs.

According to a survey made by Construtech Ventures (2019), there are 562 (five hundred and sixty-two) startups organized by segment and solution.

In the survey released by Startupi (2019), the classification of the area of operation of these companies in Brazil is presented according to Fig. 1.

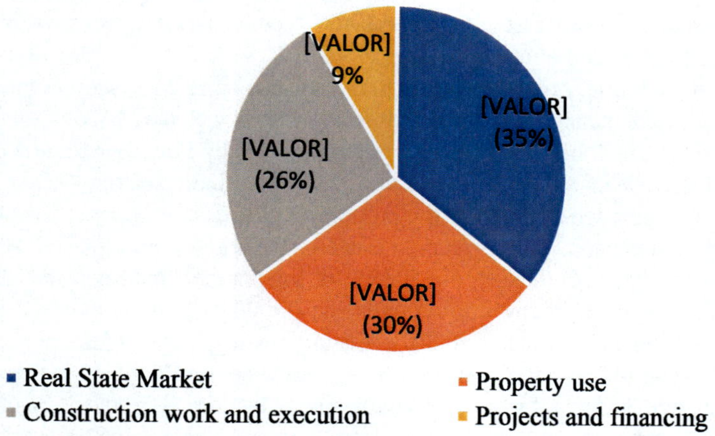

Fig. 1. Construtech companies′: area of action in Brazil

3 Research Method

The methodology used was an explanatory analysis, with the observational method, identifying the factors that determine or contribute to the occurrence of a phenomenon (Gil 2008). From the literature review, the research instrument was developed, and the sample to be researched (representatives of construtech and construction companies), and from the interviews, the data obtained were compared with the references on the subject.

For the application of the research instruments, eight companies were selected, being four construtech companies and four construction companies. Of the four construtech companies, two are from the city of Recife, Pernambuco, and two are from the city of São Paulo, the capital. Among the surveyed construction companies, two originate from Recife, one comes from the state of Minas Gerais, and the other from the state of São Paulo.

The construtech companies research instrument contemplated the following aspects:

- Data from the interviewee and the company;
- Resources and capabilities;
- Information on the product offered;
- Product development;
- Business return;
- Company strategy - vision of the future.

In the questionnaire prepared for application with the construction companies, information was requested about: the company data, resources and management, knowledge about what a construtech company is, technologies used, known by the company, and how the company promotes innovation internally".

4 Research Results

The research instrument was applied among construtech companies: Coteaqui (PE), Molegolar (PE), Instacasa (SP) and Âmbar (SP), all of them with national operations, and in the case of Molegolar, also international.

In the construtech companies surveyed we observe the emergence of the idea from the training carried out by their partners, with the "insight" of the business to be started, in the moments when they were seeking knowledge.

The presence of professionals with a higher level is evident when we observe the relationship between them and the total number of employees: Instacasa (73%), Molegolar (57%), Ambar (50%), Coteaqui (86%). The 50% rate in Ambar takes into account the employees of its plant, who mostly do not have higher education.

Sources of investment have also been diversified in these companies, as shown in Fig. 2.

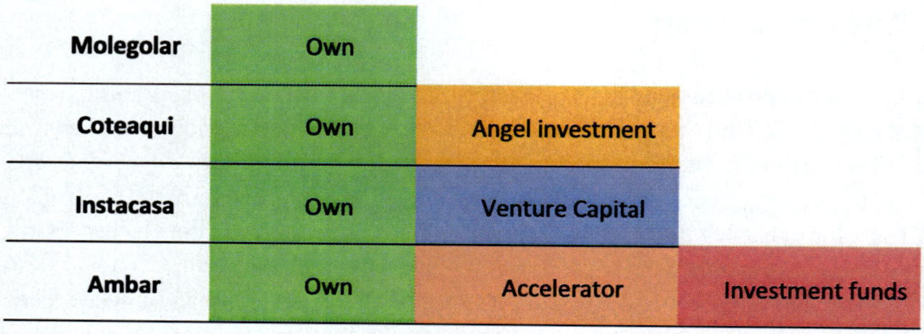

Fig. 2. Types of investment in construtech companies

According to SEBRAE (2017), the most widespread sources of capital or funding are: equity, family money and/or friends, angel investors, that is, those who are specifical-ly interested in investing in a given company; venture capital, which are groups of investors that invest capital in various companies, generally smaller and in the ex-pansion stage; and investment funds, which seek to invest in more consolidated companies with larger contributions.

The construtech companies were asked which activity of the constructor/ developer the solution was inserted for: feasibility, licensing/documentation, sales/marketing, projects, construction site, construction/production, use/maintenance, and supporting processes. Figure 3 illustrates the activities of a construction company impacted by constructech companies' solutions.

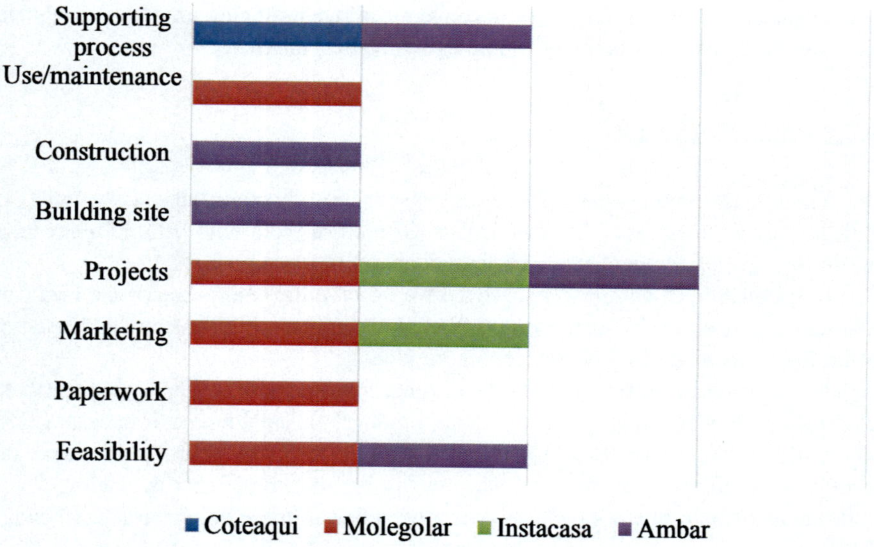

Fig. 3. Constructech companies' solutions vs. activities of a construction company

In relation to the construction companies, classified with A, B, C and D, there is a significant use of digital technologies for civil construction in the companies of the city of São Paulo (C and D), to the detriment of the companies of Recife, where there is a lower adherence to digital technologies, some of them still unknown to those interviewed.

All the construction companies have stated that they know what a construtech company is, and have shown themselves willing to hire solutions that would bring some kind of innovation to their companies. When asked which process(es), the companies would hire a solution that would bring innovation, the construction companies were unanimous in opting for the construction/ production process, i.e. the search for industrialization of processes coupled with the use of technology.

Regarding the digital technologies available today, the builders were asked which of them were already used in the companies, and the result presented in Table 1 was obtained.

Table 1. Use of technologies by construction companies*

Tecnologia	**Faz uso**			
	A	**B**	**C**	**D**
BIM			■	■
Artificial Inteligence			■	■
Cloud data	■		■	■
Drones	■	■	■	■
Augmented reality			■	■
Planning software	■	■	■	■
Project management software		■	■	■
Business management software	■	■	■	■
Digital marketing		■	■	■
3D impression				
Internet of things		■	■	
Virtual reality			■	■
Big data				■
Blockchain				■

The application of the technologies mentioned in Table 1 were: BIM in the modeling and coordination of projects; artificial intelligence allied to the use of another tech-nology, such as drones, for the mapping of pathologies in facades; augmented reality for the visualization of the parts of the projects; planning software for the execution schedule of the work; management software for the management of project activi-ties; enterprise management software (ERPs), for the integration of areas of the same kind as financial, budget and supplies; digital marketing in the dissemination of the company's products on the Internet; the Internet of things for the identification of materials on the construction site; the use of virtual reality to promote the sales of new projects; the technology of big data to analyze the historical data of the compa-ny; and the blockchain for the formalization of purchase and sale contracts. 3D printing for the production of construction components, for example, has not yet been used by any of the construction companies interviewed.

When it was asked what activities occur in companies as a way to prove an efficient model of resource management for innovation (OECD/EUROSTAT 2018), the answers were different among the surveyed Builders. While Construction Company A reported that it carries out a systematic evaluation, prioritizes new ideas and uses quantitative methods to evaluate likely returns, Construction Company B stated that it organizes innovation activities into dedicated projects with objective, budget, schedule and manager, in addition to promoting incentives for innovation activities, even informing that it does not carry out the analysis of the return on these activities.

Construction companies C and D reported that, in addition to the already mentioned activities of organization of activities and incentives for innovation, an evaluation with quantitative methods of the return of new ideas, as well as the choice of methods to allocate resources for innovation activities, based on the progress per stage.

5 Lessons Learned/Best Practices

From the data collected, we listed the best actions observed in construtech companies, which can contribute to their development and improvement:

- user research - development of Minimum Viable Product (MVP), together with the end-customer, for validation of the solution and its interfaces; the use of De-sign Thinking tools has also proven efficient in identifying user needs;
- planning - elaboration and revision of the company's strategic plan, and disclosure with the employees, as a way of engaging and always being aware of changes in the solution and in the market. This tool also proved to be effective in the search for investments, giving more credibility to the company;
- training of partners - the training, as well as the performance in the market for which they developed a solution contributed to its progress and implementation;
- training of employees - the employment of higher-level labor proved to be present in all companies, contributing to the constant improvement of the solution;
- identification of a problem to be solved - from an experience lived by each of the partners in their previous activities with the company, a need was identified, which resulted in a solution presented to the market;
- constant search for improvements and new solutions - even with the consolidation of the current solutions offered, there was a demand from these companies, either for improvement or to incorporate new solutions;

- understanding the client's needs - the development of the User Journey tool made clear the solution and user interfaces, allowing improved interaction, and meeting their needs;
- importance of the client - being always close to and listening to the client was also one of the characteristics observed during the interviews, through various communication channels (visits, meetings, applications and computerized solutions);
- collaboration - the coexistence of constructech companies in an environment that fosters the constant exchange of experiences and the expansion of the network of relationships, which occurs in spaces of interaction open to all;
- innovative environment - the performance of construtech companies in environments that encourage innovation was also observed in company visits, both in promoting the exchange of experiences and in expanding the network of relationships;
- a step forward - the search for updating has also proved to be a differential for the company in the face of the competition;
- search for investments - whether initially, as in the company's emergence, or during its performance, in order to develop new solutions throughout its operations, companies periodically participate in investment rounds to support the new products to be offered. The repercussion of the investments received is also seen in a positive way by the market, which perceives the value of the company from the investments it receives.

6 Concluding Remarks

Constructech companies are still very recent companies, with up to 5 years of operation, and have been seeking to bring solutions with the help of technology to civil construction, which is going through a peculiar moment of crisis. The impact of the emergence of these companies is already considered a favorable factor for the industry, which sees an exponential growth in the number of companies so called.

These are companies with few internal collaborators and that have developed their solutions in a relatively short space of time (between two and eight months), and that continue seeking improvements to the solutions offered.

The research identified that, although there are few professionals working in these companies, their qualification is considered important at construtech companies, since all those who work in the areas of processes and solutions, study or have higher education.

About the construction companies interviewed, different scenarios were observed depending on the regions in which they operate: despite all being large companies, and with more than three decades in the construction market, the use of digital technologies is still very little explored in the construction companies in the city of Recife, while in the construction companies based in São Paulo and Minas Gerais they use a greater number of technologies. In this context, it was also observed the little interest in knowing the existing technologies by the companies in the city of Recife, in contrast with the companies in the Southeast, which, although they use them, remain interested in giving continuity to the knowledge.

It was perceived that builders were concerned with improvements in their processes, that they seek to support the development of innovations in them, and that the process they identify as a priority for innovation is construction/production.

All the construction companies have contact with research and teaching institutions, participate in events with suppliers and entities to search for external knowledge, but there is still the possibility of using in a more comprehensive way the benefits of this interaction in the improvement of their products, processes and/or services.

References

ABDI – Agência Brasileira de Desenvolvimento Industrial. Relatório Técnico 3 – Relatório Sondagem de Inovação (2017)

Azevedo, R.C., Ensslin, L., Lacerda, R.T. de O., França, L.A., González, C.J.I., Jungles, A.E., Ensslin, S.R.: Avaliação de desempenho do processo de orçamento: estudo de caso em uma obra de construção civil. Ambiente Construído **11**(1), 85–104 (2011)

de Barros, M.M.S.B.: Metodologia para implantação de tecnologias construtivas racionalizadas na produção de edifícios. 1996. Tese de Doutorado. Universidade de São Paulo (1996)

Brockmann, C., Brezinski, H., Erbe, A.: IInnovation in construction megaprojects. J. Const. Eng. Manag. **142**(11) (2016)

Construtech Ventures. https://construtechventures.com.br/blog/proptech-construtech-contech-retech-greentech-qual-diferenca. Accessed 17 Mar 2019

FIESP. https://www.fiesp.com.br/infografico-cadeia-da-construcao/. Accessed 06 June 2019

Gil, A.C.: Métodos e técnicas de pesquisa social. 6ª ed. Editora Atlas, São Paulo (2008)

Gradvohl, R.F., de Freitas, A.A.F., Heineck, L.F.: Desenvolvimento de um modelo para análise da acumulação de capacidades tecnológicas na indústria da construção civil: subsetor de edificações. Ambiente Construído, vol. 11, jan/mar (2011)

Loreto, B.: Construtech: inovação que vem de fora do canteiro de obras, Ademi-PE, Recife, 06 jun. (2018)

Mckinsey Global Institute: Reinventing Construction Through a Productivity Revolution. McKinsey & Company (2017)

Mckinsey Global Institute. https://www.mckinsey.com/industries/capital-projects-and-infrastructure/our-insights/imagining-constructions-digital-future. Accessed 28 Sept 2018

OCDE/EUROSTAT: Oslo Manual 2018: Guidelines for Collecting, Reporting and Using Data on Innovation, 4th Edition, The Measurement of Scientific, Technological and Innovation Activities (2018)

Ozorhon, B.: Analysis of construction innovation process at project level. J. Manage. Eng. **29**, 455–463 (2013)

SEBRAE. Capital Empreendedor: estágios de investimentos. http://www.sebrae.com.br/sites/PortalSebrae/sebraeaz/capital-empreendedor-estagios-de-investimentos,464ce3796beaf410VgnVCM1000004c00210aRCRD#0. Accessed 20 Oct 2019

STARTUPI. https://startupi.com.br/2019/04/conheca-a-primeira-rede-de-investidores-anjo-com-foco-em-construtechs/. Accessed 18 June 2019

Intelligent Control of Noise and Vibrations in Building

Walid Larbi(✉)

Laboratoire de Mécanique des Structures et des Systèmes Couplés, Conservatoire National des Arts et Métiers, 292 rue Saint Martin, 75003 Paris, France
walid.larbi@cnam.fr

Abstract. Building acoustics is an important issue because it is related to human comfort and the quality of the indoor environment. The plate-like structure, such as a door, window, floor and wall, is one of the most common types of construction and building. In this work, the experimental and numerical characterization of the acoustic performances of walls and double walls found in building is investigated. First, the acoustic performances of windows in the exterior building façade are evaluated experimentally according to the standard norm ISO 10140. In the second part of this work, a finite element formulation for sound transmission through double wall panels is presented. The third part of this paper concerns the control of sound transmission through double panels using semi-passive piezoelectric shunt technique. More specifically, the system consists of two walls with an air cavity in between. The external panel has surface-mounted piezoelectric patches. The piezoelectric elements, connected with resonant shunt circuits, are used for the vibration damping of some specific resonance frequencies of the coupled system in the low-frequency range.

Keywords: Noise and vibration control · Smart materials · Building · Acoustic

1 Introduction

Double panels are often used in noise reduction since a relatively high transmission loss is achieved by introducing of an acoustic gap between the panels. Due to various effects of fluid-structure interaction, the prediction of the acoustic performance of the double wall remains very complicated in the low frequencies whatever the method used. We propose in this paper: (i) a reliable method based on experimental measurements and numerical analyzes able to predict the vibroacoustic behavior of double-glazing and (ii) to reduce the noise and vibration levels using piezoelectric shunted systems. In the literature, the article [1–3] studied experimentally the transmission loss through single, double and triple glazing. In [4], the effect of boundary conditions on this acoustic indicator of double panels was investigated. A comparison of different type of glass models can be found in [5]. For the numerical prediction approaches, several methods are available. The choice of the numerical method depends on the computational cost and the treated frequency band. The Finite Element Method (FEM) is suitable to treat problems in low frequencies. In [6], this method is applied to develop a model of a

E. Toledo Santos and S. Scheer (Eds.): ICCCBE 2020, LNCE 98, pp. 1313–1322, 2021.
https://doi.org/10.1007/978-3-030-51295-8_93

sandwich plate with viscoelastic core. The FEM is applied in [7] by the authors for the different layers of the sound barrier coupled to a variational Boundary Element Method (BEM) to account for fluid loading.

Concerning the piezoelectric shunt system, in this technology, the elastic structure is equipped with piezoelectric patches that are connected to a passive electrical circuit, called a shunt. The piezoelectric patches transform mechanical energy of the vibrating structure into electrical energy, which is then dissipated by Joule heat in the shunt circuits [8–13]. As this technique is passive (or semi-passive if some electronic components have to be powered), a critical issue is that its performances, in terms of damping efficiency, directly depend on the electromechanical coupling between the host structure and the piezoelectric elements, which has to be maximized and necessitates the development of predictive models.

The present paper is organized as follows: In the first part, the acoustic performances of windows in the exterior building façade are evaluated experimentally according to the standard norm ISO 10140. In the second part of this work, a finite element formulation of the vibro-acoustic problem is presented. The third part of this paper concerns the control of sound transmission through double panels using semi-passive piezoelectric shunt technique. In the last parts, experimental and numerical examples are presented in order to validate and analyze results calculated from the proposed formulations.

2 Experimental Measurements of Sound Reduction

The airborne sound insulation of building elements such as walls, doors, windows can be evaluated in laboratory according to standards norms such as (ISO 10140, 2010). These standards allow the determination of the acoustic performances of the structures by the measurement of the sound reduction index R (called also Sound Transmission Loss TL in the English-speaking counties).

2.1 Summary of Laboratory Measurement Method

According to different subparts of the standard ISO 10140, the laboratory test measurement takes into consideration two reverberation rooms, horizontally or vertically adjacent, separated by a common wall containing an opening in which the test element is mounted as shown in Fig. 1. The minimum volume of two chambers is 50 m^3 with a volume difference of 10%. The rooms are mounted on elastic supports and the only significant sound transmission should be through the test element. The reverberant time should not be greater than 2 s and the total absorption in the receiving chamber should be low to ensure the best possible condition of the diffuse field. During all measurements, the average relative humidity should be at least 30% and temperatures should be in the range 20 ± 3 °C.

In the source room, the sound signals should be random and have a continuous spectrum within each frequency band. For that, one or more loudspeakers can be used, which preferably should be omnidirectional to excite the sound field in the room as uniformly as possible, so the diffuse condition is approximately satisfied. This enables the acoustic pressure to be characterized by a space and frequency averaging which used to calculate the acoustic indicators in one-third octave bands within the frequency range 100 Hz to 5000 Hz as explained in the next section.

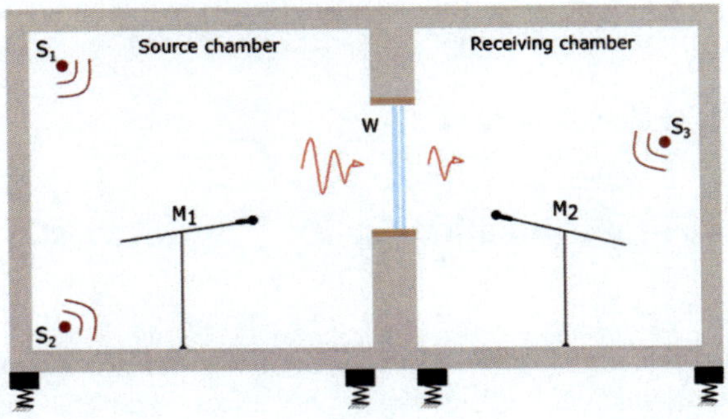

Fig. 1. Experimental set-up: Rotating microphones M_1 and M_2; Loudspeakers S_1, S_2 and S_3; Tested window W.

2.2 Acoustic Indicators

The sound reduction index R is calculated in decibels (dB) as the logarithmic inverse of the sound transmission coefficient τ

$$R = 10\log_{10}\left(\frac{1}{\tau}\right) \tag{1}$$

where τ is defined as the ratio of the sound power radiated by the test element w_r to the sound power incident w_i.

During the laboratory measurements, R is defined as the difference between the average sound pressure levels in the source room and in the receiving room, respectively L_s and L_r considering the total absorption of the receiving room A (m^2) and the area of the test specimen S (m^2):

$$R = L_s - L_r + 10\log_{10}\left(\frac{S}{A}\right) \tag{2}$$

The total absorption A is determined with Sabine's formula in which the reverberation time T_r (s) in the receiving room and its volume V (m^3) are used:

$$T_r = \frac{0.16V}{A} \tag{3}$$

L_s and L_r expressed in dB are ten times the logarithm of the ratio of the square of the sound pressure p (Pa) in the considering chamber to the square of a reference value, p_0 (Pa):

$$L_s = 10\log_{10}\frac{p^2}{p_0^2} \tag{4}$$

where the reference value p_0 is 20 μPa.

3 Finite Element Formulation for the Vibro-Acoustic Problem

Let us consider the linear vibrations of an elastic structure filled with a homogeneous, inviscid and compressible fluid, neglecting gravity effects. This configuration is chosen to model the double-glazing system. We establish in this section the local equations of the coupled problem and the corresponding finite element matrix equations.

3.1 Local Equations of the Problem

We consider an elastic structure occupying the domain Ω_{S^i} $(i = 1, 2)$ at the equilibrium. The structure is subjected to a prescribed displacement $\mathbf{u}^d$ on a part Γ_u and to surface force density $\mathbf{F}^d$ on the complementary part of its external boundary. The interior fluid domain is denoted by Ω_F and the fluid–structure interface by Σ (see Fig. 2).

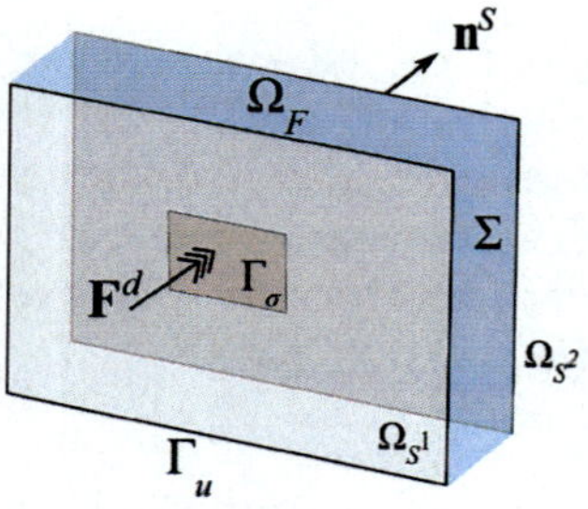

Fig. 2. Double-glazing system

We note $\boldsymbol{\sigma}$, the stress tensor and ρ_s, the density of the structure. In addition, since the compressible fluid is assumed to be inviscid, instead of describing its motion by a fluid displacement vector field, which requires an appropriate discretization of the fluid irrotationality constraint, we use the pressure scalar field p. Let us note k the wave number in the fluid ($k = \omega/c_F$) where c_F is the speed of wave propagation in the fluid,

ρ_F the mass density of the fluid, and $\mathbf{n}^s$ the normal of the structure. The local equations of the fluid-structure coupled problem described previously are [13]:

$$\text{div}\boldsymbol{\sigma}(\mathbf{u}) + \rho_s\omega^2\mathbf{u} = 0 \quad \text{in } \Omega_{S^i} \tag{5}$$

$$\boldsymbol{\sigma}(\mathbf{u})\mathbf{n}^s = \mathbf{F}^d \quad \text{on } \Gamma_\sigma \tag{6}$$

$$\mathbf{u} = \mathbf{u}^d \quad \text{on } \Gamma_u \tag{7}$$

$$\boldsymbol{\sigma}(\mathbf{u})\mathbf{n}^s = p\mathbf{n}^F \quad \text{on } \Sigma \tag{8}$$

$$\Delta p + k^2 p = 0 \quad \text{in } \Omega_F \tag{9}$$

$$\nabla p.\mathbf{n}^F = \rho_F\omega^2\mathbf{u}.\mathbf{n} \quad \text{on } \Sigma \tag{10}$$

Equation (5) corresponds to the elasto-dynamic equation in absence of body force; Eqs. (6), (7) are the prescribed mechanical boundary conditions represent the surface forces applied on Γ_σ and the displacement imposed on Γ_u, respectively; Eq. (8) results from the action of pressure forces exerted by the fluid on the structure; Eq. (9) is the Helmholtz equation; and Eq. (10) is the kinematic interface fluid–structure condition on Σ.

3.2 Finite Element Discretization

In the finite element context, the previous vibro-acoustic problem is described in terms of structure displacement u and acoustic pressure p. The discretization of the variational formulation associated to the previous Eqs. (5) to (10) leads to the following matrix system (see [6] for details):

$$\left(\begin{bmatrix} \mathbf{K}_u & -\mathbf{C}_{up} \\ 0 & \mathbf{K}_p \end{bmatrix} - \omega^2 \begin{bmatrix} \mathbf{M}_u & 0 \\ \mathbf{C}_{up}^T & \mathbf{M}_p \end{bmatrix}\right) \begin{bmatrix} \mathbf{U} \\ \mathbf{P} \end{bmatrix} = \begin{bmatrix} \mathbf{F} \\ 0 \end{bmatrix} \tag{11}$$

with $\mathbf{K}_u$ and $\mathbf{M}_u$ are the structural stiffness and mass matrices, $\mathbf{K}_p$ and $\mathbf{M}_p$ are the associated acoustic matrices, $\mathbf{C}_{up}$ is the fluid-structure coupling matrix and $\mathbf{F}$ is the structural load.

4 Noise and Vibration Reduction Using Piezoelectric Patches

In order to achieve maximum vibration dissipation and acoustic radiation attenuation of selected modes, the passive piezoelectric shunt damping technique is used. Thus, a set of P piezoelectric patches are bounded on the structure surface and connected to resistive or resonant shunt circuits. In this technology, the piezoelectric patches coverts a fraction of mechanical energy associated with the structure vibration into electrical energy, which is dissipated by heat through the resistor in the shunt circuits. Each piezoelectric patch has the shape of a plate with its upper and lower surfaces covered with very thin layer electrodes. Moreover, we denote by $R^{(p)}$ and $L^{(p)}$ the resistance and the inductance of the resonant shunt circuit connected to the pth patch (Fig. 3)

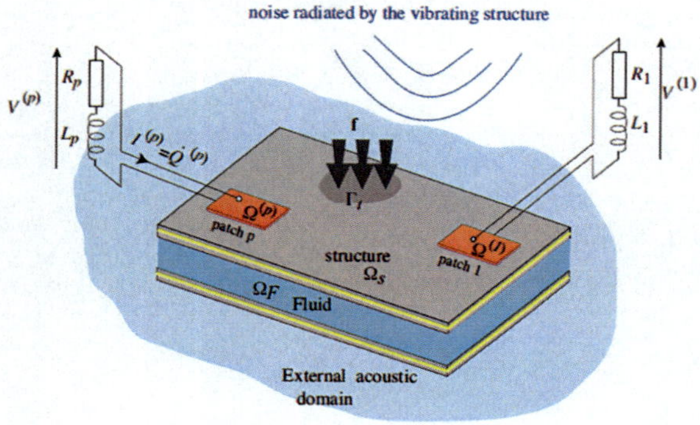

Fig. 3. Noise and vibration reduction using piezoelectric patches

For each piezoelectric patch, a set of hypotheses, which can be applied to a wide spectrum of practical applications, are formulated:

- The piezoelectric patches are thin, with a constant thickness, denoted $h^{(p)}$ for the pth patch;
- The thickness of the electrodes is much smaller than $h^{(p)}$ and is thus neglected;
- The piezoelectric patches are polarized in their transverse direction (i.e. the direction normal to the electrodes).
- Under those assumptions, the electric field vector, of components $E^{(p)}$, can be considered normal to the electrodes and uniform in the piezoelectric patch, so that for all $p \in \{1, \cdots, P\}$:

$$E_k^{(p)} = \frac{V^{(p)}}{h^{(p)}} n_k$$

where E_i is the electric field, $V^{(p)}$ is the potential difference between the upper and the lower electrode surfaces of the pth patch which is constant and n_k is the kth component of the normal unit vector to the surface of the electrodes.

The harmonic local equations of this coupled problem can be written in terms of the structure displacement u, the fluid pressure field p, the electric displacement D and the electric field E (the reader is referred to [12, 13] for more details about these equations). After variational formulation and finite element discretization, we obtain the following matrix system in frequency domain:

$$\begin{bmatrix} \mathbf{K}_u & \mathbf{C}_{uV} & -\mathbf{C}_{up} \\ -\mathbf{C}_{uV} & \mathbf{K}_V & 0 \\ 0 & 0 & \mathbf{K}_p \end{bmatrix} \begin{bmatrix} \mathbf{U} \\ \mathbf{Q} \\ \mathbf{P} \end{bmatrix} - \omega^2 \begin{bmatrix} \mathbf{M}_u & 0 & 0 \\ 0 & 0 & 0 \\ \mathbf{C}_{up}^T & 0 & \mathbf{M}_p \end{bmatrix} \begin{bmatrix} \mathbf{U} \\ \mathbf{V} \\ \mathbf{P} \end{bmatrix} = \begin{bmatrix} \mathbf{F} \\ \mathbf{Q} \\ 0 \end{bmatrix} \tag{12}$$

where U is the column vector of nodal values of mechanical displacement; $\mathbf{M}_u$ and $\mathbf{K}_u$ are the mass and stiffness matrices of the structure (elastic structure and piezoelectric patches); F is the applied mechanical force vector, P is the column vector of nodal values of acoustic pressure. Moreover, Q and V are the column vectors of electric charges and potential differences; $\mathbf{C}_{uV}$ is the electric mechanical coupled stiffness matrix; $\mathbf{K}_V$ is a diagonal matrix filled with the P capacitances of the piezoelectric patches.

The above discretized formulation equation is adapted to any elastic structure with surface-mounted piezoelectric patches. Its originality lies in the fact that the system electrical state is fully described by very few global discrete unknowns: only a couple of variables per piezoelectric patch, namely (1) the electric charge contained in the electrodes and (2) the voltage between the electrodes. Once the electrical part of the problem is fully discretized at the weak formulation step, by introducing the above cited voltage/charge variables, without any restriction on the mechanical part of the problem, any standard FE formulation can be easily modified to include the piezoelectric patches and thus the effect of an external electrical action. A second advantage of this formulation is that since global electrical variables are used, realistic electrical boundary conditions are naturally introduced. First, the equipotentiality in any of the patches electrodes is exactly satisfied when introducing the potential difference variable. Second, the use of the global charge contained in the electrodes, as the second electrical variable, is realistic since plugging an external electrical circuit to the electrodes of the patches imposes only the global charge contained in the electrodes and not a local charge surface density. Another advantage of using the global charge voltage variables is that they are intrinsically adapted to include any external electrical circuit into the electromechanical problem and to simulate the effect of shunt damping techniques. In this case, neither V nor Q is prescribed by the electrical network but the latter imposes only a relation between them [10]. For the case of a resonant shunt composed of a resistor R and an inductor L in series, connected to the pth patch, the relation writes

$$-\omega^2 L^{(p)} Q^{(p)} + i\omega R^{(p)} Q^{(p)} + V^{(p)} = 0 \tag{13}$$

Combining Eqs. (12) and (13), we finally obtain the general FE formulation of the electromechanical problem when the piezoelectric patches are shunted

$$\begin{aligned} &-\omega^2 \begin{bmatrix} \mathbf{M}_u & 0 & 0 \\ 0 & \mathbf{L} & 0 \\ \mathbf{C}_{up}^T & 0 & \mathbf{M}_p \end{bmatrix} \begin{bmatrix} \mathbf{U} \\ \mathbf{Q} \\ \mathbf{P} \end{bmatrix} + i\omega \begin{bmatrix} 0 & 0 & 0 \\ 0 & \mathbf{R} & 0 \\ 0 & 0 & 0 \end{bmatrix} \begin{bmatrix} \mathbf{U} \\ \mathbf{Q} \\ \mathbf{P} \end{bmatrix} \\ &+ \begin{bmatrix} \mathbf{K}_u + \mathbf{C}_{uV}\mathbf{K}_V^{-1}\mathbf{C}_{uV}^T & \mathbf{C}_{uV}\mathbf{K}_V^{-1} & -\mathbf{C}_{0up} \\ \mathbf{K}_V^{-1}\mathbf{C}_{uV}^T & \mathbf{K}_V^{-1} & 0 \\ 0 & 0 & \mathbf{K}_p \end{bmatrix} \begin{bmatrix} \mathbf{U} \\ \mathbf{Q} \\ \mathbf{P} \end{bmatrix} = \begin{bmatrix} \mathbf{F} \\ 0 \\ 0 \end{bmatrix} \end{aligned} \tag{14}$$

where R and L are the diagonal matrices filled with the electrical resistances and the electrical inductances of the shunt circuits.

5 Sound Transmission Loss Through Double Glazing Panel

We present in this section a comparison between experimental and numerical results of the sound transmission loss of double-glazing panel. Two panels are tested: 6/18/4 (6 mm glass panel, 18 mm argon cavity and 4 mm glass panel) and 4/20/4 (4 mm glass panel, 20 mm argon cavity and 4 mm glass panel) from 50 Hz to 500 Hz. As we can see, there is a good qualitative and quantitative agreement between numerical and experimental results for the whole frequency range. For the 6/18/4 configuration (Fig. 4. a), the maximum gap is 4.4 dB in 80 Hz and for the 4/20/4 (Fig. 4. b), the maximum gap is 3.5 dB in 160 Hz.

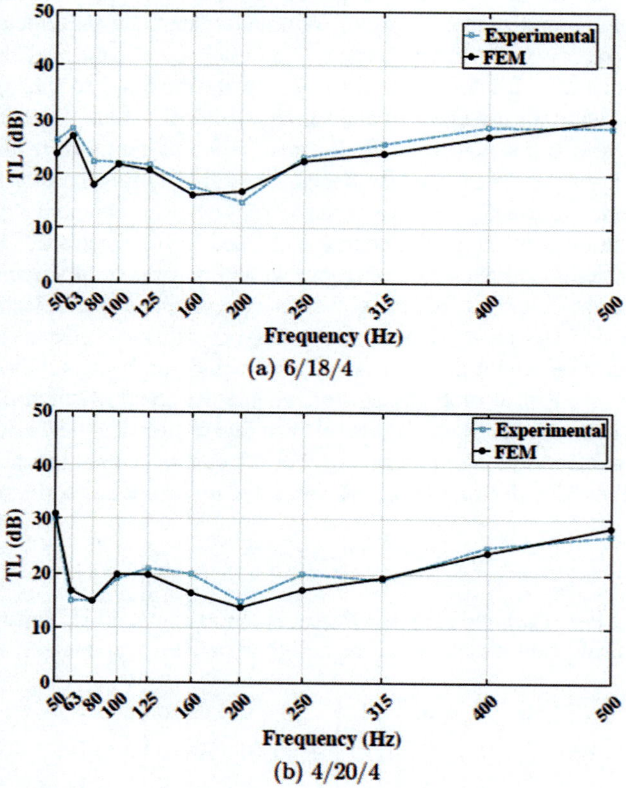

Fig. 4. Sound Transmission Loss through double glazing panel: comparison between experimental and numerical results

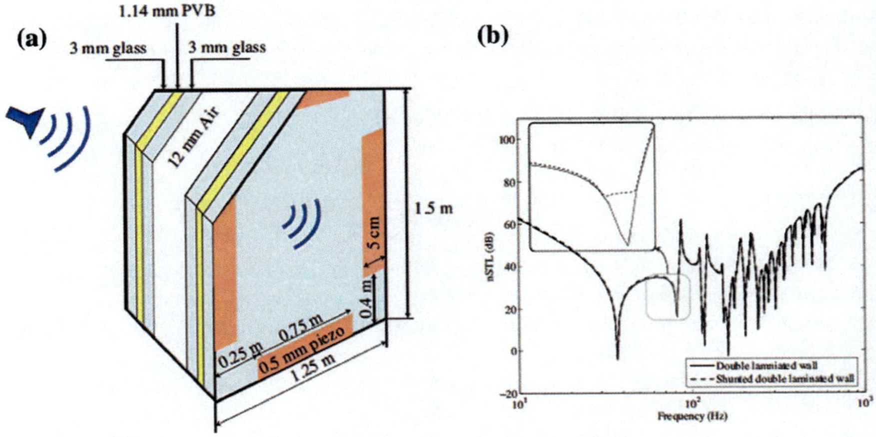

Fig. 5. (a) Double laminated glazing panels, (b) Normal sound transmission loss of laminated double panels: control of panel-air-panel resonance by shunted piezoelectric patches.

6 Example of Noise Control with Shunted Piezoelectric Patches

This example concerns the control of sound transmission through a double laminated glazing panels using shunted piezoelectric patches (Fig. 5-a). The system consists of two identical clamped laminated panels of glass separated by an air cavity. Each laminated glass is composed of two glass plates bonded together by a viscoelastic interlayer. Figure 5-b presents the normal sound transmission loss of the laminated double wall with and without piezoelectric shunt. The response is calculated with a modal reduction approach using the first 20 in vacuo structural modes and the first 20 acoustic modes of the fluid in rigid cavity with static correction. It can be seen that the resonant magnitude of the second mode (the mass-air-mass resonance) has been significantly reduced. In fact, the strain energy contained in the piezoelectric material is converted into electrical energy and hence dissipated into heat using the RL shunt device.

7 Conclusions

In this work, the vibroacoustic behavior of the double-glazing glass is studied. First, the acoustic performances of windows in the exterior building façade are evaluated experimentally according to the standard norm ISO 10140. Finite element method is then used to evaluate the acoustic performances of windows. In this approach, the window is modeled as two glazing plates with an air gap. Numerical results for two glazing configurations are obtained and compared to laboratory measurements showing a relatively good agreement between numerical and experimental sound reduction index. Finally, the control of sound transmission through double panels has be done using semi-passive piezoelectric shunt technique. In this technique, the external panel has a surface-mounted piezoelectric patches. The piezoelectric elements, connected with resonant shunt circuits, are used for the vibration damping of some specific resonance frequencies of the coupled system.

References

1. Quirt, J.D.: Sound transmission through windows I. Single and double glazing. J. Acoust. Soc. Am. **72**(3), 834–844 (1982)
2. Quirt, J.D.: Sound transmission through windows II. Double and triple glazing. J. Acoust. Soc. Am. **74**(2), 534–542 (1983)
3. Tadeu, A.J.B., Mateus, D.M.R.: Sound transmission through single, double and triple glazing. Experimental evaluation. Appl. Acoust. **62**(3), 307–325 (2001)
4. Xin, F.X., Lu, T.J.: Analytical modeling of sound transmission through clamped triple-panel partition separated by enclosed air cavities. Eur. J. Mech. - A Solids **30**(6), 770–782 (2011)
5. Miskinis, K., Dikavicius, V., Bliudzius, R., Banionis, K.: Comparison of sound insulation of windows with double glass units. Appl. Acoust. **92**(Supplement C), 42–46 (2015)
6. Larbi, W., Deü, J.-F., Ohayon, R.: Vibroacoustic analysis of double-wall sandwich panels with viscoelastic core. Comput. Struct. **174**, 92–103 (2016)
7. Sgard, F.C., Atalla, N., Nicolas, J.: A numerical model for the low frequency diffuse field sound transmission loss of double-wall sound barriers with elastic porous linings. J. Acoust. Soc. Am. **108**(6), 2865–2872 (2000)
8. Aloui, R., Larbi, W., Chouchane, M.: Global sensitivity analysis of piezoelectric energy harvesters. Compos. Struct. **228**, 111317 (2019)
9. Larbi, W., Deü, J.-F.: Reduced order finite element formulations for vibration reduction using piezoelectric shunt damping. Appl. Acoust. **147**, 111–120 (2019)
10. Larbi W., Numerical modeling of sound and vibration reduction using viscoelastic materials and shunted piezoelectric patches. Comput. Struct. (2017) available at ScienceDirect
11. Larbi, W., Pereira da Silva, L., Deü, J.-F.: An efficient FE approach for attenuation of acoustic radiation of thin structures by using passive shunted piezoelectric systems. Appl. Acoust. (2017). available at ScienceDirect
12. Larbi, W., Deü, J.-F., Ohayon, R.: Finite element formulation of smart piezoelectric composite plates coupled with acoustic fluid. Compos. Struct. **94**(2), 501–509 (2012)
13. Deü, J.-F., Larbi, W., Ohayon, R.: Piezoelectric structural acoustic problems: symmetric variational formulations and finite element results. Comput. Meth. Appl. Mech. Eng. **197**(19–20), 1715–1724 (2008)
14. Morand, H.J.-P., Ohayon, R.: Fluid-Structure Interaction". Wiley, New York (1995)

An Overview of State-of-the-Art Technologies for Data-Driven Construction

Junghoon Woo[1(✉)], Sangyun Shin[2], Ashish T. Asutosh[3], Jiaxuan Li[3], and Charles J. Kibert[1]

[1] Powell Center for Construction and Environment, Rinker School of Construction Management, University of Florida, Gainesville, USA
{jhoon3037, ckibert}@ufl.edu

[2] Center for Advanced Construction Information Modeling, Rinker School of Construction Management, University of Florida, Gainesville, USA
sang87yun@ufl.edu

[3] Powell Center for Construction and Environment, Rinker, Sr. School of Construction Management, University of Florida, Gainesville, USA
{asishasutosh, mtsdbd}@ufl.edu

Abstract. The Architecture, Engineering, and Construction (AEC) community has been implementing actions that address the significant productivity and efficiency problems that are common in on-site construction. In spite of the many advantages of introducing emerging technologies, the industry continues to rely on hard copy documents. However, the recent advent of new technologies such as AI, Blockchain, IoT, and 5G-communication is leading the built environment into a modern data-driven environment. This data-driven environment can be described as having five phases based on the data utilization criteria: data acquisition, data highway (mobile), data security, data analysis, and data realization. The AEC community can apply IoT, 3D laser scanning, and drones to the data acquisition phase. These technologies can provide the AEC community with fast, affordable, and robust data collecting methodologies. 5G-communications technologies can allow participants to communicate enormous quantities of data in seconds among the various stakeholders. Blockchain enables the AEC industry to share, secure, and store data in distributed ledgers, establishing transparency, immutability, traceability, and decentralization. It also utilizes artificial intelligence (AI) and machine learning (ML) algorithms for data analysis. AI and ML analyze big data collected from the acquisition phase and give a prediction about models or improve decision making by clustering or categorizing data. Finally, by introducing speech recognition technology, virtual and augmented reality, the designer can provide customers with a customer-centered environment. This study presents an overview of state-of-the-art technologies for data-driven environments, use cases, and the impacts of these technologies on the AEC industry productivity and efficiency.

Keywords: State-of-the-art technology · AEC industry · Data-driven construction

E. Toledo Santos and S. Scheer (Eds.): ICCCBE 2020, LNCE 98, pp. 1323–1334, 2021.
https://doi.org/10.1007/978-3-030-51295-8_94

1 Introduction

The AEC industry has been taking high impact strategies to address many of the significant productivity and efficiency problems that are common in on-site construction. The AEC needs to increase productivity annually by about 1% because it has only been growing by one-quarter of the rate of manufacturing productivity over the past 20 years. In spite of the many advantages of newly introduced strategies, it uses the same hard copy documents as it used for decades. This reluctance to adopt state-of-the-art information technologies is one of the main reasons that AEC's productivity growth has been stagnant. However, the recent advent of new technologies such as IoT, 5G-communication, Blockchain, AI, Automatic Speech Recognition (ASR), is leading the built environment into a modern data-driven environment. This data-driven environment can be described as having five phases based on the data utilization criteria: data acquisition, data highway (mobile), data security, data analysis, and data realization. Figure 1 explains the relationship between categories and technologies. This paper presents these five criteria's state-of-the-art construction technologies, use cases, and discussions.

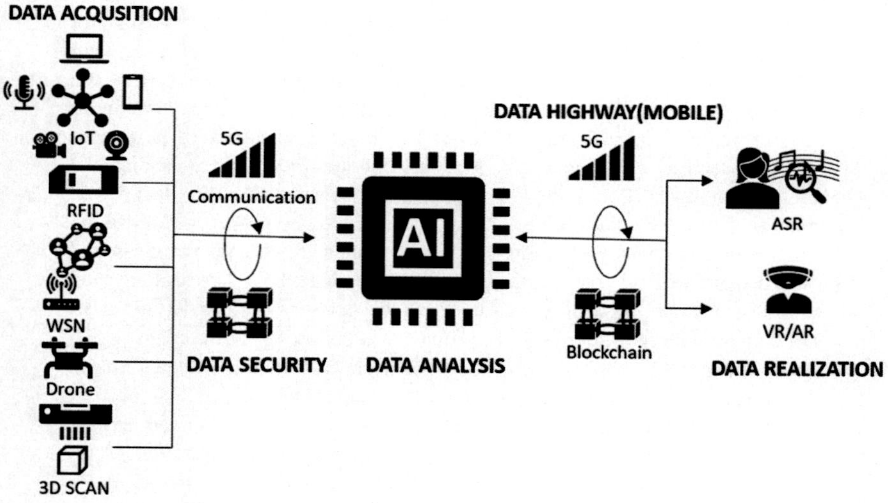

Fig. 1. An overview of state-of-the-art technologies for the AEC industry.

2 Methodology

This study conducted two tiers of analysis. One tier was general category analysis that extracted broad technological areas. The authors used "Google Scholar" and "Web of Science" to find articles related to "the state-of-the-art" and "data-driven" technology for the AEC industry. Several web and journal articles were found and general categories such as data acquisition, data analysis were extracted by the trends, clustering of technologies reported. On top of that, the other tier research conducted is to pick up specific technologies and use-cases of each technical group by googling. Since state-of-the-art technologies are nascent, just introduced to the market, the authors could get the related data mostly from web articles.

3 An Overview of State-of-the-Art Technology for Data-Driven Construction

3.1 Data Acquisition

Data acquisition of the project progress is one of the most critical managerial activities for decision making, undoubtedly relies on precise and useful monitoring systems in the architecture, engineering and construction (AEC) projects (Son et al. 2015). AEC industry has been developed data acquisition technologies for decades, and most of the advancements in data acquisition involve automated works. The purposes of the automated data acquisition are as follows: reducing human error through unmanned devices, securing large amounts of data faster, more accurately, and cheaply which was previously hard to get.

These purposes can be achieved by the following automated data acquisition technologies: Internet of Things (IoT), Radio Frequency Identification (RFID), Wireless Sensor Networks (WSN), Unmanned Aerial System (UVA, other words: Drone), 3D scanning, to name but a few. Subsequent paragraphs, however, are indented.

Internet of Things (IoT): Internet of Things (IoT) has appeared as a significant research area in construction management recently as a method of collecting data from materials manufacture to installation and inspection in building construction (Maghsoodi et al. 2020). The IoT was defined as "the interconnection of sensing and actuating devices providing the ability to share information across platforms through a unified framework, developing a common operating picture for enabling innovative applications" (Gubbi et al. 2013). In a construction project, four system blocks comprise the IoT networks: sensors, communication nodes, servers, and user interfaces (Lam et al. 2017).

Radio Frequency Identification (RFID): RFID uses electromagnetic fields to automatically identify and track tags attached to objects (Atzori et al. 2017). Two types of tags are used for different functions in data collection: passive tags collect the energy of radio waves from a nearby RFID reader, while active tags have a local power source and get energy from further RFID readers and locate more flexible (Jia et al. 2019). Tags are embedded in tracked objectives. RFID is widely used in supply chain management such as transportation tracking, and material and components assembly in construction (Gubbi et al. 2013). Near field communication chip is applied for site safety and material delivery process management (Swedberg 2015).

Wireless Sensor Networks (WSN): WSN contains a large number of different-function sensors communicate wirelessly, such as through Wi-Fi, Bluetooth, Ultra-Wideband (UWB) and ZigBee (Shen et al. 2008). WSN contains three main components namely: sensor node, communication node, and gateway (Frei et al. 2017).

Sensor node measures and storage different type of data, communication node transfers data to the gateway. Gateway receives and forwards data to the data analysis terminal wirelessly. Different kinds of sensors can be installed in an existing building in collecting data of indoor temperature, humidity, air quality, brightness, etc. (AL-Mousawi and AL-Hassani 2018). It is widely used in smart building and building energy control.

Unmanned Aerial System (UVA, Drone): UVA system is a type of WSN that any aircraft or drone that works without a human pilot onboard. It uses devices such as drones or small crafts onsite collecting 3D point cloud and transfers data through a wireless network to a user terminal. Point clouds are sets of data points in some coordinate system. The drone is widely used in data collection of site surveys, which saves cost, time and labor compare to traditional survey methods. It can also be used in building façade inspection, construction progress monitoring, and safety inspection (De Melo et al. 2017).

3D Scanning: Laser scanning and photogrammetry are two types of 3D scanning technologies. Laser scanning collects point cloud data of real-world objectively to develop an accurate 3D model on its shape and appearance. It is done by shooting laser beams followed by a distance measurement at every pointing direction. The collected data can then be used to construct digital 3D models. Photogrammetry is a science of making measurements from photographs, especially for recovering the exact positions of surface points. Point cloud technique can be used for surveying, mapping, inspection, progress monitoring, and clash detection (Kovacs et al. 2006).

As seen above, the technologies in the data acquisition phase aim at collecting big data without any human interference and errors. These technologies help to establish the foundation of data-driven circumstances.

3.2 Data Highway (Mobile)

It is evident that communication is essential in the advancement of every industry (Bailey 1953). Wireless communication is arguably the most critical segment in the global Information and Communication Technology (ICT). Notably, the data mobile highway has been in demand and growing exponentially as more people and machines use this technology to support daily activities. From the second generation 2G communication system which was introduced in 1991, it has been followed rapidly by advanced versions such as 3G and 4G and as of today, 4G LTE (Long term Evolution) has taken a shift in how data is transferred. (Wang et al. 2014). In spite of these technologies, there is a need for way faster and more effective mode of data highway which gave rise to the evolution of next-generation technology known as 5G.

5G is the next generation of mobile data highway technology that will have the capacity to replace all other mobile broadband systems. It operates on three different bands of spectrums, unlike LTE (Looper 2019). It promises to foster a more significant impact by connecting edge computing, autonomous vehicles, higher efficient industries, internet of things (IoT) and analytics infrastructure with least latency (Bayern 2019). Unlike LTE, 5G will be as fast as 20Gbps over wireless networks which is a major technological breakthrough and can revolutionize all other industries to run faster, efficient and reliable (Nordum and Clark 2017).

The advantages of 5G communication are as followings: 1) spectral efficiency with multiple frequencies allowing larger bandwidth to be extended to longer distances, 2) energy efficiency by leveraging the technological gains for both transmitters and servers, 3) enhanced mobile broadband which will allow serving densely populated metropolitan centers, 4) machine to machine type communication which enables IoT applications to be much faster and efficient, 5) ultra-reliable and low latency communication which would address the autonomous vehicle industry, 6) better satellite access and wireline coverage (Fulton 2019).

With these technological advancements on data and network management, blockchain will be able to operate on multiple networks (Bayern 2018). It has the potential to be the platform where all businesses are conducted, and 5G will be the network and data highway it will run on. It has the strength to solve many issues of the world and the way we interact with others financially (Gurnani 2019). This can be a significant revolution in the construction industry. Using smart contracts, payment methods existing in decentralized blockchain networks, the process can be much faster, transparent and much-needed automation (Johnston 2018).

A study by IHS Markit analyzed the adoption, diffusion cycle, and long-term economic contribution of 21 expected use cases. These are categorized into three main categories which are Enhanced Mobile Broadband (EMBB), Massive Internet of things (MIoT) and Mission Critical Services (MCS) (Campbell et al. 2017). Through EMBB, augmented and virtual reality can be broadly utilized. MIoT becomes an infrastructure of smart homes, cities. Finally, MCS encompass applications such as an autonomous vehicle, drone, automation, smart grid which used in the AEC industry now.

Through the 5G communication technology, people can share big data such as video footage which sizes are more than 1 GB in seconds. It means that the AEC industry provides clients with more detailed, realistic, two-way communication, attracting clients' attention, and being easy to reflect clients' ideas.

3.3 Data Security

As well known to people, Bitcoin is cyber money that is traded through the Internet in a distributed security system utilizing a publicly-opened ledger named the blockchain (Swan 2015). Blockchain is a catalog of blocks that got validated, each connecting to its precedent block and ultimately to the genesis block. A block is a data module that saves transactions for integration in the distributed ledger, the blockchain. (Antonopoulos 2014). Blockchain can be achieved even from long distant or numerous sites if the network participant has their public and private key. Since all entities can share the same copy of data in the ledger, any tampering is reported and updated in all blockchains in less than 10 min. The data varies from financial, electronic, to energy consumption, to name a but few. The data is secured cryptographically by the cyber key set (UK Government 2016).

Subsequent A smart contract utilizes codes that are autonomously carried out by the trigger of specific situations. In the perspective of the blockchain, smart contracts refer to blockchain transactions that not only trade financial assets but also possess more comprehensive specifications in them. In a more official description, a contract is an approach of utilizing blockchain platforms such as Bitcoin or Ethereum to establish agreements among people. Indeed, three characteristics of smart contracts that render them outstanding are autonomy, self-sufficiency, and decentralization (Swan 2015).

AEC industry has been trying to adopt the blockchain technology into their business processes. A supply chain is a representative area that blockchain can be an excellent solution for its complexity. Project management can apply the blockchain to secure all the data produced in the construction project. For example, HerenBouw, Amsterdam-based company, is utilizing a blockchain-embedded project management system on a commercial building construction project. The system helps project participants have in-just, precise communication, and fewer mistakes. Furthermore, in the completion of the project, blockchain can handover the project related information right away. “digital twin” blockchain system, developed by the Gardner Builders with Briq, delivers all the data such as paint colors, light bulb types, door silvers, maintenance documents, warranties, to name a few (Hegemen 2019). On top of that, the construction manager is able to get reputation data for subcontractors by blockchain. “Yeeyi,” a community platform in Australia, initiated a digital reputation ledger for a second-hand marketplace with a plethora of Chinese expatriates (Lee 2018). Life-cycle ledger is another example of blockchain use cases. Building maintenance involves an end-to-end cycle of the construction project, asking data that can be traceable when needed. “Blocxs” of OSIsoft, Amsterdam-based startup company, provides Process Intelligence (PI) system for building assets with blockchain. Blocxs shares building equipment life cycle data such as production company, logistics partners, commissioning partner, service department, and recycling partner with building owner and occupants. This application enables participants to respond to an issue from the building maintenance efficiently and reduces the cost of building operations (Mulder 2018). “Synapse” from Brickschain that located in California is an artificial intelligence and blockchain platform for the built environment. It provides the construction industry the prediction of project bidding, execution, and risk management (Talton 2018).

3.4 Data Analysis

Machine learning (ML) and pattern recognition discover the regularities automatically in data by the adoption of computational algorithms and make a decision such as data classification into different categories by using the routines found (Bishop 2006).

AEC industry has been applying ML technologies to the following areas: prediction, irregularity detection, handling system complexity. For example, the ML algorithm can forecast building energy consumption following historical weather data and occupancy details. Furthermore, it can improve building equipment’s anomaly detection, which determines whether there are issues with the facilities and responses to those issues automatically. On top of that, since building management is becoming more and more complicated, ML can help establish a better indoor quality of building systems (Primex 2020).

There are several use cases in the AEC business market. In the planning and design phase, “GenMEP” by Building System Planning, located in California, is an automated tool to create the mechanical, electrical, and plumbing (MEP) automatically, adding-on to Autodesk Revit. In the construction phase, “Building Information Modeling (BIM) 360 IQ” by Autodesk is a managerial tool that utilizes connected information and ML to forecast and focus on high-risk problems or predict subcontractor hazards. "Compact Assist by Volvo Construction Equipment (2015)" helps generate passes that the compactor draws and record asphalt mix temperature maps for the area automatically, calibrating to the final pavement requirements. "SmartTag by Smartvid.io" developed a technical photo and video platform. It involves a "SmartTag" engine that can label pictures and videos of the job field autonomously by utilizing ML, voice, and image recognition technologies. SmartTag uses a deep learning model to classify vision and voice to name the construction information, and predictively proposes safety strategies for the project participants. "SMART CONSTRUCTION" by Komatsu and NVIDIA developed an AI reinforced safety tool for job fields. It models 3D visualizations and traces the whole construction area with live interaction among workforces, equipment, and materials in the project site. Doxel, an AI-based software, argues that it improves construction productivity by adopting robots and drones with multiple sensors to track the project progress. In the operational phase, "Comfy" by Building Robotics, a startup headquartered in Oakland, is an application for building automation management. It helps users change thermostat settings indoors on-demand and apply it over time (Bharadwaj 2019a). "CogniPoint" is a building management software developed by Israel-located company, "PointGrab." CogniPoint system run by AI optimizes the lighting, HVAC for space purposes. IBM also provides the "Watson IoT Platform." It improves energy efficiency in buildings and delivers personalized/occupant-friendly spacial experiences based on data analytics. "Verdigris," a California-located company, provides software for IoT energy metering and platform for data analyzation. "Bidgely" delivers software that can help electricity businesses collect information concerning measurement and the cost of energy utilization by various home devices. The system sends customers visual data to attract attention by using the machine learning algorithm (Bharadwaj 2019b).

3.5 Data Realization

This section will deal with Automatic Speech Recognition (ASR), Augmented Reality (AR) and Virtual Reality (VR) in terms of data realization. To do that, it will also mention a definition, case studies in various fields, and the development direction for the future about automatic speech recognition.

Automatic Speech Recognition (ASR): The ASR can be determined as the autonomous, computer-conducted decoding of oral communication into a legible script instantly. Briefly, ASR is a system that enables a machine to recognize the words when a person talks into a voice recognition device, currently a smartphone, and to transform them into written scripts (Stuckless 1994).

Thanks to the popularization of Automatic Speech Recognition (ASR) with the advent of smart devices, it can be easy for people to use these systems daily. Besides, according to the 2018 Smart market report, the global automatic speech recognition market valued approximately $9.12 billion in 2017. In various areas such as finance, retail, health care, and even the AEC industry, they have tried to apply ASR systems to provide more convenient services to consumers (Bernstein et al. 2018).

In the case of the field of finance, Royal Bank of Canada (RBC) announced in 2017 that consumers can now pay their bills using Apple's Siri. Sean Amato-Gauci, vice president of the bank, said that it has become possible for RBC to provide not just more convenient solutions to clients by offering bill payments but also simple and innovative solutions by giving clients the ability to seamlessly and conveniently bank using voice commands. At the same time, the United Services Automobile Association (USAA) announced that Amazon Alexa helped USAA members to access information about account balances, transactions, and spending patterns. US Bank also announced its embracing of voice-technology. Customers can now check their account balances, hear due payment dates and the amount, obtain account transaction history, and make payments by speaking to an Amazon Alexa (Aspa 2017).

Healthcare is a sphere where speech recognition has put down deep roots as just like other areas as well. Long ago, the Healthcare Information and Management Systems Society (HIMSS) called voice recognition an "aggressively" expanding market with a 20% growth rate per year. A KLAS Enterprises report says that in spite of 50% of physicians' resistance to adopt the technology, 9 out of 10 hospitals plan to expand speech recognition use (Yelina 2017).

Augmented Reality (AR): As an emerging technology, AR integrates images of virtual objects into the real world. By inserting the virtually simulated prototypes into the real world and creating an augmented scene, AR technology could satisfy the goal of enhancing a person's perception of virtual prototyping with real entities (Whyte et al. 2000). Park et al. (2013) combined AR and BIM technologies as an inspection system that enables managers to inspect and control costs and time for more efficient job performance (Park et al. 2013). To improve the efficiency of facility maintenance management, Cheng (2016) developed a framework of a BIM-based location-aware system collaborated with AR (Cheng et al. 2016). Users can share real-time location information by Wi-fi fingerprinting and data source of BIM.

Virtual Reality (VR): The fundamental concept of virtual reality (VR) is to generate a simulated environment based on users' experience and unique insights with computers, software, and peripheral. The setting may be simulated as a real or an imaginary environment (Whyte et al. 2000, Sacks et al. 2013). The use of VR in the construction industry has expanded rapidly since the late 1980s.

VR is widely used in the AEC design in multiple aspects, including communication, safety management, information, and change tracking, and cloud-based modeling in both design and construction phases (Bouchlaghem and Liyanage 1996). Sacks et al. (2013) used VR for safety training and got a better result in safety testing data to examine, comparing to conventional classroom training (Sacks et al. 2013). Kim and Kano (2008) compared VR images with construction photographs taken by a fixed-point camera in the same direction and locations (Kim and Kano 2008). The comparison was valid for the visual assessment of construction progress in a job site. Matching VR images and construction photographs corrected the deviation angle from a 3D viewpoint.

4 Discussion

The technologies investigated in this paper improves not only construction data management but also enhances adaptability to the increasingly changing market circumstances. To gain a competitive edge in a competitive market environment, individual technologies are eliminating human involvement and are pursuing faster, more accurate, and less expensive data processing. These data-driven technologies provide critical information not only to project managers but also to clients quickly and at the right time and can help them make agile and accurate decisions. Furthermore, data-driven environments can cause a significant challenge in the AEC industry. It is the shift from a professional-oriented market to a client-oriented market.

The AEC industry has been a market that professionals have the power in decision making since, in most cases, project owners do not have professional knowledge. At best, with help from owners' representatives, owners can have leadership in the general direction of the project, and most of the details are left to the expert. However, as the state-of-the-art technologies begin to replace the realm of specialists, substitute expert decisions, and create an environment that delivers the vast amount of data needed for decision-making in real-time, the experts-oriented market quickly transforms into a client-oriented market. For example, in the stock market, AI systems have almost replaced fund managers' positions, and with the help of AI, customers can make investments with ease. The construction industry is no exception. The IT industry is taking the initiative over the traditional construction industry. Katerra, a Los Angeles-based construction company, was funded by Softbank and was founded in 2015. The company's management staff consists of mostly IT professionals, not architects, and offers customized, customer-oriented services. The AEC industry seems to need to develop a platform that accommodates market changes. It's a time of day when customers can decide a lot about the building design by using their laptops, ASRs and VRs, without any help from experts.

5 Conclusion

Industry 4.0 has become self-evident not only in information, computing, and telecommunication communities (ICT) but also in manufacturing, even in construction industries recently. Although the AEC community is slower to apply state-of-the-art technologies than other industries, it has been trying to embrace them, and its efforts are paying off in many ways. In the data acquisition phase, Sensors, chips, and networks such as IoT, RFID, WSN, Drone, and 3D scanning collect the information faster, more precise, and cheaper than ever before without human interference. The collected big data are transferred to the partner in seconds through the 5G mobile network, enhancing a fast and appropriate decision making. The blockchain system secures the data acquired in a distributed, transparent manner. The Smart contract through blockchain accelerates getting rid of intermediary in the transactions, decreasing cost and time. AI system delivers decision making, substituting experts' role. The data processed by AI are offered to clients through ASR, VR, AR vividly. The combination and collaboration among these technologies enhance the transition of the AEC industry's data-driven processes.

Despite the advance of individual technology, it is stagnant to integrate state-of-the-art technologies for the AEC industry's project and procedures. Moreover, these new technologies have a disruptive characteristic that may move the center of the AEC industry to the IT industry. And in the process, many construction experts may lose their jobs. Thus, further studies of these issues are needed.

References

5G Mobile Wireless Technology (2019). https://www.electronics-notes.com/articles/connectivity/5g-mobile-wireless-cellular/technology-basics.php. Accessed 20 Jan 2020

Al-Mousawi, A.J., AL-Hassani, H.K.: A survey in a wireless sensor network for explosives detection. Comput. Electr. Eng. **72**, 682–701 (2018)

Antonopoulos, A.M.: Mastering Bitcoin: Unlocking Digital Cryptocurrencies. O'Reilly Media Inc., Sebastopol (2014)

Aspa, J.: RBC First Bank in Canada to Enable Bill Payments Using Siri (2017). https://investingnews.com/daily/tech-investing/emerging-tech-investing/mobile-web-investing/rbc-first-bank-canada-enable-bill-payments-using-siri/. Accessed 20 Jan 2020

Atzori, L., Iera, A., Morabito, G.: Understanding the Internet of Things: definition, potentials, and societal role of a fast-evolving paradigm. Ad Hoc Netw. **56**, 122–140 (2017)

Bailey, W.K.: The importance of communication for advancement in the industry. College Compos. Commun. **4**(1), 11–13 (1953)

Bayern, M.: Why 5G, AI, cloud and blockchain will dominate the digital future of the enterprise. The Tech Republic (2018). https://www.techrepublic.com/article/why-5g-ai-cloud-and-blockchain-will-dominate-the-digital-future-of-the-enterprise/. Accessed 20 Jan 2020

Bernstein, H.M., Jones, S.A., Gudgel, J.E., Messina, F., Partyka, D.: The business value of BIM in Europe. Smart Market Report. Bedford (2018)

Bharadwaj, R.: AI Applications in Construction and Building – Current Use-Cases (2019a). https://emerj.com/ai-sector-overviews/ai-applications-construction-building/. Accessed 21 Jan 2020

Bharadwaj, R.: AI in Building Automation – Current Applications (2019b). https://emerj.com/ai-sector-overviews/ai-building-automation-current-applications/. Accessed 21 Jan 2020

Bishop, C.M.: Pattern recognition and machine learning. Springer, Cham (2006)

Boas, A.V.: "What is 5G, how fast is it, and when is it coming?" Business Insider (2019). https://www.businessinsider.com/what-is-5g-how-fast-is-it-and-when-is-it-coming-2019-1. Accessed 20 Jan 2020

Bouchlaghem, N.M., Liyanage, I.G.: Virtual reality applications in the UK's construction industry. Cib Report 89–94 (1996)

Campbell, K., Diffley, J., Flanagan, B., Morelli, B., O'Neil, B., Sideco, F.: The 5G economy: how 5G technology will contribute to the global economy. In: IHS Economics and IHS Technology. Qualcomm Technologies

Cheng, J.C.P., Chen, K., Chen, W., Ting, C.L.: A local aware augmented reality collaborative system framework for facility maintenance management. In: The 18th International Conference on Construction Applications of Virtual Reality (CONVR 2018), pp. 30–39 (2018)

Author Index

E. Toledo Santos and S. Scheer (Eds.): ICCCBE 2020, LNCE 98, pp. 1335–1338, 2021.
https://doi.org/10.1007/978-3-030-51295-8

Park, C.S., Lee, D.Y., Kwon, O.S., Wang, X.: A framework for proactive construction defect management using BIM, augmented reality and ontology-based data collection template. Autom. Constr. **33**, 61–71 (2013)
Primex: Why Machine Learning is the Future of HVAC and Building Management? (2008). https://www.primexvents.com/why-machine-learning-is-the-future-of-hvac-and-building-management. Accessed 21 Jan 2020
Sacks, R., Perlman, A., Barak, R.: Construction safety training using immersive virtual reality. Constr. Manag. Econ. **31**(9), 1005–1017 (2013)
Shen, X., Cheng, W., Lu, M.: Wireless sensor networks for resources tracking at building construction sites. Tsinghua Sci. Technol. **13**(S1), 78–83 (2008)
Son, H., Bosché, F., Kim, C.: As-built data acquisition and its use in production monitoring and automated layout of civil infrastructure: a survey. Adv. Eng. Inform. **29**(2), 172–183 (2015)
Stuckless, R.: Developments in real-time speech-to-text communication for people with impaired hearing. In: Communication access for People with Hearing Loss, pp. 197–226 (1994)
Swedberg, C.: U.K. construction company collects inspection data onsite with NFC. RFID Journel (2015)
Swan, M.: Blockchain: Blueprint for a New Economy. O'Reilly Media, Sebastopol (2015)
Talton, E.: Brickschain launches AI and machine learning platform synapse at Autodesk University (2018). https://www.newswire.com/news/brickschain-launches-ai-and-machine-learning-platform-synapse-at-20707914. Accessed 28 Jan 2020
UK Government: Distributed ledger technology: Beyond blockchain. Crown (2016)
Wang, C.X., Haider, F., Gao, X., You, X.H., Yang, Y., Yuan, D., Aggoune, H.M., Haas, H., Hepsaydir, E.: Cellular architecture and key technologies for 5G wireless communication networks. IEEE Commun. Mag. **52**(2), 122–130 (2014)
Whyte, J., Bouchlaghem, N., Thorpe, A., McCaffer, R.: From CAD to virtual reality: modelling approaches, data exchange and interactive 3D building design tools. Autom. Constr. **10**(1), 43–55 (2000)
Yamamoto, A., Yabuki, N., Fukuda, T.: Immersive virtual reality teleconference system with design change tracking and 3D editing. In: The 18th International Conference on Construction Applications of Virtual Reality (CONVR 2018), pp. 302–310 (2018)
Yelina, Y.: Speech recognition in healthcare: a significant improvement or severe headache? (2017). https://www.news-medical.net/whitepaper/20170821/Speech-Recognition-in-Healthcare-a-Significant-Improvement-or-Sev

De Melo, R.R.S., Costa, D.B., Álvares, J.S., Irizarry, J.: Applicability of unmanned aerial system (UAS) for safety inspection on construction sites. Saf. Sci. **98**, 174–185 (2017)

Fifth-Generation (5G) Telecommunications Technologies (2019). Congressional Research Service. Retrieved from https://fas.org/sgp/crs/misc/R45485.pdf. Accessed 20 Jan 2020

Frei, M., Hofer, J., Schlüter, A., Nagy, Z.: An easily-deployable wireless sensor network for building energy performance assessment. Energy Procedia **122**, 523–528 (2017)

Fulton, S.: What is 5G? All you need to know about the next generation of wireless technology. https://www.zdnet.com/article/what-is-5g-everything-you-need-to-know/. Accessed 21 Jan 21

Gubbi, J., Buyya, R., Marusic, S., Palaniswami, M.: Internet of Things (IoT): a vision, architectural elements, and future directions. Fut. Gener. Comput. Syst. **29**(7), 1645–1660 (2013)

Gurnani, C.P.: 5G and Blockchain Economics, Business Today (2019). https://www.businesstoday.in/opinion/columns/5g-and-blockchain-economics/story/313149.html. Accessed 21 Jan 2020

Hegeman, K.: Blockchian is changing the construction industry (2019). https://www.sdcexec.com/software-technology/news/21083001/blockchain-is-changing-the-construction-industry. Accessed 28 Jan 2020

Jia, M., Komeily, A., Wang, Y., Srinivasan, R.S.: Adopting Internet of Things for the development of smart buildings: a review of enabling technologies and applications. Autom. Constr. **101**, 111–126 (2019)

Johnston, P.: Prefab & any construction ripe for disruption with blockchain (2018). https://www.thefifthestate.com.au/innovation/building-construction/prefab-construction-disruption-blockchain/. Accessed 21 Jan 2020

Kim, H., Kano, N.: Comparison of construction photograph and VR image in construction progress. Autom. Constr. **17**(2), 137–143 (2008)

Kirkup, A.: Blockchain technology: could it revolutionize construction? Institute of Civil Engineers (218). https://www.ice.org.uk/news-and-insight/the-civil-engineer/december-2018/can-blockchain-transform-construction. Accessed 21 Jan 2020

Kovacs, L., Zimmermann, A., Brockmann, G., Gühring, M., Baurecht, H., Papadopulos, N.A., Schwenzer-Zimmerer, K., Sader, R., Biemer, E., Zeilhofer, H.F.: Three-dimensional recording of the human face with a 3D laser scanner. J. Plast. Reconstr. Aesthet. Surg. **59**(11), 1193–1202 (2006)

Kumar, M., Aggarwal, R.K., Leekha, G., Kumar, Y.: Ensemble feature extraction modules for improved Hindi speech recognition system. Int. J. Comput. Sci. Issues (IJCSI) **9**(3), 175 (2012)

Lam, R.C.Y., Junus, A., Cheng, W.M.Y., Li, X., Lam, L.C.H.: IoT application in construction and civil engineering works. In: 2017 International Conference on Computational Science and Computational Intelligence (CSCI), pp. 1320–1325. IEEE (2017)

Lee, S.: A Decentralized Reputation System: How Blockchain Can Restore Trust In Online Markets. https://www.forbes.com/sites/shermanlee/2018/08/13/a-decentralized-reputation-system-how-blockchain-can-restore-trust-in-online-markets/#7450d2c1481a. Accessed 28 Jan 2020

Maghsoodi, A.I., Rasoulipanah, H., López, L.M., Liao, H., Zavadskas, E.K.: Integrating interval-valued multi-granular 2-tuple linguistic BWM-CODAS approach with target-based attributes: site selection for a construction project. Comput. Ind. Eng. **139**, 106147 (2020)

Mulder, J.: Presentation: blockchain and the PI system: a case study on asset lifecycle management. https://www.osisoft.com/Presentations/Blockchain-and-the-PI-System–A-case-study-on-Asset-Lifecycle-Management/. Accessed 28 Jan 2020

Nordum, A., Clark, K.: Everything You Need to Know About 5G (2017). https://spectrum.ieee.org/video/telecom/wireless/everything-you-need-to-know-about-5g/. Accessed 21 Jan 2020